Lecture Notes in Computer Science 5065

Commenced Publication in 1973
Founding and Former Series Editors:
Gerhard Goos, Juris Hartmanis, and Jan van Leeuwen

Pierpaolo Degano   Rocco De Nicola
José Meseguer (Eds.)

Concurrency, Graphs and Models

Essays Dedicated to Ugo Montanari
on the Occasion of His 65th Birthday

Springer

Volume Editors

Pierpaolo Degano
Università di Pisa
Dipartimento di Informatica
Largo Bruno Pontecorvo 3
56127 Pisa, Italy
E-mail: degano@di.unipi.it

Rocco De Nicola
Università di Firenze
Dipartimento di Sistemi e Informatica
Viale Morgagni, 65
50143 Pisa, Italy
E-mail: denicola@dsi.unifi.it

José Meseguer
University of Illinois at Urbana-Champaign
The Thomas M. Siebel Center for Computer Science
201 N. Goodwin Ave.
Urbana, IL 61801-2302, USA
E-mail: meseguer.jose@gmail.com

The illustration appearing on the cover of this book is the work of Daniel Rozenberg
(DADARA)

Library of Congress Control Number: 2008928685

CR Subject Classification (1998): F.3, F.1, F.4, D.2.4, D.2-3, I.2.2-4, D.1, C.2

LNCS Sublibrary: SL 1 – Theoretical Computer Science and General Issues

ISSN   0302-9743
ISBN-10   3-540-68676-2 Springer Berlin Heidelberg New York
ISBN-13   978-3-540-68676-7 Springer Berlin Heidelberg New York

Springer is a part of Springer Science+Business Media

springer.com

© Springer-Verlag Berlin Heidelberg 2008
Printed in Germany

Typesetting: Camera-ready by author, data conversion by Scientific Publishing Services, Chennai, India
Printed on acid-free paper   SPIN: 12276066   06/3180   5 4 3 2 1 0

Ugo Montanari – Born 1943

Preface

This volume contains the 43 papers written by close collaborators and friends of Ugo Montanari in celebration of his 65th birthday. In some sense, the volume is a reflection, with gratitude and admiration, on Ugo's highly creative, remarkably fruitful and intellectually generous life, which is thriving as strongly as ever. It provides a snapshot of the manifold research ideas that have been deeply influenced by Ugo's work. In a sense the book gives a vantage point from which to foresee further developments to come: by Ugo himself, and by many other people encouraged and stimulated by his friendship and example.

The volume consists of seven sections, six of which are dedicated to the main research areas to which Ugo has contributed. Each of these six sections starts with a contribution by one of Ugo's closer collaborators providing an account of Ugo's contribution to the area and briefly describing the papers in the section. The six scientific sections and the respective editors are the following:

- Graph Transformation (Andrea Corradini)
- Constraint and Logic Programming (Francesca Rossi)
- Software Engineering (Stefania Gnesi)
- Concurrency (Roberto Gorrieri)
- Models of Computation (Roberto Bruni and Vladimiro Sassone)
- Software Verification (Gian-Luigi Ferrari)

The final section, edited by Fabio Gadducci, contains some *laudatio* or memories of working experiences with Ugo, as well as three more technical contributions.

All papers have undergone the scrutiny of at least two reviewers. There is one exception, the contribution by Angelo Raffaele Meo. Because of its highly technical nature and the necessities of keeping the publication deadline, we have not been able to obtain reports for it. For this reason we are not able to take full editorial responsibility for the results in the paper, which are presented as a preliminary report of work in progress to be fully developed in a later publication. Besides the three book editors and the eight section editors, we were helped by several reviewers who gave comments on the papers and made suggestions for their improvement. We would like to thank all of them for their very professional and reliable help.

This volume was presented to Ugo on the 12th of June 2008 during a one-day symposium held in Pisa at the Department of Computer Science. There were six invited talks from eminent scientists, whose friendships with Ugo date back many years. We thank Rina Dechter, Hartmut Ehrig, Robin Milner, Martin Wirsing, and Glynn Winskel for accepting our invitation.

We would also like to thank the Dipartimento di Informatica di Pisa for their support in the organization of the symposium, and the following institutions for their financial support:

- AICA: Associazione Italiana per l'Informatica ed il Calcolo Automatico,
- CINI: Consorzio Interuniversitario Nazionale per l'Informatica,
- Dipartimento di Matematica Pura ed Applicata di Padova,
- ISTI: Institute of Information Science and Technologies "A. Faedo" of CNR,
- SENSORIA project – Software Engineering for Service-Oriented Overlay Computers,
- University of Pisa.

We would like to conclude by saying that this editorial activity made us further experience how much Ugo is appreciated all around the world, and the great esteem he has in the scientific community. Congratulations Ugo!

June 2008

Rocco De Nicola
Pierpaolo Degano
José Meseguer

Organization

Additional Reviewers

Patrizia Asirelli
Paolo Baldan
Maurice ter Beek
Luca Bernardiniello
Lorenzo Bettini
Filippo Bonchi
Gerard Boudol
Roberto Bruni
Vincenzo Ciancia
Silvia Crafa
Philippe Darondeau
Cinzia Di Giusto
Alessandro Fantechi
Pierre Flener
Maria Garcia de la Banda
Roberto Guanciale
Gopal Gupta
Dan Hirsch
Eero Hyvonen
Barbara König
Ivan Lanese
Diego Latella
Francesca Levi
Alberto Lluch Lafuente
Luca Padovani
Maria Silvia Pini
G. Michele Pinna
Marco Pistore
Rosario Pugliese
Paola Quaglia
Kish Shen
Carolyn Talcott
Fabien Tarissan
Francesco Tiezzi
Emilio Tuosto
Kristen Brent Venable
Cristian Versari
Antonio Vitale

Table of Contents

Software Engineering

Concurrency Theory

Models of Computation

Software Verification

Friends

Ugo Montanari in a Nutshell

Rocco De Nicola[1], Pierpaolo Degano[2], and José Meseguer[3]

[1] Dipartimento di Sistemi e Informatica, Università di Firenze
`denicola@dsi.unifi.it`
[2] Dipartimento di Informatica, Università di Pisa
`degano@di.unipi.it`
[3] Department of Computer Science, University of Illinois, Urbana-Champaign
`meseguer@uiuc.edu`

1 Ugo's Origins

Ugo was born in Besana Brianza in 1943 where his parents had moved to escape from the Milan bombings during the Second World War. Immediately after the war he went back to Milan were he completed all his studies. Ugo got his Laurea degree in Electronic Engineering from the Politecnico di Milano in 1966, three years before the first Laurea curriculum and seventeen years before the first PhD curriculum in Computer Science started in Pisa.

At the Politecnico — *el nöster Politècnik*, as the great Milanese engineer and writer Carlo Emilio Gadda used to say — there was no need to defend a thesis, and for concluding his studies there Ugo designed and implemented (on an IBM 7040) an algorithm for tracing the *root locus* of dynamical systems with feedback, and wrote a scientific note describing it.

We cannot easily trace back who was *the* professor of Ugo, although Antonio Grasselli might be considered his "scientific father". Indeed, Ugo could take advantage of the lively cultural environment at the Politecnico, that included at that time scientists like Roberto Galimberti, Luigi Dadda, Marco Cugiani and others, who had a tremendous influence on the development of Computer Science in Italy.

A couple of years after his graduation, Ugo followed professor Antonio Grasselli, who moved to Pisa where the Computer Science curriculum[1] started in 1969. Professor Grasselli set up there an enthusiastic research group at Istituto di Elaborazione dell'Informazione (IEI) of the Italian Research Council, where in the 1950s the Calcolatrice Elettronica Pisana (CEP), the first computer designed and constructed in Italy, had been created. In 1968 Antonio organised an international school in image processing after which, following Antonio's reccomendations, Ugo went to the University of Maryland. There he worked in one

---

[1] **Pierpaolo Degano**: I was fascinated by his enthusiasm and clarity during the first introductory course on computers, so to rank Antonio as my best professor, immediately followed by Ugo, who thought me a mesmerising course at the third year, cf. the contribution by Franco Turini in this volume. Also because of these wonderful professors I had, those where my best years and I am very grateful to them and to all the great people and friends that were in Pisa at that time, cf. the contribution by Alberto Martelli in this volume.

P. Degano et al. (Eds.): Montanari Festschrift, LNCS 5065, pp. 1–8, 2008.
© Springer-Verlag Berlin Heidelberg 2008

of the leading groups in image processing, directed by professor Azriel Rosenfeld. The American experience had a great influence on Ugo, and probably Azriel Rosenfeld can be considered his second "scientific father". After Maryland, Ugo spent overall two years in the States working at Stanford, Berkeley and Carnegie Mellon Universities in a period when Computer Science was being shaped and this had a tremendous impact on his future research.

At the end of the 1970's, Ugo went back to IEI, and in 1975 moved to the Università di Pisa as a full professor. Ugo has been working there since then, but has had other important experiences abroad. Particularly important has been the one in Argentina, where together with his wife Norma Lijtmaer (a computer scientist too, with a vast and exceptional cultural richness who recently passed away) Ugo contributed to setting up ESLAI, a School for Postgraduate Studies that has produced many bright students, now working in European and American Universities. In the academic year 1986–1987, Ugo went back to the USA, this time to SRI and Stanford University, where he worked closely with Joseph Goguen, contributing key graph rewriting ideas to the model of computation and the compilation techniques of the Rewrite Rule Machine, a novel architecture that Goguen, Meseguer, and other colleagues developed with Ugo's help during those and subsequent years.[2]

2   Ugo's Research

Within the brief scope of this foreword, it would be impossible to do justice to Ugo's so rich and varied collection of seminal contributions. With around 300 scientific papers, two books, and 20 edited volumes or special journal issues on his published record, the task would require a much lengthier treatment than what is possible in this foreword. Perhaps the best we can do is to give a few, impressionistic hints by mentioning just a few papers of a seminal nature that exemplify entire areas where Ugo's work has shaped and defined the corresponding research field. It is always somewhat misleading to mention specific areas in isolation. For example, Ugo's seminal contributions to graph transformations are of a piece with his contributions to concurrency. Similarly, his first work on constraint programming was motivated by image processing applications. This rich interplay between his work in different areas must of necessity remain as an implicit *subtext* of the ineluctably brief text containing our explicit comments. However, we will make a few of these connections somewhat more explicit by mentioning some papers under more than one heading.

A general area encompassing a range of Ugo's early contributions is *image processing and artificial intelligence*. By way of example we may mention among

---

[2] **José Meseguer**: This sabbatical visit was also the time when a long-term research programme was started by Ugo and myself on categorical models of concurrency. This research programme was further advanced by my subsequent visits to Pisa and of Ugo to SRI, and was greatly boosted by a second sabbatical stay of Ugo at SRI and Stanford during the 1996–1997 academic year, and, as we briefly discuss in what follows, has contributed many important ideas to concurrency theory.

his contributions here: the first published paper on continuous skeletons in picture processing and recognition [1]; the first work on hidden line elimination for three-dimensional graphics [2]; one of the first applications of dynamic programming to picture processing [5]; and some fundamental papers on search in artificial intelligence such as [4].

Another area where Ugo has made key contributions is that of *logic and constraint programming*. The Martelli-Montanari elegant treatment as an inference system of the unification algorithm [8] has deeply influenced the entire field not only of logic programming, but also of unification theory. The paper [6] can be justly regarded as the first paper on constraint programming. The paper [14] provided a deep theoretical connection, in category theory terms, between logic programming and concurrency theory. The contributions [18,24,32,33] develop a new version of soft constraint programming, that include optimizations and probabilities.

The area of *graph transformations* has also been shaped in fundamental ways by Ugo's contributions. The paper [3] is the first journal paper on web grammars, one of the early formalisms for describing graphs. And [10] is the first paper on synchronized graph rewriting. More recent work [17,19,34] has explored in depth the use of graph transformation systems as fundamental models of concurrency.

Perhaps the broadest area where Ugo's contributions have been both numerous and seminal is *concurrency theory*. They are so many and varied, that a schematization would risk missing the point. They include: the first paper on metric spaces for fairness [9]; the already-mentioned work on synchronized graph rewriting [10] modeling distributed systems; pioneering work on the partial order and causal semantics of concurrent processes [11,12]; new categorical models of computation for concurrency in which concurrent transition systems are modeled as structured categories [13,14]; fundamental studies on the categorical semantics of Petri nets and their extensions [15,16,22,23]; the already-mentioned work on graph transformation systems as fundamental models of concurrency [17,19,34]; a new general concurrent model of computation (the tile model) based on monoidal double categories that is compositional both statically and dynamically [20,25,26]; the foundations and tool development of a finite-state verification framework for mobile calculi [27,30,31]; and a framework for defining transactions with commits and compensation for a variety of formalisms [21,28,29].

Selected Publications

1. Montanari, U.: Continuous Skeletons from Digitalized Images. Journal of the ACM 16(4), 534–549 (1969)
2. Galimberti, R., Montanari, U.: An Algorithm for Hidden Line Elimination. Communications of the ACM 12(4), 206–211 (1969)
3. Montanari, U.: Separable Graphs, Planar Graphs and Web Grammars. Information and Control 16(3), 243–267 (1970)
4. Montanari, U.: Heuristically Guided Search and Chromosome Matching. Artificial Intelligence 1, 227–245 (1970)
5. Montanari, U.: On the Optimal Detection of Curves in Noisy Pictures. Communications of the ACM 14(5) (1971)

6. Montanari, U.: Networks of Constraints: Fundamental Properties and Applications to Picture Processing. Information Sciences 7(2), 95–132 (1974)
7. Giarratana, V., Gimona, F., Montanari, U.: Observability Concepts in Abstract Data Type Specification. In: Mazurkiewicz, A. (ed.) MFCS 1976. LNCS, vol. 45, Springer, Heidelberg (1976)
8. Martelli, A., Montanari, U.: An Efficient Unification Algorithm. ACM Transactions on Programming Languages and Systems 4(2), 258–282 (1982)
9. Degano, P., Montanari, U.: Liveness properties as convergence in metric spaces. In: 16th ACM Annual Symposium on Theory of Computing, pp. 31–38 (1984)
10. Degano, P., Montanari, U.: A Model of Distributed Systems Based on Graph Rewriting. Journal of the ACM 34(2), 411–449 (1987)
11. Degano, P., De Nicola, R., Montanari, U.: A Distributed Operational Semantics For CCS Based On Condition/Event Systems. Acta Informatica 26, 59–91 (1988)
12. Degano, P., De Nicola, R., Montanari, U.: Partial Orderings Descriptions and Observations of Nondeterministic Concurrent Processes. In: de Bakker, J.W., de Roever, W.-P., Rozenberg, G. (eds.) Linear Time, Branching Time and Partial Order in Logics and Models for Concurrency. LNCS, vol. 354, pp. 438–466. Springer, Heidelberg (1989)
13. Meseguer, J., Montanari, U.: Petri Nets are Monoids. Information and Computation 88(2), 105–155 (1990)
14. Corradini, A., Montanari, U.: An Algebraic Semantics for Structured Transition Systems and its Application to Logic Programs. TCS 103, 51–106 (1992)
15. Meseguer, J., Montanari, U., Sassone, V.: Process versus Unfolding Semantics for Place/Transition Petri Nets. TCS 153(1-2), 171–210 (1996)
16. Degano, P., Meseguer, J., Montanari, U.: Axiomatizing the Algebra of Net Computations and Processes. Acta Informatica 33(7), 641–667 (1996)
17. Corradini, A., Montanari, U., Rossi, F.: Graph Processes. Fundamenta Informaticae 26(3-4), 241–265 (1996)
18. Bistarelli, S., Montanari, U., Rossi, F.: Semiring-Based Constraint Satisfaction and Optimization. Journal of the ACM 44(2), 201–236 (1997)
19. Baldan, P., Corradini, A., Montanari, U.: Unfolding and Event Structure Semantics for Graph Grammars. In: Thomas, W. (ed.) FOSSACS 1999. LNCS, vol. 1578, pp. 73–89. Springer, Heidelberg (1999)
20. Bruni, R., Montanari, U.: Cartesian Closed Double Categories, their Lambda-Notation, and the π-Calculus. In: Proc. 14th Symposium on Logic in Computer Science, pp. 246–265. IEEE Computer Society, Los Alamitos (1999)
21. Bruni, R., Montanari, U.: Zero-Safe Nets: Comparing the Collective and Individual Token Approaches. Information and Computation 156(1-2), 46–89 (2000)
22. Bruni, R., Meseguer, J., Montanari, U., Sassone, V.: Functorial Models for Petri Nets. Information and Computation 170, 207–236 (2001)
23. Baldan, P., Corradini, A., Montanari, U.: Contextual Petri Nets, Asymmetric Event Structures and Processes. Information and Computation 171, 1–49 (2001)
24. Bistarelli, S., Montanari, U., Rossi, F.: Semiring-Based Constraint Logic Programming: Syntax and Semantics. ACM Transactions on Programming Languages and Systems 23(1), 1–29 (2001)
25. Bruni, R., Meseguer, J., Montanari, U.: Symmetric and Cartesian Double Categories as a Semantic Framework for Tile Logic. Mathematical Structures of Computer Science 12, 53–90 (2002)
26. Gadducci, F., Montanari, U.: Comparing Logics for Rewriting: Rewriting Logic, Action Calculi and Tile Logic. TCS 285(2), 319–358 (2002)

27. Ferrari, G., Gnesi, S., Montanari, U., Pistore, M.: A Model Checking Verification Environment for Mobile Processes. ACM Transactions on Software Engineering and Methodology 12(4), 440–473 (2003)
28. Bruni, R., Montanari, U.: Concurrent Models for Linda with Transactions. Mathematical Structures of Computer Science 14(3), 421–468 (2004)
29. Bruni, R., Melgratti, H., Montanari, U.: Theoretical Foundations for Compensations in Flow Composition Languages. In: Abadi, M. (ed.) Proc. POPL 2005, pp. 209–220. ACM Press, New York (2005)
30. Ferrari, G., Montanari, U., Tuosto, E.: Coalgebraic Minimization of HD-Automata for the π-Calculus Using Polymorphic Types. TCS 331(2-3), 325–365 (2005)
31. Montanari, U., Pistore, M.: Structured Coalgebras and Minimal HD-Automata for the π-Calculus. TCS 340(3), 539–576 (2005)
32. Lluch Lafuente, A., Montanari, U.: Quantitative μ-Calculus and CTL Defined over Constraint Semirings. TCS 346, 135–160 (2005)
33. Bistarelli, S., Montanari, U., Rossi, F.: Soft Concurrent Constraint Programming. ACM Transactions on Computational Logic 7(3), 1–27 (2006)
34. Baldan, P., Corradini, A., Montanari, U., Ribeiro, L.: Unfolding Semantics of Graph Transformation. Information and Computation 205, 733–782 (2007)

3   Ugo's Students

Ugo Montanari has made seminal contributions to a wide range of areas in computer science. Furthermore, he has shared his enthusiasm for research with a large number of Ph.D. students, many of whom are now established researchers in various countries. More generally, his leadership, his intellectual curiosity and generosity, and his collaborative spirit have sparked very fruitful long-term collaborations with researchers and with entire research teams worldwide. Ugo's students are innumerable, if you consider also those who got their Laurea degree under his supervision — recall that there were no PhD programmes in Italy till 1983, when the first one in Computer Science started in Pisa.

Ugo is always very busy and each of his students has the impression that his work does not get sufficient attention, but when the time comes to write a paper or to discuss about scientific directions, Ugo finds always the time to sketch the work or to wisely advice the scholars[3].

Indeed, Ugo is ranked among the very first nurturers in Computer Science by a 2004 study of the Indian Institute of Science, that considers a number of authors and evaluates them both in terms of their scientific production and of the scientific production of their young students and associates (see `http://archive.csa.iisc.ernet.in/TR/2004/10/`).

Below we list those among Ugo's students that are active in universities or in research centers and have been supervised by Ugo for a *Dottorato di Ricerca*

---

[3] **Rocco De Nicola**: Ugo has had an important role in my career, it was him who convinced me to apply to Edinburgh for PhD studies when there was no PhD program in Italy. I wanted to go to lively Paris but Ugo insisted I should go to rainy Scotland. Only recently I learnt from a common friend that the reason was not only scientific; Ugo was worried that, given my attitudes, I would have got lost in the *douceur de vivre* of Paris and not concluded much. Thank you Ugo.

or for a *Laurea* when the Ph.D. program was not yet well-established in Italy. The list is ordered according to the period the degree was obtained. To give an idea of how seminal Ugo has been, for each of his students we also list their descendents. Students with joint supervision are mentioned only once and a reference is provided to the other supervisor.

1. GianFranco Prini
   (a) Luca Cardelli
2. Franco Turini
   (a) Paolo Mancarella
       i. Francesca Toni
          - Gerhard Wetzel
          - Yannis Xanthakos
       iii. Maurizio Atzori
       iv. Giacomo Terreni
   (b) Dino Pedreschi
       i. Laura Spinsanti
       ii. Francesco Bonchi (with F. Giannotti)
       iii. Mirco Nanni
       iv. Salvatore Ruggieri
       v. Mieke Massink
   (c) Fosca Giannotti
       i. Annalisa Di Deo
       ii. Giuseppe Manco
   (d) Alessandra Raffaetà (with P. Mancarella)
   (e) Chiara Renso
   (f) Andrea Bracciali (with A. Brogi)
   (g) Miriam Baglioni
   (h) Danilo Montesi
   (i) Antonio Brogi
       i. Sara Corfini
       ii. Razvan-Andrei Popescu
       iii. Simone Contiero
3. Pierpaolo Degano
   (a) Corrado Priami
       i. Linda Brodo
       ii. Davide Prandi
       iii. Claudio Eccher
       iv. Radu Mardare
       v. Paola Lecca
       vi. Federica Ciocchetta
       vii. Debora Schuch da Rosa Machado
       viii. Maria Luisa Guerriero
   (b) Chiara Bodei
   (c) Jean-Vincent Loddo
   (d) Stefano Basagni

        i. Luke Demoracski
        ii. Rituparna Ghosh
    (e) Massimo Bartoletti (with G.L. Ferrari)
    (f) Roberto Zunino
4. Stefania Gnesi
    (a) Gabriele Lenzini (with S. Etalle)
    (b) Giuseppe Lami
5. Rocco De Nicola
    (a) Luca Aceto
    (b) Michele Boreale
        i. Lucia Acciai
    (c) Rosario Pugliese
        i. Alessandro Lapadula
    (d) Flavio Corradini
        i. Diletta Romana Cacciagrano
        ii. Maria Rita Di Berardini
    (e) Roberto Segala
        i. Augusto Parma
        ii. Stefano Cattani
    (f) Michele Loreti
    (g) Lorenzo Bettini
    (h) Daniele Gorla (with R. Pugliese)
    (i) Daniele Falassi (with M. Loreti)
6. Ilaria Castellani
    (a) Ana Almeida Matos (with G. Boudol)
7. Paola Inverardi
    (a) Monica Nesi
    (b) Henry Muccini
    (c) Marco Castaldi
    (d) Massimo Tivoli
    (e) Antinisca Di Marco
    (f) Patrizio Pelliccione
    (g) Fabio Mancinelli
    (h) Leonardo Mostarda
    (i) Mauro Caporuscio
    (j) Marco Autili
    (k) Sharareh Afsharian
8. Franco Mazzanti
9. Andrea Corradini
    (a) Leila Ribeiro (with H. Ehrig)
10. Gian Luigi Ferrari
    (a) Emilio Tuosto
    (b) Simone Semprini (with C. Montangero)
    (c) Roberto Guanciale
    (d) Daniele Strollo

11. Roberto Gorrieri (with P. Degano)
    (a) Nadia Busi
    (b) Riccardo Focardi
        i. Matteo Maffei
    (c) Marco Bernardo
        i. Edoardo Bontà
    (d) Gianluigi Zavattaro
    (e) Mario Bravetti
    (f) Alessandro Aldini
    (g) Roberto Lucchi
    (h) Claudio Guidi
12. Paolo Ciancarini
    (a) Cecilia Mascolo
    (b) Davide Rossi
13. Francesca Rossi
    (a) Kristen Brent Venable
    (b) Maria Silvia Pini
14. Gioia Ristori
15. Cosimo Laneve
    (a) Samuele Carpineti
    (b) Manuel Mazzara
16. Vladimiro Sassone
    (a) Pawel Sobocinki
    (b) Damiano Macedonio (with A. Bossi)
    (c) Marco Carbone (with M. Nielsen)
17. Daniel Yankelevich
18. Fabio Gadducci
19. Stefano Bistarelli
20. Marco Pistore
21. Paola Quaglia
22. Roberto Bruni
23. Dan Hirsch (with D. Yankelevich)
24. Paolo Baldan (with A. Corradini)
25. Marzia Buscemi
26. Ivan Lanese
27. Hernan Melgratti (with R. Bruni)
28. Filippo Bonchi
29. Vincenzo Ciancia

Ugo Montanari and Graph Transformation

Andrea Corradini

Dipartimento di Informatica, Università di Pisa, Italy
andrea@di.unipi.it

Graphs are widely exploited in several fields of computer science (as well as in other disciplines) to represent in a direct and adequate way the structure of the states of a system, making it easily understandable also to a non-technical audience. In many situations, the behaviour of such systems can be specified faithfully with a rule-based approach. A graph rule describes how a state can evolve into another state by replacing a sub-state matching the left-hand side of the rule with its right-hand side.

The field of Graph Grammars (also known as Graph Rewriting Systems, or Graph Transformation Systems (GTS)) is concerned with the study of specification formalisms based on the above idea, and with their use for modelling and analysing a variety of systems emerging from several fields of computer science.

The first contributions to the field date back to the late 1960's. At that time, graph *grammars* were introduced as a generalization of string grammars, providing a finite description of a (possibly infinite) collection of graphs. Along these lines, Ugo's first paper on graph grammars ([48], published in 1970) proposed an enrichment with applicability conditions of *Web Grammars*, a formalism introduced shortly before by John Pfaltz and Azriel Rosenfeld in [53],[1] and showed that they can generate some interesting classes of graphs: this has been the first paper on graph grammars ever published in a journal.

Since then, Ugo has always been very active in the area of Graph Transformation, not only with his rich scientific production, but also taking part to the several initiatives of the "GRAGRA" research community. Ugo attended regularly since 1979 the series of quadri-annual International Workshops on Graph Grammars and Their Applications in Computer Science, and since 2002 the bi-annual International Conferences on Graph Transformations: he was Program Chair with Leila Ribeiro of the 2006 edition of the conference, a very successful event which took place in Natal, on the brazilian northern coast. Ugo and his research group participate since 1989 to a series of European projects on Graph Transformation: COMPUGRAPH I (1989-92) and II (1992-96), GET-GRATS (1997-2001, coordinated by Ugo's group), APPLIGRAPH (1997-2002), and SEGRAVIS (2002-06). Also, Ugo co-authored three chapters of the Handbooks on Graph Transformations ([27,2,49]), and is co-editor of the third volume [30] on *Concurrency, Parallelism and Distribution*.

In the 1980's Ugo started using in a systematic way Graph Transformations for the specification, modeling and analysis of concurrent and distributed systems. In joint works with Ilaria Castellani first [16] and with Pierpaolo Degano later [29] (and also with a little contribution by myself, summarized in [17], the first

---

[1] For the curious reader, *webs* are directed, node-labeled simple graphs.

P. Degano et al. (Eds.): Montanari Festschrift, LNCS 5065, pp. 9–15, 2008.

paper of my carreer), Ugo developed a formalism called *(Graph) Grammars for Distributed Systems* (GDS), where a graph represents a distributed system consisting of processes (represented by hyperedges) interacting through ports (nodes), together with its past history, represented as a partial order of events. The evolution of such a distributed system is described with an original three-level approach, which became recurrent in Ugo's work: first the evolution of each single process is specified by context-free productions which may show actions on connected ports; next a context dependent rewriting rule for a set of processes can be derived by synchronizing one rule for each process and checking that the actions shown on the same port by the connected processes are identical; finally the rewriting rule can be applied locally to the global system.

The originality of the synchronization mechanism just described has been highlighted more explicitly in recent years by the new name of the approach, *Synchronized Hyperedge Replacement* (SHR), which has been applied to the specification of Software Architecture Styles [35,36,38,37,41,42], as well as to the modeling of Wide Area Network applications [32] and Service Oriented Computing [31]. To this aim, the synchronization mechanism was extended with name-passing features [40,39,46], allowing for richer topological reconfigurations of a systems, and the use of different synchronization algebras was considered [47]. In order to prove efficiently that certain reconfigurations of Software Architectures are consistent with the corresponding style, i.e., that the resulting graphs are still derivable in the given grammar, a λ-like notation for graph derivations was introduced. A simplified and more effective approach, *Architectural Design Rewriting* (ADR) has been proposed recently [13,12], which is equipped with an implementation in the rewriting engine Maude, and where derivations are denoted by first-order terms.

A more semantic-oriented research project on which Ugo worked constantly since the early 1990's has been the development of a rich concurrent semantics for graph transformation systems. Preliminary results include the formalization in a categorical framework of existing notions of equivalences among graph derivations [18,21], and the development of event structure or partial order semantics for GTSs [51,19,20]. But more interesting results have been obtained by generalizing to GTSs several constructions and results already developed for Petri nets, an approach that perfectly fits with Ugo's attitude to relate different models of computations. The leading intuition here was the observation that P/T nets can be seen as GTSs acting on discrete graphs, a relationship that was elaborated upon, for example, in [23,7]. A milestone in this research project has been the definition of Graph Processes and the study of their properties [26,3], providing a partial-order semantics for GTSs in terms of GTSs themselves. Graph processes, in the non-deterministic version, were used in the unfolding semantics of GTSs [5] first, and later in the development of a functorial, coreflective semantics relating a category of GTS with a suitable category of event structures [4,8,9,15].

Other contributions by Ugo in the area include an axiomatization of graphs and of their derivations [22,24,25], the study of the relationships among GTSs and other computational models, including Concurrent Constraint and Logic

Programming [52,54,28,44] and the Tile Model [50], and the study of observational semantics for GTSs [6,43].

Last but not least, graph transformation systems have been used recurrently by Ugo for modeling process calculi with name passing, including the π-calculus, Mobile Ambients, and the Fusion Calculus [33,44,32,45,34,14,10,11]. There are two main advantages in providing a GTS-based representation of a nominal calculus. Firstly, in most situations the equivalence induced on agents by the graphical representation coincides with the structural equivalence of the calculus. Secondly, the graphical representation makes explicit the topology of the system and the concurrency within it, which is only implicit in the representation as process algebra term.

Papers on Graph Transformation in This Volume

Several contributions to this volume dedicated to Ugo address topics related to graph transformation.

The paper *Unfolding Graph Transformation Systems: Theory and Applications to Verification* by Paolo Baldan, Barbara König and myself presents a brief overview of the works co-authored with Ugo and other colleagues on the unfolding semantics of graph transformation systems, and on its use in the definition of a functorial semantics for GTSs and in the development of methodologies for the verification of systems modeled as finite- or infinite-state GTSs.

In the paper *Graph-Based Design and Analysis of Dynamic Software Architectures* by Roberto Bruni, Antonio Bucchiarone, Stefania Gnesi, Dan Hirsch and Alberto Lluch Lafuente, the authors compare two graph-based approaches to the specification and modeling of architectural design styles, which are aimed at validating prototypical applications before their realisation and deployment. The first approach is based on a set-theoretical variant of the standard double-pushout rewriting, effectively implemented in the Alloy system, while the second one is the ADR approach mentioned above.

The paper *Graph Transformation Units – An Overview* by Hans-Jœrg Kreowski, Sabine Kuske and Grzegorz Rozenberg presents an overview of the work done in the last decade on Graph Transformation Units, an abstract modularity framework for GTSs, independent of a specific graph transformation approach. The framework allows to encapsulate in a unit both rules and control conditions that regulate their applications, and it provides an importing mechanism suitable to structure complex specifications. Both a sequential and a parallel semantics is proposed for this composition mechanism of units.

The paper *Synchronous Multiparty Synchronizations and Transactions* by Ivan Lanese and Hernán Melgratti relates two computational models for the specification of atomic reconfigurations of complex software systems. The authors show that, under mild assumptions, each one can be mapped into the other and viceversa, preserving the operational behaviour. The first formalism is SHR, briefly mentioned above, where the distinguished feature is the multiparty synchronization needed among several hyperedges before they can evolve. The second one is a process calculus called Zero-Safe Fusion, obtained by enriching the fusion calculus, which allows for two-party synchronization, with transactional prefixes, inspired to Zero-Safe nets, a generalization of Petri nets able to describe transactions originally proposed by Ugo with Roberto Bruni.

In recent years, the definition of graph rewriting according to the algebraic, double-pushout approach has been generalized to the more abstract framework of adhesive categories, characterized by a suitable property of "well-behavedness" of pushouts with respect to pullbacks. In such categories DPO rewriting enjoys most of the results originally developed for DPO rewriting over graphs, including theorems concerning parallelism, concurrency and Church-Rosser properties. The paper *Transformations in Reconfigurable Place/Transition Systems* by Ulrike Prange, Hartmut Ehrig, Kathrin Hoffman and Julia Padberg shows that such results hold for transformation systems made of a marked Place/Transition Petri net and a set of rules describing how such nets can be transformed. This is achieved by showing that such systems, called Reconfigurable Place/Transition Systems, equipped with a suitable notion of morphism form a *weakly adhesive high-level replacement* category.

Another field where GTSs are successfully used is in the modeling of the evolution of heap based structures, arising during the execution of object oriented programs. The paper *Explicit State Model Checking for Graph Grammars* by Arend Rensink presents an overview of the GROOVE project and tool, aimed at model-checking object oriented programs through their modeling as GTSs. The paper describes a model-checking approach for graph grammars, including the definition of graph transition systems, methods for symmetry reduction (via isomorphism checking) and appropriate first-order and graph-based logics.

Linear-Ordered Graph Grammars were introduced by Ugo and Leila Ribeiro in [50] as an approach to GTS suitable for the modeling of distributed systems with mobility and object-based systems. Interestingly, also an encoding of such grammars in the Tile Model was proposed, making explicit the aspects of interactivity and compositionality of such systems. The paper *Linear-Ordered Graph Grammars: Applications to Distributed Systems Design* by Leila Ribeiro and Fernando Luís Dotti shows an application of such formalism to the description of the behaviour of distributed systems made of interacting clients and servers, in presence of faults of servers and recovering policies. The encoding as tiles is exploited to introduce a notion of *open graphs* which can be understood as graphs able to interact with the environment through a distinguished set of open nodes.

References

1. Baier, C., Hermanns, H. (eds.): CONCUR 2006. LNCS, vol. 4137. Springer, Heidelberg (2006)
2. Baldan, P., Corradini, A., Ehrig, H., Löwe, M., Montanari, U., Rossi, F.: Concurrent semantics of algebraic graph transformations. In: Ehrig, H., et al. (eds.) [30], pp. 107–187.
3. Baldan, P., Corradini, A., Montanari, U.: Concatenable graph processes: Relating processes and derivation traces. In: Larsen, K.G., Skyum, S., Winskel, G. (eds.) ICALP 1998. LNCS, vol. 1443, pp. 283–295. Springer, Heidelberg (1998)

4. Baldan, P., Corradini, A., Montanari, U.: Unfolding of double-pushout graph grammars is a coreflection. In: Ehrig, H., Engels, G., Kreowski, H.-J., Rozenberg, G. (eds.) TAGT 1998. LNCS, vol. 1764, pp. 145–163. Springer, Heidelberg (2000)

5. Baldan, P., Corradini, A., Montanari, U.: Unfolding and event structure semantics for graph grammars. In: Thomas, W. (ed.) FOSSACS 1999. LNCS, vol. 1578, pp. 73–89. Springer, Heidelberg (1999)

6. Baldan, P., Corradini, A., Montanari, U.: Bisimulation equivalences for graph grammars. In: Brauer, W., Ehrig, H., Karhumäki, J., Salomaa, A. (eds.) Formal and Natural Computing. LNCS, vol. 2300, pp. 158–190. Springer, Heidelberg (2002)

7. Baldan, P., Corradini, A., Montanari, U.: Relating SPO and DPO graph rewriting with Petri nets having read, inhibitor and reset arcs. In: Proceedings of PNGT 2004. ENTCS, vol. 127(2), pp. 5–28. Elsevier Science, Amsterdam (2005)

8. Baldan, P., Corradini, A., Montanari, U., Ribeiro, L.: Coreflective concurrent semantics for single-pushout graph grammars. In: Wirsing, M., Pattinson, D., Hennicker, R. (eds.) WADT 2002. LNCS, vol. 2755, pp. 165–184. Springer, Heidelberg (2003)

9. Baldan, P., Corradini, A., Montanari, U., Ribeiro, L.: Unfolding semantics of graph transformation. Inf. Comput. 205(5), 733–782 (2007)

10. Baldan, P., Gadducci, F., Montanari, U.: Concurrent rewriting for graphs with equivalences. In: Baier and Hermanns [1], pp. 279–294.

11. Baldan, P., Gadducci, F., Montanari, U.: Modelling calculi with name mobility using graphs with equivalences. In: Proceedings of TERMGRAPH 2006. ENTCS, vol. 176(1), pp. 85–97. Elsevier Science, Amsterdam (2007)

12. Bruni, R., Lluch Lafuente, A., Montanari, U.: Hierarchical design rewriting with Maude. In: Proceedings of WRLA 2008. ENTCS, Elsevier, Amsterdam (to appear, 2008)

13. Bruni, R., Lluch Lafuente, A., Montanari, U., Tuosto, E.: Service oriented architectural design. In: TGC 2007. LNCS, vol. 4912, pp. 186–203. Springer, Heidelberg (2007)

14. Bruni, R., Melgratti, H., Montanari, U.: Event structure semantics for nominal calculi. In: Baier, Hermanns (eds.) [1], pp. 295–309.

15. Bruni, R., Melgratti, H., Montanari, U.: Event structure semantics for dynamic graph grammars. In: Baldan, P., Ehrig, H., Padberg, J., Rozenberg, G. (eds.) Proceedings of PNGT 2006, Workshop on Petri Nets and Graph Transformation. ECE-ASST. EASST, vol. 2 (2007)

16. Castellani, I., Montanari, U.: Graph grammars for distributed systems. In: Ehrig, H., Nagl, M., Rozenberg, G. (eds.) Graph Grammars 1982. LNCS, vol. 153, pp. 20–38. Springer, Heidelberg (1983)

17. Corradini, A., Degano, P., Montanari, U.: Specifying highly concurrent data structure manipulation. In: Proceedings ACM Int. Comp. Symp, pp. 424–428. North-Holland, Amsterdam (1985)

18. Corradini, A., Ehrig, H., Löwe, M., Montanari, U., Rossi, F.: Abstract graph derivations in the double pushout approach. In: Schneider, Ehrig (eds.) [55], pp. 86–103.

19. Corradini, A., Ehrig, H., Löwe, M., Montanari, U., Rossi, F.: An event structure semantics for graph grammars with parallel productions. In: Cuny, J.E., Engels, G., Ehrig, H., Rozenberg, G. (eds.) Graph Grammars 1994. LNCS, vol. 1073, pp. 240–256. Springer, Heidelberg (1996)

20. Corradini, A., Ehrig, H., Löwe, M., Montanari, U., Rossi, F.: An event structure semantics for safe graph grammars. In: Olderog, E.-R. (ed.) PROCOMET. IFIP Transactions, vol. A-56, pp. 423–444. North-Holland, Amsterdam (1994)

21. Corradini, A., Ehrig, H., Löwe, M., Montanari, U., Rossi, F.: Note on standard representation of graphs and graph derivations. In: Schneider, Ehrig (eds.) [55], pp. 104–118.
22. Corradini, A., Montanari, U.: An algebra of graphs and graph rewriting. In: Pitt, D.H., Curien, P.-L., Abramsky, S., Pitts, A.M., Poigné, A., Rydeheard, D.E. (eds.) CTCS 1991. LNCS, vol. 530, pp. 236–260. Springer, Heidelberg (1991)
23. Corradini, A., Montanari, U.: Specification of Concurrent Systems: from Petri Nets to Graph Grammars. In: Hommel, G. (ed.) Quality of Communication-Based Systems, pp. 35–52. Kluwer Academic Publishers, Dordrecht (1995)
24. Corradini, A., Montanari, U., Rossi, F.: CHARM: Concurrency and Hiding in an Abstract Rewriting Machine. In: FGCS, pp. 887–896. IOS Press, Amsterdam (1992)
25. Corradini, A., Montanari, U., Rossi, F.: An abstract machine for concurrent modular systems: CHARM. Theor. Comput. Sci 122(1&2), 165–200 (1994)
26. Corradini, A., Montanari, U., Rossi, F.: Graph processes. Fundam. Inform. 26(3/4), 241–265 (1996)
27. Corradini, A., Montanari, U., Rossi, F., Ehrig, H., Heckel, R., Löwe, M.: Algebraic Approaches to Graph Transformation - Part I: Basic Concepts and Double Pushout Approach. In: Rozenberg, G. (ed.) Handbook of Graph Grammars and Computing by Graph Transformations: Foundations, vol. 1, pp. 163–246. World Scientific, Singapore (1997)
28. Corradini, A., Montanari, U., Rossi, F., Ehrig, H., Löwe, M.: Graph grammars and logic programming. In: Ehrig, H., Kreowski, H.-J., Rozenberg, G. (eds.) Graph Grammars 1990. LNCS, vol. 532, pp. 221–237. Springer, Heidelberg (1991)
29. Degano, P., Montanari, U.: A model for distributed systems based on graph rewriting. J. ACM 34(2), 411–449 (1987)
30. Ehrig, H., Kreowski, H.-J., Montanari, U., Rozenberg, G. (eds.): Handbook on Graph Grammars and Computing by Graph Transformation: Concurrency, Parallelism and Distribution, vol. 3. World Scientific, Singapore (1999)
31. Ferrari, G.L., Hirsch, D., Lanese, I., Montanari, U., Tuosto, E.: Synchronised hyperedge replacement as a model for service oriented computing. In: de Boer, F.S., Bonsangue, M.M., Graf, S., de Roever, W.P. (eds.) FMCO 2005. LNCS, vol. 4111, pp. 22–43. Springer, Heidelberg (2006)
32. Ferrari, G.L., Montanari, U., Tuosto, E.: Graph-based models of internetworking systems. In: Aichernig, B.K., Maibaum, T.S.E. (eds.) Formal Methods at the Crossroads. From Panacea to Foundational Support. LNCS, vol. 2757, pp. 242–266. Springer, Heidelberg (2003)
33. Gadducci, F., Montanari, U.: A concurrent graph semantics for mobile ambients. In: Proceedings of MFPS 2001. ENTCS, vol. 45, Elsevier Science, Amsterdam (2001)
34. Gadducci, F., Montanari, U.: Observing reductions in nominal calculi via a graphical encoding of processes. In: Middeldorp, A., van Oostrom, V., van Raamsdonk, F., de Vrijer, R.C. (eds.) Processes, Terms and Cycles: Steps on the Road to Infinity. LNCS, vol. 3838, pp. 106–126. Springer, Heidelberg (2005)
35. Hirsch, D., Inverardi, P., Montanari, U.: Graph grammars and constraint solving for software architecture styles. In: Magee, J., Perry, D. (eds.) Third International Software Architecture Workshop, Orlando, FL, November 1998, pp. 69–72. ACM Press, New York (1998)
36. Hirsch, D., Inverardi, P., Montanari, U.: Modeling software architectures and styles with graph grammars and constraint solving. In: Donohoe, P. (ed.) WICSA. IFIP Conference Proceedings, vol. 140, pp. 127–144. Kluwer, Dordrecht (1999)

37. Hirsch, D., Inverardi, P., Montanari, U.: Reconfiguration of software architecture styles with name mobility. In: Porto, A., Roman, G.-C. (eds.) COORDINATION 2000. LNCS, vol. 1906, pp. 148–163. Springer, Heidelberg (2000)
38. Hirsch, D., Montanari, U.: Consistent transformations for software architecture styles of distributed systems. In: Proceedings WDS 1999. ENTCS, vol. 28, Elsevier Science, Amsterdam (1999)
39. Hirsch, D., Montanari, U.: A graphical calculus for name mobility. In: Roman, G.C., Picco, G.P. (eds.) Proceedings of WSEM 2001, Toronto, Canada (2001)
40. Hirsch, D., Montanari, U.: Synchronized hyperedge replacement with name mobility. In: Larsen, K.G., Nielsen, M. (eds.) CONCUR 2001. LNCS, vol. 2154, pp. 121–136. Springer, Heidelberg (2001)
41. Hirsch, D., Montanari, U.: Two graph-based techniques for software architecture reconfiguration. In: Proceedings of GETGRATS Closing Workshop. ENTCS, vol. 51, Elsevier Science, Amsterdam (2001)
42. Hirsch, D., Montanari, U.: Shaped hierarchical architectural design. In: Proceedings of GT-VMT 2004. ENTCS, vol. 109, pp. 97–109. Elsevier Science, Amsterdam (2004)
43. König, B., Montanari, U.: Observational equivalence for synchronized graph rewriting with mobility. In: Kobayashi, N., Pierce, B.C. (eds.) TACS 2001. LNCS, vol. 2215, pp. 145–164. Springer, Heidelberg (2001)
44. Lanese, I., Montanari, U.: Software architectures, global computing and graph transformation via logic programming. In: Ribeiro, L. (ed.) Proc SBES 2002 - 16th Brazilian Symposium on Software Engineering, Anais, pp. 11–35 (2002)
45. Lanese, I., Montanari, U.: A graphical fusion calculus. In: Proceedings of the Workshop of the COMETA Project. ENTCS, vol. 104, pp. 199–215. Elsevier Science, Amsterdam (2004)
46. Lanese, I., Montanari, U.: Synchronization algebras with mobility for graph transformations. In: Proceedings of FGUC 2004. ENTCS, vol. 138(1), Elsevier Science, Amsterdam (2005)
47. Lanese, I., Montanari, U.: Hoare vs Milner: Comparing synchronizations in a graphical framework with mobility. In: Proceedings of GT-VC 2005. ENTCS, vol. 154(2), pp. 55–72. Elsevier Science, Amsterdam (2006)
48. Montanari, U.: Separable graphs, planar graphs and web grammars. Information and Control 16(3), 243–267 (1970)
49. Montanari, U., Pistore, M., Rossi, F.: Modeling concurrent, mobile and coordinated systems via graph transformations. In: Ehrig, H., et al. (eds.) [30], pp. 189–268.
50. Montanari, U., Ribeiro, L.: Linear ordered graph grammars and their algebraic foundations. In: Corradini, A., Ehrig, H., Kreowski, H.-J., Rozenberg, G. (eds.) ICGT 2002. LNCS, vol. 2505, pp. 317–333. Springer, Heidelberg (2002)
51. Montanari, U., Rossi, F.: Graph grammars as context-dependent rewriting systems: A partial ordering semantics. In: Raoult, J.-C. (ed.) CAAP 1992. LNCS, vol. 581, pp. 232–247. Springer, Heidelberg (1992)
52. Montanari, U., Rossi, F.: Graph rewriting and constraint solving for modelling distributed systems with synchronization (extended abstract). In: Ciancarini, P., Hankin, C. (eds.) COORDINATION 1996. LNCS, vol. 1061, Springer, Heidelberg (1996)
53. Pfaltz, J.L., Rosenfeld, A.: Web grammars. In: IJCAI, pp. 609–620 (1969)
54. Rossi, F., Montanari, U.: Hypergraph grammars and networks of constraints versus logic programming and metaprogramming. In: Proceedings of META 1988, pp. 531–544. MIT Press, Cambridge (1988)
55. Schneider, H.J., Ehrig, H.: Dagstuhl Seminar 1993. LNCS, vol. 776. Springer, Heidelberg (1994)

Unfolding Graph Transformation Systems: Theory and Applications to Verification

Paolo Baldan[1], Andrea Corradini[2], and Barbara König[3]

[1] Dipartimento di Matematica Pura e Applicata, Università di Padova, Italy
[2] Dipartimento di Informatica, Università di Pisa, Italy
[3] Abt. für Informatik und Ang. Kognitionswissenschaft,
Universität Duisburg-Essen, Germany
baldan@math.unipd.it, andrea@di.unipi.it,
barbara_koenig@uni-due.de

Dedicated to Ugo Montanari on the occasion of his 65th birthday

Abstract. The unfolding of a system represents in a single branching structure all its possible computations: it is the cornerstone both of semantical constructions and of efficient partial order verification techniques. In this paper we survey the contributions we elaborated in the last decade with Ugo Montanari and other colleagues, concerning the unfolding of graph transformation systems, and its use in the definition of a Winskel style functorial semantics and in the development of methodologies for the verification of finite and infinite state systems.

1 Introduction

Graph transformation systems (GTSs) [31] are recognized as an expressive specification formalism, especially suited for concurrent and distributed systems [16]: the (topo)logical distribution of a system can be represented naturally by using a graphical structure and the dynamics of the system, including the reconfigurations of its topology, can be modelled by means of graph rewriting rules. Moreover GTSs can be seen as a proper generalisation of a classical model of concurrency, i.e., Petri nets [29], since the latter are essentially rewriting systems on (multi)sets, the rewriting rules being the transitions.

In a research activity started under the guidance of Ugo Montanari the concurrent behaviour of GTSs has been thoroughly studied and a consolidated theory of concurrency for such systems is now available, including the generalisation of several semantics of Petri nets, like process and unfolding semantics (see, e.g., [13,30,7]). The unfolding construction, presented in [30] for the *single-pushout approach* and in [7] for the *double-pushout approach*, has been the basis of a functorial semantics, recently presented in [9], that generalizes to GTSs the one developed by Winskel for safe Petri nets [33]. Furthermore, building on these semantical foundations and in particular on the unfolding construction, a framework has been developed where behavioural properties of GTSs can be expressed

P. Degano et al. (Eds.): Montanari Festschrift, LNCS 5065, pp. 16–36, 2008.

and verified. As witnessed, e.g., by the approaches in [25,17] for Petri nets, truly concurrent semantics are potentially useful in the verification of finite state systems, in that they help to avoid the combinatorial explosion arising when one explores all possible interleavings of events. Such techniques have been generalized to a framework for the verification of finite state GTSs in [5]. Interestingly, several formalisms for concurrency and mobility can be encoded as GTSs, in a way that verification techniques developed for GTSs potentially carry over to such formalisms. Concurrent and mobile systems are often infinite state: in these cases we can resort to approximation techniques in order to analyze them, as proposed in [4,10,11].

In this paper we summarize a number of contributions published by the authors in collaboration with Ugo Montanari and other colleagues, developing a theory of concurrency for GTSs and a framework for the verification of systems modeled as GTSs based on such semantical foundations. We start by presenting the unfolding construction for GTSs in Section 2. Next we describe, in a succinct way due to size limitation, three frameworks where the unfolding construction plays a crucial role, namely the functorial semantics of [9] in Section 3, the finite prefix approach of [5] in Section 4, and the verification framework for infinite state GTSs based on finite over-approximations of the unfolding proposed in [4,10,11] in Section 5. Finally in Section 6 we draw some conclusions.

2   Unfolding Semantics of Graph Transformation Systems

In this section we first introduce the notion of graph rewriting used in the paper: rewriting takes place on so-called *typed graphs*, namely graphs labelled over a structure that is itself a graph [13], and it is defined according to the classical *algebraic, single-pushout approach* (see, e.g., [15]). Next we review the notion of *nondeterministic occurrence grammar*: this will be instrumental in presenting the *unfolding* of a GTS [7,30] in Section 2.3.

2.1   Graph Transformation Systems

In the sequel, given a set A we denote by A^* the set of finite strings of elements of A. Given $u \in A^*$ we write $|u|$ to indicate the length of u. If $u = a_1 \ldots a_n$ and $1 \leq i \leq n$, by $[u]_i$ we denote the i-th element a_i of u. Furthermore, if $f : A \to B$ is a function then we denote by $f^* : A^* \to B^*$ its extension to strings. When f is partial, the extension is strict, i.e., $f^*(u)$ is undefined if f is undefined on $[u]_i$ for some $i \in \{1, \ldots, |u|\}$.

Given a partial function $f : A \rightharpoonup B$ we will denote by $dom(f)$ its *domain*, i.e., the set $\{a \in A \mid f(a) \text{ is defined}\}$. Let $f, g : A \rightharpoonup B$ be two partial functions. We will write $f \leq g$ when $dom(f) \subseteq dom(g)$ and $f(x) = g(x)$ for all $x \in dom(f)$.

Definition 1 (graphs and graph morphisms). *A* (hyper)graph G *is a tuple* (V_G, E_G, c_G), *where* V_G *is a set of nodes,* E_G *is a set of edges and* $c_G : E_G \to V_G^*$ *is a connection function. A node* $v \in V_G$ *is called* isolated *if it is not connected* *to any edge. Given two graphs G, G', a partial graph morphism $f : G \rightharpoonup G'$ is a pair of partial functions $f = \langle f_V : V_G \rightharpoonup V_H, f_E : E_G \rightharpoonup E_H \rangle$ such that:*

$$c_H \circ f_E \leq f_V^* \circ c_G. \tag{1}$$

*We denote by **PGraph** the category of (unlabelled hyper-)graphs and partial graph morphisms. A morphism is called* total *if both components are total, and the corresponding subcategory of **PGraph** is denoted by **Graph**.*

Notice that, according to condition (1), if f is defined over an edge then it must be defined on all its connected nodes: this ensures that the domain of f is a well-formed graph. We will write $G_1 \simeq G_2$ if G_1 and G_2 are *isomorphic*.

Definition 2 (typed graphs). *Given a graph T, a* typed graph G over T *is a pair $\langle |G|, t_G \rangle$, where $|G|$ is a graph and $t_G : |G| \to T$ is a total morphism. A partial morphism* between T-typed graphs $f : G_1 \rightharpoonup G_2$ *is a partial graph morphisms $f : |G_1| \rightharpoonup |G_2|$ consistent with the typing, i.e., such that $t_{G_1} \geq t_{G_2} \circ f$. The category of T-typed graphs and partial typed graph morphisms is denoted by T-**PGraph**.*

A typed graph G is called injective *if the typing morphism t_G is injective. More generally, given $n \in \mathbb{N}$, the graph is called n-injective if for any item x in T it holds that $|t_G^{-1}(x)| \leq n$, namely if the number of "instances of resources" of any type x is bounded by n.*

Given a partial typed graph morphism $f : G_1 \rightharpoonup G_2$, we denote by $dom(f)$ the domain of f typed in the obvious way.

Definition 3 (graph production and direct derivation). *Given a graph T of types, a (T-typed graph)* production q *is an injective partial typed graph morphism $L_q \xrightarrow{r_q} R_q$. It is called* consuming *if the morphism is not total. A production is* node preserving *if (i) r_q is total on nodes, (ii) L_q does not contain isolated nodes, and (iii) each isolated node in R_q belongs to $r_q(L_q)$. The typed graphs L_q and R_q are called the* left-hand side *and the* right-hand side *of the production, respectively.*

A match *of a production in a graph G is a total morphism $g : L_q \to G$. A match is* valid *when for any $x, y \in |L_q|$, if $g(x) = g(y)$ then $x, y \in dom(r_q)$.*

Given a production $L_q \xrightarrow{r_q} R_q$, a typed graph G and a match $g : L_q \to G$, we say that there is a direct derivation $G \Rightarrow_q H$, if the diagram to the right is a pushout square in category T-**PGraph**.

$$
\begin{array}{ccc}
L_q & \xrightarrow{r_q} & R_q \\
g \downarrow & & \downarrow h \\
G & \xrightarrow{d} & H
\end{array}
$$

Roughly speaking, the effect of the pushout construction in T-**PGraph** is the following: graph H is obtained by first deleting from the graph G the image of the items of the left-hand side which are not in the domain of r_q, namely $g(L_q - dom(r_q))$, as well as all edges that would remain dangling, and then adding the items of the right-hand side which are not in the image of r_q, namely $R_q - r_q(dom(r_q))$. The items in the image of $dom(r_q)$ are "preserved" by the rewriting step (intuitively, they are accessed in a "read-only" manner).

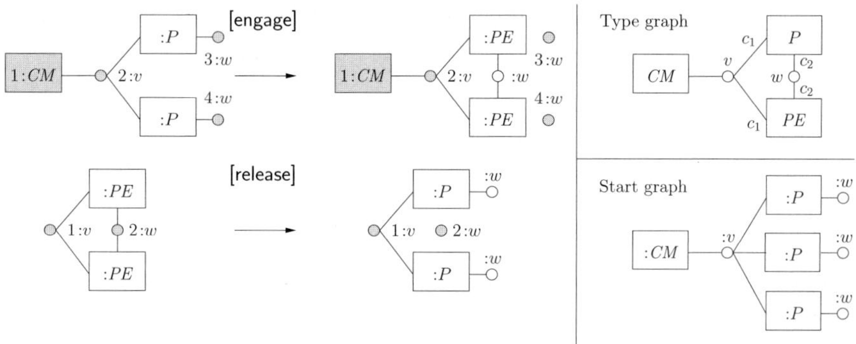

Fig. 1. The finite state GTS $\mathcal{CP}$

Definition 4 (typed GTS and derivation). *A (T-typed) SPO GTS $\mathcal{G}$ (sometimes also referred to as a (graph) grammar) is a tuple $\langle T, G_s, P, \pi \rangle$, where G_s is the (typed) start graph, P is a set of production names, and π is a function which associates a T-typed production to each name in P. We denote by $Elem(\mathcal{G})$ the set $V_T \cup E_T \cup P$.*

A derivation in $\mathcal{G}$ is a sequence of direct derivations beginning from the start graph $\rho = \{G_{i-1} \Rightarrow_{q_{i-1}} G_i\}_{i \in \{1,\dots,n\}}$, with $G_0 = G_s$: in this case we write $G_s \Rightarrow^*_{\mathcal{G}} G_n$. A T-typed graph G is reachable in $\mathcal{G}$ if $G_s \Rightarrow^*_{\mathcal{G}} G$.

We will consider only GTSs where all productions are *consuming*, and derivations where matches are *valid*. The restriction to consuming productions is standard in the framework of semantics combining concurrency and nondeterminism (see, e.g., [19,33]). On the other hand, considering valid matches only is needed to have a computational interpretation which is resource-conscious, i.e., where a resource can be consumed only once. In Sections 4 and 5 we shall further restrict to node-preserving productions, for the reasons explained there.

Example 5. Consider the GTS $\mathcal{CP}$ (a variation of the running example of [5]), modeling a system where three *processes* of type P are connected to a *communication manager* of type CM (see the start graph in Fig. 1, where edges are represented as rectangles and nodes as small circles). Two processes may establish a new connection with each other via the communication manager, becoming *processes engaged* in communication (typed PE). This transformation is modelled by the production [engage] in Fig. 1: observe that a new node connecting the two processes is created. The second production [release] terminates the communication between two partners. A typed graph G over $T_{\mathcal{CP}}$ is drawn by labeling each edge or node x of G with "$: t_G(x)$". Only when the same graphical item x belongs to both the left- and the right-hand side of a production we include its identity in the label (which becomes "$x : t_G(x)$"): in this case we also shade the item, to stress that it is preserved by the production.

2.2 Nondeterministic Occurrence Grammars

Conceptually, a *nondeterministic occurrence grammar* $\mathcal{O}$ is a structure that can be used to provide a static description of the computations of a given GTS $\mathcal{G}$: each production of $\mathcal{O}$ represents an *event*, i.e., a specific application of a production of $\mathcal{G}$, while the items of the type graph of $\mathcal{O}$ represent items of graphs reachable in derivations of $\mathcal{G}$. Analogously to what happens for Petri nets, occurrence grammars are "safe" GTSs, where the dependency relations between productions satisfy suitable acyclicity and well-foundedness requirements. The notion of safe GTS [13] generalizes the one for P/T nets which requires that each place contains at most one token in any reachable marking.

Definition 6 (safe GTS). *A GTS $\mathcal{G} = \langle T, G_s, P, \pi \rangle$ is safe if, for all H such that $G_s \Rightarrow^* H$, H is injective.*

In a safe GTS, each graph G reachable from the start graph is injectively typed, and thus we can identify it with the corresponding subgraph $t_G(|G|)$ of the type graph. With this identification, a production can be applied in G only to the subgraph of the type graph which is the image via the typing morphism of its left-hand side. Thus, according to its typing, we can safely think that a production *produces*, *preserves* or *consumes* items of the type graph. Using a net-like language, we speak of *pre-set* $^\bullet q$, *context* $\underline{q}$ and *post-set* $q^\bullet$ of a production q, defined in the obvious way. Clearly, the items of the type graph which are used by more productions may induce certain dependencies among them: this is formalized by the *causality* and *asymmetric conflict* relations introduced next, which are pivotal for the definition of occurrence grammars.

Definition 7 (causal relation). *The* causal relation *of a grammar $\mathcal{G}$ is the binary relation $<$ over $Elem(\mathcal{G})$ defined as the least transitive relation satisfying: for any node or edge x in the type graph T, and for productions $q, q' \in P$*

1. *if $x \in {}^\bullet q$ then $x < q$;*
2. *if $x \in q^\bullet$ then $q < x$;*
3. *if $q^\bullet \cap \underline{q'} \neq \emptyset$ then $q < q'$.*

As usual $\leq$ is the reflexive closure of $<$. Moreover, for $x \in Elem(\mathcal{G})$ we denote by $\lfloor x \rfloor$ the set of causes of x in P, namely $\{q \in P : q \leq x\}$.

Note that the fact that an item is preserved by q and consumed by q', i.e., $\underline{q} \cap {}^\bullet q' \neq \emptyset$, does not imply $q < q'$. Instead, such productions are in *asymmetric conflict* (see [8,28,23]): The application of q' prevents q from being applied, so that when both q and q' occur in a derivation, then q must precede q'.

Definition 8 (asymmetric conflict). *The* asymmetric conflict relation *of a grammar $\mathcal{G}$ is the binary relation $\nearrow$ over the set of productions, defined by:*

1. *if $\underline{q} \cap {}^\bullet q' \neq \emptyset$ then $q \nearrow q'$;*
2. *if ${}^\bullet q \cap {}^\bullet q' \neq \emptyset$ and $q \neq q'$ then $q \nearrow q'$;*
3. *if $q < q'$ then $q \nearrow q'$.*

Condition 1 is justified by the discussion above. Condition 2 essentially expresses the fact that the ordinary symmetric conflict is encoded, in this setting, as an asymmetric conflict in both directions. Finally, since $<$ represents a global order of execution, while $\nearrow$ determines an order of execution only locally to each computation, it is natural to impose $\nearrow$ to be an extension of $<$ (condition 3).

Definition 9 ((nondeterministic) occurrence grammar). *A (nondeterministic) occurrence grammar is a grammar $\mathcal{O} = \langle T, G_s, P, \pi \rangle$ such that*

1. *its causal relation $\leq$ is a partial order, and, for any $q \in P$, the set $\lfloor q \rfloor$ is finite and the asymmetric conflict $\nearrow$ is acyclic on $\lfloor q \rfloor$;*
2. *the start graph G_s is the set $Min(\mathcal{O})$ of minimal elements of $\langle Elem(\mathcal{O}), \leq \rangle$ (with the graphical structure inherited from T and typed by the inclusion);*
3. *any item x in T is created by at most one production in P, namely $|{}^{\bullet}x| \leq 1$;*
4. *for each $q \in P$, the typing t_{L_q} is injective on the "consumed part" $|L_q| - |dom(r_q)|$, and t_{R_q} is injective on the "produced part" $|R_q| - r_q(|dom(r_q)|)$.*

Since the start graph of an occurrence grammar $\mathcal{O}$ is determined by $Min(\mathcal{O})$, we often do not mention it explicitly. It is possible to show that, by the defining conditions, each occurrence grammar is safe. Intuitively, conditions 1–3 recast in the framework of graph grammars the analogous conditions of occurrence nets (actually of occurrence contextual nets [8]). In particular, in condition 1, the acyclicity of asymmetric conflict on $\lfloor q \rfloor$ corresponds to the requirement of irreflexivity for the conflict relation in occurrence nets. Condition 4, instead, is closely related to safety and requires that each production consumes and produces items with multiplicity one.

The finite computations of an occurrence grammar are characterized by specific subsets of productions.

Definition 10 (configuration). *Let $\mathcal{O} = \langle T, P, \pi \rangle$ be an occurrence grammar. A configuration of $\mathcal{O}$ is a finite subset of productions $C \subseteq P$ such that $\nearrow$ is acyclic on C, and for any $q \in C$, $\lfloor q \rfloor \subseteq C$. Given two configurations C and C' we write $C \sqsubseteq C'$ if $C \subseteq C'$ and for any $q \in C$, $q' \in C'$, if $q' \nearrow q$ then $q' \in C$. The set of all configurations of $\mathcal{O}$, ordered by $\sqsubseteq$, is denoted by $Conf(\mathcal{O})$.*

The intuition that a configuration represents a computation from the start state is formalised by the next result (see Proposition 6.11 of [1]), which also provides a "static" characterisation of the graph reached by such a derivation.

Proposition 11 (reachability of graphs generated by configurations). *Let $\mathcal{O}$ be an occurrence grammar, let $C \in Conf(\mathcal{O})$ be a configuration and let*

$$gr(C) = (Min(\mathcal{O}) \cup \bigcup_{q \in C} q^{\bullet}) - \bigcup_{q \in C} {}^{\bullet}q.$$

Then $gr(C)$ is reachable in $\mathcal{O}$ by applying all the productions of C in any order compatible with $\nearrow$.

As in the case of Petri nets, reachable states can be characterized in terms of a concurrency relation: this is an easy consequence of Proposition 11.

Definition 12 (concurrent graph). *Let $\mathcal{O} = \langle T, P, \pi \rangle$ be an occurrence grammar. A subgraph G of T is called* concurrent *if (1) the asymmetric conflict $\nearrow$ restricted to $\bigcup_{x \in G} \lfloor x \rfloor$ is acyclic and finitary; and (2) $\neg(x < y)$ for all $x, y \in G$.*

Proposition 13 (concurrency vs. reachability). *Let $\mathcal{O} = \langle T, P, \pi \rangle$ be an occurrence grammar and G be a subgraph of T. Then G is concurrent iff it is a subgraph of a graph reachable in $\mathcal{O}$ by applying all the productions in $\bigcup_{x \in G} \lfloor x \rfloor$ in any order compatible with $\nearrow$.*

2.3   Unfolding Construction

This section presents the unfolding construction which, applied to an SPO GTS $\mathcal{G}$, produces a nondeterministic occurrence grammar $\mathcal{U}_s(\mathcal{G})$ describing its behaviour. The idea is to begin with the start graph of the GTS, and to apply in all possible ways its productions to concurrent subgraphs, recording in the unfolding each occurrence of a production and each new graph item generated.

A basic ingredient of the construction is the *gluing* operation. It can be seen as a "partial application" of a production to a given match, in the sense that it generates the new items as specified by the production, but items that should have been deleted are not affected: intuitively, this is because such items may still be used by another production in the nondeterministic unfolding.

Definition 14 (gluing). *Let $q = r_q : L_q \rightharpoonup R_q$ be a production, G a graph and $m : L_q \to G$ a graph morphism. For any symbol z, we denote by $glue_z(q, m, G)$ the graph $\langle V, E, s, t \rangle$, where $V = V_G \cup m_z(V_{R_q})$, $E = E_G \cup m_z(E_{R_q})$, and m_z is defined by: $m_z(x) = m(x)$ if $x \in dom(r_q)$ and $m_z(x) = \langle x, z \rangle$ otherwise. The connection function and the typing are inherited from G and R_q.*

Therefore the gluing operation keeps unchanged the identity of the items already in G, and records in each newly added item from R_q the given symbol z.

The unfolding of a GTS is obtained as the union of a chain of occurrence grammars, each approximating the unfolding up to a certain causal depth.

Definition 15 (unfolding). *Let $\mathcal{G} = \langle T, G_s, P, \pi \rangle$ be a GTS. We inductively define, for each n, an occurrence grammar $\mathcal{U}_s(\mathcal{G})^{[n]} = \langle T^{[n]}, P^{[n]}, \pi^{[n]} \rangle$ and a pair of mappings $\varphi^{[n]} = \langle \varphi_T{}^{[n]} : T^{[n]} \to T, \varphi_P{}^{[n]} : P^{[n]} \to P \rangle$. Then the* unfolding $\mathcal{U}_s(\mathcal{G})$ *and the* folding morphism $\varphi_{\mathcal{G}} : \mathcal{U}_s(\mathcal{G}) \to \mathcal{G}$ *are the occurrence grammar and the morphism defined as the componentwise unions of $\mathcal{U}_s(\mathcal{G})^{[n]}$ and $\varphi^{[n]}$.*

($n = 0$) *The components of the grammar $\mathcal{U}_s(\mathcal{G})^{[0]}$ are $T^{[0]} = |G_s|$, $P^{[0]} = \pi^{[0]} = \emptyset$. Morphism $\varphi^{[0]} : \mathcal{U}_s(\mathcal{G})^{[0]} \to \mathcal{G}$ is defined by $\varphi_T{}^{[0]} = t_{G_s}$, $\varphi_P{}^{[0]} = \emptyset$.*

($n \to n + 1$) *The occurrence grammar $\mathcal{U}_s(\mathcal{G})^{[n+1]}$ is obtained by extending $\mathcal{U}_s(\mathcal{G})^{[n]}$ with all the possible production applications to concurrent subgraphs of its type graph. More precisely, let $M^{[n]}$ be the set of pairs $\langle q, m \rangle$ such that $q \in P$ is a production in $\mathcal{G}$, $m : L_q \to \langle T^{[n]}, \varphi_T{}^{[n]} \rangle$ is an injective match and $m(|L_q|)$ is a concurrent subgraph of $T^{[n]}$. Then $\mathcal{U}_s(\mathcal{G})^{[n+1]}$ is the occurrence grammar resulting after performing the following steps for each $\langle q, m \rangle \in M^{[n]}$.*

- *Add to $P^{[n]}$ the pair $\langle q, m \rangle$ as a new production name and extend $\varphi_P{}^{[n]}$ so that $\varphi_P{}^{[n]}(\langle q, m \rangle) = q$. Intuitively, $\langle q, m \rangle$ represents an occurrence of q, where the match m is needed to record the "history".*

- *Extend the type graph $T^{[n]}$ by adding to it a copy of each item generated by the application q, marked by $\langle q, m \rangle$ (in order to keep trace of the history). The morphism $\varphi_T{}^{[n]}$ is extended consequently. Formally, the T-typed graph $\langle T^{[n]}, \varphi_T{}^{[n]} \rangle$ is replaced by $\mathrm{glue}_{\langle q, m \rangle}(q, m, \langle T^{[n]}, \varphi_T{}^{[n]} \rangle)$.*

- *The production $\pi^{[n]}(\langle q, m \rangle)$ has the same untyped components as $\pi(q)$. The typing of the left-hand side is determined by m, and each item x in $|R_q| - r_q(|dom(r_q)|)$ is typed over the new item $\langle x, \langle q, m \rangle \rangle$ of the type graph.*

The most relevant property of the unfolding is the fact that it provides a compact representation of the behaviour of $\mathcal{G}$, and in particular it represents all the graphs reachable in $\mathcal{G}$, in the following sense. If T' is the type graph of the unfolding of $\mathcal{G}$, $\varphi_T : T' \to T$ is the type graph component of the folding morphism, and G is a subgraph of T', let us denote by $\varphi_T(G)$ the same graph, but typed over T by the restriction of the folding morphism, i.e., $\varphi_T(G) = \langle G, \varphi_{T|G} \rangle$. Then the next result is an easy consequence of the characterization of the unfolding as a right adjoint, shown in [9].

Theorem 16 (completeness of the unfolding). *Let $\mathcal{G} = \langle T, G_s, P, \pi \rangle$ be a* GTS. *A T-typed graph G is reachable in $\mathcal{G}$ iff there exists a configuration C in $Conf(\mathcal{U}(\mathcal{G}))$ such that $G \simeq \varphi_T(gr(C))$.*

3 Functorial Semantics: From Nets to SPO Grammars

In this section we discuss the role played by the unfolding construction in the development of a functorial semantics, first for Petri nets and then for GTSs.

3.1 A Coreflective Semantics for Petri Nets

In the theory of Petri Nets, the unfolding construction, whose generalization to SPO grammars has been presented in the previous section, is the cornerstone of a functorial semantics which has been developed by Winskel in [33], based on previous works with Nielsen and Plotkin [27]. Winskel shows that there is a chain of categorical coreflections (a special kind of adjunction), leading from the category **S-N**, having safe (marked) P/T nets as objects and suitably defined morphisms, to the category **Dom** of finitary prime algebraic domains, through the categories **O-N** of occurrence nets and **PES** of prime event structures (PESs).

$$\textbf{S-N} \; \underset{\mathcal{U}}{\overset{\mathcal{I}_{Occ}}{\rightleftarrows}} \; \textbf{O-N} \; \underset{\mathcal{E}}{\overset{\mathcal{N}}{\rightleftarrows}} \; \textbf{PES} \; \underset{\mathcal{L}}{\overset{\mathcal{P}}{\rightleftarrows}} \; \textbf{Dom}$$

The first step is the construction unwinding a safe net $N \in \textbf{S-N}$ into its unfolding $\mathcal{U}(N)$ which, as in the case of grammars, records in its branching and acyclic structure all the possible computations of the original net N. Every possible transition occurrence (*event*) is identified uniquely in the unfolding by its *history*, i.e., by the finite set of events which caused it, and events are related by the causality and *symmetric* conflict relations induced by the intersections of the pre- and post-sets: differently from the case of GTSs, in a (safe) Petri net all conflicts are symmetric because transitions do not have a context. Functor $\mathcal{U}(N) : \mathbf{S\text{-}N} \to \mathbf{O\text{-}N}$ is the right adjoint to the inclusion functor $\mathbf{O\text{-}N} \hookrightarrow \mathbf{S\text{-}N}$.

The subsequent step abstracts an occurrence net O to a PES $\mathcal{E}(O)$, which is obtained from the unfolding simply by forgetting the places, and remembering only the events and the causality and conflict relations among them. From a prime event structure E it is possible to generate freely an occurrence net $\mathcal{N}(E)$ which is the "most general" among those having E as underlying PES. Such a net is obtained by considering the events of E as transition occurrences, and introducing, among others, one fresh place for every pair of events related by causality or conflict in E, in order to enforce the same relationships in $\mathcal{N}(E)$. This construction defines a left adjoint to functor $\mathcal{E} : \mathbf{PES} \to \mathbf{O\text{-}N}$. The last step, which establishes an equivalence between the categories $\mathbf{PES}$ and $\mathbf{Dom}$, maps any event structure to its domain of configurations.

3.2   Coreflective Semantics: From Nets to SPO Grammars

During several years the first two authors cooperated with Ugo Montanari in a project aimed at generalizing the coreflective semantics of nets to graph grammars. At the beginning, most of the efforts were concentrated on the *double-pushout approach* to graph transformation, and partial results were reported in [1,7]. Only quite recently, however, a complete Winskel's style coreflective semantics has been developed successfully for the SPO approach, as reported in [9], and summarized by the following chain of adjunctions:

$$\mathbf{SPO\text{-}GG} \xleftarrow[\ \mathcal{U}_s\ ]{\perp} \mathbf{SPO\text{-}OG} \xleftarrow[\ \mathcal{E}_s\ ]{\mathcal{N}\ \ \perp} \mathbf{AES} \xleftarrow[\ \mathcal{L}_a\ ]{\mathcal{P}_a\ \ \perp} \mathbf{Dom}$$

Without delving into technical details, we summarize here the most challenging problems we had to address during our project, and the way we faced them.

Obviously, the starting point is Definition 15, describing the unfolding construction at the level of objects. To extend it to a functor $\mathcal{U}_s$ providing a right adjoint to the inclusion of the category $\mathbf{SPO\text{-}OG}$ of occurrence grammars into the category $\mathbf{SPO\text{-}GG}$ of general grammars, we first had to identify a sensible definition of grammar morphism. The chosen notion, discussed in [1], is more general than others proposed in the literature, and, unlike others, it coincides with the Petri net morphisms of [33] when restricted to graph grammars which represent Petri nets. Furthermore, inspired by corresponding results for nets in [26], we considered *semi-weighted* grammars, a class that strictly includes safe grammars, with the additional advantage of being characterized by a "static condition", not involving the behaviour but just the structure of the grammar.

Concerning the next adjunction in the chain, it is worth stressing here the presence of the category of *asymmetric* event structures ($\mathbf{AES}$) in place of $\mathbf{PES}$.

The point is that *prime* event structures, which only include the causality and *symmetric* conflict relations, are not sufficient to capture properly the dependencies among events of systems where productions may have a context, modeling the read-only access to resources. In these cases, which include SPO grammars and *contextual Petri nets*, asymmetric conflicts (see Definition 8) arise as a primitive concept. This motivated the introduction of asymmetric event structures (which are equipped with causality and *asymmetric* conflict), and the study of their category **AES**, which is shown to contain **PES** as a coreflective subcategory (see [8]). It is worth mentioning that, with the goal of providing an event structure semantics for nominal calculi, in [12] a simpler functorial semantics is presented for a restricted class of *persistent* grammars, for which category **PES** turns out to be sufficient, because asymmetric conflicts cannot arise.

Given an occurrence grammar $\mathcal{O}$, the AES $\mathcal{E}_s(\mathcal{O})$ is obtained by considering the productions as events, equipped with causality and asymmetric conflict as for Definitions 7 and 8. Moreover, a construction inspired by the work on contextual nets [8] allows one to build a canonical occurrence SPO graph grammar $\mathcal{N}(A)$ from a given asymmetric event structure A, providing a left-adjoint to functor $\mathcal{E}_s$ (technically, this works when dealing with injective matches only).

The chain of coreflection is completed by using the fact that the equivalence between **PES** and **Dom** generalizes to a coreflection between **AES** and **Dom** [8].

4   Verification of Finite State GTSs

In the approach originally proposed by McMillan for the analysis of Petri nets [25] and further developed by many authors (see, e.g., [17,18,32]) the idea is that given a finite state net, it is possible to identify a *finite* fragment of its unfolding which is *complete*, i.e., which provides full information about the system as far as reachability properties are concerned: this fragment can be characterized as the maximal prefix of the unfolding not including *cut-off events*, i.e., transitions which do not contribute to generating new markings.

In this section we summarize [5], where by exploiting the unfolding construction of Section 2.3, we have generalized McMillan's approach to SPO GTSs by introducing an original notion of "strong cut-off" (which takes into account the fact that a production can have several different histories), and we have shown how a finite complete prefix of the unfolding can be used to verify interesting properties of the graphs reachable in the GTS.

4.1   Rewriting up to Isolated Nodes

In the work on verification of graph transformation systems summarized in this and in the next section, we consider only systems consisting of node-preserving productions, as introduced in Definition 3, and rewriting *up to isolated nodes* (graphs which are isomorphic after deleting all isolated nodes are considered indistinguishable). As far as the expressive power is concerned, this is a mild restriction, since the deletion of a node can usually be modelled by leaving it isolated: Conditions (ii) and (iii) of Definition 3 guarantee that isolated nodes do not take part to the rewriting. Also, this is consistent with the fact that the logic we shall use for verification purposes (see Definition 22) is not able to distinguish graphs which are isomorphic up to isolated nodes.

In the rest of the paper, we will assume that all productions are node preserving. Moreover, given any graph G and any subset of edges $X \subseteq E_G$, we denote by $graph(X)$ the smallest subgraph of G having X as set of edges, and we say that G and G' are *isomorphic up to isolated nodes*, denoted $G \cong G'$, if $graph(E_G) \simeq graph(E_{G'})$. Finally, for a fixed $n \in \mathbb{N}$, we say that a GTS $\mathcal{G}$ is *n-bounded* if for each graph H reachable in $\mathcal{G}$ there is an n-injective graph H' such that $H' \cong H$, and a GTS is *bounded* or *finite state* if it is n-bounded for some $n \in \mathbb{N}$.

4.2   Finite Complete Prefix of Bounded GTSs

A *history* of a production in a computation is the set of all the events which must precede its application. Due to the presence of asymmetric conflicts, a production q does not have a unique history in general, because depending on the specific computation we consider, some of the productions in asymmetric conflict with q might have been applied or not before q.

Definition 17 (history). *Let $\mathcal{O}$ be an occurrence grammar, let $C \in Conf(\mathcal{O})$ be a configuration and let $q \in C$. The* history *of q in C is the set of events $C[\![q]\!] = \{q' \in C : q' \nearrow_C^* q\}$, where $\nearrow_C$ is the restriction of $\nearrow$ to C. We denote by $Hist(q)$ the set of histories of q, i.e., $Hist(q) = \{C[\![q]\!] : C \in Conf(\mathcal{O})\}$.*

Now, let $\mathcal{G} = \langle T, G_s, P, \pi \rangle$ denote a GTS, fixed throughout the section, and let $\mathcal{U}_s(\mathcal{G}) = \langle T', P', \pi' \rangle$ be its unfolding with $\varphi : \mathcal{U}_s(\mathcal{G}) \to \mathcal{G}$ the folding morphism, as in Definition 15. In order to identify a finite and complete prefix of the unfolding of a bounded GTS, the idea is to avoid useless productions in the unfolding, i.e., productions which do not contribute to generating new graphs. The definition of "cut-off event" introduced by McMillan for Petri nets in order to formalize such a notion has to be adapted to this context since, as explained above, for graph grammars a production may have different histories.

Definition 18 (cut-off). *A production $q \in P'$ of the unfolding $\mathcal{U}_s(\mathcal{G})$ is a cut-off if there exists $q' \in P'$ such that $\varphi_T(gr(\lfloor q \rfloor)) \cong \varphi_T(gr(\lfloor q' \rfloor))$ and $\|\lfloor q' \rfloor\| < \|\lfloor q \rfloor\|$.*

A production q is a strong cut-off *if for all $C_q \in Hist(q)$ there exist $q' \in P'$ and $C_{q'} \in Hist(q')$ such that $\varphi_T(gr(C_q)) \cong \varphi_T(gr(C_{q'}))$ and $|C_{q'}| < |C_q|$. The* truncation *of $\mathcal{G}$ is the greatest prefix $\mathcal{T}(\mathcal{G})$ of $\mathcal{U}_s(\mathcal{G})$ not including strong cut-offs.*

Theorem 19 (completeness and finiteness of the truncation). *The truncation $\mathcal{T}(\mathcal{G})$ is a complete prefix of the unfolding, i.e., for any reachable graph G of $\mathcal{G}$ there is a configuration C in $Conf(\mathcal{T}(\mathcal{G}))$ such that $\varphi_T(gr(C)) \cong G$. Furthermore, if $\mathcal{G}$ is bounded then the truncation $\mathcal{T}(\mathcal{G})$ is finite.*

Unfortunately, neither the statement of the above theorem nor its proof (see Appendix B of the full version of [5]) suggest a way to modify the unfolding construction of Definition 15 in order to obtain the truncation of a bounded

GTS: this is because the notion of strong cut-off refers to the set of histories of a production, that in general could be infinite. Only recently the authors proposed an algorithm (see [6]) which solves a similar problem in the simpler case of *contextual* Petri nets: we are confident that this algorithm can be adapted to GTSs, but space constraints do not allow us to elaborate on that here.

Instead, a class of GTSs can be identified for which an obvious adaptation of the unfolding construction does produce a finite complete prefix. It is characterized by a property called "read-persistence", since it appears as the graph grammar theoretical version of the notion introduced for contextual nets in [32].

Definition 20 (read-persistence). *An occurrence grammar $\mathcal{O} = \langle T, P, \pi \rangle$ is called* read-persistent *if for any $q_1, q_2 \in P$, if $q_1 \nearrow q_2$ then either q_1 is a cause of q_2, or q_1 and q_2 are in* conflict, *i.e., they cannot fire in the same derivation. A GTS $\mathcal{G}$ is called* read-persistent *if its unfolding $\mathcal{U}(\mathcal{G})$ is read-persistent.*

An adaptation of the algorithm originally proposed in [25] for ordinary nets and extended in [32] to read-persistent contextual nets, works for read-persistent GTSs as well, because in this case every production has a single history, and thus the notions of cut-off and of strong cut-off of Definition 18 coincide. Roughly, for such GTSs a complete finite prefix can be obtained by slightly modifying the inductive step of the construction of Definition 15 as follows: the production occurrences in the set $M^{[n]}$ have to be handled in increasing order according to the size of the corresponding history, and a production occurrence has to be added to the unfolding only if it is not a cut-off in the prefix computed so-far.

An obvious class of read-persistent systems consists of those GTSs where any edge preserved by productions is never consumed. For instance, the GTS $\mathcal{CP}$ in our running example is read-persistent, since CM, the only edge preserved by productions, is never consumed. Its truncation is the grammar $\mathcal{T}(\mathcal{CP})$ depicted in Fig. 2. Denote by $T_{\mathcal{T}}$ its type graph. Note that applying the production [release] to any subgraph of $T_{\mathcal{T}}$ matching its left-hand side would result in a cut-off: this is the reason why $\mathcal{T}(\mathcal{CP})$ does not include any instance of production [release]. The start graph of the truncation is isomorphic to the start graph of GTS $\mathcal{CP}$ and it is mapped injectively to the graph of types $T_{\mathcal{T}}$ in the obvious way.

4.3   Checking Properties of Reachable Graphs

Given a finite state GTS $\mathcal{G}$, a complete prefix can be used to check whether there exists at least one reachable graph satisfying a certain property F, or if a property F is an "invariant" of $\mathcal{G}$, i.e., it is satisfied by all reachable graphs. The graph properties of interest will be expressed as formulae of a quite expressive logic called $\mathcal{L}2$ (introduced below), whose induced logical equivalence on finite graphs is "isomorphism up to isolated nodes". That is, two finite graphs G and G' satisfy exactly the same formulae of $\mathcal{L}2$ if and only if $G \cong G'$.

Now, the usefulness of the truncation (or of any other finite complete prefix) resides in the fact that since for each graph G reachable in $\mathcal{G}$ there is a G' reachable in $\mathcal{T}(\mathcal{G})$ such that $\varphi_T(G') \cong G$, it is sufficient to consider the graphs

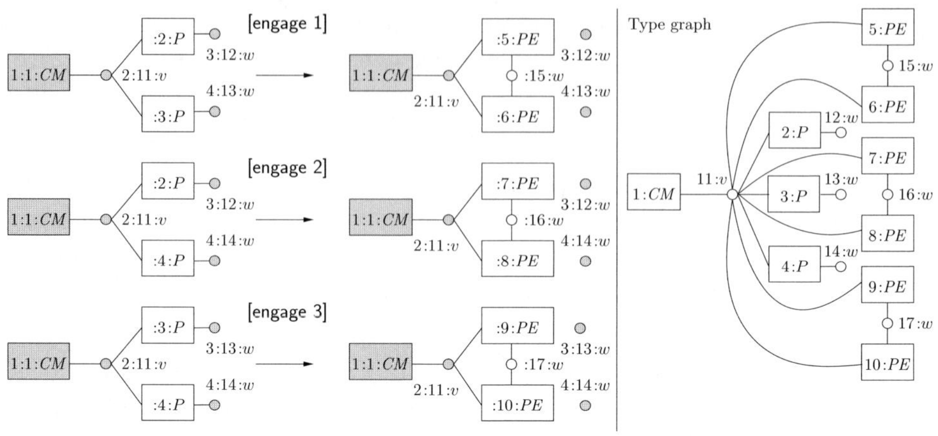

Fig. 2. The truncation $\mathcal{T}(\mathcal{CP})$ of the GTS in Fig. 1

reachable in $\mathcal{T}(\mathcal{G})$, retyped over T via $\varphi_T(\cdot)$. But we know that $\mathcal{T}(\mathcal{G})$ is an occurrence grammar, and thus the reachable graphs can be identified with subgraphs of its type graph, which in turn are uniquely identified by their sets of edges, because we rewrite up to isolated nodes.

Therefore, we can verify if F holds for all (some) reachable graphs by checking that $\varphi_T(graph(m)) \models F$ for all (some) sets of edges m reachable in $\mathcal{T}(\mathcal{G})$. This fact can be formalized in a convenient way by introducing a (safe) Petri net "underlying" the truncation, and seeing a set of edges of $\mathcal{T}(\mathcal{G})$ as a marking of such net. Furthermore, since this net is fixed and finite, it is possible to translate every formula $F \in \mathcal{L}2$ into a propositional formula over markings $M[F]$ such that, for any reachable marking m,

$$\varphi_T(graph(m)) \models F \quad \text{iff} \quad m \models M[F].$$

In this way the original problem is reduced to a verification problem of a formula over a Petri net, for which existing tools could be used.

Given an occurrence grammar $\mathcal{O}$, the underlying Petri net is an occurrence contextual net (i.e., a Petri net with read arcs, see, e.g., [8,32]).

Definition 21 (Petri net underlying an occurrence grammar). *The contextual Petri net underlying an occurrence grammar* $\mathcal{O} = \langle T', P', \pi' \rangle$, *denoted by* $\mathsf{Net}(\mathcal{O})$, *is the safe Petri net having the set of edges* $E_{T'}$ *as places and a transition for every production* $q \in P'$, *with pre-set* $^\bullet q \cap E_{T'}$, *post-set* $q^\bullet \cap E_{T'}$ *and context* $\underline{q} \cap E_{T'}$.

For instance, the Petri net $\mathsf{Net}(\mathcal{T}(\mathcal{CP}))$ underlying the truncation of $\mathcal{CP}$ (see Fig. 2) is depicted in Fig. 3. Read arcs are represented as undirected lines.

Next we define the monadic second-order logic $\mathcal{L}2$ used for specifying properties of graphs typed over a fixed graph T. Quantification is allowed over edges, but not over nodes (as in [14]).

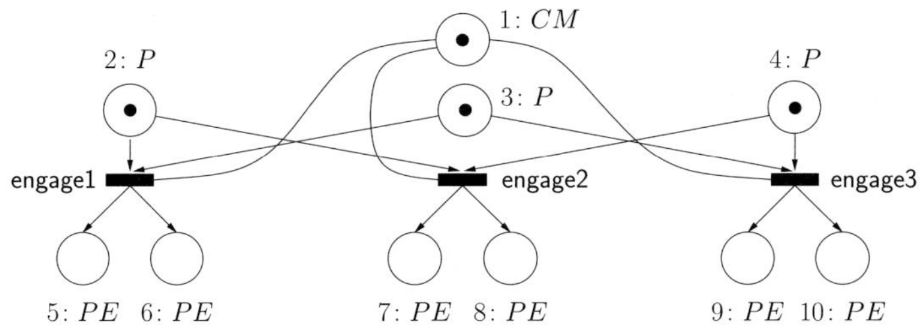

Fig. 3. The Petri net underlying the truncation $\mathcal{T}(\mathcal{CP})$ in Fig. 2

Definition 22 (graph logic). *Let $\mathcal{X}_1 = \{x, y, \ldots\}$ be a set of (first-order) edge variables and let $\mathcal{X}_2 = \{X, Y, \ldots\}$ be a set of (second-order) variables representing edge sets. The set of* graph formulae *of the logic $\mathcal{L}2$ is defined as follows, where $\ell \in E_T$, $i, j \in \mathbb{N}$:*

$$F \ ::= \ x = y \ \mid \ c_i(x) = c_j(y) \ \mid \ type(x) = \ell \ \mid \ x \in X \qquad \textit{(Predicates)}$$
$$F \vee F \ \mid \ \neg F \ \mid \ \exists x.F \ \mid \exists X.F \qquad \textit{(Connectives / Quantifiers)}$$

We denote by $free(F)$ and $Free(F)$ the sets of first-order and second-order variables, respectively, occurring free in F, defined in the obvious way.

Given a T-typed graph G, a formula F in $\mathcal{L}2$, and two valuations $\sigma : free(F) \to E_{|G|}$ and $\Sigma : Free(F) \to \mathcal{P}(E_{|G|})$, the *satisfaction relation* $G \models_{\sigma,\Sigma} F$ is defined inductively in the usual way; for instance $G \models_{\sigma,\Sigma} x = y$ iff $\sigma(x) = \sigma(y)$, $G \models_{\sigma,\Sigma} type(x) = \ell$ iff $t_G(\sigma(x)) = \ell$, and $G \models_{\sigma,\Sigma} x \in X$ iff $\sigma(x) \in \Sigma(X)$.

Interesting graph properties can be expressed in $\mathcal{L}2$, such as the existence or adjacency of edges with specific typing, and the absence of certain paths or of certain cycles. Such properties may be used to represent in the graph transformation model relevant properties of the system at hand, such as security properties or deadlock-freedom.

Recall that a *marking* of a safe net is simply a subset of its places. The syntax of the formulae over markings is

$$Q ::= e \mid \neg Q \mid Q \wedge Q \mid Q \vee Q \mid \mathsf{true} \mid \mathsf{false},$$

where $e \in E_{T'}$. These formulae are interpreted over markings of $\mathsf{Net}(\mathcal{T}(\mathcal{G}))$: $m \models e$ if $e \in m$, and logical connectives are treated as usual.

As mentioned above, given a GTS $\mathcal{G}$ and the truncation $\mathcal{T}(\mathcal{G})$ (or any other a finite complete prefix), any formula $F \in \mathcal{L}2$ can be effectively translated to a marking formula $M[F]$ such that $\mathcal{G}$ satisfies F iff the Petri net $\mathsf{Net}(\mathcal{T}(\mathcal{G}))$ underlying the prefix satisfies $M[F]$. We omit the details of the translation, which can be found in [5], and we focus on the running example $\mathcal{CP}$. Suppose, e.g., that we want to check that all graphs reachable in our sample GTS $\mathcal{CP}$ satisfy Φ, where Φ is a $\mathcal{L}2$ formula specifying that every engaged process is connected through connection c_2 to exactly one other engaged process, i.e.,

$$\Phi \equiv \forall x.(type(x) = PE \Rightarrow \exists y.(x \neq y \wedge type(y) = PE \wedge c_2(x) = c_2(y)$$
$$\wedge \ \forall z.(type(z) = PE \wedge x \neq z \wedge c_2(x) = c_2(z) \Rightarrow y = z))).$$

The encoding procedure leads to the formula $Q = M[\Phi]$

$$Q \equiv (5\colon PE \iff 6\colon PE) \land (7\colon PE \iff 8\colon PE) \land (9\colon PE \iff 10\colon PE)$$

which has to be checked for all reachable markings of the net of Fig. 3. An efficient algorithm for checking if a marking formula is satisfied by at least one reachable marking of an (occurrence) net $\mathsf{Net}(\mathcal{T}(\mathcal{G}))$ is presented in [5]: it exploits the mutual relationships between items expressed by the causality, (asymmetric) conflict and concurrency relations.

5   Verification of Infinite State GTSs

If a GTS is not finite state, obviously no finite prefix of the unfolding can be complete in the sense of Theorem 19. In this section we describe a framework, developed in [4,10,11], where behavioural properties of systems described as (possibly infinite state) GTSs can be specified and verified. Here we consider rewriting up to isolated nodes (see Section 4.1), and further we require matches to be injective on edges.

Following the guidelines of the verification technique presented in the previous section, the framework is based on finite approximations of the unfolding of a given GTS which have an underlying Petri net structure. On these structures, formulae of a suitable temporal logic interpreted over derivations of a GTS can be verified, by first translating them to a simpler logic describing computations of a fixed Petri net.

5.1   Approximating the Behaviour of GTSs

A basic ingredient of our verification framework is a technique, proposed in [4,10], for approximating the behaviour of GTSs by means of finite Petri net-like structures. More precisely, an *approximated unfolding* construction allows to generate from a given GTS $\mathcal{G}$ (which can be infinite state) suitable finite structures, called *coverings*, which provide (over-)approximations of the behaviour of $\mathcal{G}$.

The coverings of a GTS $\mathcal{G}$ are *Petri graphs over $\mathcal{G}$*, i.e., (contextual) Petri nets equipped with an additional graphical structure where the places play the role of edges, while the transitions represent applications of the productions of $\mathcal{G}$.

In the following, given a set A we denote by $A^{\oplus}$ the free commutative monoid generated by A, whose elements are finite multisets of elements of A. If $f : A \to B$ is a function, then we denote by $f^{\oplus}\colon A^{\oplus} \to B^{\oplus}$ its extension to multisets.

Definition 23 (Petri graph). *Let $\mathcal{G} = \langle T, G_0, P, \pi \rangle$ be a GTS. A Petri graph K (over $\mathcal{G}$) is a tuple $\langle G, N, m_0 \rangle$ where G is a T-typed graph and*

- *$N = \langle E_G, T_N, {}^\bullet(\,), (\,)^\bullet, (\,), p_N \rangle$ is a place/transition Petri net, where the set of places E_G is the set of edges of G, T_N is the set of transitions, ${}^\bullet(\,), (\,)^\bullet, (\,)\colon T_N \to E_G^{\oplus}$ specify the pre-set, post-set and context of each transition and $p_N\colon T_N \to P$ is the labelling function, mapping each transition to a corresponding production;*

- $m_0 \in E_G^\oplus$ is the initial marking *of the Petri graph, satisfying* $m_0 = \iota^\oplus(E_{G_0})$ *for a suitable graph morphism* $\iota : G_0 \to G$ *(i.e., m_0 must properly correspond to the initial state of $\mathcal{G}$).*

We will write $m\,[q\rangle\,m'$ if a transition labelled by $q \in P$ is enabled at marking m and its firing produces m'. A marking is called reachable (coverable) *in K if it is reachable (coverable) from the initial marking in the Petri net underlying K.*

As an example, let $\mathcal{U}_s(\mathcal{G}) = \langle T', P', \pi' \rangle$ be the unfolding of a GTS $\mathcal{G} = \langle T, G_s, P, \pi \rangle$, and let $\langle \varphi_T : T' \to T, \varphi_P : P' \to P \rangle$ be the folding morphism, as presented in Definition 15. Then it is possible to see the unfolding as a Petri graph $\langle G, N, m_0 \rangle$ for $\mathcal{G}$: the net component N is as for Definition 21, the labeling of transitions is given by φ_P, G is the T-typed graph $\langle T', \varphi_T \rangle$, and m_0 is the set of minimal edges, with respect to causality, of G.

The coverings of a GTS $\mathcal{G}$ can approximate its behaviour at different levels of accuracy. For each $k \in \mathbb{N}$, the *k-covering* of $\mathcal{G}$, denoted $\mathcal{C}^k(\mathcal{G})$, over-approximates the behaviour of $\mathcal{G}$ in the sense that every derivation sequence of $\mathcal{G}$ is mapped to a valid firing sequence of (the Petri net component of) $\mathcal{C}^k(\mathcal{G})$, and every graph reachable from the start graph of $\mathcal{G}$ can be mapped homomorphically to (the graphical component of) $\mathcal{C}^k(\mathcal{G})$, and its image is reachable in the Petri graph. Furthermore, this over-approximation is exact up to causal depth k, in the sense that each graph reachable in $\mathcal{G}$ in at most k derivation steps can be mapped *injectively* to $\mathcal{C}^k(\mathcal{G})$ (see Section 5.2).

The algorithm for the construction of the k-covering of a GTS $\mathcal{G}$ works inductively like the unfolding construction, but the classical *unfolding steps*, where the application of a production to a given match is recorded by adding to the type graph the newly generated items and to the set of productions the new production occurrence, are interleaved with suitably defined *folding steps*, which merge in the graphical part of the current Petri graph two occurrences of the left-hand side of a production, if one causally depends on the other. The termination of the algorithm is ensured by giving higher priority to folding steps, and the exactness of the approximation up to causal depth k by forbidding the application of folding steps to items of smaller depth.

We define the *depth of a transition t* (an element of $\mathbb{N} \cup \{\infty\}$) to be the length of the longest sequence $t_0 < t_1 < \ldots < t_n < t$. The depth of an edge is the maximum among the depths of transitions which contain the edge in their post-set. The k-covering $\mathcal{C}^k(\mathcal{G})$ of a GTS $\mathcal{G} = \langle T, G_0, P, \pi \rangle$ is produced by the last step of the following (terminating) algorithm which generates a sequence $K_i = \langle G_i, N_i, m_i \rangle$ of Petri graphs over $\mathcal{G}$.

1. $K_0 = \langle G_0, N_0, m_0 \rangle$, where the net N_0 contains no transitions and $m_0 = E_{G_0}$.
2. As long as one of the following steps is applicable, transform K_i into K_{i+1}, giving precedence to folding steps.

   Unfolding. Find a production $q \in P$ with $\pi(q) : L_q \rightharpoonup R_q$ and a match $n : L_q \to G_i$ such that $n^\oplus(E_{L_q})$ is coverable in K_i. Then extend K_i by gluing R_q to G_i (as described in Definition 14) and add a new transition, labelled by q, representing the application of production q.

Folding. Find a production $q \in P$ with $\pi(q) : L_q \rightharpoonup R_q$ and two matches $n, n' : L_q \rightarrow G_i$, at depth greater than or equal to k, such that
  - $n^{\oplus}(E_L)$ and $n'^{\oplus}(E_L)$ are coverable in N_i and
  - the first match has been unfolded with the introduction of a transition t and the second match causally depends on t.

Then merge the two matches, by setting $n(e) \equiv n'(e)$ for each $e \in E_{L_q}$, and factoring all components of K_i through the equivalence relation induced by $\equiv$ on edges, nodes and transitions.

For example, if we extend the GTS $\mathcal{CP}$ of Example 5 with production [fork] (see Fig. 4) that models the forking of a non-engaged process, we obtain an infinite state system: in fact graphs with an unbounded number of processes are reachable. If we compute the coarsest approximation, i.e., the 0-covering, we obtain the Petri graph shown in Fig. 4 (edge and node identities are omitted):

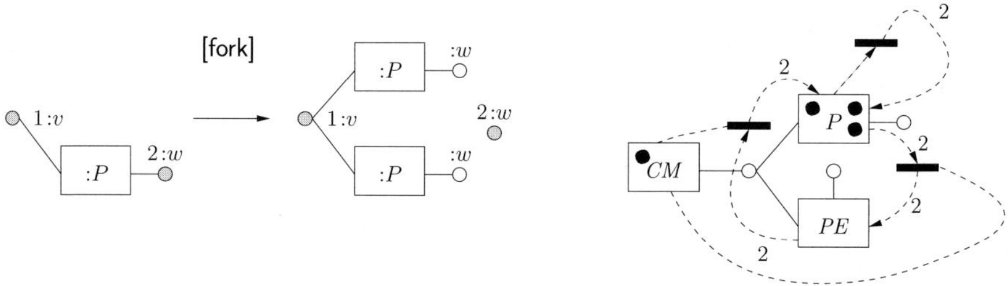

Fig. 4. Additional rule [fork] (left) and Petri graph over-approximating the GTS (right)

Several structures have been merged, for instance all engaged and all non-engaged processes. This happens because all three initial processes will be the cause of another future process, causing the fusion of all of them.

Despite the fact that the graph underlying the Petri graph is not very different from the type graph of the GTS, there are still some interesting properties of the system that can be proved by exploiting the 0-covering, using the techniques described in the next section:

  - There is always exactly one communication manager.
  - There will always be at least three processes (engaged or non-engaged).
  - The number of engaged processes is always even.
  - No engaged process is ever connected to a non-engaged process.

Due to the simplicity of the running example, these properties could easily be proved also as invariants of the transformation rules. For more complex examples we refer the reader to [4,2].

5.2   Verifying Behavioural Properties of GTSs

As mentioned above, the k-covering $\mathcal{C}^k(\mathcal{G})$ over-approximates the behaviour of the original GTS $\mathcal{G}$. In order to formalize this fact, we will first generalize the notion of a subgraph generated by a set of edges (introduced at the end of Section 4.1) to a graph generated by a marking: Let $\langle G, N, m_0 \rangle$ be a Petri graph and let $m \in E_G^\oplus$ be a marking of N. The *graph generated* by m, denoted by $graph(m)$, is the T-typed graph H without isolated nodes (which is unique up to isomorphism) such that there exists a T-typed morphism $\psi \colon H \to G$ injective on nodes with $\psi^\oplus(E_H) = m$.

Proposition 24 (simulation). *Let $\mathbf{G}$ be the set of graphs reachable from G_0 in $\mathcal{G}$ and let $\mathbf{M}$ be the set of reachable markings in $\mathcal{C}^k(\mathcal{G}) = \langle G, N, m_0 \rangle$. Then there exists a simulation $S \subseteq \mathbf{G} \times \mathbf{M}$ with the following properties:*

- *$(G_0, m_0) \in S$;*
- *whenever $(G', m') \in S$ and $G' \Rightarrow_q G''$, then there exists a marking m'' with $m' \, [q\rangle \, m''$ and $(G'', m'') \in S$;*
- *for every $(G', m') \in S$ there exists an edge-bijective graph morphism $\varphi \colon G' \to graph(m')$.*

The simulation relation just described, whose existence can be proved fairly easily by construction, allows one to exploit the finite k-covering $\mathcal{C}^k(\mathcal{G})$ to verify certain properties of the reachable graphs of GTS $\mathcal{G}$. In fact, if a given property over graphs $F \in \mathcal{L}2$ is *reflected* by edge-bijective graph morphisms (i.e., if $f : G \to G'$ is edge-bijective and $G' \models F$ then $G \models F$), then if F is satisfied by graph $graph(m)$ for all markings m reachable in $\mathcal{C}^k(\mathcal{G})$, it is also satisfied by all graphs reachable in $\mathcal{G}$.

A couple of considerations are in order here. First, unfortunately it is undecidable if a formula of $\mathcal{L}2$ is reflected by edge-bijective morphisms, but a syntactic, sufficient criterion based on a simple type system is presented in [11]. Second, the Petri net underlying the k-covering is finite, but in general it is not finite state. Nevertheless, several verification techniques and tools have been developed for the analysis of nets, and thus the possibility of reducing the verification of a property from reachable graphs of a GTS to reachable markings of a net is of high pragmatic value.

To this aim, following the guidelines we have described in Section 4.3 for finite state GTSs, first we introduced *multiset formulae* which are evaluated on markings of the k-coverings: their syntax is obtained by extending the one presented in Section 4.3 with the atomic formula $\#e \leq c$ for $e \in E_G$ and $c \in \mathbb{N}$, meaning *the number of tokens in e is smaller than or equal to c*. Next we have provided an encoding M_2 of $\mathcal{L}2$-formulae into multiset formulae, such that $graph(m) \models F \iff m \models M_2[F]$ for every reachable marking of $\mathcal{C}^k(\mathcal{G})$. This translation is a kind of quantifier elimination procedure, which is possible because the graph underlying $\mathcal{C}^k(\mathcal{G})$ is finite.

Finally, we enriched the verification framework with a temporal logic called $\mu\mathcal{L}2$, which is a propositional μ-calculus where atomic propositions are formulae of $\mathcal{L}2$. The formulae of $\mu\mathcal{L}2$ are interpreted over a *graph transition system*, i.e., a transition system where the states are graphs, and their syntax is the following:

$$M ::= \ F \mid X \mid \Diamond M \mid \Box M \mid \neg M \mid M_1 \vee M_2 \mid M_1 \wedge M_2 \mid \mu X.M \mid \nu X.M$$

where F ranges over closed formulae in $\mathcal{L}2$ and $X \in \mathcal{X}$ are proposition variables. Intuitively, an atomic proposition $F \in \mathcal{L}2$ holds in any state (graph) satisfying F according to the discussion after Definition 22. A formula $\Diamond M$ / $\Box M$ holds in a state if some / any single step leads to a state where M holds. The connectives $\neg, \vee, \wedge$ are interpreted in the usual way, and the formulae $\mu X.M$ and $\nu X.M$ represent the *least* and *greatest fixed point* over X, respectively.

Now, for suitable fragments of logic $\mu\mathcal{L}2$, e.g., the fragment $\Box\mu\mathcal{L}2$ without negation and the "possibility operator" $\Diamond$, by Proposition 24 and exploiting general results in [24], we can translate a temporal formula M over $\mathcal{G}$ where the atomic propositions are reflected by edge-bijective morphisms to a temporal formula $M_2[M]$ over markings (using for atomic propositions the encoding mentioned above), ensuring that if $\mathcal{C}^k(\mathcal{G}) \models M_2[M]$ then $\mathcal{G} \models M$, i.e., M is valid for the original GTS. We conclude by recalling that temporal state-based logics over Petri nets, i.e., logics where basic predicates have the form $\#s \leq c$, are not decidable in general, but important fragments of such logics are [20].

6   Conclusions

We presented an overview of the work on the unfolding semantics of GTSs, discussing its role for the development of a functorial concurrent semantics for GTSs and its possible applications to the verification of (infinite and finite state) systems modelled as GTSs. We used the SPO approach since due to the absence of the dangling condition it provides us with a more elegant unfolding semantics, but a large part of the theory can be equally developed for the DPO approach. For the approaches to verification, which deal with node-preserving grammars, the choice between SPO and DPO is immaterial.

The framework can appear fairly abstract and theoretical in nature. However, a prototype tool called AUGUR [21] has been implemented for computing the k-covering of a given graph transformation system. The current implementation is restricted to rules with discrete contexts. The tool can be downloaded at `http://www.ti.inf.uni-due.de/research/augur_1/`. The input and output of AUGUR is in GXL and GTXL, an XML standard for the exchange of graphs and graph transformation systems. Suitable converters have been added in order to visualize rules and Petri graphs and to extract the Petri net component of a Petri graph, which can then be analyzed with standard algorithms for nets.

Concerning the verification of finite state systems, the approach based on the construction of a finite complete prefix of the unfolding currently only applies to a special class of GTSs (read persistent GTSs). The problem of generalising the technique to the full class of GTSs is still open. An algorithm solving a similar problem in the simpler case of *contextual* Petri nets has been proposed recently [6] and we are confident that this can be adapted to GTSs.

A very stimulating direction of further research is the extension of the work on unfolding to the setting of rewriting systems over adhesive categories. Adhesive categories [22] have been recently introduced as an elegant and extremely general framework where the algebraic approaches to rewriting can be developed, encompassing rewriting on (several brands of) graphs and more general graphical structures like bigraphs or UML models. An unfolding theory for adhesive rewriting systems would thus apply uniformly to all these structures. Some promising steps have been taken in [3], which develops a concurrent semantics for adhesive rewriting systems based on deterministic processes.

References

1. Baldan, P.: Modelling concurrent computations: from contextual Petri nets to graph grammars. PhD thesis, Department of Computer Science, University of Pisa, Available as technical report n. TD-1/00 (2000)
2. Baldan, P., Corradini, A., Esparza, J., Heindel, T., König, B., Kozioura, V.: Verifying Red-Black Trees. In: Proc. of COSMICAH 2005, vol. RR-05-04, pp. 1–15. Queen Mary University, Dept. of Computer Science (2005)
3. Baldan, P., Corradini, A., Heindel, T., König, B., Sobociński, P.: Processes for Adhesive Rewriting Systems. In: Aceto, L., Ingólfsdóttir, A. (eds.) FOSSACS 2006 and ETAPS 2006. LNCS, vol. 3921, pp. 202–216. Springer, Heidelberg (2006)
4. Baldan, P., Corradini, A., König, B.: A static analysis technique for graph transformation systems. In: Larsen, K.G., Nielsen, M. (eds.) CONCUR 2001. LNCS, vol. 2154, pp. 381–395. Springer, Heidelberg (2001)
5. Baldan, P., Corradini, A., König, B.: Verifying finite-state graph grammars: an unfolding-based approach. In: Gardner, P., Yoshida, N. (eds.) CONCUR 2004. LNCS, vol. 3170, pp. 83–98. Springer, Heidelberg (2004); Full version as Tech.Rep.CS-2004-10, Dept.of Comp.Sci., University Ca' Foscari of Venice
6. Baldan, P., Corradini, A., König, B., Schwoon, S.: McMillan's Complete Prefix for Contextual Nets. In: ToPNoC - Trans. on Petri Nets and Other Models of Concurrency (to appear, 2008); Special Issue from PN 2007 Workshops and Tutorials
7. Baldan, P., Corradini, A., Montanari, U.: Unfolding and Event Structure Semantics for Graph Grammars. In: Thomas, W. (ed.) FOSSACS 1999. LNCS, vol. 1578, pp. 73–89. Springer, Heidelberg (1999)
8. Baldan, P., Corradini, A., Montanari, U.: Contextual Petri nets, asymmetric event structures and processes. Information and Computation 171(1), 1–49 (2001)
9. Baldan, P., Corradini, A., Montanari, U., Ribeiro, L.: Unfolding Semantics of Graph Transformation. Information and Computation 205, 733–782 (2007)
10. Baldan, P., König, B.: Approximating the behaviour of graph transformation systems. In: Corradini, A., Ehrig, H., Kreowski, H.-J., Rozenberg, G. (eds.) ICGT 2002. LNCS, vol. 2505, pp. 14–29. Springer, Heidelberg (2002)
11. Baldan, P., König, B., König, B.: A logic for analyzing abstractions of graph transformation systems. In: Cousot, R. (ed.) SAS 2003. LNCS, vol. 2694, pp. 255–272. Springer, Heidelberg (2003)
12. Bruni, R., Melgratti, H.C., Montanari, U.: Event structure semantics for nominal calculi. In: Baier, C., Hermanns, H. (eds.) CONCUR 2006. LNCS, vol. 4137, pp. 295–309. Springer, Heidelberg (2006)
13. Corradini, A., Montanari, U., Rossi, F.: Graph processes. Fundamenta Informaticae 26, 241–265 (1996)
14. Courcelle, B.: The expression of graph properties and graph transformations in monadic second-order logic. In: Rozenberg [31]
15. Ehrig, H., Heckel, R., Korff, M., Löwe, M., Ribeiro, L., Wagner, A., Corradini, A.: Algebraic Approaches to Graph Transformation II: Single Pushout Approach and comparison with Double Pushout Approach. In: Rozenberg [31]

16. Ehrig, H., Kreowski, H.-J., Montanari, U., Rozenberg, G.: Handbook of Graph Grammars and Computing by Graph Transformation: Concurrency, Parallelism and Distribution, vol. 3. World Scientific, Singapore (1999)
17. Esparza, J.: Model checking using net unfoldings. Science of Computer Programming 23(2–3), 151–195 (1994)
18. Esparza, J., Römer, S., Vogler, W.: An improvement of McMillan's unfolding algorithm. Formal Methods in System Design 20(20), 285–310 (2002)
19. Goltz, U., Reisig, W.: The non-sequential behaviour of Petri nets. Information and Control 57, 125–147 (1983)
20. Howell, R.R., Rosier, L.E., Yen, H.: A taxonomy of fairness and temporal logic problems for Petri nets. Theoretical Computer Science 82, 341–372 (1991)
21. König, B., Kozioura, V.: Augur 2—a new version of a tool for the analysis of graph transformation systems. In: Bruni, R., Varró, D. (eds.) Proceedings of GT-VMT 2006 (Workshop on Graph Transformation and Visual Modeling Techniques). ENTCS, Elsevier, Amsterdam (2008)
22. Lack, S., Sobociński, P.: Adhesive categories. In: Walukiewicz, I. (ed.) FOSSACS 2004. LNCS, vol. 2987, pp. 273–288. Springer, Heidelberg (2004)
23. Langerak, R.: Transformation and Semantics for LOTOS. PhD thesis, Department of Computer Science, University of Twente (1992)
24. Loiseaux, C., Graf, S., Sifakis, J., Bouajjani, A., Bensalem, S.: Property preserving abstractions for the verification of concurrent systems. Formal Methods in System Design 6, 1–35 (1995)
25. McMillan, K.L.: Symbolic Model Checking. Kluwer Academic Publishers, Dordrecht (1993)
26. Meseguer, J., Montanari, U., Sassone, V.: On the semantics of Petri nets. In: Cleaveland, W.R. (ed.) CONCUR 1992. LNCS, vol. 630, pp. 286–301. Springer, Heidelberg (1992)
27. Nielsen, M., Plotkin, G., Winskel, G.: Petri Nets, Event Structures and Domains, Part 1. Theoretical Computer Science 13, 85–108 (1981)
28. Pinna, G.M., Poigné, A.: On the nature of events: another perspective in concurrency. Theoretical Computer Science 138(2), 425–454 (1995)
29. Reisig, W.: Petri Nets: An Introduction. EACTS Monographs on Theoretical Computer Science. Springer, Heidelberg (1985)
30. Ribeiro, L.: Parallel Composition and Unfolding Semantics of Graph Grammars. PhD thesis, Technische Universität Berlin (1996)
31. Rozenberg, G. (ed.): Handbook of Graph Grammars and Computing by Graph Transformation: Foundations, vol. 1. World Scientific, Singapore (1997)
32. Vogler, W., Semenov, A., Yakovlev, A.: Unfolding and finite prefix for nets with read arcs. In: Sangiorgi, D., de Simone, R. (eds.) CONCUR 1998. LNCS, vol. 1466, pp. 501–516. Springer, Heidelberg (1998)
33. Winskel, G.: Event Structures. In: Brauer, W., Reisig, W., Rozenberg, G. (eds.) APN 1986. LNCS, vol. 255, pp. 325–392. Springer, Heidelberg (1987)

Graph-Based Design and Analysis of Dynamic Software Architectures*

Roberto Bruni[1], Antonio Bucchiarone[2,3], Stefania Gnesi[3],
Dan Hirsch[4], and Alberto Lluch Lafuente[1]

[1] Department of Computer Science, University of Pisa, Italy
`{bruni,lafuente}@di.unipi.it`
[2] IMT Alti Studi Lucca, Italy
[3] ISTI-CNR, Pisa, Italy
`{antonio.bucchiarone,stefania.gnesi}@isti.cnr.it`
[4] Argentina Software Development Center (Intel), Argentina
`dan.hirsch@intel.com`

Dedicated to Ugo Montanari in occasion of his 65th birthday

Abstract. We illustrate two ways to address the specification, modelling
and analysis of dynamic software architectures using: i) ordinary typed
graph transformation techniques implemented in Alloy; ii) a process algebraic presentation of graph transformation implemented in Maude. The
two approaches are compared by showing how different aspects can be
tackled, including representation issues, modelling phases, property specification and analysis.

1   Introduction

It is about 30 years ago when Ugo Montanari started to promote the use of graphs
and graph grammars as a multifaceted, unifying framework for the specification,
modelling and analysis of concurrent and distributed systems [12,16,17].

Since then, the way in which software artifacts are conceived and their engineering practices have evolved, but Ugo Montanari has always been able to
extend the foundations of the graph-based approach along innovative ideas so
to adapt it to emergent computational paradigms and software development
techniques.

In particular, Ugo Montanari is collaborating within the EU funded project
SENSORIA [38] to the development of a novel, comprehensive approach to the engineering of software systems for service-oriented computing. Within SENSORIA,
many efforts are devoted to the proposal and validation of sound architectural
design [14,39] principles, that focus on *architectural styles* governing the overall
structure of software systems in terms of components, their logical interrelation-
ship and their spatial distribution. Architectural styles establish the rationale

---

* Research partially supported by the EU within the FETPI Global Computing,
project IST-2005-016004 SENSORIA (*Software Engineering for Service-Oriented
Overlay Computers*) and by the Italian FIRB Project TOCAI.IT.

P. Degano et al. (Eds.): Montanari Festschrift, LNCS 5065, pp. 37–56, 2008.

for certain classes of architectures, e.g. patterns that should be fulfilled also in the presence of reconfigurations and that can even trigger and drive efficient reconfigurations.

There are many typical questions that arise during the design and analysis of software architectures. How do we represent architectures? How do we formalise architectural styles? How do we construct style conformant architectures? How do we model software architecture reconfigurations? How do we ensure style consistency? How do we express and verify architectural properties?

In this paper we compare two graph-based approaches to the specification and modelling of architectural designs. Both approaches are aimed to validate prototypical applications before their realisation and deployment.

The first approach is inspired by the tradition of algebraic approaches to graph transformation [19], an area of research in which Ugo Montanari has played a relevant role. The approach follows [5] and models dynamic software architectures using typed graph grammars (TGG).

The second approach is a recent proposal called *Architectural Design Rewriting* (ADR) [6,7,8]. ADR has been conceived by combining and reconciling apparently different formalisms, very much in the spirit that pervade many other Ugo Montanari's contributions. In particular, ADR takes inspiration of research lines in which Ugo Montanari has been deeply involved like graph-grammar based approaches to architectural styles [26], graphical models of process algebras [21] and rewriting formalisms [23].

We show the feasibility and effectiveness of TGG and ADR by implementing them using high-performant, state-of-the art formal tools. The implementation of TGG is based on Alloy [27] a light-weight approach to the modelling and analysis of software models that is raising the interest of leading researchers in the area of software architectures (see e.g. [32]). The implementation of ADR is based on Maude [13], a high-performance tool implementing Rewriting Logic [36]. We refer to the two implementations as TGG_A and ADR_M.

The main aspects on which we focus are concerned with:

Architectural Representation, i.e. convenient ways to represent a software architecture, to build it, to browse it;
Architectural Styles, i.e. convenient ways to constrain architectures under consideration to satisfy certain requirements;
Architectural Properties, i.e. convenient logical formalisms to express relevant architectural properties;
Architectural Analysis, i.e. efficient techniques and tools for verification.

We show how to tackle these aspects with both approaches. The outcome of our experience suggests that TGG is better suited for an early phase of the development, where the architectural constraints imposed by the style are defined in an iterative process of refinement of the model and style, assisted by model-finding techniques. Instead, ADR is more convenient for a more advanced phase,

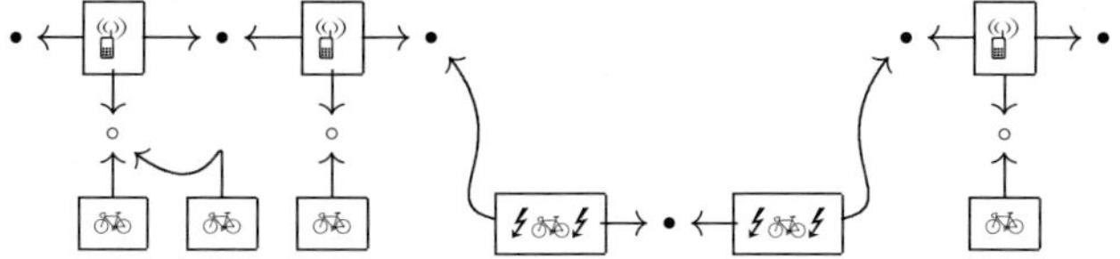

Fig. 1. The road assistance scenario

where the style is well established and structured, thus enabling flexible and powerful reconfigurations and efficient analysis via model checking.

This paper is structured as follows. Section 2 presents our running example. Section 3 offers a brief background on our design and analysis setting. Section 4 and 5 describe TGG_A and ADR_M, respectively. Section 6 compares both approaches with a special focus on analysis issues. Section 7 gives some pointers to related works. Section 8 concludes the paper and suggests future research avenues. Along with the paper we present flashes of code. For a complete vision we refer to [6,18].

2   Running Example: Road Assistance Scenario

We use as running example a simple bike scenario (see [8]), an ecological variant of the automotive case study of Sensoria. A road assistance service platform is supported by a wireless network of ad hoc stations that are situated along a road. Bikes are equipped with electronic devices that can access services as they move along the road, e.g. to request assistance in case of breakdowns. The graph in Figure 1 depicts a simple architecture of such a system. Each bike (🚲) is connected to the service access point (○) of a station (📱) which is possibly shared with other bikes. A station and its accessing bikes form a *cell*. Stations, in addition to the service access point, use two other communication points that we call chaining point (●). Such points are used to link cells in larger cell-chains. Bikes can move away from the range of the station of their current cell and enter the range of another cell. A handover protocol supports the migration of bikes to adjacent cells as in standard cellular networks. Stations can shut down, in which case their *orphan* bikes call for a repairing reconfiguration. We shall consider two shutting down situations: one in which the adjacent stations are able to bypass the connection and adopt all orphan bikes and another in which the bypassing is not possible and orphan bikes switch from their normal mode of operation to a cell mode (⚡🚲⚡), in which they become standalone stations.

The style of our example is the set of all architectures made of a connected chain of stations (or bikes in station mode) to which any bike is attached. This is a simple but very comprehensive example since it mixes features of well-known architectural styles. Indeed, the style underlying bikes and stations is basically a client-server architecture, while chains of cells resemble pipes-and-filters architectures.

3   Design and Analysis of Dynamic Software Architectures

The architecture of a software system basically consists of the structure of components and the way they are interconnected. Components are high-level computational and data entities that can range from a distributed application to a single thread, from databases to a simple data container. An architecture is *dynamic* if it can change during run-time. Typical changes, which are called *reconfigurations*, include components joining and leaving the system or changing their connections and are usually required for load balancing, fault-recovery and redimensioning software systems.

3.1   The Design of Software Architectures

The first aspect to consider in the design of a software architecture is the formalism used to describe it. Various formalisms exist ranging from the more theoretical graph-based approaches to implementation-oriented architectural programming languages (APLs) such as Java/A [28], passing through architectural description languages (ADLs) [35].

When designing an architecture, it is desirable to consider the concept of an *architectural style* [39], i.e. some set of rules or patterns indicating which components can be part of the architecture and how they can be legally interconnected. An architectural style can also be seen as a (typically infinite) set of *valid* architectures. Typical architectural styles include client-server, pipes-and-filters, layered, multi-tier and peer-to-peer.

There are basically two approaches to the definition of an architectural style. The first approach consists in defining a grammar that allows to produce all the instances of the style. This is the approach first proposed in [37] and subsequently followed in [26], a piece of work which constitutes the main inspiration of ADR. The second approach defines a set of architectural constraints that forbid or require some structural properties. Various approaches, for instance [32], use the Alloy language [27] to specify such constraints.

In this paper we adopt graphs as a suitable formalism to describe software architectures. The use of graphs as both the specification and the execution model combines the user-friendly visual representation with the formal theory for graph rewriting. One of the advantages in the combined use relies on the fact that architectural information can be encoded itself as part of the graphs, allowing for the uniform modelling of dynamic aspects such as computation, discovery, binding and reconfiguration, as graph transformations. Particularly, each architecture is represented by a hypergraph where components (resp. connectors) are modelled using hyperedges and their ports (resp. roles) by the outgoing tentacles. Components and connectors are attached together connecting the respective tentacles to the same node. Through the paper, we shall omit the prefix 'hyper' for simplicity. Ordinary directed graphs are a particular instance of hypergraph where each edge has two tentacles.

3.2   The Analysis of Architectural Properties

We consider mechanisms to express and verify the properties that we expect to be satisfied by software architectures.

Structural properties. Structural properties regard the topology of the architecture, i.e. the way components are interconnected. Note that the style definition can be considered as a special structural property. In the running example, we could consider properties regarding the number of bikes present in a cell or the effective existence of a path from the leftmost cell to the rightmost one.

Behavioural properties. Behavioural properties regard the dynamism of the architecture, i.e. the state space given by an initial architecture and its possible reconfigurations. In the running example, such properties might be the absence of deadlocks or the fact that a migration is always possible.

Among the class of properties explained above we shall focus and emphasise those that regard architectural styles. For instance, *style conformance* is a structural property that requires an architecture to be an instance of a style while *style preservation* requires all reconfigurations to preserve the style, i.e. that any reconfiguration of a style-conformant architecture results in a style-conformant architecture.

We shall consider the following analysis techniques:

Model finding. We consider the problem of analysing the state space of all possible architectures. Such analysis can serve as a computer-aided design process or as a debugging method to find out inconsistencies in the model or in its specification

Model checking. We consider the problem of verifying that a given architecture satisfies some structural or behavioural property expressed in a suitable logic.

Style matching. We consider the problem of determining whether an architecture is conformant to a certain style or whether a reconfiguration is style preserving.

4   Typed Graph Grammars with Alloy

The approach described in this section follows what discussed in [5] and it is based on the modelling of dynamic software architectures using typed graph grammars. The tool that supports this approach is Alloy [27]. Alloy provides a description language to represent software models, based on signatures and relations, which is suited for a set-theoretic presentation of graphs. Alloy also provides a logic, based on an extension of first-order logic with relational operators, to represent properties or constraints of models. We have used this logic to implement concepts like architectural styles, graph transformation rules and architectural properties. The *Alloy Analyzer* translates the model and the logical predicates into a (usually large) Boolean formula, uses efficient SAT solvers to decide satisfiability and provides a counterexample in negative case. We will show how to use these capabilities to ensure style-consistency, perform model-finding and validate architectural properties.

4.1   Designing Software Architectures

Architectures. We model software architectures using typed graphs. The three basic concepts in the implementation of graphs are *nodes*, *labels* and *edges* that we represent as Alloy signatures:

```
abstract sig Node{}
abstract sig Label{}
abstract sig Edge
{
  conn: Label->lone Node
}
```

According to the above definition, nodes and labels are atomic concepts, but edges have a relation (field) `conn` that maps each label to nodes. The multiplicity keyword `lone` in its declaration indicates that each label is mapped to at most one node. The previous signatures are marked `abstract`, which means that they have no element except those belonging to its signatures. Indeed we use subsignatures of `Node`, `Edge` and `Label` to represent concrete graph elements.

The signature `Graph` is used to define a graph as a structure made of nodes, edges and labels.

```
abstract sig Graph
{
  he: set Edge,
  n: set Node,
  l: set Label
}
```

In order to forbid ill-formed graphs we have defined some constraints, requiring for instance that the edges connect to nodes in the same graph. In Alloy the constraints of a model are organised into `paragraphs`. Constraints that are assumed always to hold are recorded as `facts`. In the following code we present some facts that we have defined to ensure a clear and efficient presentation of graphs, like requiring different graphs not to share items.

```
// two Edges  must have different set of labels (ports)
all e1,e2:Edge | e1!=e2 => (first[e1.conn]&first[e2.conn]) = none

// two Graphs have different node sets
all g1,g2:Graph | g1!=g2 => #(g1.n & g2.n)=0

// two Graphs have different edge sets
all g1,g2:Graph | g1!=g2 => #(g1.he & g2.he)=0

// two Graphs have different label sets
all g1,g2:Graph | g1!=g2 => #(g1.l & g2.l)=0
```

Architectural Styles. We represent an architectural style as a set of basic elements (modelled by a type graph plus a set of invariant constraints indicating how these elements can be legally connected). We define an Alloy module called `STYLE` that contains all these elements. It is subdivided in two parts, the first part defines basic elements of the style and the second one defines useful constraints.

Below we see the definition of each basic element using singleton extensions of `Node`, `Label` and `Edge` signatures. Note that some of them include facts in

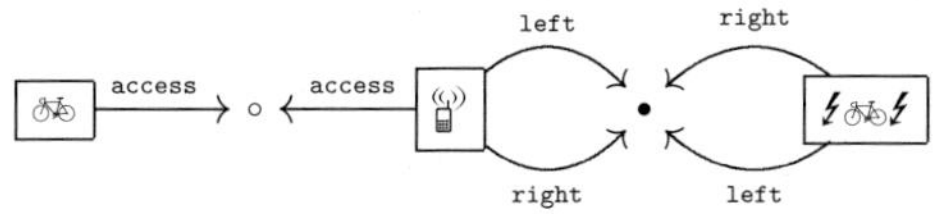

Fig. 2. Type Graph T of the running example

their body. For instance, a `Bike` component must have only one connection using an `Access` label to the `Access_Point` node. These definitions are enough to represent the type graph.

```
// Bike-Style basic elements

abstract sig Access_Point, Chain_Point extends Node{}
abstract sig Access extends Label{}
abstract sig Left extends Label{}
abstract sig Right extends Label{}
abstract sig Bike extends Edge{}
{
  #conn=1 and
  conn.univ in Access and // Projection of first column in conn
  univ.conn in Access_Point // Projection of second column in conn
}

abstract sig Bikestation extends Edge{}
{
  #conn=2 and
  conn.univ in Left+Right and
  univ.conn in Chain_Point
}

abstract sig Station extends Edge{}
{
  #conn=3 and
  conn.univ in Left+Right+Access and
  univ.conn in Chain_Point+Access_Point
}
```

The graphical representation in form of a type graph is depicted in Figure 2. Basically, a type graph offers a suitable way to type the items of a graph in a consistent manner. The idea is that the typing of a graph G over a type graph T is modelled by a total graph morphism from G to T.

The code below presents an excerpt of the additional facts needed to define our architectural style.

```
fact Style_constraints
{
  ...
  // if two stations are connected, they share one unique node
  all disj s1,s2: Station |
  attached[s1,s2]=>#(last[s1.conn]&last[s2.conn]) = 1
  // each Chain_Point node has at most two and at least one edge connected
  all cp: Chain_Point | #(conn.cp)>0 and #(conn.cp)<=2
  ...
}
```

In order get an instance of the style, generated from the previous code, we use a dummy predicate **instance** with empty parameters and body. A predicate

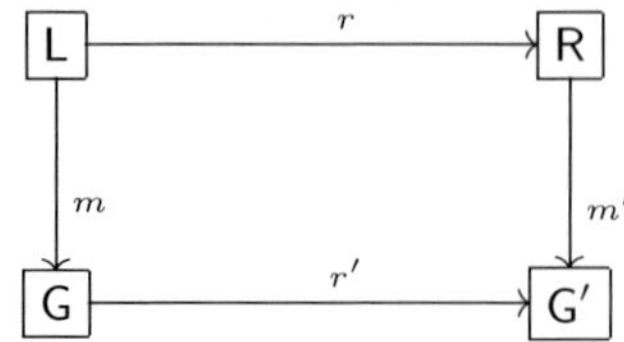

Fig. 3. SPO-based graph rewriting diagram

in Alloy is a constraint used to characterise some models of interest. Command **run** is used to find an instance of a predicate. For instance, we ask for a graph to be generated with a scope of at most two edges, one node and two labels as follows:

```
run instance for 1 Graph, 3 Station, 2 Bikestation, 4 Bike, ...
```

When we run the code above the Alloy Analyzer generates an instance and we can ask for a different one. One such instance is depicted in Figure 1 with our graphical notation.

Dynamism. The reconfiguration of a software architecture is described by a set of rewriting productions that state the possible ways in which an architecture may change. Each rule is defined as a partial, injective graph morphism $p : L \to R$, where L and R are graphs, called the *left-* and *right-hand* side. Given a graph G and a production p, a rewriting of G using p is realised using a single-pushout graph transformation approach [19] (see Fig. 3). Operationally, the rewrite is applied by finding a suitable match m (i.e. an occurrence of L in G) and the result is the graph obtained from G by removing that instance of L and releasing a fresh instance of R. Moreover, there can be items shared by L and R that are required to trigger the rewrite, but are just preserved by the transformation (some sort of interface, needed to properly attach the fresh copy of R to the existing items in G).

In order to implement what introduced above in the Alloy module called **TGG** we have defined various signatures and predicates that are used to execute a reconfiguration applying the SPO-based approach. First of all we have defined an abstract signature called **Fun** that will be used to define partial and total morphisms, matchings and productions. This signature has three fields that represent nodes, edges and labels functions mapping.

```
abstract sig Fun
{
   fN: set Node -> set Node,
   fE: set Edge -> set Edge,
   fL: set Label -> set Label
}
```

This signature can be used, for example, to define a partial morphism among two graphs. Moreover, in order to verify if a partial morphism among two graphs

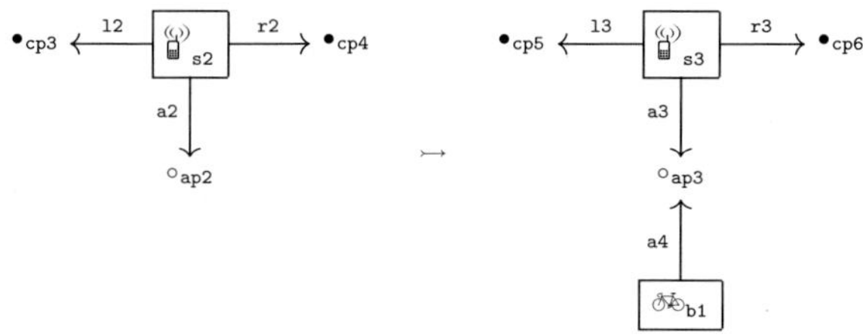

Fig. 4. Reconfiguration rule that connects a bike to a station

exists, we have defined the `isPartialMorphism` predicate that takes in input
two graphs (i.e. `G` and `H`), the respective typing functions (i.e. `t1` and `t2`) and a
mapping function (i.e. `f`). The predicate tries to find a partial morphism from `G`
to `H`, executing the set of constraints defined in its body.

```
pred isPartialMorphism [G: Graph, H: Graph, f: Fun, t1, t2: Tau]
{
  ...
  first[f.fE] = G.he
  first[f.fN] = G.n
  first[f.fL] = G.l
  all e1:G.he | all n1: G.n |
               all l1: G.l | (l1->n1) in e1.conn =>
                          f.fL[l1]->f.fN[n1] in f.fE[e1].conn &&
                          t1.tauL[l1]=t2.tauL[f.fL[l1]]
  ...
}
```

A `production` is defined using the signature `Production` that consists of a
left- and right-hand side graphs and the morphism indicating the items being
preserved.

```
abstract sig Production
{
  lhs: Graph,
  rhs: Graph,
  p: Fun
}
```

Using Alloy, we declare the signature of the previous production as a singleton
extension of `Prod` and define facts that characterise the left- and right-hand side
graphs. The rule that specifies the connection of a bike to a station is shown
in Figure 4 and it is defined by the code below (where the obvious definition of
`rhs1` is omitted for simplicity).

```
one sig lhs1 extends Graph{}
{
  he = s2
  n = cp3 + cp4 + ap2
  l = 12 + r2 + a2
  s2.conn = 12->cp3 + r2->cp4 + a2->ap2
}

one sig rhs1 extends Graph{} { ... }
```

```
one sig p1 extends Fun{}
{
  p.fN = cp3->cp5 + cp4->cp6 + ap2->ap3
  p.fE = s2->S3
  p.fL = l1->l3 + r2->r3 + a2->a3
}

one sig connect_bike extends Production{}
{
  lhs = lhs1, rhs = rhs1, p = p1
}
```

A single reconfiguration step is implemented using two distinct predicates (i.e., **rwStepPre** and **rwStepPost**. They are used to verify conditions that must hold in the host and target graph. The predicate **rwStepPre** checks the well formedness of the production **Pr** and the validity of matching **m1** of the left-side of the production (i.e., **Pr.lhs**) in the host graph **G**.

```
pred rwStepPre[G1:Graph, Pr: Production, m1: Fun, t1:Tau, t2:Tau, t3:Tau, t4: Tau ]
{
  // t1 defines types for G1, t2 defines typed for Pr.lhs
  isProd[Pr,Pr.p, t2,t3]
  // a production rule is applicable to a graph G1 if there is a matching of lhs into G1
  isMatch[Pr.lhs,G1,M1,t2,t1]
}
```

rwStepPost is responsible to execute a single SPO-based rewriting approach generating two morphisms **m2:Pr.rhs** $\rightarrow$ **G2** and **r2:G1** $\rightarrow$ **G2** and the target graph **G2**.

```
pred rwStepPost[G1:Graph, G2:Graph, Pr: Production, m1:Fun, m2:Fun, r1:Fun,
             r2:Fun,t1:Tau, t2:Tau,t3:Tau,t4:Tau]
{
  // m1 : L->G1
  isMatch[Pr.lhs,G1,m1,t2,t1]
  //m2: R->G2
  isTotalTGM[Pr.rhs,G2,m2]
  // r2: G1->G2
  isPartialMorphism[G1,G2,r2]
}
```

4.2   Analysis in Alloy

Structural Properties. In the Alloy language, assertions are constraints that should follow from the facts and must be checked. Using the Alloy Analyzer, it is possible to validate assertions, by searching for possible (finite) counterexamples for them, under the constraints imposed in the specification of the system. It is hence possible to specify that a given property P is invariant under sequences of applications of some operations. In our case this operation is the rewriting step that from an initial graph **G** and a production **P** generated a new graph **G'**. A technique useful for stating the invariance of a property P consists of specifying that P holds in the initial graph, and that for every non initial graph and rewriting operation, the following holds: **P(G)** and **rwStep(G,G')** $\rightarrow$ **P(G')**.

For this objective we have defined a set of properties that each architecture, after a rewriting step must satisfy. This set is memorised in an Alloy module called `PROPERTIES`. Structural properties are specified in Alloy defining functions and logical predicates. For instance, we have defined a predicate to express the property that there exists an acyclic path formed by stations that connects the left- and right-most stations. Functions `leftmost` and `rightmost` in the code below identify the left- and right-most stations of an architecture. In order to check this property we use the transitive closure of relation `next`.

```
fun leftmost[g:Graph,t:Tau]:one Edge {let e={e:g.he | t.tauE[e]=S and none= e.~next}| e }

pred Property1 [G:Graph,t:Tau]{ rightmost[G,t] in leftmost[G,t].^next }
```

Behavioural properties. Behavioural properties such as style preservation are suitably specified using DynAlloy [22], an extension of the Alloy language with syntactic sugar to define `actions` as a model of state changes. An action is the means by which the Analyzer transforms the system state after its execution. Regular expressions over actions and predicates can then be combined to express to express complex behavioural properties [22]. This issue is ongoing work [9].

5   Architectural Design Rewriting with Maude

The approach presented in this section is based on ADR [8], a formal framework for the development and reconfiguration of software architectures based on term-rewriting. An architectural style in ADR consists of a set of architectural elements and operations called *design productions* which define the well-formed compositions of architectures. Roughly, a term built out of such ingredients constitutes the proof that a design was constructed according to the style, and the value of the term, called *design*, is the constructed software architecture.

The tool support for the approach is based on Maude [13]. Maude naturally supports most of the features of ADR and also provides a powerful set of tools in the same framework including a model checker and a theorem prover.

5.1   Designing Software Architectures with Maude

Architectures. Software architectures and their constituents are represented uniformly in ADR by suitable graphs that we call *designs*. One difference with the TGG_A approach is that designs add an interface to the graph representing the architecture. The interface of a design is represented by an edge. The internal structure (called *body*) of a design is the architecture graph.

We have implemented various modules named `GRAPH-*` with the necessary machinery to construct graphs. Designs are implemented in a functional module called `DESIGN`, which includes the definition of the sort `Design`, a constructor operation `design` and an operation to replace a *hole* in a design (representing an unspecified part of the architecture) by a design whose interface is compatible with the hole.

Below we include an excerpt of module `DESIGN`. First, we see the declaration of sort `Design`. In a next line, the type of operation `design` is defined: it builds a

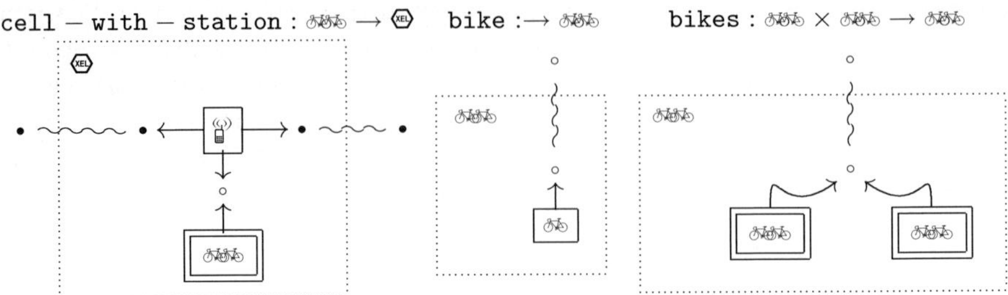

Fig. 5. Some design productions

design from a typed graph (here implemented as graph morphisms) representing the interface, a typed graph representing the body, a mapping relating the nodes in the interface with those in the body, and a list of edges representing holes. Membership equations not shown here take care of the well formedness of the design. Last, the type of operation `apply` is defined: it takes two arguments of type `Design` (the first intended to have a hole and the second being the design with proper interface to be placed in the hole) and returns the `Design` after the replacement. This complex operation basically implements the concept of type-consistent hyper-edge replacement [24].

```
sort Design .
op design : GraphMorphism GraphMorphism Map{Node,Node} List{Edge} -> Design .
op apply : Design Design -> Design .
```

Architectural Styles. The principle of ADR consists in defining an architectural style as a suitable algebra over designs. This approach can be seen as an algebraic recasting of context-free graph grammar-based approaches to architectural styles [37]. So, while in TGGs a style is given by logical predicates that forbid illegal architectures, in ADR the style is given by a generative mechanism that allows us to construct legal architectures only.

In Maude we define an architectural style as a set of sorts (the architectural types) which are subsorts of `Design`, a set of operations, called *design productions*, on such sorts (the legal ways of composing designs) and an interpretation of terms as designs.

Below we show an excerpt of module BIKE-STYLE which implements the architectural style in an abstract manner (i.e. no interpretation over designs is yet defined). Sorts `Cells` and `Bikes` are the types of the whole system (a chain of cells) and the type of bikes allocated in the cells, respectively. We see constructors `nocell` and `bikestation`, respectively representing an empty cell and a cell formed by one bike in station mode. Operation `cell-with-station` allows to construct from a collection of bikes (the inclusion of a fresh station is embedded in the operation), while `chain` builds a chain of cells by concatenating two chains of cells. Observe that one can annotate operations with axioms such as the associativity of `chain` or the fact that it has `nocell` as identity. Some of these productions are graphically represented in Fig. 5. Roughly, double boxes correspond to

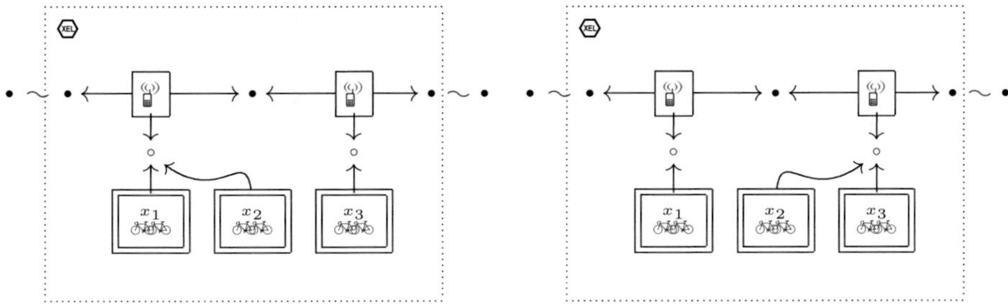

Fig. 6. Rule `migrate − right`

the parameters, the dotted boxes enclose the result of the productions, with the corresponding codomain type in its left corner, while waved lines indicate how internal nodes are exposed in the interface.

```
sort Cells . *** Chain of cells
sort Bikes . *** Collection of bikes
...
op nobike : -> Bikes .
op bike : -> Bike.
op bikes : Bikes Bikes -> Bikes [assoc comm id: nobike] .
op nocell : -> Cells .
op bikestation : -> Cells .
op cell-with-station : Bikes -> Cells .
op chain : Cells Cells -> Cells [assoc id: nocell] .
```

Dynamism. Contrary to TGG, reconfiguration rules in ADR are defined as rewrite rules over design terms, instead of over plain graphs or designs. In addition, ADR rules can be conditional, labelled and inductively defined and can be used to define complex behaviours and reconfigurations.

Below we see rule `migrate-right` (see Fig. 6). Note that because `bikes` is associative, commutative and has `nobike` as identity element, the rule can be matched to perform the migration of any number of bikes, while in TGG_A we have a rule able to migrate one bike at each step, only.

```
rl [migrate-right] : chain(cell-with-station(bikes(x1,x2)),cell-with-station(x3))
                  => chain(cell-with-station(x1),cell-with-station(bikes(x2,x3))) .
```

The module also includes the more complex reconfiguration that deals with shutdown by repairing the connection with an ad-hoc chain of bikes in station mode. Rule `bike2cell` allows a bike to reconfigure itself into a bike in station mode using label `'tocell`. Rule `bikes2cell` allows to propagate such reconfigurations inside collections of bikes. Finally, rule `cell2chain` allows reconfigure a station whenever the corresponding bikes are reconfigured into a chain of cells.

```
rl [bike2cell] : bike  => {'tocell}bikestation .

crl [bikes2cell] : bikes(x1,x2) => {'tocell}chain(y1,y2)
  if x1 =/= nobike /\ x2 =/= nobike /\ x1 => {'tocell}y1 /\ x2 => {'tocell}y2 .

crl [cell2chain] : cell-with-station(x1) => y1
  if x1 => {'tocell}y1 .
```

5.2   Analysis with Maude

The property specification mechanisms of our approach is based on the use of suitable logics to reason about structural and dynamic aspects of software architectures.

Structural Properties. Recall that contrary to TGG_A, ADR_M considers two aspects of the structural information of a software architecture: the design term, which can be seen as an abstraction, a proof of style conformance or an encoding of the construction process, and the design itself. We use different logics to reason about each aspect.

A natural and structured way to reason about design terms is the use of a suitable spatial logic (e.g. [11,34]). Basically, for each design production `f` used to compose designs the logic incorporates a spatial operator `f-so` to decompose a design. For instance, formula `chain-so(phi1,phi2)` is satisfied by all those designs of the form `chain(x1,x2)`, where design `x1` satisfies formula `phi1` and design `x2` satisfies formula `phi2`.

As an example consider the property that states a collection of bikes has at least n bikes. We see that this property can be inductively defined as below. For n equal to zero the formula always holds. For $n + 1$ (we use the successor constructor `s` below) the formula holds whenever the term is decomposable as the composition via operation `bikes` of one bike (`bike-so`) and a term with at least n bikes.

```
op at-least-k-bikes : Nat -> Prop .
vars n : Nat .
eq at-least-k-bikes(0) = True .
eq at-least-k-bikes( (s n )) = bikes-so(bike-so,at-least-k-bikes(n)) .
```

Our implementation also includes Courcelle's Monadic Second-Order (MSO) logic [15]. The rough idea is that the logic allows us to quantify over sets of nodes and edges. This allows us to reason about the complex interpreted structure of graphs that is hidden at the level of design terms.

Behavioural Properties. The dynamic aspects are expressed using the Linear-time Temporal Logic (LTL) for which Maude provides a module where only the state predicates have to be defined. Roughly, one is able to reason about infinite sequences of reconfigurations, by expressing properties on the ordering of state observations. Such observations are state predicates given by closed structural formulae.

As an example we state that it is always true that a collection of bikes has at least 2 bikes as the formula `[] at-least-k-bikes(2)`, where `[]` denotes the *always* temporal operator.

6   Comparison

We overview a brief comparison of the presented approaches w.r.t. the issues discussed in Section 3. Figure 7 summarises the main aspects and advantages of each approach.

ADR$_M$	**Aspect**	**TGG$_A$**
ADR (SOS, HR) *More flexible*	Formal Support	SPO Graph Rewriting *Well-studied, resource conscious*
Maude *Expressive, simple, performant*	Tool	Alloy *Light-weight, FO/SAT-based*
Hierarchical, Interfaced Graphs *Structured view, reusability*	Architectures	Flat Graphs *Concrete view*
Constructive (Algebra) *Style consistent*	Styles	Declarative (Alloy Logic) *Prototypical*
Term rewrites *Inductive, conditional*	Reconfigurations	Graph rewrites *Item trace*
Ad-hoc, VLRL, MSO *Expressive, flexible*	Structural properties	Alloy Logic *Simple, based on FO*
Ad-hoc, LTL *Expressive, well-studied*	Behavioural properties	Pre/Post, DynAlloy *Regular expressions, ongoing work*
Ad-hoc *Flexible, ongoing work*	Style Matching	Alloy Analyzer *Built-in, based on SAT*
By construction *Must not be proved*	Style Preservation	Alloy Analyzer *Performant*
Search strategies *Built-in, flexible*	Model Finding	Alloy Analyzer *Built-in*
LTL Model Checker *Efficient, flexible*	Model Checking	Alloy Analyzer *Bounded*

Fig. 7. Main aspects and advantages of each approach

6.1 Designing Software Architectures

Architectures. Both approaches represent architectures as suitable graphs. The main differences are that TGG represents architectures by flat graphs while ADR considers additional structural information. First, graphs have an interface which supports the compositionality of architectures. Second, graphs are equipped with a design term which serves various purposes: it is a proof of style consistency, it is a witness of the design process, it can be used to offer a hierarchical view at a suitable level of abstraction.

Architectural Styles. Each approach follows a different tradition. TGG$_A$ uses explicit structural constraints by means of logic predicates. This approach is more suited to a reactive modelling process: the software architect constructs a model and the system reacts reporting style inconsistencies. ADR$_M$ uses an implicit generative mechanism inspired by context-free graph grammars. This approach is more suited to a proactive modelling process: the software architect performs a decision procedure (i.e. refinements and compositions) that is guaranteed to be style consistent and not faulty.

Representing dynamism. Each approach defines dynamism at different levels of abstraction. TGG defines dynamism by means of local rewriting rules on flat graphs, while ADR defines rules on design terms. ADR rules are more abstract and also more expressive, because conditional labelled rules are also allowed. In addition, ADR rules guarantee style preservation by construction. On the other hand, ADR has still no mechanism to keep trace of reconfigured items, while this is done in TGG by the notion of trace morphism.

6.2   Specifying Architectural Properties

Structural properties. Structural properties are expressed in TGG_A by means of the same formalism used to define architectural styles, namely the Alloy logic. Maude does not offer a built-in structural logic but spatial logics such as the Verification Logic for Rewriting Logic [34] arise naturally and we can adjust the property specification mechanism at will. Indeed, we implemented property specification mechanisms tailored for the abstract view of design terms and the more concrete view of designs.

Behavioural properties. Alloy does not offer a standard behavioural logic and one has to rely on ad-hoc mechanisms or recent extensions such as DynAlloy. Maude, instead, comes with a built-in implementation of a Linear-time Temporal Logic (LTL) module, for which the user must provide just the state predicates.

6.3   Analysing Software Architectures

Style Matching. Checking style-consistency in TGG_A is done via the same mechanism as the verification of structural properties, i.e. it amounts to verify whether an architecture satisfies the predicate characterising the architectural style. Instead, in ADR_M we need a parsing mechanism, able to determine for given graph g, whether there is a design term d such that the design corresponding to d has a body graph isomorphic to g. This mechanism is under implementation.

Model Finding. Model finding is the main analysis capability offered by Alloy. The Alloy Analyzer basically explores (a bounded fragment) of the state space of all possible models and is able to show example instances satisfying or violating the desired properties. For instance, we can easily use the Alloy Analyzer to construct initial architectures: we need to ask for an instance graph satisfying the style predicates and having a certain number of bikes and stations. Model finding can also serve to the purpose of analysis. For instance, to validate if the style predicates really define what the software architect means. The use of bounds is justified by Alloy's small scope hypothesis that states that "most bugs have small counterexamples" [27]. This means that examining small architectures is often enough to detect possible major flaws.

Model finding is not provided by Maude directly. In order to perform model finding with Maude we need basically two mechanisms: one to generate a state space of models and one to explore it. We have defined a rewrite theory that simulates a design-by-refinement process, roughly consisting of the context-free graph grammar obtained by a left-to-right reading the design productions. Another generative mechanism that we can implement produces random graphs by adding items iteratively. The resulting graph could be used as the body of a design. In order to explore such state spaces we can use the LTL model checker, the `search` command or rewriting strategies.

Model Checking. Alloy does not have traditional model checking capabilities but its model finding features can be used instead to encode some (bounded) model checking problems. Basically, models and predicates are translated into a first-order formula and solved with efficient SAT solvers. Instead, model checking in Maude is supported by an efficient built-in LTL model checker. We have also implemented the satisfaction relation for the above mentioned MSO and spatial logics.

An alternative to model checking is the use of manual or computer-assisted theorem proving, which is highly facilitated in ADR by the hierarchical nature of *design productions*, which allows for proof based on structural induction. This issue is currently being investigated.

7   Related Work

An exhaustive enumeration and a deep comparison with related work is out of the scope of this paper. We mention, however some of the works that have inspired or share concepts with our approaches.

Many research works are focusing on the design and analysis of dynamic software architectures, applying formal methods such as graphs [37,25,3], logics [20,1] and process algebras [2,10].

As a matter of fact ADR has taken inspiration from initiatives that promote the conciliation of software architectures and process calculi by means of graphical methods [33].TGG instead, follows the tradition of graph grammars and applied to software architectures [37] and combines them with logical approach of Alloy.

Our approaches also shares concepts with various approaches ranging from process calculi that deal with reconfigurable component based architectures (e.g. [1]) to graphical representation of concurrent systems such as those based on Synchronized Hyperedge Replacement [21] or Bigraphs [29].

Maude and Alloy have been already used for the design and analysis of software architectures or graph transformation systems. For instance, in [30] Maude is used to model and verify software architectures given in LfP, a system description language with hierarchical behaviour. On the other hand, the work presented in [4] proposes a methodology to analyse transformation systems by means of Alloy and its supporting tool. They encode graph transformation systems into Alloy and verify properties such as reachability of given configurations through a finite sequence of steps or whether given sequences of rules can be obtained on an initial graph, and show all the configurations that can be obtained by applying a bounded sequence of rule instances on a given initial graph.

Another source of related work regard Architectural Description Languages (ADLs) to model and analyse software architectures [35] or UML-based approaches to model dynamic architectures [31].

8   Conclusion

In our experience, TGG_A and ADR_M are both flexible formal methods for the design and analysis of dynamic software architectures. Moreover, we promote their synergic application as each of them can be focused on the development phases where they are more effective.

In the early phase of the development, the architectural style is typically not well understood and TGG looks more appealing since the software architect can follow an incremental approach, adding, removing or refining architectural constraints with the help of the inconsistencies shown by the Alloy Analyzer.

Later, when the architectural style is well understood and stable, the software architect can define the style in a generative manner in the form of an ADR style. The benefits of an ADR style arise then from its hierarchical and structured nature. First, complex reconfigurations can be inductively defined on the structure of designs. Second, properties can be defined again inductively and at an abstract level. Last, but not least, model checking and theorem proving can be applied to exploit the hierarchical structure.

This separation of phases also matches the choice of the tools. Indeed, TGG could have been easily implemented in Maude, but Maude lacks of a built-in, efficient mechanism to perform model finding. On the other hand, implementing ADR in Alloy would have required a lot more efforts to encode all the mechanisms that are native in Maude such as normal forms, memberships or conditional rewrite rules, and built-in tools such as the LTL model checker.

Current work is devoted to validate our ideas by enriching our experience with more realistic examples, taken from global computing areas like service oriented computing and grid computing. In future work we plan to extend our comparative analysis to further interesting aspects such as the ordinary behaviour of software components, the analysis of architectural styles and run-time implementation. Another interesting perspective is to investigate other ADR suited interpreted algebras, other than graphs. Constraints systems, for instance, seem well suited and appealing to deal with non functional aspects like quality of service.

References

1. Aguirre, N., Maibaum, T.S.E.: Hierarchical temporal specifications of dynamically reconfigurable component based systems. Electr. Notes Theor. Comput. Sci. 108, 69–81 (2004)
2. Allen, R., Douence, R., Garlan, D.: Specifying and analyzing dynamic software architectures. In: Astesiano, E. (ed.) FASE 1998. LNCS, vol. 1382, Springer, Heidelberg (1998)
3. Baresi, L., Heckel, R., Thöne, S., Varró, D.: Style-based refinement of dynamic software architectures. In: Proceedings of WICSA 2004, pp. 155–166. IEEE Computer Society Press, Los Alamitos (2004)
4. Baresi, L., Spoletini, P.: On the use of Alloy to analyze graph transformation systems. In: Corradini, A., Ehrig, H., Montanari, U., Ribeiro, L., Rozenberg, G. (eds.) ICGT 2006. LNCS, vol. 4178, pp. 306–320. Springer, Heidelberg (2006)

5. Bruni, R., Bucchiarone, A., Gnesi, S., Melgratti, H.: Modelling dynamic software architectures using typed graph grammars. In: Proceedings of GTVC 2007. ENTCS, Elsevier, Amsterdam (to appear)
6. Bruni, R., Lluch Lafuente, A., Montanari, U.: Hierarchical design rewriting with Maude. In: Proceedings of WRLA 2008, Elsevier, Amsterdam (to appear, 2008)
7. Bruni, R., Lluch Lafuente, A., Montanari, U., Tuosto, E.: Service oriented architectural design. In: TGC 2007. LNCS, vol. 4912, pp. 186–203. Springer, Heidelberg (2007)
8. Bruni, R., Lluch Lafuente, A., Montanari, U., Tuosto, E.: Style based reconfigurations of software architectures. Technical Report TR-07-17, Dipartimento di Informatica, Università di Pisa (2007)
9. Bucchiarone, A., Galeotti, J.P.: Dynamic software architectures verification using DynAlloy. In: Proceedings of GT-VMT 2008 (to appear, 2008)
10. Canal, C., Pimentel, E., Troya, J.M.: Specification and refinement of dynamic software architectures. In: Proceedings of WICSA 1999. IFIP Conference Proceedings, vol. 140, pp. 107–126. Kluwer, Dordrecht (1999)
11. Cardelli, L., Gardner, P., Ghelli, G.: A spatial logic for querying graphs. In: Widmayer, P., Triguero, F., Morales, R., Hennessy, M., Eidenbenz, S., Conejo, R. (eds.) ICALP 2002. LNCS, vol. 2380, pp. 597–610. Springer, Heidelberg (2002)
12. Castellani, I., Montanari, U.: Graph grammars for distributed systems. In: Ehrig, H., Nagl, M., Rozenberg, G. (eds.) Graph Grammars 1982. LNCS, vol. 153, pp. 20–38. Springer, Heidelberg (1983)
13. Clavel, M., Durán, F., Eker, S., Lincoln, P., Martí-Oliet, N., Meseguer, J., Talcott, C.: All About Maude - A High-Performance Logical Framework. LNCS, vol. 4350. Springer, Heidelberg (2007)
14. Clements, P., Garlan, D., Bass, L., Stafford, J., Nord, R., Ivers, J., Little, R.: Documenting Software Architectures: Views and Beyond. Pearson Education, London (2002)
15. Courcelle, B.: The expression of graph properties and graph transformations in monadic second-order logic. In: Handbook of Graph Grammars and Computing by Graph Transformation, pp. 313–400. World Scientific, Singapore (1997)
16. Degano, P., Montanari, U.: Concurrent histories: A basis for observing distributed systems. J. Comput. Syst. Sci. 34(2/3), 422–461 (1987)
17. Degano, P., Montanari, U.: A model for distributed systems based on graph rewriting. J. ACM 34(2), 411–449 (1987)
18. Dynamic Software Architectures for Global Computing Systems, `http://fmt.isti.cnr/~antonio`
19. Ehrig, H., Heckel, R., Korff, M., Löwe, M., Ribeiro, L., Wagner, A., Corradini, A.: Algebraic approaches to graph transformation - Part II: Single pushout approach and comparison with double pushout approach. In: Handbook of Graph Grammars, pp. 247–312. World Scientific, Singapore (1997)
20. Endler, M.: A language for implementing generic dynamic reconfigurations of distributed programs. In: Proceedings of BSCN 1994, pp. 175–187 (1994)
21. Ferrari, G.L., Hirsch, D., Lanese, I., Montanari, U., Tuosto, E.: Synchronised hyperedge replacement as a model for service oriented computing. In: de Boer, F.S., Bonsangue, M.M., Graf, S., de Roever, W.-P. (eds.) FMCO 2005. LNCS, vol. 4111, pp. 22–43. Springer, Heidelberg (2006)
22. Frias, M., Galeotti, J., Pombo, C.L., Aguirre, N.: DynAlloy: Upgrading Alloy with actions. In: ICSE 2005, pp. 442–450. ACM Press, New York (2005)
23. Gadducci, F., Montanari, U.: Comparing logics for rewriting: rewriting logic, action calculi and tile logic. Theor. Comput. Sci. 285(2), 319–358 (2002)

24. Habel, A.: Hyperedge Replacement: Grammars and Languages. Springer-Verlag New York, Inc., Secaucus, NJ, USA (1992)
25. Hirsch, D., Inverardi, P., Montanari, U.: Modeling software architectures and styles with graph grammars and constraint solving. In: WICSA 1999. IFIP Conference Proceedings, vol. 140, Kluwer Academic Publishers, Dordrecht (1999)
26. Hirsch, D., Montanari, U.: Shaped hierarchical architectural design. Electronic Notes on Theoretical Computer Science 109, 97–109 (2004)
27. Jackson, D.: Software Abstractions: Logic, Language and Analysis. MIT Press, Cambridge (2006)
28. Java/A, http://www.pst.ifi.lmu.de/Research/current-projects/java-a/
29. Jensen, O.H., Milner, R.: Bigraphs and mobile processes. Technical Report 570, Computer Laboratory, University of Cambridge (2003)
30. Jerad, C., Barkaoui, K., Grissa-Touzi, A.: Hierarchical verification in Maude of LfP software architectures. In: Oquendo, F. (ed.) ECSA 2007. LNCS, vol. 4758, pp. 156–170. Springer, Heidelberg (2007)
31. Kacem, M.H., Kacem, A.H., Jmaiel, M., Drira, K.: Describing dynamic software architectures using an extended UML model. In: Proceedings of SAC 2006, pp. 1245–1249. ACM, New York (2006)
32. Kim, J.S., Garlan, D.: Analyzing architectural styles with Alloy. In: Proceedings of ROSATEA 2006, pp. 70–80. ACM Press, New York (2006)
33. König, B., Montanari, U., Gardner, P. (eds.): Graph Transformations and Process Algebras for Modeling Distributed and Mobile Systems, IBFI, Schloss Dagstuhl, Germany, June 6-11, 2004. Dagstuhl Seminar Proceedings, vol. 04241 (2004), http://www.dagstuhl.de/04241/
34. Martí-Oliet, N., Pita, I., Fiadeiro, J.L., Meseguer, J., Maibaum, T.S.E.: A verification logic for rewriting logic. J. Log. Comput. 15(3), 317–352 (2005)
35. Medvidovic, N., Taylor, R.N.: A classification and comparison framework for software architecture description languages. IEEE Transactions on Software Engineering 26(1), 70–93 (2000)
36. Meseguer, J., Rosu, G.: The rewriting logic semantics project. TCS 373(3), 213–237 (2007)
37. Métayer, D.L.: Describing software architecture styles using graph grammars. IEEE Trans. Software Eng. 24(7), 521–533 (1998)
38. Sensoria Project, http://sensoria.fast.de/
39. Shaw, M., Garlan, D.: Software Architectures: Perspectives on an emerging discipline. Prentice-Hall, Englewood Cliffs (1996)

Graph Transformation Units – An Overview*

Hans-Jörg Kreowski[1], Sabine Kuske[1], and Grzegorz Rozenberg[2]

[1] University of Bremen, Department of Computer Science
P.O. Box 33 04 40, 28334 Bremen, Germany
`{kreo,kuske}@informatik.uni-bremen.de`
[2] Leiden University,
Leiden Institute for Advanced Computer Science (LIACS)
2333 CA Leiden, The Netherlands
`rozenber@liacs.nl`

Abstract. In this paper, we give an overview of the framework of graph transformation units which provides syntactic and semantic means for analyzing, modeling, and structuring all kinds of graph processing and graph transformation.

1 Introduction

Graphs are used in computer science in many forms and contexts, and for many purposes. Well-known examples are Petri nets, entity-relationship diagrams, UML diagrams, and state graphs of finite automata. In many applications, graphs are not of interest as singular entities, but as members of graph and diagram languages, as states of processes and systems, as structures underlying algorithms, as networks underlying communication, tour planning, production planning, etc. Therefore, there is genuine need to generate, recognize, process, and transform graphs. The development and investigation of means and methods to achieve this goal is the focus of the area of graph transformation. (See the Handbook of Graph Grammars and Computing by Graph Transformation [1–3].)

In this paper, we give an overview of the framework of graph transformation units which provides syntactic and semantic means for analyzing, modeling, and structuring all kinds of graph processing and graph transformation. In particular, graph transformation units can be used to generate graph languages, to specify graph algorithms, to transform models whenever they are represented by graphs, and to define the operational semantics of data processing systems whenever the data and system states are given as graphs.

Graph transformation units encapsulate rules and control conditions that regulate the application of rules to graphs including the specification of initial and

---

* The first two authors would like to acknowledge that their research is partially supported by the Collaborative Research Centre 637 (Autonomous Cooperating Logistic Processes: A Paradigm Shift and Its Limitations) funded by the German Research Foundation (DFG).

P. Degano et al. (Eds.): Montanari Festschrift, LNCS 5065, pp. 57–75, 2008.

terminal graphs. To support the re-use and the stepwise development of graph transformation units, also a structuring principle is introduced that allows the import of units by units. The operational semantics of a unit transforms initial graphs into terminal graphs by interleaving rule applications and calls of imported units so that the control condition is fulfilled.

An important aspect of this framework is its approach independence, meaning that all introduced concepts work for all kinds of graphs, rules, rule applications, and control conditions. In particular, one can build new approaches from given ones through the product of approaches. In this way, graph transformation units can be combined in such a way that tuples of graphs are transformed component-wise. As a consequence, the interleaving semantics covers computable relations on graphs with m input graphs and n output graphs for arbitrary numbers m, n rather than binary relations between initial and terminal graphs.

This overview is organized in the following way. In Section 2, we recall the basic notions of graphs and of rule-based graph transformation. In Section 3, simple graph transformation units encompassing rules and control mechanisms are introduced, while a structuring principle by means of import is added in Section 4. Section 5 addresses the idea of approach independence and the product type. In the last section we discuss some related work and point out some further aspects such as the iterated interleaving semantics in case of cyclic import, the interlinking semantics that supplements the sequential interleaving semantics by parallel and concurrent elements, and autonomous units which interact in a common graph environment.

2   Graphs and Rule-Based Graph Transformation

In this section, we recall the basic notion of graphs, matches, rules, and rule application as they are needed to model the examples of this paper.

Graphs are quite generic structures which are encountered in the literature in many variants: directed and undirected, labeled and unlabeled, simple and multiple, with binary edges and hyperedges, etc. In this survey, we focus on directed, edge-labeled, and multiple graphs with binary edges (see 2.1). As a running example, we consider the generation of simple paths and related problems including the search for Hamiltonian paths. Please note that the search for simple paths of certain lengths and in particular the search for Hamiltonian paths are NP-complete problems.

Graphs are often considered as inputs of algorithms and processes so that methods are needed to search and manipulate graphs. Graphs may also represent states of systems so that methods for updates and state transitions are needed. Also, graphs may often be used to specify the structure of all the data of interest, e.g., the set of all connected and planar graphs. Like in the case of string languages, one needs then mechanisms to generate and recognize graph languages. To meet all these needs, rule-based graph transformation is defined in 2.4 to 2.6.

2.1   Graphs

Let Σ be a set of labels. A *graph* over Σ is a system $G = (V, E, s, t, l)$ where V is a finite set of *nodes*, E is a finite set of *edges*, $s, t\colon E \to V$ are mappings assigning a *source* $s(e)$ and a *target* $t(e)$ to every edge in E, and $l\colon E \to \Sigma$ is a mapping assigning a label to every edge in E. An edge e with $s(e) = t(e)$ is also called a *loop*. The components V, E, s, t, and l of G are also denoted by V_G, E_G, s_G, t_G, and l_G, respectively. The set of all graphs over Σ is denoted by $\mathcal{G}_\Sigma$. We reserve a specific label $*$ which is omitted in drawings of graphs. In this way, graphs where all edges are labeled with $*$ may be seen as *unlabeled graphs*.

Example 1. Consider the graphs G_0, G_1, G_{12}, G_{123}, and G_{1234} in Figure 1.

A box $\square$ represents a node with an unlabeled loop. Therefore, G_0 has four nodes, four loops and five additional unlabeled edges. The other graphs are variants of G_0. We use $\circledcirc$ to represent a *begin*-node which is a node with a loop labeled with *begin*. Analogously, $\bigcirc$ represents an *end*-node, and $\circledcirc$ represents a node with a *begin*-loop and an *end*-loop. If one starts in the *begin*-node and follows the p-labeled edges, one reaches the *end*-node in the graphs G_{12}, G_{123}, and G_{1234}. In each case, the sequence of p-edges defines a simple path of G_0, where the intermediate nodes have no loops. In G_1, the *begin*-node and the *end*-node are identical which means that the corresponding simple path has length 0.

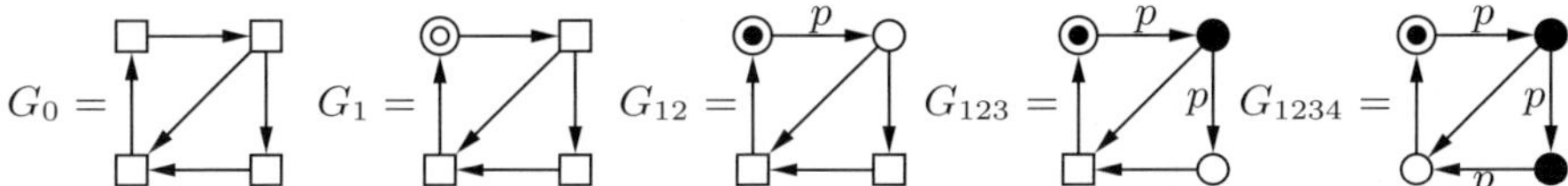

Fig. 1. G_0 with some of its simple paths

If one numbers the nodes of G_0 clockwise by 1 to 4 starting in the upper left-most corner, then the node sequences 1, 12, 123, and 1234 define simple paths of G_0 that correspond to the simple paths in G_1, G_{12}, G_{123}, and G_{1234}, resp. In this way, every simple path s of G_0 can be represented by a graph G_s. The set of all those graphs is denoted by $SP(G_0)$. The next subsection describes paths more formally.

2.2   Paths

One of the most important concepts concerning graphs is the notion of a path that is the subject of many research problems and applications of graphs such as connectivity, shortest paths, long simple paths, Eulerian paths, Hamiltonian paths, traveling salesperson problem, etc.

Given a graph $G = (V, E, s, t, l)$, a path from node v to node v' is a sequence of edges $p = e_1 \ldots e_n$ with $n \geq 1$, $s(e_1) = v$, $s(e_i) = t(e_{i-1})$ for $i = 2, \ldots, n$, and $t(e_n) = v'$. The *length* of p is n. Moreover, the empty sequence λ is considered to be a path from v to v of length 0 for each $v \in V$. A path $p = e_1 \ldots e_n$ from v

to v' *visits* the nodes $V(p) = \{v\} \cup \{t(e_i) \mid i = 1, \ldots, n\}$. A path p is *simple* if it visits no node twice, i.e., $\#V(p) = length(p) + 1$. [1] A simple path is *Hamiltonian* if it visits all nodes, i.e., $\#V(p) = \#V$.

2.3   Graph Morphisms, Subgraphs, and Matches

For graphs $G, H \in \mathcal{G}_\Sigma$, a *graph morphism* $g\colon G \to H$ is a pair of mappings $g_V\colon V_G \to V_H$ and $g_E\colon E_G \to E_H$ that are structure-preserving, i.e., $g_V(s_G(e)) = s_H(g_E(e))$, $g_V(t_G(e)) = t_H(g_E(e))$, and $l_H(g_E(e)) = l_G(e)$ for all $e \in E_G$.

If the mappings g_V and g_E are bijective, then g is an *isomorphism*, and G and H are called *isomorphic*.

If the mappings g_V and g_E are inclusions, then G is called a *subgraph* of H, denoted by $G \subseteq H$.

For a graph morphism $g\colon G \to H$, the image of G in H is called a *match* of G in H, i.e., the match of G with respect to the morphism g is the subgraph $g(G) \subseteq H$. If the mappings g_V and g_E are injective, the match $g(G)$ is also called *injective*. In this case, G and $g(G)$ are isomorphic.

Example 2. There is a graph morphism from the graph $L_{run} = \!\!$○────▶□ into some $G_s \in SP(G_0)$ whenever G_s has a subgraph isomorphic to L_{run}. Hence, there are two graph morphisms into G_{12}, there is one graph morphism into G_1 and one into G_{123}, but no graph morphism into G_{1234}.

Consider the injective match of L_{run} in G_{12} given by the right-most vertical edge. The removal of the edges of this match yields the subgraph

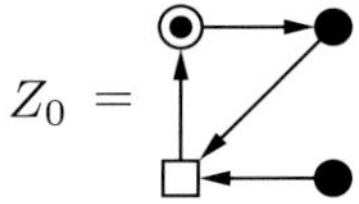

2.4   Graph Transformation Rule

The idea of a graph transformation rule is to express which part of a graph is to be replaced by another graph. Unlike strings, a subgraph to be replaced can be linked in many ways (i.e., by many edges) with the rest of the graph. Consequently, a rule also has to specify which kind of links are allowed. This is done with the help of a third graph that is common to the replaced and the replacing graph.

Formally, a *rule* $r = (L \supseteq K \subseteq R)$ consists of three graphs $L, K, R \in \mathcal{G}_\Sigma$ such that K is a subgraph of L and R. The components L, K, and R of r are called *left-hand side*, *gluing graph*, and *right-hand side*, respectively.

Example 3. Consider the following two rules.

---

[1] $\#X$ denotes the number of elements of a finite set X.

$$start \quad = \quad \square \quad \supseteq \quad \bullet \quad \subseteq \quad \circledcirc$$

$$run \quad = \quad \circ \!\longrightarrow\! \square \ \supseteq \ \bullet \qquad \bullet \ \subseteq \ \bullet \overset{p}{\longrightarrow} \circ$$

The rule *start* describes the removal of an unlabeled loop and the addition of a *begin*-loop and an *end*-loop at the same node. The rule *run* replaces an unlabeled edge by a *p*-edge removing the two loops of the left-hand side and adding an *end*-loop at the target node of the right-hand side. The identity of the nodes is chosen in such a way that the direction of the edge is preserved, i.e., the two sources are equal and the two targets are equal.

2.5 Application of a Graph Transformation Rule

The application of a graph transformation rule to a graph G consists of replacing an injective match of the left-hand side in G by the right-hand side in such a way that the match of the gluing graph is kept. Hence, the application of $r = (L \supseteq K \subseteq R)$ to a graph $G = (V, E, s, t, l)$ consists of the following three steps.

1. An injective match $g(L)$ of L in G is chosen.
2. Now the nodes of $g_V(V_L - V_K)$ are removed, and the edges of $g_E(E_L - E_K)$ as well as the edges incident to removed nodes are removed yielding the *intermediate graph* $Z \subseteq G$.
3. Afterwards the right-hand side R is added to Z by gluing Z with R in $g(K)$ yielding the graph $H = Z + (R - K)$ with $V_H = V_Z + (V_R - V_K)$ [2] and $E_H = E_Z + (E_R - E_K)$. The edges of Z keep their labels, sources, and targets so that $Z \subseteq H$. The edges of R keep their labels. They keep their sources and targets provided that those belong to $V_R - V_K$. Otherwise, $s_H(e) = g(s_R(e))$ for $e \in E_R - E_K$ with $s_R(e) \in V_K$, and $t_H(e) = g(t_R(e))$ for $e \in E_R - E_K$ with $t_R(e) \in V_K$.

The application of a rule r to a graph G is denoted by $G \underset{r}{\Longrightarrow} H$, where H is the graph resulting from the application of r to G. A rule application is called a *direct derivation*. The subscript r may be omitted if it is clear from the context.

Given a finite set of rules and a finite graph G, the number of injective matches is bounded by a polynomial in the size of G because the sizes of left-hand sides of rules are bounded by a constant. Given an injective match, the construction of the directly derived graph is linear in the size of G. Therefore, it needs polynomial time to find a match and to construct a direct derivation.

Example 4. The rule *run* of Example 3 can be applied to the graph G_{12} in two ways. One injective match of L_{run} is given by the right vertical edge of G_{12}. The intermediate graph is Z_0 as constructed in Example 2. And the derived graph is G_{123}. Consequently, one gets a direct derivation $G_{12} \underset{run}{\Longrightarrow} G_{123}$.

---

[2] Given sets X and Y, $X + Y$ denotes the disjoint union of X and Y.

Accordingly, the rule *run* can be applied to other graphs in $SP(G_0)$ yielding a graph in $SP(G_0)$ in each case. Here is the complete list of direct derivations by applying *run*:

$$G_1 \Longrightarrow G_{12}, \; G_2 \Longrightarrow G_{23}, \; G_2 \Longrightarrow G_{24}, \; G_3 \Longrightarrow G_{34}, \; G_4 \Longrightarrow G_{41},$$
$$G_{12} \Longrightarrow G_{123}, G_{12} \Longrightarrow G_{124}, \; G_{23} \Longrightarrow G_{234}, \; G_{24} \Longrightarrow G_{241}, \; G_{34} \Longrightarrow G_{341},$$
$$G_{41} \Longrightarrow G_{412}, G_{123} \Longrightarrow G_{1234}, G_{234} \Longrightarrow G_{2341}, G_{341} \Longrightarrow G_{3412}, G_{412} \Longrightarrow G_{4123}.$$

Moreover, the rule *start* can be applied to G_0 in four ways deriving G_1, G_2, G_3 and G_4.

2.6 Derivation and Application Sequence

The sequential composition of direct derivations $d = G_0 \underset{r_1}{\Longrightarrow} G_1 \underset{r_2}{\Longrightarrow} \cdots \underset{r_n}{\Longrightarrow} G_n$ ($n \in \mathbb{N}$) is called a *derivation* from G_0 to G_n. As usual, the derivation from G_0 to G_n can also be denoted by $G_0 \underset{P}{\overset{n}{\Longrightarrow}} G_n$ where $\{r_1, \ldots, r_n\} \subseteq P$, or just by $G_0 \underset{P}{\overset{*}{\Longrightarrow}} G_n$. The string $r_1 \cdots r_n$ is the *application sequence* of the derivation d.

Example 5. The direct derivations in Example 3 can be composed into the following derivations:

$$G_0 \Longrightarrow G_1 \Longrightarrow G_{12} \Longrightarrow G_{123} \Longrightarrow G_{1234},$$
$$G_0 \Longrightarrow G_1 \Longrightarrow G_{12} \Longrightarrow G_{124},$$
$$G_0 \Longrightarrow G_2 \Longrightarrow G_{23} \Longrightarrow G_{234} \Longrightarrow G_{2341},$$
$$G_0 \Longrightarrow G_2 \Longrightarrow G_{24} \Longrightarrow G_{241},$$
$$G_0 \Longrightarrow G_3 \Longrightarrow G_{34} \Longrightarrow G_{341} \Longrightarrow G_{3412},$$
$$G_0 \Longrightarrow G_4 \Longrightarrow G_{41} \Longrightarrow G_{412} \Longrightarrow G_{4123}.$$

Altogether, these derivations show that exactly the graphs in $SP(G_0)$ can be derived from G_0 by applying the rule *start* once and then the rule *run* k-times for $k \in \{0, 1, 2, 3\}$.

It is not difficult to see that one can generate the set of all simple paths of every unlabeled graph if each of its node has a simple unlabeled loop and if *start* is only applied in the first derivation step. Moreover, the length of each such derivation equals the number of nodes visited by the derived path. In particular, the length is bounded by the size of the initial graph. The proof can be done by induction on the lengths of derivations on one hand and on the lengths of simple paths on the other hand.

The six derived graphs above correspond to the dead-ended simple paths of G_0 being those that cannot be prolonged by a further edge. The four graphs $G_{1234}, G_{2341}, G_{3421}$, and G_{4123} represent the Hamiltonian paths of G_0.

3 Simple Graph Transformation Units

A rule yields a binary relation on graphs and a set of rules a set of derivations. The example of simple paths shows (like many other examples would show) that more features are needed to model processes on graphs in a proper way, in particular one needs initial graphs to start the derivation process, terminal graphs to stop it, and some control conditions to regulate it. This leads to the concept of a simple graph transformation unit. The well-known notion of a graph grammar is an important special case.

Analogously to Chomsky grammars in formal language theory, graph transformation can be used to generate graph languages. A graph grammar consists of a set of rules, a start graph, and a terminal expression fixing the set of terminal graphs. This terminal expression is a set $\Delta \subseteq \Sigma$ of terminal labels admitting all graphs that are labeled over Δ.

3.1   Graph Grammar

A *graph grammar* is a system $GG = (S, P, \Delta)$, where $S \in \mathcal{G}_\Sigma$ is the *initial graph* of GG, P is a finite set of graph transformation rules, and $\Delta \subseteq \Sigma$ is a set of *terminal symbols*. The *generated language* of GG consists of all graphs $G \in \mathcal{G}_\Sigma$ that are labeled over Δ and that are derivable from the initial graph S via successive application of the rules in P, i.e., $L(GG) = \{G \in \mathcal{G}_\Delta \mid S \overset{*}{\underset{P}{\Longrightarrow}} G\}$.

Example 6. The following graph grammar

> *unlabeled graphs*
>   initial: *empty*
>   rules: *new-node* =   *empty* $\supseteq$ *empty* $\subseteq$ $\square$
>
>   *new-edge* = $\bullet$ $\bullet \supseteq \bullet$ $\bullet \subseteq \bullet\!\longrightarrow\!\bullet$
>   terminal: $\{*\}$

generates the unlabeled graphs with a single unlabeled loop at each node. The start graph is the empty graph. The rule *new-node* adds a node with an unlabeled loop in each application. The rule *new-edge* adds an edge between two nodes. Due to the fact that we consider only injective matches, no two nodes can be identified in a match of the left-hand side of the rule *new-edge*. This guarantees that no new loops are generated. The terminal expression $\{*\}$ specifies all unlabeled graphs. Since the given rules transform unlabeled graphs into unlabeled ones, all derived graphs belong to the generated language. It should be noted that the grammar *unlabeled graphs* generates exactly the graphs that are used as initial graphs to obtain their simple paths in Example 5.

As for formal string languages, one does not only want to generate languages, but also to recognize them or to verify certain properties. Moreover, for modeling and specification aspects one wants to have additional features like the possibility to cut down the non-determinism inherent in rule-based graph transformation. This can be achieved through the concept of transformation units (see, e.g., [4–6]), which generalize standard graph grammars in the following ways:

- Transformation units allow a set of initial graphs instead of a single one.
- The class of terminal graphs can be specified in a more general way.
- The derivation process can be controlled.

The first two points are achieved by replacing the initial graph and the terminal alphabet of a graph grammar by a graph class expression specifying sets of initial and terminal graphs. The regulation of rule application is obtained by means of so-called control conditions.

3.2   Graph Class Expressions

A *graph class expression* may be any syntactic entity X that specifies a class of graphs $SEM(X) \subseteq \mathcal{G}_\Sigma$. A typical example is the above-mentioned subset $\Delta \subseteq \Sigma$ with $SEM(\Delta) = \mathcal{G}_\Delta \subseteq \mathcal{G}_\Sigma$. Forbidden structures are also frequently used. Let F be a graph, then $SEM(forbidden(F))$ contains all graphs G such that there is no graph morphism $f : F \to G$. Another useful type of graph class expressions is given by sets of rules. More precisely, for a set P of rules, $SEM(reduced(P))$ contains all *P-reduced* graphs, i.e., graphs to which none of the rules in P can be applied. Finally, it is worth noting that a graph grammar GG itself may serve as a graph class expression with $SEM(GG) = L(GG)$.

3.3   Control Conditions

A *control condition* is any syntactic entity that cuts down the non-determinism of the derivation process. A typical example is a regular expression over a set of rules (or any other string-language-defining device). Let C be a regular expression specifying the language $L(C)$. Then a derivation with application sequence s is *permitted* by C if $s \in L(C)$. As a special case of this type of control condition, the condition *true* allows every application sequence, i.e., $L(C) = P^*$, where P is the set of underlying graph transformation rules. Another useful control condition is *as long as possible*, which requires that all rules be applied as long as possible. More precisely, let P be the set of underlying rules. Then $SEM(as\ long\ as\ possible)$ allows all derivations $G \underset{P}{\Longrightarrow} G'$ such that no rule of P is applicable to G'. Hence, this control condition is similar to the graph class expression $reduced(P)$ introduced above. Also similar to *as long as possible* are *priorities* on rules, which are partial orders on rules such that if $p_1 > p_2$, then p_1 must be applied as long as possible before any application of p_2. More details on control conditions for transformation units can be found in [7].

By now, we have collected all components for defining simple graph transformation units.

3.4   Simple Graph Transformation Units

A *simple graph transformation unit* is a system $tu = (I, P, C, T)$, where I and T are graph class expressions to specify the *initial* and the *terminal* graphs respectively, P is a set of rules, and C is a control condition.

Each such transformation unit *tu* specifies a binary relation $SEM(tu) \subseteq SEM(I) \times SEM(T)$ that contains a pair (G, H) of graphs if and only if there is a derivation $G \overset{*}{\underset{P}{\Longrightarrow}} H$ permitted by C. The semantic relation $SEM(tu)$ may also be denoted by $RA(tu)$ if the aspect is stressed that it is based on rule application.

Example 7. The constructions from Examples 1 through 6 can be summarized by the following simple graph transformation unit:

> *simple paths*
>   initial: *unlabeled graphs*
>   rules: *start, run*
>   control: *start ; run**
>   terminal: *all*

The initial graphs are unlabeled graphs with a single loop at each node as generated by the graph grammar *unlabeled graphs* in Example 6. The rules *start* and *run* are given in Example 3, and the control condition is a regular expression over the set of rules with the sequential composition ; and the Kleene star * (specifying that a single application of *start* can be followed by an arbitrary sequence of applications of *run*). All graphs derived in this way from initial graphs are accepted as terminal. This is expressed by the graph class expression *all*. As discussed in Example 5, this unit generates all simple paths of each initial graph.

Analoguously, one can model *dead-ended simple paths* and *Hamiltonian paths* by replacing the terminal graph expression *all* by the expressions *reduced*({*run*}) and *forbidden*($\square$), resp.

4   Graph Transformation Units with Structuring

Simple graph transformation units allow one to model computational processes on graphs in the small. For modeling in the large, structuring concepts are needed for several reasons:

(1) To describe and solve a practical problem, one may need hundreds, thousands, or even millions of rules. Whereas a single set of rules of such a size would be hard to understand, the division into small components could help.
(2) Many problems can be solved by using the solutions of other problems. For example, most kinds of tour planning require a shortest path algorithm. Therefore, it would be unreasonable to model everything from scratch rather than to re-use available and working components.
(3) The modeling of a large system requires the subdivision into smaller pieces and the distribution of subtasks. But then it is necessary to know how components interact with each other.

Graph transformation units can be provided with a structuring principle by the *import* and *use* concept. The basic observation behind this concept is that the semantics of a simple graph transformation unit is a binary relation on graphs, like the rule application relation. Therefore, a unit (maybe with many rules) can play the role of a single rule. Units may use or import entities that describe binary relations on graphs. In particular, units may be used. This allows re-use as well as the distribution of tasks. See, e.g., [6, 8] for more details.

4.1   Units with Import and Interleaving Semantics

A *graph transformation unit with import* is a system $tu = (I, U, P, C, T)$, where (I, P, C, T) is a simple graph transformation unit, and the *use* component U is a set of identifiers (which is disjoint of P).

If each $u \in U$ defines a relation $SEM(u) \in \mathcal{G} \times \mathcal{G}$, then tu specifies a binary relation on graphs $INTER_{SEM}(tu)$ defined as follows:

$$(G, G') \in INTER_{SEM}(tu) \quad \text{if} \quad G \in SEM(I), G' \in SEM(T),$$

and there is a sequence $G_0, \ldots, G_n$ with $G = G_0, G_n = G'$, and, for $i = 1, \ldots, n, G_{i-1} \underset{r}{\Longrightarrow} G_i$ for some $r \in P$ or $(G_{i-1}, G_i) \in SEM(u)$ for some $u \in U$. Moreover, (G, G') must be accepted by the control condition C.

This relation is called *interleaving semantics* because the computation interleaves rule applications and calls of the imported relations. It should be noted that each choice of used relations defines an interleaving relation.

4.2   Networks of Units

The definition of structuring by import allows the use of any relation whenever it has an identifier. There may be some library that offers standard relations. Or one may use another framework that supports the modeling of relations of graphs. But the most obvious choice is the import of units as use component. This leads to sets of units that are closed under import:

Let V be a set of identifiers and $tu(v) = (I(v), U(v), P(v), C(v), T(v))$ a graph transformation unit with import for each $v \in V$. Then the set $\{tu(v) \mid v \in V\}$ is *closed under import* if $U(v) \subseteq V$ for all $v \in V$.

If one considers V as a set of nodes, the pairs (v, v') for $v' \in U(v)$ as edges with the projections as source and target, and the mapping tu as labeling, then one gets a *network* $(V, \{(v, v') \mid v' \in U(v), v \in V\}, tu)$.

If this network is finite and acyclic, then each of its units can be assigned with an import level that is used to define the interleaving semantics of networks in the next subsection:

$$level(v) = \begin{cases} 0 & \text{if } U(v) = \emptyset, \\ n + 1 & \text{if } n = \max\{level(v') \mid v' \in U(v)\}. \end{cases}$$

4.3   Interleaving Semantics of Acyclic Networks

The interleaving semantics of a unit with import requires that the used relations are predefined. This can be guaranteed in finite and acyclic networks of units level-wise so that the semantics can be defined inductively. The units of level 0 are simple graph transformation units such that the rule application semantics is defined. If the semantic relations up to level n are defined by induction hypothesis, then the interleaving semantics of units on level $n + 1$ is defined.

Let (V, E, tu) be a finite and acyclic network of units. Then the interleaving semantics $INTER(tu(v))$ for each $v \in V$ is inductively defined by:

- $INTER(tu(v)) = RA(tu(v))$ for $v \in V$ with $level(v) = 0$;
- assume that $INTER(tu(v)) \subseteq \mathcal{G} \times \mathcal{G}$ is defined for all $v \in V$ with $level(v) \leq n$ for some $n \in \mathbb{N}$; then $INTER(tu(v))$ is equal to $INTER_{SEM}(tu(v))$ with $SEM(v') = INTER(tu(v'))$ for $v' \in U_{tu(v)}$ for $v \in V$ with $level(v) = n + 1$.

Example 8. As shown in Example 7, the generation of dead-ended paths and Hamiltonian paths coincides with the generation of simple paths except for stronger terminal graph expressions. Hence, they can be modeled by the import and re-use of *simple paths*.

<table>
<tr><td>*dead-ended simple paths*</td><td>*Hamiltonian paths*</td></tr>
<tr><td>uses: *simple paths*</td><td>uses: *simple paths*</td></tr>
<tr><td>control : *simple paths*</td><td>control : *simple paths*</td></tr>
<tr><td>terminal *reduced*(*run*)</td><td>terminal *forbidden*($\square$)</td></tr>
</table>

5   Approach Independence and Product Type

Graph class expressions and control conditions are introduced as generic concepts that can be chosen out of a spectrum of possibilities. On the other hand, the graphs, the rules and their application are fixed in a specific way in the preceding sections. This is done to be able to illustrate the concepts by explicit examples. The notions of graph transformation units and the interleaving semantics are independent of the particular kind of graphs and rules one assumes. This is made precise by the generic notion of a graph transformation approach.

5.1   Graph Transformation Approach

A *graph transformation approach* $\mathcal{A} = (\mathcal{G}, \mathcal{R} \Longrightarrow, \mathcal{X}, \mathcal{C})$ consists of a class of graphs $\mathcal{G}$, a class of rules $\mathcal{R}$, a rule application operator $\Longrightarrow$ that provides a binary relation on graphs $\underset{r}{\Longrightarrow} \subseteq \mathcal{G} \times \mathcal{G}$ for every $r \in \mathcal{R}$, a class of graph class expressions $\mathcal{X}$ with $SEM(x) \subseteq \mathcal{G}$ for every $x \in \mathcal{X}$, and a class of control conditions $\mathcal{C}$ with $SEM(c) \subseteq \mathcal{G} \times \mathcal{G}$ for every $c \in \mathcal{C}$.

All the notions of Sections 3 and 4 remain valid over such an approach $\mathcal{A}$ if one replaces the class of graphs $\mathcal{G}_\Sigma$ by $\mathcal{G}$, the class of rules by $\mathcal{R}$, and each rule application by its abstract counterpart. In this sense, the modeling by graph transformation units is approach-independent because it works independently of a particular approach.

The approach independence is meaningful in at least two ways. On the one hand, it means that everybody can use and choose his or her favorite kinds of graphs, rules, rule applications, graph class expressions, and control conditions. On the other hand, it allows one to adapt given approaches to particular applications or to build new approaches out of given ones without the need to change the modeling concepts.

5.2   Restriction

Given a graph transformation approach $\mathcal{A} = (\mathcal{G}, \mathcal{R}, \Longrightarrow, \mathcal{X}, \mathcal{C})$, each subclass $\mathcal{G}' \subseteq \mathcal{G}$ induces a restricted approach $restrict(\mathcal{A}, \mathcal{G}') = (\mathcal{G}', \mathcal{R}, \Longrightarrow, \mathcal{X}, \mathcal{C})$, where the semantics of $\Longrightarrow, \mathcal{X}$, and $\mathcal{C}$ are taken from $\mathcal{A}$, but restricted to the graphs of $\mathcal{G}'$.

Similarly, all other components of $\mathcal{A}$ may be restricted – while this can be done for $\mathcal{X}$ and $\mathcal{C}$ without side effects, the restriction of rules and rule application may influence the semantics of $\mathcal{X}$ and $\mathcal{C}$, because graph class expressions and control conditions may refer to rules.

With respect to the class of graphs $\mathcal{G}_\Sigma$ as considered in the preceding sections, various restrictions are possible: unlabeled graphs, connected graphs, planar graphs, etc. A rule application to such a sort of graph would only be accepted if the derived graph is of the same type. An undirected graph can be represented by a directed graph if each undirected edge is replaced by two directed edges in opposite direction. In this way, undirected graphs can be handled in the given framework as a restriction of the introduced approach.

An example of a restricted kind of rules is a rule the left-hand side of which coincides with the gluing graph so that no removal takes place if the rule is applied.

A typical restriction concerning the rule application is the requirement of injective matching as used in our sample approach. This is requested in many graph transformation approaches in the literature.

5.3   Product Type

Another approach-building operator is the product of approaches. The idea is to consider tuples of graphs and rules where a tuple of rules is applied to a tuple of graphs by applying the component rules to the component graphs simultaneously. It is not always convenient to apply a rule to each component graph. Therefore, we add the symbol $-$ to each class of rules. Whenever $-$ is a component of a tuple of rules, the corresponding component graph remains unchanged.

Let $\mathcal{A}_i = (\mathcal{G}_i, \mathcal{R}_i, \Longrightarrow_i, \mathcal{X}_i, \mathcal{C}_i)$ for $i = 1, \ldots, n$ for some $n \in \mathbb{N}$ be a graph transformation approach. Then the product approach is defined by

$$\prod_{i=1}^{n} \mathcal{A}_i = (\prod_{i=1}^{n} \mathcal{G}_i, \prod_{i=1}^{n} (\mathcal{R}_i \cup \{-\}), \Longrightarrow, \prod_{i=1}^{n} \mathcal{X}_i, \prod_{i=1}^{n} \mathcal{C}_i)$$

where

- $(G_1, \ldots, G_n) \Longrightarrow (G_1', \ldots, G_n')$ for $(r_1, \ldots, r_n) \in \prod_{i=1}^{n}(\mathcal{R}_i \cup \{-\})$ if, for $i = 1, \ldots, n, G_i \underset{r_i}{\Longrightarrow} G_i'$ for $r_i \in \mathcal{R}_i$ and $G_i = G_i'$ for $r_i = -$,

- $(G_1, \ldots, G_n) \in SEM(x_1, \ldots, x_n)$ for $(x_1, \ldots, x_n) \in \prod_{i=1}^{n} \mathcal{X}_i$ if $G_i \in SEM(x_i)$ for $i = 1, \ldots, n$,

- $((G_1, \ldots, G_n), (G_1', \ldots, G_n')) \in SEM(c_1, \ldots, c_n)$ for $(c_1, \ldots, c_n) \in \prod_{i=1}^{n} \mathcal{C}_i$ if $(G_i, G_i') \in SEM(c_i)$ for $i = 1, \ldots, n$.

5.4 Tuples of Graph Transformation Units

Analoguously to the tupling of graphs, rules, rule applications, graph class expressions and control conditions in the product of graph transformation approaches, graph transformation units over the approaches can be tupled.

Let $tu_i = (I_i, U_i, P_i, C_i, T_i)$, for each $i = 1, \ldots, n$ for some $n \in \mathbb{N}$, be a graph transformation unit over the graph transformation approach $\mathcal{A}_i = (\mathcal{G}_i, \mathcal{R}_i, \Longrightarrow_i, \mathcal{X}_i, \mathcal{C}_i)$. Then the tuple

$$(tu_1, \ldots, tu_n) = ((I_1, \ldots, I_n), \prod_{i=1}^{n} U_i, \prod_{i=1}^{n} P_i, (C_1, \ldots, C_n), (T_1, \ldots, T_n))$$

is a graph transformation unit over the product approach $\prod_{i=1}^{n} \mathcal{A}_i$.

The important aspect of the tupling of units is that the choice of the components of the single units induces automatically all the components of the tuple of units by tupling the graph class expressions for initial and terminal graphs resp. and the control conditions as well as by the products of the import and rule sets.

As the tuple of units is defined component-wise, its semantics is the product of the semantic relations of the components up to a reordering of components. Let $SEM_i(u) \subseteq \mathcal{G}_i \times \mathcal{G}_i$ be a binary relation on graphs for each $u \in U_i, i = 1, \ldots, n$. Let $SEM_i(u_1, \ldots, u_n) \subseteq \prod_{i=1}^{n} \mathcal{G}_i \times \prod_{i=1}^{n} \mathcal{G}_i$ be defined for $(u_1, \ldots u_n) \in \prod_{i=1}^{n} U_i$ by $((G_1, \ldots, G_n), (G_1', \ldots, G_n')) \in SEM((u_1, \ldots, u_n))$ if and only if $(G_i, G_i') \in SEM_i(u_i)$ for $i = 1, \ldots, n$. Then the following holds: $((G_1, \ldots, G_n), (G_1', \ldots, G_n')) \in INTER_{SEM}(tu)$ if and only if $(G_i, G_i') \in INTER_{SEM_i}(tu_i)$ for $i = 1, \ldots, n$.

As the interleaving semantics of a tuple of units is given by the product of the interleaving semantics, the tuple of units tu may also be denoted by $tu_1 \times \cdots \times tu_n$ to stress the meaning already on the syntactic level.

Example 9. Consider the tuple of units *simple paths* $\times$ *nat* $\times$ *bool* where the first component is given in Example 7 and *nat* and *bool* are modeled as follows:

nat

    initial: *nat0*

    rules: $pred =$ ●——▸○ $\supseteq$ ● $\subseteq$ ○ [3]
                 1   1   1

         is 0 $=$ ◎ $\supseteq$ ◎ $\subseteq$ ◎

bool

    initial: *empty*

    rules: *set true* $= empty \supseteq empty \subseteq$ (node with *true*-loop)

         terminal: (node with *true*-loop) $= TRUE$

nat0

    initial: ◎ $= 0$

    rules: $succ =$ ○ $\supseteq$ ● $\subseteq$ ●——▸○
                 1   1   1

The start graph of *nat0* is a node with a *begin*-loop and an *end*-loop which may be seen as the number 0. The application of *succ* adds an edge to the *end*-node while the added target becomes the new *end*. Therefore, the derived graphs are simple paths of the form

 $= n$

with n edges, for some $n \in \mathbb{N}$, representing the number n. There is no control, and all graphs are terminal. In other words, *nat0* generates the natural numbers being the initial graphs of *nat*. The rule *pred* is inverse to *succ* so that its application transforms the graph $n + 1$ into the graph n. The rule *is 0* is applicable to a graph n if and only if $n = 0$ such that the rule provides a 0-test. Altogether, the unit *nat* can count down a given number and test whether 0 is reached.

The unit *bool* is extremely simple. The start graph is empty. The only rule adds a node with a *true*-loop whenever applied. But after one application, the terminal graph is reached already. This graph can be seen as the truth value *TRUE*.

According to the definitions of the components, an initial graph of the tuple of units *simple paths* $\times$ *nat* $\times$ *bool* has the form $(G, n, empty)$, where G is an unlabeled graph with an unlabeled loop at each node. Consider, in particular, $(G_0, 3, empty)$. To such an initial graph, one may apply the triple rule $(start, -, -)$ replacing an unlabeled loop of G by a *begin*- and an *end*-loop and keeping n and *empty* unchanged. For example, one can derive $(G_1, 3, empty)$ from $(G_0, 3, empty)$. Now one may apply the triple rule $(run, pred, -)$ repeatedly. This builds a simple path in G edge by edge while n is decreased 1 by 1. If G has a simple path of length n, one can derive $(G', 0, empty)$ in this way. For example, one can derive $(G_{1234}, 0, empty)$ from $(G_1, 3, empty)$. Finally, the triple rule $(-, is\ 0, set\ true)$ becomes applicable deriving $(G', 0, TRUE)$ and $(G_{1234}, 0, TRUE)$ in particular.

---

[3] The number 1 identifies the identical nodes to make the inclusions unambiguous.

Altogether, this models a test whether a graph has a simple path of a certain length.

> *simple paths of some lengths*
>   initial: : *simple paths* × *nat*
>   uses : *simple paths* × *nat* × *bool*
>   control : $(start, -, -); (run, pred, -)^*; (-, is\ 0, set\ true)$
>   terminal : *bool*

Due to the import, this unit is based on the tuple of units considered above. This provides the tuple rules in the control condition, which is just a regular expression over the set of rules. But the graph class expressions are of a new type related to the product. The initial expression means that the first two components can be chosen freely as inputs of the modeled test. The third component is always the empty graph so that it can be added by default. The terminal expression means the projection to the third component as output of the test.

The semantic analysis shows that the unit *simple paths of some lengths* relates a graph G and a number n to the truth value $TRUE$ if and only if G has a simple path of length n. Moreover, the length of every derivation is bounded by $n + 2$ because n can be decreased by 1 n times at most. It is also bounded by the number m of nodes of G plus 1 because simple paths are shorter than m. Because the number of matches for the rules is also bounded polynomially, the unit proves that the test for simple paths of certain lengths is in the class NP. This is a well-known fact in this case, but illustrates that the introduced framework supports proofs like this.

If the input length is chosen as the number of nodes of the input graph minus 1, then the unit yields $TRUE$ if and only if the input graph has a Hamiltonian path. In this way, the test for simple paths of certain lengths turns out to be NP-complete because the Hamiltonian-path problem is NP-complete and a special case.

5.5   Typing of Units

A graph transformation unit models a relation between initial and terminal graphs. Hence one may say that the type of a unit $tu = (I, P, C, T)$ is $I \rightarrow T$. The introduced product type allows a more sophisticated typing of the form $I_1 \times \cdots \times I_m \rightarrow T_1 \times \cdots \times T_n$. This works as follows. Let $tu_1 \times \cdots \times tu_p$ be a tuple of units, let $input : \{1, \ldots, m\} \rightarrow \{1, \ldots, p\}$ be an injective mapping with $I_i = I(tu_{input(i)})$ for $i = 1, \ldots, m$, and let $output : \{1, \ldots, n, \} \rightarrow \{1, \ldots, p\}$ be a mapping with $T_j = T(tu_{output(j)})$. Moreover, there may be some extra control condition c for the tuple of units. Then this defines a unit of type $I_1 \times \cdots \times I_m \rightarrow T_1 \times \cdots \times T_n$ by

> *typed unit*
>   initial: *input*
>   uses: $tu_1 \times \cdots \times tu_p$
>   control: c
>   terminal: *output*

This unit relates graph tuple $(G_1, \ldots, G_m) \in \prod_{i=1}^{m} SEM(I_i)$ with the graph tuple $(H_1, \ldots, H_n) \in \prod_{i=1}^{n} SEM(T_j)$ if there are graphs $((G'_1, \ldots, G'_p), (H'_1, \ldots, H'_p))$ belonging to the interleaving semantics of $tu_1 \times \cdots \times tu_p$, fulfilling in addition the control condition c, and $G_i = G'_{input(i)}$ for $i = 1, \ldots, m$ as well as $H_j = H'_{output(j)}$ for $j = 1, \ldots, n$, where the graphs G_i with $i \notin input(\{1, \ldots, m\})$ can be chosen arbitrarily. Some of the components of the product are inputs, some outputs, some may be auxiliary. Initially, the input components are given. The other components must be chosen which is meaningful if there are unique initial graphs that can serve as defaults. Then the product is running component-wise according to the component rules and control conditions reflecting its extra control condition. If all components are terminal, the output components are taken as results.

A more detailed investigation of the product type can be found in [9, 10].

6   Further Research and Related Work

In the preceding sections, we have given an overview of graph transformation units as devices to model algorithms, processes, and relations on graphs. Such units consist of rules together with specifications of initial and terminal graphs as well as control conditions to cut down the nondeterminism of rule applications. Moreover, units can import other units (or other relations on graphs) providing in this way possibilities of re-use and of structuring. The operational semantics of graph transformation units is given by the interleavings of rule applications and calls of imported relations; it yields a relation between initial and terminal graphs. This interleaving semantics is well-defined if the import structure is acyclic. All the considered concepts work for arbitrary graph transformation approaches, where an approach is the computational base underlying graph transformation units. Such an approach consists of classes of graphs, rules, graph class expressions, and of control conditions as well as a notion of rule application. Every component of an approach can be chosen out of a wide spectrum of graph transformation concepts one encounters in the literature. An interesting aspect of this kind of approach independence is the possibility to construct new approaches from given ones. For example, this allows one to transform undirected graphs as a restriction of an approach for directed graphs. This also includes the product of approaches which handles tuples of graphs and provides a quite general typing of semantic relations.

With respect to the modeling of graph algorithms, Plump's and Steinert's concept of graph programs [11] is closely related to transformation units (cf. also Mosbah and Ossamy [12]). The major difference is that graph programs are based on a particular graph transformation approach. In contrast to this, approach independence allows the modelers to choose their favorite approaches of which the area of graph transformation offers quite a spectrum (see, e.g., [1] for the most frequently used approaches and Corradini et al. [13], Drewes et al. [14], and Klempien-Hinrichs [15] as examples of newer approaches). In this context, one may note that the categorial framework of adhesive categories provides a kind of approach independence. While rules and rule application are fixed by the use of pushouts, one can enjoy quite a variety of classes of graphs due to the possible choices of the underlying category (see, e.g., [16, 17]).

As a structuring principle, transformation units are closely related to other module concepts for graph transformation systems like the ones introduced by Ehrig and Engels [18], by Taentzer and Schürr [19], by Große-Rhode, Parisi-Presicce and Simeoni [20, 21] as well as Schürr's and Winter's package concept [22]. Heckel et al. [23] classify and compare all these concepts including transformation units in some detail.

We point out now some further possible directions of the investigation of graph transformation units:

1. If one permits cyclic import, meaning that units can use and help each other in a recursive way, then the interleaving semantics as defined in Section 4 is no longer meaningful because the imported relations cannot be assumed to be defined already. In this case, the infinite iteration of the interleaving construction works. (see, e.g.,[24]).
2. The interleaving semantics is based on the iterated sequential composition of the relations given by rule application and the imported relations. Therefore, it is a purely sequential semantics. But one may replace the sequential composition by other operations on relations like, for example, parallel composition. Every choice of such operations defines an interlinking semantics of graph transformation units that, in particular, covers modes of parallel and concurrent processing (see, e.g.,[25]).
3. The semantic relations considered so far are associated to single units which control the computations and call the service of other units. In this sense, a graph transformation unit is a centralized computational entity. The concept of autonomuous units (see, e.g.,[26–28]) is a generalization to a decentralized processing of graphs. The autonomuous units in a community act, interact, and communicate in a common environment, with each of them controling its own activities autonomuously.

Acknowledgement. We are grateful to Andrea Corradini for valuable comments on an earlier version of the overview.

References

1. Rozenberg, G. (ed.): Handbook of Graph Grammars and Computing by Graph Transformation. Foundations, vol. 1. World Scientific, Singapore (1997)
2. Ehrig, H., Engels, G., Kreowski, H.J., Rozenberg, G. (eds.): Handbook of Graph Grammars and Computing by Graph Transformation. Applications, Languages and Tools, vol. 2. World Scientific, Singapore (1999)
3. Ehrig, H., Kreowski, H.J., Montanari, U., Rozenberg, G. (eds.): Handbook of Graph Grammars and Computing by Graph Transformation. Concurrency, Parallelism, and Distribution, vol. 3. World Scientific, Singapore (1999)

4. Andries, M., Engels, G., Habel, A., Hoffmann, B., Kreowski, H.J., Kuske, S., Plump, D., Schürr, A., Taentzer, G.: Graph transformation for specification and programming. Science of Computer Programming 34(1), 1–54 (1999)

5. Kreowski, H.J., Kuske, S.: Graph transformation units and modules. In: [2], pp. 607–638.

6. Kreowski, H.J., Kuske, S.: Graph transformation units with interleaving semantics. Formal Aspects of Computing 11(6), 690–723 (1999)

7. Kuske, S.: More about control conditions for transformation units. In: Ehrig, H., Engels, G., Kreowski, H.-J., Rozenberg, G. (eds.) TAGT 1998. LNCS, vol. 1764, pp. 323–337. Springer, Heidelberg (2000)

8. Kuske, S.: Transformation Units—A Structuring Principle for Graph Transformation Systems. PhD thesis, University of Bremen (2000)

9. Klempien-Hinrichs, R., Kreowski, H.J., Kuske, S.: Rule-based transformation of graphs and the product type. In: van Bommel, P. (ed.) Transformation of Knowledge, Information, and Data: Theory and Applications, pp. 29–51. Idea Group Publishing, Hershey, Pennsylvania (2005)

10. Klempien-Hinrichs, R., Kreowski, H.J., Kuske, S.: Typing of graph transformation units. In: Ehrig, H., Engels, G., Parisi-Presicce, F., Rozenberg, G. (eds.) ICGT 2004. LNCS, vol. 3256, pp. 112–127. Springer, Heidelberg (2004)

11. Plump, D., Steinert, S.: Towards graph programs for graph algorithms. In: Ehrig, H., Engels, G., Parisi-Presicce, F., Rozenberg, G. (eds.) ICGT 2004. LNCS, vol. 3256, pp. 128–143. Springer, Heidelberg (2004)

12. Mosbah, M., Ossamy, R.: A programming language for local computations in graphs: Computational completeness. In: 5th Mexican International Conference on Computer Science (ENC 2004), pp. 12–19. IEEE Computer Society, Los Alamitos (2004)

13. Corradini, A., Heindel, T., Hermann, F., König, B.: Sesqui-pushout rewriting. In: Corradini, A., Ehrig, H., Montanari, U., Ribeiro, L., Rozenberg, G. (eds.) ICGT 2006. LNCS, vol. 4178, pp. 30–45. Springer, Heidelberg (2006)

14. Drewes, F., Hoffmann, B., Janssens, D., Minas, M., Eetvelde, N.V.: Adaptive star grammars. In: Corradini, A., Ehrig, H., Montanari, U., Ribeiro, L., Rozenberg, G. (eds.) ICGT 2006. LNCS, vol. 4178, pp. 77–91. Springer, Heidelberg (2006)

15. Klempien-Hinrichs, R.: Hyperedge substitution in basic atom-replacement languages. In: Corradini, A., Ehrig, H., Kreowski, H.-J., Rozenberg, G. (eds.) ICGT 2002. LNCS, vol. 2505, pp. 192–206. Springer, Heidelberg (2002)

16. Ehrig, H., Habel, A., Padberg, J., Prange, U.: Adhesive high-level replacement systems: A new categorical framework for graph transformation. Fundamenta Informaticae 74(1), 1–29 (2006)

17. Ehrig, H., Ehrig, K., Prange, U., Taentzer, G. (eds.): Fundamentals of Algebraic Graph Transformation. Springer, Heidelberg (2006)

18. Ehrig, H., Engels, G.: Pragmatic and semantic aspects of a module concept for graph transformation systems. In: Cuny, J., Engels, G., Ehrig, H., Rozenberg, G. (eds.) Graph Grammars 1994. LNCS, vol. 1073, pp. 137–154. Springer, Heidelberg (1996)

19. Taentzer, G., Schürr, A.: DIEGO, another step towards a module concept for graph transformation systems. In: Corradini, A., Montanari, U. (eds.) SEGRAGRA 1995, Joint COMPUGRAPH/SEMAGRAPH Workshop on Graph Rewriting and Computation. Electronic Notes in Theoretical Computer Science, vol. 2, Elsevier, Amsterdam (1995)

20. Grosse-Rhode, M., Parisi Presicce, F., Simeoni, M.: Refinements and modules for typed graph transformation systems. In: Fiadeiro, J.L. (ed.) WADT 1998. LNCS, vol. 1589, pp. 137–151. Springer, Heidelberg (1999)
21. Große-Rhode, M., Parisi-Presicce, F., Simeoni, M.: Formal software specification with refinements and modules of typed graph transformation systems. Journal of Computer and System Sciences 64(2), 171–218 (2002)
22. Schürr, A., Winter, A.J.: UML packages for PROgrammed Graph REwriting Systems. In: Ehrig, H., Engels, G., Kreowski, H.-J., Rozenberg, G. (eds.) TAGT 1998. LNCS, vol. 1764, pp. 396–409. Springer, Heidelberg (2000)
23. Heckel, R., Engels, G., Ehrig, H., Taentzer, G.: Classification and comparison of module concepts for graph transformation systems. In: [2], pp. 639–689.
24. Kreowski, H.J., Kuske, S., Schürr, A.: Nested graph transformation units. International Journal on Software Engineering and Knowledge Engineering 7(4), 479–502 (1997)
25. Janssens, D., Kreowski, H.J., Rozenberg, G.: Main concepts of networks of transformation units with interlinking semantics. In: Kreowski, H.J., Montanari, U., Orejas, F., Rozenberg, G., Taentzer, G. (eds.) Formal Methods in Software and Systems Modeling. LNCS, vol. 3393, pp. 325–342. Springer, Heidelberg (2005)
26. Hölscher, K., Kreowski, H.-J., Kuske, S.: Autonomous units and their semantics — the sequential case. In: Corradini, A., Ehrig, H., Montanari, U., Ribeiro, L., Rozenberg, G. (eds.) ICGT 2006. LNCS, vol. 4178, pp. 245–259. Springer, Heidelberg (2006)
27. Hölscher, K., Klempien-Hinrichs, R., Knirsch, P., Kreowski, H.J., Kuske, S.: Autonomous units: Basic concepts and semantic foundation. In: Hülsmann, M., Windt, K. (eds.) Understanding Autonomous Cooperation and Control in Logistics. The Impact on Management, Information and Communication and Material Flow, Springer, Berlin, Heidelberg, New York (2007)
28. Kreowski, H.J., Kuske, S.: Autonomous units and their semantics - the parallel case. In: Fiadeiro, J.L., Schobbens, P.-Y. (eds.) WADT 2006. LNCS, vol. 4409, pp. 56–73. Springer, Heidelberg (2007)

Synchronous Multiparty Synchronizations and Transactions[*]

Ivan Lanese[1] and Hernán Melgratti[2]

[1] Dipartimento di Scienze dell'Informazione, Università di Bologna,
`lanese@cs.unibo.it`
[2] Departamento de Computación, FCEyN, Universidad de Buenos Aires,
`hmelgra@dc.uba.ar`

*Dedicated to Ugo Montanari
on the occasion of his 65th birthday*[**]

In this paper we analyze how a powerful synchronization mechanism such as synchronous multiparty synchronizations, which is able to specify atomic reconfigurations of large systems, can be implemented using binary synchronizations combined with a transactional mechanism. To this aim we show a mapping from SHR, a graph transformation framework allowing multiparty synchronizations, to a generalization of Fusion Calculus featuring a transactional mechanism inspired by the Zero-Safe Petri nets. To complete the correspondence between the two formalisms we also present a mapping in the opposite direction.

1 Introduction

A key aspect when specifying complex software systems is to find the right level of abstraction, and in particular the kind of primitives to be used. On the one side, one would expect to rely on very powerful primitives that enable for a simple and compact specification of complex behaviours. On the other side, one would prefer a small number of easily implementable primitives to shorten the gap between modeling and implementation.

In this paper we concentrate on synchronous multiparty synchronizations, which are a powerful tool for defining complex reconfigurations that allow the propagation of synchronizations through a chain of pairwise synchronized components. As an instance of this approach we have chosen Synchronized Hyperedge Replacement (SHR) [1,2,3], a graph transformation framework in which complex transformations are specified in terms of simple rules that define the behavior of single edges (or, more precisely, hyperedges). Although this makes SHR specifications quite compact, relying on a synchronous multiparty synchronization mechanism has always been considered as a drawback, mainly from the implementation point of view. We address this problem by showing how synchronous

---

[*] Research partially funded by EU Integrated Project Sensoria, contract n. 016004.

[**] We decided to work on highlighting the relationships among three different computational models (namely, the Fusion Calculus, Synchronized Hyperedge Replacement and Zero-Safe nets) since looking at problems from different perspectives is one of the lessons we have learnt from Ugo.

P. Degano et al. (Eds.): Montanari Festschrift, LNCS 5065, pp. 76–95, 2008.

Prefixes:
$$\alpha ::= \quad u\vec{x} \quad (\text{Input}) \quad | \quad \overline{u}\vec{x} \quad (\text{Output}) \quad | \quad \phi \quad (\text{Fusion})$$

Agents:

$$S ::= \sum_i \alpha_i.P_i \ (\text{Guarded sum})$$

$$P ::= \quad S \quad (\text{Sequential Agent}) \qquad\qquad P_1|P_2 \ (\text{Composition})$$
$$(x)P \ (\text{Scope}) \qquad\qquad\qquad C[\vec{v}] \ (\text{Constant application})$$

Fig. 1. Fusion Calculus syntax

multiparty synchronizations can be implemented by using normal binary synchronizations combined with a transactional mechanism. To this end we consider Fusion Calculus [4], a variant of π-calculus [5] whose close relation with SHR has been first shown in [6], and extend it with a transactional mechanism inspired by Zero-Safe Petri nets [7]. We call the resulting calculus Zero-Safe Fusion. While it is clear from [6] that Fusion Calculus is less expressive than SHR, we show that Zero-Safe Fusion is able to model the full SHR. We also show how to extend the mapping from Fusion to SHR in [6] to deal with Zero-Safe Fusion, thus proving that the two frameworks have the same expressive power.

Structure of the paper: Sections 2 and 3 present the background on Fusion Calculus and SHR respectively. Section 4 introduces Zero-Safe Fusion Calculus. Then, the mappings from SHR to Zero-Safe Fusion is presented in Section 5, while Section 6 discusses the assumptions made to simplify the mapping. Section 7 shows the inverse mapping. Finally, Section 8 draws some conclusions. For space limitation, we omit here the proofs of main results and we refer the interested reader to the full version of this paper [8].

2   Fusion Calculus

Fusion [4,9] has been proposed as a calculus for modeling distributed and mobile systems, and it is based on binary synchronization and name fusion. Interestingly, it both simplifies and generalizes the π-calculus [5]. This paper will consider Fusion Calculus with guarded summation, constant definitions to model infinite behaviors, and without match and mismatch operators. This section reports on its original syntax and structural congruence, while we refer to [9] for the original presentation of its operational semantics. Section 4 defines the operational semantics of a generalization of Fusion Calculus that accounts for transactions, which subsumes the original calculus.

We rely on an infinite set $\mathcal{N}$ of channel names ranged over by $u, v, \ldots, z$. We write $\vec{x}$ for indicating a vector of names, and ϕ for denoting an equivalence relation over $\mathcal{N}$ that can be represented by a finite set of equalities (that is always considered closed under reflexivity, symmetry and transitivity).

Definition 1 (Fusion syntax). *The sets of Fusion prefixes, ranged over by α, and Fusion agents, ranged over by $P, Q, \ldots$ are defined in Figure 1. Constants, ranged over by C are defined by equalities like $C[\vec{x}] \triangleq P$. We require the occurrence of constants to be prefix-guarded inside constant definitions.*

Note that we syntactically distinguish sequential agents S, i.e., agents whose topmost operator is a summation, from general agents P. We define the inactive process 0 as the empty summation, and we will omit trailing occurrences of 0. We use + to denote binary summation.

The scope operator acts as a binder for names, thus x is bound in $(x)P$. Given an agent P, $\mathrm{fn}(P)$ denotes the set of all names occurring free in P. For constant definitions $C[\vec{x}] \triangleq P$ we require $\mathrm{fn}(P) \subseteq \vec{x}$. Processes are equivalence classes of agents defined by the structural congruence introduced below.

Definition 2. *The structural congruence $\equiv$ between agents is the least congruence satisfying the α-conversion law, the abelian monoid laws for summation and composition (associativity, commutativity and 0 as identity), the scope laws $(x)0 \equiv 0$, $(x)(y)P \equiv (y)(x)P$, the scope extension law $P|(z)Q \equiv (z)(P|Q)$ when $z \notin \mathrm{fn}(P)$, and the law for constants $C[\vec{v}] \equiv P\{\vec{v}/\vec{x}\}$ if $C[\vec{x}] \triangleq P$.*

Note that fn can be trivially extended from agents to processes.

3   Synchronized Hyperedge Replacement

Synchronized Hyperedge Replacement (SHR) is a graph transformation framework in which the evolution of a whole graph is defined in terms of synchronizing rules, called productions. Each production describes the evolution of a single edge. SHR has been introduced in [10], and several variants studying alternative mechanisms for handling synchronization and mobility have been proposed in the literature [11,12,13]. In this paper we concentrate on Milner SHR [11] as presented in [12]: the synchronization model has been inspired by Milner's π-calculus [5], and the mobility model follows the Fusion style.

Our graphs (or, more precisely, hypergraphs) are composed by edges and nodes. Each edge has a label L and it is attached to n nodes, where n is the rank $\mathrm{rank}(L)$ of its label. A graph is connected to its environment by an interface, which is a subset of the nodes of the graph. Nodes in the interface are free, while the others are bound. We give an algebraic definition of graphs.

Definition 3 (Graphs). *Let $\mathcal{N}$ be a fixed infinite set of nodes and LE a ranked set of labels. A graph has the form $\Gamma \vdash G$ where:*

1. *$\Gamma \subseteq \mathcal{N}$ is a finite set of nodes (the free nodes of the graph);*
2. *G is a graph term generated by the grammar*
   $G ::= L(\vec{x}) \mid G_1|G_2 \mid \nu y\, G \mid \mathrm{nil}$
   where $\vec{x} \in \mathcal{N}^*$ is a vector of nodes, $L \in LE$ is an edge label with $\mathrm{rank}(L) = |\vec{x}|$ and $y \in \mathcal{N}$ is a node.

Table 1. Structural congruence for graphs

$$(AG1)\ (G_1|G_2)|G_3 \equiv G_1|(G_2|G_3) \quad (AG2)\ G_1|G_2 \equiv G_2|G_1 \quad (AG3)\ G|nil \equiv G$$
$$(AG4)\ \nu x\ \nu y\ G \equiv \nu y\ \nu x\ G \quad (AG5)\ \nu x\ G \equiv G \text{ if } x \notin \mathrm{fn}(G)$$
$$(AG6)\ \nu x\ G \equiv \nu y\ G\{y/x\} \text{ if } y \notin \mathrm{fn}(G)$$
$$(AG7)\ \nu x\ (G_1|G_2) \equiv G_1|\nu x\ G_2 \text{ if } x \notin \mathrm{fn}(G_1)$$

The restriction operator ν is a binder, and we denote with $\mathrm{fn}(G)$ the set of free nodes in G. We demand that $\mathrm{fn}(G) \subseteq \Gamma$.

Graph terms and graphs are considered up to the structural congruence $\equiv$ given by the axioms in Table 1.

Example 1. A ring-shaped graph composed by three edges with label R of rank 2 can be written as:

$$x \vdash \nu y, z\ R(x,y)|R(y,z)|R(z,x)$$

Notice that here nodes y and z are bound, while x is free.

Graphs interact by performing actions on nodes, thus we assume a set of actions *Act*. Since we are interested in Milner synchronization, we assume that $Act = In \uplus Out \uplus \{\tau\}$ where In is a set of input actions, Out a set of output actions, and τ a special action denoting a complete (internal) synchronization. Each action a has an arity $\mathrm{arity}(a)$, denoting the number of parameters. We assume a bijection $\bar{\ }$ between In and Out (and we use the same notation to denote the inverse) such that $\mathrm{arity}(\bar{a}) = \mathrm{arity}(a)$. We also assume $\mathrm{arity}(\tau) = 0$. Then, the transitions describing the evolution of a graph are as follows.

Definition 4 (SHR transitions). *A SHR transition is a relation of the form:*

$$\Gamma \vdash G \xrightarrow{\Lambda, \pi} \Phi \vdash G'$$

where $\Gamma \vdash G$ and $\Phi \vdash G'$ are graphs, $\Lambda : \Gamma \to (Act \times \mathcal{N}^*)$ is a partial function and $\pi : \Gamma \to \Gamma$ is an idempotent renaming. Function Λ assigns to each node x the action a and the vector $\vec{y}$ of node references communicated to x. If $\Lambda(x) = (a, \vec{y})$ then we define $\mathrm{act}_\Lambda(x) = a$ and $\mathrm{n}_\Lambda(x) = \vec{y}$. We require that $\mathrm{arity}(\mathrm{act}_\Lambda(x)) = |\,\mathrm{n}_\Lambda(x)|$.
    We define:

- $\mathrm{n}(\Lambda) = \{z | \exists x.z \in \mathrm{n}_\Lambda(x)\}$ *set of communicated nodes;*
- $\Gamma_\Lambda = \mathrm{n}(\Lambda) \setminus \Gamma$ *set of communicated fresh nodes.*

Renaming π allows to merge nodes. Since π is idempotent, it maps every node into a standard representative of its equivalence class. We require that $\forall x \in \mathrm{n}(\Lambda).x\pi = x$, i.e., only references to representatives can be communicated. Furthermore, we require $\Phi = \Gamma\pi \cup \Gamma_\Lambda$, namely free nodes can be merged but never erased ($\supseteq$) and new nodes are bound unless communicated ($\subseteq$).

We will not write π when it is the identity.

Example 2 (Ring to star). The transition modeling the classical [1] ring-to-star reconfiguration for the ring graph in Example 1 is:

$$x \vdash \nu y, z \ R(x,y)|R(y,z)|R(z,x) \xrightarrow{(x,\tau,\langle\rangle)} x \vdash \nu y, z, w \ S(w,x)|S(w,y)|S(w,z)$$

The graph on the right-hand-side is composed by edges with label S of arity 2. Notice that actions are only observed on the free nodes, i.e. on x.

We will use this example as running example throughout the paper.

Definition 5 (Productions). *A production is an SHR transition of the form:*

$$x_1, \ldots, x_n \vdash L(x_1, \ldots, x_n) \xrightarrow{\Lambda, \pi} \Phi \vdash G$$

where $x_1, \ldots, x_n$ are all distinct.

For each edge label L of rank n we assume an idle production $x_1, \ldots, x_n \vdash$ $L(x_1, \ldots, x_n) \xrightarrow{\Lambda_\perp, id} x_1, \ldots, x_n \vdash L(x_1, \ldots, x_n)$ where $\Lambda_\perp$ is always undefined. Idle productions are included in all sets of productions.

Given a set *Prod* of productions, valid SHR transitions are derived from the productions in *Prod* via the inference rules below.

Definition 6 (Inference rules for Milner SHR)

$$(\textit{prod-M}) \quad \frac{x_1, \ldots, x_n \vdash L(x_1, \ldots, x_n) \xrightarrow{\Lambda, \pi} \Phi \vdash G \in Prod}{\left(x_1, \ldots, x_n \vdash L(x_1, \ldots, x_n) \xrightarrow{\Lambda, \pi} \Phi \vdash G\right)\sigma}$$

where σ is a bijective renaming and it is applied to all the parts of the production.

$$(\textit{par-M}) \quad \frac{\Gamma \vdash G_1 \xrightarrow{\Lambda, \pi} \Phi \vdash G_2 \qquad \Gamma' \vdash G_1' \xrightarrow{\Lambda', \pi'} \Phi' \vdash G_2'}{\Gamma, \Gamma' \vdash G_1|G_1' \xrightarrow{\Lambda \cup \Lambda', \pi \cup \pi'} \Phi, \Phi' \vdash G_2|G_2'}$$

where $(\Gamma \cup \Phi) \cap (\Gamma' \cup \Phi') = \emptyset$.

$$(\textit{merge-M}) \quad \frac{\Gamma \vdash G_1 \xrightarrow{\Lambda, \pi} \Phi \vdash G_2}{\Gamma\sigma \vdash G_1\sigma \xrightarrow{\Lambda', \pi'} \Phi' \vdash \nu U \ G_2\sigma\rho}$$

where $\sigma = \{x/y\}$ is a renaming and:

1.a $x, y \in \text{dom}(\Lambda) \Rightarrow \text{act}_\Lambda(x) = a \wedge \text{act}_\Lambda(y) = \bar{a} \wedge a \neq \tau$
1.b $\rho = \text{mgu}(\{(\mathsf{n}_\Lambda(x))\sigma = (\mathsf{n}_\Lambda(y))\sigma\} \cup \{z\sigma = w\sigma | z\pi = w\pi\})$ where ρ maps names to representatives in $\Gamma\sigma$ whenever possible
1.c $\Lambda'(z) = \begin{cases} (\tau, \langle\rangle) & \text{if } z = x \wedge x, y \in \text{dom}(\Lambda) \\ (\Lambda(w))\sigma\rho & \text{if } \exists w \in \text{dom}(\Lambda).w\sigma = z \\ \text{undefined} & \text{otherwise} \end{cases}$
1.d $\pi' = \rho|_{\Gamma\sigma}$

1.e $U = (\varPhi\sigma\rho) \setminus \varPhi'$

$$(\textit{res-M}) \quad \frac{\Gamma, x \vdash G_1 \xrightarrow{\Lambda,\pi} \varPhi \vdash G_2}{\Gamma \vdash \nu x\ G_1 \xrightarrow{\Lambda|_\Gamma,\pi|_\Gamma} \varPhi' \vdash \nu Z\ G_2}$$

where:

2.a $(\exists y \in \Gamma.x\pi = y\pi) \Rightarrow x\pi \neq x$
2.b $\mathrm{act}_\Lambda(x)$ undefined $\vee\ \mathrm{act}_\Lambda(x) = \tau$
2.c $Z = \{x\}$ if $x \notin \mathrm{n}(\Lambda|_\Gamma), Z = \emptyset$ otherwise

$$(\textit{new-M}) \quad \frac{\Gamma \vdash G_1 \xrightarrow{\Lambda,\pi} \varPhi \vdash G_2}{\Gamma, x \vdash G_1 \xrightarrow{\Lambda,\pi} \varPhi, x \vdash G_2}$$

where $x \notin \Gamma \cup \varPhi$.

Rule (prod-M) says that productions are the basic transitions, and it allows them to be bijectively renamed. Rule (par-M) accounts for the parallel evolution of two disjoint components of a graph. Rule (merge-M) deals with synchronization and allows to merge two nodes x and y into a unique node x (represented by the substitution $\sigma = \{x/y\}$) provided that actions performed on x and y are complementary (if they are both defined). The result of such an interaction is represented by the action τ over the node x (see the definition of Λ' in side-condition *1.c*). The effect of merging two nodes is also reflected by the renaming π'. Note that in addition to the original renaming π, π' should reflect that (i) x and y have been merged (second operand of the set union expression in side-condition *1.b*) and (ii) the names communicated by the action performed on x are fused with the names communicated by the action on y (first operand of the set union expression in side-condition *1.b*). Then, the obtained ρ and π' are applied to the names in the transition label Λ, to the obtained graph G_2, and to the set $\varPhi$ of free names of G_2. Moreover, νU binds the names that occur free in G_2 but are not in $\varPhi'$. This is analogous to the rule *close* of π-calculus. Rule (res-M) allows to bind a node x whenever no action is performed on it or a τ action is performed. Finally, rule (new-M) allows to introduce a free isolated node x to both the left- and right-hand-side graphs of the transition.

Example 3. The ring-to-star reconfiguration transition in Example 2 can be derived from the production:

$$x, y \vdash R(x, y) \xrightarrow{(x,in_1,\langle w\rangle),(y,out_1,\langle w\rangle)} x, y, w \vdash S(w, x)$$

using the inference rules assuming $\overline{in_1} = out_1$. For each node, we will have a synchronization between the actions in_1 and out_1, which will also merge the nodes w belonging to different productions (actually, nodes w are renamed by (prod-M) in order to make rule (par-M) applicable).

The following lemma shows that derivations of SHR transitions can be normalized to some extent.

Lemma 1. *Suppose to have a derivation for an SHR transition. Then the same transition can be derived using a normalized derivation where no rule different from (res-M) is applied after any application of rule (res-M), i.e. all applications of (res-M) are at the end of the derivation.*

Proof. This is a consequence of the fact that structurally equivalent graphs have the same transitions, see e.g. [14].

4   Zero-Safe Fusion Calculus

In this section we introduce *Zero-Safe Fusion* (ZS Fusion), which extends Fusion Calculus [4,9] with a transaction mechanism inspired by the Zero-Safe Petri nets [7]. At the syntactic level we introduce a new form of prefixes, called *transactional prefixes*, which are composed by several standard prefixes that are executed as a unique action. Transactional prefixes are a general version of the join prefixes introduced by the Join calculus [15] and adopted by the *General Rendezvous Calculus* (GRC) [16]. The main difference is that our proposal allows for output and fusion actions inside join prefixes, while Join and GRC do not. Moreover, all actions occurring in a transactional prefix can be executed concurrently. This is the main difference of ZS Fusion with respect to other proposals appeared in the literature [17,18], in which transactional prefixes are sequences of basic actions.

4.1   Syntax

ZS Fusion agents are analogous to Fusion agents introduced by Definition 1, while sequential agents have the following form:

$$S ::= \sum_i \beta_i.P_i \quad \text{(Guarded sum)} \qquad \text{where } \beta \text{ is} \quad \beta ::= \wedge_i \alpha_i \quad \text{(Transaction)}$$

The prefix $\wedge_i \alpha_i$ stands for the transactional execution of several standard prefixes α_i. Normal prefixes are recovered as transactional prefixes containing just one component. Thus, we use $\alpha_1.P$ as a shorthand for $\wedge_{i \in \{1\}} \alpha_i.P$. We also use $\wedge.P$ as a shorthand for $\wedge_{i \in \emptyset} \alpha_i.P$.

In addition to the static syntax, we use the production below as run-time syntax to provide the operational semantics of the calculus.

$$S ::= \langle \beta \rangle.P \qquad \text{(Partial transaction)}$$

where the prefix $\langle \beta \rangle$ denotes a started transaction that has to execute β to complete. We write $\langle \rangle.P$ as a shorthand for $\langle \wedge_{i \in \emptyset} \alpha_i \rangle.P$.

Processes are agents up-to the structural congruence in Definition 2.

A sequential process S is *stable* if it has the form $\sum_i \beta_i.P_i$ or $\langle\rangle.P$, i.e., all partial transactional prefixes are empty. A sequential process S is *static* if it has the form $\sum_i \beta_i.P_i$. A process P is stable (resp. static) when all its sequential subprocesses are stable (resp. static).

Given a stable process P, $\widetilde{P}$ denotes the static process obtained from P by removing all prefixes of the form $\langle\rangle$. Formally,

$$\widetilde{\langle\rangle.P} = P \qquad \widetilde{P_1|P_2} = \widetilde{P_1}|\widetilde{P_2} \qquad \widetilde{(x)P} = (x)\widetilde{P} \qquad \widetilde{P} = P \; otherwise$$

4.2 Operational Semantics

We define the operational semantics of the ZS Fusion Calculus in two steps: first, we define a small-step behavior, and then we provide the transactional behavior of the calculus in terms of the small-steps.

Transactions require a few constraints to be satisfied:

1. the starting and final processes of each transaction must be static: this ensures that each transactional prefix has been completely executed;
2. there can be at most one communication/synchronization on each channel during each transaction: intuitively this means that channels are resources that cannot be shared, thus during each transaction only one process or a pair of interacting processes can use them;
3. fusions produced inside a transaction should be applied only when the transaction ends: on one side, this corresponds to the fact that tokens produced by a Zero-Safe transaction are available only at the end of the transaction and, on the other side, this ensures that the order in which elementary actions are executed inside a transaction is irrelevant.

In order to satisfy conditions 2 and 3, we define the small-step semantics only for processes without restriction, while we take restrictions into account when computing transactions. This is necessary since restrictions force the application of fusions, which is against condition 3. Notice also that α-conversion of bound names would make the checking of condition 2 more complex.

The small-step semantics exploits the labels below:

$$\gamma ::= u\vec{x} \mid \overline{u}\vec{x} \mid \{\vec{x} = \vec{y}\} \mid \gamma \wedge \gamma$$

We write $\emptyset$ for the empty substitution and we consider $\emptyset$ as the neutral element for $\wedge$.

The first three labels are standard Fusion labels for input, output and fusions, while the last one is the label for transactions. Given an action γ, $subj(\gamma)$ is the set of channels over which the actions in γ are taking place. Similarly $obj(\gamma)$ is the set of objects. They are defined as follows, where $\mathrm{Set}(\vec{x})$ is the set of names in $\vec{x}$.

$$subj(u\vec{x}) = \{u\} \qquad\qquad subj(\overline{u}\vec{x}) = \{u\}$$
$$subj(\{\vec{x} = \vec{y}\}) = \emptyset \qquad subj(\gamma_1 \wedge \gamma_2) = subj(\gamma_1) \cup subj(\gamma_2)$$

$$obj(u\vec{x}) = \mathrm{Set}(\vec{x}) \qquad\qquad obj(\overline{u}\vec{x}) = \mathrm{Set}(\vec{x})$$
$$obj(\{\vec{x} = \vec{y}\}) = \emptyset \qquad obj(\gamma_1 \wedge \gamma_2) = obj(\gamma_1) \cup obj(\gamma_2)$$

(START)
$$subj(\alpha_i) \vdash \alpha_1 \wedge \ldots \wedge \alpha_n.P \xrightarrow{\alpha_i} \langle \alpha_1 \wedge \ldots \wedge \alpha_{i-1} \wedge \alpha_{i+1} \wedge \ldots \wedge \alpha_n \rangle.P$$

(PREF)
$$subj(\alpha_i) \vdash \langle \alpha_1 \wedge \ldots \wedge \alpha_n \rangle.P \xrightarrow{\alpha_i} \langle \alpha_1 \wedge \ldots \wedge \alpha_{i-1} \wedge \alpha_{i+1} \wedge \ldots \wedge \alpha_n \rangle.P$$

(EMPTY)
$$\emptyset \vdash \wedge.P \xrightarrow{\emptyset} \langle \rangle.P$$

(SUM)
$$\frac{\Gamma \vdash P \xrightarrow{\gamma} P'}{\Gamma \vdash P + Q \xrightarrow{\gamma} P'}$$

(PAR)
$$\frac{\Gamma \vdash P \xrightarrow{\gamma} P'}{\Gamma \vdash P|Q \xrightarrow{\gamma} P'|Q}$$

(COM)
$$\frac{\Gamma \vdash P \xrightarrow{u\vec{x}} P' \quad \Gamma \vdash Q \xrightarrow{\overline{u}\vec{y}} Q' \quad |\vec{x}| = |\vec{y}|}{\Gamma \vdash P|Q \xrightarrow{\{\vec{x}=\vec{y}\}} P'|Q'}$$

(SEQ)
$$\frac{\Gamma \vdash P \xrightarrow{\gamma_1} P'' \quad \Gamma' \vdash P'' \xrightarrow{\gamma_2} P' \quad \Gamma \cap \Gamma' = \emptyset}{\Gamma \cup \Gamma' \vdash P \xrightarrow{\gamma_1 \wedge \gamma_2} P'}$$

Fig. 2. Small-step semantics for Zero-Safe Fusion

We also define the following auxiliary functions to work with labels:

$$\mathrm{subst}(u\vec{x}) = \emptyset \qquad \mathrm{subst}(\overline{u}\vec{x}) = \emptyset$$
$$\mathrm{subst}(\{\vec{x} = \vec{y}\}) = \{\vec{x} = \vec{y}\} \qquad \mathrm{subst}(\gamma_1 \wedge \gamma_2) = \mathrm{subst}(\gamma_1) \cup \mathrm{subst}(\gamma_2)$$

$$\mathrm{comm}(u\vec{x}) = \{u\vec{x}\} \qquad \mathrm{comm}(\overline{u}\vec{x}) = \{\overline{u}\vec{x}\}$$
$$\mathrm{comm}(\{\vec{x} = \vec{y}\}) = \emptyset \qquad \mathrm{comm}(\gamma_1 \wedge \gamma_2) = \mathrm{comm}(\gamma_1) \cup \mathrm{comm}(\gamma_2)$$

The small-step semantics is defined by the labeled transition system in Figure 2. Rules are quite standard. The key point is that the set Γ records channels involved in the action γ. Note that sequential composition is allowed only when the actions take place over different channels. Rule (EMPTY) allows to execute empty transactions.

Labels for transactional semantics have the form $(Y)S, \phi$ where Y is the set of extruded names, S is a set of input/output actions $x\vec{y}$ or $\overline{x}\vec{y}$ and ϕ is a fusion. We drop (Y) if Y is empty.

Given a fusion ϕ we define $\phi \setminus z$ as $(\phi \cap ((\mathcal{N} \setminus \{z\}) \times (\mathcal{N} \setminus \{z\}))) \cup \{(z, z)\}$. We extend to labels the notions of free and bound names (the names in a fusion are those names belonging to non-singleton equivalence classes).

Transactional semantics is defined by the labeled transition system in Figure 3. Rule (CMT) states that a transactional step is a small-step among two stable processes. Rule (SCOPE) allows to bind a name z belonging to a non-trivial equivalence class induced by ϕ (and not occurring as a subject in S). Name z is renamed by an equivalent name x both in the label and in the resulting process. The remaining rules are the usual ones.

The following definition introduces a notion of decomposition of Fusion processes that will be used in the following sections.

Definition 7. *The standard decomposition of a process P is defined as:*

$$P = \hat{P}\sigma_P$$

(CMT)
$$\frac{\Gamma \vdash P \xrightarrow{\gamma} P' \qquad P \text{ and } P' \text{ stable}}{P \xrightarrow{(\mathrm{comm}(\gamma),\mathrm{subst}(\gamma))} \widetilde{P'}}$$

(SCOPE)
$$\frac{P \xrightarrow{(Y)S,\phi} P', z\phi x, z \neq x, z \notin subj(S)}{(z)P \xrightarrow{(Y)S\{x/z\},\phi\backslash z} P'\{x/z\}}$$

(PASS)
$$\frac{P \xrightarrow{\lambda} P', z \notin \mathrm{fn}(\lambda)}{(z)P \xrightarrow{\lambda} (z)P'}$$

(OPEN)
$$\frac{P \xrightarrow{(Y)S,\phi} P', z \in obj(S) \setminus Y, z \notin subj(S), z \notin \mathrm{fn}(\phi)}{(z)P \xrightarrow{(Y\cup\{z\})S,\phi} P'}$$

(STRUCT)
$$\frac{P \xrightarrow{\lambda} P' \quad P \equiv Q \quad P' \equiv Q'}{Q \xrightarrow{\lambda} Q'}$$

Fig. 3. Transactions for Zero-Safe Fusion

where σ_P is the standard substitution and $\hat{P}$ is the standard process of P. This decomposition satisfies:

$$P = Q\sigma \text{ implies } \hat{P} = \hat{Q} \wedge \sigma_P = \sigma_Q\sigma$$

We denote with $fnarray(P)$ the array of the free name occurrences in P, ordered according to some fixed order dictated by the structure of (a suitable normal form w.r.t $\equiv$ of) P. In particular $\sigma_P = \{fnarray(P)/fnarray(\hat{P})\}$.

The following propositions about small-steps will be used for studying properties of transactions in ZS Fusion.

Proposition 1. *If $\Gamma \vdash P \xrightarrow{\gamma_1 \wedge \gamma_2} P''$ then there exists P' such that $\Gamma' \vdash P \xrightarrow{\gamma_1} P'$ and $\Gamma'' \vdash P' \xrightarrow{\gamma_2} P''$ with $\Gamma' \cup \Gamma'' = \Gamma$.*

Proposition 2. *If $\Gamma \vdash P \xrightarrow{\gamma_1 \wedge \gamma_2} P'$ then $\Gamma \vdash P \xrightarrow{\gamma_2 \wedge \gamma_1} P'$.*

Proposition 3. *If $\Gamma \vdash P \xrightarrow{\gamma_1 \wedge (\gamma_2 \wedge \gamma_3)} P'$ then $\Gamma \vdash P \xrightarrow{(\gamma_1 \wedge \gamma_2) \wedge \gamma_3} P'$.*

Because of the propositions above from now on we will consider labels up to associativity and commutativity of $\wedge$.

Proposition 4. *Given a process $P = \langle\beta\rangle.Q$ or $P = \beta.Q$. If $\Gamma \vdash P \xrightarrow{\gamma} P'$ with P' stable, then*

1. *$\gamma = \beta$ and $P' = \langle\rangle.Q$;*
2. *$\Gamma = \{subj(\alpha)|\alpha \in comm(\gamma)\}$.*

Proposition 5. *If $\Gamma \vdash P \xrightarrow{\gamma} Q$, then*

1. *$subj(\alpha_i) \neq subj(\alpha_j)$ for all $\alpha_i, \alpha_j \in comm(\gamma)$ with $i \neq j$.*
2. *$\{subj(\alpha)|\alpha \in comm(\gamma)\} \subseteq \Gamma$.*

Notice that in the proposition above we may have $\{subj(\alpha)|\alpha \in comm(\gamma)\} \subsetneq \Gamma$ since names on which a synchronization is performed are in the right-hand-side but not in the left-hand-side.

In order to state the proposition below we introduce the following notation: $\Gamma \vdash P \xrightarrow{\gamma}_\epsilon P'$ stands for either $\Gamma \vdash P \xrightarrow{\gamma} P'$ or $P = P'$, with empty Γ and γ.

Proposition 6. *If $\Gamma \vdash P_1|P_2 \xrightarrow{\gamma}_\epsilon Q$ with Q stable, then $\Gamma_1 \vdash P_1 \xrightarrow{\gamma_1}_\epsilon Q_1$ and $\Gamma_2 \vdash P_2 \xrightarrow{\gamma_2}_\epsilon Q_2$ with:*

1. *$\Gamma = \Gamma_1 \cup \Gamma_2$, $\Gamma_1 \cap \Gamma_2 = subj(\gamma_1) \cap subj(\gamma_2)$;*
2. *$\gamma = \gamma_1|_D \wedge \gamma_2|_D \wedge F_{\gamma_1|\gamma_2}$ where*
   $D = \Gamma \setminus (subj(\gamma_1) \cap subj(\gamma_2))$, $\gamma|_D$ denotes the conjunction of all actions in γ whose subjects are in D, and $F_{\gamma_1|\gamma_2}$ denotes the set of fusions $\vec{x} = \vec{y}$ such that $a\vec{x} \in \gamma_1$ and $\overline{a}\vec{y} \in \gamma_2$;
3. *$Q \equiv Q_1|Q_2$.*

Proof. If follows by induction on the derivation $\Gamma \vdash P_1|P_2 \xrightarrow{\gamma} Q$. ☐

The following proposition shows that ZS Fusion transactions are fully determined by their set of consumed prefixes. We define prefixes β_i consumed by a transaction $P \overset{\lambda}{\Longrightarrow} P'$ as the ones occurring in P but not in P'. To this end, one needs to distinguish different occurrences of the same prefix. We will not formalize this idea, but, roughly, given a process one can assign unique tags to each prefix, and make transitions to preserve the tag of non-consumed prefixes.

Proposition 7. *Let P be a process. If $P \overset{\lambda}{\Longrightarrow} Q$ and $P \overset{\lambda'}{\Longrightarrow} Q'$ are two transactions consuming the same prefixes (i.e., prefixes with the same tags), then $\lambda = \lambda'$ and $Q \equiv Q'$.*

The following lemma characterizes sets of compatible prefixes, i.e., prefixes that can be consumed together in a transaction.

Lemma 2. *Let P be a process, and consider a transaction that consumes a set of prefixes $S = \{\beta_i\}_{i \in I}$ in P. Then for each name x in $subj(S)$ there are two possibilities:*

- *x is not bound in P, and it is the subject of exactly one action in S;*
- *x is the subject of two complementary actions in S.*

5   From SHR to Zero-Safe Fusion

This section presents a translation from SHR to ZS Fusion Calculus, i.e., given a graph $\Gamma \vdash G$ and a set of productions $Prod$ describing its behavior, we translate them into a process $[\![\Gamma \vdash G]\!]_{g2p}$ together with a set of constant definitions $[\![Prod]\!]_{p2c}$. We also provide an operational correspondence result relating the original SHR model and its encoding into ZS Fusion.

Table 2. Translations

Translation	From	To	Section
$[\![\bullet]\!]_{g2p}$	Graphs	Processes	5
$[\![\bullet]\!]_{p2c}$	Production sets	Constant definitions	5
$[\![\bullet]\!]_{a2p}$	SHR actions	Prefixes/labels	5
$[\![\bullet]\!]_{p2p}$	Productions	Processes	5
$[\![\bullet]\!]_{a2g}$	Agents	Graph terms	7
$[\![\bullet]\!]_{p2a}$	Prefixes	SHR actions	7

In order to define the mapping above and the inverse mapping presented in Section 7, we exploit different translation functions, denoted by $[\![\bullet]\!]_\pi$, distinguished by their subscript π. To make the presentation clearer we summarize the used translation functions in Table 2.

In this section we only consider SHR models whose set Act of SHR actions is such that In contains at most one action for each arity. Consequently, we use in_n to denote the input action of arity n, and out_n for the corresponding output action. In addition, we consider only productions $\Gamma \vdash G \xrightarrow{\Lambda,\pi} \Phi \vdash G'$ where $\pi = id$ and action τ is never used in Λ. We also assume that each SHR edge is attached to any chosen node at most once. Section 6 shows that these assumptions are not restrictive.

We assume two bijective mappings: one between SHR nodes and ZS Fusion names, and the other between SHR edge labels and Fusion constant names. For simplicity, we will not write these mappings explicitly.

Definition 8 (Translation from graphs to processes). *We define the translation $[\![\Gamma \vdash G]\!]_{g2p}$ of a graph $\Gamma \vdash G$ by induction on the structure of G (we drop Γ because it is not relevant):*

$$[\![L(\vec{x})]\!]_{g2p} = L[\vec{x}] \qquad [\![G_1|G_2]\!]_{g2p} = [\![G_1]\!]_{g2p}\,|\,[\![G_2]\!]_{g2p}$$

$$[\![\nu y\ G]\!]_{g2p} = (y)[\![G]\!]_{g2p} \qquad [\![\mathrm{nil}]\!]_{g2p} = 0$$

Lemma 3. *If $G \equiv G'$ then $[\![G]\!]_{g2p} \equiv [\![G']\!]_{g2p}$.*

We also need a translation from SHR actions to ZS Fusion prefixes. We overload the notation using the same translation function also from SHR synchronization functions Λ to ZS Fusion labels, since it is a straightforward generalization of the previous one.

Definition 9 (Translation for actions). *Basic actions are translated as follows:*

$[\![(x, in_n, \vec{y})]\!]_{a2p} = x\vec{y}$

$[\![(x, out_n, \vec{y})]\!]_{a2p} = \overline{x}\vec{y}$

The translation of a synchronization function Λ is the set S obtained by translating all the basic actions in Λ but the ones of the form $(x, \tau, \langle\rangle)$, which are deleted. With $[\![\Lambda]\!]_{a2p}$ we also denote $\bigwedge_{\alpha \in S} \alpha$.

As already mentioned, constant definitions are obtained from productions. In particular, we produce a constant definition for each edge label L.

Definition 10 (Translation from productions to constants). *Let L be a label and let $Prod_L$ be the set of non-idle productions for L. We can assume that all the productions in $Prod_L$ have the same left-hand-side $L(x_1,\ldots,x_n)$ (this can be guaranteed up to α-conversion). Then, the translation of $Prod_L$ gives the following definition:*

$$L[x_1,\ldots,x_n] \triangleq \sum_{Pr \in Prod_L} [\![Pr]\!]_{p2p}$$

where the translation of a single production is:

$$[\![\Gamma \vdash L(x_1,\ldots,x_n) \xrightarrow{\Lambda} \Gamma' \vdash G]\!]_{p2p} = (Y')(\bigwedge_{l \in \Lambda} [\![l]\!]_{a2p}).[\![\Gamma' \vdash G]\!]_{g2p}$$

where $Y' = \Gamma' \setminus \Gamma$. To be precise the translation above does not produce ZS Fusion processes, since one can have restrictions just below sums, but one can easily move them outside to have guarded sums.

Example 4. The production

$$x,y \vdash R(x,y) \xrightarrow{(x,in_1,\langle w\rangle),(y,out_1,\langle w\rangle)} x,y,w \vdash S(w,x)$$

from Example 3 gives rise to the following constant definition:

$$R[x,y] \triangleq \nu w\ (xw \wedge \overline{y}w).S[w,x]$$

If we consider also a production:

$$x,y \vdash R(x,y) \rightarrow x,y \vdash \nu z\ R(x,z)|R(z,y)$$

then the constant definition becomes:

$$R[x,y] \triangleq \nu w,z\ (xw \wedge \overline{y}w).S[w,x] + \wedge.(R(x,z)|R(z,y))$$

We describe now the relation between the possible evolutions of a graph $\Gamma \vdash G$ and the computations of $[\![\Gamma \vdash G]\!]_{g2p}$. Roughly, there is a bijective correspondence between transitions of $\Gamma \vdash G$ and transactions of $[\![\Gamma \vdash G]\!]_{g2p}$. Nevertheless, the two models handle fusions differently: transactions in ZS Fusion apply the computed fusion ϕ to restricted names only, and propagate the remaining part, while SHR transitions apply the substitution π (i.e., an mgu of ϕ) immediately to both the obtained graph and the objects of performed actions. We introduce the notation $S \triangleleft \pi$ to indicate that the substitution π is applied only to the objects of the actions in S. We also write open(P) to denote the set of processes obtained from P by removing all restrictions (this is a set since bound names are α-convertible; we assume however that distinct names are kept distinct).

We start the proof of operational correspondence by stating the following lemma that deals with small-step transitions.

Lemma 4. *For each graph $\Gamma \vdash G$ without restriction operator and each set of productions Prod, if there is an SHR transition $\Gamma \vdash G \xrightarrow{\Lambda,\pi} \Phi \vdash G'$ then for each $P' \in \mathrm{open}(\llbracket G \rrbracket_{g2p})$ there is a small-step $P' \xrightarrow{\gamma} P$ such that π is an mgu of $\mathrm{subst}(\gamma)$, $\llbracket \Lambda \rrbracket_{a2p} = \mathrm{comm}(\gamma) \lhd \pi$ and $\widetilde{P}\pi \in \mathrm{open}(\llbracket G' \rrbracket_{g2p})$.*

Proof. The proof is by induction on the derivation of the SHR transition.

Theorem 1. *For each graph $\Gamma \vdash G$ and each set of productions Prod, there is an SHR transition $\Gamma \vdash G \xrightarrow{\Lambda,\pi} \Phi \vdash G'$ iff there is a ZS Fusion transaction $\llbracket G \rrbracket_{g2p} \xRightarrow{(Y)S,\phi} P$ such that π is an mgu of ϕ, $S \lhd \pi = \llbracket \Lambda \rrbracket_{a2p}$, $\llbracket G' \rrbracket_{g2p} = P\pi$ and $Y = (\Phi \setminus \Gamma)$.*

Proof. The *only if* part is by induction on the derivation of the SHR transition exploiting Lemma 4. The *if* part exploits Lemma 2 and Lemma 7.

Example 5 (Ring to star). We can now apply the translation to the ring-to-star reconfiguration in Example 2. We remember that the final transition is:

$$x \vdash \nu y,z \; R(x,y)|R(y,z)|R(z,x) \xrightarrow{(x,\tau,\langle\rangle)} x \vdash \nu x,y,w \; S(w,x)|S(w,y)|S(w,z)$$

In the ZS Fusion setting we have:

$$\llbracket x \vdash \nu y,z \; R(x,y)|R(y,z)|R(z,x) \rrbracket_{g2p} = (yz) \; R[x,y]|R[y,z]|R[z,x]$$

We also have the following constant definition:

$$R[x,y] \triangleq \nu w \; (xw \wedge \overline{y}w).S[w,x]$$

Thus, the starting process is structural congruent to:

$$(w_1 w_2 w_3 yz) \; (xw_1 \wedge \overline{y}w_1).S[w_1,x]|(yw_2 \wedge \overline{z}w_2).S[w_2,y]|(zw_3 \wedge \overline{x}w_3).S[w_3,z]$$

The only small step that can be lifted to a transaction is:

$$(xw_1 \wedge \overline{y}w_1).S[w_1,x]|(yw_2 \wedge \overline{z}w_2).S[w_2,y]|(zw_3 \wedge \overline{x}w_3).S[w_3,z]$$
$$\xrightarrow{\{w_1=w_2=w_3\}} \langle\rangle.S[w_1,x]|\langle\rangle.S[w_2,y]|\langle\rangle.S[w_3,z]$$

By applying rule (Cmt) and rule (Pass) twice, we derive:

$$(yz)(xw_1 \wedge \overline{y}w_1).S[w_1,x]|(yw_2 \wedge \overline{z}w_2).S[w_2,y]|(zw_3 \wedge \overline{x}w_3).S[w_3,z]$$
$$\xRightarrow{\{w_1=w_2=w_3\}} (yz)S[w_1,x]|S[w_2,y]|S[w_3,z]$$

Finally, we can apply rule (Scope) twice and rule (Pass) for deriving:

$$(w_1 w_2 w_3 yz)(xw_1 \wedge \overline{y}w_1).S[w_1,x]|(yw_2 \wedge \overline{z}w_2).S[w_2,y]|(zw_3 \wedge \overline{x}w_3).S[w_3,z]$$
$$\xRightarrow{\emptyset} (w_1 yz)S[w_1,x]|S[w_1,y]|S[w_1,z]$$

which is the desired transaction.

6   About Assumptions

This section is devoted to discuss the assumptions made in the previous section, in particular:

1. there is at most one SHR input action for each arity;
2. each edge is attached to any chosen node at most once;
3. in all SHR productions, the substitution π is the identity and action τ is never used.

The discussions below, and the first one in particular, besides justifying the assumptions made to simplify the mapping, highlight some properties of SHR.

6.1   Simulating Multiple Actions

SHR, contrary to common nominal calculi that have prefixes of the form $x\vec{y}$, featuring only the subject and the objects, has prefixes of the form $xa\vec{y}$ including also an action a. We show below that while having actions is very handy from the specification point of view since it allows for much more compact specifications, it does not change the expressive power. Notice that the simulation of many actions exploits multiparty synchronizations, and could not be done in such a way e.g. in Fusion Calculus. Notice also that prefixes of different arities behave in different ways also in Fusion Calculus, thus having no actions actually corresponds in SHR to have at most one action for each arity.

To show that the introduction of multiple actions does not change the expressive power, we show that any Milner SHR model can be mapped into a Milner SHR model where any two input actions have different arities preserving the original behavior.

The mapping works as follows. Suppose we have n input actions denoted by a_i with $i \in \{1, \ldots, n\}$, and their corresponding output actions. We consider a mapping that translates any node x into a tuple of $n+1$ nodes written $x, x_1, \ldots, x_n$. For instance, if we have two input actions a_1 and a_2 with arity 1 then the graph $x, y \vdash L(x, y)$ is translated into $x, x_1, x_2, y, y_1, y_2 \vdash L(x, x_1, x_2, y, y_1, y_2)$. Then, performing an action a_n over the node x is translated as the execution of the unique action a on both x_n and x, where x is used for mutual exclusion. The same parameters, that are the translation of the original ones, are sent on both channels.

We denote the translation function defined above by $[\![\bullet]\!]_{n21}$, and we consider its trivial extension to SHR productions.

Example 6. The transition $x, y \vdash L(x, y) \xrightarrow{(x, a_1, \langle y \rangle)} x, y \vdash L(x, x)$ is translated into:

$$x, x_1, x_2, y, y_1, y_2 \vdash L(x, x_1, x_2, y, y_1, y_2) \xrightarrow{(x, a, \langle y, y_1, y_2 \rangle), (x_1, a, \langle y, y_1, y_2 \rangle)}$$
$$x, x_1, x_2, y, y_1, y_2 \vdash L(x, x_1, x_2, x, x_1, x_2)$$

The following theorem states the behavioral correspondence between the two models.

Theorem 2. *The transition* $\Gamma \vdash G \xrightarrow{\Lambda,\pi} \Phi \vdash G'$ *can be derived using productions in* Prod *iff* $[\![\Gamma \vdash G]\!]_{n21} \xrightarrow{[\![\Lambda]\!]_{n21},[\![\pi]\!]_{n21}} [\![\Phi \vdash G']\!]_{n21}$ *can be derived using productions in* $[\![Prod]\!]_{n21}$.

Proof. By induction on the derivation $\Gamma \vdash G \xrightarrow{\Lambda,\pi} \Phi \vdash G'$.

6.2   Each Edge is Attached to Any Chosen Node at Most Once

This is the most important assumption that we have made. Notice that such a restriction is not preserved by general SHR transitions. Nevertheless, we require this condition in order to simplify the definition of the encoding presented in Section 5. This restriction avoids the following mismatch: actions belonging to the same prefix in a ZS Fusion process are executed sequentially and, hence, they cannot interact with each other, while two actions in the same SHR production can interact. For instance, the production $x,y \vdash L(x,y) \xrightarrow{(x,a,\langle\rangle),(y,\bar{a},\langle\rangle)} x,y \vdash L(x,y)$ may be applied to the edge $L(x,x)$ for deriving the transition $x \vdash L(x,x) \xrightarrow{(x,\tau,\langle\rangle)} x \vdash L(x,x)$.

Nevertheless, we can model the above system by avoiding non linear attachments as follows. We represent the edge $L(x,y)$ by using two different edges $L_1(x,w)$ and $L_2(y,w)$, whose corresponding productions are:

$$x,w \vdash L_1(x,w) \xrightarrow{(x,a,\langle\rangle),(w,b,\langle\rangle)} x,y \vdash L_1(x,w)$$
$$y,w \vdash L_2(y,w) \xrightarrow{(y,\bar{a},\langle\rangle),(w,\bar{b},\langle\rangle)} x,y \vdash L_2(y,w)$$

Note that we use a new auxiliary action b for synchronizing the rewritings of L_1 and L_2. Then, the initial graph $x,y \vdash L(x,y)$ can be simulated by the graph $x,y \vdash \nu w \ (L_1(x,w)|L_2(y,w))$. Consequently, we can simulate the transition $x \vdash L(x,x) \xrightarrow{(x,\tau,\langle\rangle)} x \vdash L(x,x)$ with the following one:

$$x \vdash \nu w \ (L_1(x,w)|L_2(x,w)) \xrightarrow{(x,\tau,\langle\rangle)} x \vdash \nu w \ (L_1(x,w)|L_2(x,w))$$

6.3   No Non-trivial Renamings Nor τ Actions in SHR Productions

Requiring substitutions appearing in SHR productions to be the identity does not affect the expressive power, since it is almost folklore that in SHR arbitrary idempotent substitutions can be generated by adding a synchronization on a private node z. Analogously, a τ action on a node can be generated by performing two complementary actions on the node itself. Thus one can use a simple encoding to map SHR models with arbitrary substitutions and τ actions onto models without them.

7   From Zero-Safe Fusion to SHR

We extend the approach followed in [6] for translating Fusion processes into SHR graphs. Basically, every sequential agent is encoded as a graph containing exactly one edge, whose label corresponds to the sequential agent renamed so to use only standard names. To this end, we exploit the notion of standard decomposition introduced by Definition 7.

Definition 11 (Translation of ZS Fusion agents). *The translation $[\![\_]\!]_{a2g}$ from Fusion agents to graph terms is defined as follows:*

- *$[\![0]\!]_{a2g} = nil$*
- *$[\![S]\!]_{a2g} = L_{\hat{P}}(fnarray(P))$, where P is the process corresponding to the sequential agent S.*
- *$[\![P_1|P_2]\!]_{a2g} = [\![P_1]\!]_{a2g}|[\![P_2]\!]_{a2g}$*
- *$[\![(x)P]\!]_{a2g} = \nu x\ [\![P]\!]_{a2g}$*
- *$[\![C[\vec{v}]]\!]_{a2g} = [\![P\{\vec{v}/\vec{x}\}]\!]_{a2g}$, if $C[\vec{x}] \triangleq P$*

Lemma 5. $P_1 \equiv P_2 \Rightarrow [\![P_1]\!]_{a2g} \equiv [\![P_2]\!]_{a2g}$.

Note that a sequential agent $\sum_i \beta_i.P_i$ is translated as a graph $[\![\sum_i \beta_i.P_i]\!]_{a2g}$, containing just one edge. Then, we define the set of SHR productions corresponding to a ZS Fusion process by defining the behavior of each edge. The following auxiliary definition provides a mapping for communication actions.

Definition 12 (Translation of communication actions). *Communication actions are translated as follows:*

- *$[\![u\vec{x}]\!]_{p2a} = (in_n, \vec{x})$ where $n = |\vec{x}|$*
- *$[\![\bar{u}\vec{x}]\!]_{p2a} = (out_n, \vec{x})$ where $n = |\vec{x}|$*

The productions corresponding to a particular sequential agent are defined as follows.

Definition 13 (Productions for a ZS Fusion agent). *Let $S = \sum_i \beta_i.P_i$ be a sequential agent and $\Gamma = fn(S)$ the set of free names in S. The set of productions $\mathcal{P}_S$ associated with S contains the following rules (one for each possible i).*

$$\Gamma \vdash [\![\sum_i \beta_i.P_i]\!]_{a2g} \xrightarrow{\Lambda_{\beta_i} \lhd \pi_{\beta_i}, \pi_{\beta_i}} \Gamma\pi_{\beta_i} \vdash [\![P_i\pi_{\beta_i}]\!]_{a2g}$$

where

$$\Lambda_{\beta_i}(x) = \begin{cases} [\![\alpha]\!]_{p2a} & \text{if } \alpha \in comm(\beta_i) \text{ and } subj(\alpha) = x \\ undefined & otherwise \end{cases}$$

and π_{β_i} is an mgu of $subst(\beta_i)$.

Lemma 6. *Let P be a ZS Fusion process and σ a renaming. Then $[\![P]\!]_{a2g}\sigma = [\![P\sigma]\!]_{a2g}$.*

The remaining of this section is devoted to show the completeness and soundness of the encoding. The following lemma states the completeness of the encoding for small-steps.

Lemma 7. *Let P be a static ZS Fusion process. If $\Delta \vdash P \xrightarrow{\gamma_1} P''$ and $\Delta' \vdash P'' \xrightarrow{\gamma_2}_\epsilon P'$ where P' is stable and $\Delta \cap \Delta' = \emptyset$, then for each $\Gamma \supseteq \mathrm{fn}(P)$ and for each mgu π of $\mathrm{subst}(\gamma_1 \wedge \gamma_2)$:*

$$\Gamma \vdash [\![P]\!]_{a2g} \xrightarrow{\Lambda \lhd \pi, \pi} \Gamma\pi \vdash [\![\widetilde{P'}\pi]\!]_{a2g}$$

where

$$\Lambda(x) = \begin{cases} [\![\alpha]\!]_{p2a} & \text{if } \alpha \in comm(\gamma_1 \wedge \gamma_2) \text{ and } subj(\alpha) = x \\ (\tau, \langle \rangle) & \text{if } x \in \Delta \cup \Delta' \text{ and } \not\exists \alpha \in comm(\gamma_1 \wedge \gamma_2) \text{ with } subj(\alpha) = x \\ \text{undefined} & \text{otherwise} \end{cases}$$

Proof. By induction on the derivation $\Delta \vdash P \xrightarrow{\gamma_1} P''$.

The completeness of the encoding considering ZS transactions and SHR transitions is stated by the following theorem.

Theorem 3. *Let P be a static ZS Fusion process. If $P \xrightarrow{(Y)S,\phi} P'$, then for each $\Gamma \supseteq \mathrm{fn}(P)$ and for each π mgu of ϕ:*

$$\Gamma \vdash [\![P]\!]_{a2g} \xrightarrow{\Lambda \lhd \pi, \pi} \Gamma\pi \vdash [\![\widetilde{P'}\pi]\!]_{a2g}$$

where

$$\Lambda(x) = \begin{cases} [\![\alpha]\!]_{p2a} & \text{if } \alpha \in S \text{ and } subj(\alpha) = x \\ \text{undefined or } (\tau, \langle \rangle) & \text{otherwise} \end{cases}$$

Proof. By induction on the derivation $P \xrightarrow{(Y)S,\phi} P'$, using Lemma 7.

The following results show the soundness of the encoding.

Lemma 8. *Let P be a static ZS Fusion process without restrictions. If $\Gamma \vdash [\![P]\!]_{a2g} \xrightarrow{\Lambda, \pi} \Phi \vdash G$ with $fn(P) \subseteq \Gamma$, then $\Delta \vdash P \xrightarrow{\gamma} P'$, with $G = [\![\widetilde{P'}\pi]\!]_{a2g}$, $\Phi = \Gamma\pi$, π is an mgu of $\mathrm{subst}(\gamma)$, and $\Lambda = \Lambda' \lhd \pi$ where*

$$\Lambda'(x) = \begin{cases} [\![\alpha]\!]_{p2a} & \text{if } \alpha \in comm(\gamma) \text{ and } subj(\alpha) = x \\ (\tau, \langle \rangle) & \text{if } x \in \Delta \text{ and } \not\exists \alpha \in comm(\gamma) \text{ with } subj(\alpha) = x \\ \text{undefined} & \text{otherwise} \end{cases}$$

Proof. By induction on the derivation $\Gamma \vdash [\![P]\!]_{a2g} \xrightarrow{\Lambda, \pi} \Phi \vdash G$.

Theorem 4. *Let P be a static ZS Fusion process. If $\Gamma \vdash [\![P]\!]_{a2g} \xrightarrow{\Lambda,\pi} \Phi \vdash G$ with $fn(P) \subseteq \Gamma$, then $P \xRightarrow{(Y)S,\phi} P'$, with $G = [\![\widetilde{P'\pi}]\!]_{a2g}$, $\Phi = (\Gamma\pi \cup \{Y\})$, π is an mgu of ϕ, and $\Lambda = \Lambda' \lhd \pi$ where*

$$\Lambda'(x) = \begin{cases} [\![\alpha]\!]_{p2a} & \text{if } \alpha \in S \text{ and } subj(\alpha) = x \\ undefined \ or \ (\tau,\langle\rangle) & otherwise \end{cases}$$

Proof. By induction on $\Gamma \vdash [\![P]\!]_{a2g} \xrightarrow{\Lambda,\pi} \Phi \vdash G$, using Lemma 8.

8   Conclusion

We have shown the strict relationship between SHR and ZS Fusion Calculus, in particular highlighting how synchronous multiparty synchronizations can be implemented using binary synchronizations and a transactional mechanism. This can be used as a starting point to provide a distributed implementation of SHR. We leave for future work the analysis of the relationships between SHR and ZS Fusion from an observational point of view, instead of the operational one considered here. Actually, there are some differences between the observations that can be done on SHR and on ZS Fusion, e.g. because of the different treatment of substitutions and because of τ actions, thus the observational relation between the two models is not straightforward.

Other possible lines for future work are the analysis of the expressive power and usability of ZS Fusion calculus, and its comparison with other transactional frameworks such as, e.g., CJoin [19], πt [20], and Webπ [21]. The transactional mechanism of ZS Fusion is quite minimal and more in the ACID transactions style, thus it would be interesting to compare it with the calculi above that exploit the compensation approach.

References

1. Hirsch, D., Inverardi, P., Montanari, U.: Reconfiguration of software architecture styles with name mobility. In: Porto, A., Roman, G.-C. (eds.) COORDINATION 2000. LNCS, vol. 1906, pp. 148–163. Springer, Heidelberg (2000)
2. Hirsch, D.: Graph transformation models for software architecture styles. PhD thesis, Departamento de Computación, Facultad de Ciencias Exactas y Naturales, U.B.A (2003)
3. Lanese, I., Montanari, U.: Hoare vs Milner: Comparing synchronizations in a graphical framework with mobility. In: Proc. of GT-VC 2005. Elect.Notes in Th.Comput.Sci., vol. 154(2), pp. 55–72. Elsevier Science, Amsterdam (2005)
4. Parrow, J., Victor, B.: The fusion calculus: Expressiveness and symmetry in mobile processes. In: Proc. of LICS 1998. IEEE Computer Society Press, Los Alamitos (1998)
5. Milner, R., Parrow, J., Walker, D.: A calculus of mobile processes. Inform.and Comput. 100, 1–77 (1992)

6. Lanese, I., Montanari, U.: A graphical fusion calculus. In: Proc. of the Workshop of the COMETA Project. Elect.Notes in Th.Comput.Sci., vol. 104, pp. 199–215. Elsevier Science, Amsterdam (2004)

7. Bruni, R., Montanari, U.: Zero-safe nets: Comparing the collective and individual token approaches. Inform.and Comput. 156(1-2), 46–89 (2000)

8. Lanese, I., Melgratti, H.: Synchronous multiparty synchronizations and transactions, http://cs.unibo.it/~lanese/work/ugo65-TR.pdf

9. Victor, B.: The fusion calculus: Expressiveness and symmetry in mobile processes. PhD thesis, Dept.of Computer Systems, Uppsala University, Sweden (1998)

10. Castellani, I., Montanari, U.: Graph grammars for distributed systems. In: Ehrig, H., Nagl, M., Rozenberg, G. (eds.) Graph Grammars 1982. LNCS, vol. 153, pp. 20–38. Springer, Heidelberg (1983)

11. Hirsch, D., Montanari, U.: Synchronized hyperedge replacement with name mobility. In: Larsen, K.G., Nielsen, M. (eds.) CONCUR 2001. LNCS, vol. 2154, pp. 121–136. Springer, Heidelberg (2001)

12. Ferrari, G., Montanari, U., Tuosto, E.: A LTS semantics of ambients via graph synchronization with mobility. In: Restivo, A., Ronchi Della Rocca, S., Roversi, L. (eds.) ICTCS 2001. LNCS, vol. 2202, pp. 1–16. Springer, Heidelberg (2001)

13. Lanese, I., Montanari, U.: Synchronization algebras with mobility for graph transformations. In: Proc. of FGUC 2004. Elect.Notes in Th.Comput.Sci., vol. 138, pp. 43–60. Elsevier, Amsterdam (2004)

14. Lanese, I.: Synchronization strategies for global computing models. PhD thesis, Computer Science Department, University of Pisa, Pisa, Italy (2006)

15. Fournet, C., Gonthier, G.: The reflexive CHAM and the join-calculus. In: Proc. of POPL 1996, pp. 372–385. ACM Press, New York (1996)

16. Bocchi, L., Wischik, L.: A process calculus of atomic commit. In: Proc. of WS-FM 2004. Elect.Notes in Th.Comput.Sci., vol. 105, pp. 119–132. Elsevier, Amsterdam (2004)

17. Bruni, R., Montanari, U.: Concurrent models for linda with transactions. Math.Struct.in Comput.Sci. 14(3), 421–468 (2004)

18. Gorrieri, R., Marchetti, S., Montanari, U.: A2CCS: Atomic actions for CCS. Theoret.Comput.Sci. 72(2&3), 203–223 (1990)

19. Bruni, R., Melgratti, H., Montanari, U.: Nested commits for mobile calculi: extending Join. In: Proc. of IFIP-TCS 2004, pp. 569–582. Kluwer Academic, Dordrecht (2004)

20. Bocchi, L., Laneve, C., Zavattaro, G.: A calculus for long-running transactions. In: Najm, E., Nestmann, U., Stevens, P. (eds.) FMOODS 2003. LNCS, vol. 2884, pp. 124–138. Springer, Heidelberg (2003)

21. Laneve, C., Zavattaro, G.: Foundations of web transactions. In: Sassone, V. (ed.) FOSSACS 2005. LNCS, vol. 3441, pp. 282–298. Springer, Heidelberg (2005)

Transformations in
Reconfigurable Place/Transition Systems[*]

Ulrike Prange, Hartmut Ehrig, Kathrin Hoffmann, and Julia Padberg

Technische Universität Berlin, Germany
`{uprange,ehrig,hoffmann,padberg}@cs.tu-berlin.de`

Abstract. Reconfigurable place/transition systems are Petri nets with initial markings and a set of rules which allow the modification of the net during runtime in order to adapt the net to new requirements. For the transformation of Petri nets in the double pushout approach, the categorical framework of adhesive high-level replacement systems has been instantiated to Petri nets. In this paper, we show that also place/transition systems form a weak adhesive high-level replacement category. This allows us to apply the developed theory also to tranformations within reconfigurable place/transition systems.

1 Introduction

Petri nets are an important modeling technique to describe discrete distributed systems. Their nondeterministic firing steps are well-suited for modeling the concurrent behavior of such systems. The formal treatment of Petri nets as monoids by Meseguer and Montanari in [1] has been an important step for a rigorous algebraic treatment and analysis of Petri nets which is also used in this paper.

As the adaptation of a system to a changing environment gets more and more important, Petri nets that can be transformed during runtime have become a significant topic in recent years. Application areas cover e.g. computer supported cooperative work, multi agent systems, dynamic process mining and mobile networks. Moreover, this approach increases the expressiveness of Petri nets and allows for a formal description of dynamic changes.

In [2], the concept of reconfigurable place/transition (P/T) systems was introduced for modeling changes of the net structure while the system is kept running. In detail, a reconfigurable P/T system consists of a P/T system and a set of rules, so that not only the follower marking can be computed but also the net structure can be changed by rule application. So, a new P/T system is obtained that is more appropriate with respect to some requirements of the environment. Moreover, these activities can be interleaved. In [3], the conflict situation of transformation and token firing has been dealt with. In this paper, we give the formal foundation for transformations of P/T systems.

For rule-based transformations of P/T systems we use the framework of adhesive high-level replacement (HLR) systems [4, 5] that is inspired by graph transformation systems [6]. Adhesive HLR systems have been recently introduced as a new categorical framework for graph transformation in the double pushout approach [4, 5]. They

---

* This work has been partly funded by the research project *for* MA*ℓ*NET of the German Research Council (see http://tfs.cs.tu-berlin.de/formalnet/).

combine the well-known framework of HLR systems with the framework of adhesive categories introduced by Lack and Sobociński [7]. The main concept behind adhesive categories are the so-called van Kampen squares. These ensure that pushouts along monomorphisms are stable under pullbacks and, vice versa, that pullbacks are stable under combined pushouts and pullbacks. In the case of weak adhesive HLR categories, the class of all monomorphisms is replaced by a subclass $\mathcal{M}$ of monomorphisms closed under composition and decomposition, and for the van Kampen properties certain morphisms have to be additionally $\mathcal{M}$-morphisms.

In this paper, we present the formal foundations for transformations of nets with markings. We show that the category of P/T systems is a weak adhesive HLR category which allows the application of the developed theory also to tranformations within reconfigurable P/T systems. This theory comprises many results concerning the applicability of rules, the embedding and extension of transformations, parallel and sequential dependence and independence, and concurrency of rule applications, and hence gives precise notions for concurrent or conflicting situations in reconfigurable P/T systems. Our work is illustrated by an example in the area of mobile emergency scenarios.

This paper is organized as follows. In Section 2, we introduce weak adhesive HLR categories and adhesive HLR systems. The notion of reconfigurable P/T systems is presented in Section 3. In Section 4, we show that the category **PTSys** used for reconfigurable P/T systems is a weak adhesive HLR category. Finally, we give a conclusion and outline related and future work in Section 5.

2 Adhesive HLR Categories and Systems

In this section, we give a short introduction to weak adhesive HLR categories and summarize some important results for adhesive HLR systems (see [4]) which are based on adhesive categories introduced in [7].

The intuitive idea of an adhesive or (weak) adhesive HLR category is a category with suitable pushouts and pullbacks which are compatible with each other. More precisely, the definition is based on so-called van Kampen squares.

The idea of a van Kampen (VK) square is that of a pushout which is stable under pullbacks, and vice versa that pullbacks are stable under combined pushouts and pullbacks.

Definition 1 (van Kampen square). *A pushout (1) is a* van Kampen square *if for any commutative cube (2) with (1) in the bottom and the back faces being pullbacks it holds that: the top face is a pushout if and only if the front faces are pullbacks.*

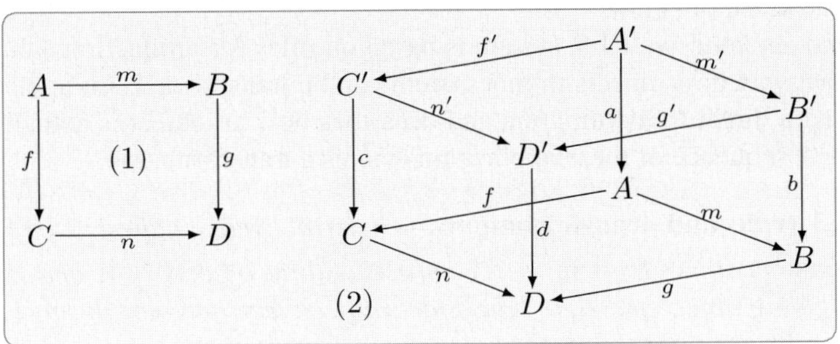

Not even in the category **Sets** of sets and functions each pushout is a van Kampen square. Therefore, in (weak) adhesive HLR categories only those VK squares of Def. 1 are considered where m is in a class $\mathcal{M}$ of monomorphisms. A pushout (1) with $m \in \mathcal{M}$ and arbitrary f is called a pushout along $\mathcal{M}$.

The main difference between (weak) adhesive HLR categories as described in [4, 5] and adhesive categories introduced in [7] is that a distinguished class $\mathcal{M}$ of monomorphisms is considered instead of all monomorphisms, so that only pushouts along $\mathcal{M}$-morphisms have to be VK squares. In the weak case, only special cubes are considered for the VK square property.

Definition 2 ((weak) adhesive HLR category). *A category* **C** *with a morphism class* $\mathcal{M}$ *is a* (weak) adhesive HLR category, *if*

1. $\mathcal{M}$ *is a class of monomorphisms closed under isomorphisms, composition* $(f : A \to B \in \mathcal{M}, g : B \to C \in \mathcal{M} \Rightarrow g \circ f \in \mathcal{M})$ *and decomposition* $(g \circ f \in \mathcal{M}, g \in \mathcal{M} \Rightarrow f \in \mathcal{M})$,
2. **C** *has pushouts and pullbacks along* $\mathcal{M}$-*morphisms and* $\mathcal{M}$-*morphisms are closed under pushouts and pullbacks,*
3. *pushouts in* **C** *along* $\mathcal{M}$-*morphisms are (weak) VK squares.*

For a weak VK square, the VK square property holds for all commutative cubes with $m \in \mathcal{M}$ *and* $(f \in \mathcal{M}$ *or* $b, c, d \in \mathcal{M})$ *(see Def. 1).*

Remark 1. $\mathcal{M}$-morphisms closed under pushouts means that given a pushout (1) in Def. 1 with $m \in \mathcal{M}$ it follows that $n \in \mathcal{M}$. Analogously, $n \in \mathcal{M}$ implies $m \in \mathcal{M}$ for pullbacks.

The categories **Sets** of sets and functions and **Graphs** of graphs and graph morphisms are adhesive HLR categories for the class $\mathcal{M}$ of all monomorphisms. The categories **ElemNets** of elementary nets and **PTNet** of place/transition nets with the class $\mathcal{M}$ of all corresponding monomorphisms fail to be adhesive HLR categories, but they are weak adhesive HLR categories (see [8]). Elementary Petri nets, also called condition/event nets, have a weight restricted to one, while place/transition nets allow arbitrary finite arc weights. Instead of the original set theoretical notations used in [9, 10] we have used in [4] a more algebraic version based on power set or monoid constructions as introduced in [1].

Now we are able to generalize graph transformation systems, grammars and languages in the sense of [11, 4].

In general, an adhesive HLR system is based on rules (or productions) that describe in an abstract way how objects in this system can be transformed. An application of a rule is called a direct transformation and describes how an object is actually changed by the rule. A sequence of these applications yields a transformation.

Definition 3 (rule and transformation). *Given a (weak) adhesive HLR category* $(\mathbf{C}, \mathcal{M})$, *a rule* $prod = (L \xleftarrow{l} K \xrightarrow{r} R)$ *consists of three objects* L, K *and* R *called left hand side, gluing object and right hand side, respectively, and morphisms* $l : K \to L$, $r : K \to R$ *with* $l, r \in \mathcal{M}$.

Given a rule prod $= (L \xleftarrow{l} K \xrightarrow{r} R)$ *and an object G with a morphism* $m : L \to G$, *called match, a* direct transformation $G \stackrel{prod,m}{\Longrightarrow} H$ *from G to an object H is given by the following diagram, where (1) and (2) are pushouts. A sequence* $G_0 \Longrightarrow G_1 \Longrightarrow ... \Longrightarrow G_n$ *of direct transformations is called a* transformation *and is denoted as* $G_0 \stackrel{*}{\Longrightarrow} G_n$.

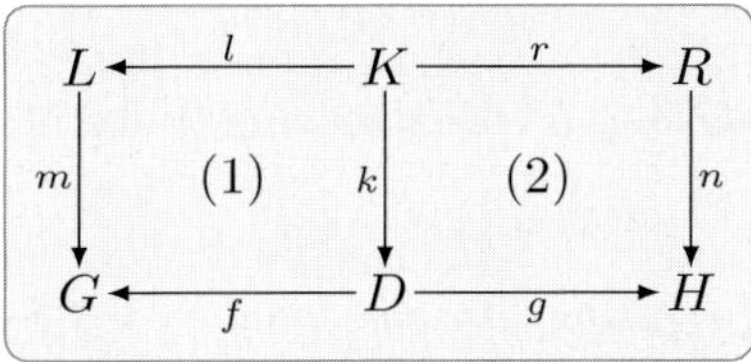

An adhesive HLR system $AHS = (\mathbf{C}, \mathcal{M}, RULES)$ *consists of a (weak) adhesive HLR category* $(\mathbf{C}, \mathcal{M})$ *and a set of rules RULES.*

Remark 2. Note that given a rule *prod* and a match m pushout (1) is constructed as the pushout complement, which requires a certain gluing condition to be fulfilled.

3 Reconfigurable P/T Systems

In this section, we formalize reconfigurable P/T systems as introduced in [2]. As net formalism we use P/T systems following the notation of "Petri nets are Monoids" in [1].

Definition 4 (P/T system). *A P/T net is given by* $PN = (P, T, pre, post)$ *with places P, transitions T, and pre and post domain functions* $pre, post : T \to P^{\oplus}$.
  A P/T system $PS = (PN, M)$ *is a P/T net PN with marking* $M \in P^{\oplus}$.

$P^{\oplus}$ is the free commutative monoid over P. The binary operation $\oplus$ leads to the monoid notation, e.g. $M = 2p_1 \oplus 3p_2$ means that we have two tokens on place p_1 and three tokens on p_2. Note that M can also be considered as a function $M : P \to \mathbf{N}$, where only for a finite set $P' \subseteq P$ we have $M(p) \geq 1$ with $p \in P'$. We can switch between these notations by defining $\sum_{p \in P} M(p) \cdot p = M \in P^{\oplus}$. Moreover, for $M_1, M_2 \in P^{\oplus}$ we have $M_1 \leq M_2$ if $M_1(p) \leq M_2(p)$ for all $p \in P$. A transition $t \in T$ is M-enabled for a marking $M \in P^{\oplus}$ if we have $pre(t) \leq M$, and in this case the follower marking M' is given by $M' = M \ominus pre(t) \oplus post(t)$ and $(PN, M) \xrightarrow{t} (PN, M')$ is called a firing step. Note that $\ominus$ is the inverse of $\oplus$, and $M_1 \ominus M_2$ is only defined if we have $M_2 \leq M_1$.

  In order to define rules and transformations of P/T systems we introduce P/T morphisms which preserve firing steps by Condition (1) below. Additionally they require that the initial marking at corresponding places is increasing (Condition (2)) or equal (Condition (3)).

Definition 5 (P/T Morphism). *Given P/T systems* $PS_i = (PN_i, M_i)$ *with* $PN_i = (P_i, T_i, pre_i, post_i)$ *for* $i = 1, 2$, *a P/T morphism* $f : (PN_1, M_1) \to (PN_2, M_2)$ *is given by* $f = (f_P, f_T)$ *with functions* $f_P : P_1 \to P_2$ *and* $f_T : T_1 \to T_2$ *satisfying*

(1) $f_P^{\oplus} \circ pre_1 = pre_2 \circ f_T$ and $f_P^{\oplus} \circ post_1 = post_2 \circ f_T$,
(2) $M_1(p) \leq M_2(f_P(p))$ for all $p \in P_1$.

Note that the extension $f_P^{\oplus} : P_1^{\oplus} \to P_2^{\oplus}$ of $f_P : P_1 \to P_2$ is defined by $f_P^{\oplus}(\sum_{i=1}^n k_i \cdot p_i) = \sum_{i=1}^n k_i \cdot f_P(p_i)$. (1) means that f is compatible with pre and post domains, and (2) that the initial marking of PN_1 at place p is smaller or equal to that of PN_2 at $f_P(p)$.

Moreover, the P/T morphism f is called strict *if f_P and f_T are injective and*

(3) $M_1(p) = M_2(f_P(p))$ for all $p \in P_1$.

P/T systems and P/T morphisms form the category **PTSys**, *where the composition of P/T morphisms is defined componentwise for places and transitions.*

Remark 3. For our morphisms we do not always have $f_P^{\oplus}(M_1) \leq M_2$. E.g., $M_1 = p_1 \oplus p_2, M_2 = p$ and $f_P(p_1) = f_P(p_2) = p$ implies $f_P^{\oplus}(M_1) = 2p > p = M_2$, but $M_1(p_1) = M_1(p_2) = 1 = M_2(p)$.

P/T Nets and morphisms satisfying (1) form the category **PTNet**.

Based on the category **PTSys** and the morphism class $\mathcal{M}_{strict}$ of all strict P/T morphisms we are now able to define reconfigurable P/T systems. They allow the modification of the net structure using rules and net transformations of P/T systems, which are instantiations of the corresponding categorical concepts defined in Section 2.

Definition 6 (Reconfigurable P/T System). *Given a P/T system (PN, M) and a set RULES of rules, a* reconfigurable P/T system *is defined by $((PN, M), RULES)$.*

Example 1. We will illustrate the main idea of reconfigurable P/T systems in the area of a mobile scenario. This work is part of a collaboration with some research projects where the main focus is on an adaptive workflow management system for mobile ad-hoc networks, specifically targeted to emergency scenarios [1].

Our scenario takes place in an archaeological disaster/recovery mission: after an earthquake, a team (led by a team leader) is equipped with mobile devices (laptops and PDAs) and sent to the affected area to evaluate the state of archaeological sites and the state of precarious buildings. The goal is to draw a situation map in order to schedule restructuring jobs. The team is considered as an overall mobile ad-hoc network in which the team leader's device coordinates the other team members' devices by providing suitable information (e.g. maps, sensible objects, etc.) and assigning activities. For our example, we assume a team consisting of a team leader as picture store device and two team members as camera device and bridge device, respectively. A typical cooperative process to be enacted by a team is shown in Fig. 1 as P/T system (PN_1, M_1), where only the team leader and one of the team members are yet involved in activities.

The work of the team is modeled by firing steps. So to start the activities of the camera device the follower marking of the P/T system (PN_1, M_1) is computed by firing the transition *Select Building*, then the task *Go to Destination* can be executed etc.

As a reaction to changing requirements, rules can be applied to the net. A rule $prod = ((L, M_L) \xleftarrow{l} (K, M_K) \xrightarrow{r} (R, M_R))$ is given by three P/T systems and a span of two

---

[1] IST FP6 WORKPAD: `http://www.workpad-project.eu`

$$(PN_1, M_1)$$

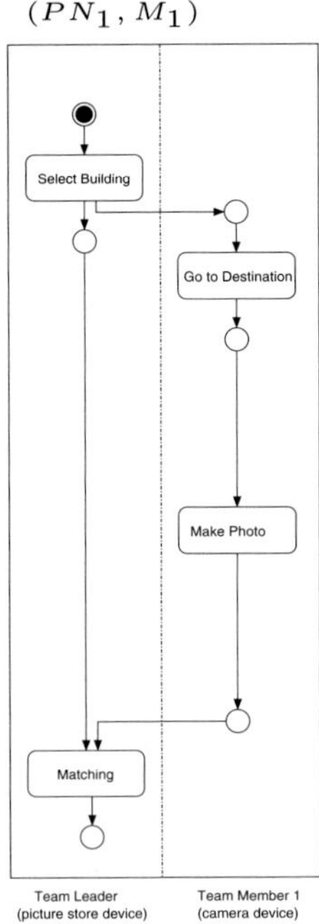

Fig. 1. Cooperative process of the team

strict P/T morphisms l and r (see Def. 3). For the application of the rule to the P/T system (PN_1, M_1), we additionally need a match morphism m that identifies the left-hand side L in PN_1.

The activity of taking a picture can be refined into single steps by the rule $prod_{photo}$, which is depicted in the top row of Fig. 2. The application of this rule to the net (PN_1, M_1) leading to the transformation $(PN_1, M_1) \overset{prod_{photo}, m}{\Longrightarrow} (PN_2, M_2)$ is shown in Fig. 2.

To predict a situation of disconnection, a movement activity of the bridge device has to be introduced in our system. In more detail, the workflow has to be extended by a task to follow the camera device. For this reason we provide the rule $prod_{follow}$ depicted in the upper row in Fig. 3. Then the transformation step $(PN_2, M_2) \overset{prod_{follow}, m'}{\Longrightarrow} (PN_3, M_3)$ is shown in Fig. 3.

Summarizing, our reconfigurable P/T system $((PN_1, M_1), \{prod_{photo}, prod_{follow}\})$ consists of the P/T system (PN_1, M_1) and the set of rules $\{prod_{photo}, prod_{follow}\}$ as described above.

Conflicts in Reconfigurable P/T Systems

The traditional concurrency situation in P/T systems without capacities is that two transitions with overlapping pre domain are both enabled and together require more tokens than available in the current marking. As the P/T system can evolve in two different ways, the notions of conflict and concurrency become more complex. We illustrate the situation in Fig. 4, where we have a P/T system (PN_0, M_0) and two transitions that are

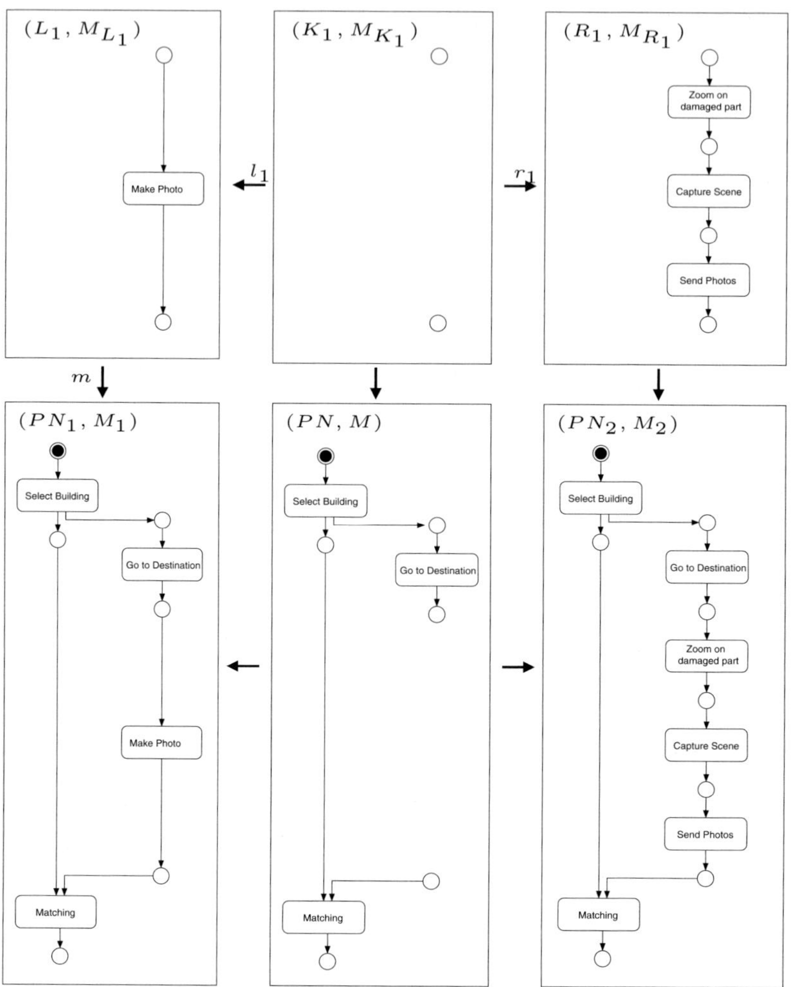

Fig. 2. Transformation step $(PN_1, M_1) \stackrel{prod_{photo},m}{\Longrightarrow} (PN_2, M_2)$

both enabled leading to firing steps $(PN_0, M_0) \xrightarrow{t_1} (PN_0, M_0')$ and $(PN_0, M_0) \xrightarrow{t_2} (PN_0, M_0'')$, and two transformations $(PN_0, M_0) \stackrel{prod_1,m_1}{\Longrightarrow} (PN_1, M_1)$ and $(PN_0, M_0) \stackrel{prod_2,m_2}{\Longrightarrow} (PN_2, M_2)$ via the corresponding rules and matches.

The squares $(1) \ldots (4)$ can be obtained under the following conditions:

For square (1), we have the usual condition for P/T systems that t_1 and t_2 need to be conflict free, so that both can fire in arbitrary order or in parallel yielding the same marking.

For squares (2) and (3), we require parallel independence as introduced in [3]. Parallel independence allows the execution of the transformation step and the firing step in arbitrary order leading to the same P/T system. Parallel independence of a transi-

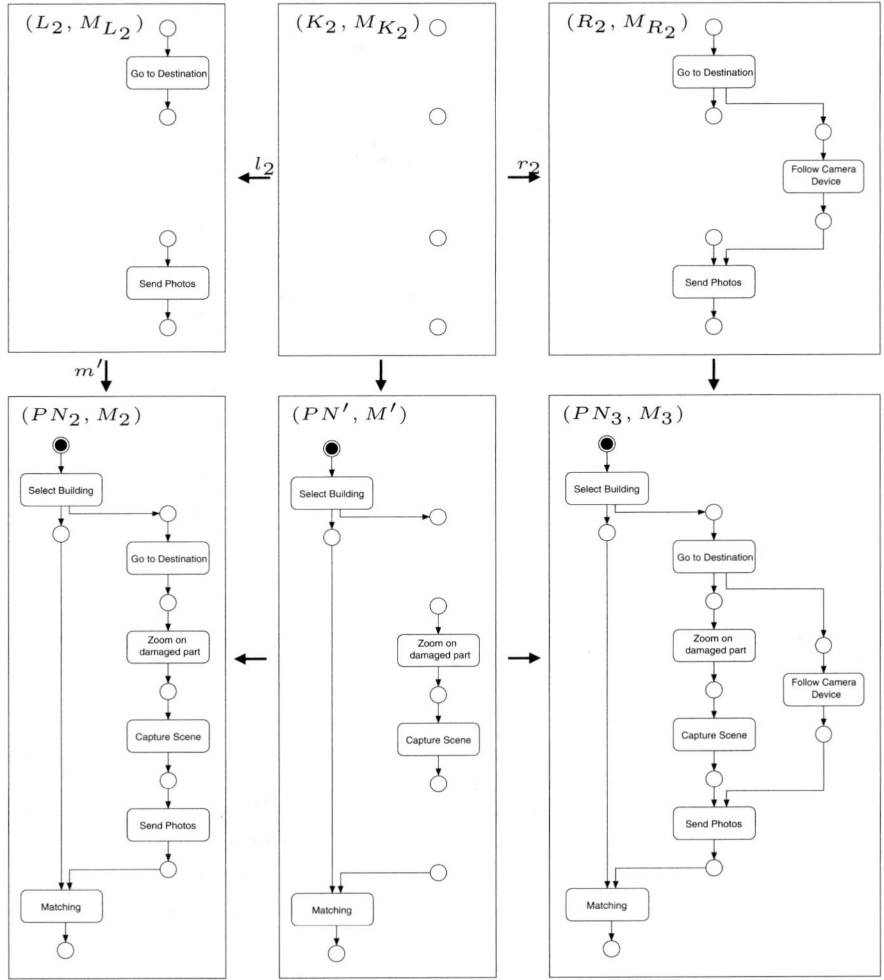

Fig. 3. Transformation step $(PN_2, M_2) \overset{prod_{follow}, m'}{\Longrightarrow} (PN_3, M_3)$

tion and a transformation is given – roughly stated – if the corresponding transition is not deleted by the transformation and the follower marking is still sufficient for the match of the transformation. A detailed formal presentation and analysis of this case is given in [3].

For square (4), we have up to now no conditions to ensure parallel or sequential application of both rules. In this paper, we give these conditions by using results for adhesive HLR systems (see Section 2).

Note that in our framework it is not possible to reduce the conflicts to the case of square (4) by implementing the firing steps by rules. This is due to the fact that the rule morphisms have to be marking strict. Moreover, not only rules but rule schemas would

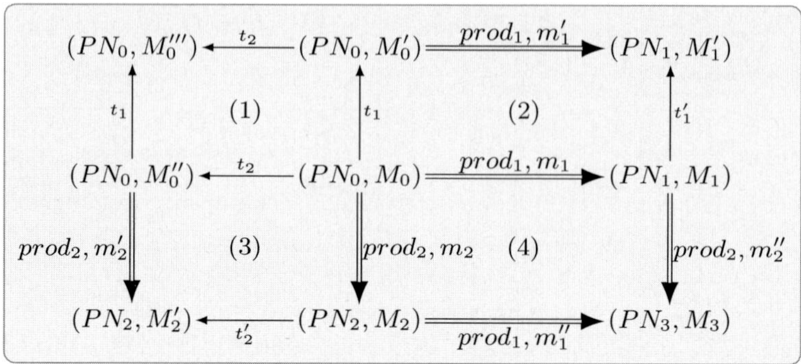

Fig. 4. Concurrency in reconfigurable P/T systems

be needed leading to one rule for each kind of transition with n ingoing and m outgoing arcs.

In [4], the following main results for adhesive HLR systems are shown for weak adhesive HLR categories:

1. Local Church-Rosser Theorem,
2. Parallelism Theorem,
3. Concurrency Theorem.

The Local Church-Rosser Theorem allows one to apply two graph transformations $G \implies H_1$ via $prod_1$ and $G \implies H_2$ via $prod_2$ in an arbitrary order leading to the same result H, provided that they are parallel independent. In this case, both rules can also be applied in parallel, leading to a parallel graph transformation $G \implies H$ via the parallel rule $prod_1 + prod_2$. This second main result is called the Parallelism Theorem and requires binary coproducts together with compatibility with $\mathcal{M}$ (i.e. $f, g \in \mathcal{M} \Rightarrow f + g \in \mathcal{M}$). The Concurrency Theorem is concerned with the simultaneous execution of causally dependent transformations, where a concurrent rule $prod_1 * prod_2$ can be constructed leading to a direct transformation $G \implies H$ via $prod_1 * prod_2$ (see Ex. 2 in Section 4).

4   P/T Systems as Weak Adhesive HLR Category

In this section, we show that the category **PTSys** used for reconfigurable P/T systems together with the class $\mathcal{M}_{strict}$ of strict P/T morphisms is a weak adhesive HLR category. Therefore, we have to verify the properties of Def. 2.

First we shall show that pushouts along $\mathcal{M}_{strict}$-morphisms exist and preserve $\mathcal{M}_{strict}$-morphisms.

Theorem 1. *Pushouts in* **PTSys** *along* $\mathcal{M}_{strict}$ *exist and preserve* $\mathcal{M}_{strict}$-*morphisms, i.e. given P/T morphisms f and m with m strict, then the pushout (PO) exists and n is also a strict P/T morphism.*

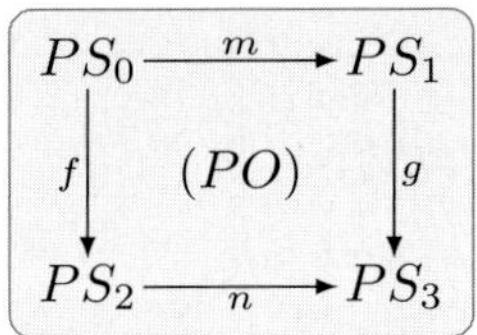

Construction.   Given P/T systems $PS_i = (PN_i, M_i)$ for $i = 0, 1, 2$ and $f, m \in$ **PTSys** with $m \in \mathcal{M}_{strict}$ we construct PN_3 as pushout in **PTNet**, i.e. componentwise in **Sets** on places and transitions. The marking M_3 leading to the P/T system $PS_3 = (PN_3, M_3)$ is defined by

(1) $\forall p_1 \in P_1 \backslash m(P_0)$: $M_3(g(p_1)) = M_1(p_1)$

(2) $\forall p_2 \in P_2 \backslash f(P_0)$: $M_3(n(p_2)) = M_2(p_2)$

(3) $\forall p_0 \in P_0$: $M_3(n \circ f(p_0)) = M_2(f(p_0))$

Remark 4. Actually, we have $M_3 = g^{\oplus}(M_1 \ominus m^{\oplus}(M_0)) \oplus n^{\oplus}(M_2)$. (2) and (3) can be integrated, i.e. it is sufficient to define $\forall p_2 \in P_2$: $M_3(n(p_2)) = M_2(p_2)$.

Proof. Since PN_3 is a pushout in **PTNet** with g, n jointly surjective we construct a marking for all places $p_3 \in P_3$. (1) and (2) are well-defined because g and n are injective on $P_1 \backslash m(P_0)$ and $P_2 \backslash f(P_0)$, respectively. (3) is well-defined because for $n(f(p_0)) = n(f(p_0'))$, n being injective implies $f(p_0) = f(p_0')$ and hence $M_2(f(p_0)) = M_2(f(p_0'))$.

First we shall show that g, n are P/T morphisms and n is strict.

1. $\forall p_1 \in P_1$ we have:

    1. $p_1 \in P_1 \backslash m(P_0)$ and $M_1(p_1) \stackrel{(1)}{=} M_3(g(p_1))$ or

    2. $\exists p_0 \in P_0$ with $p_1 = m(p_0)$ and $M_1(p_1) = M_1(m(p_0)) \stackrel{m \text{ strict}}{=}$
    $M_0(p_0) \stackrel{f \in \textbf{PTSys}}{\leq} M_2(f(p_0)) \stackrel{(3)}{=} M_3(n(f(p_0))) = M_3(g(m(p_0))) = M_3(g(p_1))$.
    This means $g \in$ **PTSys**.

2. $\forall p_2 \in P_2$ we have:

    1. $p_2 \in P_2 \backslash f(P_0)$ and $M_2(p_2) \stackrel{(2)}{=} M_3(n(p_2))$ or

    2. $\exists p_0 \in P_0$ with $p_2 = f(p_0)$ and $M_2(p_2) = M_2(f(p_0)) \stackrel{(3)}{=} M_3(n(f(p_0))) = M_3(n(p_2))$.
    This means $n \in$ **PTSys** and n is strict.

It remains to show the universal property of the pushout.

Given morphisms $h, k \in$ **PTSys** with $h \circ f = k \circ m$, we have a unique induced morphism x in **PTNet** with $x \circ n = h$ and $x \circ g = k$. We shall show that $x \in$ **PTSys**, i.e. $M_3(p_3) \leq M_4(x(p_3))$ for all $p_3 \in P_3$.

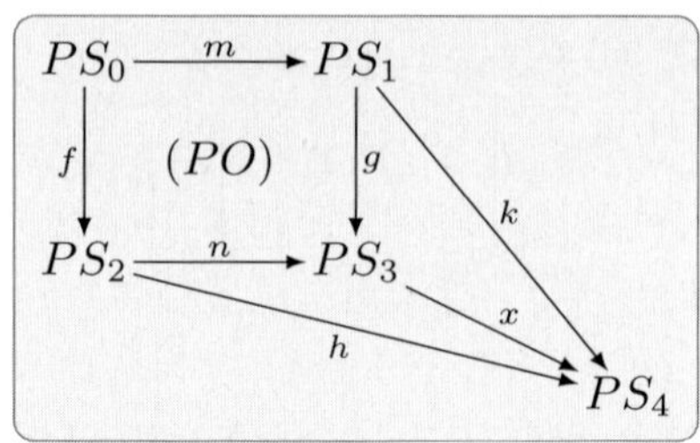

1. For $p_3 = g(p_1)$ with $p_1 \in P_1 \backslash m(P_0)$ we have $M_3(p_3) = M_3(g(p_1)) \overset{(1)}{=}$
$M_1(p_1) \overset{k \in \mathbf{PTSys}}{\leq} M_4(k(p_1)) = M_4(x(g(p_1))) = M_4(x(p_3))$.

2. For $p_3 = n(p_2)$ with $p_2 \in P_2$ we have $M_3(p_3) = M_3(n(p_2)) \overset{(2) \text{ or } (3)}{=}$
$M_2(p_2) \overset{h \in \mathbf{PTSys}}{\leq} M_4(h(p_2)) = M_4(x(n(p_2))) = M_4(x(p_3))$. $\qquad\square$

As next property, we shall show that pullbacks along $\mathcal{M}_{strict}$-morphisms exist and preserve $\mathcal{M}_{strict}$-morphisms.

Theorem 2. *Pullbacks in* **PTSys** *along* $\mathcal{M}_{strict}$ *exist and preserve* $\mathcal{M}_{strict}$*-morphisms, i.e. given P/T morphisms g and n with n strict, then the pullback (PB) exists and m is also a strict P/T morphism.*

$$PS_0 \xrightarrow{\ m\ } PS_1$$
$$f \downarrow \quad (PB) \quad \downarrow g$$
$$PS_2 \xrightarrow{\ n\ } PS_3$$

Construction. Given P/T systems $PS_i = (PN_i, M_i)$ for $i = 1, 2, 3$ and $g, n \in$ **PTSys** with $n \in \mathcal{M}_{strict}$ we construct PN_0 as pullback in **PTNet**, i.e. componentwise in **Sets** on places and transitions. The marking M_0 leading to the P/T system $PS_0 = (PN_0, M_0)$ is defined by

$$(*)\ \ \forall p_0 \in P_0 : M_0(p_0) = M_1(m(p_0)).$$

Proof. Obviously, M_0 is a well-defined marking. We have to show that f, m are P/T morphisms and m is strict.

1. $\forall p_0 \in P_0$ we have: $M_0(p_0) \overset{(*)}{=} M_1(m(p_0)) \overset{g \in \mathbf{PTSys}}{\leq} M_3(g(m(p_0))) = M_3(n(f(p_0))) \overset{n \text{ strict}}{=} M_2(f(p_0))$. This means $f \in$ **PTSys**.

2. $\forall p_0 \in P_0$ we have: $M_0(p_0) \overset{(*)}{=} M_1(m(p_0))$, this means $m \in$ **PTSys** and m is strict.

It remains to show the universal property of the pullback.

Given morphisms $h, k \in$ **PTSys** with $n \circ h = g \circ k$, we have a unique induced morphism x in **PTNet** with $f \circ x = h$ and $m \circ x = k$. We shall show that $x \in$ **PTSys**, i.e. $M_4(p_4) \leq M_0(x(p_4))$ for all $p_4 \in P_4$.

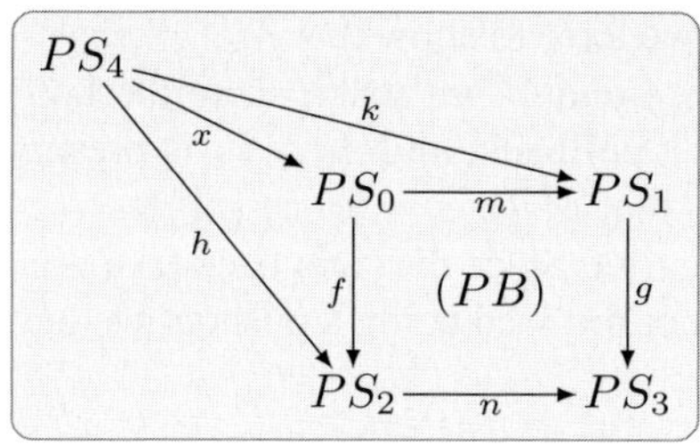

For $p_4 \in P_4$ we have $M_4(p_4) \overset{k \in \mathbf{PTSys}}{\leq} M_1(k(p_4)) = M_1(m(x(p_4))) \overset{m \text{ strict}}{=}$
$M_0(x(p_4))$. $\qquad\qquad\qquad\qquad\qquad\qquad\qquad\qquad\qquad\qquad\qquad\qquad\qquad\qquad\qquad\quad\square$

It remains to show the weak VK property for P/T systems. We know that $(\mathbf{PTNet}, \mathcal{M})$ is a weak adhesive HLR category for the class $\mathcal{M}$ of injective morphisms [4, 8], hence pushouts in $\mathbf{PTNet}$ along injective morphisms are van Kampen squares. But we have to give an explicit proof for the markings in $\mathbf{PTSys}$, because diagrams in $\mathbf{PTSys}$ as in Thm. 1 with $m, n \in \mathcal{M}_{strict}$, which are componentwise pushouts in the P- and T-component, are not necessarily pushouts in $\mathbf{PTSys}$, since we may have $M_3(g(p_1)) > M_1(p_1)$ for some $p_1 \in P_1 \backslash m(P_0)$.

Theorem 3. *Pushouts in $\mathbf{PTSys}$ along $\mathcal{M}_{strict}$-morphisms are weak van Kampen squares.*

Proof. Given the following commutative cube (C) with $m \in \mathcal{M}_{strict}$ and ($f \in \mathcal{M}_{strict}$ or $b, c, d \in \mathcal{M}_{strict}$), where the bottom face is a pushout and the back faces are pullbacks, we have to show that the top face is a pushout if and only if the front faces are pullbacks.

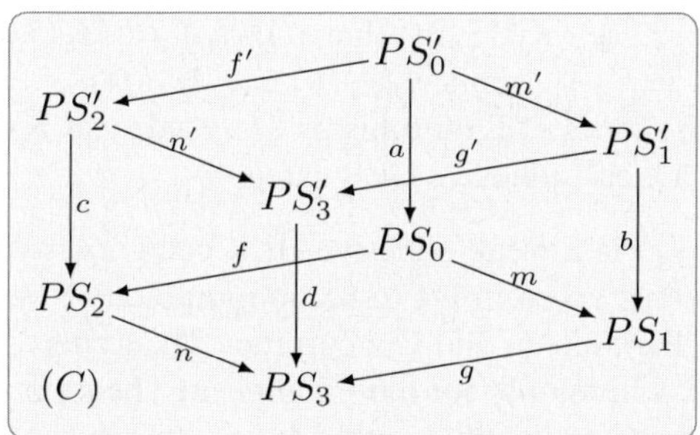

"$\Rightarrow$" If the top face is a pushout then the front faces are pullbacks in $\mathbf{PTNet}$, since all squares are pushouts or pullbacks in $\mathbf{PTNet}$, respectively, where the weak VK property holds. For pullbacks as in Thm. 2 with $m, n \in \mathcal{M}_{strict}$, the marking M_0 of PN_0 is completely determined by the fact that $m \in \mathcal{M}_{strict}$. Hence a diagram in $\mathbf{PTSys}$ with $m, n \in \mathcal{M}_{strict}$ is a pullback in $\mathbf{PTSys}$ if and only if it is a pullback in $\mathbf{PTNet}$ if and only if it is a componentwise pullback in $\mathbf{Sets}$. This means, the front faces are also pullbacks in $\mathbf{PTSys}$.

"$\Leftarrow$" If the front faces are pullbacks we know that the top face is a pushout in $\mathbf{PTNet}$. To show that it is also a pushout in $\mathbf{PTSys}$ we have to verify the conditions (1)-(3) from the construction in Thm. 1.

(1) For $p_1' \in P_1' \setminus m'(P_0')$ we have to show that $M_3'(g'(p_1')) = M_1'(p_1')$.

If f is strict then also g and g' are strict, since the bottom face is a pushout and the right front face is a pullback, and $\mathcal{M}_{strict}$ is preserved by both pushouts and pullbacks. This means that $M_1'(p_1') = M_3'(g'(p_1'))$.

Otherwise b and d are strict. Since the right back face is a pullback we have $b(p_1') \in P_1 \setminus m(P_0)$. With the bottom face being a pushout we have

$$(a)\ M_3(g(b(p_1'))) \overset{(1)}{=} M_1(b(p_1')).$$

It follows that $M_3'(g'(p_1')) \overset{d\ \text{strict}}{=} M_3(d(g'(p_1'))) = M_3(g(b(p_1'))) \overset{(a)}{=} M_1(b(p_1')) \overset{b\ \text{strict}}{=} M_1'(p_1')$.

(2) and (3) For $p_2' \in P_2'$ we have to show that $M_3'(n'(p_2')) = M_2'(p_2')$.

With m being strict also n and n' are strict, since the bottom face is a pushout and the left front face is a pullback, and $\mathcal{M}_{strict}$ is preserved by both pushouts and pullbacks. This means that $M_2'(p_2') = M_3'(n'(p_2'))$.

$\square$

We are now ready to show that the category of P/T systems with the class $\mathcal{M}_{strict}$ of strict P/T morphisms is a weak adhesive HLR category.

Theorem 4. *The category* $(\mathbf{PTSys}, \mathcal{M}_{strict})$ *is a weak adhesive HLR category.*

Proof. By Thm. 1 and Thm. 2, we have pushouts and pullbacks along $\mathcal{M}_{strict}$-morphisms in $\mathbf{PTSys}$, and $\mathcal{M}_{strict}$ is closed under pushouts and pullbacks. Moreover, $\mathcal{M}_{strict}$ is closed under composition and decomposition, because for strict morphisms $f : PS_1 \to PS_2$, $g : PS_2 \to PS_3$ we have $M_1(p) = M_2(f(p)) = M_3(g \circ f(p))$ and $M_1(p) = M_3(g \circ f(p))$ implies $M_1(p) = M_2(f(p)) = M_3(g \circ f(p))$. By Thm. 3, pushouts along strict P/T morphisms are weak van Kampen squares, hence $(\mathbf{PTSys}, \mathcal{M}_{strict})$ is a weak adhesive HLR category. $\square$

Since $(\mathbf{PTSys}, \mathcal{M}_{strict})$ is a weak adhesive HLR category, we can apply the results for adhesive HLR systems given in [4] to reconfigurable P/T systems. Especially, the Local Church-Rosser, Parallelism and Concurrency Theorems as discussed in Section 2 are valid in $\mathbf{PTSys}$, where only for the Parallelism Theorem we need as additional property binary coproducts compatible with $\mathcal{M}_{strict}$, which can be easily verified.

Example 2. If we analyze the two transformations from Ex. 1 in Section 3 depicted in Figs. 2 and 3 we find out that they are sequentially dependent, since $prod_{photo}$ creates the transition *Send Photos* which is used in the match of the transformation $(PN_2, M_2) \overset{prod_{follow}, m'}{\Longrightarrow} (PN_3, M_3)$. In this case, we can apply the Concurrency Theorem and construct a concurrent rule $prod_{conc} = prod_{photo} * prod_{follow}$ that describes the concurrent changes of the net done by the transformations. This rule is depicted in the top row of Fig. 5 and leads to the direct transformation $(PN_1, M_1) \overset{prod_{conc}, m''}{\Longrightarrow} (PN_3, M_3)$, integrating the effects of the two single transformations into one direct one.

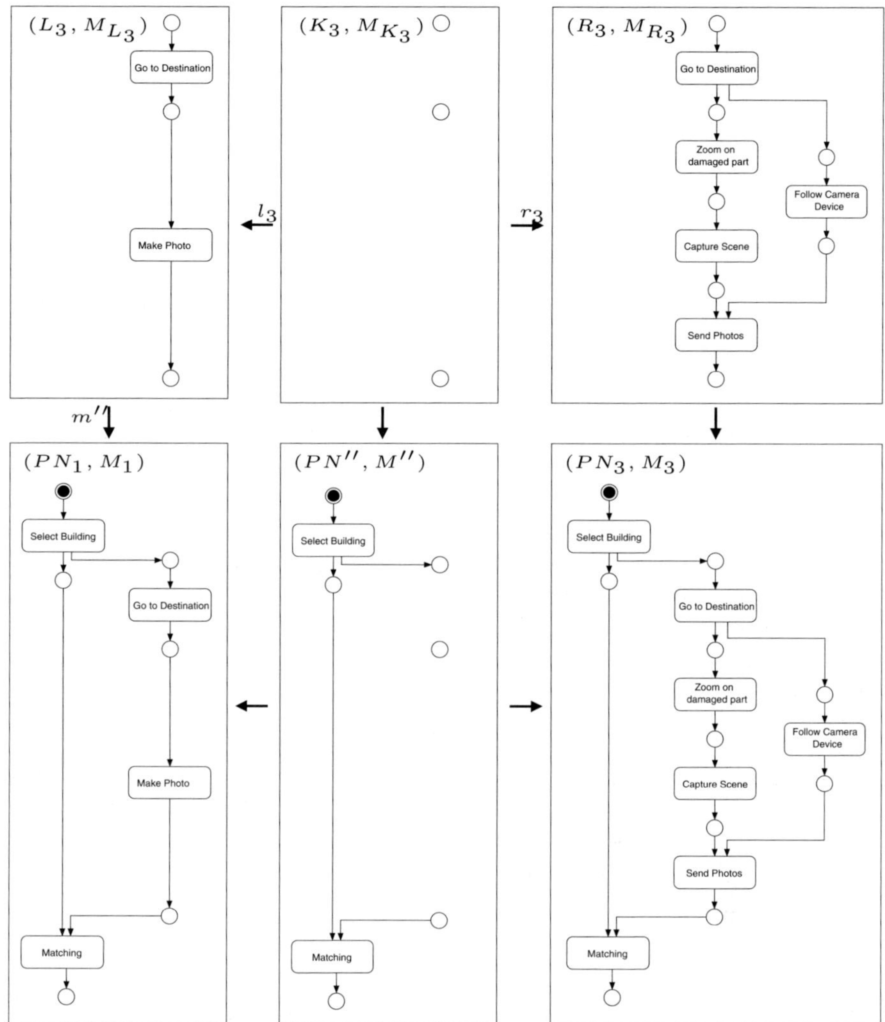

Fig. 5. Direct transformation of (PN_1, M_1) via the concurrent rule $prod_{conc}$

5   Conclusion

In this paper, we have shown that the category **PTSys** of P/T systems, i.e. place/transition nets with markings, is a weak adhesive HLR category for the class $\mathcal{M}_{strict}$ of strict P/T morphisms. This allows the application of the rich theory for adhesive HLR systems like the Local Church-Rosser, Parallelismus and Concurrency Theorems to net transformations within reconfigurable P/T systems.

Related Work

Transformations of nets can be considered in various ways. Transformations of Petri nets to a different Petri net class (e.g. in [12, 13, 14]), to another modeling formalism or vice versa (e.g in [15, 16, 17, 18, 19, 20]) are well examined and have yielded many important results. Transformation of one net into another without changing the net class is often used for purposes of forming a hierarchy in terms of reductions or abstraction (e.g. in [21, 22, 23, 24, 25]), or transformations are used to detect specific properties of nets (e.g. in [26, 27, 28, 29]). For the relationship of Petri nets with process algebras and applications to workflow management we refer to [30] and [31], respectively.

Net transformations that aim directly at changing the net in arbitrary ways as known from graph transformations were developed as a special case of HLR systems e.g. in [4]. The general approach can be restricted to transformations that preserve specific properties as safety or liveness (see [14, 32, 33]). Closely related are those approaches that propose changing nets in specific ways in order to preserve specific semantic properties, as behaviour-preserving reconfigurations of open Petri nets (e.g. in [34]), equivalent (I/O-) behavior (e.g in [35, 36]), invariants (e.g. in [37]) or liveness (e.g. in [38, 31]).

In [2], the concept of "nets and rules as tokens" has been introduced that is most important to model changes of the net structure while the system is kept running, while [3] continues our work by transferring the results of local Church-Rosser, which are well known for term rewriting and graph transformations, to the consecutive evolution of a P/T system by token firing and rule applications. The concept of "nets and rules as tokens" has been used in [39] for a layered architecture for modeling workflows in mobile ad-hoc networks, so that changes given by net transformation are taken into account and the way consistency is maintained is realized by the way rules are applied.

In [40], rewriting of Petri nets in terms of graph grammars are used for the reconfiguration of nets as well, but this approach lacks the "nets as tokens"-paradigm.

Future Work

Ongoing work concerns a prototype system for the editing and simulation of such distributed workflows. For the application of net transformation rules, this tool will provide an export to AGG [41], a graph transformation engine as well as a tool for the analysis of graph transformation properties like termination and rule independence. Furthermore, the token net properties could be analyzed using the Petri Net Kernel [42], a tool infrastructure for Petri nets of different net classes.

On the theoretical side, there are other relevant results in the context of adhesive HLR systems which could be interesting to apply within reconfigurable P/T systems. One of them is the Embedding and Extension Theorem, which deals with the embedding of a transformation into a larger context. Another one is the Local Confluence Theorem, also called Critical Pair Lemma, which gives a criterion when two direct transformations are locally confluent. Moreover, it would be interesting to integrate these aspects with those of property preserving transformations, like lifeness and safety, studied in [14, 32, 33]. As future work, it would be important to verify the additional properties necessary for these results.

Another extension will be to consider rules with negative application conditions, which restrict the applicability of a rule by defining structures that are not allowed to exist. In [43], a theory of adhesive HLR systems with negative application conditions is developed, which should be applied and extended to reconfigurable P/T systems.

For the modeling of complex systems, often not only low-level but also high-level Petri nets are used, that combine Petri nets with some data specification [44]. In [8, 45], it is shown that different kinds of algebraic high-level (AHL) nets and systems form weak adhesive HLR categories. More theory for reconfigurable Petri systems based on high-level nets is needed, since the integration of data and data dependencies leads to more appropriate models for many practical problems.

References

[1] Meseguer, J., Montanari, U.: Petri Nets are Monoids. Information and Computation 88(2), 105–155 (1990)

[2] Hoffmann, K., Ehrig, H., Mossakowski, T.: High-Level Nets with Nets and Rules as Tokens. In: Ciardo, G., Darondeau, P. (eds.) ICATPN 2005. LNCS, vol. 3536, pp. 268–288. Springer, Heidelberg (2005)

[3] Ehrig, H., Hoffmann, K., Padberg, J., Prange, U., Ermel, C.: Independence of Net Transformations and Token Firing in Reconfigurable Place/Transition Systems. In: Kleijn, J., Yakovlev, A. (eds.) ICATPN 2007. LNCS, vol. 4546, pp. 104–123. Springer, Heidelberg (2007)

[4] Ehrig, H., Ehrig, K., Prange, U., Taentzer, G.: Fundamentals of Algebraic Graph Transformation. EATCS Monographs. Springer, Heidelberg (2006)

[5] Ehrig, H., Habel, A., Padberg, J., Prange, U.: Adhesive High-Level Replacement Systems: A New Categorical Framework for Graph Transformation. Fundamenta Informaticae 74(1), 1–29 (2006)

[6] Rozenberg, G. (ed.): Handbook of Graph Grammars and Computing by Graph Transformation: Foundations, vol. 1. World Scientific, Singapore (1997)

[7] Lack, S., Sobociński, P.: Adhesive and Quasiadhesive Categories. Theoretical Informatics and Applications 39(3), 511–546 (2005)

[8] Prange, U.: Algebraic High-Level Nets as Weak Adhesive HLR Categories. Electronic Communications of the EASST 2, 1–13 (2007)

[9] Reisig, W.: Petri Nets. EATCS Monographs on Theoretical Computer Science, vol. 4. Springer, Berlin (1985)

[10] Nielsen, M., Rozenberg, G., Thiagarajan, P.S.: Elementary Transition Systems. Theoretical Computer Science 96(1), 3–33 (1992)

[11] Ehrig, H.: Introduction to the Algebraic Theory of Graph Grammars (A Survey). In: Ng, E.W., Ehrig, H., Rozenberg, G. (eds.) Graph Grammars 1978. LNCS, vol. 73, pp. 1–69. Springer, Heidelberg (1979)

[12] Billington, J.: Extensions to Coloured Petri Nets. In: Proceedings of PNPM 1989, pp. 61–70. IEEE, Los Alamitos (1989)

[13] Campos, J., Sánchez, B., Silva, M.: Throughput Lower Bounds for Markovian Petri Nets: Transformation Techniques. In: Proceedings of PNPM 1991, pp. 322–331. IEEE, Los Alamitos (1991)

[14] Urbášek, M.: Categorical Net Transformations for Petri Net Technology. PhD thesis, TU Berlin (2003)

[15] Belli, F., Dreyer, J.: Systems Modelling and Simulation by Means of Predicate/Transition Nets and Logic Programming. In: Proceedings of IEA/AIE 1994, pp. 465–474 (1994)

[16] Bessey, T., Becker, M.: Comparison of the Modeling Power of Fluid Stochastic Petri Nets (FSPN) and Hybrid Petri Nets (HPN). In: Proceedings of SMC 2002, vol. 2, pp. 354–358. IEEE, Los Alamitos (2002)

[17] de Lara, J., Vangheluwe, H.: Computer Aided Multi-Paradigm Modelling to Process Petri-Nets and Statecharts. In: Corradini, A., Ehrig, H., Kreowski, H.-J., Rozenberg, G. (eds.) ICGT 2002. LNCS, vol. 2505, pp. 239–253. Springer, Heidelberg (2002)

[18] Kluge, O.: Modelling a Railway Crossing with Message Sequence Charts and Petri Nets. In: Ehrig, H., Reisig, W., Rozenberg, G., Weber, H. (eds.) Petri Net Technology for Communication-Based Systems. LNCS, vol. 2472, pp. 197–218. Springer, Heidelberg (2003)

[19] Parisi-Presicce, F.: A Formal Framework for Petri Net Class Transformations. In: Ehrig, H., Reisig, W., Rozenberg, G., Weber, H. (eds.) Petri Net Technology for Communication-Based Systems. LNCS, vol. 2472, pp. 409–430. Springer, Heidelberg (2003)

[20] Corts, L., Eles, P., Peng, Z.: Modeling and Formal Verification of Embedded Systems Based on a Petri Net Representation. Journal of Systems Architecture 49(12-15), 571–598 (2003)

[21] Haddad, S.: A Reduction Theory for Coloured Nets. In: Rozenberg, G. (ed.) APN 1989. LNCS, vol. 424, pp. 209–235. Springer, Heidelberg (1990)

[22] Desel, J.: On Abstraction of Nets. In: Rozenberg, G. (ed.) APN 1991. LNCS, vol. 524, pp. 78–92. Springer, Heidelberg (1991)

[23] Esparza, J., Silva, M.: On the Analysis and Synthesis of Free Choice Systems. In: Rozenberg, G. (ed.) APN 1990. LNCS, vol. 483, pp. 243–286. Springer, Heidelberg (1991)

[24] Chehaibar, G.: Replacement of Open Interface Subnets and Stable State Transformation Equivalence. In: Rozenberg, G. (ed.) APN 1993. LNCS, vol. 674, pp. 1–25. Springer, Heidelberg (1993)

[25] Bonhomme, P., Aygalinc, P., Berthelot, G., Calvez, S.: Hierarchical Control of Time Petri Nets by Means of Transformations. In: Proceedings of SMC 2002, vol. 4, pp. 6–11. IEEE Computer Society Press, Los Alamitos (2002)

[26] Berthelot, G.: Checking Properties of Nets Using Transformation. In: Rozenberg, G. (ed.) APN 1985. LNCS, vol. 222, pp. 19–40. Springer, Heidelberg (1986)

[27] Berthelot, G.: Transformations and Decompositions of Nets. In: Brauer, W., Reisig, W., Rozenberg, G. (eds.) APN 1986. LNCS, vol. 254, pp. 359–376. Springer, Heidelberg (1987)

[28] Best, E., Thielke, T.: Orthogonal Transformations for Coloured Petri Nets. In: Azéma, P., Balbo, G. (eds.) ICATPN 1997. LNCS, vol. 1248, pp. 447–466. Springer, Heidelberg (1997)

[29] Murata, T.: Petri Nets: Properties, Analysis and Applications. In: Proceedings of the IEEE, vol. 77, pp. 541–580. IEEE Computer Society Press, Los Alamitos (1989)

[30] Best, E., Devillers, R., Koutny, M.: The Box Algebra = Petri Nets + Process Expressions. Information and Computation 178(1), 44–100 (2002)

[31] van der Aalst, W.: The Application of Petri Nets to Workflow Management. Journal of Circuits, Systems and Computers 8(1), 21–66 (1998)

[32] Padberg, J., Gajewsky, M., Ermel, C.: Rule-based Refinement of High-Level Nets Preserving Safety Properties. Science of Computer Programming 40(1), 97–118 (2001)

[33] Padberg, J., Urbášek, M.: Rule-Based Refinement of Petri Nets: A Survey. In: Ehrig, H., Reisig, W., Rozenberg, G., Weber, H. (eds.) Petri Net Technology for Communication-Based Systems. LNCS, vol. 2472, pp. 161–196. Springer, Heidelberg (2003)

[34] Baldan, P., Corradini, A., Ehrig, H., Heckel, R., König, B.: Bisimilarity and Behaviour-Preserving Reconfigurations of Open Petri Nets. In: Mossakowski, T., Montanari, U., Haveraaen, M. (eds.) CALCO 2007. LNCS, vol. 4624, pp. 126–142. Springer, Heidelberg (2007)

[35] Balbo, G., Bruell, S., Sereno, M.: Product Form Solution for Generalized Stochastic Petri Nets. IEEE Transactions on Software Engineering 28(10), 915–932 (2002)

[36] Carmona, J., Cortadella, J.: Input/Output Compatibility of Reactive Systems. In: Aagaard, M.D., O'Leary, J.W. (eds.) FMCAD 2002. LNCS, vol. 2517, pp. 360–377. Springer, Heidelberg (2002)

[37] Cheung, T., Lu, Y.: Five Classes of Invariant-Preserving Transformations on Colored Petri Nets. In: Donatelli, S., Kleijn, J. (eds.) ICATPN 1999. LNCS, vol. 1639, pp. 384–403. Springer, Heidelberg (1999)

[38] Esparza, J.: Model Checking Using Net Unfoldings. Science of Computer Programming 23(2-3), 151–195 (1994)

[39] Padberg, J., Hoffmann, K., Ehrig, H., Modica, T., Biermann, E., Ermel, C.: Maintaining Consistency in Layered Architectures of Mobile Ad-hoc Networks. In: Dwyer, M.B., Lopes, A. (eds.) FASE 2007. LNCS, vol. 4422, pp. 383–397. Springer, Heidelberg (2007)

[40] Llorens, M., Oliver, J.: Structural and Dynamic Changes in Concurrent Systems: Reconfigurable Petri Nets. IEEE Transactions on Computers 53(9), 1147–1158 (2004)

[41] AGG Homepage (2007), http://tfs.cs.tu-berlin.de/agg

[42] Kindler, E., Weber, M.: The Petri Net Kernel - An Infrastructure for Building Petri Net Tools. Software Tools for Technology Transfer 3(4), 486–497 (2001)

[43] Lambers, L., Ehrig, H., Prange, U., Orejas, F.: Parallelism and Concurrency in Adhesive High-Level Replacement Systems with Negative Application Conditions. Electronic Notes in Theoretical Computer Science (to appear 2008)

[44] Padberg, J., Ehrig, H., Ribeiro, L.: Algebraic High-Level Net Transformation Systems. Mathematical Structures in Computer Science 5(2), 217–256 (1995)

[45] Prange, U.: Algebraic High-Level Systems as Weak Adhesive HLR Categories. Electronic Notes in Theoretical Computer Science (to appear, 2008)

Explicit State Model Checking for Graph Grammars

Arend Rensink

Department of Computer Science, University of Twente, The Netherlands
`rensink@cs.utwente.nl`

Abstract. In this paper we present the philosophy behind the GROOVE project, in which graph transformation is used as a modelling formalism on top of which a model checking approach to software verification is being built. We describe the basic formalism, the current state of the project, and (current and future) challenges.

1   Introduction

Our primary interest in this paper is software model checking, in particular of object-oriented programs. Model checking has been quite successful as a hardware verification technique and its potential application to software is receiving wide research interest. Indeed, software model checkers are being developed and applied at several research institutes; we mention Bogor [32] and Java Pathfinder [17] as two well-known examples of model checkers for Java.

Despite these developments, we claim that there is an aspect of software that does not occur in this form in hardware, and which is only poorly covered by existing model checking theory: dynamic (de)allocation, both on the heap (due to object creation and garbage collection) and on the stack (due to mutual and recursive method calls and returns). Classical model checking approaches are based on propositional logic with a fixed number of propositions; this does not allow a straightforward representation of systems that may involve variable, possibly unbounded numbers of objects. Although there exist workarounds for this (as evidenced by the fact that, as we have already seen, there are working model checkers for Java) we strongly feel that a better theoretical understanding of the issues involved is needed.

Graphs are an obvious choice for modelling the structures involved, at least informally; direct evidence of this can be found in the fact that any textbook of object-oriented programming uses graphs (of some form) for illustrative purposes. Indeed, a graph model is a very straightforward way to visualise and reason about heap and stack structures, at least when they are of restricted size. In fact, there is no *a priori* reason why this connection cannot be exploited beyond the informal, given the existence of a rich theory of (in particular) graph transformation — see for instance the handbook [33], or the more recent textbook [8]. By adopting graph transformation, one can model the computation steps of object-oriented systems through rules working directly on the graphs, rather than through some intermediate modelling language, such as a process algebra.

This insight has been the inspiration for the GROOVE project and tool.[1] Though the idea is in itself not revolutionary or unique, the approach we have followed differs

---

[1] GROOVE stands for "GRaphs for Object-Oriented VErification.

P. Degano et al. (Eds.): Montanari Festschrift, LNCS 5065, pp. 114–132, 2008.

from others in the fact that it is based on state space generation directly from graph grammars; hence, neither do we use the input language of an existing model checker to translate the graph rules to, like in [35,12], nor do we attempt to prove properties on the level of graph grammars, like in [22,14,21]. In this paper, we present the elements of this approach, as well as the current state of the research. It is thus essentially the successor [25], where we first outlined the approach.

The paper is structured as follows: in Sect. 2 we introduce the formal notion of graphs and transformations, in a constructive way rather than relying on category theoretical notions. In Sect. 3 we define automata, so-called *graph transition systems*, on top of graph grammars; we also define bisimilarity and show that there exist minimal (reduced) automata. (This is a new result, achieved by abstracting away from symmetries in a somewhat different way than by Montanari and Pistore in [24].) In Sect. 4 we define first-order temporal logic on top of graph transition systems; we also present an equivalent temporal logic based on graph morphisms as core elements, along the lines of [28,6]. Finally, in Sect. 5 we give an evaluation and outlook.

2   Transformation of Simple Graphs

We model system states as graphs. Immediately, we are faced with the choice of graph formalism. In order to make optimal use of the existing theory on graph transformation, in particular the algebraic approach [8], it is preferable to select a definition that gives rise to a (weakly) adhesive HLR category (see [20,10]), such as multi-sorted graphs (with separate sorts for nodes and edges, and explicit source and target functions), or attributed graphs [7] built on top of those. On the other hand, in the GROOVE project and tool [27], in which the approach described in this paper was developed, we have chosen to use simple graphs, which do not fulfill these criteria (unless one also restricts the rules to regular left morphisms, which we have not done), and single-pushout transformation, as first defined by Löwe in [23]. There were two main reasons for this choice:

- In the envisaged domain of application (operational semantics of object-oriented systems) there is little use for edges with identities (see, e.g., [18]);
- The most straightforward connection to first-order logic is to interpret edges as binary predicates (see, e.g., [28]); again, this ignores edge identities.

In this paper, we stick to this choice and present the approach based on simple graphs; in Sect. 5 we will come back to this issue.

Throughout this paper we will assume the existence of a universe of labels Label, and a universe of node identities Node.

Definition 1 (simple graph)

- *A simple graph is a tuple $\langle V, E \rangle$, where $V \subseteq$ Node is a set of nodes and $E \subseteq V \times$ Label $\times V$ a set of edges. Given $e = (v, a, w) \in E_G$, we denote $src(e) = v$, $lab(e) = a$ and $tgt(e) = w$ for its source, label and target, respectively.*

- *Given two simple graphs G, H, a (partial) graph morphism $f: G \to H$ is a pair of partial functions $f_V: V_G \rightharpoonup V_H$ and $f_E: E_G \rightharpoonup E_H$, such that for all $e \in dom(f_E)$, $f_E(e) = (f_V(src(e)), lab(e), f_V(tgt(e)))$.*

Some notation and terminology.

- A morphism f is called *total* if f_V and f_E are total functions, *injective* if they are injective functions, and an *isomorphism* if they are bijective functions. The total, injective morphisms are sometimes called *monos*.[2]
- For $f: G \to H$, we call G and H the *source* and *target* of f, denoted $src(f)$ and $tgt(f)$, respectively. A pair of morphisms with a common source is called a *span*; with a common target, a *co-span*.
- We write $G \cong H$ to denote that there is an isomorphism from G to H, and $\phi: G \cong H$ to denote that ϕ is such an isomorphism. This is extended to the individual morphisms of spans $\xleftarrow{f} \xrightarrow{g}$: we write $f \cong g$ for such morphisms if there is an isomorphism $\phi: tgt(f) \to tgt(g)$ such that $g = \phi \circ f$.
- We use Morph to denote the set of all (partial) graph morphisms.

Some example graphs will follow below. As a further convention, in figures we will use node labels to represent self-edges; for instance, in Fig. 1, the labels Buffer, Cell and Object are used in this way.

Graph morphisms are used for many different purposes, but the following uses stand out in the context of graph transformation:

- Isomorphisms are used to capture the fact that two graphs are essentially the same. The whole theory of graph transformation is set up to be insensitive to isomorphism, meaning that it is all right to pick the most convenient isomorphic representative.
- Total morphisms describe an embedding of one graph into another, possibly while merging nodes. If a total morphism is also injective, then there is no merging, and the source graph is sometimes called an (isomorphic) *subgraph* of the target.
- Arbitrary (partial) morphisms are used to capture the *difference* between graphs, in terms of the exact change from the source graph to the target graph. To be precise, the change consists of deletion of those source graph elements on which the morphism is not defined, merging of those elements on which it is not injective, and addition of those target graph elements that are not in the image of the morphism.

A core construction in graph transformation is the so-called *pushout*. This is used as a way to combine, or glue together, different changes to the same graph; or in particular, an embedding on the one hand and a change on the other — where both the embedding and the change are captured by morphisms, as in the second and third items above. In the following, we define pushouts constructively.

Definition 2 (pushout). *For $i = 1, 2$, let $f_i: G \to H_i$ be morphisms in Morph such that $H_i = \langle V_i, E_i \rangle$ with $V_1 \cap V_2 = \emptyset$; let $*$ be arbitrary such that $* \notin V_1 \cup V_2$.*

---

[2] This name stems from category theory, where monos are arrows satisfying a particular decomposition property. We do not elaborate on this issue here.

1. *For $i = 1, 2$, let $\bar{f}_i \colon G \to \bar{H}_i$ be a total extension of f_i, defined by adding a distinct fresh node v' to H_i and setting $\bar{f}_i(v) = v'$ for each $v \in V_G \setminus dom(f_{V,i})$. Hence, $\bar{H}_i = \langle \bar{V}_i, \bar{E}_i \rangle$ such that $\bar{V}_i$ extends V_i with the fresh nodes, and $\bar{E}_i$ extends E_i with the fresh edges implied by the totality of $\bar{f}_i$.*
2. *Let $\bar{V} = \bar{V}_1 \cup \bar{V}_2$ be the union of the extended node sets, and $\bar{E} = \bar{E}_1 \cup \bar{E}_2$ that of the extended edge sets. Let $\simeq \,\subseteq \bar{V} \times \bar{V}$ be the smallest equivalence relation such that $\bar{f}_{V,1}(v) \simeq \bar{f}_{V,2}(v)$ for all $v \in V_G$; and likewise for edges.*
3. *Let $W = \{X \in \bar{V}/\!\simeq \; | \; X \subseteq V_1 \cup V_2\}$, and for $i = 1, 2$, define $g_{V,i} \colon V_i \to \bar{W}$ such that for all $v \in V_i$*

$$g_{V,i} \colon v \mapsto [v]_{\simeq} \quad \text{if } [v]_{\simeq} \subseteq V_1 \cup V_2 \ .$$

Let $F = \{([v]_{\simeq}, a, [w]_{\simeq}) \mid [(v, a, w)]_{\simeq} \subseteq E_1 \cup E_2\}$; moreover, for $i = 1, 2$, define $g_{E,i} \colon E_i \to F$ such that for all $(v, a, w) \in E_i$

$$g_{E,i} \colon (v, a, w) \mapsto ([v]_{\simeq}, a, [w]_{\simeq}) \quad \text{if } [(v, a, w)]_{\simeq} \subseteq E_1 \cup E_2 \ .$$

4. *Let $K = \langle W, F \rangle$; then $g_i \colon H_i \to K$ are morphisms for $i = 1, 2$.*

K *together with the morphisms* g_i *is called the* pushout *of the span* f_1, f_2*; together they form the following* pushout diagram.

$$
\begin{array}{ccc}
G & \xrightarrow{\;f_1\;} & H_1 \\[4pt]
{\scriptstyle f_2}\big\downarrow & & \big\downarrow{\scriptstyle g_1} \\[4pt]
H_2 & \xrightarrow{\;g_2\;} & K
\end{array}
$$

The intuition behind the construction is as follows: first (step 1) we (essentially) construct the disjoint union of the two target graphs H_1 and H_2, augmented with images for those elements of G for which the morphisms f_1 resp. f_2 were originally undefined. These fresh images later work like little "time bombs" that obliterate themselves, together with all associated (i.e., equivalent) elements. In the resulting extended disjoint union, we call two elements equivalent (step 2) if they have a common source in G, under the (extended) morphisms $\bar{f}_1$ or $\bar{f}_2$. Then (step 3), we construct the quotient with respect to this equivalence, omitting however the (equivalence classes of the) fresh nodes and edges added earlier — in other words, this is when the bombs go off.

The name "pushout" for the object constructed in the previous definition is justified by the fact that it satisfies a particular co-limit property, stated formally in the following proposition.

Proposition 1 (pushout property). *Given morphisms* f_i *as in Def. 2, the pushout is a co-limit of the diagram consisting of the span* f_1, f_2*, meaning that*

- $g_1 \circ f_1 = g_2 \circ f_2$*;*
- *Given any* $h_i \colon H_i \to L$ *for* $i = 1, 2$ *such that* $h_1 \circ f_1 = h_2 \circ f_2$*, there is a unique morphism* $k \colon K \to L$ *such that* $h_1 = k \circ g_1$ *and* $h_2 = k \circ g_2$*; in other words, such that the following diagram commutes*

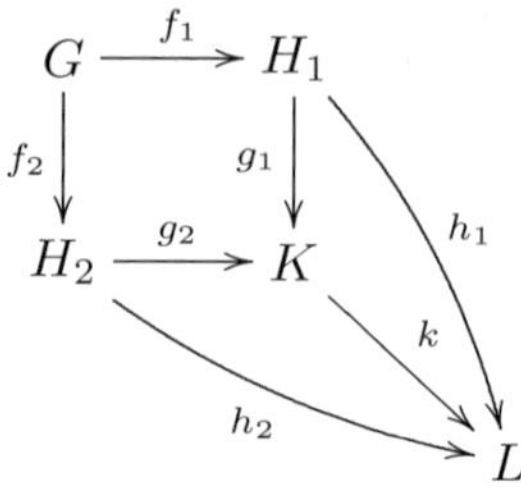

Proof (sketch). We construct k. For a given $W \in V_K$, let i be such that $W \cap V_i \neq \emptyset$; let $k_V(W) = h_i(w)$ for $w \in W \cap V_i$. This is well-defined due to the fact that W is a $\simeq$-equivalence class, in combination with the confluence of the f_1g_1/f_2g_2- and f_1k_1/f_2k_2-squares of the diagram. k_E is defined likewise. k satisfies the necessary commutation properties by construction. Its uniqueness in this regard can be established by observing that no other image for any of the nodes or edges of K will make the pushout diagram commute.

The mechanism we use for generating state spaces is based on graph grammars, consisting of a set of graph production rules and a start graph. The necessary ingredients are given by the following definition.

Definition 3 (graph grammar)

- *A graph production rule is a tuple* $r = \langle p\colon L \to R, ac \rangle$, *where* $p \in$ Morph, *and* ac *is an* application condition *on total graph morphisms* $m\colon L \to G$ *(for arbitrary G) that is well-defined up to isomorphism of G. We write* $m \models ac$ *to denote that m satisfies ac. (Well-definedness up to isomorphism of G means that* $m \models ac$ *if and only if* $\phi \circ m \models ac$ *for all graph isomorphisms* $\phi\colon G \to H$.)

  A graph grammar is a tuple $\mathcal{G} = \langle \mathcal{R}, I \rangle$, *where* $\mathcal{R}$ *is a set of production rules and I is an initial graph.*

- *Given a graph production rule r, an r-derivation is a four-tuple* (G, r, m, H), *typically denoted* $G \overset{r,m}{\Longrightarrow} H$, *such that* $m\colon L_r \to G \models ac_r$ *and H is isomorphic to the pushout graph; i.e., the following square is a pushout:*

$$
\begin{array}{ccc}
L_r & \overset{p_r}{\longrightarrow} & R_r \\
\downarrow{\scriptstyle m} & & \downarrow{\scriptstyle m'} \\
G & \underset{f}{\longrightarrow} & K \cong H
\end{array}
$$

  A $\mathcal{G}$-derivation ($\mathcal{G}$ a graph grammar) is an r-derivation for some $r \in \mathcal{R}_\mathcal{G}$.

The definition is slightly sloppy in that our pushout construction is only defined if the right hand side R_r and the host graph G have disjoint node sets. This is in practice not a problem because we are free to take isomorphic representatives where required; in particular, we can make sure that the derived graphs have nodes that are distinct from all right hand side graphs.

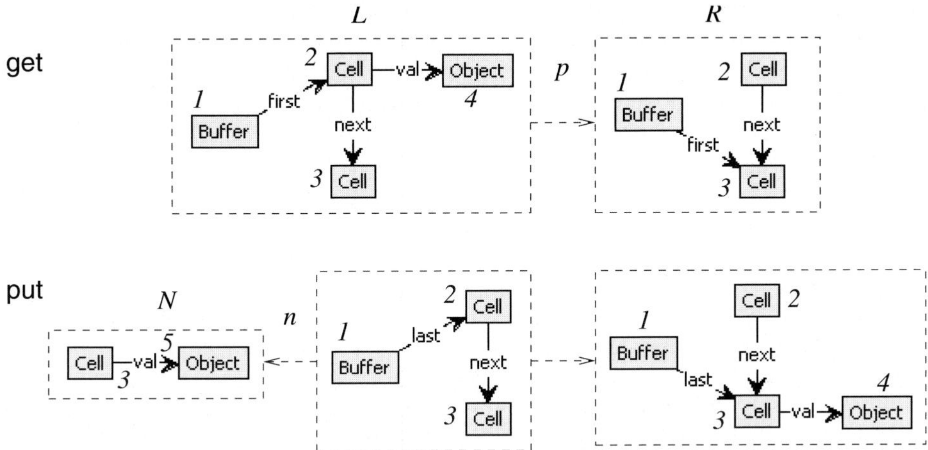

Fig. 1. Graph production rules for a circular buffer

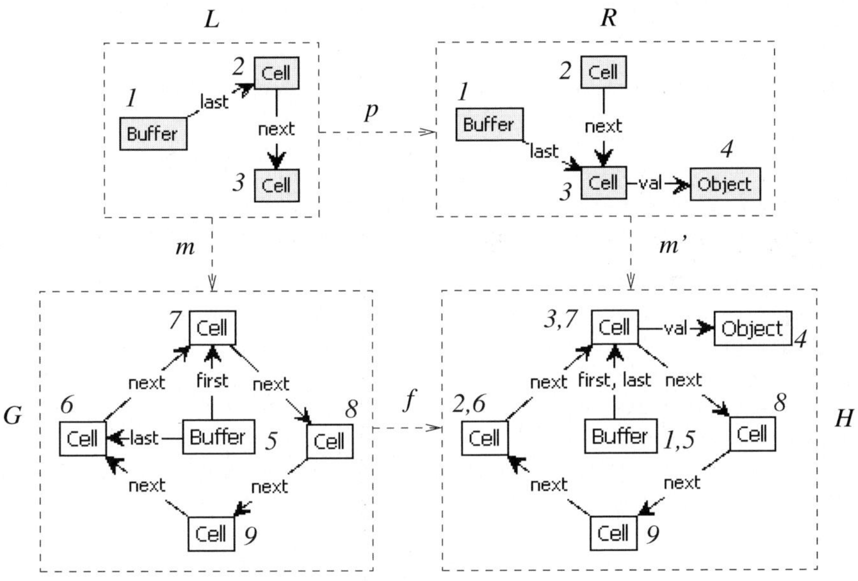

Fig. 2. Example derivation using the rule get from Fig. 1

As a running example, we use the graph grammar consisting of the two rules in Fig. 1, which retrieve, resp. insert, objects in a circular buffer. The rule put has a so-called *negative application condition* (see [13]), in the form of a morphism $n: L \to N$ from the left hand side of the rule; the satisfaction of such a condition is defined by

$$m: L \to G \models n \quad :\Leftrightarrow \quad \nexists f: N \to G : f \circ n = m \upharpoonright dom(n)$$

(where $dom(n)$ is the sub-graph of L on which n is defined, and $m \upharpoonright H$ stands for the restriction of the morphism m to the graph H). The morphisms are indicated by the numbers in the figure: nodes are mapped to equally labelled nodes in the target graph (if such a target node exists, elsewhere the morphism is undefined), and edges are mapped accordingly. An example derivation is shown in Fig. 2, given a match $m = \{(1,5),(2,6),(3,7)\}$. In terms of Def. 2, this gives rise to

$$\bar{V} = \{1,2,3,4,5,6,7,8,9\}$$
$$W = \{\{1,5\},\{2,6\},\{3,7\},\{4\},\{8\},\{9\}\} \ .$$

3   Graph Transition Systems

As related in the introduction above, the core of our approach is explicit state space generation, where the states are essentially graphs. Rather than completely identifying states with graphs (like we did in the original [25]), in this paper we follow [2] by merely requiring that every state *has* an associated graph. This leaves room for cases where there is more information in a state than just the underlying graph.

Definition 4 (graph transition system). *A* graph transition system *(GTS) is a tuple* $S = \langle Q, T, q_0 \rangle$ *where*

- Q *is a set of states with, for every* $q \in Q$, *an associated graph* $G_q \in$ Graph;
- $T \subseteq Q \times$ Morph $\times Q$ *is a set of transitions, such that* $src(\alpha) = G_q$ *and* $tgt(\alpha) \cong G_{q'}$ *for all* $(q, \alpha, q') \in T$. *As usual, we write* $q \xrightarrow{\alpha} q'$ *as equivalent to* $(q, \alpha, q') \in T$.
- $q_0 \in Q$ *is the initial state.*

S *is called* symmetric *if* $(q, \alpha, q') \in T$ *implies* $(q, \alpha \circ \phi, q') \in T$ *for all* $\phi \colon G_q \cong G_q$.

We write q_t, α_t, q'_t for the source state, morphism and target state of a transition t, and q_i etc. for the components of t_i. Note that the target graphs of the morphisms associated with the transitions are only required to be isomorphic, rather than identical, to the graphs associated with their target states. Obviously, this only makes a difference for graphs having non-trivial symmetries, since otherwise the isomorphisms are unique and might as well be appended to the transition morphisms. Using the definition given here, we avoid to distinguish between symmetric cases, and hence it is possible to minimise with respect to bisimilarity — see below.

An example symmetric GTS is shown in Fig. 3. The morphisms associated with the transitions are indicated by node mappings at the arrows; all the morphisms have empty edge mappings. The two left-to-right transitions are essentially the same, since their associated morphisms are "isomorphic;" that is, there is an isomorphism between their target graphs that equalises them —namely, based on node mapping $(3,4),(4,3)$. On the other hand, this is not true for the right-to-left transitions: the node mapping $(1,2),(2,1)$ is not an isomorphism of the left hand side graph. Indeed, by symmetry, the presence of each of the right-to-left transition implies the presence of the other.

We can now understand a GTS as *being generated by* a graph grammar $\mathcal{G}$, if the start state's associated graph is isomorphic to the start graph of $\mathcal{G}$, and there are transitions corresponding to all the derivations of $\mathcal{G}$ (modulo isomorphism). That is, we call a GTS S generated by $\mathcal{G}$ if the following conditions hold:

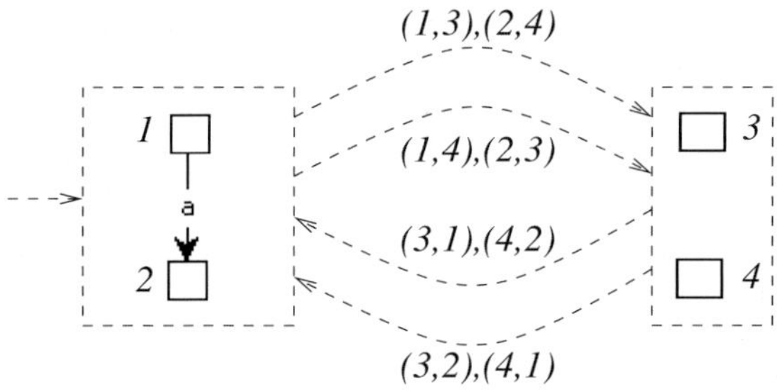

Fig. 3. Example symmetric GTS, removing and re-adding an edge

- G_{q_0} is isomorphic to I;
- For any $q \in Q$ and any $\mathcal{G}$-derivation $G_q \xrightarrow{r,m} H$, S has a transition $q \xrightarrow{\alpha} q'$ such that $\alpha \cong p_r \uparrow m$.
- Likewise, for any transition $q \xrightarrow{\alpha} q'$, there is a $\mathcal{G}$-derivation $G_q \xrightarrow{r,m} H$ such that $\alpha \cong p_r \uparrow m$.

For instance, Fig. 4 shows a GTS generated using the rules in Fig. 1, taking G from Fig. 2 as a start graph.

Grammar-generated GTSs are close to the history-dependent automaton (HDA) of Montanari and Pistore (see [24]). There, states have associated sets of names, which are "published" through labelled transitions, the labels also having names and the transitions carrying triple co-spans of total injective name functions, from the source state, target state and label to a common set of names associated with the transition.

If we limit our rules to injective morphisms, then the derivation morphisms will be injective, too. Injective partial morphisms $\alpha: G \to H$ are in fact equivalent to

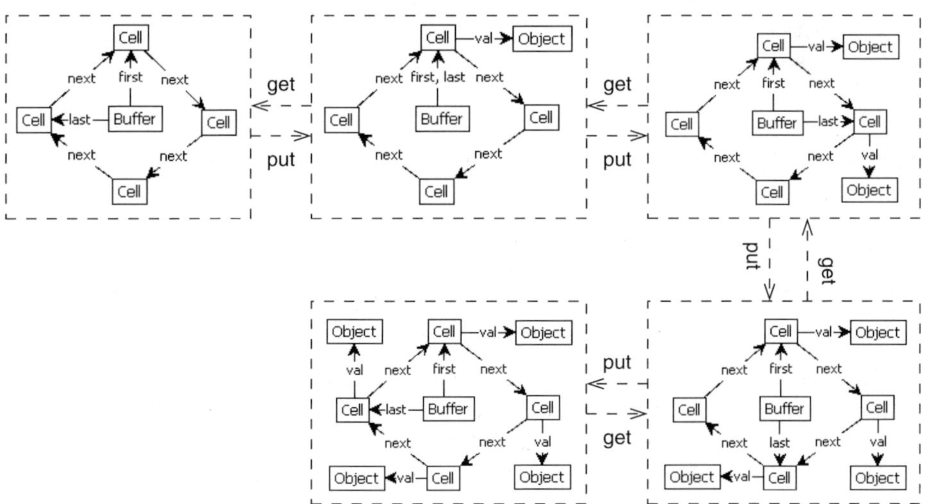

Fig. 4. GTS generated from the circular buffer rules

co-spans of monos $\alpha_L\colon G \to U, \alpha_R\colon H \to U$, where U is the "union" of G and H while gluing together the elements mapped onto each other by α. If, furthermore, we require all matchings to be injective as well (through the application condition ac), any $\mathcal{G}$-generated GTS gives rise to a HDA, where the names are given by node identities. The only catch is that, as mentioned before, the actual isomorphism from $tgt(\alpha)$ to $G_{q'}$ in a transition $q \xrightarrow{\alpha} q'$ is not part of the GTS, whereas in HDA the name mappings are precise. On the other hand, this information is abstracted away in *HDA with symmetries* (see [24]). We conclude:

Proposition 2. *If $\mathcal{G}$ is such that all rules in $\mathcal{R}$ are injective, and $m \models ac$ only if m is injective, then every $\mathcal{G}$-generated GTS uniquely gives rise to a HDA with symmetries, in which the transition labels are tuples of rules and matchings.*

In contrast to HDAs, however, GTS transitions are reductions and not reactions. In other words, they do not reflect communications with the "outside world". In fact, the behaviour modelled by a GTS is not primarily captured by the transition labels but by the structure of the states; as we will see, the logic we use to express GTS properties can look inside the states. For that reason, although we can indeed define a notion of bisimilarity —inspired by HDA bisimilarity— which abstracts away to some degree from the branching structure, the relation needs to be very discriminating on states.[3]

Definition 5 (bisimilarity). *Given two GTSs S_1, S_2, a bisimulation between S_1 and S_2 is an isomorphism-indexed relation $(\rho_\phi)_\phi \subseteq (Q_1 \times Q_2) \cup (T_1 \times T_2)$ such that*

- *For all $q_1 \, \rho_\phi \, q_2$, the following hold:*
    - $\phi\colon G_{q_1} \cong G_{q_2}$;
    - *For all $q_1 \xrightarrow{\alpha_1} q_1'$, there is a $q_2 \xrightarrow{\alpha_2} q_2'$ such that $(q_1, \alpha_1, q_1') \, \rho_\phi \, (q_2, \alpha_2, q_2')$;*
    - *For all $q_2 \xrightarrow{\alpha_2} q_2'$, there is a $q_1 \xrightarrow{\alpha_1} q_1'$ such that $(q_1, \alpha_1, q_1') \, \rho_\phi \, (q_2, \alpha_2, q_2')$;*
- *For all $t_1 \, \rho_\phi \, t_2$: $q_1 \, \rho_\phi \, q_2$, $\alpha_1 \cong \alpha_2 \circ \phi$, and there is a ψ such that $q_1' \, \rho_\psi \, q_2'$;*
- *$q_{0,1} \, \rho_\phi \, q_{0,2}$ for some ϕ.*

S_1 and S_2 are said to be bisimilar, *denoted $S_1 \sim S_2$, if there exists a bisimulation between them.*

For instance, although the GTS generated by a graph grammar is not unique, it is unique modulo bisimilarity.

Theorem 1. *If S_1 and S_2 are both generated by a graph grammar $\mathcal{G}$, then $S_1 \sim S_2$.*

Thus, bisimulation establishes binary relations between the states and transitions of two GTSs. As usual, this can be used to *reduce* GTSs, as follows:

- Between any pair of GTSs there exists a *largest* bisimulation, which can be defined as the union of all bisimulations (pointwise along the ϕ). (The proof that this is indeed again a bisimulation is straightforward.)
- If S_1 and S_2 are the same GTS (call it S), then the largest bisimulation gives rise to an equivalence relation ρ over the states and transitions of S.

---

[3] In this paper, we use bisimilarity only to *minimise* GTSs, not so much to establish a theory of equivalence.

- Given the equivalence ρ, pick a representative from every equivalence class of states Q/ρ. For any $q \in Q$, let $\hat{q}$ denote the representative from $[q]_\rho$.
- Similarly, for every equivalence class of transitions $U \in T/\rho$, pick a representative with as source the (uniquely defined) state $\hat{q}_t$ for $t \in U$. For any t, let $\hat{t}$ denote the selected representative from $[t]_\rho$. (Note that this means $q_{\hat{t}}$ is the representative state selected from $[q_t]_\rho$, but $q'_{\hat{t}}$ is not necessarily the representative for $[q'_t]_\rho$.)
- Define $\hat{S} = \langle \hat{Q}, \hat{T}, \hat{q}_0 \rangle$ where

$$\hat{Q} = \{\hat{q} \mid q \in Q\}$$
$$\hat{T} = \{(q_{\hat{t}}, \alpha_{\hat{t}}, \hat{q}'_t) \mid t \in T\} \ .$$

Thus, this construction collapses states with isomorphic graphs and transitions with isomorphic graph changes. For instance, in Fig. 3, only one of the two left-to-right transitions remains in the reduced transition system. The transition system in Fig. 4 cannot be reduced further (there are no non-trivial isomorphisms); in fact, it has already been reduced up to symmetry, since (for instance) the precise graph reached after applying put $\cdot$ get from the initial state is not identical to the start graph; rather, it can be thought of as isomorphically rotated clockwise by $90°$. Indeed, reduction with respect to bisimilarity exactly corresponds to symmetry reduction for model checking (see, e.g., [11]).

It may actually not be clear that $\hat{S}$ is well-defined, since it relies on the choice of representatives from the equivalence classes Q/ρ and T/ρ. To show well-definedness we must first define *isomorphism* of GTSs.

Definition 6 (GTS isomorphism). *Two GTSs S_1, S_2 are* isomorphic *if there exists a pair of mappings $\phi_Q \colon Q_1 \to (\mathrm{Morph} \times Q_2)$ and $\phi_T \colon T_1 \to T_2$ such that*

- *For all $q_1 \in Q_1$, $\phi_Q(q_1) = (\psi, q_2)$ such that $\psi \colon G_{q_1} \cong G_{q_2}$;*
- *For all $t_1 \in T_1$, if $\phi_Q(q_1) = (\psi, q_2)$ and $\phi_Q(q'_1) = (\psi', q'_2)$ then $\phi_T(t_1) = (q_2, \alpha_2, q'_2)$ such that $\alpha_2 \circ \psi \cong \alpha_1$;*
- *$\phi_Q(q_{1,0}) = (\psi, q_{2,0})$ for some ψ.*

The following theorem states the crucial properties of the GTS reduction.

Theorem 2. *Given any GTS S, the reduced GTS $\hat{S}$ is well-defined up to isomorphism and satisfies $\hat{S} \sim S$. Furthermore, for any bisimulation $(\hat{\rho}_\phi)_\phi$ between $\hat{S}$ and itself, $\hat{q}_1 \ \hat{\rho}_\phi \ \hat{q}_2$ implies $\hat{q}_1 = \hat{q}_2$ for all $\hat{q}_1, \hat{q}_2 \in \hat{Q}$.*

Proof (sketch).

- It is straightforward to check that any choice of representatives from the ρ-induced equivalence classes of states and transitions gives rise to an isomorphic GTS. For instance, in Fig. 3 it is not important which of the two left-to-right transition is chosen. (This indifference crucially depends on the condition $\alpha_2 \circ \psi \cong \alpha_1$ in the isomorphism condition for transitions; if we would require $\alpha_2 \circ \psi = \psi' \circ \alpha_1$ then the uniqueness up to isomorphism would break down.)
- Let $(\rho_\phi)_\phi$ be the largest bisimulation over S, used in the construction of $\hat{S}$. $\hat{S} \sim S$ is then immediate, using as bisimilarity the restriction of $(\rho_\phi)_\phi$ on the left hand side to states of $\hat{S}$.

- It can be proved that $\hat{q}_1 \; \hat{\rho}_\phi \; \hat{q}_2$ implies that also $\hat{q}_1 \; \rho_\phi \; \hat{q}_2$ in the original GTS S. But then $\hat{q}_1$ and $\hat{q}_2$ are both representatives of the same ρ-equivalence class of states in S, implying they are the same.

(Note that it is *not* the case that $\hat{S}$ has only trivial bisimulations, i.e., such that $q \; \rho_\phi \; q$ implies that ϕ is the identity, since in contrast to [24] we are not abstracting graphs up to symmetry.) The following is immediate.

Corollary 1 (canonical GTS). *Given a graph grammar $\mathcal{G}$, there is a smallest GTS (unique up to isomorphism) generated by $\mathcal{G}$. We call this the* canonical GTS of $\mathcal{G}$, *and denote it $S_{\mathcal{G}}$.*

The following property of the canonical GTS is a consequence of the definition of derivations and the assumption that ac is well-defined up to isomorphism.

Proposition 3 (symmetry of canonical GTSs). *For any graph grammar $\mathcal{G}$, the canonical GTS $S_{\mathcal{G}}$ is symmetric.*

4   First-Order Temporal Logic

Besides providing a notion of symmetry, the transition morphisms of GTSs also keep track of the identity of entities. For instance, Fig. 4 contains all the information necessary to check that entities are retrieved in the order they are inserted and that no entity is inserted without eventually being retrieved. All this is established through the node identities of the val-labelled nodes; no data values need be introduced. Such properties can be expressed formally as formulae generated in a special temporal logic.

4.1   First-Order Linear Temporal Logic

The usual temporal logics are *propositional*, meaning that their smallest building blocks are "atomic" propositions, whose truth is a priori known for every state.[4] For expressing properties that trace the evolution of entities over transitions, however, we need variables that exist, and remain bound to the same value, *outside* the temporal modalities. An example of a logic that has this feature is *first-order linear temporal logic* (FOLTL), generated by the following grammar:

$$\Phi ::= x \mid a(x, y) \mid \mathbf{tt} \mid \neg\Phi \mid \Phi \vee \Psi \mid \exists x.\Phi \mid \mathsf{X}\Phi \mid \Phi \, \mathsf{U} \, \Psi \, .$$

The meaning of the predicates is:

- x expresses that the first-order variable x has a definite value (which is taken from Node). As we will see, this is not always the case: x may be undefined.
- $a(x, y)$ expresses that there is an a-labelled edge from node x to node y.
- $\mathbf{tt}$ (true), $\neg\Phi$ and $\Phi \vee \Psi$ have their standard meaning.

---

[4] In practice, such propositions may themselves well be (closed) first-order formulae, evaluated over each state.

- $\mathsf{X}\Phi$ and $\Phi \,\mathsf{U}\, \Psi$ are the usual linear temporal logic operators: $\mathsf{X}\Phi$ expresses that Φ is true in the next state, whereas $\Phi \,\mathsf{U}\, \Psi$ expresses that Ψ is true at some state in the future, and until that time, Φ is true.

In addition, we use the common auxiliary propositional operators $\wedge$, $\Rightarrow$ etc., as well as the temporal operators G (for "Globally") and F (for "in the Future"). Furthermore, we use $\exists x : a.\Phi$ with $a \in$ Label as abbreviation for $\exists x.a(x, x) \Rightarrow \Phi$. Some example formulae which can be interpreted over the circular buffer system of Fig. 4 are:

1. $\forall c :$ Cell.$\nexists v.$val(c, v) is a non-temporal formula expressing that in the current state there is no val-edge.
2. $\forall c :$ Cell. $\mathsf{F}\ \exists v.$val(c, v) is a temporal formula expressing that all currently existing cells will eventually be filled (though maybe not all at the same time).
3. $\mathsf{F}\ \forall c :$ Cell.$\exists v.$val(c, v) expresses that eventually all cells will be filled (at the same time).
4. $\exists o :$ Object.$\mathsf{X}\ \neg o$ expresses that there exists an Object-node that will be gone in the next state.
5. $\forall b :$ Buffer.$(\exists c, o.$first$(b, c) \wedge$val$(c, o)) \,\mathsf{U}\, (\nexists c, o.val(c, o))$ expresses that eventually the buffer is empty, and until that time, the first cell has a value.
6. $(\mathsf{X}\,\exists o.$val$(c, o)) \wedge (\nexists o.\mathsf{X}\,val(c, o))$ expresses that, although in the next state the cell c will have a value, that value does not already exist in the current state. This implies that that value is freshly created in the next state.

Formulae are interpreted over infinite sequences of graph morphisms, in combination with a valuation of the free variables. The definition requires some auxiliary concepts and notation.

- A *path* is an infinite sequence of consecutive morphisms $m_1\, m_2 \cdots$, i.e., such that $tgt(m_i) = src(m_{i+1})$ for all $i \geq 1$. We let σ range over paths. For all $1 \leq i \leq |\sigma|$, σ_i denotes the i'th element of σ (i.e., m_i), and, σ^i the tail of σ starting at the i'th element (i.e., $m_i\, m_{i+1} \cdots$).
- A *run* of a GTS S is an infinite sequence of pairs $\rho = (t_1, \phi_1)(t_2, \phi_2) \cdots$ such that $q_1 = q_0$, and for all $i \geq 1$
  - $q_i' = q_{i+1}$;
  - Either $t_i \in T$ or $t_i = (q, id_{G_q}, q)$ where $\nexists t \in T : q_t = q$ (so we stutter upon reaching a final state);
  - $\phi_i\colon tgt(\alpha_i) \cong src(\alpha_{i+1})$.

  Given a run ρ, the path σ_ρ generated by ρ is the sequence of morphisms $(\phi_1 \circ \alpha_1)(\phi_2 \circ \alpha_2) \cdots$.
- θ is a partial valuation of variables to elements of Node. If f is a graph morphism, then $f \circ \theta$ is a new valuation defined by concatenating f_V with θ. We use $\theta\{v/x\}$ (with $v \in$ Node) to denote a derived valuation that maps x to v and all other variables to their θ-images.

Satisfaction of a formula is expressed through a predicate of the form $G, \sigma, \theta \models \phi$, where $G = src(\sigma_1)$. The following set of rules defines this predicate inductively. The main point to notice is the modification of the valuation θ in the rule for $\mathsf{X}\phi$. Here the effect of the transformation morphism is brought to bear. For one thing, it is possible that a variable becomes undefined, if the node that it was referring to is deleted by the transformation.

$$
\begin{aligned}
&G, \sigma, \theta \models x && \text{iff } \theta(x) \text{ is defined} \\
&G, \sigma, \theta \models a(x, y) && \text{iff } (\theta(x), a, \theta(y)) \in E_G \\
&G, \sigma, \theta \models \mathbf{tt} && \text{always} \\
&G, \sigma, \theta \models \neg\Phi && \text{iff not } G, \sigma, \theta \models \Phi \\
&G, \sigma, \theta \models \Phi_1 \vee \Phi_2 && \text{iff } G, \sigma, \theta \models \Phi_1 \text{ or } G, \sigma, \theta \models \Phi_2 \\
&G, \sigma, \theta \models \exists x : \Phi && \text{iff } G, \sigma, \theta\{v/x\} \models \Phi \text{ for some } v \in N_G \\
&G, \sigma, \theta \models \mathsf{X}\Phi && \text{iff } src(\sigma_2), \sigma^2, \sigma_1 \circ \theta \vdash \Phi \\
&G, \sigma, \theta \models \Phi_1 \cup \Phi_2 && \text{iff } \exists i \geq 0 : G, \sigma, \theta \models \mathsf{X}^i\Phi_2 \text{ and } \forall 0 \leq j < i : G, \sigma, \theta \models \mathsf{X}^j\Phi_1
\end{aligned}
$$

We define the *validity* of a formula on a GTS S as follows:

$$S \models \Phi \quad \text{if for all runs } \rho \text{ of } S \text{ and all valuations } \theta : \quad G_{q_0}, \sigma_\rho, \theta \models \Phi$$

For instance, of the example formulae presented above, nrs. 1, 2, 5 and 6 (provided c is mapped to the cell pointed to by first) are valid on the GTS of Fig. 4, whereas the others are not. Property 2, for instance, holds because the morphisms associated with the transitions are such that after a finite number of transitions, each cell is mapped onto a cell with an outgoing val-edge. Property 5, on the other hand, is trivially valid since the start state is already empty; however, when another state is picked as start state it becomes invalid, since although it would hold for some paths of that modified GTS, there are runs that never enter the state where the buffer is empty.

The following is an important property, since it shows that bisimilarity minimisation does not change the validity of FOLTL properties.

Theorem 3. *If S_1, S_2 are GTSs, then $S_1 \sim S_2$ implies $S_1 \models \Phi$ iff $S_2 \models \Phi$ for all $\Phi \in$ FOLTL.*

We can now finally formulate the model checking question:

Model checking problem: Given a graph grammar $\mathcal{G}$ and a formula Φ, does $S_{\mathcal{G}} \models \Phi$ hold?

In general, this question is certainly undecidable. In cases where $S_{\mathcal{G}}$ is finite, however, we can use the following (which can be shown for instance by a variation on [29]):

Theorem 4. *Given a finite GTS S and a formula Φ, the property $S \models \Phi$ is decidable, with worst-case time complexity exponential in the number of variables in Φ.*

4.2 Graph-Based Linear Temporal Logic

We now present an alternative logic, which we call GLTL, based only on graphs (rather than predicates and variables) but equivalent (for simple graphs) to FOLTL as presented above. The ideas are based on our own work in [28], originally conceived as an extension to negative application conditions [13]; the same basic ideas were later worked out in a slightly different form in [6]. The extension of this principle to temporal logic is new here.

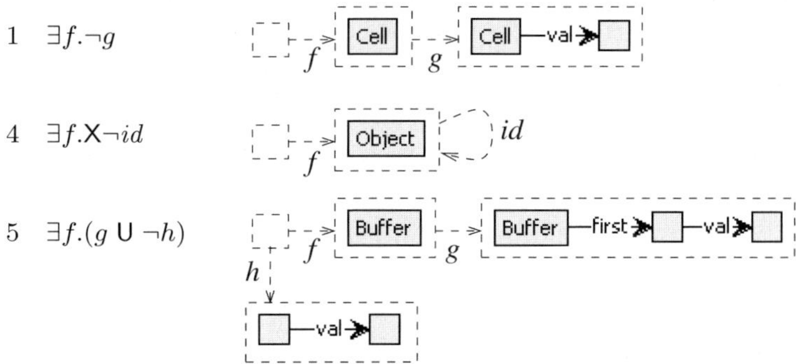

Fig. 5. Formulae in GLTL, corresponding to FOLTL nrs. 1, 4 and 5

The basic idea is to use morphisms as core elements of formulae. Thus, a GLTL formula is generated by the following grammar:

$$\Omega ::= \mathbf{tt} \mid \neg\Omega \mid \Omega \vee \Omega \mid \exists f.\Omega \mid \mathsf{X}\Omega \mid \Omega \mathrel{\mathsf{U}} \Omega \ .$$

As a convenient notation we use the morphism f on its own as equivalent to $\exists f.\mathbf{tt}$.

A formula of the form $\exists f.\Omega$ is evaluated under an existing total matching θ of $src(f)$ to the current graph; the formula is satisfied if θ can be factored through f, i.e., there exists an η from $tgt(f)$ such that $\theta = \eta \circ f$. In fact, $src(f)$ acts as a "type" of $\exists f.\Omega$, and the sub-formula Ω is typed by $tgt(f)$, meaning that its evaluation can assume the existence of η. This notion of "type" replaces the notion of free variables of a (traditional) first-order formula. Types can be computed as follows:

$$type(\mathbf{tt}) = \langle \emptyset, \emptyset \rangle$$
$$type(\neg\Omega) = type(\Omega)$$
$$type(\Omega_1 \vee \Omega_2) = type(\Omega_1) \cup type(\Omega_2)$$
$$type(\exists f.\Omega) = src(f) \quad \text{if } type(\Omega) \subseteq tgt(f)$$
$$type(\mathsf{X}\Omega) = type(\Omega)$$
$$type(\Omega_1 \mathrel{\mathsf{U}} \Omega_2) = type(\Omega_1) \cup type(\Omega_2) \ .$$

Here, the union of two graphs is the (ordinary, not disjoint) union of the node and edge sets, and the sub-graph relation is likewise defined pointwise. (In fact, the types Ω_i of the operands of $\vee$ and U are regarded as sub-types of the type of the composed formula.) The side condition in the type definition for $\exists f.\Omega$ implies that the type can be undefined, namely if the type of Ω is not a sub-graph of $tgt(f)$. We only consider typable formulae.

For example, Fig. 5 shows some GLTL formulae that are equivalent to FOLTL formulae given earlier.

The semantics of GLTL is a relatively straightforward modification of FOLTL. The valuation θ is now a total graph morphism from the type of the formula to the graph. Due to the type definition, this means that in the evaluation we sometimes have to restrict

θ to sub-types. The rules that are different from FOLTL are given below; for the other operators, the semantics is precisely as defined above.

$$G, \sigma, \theta \models \Omega_1 \vee \Omega_2 \text{ iff } G, \sigma, \theta \upharpoonright type(\Omega_1) \models \Omega_1 \text{ or } G, \sigma, \theta \upharpoonright type(\Omega_2) \models \Omega_2$$
$$G, \sigma, \theta \models \exists f : \Omega \quad \text{iff there is a } \eta: tgt(f) \to G \text{ such that } \theta = \eta \circ f \text{ and } G, \sigma, \eta \models \Omega$$
$$G, \sigma, \theta \models \Omega_1 \cup \Omega_2 \text{ iff } \exists i \geq 0 : G, \sigma, \theta \upharpoonright type(\Omega_2) \models \mathsf{X}^i \Omega_2$$
$$\text{and } \forall 0 \leq j < i : G, \sigma, \theta \upharpoonright type(\Omega_1) \models \mathsf{X}^j \Omega_1$$

There exists a relatively straightforward translation back and forth between FOLTL and GLTL. We explain the principles here on an intuitive level; see Fig. 5 for some concrete examples.

- From FOLTL to GLTL, formulae of the form $a(x, y)$ are translated to morphisms f with $src(f) = \langle \{x, y\}, \emptyset \rangle$ and $tgt(f) = \langle \{x, y\}, \{(x, a, y)\} \rangle$; formulae $x = y$ to non-injective morphisms mapping a two-node discrete graph to a one-node discrete graph while merging the nodes; and formulae $\exists x.\Phi$ to $\exists f.\Omega$ where f adds a single, unconnected node to its source.
- From GLTL to FOLTL, $\exists f.\Omega$ is translated to

$$\exists z_1, \ldots, z_n. \bigwedge_i a_i(x_i, y_i) \wedge \bigwedge_j (x_j = y_j) \wedge \Phi \ ,$$

  where the z_k are variables representing the nodes that are new in $tgt(f)$ (i.e., not used as images by f), the $a_i(x_i, y_i)$ are edges that are new in $tgt(f)$, the $x_i = y_i$ equate nodes on which f is non-injective (in both cases, the x_i and y_i correspond to some z_j), and Φ is the translation of Ω.

We state the following result without proof.

Theorem 5. *There exist translations gltl*: FOLTL $\to$ GLTL *and foltl*: GLTL $\to$ FOLTL *such that for all GTSs S:*

$$S \models \Phi \iff S \models gltl(\Phi)$$
$$S \models \Omega \iff S \models foltl(\Omega) \ .$$

5   Evaluation and Future Work

In this section we evaluate some issues regarding choices made in the approach, as well as possible extensions and future challenges. In the course of this we will also touch upon related work, insofar not already discussed.

Graph formalism. In Sect. 3, we have discussed our choice of graph formalism, in the light of the existing algebraic theory surrounding DPO rewriting, only little of which has been successfully transferred to SPO. Let us briefly investigate what has to be done to lift our approach to a general DPO setting; that is, to a category of graphs that is an adhesive HLR category, with a set $\mathcal{M}$ of monos.

- In Sect. 3, our graph transitions carry partial morphisms. In a DPO setting, this should be turned into a span of arrows, of which the left arrow (pointing to the source graph of the transition) should be in $\mathcal{M}$. The corresponding notions of bisimilarity and isomorphism will become slightly more complicated, but we expect that the same results will still hold.

- In Sect. 4, it is not clear how to interpret FOLTL in an arbitrary HLR category. On the other hand, GLTL as defined in Sect. 4.2 can easily be generalised. For this purpose, the construction of the type of a formula (which now relies on subgraphs and graph union, neither of which can be generalised directly to categorical constructions) should be revised.

  A straightforward solution is to provide explicit monos with the binary operators to generalise the sub-graph relation; i.e., the formulae become $\Omega_1 \vee_{\iota,\kappa} \Omega_2$ and $\Omega_1 \cup_{\iota,\kappa} \Omega_2$, where ι, κ is a co-span of monos (in $\mathcal{M}$) such that $src(\iota) = type(\Omega_1)$, $src(\kappa) = type(\Omega_2)$ and $tgt(\iota) = tgt(\kappa)$ equals the type of the formula as a whole. Furthermore, the morphism f in $\exists f.\Omega$ should be replaced by a span $\xleftarrow{\iota}\xrightarrow{f}$, where $type(\Omega) = tgt(f)$ and $tgt(\iota)$ is the type of the whole formula.

Existing model checkers. In the last decades, model checking has given rise to a large number of successful academic and commercial tools, such as SPIN [16], BLAST [4], JPF [17], Murphi [5] or Bogor [32]. Many of these tools share the aims of the GROOVE project, viz., verifying actual (object-oriented) code. It is, therefore, justified to ask what we can hope to add to this field, given the inherent complexities of the graph transformation approach. In fact, there are two distinct issues involved:

- *Graph transformation as a specification paradigm.* In our approach, we essentially propose to use graph transformations as a language to specify the semantics of programming languages. Existing tools use textual modelling languages for this purpose, such as Promela for SPIN or BIR for Bogor, or rely on the available compilation to byte code, as in the case of JPF.

  We believe graph transformations to be a viable alternative, for the following reasons:
    - Graphs provide a syntax which is very close to, if not coincides with, the intuitive understanding of object-oriented data structures (or even heap structures in general). Thus, graph-based operational semantics is easy to understand (see also [18]).
    - Graphs are also very close to diagram models as used in visual languages, and so provide an integrated view along the software engineering process.
    - As numerous case studies have shown, graph transformation can alternatively be used as a specification formalism in its own right. Verification techniques based on this paradigm can therefore also be used outside the context of software model checking.
- *Graph transition systems as verification models.* Even if one accepts the arguments given above in favour of graph transformation as a specification paradigm, this does not immediately imply creating a new model checker. Instead, it is perfectly thinkable to encode graph derivations in terms of the input languages of one of the existing tools, and so avoid re-inventing (or re-implementing) the wheel. Verification approaches based on this idea are, for instance, [35,12].

  We believe that it is nevertheless worthwhile to implement model checking directly on top of graph transition systems, for the following reasons:

- *Symmetry reduction.* In graphs, symmetry is equivalent to isomorphism, and collapsing the state space modulo isomorphism is an immediate method for (non-trivial) symmetry reduction (see also [30]). This connection is lost when graphs are encoded in some other language, since such an encoding invariably involves breaking symmetries.
- *Unboundedness.* Many of the existing model checkers rely on fixed-size bit vectors to represent states. Graphs, however, are not a priori bounded. In order to perform the encoding, it is therefore necessary to choose an upper bound for the state size, and to increase this if it turns out to have been set too low — which involves repeating the encoding step.
- *Encoding.* The complexity of finding acceptable matchings for rules does not suddenly disappear when the problem is encoded in another language. Instead, the encoding itself typically involves predicting or checking all possible matches; so the complexity of the graph transformation paradigm is (partially) shifted from the actual model checking to the encoding process.

Alternative approaches. The explicit state model checking approach presented here is probably the most straightforward way to define and implement verification for graph grammars. A promising alternative is the Petri graph method proposed by König et al.; see, e.g., [3,1,19]. This approach uses *unfolding* techniques developed originally for Petri nets, and offers good abstractions (identified below as one of the more important future work items in our approach); thus, the aproach can yield answers for arbitrary graph grammars. On the other hand, the logic supported is more limited, and it is not clear if symmetry reduction is possible.

Future work. Finally, we identify the following major challenges to be addressed before explicit-state model checking for graph grammars can really take off.

- *Partial order reduction.* This refers to a technique for only generating part of the state space, on which nevertheless a given fragment of the logic (typically, X-free LTL) can still be checked. Traditional techniques for partial order reduction do not apply directly to the setting of graph grammars, since the number of entities is not *a priori* known. (Note that the unfolding approach of [3,1] is in fact also a partial order reduction.)
- *Abstraction.* Instead of taking concrete graphs, which can grow unboundedly large, one may define graph abstractions, which collapse and combine parts of the graphs that are sufficiently similar. Inspired by *shape analysis* [34], we have investigated possible abstractions in [26,31]. This is still ongoing research; no implementation is yet available.
- *Compositionality.* In Sect. 3 we have pointed out the analogy of GTSs with History-Dependent automata, a long-established model for communicating systems, sharing some of the dynamic nature of graphs. This analogy can be turned around to inspire a notion of communication between graph transition systems where (parts of) graphs are echanged, leading to a theory of compositionality for graph grammars. The only work we know of in this direction is König and Ehrig's *borrowed context* [9], and the *synchronised hyperedge replacement* by Hirsch, Montanari and others [15].

References

1. Baldan, P., Corradini, A., König, B.: Verifying finite-state graph grammars: An unfolding-based approach. In: Gardner, P., Yoshida, N. (eds.) CONCUR 2004. LNCS, vol. 3170, pp. 83–98. Springer, Heidelberg (2004)
2. Baldan, P., Corradini, A., König, B., Lluch-Lafuente, A.: A temporal graph logic for verification of graph transformation systems. In: Fiadeiro, J.L., Schobbens, P.-Y. (eds.) WADT 2006. LNCS, vol. 4409, pp. 1–20. Springer, Heidelberg (2007)
3. Baldan, P., König, B.: Approximating the behaviour of graph transformation systems. In: Corradini, A., Ehrig, H., Kreowski, H.-J., Rozenberg, G. (eds.) ICGT 2002. LNCS, vol. 2505, pp. 14–29. Springer, Heidelberg (2002)
4. Beyer, D., Henzinger, T.A., Jhala, R., Majumdar, R.: The software model checker Blast. STTT 9(5-6), 505–525 (2007)
5. Dill, D.L.: The murϕ verification system. In: Alur, R., Henzinger, T.A. (eds.) CAV 1996. LNCS, vol. 1102, pp. 390–393. Springer, Heidelberg (1996)
6. Ehrig, H., Ehrig, K., Habel, A., Pennemann, K.H.: Theory of constraints and application conditions: From graphs to high-level structures. Fundam. Inform. 74(1), 135–166 (2006)
7. Ehrig, H., Ehrig, K., Prange, U., Taentzer, G.: Fundamental theory for typed attributed graphs and graph transformation based on adhesive HLR categories. Fundam. Inform. 74(1), 31–61 (2006)
8. Ehrig, H., Ehrig, K., Prange, U., Taentzer, G.: Fundamentals of Algebraic Graph Transformation. Springer, Heidelberg (2006)
9. Ehrig, H., König, B.: Deriving bisimulation congruences in the dpo approach to graph rewriting. In: Walukiewicz, I. (ed.) FOSSACS 2004. LNCS, vol. 2987, pp. 151–166. Springer, Heidelberg (2004)
10. Ehrig, H., Padberg, J., Prange, U., Habel, A.: Adhesive high-level replacement systems: A new categorical framework for graph transformation. Fundam. Inform. 74(1), 1–29 (2006)
11. Emerson, E.A., Sistla, A.P.: Symmetry and model checking. Formal Methods in System Design 9(1/2), 105–131 (1996)
12. Ferreira, A.P.L., Foss, L., Ribeiro, L.: Formal verification of object-oriented graph grammars specifications. In: Rensink, A., Heckel, R., König, B. (eds.) Graph Transformation for Concurrency and Verification (GT-VC). Electr. Notes Theor. Comput. Sci, vol. 175, pp. 101–114 (2007)
13. Habel, A., Heckel, R., Taentzer, G.: Graph grammars with negative application conditions. Fundam. Inform. 26(3/4), 287–313 (1996)
14. Habel, A., Pennemann, K.H.: Satisfiability of high-level conditions. In: Corradini, A., Ehrig, H., Montanari, U., Ribeiro, L., Rozenberg, G. (eds.) ICGT 2006. LNCS, vol. 4178, pp. 430–444. Springer, Heidelberg (2006)
15. Hirsch, D., Montanari, U.: Synchronized hyperedge replacement with name mobility. In: Larsen, K.G., Nielsen, M. (eds.) CONCUR 2001. LNCS, vol. 2154, pp. 121–136. Springer, Heidelberg (2001)
16. Holzmann, G.J.: The model checker SPIN. IEEE Trans. Software Eng. 23(5), 279–295 (1997)
17. Java PathFinder – A Formal Methods Tool for Java,
    `http://ase.arc.nasa.gov/people/havelund/jpf.html`
18. Kastenberg, H., Kleppe, A.G., Rensink, A.: Defining object-oriented execution semantics using graph transformations. In: Gorrieri, R., Wehrheim, H. (eds.) FMOODS 2006. LNCS, vol. 4037, pp. 186–201. Springer, Heidelberg (2006)
19. König, B., Kozioura, V.: Counterexample-guided abstraction refinement for the analysis of graph transformation systems. In: Hermanns, H., Palsberg, J. (eds.) TACAS 2006. LNCS, vol. 3920, pp. 197–211. Springer, Heidelberg (2006)

20. Lack, S., Sobocinski, P.: Adhesive categories. In: Walukiewicz, I. (ed.) FOSSACS 2004. LNCS, vol. 2987, pp. 273–288. Springer, Heidelberg (2004)
21. Lambers, L., Ehrig, H., Orejas, F.: Efficient detection of conflicts in graph-based model transformation. Electr. Notes Theor. Comput. Sci. 152, 97–109 (2006)
22. Levendovszky, T., Prange, U., Ehrig, H.: Termination criteria for dpo transformations with injective matches. Electron. Notes Theor. Comput. Sci. 175(4), 87–100 (2007)
23. Löwe, M.: Algebraic approach to single-pushout graph transformation. Theoretical Computer Science 109(1–2), 181–224 (1993)
24. Montanari, U., Pistore, M.: History-dependent automata: An introduction. In: Bernardo, M., Bogliolo, A. (eds.) SFM-Moby 2005. LNCS, vol. 3465, pp. 1–28. Springer, Heidelberg (2005)
25. Rensink, A.: Towards model checking graph grammars. In: Gruner, S., Presti, S.L., eds.: Workshop on Automated Verification of Critical Systems (AVoCS), Southampton, UK. Volume DSSE-TR-, -02 of Technical Report., University of Southampton (2003) 150–160 (2003)
26. Rensink, A.: Canonical graph shapes. In: Schmidt, D. (ed.) ESOP 2004. LNCS, vol. 2986, pp. 401–415. Springer, Heidelberg (2004)
27. Rensink, A.: The GROOVE simulator: A tool for state space generation. In: Pfaltz, J.L., Nagl, M., Böhlen, B. (eds.) AGTIVE 2003. LNCS, vol. 3062, pp. 479–485. Springer, Heidelberg (2004)
28. Rensink, A.: Representing first-order logic using graphs. In: Ehrig, H., Engels, G., Parisi-Presicce, F., Rozenberg, G. (eds.) ICGT 2004. LNCS, vol. 3256, pp. 319–335. Springer, Heidelberg (2004)
29. Rensink, A.: Model checking quantified computation tree logic. In: Baier, C., Hermanns, H. (eds.) CONCUR 2006. LNCS, vol. 4137, pp. 110–125. Springer, Heidelberg (2006)
30. Rensink, A.: Isomorphism checking in GROOVE. In: Zündorf, A., Varró, D. (eds.) Graph-Based Tools (GraBaTs). Electronic Communications of the EASST, European Association of Software Science and Technology, vol. 1, Natal, Brazil (September 2007)
31. Rensink, A., Distefano, D.: Abstract graph transformation. Electr. Notes Theor. Comput. Sci. 157(1), 39–59 (2006)
32. Robby, D.M.B., Hatcliff, J.: Bogor: A flexible framework for creating software model checkers. In: McMinn, P. (ed.) Testing: Academia and Industry Conference - Practice And Research Techniques (TAIC PART), pp. 3–22. IEEE Computer Society, Los Alamitos (2006)
33. Rozenberg, G. (ed.): Handbook of Graph Grammars and Computing by Graph Transformation. Foundations, vol. I. World Scientific, Singapore (1997)
34. Sagiv, S., Reps, T.W., Wilhelm, R.: Parametric shape analysis via 3-valued logic. ACM Trans. Program. Lang. Syst. 24(3), 217–298 (2002)
35. Varró, D.: Automated formal verification of visual modeling languages by model checking. Software and System Modeling 3(2), 85–113 (2004)

Linear-Ordered Graph Grammars: Applications to Distributed Systems Design

Leila Ribeiro[1,*] and Fernando Luís Dotti[2]

[1] Instituto de Informática, Universidade Federal do Rio Grande do Sul,
Porto Alegre, Brazil
`leila@inf.ufrgs.br`
[2] Faculdade de Informática, Pontifícia Universidade Católica do Rio Grande do Sul
Porto Alegre, Brazil
`fldotti@inf.pucrs.br`

Abstract. Linear-ordered graph grammars (LOGGs) are a special kind of graph grammars that were inspired by the general definitions of graph grammars and by tile systems. In this paper we show that this kind of grammar is particularly suited for the specification of distributed systems. Moreover, we discuss a simple extension of LOGGs inspired by the representation using tiles, leading to a notion of open graphs that can be very useful in a wider range of applications.

1   Introduction

Distributed systems may be spread over several computational nodes and cross different communication links, may have mobile components, and engage in communication with a priori unknown other components. In this scenario, it becomes even more important to be able to specify the systems behavior without ambiguity as well as to assure functional and non-functional properties of the system as early as possible during system construction. Abstractions that capture the appropriate aspects of the system under construction are thus highly desirable.

Graph grammars [Roz97] is a formalism that has been used to model many features of computational systems. Systems that involve concurrency and distribution can take advandage from the modeling of states as graphs, and of state changes as rule applications (that can be performed in parallel in many parts of the graph representing the state). A large part of the theory developed in the algebraic approach to graph grammars [Ehr79] explains the behavior of concurrent systems (see [EKMR99] for an overview of main results). However, most of this theory was developed for restricted kinds of graph grammars. One of the most used restrictions is that rules are not allowed to glue nodes (or arcs). Although in many applications this restriction might be adequate, in some of them this poses a severe restriction on the specification language. For instance, whenever we want to express the fact that two servers, known by many client

* The work of this author was partially supported by CNPq.

P. Degano et al. (Eds.): Montanari Festschrift, LNCS 5065, pp. 133–150, 2008.

processes, will be joined, such a glue operation would be needed to allows a natural representation.

Linear-ordered graph grammars were proposed in [MR02], and the definition raised as a collaboration between the University of Pisa and the Federal University of Rio Grande do Sul. The topic was mainly developed during the visits of Ugo Montanari to Brazil within the scope of the cooperation project IQ-Mobile. We were searching for a way to describe graph grammars (in particular, algebraic graph grammars), and their concurrent behavior as tile rewriting systems [GM00, GM02, FM00]. One of the goals we had in mind was that the resulting formalism would be suited for the description of mobile and distributed systems. The intuitive idea was to model persistent components of a system (like objects, communication channels or places) as graph nodes, and resources that are generated/consumed as arcs. Persistent items may not be deleted in computations, but may be glued (if this is interesting from the application point of view). Resources may not be glued, and the approach is resource conscious (the number of resources necessary to perform a computation is important and may not be abstracted). In [MR02], LOGGs were introduced, but there was no more pratical example to illustrate their features. Here we recall these definitions and illustrate them using an example of fault recovery of distributed systems.

The description of graphs introduced in [MR02] gave a hint on other classes of graphs that could be used to describe distributed systems. Although LOGGs were developed only for usual graphs, a more general class of graphs called *open graphs* was defined. Here we explore this definition, discussing the advantages of this concept for specification, and also the modifications that would be necessary in the theory to suitably describe computations using open graphs. One of the most interesting features of this approach is the possibility to model the interplay between the horizontal composition of graphs (viewed as tiles) and the vertical composition (used to model derivations).

The structure of this paper is as follows: Section 2 describes Linear-Ordered Graph Grammars; Section 3 presents an example application of a distributed system using LOGGs; Section 4 describes LOGGs as Tile Systems; Section 5 presents an extension of LOGGs using Open Graphs and Section 6 concludes the paper.

2   Linear-Ordered Graph Grammars

In this section we recall the basic concepts of graph rewriting in the single pushout (SPO) approach [Low93, EHK$^+$97] for a special kind of graph and rules. This kind of grammars, called linear-ordered graph grammars, short LOGGs, were introduced in [MR02] to enable a natural description of some features of distributed systems that were difficult to describe using other approaches. Section 3 illustrates the definitions of this section with an application. Instead of the standard definitions of partial morphisms based on total morphisms from subgraphs, we will explicitly use partial functions and a weak commutativity requirement. In [Kor95] it was shown that the two definitions are equivalent.

Definition 1 (Weak Commutativity). *For a (partial) function $f : A \to B$ with domain $dom(f)$, let $f? : A \leftarrow dom(f)$ and $f! : dom(f) \to B$ denote the corresponding domain inclusion and the domain restriction. Given functions $a : A \to A'$, $b : B \to B'$ and $f' : A' \to B'$, where a and b are total, we write $f' \circ a \geq b \circ f$ and say that the diagram **commutes weakly** iff $f' \circ a \circ f? = b \circ f!$.*

If f and f' are total, weak commutativity coincides with commutativity. The commutativity condition introduced above means that the partial functions $b \circ f$ and $f' \circ a$ must agree on all the elements of A on which f is defined. The term "weak" is used because on elements of A on which f is undefined, the two functions can behave differently, i.e., $f' \circ a$ can be defined.

In this paper we use undirected hypergraphs, and therefore we will only define a *source* function connecting each arc with the list of nodes it is connected to. We use the notation S^* to denote the set of all (finite) lists of elements of S.

Definition 2 ((Hyper)Graph, (Hyper)Graph Morphism). *A **(hyper) graph** $G = (N_G, A_G, source^G)$ consists of a set of nodes N_G, a set of arcs A_G and a total function $source^G : A_G \to N_G^*$, assigning to each arc a list of nodes.*

*A **(partial) graph morphism** $g : G \to H$ from a graph G to a graph H is a pair of partial functions $g_N : N_G \to N_H$ and $g_A : A_G \to A_H$ which are weakly homomorphic, i.e., $g_N^* \circ source^G \geq source^H \circ g_A$ (g_N^* is the extension of g_N to lists of nodes). A morphism is called total if both components are total. The category of hypergraphs and partial hypergraph morphisms is denoted by* **HGraphP** *(identities and composition are defined componentwise).*

To distinguish different kinds of nodes and arcs, we will use a notion of typed hypergraphs, analogous to typed graphs [CMR96][1].

Definition 3 (Typed Hypergraphs). *A typed hypergraph is a tuple $HG^{TG} = (HG, type^{HG}, TG)$ where HG and TG are hypergraphs, called instance and type graph, respectively, and $type^{HG} : HG \to TG$ is a total hypergraph morphism, called typing morphism.*

*A morphism between typed hypergraphs $HG1^{TG}$ and $HG2^{TG}$ is a partial graph morphism $f : HG1 \to HG2$ such that $type1^{HG1} \geq type2^{HG2} \circ f$. The category of all hypergraphs typed over a type graph TG, denoted by **THGraphP(TG)**, consists of all hypergraphs over TG as objects and all morphisms between typed hypergraphs (identities and composition are the identities and composition of partial functions).*

For the well-definedness of the categories above and the proof that these categories have pushouts we refer to [Kor95] (these categories can be constructed as generalized graph structures categories). The construction of pushouts in such categories is done componentwise, followed by a free construction to make the

---

[1] Note that, due to the use of partial morphisms, this is not just a comma category construction: the morphism *type* is total whereas morphisms among graphs are partial, and we need weak commutativity instead of commutativity. In [Kor95] a way to construct such categories was defined.

source and *type* components total (actually, as types are never changed by morphisms, this free construction is not needed in the case of typed graphs). In the rest of the paper, we will usually denote typed hypergraphs just by graphs.

Now we recall the definition of linear-ordered graph grammar as defined in [MR02]: the nodes and arcs of each graph are ordered, the rules do not delete nodes and do not collapse arcs, and the matches do not collapse arcs. With this kind of restriction, the derivation steps can be obtained componentwise as a pushout of nodes and pushout of arcs (because no node is deleted).

Definition 4 (Linear Ordered Graph Grammars (LOGGs)). *A **linear ordered graph** over a type graph $TG = (\{s\}, A_{TG}, source^{TG})$ is a graph $HT^{TG} = (HG, type^{HG}, TG)$ with $HG = (N, A_{HG}, source^{HG})$ where $N = \{0, \ldots, n\}$, with $n \in \mathbb{N}$, and $A_{HG} = \bigcup_{a \in A_{TG}} A_{HG}^a$, with $A_{HG}^a = \{a_i \mid i \in \{0, \ldots, m-1\}\}$, where m is the cardinality of $\{ha \in A_{HG} \mid type^{HG}(ha) = a\}$. A morphism between linear ordered graphs is simply a typed graph morphism: no additional requirements are imposed.*

*A **linear ordered graph grammar** is a grammar $GG = (TG, IG, Rules)$ where TG is a type graph having a single node, IG is a graph typed over TG called the start graph, $Rules$ is a set of typed graph morphisms and for each rule $r = (r_N, r_A) : L \to R$ of Rules, r_A is injective and r_N is total. A **linear ordered match** $m = (m_N, m_A) : L \to G$ is a total typed graph morphism where the m_A component is injective.*

The semantics of a linear ordered graph grammar is defined as for a usual graph grammars, taking into account that only linear ordered matches are allowed, that is, a match is resource conscious on arcs, but may identify nodes.

Definition 5 (Derivation Step, Sequential Semantics of a LO-Graph Grammar). *Given a linear-ordered graph grammar $GG = (TG, IG, Rules)$, a rule $r : L \to R \in Rules$, and a graph $G1$ typed over TG, and a linear ordered match $m : L \to G1$, a **derivation step** $G1 \xRightarrow{r,m} G2$ using rule r and match m is a pushout in the category **THGraphP(TG)**.*

$$
\begin{array}{ccc}
L & \xrightarrow{\;r\;} & R \\
{\scriptstyle m}\downarrow & (PO) & \downarrow{\scriptstyle m'} \\
G1 & \xrightarrow{\;r'\;} & G2
\end{array}
$$

*A derivation sequence of GG is a sequence of derivation steps $G_i \xRightarrow{r_i, m_i} G_{i+1}$, $i \in \{0, \ldots, n\}$, $n \in \mathbb{N}$, where $G_0 = IG$, $r_i \in Rules$. The **sequential semantics** of GG is the set of all sequential derivations of GG.*

3   Example: Distributed Systems in Presence of Faults

Distributed systems may be spread over several computational nodes connected through wide area networks, where reliability and availability levels may vary. Therefore, while modeling distributed systems one important aspect to take into consideration is the possibility of fault occurrences. In [DRS04, DMS05] Graph Transformation Systems where employed to model and analyze distributed systems in presence of some kinds of faults.

There are several kinds of faults in distributed systems. Here we concentrate on "crash" faults, which are very often considered while developing distributed systems. According to the crash fault, a process fails by halting and the processes that communicate with the halted process are not warned about the fault. If a process that offers a service crashes, it has to be replaced by another process and the service continued. This may be performed in several ways, depending on the kind of system. Very often, a server process has one or more backup processes which are kept synchronized and in case the server crashes, the primary backup process assumes the service. In this situation, the other processes have to update references to the new server. With the appropriate usage of an underlying communication platform, the communication is kept consistent from the point of view of the user processes, even if the server is replaced.

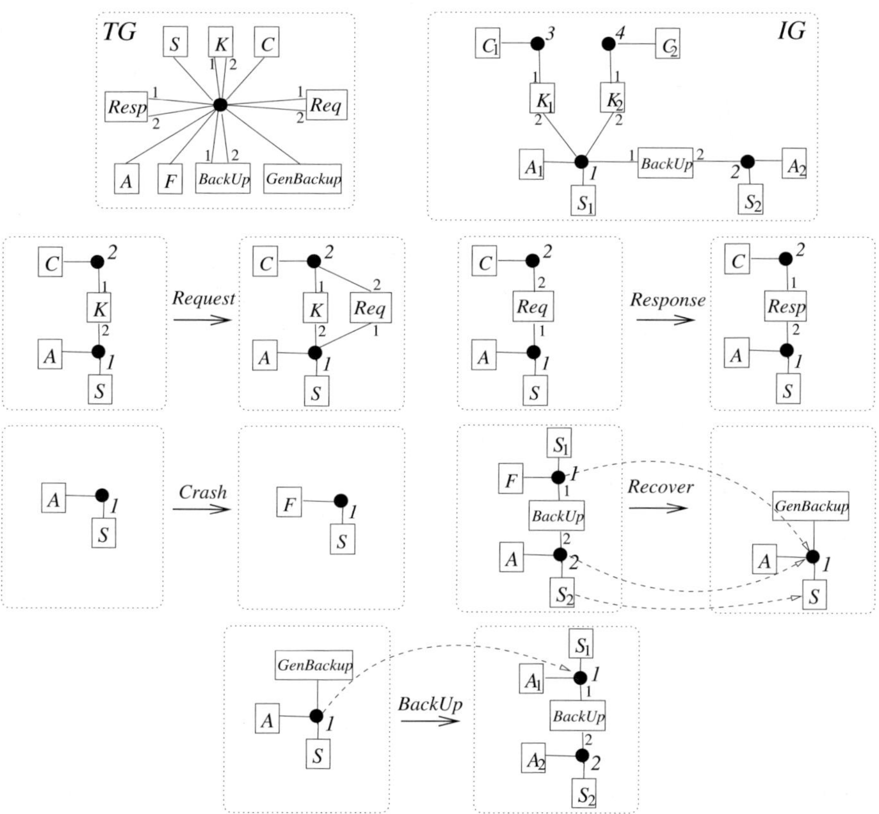

Fig. 1. Fault/Recovery System

This abstraction is well captured using linear-ordered graph grammars. The corresponding grammar is shown in Figure 1. Arcs are drawn as squares and nodes as bullets. The tentacles of the arcs describe the *source* function and are numbered to indicate their order. Indices on nodes and arcs here just indicate different instances of items with same types (when there is only one instance, we may omit the index 1). Nodes and arcs with same indices in left- and rigth-hand sides of rules denote preserved items, unless explicitly stated otherwise by connecting left- and right-hand side items by a dashed arrow. The type graph has one kind of node (to represent communication channels), and many kinds of arcs: S and C arcs mark which nodes represent server and client processes, respectively; A and F arcs represent that servers are active or crashed, respectively; K arcs are used to model that a process knows a server; Req and Resp arcs describe requests and responses; BackUp arcs relate servers that are synchronized (one is backup of the other) and GenBackup arcs are used to warn that a new backup server shall be created (because one server has crashed). A client process C may use the service of an active (A) server S through requests Req (rule *Request*). In normal operation, the service answers the requisitions back to the process via Resp (rule *Response*). However, a server process may crash (or fail), switching to F mode (rule *Crash*). As discussed above, the detection of a failed server will fire the server replacement behavior which, in this case, means that a back-up server will be activated and resume the operation assuming the same state as the crashed server before the fault arises. This is represented by collapsing the communication channels of the involved servers in rule *Recover*. In addition, this rule generated a warning in the system to inform that a new backup server is needed. The new back-up server process is created in rule *BackUp*. Note that collapsing nodes in this case provides a very useful abstraction for several aspects: (i) the state of the service is not lost; (ii) references (from user processes) are updated consistently; (iii) ongoing messages in communication channels are not lost. This effect is very difficult to achieve (if not impossible at all) in approaches that do not allow non injective rules.

An example of rule application is shown in Figure 2, in which a crashed server known by two processes is replaced by its backup. All requests were moved to the new server, and it is transparent for the client processes that this change of servers has taken place (as required by an adequate recovery procedure).

4   Tile Semantics of LOGG

Here we will reproduce the main ideas presented in [MR02], showing that the semantics of LOGGs can be suitably described by tile systems. First, we review the main concepts of tile systems and theories that are relevant to model graph grammars (sect. 4.1) and then discuss how LOGGs and their semantics can be defined in terms of tile systems (sect. 4.2).

4.1   Tile Systems for Graph Grammars

A tile can be seen as a square consisting of four arrows: the horizontal ones describe states, and the vertical ones observations [GM02, FM00]. Tile systems exhibit an algebraic flavor (tiles have a straightforward axiomatization as monoidal

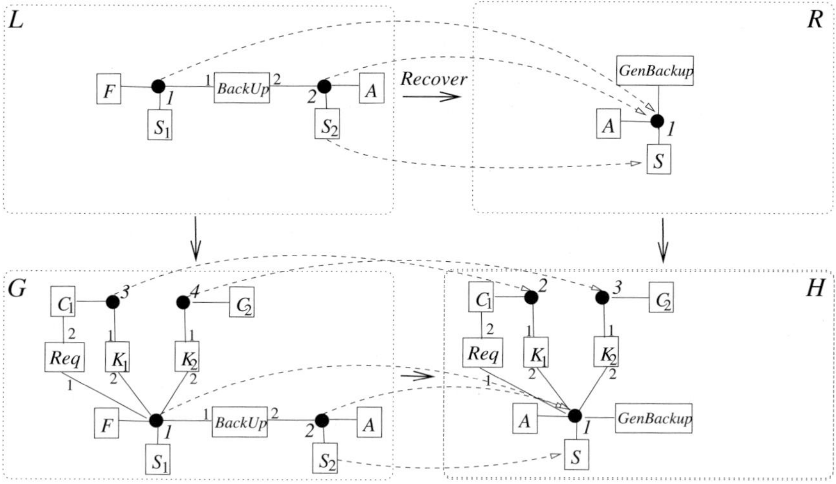

Fig. 2. Derivation using rule *Recover* at match m

double categories) which may allow for universal constructions, compositionality and refinement in the classical style of algebraic semantics [MM98]. The first step to model graph grammars as tile system is to describe how to model graphs and graph morphisms as tiles. Actually, graphs will be modeled as arrows of a suitable theory and graph morphisms as tiles.

Definition 6 (One-sorted (hyper-)signatures). *A **hyper-signature** $\Sigma = (S, OP)$ consists of a singleton $S = \{s\}$ and a family $OP = \{OP^n\}_{n \in \mathbb{N}}$ of sets of operators with n arguments and no result (no target sort), where n is a natural number. Given an operation $op \in OP^n$, n is the arity of op, and the domain of op is denoted by s^n.*

Now we define an extension of a signature Σ that will add as sorts all operation names in Σ and as target of each operation the corresponding sort.

Definition 7 (Extended (hyper-)signatures). *Given a signature $\Sigma = (S, OP)$, its **extension** is a signature $\Sigma^E = (S_E, OP_E)$, where $S_E = \{\underline{s}\} \cup \{\underline{op} \mid op : s^n \to 0 \in OP^n, n \in \mathbb{N}\}$ and $OP_E = \{op^E : \underline{s}^n \to \underline{op} \mid op : s^n \to 0 \in OP, n \in \mathbb{N}\}$.*

An extended version of the gs-monoidal theory defined in [CG99] is used to model (linear-ordered hyper) graphs, and a theory called pgm was defined in [MR02] to model partial graph morphisms. We now review the necessary concepts to define these theories.

Definition 8 (Connection diagrams). *A connection diagram G is a 4-tuple $\langle O_G, A_G, \delta_0, \delta_1 \rangle$: O_G, A_G are sets whose elements are called respectively objects and arrows, and $\delta_0, \delta_1 : A_G \to O_G$ are functions, called respectively source and target. A connection diagram G is reflexive if there exists an identity function*

140   L. Ribeiro and F.L. Dotti

$id_- : O_G \rightarrow A_G$ *such that* $\delta_0(id_a) = \delta_1(id_a) = a$ *for all* $a \in O_G$; *it is* with pairing *if its class* O_G *of objects forms a monoid; it is* monoidal *if it is reflexive with pairing and also its class of arrows forms a monoid, such that id,* δ_0 *and* δ_1 *respect the neutral element and the monoidal operation.*

In the following we will use the free commutative monoid construction over a set S, denoted by $(S^\otimes, \otimes, \underline{0})$. The elements of $S^\otimes$ can be seen as finite multisets of elements of S, and will be referred to using underlined variables. If S is a singleton set, we will denote elements of $S^\otimes$ by underlined natural numbers (indicating the number of times the only element of S appears in the list, for example $\underline{s} \otimes \underline{s} \otimes \underline{s}$ will be denoted by $\underline{3}$).

Gs-moniodal theories define a set of arrows that can be used to characterize (linear-ordered) graphs. The following inductive definition describes the basic arrows and composition operators, together with a set of axioms that arrows have to satisfy. Given a (hyper)signature $\Sigma = (S, OP)$, basic arrows are created by the axioms *identities* (identities on nodes/arcs are gs-monoidal arrows), *generators* (for each hyperarc $op : S^n \rightarrow 0$ of the signature, an arrow with exaclty n nodes as source and op as target is in the theory), *duplicators* (arrows that duplicate the number of nodes are allowed), *permutations* (arrows may change the order of nodes/arcs), and *dischargers* (arrows may delete nodes). To build further arrows, composition operators are defined by the inference rules *sum* (the parallel - monoidal - composition of arrows is an arrow), and *composition* (arrows may be sequentially composed to generate new arrows). Graphs will be modeled by special gs-monoidal arrows.

Definition 9 (gs-monoidal theories). *Given a (hyper)signature* $\Sigma = (S, OP)$ *and its extension* $\Sigma^E = (S_E, OP_E)$, *the associated* **gs-monoidal theory GS(Σ)** *is the monoidal connection diagram with objects the elements of the free commutative monoid over* S_E $((S_E)^\otimes, \otimes, \underline{0})$ *and arrows generated by the following inference rules:* generators, sum, identities, composition, duplicators, dischargers *and* permutations *in Table 1.*

Table 1. Inference Rules for gs-monoidal Theories

$$(identities) \quad \frac{\underline{x} \in S_E^\otimes}{id_{\underline{x}} : \underline{x} \rightarrow \underline{x} \in \mathbf{GS}(\Sigma)} \qquad (generators) \quad \frac{f \in \Sigma_n}{f : \underline{n} \rightarrow \underline{f} \in \mathbf{GS}(\Sigma)}$$

$$(duplicators) \quad \frac{\underline{n} \in S^\otimes}{\nabla_{\underline{n}} : \underline{n} \rightarrow \underline{n} \otimes \underline{n} \in \mathbf{GS}(\Sigma)} \qquad (dischargers) \quad \frac{\underline{n} \in S^\otimes}{!_{\underline{n}} : \underline{n} \rightarrow \underline{0} \in \mathbf{GS}(\Sigma)}$$

$$(permutations) \quad \frac{\underline{x}, \underline{y} \in S_E^\otimes}{\rho_{\underline{x},\underline{y}} : \underline{x} \otimes \underline{y} \rightarrow \underline{y} \otimes \underline{x} \in \mathbf{GS}(\Sigma)}$$

$$(sum) \quad \frac{s : \underline{x} \rightarrow \underline{y}, t : \underline{x}' \rightarrow \underline{y}'}{s \otimes t : \underline{x} \otimes \underline{x}' \rightarrow \underline{y} \otimes \underline{y}' \in \mathbf{GS}(\Sigma)} \qquad (composition) \quad \frac{s : \underline{x} \rightarrow \underline{y}, t : \underline{y} \rightarrow \underline{z}}{s; t : \underline{x} \rightarrow \underline{z} \in \mathbf{GS}(\Sigma)}$$

The composition operator $;$ is associative, and the monoid of arrows satisfies

(i) the functoriality axiom: for all arrows $s, s't, t' \in \mathbf{GS}(\Sigma)$

$$(s \otimes t); (s' \otimes t') = (s; s') \otimes (t; t') \qquad (\textit{whenever both sides are defined})$$

(ii) the identity axiom: for all $s : \underline{x} \to \underline{y}$

$$id_{\underline{x}}; s = s = s; id_{\underline{y}}$$

(iii) the monoidality axioms: for all $\underline{n}, \underline{m} \in S^{\otimes}$ and $\underline{x}, \underline{y}, \underline{z} \in S_E^{\otimes}$

$$\rho_{\underline{x}\otimes \underline{y},\underline{z}} = (id_{\underline{x}}\otimes\rho_{\underline{y},\underline{z}}); (\rho_{\underline{x},\underline{z}}\otimes id_{\underline{y}}) \qquad !_{\underline{0}} = \nabla_{\underline{0}} = \rho_{\underline{0},\underline{0}} = id_{\underline{0}} \qquad \rho_{\underline{0},\underline{x}} = \rho_{\underline{x},\underline{0}} = id_{\underline{x}}$$

$$id_{\underline{x}\otimes \underline{y}} = id_{\underline{x}} \otimes id_{\underline{y}} \qquad !_{\underline{x}\otimes \underline{y}} = !_{\underline{x}}\otimes !_{\underline{y}} \qquad \nabla_{\underline{n}\otimes \underline{m}} = (\nabla_{\underline{n}} \otimes \nabla_{\underline{m}}); (id_{\underline{n}} \otimes \rho_{\underline{n},\underline{m}} \otimes id_{\underline{m}})$$

(iv) the coherence axioms: for all $\underline{n} \in S^{\otimes}$ and $\underline{x}, \underline{y} \in S_E^{\otimes}$

$$\nabla_{\underline{n}}; (id_{\underline{n}} \otimes \nabla_{\underline{n}}) = \nabla_{\underline{n}}; (\nabla_{\underline{n}} \otimes id_{\underline{n}}) \qquad\qquad \nabla_{\underline{n}}; \rho_{\underline{n},\underline{n}} = \nabla_{\underline{n}}$$

$$\nabla_{\underline{n}}; (id_{\underline{n}}\otimes !_{\underline{n}}) = id_{\underline{n}} \qquad\qquad \rho_{\underline{x},\underline{y}}; \rho_{\underline{x},\underline{y}} = id_{\underline{x}} \otimes id_{\underline{y}}$$

(v) and the naturality axiom: for all $s : \underline{x} \to \underline{y}, t : \underline{z} \to \underline{w}$

$$(s \otimes t); \rho_{\underline{y},\underline{w}} = \rho_{\underline{x},\underline{z}}; (t \otimes s)$$

Arrows of gs-monoidal theories that are constructed without the generators *axiom are called* **basic gs-monoidal arrows**. *The theory obtained only with basic arrows is denoted by* **bGS**(Σ).

For example, consider graph L of Figure 2. Assuming that the sort of nodes is s, this graph can be modeled as the following gs-monoidal arrow: $(\nabla_1 \otimes \nabla_2); (F \otimes \nabla_1 \otimes \nabla_2 \otimes A); (id_F \otimes S \otimes Backup \otimes S \otimes id_A): \underline{s} \otimes \underline{s} \to \underline{F} \otimes \underline{S} \otimes \underline{Backup} \otimes \underline{S} \otimes \underline{A}$. Graphically, this arrow is represented in Figure 3 (1 and 2 are the first and second occurrences of sort s in $\underline{s} \otimes \underline{s}$).

Notation: In the pictures, the operator $\otimes$ and the underlines in the domain and codomain of the arrows will be omitted. Moreover, we will draw indexed bullets ($\bullet$) in the graphical representation to describe the instances of sorts.

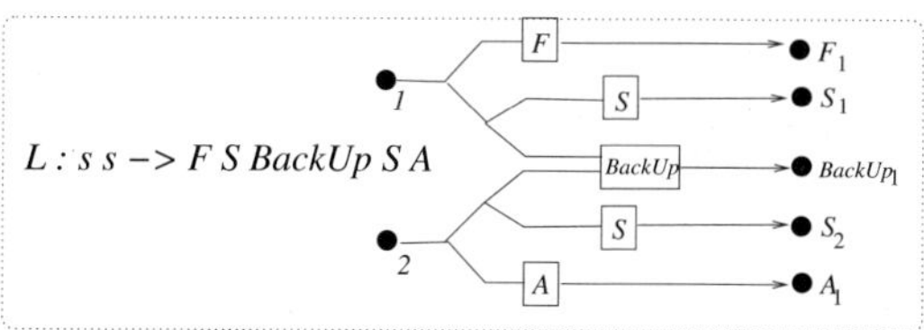

Fig. 3. Arrow of a gs-monoidal theory

The theory defined below is used to model a special kind of graph morphism: morphisms that are total on nodes (and may identify nodes), and are partial on arcs (and may not identify arcs). Identification of nodes is achieved by the ∇ operator, as in gs-monoidal theories. The possibility of deleting arcs is achieved by the $!$ operator, yielding partial functions (only for arcs).

Definition 10 (pgm-monoidal theories). *Given a (hyper)signature $\Sigma = (S, OP)$ and its extension $\Sigma^E = (S_E, OP_E)$, the associated* **pgm-monoidal theory** $\mathbf{PGM}(\Sigma)$ *is the monoidal connection diagram with objects the elements of the free commutative monoid over S_E $((S_E)^\otimes, \otimes, \underline{0})$ and arrows generated by the following inference rules:* sum, identities, composition *and* permutations *(analogous to the rules in Def. 9), and the rules in Table 2.*

Table 2. Additional Inference Rules for pgm-monoidal Theories

$$(new) \quad \frac{\underline{x} \in (S_E - S)^\otimes}{\mathsf{i}_{\underline{x}} : \underline{0} \to \underline{x} \in \mathbf{PGM}(\Sigma)}$$

$$(duplicators) \quad \frac{\underline{n} \in S^\otimes}{\nabla_{\underline{n}} : \underline{n} \to \underline{n} \otimes \underline{n} \in \mathbf{PGM}(\Sigma)} \qquad (dischargers) \quad \frac{\underline{x} \in S_E^\otimes}{!_{\underline{x}} : \underline{x} \to \underline{0} \in \mathbf{PGM}(\Sigma)}$$

The composition operator ; is associative and the monoid of arrows satisfies the functoriality, identity, coherence *and* naturality *axioms (Def. 9); as well as the* monoidality *axioms (all of Def. 9 plus $!_{\underline{0}} = \mathsf{i}_{\underline{0}} = \rho_{\underline{0},\underline{0}} = id_{\underline{0}}$ and $\mathsf{i}_{\underline{x} \otimes \underline{y}} = \mathsf{i}_{\underline{x}} \otimes \mathsf{i}_{\underline{y}}$, for all $\underline{x}, \underline{y}, \underline{z} \in S_E^\otimes$).*
The theory obtained using the same objects and arrows generated by the rules above, except the composition *is called* **short pgm-graph theory**, *denoted by* $\mathbf{sPGM}(\Sigma)$. *The theory obtained using all rules except* new *and with* discharger *only for $\underline{x} \in S^\otimes$ is called* **basic pgm-graph theory**, *denoted by* $\mathbf{bPGM}(\Sigma)$.

Note that in pgm-monoidal theories the *generators* axiom was not used. This has the effect that arrows of these theories do not correspond to graphs as gs-monoidal theories (because arcs are not allowed), they rather describe relationships between the objects (that are lists of nodes and arc labels). Moreover, an arrow of $\mathbf{bPGM}(\Sigma)$ is also an arrow of $\mathbf{bGS}(\Sigma)$.

The following definition relies on the fact that basic gs-monoidal arrows of one sorted signatures correspond to total functions in the inverse direction, that is, each gs-monoidal arrow $\underline{n} \to \underline{m}$ corresponds to a total function $\underline{m} \to \underline{n}$ (see [CG99] for the proof). As basic pgm-monoidal arrows are also basic gs-monoidal arrows, they also correspond to total functions. This will be used to construct pullback squares of such arrows based on pushouts of functions. These pushouts will model the node-component morphisms of rule applications.

Definition 11 (Derivation Pair). *Given one-sorted signatures Σ_h and Σ_v, and arrows $s : \underline{n} \to \underline{m} \in \mathbf{bGS}(\Sigma_h)$ and $t : \underline{q} \to \underline{m} \in \mathbf{bPGM}(\Sigma_v)$. The* **derivation pair** *of s and t, denoted by $deriPair(s,t)$, is a pair of a basic gs-monoidal and a basic pgm-monoidal arrows $(s' : \underline{p} \to \underline{q}, t' : \underline{p} \to \underline{n})$ such that the inverse square is a pushout in the category of sets and total functions.*

Now we define a tile rewrite system that can be used to model graph rewriting.

Definition 12 (Tile sequent, Tile rewrite system). *Let Σ_h and Σ_v be two (one sorted) signatures, called the* horizontal *and the* vertical *signature respectively. A Σ_h-Σ_v tile sequent is a quadruple $l \xrightarrow[b]{a} r$, where $l : \underline{x} \to \underline{y}$ and $r : \underline{p} \to \underline{q}$ are arrows of $\mathbf{GS}(\Sigma_h)$ while $a : \underline{x} \to \underline{p}$ and $b : \underline{y} \to \underline{q}$ are arrows of $\mathbf{PGM}(\Sigma_v)$. Arrows l, r, a and b are called respectively* initial configuration, final configuration, trigger *and* effect *of the tile. Trigger and effect are called* observations. *Underlined strings $\underline{x}$, $\underline{y}$, $\underline{p}$ and $\underline{q}$ are called respectively* initial input interface, initial output interface, final input interface *and* final output interface.

A short tile sequent is a tile $l \xrightarrow[b]{a} r$ *where observations a and b are arrows of the short pgm-graph theory $\mathbf{sPGM}(\Sigma_v)$ (i.e. no sequential composition is allowed to build them).*

A tile rewrite system (TRS) $\mathcal{R}$ *is a triple $\langle \Sigma_h, \Sigma_v, R \rangle$, where R is a set of Σ_h-Σ_v short tile sequents called* rewrite rules.

A TRS $\mathcal{R}$ can be considered as a logical theory, and new sequents can be derived from it via certain inference rules.

Definition 13 (pgm-monoidal tile logic). *Let $\mathcal{R} = \langle \Sigma_h, \Sigma_v, R \rangle$ be a TRS. Then we say that $\mathcal{R}$ entails the class $\mathbf{R}$ of the tile sequents $s \xrightarrow[b]{a} t$ obtained by finitely many applications of the inference rules depicted in Table 3.*

Table 3. pgm-monoidal Tile Logic

$$
(generators)\ \frac{s \xrightarrow[b]{a} t \in \mathcal{R}}{s \xrightarrow[b]{a} t \in \mathbf{R}}
\qquad
(h\text{-}refl)\ \frac{s : \underline{x} \to \underline{y} \in \mathbf{GS}(\Sigma_h)}{id_s = s \xrightarrow[id_{\underline{y}}]{id_{\underline{x}}} s \in \mathbf{R}}
\qquad
(v\text{-}refl)\ \frac{a : \underline{x} \to \underline{y} \in \mathbf{PGM}(\Sigma_v)}{id_a = id_{\underline{x}} \xrightarrow{a}_{a} id_{\underline{y}} \in \mathbf{R}} ;
$$

$$
(p\text{-}comp)\ \frac{\alpha = s \xrightarrow[b]{a} t,\ \alpha' = s' \xrightarrow[b']{a'} t' \in \mathbf{R}}{\alpha \otimes \alpha' = s \otimes s' \xrightarrow[b \otimes b']{a \otimes a'} t \otimes t' \in \mathbf{R}}
$$

$$
(h\text{-}comp)\ \frac{\alpha = s \xrightarrow[c]{a} t,\ \alpha' = s' \xrightarrow[b]{c} t' \in \mathbf{R}}{\alpha * \alpha' = s;s' \xrightarrow[b]{a} t;t' \in \mathbf{R}}
\qquad
(v\text{-}comp)\ \frac{\alpha = s \xrightarrow[b]{a} u,\ \alpha' = u \xrightarrow[b']{a'} t \in \mathbf{R}}{\alpha \cdot \alpha' = s \xrightarrow[b;b']{a;a'} t \in \mathbf{R}} ;
$$

$$
(perm)\ \frac{a : \underline{x} \to \underline{y},\ b : \underline{x'} \to \underline{y'} \in \mathbf{PGM}(\Sigma_v)}{\rho_{a,b} = \rho_{\underline{x},\underline{x'}} \xrightarrow[b \otimes a]{a \otimes b} \rho_{\underline{y},\underline{y'}} \in \mathbf{R}}
$$

$$
(PBnodes)\ \frac{s : \underline{n} \to \underline{m} \in \mathbf{bGS}(\Sigma_h),\ a : \underline{q} \to \underline{m} \in \mathbf{bPGM}(\Sigma_v)}{\underline{s} \xrightarrow[a]{a'} \underline{s'} \in \mathbf{R}} (s', a') = deriPair(s, a)
$$

4.2   Interpretation of LOGGs as Tile Rewriting Systems

The idea introduced in [MR02] of representing graphs and rules as tiles is the following:

Graphs: Graphs will be modeled as the horizontal components of the tiles. Let $TG = (\{s\}, A, source^{TG})$ be a type graph. Based on this graph we can build a signature $\Sigma_h = (\{s\}, A_\Sigma)$, where A_Σ contains all arcs of A as operations (the information about the arity is given by the $source^{TG}$ function). A graph will be then an arrow $\underline{x} \to \underline{y}$ of the corresponding gs-monoidal theory, where $\underline{x}, \underline{y} \in (\{s\} \uplus A)^\otimes$. This means that the mapping from $\underline{x}$ to $\underline{y}$ is constructed by using, besides the arcs of the graph, the operations of identity and permutation for all sorts and, for elements of sort s (nodes), duplication and discharging may also be used allowing, respectively, the same node to be source of more than one hyperarcs, and that there are nodes that are not source of any hyperarc. For the hyperarcs of the graph we include a target function that assigns to each arc an occurrence of its type (indexed by a natural number). Usual (closed) graphs correspond to the special case of $G : \underline{x} \to \underline{y}$ when we have $\underline{x} \in \{s\}^\otimes$ and $\underline{y} \in A^\otimes$. Other graphs are called *open graphs*, and will be used as auxiliary components to allow the modeling of the direct derivations.

Rules: A rule $r : L \to R$ is a (partial) graph morphism $r = (r^N, r^A)$. Such a rule can be represented as a tile having as horizontal arrows the graphs L and R, and as vertical arrows mappings that allow to glue nodes, delete arcs, preserve and create arcs and nodes. These vertical arrows are arrows of the pgm-monoidal theory for the signature $\Sigma_v = \Sigma_h$. Rule *Recover* of Figure 1 corresponds to the tile of Figure 4 (vertical mapping is shown as dashed arrows, creation - new - is denoted by the ^{i}, and deletion - bang - is denoted by !). Note that at the west side of the tile *Recover* we have modeled the component $Recover^N$ of the rule, whereas in the east side we model the component $Recover^A$.

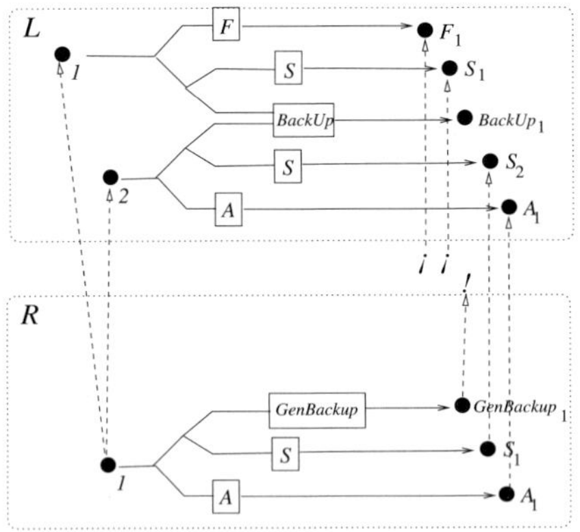

Fig. 4. Modeling a rule as a tile

Definition 14 (Signature of a Type Graph). *Given a type graph $TG = (N, A, source^{TG})$, the corresponding hypergraph signature is $\Sigma_{TG} = (N, A_{TG})$, where $A_{TG} = \{a : source^{TG}(a) \to \underline{0} | a \in A\}$.*

In [MR02] it was shown that the graphs typed over $TG = (N, A, source^{TG})$ with N being a singleton are the arrows of kind $\underline{n} \to \underline{a}$ of a gs-monoidal theory with signature Σ_{TG}, where $\underline{n} \in N^{\otimes}$ and $\underline{a} \in (N^E - N)^{\otimes}$, with $\Sigma_{TG}^E = (N^E, A^E)$. Morevover, graph morphisms typed over TG are tiles having as horizontal and vertical signature Σ_{TG} where the north and south sides are the arrows corresponding to graphs (the left- and right-hand sides of the rule, respectively), and the vertical arrows are arrows of $\mathbf{PGM}(\Sigma_{TG})$. A direct derivation of a grammar GG can be modeled by a suitable composition of tiles of the pgm-monoidal tile logic $\mathbf{R}$ obtained by the TRS $\langle \Sigma_h, \Sigma_v, R \rangle$, where $\Sigma_h = \Sigma_v = \Sigma_{TG}$ (as discussed above), and R is the set of tiles representing the rules of GG. A derivation $G \stackrel{r,m}{\Longrightarrow} H$ using rule $r : L \to R$ at match $m : L \to G$ can be obtained as a composition of tiles that will give raise to a tile $r' : G \to H$. This can be done in 4 steps (see [MR02] for detailed explanations):

1. **Context Graph**: Construct the (context) graph G', that contains all nodes of G and all arcs that are in G and are not in the image of the match m (that is, G' is G after removing the deleted and preserved arcs).
2. **Tile 1**: Construct the tile $r \otimes id_{G'}$, that is the parallel composition of the tile corresponding to rule r and the identity tile of graph G'.
3. **Tile 2**: Construct the tile corresponding to the match and derivation on the node component: the north side of this tile corresponds to m^N (the component of the match morphism that maps nodes), the east side is the west side of the tile obtained in step 2, and the remaining sides will be mappings of nodes such that the resulting square commutes and has no nodes that are not in the north or east sides already.
4. **Resulting Tile**: The result of the application of rule r at match m is then given by the sequential composition of the tiles obtained in the last 2 steps: $Tile\ 1 * Tile\ 2$ (obtained by the application of rule $h\text{-}comp$ of $\mathbf{R}$).

5   Open Graphs

Although open graphs were defined in [MR02], they were just used to build auxiliary tiles to enable the description of the derivation as tile rewrite systems. Since LOGGs were defined only for closed graphs (the initial graph and all rules were only allowed to have this kind of graphs), it was not possible to use the feature of horizontal composition of tiles to model composition of parts of a system, we could only use the vertical dimension to model the evolution of the system via the application of rules. In the sequel we will define a special kind of open graph (in which only nodes may be shared), and show that small changes in the definitions of Sect. 4 provide a much more interesting specification formalism.

Definition 15. *A (node-)open graph typed over $TG = (N, A, source^{TG})$, with N being a singleton, is a tuple $OG^{TG} = (G, ON)$, where $G = (OG, type^{OG}, TG)$ is a typed graph and ON is a subset of nodes of OG, called the* open nodes.

According to this definition, an open graph is just a typed graph with a distinguished set of nodes that are called *open*, meaning that these are the points to which other graphs may be attached when composing this graph with another one. Graphs with no open nodes will be called closed graphs (this is the case of all graphs defined in the previous sections).

In terms of gs-monoidal theories, open graph can be characterized as arrows of the kind $\underline{n} \to \underline{na}$, where $\underline{n}$ is a multiset of nodes (like in closed graphs) and $\underline{na}$ may contain nodes and arcs (in closed graphs, this component consisted only of arcs). The nodes that appear in $\underline{na}$ are the open nodes. An additional requirement is necessary to assure compatibility with Def. 15: due to the duplication operation, it could be possible to have a gs-monoidal arrow with, for example, two nodes in $\underline{na}$ corresponding to the same node in $\underline{n}$ – but such an arrow should not correspond to an open graph.

Characterization 16. *An arrow $\underline{n} \to \underline{na}$ of a gs-monoidal theory with signature Σ_{TG}, with $\Sigma_{TG}^E = (N^E, A^E)$, represents an (node-)open graph if $\underline{n} \in N^\otimes$, $\underline{na} \in (N^E)^\otimes$, and each node of $\underline{n}$ is connected to at most one node node of $\underline{na}$.*

In the rest of this paper, we refer to open graphs instead of node-open graphs. More general forms of graphs (in which $\underline{n} \in (N^E)^\otimes$) will be called *general graphs*.

5.1   LOGGs Using Open Graphs

We can rephrase the definition of LOGGs now by using open graphs instead of graphs, and the morphisms will be tiles having open graphs as horizontal arrows and arrows of $\mathbf{PGM}(\Sigma_{TG})$ as vertical arrows. As an example, we will consider the specification of client/server systems. While working with distributed systems it is highly desirable to be able to compose systems from separate modules via well defined interfaces, as well as to specify open systems. Openess is used in this context in the sense that parts of the system are open to engage in collaboration with previously unknown other modules. The collaboration may take place through the exchange of well defined messages using appropriate channels.

Therefore, while specifying one part of the system it is important to have suitable means to make explicit which of its parts are open for collaboration. On the other hand, it is also important to state clearly what is assumed from other modules that may engage in communication during the life-time of the system. Using LOGGs, we can identify approapriate abstractions for such cases.

The example depicted in Figure 5 shows a generic situation where a server S is open to communicate with clients C. Although it would be possible to represent graphs and rules as arrows/tiles, we will stick to the usual graphical representation, marking open nodes (nodes that are present in the left- and right-hand sides of gs-monoidal arrows) with a circle. The type graph was not depicted, since it is straightforward (has one node and all types of arcs that appear in the specification). Client and server communicate via a communication channel. A client may send a request Req to the server. The server, in order to accomplish the request, may have to allocate internal resources R. This situationis quite common in distributed systems: for instance, TCP connections are opened depending on available memory needed to store incomming data through the connection; or

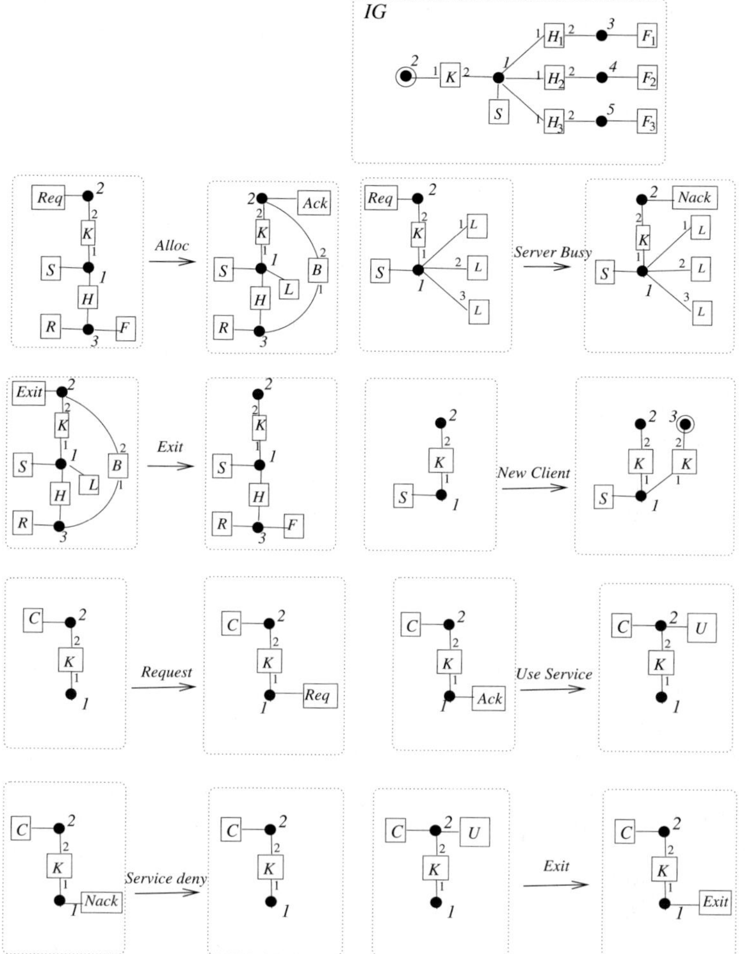

Fig. 5. Client/Server System

even the service for a specific client may need a new thread. The resource is then assigned to the client (L) and the request is positively responded with `Ack`. If no resources are avaliable, the server may deny the service through `Nack`.

The client may eventualy cease service usage (`Exit`), leading the resource to be freed (F). Note that when a client and a server engage in a communication, the client and server share a private communication channel (a closed node). In order to allow the possibility of new clients to communicate with the server, an open communication channel node has to be created, thus the rule for open channel creation is needed (rule *NewClient*). This situation is analogous to the use of the replication operator in process algebra, and is extremely useful in the design of open systems, since a server does not know a priori how many other clients may be using it, and the fact that we may have open nodes provides the necessary abstraction to describe such a situation: the semantics will consider all possible compositions of a server with any number of clients.

To be able to define the desired behavior as a tile systems, we need additional tiles: after performing the 4 steps described in section 4, we must compose the resulting tile $T3$ with an additional tile to glue the open nodes according to the way defined in the rule/match. This additional tile can be obtained as the parallel composition of a tile that is the horizontal identity of the arcs involved in the west arrow of $T3$ and a tile that has as east side the node component of the west arrow of $T3$, and whose north arrow describe the gluing of open nodes specified by the derivation. This tile corresponds to a pushout of total functions, and to obtain this effect, we must take the north and south arrows from a theory that allows to glue nodes (for example, a $\mathbf{coGS}(\Sigma_{TG})$ [BM02]). Like the rule $PBnodes$ of Table 3, a corresponding rule should be included in the pgm-monoidal theory to generate the needed tiles. Moreover, the semantics should take into account the composition of a server with any number of clients, since the behavior of a server without clients would be empty. This will be discussed in the following section.

5.2   Composition of Open Graphs

An important feature of this new model is the possibility to merge partial states of a system via interface items (open nodes). But, since we still do not allow arcs in the interface, a more involved construction is needed to build the composition. Given two open graphs $G1 : \underline{n} \to \underline{me}$ and $G2 : \underline{n'} \to \underline{m'e'}$, we need to make the target of $G1$ equal to the source of $G2$ to enable the composition (we assume that $G2$ does not necessarily have the same number of nodes as open nodes in $G1$, since in practical application this would be a strong requirement). The idea is to define that the first m nodes of $G2$ will be glued to corresponding open nodes of $G1$, and the rest of the nodes will remain in the resulting graph. To obtain this effect, we need to to build the following arrow $(G1 \otimes id_{NG2}); (G2 \otimes id_{AG1}) :$ $n'' \Rightarrow m''e''$, where id_{NG2} is the identity arrow on the nodes of $G2$ that will not be matched to open nodes of $G1$, id_{AG1} is the identity arrow on arcs of $G1$. This way, n'' contains all nodes of $G1$ and all nodes of $G2$ that were not merged with nodes of $G1$, m'' contains all open nodes of $G2$ and e'' contains all arcs of $G1$ and $G2$. This construction is illustrated in Figure 6. Note that this composition is not commutative: we compose the open nodes of the first graph with nodes of the second (and the latter may be open or closed nodes).

To enable that composition takes place during the evolution of the system via rules, we need to enrich further the tile system with (i) tiles in which the north and south arrows consist only of nodes and are identity arrows, and the east and west arrows denote the same total function; (ii) tiles in which the north and south arrows consist only of arcs and are identities and east and west arrows describe the same partial injective function. Note that the whole functionality of the system only takes place when the interplay between horizontal and vertical

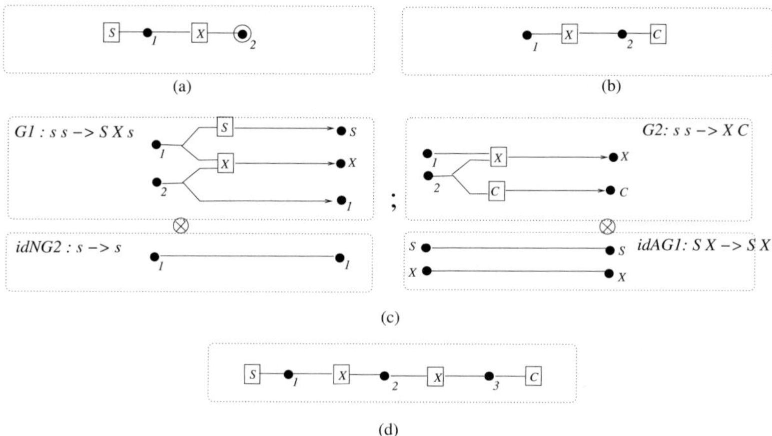

Fig. 6. (a) Graph $G1$ (b) Graph $G2$ (c) Construction of Composition (d) Result

composition is allowed: in the example, the system starts with an open communication channel, and in this situation, no rules may be applied; when a client composes with this system (this may happen since all graphs are available as tiles in the tile system of a LOGG), this open channel is restricted for the communication with this client, and the rules of the server become applicable (including the rule that creates a new communication channel). This very interesting view on the behavior of open systems is inspired by the tile modeling.

6   Conclusion and Future Work

In this paper we reviewed the concepts of Linear-Ordered Graph Grammars (LOGGs) as presented in [MR02]. We showed an example of application of this kind of grammars, taking advantages of its features to model distributed systems. We then discussed how the notion of *open graph* can be exploited to model open systems. In this framework, the behavior is defined not only by the rules that specify a system, but also by the different ways in which this system may be composed with other systems. The cooperation between horizontal and vertical compositions for a system can be naturally described in the tile systems setting.

This contribution makes evident that LOGGs provide a link between the area of graph grammars and tile systems, showing that both areas of research can benefit from each other. In particular, we showed that by modeling graph grammars as tile systems, we immediately got ideas on novel kinds of graphs, on how composition of these new kinds of graphs may be defined, and on the relationships between composition and evolution (through derivations) of a system.

Since it was not the aim of this paper to show all formal definitions, but to provide insights on possible applications and extensions of linear-ordered graph grammars, a lot of work still has to be done to formalize the presented ideas.

References

[BM02]   Bruni, R., Montanari, U.: Dynamic connectors for concurrency. Theor. Comput. Sci. 281(1–2), 131–176 (2002)

[CG99]   Corradini, A., Gadducci, F.: An algebraic presentation of term graphs via gs-monoidal categories. Applied Categorical Structures 7, 299–331 (1999)

[CMR96]   Corradini, A., Montanari, U., Rossi, F.: Graph processes. Fundamenta Informaticae 26(3–4), 241–265 (1996)

[DMS05]   Dotti, F., Mendizabal, O., Santos, O.: Verifying fault-tolerant distributed systems using object-based graph grammars. In: Maziero, C.A., Gabriel Silva, J., Andrade, A.M.S., Assis Silva, F.M.d. (eds.) LADC 2005. LNCS, vol. 3747, pp. 80–100. Springer, Heidelberg (2005)

[DRS04]   Dotti, F., Ribeiro, L., Santos, O.: Specification and analysis of fault behaviours using graph grammars. In: Pfaltz, J.L., Nagl, M., Böhlen, B. (eds.) AGTIVE 2003. LNCS, vol. 3062, pp. 120–133. Springer, Heidelberg (2004)

[EHK+97]   Ehrig, H., Heckel, R., Korff, M., Lowë, M., Ribeiro, L., Wagner, A., Corradini, A.: Algebraic approaches to graph transformation II: Single pushout approach and comparison with double pushout approach. In: Handbook of graph grammars and computing by graph transformation, vol. 2, pp. 247–312. World Scientific, Singapore (1997)

[Ehr79]   Ehrig, H.: Introduction to the algebraic theory of graph grammars. In: Ng, E.W., Ehrig, H., Rozenberg, G. (eds.) Graph Grammars 1978. LNCS, vol. 73, pp. 1–69. Springer, Heidelberg (1979)

[EKMR99]   Ehrig, H., Kreowski, H.-J., Montanari, U., Rozenberg, G. (eds.): Handbook of graph grammars and computing by graph transformation, vol. 3: Concurrency, parallelism and distribution. World Scientific, Singapore (1999)

[FM00]   Ferrari, G., Montanari, U.: Tile formats for located and mobile systems. Information and Computation 156(1/2), 173–235 (2000)

[GM00]   Gadducci, F., Montanari, U.: The tile model. In: Plotkin, G., Stirling, C., Tofte, M. (eds.) Proof, Language and Interaction: Essays in Honour of Robin Milner, MIT Press, Cambridge (2000)

[GM02]   Gadducci, F., Montanari, U.: Comparing logics for rewriting: Rewriting logic, action calculi and tile logic. Theor. Comput. Sci. 285(2), 319–358 (2002)

[Kor95]   Korff, M.: Generalized graph structures with application to concurrent object-oriented systems, Ph.D. thesis, Technische Universität Berlin (1995)

[Low93]   Lowë, M.: Algebraic approach to single-pushout graph transformation. Theor. Comput. Sci. 109, 181–224 (1993)

[MM98]   Meseguer, J., Montanari, U.: Mapping tile logic into rewriting logic. In: Parisi-Presicce, F. (ed.) WADT 1997. LNCS, vol. 1376, pp. 62–91. Springer, Heidelberg (1998)

[MR02]   Montanari, U., Ribeiro, L.: Linear ordered graph grammars and their algebraic foundations. In: Corradini, A., Ehrig, H., Kreowski, H.-J., Rozenberg, G. (eds.) ICGT 2002. LNCS, vol. 2505, pp. 317–333. Springer, Heidelberg (2002)

[Roz97]   Rozenberg, G.: Handbook of graph grammars and computing by graph transformation, vol. 1: Foundations. World Scientific, Singapore (1997)

Constraint and Logic Programming:
Ugo Montanari's Main Contributions and
Introduction to the Volume Section

Francesca Rossi

Dipartimento di Matematica Pura e Applicata, Università di Padova, Italy
`frossi@math.unipd.it`

Abstract. This short note introduces the area of constraint and logic programming. I will first briefly describe the main contributions of Ugo Montanari to these disciplines. Then I will summarize the papers contained in this volume, and concerned with either constraint or logic programming, that have been written in honour of Ugo Montanari.

1   Constraint Programming

Constraint programming (CP) [1,8,11,15,14] is a powerful paradigm for solving combinatorial search problems. CP was born as a multi-disciplinary research area that embeds techniques and notions coming from many other areas, among which artificial intelligence, computer science, databases, programming languages, and operations research. Constraint programming is currently applied with success to many domains, such as scheduling, planning, vehicle routing, configuration, networks, and bioinformatics.

Constraint solvers take a real-world problem, represented in terms of decision variables and constraints, and find assignments to all the variables that satisfy all the constraints. Constraint solvers search the solution space either systematically, as with backtracking or branch and bound algorithms, or use forms of local search which may be incomplete. Systematic methods often interleave search and inference, where inference consists of propagating the information contained in one constraint to the neighboring constraints. Such inference, called constraint propagation, is usually very useful, since it may greatly reduce the parts of the search space that need to be visited. Constraints that are often used in real-life problems, called global constraints, come with their own ad-hoc efficient propagation mechanisms, that make them especially efficient to use.

The initial ideas underlying the whole constraint programming research area emerged in the 70's with several pioneering papers on constraint propagation. Among them, it is no doubt that the most influential of all has been the 1974 paper by Ugo Montanari titled "Networks of Constraints: Fundamental Properties and Applications to Picture Processing" (Information Science, vol. 7, p. 95–132, 1974) [13], where, for the first time, a form of constraint propagation called path consistency was defined and studied in depth. This paper is one of the most cited and influential in the field of constraint programming. For the

P. Degano et al. (Eds.): Montanari Festschrift, LNCS 5065, pp. 151–154, 2008.

first time, there was a clear and formal study of a new notion of local consistency, as well as of many other concepts such as the notion of decomposable networks of constraints. The influence was both in terms of the results contained in the paper, as well as in terms of a very formal and clear style of presentation, typical of all Ugo's papers, that was very rare in Artificial Intelligence at that time.

In 1995, when CP was already well developed and with many research lines, Ugo made another important contribution to the field, by introducing, together with S. Bistarelli and myself, the notion of soft constraints, that is, constraints that can have several levels of satisfiability [3,2]. To do this, algebraic concepts were used, such as the notion of semiring, that was used to model the set of satisfiability levels. The main contribution was the introduction of a very general framework where several classes of soft constraints could be modelled, and where properties could be proven once and for all, and then inhertited by all the instances. Since then, the area of soft constraints has evolved greatly, with other modelling formalisms, solving techniques, applications, and theoretical results. Further contributions of Ugo along this line include the embedding of soft constraints in programming paradigms, such as constraint logic programming [4,5] and concurrent concurrent programming [6,7].

2   Logic Programming

Logic programming [10] is a declarative programming paradigm where programs are not made of commands nor of functions, but of logical implications (called clauses) between collections of predicates. Executing a logic program means asking whether a certain statement (called the goal) is true under the logical theory modelled by the clauses. To answer this question, the current goal is recursively "unified" with the conclusion (also called the head) of a clause. Unification is therefore a crucial mechanism to execute a logic program such as those written in Prolog. However, it can be very expensive to compute.

Ugo Montanari, together with Alberto Martelli, gave an essential contribution to logic programming with the paper "An Efficient Unification Algorithm" (ACM TOPLAS, 1982) [12], by introducing an efficient way to unify two terms. In this paper, the problem of unifying two terms is seen as the problem of finding the solution of a set of equations, and it is shown that this can be done very efficiently. For a historical view of how this result came about, see also the paper by Alberto Martelli in this volume.

3   Papers About Constraint and Logic Programming in This Volume

The paper "Semiring-based soft constraints", by Stefano Bistarelli and Francesca Rossi, describes the work done together with Ugo Montanari for the introduction of semiring-based soft constraints, as well as the main lines of research that have been followed since then to extend, use, and make soft constraints more applicable.

Constraint programming is a declarative programming framework, where, ideally, a user should state its constraints, and the underlying solver should solve them. There are many other declarative programming paradigms, and most of them, included CP, suffer from the lack of debugging tools. This is inherently related to the declarative nature of such paradigms. The paper "Declarative Debugging of Membership Equational Logic Specifications", by Rafael Caballero, Narciso Marti-Oliet, Adrian Riesco, and Alberto Verdejo, shows how to use declarative (or algorithmic debugging) in the context of Maude, a high-level declarative language supporting equational and rewriting logic computations.

Classical spreadsheet software, such as Excel, embed a very simple form of constraint programming, since it is possible to relate the value of a cell to the value of other cells via some function. By extending these functions to be generic constraints, we greatly enlarge the applicability of this software. The paper "SPREADSPACES: Mathematically-Intelligent Graphical Spreadsheets", by Nachum Dershowitz and Claude Kirchner, shows how to enhance classical spreadsheets to obtain systems where spreadsheet computation, as well as constraint solving and optimization, can be carried on in a graphical environment.

Constraints have been embedded in many programming paradigms, but the one that has shown to be the most suitable is the logic programming paradigm. The paper "An Overview of The Ciao Multiparadigm Language and Program Development Environment, and its Design Philosophy", by Manuel Hermenegildo, Francisco Bueno, Manuel Carro, Pedro Lopez-Garcia, and German Puebla, is a brief overview of the CIAO programming language, that started as a constraint logic programming language [9], and that now supports functional, logic, constraint, and object-oriented programming, in an environment with concurrent, parallel, and distributed executions, as well as many auxiliary tools such as debuggers, verifiers, and visualizers.

Constraint programming problems are usually solved via systematic search, which traverses a search tree via depth-first backtracking search. Search trees have only OR-nodes, that model branching via variable instantiation. By augmenting them with AND-nodes, modelling problem decomposition, it is possible to exploit variable independency during search. This can lead to exponential speed-up. Binary decision diagrams are also widely and effectively used to model Boolean functions. The paper "AND/OR Multi-Valued Decision Diagrams for Constraint Networks", by Robert Mateescu and Rina Dechter, shows how to combine the idea of decision diagrams and of AND/OR search structures to model compactly and solve more efficiently constraint problems, by providing a survey of recent results in this line of research.

References

1. Apt, K.R.: Principles of Constraint Programming. Cambridge University Press, Cambridge (2003)
2. Bistarelli, S., Montanari, U., Rossi, F.: Semiring-based Constraint Solving and Optimization. Journal of the ACM 44(2), 201–236 (1997)

3. Bistarelli, S., Montanari, U., Rossi, F.: Constraint Solving over Semirings. In: Proc. IJCAI 1995, Morgan Kaufmann, San Francisco (1995)
4. Bistarelli, S., Montanari, U., Rossi, F.: Semiring-based Constraint Logic Programming. In: Proc. 15th International Joint Conference on Artificial Intelligence (IJCAI 1997), pp. 352–357. Morgan Kaufmann, San Francisco (1997)
5. Bistarelli, S., Montanari, U., Rossi, F.: Semiring-based Constraint Logic Programming: Syntax and Semantics. ACM Transactions on Programming Languages and System (TOPLAS) 23(1), 1–29 (2001)
6. Bistarelli, S., Montanari, U., Rossi, F.: Soft Concurrent Constraint Programming. In: Le Métayer, D. (ed.) ESOP 2002. LNCS, vol. 2305, Springer, Heidelberg (2002)
7. Bistarelli, S., Montanari, U., Rossi, F.: Soft Concurrent Constraint Programming. ACM Transactions on Computational Logic (TOCL) 7(3), 1–27 (2006)
8. Dechter, R.: Constraint Processing. Morgan Kaufmann, San Francisco (2003)
9. Jaffar, J., Lassez, J.L.: Constraint Logic Programming. In: Proc. POPL 1987, ACM, New York (1987)
10. Lloyd, J.: Foundations of logic programming. Springer, Heidelberg (1984)
11. Marriott, K., Stuckey, P.J.: Programming with Constraints. MIT Press, Cambridge (1998)
12. Martelli, A., Montanari, U.: An efficient unification algorithm. ACM Transactions on Programming Languages and Systems 4(2), 258–282 (1982)
13. Montanari, U.: Networks of Constraints: Fundamental Properties and Applications to Picture Processing. Information Science 7, 95–132 (1974)
14. Rossi, F., Van Beek, P., Walsh, T. (eds.): Handbook of constraint programming. Elsevier, Amsterdam (2006)
15. Tsang, E.P.K.: Foundations of Constraint Satisfaction. Academic Press, London (1993)

Semiring-Based Soft Constraints

Stefano Bistarelli[1,2] and Francesca Rossi[3]

[1] Dipartimento di Scienze, Università "G. d'Annunzio" di Chieti-Pescara, Italy,
`bista@sci.unich.it`,
[2] IIT-CNR, Pisa, Italy
`stefano.bistarelli@iit.cnr.it`
[3] Dipartimento di Matematica Pura e Applicata, Università di Padova, Italy
`frossi@math.unipd.it`

Abstract. The semiring-based formalism to model soft constraint has been introduced in 1995 by Ugo Montanari and the authors of this paper. The idea was to make constraint programming more flexible and widely applicable. We also wanted to define the extension via a general formalism, so that all its instances could inherit its properties and be easily compared. Since then, much work has been done to study, extend, and apply this formalism. This papers gives a brief summary of some of these research activities.

1   Before Soft Constraints: A Brief Introduction to Constraint Programming

Constraint programming [1,42,60,74,68] is a powerful paradigm for solving combinatorial search problems that draws on a wide range of techniques from artificial intelligence, computer science, databases, programming languages, and operations research. Constraint programming is currently applied with success to many domains, such as scheduling, planning, vehicle routing, configuration, networks, and bioinformatics.

The basic idea in constraint programming is that the user states the constraints and a general purpose constraint solver solves them. Constraints are just relations, and a constraint satisfaction problem (CSP) states which relations should hold among the given decision variables. For example, in scheduling activities in a company, the decision variables might be the starting times and the durations of the activities, as well as the resources needed to perform them, and the constraints might be on the availability of the resources and on their use for a limited number of activities at a time.

Constraint solvers take a real-world problem, represented in terms of decision variables and constraints, and find an assignment of values to all the variables that satisfies all the constraints. Constraint solvers search the solution space either systematically, as with backtracking or branch and bound algorithms, or use forms of local search which may be incomplete. Systematic methods often interleave search and inference, where inference consists of propagating the

P. Degano et al. (Eds.): Montanari Festschrift, LNCS 5065, pp. 155–173, 2008.

information contained in one constraint to the neighboring constraints. Such inference, usually called *constraint propagation*, may reduce the parts of the search space that need to be visited.

Rather than trying to satisfy a set of constraints, sometimes people want to optimize them. This means that there is an objective function that tells us the quality of each solution, and the aim is to find a solution with optimal quality. To solve such problems, techniques such as branch and bound are usually used.

The initial ideas underlying the whole constraint programming research area emerged in the '70s with several pioneering papers on local consistency, among which the 1974 paper by Ugo Montanari [63], where for the first time a form of constraint propagation, called *path consistency*, was defined and studied in depth. Since then, the field has evolved greatly, and theoretical study has been coupled with application work, that has shown the need for several extensions of the classical constraint formalism. The introduction of semiring-based soft constraints lies within this evolution thread.

In the classical notion of constraint programming, constraints are relations. Thus a constraint can either be satisfied or violated. In the early '90s, some attempts had been made to generalize the notion of constraint to an object with more than just two levels of satisfiability.

For example, fuzzy constraints [46,69] allow for the whole range of satisfiability levels between 0 and 1. Then, the quality of a solution is the minimum level of satisfiability of the constraints for that solution. The aim is then to find a solution whose quality is highest.

Because fuzzy constraints suffer from the so-called "drowning effect" (where the worst level of satisfiability "drowns" all the others), lexicographic constraints were introduced [49], to obtain a more discriminating ordering of the solutions, where also solutions with the same worst level can be distinguished.

Another extension to classical constraints are the so-called probabilistic constraints [48], where, in the context of an uncertain model of the real world, each constraint is associated to the probability of being present in the real problem. Solutions are then associated to their conjoint probability (assuming independence of the constraints), and the aim is to find a solution with the highest probability.

In weighted constraints, instead, each constraint is given a weight, and the aim is to find a solution for which the sum of the weights of the satisfied constraints is maximal. A very useful instance of weighted constraints are MaxCSPs, where weights are just 0 or 1 (0 if the constraint is violated and 1 if it is satisfied). In this case, we therefore want to satisfy as many constraints as possible.

While fuzzy, lexicographic, and probabilistic constraints were defined for modeling purposes, that is, to model real-life situations that could not be faithfully modeled via classical constraints, weighted constraints and MaxCSPs were mainly addressing over-constrained problems, where there are so many constraints that the problem has no solution. In fact, the aim is to satisfy as many constraints as possible, possibly using priorities (modeled by the weights) to have more discriminating power.

This second line of reasoning lead also to the definition of the first general framework to extend classical constraints, called partial constraint satisfaction [51]. In partial CSPs, over-constrained problems are addressed by defining a metric over constraint problems, and by trying to find a solution of a problem which is as close as possible to the given one according to the chosen metric. MaxCSPs are then just an instance of partial CSPs, where the metric is based on the number of satisfied constraints.

2 Semiring-Based Soft Constraints: Main Idea and Properties

The idea of the semiring-based formalism [26,27,61,7] was to further extend the classical constraint notion, and to do it with a formalism that could encompass most of the existing extensions, as well as other ones not yet defined, with the aim to provide a single environment where properties could be proven once and for all, and inherited by all the instances.

At the technical level, this was done by adding to the usual notion of a CSP the concept of a structure representing the levels of satisfiability of the constraints. Such a structure is a set with two operations: one (written $+$) is used to generate an ordering over the levels, while the other one ($\times$) is used to define how two levels can be combined and which level is the result of such combination. Because of the properties required on such operations, this structure is similar to a semiring: from here the terminology of "semiring-based soft constraints", that is, constraints with several levels of satisfiability, and whose levels are (totally or partially) ordered according to the semiring structure. Fuzzy, lexicographic, probabilistic, weighted, and MaxCSPs are all instances of the semiring-based framework. In general, problems defined according to the semiring-based framework are called *soft constraint satisfaction problems* (SCSPs).

Figure 1 shows a weighted CSP as a graph. Variables and constraints are represented respectively by nodes and by undirected arcs (unary for c_1 and c_3, and binary for c_2), and semiring values are written to the right of each tuple. Here we assume that the domain of the variables contains only elements a, b and c. An optimal solution of this problem is $(X = b, Y = c)$, that has weight 7.

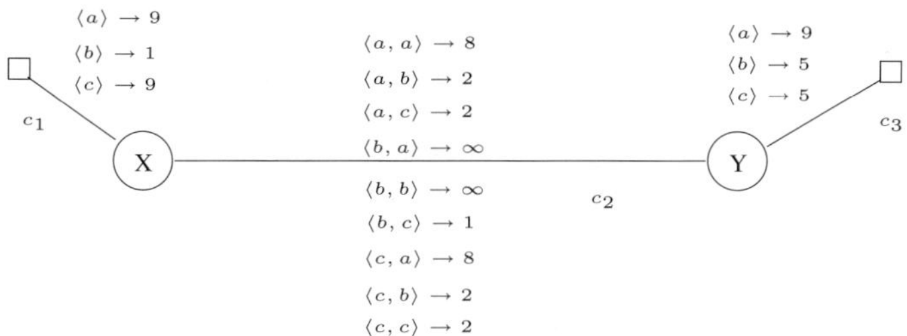

Fig. 1. A weighted CSP

In the same year in which semiring-based soft constraints were introduced (1995), another general formalism to model constraints with several levels of satisfiability was proposed: the so-called "valued" constraints [72]. Valued constraints are very similar to semiring-based soft constraints, except that their levels of satisfiability cannot be partially ordered [13].

The possibility of partially ordered set of levels of satisfiability can be useful in several scenarios. When the levels are the result of the combination of several optimization criterion, it is natural to have a Pareto-like approach in combining such criteria, that naturally leads to a partial order. Also, even if we have just one optimization criteria, we may want to insist in declaring some levels as incomparable, because of what they model. In fact, the elements of the semiring structure do not need to be numbers, but can be any objects that we want to associate to a way of giving values of the variables of a constraints. If, for example, the objects are all the subsets of a certain set, then we have a partial order under subset inclusion.

One of the strengths of constraint programming is the ability to remove local inconsistencies via constraint propagation. This techniques can be extended to soft constraints. If some properties of the semiring structure (mainly the idempotence of the combination operator) hold, this extension has the same desirable properties as the classical notion. That is, it terminates, it returns an equivalent problem, and it is independent on the order of the application over constraints [22]. Otherwise, a different notion of constraint propagation can be defined, which enjoys some of the properties but not all (for example, independence does not hold any longer) [39,40,71,21].

Some real-life situations cannot be modeled via soft constraints with idempotent operators. For this reason, a more general notion that does not assume this property has been introduced in [75]. In this more general setting, semiring valuations are useful, for example, when counting the number of solutions.

Another extension of the semiring-based framework has been proposed in [54], where a metric space has been combined with semiring-based constraints to capture distances between preference levels. In [59] the distance is then used to define a notion of constraint relaxation.

It is known that non-binary classical constraints can always be modeled by binary constraints, if enough new variables are introduced (primal representation) or if we use variables with tuple domains (dual representation). In [58] this issue has been considered in the context of soft constraints, and it was shown that any set of semiring-based soft constraints can be modeled via unary soft constraints plus classical binary constraints.

3   Embedding Soft Constraints in Programming Paradigms

Classical constraints have been embedded in several programming paradigms, such as logic programming and concurrent programming. This has lead to Constraint

Logic Programming (CLP) [60,55,56] and Concurrent Constraint (cc) programming [70]. Similar attempts have been done with semiring-based soft constraints.

3.1 Soft CLP

To build applications with constraints, a language is needed where such constraints are easily embedded and handled as first class objects. This is why soft constraints have been embedded in the Constraint Logic Programming (CLP) [56] formalism. The resulting paradigm, called SCLP (for Semiring-based CLP, or also Soft CLP) [29,30], has the advantage of treating in a uniform way, and with the same underlying machinery, all constraints that can be seen as instances of the semiring-based approach. This leads to a high-level declarative programming formalism where real-life problems involving constraints of all these kinds can be easily modeled and solved.

As usual for logic programming languages, three equivalent semantics have been defined for SCLP: model-theoretic, fix-point, and operational, which are conservative extensions of the corresponding ones for LP. Additionally, the decidability of the semantics of SCLP programs have been investigated: if a goal has a semiring value greater than or equal to a certain value in the semiring, then we can discover this in finite time. Moreover, for SCLP programs without functions, the problem is completely decidable: the semantics of a goal can be computed in finite and bounded time.

The SCLP framework has been implemented [53] on top of an existing CLP(FD) language. The resulting language, called CLP(FD,S), is parametric with respect to the semiring S, and can handle semiring-based soft constraints over S in problems where variables have finite domains.

3.2 Soft cc

Semiring-based soft constraints have also been embedded in concurrent languages. The framework proposed in [31,25] (called scc) extends the cc programming framework [70] by using soft constraints instead of classical ones.

In cc programming, a set of agents share a store which contains constraints. An agent can *ask* if a constraint is entailed by the store, or can *tell* (that is, add) a new constraint to the store. In scc, the notions of ask and tell are parameterized with respect to the level of consistency of the store or the semiring level of each instance of the constraints present in the store. In this way, each tell and ask agent is equipped with a preference (or consistency) threshold which is used to determine their success, failure, or suspension, as well as to prune the search.

Scc programming has been also extended to deal with timed [20] and non-monotonic [35] issues. The timed extension is based on the hypothesis of *bounded asynchrony*: computation takes a bounded period of time and is measured by a discrete global clock. Action prefixing is then considered as the syntactic marker which distinguishes a time instant from the next one. In non-monotonic scc some new actions provide the user with explicit non-monotonic operations: retract(c), to remove constraint c from the current store; updateX(c), to transactionally relax all the constraints of the store that deal with variables in set X, and then add a constraint c; and nask(c), to test if constraint c is not entailed by the store.

The framework has been implemented [43]. In particular, the soft constraint constructs were adapted to and integrated within the propagation process of Mozart [47].

Soft constraints have been embedded also in the Constraint Handling Rule (CHR) framework, a formalism to specify constraint solvers and constraint propagation algorithms via a set of rewriting rules [24,19]. The obtained system allows one to design and specify naturally soft constraint solvers, including soft propagation algorithms.

4   Extending Soft Constraints to Model Other Kinds of Preferences

Semiring-based soft constraints are a way to model preferences. However, preferences can be of various kinds, and semiring-based soft constraints, as originally defined, are good at modeling only some of them.

For example, preferences can be quantitative or qualitative (e.g., "I like this at level 10" versus "I like this more than that"). They can also be conditional (e.g., "If the main dish is fish, I prefer white wine to red wine") or bipolar (e.g., "I like fish a lot, and I slightly dislike meat"). Soft constraints can model directly and naturally quantitative preferences, but are not as good at modeling qualitative, conditional, or bipolar preferences. We will now summarize some approaches to model these other kinds of preferences via extensions or adaptations of semiring-based soft constraints.

4.1   Bipolar Preferences

Bipolarity is an important topic in several fields, such as psychology and multi-criteria decision making, and it has recently attracted interest in the AI community, especially in argumentation and qualitative reasoning. Bipolarity in preference reasoning can be seen as the possibility to stating both degrees of satisfaction (that is, positive preferences) and degrees of rejection (that is, negative preferences).

Positive and negative preferences can be thought as two symmetric concepts, and thus one can think that they can be dealt with via the same operators. However, this may not model what one usually expects in real scenarios. For example, when we have a scenario with two objects A and B, if we like both A and B, then having both A and B should be more preferred than having just A or B alone. On the other hand, if we don't like A nor B, then the preference of A and B together should be smaller than the preferences of A or B alone. That is, the combination of positive preferences should produce a higher (positive) preference, while the combination of negative preferences should give us a lower (negative) preference.

When dealing with both kinds of preferences, it is natural to express also *indifference*, which means that we express neither a positive nor a negative preference over an object. For example, we may say that we like peaches, we don't like bananas, and we are indifferent to apples. Then, a desired behavior of indifference is that, when combined with any preference (either positive or negative), it should not influence the overall preference. For example, if we like peaches and we are indifferent to apples, a dish with peaches and apples should have overall a positive preference.

Finally, besides combining preferences of the same type, we also want to be able to combine positive with negative preferences. The most natural and intuitive way to do so is to allow for *compensation*. Comparing positive against negative aspects and compensating them with respect to their strength is one of the core features of decision-making processes, and it is, undoubtedly, a tactic universally applied to solve many real life problems. For example, if we have a meal with meat (that we like very much) and wine (that we don't like), then what should be the preference of the meal? To know that, we should be able to compensate the positive preference given to meat with the negative preference given to wine. The expected result is a preference which is between the two, and which should be positive if the positive preference is "stronger" than the negative one.

Semiring-based soft constraints can only model negative preferences, since in this framework preference combination returns lower preferences. However, this framework can be generalized to model also positive preferences. In [65] this is done by defining a new algebraic structure to model positive preferences. The two structures are then linked by setting the highest negative preference to coincide with the lowest positive preference, to model indifference. Then, a combination operator between positive and negative preferences is defined to model preference compensation. To find optimal solutions of bipolar problems, it is possible to adapt usual soft constraint propagation and branch and bound.

4.2   Conditional Qualitative Preferences

While soft constraints cannot model conditional qualitative preferences directly, CP-nets [37] (Conditional Preference networks) can. CP-nets exploit conditional preferential independence by structuring a user's possibly complex preference ordering with the ceteris paribus (that is, "all else being equal") assumption. CP-nets are sets of conditional ceteris paribus preference statements (cp-statements). For instance, the statement "I prefer red wine to white wine if meat is served" asserts that, given two meals that differ only in the kind of wine served, and both containing meat, the meal with red wine is preferable to the meal with white wine.

If we compare the expressive power of CP-nets and soft constraints, we may see that classical constraints are at least as expressive as CP-nets in terms of optimal solutions. In fact, it is possible to show that, given any CP-net, we can obtain in polynomial time a set of classical constraints whose solutions are the optimal outcomes of the CP-net [36]. However, if we are interested not just in the optimal solutions, but in the whole solution ordering, CP-nets and soft constraints are incomparable.

However, it is possible to approximate a CP-net ordering via soft constraints, achieving tractability while sacrificing precision to some degree [45]. Different approximations can be characterized by how much of the original ordering they preserve, the time complexity of generating the approximation, and the time complexity of comparing outcomes in the approximation. It is vital that such approximations are information preserving; that is, what is ordered in the CP-net is also ordered in the same way in the soft constraint problem. Another desirable property of approximations is that they preserve the ceteris paribus property.

The possibility of approximating CP-nets via soft constraints means that, with only a soft constraint solver at hand, we can model and solve real-life problems containing either qualitative and quantitative preferences.

5   Mastering the Complexity of Modeling and Solving Soft Constraint Problems

In constraint problems we look for a solution, while in soft constraint problems we look for an optimal solution. Thus, soft constraint problems are more difficult to handle by a solver. To ease this difficulty, several AI techniques have been exploited. Here we cite just three of them: abstraction, symmetry breaking, and explanation generation.

Abstraction is used to work on a simplified version of the given problem, thus hoping to have a significantly smaller search space, while explanation generation is used to ease the understanding of the behavior of the solver. For example, it is not always easy for a user to understand why there are no better solution than the one returned. Symmetry breaking, instead, aims at simplifying the problem by pruning part of the search space, via the elimination of symmetric (or interchangeable) assignments.

An added difficulty in dealing with soft constraints comes also in the modeling phase, where a user has to understand how to model faithfully his real-life problem via soft constraints. Since soft constraints require the specification of all the preferences inside the constraints, it may be too tedious for a user to do this. Also, some users may prefer to not reveal all their preferences because of privacy reasons. In both cases, we end up with a soft constraint problem where some preferences are missing. To reason in this scenario, we may use techniques like machine learning to complete the problem, or we may try to find an optimal solution without completing the problem, or by eliciting only a small number of missing preferences.

5.1   Abstraction

Although it is obvious that SCSPs are much more expressive than classical CSPs, they are also more difficult to process and to solve. Therefore, sometimes it may be too costly to find all, or even only one, optimal solution. Also, although classical propagation techniques like arc-consistency can be extended to SCSPs, even such techniques can be too costly to be used, depending on the size and structure of the partial order associated to the SCSP. Finally, sometimes we may not have a solver for the class of soft constraints we need to solve, while we may have a solver for a "simpler" class of soft constraints.

For these reasons, it may be reasonable to work on a simplified version of the given soft constraint problem, trying however to not loose too much information. Such a simplified version can be defined by means of the notion of abstraction, which takes an SCSP and returns a new one which is simpler to solve. Here, as in many other works on abstraction, "simpler" may have several meanings, like the fact that a certain solution algorithm finds a solution, or an optimal solution, in a fewer number of steps, or also that the abstracted problem can be processed by a machinery which is not available in the concrete context.

To define an abstraction, we may use for example the theory of Galois insertions [41]: given an SCSP (the *concrete* one), we may get an abstract SCSP by just changing the associated semiring, and relating the two structures (the concrete and the abstract one) via a *Galois insertion*. Note that this way of abstracting constraint problems does not change the structure of the problem (the set of variables remains the same, as well as the set of constraints), but just the semiring values to be associated to the tuples of values for the variables in each constraint [10].

Once we reason on the abstracted version of a given problem, we can bring back to the original problem some (or possibly all) of the information derived in the abstract context, and then continue the solution process on the transformed problem, which is a concrete problem equivalent to the given one. The hope is that, by following this route, we get to the final goal faster than just solving the original problem.

Given any optimal solution of the abstract problem, we can find upper and lower bounds for an optimal solution for the concrete problem. It is also possible to define iterative hybrid algorithms which can approximate an optimal solution of a soft constraint problem by solving a series of problems which abstract, in different ways, the original problem. These are anytime algorithms since they can be stopped at any phase, giving better and better approximations of an optimal solution.

Experimental results show that this line is promising, for example when we want to solve a fuzzy CSP but we just have a solver for classical constraints [33].

5.2   Symmetry Breaking

The existence of symmetries in a problem has the effect of artificially increasing the size of the search space that is explored by search algorithms. Therefore, a typical approach is to break the symmetries in the problem so that only unique solutions are returned. The significant advantage is that not only do we return fewer solutions, but we also reduce the search effort required to find these solutions by eliminating symmetric branches of the search tree.

In [23,73] an approach is presented to deal with symmetry in the semiring-based framework for soft constraints, and it is shown that breaking symmetries in soft constraint satisfaction problems improves the efficiency of search. In [11,12] interchangeability has been extended to the soft CSP framework by adding a notion of threshold and degradation. With these extensions, values are considered interchangeable only by checking solutions with a semiring level better than a given threshold (thus disregarding differences among solutions that are not sufficiently good), or solutions whose exchange cannot degrade the current solution by more than a given factor.

5.3   Explanations

One of the most important features of problem solving in an interactive setting is the capacity of the system to provide the user with justifications, or explanations, for its operations. Such justifications are especially useful when the user is interested in what happens at any time during search, because he/she can alter features of the problem to facilitate the problem solving process.

Basically, the aim of an explanation is to show clearly why a system acted in a certain way after certain events. In the context of constraint problems, explanations can be viewed as answers to user's questions like the following: Why isn't it possible to obtain a solution? Why is there a conflict between these values for these variables? Why did the system select this value for this variable? In soft constraint problems, explanation should certainly take preferences into account.

In addition to providing explanations, interactive systems should be able to show the consequences, or implications, of an action to the user, which may be useful in deciding which choice to make next. In this way, they can provide a sort of "what-if" kind of reasoning, which guides the user towards good future choices. Implications can be viewed as answers to questions like the following: What would happen if this variable could only take on these values? What would happen if this value were added to the domain of this variable? Fortunately, in soft constraint problems this capacity can be implemented with the same machinery that is used to give explanations.

A typical example of an interactive system where constraints and preferences may be used, and where explanations can be very useful, are configurators. A configurator is a system which interacts with a user to help him/her to configure a product. A product can be seen as a set of component types, where each type corresponds to a certain finite number of concrete components, and a set of compatibility constraints among subsets of the component types. A user configures a product by choosing a concrete component for each component type, such that all the compatibility constraints as well as personal preferences are satisfied. For example, in a car configuration problem, a user may prefer red cars, but may also not want to completely rule out other colors. Thus red will get a higher preference with respect to the other colors.

Constraint-based technology is currently used in many configurators to both model and solve configuration problems: component types are represented by variables, having as many values as the concrete components, and both compatibility and personal constraints are represented as constraints (or soft constraints) over subsets of such variables. At present, user choices during the interaction with the configurator are usually restricted to specifying unary constraints, in which a certain value is selected for a variable.

Whenever a choice is made, the corresponding (unary) constraint is added to existing compatibility and personal constraints, and some constraint propagation notion is enforced, for example arc-consistency (AC) [68], to rule out (some of the) future choices that are not compatible with the current choice. While providing justifications based on search is difficult, arc-consistency enforcing has been used as a source of guidance for justifications, and it has been exploited to help the users in some of the scenarios mentioned above. For example, in [50], it is shown how AC enforcement can be used to provide both justifications for choice elimination, and also guidance for conflict resolution.

The same approach can be used also for configurators with preferences [64], using a generalized version of arc-consistency, whose application may decrease the preferences in some constraints. When a user makes a choice for a specific feature of a configuration problem, the preferences of the other features are automatically lowered to the minimum value. This triggers arc-consistency, which in turn lowers other preferences for features that are correlated to the chosen one. If all the values of a feature have minimal preference, that feature cannot be instantiated to any value. Thus the sequence of choices already made cannot lead to a complete configuration. Otherwise, the configuration process may continue, but the lowered preference values can help in guiding towards the best complete configurations. It is therefore possible to compute explanations describing why the preferences for some values decrease, and suggesting at the same time which assignment has to be chosen, or retracted, in order to maximize the quality of a complete configuration.

5.4   Learning

In a soft constraint problem, sometimes one may know his/her preferences over some of the solutions, but have no idea on how to code this knowledge into the constraint problem in terms of local preferences. That is, one has a global idea about the *goodness* of a solution, but does not know the contribution of each single constraint to such a measure. In such a situation, it is difficult both to associate a preference to the other solutions in a compatible way, and to understand the importance of each tuple and/or constraint. In other situations, one may have just a rough estimate of the local preferences, either for the tuples or for the constraints.

In [66], this scenario is theoretically addressed by proposing to use learning techniques based on gradient descent. More precisely, it is assumed that the level of preference for some solutions (that is, the *examples*) is known, and a suitable learning technique is defined to learn, from these examples, values to be associated with each constraint tuple, in a way that is compatible with the examples. In [6] the theoretical framework proposed in [66] is implemented, and results of several experiments are shown.

Soft constraint learning has been embedded also in a general interactive constraint framework, where users can state both usual preferences over constraints and also preferences over solutions proposed by the system [67]. In this way, the modeling and the solving process are heavily interleaved. Moreover, the two tasks can be done also incrementally, by specifying some preferences at each step, and obtaining better and better solutions at each step. In this way, the examples needed are much less, since they are not given by the user all at the beginning of the solving process, but are guided by the solver, that proposes the next best solutions and asks the user to give a feedback on them.

5.5   Incompleteness and Elicitation

Sometimes the task of specifying a whole soft CSP may be so heavy that a user may be unwilling to provide a complete specification. For example, some preferences may be omitted. Preference omission may have several reasons, such as privacy concerns, or timing issues among several users concurring in the specification of a soft CSP.

Even if a soft CSP has some missing preferences, it could still be feasible to find an optimal solution. However, there is more than one notion of "optimality". Two extreme notions of optimal solutions are the following: *possibly optimal* solutions are assignments to all the variables that are optimal in *at least one way* currently unspecified preferences can be revealed, while *necessarily optimal* solutions are assignments to all the variables that are optimal in *all ways* in which currently unspecified preferences can be revealed.

Given an incomplete soft CSP, its set of possibly optimal solutions is never empty, while the set of necessarily optimal solutions can be empty. Of course what we would like to find is a necessarily optimal solution, to be on the safe side: such solutions are optimal regardless of how the missing preferences would be specified. However, if this set is empty, we can interleave search and preference elicitation. More precisely, we can ask the user to provide some of the missing preferences and try to find, if any, a necessarily optimal solution of the new incomplete soft CSP. Then we can repeat the process until the current problem has at least one necessarily optimal solution. Experimental results show that this process ends after eliciting a very small percentage of the missing preferences [52].

6   Applying Soft Constraints

Soft constraints have been applied to several scenarios. Here we will review some of them. Others can be found in [61].

6.1   Temporal Reasoning

Reasoning about time is a core issue in many real life problems, such as planning and scheduling for production plants, transportation, and space missions. Several approaches have been proposed to reason about temporal information. Temporal constraints have been among the most successful in practice.

In temporal constraint problems, variables either represent instantaneous events or temporal intervals. Temporal constraints allow one to put temporal restrictions either on when a given event should occur, or on how long a given activity should last. Several qualitative and quantitative constraint-based temporal formalisms have been proposed. A qualitative temporal constraint defines which temporal relations, e.g. *before, after, during*, are allowed between two temporal intervals representing two activities. A quantitative temporal constraint instead defines restrictions between the start and end times of some activities. Once the constraints have been stated, the goal is to find an occurrence time, or duration, for all the events, such that all temporal constraints are respected.

The expressive power of classical temporal constraints may however be insufficient to model faithfully all the aspects of the problem. For example, it may be that some durations are more preferable than others, such as in "I can have lunch between 11:30am and 2pm, but I prefer to have it at 1pm", or in "I can meet you between 9am and 10am, but the earlier the better". For this reason, both qualitative and quantitative temporal reasoning formalisms have been extended with quantitative preferences to allow for the specification of such a kind of statements.

In particular, the qualitative approach has been augmented with fuzzy preferences, that are associated with the relations among temporal intervals allowed by the constraints. Higher values represent a more preferred relation. Once such constraints have been stated, the goal is to find a temporal assignment to all the variables with the highest "lower" preference on any constraint.

In [57] the semiring-based formalism has been combined with temporal quantitative constraints. The result are soft temporal constraints where each allowed duration or occurrence time for a temporal event is associated to a (fuzzy) preference representing the desirability of that specific time. Tractability results have been shown for some classes of these problems, characterized by the absence of disjunctions in the temporal constraints (as in the classical case) and by the shape of the preference functions.

6.2   Security

The semiring-based framework has been used to tackle several security problems, such as protocols, policies, and system/network security. The basic idea used is to consider security not as a Boolean predicate but as a notion suitable to be represented via different levels. So, instead of just having secure and non secure protocols, we have protocols that satisfy the confidentiality and authentication goals with a certain security level [2,3]. By considering such security levels (for instance public, confidential, secret, and top-secret), protocols can be better analyzed and sometimes new flaws [4,5] can be found.

The security of systems and networks have been analyzed by considering respectively integrity policies and trusts among nodes of a network. The integrity of a system can then be evaluated by checking how much it satisfies some specified (soft) constraints [14,15]. In a similar manner, by adding constraints on the type of flow permitted or denied among the nodes of a network, a flow analysis can be executed, revealing inter-operation [16] or cascading [17,18] problems.

Soft constraints and logic programming together have been also used [34] to represent the concept of multi-trust, which is aimed at computing trust by collectively involving a group of trustees at the same time: the trustor needs the concurrent support of multiple individuals to accomplish its task.

6.3   Routing and Quality of Service

Semirings are also used for Routing and Quality of Service (QoS). For instance, in [62] the authors give a generic algorithm for finding singlesource shortest distances in a weighted directed graph when the weights are elements of a semiring. The same algorithm can also be used to solve efficiently classical shortest paths problems or to find the kshortest distances in a directed graph. An interesting foundational model has been instead introduced in [44]. The model handles QoS attributes as semantic constraints within a graphical calculus for mobility. The QoS attributes are related to the programming abstractions and are exploited to select, configure, and dynamically modify the underlying system oriented QoS mechanisms.

Semirings and constraints together are instead used in [28,32] where a formal model to represent and solve routing and multicast routing problems in multicast networks with QoS has been suggested. And-or graphs have been used to represent the network and SCLP programs are used to compute the best path (or the best tree when multicast is considered), according to QoS criteria. Another approach extends instead cc programming [38]. In the resulting framework, Service Level Agreement requirements are (soft) constraints that can be generated either by a single party or by the synchronization of two agents.

6.4   Data Mining

The paradigm of pattern discovery based on constraints was introduced with the aim of providing the user with a tool to drive the discovery process towards potentially interesting patterns, with the positive side effect of achieving a more efficient computation.

In classical constraint-based mining, a constraint is a Boolean function which returns either true or false. In [8,9] a new paradigm of pattern discovery based on soft constraints has been introduced. This provides a rigorous theoretical framework, which is very general (having the classical paradigm as a particular instance), and able to measure the level of interest of a pattern.

7   Conclusions and Future Scenarios

Ugo Montanari and the authors of this paper introduced twelve years ago the notion of semiring-based soft constraints. Since then, the semiring-based framework has been further studied, extended, and applied to several scenarios. In this paper we have described the extension of this framework to deal with both positive and negative preferences, as well as its adaption to deal with conditional qualitative preferences. Moreover, we have briefly outlined how this kind of constraints have been embedded in constraint-based languages, both sequential and concurrent. Furthermore, we have described how the complexity of modeling and solving a problem with soft constraints has been mitigated via the use of techniques such as abstraction, explanation generation, symmetry breaking, machine learning, and preference elicitation. Finally, we have mentioned some applications to the fields of security, QoS, data mining, and temporal reasoning.

Although the initial idea of semiring-based soft constraints generated a huge amount of research, both on the theoretical and on the application side, much more can be done to make soft constraints more useful and widely applicable. For example, many tractability results, developed for classes of classical constraints, can be studied and adapted to the setting of soft constraints. Also, specialized solvers for specific classes of soft constraints, as well as general solvers for the whole class, should be developed. We also believe that the generalization started with the introduction of the soft constraint formalism should be continued to achieve a single framework where many kinds of preferences should be easily modeled and solved. We also believe that uncertainty issues should be taken into consideration, as well as multi-agent scenarios where several agents express their preferences over a common set of objects and want to agree over the choice of one or more objects which are highly preferred by all of them. On the application side, the semiring idea could be used in reputation logic to give a quantitative and qualitative measure to the notion of trust among users. The field of QoS seems also a field that deserve further study.

8   A Special Thank

We would really like to thank Ugo Montanari for the very interesting and exciting time spent working together on the subject of this paper and on other research issues. His knowledge, skills, generosity, and passion for research has always been for us an inspiring example that drives our work and fills us with pride for having shared some research activities with him.

References

1. Apt, K.R.: Principles of Constraint Programming. Cambridge University Press, Cambridge (2003)
2. Bella, G., Bistarelli, S.: Soft Constraints for Security Protocol Analysis: Confidentiality. In: Ramakrishnan, I.V. (ed.) PADL 2001. LNCS, vol. 1990. Springer, Heidelberg (2001)
3. Bella, G., Bistarelli, S.: Soft Constraint Programming to Analysing Security Protocols. In: Theory and Practice of Logic Programming (TPLP), special Issue on Verification and Computational Logic, vol. 4(5), pp. 1–28. Cambridge University Press, Cambridge (2004)

4. Bella, G., Bistarelli, S.: Confidentiality levels and deliberate/indeliberate protocol attacks. In: Christianson, B., Crispo, B., Malcolm, J.A., Roe, M. (eds.) Security Protocols 2002. LNCS, vol. 2845, pp. 104–119. Springer, Heidelberg (2004)
5. Bella, G., Bistarelli, S.: Information Assurance for Security Protocols. Computers & Security 24(4), 322–333 (2005)
6. Biso, A., Rossi, F., Sperduti, A.: Experimental results on Learning Soft Constraints. In: Proc. KR 2000 (7th Int. Conf. on Principles of Knowledge Representation and Reasoning) (2000)
7. Bistarelli, S.: Semirings for Soft Constraint Solving and Programming. LNCS, vol. 2962. Springer, Heidelberg (2004)
8. Bistarelli, S., Bonchi, F.: Soft Constraint Based Pattern Mining. Data & Knowledge Engineering 62(1) (2007)
9. Bistarelli, S., Bonchi, F.: Extending the Soft Constraint Based Mining Paradigm. In: Džeroski, S., Struyf, J. (eds.) KDID 2006. LNCS, vol. 4747, Springer, Heidelberg (2007)
10. Bistarelli, S., Codognet, P., Rossi, F.: Abstracting Soft Constraints: Framework, Properties, Examples. Artificial Intelligence Journal 139, 175–211 (2002)
11. Bistarelli, S., Faltings, B., Neagu, N.: Interchangeability in Soft CSPs. In: Van Hentenryck, P. (ed.) CP 2002. LNCS, vol. 2470, Springer, Heidelberg (2002)
12. Bistarelli, S., Faltings, B., Neagu, N.: Experimental Evaluation of Interchangeability in Soft CSPs. In: Apt, K.R., Fages, F., Rossi, F., Szeredi, P., Váncza, J. (eds.) CSCLP 2003. LNCS (LNAI), vol. 3010, Springer, Heidelberg (2004)
13. Bistarelli, S., Fargier, H., Montanari, U., Rossi, F., Schiex, T., Verfaillie, G.: Semiring-Based CSPs and Valued CSPs: Frameworks, Properties, and Comparison. In: Constraints, vol. 4(3), Kluwer, Dordrecht (1999)
14. Bistarelli, S., Foley, S.N.: Analysis of Integrity Policies using Soft Constraints. In: Proc. IEEE POLICY 2003 (2003)
15. Bistarelli, S., Foley, S.N.: A Constraint Based Framework for Dependability Goals: Integrity. In: Anderson, S., Felici, M., Littlewood, B. (eds.) SAFECOMP 2003. LNCS, vol. 2788. Springer, Heidelberg (2003)
16. Bistarelli, S., Foley, S.N., O'Sullivan, B.: Reasoning about Secure Interoperation using Soft Constraints. In: Proc. IFIP TC1 WG1.7 Workshop on Formal Aspects in Security and Trust (FAST), Kluwer, Dordrecht (2005)
17. Bistarelli, S., Foley, S.N., O'Sullivan, B.: Detecting and Eliminating the Cascade Vulnerability Problem from Multi-level Security Networks using Soft Constraints. In: Proc. IAAI 2004, AAAI Press, Menlo Park (2004)
18. Bistarelli, S., Foley, S.N., O'Sullivan, B.: A soft constraint-based approach to the cascade vulnerability problem. Journal of Computer Security 13(5), 699–720; Special Issue: Security Track at ACM Symposium on Applied Computing (2004)
19. Bistarelli, S., Fruehwirth, T., Marte, M., Rossi, F.: Soft Constraint Propagation and Solving in Constraint Handling Rules. Special issue of Computational Intelligence on Preferences in AI and CP 20(2), 287–307 (2004)
20. Bistarelli, S., Gabbrielli, M., Meo, M.C., Santini, F.: Timed Soft Concurrent Constraint Programs. In: Proc. CP2007 Doctoral Program, Providence RI - USA (2007)
21. Bistarelli, S., Gadducci, F.: Enhancing constraints manipulation in semiring-based formalisms. In: Proc. ECAI 2006, pp. 63–67. IOS Press, Amsterdam (2006)
22. Bistarelli, S., Gennari, R., Rossi, F.: General Properties and Termination Conditions for Soft Constraint Propagation. Constraints: An International Journal, Constraints 8(1) (2003)

23. Bistarelli, S., Keleher, J., O'Sullivan, B.: Symmetry Breaking in Soft CSPs. In: Proc. AI-2003, BCS Conference Series Research and Development in Intelligent Systems xx. Springer, Heidelberg (2004)
24. Bistarelli, S., Marte, M., Fruhwirth, T., Rossi, F.: Soft Constraint propagation and Solving with CHRs. In: Proc. SAC 2002 (ACM Symposium on Applied Computing), Madrid (March 2002)
25. Bistarelli, S., Montanari, U., Rossi, F.: Soft Concurrent Constraint Programming. ACM Transactions on Computational Logic (TOCL) 7(3), 1–27 (2006)
26. Bistarelli, S., Montanari, U., Rossi, F.: Semiring-based Constraint Solving and Optimization. Journal of the ACM 44(2), 201–236 (1997)
27. Bistarelli, S., Montanari, U., Rossi, F.: Constraint Solving over Semirings. In: Proc. IJCAI 1995, Morgan Kaufmann, San Francisco (1995)
28. Bistarelli, S., Montanari, U., Rossi, F.: Soft Constraint Logic Programming and Generalized Shortest Path Problems. Journal of Heuristics 8(1) (2002)
29. Bistarelli, S., Montanari, U., Rossi, F.: Semiring-based Constraint Logic Programming. In: Proc.15th International Joint Conference on Artificial Intelligence (IJCAI 1997), pp. 352–357. Morgan Kaufman, San Francisco (1997)
30. Bistarelli, S., Montanari, U., Rossi, F.: Semiring-based Constraint Logic Programming: Syntax and Semantics. ACM Transactions on Programming Languages and System (TOPLAS) 23(1), 1–29 (2001)
31. Bistarelli, S., Montanari, U., Rossi, F.: Soft Concurrent Constraint Programming. In: Le Métayer, D. (ed.) ESOP 2002. LNCS, vol. 2305, Springer, Heidelberg (2002)
32. Bistarelli, S., Montanari, U., Rossi, F., Santini, F.: Modelling Multicast Qos Routing by using Best-Tree Search in AND-OR Graphs and Soft Constraint Logic Programming. In: Proc. Fifth Workshop on Quantitative Aspects of Programming Languages QAPL 2007, ENTCS (2007)
33. Bistarelli, S., Pilan, I., Rossi, F.: Abstracting Soft Constraints: Some Experimental Results. In: Apt, K.R., Fages, F., Rossi, F., Szeredi, P., Váncza, J. (eds.) CSCLP 2003. LNCS (LNAI), vol. 3010, Springer, Heidelberg (2004)
34. Bistarelli, S., Santini, F.: Propagating Multitrust within Trust Networks. In: Proc. SAC 2008 (to appear, 2008)
35. Bistarelli, S., Santini, F.: Non monotonic soft cc. IIT TR-15/2007 (2007)
36. Brafman, R., Dimopoulos, Y.: Preference-based constraint optimization. Computational Intelligence 22(2), 218–245 (2004)
37. Boutilier, C., Brafman, R.I., Domshlak, C., Hoos, H.H., Poole, D.: CP-nets: A tool for representing and reasoning with conditional ceteris paribus preference statements. Journal of Artificial Intelligence Research 21, 135–191 (2004)
38. Buscemi, M., Montanari, U.: CC-Pi: A Constraint-Based Language for Specifying Service Level Agreements. In: De Nicola, R. (ed.) ESOP 2007. LNCS, vol. 4421, Springer, Heidelberg (2007)
39. Cooper, M.C.: Reduction operations in fuzzy or valued constraint satisfaction. Fuzzy Sets and Systems 134(3), 311–342 (2003)
40. Cooper, M.C., Schiex, T.: Arc Consistency for Soft Constraints. Artificial Intelligence 154(1-2), 199–227 (2004)
41. Cousot, P., Cousot, R.: Abstract Interpretation: A Unified lattice Model for static Analysis of Programs by Construction or Approximation of Fixpoints. In: Proc.4th ACM Symposium on Principles of Programming Languages (POPL 1977). ACM, New York (1977)
42. Dechter, R.: Constraint Processing. Morgan Kaufmann, San Francisco (2003)

Declarative Debugging of Membership Equational Logic Specifications[*]

Rafael Caballero, Narciso Martí-Oliet, Adrián Riesco, and Alberto Verdejo

Facultad de Informática, Universidad Complutense de Madrid, Spain

Abstract. Algorithmic debugging has been applied to many declarative programming paradigms; in this paper, it is applied to the rewriting paradigm embodied in Maude. We introduce a declarative debugger for executable specifications in membership equational logic which correspond to Maude functional modules. Declarative debugging is based on the construction and navigation of a debugging tree which logically represents the computation steps. We describe the construction of appropriate debugging trees for oriented equational and membership inferences. These trees are obtained as the result of collapsing in proof trees all those nodes whose correctness does not need any justification. We use an extended example to illustrate the use of the declarative debugger and its main features, such as two different strategies to traverse the debugging tree, use of a correct module to reduce the number of questions asked to the user, and selection of trusted vs. suspicious statements by means of labels. The reflective features of Maude have been extensively used to develop a prototype implementation of the declarative debugger for Maude functional modules using Maude itself.

Keywords: declarative debugging, membership equational logic, Maude, functional modules.

1 Introduction

As argued in [23], the application of declarative languages out of the academic world is inhibited by the lack of convenient auxiliary tools such as *debuggers*. The traditional separation between the problem logic (defining *what* is expected to be computed) and control (*how* computations are carried out actually) is a major advantage of these languages, but it also becomes a severe complication when considering the task of debugging erroneous computations. Indeed, the involved execution mechanisms associated to the control make it difficult to apply the typical techniques employed in imperative languages based on step-by-step trace debuggers.

Consequently, new debugging approaches based on the language's semantics have been introduced in the field of declarative languages, such as *abstract diagnosis*, which formulates a debugging methodology based on abstract interpretation [9,1], or *declarative debugging*, also known as *algorithmic debugging*, which was first introduced by E. Y. Shapiro [19] and that constitutes the framework of this work. Declarative debugging

---

[*] Research supported by MEC Spanish projects *DESAFIOS* (TIN2006-15660-C02-01) and *MERIT-FORMS* (TIN2005-09027-C03-03), and Comunidad de Madrid program *PROMESAS* (S-0505/TIC/0407).

P. Degano et al. (Eds.): Montanari Festschrift, LNCS 5065, pp. 174–193, 2008.

has been widely employed in the logic [10,14,22], functional [21,17,16,18], and multiparadigm [5,3,11] programming languages. Declarative debugging is a semi-automatic technique that starts from a computation considered incorrect by the user (error symptom) and locates a program fragment responsible for the error. The declarative debugging scheme [15] uses a *debugging tree* as a logical representation of the computation. Each node in the tree represents the result of a computation step, which must follow from the results of its child nodes by some logical inference. Diagnosis proceeds by traversing the debugging tree, asking questions to an external oracle (generally the user) until a so-called *buggy node* is found. A buggy node is a node containing an erroneous result, but whose children all have correct results. Hence, a buggy node has produced an erroneous output from correct inputs and corresponds to an erroneous fragment of code, which is pointed out as an error.

During the debugging process, the user does not need to understand the computation operationally. Any buggy node represents an erroneous computation step, and the debugger can display the program fragment responsible for it. From an explanatory point of view, declarative debugging can be described as consisting of two stages, namely the debugging tree *generation* and its *navigation* following some suitable strategy [20].

In this paper we present a declarative debugger for *Maude functional modules* [7, Chap. 4]. Maude is a high-level language and high-performance system supporting both equational and rewriting logic computation for a wide range of applications. It is a declarative language because Maude modules correspond in general to specifications in rewriting logic [12], a simple and expressive logic which allows the representation of many models of concurrent and distributed systems. This logic is an extension of equational logic; in particular, Maude functional modules correspond to specifications in *membership equational logic* [2,13], which, in addition to equations, allows the statement of *membership assertions* characterizing the elements of a sort. In this way, Maude makes possible the faithful specification of data types (like sorted lists or search trees) whose data are defined not only by means of constructors, but also by the satisfaction of additional properties.

For a specification in rewriting or membership equational logic to be executable in Maude, it must satisfy some executability requirements. In particular, Maude functional modules are assumed to be confluent, terminating, and sort-decreasing[1] [7], so that, by orienting the equations from left to right, each term can be reduced to a unique canonical form, and semantic equality of two terms can be checked by reducing both of them to their respective canonical forms and checking that they coincide. Since we intend to debug functional modules, we will assume throughout the paper that our membership equational logic specifications satisfy these executability requirements.

The Maude system supports several approaches for debugging Maude programs: tracing, term coloring, and using an internal debugger [7, Chap. 22]. The tracing facilities allow us to follow the execution on a specification, that is, the sequence of rewrites that take place. Term coloring consists in printing with different colors the operators used to build a term that does not fully reduce. The Maude debugger allows the user to define break points in the execution by selecting some operators or statements. When

---

[1] All these requirements must be understood *modulo* some axioms such as associativiy and commutativity that are associated to some binary operations.

a break point is found the debugger is entered. There, we can see the current term and execute the next rewrite with tracing turned on.

The Maude debugger has as a disadvantage that, since it is based on the trace, it shows to the user every small step obtained by using a single statement. Thus the user can loose the general view of the *proof* of the incorrect inference that produced the wrong result. That is, when the user detects an unexpected statement application it is difficult to know where the incorrect inference started.

Here we present a different approach based on declarative debugging that solves this problem for functional modules. The debugging process starts with an incorrect transition from the initial term to a fully reduced unexpected one. Our debugger, after building a proof tree for that inference, will present to the user questions of the following form: "Is it correct that T fully reduces to T'?", which in general are easy to answer. Moreover, since the questions are located in the proof tree, the answer allows the debugger to discard a subset of the questions, leading and shortening the debugging process.

The current version of the tool has the following characteristics:

- It supports all kinds of functional modules: operators can be declared with any combination of axiom attributes (except for the attribute `strat`, that allows to specify an evaluation strategy); equations can be defined with the `otherwise` attribute; and modules can be parameterized.[2]
- It provides two strategies to traverse the debugging tree: *top-down*, that traverses the tree from the root asking each time for the correctness of all the children of the current node, and then continues with one of the incorrect children; and *divide and query*, that each time selects the node whose subtree's size is the closest one to half the size of the whole tree, keeping only this subtree if its root is incorrect, and deleting the whole subtree otherwise.
- Before starting the debugging process, the user can select a module containing only correct statements. By checking the correctness of the inferences with respect to this module (i.e., using this module as oracle) the debugger can reduce the number of questions asked to the user.
- It allows the user to debug Maude functional modules where some equations and memberships are suspicious and have been labeled (each one with a different label). Only these labeled statements generate nodes in the proof tree, while the unlabeled ones are considered correct. The user is in charge of this labeling. Moreover, the user can answer that he trusts the statement associated with the currently questioned inference; that is, statements can be trusted "on the fly."

Detailed proofs of the results, additional examples, and much more information about the implementation can be found in the technical report [4], which, together with the Maude source files for the debugger, is available from the webpage `http://maude.sip.ucm.es/debugging`.

---

[2] For the sake of simplicity, our running example will be unparameterized, but it can easily be parameterized, as shown in [4].

2 Maude Functional Modules

Maude uses a very expressive version of equational logic, namely membership equational logic (*MEL*) [2,13], which, in addition to equations, allows the statement of membership assertions characterizing the elements of a sort. Below we present the logic and how its specifications are represented as Maude functional modules.

2.1 Membership Equational Logic

A *signature* in *MEL* is a triple (K, Σ, S) (just Σ in the following), with K a set of *kinds*, $\Sigma = \{\Sigma_{k_1 \ldots k_n, k}\}_{(k_1 \ldots k_n, k) \in K^* \times K}$ a many-kinded signature, and $S = \{S_k\}_{k \in K}$ a pairwise disjoint K-kinded family of sets of *sorts*. The kind of a sort s is denoted by $[s]$. We write $T_{\Sigma,k}$ and $T_{\Sigma,k}(X)$ to denote respectively the set of ground Σ-terms with kind k and of Σ-terms with kind k over variables in X, where $X = \{x_1 : k_1, \ldots, x_n : k_n\}$ is a set of K-kinded variables. Intuitively, terms with a kind but without a sort represent undefined or error elements. *MEL* atomic formulas are either *equations* $t = t'$, where t and t' are Σ-terms of the same kind, or *membership assertions* of the form $t : s$, where the term t has kind k and $s \in S_k$. *Sentences* are universally-quantified Horn clauses of the form $(\forall X) A_0 \Leftarrow A_1 \wedge \ldots \wedge A_n$, where each A_i is either an equation or a membership assertion, and X is a set of K-kinded variables containing all the variables in the A_i. Order-sorted notation $s_1 < s_2$ (with $s_1, s_2 \in S_k$ for some kind k) can be used to abbreviate the conditional membership $(\forall x : k)\, x : s_2 \Leftarrow x : s_1$. A *specification* is a pair (Σ, E), where E is a set of sentences in *MEL* over the signature Σ.

Models of *MEL* specifications are called algebras. A Σ-*algebra* $\mathcal{A}$ consists of a set A_k for each kind $k \in K$, a function $A_f : A_{k_1} \times \cdots \times A_{k_n} \longrightarrow A_k$ for each operator $f \in \Sigma_{k_1 \ldots k_n, k}$, and a subset $A_s \subseteq A_k$ for each sort $s \in S_k$, with the meaning that the elements in sorts are well-defined, whereas elements in a kind not having a sort are undefined or error elements. The meaning $[\![t]\!]_{\mathcal{A}}$ of a term t in an algebra $\mathcal{A}$ is inductively defined as usual. Then, an algebra $\mathcal{A}$ satisfies an equation $t = t'$ (or the equation holds in the algebra), denoted $\mathcal{A} \models t = t'$, when both terms have the same meaning: $[\![t]\!]_{\mathcal{A}} = [\![t']\!]_{\mathcal{A}}$. In the same way, satisfaction of a membership is defined as: $\mathcal{A} \models t : s$ when $[\![t]\!]_{\mathcal{A}} \in A_s$.

A specification (Σ, E) has an initial model $\mathcal{T}_{\Sigma/E}$ whose elements are E-equivalence classes of terms $[t]$. We refer to [2,13] for a detailed presentation of (Σ, E)-algebras, sound and complete deduction rules, initial algebras, and specification morphisms.

Since the *MEL* specifications that we consider are assumed to satisfy the executability requirements of confluence, termination, and sort-decreasingness, their equations $t = t'$ can be oriented from left to right, $t \to t'$. Such a statement holds in an algebra, denoted $\mathcal{A} \models t \to t'$, exactly when $\mathcal{A} \models t = t'$, i.e., when $[\![t]\!]_{\mathcal{A}} = [\![t']\!]_{\mathcal{A}}$. Moreover, under those assumptions an equational condition $u = v$ in a conditional equation can be checked by finding a common term t such that $u \to t$ and $v \to t$. This is the notation we will use in the inference rules and debugging trees studied in Sect. 3.

2.2 Representation in Maude

Maude functional modules, introduced with syntax `fmod ... endfm`, are executable *MEL* specifications and their semantics is given by the corresponding initial membership algebra in the class of algebras satisfying the specification.

In a functional module we can declare sorts (by means of keyword `sort(s)`); subsort relations between sorts (`subsort`); operators (`op`) for building values of these sorts, giving the sorts of their arguments and result, and which may have attributes such as being associative (`assoc`) or commutative (`comm`), for example; memberships (`mb`) asserting that a term has a sort; and equations (`eq`) identifying terms. Both memberships and equations can be conditional (`cmb` and `ceq`).

Maude does automatic kind inference from the sorts declared by the user and their subsort relations. Kinds are *not* declared explicitly, and correspond to the connected components of the subsort relation. The kind corresponding to a sort s is denoted [s]. For example, if we have sorts `Nat` for natural numbers and `NzNat` for nonzero natural numbers with a subsort `NzNat` < `Nat`, then [NzNat] = [Nat].

An operator declaration like[3]

```
op _div_ : Nat NzNat -> Nat .
```

is logically understood as a declaration at the kind level

```
op _div_ : [Nat] [Nat] -> [Nat] .
```

together with the conditional membership axiom

```
cmb N div M : Nat if N : Nat and M : NzNat .
```

2.3   A Buggy Example: Non-empty Sorted Lists

Let us see a simple example showing how to specify sorted lists of natural numbers in Maude. The following module includes the predefined module `NAT` defining the natural numbers.

```
(fmod SORTED-NAT-LIST is
  pr NAT .
```

We introduce sorts for non-empty lists and sorted lists. We identify a natural number with a sorted list with a single element by means of a subsort declaration.

```
sorts NatList SortedNatList .
subsorts Nat < SortedNatList < NatList .
```

The lists that have more than one element are built by means of the associative juxtaposition operator __.

```
op __ : NatList NatList -> NatList [ctor assoc] .
```

We define now when a list (with more than one element) is sorted by means of a membership assertion. It states that the first number must be smaller than the first of the rest of the list, and that the rest of the list must also be sorted.

---

[3] The underscores indicate the places where the arguments appear in mixfix syntax.

```
vars E E' : Nat .    var L : NatList .    var OL : SortedNatList .
cmb [olist] : E L : SortedNatList if E <= head(L) /\ L : SortedNatList .
```

The definition of the `head` function distinguishes between lists with a single element and longer ones.

```
op head : NatList -> Nat .
eq [hd1] : head(E) = E .
eq [hd2] : head(L E) = E .
```

We also define a sort function which sorts a list by successively inserting each natural number in the appropriate position in the sorted sublist formed by the numbers previously considered.

```
op insertion-sort : NatList -> SortedNatList .
op insert-list : SortedNatList Nat -> SortedNatList .
eq [is1] : insertion-sort(E) = E .
eq [is2] : insertion-sort(E L) = insert-list(insertion-sort(L), E) .
```

The function `insert-list` distinguishes several cases. If the list has only one number, the function checks if it is bigger than the number being inserted, and returns the sorted list. If the list has more than one element, the function checks that the list is previously sorted; if the number being inserted is smaller than the first of the list, it is located as the (new) first element, while if it is bigger we keep the first element and recursively insert the element in the rest of the list.

```
ceq [il1] : insert-list(E, E') = E' E if E' < E .
eq [il2] : insert-list(E, E') = E E' [owise] .
ceq [il3] : insert-list(E OL, E') = E E' OL
 if E' <= E /\ E OL : SortedNatList .
ceq [il4] : insert-list(E OL, E') = E insert-list(OL, E')
 if E OL : SortedNatList [owise] .
endfm)
```

Now, we can reduce a term in this module. For example, we can try to sort the list 3 4 7 6 with

```
Maude> (red insertion-sort(3 4 7 6) .)
result SortedNatList : 6 3 4 7
```

But... the list obtained *is not sorted!* Moreover, Maude infers that *it is sorted*. Did you notice the bugs? We will show how to use the debugger in Sect. 4.3 to detect them.

3   Declarative Debugging of Maude Functional Modules

We describe how to build the debugging trees for *MEL* specifications. Detailed proofs can be found in [4].

(Reflexivity)
$$\frac{}{e \to e} \; (Rf)$$

(Transitivity)
$$\frac{e_1 \to e' \quad e' \to e_2}{e_1 \to e_2} \; (Tr)$$

(Congruence)
$$\frac{e_1 \to e'_1 \quad \cdots \quad e_n \to e'_n}{f(e_1, \ldots, e_n) \to f(e'_1, \ldots, e'_n)} \; (Cong)$$

(Subject Reduction)
$$\frac{e \to e' \quad e' : s}{e : s} \; (SRed)$$

(Membership)
$$\frac{\{\theta(u_i) \to t_i \leftarrow \theta(u'_i)\}_{1 \leq i \leq n} \quad \{\theta(v_j) : s_j\}_{1 \leq j \leq m}}{\theta(e) : s} \; (Mb)$$
$$\text{if } e : s \Leftarrow u_1 = u'_1 \wedge \cdots \wedge u_n = u'_n \wedge v_1 : s_1 \wedge \cdots \wedge v_m : s_m$$

(Replacement)
$$\frac{\{\theta(u_i) \to t_i \leftarrow \theta(u'_i)\}_{1 \leq i \leq n} \quad \{\theta(v_j) : s_j\}_{1 \leq j \leq m}}{\theta(e) \to \theta(e')} \; (Rep)$$
$$\text{if } e \to e' \Leftarrow u_1 = u'_1 \wedge \cdots \wedge u_n = u'_n \wedge v_1 : s_1 \wedge \cdots \wedge v_m : s_m$$

Fig. 1. Semantic calculus for Maude functional modules

3.1   Proof Trees

Before defining the debugging trees employed in our declarative debugging framework we need to introduce the semantic rules defining the specification semantics. The inference rules of the calculus can be found in Fig. 1, where θ denotes a substitution.

They are an adaptation to the case of Maude functional modules of the deduction rules for *MEL* presented in [13]. The notation $\theta(u_i) \to t_i \leftarrow \theta(u'_i)$ must be understood as a shortcut for $\theta(u_i) \to t_i, \theta(u'_i) \to t_i$. We assume the existence of an *intended interpretation I* of the specification, which is a Σ-algebra corresponding to the model that the user had in mind while writing the statements E, i.e., the user expects that $I \models e \to e'$, $I \models e : s$ for each reduction $e \to e'$ and membership $e : s$ computed w.r.t. the specification (Σ, E). As a Σ-algebra, I must satisfy the following proposition:

Proposition 1. *Let $S = (\Sigma, E)$ be a MEL specification and let $\mathcal{A}$ be any Σ-algebra. If $e \to e'$ (respectively $e : s$) can be deduced using the semantic calculus rules* reflexivity, transitivity, congruence, *or* subject reduction *using premises that hold in $\mathcal{A}$, then $\mathcal{A} \models e \to e'$ (respectively $\mathcal{A} \models e : s$).*

Observe that this proposition cannot be extended to the *membership* and *replacement* inference rules, where the correctness of the conclusion depends not only on the calculus but also on the associated specification statement, which could be wrong.

We will say that $e \to e'$ (respectively $e : s$) is *valid* when it holds in I, and *invalid* otherwise. Declarative debuggers rely on some external oracle, normally the user, in order to obtain information about the validity of some nodes in the debugging tree. The concept of validity can be extended to distinguish *wrong equations* and *wrong membership axioms*, which are those specification pieces that can deduce something invalid from valid information.

Definition 1. *Let* $R \equiv (af \Leftarrow u_1 = u'_1 \wedge \cdots \wedge u_n = u'_n \wedge v_1 : s_1 \wedge \cdots \wedge v_m : s_m)$, *where af denotes an atomic formula, that is, either an oriented equation or a membership axiom in a specification S. Then:*

- $\theta(R)$ *is a* wrong equation instance *(respectively, a* wrong membership axiom instance*) w.r.t. an intended interpretation I when*
  - *There exist $t_1, \ldots, t_n$ such that $I \models \theta(u_i) \rightarrow t_i$, $I \models \theta(u'_i) \rightarrow t_i$ for $i = 1 \ldots n$.*
  - $I \models \theta(v_j) : s_j$ *for $j = 1 \ldots m$.*
  - $\theta(af)$ *does not hold in I.*
- R *is a* wrong equation *(respectively, a* wrong membership axiom*) if it admits some wrong instance.*

It will be convenient to represent deductions in the calculus as *proof trees*, where the premises are the child nodes of the conclusion at each inference step. For example, the proof tree depicted in Fig. 2 corresponds to the result of the reduction in the specification for sorted lists described at the end of Sect. 2.3. For obvious reasons, the operation names have been abbreviated in a self-explanatory way; furthermore, each node corresponding to an instance of the *replacement* or *membership* inference rules has been labelled with the label of the equation or membership statement which is being applied.

In declarative debugging we are specially interested in *buggy nodes* which are invalid nodes with all its children valid. The following proposition characterizes buggy nodes in our setting.

Proposition 2. *Let N by a buggy node in some proof tree in the calculus of Fig. 1 w.r.t. an intended interpretation I. Then:*

1. *N is the consequence of either a* membership *or a* replacement *inference step.*
2. *The equation associated to N is a wrong equation or a wrong membership axiom.*

3.2   Abbreviated Proof Trees

Our goal is to find a buggy node in any proof tree T rooted by the initial error symptom detected by the user. This could be done simply by asking questions to the user about the validity of the nodes in the tree according to the following *top-down* strategy:

> **Input:** A tree T with an invalid root.
> **Output:** A buggy node in T.
> **Description:** Consider the root N of T. There are two possibilities:
>> – If all the children of N are valid, then finish identifying N as buggy.
>> – Otherwise, select the subtree rooted by any invalid child and use recursively the same strategy to find the buggy node.

Proving that this strategy is complete is straightforward by using induction on the height of T. As an easy consequence, the following result holds:

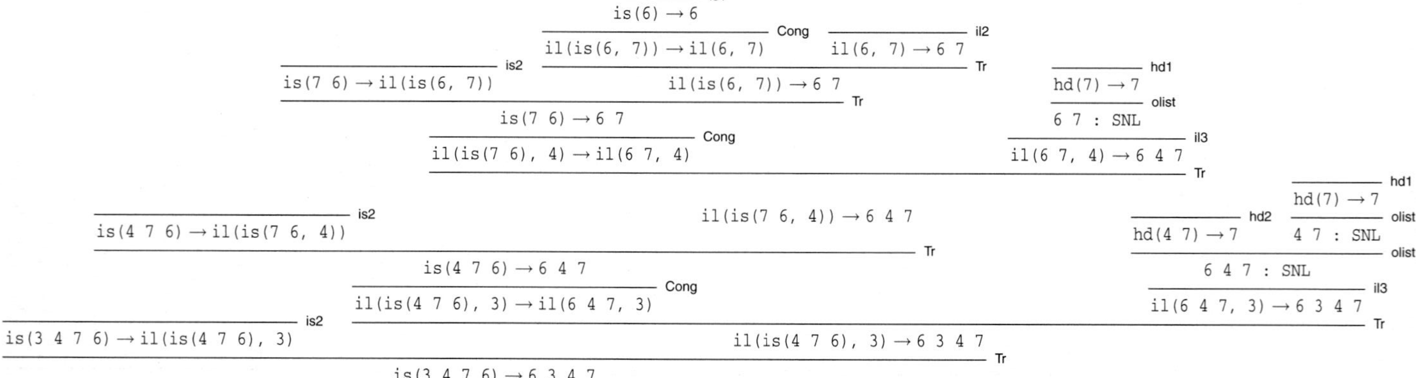

Fig. 2. Proof tree for the sorted lists example

Proposition 3. *Let T be a proof tree with an invalid root. Then there exists a buggy node $N \in T$ such that all the ancestors of N are invalid.*

However, we will not use the proof tree T as debugging tree, but a suitable abbreviation which we denote by $APT(T)$ (from *Abbreviated Proof Tree*), or simply APT if the proof tree T is clear from the context. The reason for preferring the APT to the original proof tree is that it reduces and simplifies the questions that will be asked to the user while keeping the soundness and completeness of the technique. In particular the APT contains only nodes related to the *replacement* and *membership* inferences using statements included in the specification, which are the only possible buggy nodes as Proposition 2 indicates. Fig. 3 shows the definition of $APT(T)$. The T_i represent proof trees corresponding to the premises in some inferences.

The rule APT_1 keeps the root unaltered and employs the auxiliary function APT' to produce the children subtrees. APT' is defined in rules $APT_2 \ldots APT_8$. It takes a proof tree as input parameter and returns a forest $\{T_1, \ldots, T_n\}$ of APTs as result. The rules for APT' are assumed to be tried top-down, in particular APT_4 must not be applied if APT_3 is also applicable. It is easy to check that every node $N \in T$ that is the conclusion of a *replacement* or *membership* inference has its corresponding node $N' \in APT(T)$ labeled with the same abbreviation, and conversely, that for each $N' \in APT(T)$ different from the root, there is a node $N \in T$, which is the conclusion of a *replacement* or *membership* inference. In particular the node associated to $e_1 \to e_2$ in the righthand side of APT_3 is the node $e_1 \to e'$ of the proof tree T, which is not included in the $APT(T)$. We have chosen to introduce $e_1 \to e_2$ instead of simply $e_1 \to e'$ in the $APT(T)$ as a pragmatic way of simplifying the structure of the APTs, since e_2 is obtained from e' and hence likely simpler (the root of the tree T' in APT_3 must be necessarily of the form $e' \to e_2$ by the structure of the inference rule for transitivity in Fig. 1). We will formally state below (Theorem 1) that skipping $e_1 \to e'$ and introducing instead $e_1 \to e_2$ is safe from the point of view of the debugger.

Although $APT(T)$ is no longer a proof tree we keep the inference labels (Rep) and (Mb), assuming implicitly that they contain a reference to the equation or membership axiom used at the corresponding step in the original proof trees. It will be used by the debugger in order to single out the incorrect fragment of specification code.

Before proving the correctness and completeness of the debugging technique we need some auxiliary results. The first one indicates that APT' transforms a tree with invalid root into a set of trees such that at least one has an invalid root. We denote the root of a tree T as $root(T)$.

Lemma 1. *Let T be a proof tree such that $root(T)$ is invalid w.r.t. an intended interpretation I. Then there is some $T' \in APT'(T)$ such that $root(T')$ is invalid w.r.t. I.*

An immediate consequence of this result is the following:

Lemma 2. *Let T be a proof tree and $APT(T)$ its abbreviated proof tree. Then the root of $APT(T)$ cannot be buggy.*

The next theorem guarantees the correctness and completeness of the debugging technique based on APTs:

$$(\textbf{APT}_1) \; APT\left(\frac{T_1 \ldots T_n}{af}(R)\right) = \frac{APT'\left(\frac{T_1 \ldots T_n}{af}(R)\right)}{af} \quad \text{(with } (R) \text{ any inference rule)}$$

$$(\textbf{APT}_2) \; APT'\left(\frac{}{e \to e}(Rf)\right) \qquad = \emptyset$$

$$(\textbf{APT}_3) \; APT'\left(\frac{\dfrac{T_1 \ldots T_n}{e_1 \to e'}(Rep) \quad T'}{e_1 \to e_2}(Tr)\right) = \left\{\frac{APT'(T_1) \ldots APT'(T_n) \; APT'(T')}{e_1 \to e_2}(Rep)\right\}$$

$$(\textbf{APT}_4) \; APT'\left(\frac{T_1 \quad T_2}{e_1 \to e_2}(Tr)\right) \qquad = \{APT'(T_1), \; APT'(T_2)\}$$

$$(\textbf{APT}_5) \; APT'\left(\frac{T_1 \ldots T_n}{e_1 \to e_2}(Cong)\right) \qquad = \{APT'(T_1), \ldots, APT'(T_n)\}$$

$$(\textbf{APT}_6) \; APT'\left(\frac{T_1 \quad T_2}{e : s}(SRed)\right) \qquad = \{APT'(T_1), \; APT'(T_2)\}$$

$$(\textbf{APT}_7) \; APT'\left(\frac{T_1 \ldots T_n}{e : s}(Mb)\right) \qquad = \left\{\frac{APT'(T_1) \ldots APT'(T_n)}{e : s}(Mb)\right\}$$

$$(\textbf{APT}_8) \; APT'\left(\frac{T_1 \ldots T_n}{e_1 \to e_2}(Rep)\right) \qquad = \left\{\frac{APT'(T_1) \ldots APT'(T_n)}{e_1 \to e_2}(Rep)\right\}$$

Fig. 3. Transforming rules for obtaining abbreviated proof trees

Theorem 1. *Let S be a specification, I its intended interpretation, and T a finite proof tree with invalid root. Then:*

- *$APT(T)$ contains at least one buggy node (completeness).*
- *Any buggy node in $APT(T)$ has an associated wrong equation in S.*

The theorem states that we can safely employ the abbreviated proof tree as a basis for the declarative debugging of Maude functional modules: the technique will find a buggy node starting from any initial symptom detected by the user. Of course, these results assume that the user answers correctly all the questions about the validity of the APT nodes asked by the debugger (see Sect. 4.1).

The tree depicted in Fig. 4 is the abbreviated proof tree corresponding to the proof tree in Fig. 2, using the same conventions w.r.t. abbreviating the operation names. The debugging example described in Sect. 4.3 will be based on this abbreviated proof tree.

4   Using the Debugger

Before describing the basics of the user interaction with the debugger, we make explicit what is assumed about the modules introduced by the user; then we present the available commands and how to use them to debug the buggy example introduced in Sect. 2.3.

```
                                                        ———————— hd1
                                                        hd(7) → 7                                             ———————— hd1
                                                        ———————— olist                    ———————— hd2       hd(7) → 7
——————————— is1   ———————————————— il2                  6 7 : SNL                         hd(4 7) → 7         ———————— olist
is(6) → 6         il(6, 7) → 6 7                 ————————————————————— il3    —————————————————————————————— 4 7 : SNL
——————————————————————————————————— is2         il(6 7, 4) → 6 4 7                         6 4 7 : SNL        ———————— olist
          is(7 6) → 6 7                          —————————————————————— is2    ——————————————————————————————————————— il3
——————————————————————————————————————          is(4 7 6) → 6 4 7              il(6 4 7, 3) → 6 3 4 7
          is(4 7 6) → 6 4 7                                                     ——————————————————————————————————————— is2
                          —————————————————————————————————————————————————————————————————————————————————————————————
                                                   is(3 4 7 6) → 6 3 4 7
```

Fig. 4. Abbreviated proof tree for the sorted lists example

4.1 Assumptions

Since we are debugging Maude functional modules, they are expected to satisfy the appropriate executability requirements, namely, the specifications have to be terminating, confluent, and sort-decreasing.

One interesting feature of our tool is that the user is allowed to trust some statements, by means of labels applied to the suspicious statements. This means that the unlabeled statements are assumed to be correct. A trusted statement is treated in the implementation as the *reflexivity*, *transitivity*, and *congruence* rules are treated in the *APT* transformation described in Fig. 3; more specifically, an instance of the *membership* or *replacement* inference rules corresponding to a trusted statement is collapsed in the abbreviated proof tree.

In order to obtain a nonempty abbreviated proof tree, the user must have labeled some statements (all with different labels); otherwise, everything is assumed to be correct. In particular, the buggy statement must be labeled in order to be found. When not all the statements are labeled, the correctness and completeness results shown in Sect. 3 are conditioned by the goodness of the labeling for which the user is responsible.

Although the user can introduce a module importing other modules, the debugging process takes place in the *flattened* module. However, the debugger allows the user to trust a whole imported module.

As mentioned in the introduction, navigation of the debugging tree takes place by asking questions to an external oracle, which in our case is either the user or another module introduced by the user. In both cases the answers are assumed to be correct. If either the module is not really correct or the user provides an incorrect answer, the result is unpredictable. Notice that the information provided by the correct module need not be complete, in the sense that some functions can be only partially defined. In the same way, the signature of the correct module need not coincide with the signature of the module being debugged. If the correct module cannot help in answering a question, the user may have to answer it.

4.2 Commands

The debugger is initiated in Maude by loading the file dd.maude (available from http://maude.sip.ucm.es/debugging), which starts an input/output loop that allows the user to interact with the tool.

As we said in the introduction, the generated proof tree can be navigated by using two different strategies, namely, *top-down* and *divide and query*, the latter being the default one. The user can switch between them by using the commands (top-down strategy .) and (divide-query strategy .). If a module with correct definitions is used to reduce the number of questions, it must be indicated before starting the debugging process with the command (correct module MODULE-NAME .). Moreover, the user can trust all the statements in several modules with the command (trust[*] MODULE-NAMES-LIST .) where * means that modules are considered flattened.

Once we have selected the strategy and, optionally, the module above, we start the debugging process with the command[4]

```
(debug [in MODULE-NAME :] INITIAL-TERM -> WRONG-TERM .)
```

If we want to debug only with a subset of the labeled statements, we use the command

```
(debug [in MODULE-NAME :] INITIAL-TERM -> WRONG-TERM with LABELS .)
```

where LABELS is the set of suspicious equation and membership axiom labels that must be taken into account when generating the debugging tree.

In the same way, we can debug a membership inference with the commands

```
(debug [in MODULE-NAME :] INITIAL-TERM : WRONG-SORT .)
(debug [in MODULE-NAME :] INITIAL-TERM : WRONG-SORT with LABELS .)
```

How the process continues depends on the selected strategy. In case the top-down strategy is selected, several nodes will be displayed in each question. If there is an invalid node, we must select one of them with the command (node N .), where N is the identifier of this wrong node. If all the nodes are correct, we type (all valid .).

In the divide and query strategy, each question refers to one inference that can be either correct or wrong. The different answers are transmitted to the debugger with the commands (yes .) and (no .). Instead of just answering yes, we can also *trust* some statements on the fly if, once the process has started, we decide the bug is not there. To trust the current statement we type the command (trust .).

Finally, we can return to the previous state in both strategies by using the command (undo .).

4.3   Sorted Lists Revisited

We recall from Sect. 2.3 that if we try to sort the list 3 4 7 6, we obtain the strange result

```
Maude> (red insertion-sort(3 4 7 6) .)
result SortedNatList : 6 3 4 7
```

That is, the function returns an unsorted list, but Maude infers it is sorted. We can debug the buggy specification by using the command

```
Maude> (debug in SORTED-NAT-LIST : insertion-sort(3 4 7 6) -> 6 3 4 7 .)
```

---

[4] If no module name is given, the current module is used by default.

$$\cfrac{\cfrac{}{\mathtt{is(6) \to 6}}\ is1 \quad \cfrac{}{\mathtt{il(6,\ 7) \to 6\ 7}}\ il2}{\mathtt{is(7\ 6) \to 6\ 7}}\ is2 \qquad \cfrac{\cfrac{\cfrac{\cfrac{}{\mathtt{hd(7) \to 7}}\ hd1}{\mathtt{6\ 7\ :\ SNL}}\ olist}{\mathtt{il(6\ 7,\ 4) \to 6\ 4\ 7}}\ il3}{}\ is2$$

$$\cfrac{}{\mathtt{is(4\ 7\ 6) \to 6\ 4\ 7}}$$

Fig. 5. Abbreviated proof tree after the first question

With this command the debugger computes the tree shown in Fig. 4. Since the default navigation strategy is divide and query, the first question is

```
Is this transition (associated with the equation is2) correct?
insertion-sort(4 7 6) -> 6 4 7
Maude> (no .)
```

We expect `insertion-sort` to order the list, so we answer negatively and the subtree in Fig. 5 is selected to continue the debugging. The next question is

```
Is this transition (associated with the equation is2) correct?
insertion-sort(7 6) -> 6 7
Maude> (yes .)
```

Since the list is sorted, we answer `yes`, so this subtree is deleted (Fig. 6 left). The debugger asks now the question

```
Is this membership (associated with the membership olist) correct?
6 7 : SortedNatList
Maude> (yes .)
```

This sort is correct, so this subtree is also deleted (Fig. 6 right) and the next question is prompted.

```
Is this transition (associated with the equation il3) correct?
insert-list(6 7, 4) -> 6 4 7
Maude> (no .)
```

With this information the debugger selects the subtree and, since it is a leaf, it concludes that the node is associated with the buggy equation.

```
The buggy node is:
insert-list(6 7, 4) -> 6 4 7
With the associated equation: il3
```

Fig. 6. Abbreviated proof trees after the second and third questions

That is, the debugger points to the equation i13 as buggy. If we examine it

```
ceq [i13] : insert-list(E OL, E') = E E' OL
  if E' <= E /\ E OL : SortedNatList .
```

we can see that the order of E and E' in the righthand side is wrong and we can proceed to fix it appropriately.

We can check the fixed function by sorting again the list 3 4 7 6. We obtain now the sorted list 3 4 6 7. Then, we have solved one problem, but if we reduce the unsorted list 6 3 4 7

```
Maude> (red 6 3 4 7 .)
result SortedNatList : 6 3 4 7
```

we can see that Maude continues assigning to it an incorrect sort.

We can check this inference by using the command

```
Maude> (debug 6 3 4 7 : SortedNatList .)
```

The first question the debugger prompts is

```
Is this membership (associated with the membership olist) correct?
3 4 7 : SortedNatList
Maude> (yes .)
```

Of course, this list is sorted. The following question is

```
Is this transition (associated with the equation hd2) correct?
head(3 4 7) -> 7
Maude> (no .)
```

But the head of a list should be the first element, not the last one, so we answer no. With only these two questions the debugger prints

```
The buggy node is:
head(3 4 7) -> 7
With the associated equation: hd2
```

If we check the equation hd2, we can see that we take the element from the wrong side.

```
eq [hd2] : head(L E) = E .
```

To debug this module we have used the default divide and query strategy. Let us check it now with the top-down strategy. We debug again the inference

```
insertion-sort(3 4 7 6) -> 6 3 4 7
```

in the initial module with the two errors. The first question asked in this case is

```
Is any of these nodes wrong?
Node 0 : insertion-sort(4 7 6) -> 6 4 7
Node 1 : insert-list(6 4 7, 3) -> 6 3 4 7
Maude> (node 0 .)
```

Both nodes are wrong, so we select, for example, the first one. The next question is

```
Is any of these nodes wrong?
Node 0 : insertion-sort(7 6) -> 6 7
Node 1 : insert-list(6 7, 4) -> 6 4 7
Maude> (node 1 .)
```

This time, only the second node is wrong, so we select it. The debugger prints now

```
Is any of these nodes wrong?
Node 0 : 6 7 : SortedNatList
Maude> (all valid .)
```

There is only one node, and it is correct, so we give this information to the debugger, and it detects the wrong equation.

```
The buggy node is:
insert-list(6 7, 4) -> 6 4 7
With the associated equation: il3
```

But remember that we chose a node randomly when the debugger showed two wrong nodes. What happens if we select the other one? The following question is printed.

```
Is any of these nodes wrong?
Node 0 : 6 4 7 : SortedNatList
Maude> (node 0 .)
```

Since this single node is wrong, we choose it and the debugger asks

```
Is any of these nodes wrong?
Node 0 : head(4 7) -> 7
Node 1 : 4 7 : SortedNatList
Maude> (node 0 .)
```

The first node is the only one erroneous, so we select it. With this information, the debugger prints

```
The buggy node is:
head(4 7) -> 7
With the associated equation: hd2
```

That is, the second path finds the other bug. In general, this strategy finds different bugs if the user selects different wrong nodes.

In order to prune the debugging tree, we can define a module specifying the sorting function `sort` in a correct, but inefficient, way. This module will define the functions `insertion-sort` and `insert-list` by means of `sort`.

```
(fmod CORRECT-SORTING is
  pr NAT .
  sorts NatList SortedNatList .
  subsorts Nat < SortedNatList < NatList .
  vars E E' : Nat .  vars L L' : NatList .  var OL : SortedNatList .
  op __ : NatList NatList -> NatList [ctor assoc] .
  cmb E E' : SortedNatList if E <= E' .
  cmb E E' L : SortedNatList if E <= E' /\ E' L : SortedNatList .
```

The `sort` function is defined by switching unsorted adjacent elements in all the possible cases for lists.

```
  op sort : NatList -> SortedNatList .
  ceq sort(L E E' L') = sort(L E' E L') if E' < E .
  ceq sort(L E E') = sort(L E' E) if E' < E .
  ceq sort(E E' L) = sort(E' E L) if E' < E .
  ceq sort(E E') = E' E if E' < E .
  eq sort(L) = L [owise] .
```

We now use `sort` to implement `insertion-sort` and `insert-list`.

```
  op insertion-sort : NatList -> SortedNatList .
  op insert-list : SortedNatList Nat -> SortedNatList .
  eq insertion-sort(L) = sort(L) .
  eq insert-list(OL, E) = sort(E OL) .
endfm)
```

We can use this module to prune the debugging trees built by the `debug` commands if we previously introduce the command

```
Maude> (correct module CORRECT-SORTING .)
```

Now, we try to debug the initial module (with two errors) again. In this example, all the questions about correct inferences have been pruned, so all the answers are negative. In general, the correct module does not have to be complete, so some correct inferences could be presented to the user.

```
Maude> (debug in SORTED-NAT-LIST : insertion-sort(3 4 7 6) -> 6 3 4 7 .)

Is this transition (associated with the equation il3) correct?
insert-list(6 4 7, 3) -> 6 3 4 7
Maude> (no .)

Is this membership (associated with the membership olist) correct?
6 4 7  : SortedNatList
Maude> (no .)

Is this transition (associated with the equation hd2) correct?
head(4 7) -> 7
Maude> (no .)
```

```
The buggy node is:
head(4 7) -> 7
With the associated equation: hd2
```

The correct module also improves the debugging of the membership. With only one question we obtain the buggy equation.

```
Maude> (debug in SORTED-NAT-LIST : 6 3 4 7 : SortedNatList .)

Is this transition (associated with the equation hd2) correct?
head(3 4 7) -> 7
Maude> (no .)

The buggy node is:
head(3 4 7) -> 7
With the associated equation: hd2
```

4.4   Implementation

Exploiting the fact that rewriting logic is reflective [6,8], a key distinguishing feature of Maude is its systematic and efficient use of reflection through its predefined META-LEVEL module [7, Chap. 14], a feature that makes Maude remarkably extensible and that allows many advanced metaprogramming and metalanguage applications. This powerful feature allows access to metalevel entities such as specifications or computations as data. Therefore, we are able to generate and navigate the debugging tree of a Maude computation using operations in Maude itself. In addition, the Maude system provides another module, LOOP-MODE [7, Chap. 17], which can be used to specify input/output interactions with the user. Thus, our declarative debugger for Maude functional modules, including its user interactions, is implemented in Maude itself, as an extension of Full Maude [7, Chap. 18]. As far as we know, this is the first declarative debugger implemented using such reflective techniques.

The implementation takes care of the two stages of generating and navigating the debugging tree. Since navigation is done by asking questions to the user, this stage has to handle the navigation strategy together with the input/output interaction with the user.

To build the debugging tree we use the facts that the equations defined in Maude functional modules are both *terminating* and *confluent*. Instead of creating the complete proof tree and then abbreviating it, we build the abbreviated proof tree directly.

The main function in the implementation of the debugging tree generation is called createTree. It receives the module where a wrong inference took place, a correct module (or the constant noModule when no such module is provided) to prune the tree, the term initially reduced, the (erroneous) result obtained, and the set of suspicious statement labels. It keeps the initial inference as the root of the tree and uses an auxiliary function createForest that, in addition to the arguments received by createTree, receives the module "cleaned" of suspicious statements, and generates the abbreviated forest corresponding to the reduction between the two terms passed as arguments. This transformed module is used to improve the efficiency of the tree construction, because we can use it to check if a term reaches its final form only using trusted equations, thus avoiding to build a tree that will be finally empty.

Regarding the navigation of the debugging tree, we have implemented two strategies. In the top-down strategy the selection of the next node of the debugging tree is done by the user, thus we do not need any function to compute it. The divide and query strategy used to traverse the debugging tree selects each time the node whose subtree's size is the closest one to half the size of the whole tree, keeping only this subtree if its root is incorrect, and deleting the whole subtree otherwise. The function `searchBestNode` calculates this best node by searching for a subtree minimizing an appropriate function.

The technical report [4] provides a full explanation of this implementation, including the user interaction.

5   Conclusions and Future Work

In this paper we have developed the foundations of declarative debugging of executable *MEL* specifications, and we have applied them to implement a debugger for Maude functional modules. As far as we know, this is the first debugger implemented in the same language it debugs. This has been made possible by the reflective features of Maude. In our opinion, this debugger provides a complement to existing debugging techniques for Maude, such as tracing and term coloring. An important contribution of our debugger is the help provided by the tool in locating the buggy statements, assuming the user answers correctly the corresponding questions. The debugger keeps track of the questions already answered, in order to avoid asking the same question twice.

We want to improve the interaction with the user by providing a complementary graphical interface that allows the user to navigate the tree with more freedom. We are also studying how to handle the `strat` operator attribute, that allows the specifier to define an evaluation strategy. This can be used to represent some kind of laziness.

We plan to extend our framework by studying how to debug *system modules*, which correspond to rewriting logic specifications and have rules in addition to memberships and equations. These rules can be non-terminating and non-confluent, and thus behave very differently from the statements in the specifications we handle here. In this context, we also plan to study how to debug *missing answers* [14] in addition to the wrong answers we have treated thus far.

References

1. Alpuente, M., Comini, M., Escobar, S., Falaschi, M., Lucas, S.: Abstract diagnosis of functional programs. In: Leuschel, M. (ed.) LOPSTR 2002. LNCS, vol. 2664, pp. 1–16. Springer, Heidelberg (2003)
2. Bouhoula, A., Jouannaud, J.-P., Meseguer, J.: Specification and proof in membership equational logic. Theoretical Computer Science 236, 35–132 (2000)
3. Caballero, R.: A declarative debugger of incorrect answers for constraint functional-logic programs. In: WCFLP 2005: Proceedings of the 2005 ACM SIGPLAN Workshop on Curry and Functional Logic Programming, pp. 8–13. ACM Press, New York (2005)

4. Caballero, R., Martí-Oliet, N., Riesco, A., Verdejo, A.: Declarative debugging of Maude functional modules. Technical Report 4/07, Dpto.Sistemas Informáticos y Computación, Universidad Complutense de Madrid (2007), http://maude.sip.ucm.es/debugging
5. Caballero, R., Rodríguez-Artalejo, M.: DDT: A declarative debugging tool for functional-logic languages. In: Kameyama, Y., Stuckey, P.J. (eds.) FLOPS 2004. LNCS, vol. 2998, pp. 70–84. Springer, Heidelberg (2004)
6. Clavel, M.: Reflection in Rewriting Logic: Metalogical Foundations and Metaprogramming Applications. CSLI Publications, Stanford University (2000)
7. Clavel, M., Durán, F., Eker, S., Lincoln, P., Martí-Oliet, N., Meseguer, J., Talcott, C.: All About Maude - A High-Performance Logical Framework. LNCS, vol. 4350. Springer, Heidelberg (2007)
8. Clavel, M., Meseguer, J.: Reflection in conditional rewriting logic. Theoretical Computer Science 285(2), 245–288 (2002)
9. Comini, M., Levi, G., Meo, M.C., Vitiello, G.: Abstract diagnosis. Journal of Logic Programming 39(1-3), 43–93 (1999)
10. Lloyd, J.W.: Declarative error diagnosis. New Generation Computing 5(2), 133–154 (1987)
11. MacLarty, I.: Practical Declarative Debugging of Mercury Programs. PhD thesis, University of Melbourne (2005)
12. Meseguer, J.: Conditional rewriting logic as a unified model of concurrency. Theoretical Computer Science 96(1), 73–155 (1992)
13. Meseguer, J.: Membership algebra as a logical framework for equational specification. In: Parisi-Presicce, F. (ed.) WADT 1997. LNCS, vol. 1376, pp. 18–61. Springer, Heidelberg (1998)
14. Naish, L.: Declarative diagnosis of missing answers. New Generation Computing 10(3), 255–286 (1992)
15. Naish, L.: A declarative debugging scheme. Journal of Functional and Logic Programming 1997(3), 1–27 (1997)
16. Nilsson, H.: How to look busy while being as lazy as ever: the implementation of a lazy functional debugger. Journal of Functional Programming 11(6), 629–671 (2001)
17. Nilsson, H., Fritzson, P.: Algorithmic debugging of lazy functional languages. Journal of Functional Programming 4(3), 337–370 (1994)
18. Pope, B.: Declarative debugging with Buddha. In: Vene, V., Uustalu, T. (eds.) AFP 2004. LNCS, vol. 3622, pp. 273–308. Springer, Heidelberg (2005)
19. Shapiro, E.Y.: Algorithmic Program Debugging. ACM Distinguished Dissertation. MIT Press, Cambridge (1983)
20. Silva, J.: A comparative study of algorithmic debugging strategies. In: Puebla, G. (ed.) LOPSTR 2006. LNCS, vol. 4407, pp. 143–159. Springer, Heidelberg (2007)
21. Takahashi, N., Ono, S.: DDS: A declarative debugging system for functional programs. Systems and Computers in Japan 21(11), 21–32 (1990)
22. Tessier, A., Ferrand, G.: Declarative diagnosis in the CLP scheme. In: Analysis and Visualization Tools for Constraint Programming, Constraint Debugging (DiSCiPl project), pp. 151–174. Springer, Heidelberg (2000)
23. Wadler, P.: Why no one uses functional languages. SIGPLAN Not. 33(8), 23–27 (1998)

SPREADSPACES:
Mathematically-Intelligent
Graphical Spreadsheets

Nachum Dershowitz[1,*] and Claude Kirchner[2]

[1] School of Computer Science, Tel Aviv University, Ramat Aviv 69978, Israel
`Nachum.Dershowitz@cs.tau.ac.il`
[2] INRIA Bordeaux – Sud-Ouest, Talence, France
`Claude.Kirchner@inria.fr`

Abstract. Starting from existing spreadsheet software, like Lotus 1-2-3®, Excel®, or Spreadsheet 2000®, we propose a sequence of enhancements to fully integrate constraint-based reasoning, culminating in a system for reactive, graphical, mathematical constructions. This is driven by our view of constraints as the essence of (spreadsheet) computation, rather than as an add-on tool for expert users. We call this extended computational metaphor, *spreadspaces*.

We believe that research towards more general and realistic constraint solving frameworks has to go on in parallel with the effort to make fewer and fewer requests to the user. In other words, users should be asked only for as much as they want to give the system. This amount of information (decided by users but with a minimum set by the system below which most precision is lost) is then used by the system to construct the whole constraint problem.

—Ugo Montanari and Francesca Rossi [18]

1 Overview

Our ultimate goal in this work is the design of a graphical environment for spreadsheet-like computations, including solving and optimization, wherein the graphical interface serves as an input medium, in addition to its traditional output rôle. Changing a displayed value, be it graphical or textual, results immediately in the appropriate changes to values it depends on. This integrated system, with its transparent graphical mode of interaction, will dramatically extend the capabilities of existing commercial products, providing sophisticated mathematical intelligence for the computationally naïve.

Spreadspaces do not have the look or feel of spreadsheets [20], or even of graphical spreadsheets, but rather that of a graphical user interface. At the

---

* This author's research was performed while on leave.

P. Degano et al. (Eds.): Montanari Festschrift, LNCS 5065, pp. 194–208, 2008.

design level, the system serves as a graphical design environment with definable and extensible graphical objects. Thus, educators, for example, can design spreadspaces within which schoolchildren can solve problems and investigate variants.

Target users are everyday users of personal computers. Examples of spreadspace applications include:

- Exploring mathematical relations (see the baguette example below).
- Simulating physical devices (such as a pendulum).
- High school problem solving.
- Contingent ("what-if") financial calculations (for example, home loan planning, tax computations).
- Logical and mathematical puzzles (for example, map coloring, crypto-arithmetic).

Today's spreadsheets provide ad-hoc constraint solving [8,16,7], mainly via linear programming, and incorporate sophisticated graphical output. But these features are patched on top of the basic spreadsheet, making the interface difficult and limiting its general use.

The following sections lead us from minor cosmetic enhancements of current spreadsheets, through sophisticated tools for the incorporation of mathematical intelligence, to user-friendly graphical spreadspaces.

2   Cosmetic Constraints

The goal was to give the user a conceptual model which was unsurprising – it was called the principle of least surprise. We were illusionists synthesizing an experience. Our model was the spreadsheet – a simple paper grid that would be laid out on a table. The paper grid provided an organizing metaphor for a working with series of numbers. While the spreadsheet is organized we also had the back-of-envelope model which treated any surface as a scratch pad for working out ideas.

—Bob Frankston (coinventor of Visicalc)

We begin with some simple "cosmetic" improvements to modern-day spreadsheets. The central notion is that of *constraints*, which are boolean (true/false) formulæ, involving comparisons and conditionals, that are required to evaluate to `true`. Constraints extend ordinary formulæ by allowing the user to specify more general relations between variables. Guaranteeing the truth of a constraint forces the variables it involves to take on appropriate values. Values that make a constraint true are called a *solution*. Typical constraints involve inequalities (such as `Years` < 80), type information (`Years: Integer`), and logical combinations (`(Years=62 and Gender=Female) or (Years=65 and Gender=Male)`). To satisfy a constraint, the variables (like `Years`) appearing in it are set by the system to appropriate values by computation and solving mechanisms.

To incorporate constraints into the spreadsheet paradigm, we do the following:

1. Add a new kind of cell for constraints. One should be able to switch the type of a cell from boolean to constraint and back again easily. This facilitates debugging a set of constraints.
2. Only cells with an empty value are considered to be variables. Special flags can be used to indicate preference for maximal or minimal possible solutions. Cells that contain user-supplied values, like X3 having the value 4, would be interpreted as an implicit constraint, viz. X3 = 4. This is in contrast with current systems which allow the solver to modify cells containing user-supplied values.
3. As in today's spreadsheets, symbolic names (like `Current_Price`) can be given to cells instead of their Cartesian name (e.g. `G13`), and these could be used in expressing constraints.
4. It would be nice to allow cells to contain interval values (like 0..100) to express ranges of possible inputs or outputs.

As a simple example, consider the problem of graphing the price of a (nice, fresh and crusty) baguette under variable inflation rates.

Using today's spreadsheets, one can answer the question, "What is the lowest inflation rate such that the price of a baguette will increase tenfold in the span of a person's lifetime?", by using the following spreadsheet, where the constraints are specified and solved via the integrated solver:

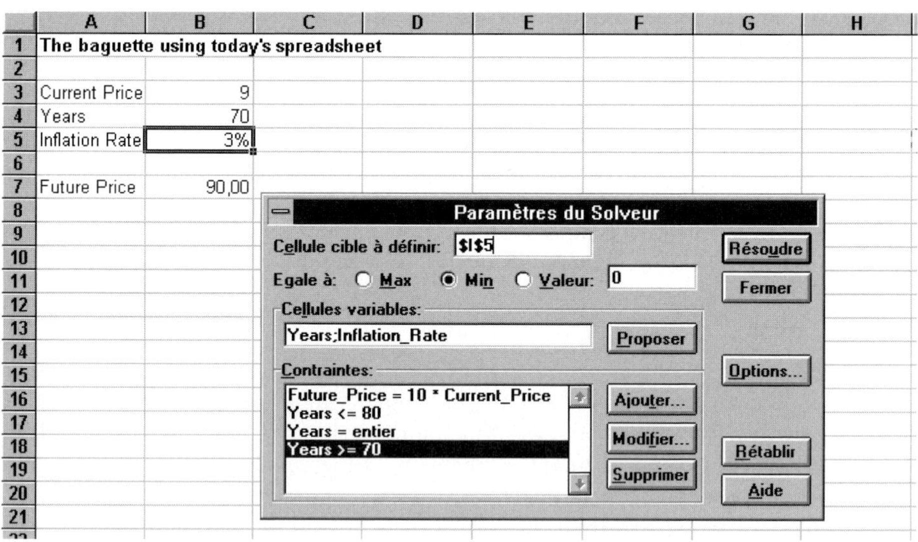

With the same example, using a constraint spreadsheet, a user will input the following spreadsheet containing two constraints: a type constraint in C2 and an inequality constraint in C3. The variable to be maximized is B2:

	A	B	C	D
1	Current_Price	9		
2	Years		:70$\leq$B2$\leq$80	:B2 integer
3	Inflation_Rate	!min		
4				
5	Future_Price	B1*(1+B3)^B2	:B5=10*B1	

The system is designed to automatically solve the constraint and display the following values:

	A	B	C	D
1	Current_Price	9		
2	Years	70	*solved*	*solved*
3	Inflation_Rate	3.34%		
4				
5	Future_Price	90	*solved*	

The value 3.34% is assigned to the cell B3 (better known as Inflation_Rate) since it is the smaller number that allows the constraints given in the spreadsheet to be satisfied. Constraints are written in cells like C2 and D2 to specify that Years should be an integer in between 70 and 80. In its solved form the constraint spreadsheet displays the result of the computation in cells that contain formulæ (like B5), and either *solved*, when the constraint is satisfied, or false, if no solution has been found, in cells that contain constraints (like C5) .

Formally, a *spreadspace* is a finite set of constraints (which are finite or infinite relations) on values of cells. At any given moment, each cell is either "protected" (user-supplied input or derived therefrom) or "variable". Relations defined in constraints are not directional: whether a constraint X3*3=2*Y3 would cause X3 to be calculated from a known value of Y3 or vice-versa would depend on the context. A variable can be determined, not only by fixing related rigid values and calculating functional dependencies (as in backsolving a value for X3 from the equation Y3=3*X3 and a fixed value for Y3), but also by solving several inequalities (like Y3=X3*X3, X3>0, and 10<Y3<20, for *integer* X3 and Y3), depending on the sophistication of the available solver routines.

Thus, processing spreadsheets with constraint cells involves the following steps:

1. Extract the set of constraints from the spreadsheet.
2. Choose which variable cell(s) to solve for.
3. Attempt to solve the constraints using constraint solvers.
4. If the set of solutions is non-empty, determine which solution should be fed back to the appropriate cells.

Today's systems have capabilities to "backsolve" single constraints, optimize by linear programming, and solve some non-linear equations using Newton's method and the like. Only rudimentary solving capabilities for integers are available. As we outline in the next section, more powerful tools are in fact available.

3   Constrained Spreadsheets

> *What was important were the features we'd left. We'd already discussed wall-sized interactive displays with live graphics but the systems weren't up to it. More important, the grid provided the simplifying structure that made it a spreadsheet as opposed to a more general surface.*
>
> —*Bob Frankston (letter to D. J. Power, 15 April 1999)*

Starting from the cosmetically improved spreadsheet of the prior section, we aim to add more sophisticated solvers. The driving engine is a cooperating set of numerical and symbolic constraint-solving modules. They transform the extracted set of constraints into a "solved form".

The system should incorporate as much mathematical ability as possible. These could include facilities for:

- interval arithmetic,
- finite domains,
- finite sets,
- propositional calculus,
- algebraic identities (associativity, commutativity, etc.),
- polynomials.

Such capabilities exist in computer algebra systems designed to solve elaborated constraints like: Axiom, Maple, Mathematica®, MuPAD®, Numerica. Finite domain solvers can be solved using ILOG® Solver ([11]) or GNU Prolog ([21]) or specialized solvers written in general-purpose or rule based languages like ELAN ([5,2]). In some cases, searching for solutions might be necessary. Several works stemming from the declarative programming community extend classical spreadsheets with constraints, including, among others, instance [15,10,23,3,12].

To achieve the kind of capabilities we envision, a blackboard architecture, with component solvers contributing partial solutions to the listed constraints, is indicated.

4   Constrained Graphics

Modern spreadsheets provide tools for generating graphical representation of spreadsheet data. The resultant graphs can be sized, placed, and annotated, as desired. For instance, TK!solver [13,14], Spreadsheet 2000® [24] and iWork® Numbers [4] for the Mac provide nice interfaces that are more graphical and less tabular than standard spreadsheets. The relationship between graphics and constraints has a very long history to which Ugo Montanari has contributed greatly [17]. The relationship with spreadsheets has also been developed by numerous authors, including, for example, [6,9].

The parameters of the graphics (position, color, spacing, etc.) should be linkable to the spreadsheet itself. Furthermore, constraint solving could be employed to determine their value.

Graphical objects should include dials, meters, switches, etc. The baguette spreadsheet could be portrayed in the following manner:

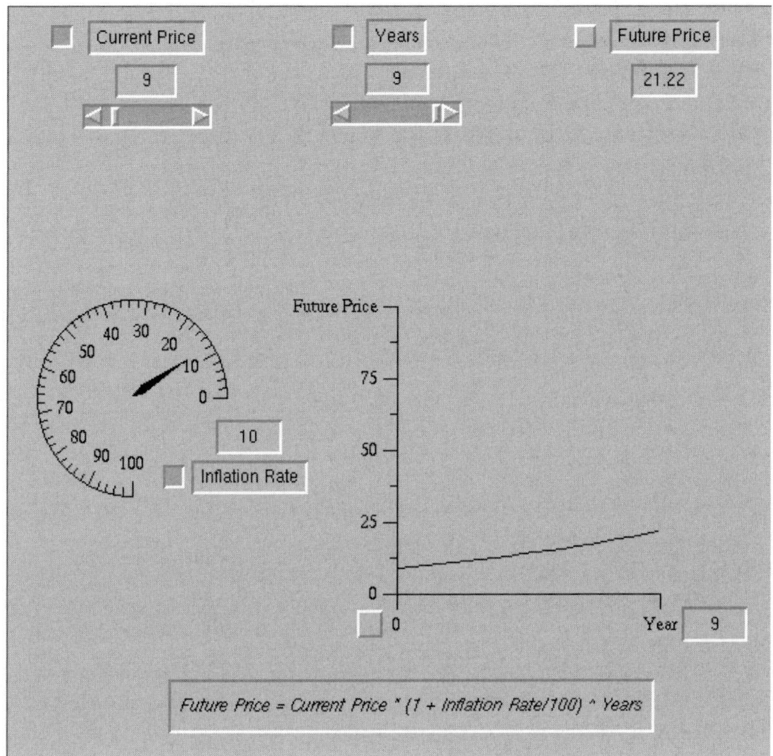

Importantly, it is not hard to express graphical objects themselves as sets of relatively simple constraints. Thus constraint-solving could be used to calculate the graphical representations. This adds a lot more expressivity.

5   Active Graphics

Once we have graphics expressed as first-class constraints, the only difference being that results are displayed on a screen, it is possible to allow the user to directly manipulate the graphical objects, causing constraints to be solved and other displays to change accordingly. At this stage, the system would no longer bear any external resemblance to spreadsheets.

Virtually all interaction becomes graphical. Graphical output would not be an add-on that sits atop arithmetic computations, as in today's systems, but would be fully integrated with the calculations and constraint solving. By representing graphical objects and their properties (value, size, color, etc.) in this way, changes the user makes to the graphical objects will immediately result in new values that drive other parts of the spreadspace.

Returning to the baguette example, the same spreadspace as above can be used to gracefully solve all the following queries:

1. *What will the price of a baguette be in 9 years if the current price is 9 pesos, and the inflation rate is 10%?*
   To indicate what the input values are, the user pushes the buttons alongside `Current_Price`, `Years` and `Inflation_Rate`. Then the user sets the current price to 9 and the rate to 10. The answer is graphed and also displayed in the `Future_Price` cell as shown in the figure just above.
2. *What happens if the inflation rate rises to 20% (35%)?*
   The user just uses the mouse to move the dial to the appropriate values.

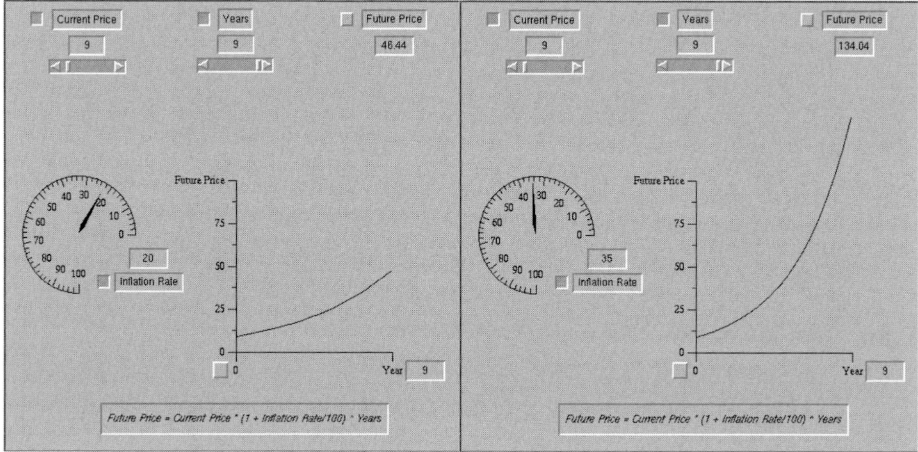

3. *What is the lowest inflation rate such that the price of a baguette will increase tenfold in the span of a person's lifetime?*
   This time, the user pushes `Current_Price`, `Future_Price` and `Inflation_Rate`, and sets the final price to 10 times as much (90). Lastly he/she turns the inflation rate dial until a satisfactory value appears in the `Years` cell: graphics become active.
4. *What inflation rate causes the price to increase tenfold in only 4 years?*
   Starting with the previous state, the user reverses the statuses of `Years` and `Inflation_Rate` by toggling their buttons, and then sets `Years` to 4. The inflation rate is displayed and the graph is updated.

This spreadspace is simply constructed by choosing the graphical elements from menus, placing and sizing them with the mouse, changing some of the default values to better ones. Entering the formula either textually or in a menu-driven manner relates the various entities mathematically. Dragging cell names or values can streamline the construction of the formula.

To give a deeper intuition of the way it works, let us write a script for some of the actions needed to create the baguette spreadspace:

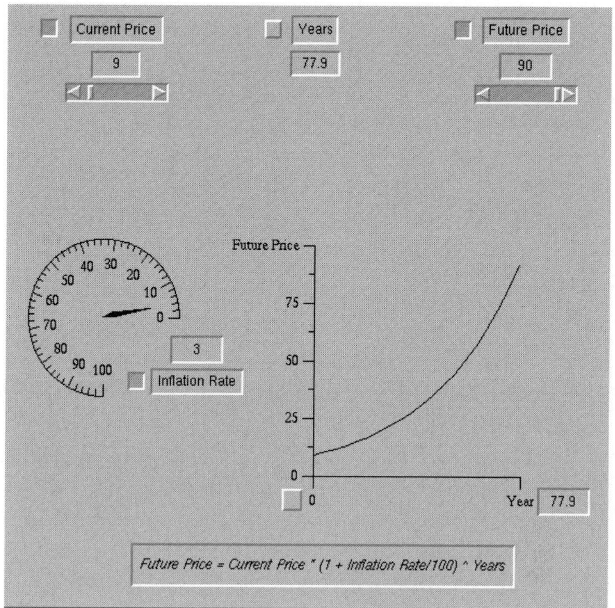

1. I pull down the **device** menu and choose a **dial**. I place it where I want on the screen and stretch it to the desired size.

   An instance of the object dial is created and therefore the following constraints are added to the currently empty constraint store. We assume the dial to have its center at coordinates (a, b) and to be of radius r (all specified indeed graphically by the action of the user who drags the dial on the working space):

   $\mathtt{dial.value} = 0$ ⟶ the default value presented by the dial

   $\mathtt{dial.center} = (a, b)$

   $\mathtt{dial.radius} = r$

   $\mathtt{dial.shape} = ((x - a)^2 + (y - b)^2 = r^2)$

   $\mathtt{dial.min} = 0$

   $\mathtt{dial.max} = \mathtt{dial.min} + 100$

   $\mathtt{dial.marking}[\mathtt{dial.min} : \mathtt{dial.max}].\mathtt{color} = \mathtt{black}$

   $\mathtt{dial.foreground} = \mathtt{black}$

   Finally, since by default the value of this object can be either set by the user interactively (in which case **dial.in** is **true**, value by default) or set by the constraint solver (in which case **dial.out** is **true**), the following constraints are added:

   $\mathtt{dial.readWrite} = (\mathtt{dial.in} \neq \mathtt{dial.out})$

   $\mathtt{dial.in} = \mathtt{true}$

2. I change the *high value* from the default 100 to 200, and all the intermediate values change to match.

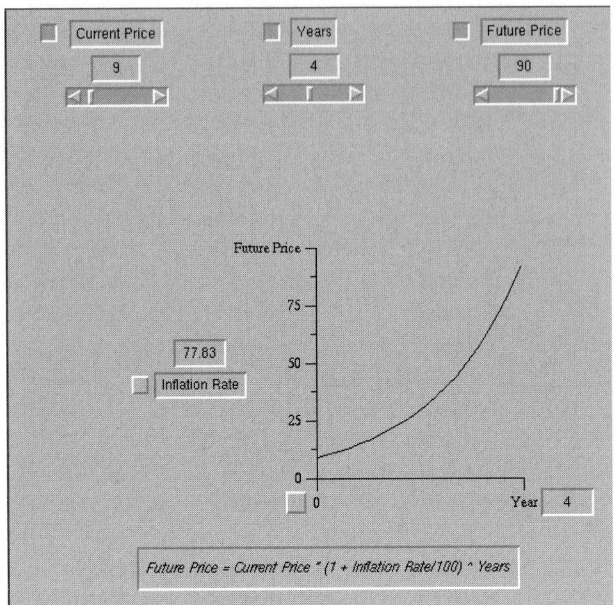

The constraint store now contains:
dial.value $= 0$
dial.center $= (a, b)$
dial.radius $= r$
dial.shape $= ((x - a)^2 + (y - b)^2 = r^2)$
dial.min $= 0$
dial.max $= 200$
dial.marking[dial.min : dial.max].color $=$ black
dial.foreground $=$ black

3. I change the color of the markings in the range [100 : 200] to red, by catching
   all of them, holding the mouse button down to get the list of attributes, and
   then choosing the foreground color item, which gives a palette from which
   I choose a dark blue. I leave the default low value of 0 and default interval
   markings.

   So now we have the following:
   dial.value $= 0$
   dial.center $= (a, b)$
   dial.radius $= r$
   dial.shape $= ((x - a)^2 + (y - b)^2 = r^2)$
   dial.min $= 0$
   dial.max $= 200$
   dial.marking[dial.min : 100[.color $=$ black
   dial.marking[100 : dial.max].color $=$ red
   dial.foreground $=$ darkBlue

4. Then I choose a simple rectangular display from the menu, placing it near
   the dial.
   This has the effect to add to the previous constraint store the following:

   $\texttt{rectDisplay1.position} = (c, d)$ determined by the user action
   $\texttt{rectDisplay1.height} = 5$ default value
   $\texttt{rectDisplay1.length} = \texttt{rectDisplay1.height} * 5$ default value
   $\texttt{rectDisplay1.type} = \texttt{real}$
   $\texttt{rectDisplay1.value} = 0$
   $\texttt{rectDisplay1.min} = 0$
   $\texttt{rectDisplay1.max} = \texttt{rectDisplay1.min} + 100$
   $\texttt{rectDisplay1.foreground} = \texttt{black}$

   etc.

As one can see, an explicit set of constraints is built using a graphical interactive interface. It represents exactly all the behavioral knowledge the user wants to put into his or her model.

6   Examples

Here are a few simple examples highlighting some of the original features of spreadspaces.

6.1   Color-Changing Rectangle

Context: The user is resizing a rectangle by dragging one of the corners or sides of the rectangle with her mouse.

Constraint: The designer of the current spreadspace has written the following constraints, where P identifies the perimeter of the rectangle.

$$P > 3 \;\Rightarrow\; \texttt{rectangle.backGroundColor} = \texttt{red}$$
$$P \leq 3 \;\Rightarrow\; \texttt{rectangle.backGroundColor} = \texttt{blue}$$

Behavior: When the user is in-playing the size of the rectangle with her mouse, the color of the rectangle is displayed in red when the perimeter of the rectangle is larger than 3 in the current length unit. Otherwise, it is shown in blue.

6.2   Standard Spreadsheet

Context: The user uses a spreadsheet to understand the relationship between the total amount of money "available", the amount "allocated" (earmarked) and the amount still available. With a standard spreadsheet, depending on the quantity one wants to compute, one has to make three different computations expressed in three different spreadsheets as illustrated in Fig. 1.

What is the total	
Already allocated	34000
Still available	16000
Total available	50000

What is available	
Already allocated	34000
Still available	16000
Total available	50000

What is allocated	
Already allocated	34000
Still available	16000
Total available	50000

Fig. 1. Spreadsheet examples

Constraint: The designer of that spreadspace simply writes the following constraint:

$$\texttt{Already\_allocated} + \texttt{Still\_available} = \texttt{Total\_available}$$

Behavior: As soon as the value of 2 of the above variables are known, the third is fulfilled automatically.

6.3   Red First

Context: The user sees three circles and can in-play the color of them using a menu poping up when (s)he clicks right on one of the circles. At the beginning the circles have the same background color than the overall background (which is assumed not to be red!).

Constraint: The designer of that spreadspace has written the following constraint:

$$(\texttt{C1.backGroundColor} = \textbf{red})$$
$$\oplus \, (\texttt{C2.backGroundColor} = \textbf{red})$$
$$\oplus \, (\texttt{C3.backGroundColor} = \textbf{red})$$
$$= 1$$

where $\oplus$ denotes exclusive or.

Behavior: The only valid in-play of the user will be to enter the first one to be red, and the other not to be red. This will therefore force the user to behaves accordingly.

7 S_p^2 System Organization

The diagram in Fig. 2 exemplifies how a user of the S_p^2 spreadspace system we are describing fills in values for the fields of the cell just positioned on the spreadspace.

1. The user clicks on the `Name` field, types "Already allocated", and hits the return key.
2. The INTERACTION MANAGER
   (a) identifies `C1.Name` as the field modified, and

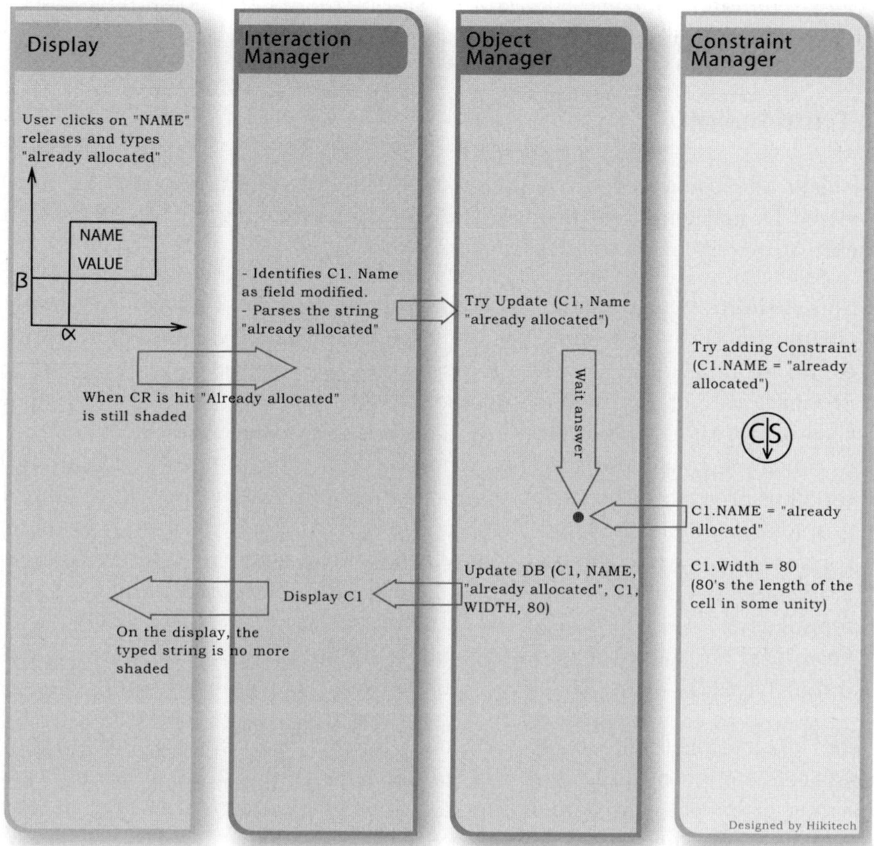

Fig. 2. Information flow from user to constraints

(b) parses the string "Already allocated".

3. The OBJECT MANAGER tries now to update the cell and – for this purpose – initiates the solver and waits for an answer.

4. The CONSTRAINT MANAGER tries to add the new constraint `C1.Name =` "Already allocated".

 The solver detects no contradiction, but has computed (by constraint propagation, based on the name and default font) that the width of the cell should be 80, and thus `C1.Width = 80` is added as a constraint, and that value is also passed to the OBJECT MANAGER.

5. The OBJECT MANAGER updates the `Name` and `Width` fields of cell `C1` in the database.

6. The INTERACTION MANAGER updates the display with the new values. No values are shaded at this point on the display.

 In this context, one can see the importance of the constraint satisfier and/or solver. Of course, all the work on constraint solving, combination, propagation and clever handling of constraint stores shall be reused and possibly adapted to handle huge numbers of heterogeneous constraints.

8   Conclusion

It is clear that people use computers to do many computations that can be expressed as mathematical problem solving. However, many of these tasks are difficult or inconvenient with current software.

- Spreadsheets may provide constraint-solving, but only as an afterthought; though relatively powerful, the interface is not intuitive, and the computational meaning of constraints is obscured. Most aspects of the layout and graphics are not integrated into the spreadsheet, even in graphical spreadsheets, and the graphics certainly have no connection with the solver.
- Commercial constraint solvers are designed for programmers, and cannot be used by spreadsheet users and their kin. Symbolic systems may have powerful graphing capabilities, but solving requires mathematical and programming sophistication, since solving usually necessitates heavy user-interaction. Existing solvers cannot cooperate, and future improvements cannot be added modularly.
- Graphical interface design systems allow one to construct the kind of objects in our baguette example, but all calculations must be hard-wired.

The beauty and value of spreadspaces lie in their seamless integration of spreadsheet computations, constraint solving, and optimization, in an active and appealing graphical environment. As such, it contributes to the large research interest in spreadsheets, both with regard to their deductive extensions [1,22] and from the risk point of view [19].

Acknowledgements

Many thanks to Carlos Castro who, a long time ago, coded the baguette spreadspace into the ILOG view and solver environments and to Franz Kirchner for his graphical design of the information flow presented in Fig. 2. Thanks also to the ILOG company for letting us freely use their software.

References

1. Burnett, M., Yang, S., Summet, J.: A scalable method for deductive generalization in the spreadsheet paradigm. Interactions 9(5), 9–11 (2002)
2. Castro, C.: Building Constraint Satisfaction Problem Solvers Using Rewrite Rules and Strategies. Fundamenta Informaticae 34, 263–293 (1998)
3. Chitnis, S., Yennamani, M., Gupta, G.: Exsched: Solving constraint satisfaction problems with the spreadsheet paradigm. The Computing Research Repository (CoRR), abs/cs/0701109 (2007)

4. Apple Corp. Numbers (2008), `http://www.apple.com/iwork/numbers`
5. ELAN, `http://elan.loria.fr`
6. Gupta, G., Akhter, S.F.: Knowledgesheet: A graphical spreadsheet interface for interactively developing a class of constraint programs. In: Pontelli, E., Santos Costa, V. (eds.) PADL 2000. LNCS, vol. 1753, pp. 308–323. Springer, Heidelberg (2000)
7. Horn, B.: Constraint patterns as a basis for object-oriented programming. In: Proceedings of the SIGPLAN International Conference on Object-Oriented Programming, Systems, Languages, and Applications (OOPSLA), pp. 218–233. ACM Press, New York (1992)
8. Hower, W., Graf, W.: A bibliographical survey of constraint-based approaches to CAD, graphics, layout, visualization, and related topics. Knowl.-Based Syst. 9(7), 449–464 (1996)
9. Chi, E.H.H., Riedl, J., Barry, P., Konstan, J.: Principles for information visualization spreadsheets. IEEE Comput. Graph. Appl. 18(4), 30–38 (1998)
10. Hyvönen, E., Pascale, S.D.: A new basis for spreadsheet computing: Interval solver for Microsoft Excel. In: Proceedings of the Sixteenth National Conference on Artificial intelligence and the Eleventh Innovative Applications of Artificial Intelligence Conference (AAAI 1999/ IAAI 1999), Menlo Park, CA, pp. 799–806. American Association for Artificial Intelligence (1999)
11. ILOG, `http://www.ilog.fr`
12. Jayaraman, B., Tambay, P.: Modeling engineering structures with constrained objects. In: Krishnamurthi, S., Ramakrishnan, C.R. (eds.) PADL 2002. LNCS, vol. 2257, pp. 28–46. Springer, Heidelberg (2002)
13. Konopasek, M., Jayaraman, S.: The TK! Solver Book: A Guide to Problem-Solving in Science, Engineering, Business, and Education. McGraw-Hill, Osborne (1984)
14. Konopasek, M., Jayaraman, S.: Constraint and declarative languages for engineering applications: The TK!Solver contribution. Proceedings of the IEEE 73(12), 1791–1806 (1985)
15. Kunstmann, T., Frisch, M., Muller, R.: A declarative programming environment based on constraints. In: Proceedings of the 11th International IEEE Symposium on Visual Languages (VL 1995), Washington DC, September 1995, pp. 120–121. IEEE Computer Society Press, Los Alamitos (1995)
16. Kwaiter, G., Gaildrat, V., Caubet, R.: Modelling with constraints: A bibliographical survey. In: Proceedings of the Second International Conference on Information Visualisation (IV 1998), pp. 211–220. IEEE Computer Society Press, Los Alamitos (1998)
17. Montanari, U.: Networks of constraints: Fundamental properties and applications to picture processing. Inf. Sci. 7, 95–132 (1974)
18. Montanari, U., Rossi, F.: Constraint solving and programming: What's next? ACM Comput. Surv., 70 (1996)
19. Panko, R.R., Halverson Jr., R.P.: Spreadsheets on trial: A survey of research on spreadsheet risks. In: Proceedings of the 29th Hawaii International Conference on System Sciences (HICSS): Decision Support and Knowledge-Based Systems, January 1996, vol. 2, pp. 326–335. IEEE Computer Society Press, Los Alamitos (1996)
20. Power, D.J.: A brief history of spreadsheets (August 2004), `http://dssresources.com/history/sshistory.html`
21. GNU Prolog, `http://www.gprolog.org`

22. Ramakrishnan, C.R., Ramakrishnan, I.V., Warren, D.S.: Deductive spreadsheets using tabled logic programming. In: Etalle, S., Truszczyński, M. (eds.) ICLP 2006. LNCS, vol. 4079, pp. 391–405. Springer, Heidelberg (2006)
23. Stadelmann, M.: A spreadsheet based on constraints. In: Proceedings of the ACM Symposium on User Interface Software and Technology, pp. 217–224 (1993)
24. Wilson, S.: Visual programming: Building a visual programming language. Mac. Tech. 13(4) (2000)

An Overview of the Ciao Multiparadigm Language and Program Development Environment and Its Design Philosophy

M.V. Hermenegildo, F. Bueno, M. Carro, P. López, J.F. Morales,
and G. Puebla

IMDEA Institute for Software Development Technology
Universidad Politécnica de Madrid
University of New Mexico
Universidad Complutense de Madrid

Abstract. We describe some of the novel aspects and motivations behind the design and implementation of the Ciao multiparadigm programming system. An important aspect of Ciao is that it provides the programmer with a large number of useful features from different programming paradigms and styles, and that the use of each of these features can be turned on and off at will for each program module. Thus, a given module may be using e.g. higher order functions and constraints, while another module may be using objects, predicates, and concurrency. Furthermore, the language is designed to be extensible in a simple and modular way. Another important aspect of Ciao is its programming environment, which provides a powerful preprocessor (with an associated assertion language) capable of statically finding non-trivial bugs, verifying that programs comply with specifications, and performing many types of program optimizations. Such optimizations produce code that is highly competitive with other dynamic languages or, when the highest levels of optimization are used, even that of static languages, all while retaining the interactive development environment of a dynamic language. The environment also includes a powerful auto-documenter. The paper provides an informal overview of the language and program development environment. It aims at illustrating the design philosophy rather than at being exhaustive, which would be impossible in the format of a paper, pointing instead to the existing literature on the system.

1 Origins and Initial Motivations

Ciao [50,48,5,25,49] is a modern, multiparadigm programming language with an advanced programming environment. The ultimate motivation behind the system is to develop a combination of programming language and development tools that together help programmers produce in less time and with less effort code which has fewer or no bugs and which also performs very efficiently on platforms from small embedded processors to powerful multicore architectures.

P. Degano et al. (Eds.): Montanari Festschrift, LNCS 5065, pp. 209–237, 2008.

Ciao has its main roots in the &-Prolog language and system [51]. &-Prolog's design was aimed at achieving higher performance than state of the art sequential logic programming systems by exploiting (and-)parallelism. This required the development of a specialized abstract machine capable of running in parallel and very efficiently a large number of (possibly non-deterministic) goals [44,51] (the abstract machine was derived from early versions of SICStus Prolog). It also required extending the source language to allow expressing parallelism and concurrency in programs. This made it possible for the user to parallelize programs manually in a relatively simple way. The system was later extended to support constraint programming, including the concurrent and parallel execution of such programs [30]. Significant work was done with Ugo Montanari and Francesca Rossi in this context through the development of a true concurrency semantics that implied the possibility of exploiting truly maximum parallelism in the execution of constraint programs [68,10,11]. Additional work was also performed to extend the system to support other computation rules, such as the Andorra principle [90,7,79,9], other sublanguages, etc.[1]

The experience of this process of gradual extension of the capabilities of the &-Prolog system inspired some of the fundamental concepts underlying the Ciao system. In particular, while each of the functionalities mentioned above (Andorra execution, constraint programming, concurrent programming, etc.) was typically implemented up to that time by a separate (and complex) system comprising compiler, abstract machine, etc. we observed that all such extensions could in fact be supported efficiently within the same system provided the underlying machinery implemented a relatively limited set of basic constructs (a *kernel language*) [50,48], coupled with an easily programmable and modular way of defining new syntax and giving semantics to it in terms of that kernel language. This approach is, of course, not exclusive to Ciao, but in Ciao the facilities that enable building from a simple kernel are very explicitly available (and their use encouraged) from the system programmer level to the application programmer level.

---

[1] It is interesting to note that a great deal of this initial work on the design and implementation of Ciao occurred within the ACCLAIM EU project, in which the Andorra Kernel Language (AKL) and its successor, the Oz language [42], were also developed. The great group of people involved in the project, including Ugo Montanari, Seif Haridi, Gert Smolka, Peter VanRoy, David Warren, and many others resulted in a very fruitful collaboration that effectively gave birth to modern multiparadigm languages. Within this collaborative context, Ciao took different paths to AKL and Oz in many aspects, including for example the use of assertions and global analysis support, the fact that in Ciao non-determinism (backtracking) is implicit, or the use of a (Prolog-derived) syntax aimed at easily supporting meta-programming. As another example, in Ciao the language has always been sequential by default, i.e., concurrency has to be added explicitly by the user (or the parallelizer), whereas in the original Oz and AKL designs the language was concurrent by default (although this has been changed in later Oz designs). Interesting aspects of Oz include for example the extensive development and use of computational spaces as first-level constructs.

This is one of the fundamental capabilities of the Ciao system, which effectively allows Ciao to support multiple programming paradigms and styles. In Ciao all operators, "builtins," and most other syntactic and semantic language constructs such as conditionals or loops are not part of a predefined "language." Instead they are user-modifiable constructs living in *libraries* which can be loaded or unloaded at will thanks to the notion of *packages* [15]. This is the mechanism which allows adding new syntax to the language and giving semantics to this syntax. Most importantly, such packages, and thus the restrictions and extensions to the language that they provide, can be activated or deactivated separately on a per-module/class basis without interfering with each other.[2] The different source-level constructs (and sub-languages / DSLs) are supported by a compilation process defined within the corresponding package, typically via a set of rules defining *source-to-source* transformation into the kernel language, with the (rather infrequent) help of modules or classes written in an external language using one of the several interfaces provided. The approach of compiling to a common kernel implies that the programming styles that Ciao implements share much at both the semantic and implementation levels, and they naturally reuse significant portions of the compiler, documenter, abstract machine, etc.

Another fundamental characteristic of the Ciao system is that it provides a powerful preprocessor, called CiaoPP [8,52,53], which is capable of statically finding non-trivial bugs, verifying that the program complies with specifications, and performing many types of program optimizations. A key ingredient for the above task is the Ciao assertion language [82]. While not strictly required for developing or compiling programs, the preprocessor and assertion language are important and distinctive components of the Ciao design and they also have their origin in earlier work stemming from &-Prolog. In particular, the &-Prolog compiler included a parallelizer which was capable of automatically annotating programs for parallel execution [51,74,71]. This required developing advanced program analysis technology based on abstract interpretation [27] (leading to the development of the PLAI analyzer [91,73,55,76]) which allowed inferring program properties such as independence among program variables [73,75], absence of side effects [72], non-failure [36], data structure shape and instantiation state ("moded types") [87], or even being able to infer upper and lower bounds on the sizes of data structures and the cost of procedures [33,32,37,34], which was instrumental for performing automatic granularity control [33,66,65]. Also, in addition to automatic parallelization the &-Prolog compiler performed other optimizations such as multiple (abstract) specialization [84]. While the &-Prolog inference technology was aimed at performing program optimizations to maximize execution speed and minimize resource consumption, interacting with the system it soon became clear that the wealth of information inferred by the analyzers would also be very useful as an aid in the program development process, and this led to the idea of the Ciao assertion language and preprocessor, as we will discuss later.

---

[2] In fact, some Ciao packages are intended to be portable so that they can be used with little modification in other logic and constraint logic programming systems.

```
1   :- module(_, _, [functional, lazy]).
2
3   nrev([])      := [].
4   nrev([H|T])  := ~conc(nrev(T), [H]).
5
6   conc([],     L)  := L.
7   conc([H|T], K)  := [H | conc(T, K)].
8
9   fact(N)  := N=0 ? 1
10        |   N>0 ? N * fact(--N).
11
12  :- lazy fun_eval nums_from/1.
13  nums_from(X)  := [X | nums_from(X+1)].
14
15  :- use_module(library('lazy/lazy_lib'), [take/3]).
16  nums(N)  := ~take(N, nums_from(0)).
```

Fig. 1. Some examples in Ciao functional notation

2 Supporting Multiple Paradigms

We will now show some examples of how the extensibility of the kernel language mentioned before allows Ciao to incorporate the fundamental constructs from a number of programming paradigms. In particular, the system currently offers, as a combination of syntactic and semantic extensions, the following programming models:

- *Functional Programming:* functional notation is provided by a set of packages which, besides a convenient syntax to define functions (or predicates using a function-like layout), gives support for semantic extensions which include higher-order facilities (e.g., function abstractions and applications thereof) and, if so required, lazy evaluation. For illustration, Figure 1 lists a number of examples using the Ciao functional notation. nrev and conc are written by using multiple :=/2 definitions. fact is written using a disjunction of guards (which actually commits the system to the first matching choice). The ~ prefix (eval, which can often be omitted) is the opposite of quote and states that its argument is a call (as opposed to a data structure to unify with). All of this syntax is defined in the functional package, which is loaded into the module (line 1). nums_from is declared lazy, which is possible thanks to the lazy package, also loaded into the module. An important point is that these packages only modify the syntax and semantics of this module, and other modules can use any other packages. Finally, nums uses take from the library of lazy functions/predicates.

   In general, functional notation is just syntax and thus the following query (loading the functional package in the top level allows using functional notation –the top level behaves in this sense exactly as a module):

```
1   :- module(_, _, [functional, hiord, bfall]).
2
3   color := red | blue | green.
4
5   list := [] | [_ | list].
6
7   list_of(T) := [] | [~T | list_of(T)].
```

```
8   :- module(_, _, [hiord, bfall]).
9
10  color(red). color(blue). color(green).
11
12  list([]).
13  list([_|T]) :- list(T).
14
15  list_of(_, []).
16  list_of(T, [X|Xs]) :- T(X), list_of(T, Xs).
```

Fig. 2. Examples in Ciao functional notation and state of translation after applying the `functional` package

```
?- use_package(functional).
?- [3,2,1] = ~nrev(X).
```

produces the answer:

```
X = [1,2,3]
```

As mentioned before other constructs such as conditionals do commit the system to the first matching case. More strictly "functional" behavior (e.g., being single moded, in the sense that a fixed set of inputs must always be ground and for them a single output is produced, etc.) can be enforced using *assertions*, to be discussed later. Figure 2 lists more examples using `functional` and other packages, and the result after applying just the transformations implied by the `functional` package. Note the use of higher order in `list_of`. More details on Ciao's functional notation can be found in [21].

– *Logic Programming Flavors:* a set of packages (which are loaded by default when a Prolog module is read in) provide support for full ISO-Prolog and a number of other classical "builtins" expected by users to be provided by Prolog systems —except that of course in Ciao rather than builtins all of them are optional features brought in from the libraries.[3] This is signaled by

---

[3] The support of Prolog is done in such a way that Prolog code runs without modification, and the system top level comes up by default in Prolog mode. As a result, many Ciao users which come to the system looking for a good Prolog implementation do get what they expect and, if they do not poke further into the menus and manuals, may never realize that Ciao is in fact quite a different beast.

simply not using the third argument (the one devoted to listing packages) in module declarations.

Alternatively, by avoiding the loading of the Prolog packages the user can restrict the module to use only pure logic programming, without any of Prolog's impure features. For example, the second listing in Figure 2 is a pure logic programming module. If a call to a Prolog builtin such as `assert` were to appear within the module it would be signaled by the compiler as calling an undefined predicate. Features such as, for example, declarative I/O, can be added to such pure modules, by loading additional libraries. This also allows adding individual features of Prolog on a needed basis.

Higher-order logic programming with predicate abstractions is supported through the `hiord` package. This is also illustrated in the second listing in Figure 2. As a further example of the capabilities of the `hiord` package consider the queries:

```
?- use_package(hiord), use_module(library(hiordlib)).
?- P = ( _(X,Y) :- Y = f(X) ),    map([1, 3, 2], P, R).
```

where, after loading the higher-order package `hiord` and instantiating P to the predicate abstraction `_(X,Y)` `:-` `Y = f(X)`, `map([1, 3, 2], P, R)` is applied to P producing:

```
R = [f(1), f(3), f(2)]
```

The (reversed) query:

```
?- P = ( _(X,Y) :- Y = f(X) ),    map(M, P, [f(1), f(3), f(2)]).
```

produces:

```
M = [1, 3, 2]
```

- *Additional Computation Rules:* In addition to the usual depth-first, left-to-right execution regime of Prolog, and again by loading suitable packages, other computation rules such as breadth-first, iterative deepening, Andorra model, etc. are available. As an example in the second listing in Figure 2 any calls to `color`, `list`, and `list_of` will be executed breadth-first.[4] Tabling [24]

---

[4] The possibility of using different control rules has shown useful not only in applications but also (and very specially) when teaching logic programming. In our experience, it is cumbersome to make the first introductory lectures to logic programming using Prolog since the particular (albeit often practically useful) quirks and the subsequent non-termination of Prolog get in the way of teaching the fundamental concepts of logic programming. We have found that it makes perfect sense to start with a purer logic language, with better termination and fairness characteristics. The Ciao breadth-first mode has proved quite useful for this (see `http://www.cliplab.org/logalg` for the slides of our course based on this approach). Once the beauty of pure logic programming is experienced the student is then introduced to the practical and powerful choices made in the design of Prolog, and later to topics and functionality beyond Prolog, such as those outlined in this document, all within the same system.

```
1   :- module(_,_,[fsyntax,clpqf]).
2
3   fact(.=. 0)   :=  .=.  1.
4   fact(N)       :=  .=.  N*fact(.=. N-1)  :- N .>. 0.
5
6   sorted := [] | [_].
7   sorted([X,Y|Z])  :-  X .<. Y, sorted([Y|Z]).
```

Fig. 3. Ciao constraints (combined with functional notation)

is currently being added using an approach which relies mostly on a source-to-source program transformation (which is performed using a package) and an external C library which is accessed using one of the available foreign interfaces [29]. The underlying abstract machine did not have to be changed, and therefore sequential execution is left essentially untouched.

— *Constraint Programming:* several constraint solvers and classes of constraints using these solvers are supported including CLP($\mathcal{Q}$), CLP($\mathcal{R}$) (a derivation of [57]), and CLP($\mathcal{FD}$). The constraint languages and solvers, which are built on more basic blocks such as attributed variables [56] and/or the higher-level Constraint Handling Rules (CHR) [39,88], are also extensible at the user level. As an example, Figure 3 provides two examples using Ciao CLP(Q) constraints, combined with functional notation. For example, line 3 can be read as: if the argument of `fact` is constrained to 0 then the "output" argument is constrained to 1. In the next line if the argument of `fact` is constrained to be greater than 0 then the "output" is constrained to be equal to N*fact(.=. N-1). The two definitions (`fact` and `sorted`) can be called with their arguments in any state of instantiation. For example, the query

```
?- sorted(X).
```

returns:

```
X = [] ? ;
```

```
X = [_] ? ;
```

```
X = [_A, _B],
_A .<. _B ? ;
```

```
X = [_A, _B, _C],
_B .<. _C,
_A .<. _B ? ;
```

```
X = [_A, _B, _C, _D],
```

```
_C .<. _D,
_B .<. _C,
_A .<. _B ?
```

etc.

- *Object-Oriented Programming:* object-oriented programming is provided by the O'Ciao `class` and `object` packages [80]. These packages provide capabilities for class definition, object instantiation, encapsulation and replication of state, inheritance, interfaces, etc. These features are designed to be natural extensions of the underlying module system.
- *Concurrency, Parallelism, and Distributed Execution:* other packages bring in extensive capabilities for expressing concurrency (including a concurrent, shared version of the internal fact database which can be used for synchronization), distribution, and parallel execution [14,11,23]. A notion of "active objects" also allows compiling objects so that they are ultimately mapped to a standalone process, which can then be transparently accessed by the rest of an application. This provides simple ways to implement servers and services in general.

In addition to characteristics that are specific to certain programming paradigms, many other additional features are available through libraries such as, e.g.:

- Structures with named arguments (feature terms), a trimmed-down version of ψ-terms [2] which makes it possible to compile statically all structure unifications to Prolog unifications, which ensures that using them adds no overhead to the execution.
- Persistence, which makes it possible to transparently save and restore the state of selected facts of the dynamic database of a program on exit and startup. This is the base of a high-level interface with databases [26].
- Answer set programming [38], an alternative logic programming model based on the *stable model semantics* [40].
- WWW programming, where Ciao provides libraries to easily establish a mapping between HTML / XML and Herbrand terms, to easily handle (generate / transform / inspect / ...) them in order to for example, write CGIs (or complete web sites) quite easily Ciao [17].

Again, all of these can be activated or deactivated on a *per-module / class* basis.

3   The Ciao Approach to Assertions

Many languages (e.g. Mercury [89,43] or Haskell [59,58], to cite some modern, well-known examples from the logic and functional programming communities) impose certain type-related requirements, e.g., all types (and, when relevant, modes) used have to be defined explicitly or all procedures have to be "well-typed" and "well-moded".

One argument in favor of such declarations and restrictions is that they can be useful to clarify interfaces and meanings, and in general to make large programs more maintainable and well documented, facilitating "programming in the large." Besides, the compiler may use them to generate more specific code, which can be better in several ways (e.g., performance-wise).

We certainly agree with this! But at the same time we also wanted Ciao to be useful (as, say, Prolog, Scheme, or, more recently, Python) for "programming in the small," prototyping, developing simple scripts, or simply experimenting while trying to find a solution to a problem, ... and for this we feel type and mode declarations and other related restrictions can sometimes get in the way.

Fortunately, we came up with a unique solution to this apparent conundrum: Ciao includes a very rich assertion language (and a methodology for dealing with such assertions) [52,82] which allows expressing not only classical types, but also a much wider variety of properties (modes, determinacy, non-failure, cost, ...), but in Ciao *these assertions are optional.* This solution makes Ciao very useful both for programming in the small and in the large. We believe that Ciao's solution to the issue of assertions combines effectively the advantages of the strongly typed and untyped language approaches, bringing the best of both worlds to the programmer, but within a broader scope which, as we will see, makes it possible to use a uniform language to express more program properties (and, therefore, to interact with a tool able to check or infer / reconstruct them). This is, in some sense, related to the *soft typing* approach, pioneered in [20], but it differs from it in that it is not restricted to types. Instead, the framework is open regarding the kind of properties that can be expressed in the corresponding assertions. As an example, systems which aim at performing automatic compile-time checking are often rather strict about the properties which the user can write in assertions. This is understandable because otherwise, the underlying static analyses are of little use for proving the assertions. In our case, we use the same assertion language for different purposes, including run-time checking.

```
 1  :- module(_, [nrev/2], [assertions, fsyntax, nativeprops]).
 2  :- entry nrev/2 : {list, ground} * var.
 3
 4        :- pred nrev(A, B)   : list(A) => list(B).
 5        :- success nrev(A, B) => size_o(B, length(A)).
 6        :- comp nrev(_, _)   + (not_fails, is_det).
 7        :- comp nrev(A, _)   + steps_o(length(A)).
 8
 9
10  nrev([])      := [].
11  nrev([H|L])   := ~conc(~nrev(L),[H]).
12
13        :- comp conc(_, _, _) + (terminates, non_det).
14        :- comp conc(A, _, _) + steps_o(length(A)).
15
16  conc([],     L) := L.
17  conc([H|L],  K) := [ H | ~conc(L,K) ].
```

Fig. 4. Naive reverse with some –partially erroneous– assertions

Therefore, the user may use properties which go beyond those which the static analyses in the system can prove. Of course, even though such assertions may sometimes be useful at compile-time for certain purposes, the user cannot expect CiaoPP to automatically always be able to verify such assertions statically.

As an example, Figure 4 includes the definitions of `nrev` and `conc` (similarly to Figure 1) but also states some program properties expressed using the Ciao assertion language (whose syntax and semantics are made available to the module by means of the `assertions` package). For example, the assertion in line 4 expresses that when `nrev` is called (`:`) the first argument should be a list, and the second one should be a list on success (`=>`). The `+` field in `comp` assertions can contain a conjunction of global properties of the *computation* of the predicate (as the one in line 7). The predicates which are used in such assertions can be in libraries (such as the `nativeprops` library used in the figure) or defined by the user. For example, the definitions in Figure 2 can be used as types in assertions (e.g., line 4 in Figure 4).

4   Program Documentation, Static Debugging, and Verification

One of the most useful characteristics of the assertions used in Ciao is that they are designed to serve many purposes. First, any assertions present in programs can be processed by an autodocumenter (`lpdoc` [46]) in order to generate useful documentation. Also, assertions are analyzed interactively during program development by the system preprocessor (CiaoPP) which can *find non-trivial bugs* statically, *verify* that the program complies with the assertions, or even generate automatically proofs of correctness that can be shipped with programs and checked easily at the receiving end (using the *proof/abstraction carrying code* approach [3]). Even if a program contains no user-provided assertions, Ciao can check the program against the assertions contained in the libraries used by the program, thus potentially catching additional bugs at compile time. If the system cannot prove nor disprove some property at compile time, the system can (optionally again) introduce a run-time check for such property in the executable. For homogeneity, and to ease information exchange among the autodocumenter and the different checkers and analyzers, analysis results are reported using also the assertion language —which, since it is readable by humans, can be inspected by a programmer, for example to make sure that the results of the analyses agree with the intended meaning of the program.

Interestingly, the same underlying technology (global analysis based on abstract interpretation) that allows the system to obtain useful results even when assertions are not present for all predicates, also allows dealing with complex properties, beyond classical types, in a safe way. As a result, for example, the programmer has the possibility of stating assertions about the efficiency of the program (lower and/or upper bounds on the computational cost of procedures [37,35]) which the system will try to verify or falsify, thus performing automatic debugging and validation of the *performance* of programs. Many

```
1    :- module(qsort, [qsort/2], [assertions, fsyntax]).
2    :- use_module(compare, [geq/2, lt/2]).
3    :- entry qsort/2 : {list(num), ground} * var.
4
5    qsort([])       := [].
6    qsort([X|L])  := ~append(~qsort(L1), [X|~qsort(L2)])
7                  :- partition(L, X, L1, L2).
8
9    append([],      X) := X.
10   append([H|X], Y)  := [H | ~append(X,Y)].
11
12   partition([],_B,[],[]).
13   partition([E|R],C,[E|Left1],Right)  :-
14           lt(E,C),   partition(R,C,Left1,Right).
15   partition([E|R],C,Left,[E|Right1])  :-
16           geq(E,C),  partition(R,C,Left,Right1).
```

Fig. 5. A modular `qsort` program

other interesting properties of the predicates and literals of the program can be
handled, such as data structure shape (including pointer sharing), bounds on
data structure sizes, and other operational variable instantiation properties, as
well as procedure-level properties such as determinacy [63], non-failure [12,36],
termination, and bounds on the execution time [67], as well as on the consumption a large class user-defined resources [77].

Assertions also allow programmers to describe the relevant properties of modules or classes which are not yet written or are written in other languages. This
is also done in other languages but often using different types of assertions for
each purpose. In contrast in Ciao the same assertion language is used again for
this task. This, interestingly, makes it possible to run checkers / verifiers / docu-
menters against code which is only partially developed: the traditional "stubs",
which have to be changed later on for a working version, can be replaced by an
assertion declaring how the predicate should behave, with the advantage that
this declared behavior can effectively be checked against its uses.

We will now present some examples which illustrate how these capabilities
are used in practice, and which also help introduce some aspects of the assertion
language. The first example will illustrate automatic inference of non-trivial code
properties while the second will focus on the use of assertions in verification and
debugging, and more specifically to detect problems in the expected performance
of a program.

As mentioned before, CiaoPP includes a non-failure analysis, based on [36]
and [12], which can detect procedures and goals that can be guaranteed not
to fail, i.e., to produce at least one solution or not to terminate. It also can
detect predicates that are "covered", i.e., such that for any input (included in
the calling type of the predicate), there is at least one clause whose "test" (head
unification and body builtins) succeeds. CiaoPP also includes a determinacy
analysis based on [63], which can detect predicates which produce at most one

solution, or predicates whose clause tests are mutually exclusive, even if they are not deterministic (because they call other predicates that can produce more than one solution). We will use the code in Figure 5. The aforementioned analyses infer different types of information which include, among others, that expressed by the following assertion:

```
:- true pred qsort(A,B)
        : ( list(A,num), var(B) ) => ( list(A,num), list(B,num) )
        + ( not_fails, covered, is_det, mut_exclusive ).
```

which expresses that, if `qsort(A, B)` is called with a list of numbers in A and a variable in B, then B will on exit be a list of numbers and the predicate will not fail, will give at most one solution, and will not perform backtracking at the level of its clauses (the + field in `pred` assertions can contain a conjunction of global properties of the *computation* of the predicate.)

CiaoPP can also infer lower and upper bounds on the sizes of terms and the computational cost of predicates [37,35]. The cost bounds are expressed as functions on the sizes of the input arguments and yield the number of resolution steps. Various measures can be used for the "size" of an input, such as list length, term size, term depth, integer value, etc. Note that obtaining a finite upper bound on cost also implies proving *termination* of the predicate. As an example, the following assertion is part of the output of the upper bounds analysis:

```
:- true pred append(A,B,C)
         : ( list(A,num), list1(B,num), var(C) )
        => ( list(A,num), list1(B,num), list1(C,num),
             size_ub(A,length(A)), size_ub(B,length(B)),
             size_ub(C,length(B)+length(A)) )
         + steps_ub(length(A)+1).
```

Note that in this example the size measure used is list length. The property `size_ub(C,length(B)+length(A)` means that an (upper) bound on the size of the third argument of `append/3` is the sum of the sizes of the first and second arguments. The inferred upper bound on computational steps is the length of the first argument of `append/3`.

The following is the output of the lower-bounds analysis:

```
:- true pred append(A,B,C)
         : ( list(A,num), list1(B,num), var(C) )
        => ( list(A,num), list1(B,num), list1(C,num),
             size_lb(A,length(A)), size_lb(B,length(B)),
             size_lb(C,length(B)+length(A)) )
         + ( not_fails, covered, steps_lb(length(A)+1) ).
```

The lower-bounds analysis uses information from the non-failure analysis, without which a trivial lower bound of 0 would be derived.

As a second example, we illustrate how in CiaoPP it is possible to state assertions about the efficiency of the program which the system will try to verify or falsify, thus implementing a form of performance debugging and validation.

This is done by specifying lower and/or upper bounds on the computational cost of predicates (given in number of execution steps). Consider for example again the naive reverse program in Figure 4. The assertion in line 7 states that `nrev` should be linear in the length of the (input) argument A. CiaoPP can be used to verify (or disprove) this assumption by running the analyzer (as before) to infer bounds on costs and then comparing them with the assertion. In fact, `nrev` is of course quadratic. With compile-time error checking turned on, and mode, type, non-failure and lower-bound cost analysis selected, CiaoPP issues the following error message:

```
ERROR: false comp assertion:
          :- comp nrev(A,B) : true => steps_o(length(A))
       because in the computation the following holds:
          steps_lb(0.5*exp(length(A),2)+1.5*length(A)+1)
```

This message states that `nrev` will take at least $\frac{length(A)^2 + 3\ length(A)}{2} + 1$ resolution steps (which is the cost analysis output), while the assertion requires the cost to be in $O(length(A))$ resolution steps. As a result, the worst case asymptotic complexity stated in the user-provided assertion is proved wrong by the lower bound cost assertion inferred by the analysis. This allows detecting the inconsistency and proving that the program does not satisfy the efficiency requirements imposed. Note that upper-bound cost assertions can be proved to hold by means of upper-bound cost analysis if the bound computed by analysis is lower or equal than the upper bound stated by the user in the assertion. The converse holds for lower-bound cost assertions [8]. Thanks to this functionality, CiaoPP can also certify programs with resource consumption assurances as well as efficiently checking such certificates [47].

5   High Performance with Less Effort

A potential benefit of strongly typed languages is performance: the compiler can generate more efficient code with the additional type and mode information that the user provides. Performance is a good thing, of course. However, we do not want to put the burden of efficient compilation on the user by requiring the presence of many program declarations: the compiler should certainly take advantage of any information given by the user, but if the information is not available it should do the work of inferring such program properties. This is the approach taken in Ciao: as we have seen before, when assertions are not present in the program Ciao's analyzers try to *infer* them. Most of these analyses are performed at the kernel language level, so that the same analyzers are used for several of the supported programming models. The information inferred by the global analyzers is used to perform optimizations, including multiple abstract specialization [85], partial evaluation [81], dead code removal, goal reordering, reduction of concurrency / dynamic scheduling [83], low-level optimization (including optimized compilation to native code via C), and others [53].

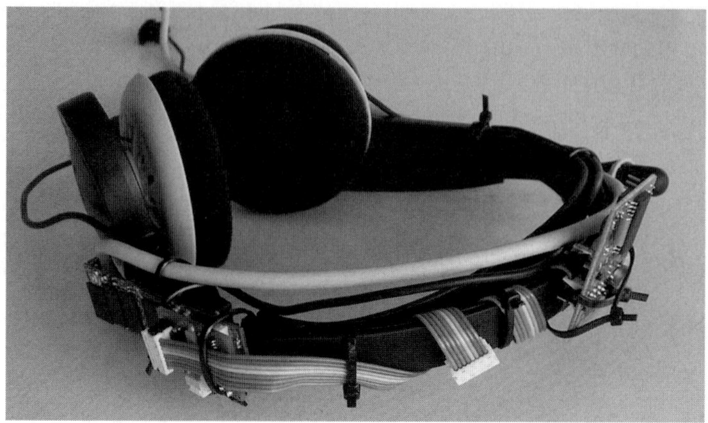

Fig. 6. Headset with a Gumstix processor (left) and 3-D compass (right)

The objective is again to achieve the best of both worlds: with no assertions or analysis information the low-level Ciao compiler (`ciaoc` [16]) generates code which is highly competitive in speed and size with the best dynamically typed systems. And then, when useful information is present (either coming from the user or *inferred by the system analyzers*) the optimizing compiler (see, e.g., [69] for an early description) can produce code that is competitive with that coming from strongly-typed systems. Ciao's highly optimized compilation has been successfully tested for example in applications with tight resource usage constraints (including real-time) [19], obtaining a 7-fold speed-up w.r.t. the default byte-code compilation (the performance of which is similar to that of state of the art abstract machine-based systems). The application in hand was real-time spacial placement of sound sources for a virtual reality suit and ran in a small ("Gumstix") processor embedded within a headset (Figure 6). It is interesting to note that this level of performance is only around 20-40% slower than a comparable implementation in C of the same application.

A particularly interesting optimization performed by CiaoPP, and which is inherited from the &-Prolog system, is *automatic parallelization* [45,41]. This is specially relevant nowadays given that the wide availability of multicore processors has made parallel computers mainstream. We illustrate this by means of a simple example using goal-level program parallelization [6,22]. This optimization is performed as a source-to-source transformation, in which the input program is *annotated* with parallel expressions. The parallelization algorithms, or annotators [71], exploit parallelism under certain *independence* conditions, which allow guaranteeing interesting correctness and no-slowdown properties for the parallelized programs [54,31]. This process is made more complex by the presence of variables shared among goals and pointers among data structures at runtime. Let us consider again the program in Figure 5. A possible parallelization (obtained in this case with the "MEL" annotator [71]) is:

```
qsort([X|L],R) :-
      partition(L,X,L1,L2),
      (
          indep(L1, L2) ->
          qsort(L2,R2) & qsort(L1,R1)
      ;
          qsort(L2,R2), qsort(L1,R1)
      ),
      append(R1,[X|R2],R).
```

which indicates that, provided that L1 and L2 do not have variables in common at run-time, then the recursive calls to qsort can be run in parallel. Assuming that lt/2 and geq/2 in Figure 5 need their arguments to be ground (note that this may be either inferred by analyzing the implementation of lt/2 and geq/2 or by stated by the user using suitable assertions), the information inferred by the abstract interpreter using, e.g., mode and sharing/freeness analysis, can determine that L1 and L2 are ground after partition, and therefore they do not have variables to share. As a result, the independence test and the corresponding conditional is simplified via abstract executability and the annotator yields instead the following code:

```
qsort([X|L],R) :-
      partition(L,X,L1,L2),
      qsort(L2,R2) & qsort(L1,R1),
      append(R1,[X|R2],R).
```

which is much more efficient since it has no run-time test. This test simplification process is described in detail in [6] where the impact of abstract interpretation in the effectiveness of the resulting parallel expressions is also studied.

The tests in the above example aim at *strict* independent and-parallelism. However, the annotators are parametrized on the notion of independence. Different tests can be used for different independence notions: non-strict independence [13], constraint-based independence [31], etc. Moreover, all forms of and-parallelism in logic programs can be seen as independent and-parallelism, provided the definition of independence is applied at the appropriate granularity level.[5]

The information produced by the CiaoPP cost analyzers is also used to perform combined compile–time/run–time resource control. An example of this is *task granularity control* [65] of parallelized code. Such parallel code can be the output of the process mentioned above or code parallelized manually. In general, this run-time granularity control process involves computing sizes of terms involved in granularity control, evaluating cost functions, and comparing the result with a threshold to decide between parallel and sequential execution. Optimizations to this general process include cost function simplification and improved term size computation [64].

---

[5] For example, stream and-parallelism can be seen as independent and-parallelism if the independence of "bindings" rather than goals is considered.

6 Incremental Compilation: Other Support for Programming in the Small and in the Large

In addition to all the functionality provided by the preprocessor and assertions, programming in the large is further supported again by the module/class system [15,80]. This design is the real enabler of Ciao's modular program development support tools, effective global program analysis, modular static debugging, and module-based automatic incremental compilation and optimization. The analyzers and compiler take advantage of the module system and module dependencies to reanalyze / recompile only part of the application modules after one of them is changed, without any need to define "makefiles" or similar dependency-related additional files.

Application deployment is enhanced beyond the traditional Prolog top-level, since the system offers a full-featured interpreter but also supports the use of Ciao as a scripting language and a compiled language. Several types of executables can be easily built, from multiarchitecture *bytecode* executables to single-architecture, standalone executables. Multiple platforms are supported, including Windows, Linux, Mac Os X, and many other Un*x-based OSs.

Modular distribution of user and system code in Ciao, coupled with modular analysis, allows the generation of stripped executables, with only those builtins and libraries used by the program. Those reduced-size executables permit programming in the small when strict space constraints are present. Flexible development of applications and libraries that use components written in several languages is also allowed, by means of compiler and abstract machine support for multiple bidirectional foreign interfaces to C/C++, Java, Tcl/Tk, SQL databases (with a notion of predicate persistence), etc. The interfaces are expressed in (and any necessary glue code automatically generated from) descriptions written in the assertion language, as previously stated.

7 An Advanced Integrated Development Environment

Another design objective of Ciao has been to provide a truly productive program development environment that integrates all of the tools mentioned before in order to fulfill the objective of allowing the development of correct and efficient programs in as little time and with as little effort as possible. This includes a *rich graphical development interface*, based on the latest, graphical versions of Emacs and offering menu and widget-based interfaces with direct access to the top-level/debugger, preprocessor, and autodocumenter, as well as an embeddable source-level debugger with breakpoints, and several execution visualization tools. In addition, a plugin with very similar functionality is also available for the Eclipse environment.

The programming environment makes it possible to start the top-level, the debugger, or the preprocessor, and to load the current module within them by pressing a button or via a pair of keystrokes. Figure 7 shows a source file (with syntax highlighting, top level, menus, buttons, etc.). Tracing the execution in

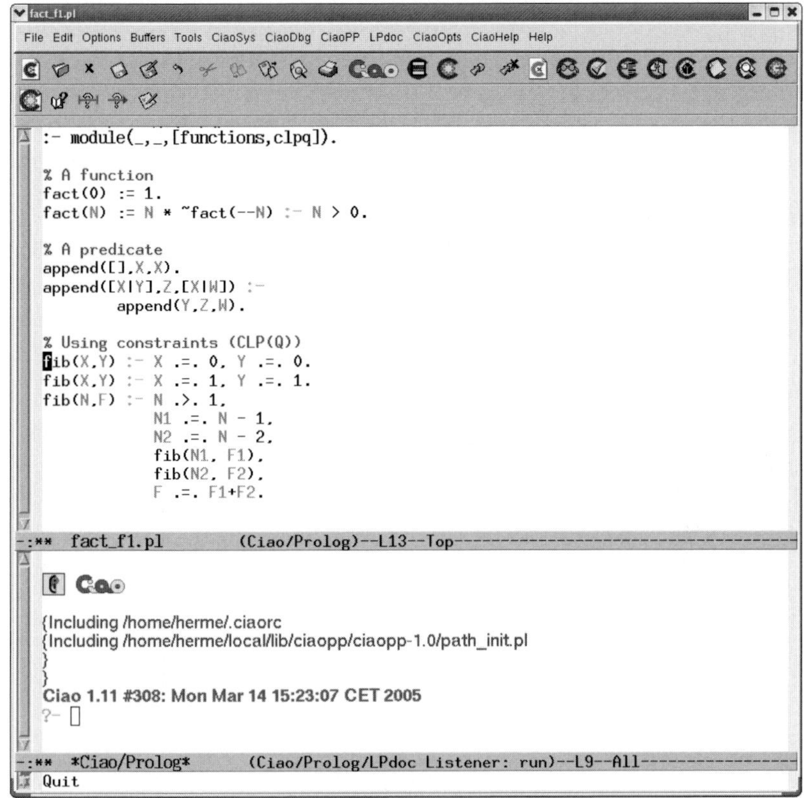

Fig. 7. A program file and the top-level interpreter

the debugger makes the current statement in the program be highlighted in an
additional buffer containing the debugged file (Figure 8).

The environment provides also automated access to the documentation, extensive syntax highlighting, auto-completion, or auto-location of errors in the
source, and is highly customizable (to set, for example, alternative installation
directories or the location of some binaries). Figure 9 shows the location in the
source of a simple syntactic error. The direct access to the preprocessor allows
interactive control of all the static debugging, verification, and program transformation facilities. As an example, Figure 10 shows CiaoPP signaling in the
source a semantic error (in the same way as the previous simple syntactic error). In particular, it is the cost-related error discussed previously in which the
compiler detects (statically!) that the definition of **nrev** does not comply with
the assertion requiring it to be of linear complexity. The direct access to the
auto-documentation facilities [46] allows using a pair of keystrokes to generate
human-readable program documentation from the current file in a variety of
formats from the assertions, directives, and machine-readable comments present
in the program being developed or in the system's libraries, as well as all other

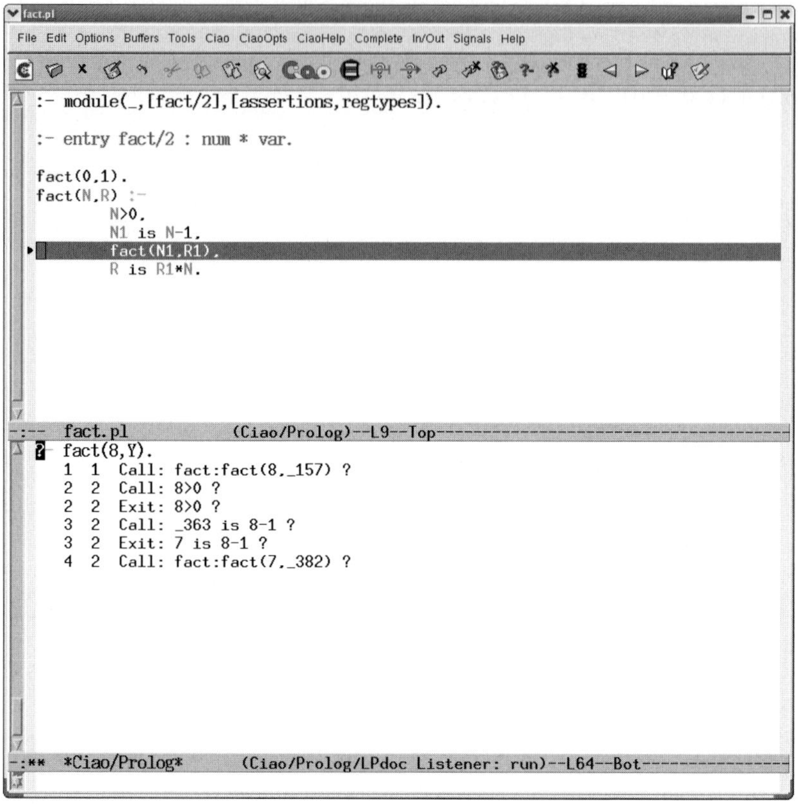

Fig. 8. The source debugger in action

program information available to the compiler. This direct access to the documenter and on a per-module basis is very useful in practice in order to build documentation incrementally and to make sure that, for example, cross references between files are are well resolved and that the documentation itself is well structured and formatted. As a further example of the different components and capabilities of the environment, Figure 11 shows a VisAndOr [18] depiction of an and-parallel execution.

8   Some Final Thoughts on Parallelism, Dynamic Languages, and Mainstream Programming

Interestingly many of the motivations behind the development of Ciao over the years have acquired presently even more crucial importance. Parallelism capabilities are becoming ubiquitous thanks to the widespread use of multi-core processors. Indeed, most laptops on the market contain two cores (typically capable of running up to four threads simultaneously) and single-chip, 8-core servers are

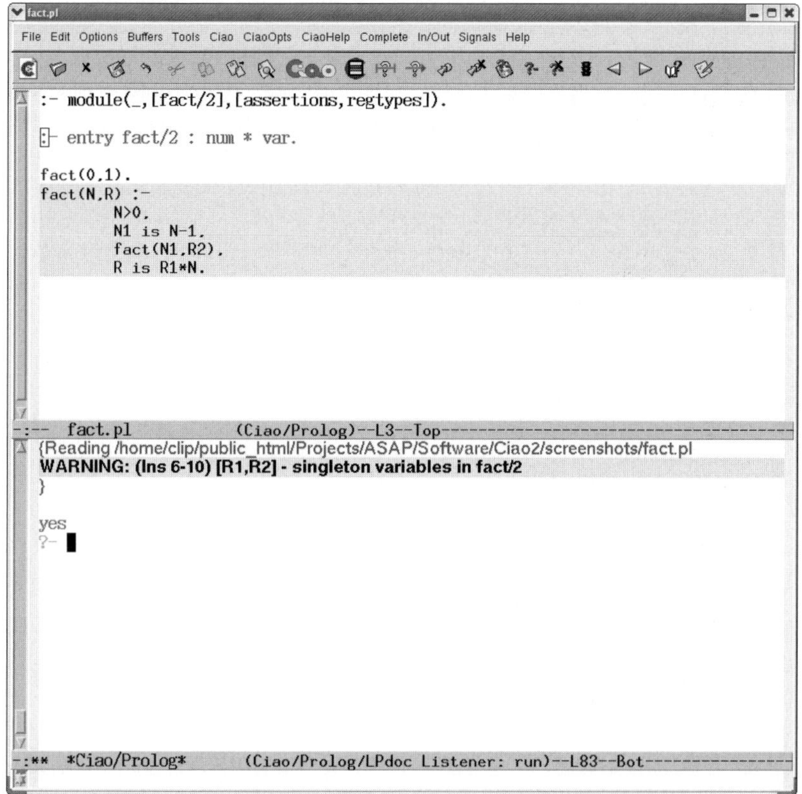

Fig. 9. Error location in the source –a simple syntactic error

now in widespread use. Furthermore, the trend is that the number of on-chip cores will double with each processor generation. In this context, being able to exploit such parallel execution capabilities in programs as easily as possible becomes more and more a necessity. However, it is well-known [61] that parallelizing programs is a hard challenge. This has renewed interest in language-related designs and tools which can simplify the task of producing parallel programs.

At the same time, the environment in which much software needs to be developed nowadays (decoupled software development, use of components and services, increased interoperability constraints, need for dynamic update or self-reconfiguration, mash-ups) is posing requirements which align with the classical arguments for dynamic languages but which in fact go beyond them. Examples of often required dynamic features include making it possible to (partially) test and verify applications which are partially developed, and which will never be "complete" or "final", or which need to have flexibility in their APIs because they need to have a variable number of arguments or their "entry points" evolve over time in an asynchronous, decentralized fashion (e.g., services,

```
emacs@gazelle

File  Edit  Options  Buffers  Tools  Ciao  CiaoOpts  CiaoHelp  Complete  In/Out  Si

:- module(_, [nrev/2], [assertions,functional,regtypes,nativeprops]).

:- entry nrev/2 : {list, ground} * var.


   :- pred nrev(A,B)   : list(A) => num(B).
   :- success nrev(A,B) => size_o(B,length(A)).
   :- comp nrev(_,_)   + ( not_fails, is_det ).
   :- comp nrev(A,_)   + steps_o( length(A) ).

nrev( [] )      := [].
nrev( [H|L] ) := ~conc( nrev(L),[H] ).


   :- comp conc(_,_,_) + ( terminates, non_det ).
   :- comp conc(A,_,_) + steps_o(length(A)).

conc( [],      L ) := L.
--:---    revf_assrt_bug.pl    Top L1        (Ciao)-------------------------------
{ERROR (ctchecks_pred_messages): (lns 9-9) False assertion:
:- check comp nrev(A,_1)
          + steps_o(length(A)).

because
on comp revf_assrt_bug:nrev(A,_) :

[generic_comp] : steps_ub(0.5*exp(length(A),2)+1.5*length(A)+1),
 steps_lb(0.5*exp(length(A),2)+1.5*length(A)+1),not_fails,covered,
 mut_exclusive,is_det

}
{ERROR (ctchecks_pred_messages): (lns 5-6) False assertion:
:- check success nrev(A,B)
          : list(A)
        => num(B).

because
-U:**-  *Ciao-Preprocessor*    62% L40     (Ciao/CiaoPP/LPdoc Listener: run)-------
```

Fig. 10. Error location in the source –a cost error

including web services). These requirements, coupled with their intrinsic agility
in development, have made dynamic programming languages (such as Python,
Ruby, Lua, JavaScript, Perl, PHP, etc.) a very attractive option in recent years
for a number of purposes that go well beyond simple scripting. Parts written in
these languages often become essential components (if not the central implementation vehicle) of mainstream applications. The practical relevance of dynamic
features is also illustrated by the many successful languages and frameworks
which aim at bringing together ideas of both worlds. For example, Objective-
C [28], which mixes C, object orientation, and the possibility of having dynamically typed variables and messages, is currently used as the base of Mac OS X,
and it was used before in NextStep. Other frameworks, such as Java and .NET,
are intensely working on ensuring and improving the interoperability among dynamic and static languages by including support for dynamicity in their virtual
machines. Another example is the future fourth revision of ECMAScript [1] on
which the JavaScript and ActionScript languages are based, that will include
optional (soft-)type declarations to allow the compiler to generate more efficient

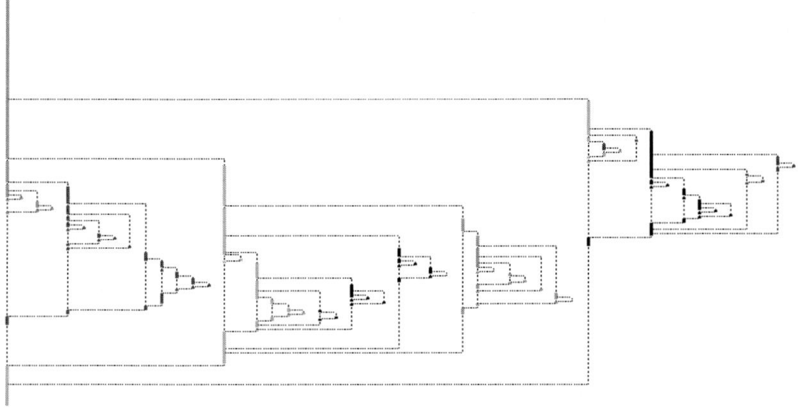

Fig. 11. VisAndOr depiction of an and-parallel execution of QuickSort

code and detect more errors. The Tamarin project [70] intends to use this additional information to generate faster code. For Python, the PyPy project [86] designed a language, RPython [4] that imposes constraints on the programs to ensure that they can be statically typed. RPython is moving forward as a general purpose language.

At the same, detecting errors at compile-time and inferring properties required to optimize programs, are still important issues in real-world applications. This has also brought the development of safe versions of traditional languages, such as, e.g., CCured [78] or Cyclone [60] for C, as well as of systems that offer capabilities similar to those of the Ciao assertion preprocessor, such as Necula et al.'s Deputy [6] or Leino et al.'s Spec# [62].

We believe that Ciao has pioneered and is continuing to push the state of the art in these currently very relevant and challenging areas, and offers a unique combination of features which directly address many of these challenges. The Ciao approach to exploiting parallelism provides powerful parallelizers and at the same time allows programmer and parallelizer to cooperate. Programmers can choose between expressing manually the parallelism with high-level constructs, letting the compiler discover the parallelism, or a combination of both. Parts of a program can be parallelized by hand and other parts automatically. Furthermore, the parallelizer also checks manual parallelizations for correctness. Finally, the output of the parallelizer is expressed in the same high level language, which means that programmers can easily inspect (and improve) the parallelizations produced by the compiler. At the heart of these capabilities are CiaoPP's powerful, modular, and incremental abstract interpretation-based program analyzers. The use of this technology was pioneered by &-Prolog and Ciao (it was arguably the first use of abstract interpretation in a real compiler) and we continue to believe it is the most promising nowadays, and they are being

---

[6] http://deputy.cs.berkeley.edu/

adopted or will be adopted by many systems (see, e.g., [45] for further discussion of this topic).

Regarding the conundrum between statically and dynamically checked languages, Ciao has also pioneered and continues to push the state of the art of what we believe is the most promising approach in order to be able to obtain the best of both worlds: the combination of a flexible, multi-purpose assertion language with sophisticated assertion processing based on strong program analysis technology. This allows support for dynamic language features while at the same time having the capability of achieving the performance and efficiency of static systems. It also allows being able to work in a seamless way with a large class of properties, some of them even user-defined, and which go well beyond traditional types. Again, at the heart of these capabilities are CiaoPP's abstract interpretation-based analyzers.

Finally, we also believe that Ciao's language design offers unique possibilities due to its simple and powerful extensibility features, which not only allow to selectively bring in the constructs of multiple programming paradigms, but also make it possible for the programmer to easily extend (and restrict) the language as needed, syntactically and semantically, and to quickly design domain-specific languages.

Probing Further

The reader is encouraged to explore the system, its documentation, and the tutorial papers that have been published on it. We are currently working on the new 1.14 system version which includes significant enhancements with respect to the previous version (1.10). In addition to the autodocumenter, we plan to include a beta version of the preprocessor in the default Ciao distribution (up to now, CiaoPP was only distributed on demand and installed separately). Ciao 1.14 is available already on demand from the Ciao subversion repository.

But, Why Is it Called Ciao?

After reading the previous paragraphs the reader may have already seen the logic behind the "Ciao Prolog" phrase. Ciao is an interesting word which is used both to say *hello* and *goodbye*. Ciao intends to be a truly excellent, high-performance, and freely available ISO-Prolog system which can be used as a classical Prolog, in both academic and industrial environments (and, in particular, to introduce users to Prolog and to constraint and logic programming –the *hello* Prolog part). But Ciao is also a new-generation, multiparadigm programming language and sophisticated program development environment for large, complex applications which goes well beyond Prolog and other classical logic programming languages –the *goodbye* Prolog part. And it has the advantage (when compared to other modern systems that support different forms of logic programming) that it does so while keeping full Prolog compatibility when desired.

Contact / download info

```
http://www.ciaohome.org
http://www.cliplab.org
ciao@clip.dia.fi.upm.es
```
The Ciao Development Team
Technical U. of Madrid, Spain
U. of New Mexico, USA
U. Complutense de Madrid, Spain
IMDEA-Institute for Software Development
Technology

Ciao is free software protected to remain so by the GNU LGPL license. It can be used freely to develop both free and commercial applications.

Acknowledgments

The Ciao system is in continuous and very active development through the collaborative effort of numerous members of several institutions, including UPM, UNM, UCM, Roskilde U., U. of Melbourne, Monash U., U. Arizona, Linköping U., NMSU, K. U. Leuven, Bristol U., Ben-Gurion U., INRIA, as well as many others. The development of the Ciao system has been supported by a number of European, Spanish, and other international projects (currently by EU IST-15905 *MOBIUS* and *S-CUBE* projects, the Spanish TIN-2005-09207 *MERIT* project, and the CAM *PROMESAS* program). Manuel Hermenegildo is also supported by the IST Prince of Asturias Chair at the University of New Mexico and IMDEA-Software, the Madrid Institute for Software Development Technology. The system documentation and related publications contain more specific credits to the many contributors to the system. We would also like to thank the anonymous reviewers for providing very constructive and useful comments which have contributed to improving the paper.

References

1. ECMA 2008: ECMAscript Edition 4 Specification Wiki (2008), http://wiki.ecmascript.org
2. Aït-Kaci, H.: An Introduction to LIFE – Programming with Logic, Inheritance, Functions and Equations. In: Proceedings of the 1993 International Symposium on Logic Programming (1993)
3. Albert, E., Puebla, G., Hermenegildo, M.: Abstraction-Carrying Code. In: Baader, F., Voronkov, A. (eds.) LPAR 2004. LNCS (LNAI), vol. 3452, Springer, Heidelberg (2005)
4. Ancona, D., Ancona, M., Cuni, A., Matsakis, N.D.: RPython: a Step towards Reconciling Dynamically and Statically Typed OO Languages. In: DLS 2007: Proceedings of the 2007 Symposium on Dynamic Languages, pp. 53–64. ACM Press, New York (2007)
5. Bueno, F., Cabeza, D., Carro, M., Hermenegildo, M., López-García, P., Puebla, G. (eds.): The Ciao System. Ref. Manual (v1.13). Technical report, C. S. School (UPM) (2006), http://www.ciaohome.org

6. Bueno, F., García de la Banda, M., Hermenegildo, M.: Effectiveness of Abstract Interpretation in Automatic Parallelization: A Case Study in Logic Programming. ACM TOPLAS 21(2), 189–238 (1999)
7. Bueno, F., Debray, S.K., García de la Banda, M., Hermenegildo, M.: Transformation-based Implementation and Optimization of Programs Exploiting the Basic Andorra Model. Technical Report CLIP11/95.0, Facultad de Informática, UPM (May 1995)
8. Bueno, F., Deransart, P., Drabent, W., Ferrand, G., Hermenegildo, M., Maluszynski, J., Puebla, G.: On the Role of Semantic Approximations in Validation and Diagnosis of Constraint Logic Programs. In: Proc. of the 3rd. Int'l Workshop on Automated Debugging–AADEBUG 1997, Linköping, Sweden, May 1997, pp. 155–170. U.of Linköping Press (1997)
9. Bueno, F., Hermenegildo, M.: An Automatic Translation Scheme from Prolog to the Andorra Kernel Language. In: Proc. of the 1992 International Conference on Fifth Generation Computer Systems, Institute for New Generation Computer Technology (ICOT), vol. 2, pp. 759–769 (June 1992)
10. Bueno, F., Hermenegildo, M., Montanari, U., Rossi, F.: From Eventual to Atomic and Locally Atomic CC Programs: A Concurrent Semantics. In: Rodríguez-Artalejo, M., Levi, G. (eds.) ALP 1994. LNCS, vol. 850, pp. 114–132. Springer, Heidelberg (1994)
11. Bueno, F., Hermenegildo, M., Montanari, U., Rossi, F.: Partial Order and Contextual Net Semantics for Atomic and Locally Atomic CC Programs. Science of Computer Programming 30, 51–82 (1998); Special CCP 1995 Workshop issue
12. Bueno, F., López-García, P., Hermenegildo, M.: Multivariant Non-Failure Analysis via Standard Abstract Interpretation. In: Kameyama, Y., Stuckey, P.J. (eds.) FLOPS 2004. LNCS, vol. 2998, pp. 100–116. Springer, Heidelberg (2004)
13. Cabeza, D., Hermenegildo, M.: Extracting Non-strict Independent And-parallelism Using Sharing and Freeness Information. In: LeCharlier, B. (ed.) SAS 1994. LNCS, vol. 864, pp. 297–313. Springer, Heidelberg (1994)
14. Cabeza, D., Hermenegildo, M.: Distributed Concurrent Constraint Execution in the CIAO System. In: Proc. of the 1995 COMPULOG-NET Workshop on Parallelism and Implementation Technologies, Utrecht, NL, September 1995, U. Utrecht / T.U. Madrid (1995), http://www.cliplab.org/
15. Cabeza, D., Hermenegildo, M.: A New Module System for Prolog. In: Palamidessi, C., Moniz Pereira, L., Lloyd, J.W., Dahl, V., Furbach, U., Kerber, M., Lau, K.-K., Sagiv, Y., Stuckey, P.J. (eds.) CL 2000. LNCS (LNAI), vol. 1861, pp. 131–148. Springer, Heidelberg (2000)
16. Cabeza, D., Hermenegildo, M.: The Ciao Modular, Standalone Compiler and Its Generic Program Processing Library. Special Issue on Parallelism and Implementation of (C)LP Systems 30(3); ENTCS. Elsevier - North Holland (March 2000)
17. Cabeza, D., Hermenegildo, M.: Distributed WWW Programming using (Ciao-)Prolog and the PiLLoW Library. Theory and Practice of Logic Programming 1(3), 251–282 (2001)
18. Carro, M., Gómez, L., Hermenegildo, M.: Some Paradigms for Visualizing Parallel Execution of Logic Programs. In: 1993 International Conference on Logic Programming, pp. 184–201. MIT Press, Cambridge (June 1993)
19. Carro, M., Morales, J., Muller, H.L., Puebla, G., Hermenegildo, M.: High-Level Languages for Small Devices: A Case Study. In: Flautner, K., Kim, T. (eds.) Compilers, Architecture, and Synthesis for Embedded Systems, pp. 271–281. ACM Press, Sheridan (October 2006)

20. Cartwright, R., Fagan, M.: Soft Typing. In: Programming Language Design and Implementation (PLDI 1991), SIGPLAN, pp. 278–292. ACM, New York (1991)
21. Casas, A., Cabeza, D., Hermenegildo, M.: A Syntactic Approach to Combining Functional Notation, Lazy Evaluation and Higher-Order in LP Systems. In: Hagiya, M., Wadler, P. (eds.) FLOPS 2006. LNCS, vol. 3945. Springer, Heidelberg (2006)
22. Casas, A., Carro, M., Hermenegildo, M.: Annotation Algorithms for Unrestricted Independent And-Parallelism in Logic Programs. In: 17th International Symposium on Logic-based Program Synthesis and Transformation (LOPSTR 2007), The Technical University of Denmark, August 2007. LNCS, vol. 4915, pp. 138–153. Springer, Heidelberg (2007)
23. Casas, A., Carro, M., Hermenegildo, M.: Towards a High-Level Implementation of Execution Primitives for Non-restricted, Independent And-parallelism. In: Warren, D.S., Hudak, P. (eds.) 10th International Symposium on Practical Aspects of Declarative Languages (PADL 2008). LNCS, vol. 4902, pp. 230–247. Springer, Heidelberg (2008)
24. Chen, W., Warren, D.S.: Tabled Evaluation with Delaying for General Logic Programs. Journal of the ACM 43(1), 20–74 (1996)
25. The Ciao Development Team. The Ciao Multiparadigm Language and Program Development Environment, The ALP Newsletter 19(3). The Association for Logic Programming (November 2006), http://www.logicprogramming.org/newsletter/nov06/index.html
26. Correas, J., Gomez, J.M., Carro, M., Cabeza, D., Hermenegildo, M.: A Generic Persistence Model for CLP Systems (And Two Useful Implementations). In: Jayaraman, B. (ed.) PADL 2004. LNCS, vol. 3057, pp. 104–119. Springer, Heidelberg (2004)
27. Cousot, P., Cousot, R.: Abstract Interpretation: a Unified Lattice Model for Static Analysis of Programs by Construction or Approximation of Fixpoints. In: Proc. of POPL 1977, pp. 238–252 (1977)
28. Cox, B.J.: Object Oriented Programming: An Evolutionary Approach. Addison Wesley (1991), http://developer.apple.com/documentation/Cocoa/Conceptual/ObjectiveC/ObjC.pdf
29. de Guzmán, P.C., Carro, M., Hermenegildo, M., Silva, C., Rocha, R.: An Improved Continuation Call-Based Implementation of Tabling. In: Warren, D.S., Hudak, P. (eds.) 10th International Symposium on Practical Aspects of Declarative Languages (PADL 2008). LNCS, vol. 4902, pp. 198–213. Springer, Heidelberg (2008)
30. García de la Banda, M., Bueno, F., Hermenegildo, M.: Towards Independent And-Parallelism in CLP. In: Kuchen, H., Swierstra, S.D. (eds.) PLILP 1996. LNCS, vol. 1140, pp. 77–91. Springer, Heidelberg (1996)
31. García de la Banda, M., Hermenegildo, M., Marriott, K.: Independence in CLP Languages. ACM Transactions on Programming Languages and Systems 22(2), 269–339 (2000)
32. Debray, S.K., Lin, N.W.: Cost analysis of logic programs. TOPLAS 15(5) (1993)
33. Debray, S.K., Lin, N.-W., Hermenegildo, M.: Task Granularity Analysis in Logic Programs. In: Proc. PLDI 1990, June 1990, pp. 174–188. ACM Press, New York (1990)
34. Debray, S.K., López-García, P., Hermenegildo, M., Lin, N.-W.: Lower Bound Cost Estimation for Logic Programs. In: 1997 International Logic Programming Symposium, pp. 291–305. MIT Press, Cambridge (1997)
35. Debray, S.K., López-García, P., Hermenegildo, M., Lin, N.-W.: Lower Bound Cost Estimation for Logic Programs. In: ILPS 1997. MIT Press, Cambridge (1997)

36. Debray, S.K., López-García, P., Hermenegildo, M.: Non-Failure Analysis for Logic Programs. In: ICLP 1997, pp. 48–62. MIT Press, Cambridge (1997)
37. Debray, S.K., López-García, P., Hermenegildo, M., Lin, N.-W.: Estimating the Computational Cost of Logic Programs. In: LeCharlier, B. (ed.) SAS 1994. LNCS, vol. 864, pp. 255–265. Springer, Heidelberg (1994)
38. El-Khatib, O., Pontelli, E., Son, T.C.: Integrating an Answer Set Solver into Prolog: ASP-PROLOG. In: LPNMR, pp. 399–404 (2005)
39. Frühwirth, T.: Theory and Practice of Constraint Handling Rules. Journal of Logic Programming, Special Issue on Constraint Logic Programming 37(1-3) (October 1998)
40. Gelfond, M., Lifschitz, V.: The Stable Model Semantics for Logic Programming. In: International Conference on Logic Programming 1988, pp. 1070–1080 (1988)
41. Gupta, G., Pontelli, E., Ali, K., Carlsson, M., Hermenegildo, M.: Parallel Execution of Prolog Programs: a Survey. ACM Transactions on Programming Languages and Systems 23(4), 472–602 (2001)
42. Haridi, S., Franzén, N.: The Oz Tutorial. DFKI (February 2000), http://www.mozart-oz.org
43. Henderson, F., Somogyi, Z., Conway, T.: Determinism Analysis in the Mercury Compiler. In: Proc. Australian Computer Science Conference, Melbourne, Australia, pp. 337–346 (January 1996)
44. Hermenegildo, M.: An Abstract Machine for Restricted AND-parallel Execution of Logic Programs. In: Shapiro, E. (ed.) ICLP 1986. LNCS, vol. 225, pp. 25–40. Springer, Heidelberg (1986)
45. Hermenegildo, M.: Automatic Parallelization of Irregular and Pointer-Based Computations: Perspectives from Logic and Constraint Programming. In: Lengauer, C., Griebl, M., Gorlatch, S. (eds.) Euro-Par 1997. LNCS, vol. 1300, pp. 31–46. Springer, Heidelberg (1997)
46. Hermenegildo, M.: A Documentation Generator for (C)LP Systems. In: Palamidessi, C., Moniz Pereira, L., Lloyd, J.W., Dahl, V., Furbach, U., Kerber, M., Lau, K.-K., Sagiv, Y., Stuckey, P.J. (eds.) CL 2000. LNCS (LNAI), vol. 1861, pp. 1345–1361. Springer, Heidelberg (2000)
47. Hermenegildo, M., Albert, E., López-García, P., Puebla, G.: Some Techniques for Automated, Resource-Aware Distributed and Mobile Computing in a Multi-Paradigm Programming System. In: Danelutto, M., Vanneschi, M., Laforenza, D. (eds.) Euro-Par 2004. LNCS, vol. 3149, pp. 21–37. Springer, Heidelberg (2004)
48. Hermenegildo, M., Bueno, F., Cabeza, D., Carro, M., García de la Banda, M., López-García, P., Puebla, G.: The CIAO Multi-Dialect Compiler and System: An Experimentation Workbench for Future (C)LP Systems. In: Parallelism and Implementation of Logic and Constraint Logic Programming, Nova Science, Commack, NY, USA, pp. 65–85 (April 1999)
49. Hermenegildo, M.: The Ciao Development Team. Why Ciao? –An Overview of the Ciao System's Design Philosophy. Technical Report CLIP7/2006.0, Technical University of Madrid (UPM), School of Computer Science, UPM (December 2006), http://cliplab.org/papers/ciao-philosophy-note-tr.pdf
50. Hermenegildo, M.: The CLIP Group. In: Borning, A. (ed.) PPCP 1994. LNCS, vol. 874, pp. 123–133. Springer, Heidelberg (1994)
51. Hermenegildo, M., Greene, K.: The &-Prolog System: Exploiting Independent And-Parallelism. New Generation Computing 9(3,4), 233–257 (1991)

52. Hermenegildo, M., Puebla, G., Bueno, F.: Using Global Analysis, Partial Specifications, and an Extensible Assertion Language for Program Validation and Debugging. In: The Logic Programming Paradigm: a 25-Year Perspective, pp. 161–192. Springer, Heidelberg (1999)

53. Hermenegildo, M., Puebla, G., Bueno, F., López-García, P.: Integrated Program Debugging, Verification, and Optimization Using Abstract Interpretation (and The Ciao System Preprocessor). Science of Computer Programming 58(1–2), 115–140 (2005)

54. Hermenegildo, M., Rossi, F.: Strict and Non-Strict Independent And-Parallelism in Logic Programs: Correctness, Efficiency, and Compile-Time Conditions. Journal of Logic Programming 22(1), 1–45 (1995)

55. Hermenegildo, M., Warren, R., Debray, S.K.: Global Flow Analysis as a Practical Compilation Tool. Journal of Logic Programming 13(4), 349–367 (1992)

56. Holzbaur, C.: Metastructures vs. Attributed Variables in the Context of Extensible Unification. In: Bruynooghe, M., Wirsing, M. (eds.) PLILP 1992. LNCS, vol. 631, pp. 260–268. Springer, Heidelberg (1992)

57. Holzbaur, C.: SICStus 2.1/DMCAI Clp 2.1.1 User's Manual. University of Vienna (1994)

58. Hudak, P., Peyton-Jones, S., Wadler, P., Boutel, B., Fairbairn, J., Fasel, J., Guzman, M.M., Hammond, K., Hughes, J., Johnsson, T., Kieburtz, D., Nikhil, R., Partain, W., Peterson, J.: Report on the Programming Language Haskell. Haskell Special Issue, ACM Sigplan Notices 27(5) (1992)

59. Hudak, P., Hughes, J., Jones, S.P., Wadler, P.: A History of Haskell: Being Lazy with Class. In: HOPL III: Proceedings of the third ACM SIGPLAN conference on History of programming languages, pp. 1–55. ACM Press, New York (2007)

60. Jim, T., Morrisett, J.G., Grossman, D., Hicks, M.W., Cheney, J., Wang, Y.: Cyclone: A safe dialect of c. In: Ellis, C.S. (ed.) USENIX Annual Technical Conference, General Track, pp. 275–288. USENIX (2002)

61. Karp, A.H., Babb, R.C.: A Comparison of 12 Parallel Fortran Dialects. IEEE Software (September 1988)

62. Leavens, G.T., Leino, K.R.M., Müller, P.: Specification and verification challenges for sequential object-oriented programs. Formal Asp. Comput. 19(2), 159–189 (2007)

63. López-García, P., Bueno, F., Hermenegildo, M.: Determinacy Analysis for Logic Programs Using Mode and Type Information. In: Bruynooghe, M. (ed.) LOPSTR 2004. LNCS, vol. 3018, pp. 19–35. Springer, Heidelberg (2004)

64. López-García, P., Hermenegildo, M.: Efficient Term Size Computation for Granularity Control. In: Proc. of ICLP 1995 (1995)

65. López-García, P., Hermenegildo, M., Debray, S.K.: A Methodology for Granularity Based Control of Parallelism in Logic Programs. J. of Symbolic Computation, Special Issue on Parallel Symbolic Computation 21, 715–734 (1996)

66. López-García, P., Hermenegildo, M., Debray, S.K.: Towards Granularity Based Control of Parallelism in Logic Programs. In: Proc. of First International Symposium on Parallel Symbolic Computation, PASCO 1994 (1994)

67. Mera, E., López-García, P., Puebla, G., Carro, M., Hermenegildo, M.: Combining Static Analysis and Profiling for Estimating Execution Times. In: Hanus, M. (ed.) PADL 2007. LNCS, vol. 4354, Springer, Heidelberg (2006)

68. Montanari, U., Rossi, F., Bueno, F., García de la Banda, M., Hermenegildo, M.: Towards a Concurrent Semantics-based Analysis of CC and CLP. In: Principles and Practice of Constraint Programming, vol. 874, pp. 151–161. Springer, Heidelberg (1994)

69. Morales, J., Carro, M., Hermenegildo, M.: Improving the Compilation of Prolog to C Using Moded Types and Determinism Information. In: Jayaraman, B. (ed.) PADL 2004. LNCS, vol. 3057, pp. 86–103. Springer, Heidelberg (2004)
70. Mozilla. Tamarin Project (2008), http://www.mozilla.org/projects/tamarin/
71. Muthukumar, K., Bueno, F., García de la Banda, M., Hermenegildo, M.: Automatic Compile-time Parallelization of Logic Programs for Restricted, Goal-level, Independent And-parallelism. Journal of Logic Programming 38(2), 165–218 (1999)
72. Muthukumar, K., Hermenegildo, M.: Complete and Efficient Methods for Supporting Side Effects in Independent/Restricted And-parallelism. In: 1989 International Conference on Logic Programming, pp. 80–101. MIT Press, Cambridge (1989)
73. Muthukumar, K., Hermenegildo, M.: Determination of Variable Dependence Information at Compile-Time Through Abstract Interpretation. In: 1989 North American Conference on Logic Programming, pp. 166–189. MIT Press, Cambridge (1989)
74. Muthukumar, K., Hermenegildo, M.: The CDG, UDG, and MEL Methods for Automatic Compile-time Parallelization of Logic Programs for Independent And-parallelism. In: Int'l. Conference on Logic Programming, pp. 221–237. MIT Press, Cambridge (1990)
75. Muthukumar, K., Hermenegildo, M.: Combined Determination of Sharing and Freeness of Program Variables Through Abstract Interpretation. In: ICLP (1991)
76. Muthukumar, K., Hermenegildo, M.: Compile-time Derivation of Variable Dependency Using Abstract Interpretation. JLP 13(2/3), 315–347 (1992)
77. Navas, J., Mera, E., López-García, P., Hermenegildo, M.: User-definable resource bounds analysis for logic programs. In: Dahl, V., Niemelä, I. (eds.) ICLP 2007. LNCS, vol. 4670. Springer, Heidelberg (2007)
78. Necula, G.C., Condit, J., Harren, M., McPeak, S., Weimer, W.: CCured: type-safe retrofitting of legacy software. ACM Trans. Program. Lang. Syst. 27(3), 477–526 (2005)
79. Olmedilla, M., Bueno, F., Hermenegildo, M.: Automatic Exploitation of Non-Determinate Independent And-Parallelism in the Basic Andorra Model. In: Logic Program Synthesis and Transformation, 1993. Workshops in Computing, pp. 177–195. Springer, Heidelberg (1993)
80. Pineda, A., Bueno, F.: The O'Ciao Approach to Object Oriented Logic Programming. In: Colloquium on Implementation of Constraint and LOgic Programming Systems (ICLP associated workshop), Copenhagen (July 2002)
81. Puebla, G., Albert, E., Hermenegildo, M.: Abstract Interpretation with Specialized Definitions. In: Yi, K. (ed.) SAS 2006. LNCS, vol. 4134. Springer, Heidelberg (2006)
82. Puebla, G., Bueno, F., Hermenegildo, M.: An Assertion Language for Constraint Logic Programs. In: Deransart, P., Małuszyński, J. (eds.) DiSCiPl 1999. LNCS, vol. 1870, pp. 23–61. Springer, Heidelberg (2000)
83. Puebla, G., García de la Banda, M., Marriott, K., Stuckey, P.: Optimization of Logic Programs with Dynamic Scheduling. In: 1997 International Conference on Logic Programming, June 1997, pp. 93–107. MIT Press, Cambridge (1997)
84. Puebla, G., Hermenegildo, M.: Implementation of Multiple Specialization in Logic Programs. In: Proc. of PEPM 1995, pp. 77–87. ACM Press, New York (1995)
85. Puebla, G., Hermenegildo, M.: Abstract Multiple Specialization and its Application to Program Parallelization. JLP 41(2&3), 279–316 (1999)
86. Rigo, A., Pedroni, S.: PyPy's Approach to Virtual Machine Construction. In: Dynamic Languages Symposium 2006. ACM Press, New York (2006)
87. Saglam, H., Gallagher, J.: Approximating constraint logic programs using polymorphic types and regular descriptions. Technical Report CSTR-95-17, Department of Computer Science, University of Bristol, Bristol BS8 1TR (1995)

88. Schrijvers, T.: Analyses, Optimizations and Extensions of Constraint Handling Rules. PhD thesis, K.U.Leuven, Belgium (June 2005)
89. Somogyi, Z., Henderson, F., Conway, T.: The Execution Algorithm of Mercury: an Efficient Purely Declarative Logic Programming Language. JLP 29(1–3) (October 1996)
90. Warren, D.H.D.: Logic Programming Languages, Parallel Implementations, and the Andorra Model. Invited talk, slides presented at ICLP 1993 (1993)
91. Warren, R., Hermenegildo, M., Debray, S.K.: On the Practicality of Global Flow Analysis of Logic Programs. In: Fifth International Conference and Symposium on Logic Programming, pp. 684–699. MIT Press, Cambridge (1988)

AND/OR Multi-valued Decision Diagrams
for Constraint Networks

Robert Mateescu[1,*] and Rina Dechter[2]

[1] Electrical Engineering Department, California Institute of Technology
Pasadena, CA 91125
mateescu@paradise.caltech.edu

[2] Donald Bren School of Information and Computer Science, University of California, Irvine
Irvine, CA 92697
dechter@ics.uci.edu

Abstract. The paper is an overview of a recently developed compilation data structure for graphical models, with specific application to constraint networks. The AND/OR Multi-Valued Decision Diagram (AOMDD) augments well known decision diagrams (OBDDs, MDDs) with AND nodes, in order to capture function decomposition structure. The AOMDD is based on a pseudo tree of the network, rather than a linear ordering of its variables. The AOMDD of a constraint network is a canonical form given a pseudo tree. We describe two main approaches for compiling the AOMDD of a constraint network. The first is a top down, search-based procedure, that works by applying reduction rules to the trace of the memory intensive AND/OR search algorithm. The second is a bottom up, inference-based procedure, that uses a Bucket Elimination schedule. For both algorithms, the compilation time and the size of the AOMDD are, in the worst case, exponential in the *treewidth* of the constraint graph, rather than *pathwidth* as is known for ordered binary decision diagrams (OBDDs).

1 Introduction

The paper is an overview of AND/OR Multi-Valued Decision Diagrams (AOMDDs) as a compiled data structure for constraint networks. We present here an extension of the work in [1], while still maintaining the focus on constraint networks. AOMDDs for weighted graphical models and for constraint optimization were presented in [2,3].

The AOMDD is based on two existing frameworks: (1) AND/OR search spaces for graphical models; (2) binary decision diagrams (BDDs). AND/OR search spaces [4,5,6] have proven to be a unifying framework for various classes of search algorithms for graphical models. The main novelty is the exploitation of independencies between variables during search, which can provide exponential speedups over traditional search methods that can be viewed as traversing an OR structure. The AND nodes capture problem decomposition into *independent subproblems*, and the OR nodes represent branching according to variable values.

Decision diagrams are widely used in many areas of research, especially in software and hardware verification [7,8]. A BDD represents a Boolean function by a directed

---

* This work was done while at the University of California, Irvine.

P. Degano et al. (Eds.): Montanari Festschrift, LNCS 5065, pp. 238–257, 2008.

acyclic graph with two sink nodes (labeled 0 and 1). Every internal node is labeled with a variable and has exactly two children: *low* for 0 and *high* for 1. If isomorphic nodes (i.e., with the same label and identical children) are not merged, we have the full search *tree* explored by the backtracking algorithm. The tree is ordered if variables are encountered in the same order along every branch. The tree can then be compressed by merging isomorphic nodes, and by eliminating redundant nodes (i.e., whose *low* and *high* children are identical). The result is the celebrated *reduced ordered binary decision diagram*, or OBDD for short, introduced by Bryant [9]. However, the underlying structure is OR (i.e., linear structure rather than tree). If AND/OR search trees are reduced by node merging and redundant nodes elimination we get a compact search graph that can be viewed as a BDD representation augmented with AND nodes.

This paper shows how to combine the two ideas, creating a decision diagram that has an AND/OR structure, thus exploiting problem decomposition. As a detail, the number of values of variables is also increased from two to any constant. In the context of constraint networks, decision diagrams can be used to represent the whole set of solutions, facilitating solutions count, solution enumeration and queries on equivalence of constraint networks. The benefit of moving from OR structure to AND/OR is in a lower complexity of the algorithms and size of the compiled structure. It typically moves from being bounded exponentially in *pathwidth* pw^*, which is characteristic to chain decompositions or linear structures, to being exponentially bounded in *treewidth* w^*, which is characteristic of tree structures (it always holds that $w^* \leq pw^*$ and $pw^* \leq w^* \cdot \log n$). In both cases, the compiled structure achieved in practice is often far smaller than what the bounds suggest.

A decision diagram offers a compilation of a propositional knowledge-base. A multi-valued AND/OR decision diagram extends compilation to general graphical models. The *knowledge compilation* approach has become an important research direction in automated reasoning in the past decades [10,11,12]. Typically, a knowledge representation language is compiled into a compact data structure on which various queries can be answered quickly. Accordingly, the computational effort can be divided between an *offline* and an *online* phase where most of the work is pushed offline. Compilation can also be used to generate compact building blocks to be used by online algorithms multiple times. Macro-operators compiled during or prior to search can be viewed in this light [13], while in graphical models the building blocks are the functions whose compact, compiled, representation can be used effectively across many tasks.

As one example, consider product configuration tasks and imagine a user that chooses sequential options to configure a product. In a naive system, the user would be allowed to choose any valid option at the current level based only on the initial constraints, until either the product is configured, or else, when a dead-end is encountered, the system would backtrack to some previous state and continue from there. This would in fact be a search through the space of possible partial configurations. Needless to say, it would be very unpractical, and would offer the user no time guarantee. A system based on compilation would actually build the *backtrack-free* search space in the offline phase, and represent it as compactly as possible. In the online phase, only valid partial configurations (i.e., that can be extended to a full valid configuration) are allowed, and depending on the query type, response time guarantees can be offered in terms of the size of the compiled structure.

Numerous other examples, such as diagnosis and planning problems can be formulated as graphical models, and for which compilation would be useful. Compilation in diagnosis can facilitate fast detection of possible faults or explanations for some unusual behavior. In planning, compilation would allow swift adjustments according to changes in the environment. Probabilistic models are one of the most used types of graphical models, and the basic query is to compute conditional probabilities of some variables given the evidence. A compact compilation of a probabilistic model would allow fast response for any change in the evidence along time. Formal verification is another example where compilation is heavily used to compare equivalence of circuit design, or to check the behavior of a circuit. *Binary Decision Diagram* (BDD) [9] are arguably the most widely known and used compiled structure.

Our AOMDD proposal is related to two earlier research lines within the BDD literature. The first is the work on Disjoint Support Decompositions (DSD) [14], investigated within the area of design automation [15], that were proposed as enhancements for BDDs aimed at exploiting function decomposition. The second is the work on BDD trees [16]. Another related proposal is the recent work by Fargier and Vilarem [17] on compiling CSPs into tree-driven automata. We will comment more on the relationship between these work and AOMDD in the related work section.

The structure of the paper is as follows. Section 2 provides the preliminaries. Section 3 gives and overview of AND/OR search spaces. Section 4 introduces the AOMDD and Section 5 shows that it is a canonical form for constraint networks. Section 6 describes a search based algorithm for compiling the AOMDD. Section 7 presents a compilation algorithm based on a Bucket Elimination schedule and the APPLY operation. Section 8 presents related work and Section 9 concludes.

2   Preliminaries

A constraint network and its associated graph are defined in the usual way:

Definition 1 (constraint network). *A constraint network is a 3-tuple* $\mathcal{R} = \langle \mathbf{X}, \mathbf{D}, \mathbf{C} \rangle$, *where:* $\mathbf{X} = \{X_1, \ldots, X_n\}$ *is a set of variables;* $\mathbf{D} = \{D_1, \ldots, D_n\}$ *is the set of their finite domains of values, with cardinalities* $k_i = |D_i|$ *and* $k = \max_{i=1}^{n} k_i$ *;* $\mathbf{C} = \{C_1, \ldots, C_r\}$ *is a set of constraints over subsets of* $\mathbf{X}$. *Each constraint is defined as* $C = (S_i, R_i)$, *where* S_i *is the set of variables on which the constraint is defined, called its scope, and* R_i *is the relation defined on* S_i.

Definition 2 (constraint graph). *The* constraint graph *(or primal graph) of a constraint network is an undirected graph,* $G = (\mathbf{X}, E)$, *that has variables as its vertices and an edge connecting any two variables that appear in the scope (set of arguments) of the same constraint.*

A pseudo tree resembles the tree rearrangements introduced in [18]:

Definition 3 (pseudo tree). *A pseudo tree of a graph* $G = (\mathbf{X}, E)$ *is a rooted tree* $\mathcal{T}$ *having the same set of nodes* $\mathbf{X}$, *such that every arc in* E *is a backarc in* $\mathcal{T}$ *(i.e., it connects nodes on the same path from root).*

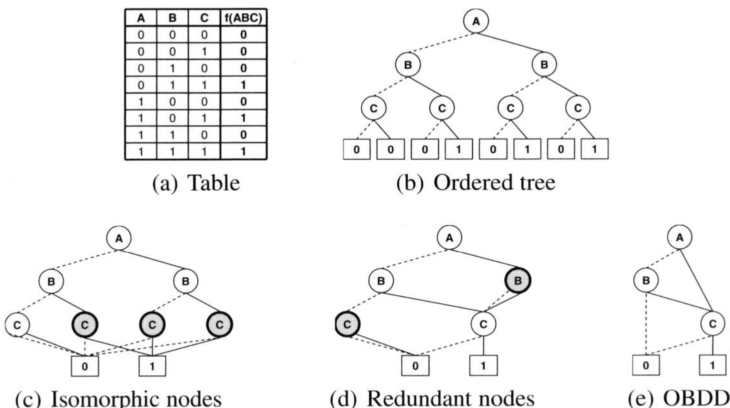

(a) Table   (b) Ordered tree

(c) Isomorphic nodes   (d) Redundant nodes   (e) OBDD

Fig. 1. Boolean function representation and reduction rules

Definition 4 (induced graph, induced width, treewidth, pathwidth). *An ordered graph is a pair (G, d), where G is an undirected graph, and $d = (X_1, ..., X_n)$ is an ordering of the nodes. The width of a node in an ordered graph is the number of neighbors that precede it in the ordering. The width of an ordering d, denoted $w(d)$, is the maximum width over all nodes. The induced width of an ordered graph, $w^*(d)$, is the width of the induced ordered graph obtained as follows: for each node, from last to first in d, its preceding neighbors are connected in a clique. The induced width of a graph, w^*, is the minimal induced width over all orderings. The induced width is also equal to the treewidth of a graph. The pathwidth pw^* of a graph is the treewidth over the restricted class of orderings that correspond to chain decompositions.*

2.1 Binary Decision Diagrams Review

Decision diagrams are widely used in many areas of research to represent decision processes. In particular, they can be used to represent functions. Due to the fundamental importance of Boolean functions, a lot of effort has been dedicated to the study of *Binary Decision Diagrams* (BDDs), which are extensively used in software and hardware verification [7,8]. Bryant [9] introduced the *Ordered Binary Decision Diagram* (OBDD). The order of variables along any path of an OBDD is the same. OBBDs provide a compact representation and efficient operations (the *apply* procedure, that combines two OBDDs by an operation is at most quadratic in the sizes of the input diagrams).

Example 1. Figure 1(a) shows a table representation of a Boolean function. A binary tree representation is shown in Figure 1(b). The internal round nodes represent the variables, the solid edges are the 1 (or high) value, and the dotted edges are the 0 (or low) value. The leaf square nodes show the value of the function for each assignment along a path. The tree is ordered, because variables appear in the same order along each path.

There are two reduction rules that transform a decision diagram into an equivalent one: *(1) isomorphism:* merge nodes that have the same label and the same children; *(2) redundancy:* eliminate nodes whose low and high edges point to the same node, and

complete, which means that there may still exist unifiable nodes in the search graph that do not have identical contexts. Moreover, some of the nodes in the context minimal AND/OR graph may be redundant, for example when the set of solutions rooted at variable X_i does not depend on the specific value assigned to X_i (this situation is not detectable based on context). This is sometimes termed as "interchangeable values" or "symmetrical values".

We propose to augment the minimal AND/OR search graph with removing redundant variables as is common in OBDD representation. This yields a data structure that we call AND/OR BDD, that exploits decomposition by using AND nodes. We present the extension over multi-valued variables yielding AND/OR MDD or AOMDD. Subsequently we present two algorithms for compiling the canonical AOMDD of a constraint network: the first is search based, and uses the memory intensive AND/OR graph search to generate the context minimal AND/OR graph, and then reduces it bottom up by applying reduction rules; the second is inference based, and uses a Bucket Elimination schedule to combine the AOMDDs of initial functions by APPLY operations (similar to the *apply* for OBDDs). Both approaches have the same worst case complexity as the AND/OR graph search with context based caching, and also the same complexity as Bucket Elimination, namely time and space exponential in the treewidth of the problem, $O(n\ k^{w^*})$.

4.1 From AND/OR Search Graphs to Decision Diagrams

We will now show how we can process an AND/OR search graph by reduction rules similar to the case of OBDDs, in order to obtain a representation of minimal size. In the case of OBDDs, a node is labeled with a variable name, for example A, and the *low* (dotted line) and *high* (solid line) outgoing arcs capture the restriction of the function to the assignments $A = 0$ or $A = 1$. To determine the value of the function, one needs to follow either one or the other (but not both) of the outgoing arcs from A. The straightforward extension of OBDDs to multi-valued variables (multi-valued decision diagrams, or MDDs) was presented in [22].

We generalize the OBDD and MDD representations by allowing each outgoing arc to be an AND arc. An AND arc connects a node to a set of nodes, and captures the decomposition of the problem into independent components.

We define the AND/OR Decision Diagram representation based on AND/OR search graphs. We find it useful to introduce the *meta-node* data structure, which defines small portions of any AND/OR graph, based on an OR node and its AND children:

Definition 11 (meta-node). *A meta-node u in an AND/OR search graph of a constraint network consists of an OR node labeled X (therefore $var(u) = X$) and its k AND children labeled $x_1, \ldots, x_k$ that correspond to the value assignments of X. Each AND node labeled x_i stores a list of pointers to child meta-nodes, denoted by $u.children_i$.*

We also define two special meta-nodes, that will play the role of the terminal nodes in OBDDs. The terminal meta-node **0** indicates the inconsistent assignments, while the terminal meta-node **1** indicates the consistent ones.

Any AND/OR search graph can now be viewed as a diagram of meta-nodes, simply by grouping OR nodes with their AND children, and adding the terminal meta-nodes appropriately.

It is now easy to see when a variable is redundant with respect to the outcome of the function based on the current partial assignment. Intuitively, any assignment to a redundant variable should lead to the same set of solutions.

Definition 12 (redundant meta-node). *Given an AND/OR search graph $\mathcal{G}$ represented with meta-nodes, a meta-node u with $var(u) = X$ and $|D(X)| = k$ is redundant iff $u.children_1 = \ldots = u.children_k$.*

An AND/OR graph $\mathcal{G}$, that contains a redundant meta-node u, can be transformed into an equivalent graph $\mathcal{G}'$ by replacing any incoming arc into u with its common list of children $u.children_1$ (joined in an AND arc), and then removing u and its outgoing arcs from $\mathcal{G}$.

The notion of isomorphism is extended naturally from AND/OR graphs to meta-nodes.

Definition 13 (isomorphic meta-nodes). *Given an AND/OR search graph $\mathcal{G}$ represented with meta-nodes, two meta-nodes u and v having $var(u) = var(v) = X$ and $|D(X)| = k$ are isomorphic iff $u.children_i = v.children_i$, $\forall i \in \{1, \ldots, k\}$.*

Naturally, the AND/OR graph obtained by merging isomorphic meta-nodes is equivalent to the original one. We can now define the AND/OR Multi-Valued Decision Diagram:

Definition 14 (AOMDD). *An AND/OR Multi-Valued Decision Diagram (AOMDD) is a weighted AND/OR search graph that is completely reduced by isomorphic merging and redundancy removal, namely:*

(1) it contains no isomorphic meta-nodes; and

(2) it contains no redundant meta-nodes.

5   AOMDDs for Constraint Networks Are Canonical Forms

It is well known that OBDDs are canonical representations of Boolean functions given an ordering of the variables [9], and this property extends to MDDs [22]. In the case of AOBDDs and AOMDDs, the canonicity is with respect to a pseudo tree, following the transition from total orders (that correspond to linear orderings) to partial orders (that correspond to pseudo tree orderings).

Proposition 1. *Let f be a function, not always zero, defined by a constraint network over $\mathbf{X}$. Given a partition $\{\mathbf{X}^1, \ldots, \mathbf{X}^m\}$ of the set of variables $\mathbf{X}$ (namely, $\mathbf{X}^i \cap \mathbf{X}^j = \phi, \forall\, i \neq j$, and $\mathbf{X} = \cup_{i=1}^{m} \mathbf{X}^i$), if $f = f_1 \otimes \ldots \otimes f_m$ and $f = g_1 \otimes \ldots \otimes g_m$, such that $scope(f_i) = scope(g_i) = \mathbf{X}^i$ for all $i \in \{1, \ldots, m\}$, then $f_i = g_i$ for all $i \in \{1, \ldots, m\}$. Namely, if f can be decomposed over the given partition, then the decomposition is unique.*

Based on the previous proposition, in can be shown that AOMDDs for constraint networks are canonical representations given a pseudo tree.

Theorem 4 (AOMDDs are canonical for a given pseudo tree). *Given a constraint network, and a pseudo tree T of its constraint graph, there is a unique (up to isomorphism) AOMDD that represents it, and it has the minimal number of meta-nodes.*

The proof, omitted for space reasons, is by structural induction over the depth of the pseudo tree.

A constraint network is defined by its relations (or functions). There exist equivalent constraint networks that are defined by different sets of functions, even having different scope signatures. However, equivalent constraint networks define the same function, and we can ask if the AOMDD of different equivalent constraint networks is the same. The following theorem can be derived immediately from Theorem 4.

Theorem 5. *Two equivalent constraint networks that admit the same pseudo tree T have the same AOMDD based on T.*

6   Using AND/OR Search to Generate AOMDDs

In Section 4.1 we described how we can transform an AND/OR graph into an AOMDD by applying reduction rules. In Section 6.1 we describe the explicit algorithm that takes as input a constraint network, performs AND/OR search with context-based caching to obtain the context minimal AND/OR graph, and in Section 6.2 we give the procedure that applies the reduction rules bottom up to obtain the AOMDD. The reduction procedure can actually be incorporated in the search algorithm, but we present it separately for clarity.

6.1   Algorithm AND/OR-SEARCH-AOMDD

Algorithm 1, called AND/OR-SEARCH-AOMDD, compiles a constraint network into an AOMDD. A memory intensive (with context-based caching) AND/OR search is used to create the context minimal AND/OR graph (see Definition 10). The input to AND/OR-SEARCH-AOMDD is a constraint network R and a pseudo tree T, that also defines the OR-context of each variable.

Each variable X_i has an associated cache table, whose scope is the context of X_i in T. This ensures that the trace of the search is the context minimal AND/OR graph. A list denoted by L^{X_i} (see line 34), is used for each variable X_i to save pointers to meta-nodes labeled with X_i. These lists are used by the procedure that performs the bottom up reduction, per layers of the AND/OR graph (one layer contains all the nodes labeled with one given variable). The fringe of the search is maintained on a stack called OPEN. The current node (either OR or AND node) is denoted by n, its parent by p, and the current path by π_n. The children of the current node are denoted by $successors(\text{n})$. For each node n, the Boolean attribute $consistent(\text{n})$ indicates if the current path can be extended to a solution. This information is useful for pruning the search space.

Algorithm 1: AND/OR SEARCH - AOMDD

> **input**   : $\mathcal{R} = \langle \mathbf{X}, \mathbf{D}, \mathbf{C} \rangle$; pseudo tree $\mathcal{T}$ rooted at X_1; parents pa_i (OR-context) for every variable X_i.
> **output** : AOMDD of $\mathcal{R}$.

1 **forall** $X_i \in \mathbf{X}$ **do**
2   Initialize context-based cache table $Cache_{X_i}(pa_i)$ with $\mathtt{null}$ entries

3 Create new OR node t, labeled with X_i; $consistent(\mathrm{t}) \leftarrow true$; push t on top of $\mathtt{OPEN}$
4 **while** $\mathtt{OPEN} \neq \phi$ **do**
5   n $\leftarrow top(\mathtt{OPEN})$; remove n from $\mathtt{OPEN}$ // **Forward**
6      $successors(\mathrm{n}) \leftarrow \phi$
7   **if** n *is an OR node labeled with* X_i **then**   // OR-expand
8   **if** $Cache_{X_i}(asgn(\pi_n)[pa_i]) \neq \mathtt{null}$ **then**
9   Connect parent of n to $Cache_{X_i}(asgn(\pi_n)[pa_i])$ // Use the cached pointer
10   **else**
11   **forall** $x_i \in D_i$ **do**
12   Create new AND node t, labeled with $\langle X_i, x_i \rangle$
13   **if** $\langle X_i, x_i \rangle$ **is consistent** *with* π_n **then**   // Constraint Propagation
14                   $consistent(\mathrm{t}) \leftarrow true$
15   add t to $successors(\mathrm{n})$
16   **else**
17                   $consistent(\mathrm{t}) \leftarrow false$
18   make terminal **0** the only child of t

19   **if** n *is an AND node labeled with* $\langle X_i, x_i \rangle$ **then**   // AND-expand
20   **if** $children_{\mathcal{T}}(X_i) == \phi$ **then**
21   make terminal **1** the only child of n
22   **else**
23   **forall** $Y \in children_{\mathcal{T}}(X_i)$ **do**
24   Create new OR node t, labeled with Y
25                $consistent(\mathrm{t}) \leftarrow false$
26   add t to $successors(\mathrm{n})$

27   Add $successors(\mathrm{n})$ to top of $\mathtt{OPEN}$
28   **while** $successors(\mathrm{n}) == \phi$ **do**   // **Backtrack**
29   let p be the parent of n
30   **if** n *is an OR node labeled with* X_i **then**
31   **if** $X_i == X_1$ **then**   // Search is complete
32   BottomUpReduction   // begin reduction to AOMDD
33             $Cache(asgn(\pi_n)[pa_i]) \leftarrow$ n   // Save in cache
34   Add meta-node of n to the list L^{X_i}
35             $consistent(\mathrm{p}) \leftarrow consistent(\mathrm{p}) \wedge consistent(\mathrm{n})$
36   **if** $consistent(\mathrm{p}) == false$ **then**   // Check if p is dead-end
37   remove $successors(\mathrm{p})$ from $\mathtt{OPEN}$
38                $successors(\mathrm{p}) \leftarrow \phi$

39   **if** n *is an AND node labeled with* $\langle X_i, x_i \rangle$ **then**
40             $consistent(\mathrm{p}) \leftarrow consistent(\mathrm{p}) \vee consistent(\mathrm{n})$;

41   remove n from $successors(\mathrm{p})$
42   n $\leftarrow$ p

The algorithm is based on two mutually recursive steps: **Forward** (beginning at line 5) and **Backtrack** (beginning at line 28), which call each other (or themselves) until the search terminates. In the forward phase, the AND/OR graph is expanded top down. The two types of nodes, AND and OR, are treated differently according to their semantics.

Before an OR node is expanded, the cache table of its variable is checked (line 8). If the entry is not $\mathtt{null}$, a link is created to the already existing OR node that roots the graph equivalent to the current subproblem. Otherwise, the OR node is expanded by generating its AND descendants. Each value x_i of X_i is checked for consistency (line 13). Any level of constraint propagation can be performed in this step (e.g., look ahead, arc consistency, path consistency, i-consistency etc.). The computational overhead can increase, in the hope of pruning the search space more aggressively. We should note that constraint propagation is not crucial for the algorithm, and the complexity guarantees are maintained even without it. The consistent AND nodes are added to the list of successors of n (line 15), while the inconsistent ones are linked to the terminal **0** meta-node (line 18).

An AND node n labeled with $\langle X_i, x_i \rangle$ is expanded (line 19) based on the structure of the pseudo tree. If X_i is a leaf in $\mathcal{T}$, then n is linked to the terminal **1** meta-node (line 21). Otherwise, an OR node is created for each child of X_i in $\mathcal{T}$ (line 23).

The forward step continues as long as the current node is not a dead-end and still has unevaluated successors. The backtrack phase is triggered when a node has an empty set of successors (line 28). Note that, as each successor is processed, it is removed from the set of successors in line 41. When the backtrack reaches the root (line 31), the search is complete, the context minimal AND/OR graph has been generated, and the Procedure BOTTOMUPREDUCTION is called.

When the backtrack step processes an OR node (line 30), it saves a pointer to it in cache, and also adds a pointer to the corresponding meta-node to the list L^{X_i}. The *consistent* attribute of the AND parent p is updated by conjunction with $consistent(\text{n})$. If the AND parent p becomes inconsistent, it is not necessary to check its remaining OR successors (line 37). When the backtrack step processes an AND node (line 39), the *consistent* attribute of the OR parent p is updated by disjunction with $consistent(\text{n})$.

The AND/OR search algorithm usually maintains a value for each node, corresponding to a task that is solved (e.g., counting solutions, or cost of the optimal solution). We did not include values in our description because an AOMDD is just an equivalent representation of the original constraint network $\mathcal{R}$. Any task over $\mathcal{R}$ can be solved by a traversal of the AOMDD. It is however up to the user to include more information in the meta-nodes (e.g., number of solutions for a subproblem).

6.2 Reducing the Context Minimal AND/OR Graph to an AOMDD

Procedure `BottomUpReduction` processes the variables bottom up relative to the pseudo tree $\mathcal{T}$. We use the depth first traversal ordering of $\mathcal{T}$ (line 1), but any other bottom up ordering is as good. The outer *for* loop (starting at line 11) goes through each level of the context minimal AND/OR graph (where a level contains all the OR and AND nodes labeled with the same variable, in other words it contains all the meta-nodes of that variable). For efficiency, and to ensure the complexity guarantees, a hash table, initially empty, is used for each level. The inner *for* loop (starting at line 11) goes through all the meta-nodes of a level, that are also saved (or pointers to them are saved) in the list L^{X_i}. For each new meta-node n in the list L^{X_i}, in line 6 the hash table H is checked to verify if a node isomorphic with n already exists. If the hash table H already contains a node p corresponding to the hash key $(X_i, n.children_1, \ldots, n.children_{k_i})$, then p and n are isomorphic and should be merged. Otherwise, if the new meta-node n

---

Procedure `BottomUpReduction`

> **input**   : A constraint network $\mathcal{R} = \langle \mathbf{X}, \mathbf{D}, \mathbf{C} \rangle$; a pseudo tree $\mathcal{T}$ of the primal graph, rooted at X_1;
>   Context minimal AND/OR graph, and lists L^{X_i} of meta-nodes for each level X_i.
> **output** : AOMDD of $\mathcal{R}$.

1 Let $d = \{X_1, \ldots, X_n\}$ be the depth first traversal ordering of $\mathcal{T}$
2 **for** $i \leftarrow n$ **down to** 1 **do**
3   Let H be a hash table, initially empty
4   **forall** *meta-nodes* n *in* L^{X_i} **do**
5   **if** $H(X_i, n.children_1, \ldots, n.children_{k_i})$ *returns a meta-node* p **then**
6   merge n with p in the AND/OR graph

7   **else if** n *is redundant* **then**
8   eliminate n from the AND/OR graph

9   **else**
10   hash n into the table H:
11 $H(X_i, n.children_1, \ldots, n.children_{k_i}) \leftarrow$ n

12 **return** *reduced AND/OR graph*

---

is redundant, then it is eliminated from the AND/OR graph. If none of the previous two conditions is met, then the new meta-node n is hashed into the table H.

Proposition 2. *The output of Procedure* `BottomUpReduction` *is the AOMDD of* $\mathcal{R}$ *along the pseudo tree* $\mathcal{T}$*, namely the resulting AND/OR graph is completely reduced.*

Note that we explicated Procedure `BottomUpReduction` separately only for clarity. In practice, it can actually be included in Algorithm AND/OR-SEARCH-AOMDD, and the reduction rules can be applied whenever the search backtracks. We can maintain a hash table for each variable, during the AND/OR search, to store pointers to meta-nodes. When the search backtracks out of an OR node, it can already check the redundancy of that meta-node, and also look up in the hash table to check for isomorphism. Therefore, the reduction of the AND/OR graph can be done during the AND/OR search, and the output will be the AOMDD of $\mathcal{R}$.

From Theorem 3 and Proposition 2 we can conclude:

Theorem 6. *Given a constraint network* $\mathcal{R}$ *and a pseudo tree* $\mathcal{T}$ *of its constraint graph* G*, the AOMDD of* $\mathcal{R}$ *corresponding to* $\mathcal{T}$ *has size bounded by* $O(n \; k^{w_{\mathcal{T}}^*(G)})$ *and it can be computed by Algorithm* AND/OR-SEARCH-AOMDD *in time* $O(n \; k^{w_{\mathcal{T}}^*(G)})$*, where* $w_{\mathcal{T}}^*(G)$ *is the induced width of* G *over the depth first traversal of* $\mathcal{T}$*, and* k *bounds the domain size.*

7   Using Bucket Elimination to Generate AOMDDs

In this section we propose to use a Bucket Elimination (**BE**) type algorithm to guide the compilation of a constraint network into an AOMDD. The idea is to express the constraints as AOMDDs, and then combine them via the APPLY operator by following a **BE** schedule. The APPLY is a procedure very similar to that from OBDDs [9], but it is adapted to AND/OR search graphs. It takes as input two constraints represented as AOMDDs based on the same pseudo tree, and outputs their join, also represented as an AOMDD based on the same pseudo tree.

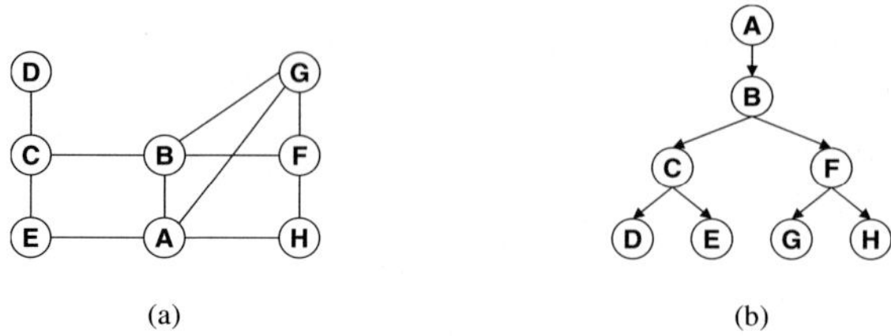

Fig. 4. (a) Constraint graph for $\mathbf{C} = \{C_1, \ldots, C_9\}$, where $C_1 = F \vee H$, $C_2 = A \vee \neg H$, $C_3 = A \oplus B \oplus G$, $C_4 = F \vee G$, $C_5 = B \vee F$, $C_6 = A \vee E$, $C_7 = C \vee E$, $C_8 = C \oplus D$, $C_9 = B \vee C$; (b) Pseudo tree (bucket tree) for ordering $d = (A, B, C, D, E, F, G, H)$

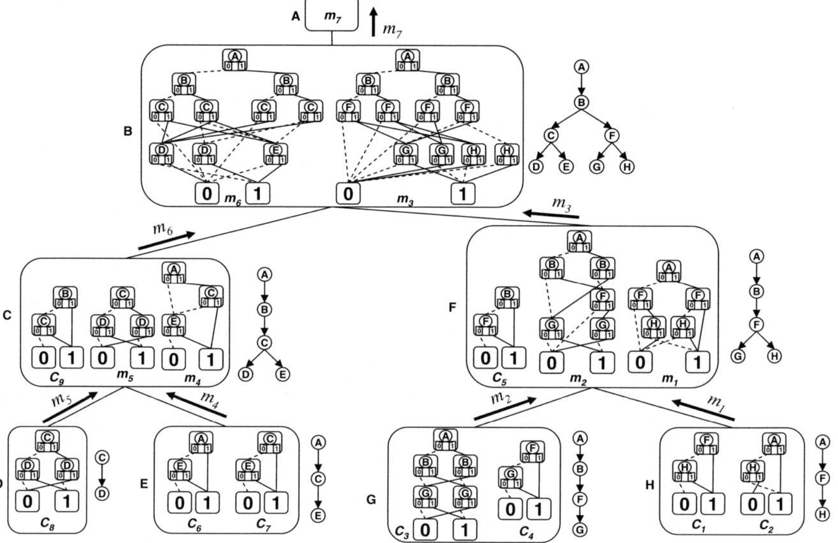

Fig. 5. Execution of BE with AOMDDs

Example 6. Consider the constraint network defined by $\mathbf{X} = \{A, B, \ldots, H\}$, $D_A = \ldots = D_H = \{0, 1\}$ and the constraints (where $\oplus$ denotes XOR): $C_1 = F \vee H$, $C_2 = A \vee \neg H$, $C_3 = A \oplus B \oplus G$, $C_4 = F \vee G$, $C_5 = B \vee F$, $C_6 = A \vee E$, $C_7 = C \vee E$, $C_8 = C \oplus D$, $C_9 = B \vee C$. The constraint graph is shown in Figure 4(a). Consider the ordering $d = (A, B, C, D, E, F, G, H)$. The pseudo tree (or bucket tree) induced by d is given in Fig. 4(b). Figure 5 shows the execution of **BE** with AOMDDs along ordering d. Initially, the constraints C_1 through C_9 are represented as AOMDDs and placed in the bucket of their latest variable in d. The scope of any original constraint always appears on a path from root to a leaf in the pseudo tree. Therefore, each *original* constraint is represented by an AOMDD based on a chain. (i.e., there is no branching into independent components at any point). The chain is just the scope of the constraint, ordered according to d. For bi-valued variables, the original constraints are represented by OBDDs, for multiple-valued variables they are MDDs. Note that we depict meta-nodes:

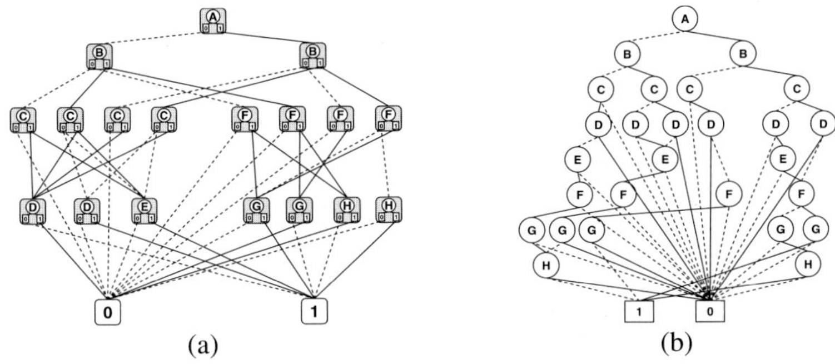

Fig. 6. (a) The final AOMDD; (b) The OBDD corresponding to d

one OR node and its two AND children, that appear inside each gray node. The dotted edge corresponds to the 0 value (the *low* edge in OBDDs), the solid edge to the 1 value (the *high* edge). We have some redundancy in our notation, keeping both AND value nodes and arc-types (doted arcs from "0" and solid arcs from "1").

The **BE** scheduling is used to process the buckets in reverse order of d. A bucket is processed by *joining* all the AOMDDs inside it, using the APPLY operator. However, the step of elimination of the bucket variable is omitted because we want to generate the full AOMDD. In our example, the messages $m_1 = C_1 \bowtie C_2$ and $m_2 = C_3 \bowtie C_4$ are still based on chains, so they are still OBDDs. Note that they still contain the variables H and G, which have not been eliminated. However, the message $m_3 = C_5 \bowtie m_1 \bowtie m_2$ is not an OBDD anymore. We can see that it follows the structure of the pseudo tree, where F has two children, G and H. Some of the nodes corresponding to F have two outgoing edges for value 1.

The processing continues in the same manner. The final output of the algorithm, which coincides with m_7, is shown in Figure 6(a). The OBDD based on the same ordering d is shown in Fig. 6(b). Notice that the AOMDD has 18 nonterminal nodes and 47 edges, while the OBDD has 27 nonterminal nodes and 54 edges.

7.1   Algorithm BE-AOMDD

Algorithm 2, called BE-AOMDD, creates the AOMDD of a constraint network by using a **BE** schedule for APPLY operations. Given an order d of the variables, a pseudo tree is created based on the constraint graph (this is just the bucket tree, or elimination tree of d). Each initial constraint C_i is then represented as an AOMDD, denoted by $\mathcal{G}_{C_i}^{aomdd}$, and placed in its bucket. To obtain the AOMDD of a constraint, its scope is ordered according to d, a search tree (based on a chain) that represents C_i is generated, and then reduced by Procedure `BottomUpReduction`. Then, the algorithm proceeds exactly like **BE**, with the only difference that the join of constraints (represented as AOMDDs) is realized by the APPLY algorithm, and variables are not eliminated but carried around to the destination bucket. The messages between buckets are initialized with the dummy AOMDD of **1**, denoted by $\mathcal{G}_1^{aomdd}$, which is neutral for join.

Algorithm 2: BE-AOMDD

> **input**   : Constraint network $\mathcal{R} = \langle \mathbf{X}, \mathbf{D}, \mathbf{C} \rangle$, where $\mathbf{X} = \{X_1, \ldots, X_n\}$, $\mathbf{C} = \{C_1, \ldots, C_r\}$;
> order $d = (X_1, \ldots, X_n)$
> **output**   : AOMDD representing $\bowtie_{i \in \mathbf{F}} C_i$
>
> **1** Let $\mathcal{T}$ be the pseudo tree (bucket tree) corresponding to d; **for** $i \leftarrow 1$ **to** r **do**   `// place constraints in buckets`
> **2**   place $\mathcal{G}_{C_i}^{aomdd}$ in the bucket of its latest variable in d
>
> **3** **for** $i \leftarrow n$ **down to** 1 **do**   `// process buckets`
> **4** $message(X_i) \leftarrow \mathcal{G}_{\mathbf{1}}^{aomdd}$ `// initialize with AOMDD of 1;`
> **5**   **while** $bucket(X_i) \neq \phi$ **do**   `// combine AOMDDs in bucket of $X_i$`
> **6**   pick $\mathcal{G}_f^{aomdd}$ from $bucket(X_i)$;
> **7** $bucket(X_i) \leftarrow bucket(X_i) \setminus \{\mathcal{G}_f^{aomdd}\}$;
> **8** $message(X_i) \leftarrow \text{APPLY}(message(X_i), \mathcal{G}_f^{aomdd})$
> **9**   add $message(X_i)$ to the bucket of the parent of X_i in $\mathcal{T}$
>
> **10** **return** $message(X_1)$

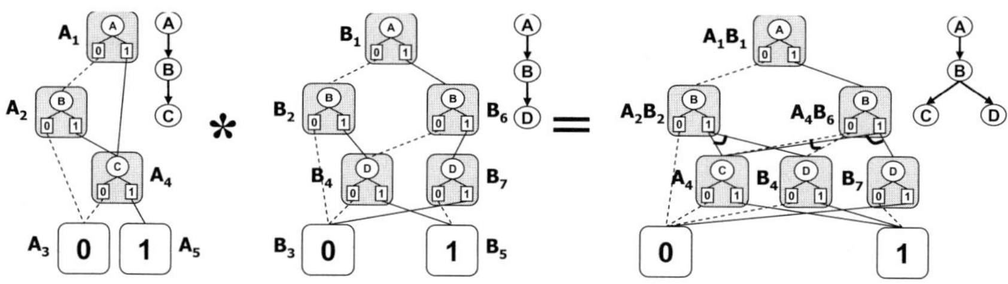

Fig. 7. Example of APPLY operation

7.2   The AOMDD APPLY Operation

The apply operator takes as input two AOMDDs representing constraints C_1 and C_2 and returns an AOMDD representing their join $C_1 \bowtie C_2$. In OBDDs the *apply* operator combines two input diagrams based on the same variable ordering. Likewise, in order to combine two AOMDDs we assume that they are based on the same pseudo tree. This restriction is satisfied when we use APPLY to combine the constraints in the same bucket of the **BE** based algorithm. There are also more relaxed version of APPLY, when the pseudo trees of the two input constraints need only be *compatible*, rather than identical. Intuitively, this means that the pseudo trees generate partial orders (based on descendance relation) that are not in conflict. For space reasons, we will not present the details of the APPLY algorithm here, but refer the reader to [1]. We will just briefly describe it by an example.

Example 7. Figure 7 shows the result of combining two Boolean functions by an AND operation (or product). The input functions f and g are represented by AOMDDs based on chain pseudo trees, while the results is based on the pseudo tree that expresses the decomposition after variables A and B are instantiated. The APPLY operator performs a depth first traversal of the two input AOMDDs, and generates the resulting AOMDD based on the output pseudo tree. Similar to the case of OBDDs, a function or an AOMDD can be identified by its root meta-node. In this example the input meta-nodes have labels ($A_1, A_2, B_1, B_2,$ *etc.*). The output meta-node labeled by A_2B_2 is the root of a diagram that represents the function obtained by combining the functions rooted by A_2 and B_2.

The complexity of BE-AOMDD and the output size are similar to those of the search based algorithm:

Theorem 7. *The space complexity of* BE-AOMDD *and the size of the output AOMDD are* $O(n\ k^{w^*})$, *where* n *is the number of variables,* k *is the maximum domain size and* w^* *is the treewidth of the bucket tree. The time complexity is bounded by* $O(r\ k^{w^*})$, *where* r *is the number of initial functions.*

8   Related Work

There are various lines of related research. The formal verification literature, beginning with [9] contains a very large number of papers dedicated to the study of BDDs. However, BDDs are in fact OR structures (the underlying pseudo tree is a chain) and do not take advantage of the problem decomposition in an explicit way. The complexity bounds for OBDDs are based on *pathwidth* rather than *treewidth*.

As noted earlier, the work on Disjoint Support Decomposition (DSD) is related to AND/OR BDDs in various ways [14]. The main common aspect is that both approaches show how structure decomposition can be exploited in a BDD-like representation. DSD is focused on Boolean functions and can exploit more refined structural information that is inherent to Boolean functions. In contrast, AND/OR BDDs assumes only the structure conveyed in the constraint graph. They are therefore more broadly applicable to any constraint expression and also to graphical models in general. They allow a simpler and higher level exposition that yields graph-based bounds on the overall size of the generated AOMDD.

McMillan introduced the BDD trees [16], along with the operations for combining them. For circuits of bounded tree width, BDD trees have linear space upper bound $O(|g|2^{w2^{2w}})$, where $|g|$ is the size of the circuit g (typically linear in the number of variables) and w is the treewidth. This bound hides some very large constants to claim the linear dependence on $|g|$ when w is bounded. However, McMillan maintains that when the input function is a CNF expression BDD-trees have the same bounds as AND/OR BDDs, namely they are exponential in the treewidth only.

The AND/OR structure restricted to propositional theories is very similar to deterministic decomposable negation normal form (d-DNNF) [11]. More recently, in [23], the trace of the DPLL algorithm is used to generate an OBDD, and compared with the typical formal verification approach of combining the OBDDs of the input function according to some schedule. The structures that were investigated are still OR.

McAllester [24] introduced the case factor diagrams (CFD) which subsume Markov random fields of bounded tree width and probabilistic context free grammars (PCFG). CFDs are very much related to the AND/OR graphs. The CFDs target the minimal representation, by exploiting decomposition (similar to AND nodes) but also by exploiting context sensitive information and allowing dynamic ordering of variables based on context. CFDs do not eliminate the redundant nodes, and part of the cause is that they use zero suppression. There is no claim about CFDs being a canonical form, and also there is no description of how to combine two CFDs.

More recently, independently and in parallel to our work on AND/OR graphs [4,5], Fargier and Vilarem [17] proposed the compilation of CSPs into tree-driven automata, which have many similarities to our work. Their main focus is the transition from linear automata to tree automata (similar to that from OR to AND/OR), and the possible savings for tree-structured networks and hyper-trees of constraints due to decomposition. Their compilation approach is guided by a tree-decomposition while ours is guided by a variable-elimination based algorithms. And, it is well known that Bucket Elimination and cluster-tree decomposition are in principle, the same [25].

9   Conclusion

This paper gives an overview of a new compilation data structure for constraint networks. The AND/OR Multi-valued Decision Diagram (AOMDD) [1,2,3] emerges from the study of AND/OR search spaces for graphical models [4,5,26,6] and ordered binary decision diagrams (OBDDs) [9]. Graphical models algorithms that are search-based and compiled data-structures such as BDDs differ primarily by their choices of time vs memory. When we move from regular OR to an AND/OR search space, the spectrum of algorithms available is improved. We believe that the AND/OR search space clarifies the available choices and helps guide the user into making an informed selection of the algorithm that would fit best the particular query asked, the specific input function and the available computational resources.

We presented the two main algorithmic approaches for compiling an AOMDD for constraint networks. The first is a top down procedure, that uses memory intensive AND/OR search, and applies reduction rules to the trace of the search. The second is a bottom up procedure that uses a Bucket Elimination schedule to combine the constraints via the APPLY operator.

As part of our current and future work, we are implementing, and experimenting with, the algorithms described here in order to provide an empirical evaluation.

Acknowledgments

This work was supported in part by the NSF grants IIS-0412854 and IIS-0713118.

References

1. Mateescu, R., Dechter, R.: Compiling constraint networks into AND/OR multi-valued decision diagrams (AOMDDs). In: Benhamou, F. (ed.) CP 2006. LNCS, vol. 4204, pp. 329–343. Springer, Heidelberg (2006)
2. Mateescu, R., Dechter, R.: And/or multi-valued decision diagrams (aomdds) for weighted graphical models. In: Proceedings of the 23rd Conference on Uncertainty in Artificial Intelligence, UAI 2007 (2007)
3. Mateescu, R., Marinescu, R., Dechter, R.: AND/OR multi-valued decision diagrams for constraint optimization. In: Bessière, C. (ed.) CP 2007. LNCS, vol. 4741, pp. 498–513. Springer, Heidelberg (2007)

4. Dechter, R., Mateescu, R.: Mixtures of deterministic-probabilistic networks and their AND/OR search space. In: Proceedings of the 20th Conference on Uncertainty in Artificial Intelligence, UAI 2004, pp. 120–129 (2004)
5. Dechter, R., Mateescu, R.: The impact of AND/OR search spaces on constraint satisfaction and counting. In: Wallace, M. (ed.) CP 2004. LNCS, vol. 3258, pp. 731–736. Springer, Heidelberg (2004)
6. Dechter, R., Mateescu, R.: AND/OR search spaces for graphical models. Artificial Intelligence 171, 73–106 (2007)
7. Clarke, E., Grumberg, O., Peled, D.: Model Checking. MIT Press, Cambridge (1999)
8. McMillan, K.L.: Symbolic Model Checking. Kluwer Academic Publishers, Dordrecht (1993)
9. Bryant, R.E.: Graph-based algorithms for boolean function manipulation. IEEE Transactions on Computers 35, 677–691 (1986)
10. Selman, B., Kautz, H.: Knowledge compilation and theory approximation. Journal of the ACM 43, 193–224 (1996)
11. Darwiche, A., Marquis, P.: A knowledge compilation map. Journal of Artificial Intelligence Research (JAIR) 17, 229–264 (2002)
12. Cadoli, M., Donini, F.M.: A survey on knowledge compilation. AI Communications 10, 137–150 (1997)
13. Korf, R., Felner, A.: Disjoint pattern database heuristics. Artificial Intelligence 134, 9–22 (2002)
14. Bertacco, V., Damiani, M.: The disjunctive decomposition of logic functions. In: International Conference on Computer-Aided Design (ICCAD), pp. 78–82 (1997)
15. Brayton, R., McMullen, C.: The decomposition and factorization of boolean expressions. In: ISCAS, Proceedings of the International Symposium on Circuits and Systems, pp. 49–54 (1982)
16. McMillan, K.L.: Hierarchical representation of discrete functions with application to model checking. In: Computer Aided Verification, pp. 41–54 (1994)
17. Fargier, H., Vilarem, M.: Compiling csps into tree-driven automata for interactive solving. Constraints 9, 263–287 (2004)
18. Freuder, E.C., Quinn, M.J.: Taking advantage of stable sets of variables in constraint satisfaction problems. In: Proceedings of the Ninth International Joint Conference on Artificial Intelligence (IJCAI 1985), pp. 1076–1078 (1985)
19. Dechter, R.: Bucket elimination: A unifying framework for reasoning. Artificial Intelligence 113, 41–85 (1999)
20. Bayardo, R., Miranker, D.: A complexity analysis of space-bound learning algorithms for the constraint satisfaction problem. In: Proceedings of the Thirteenth National Conference on Artificial Intelligence (AAAI 1996), pp. 298–304 (1996)
21. Darwiche, A.: Recursive conditioning. Artificial Intelligence 125, 5–41 (2001)
22. Srinivasan, A., Kam, T., Malik, S., Brayton, R.K.: Algorithms for discrete function manipulation. In: International Conference on Computer-Aided Design (ICCAD), pp. 92–95 (1990)
23. Huang, J., Darwiche, A.: Dpll with a trace: From sat to knowledge compilation. In: Proceedings of the Nineteenth International Joint Conference on Artificial Intelligence (IJCAI 2005), pp. 156–162 (2005)
24. McAllester, D., Collins, M., Pereira, F.: Case-factor diagrams for structured probabilistic modeling. In: Proceedings of the 20th Conference on Uncertainty in Artificial Intelligence, UAI 2004, pp. 382–391 (2004)
25. Dechter, R., Pearl, J.: Tree clustering for constraint networks. Artificial Intelligence 38, 353–366 (1989)
26. Mateescu, R., Dechter, R.: The relationship between AND/OR search and variable elimination. In: Proceedings of the 21st Conference on Uncertainty in Artificial Intelligence, UAI 2005, pp. 380–387 (2005)

Software Engineering: Ugo Montanari's Main Contributions and Introduction to the Section

Stefania Gnesi

Istituto di Scienza e Tecnologie dell'Informazione "A. Faedo", ISTI - CNR, Pisa

Ugo Montanari began to work in software engineering related topics in the early eighties when he started to promote the use of formal techniques in industries. In that period Ugo established a strong cooperation mainly with Olivetti, that was for more than a decade a very active italian ICT company.

Those were also the years when the first Progetto Finalizzato Informatica started and in this context Ugo chaired the P1 subproject, "Industria Nazionale del settore: Architettura e Struttura dei Sistemi di Elaborazione" whose aims were: i) the development of a prototypical local network meant also to be used to develop software products for the public administration; ii) the realization of a prototypical micro processor with dependability, availability and reconfigurability characteristics to be employed for industrial automation; iii) the development of methods and software programs to be used as basis for the software production.

Continuing the involvement of Olivetti a project was funded by the European Community under the multi-annual programme CEC MAP for the development of an Ada Compiler and Programming Systems. Few years later another CEC MAP project on the Formal Definition of Ada was funded and very successfully carried out by a consortium that included Università di Genova with Prof. Egidio Astesiano and University of Copenhagen with Prof. Dines Bjørner. As a follow up of these projects an interesting workshop on Software Factories and Ada was organized [1] by Ugo and Nico Habermann with the aims of presenting advances in software engineering and their relations with the development of the Ada language. In the paper [2] presented at the 1st European Software Engineering Conference, Ugo proposed an execution environment for the formal definition of Ada based on a logic programming approach. The aim of this paper was to evaluate the feasibility and effectiveness of an interpreter based on the formal definition of Ada.

The new directions in software development became a very important forum of discussions as the basis for technological and theoretical advances with the purpose of making the software production more rigorous. In 1987, Ugo was one of the organizers of the Advanced Seminar on Foundations of Innovative Software Development as part of the International Joint Conference on Theory and Practice of Software Development (TAPSOFT) [3,4].

Based also on these project experiences Ugo has always maintained strong contacts with the Italian ICT industry, establishing cooperation with companies such as SELENIA, (nowadays Selex), the research branch of Telecom Italia CSELT (nowadays TILAB) and many others. Cooperations that have been pursued within national and international research projects in the software engineering area, such as for example the recent participation to the EU projects Agile: Architectures for Mobility Information

P. Degano et al. (Eds.): Montanari Festschrift, LNCS 5065, pp. 258–260, 2008.

Societies Technology [5] and SENSORIA: Software Engineering for Service-Oriented Overlay Computers, both chaired by Prof. Martin Wirsing; to the Italian MUR projects "Architetture Software ad Alta Qualità di Servizio per Global Computing su Cooperative Wide Area Networks" and Tocai "Tecnologie Orientate alla Conoscenza per Aggregazioni di Imprese per Internet ".

From the research point of view, the activity of Ugo in the software engineering area was mainly in topics related to Software Architectures. An important issue in the area of software architecture is the specification of reconfiguration and mobility of systems proposing the use of graphs as modeling framework [6,7,8,9]. The architectural design of systems deals with the high level structuring of configurations. Checking that a system belongs to an architectural style (or shape) implies that the architecture is an instance of a structurally defined class. In [10] an approach for representing hierarchical software architecture shapes using types was proposed.

Organization of the Software Engineering Section

Seven papers belong to this section, most of them are from friends that worked with Ugo in the projects we have mentioned above.

The first one is by Egidio Astesiano, Gianna Reggio and Filippo Ricca: "Modeling Business within a UML-based Rigorous Software Development Approach". In this paper the authors provided an attempt at showing that software system development and business modeling can be aligned and that this means not only that there must be a strict correlation between the two, but that it could be necessary to adopt for both the same conceptual frame and notation. To that end, the authors present an approach, fully integrated within a UML-based rigorous model-driven method.

The second one is by Dines Bjørner: "From Domain to Requirements". In this paper a view of requirement engineering and its relation with domain engineering is shown, presenting a summary of the essentials of domain engineering and then the essence of two aspects of requirements: the domain requirements and the interface requirements prescriptions as they relate to domain descriptions.

A framework for organizational knowledge management based on Business Process Modeling (BPM), that is the main modeling practice connecting the management and engineering disciplines in software development, is presented in the paper "Business Process Modeling for Organizational Knowledge Management" by Luca Abeti, Paolo Ciancarini and Rocco Moretti.

In the paper "Event-based Service Coordination" by Gianluigi Ferrari, Roberto Guanciale, Daniele Strollo and Emilio Tuosto, the problem of designing and implementing a framework for programming service coordination policies is tackled presenting the prototype implementation of Java Signal Core Layer (JSCL), a coordination middleware for services based on the event notification paradigm.

The main motivations that lead to the present need for supporting continuous software evolution have been analyzed by Carlo Ghezzi, Paola Inverardi, and Carlo Montangero in the paper "Dynamically Evolvable Dependable Software From Oxymoron To Reality"

In "The Temporal Logic of Rewriting: A Gentle Introduction" by José Meseguer the temporal logic of rewriting TLR* is presented. It extends the CTL* logic adding action atoms, in the form of spatial action patterns. Semantically and pragmatically, however, when used together with rewriting logic as a 'tandem' of system specification and property specification logics, it has substantially more expressive power than purely state-based logics like CTL* or purely action-based logics like ACTL*.

The last paper of this section is "A Heterogeneous Approach to UML Semantics" by Mara Victoria Cengarle, Alexander Knapp, Andrzej Tarlecki and Martin Wirsing. In this paper a heterogeneous approach to the semantics of UML is proposed, where each diagram type can be described in its 'natural' semantics, and the relations between diagram types are expressed by appropriate translations.

References

1. Habermann, A.N., Montanari, U. (eds.): System Development and Ada. LNCS, vol. 275. Springer, Heidelberg (1987)
2. Fantechi, A., Gnesi, S., Inverardi, P., Montanari, U.: An Executon Environment for the Formal Definiton of Ada. In: Nichols, H.K., Simpson, D. (eds.) ESEC 1987. LNCS, vol. 289, pp. 327–335. Springer, Heidelberg (1987)
3. Ehrig, H., Kowalski, R.A., Levi, G., Montanari, U.: CAAP 1987 and TAPSOFT 1987. LNCS, vol. 249. Springer, Heidelberg (1987)
4. Ehrig, H., Kowalski, R.A., Levi, G., Montanari, U.: TAPSOFT 1987 and CFLP 1987. LNCS, vol. 250. Springer, Heidelberg (1987)
5. Andrade, L.F., et al.: AGILE: Software Architecture for Mobility. In: Wirsing, M., Pattinson, D., Hennicker, R. (eds.) WADT 2003. LNCS, vol. 2755, pp. 1–33. Springer, Heidelberg (2003)
6. Hirsch, D., Montanari, U.: Consistent transformations for software architecture styles of distributed systems. Electr. Notes Theor. Comput. Sci. 28 (1999)
7. Hirsch, D., Inverardi, P., Montanari, U.: Modeling Software Architecutes and Styles with Graph Grammars and Constraint Solving. In: WICSA 1999 Software Architecture, TC2 First Working IFIP Conference on Software Architecture (WICSA1). IFIP Conference Proceedings, San Antonio, Texas, USA, 22-24 February 1999, pp. 127–144 (1999)
8. Hirsch, D., Inverardi, P., Montanari, U.: Reconfiguration of Software Architecture Styles with Name Mobility. In: Porto, A., Roman, G.-C. (eds.) COORDINATION 2000. LNCS, vol. 1906, pp. 148–163. Springer, Heidelberg (2000)
9. Hirsch, D., Montanari, U.: Two Graph-Based Techniques for Software Architecture Reconfiguration. Electr. Notes Theor. Comput. Sci. 51 (2001)
10. Hirsch, D., Montanari, U.: Shaped Hierarchical Architectural Design. Electr. Notes Theor. Comput. Sci. 109, 97–109 (2004)

Modeling Business within a UML-Based Rigorous Software Development Approach

Egidio Astesiano[1], Gianna Reggio[1], and Filippo Ricca[2]

[1] DISI, Università di Genova, Italy
{astes, gianna.reggio}@disi.unige.it
[2] Unità CINI at DISI*, Genova, Italy
filippo.ricca@disi.unige.it

Abstract. We share the view that software system development and business modeling have to be aligned. In our opinion that means not only that there must be a strict correlation between the two, but that we should adopt both the same conceptual frame and notation, enforcing a seamless activity flow between them. In this paper we offer an attempt at showing that such an easy bridge can be provided. To that end we present an approach, fully integrated within a UML-based rigorous model-driven method, where business processes are viewed in the context of the overall business structure and may be modeled hierarchically at various levels of detail.

1 Introduction

The term Business Modeling (BM) has been used in the years with different meanings. For example, at the beginning of the years 2000 BM has been introduced in the well-known RUP (Rational Unified Process Model) model for software development as a "discipline", alongside Requirements and Analysis & Design, to mean what was then usually called Domain Model, but with some emphasis for its use in the development of business applications. Nowadays, BM still keeps a close relationship with domain modeling, but its link with business organization and goals is more stringent.

To put our work into perspective, we may distinguish different purposes, viewpoints, and trends in the current treatment of this subject. The first distinction comes from the focus: either more on the business management side or on the side of the information systems supporting the business. In the domain of Business Process Re-engineering and Enterprise Architecture, BM is focused on the aim of analyzing, organizing and managing the business activities. On the other side, BM is viewed as preliminary to and integrated with the development of information systems. For example in [1] "the business model is used to express the part played by the product (system or component) being developed in the context of the business that will fund its development (or purchase it) and use it"; and [2] is concerned with "the alignment of business processes and IT".

A further distinction on both sides is on the main concern and the granularity level of the modeling. Some approaches are only concerned with the process

---

* Laboratorio Iniziativa Software FINMECCANICA/ELSAG spa - CINI.

P. Degano et al. (Eds.): Montanari Festschrift, LNCS 5065, pp. 261–277, 2008.

flow and there BM is in essence the description of the processes composing the business. Others are instead more comprehensive, paying attention to the business context; among those, there is a varying degree of concern for the economic aspects, including risk analysis and heuristics for process identification, or for the operational aspects, such as the process activities.

In all the approaches there seems to be a consensus for the need or the relevance, at least, of a supporting notation. Still such consensus has not been reached about the choice of a notation, though currently BPMN (Business Process Modeling Notation, see [3]), and (part of) the UML are clearly the favorite. In particular, it has been shown how BPMN can be replaced by the UML Activity Diagrams (see [4]) at the process definition level, while, at a more or less "comprehensive level", various UML Profiles for Business Modeling have been proposed (see, e.g., for UML 1.4 [5] and for UML 2 [2], also for references).

Also on the basis of some experience on the industry side, we very much agree with the point put forward in [2] that "most software developers are not aware of business processes or are not able to read the models". Thus we are convinced of the need of building a bridge between the business and the system development side, or better the need of providing methodological tools for a seamless flow of activity, from business to systems. This is a point much advocated in [6], where the contiguity, and even a substantial overlapping of BM and system development, especially according to the SOA paradigm [7], is argued. For us too, a motivation of our work is to pave the way to an efficient use for the SOA paradigm. Indeed we believe that, in order to integrate Business Analysis and SOA software development, we first need means to treat relevant services and business processes within the same notation frame and with a clearly indicated correlation.

To that end, we devote this paper to present the integration of Business Modeling, including Process Modeling, within *MARS*, a UML based Model-driven Adaptively Rigorous approach to Software development [8,9,10]. *MARS* falls into the class of what we have called well-founded methods [11], namely it enforces a tighter and more precise structuring of the artifacts for the different phases of the software development process, than required by most MDA [12] compliant methods. That characteristic helps inexperienced developers speed-up the process and at the same time facilitates the consistency checks among the various artifacts, and hence their final quality. Moreover, *MARS* strives to balance formalism and easiness of use: the formal background provides the foundational rigor but is kept hidden from the developer.

Here we summarize the key ideas of the work presented in this paper:

- adopting as visual notation UML 2.0 and following a multiview approach;
- modeling a business process not in isolation, but within the context of the various processes composing the business, and of their mutual relationships (inclusion, specialization, dependency, ...);
- modeling all the entities involved in the business or at the business boundary (i.e., part of the domain in which the business operates) and their various aspects (e.g., organizational aspects);
- precise modeling of the activities that build up the business processes.

There are, we believe, two main reasons for choosing a fully UML notation, including the process description: first it gives the possibility of defining the Business Process Modeling in its context (the mentioned comprehensive approach, as in [2]); moreover it allows to keep BM into the same frame of the software development, not running the risk of facing a different paradigm and even different meanings of terms, as it would be if we used for example BPMN for the process description (e.g., BPMN is not object-oriented). Furthermore, *MARS* tries to propose consistent ways to handle common aspects in the various development phases; for example a **Data View** is part of the Requirement and of the Design model (since *MARS* follows the principle of not confusing data types and objects), and so also the Business Model will have a **Data View**.

We illustrate our proposal by means of a running case study **ASSOC** where "the business" consists in organizing and running an association of persons, not further detailed. In the following sections we present first the Overall Structure of the BM, then the Business Context and, separately, the Business Processes. A section is then devoted to discuss and compare some related work, before providing our conclusions. For readers non-expert in UML we have added in the footnotes some explanations for the less common constructs.

2   Business Modelling: The Overall Structure

The overall structure of a **Business Model** following the *MARS* method [8,9,10] is shown in Fig. 1 by means of a UML class diagram.

We propose to model all the entities that are part of the business by the **Static View**, which is essentially a UML class diagram. Such entities are then classified into business objects, business workers, and external. Further orthogonal classifications may be introduced (e.g., autonomous versus passive, human versus hardware/software system). The way the business workers are organized may be also modelled by means of the **Organization View**. The **Business Process View** then presents the business processes that define how the business is carried out. It consists of the **Business Process Overview Diagram**, that is a summary of the various processes and of their mutual relationships, together with the descriptions of all the processes, the **Business Process Description**. The description of a business process mainly presents the business entities taking part in it, and the activities carried out. The **Data View** lists and makes precise all the data types used to type attributes and operations appearing in the other views.

In the paper we use as a running example the case study **ASSOC**, where "the business" is managing an association of persons, not further detailed. The association is established at some point and may terminate subsequently. The persons may join and leave the association at any time. The association is run by a board elected by all the associates and chaired by one of its members. A secretary, which is not an associate, will take care of the administrative tasks.

In the following sections we present the various parts of the **Business Model**, illustrating them on the **ASSOC** case study.

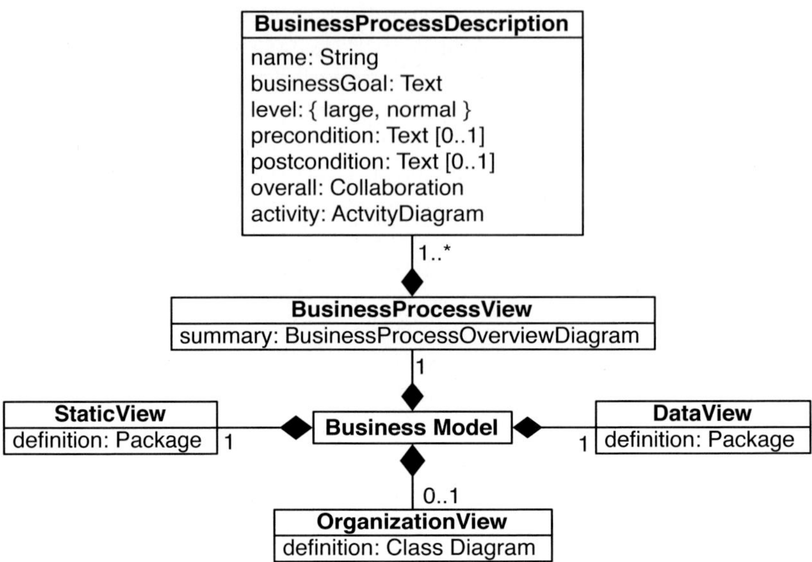

Fig. 1. Business Model Structure

3   Business Modeling: The Context

The context of the business, in our approach, consists of the involved business
entities and of how they are structured and organized.

3.1   Data View

The Data View introduces all the non-predefined data types needed in the other
views. It is useful to guarantee a better coherence among the various views
introducing a common basic vocabulary, and help distinguish in the Static View
a business object and the data used to describe it. Consider, for example, the case
of a business object Order that has an attribute deliverAddress of class Address:
the class Address will be given in the Data View, whereas the class Order will
appear in the Static View; thus there will be no doubt whether the address is a
business object or not.

Technically, the Data View is a UML package containing a class diagram,
containing only data types[1] and where the relationships are either specialization
or aggregation or composition.

In Fig. 2 we present the Data View of the Business Model relative to the ASSOC
case study. It defines six data structures: the votes (notice that a vote may also

---

[1] A UML data type is a classifier whose instances are pure values, i.e., they have no
identity and their state cannot be changed; thus the operations of a data type are
all queries, i.e., pure functions.

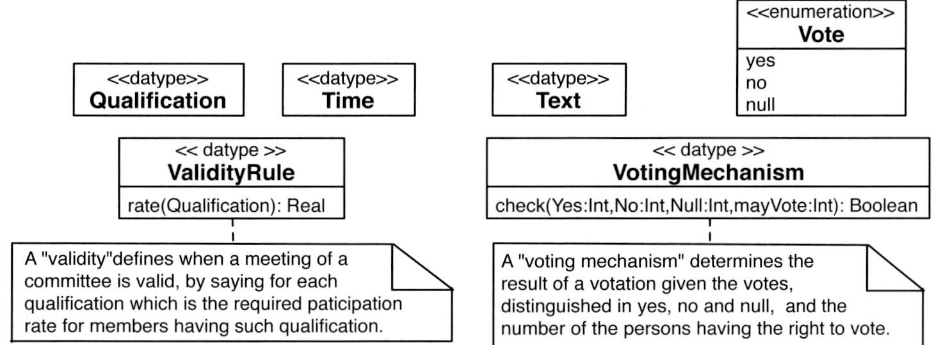

Fig. 2. ASSOC Business Model: Data View

be null, i.e., an abstention), the qualification of the associates having different rights and duties, the voting mechanism (how to vote on an item on the agenda of a meeting), the validity rules (defining when a meeting is valid, depending on the number of participants of the various qualifications), the time and text. Notice how the flexibility of the UML notation allow the modeler to choose the preferred degree of precision; for example, some data types, as Qualification and Text, are not detailed at all, and others, as VotingMechanism, are abstractly modeled without detailing their internal structure.

3.2   Static View

The Static View, a UML package containing at least one class diagram, makes explicit which are the entities appearing in the business (modeled by objects whose classes appear in such diagram) and their mutual relationships (modeled by associations among the corresponding classes), if any. The other elements of the class diagram, such as class attributes, operations and constraints, and the other diagrams in the package (such as state machines defining the behaviour of some classes) may be used to model relevant aspects of such entities. In the Static View we may use the following stereotypes to classify the business entities:

≪**bw**≫   business workers, those entities performing the basic actions of the business;

≪**bo**≫   business objects, those entities over which the basic actions are performed;

≪**ext**≫   external entities, those entities outside the responsibility of the business, but with which the business needs to interact.

The above three stereotypes are mutually exclusive, and each class appearing in the Static View should be decorated with one of them; they represent the role of the entities w.r.t. the business.

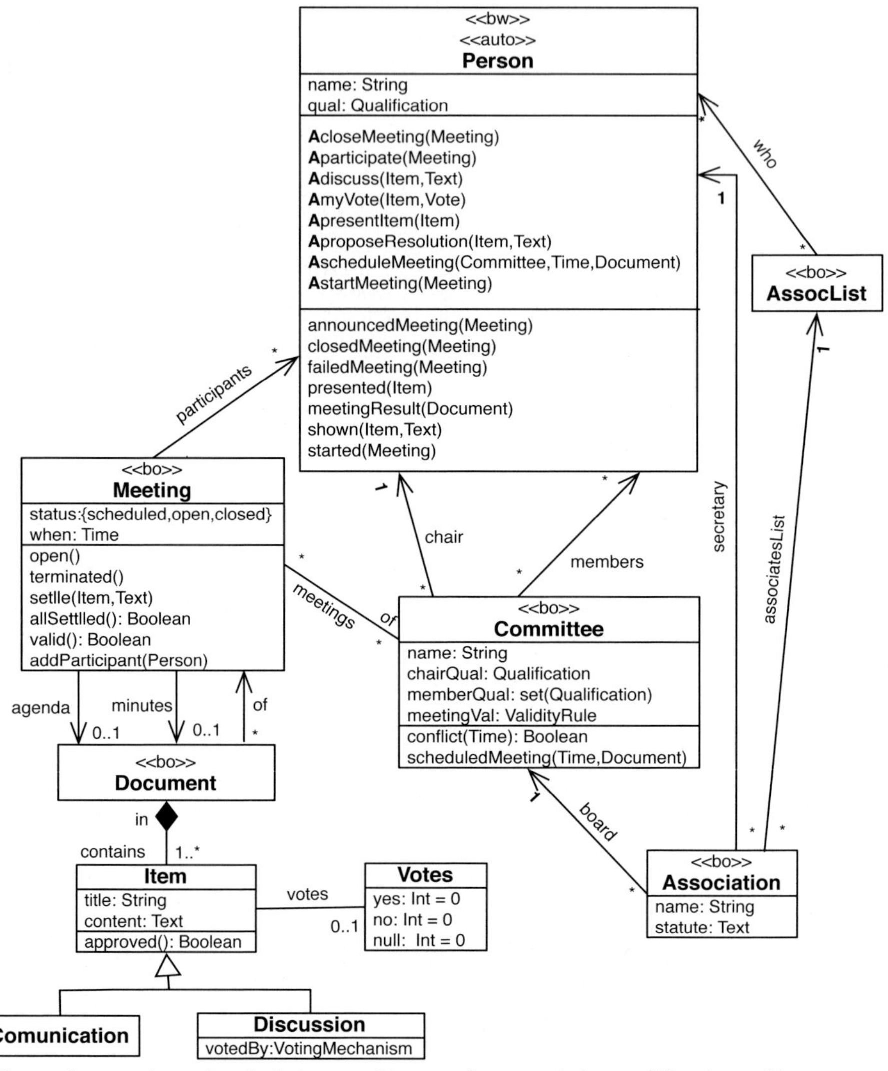

For each committee, its chair is one of its members, and the qualifications of its chair and of its members are ok.

context C: Committee inv:
    C.members->includes(C.chair) and C.chair.qual = C.chairQual and
    C.memberQual->includesAll(C.members.qual)

The scheduled meetings do not have minutes.

context M:Meeting inv: (M.status = scheduled) <=> M.minutes.size()= 0

The secretary of an association cannot be a member of the same.

context A: Association inv: (A.associatesList.who->excludes(A.secretary)

Fig. 3. ASSOC Business Model: Static View

It may be useful also to model the fact that an entity be able to perform autonomous behaviour. In our approach we use the active class stereotype ≪auto≫ to indicate those business entities capable of autonomous behaviour, i.e., those not just reacting to external stimuli. An autonomous act of one such entity is modeled by sending to itself a special kind of UML *signals*[2]. The signals modeling the autonomous acts are stereotyped by ≪A≫ (visually denoted by identifiers starting with a bold capital **A**). We collect the ≪A≫ signals that the instances of a class are able to (self) send, by listing all their receptions in a specific compartment of the class icon, placed between the attribute and the operation compartments (see for example the class **Person** in Fig. 3); and we represent the signal attributes as operation arguments[3].

In Fig. 3 we present the **Static View** of the **ASSOC** case study. All business workers are persons that will play the various roles required by the business, as associates, members of the board, secretary and so on, as will be presented later in the **Business Process View**. The business objects are the meetings, committees, associations, documents and the list of the associates. A meeting can have an agenda and/or a minute: both are of type **Document**. A document is composed of items[4], specialized in **Communication** and **Discussion** (the difference is that the former is concluded without a vote while the latter with a vote). Notice how the use of the UML constraints, expressed using OCL, allows to model quite precisely the relationships among the various business entities (see bottom of Fig. 3).

3.3   Organization View

The **Organization View** consists of a class diagram where the organizational structure of the business is modeled. The classes appearing in that diagram are classes with stereotype either ≪bw≫ or ≪bo≫ already appearing in the **Static View**[5], or new classes. Some classes may be stereotyped by ≪ou≫ (visually depicted using a thick line), and will represent the organizational units[6]. The aggregation and composition associations will depict the hierarchical structure among the various units, and the membership relation between units and workers, also clarifying their roles inside the units. Constraints may be added by stating properties on

---

[2] From [13]: "A signal triggers a reaction in the receiver in an asynchronous way and without a reply. By declaring a reception associated to a given signal, a classifier specifies that its instances will be able to receive that signal, and will respond to it with the designated behavior."

[3] The UML requires to depict each signal similarly to a class with an attribute corresponding to each signal argument, but this will excessively clutter the **Static View**.

[4] Noting that **Item** is represented in Fig. 3 without ≪bo≫, because parts inherit the stereotypes of the "whole"

[5] Recall that in the UML the same class may appear in several class diagrams of the same model; furthermore only part of the class features may be shown in a diagram.

[6] In UML a class may have several stereotypes, and thus a class having stereotype ≪ou≫ may have also stereotype ≪bo≫; clearly the ≪ou≫ will be shown only in this view.

the associations and on the class instances, and thus on the way the business entities are organized inside the business.

Whenever needed it is also possible to add some object diagram, where the instances are typed using the classes appearing in this class diagram, getting a classical "organizational chart".

When modeling the business processes by means of UML activity diagrams, see Sect. 4, it will be possible to see what is done by the various organization units using the swimlanes mechanism.

In Fig. 4 we present the **Organization View** of the ASSOC case study. The organization units are the association itself, the board and the staff, made by the secretary and by the chair of the board.

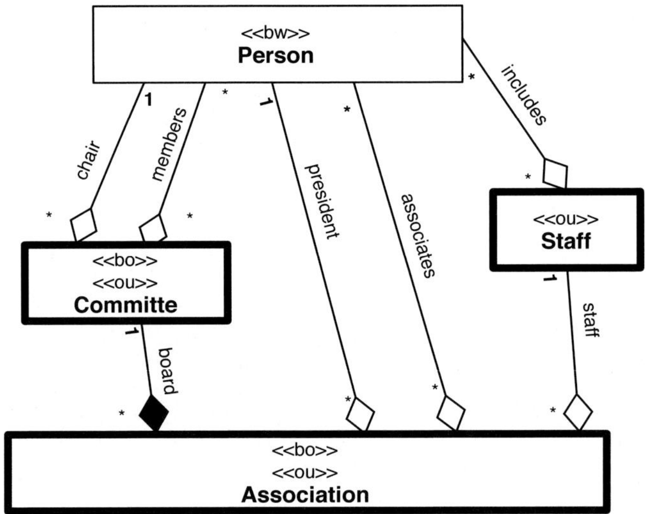

The staff consists of the secretary and the chair of the board, and the
associates are defined by the association list, introduced in the **Static View** (see Fig. 3).
context AS: Association inv:
   AS.staff.includes = { AS.secretary, AS.board.chair } and
     AS.associates = AS.associatesList.who

Fig. 4. ASSOC Business Model: Organization View

4   Business Modeling: The Business Processes

We assume that in a business there are many different processes, thus our approach requires first to give an overview of them and of their mutual relationships (Business Process Overview Diagram), and after to describe the various processes (Business Process Description).

4.1   The Business Process Overview

The Business Process Overview Diagram visually summarizes the processes of the modeled business together with the business entities taking part in them, and their mutual relationships.

We use a stereotype $\ll$BP$\gg$ of the UML model element "use case" to represent a business process, whereas [5] uses "business use case". A use stereotyped by $\ll$BP$\gg$ is visually depicted by an oval with a thick line. A Business Process Overview Diagram is then a UML use case diagram, where the use cases are stereotyped with $\ll$BP$\gg$, the actors are the business entities taking part in them, and the lines between ovals and actors show the participation. It is important to note that, the box showing the boundary of the system is lacking: in modeling the business there is no need to show "THE" system (that will be considered when developing a software system to support the business). In this diagram it is also possible to depict:

- inclusion relationships among the business processes (using the corresponding relationship between use cases of the UML, visually represented by a dashed arrow[7].
- dependency relationships (with the standard meaning in the UML: BP_1 depends on BP_2 whenever any modification in BP_2 may affect BP_1);
- specializations (recall that use cases, and thus business processes, are UML classifiers).

The actors are typed using the classes appearing in the Static View and in the Organization View. Moreover, actors may have a multiplicity marking: that means that several instances of a class take part in the business process.

In Fig. 5 we show the Business Process Overview Diagram relative to the ASSOC case study. The diagram shows that the main business process Managing-Association is built out of other processes (it includes them) as Establishing Association, Hold Board Meeting. Moreover both business workers – like SECT, the secretary, and CHAIR, the chair of the board – and business objects – like BD, the board, and AL, the list of associates – take part in the various processes. The stereotype $\ll$large$\gg$ and the icons used for the various kinds of actors are introduced in the following section.

4.2   The Business Process Descriptions

The description of a business process (see Fig. 1) follows a template presented below by listing all its parts.

- "Name": it identifies the business process.
- "Business goal": a natural language text, it describes the goal (intention) of the business process;

---

[7] For simplicity we drop the decoration $\ll$inclusion$\gg$ over the dashed line of the inclusion, since there is no possibility to confuse it with an extension relation (not used in this case).

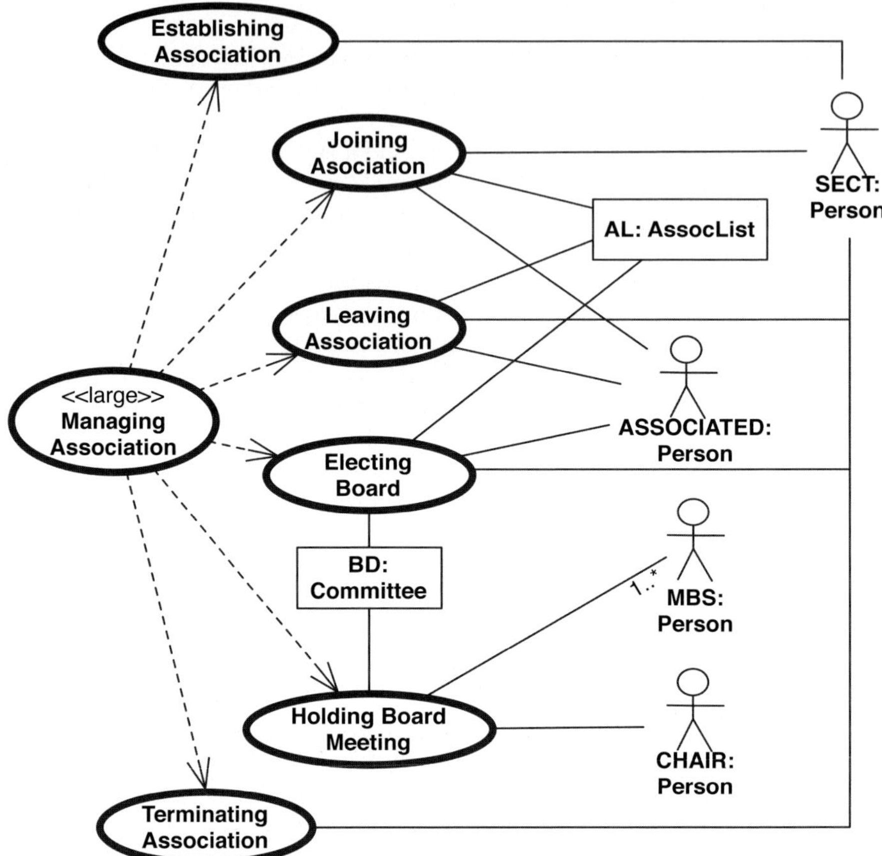

Fig. 5. ASSOC Business Model: Business Process Overview Diagram

- "Precondition" and "Postcondition" (both are optional): natural language text, they state what we assume about the current state of the business before/after the execution of the business process.
- "Level", i.e., the granularity of the process: – *large* (a rather large process usually made up by composing many business processes, corresponding to a reasonably high level business goal and to a manager point of view), – *normal* (what is usually intended as a business process from a business worker point of view: these are the processes that the people involved in the business may discuss). The large business processes are represented by the stereotype ≪large≫.
- "Overall view" (see the detail in Sect. 4.3): it summarizes the business entities taking part in the process, the parameters and their mutual relationships.
- "Activity": a presentation of the activities, together with their causal/ temporal relationships, that build up the process (i.e., the workflow view of the process).

4.3   Overall View of a Business Process

The *overall view* of a business process is a, possibly parameterized, UML collaboration[8], where the roles, typed using classes appearing in the Static View or in the Organization View, correspond to all the business entities taking part in the business process; usually no connectors among them are shown, since we are not interested in the message exchange among them. We choose to represent the collaboration in the following way (the UML offers two different ways to depict the collaboration, see [13, Sect. 9.3.3]):

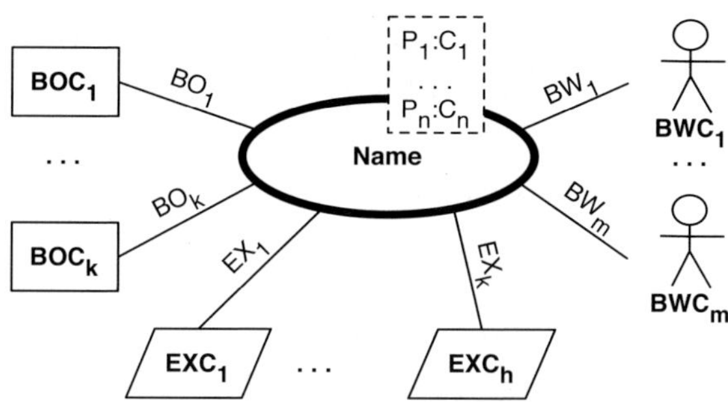

where Name is the name of the business process, $P_1, \ldots, P_n$ are the parameters, $BO_1, \ldots, BO_k, BW_1, \ldots BW_m, EX_1, \ldots, EX_h$ are the business entity (roles to be played by business entities) taking part in the process; $C_1, \ldots, C_n, BOC_1, \ldots, BOC_k, BWC_1, \ldots, BWC_m, EXC_1, \ldots, EXC_h$ are classes appearing in the Static View or in the Organization View. We use different icons for the classes depending on their stereotype (box, sticky man and parallelogram[9]). Since we can attach a constraint to the collaboration, we can state, if any, which conditions the business entities and the parameters must satisfy to take part in the business process. Fig. 6 (up) shows the overall view of the business process Holding Board Meeting.

4.4   Activity of a Business Process

The **Activity** of a business process is described by means of a UML activity diagram. The actions appearing in that diagram can be only calls of the operations/signals of the business entities taking part in it, which have been introduced in the Static View.

To keep the presentations of the activity of the business processes simple and quite readable, we strongly suggest to define appropriate auxiliary sub-activities, and then to reuse them taking advantage of the UML 2.0 construct "rake" (see

---

[8] From [13]: collaboration is a kind of classifier and defines a set of cooperating entities to be played by instances (its roles), as well as a set of connectors that define communication paths between the participating instances.

[9] The last one inspired by the icon used for external entities in the old fashion flow chart.

[13, Sect. 12.3.14]). If a business process BP_1 is included into the business process BP_2, then the behaviour of BP_2 will be modeled by an activity diagram invoking the activity modeling the behaviour of BP_1, using again the rake construct.

In Fig. 6 we present the description of the process **Holding Board Meeting**, the more detailed definitions of some of the activities appearing in that figure are reported in Fig. 7, whereas the remaining auxiliary activities and the descriptions of the other business processes appearing in Fig. 5 can be found in [14]. The UML 2.0 "rake" construct indicates that the activities are further on described by means of auxiliary activity diagrams given apart. Notice in Fig. 6 how the use of the auxiliary sub-activities, which can be then detailed apart, allows also to get a less precise but quite expressive and useful presentation of the processes activity. The level of abstraction of this view is hence comparable to the one of BPMN [3].

5   Related Works

The literature reports several proposals for business modeling having the purpose of representing various aspects of a business, such as operational processes, organizational structures, goals, and risk analysis. In this section, we focus only on related works in business modeling which are based or extend UML.

The UML profile for business modeling of the OMG is defined in the chapter 4 of the UML 1.4 specification [15]. The model consists of two views, one external (described by the use case model) and one internal view (described by the object model). As already noted in [2], the model lacks: (i) a description of a detailed process flow as sequence of activities, but also (ii) the business context.

The profile for business modeling proposed by Johnston [5], based on prior work by Rational Software and Objectory, extends the OMG proposal in several directions. It adds new stereotypes such as business actor, business entity, business goal, business worker and substantially organizes the business model into two parts: business use case model and business analysis model. Moreover it introduces goals, events and activity diagrams to model the flow of the business processes (i.e., the dynamic behaviour). This profile is a component of the RUP©(Rational Unified Process). While we share with Johnston [5] the use of UML stereotypes, events and activity diagram and we have taken from him several fundamental ideas (for example our **Business Process Overview Diagram** is similar to the Business Use Case Model proposed in [5]) our **Business Model** multi-view structure is richer. Most important, in [5], the actions of the activity diagram are expressed by means of natural language instead in our proposal they are operations in the "Object's World".

COMET [1] is a use-case driven and MDA [12] based methodology aimed at supporting the process of developing and maintaining systems, components, products and product families. That methodology, among other things, includes also a business modeling method. Even if, some ingredients of the business modeling method of COMET are similar to ours (for example the use of UML activity diagrams), the point of view and the aim of the business models is different. In

Business Process: Holding Board Meeting
Goal: The functioning of the association is managed by its board, that discusses and decides during its meeting. The goal of this process is to have the meeting of the association board.
Level: Normal

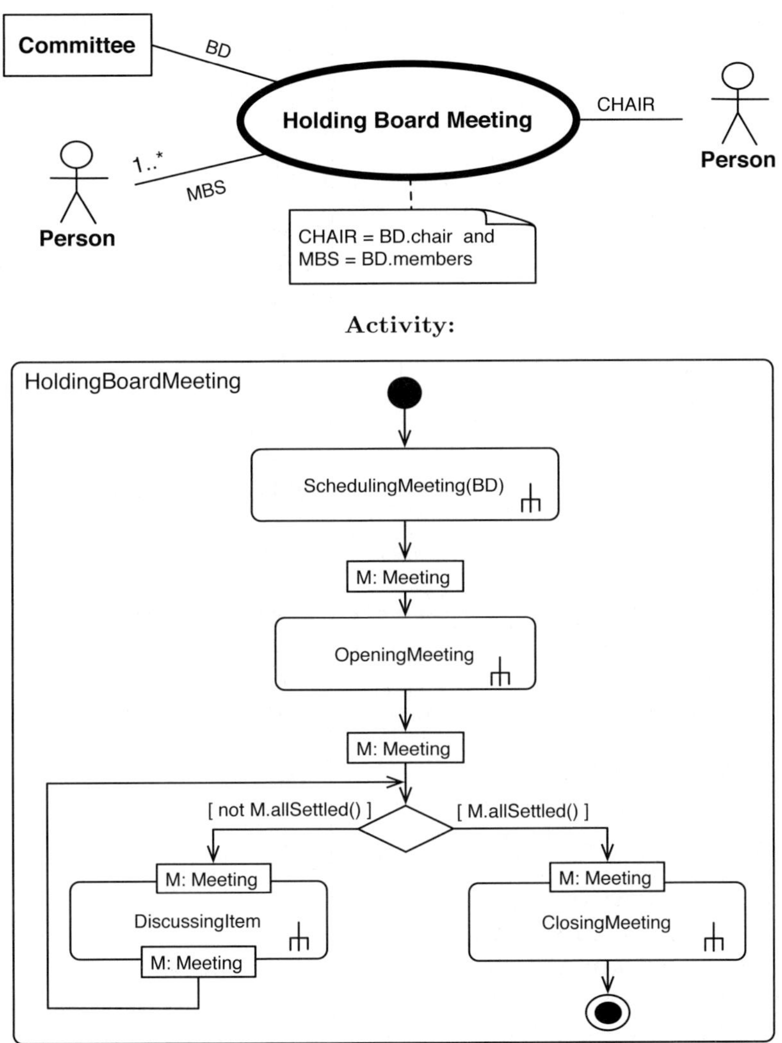

Fig. 6. Description of the business process Holding Board Meeting

[1] the business model is used "to express the part played by the product (system or component) being developed in the context of the business that will fund its development (or purchase it) and use it". We use instead the model to describe the business, to understand it better and, possibly, to re-engineer it.

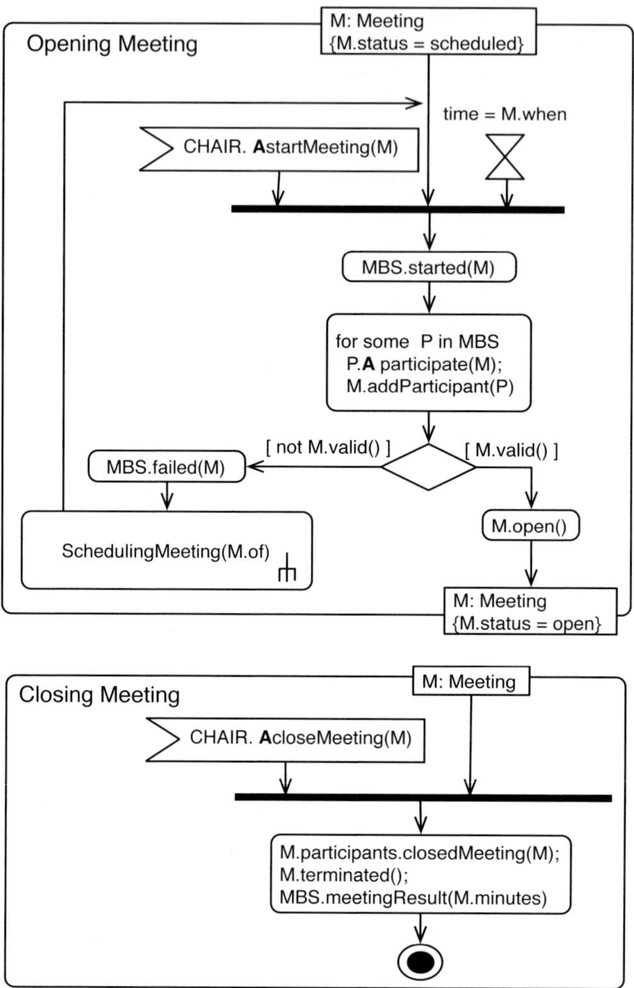

Fig. 7. Two auxiliary sub-activities used in **Holding Board Meeting** (UML signal receptions, as **A**StartMeeting, are represented with the appropriate icon)

One of the first UML 2 profiles for business process modeling has been proposed by List and Korherr in [2]. That profile provides two complementary perspectives, the business and the sequence perspective. The first, described substantially using stereotypes, presents the business processes integrating together several aspects like goals, measures, customers, deliverables, process types etc. The sequence perspective refines the business perspective and describes the detailed flow of the process (i.e., the workflow). The main differences between our and their proposal are that: (i) we use more views (data, static, organization, business processes overview) to represent the static part of the processes, while List and Korherr [2] put all of them together in a unique view and (ii) we chose the activity diagrams to represent/describe the flow of the process, while they prefer leaving this decision to the process modeler, who is free to choose among activity diagrams, BPMN [3] and other means.

In [16], Brambilla et alt. adopt a formalized model-driven development process [12] for workflow-based applications and argue for the automatic integration of the workflow model within the domain model. That is an alternative way to implement workflow models. Usually they are implemented with the help of dedicated workflow management systems which have the drawback of being heavy-weight applications. The initial workflow model is expressed in BPMN while the domain model is represented with a UML class diagram. Then, model-driven development methods may use the workflow extended domain model, produced combining an OCL representation of the given workflow model and the domain model, to automatically generate an implementation of the system enforcing the business processes in the final technology platform. Even if both our purpose and technical treatment is completely different, we give the same importance to domain modeling (for us the domain is represented by means of the static view) and to the integration of the workflow models within domain models.

Sensoria[10] is an integrated project (6th Framework Programme) funded by the EU which aim is to develop a novel comprehensive approach to the engineering of Service-Oriented Systems, where foundational theories, techniques and methods are integrated in a pragmatic software engineering approach. The research themes range across the whole lifecycle of software development from requirements to deployment including coding and maintenance. The main research topics of Sensoria are: Modelling Service-Oriented Software, Formal analysis of Service-Oriented Software and Deployment and Runtime issues. In several of their works [17,18] it seems that the conception of business process is different from ours. They have a view more oriented to SOA, i.e., they consider a business process as an orchestration of services while we have a more abstract view: for us, a business process is a collection of interrelated tasks, which solve a particular issue. Another difference is that we consider the business modeling as a first step in the development of SOA based applications, i.e., the services are derived directly starting from the business model. Instead, they, for the same aim, use a Use case model (see [18] for an example) conveniently tailored for SOA based applications. A workflow based approach to business process modeling (named StPowla) that integrates a simple graphical notation, a natural policy language — Appel — and the Service Oriented Architecture, to assemble and orchestrate available services in the business process is presented in [19].

6   Conclusions and Future Work

With the aim, shared by many other colleagues, of filling the gap between Business Modeling (BM) and software development, and also building over some previous work by us and others, as mentioned in the Introduction and in Sect. 5,

---

[10] http://www.sensoria-ist.eu/

we have proposed to treat the BM following a UML-based model-driven rigorous method, in much the same way we handle Domain Modeling, but with a special concern for the business context and the description of the business processes within that context. In that way we may benefit of all the facilities provided by the UML, such as the reuse of artifacts, possibly with the help of specialization both for business processes and the various entities and actors involved in the business.

A most important feature of our approach is that we adopt the same conceptual and terminological framework for BM and software system development. However, starting with the direct modeling of the business processes without requiring a preliminary alternative representation (say, in BPMN or other workflow notation), we may extract those other representations at various levels of abstraction/detail. A further step in our work will consist in providing some automatic way to recover those representations. Also we have not insisted here on the adaptive aspect of the MARS approach that allows to pass from an extremely precise UML description of every step, to a variety of "light" descriptions, including textual parts, but still well-founded, because based on a rigorous underlying model, with which they may be confronted.

We are currently experimenting our way to model business in two specific projects in cooperation with some companies. On one side we want to derive from the business model a realization of a supporting application using state-of-the-art workflow engines and content management systems. On the other side, we are looking for a viable method to discover and orchestrate the services needed to execute the modeled processes, following a SOA approach.

References

1. COMET Task Force: COMET methodology. Website, last access 13 December 2007 (2007), `http://www.modelbased.net/comet/index.html`
2. List, B., Korherr, B.: A UML 2 Profile for Business Process Modelling. In: Akoka, J., Liddle, S.W., Song, I.-Y., Bertolotto, M., Comyn-Wattiau, I., van den Heuvel, W.-J., Kolp, M., Trujillo, J., Kop, C., Mayr, H.C. (eds.) ER Workshops 2005. LNCS, vol. 3770. Springer, Heidelberg (2005)
3. Business Process Management Initiative (BPMI): Business Process Modeling Notation (BPMN) (2004) (last access December 13, 2007), `www.bpmn.org/Documents/ BPMN%20V1-0%20May%203%202004.pdf`
4. White, A.: Process Modeling Notations and Workflow Patterns (2004) (last access December 13, 2007) (2004), `www.bpmn.org/Documents/Notations% 20and%20Workflow%20Patterns.pdf`
5. Johnston, S.: Rational UML Profile for business modelling (last access December 13, 2007) (2004), `www.ibm.com/developerworks/rational/library/5167.html`
6. Zimmermann O., Krogdahl P., and Gee C.: Elements of Service-Oriented Analysis and Design (last access December 13, 2007) (2004), `www.ibm.com/developerworks/ library/ws-soad1/`
7. Erl, T.: SOA: Principles of Service Design. The Prentice Hall Service-Oriented Computing Series from Thomas Erl (2005)

8. Astesiano, E., Reggio, G.: Tight structuring for precise UML-based requirement specifications. In: Wirsing, M., Knapp, A., Balsamo, S. (eds.) RISSEF 2002. LNCS, vol. 2941. Springer, Heidelberg (2004)

9. Astesiano, E., Reggio, G.: Towards a well-founded UML-based development method. In: Conference on Software Engineering and Formal Methods, SEFM 2003, September 22-27 (2003)

10. Astesiano, E., Reggio, G.: MARS: Model-based Adaptively Rigorous Software development. Technical Report DISI–TR–2007–12, DISI – Università di Genova, Italy (2007), `ftp://ftp.disi.unige.it/person/ReggioG/MARS01.pdf`

11. Astesiano, E., Reggio, G., Cerioli, M.: From formal techniques to well-founded software development methods. In: Aichernig, B.K., Maibaum, T.S.E. (eds.) Formal Methods at the Crossroads. From Panacea to Foundational Support. LNCS, vol. 2757. Springer, Heidelberg (2003)

12. Mellor, S.J., Scott, K., Uhl, A., Weise, D.: MDA Distilled. Addison-Wesley Object Technology Series (2004)

13. UML Revision Task Force: OMG UML Specification 2.0. (2004) `www.omg.org/docs/formal/05-07-04.pdf`.

14. Astesiano, E., Reggio, G., Ricca, F.: Modeling Business within a UML-based Rigorous Software Development Approach (Complete Version). Technical Report DISI–TR–2007–20, DISI – Università di Genova, Italy (2007), `ftp://ftp.disi.unige.it/person/ReggioG/BPComplete.pdf`

15. UML Revision Task Force: OMG UML Specification 1.4. (2001), `www.omg.org/docs/formal/01-09-75.pdf`

16. Brambilla, M., Cabot, J., Comai, S.: Automatic generation of workflow-extended domain models. In: Engels, G., Opdyke, B., Schmidt, D.C., Weil, F. (eds.) MODELS 2007. LNCS, vol. 4735. Springer, Heidelberg (2007)

17. Koch, N., Mayer, P., Heckel, R., Gonczy, L., Montangero, C.: UML for Service-Oriented Systems. Technical Report D1.4a, Sensoria, Munich, Germany (2007), `www.pst.ifi.lmu.de/projekte/Sensoria/del_24/D1.4.a.pdf`

18. Koch, N., Berndl, D.: Automotive Case Study: Requirements Specification and Modelling of Selected Scenarios. Technical Report D8.2a, Sensoria, Munich, Germany (2007), `www.pst.informatik.uni-muenchen.de/projekte/Sensoria/del_24/D8.2.a.pdf`

19. Gorton, S., Montangero, C., Reiff-Marganiec, S., Semini, L.: StPowla: SOA, Policies and Workflows. In: Proc. 3rd Int. Workshop on Engineering Service-Oriented Applications: Analysis, Design, and Composition, Austria (2007), `www.cs.le.ac.uk/people/smg24/papers/wesoa07.pdf`

From Domain to Requirements

Dines Bjørner

[1] Faculté des Sciences, Bureau 266, LORIA & Université Henri Poincaré Nancy 1,
BP 239, F-54506 Vandœuvre lès Nancy, France[*]
[2] Professor emeritus, DTU Informatics, Bldg. 325, Technical University of Denmark,
DK-2800 Kgs. Lyngby, Denmark
[3] Fredsvej 11, DK-2840 Holte, Danmark
bjorner@gmail.com,
www.imm.dtu.dk/~db

Abstract. We first present a summary of essentials of domain engineering, its motivation, and its modelling of abstractions of domains through the modelling of the intrinsics, support technologies, management and organisation, rules and regulations, scripts, and human behaviour of whichever domain is being described.

Then we present the essence of two (of three) aspects of requirements: the domain requirements and the interface requirements prescriptions as they relate to domain descriptions and we survey the basic operations that "turn" a domain description into a domain requirements prescription: projection, instantiation, determination, extension and fitting. An essence of interface requirements is also presented: the "merging" of shared entities, operations, events and behaviours of the domain with those of the machine (i.e., the hardware and software to be designed).

1 Introduction

This paper presents a model of early stages of software development that is not conventional. The model is presented in two alternating ways: (i) we present some of the principles and techniques of that unconventional software development method, and (ii) we present — what in the end, that is, taken across the paper, amounts to a relatively large example.

In summary: the objective of the present paper is to relate domain engineering to requirements engineering and to show that one can obtain an altogether different basis for requirements engineering.

2 The Triptych Principle of Software Engineering

We start, unconventionally, by enunciating a principle. The principle expresses how we see software development as centrally consisting of three "programming-like" phases based on the following observation: before software can be designed

---

[*] This paper was written with the financial support of Université Henri Poincaré and INRIA during the author's two month visit: October 15 – December 14, 2007.

P. Degano et al. (Eds.): Montanari Festschrift, LNCS 5065, pp. 278–300, 2008.

we must understand its requirements, and before requirements can be prescribed
we must understand the application domain. We therefore see software development proceeding, ideally, in three phases: a first phase of domain engineering, a
second phase of requirements engineering, and a third phase of software design.

The first paragraphs of Sects. 3 and 4 explain what the objectives of domain
engineering and requirements engineering are. The sections otherwise outline
major development stages and steps of these two phases.

3 Domain Engineering

The objective of domain engineering is to create a domain description. A domain
description specifies entities, functions, events and behaviours of the domain such
as the domain stakeholders think they are. A domain description thus (indica-
tively) expresses what there is. A domain description expresses no requirements
let alone anything about the possibly desired (required) software.

3.1 Stages of Domain Engineering

To develop a proper domain description necessitates a number of development
stages: (i) identification of stakeholders, (ii) domain knowledge acquisition, (iii)
business process rough-sketching, (iv) domain analysis, (v) domain modelling:
developing abstractions and verifying properties, (vi) domain validation and (vii)
domain theory building.

Business process (BP) rough-sketching amount to rough, narrative outlines of
the set of business processes as experienced by each of the stakeholder groups.
BP engineering is in contrast to BR re-engineering (BPR) which we shall cover
later, but briefly in Sect. 4.2.

We shall only cover domain modelling.

3.2 First Example of a Domain Description

We exemplify a transportation domain. By transportation we shall mean *the
movement of vehicles from hubs to hubs along the links of a net.*

Rough Sketching — Business Processes. The basic *entities* of the transportation "business" are the (i) *net*s with their (ii) *hub*s and (iii) *link*s, the (iv)
*vehicle*s, and the (v) *traffic* (of vehicles on the net). The basic *functions* are those
of (vi) vehicles entering and leaving the net (here simplified to entering and leaving at hubs), (vii) for vehicles to make movement transitions along the net, and
(viii) for inserting and removing links (and associated hubs) into and from the
net. The basic *events* are those of (ix) the appearance and disappearance of
vehicles, and (x) the breakdown of links. And, finally, the basic behaviours of
the transportation business are those of (xi) vehicle journey through the net and
(xii) net development & maintenance including insertion into and removal from
the net of links (and hubs).

Narrative — Entities. By an *entity* we mean *something we can point to, i.e., something manifest, or a concept abstracted from, such a phenomenon or concept thereof.*

Among the many entities of transportation we start with nets, hubs, and links.

A transportation net consists of hubs and links. Hubs and links are different kinds of entities. Conceptually hubs (links) can be uniquely identified. From a link one can observe the identities of the two distinct hubs it links. From a hub one can observe the identities of the one or more distinct links it connects.

Other entities such as vehicles and traffic could as well be described. Please think of these descriptions of entities as descriptions of the real phenomena and (at least postulated) concepts of an actual domain.

Formalisation — Entities

type H, HI, L, LI, N = H-**set** × L-**set**
value obs_HI: H→HI, obs_LI: L→LI, obs_HIs: L→HI-**set**,obs_LIs: H→LI-**set**
axiom
   ∀ (hs,ls):N •
      card hs≥2 ∧ **card** ls≥1 ∧ ∀ h:H • h ∈ hs ⇒
        ∀ li:LI • li ∈ obs_LIs(h) ⇒
          ∃ l′:L • l′ ∈ ls∧li=obs_LI()∧obs_HI(h) ∈ obs_HIs(l′)∧
     ∀ l:L • l ∈ ls ⇒
        ∃ h′,h″:H • {h′,h″}⊆hs∧obs_HIs(l)={obs_HI(h′),obs_HI(h″)}
value xtr_HIs: N → HI-**set**,xtr_LIs: N → LI-**set**

Narrative — Operations. By an *operation* (of a domain) we mean *a function that applies to entities of the domain and yield entities of that domain — whether these entities are actual phenomena or concepts of these or of other phenomena.*

Actions (by domain stakeholders) amount to the execution of operations.

Among the many operations performed in connection with transportation we illustrate some on nets. To a net one can join new links in either of three ways: The new link connects two new hubs — so these must also be joined , or The new link connects a new hub with an existing hub — so it must also be joined, or The new link connects two existing hubs. In any case we must either provide the new hubs or identify the existing hubs.

From a net one can remove a link. Three possibilities now exists: The removed link would leave its two connected hubs isolated unless they are also removed — so they are; The removed link would leave one of its connected hubs isolated unless it is also removed — so it is; or The removed link connects two hubs into both of which other links are connected — so all is OK. (Note our concern for net invariance.) Please think of these descriptions of operations as descriptions of the real phenomena and (at least postulated) concepts of an actual domain. (Thus

they are not prescriptions of requirements to software let alone specifications of
software operations.)

Formalisation — Operations

type
   NetOp = InsLnk | RemLnk
   InsLnk == 2Hs(h1:H,l:L,h2:H)|1H(hi:HI,l:L,h:H)|0H(hi1:HI,l:L,hi2:HI)
   RemLnk == RmvL(li:LI)
value
   int_NetOp: NetOp $\to$ N $\xrightarrow{\sim}$ N
   pre int_NetOp(op)(hs,ls) $\equiv$
      case op **of**
         2Hs(h1,l,h2) $\to$
            {h1,h2}$\cap$ hs={}$\land$l$\notin$ ls$\land$
            obs_HIs(l)={obs_HI(h1),obs_HI(h2)}$\land$
               {obs_HI(h1),obs_HI(h2)}$\cap$ xtr_HIs(hs)={}$\land$
               obs_LIs(h1)={li}$\land$obs_LIs(h2)={li},
         1H(hi,l,h) $\to$ ...,
         0H(hi1,l,hi2) $\to$ ...
      end

   int_NetOp(op)(hs,ls) $\equiv$
      case op **of**
         2Hs(h1,l,h2) $\to$
            (hs $\cup$ {h1,h2},ls $\cup$ {l}),
         1H(hi,l,h) $\to$
            (hs\{xtr_H(hi,hs)}$\cup${h,aLI(xtr_H(hi,hs),obs_LI(l))},ls $\cup$ {l}),
         0H(hi1,l,hi2) $\to$ ...,
         RmvL(li) $\to$ ...
      end

   xtr_H: HI $\times$ H-**set** $\xrightarrow{\sim}$ H
   xtr_H(hi,hs) $\equiv$ **let** h:H $\bullet$ h $\in$ hs $\land$ obs_HI(h)=hi **in** h **end**
   pre $\exists$ h:H $\bullet$ h $\in$ hs $\land$ obs_HI(h)=hi

   aLI: H $\times$ LI $\to$ H, sLI: H $\times$ LI $\to$ H
   aLI(h,li) **as** h$'$,
      pre li $\notin$ obs_LIs(h), **post** obs_LIs(h$'$)={li} $\cup$ obs_LIs(h)$\land$ ...
   sLI(h$'$,li) **as** h,
      pre li $\in$ obs_LIs(h$'$), **post** obs_LIs(h)=obs_LIs(h$'$)\{li}$\land$ ...

The ellipses, ..., shall indicate that previous properties of h holds for h$'$.

Narrative — Events. By an *event* of a domain we shall here mean *an instantaneous change of domain state (here, for example, "the" net state) not directly brought about by some willed action of the domain but either by "external" forces or implicitly, as an unintended result of a willed action.*

Among the "zillions" of events that may occur in transportation we single out just one. A link of a net ceases to exist as a link.[1]

In order to model transportation events we — ad hoc — introduce a transportation state notion of a net paired with some — ad hoc — "conglomerate" of remaining state concepts referred to as $\omega : \Omega$.

Formalisation — Events

type
   Link_Disruption == LiDi(li:LI)
channel
   x:(Link_Disruption|...)
value
   transportation_transition: $(N \times \Omega) \to$ **in** x $(N \times \Omega)$
   transportation_transition(n,ω) $\equiv$
     ... $\lceil$ **let** xv = x? **in**
          case xv **of**
             LiDi(li) $\to$ (int_NetOp(RmvL(li))(hs,ls),line_dis(ω))
             ... end end $\lceil$ ...
   line_dis: $\Omega \to \Omega$

Narrative — Behaviours. By a *behaviour* we mean *a possibly infinite sequence of zero, one or more actions and events.*

We illustrate just one of very many possible transportation behaviours.

A net behaviour is a sequence of zero, one or more executed net operations: the openings (insertions) of new links (and implied hubs) and the closing (removals) of existing links (and implied hubs), and occurrences of external events (limited here to link disruptions).

Formalisation — Behaviours

channel
   x:...
value
   transportation_transition: $(N \times \Omega) \to$ **in** x $(N \times \Omega)$
   transportation_transition(n,ω) $\equiv$
     ... $\lceil$ **let** xv = x? **in case** xv **of** ... **end end**

> ... $\sqcap$ **let** op:NetOp • **pre** IntNetOp(op)(n) **in** IntNetOp(op)(n) **end** ...
> transportation: $(N \times \Omega) \rightarrow$ **in** x **Unit**
> transportation(n,ω) $\equiv$
>   **let** (n$'$,ω') = transportation_transition(n,ω) **in**
>   transportation (n$'$,ω') **end**

3.3 Domain Modelling: Describing Facets

Domain modelling, as we shall see, entails modelling a number of domain facets.

By a *domain facet* we mean *one amongst a finite set of generic ways of analysing a domain: a view of the domain, such that the different facets cover conceptually different views, and such that these views together cover the domain.*

These are the facets that we find "span" a domain in a pragmatically sound way: intrinsics, support technology, management & organisation, rules & regulations, scripts and human behaviour: We shall now survey these facets.

Domain Intrinsics. By *domain intrinsics* we mean *those phenomena and concepts of a domain which are basic to any of the other facets (listed earlier and treated, in some detail, below), with such domain intrinsics initially covering at least one specific, hence named, stakeholder view.*

For the large example of Sect. 3.2, we claim that the net, hubs and links were intrinsic phenomena of the transportation domain; and that the operations of joining and removing links were not: one can explain transportation without these operations. We will now augment the domain description of Sect. 3.2 with an intrinsic concept, namely that of the states of hubs and links: where these states indicate desirable directions of flow of movement.

A Transportation Intrinsics — Narrative. With a hub we can associate a concept of hub state. The pragmatics of a hub state is that it indicates desirable directions of flow of vehicle movement from (incoming) links to (outgoing) links. The syntax of indicating a hub state is (therefore) that of a possibly empty set of triples of two link identifiers and one hub identifier where the link identifiers are those observable from the identified hub.

With a link we can associate a concept of link state. The pragmatics of a link state is that it indicates desirable directions of flow of vehicle movement from (incoming, identified) hubs to (outgoing, identified) hubs along an identified link. The syntax of indicating a link state is (therefore) that of a possibly empty set of triples of pairs of identifiers of link connected hub and a link identifier where the hub identifiers are those observable from the identified link.

A Transportation Intrinsics — Formalisation.

type
  X = LI×HI×LI [crossings **of** a hub], P = HI×LI×HI [paths **of** a link]
  HΣ = X-**set** [hub states],   LΣ = P-**set** [link states]
value
  obs_HΣ: H → HΣ, obs_LΣ: L → LΣ,
  xtr_Xs: H → X-**set**, xtr_Ps: L → P-**set**
  xtr_Xs(h) ≡
     {(li,hi,li′)|li,li′:LI,hi:HI•{li,li′}⊆obs_LIs(h)∧hi=obs_HI(h)}
  xtr_Ps(l) ≡
     {(hi,li,hi′)|hi,hi′:HI,li:LI•{hi,hi′}=obs_HIs(l)∧li=obs_LI(l)}
axiom ∀ n:N,h:H;l:L • h ∈ obs_Hs(n)∧l ∈ obs_Ls(n) ⇒
     obs_HΣ(h)⊆xtr_Xs(h) ∧ obs_LΣ(l)⊆xtr_Ps(l)

Domain Support Technologies. By *domain support technologies* we mean
*ways and means of implementing certain observed phenomena or certain con-
ceived concepts.*

A Transportation Support Technology Facet — Narrative, 1. Earlier we claimed
that the concept of hub and link states was an intrinsics facet of transport nets.
But we did not describe how hubs or links might change state, yet hub and link
state changes should also be considered intrinsic facets. We there introduce the
notions of hub and link state spaces and hub and link state changing operations.
A hub (link) state space is the set of all states that the hub (link) may be in. A
hub (link) state changing operation can be designated by the hub and a possibly
new hub state (the link and a possibly new link state).

A Transportation Support Technology Facet — Formalisation, 1.

type HΩ = HΣ-**set**, LΩ = LΣ-**set**
value obs_HΩ: H → HΩ, obs_LΩ: L → LΩ
axiom ∀ h:H • obs_HΣ(h) ∈ obs_HΩ(h) ∧ ∀ l:L • obs_LΣ(l) ∈ obs_LΩ(l)
value
  chg_HΣ: H × HΣ → H, chg_LΣ: L × LΣ → L
  chg_HΣ(h,hσ) **as** h′, **pre** hσ ∈ obs_HΩ(h), **post** obs_HΣ(h′)=hσ
  chg_LΣ(l,lσ) **as** l′, **pre** lσ ∈ obs_LΩ(h), **post** obs_HΣ(l′)=lσ

A Transportation Support Technology Facet — Narrative, 2. Well, so far we
have indicated that there is an operation that can change hub and link states.
But one may debate whether those operations shown are really examples of

a support technology. (That is, one could equally well claim that they remain examples of intrinsic facets.) We may accept that and then ask the question: How to effect the described state changing functions ? In a simple street crossing a semaphore does not instantaneously change from red to green in one direction while changing from green to red in the cross direction. Rather there is are intermediate sequences of green/yellow/red and red/yellow/green states to help avoid vehicle crashes and to prepare vehicle drivers. Our "solution" is to modify the hub state notion.

A Transportation Support Technology Facet — Formalisation, 2.

type
    Colour == red | yellow | green
    X = LI×HI×LI×Colour [crossings **of** a hub]
    HΣ = X-**set** [hub states]
value
    obs_HΣ: H $\to$ HΣ, xtr_Xs: H $\to$ X-**set**
    xtr_Xs(h) $\equiv$
        $\{$(li,hi,li$'$,c)|li,li$'$:LI,hi:HI,c:Colour•$\{$li,li$'\}\subseteq$obs_LIs(h)$\wedge$hi=obs_HI(h)$\}$
axiom
    $\forall$ n:N,h:H • h $\in$ obs_Hs(n) $\Rightarrow$ obs_HΣ(h)$\subseteq$xtr_Xs(h) $\wedge$
        $\forall$ (li1,hi2,li3,c),(li4,hi5,li6,c$'$):X •
            $\{$(li1,hi2,li3,c),(li4,hi5,li6,c$'$)$\}\subseteq$obs_HΣ(h) $\wedge$
            li1=li4 $\wedge$ hi2=hi5 $\wedge$ li3=li6 $\Rightarrow$ c=c$'$

A Transportation Support Technology Facet — Narrative, 3. We consider the colouring, or any such scheme, an aspect of a support technology facet. There remains, however, a description of how the technology that supports the intermediate sequences of colour changing hub states.

We can think of each hub being provided with a mapping from pairs of "stable" (that is non-yellow coloured) hub states (hσ_i,hσ_f) to well-ordered sequences of intermediate "un-stable' (that is yellow coloured) hub states paired with some time interval information $\langle(h\sigma', t\delta'), (h\sigma'', t\delta''), \ldots, (h\sigma'^{\cdots'}, t\delta'^{\cdots'})\rangle$ and so that each of these intermediate states can be set, according to the time interval information,[2] before the final hub state (hσ_f) is set.

A Transportation Support Technology Facet — Formalisation, 3.

type
    TI [time interval]
    Signalling = (HΣ × TI)*
    Sema = (HΣ × HΣ) $\overrightarrow{m}$ Signalling
value
    obs_Sema: H $\to$ Sema,
    chg_HΣ: H × HΣ $\to$ H,

```
chg_HΣ_Seq: H × HΣ → H
chg_HΣ(h,hσ) as h'
    pre hσ ∈ obs_HΩ(h) post obs_HΣ(h')=hσ
chg_HΣ_Seq(h,hσ) ≡
    let sigseq = (obs_Sema(h))(obs_Σ(h),hσ) in sig_seq(h)(sigseq) end
sig_seq: H → Signalling → H
sig_seq(h)(sigseq) ≡
    if sigseq=⟨⟩ then h else
    let (hσ,tδ) = hd sigseq in let h' = chg_HΣ(h,hσ);
    wait tδ;
    sig_seq(h')(tl sigseq) end end end
```

Domain Management & Organisation. By *domain management* we mean
people (such decisions) (i) who (which) determine, formulate and thus set stan-
dards (cf. rules and regulations, a later lecture topic) concerning strategic, tac-
tical and operational decisions; (ii) who ensure that these decisions are passed
on to (lower) levels of management, and to "floor" staff; (iii) who make sure
that such orders, as they were, are indeed carried out; (iv) who handle undesir-
able deviations in the carrying out of these orders cum decisions; and (v) who
"backstop" complaints from lower management levels and from floor staff.

We use the connective '&' (ampersand) in lieu of the connective 'and' in order
to emphasise that the joined concepts (A & B) hang so tightly together that it
does not make sense to discuss one without discussing the other.

By *domain organisation* we mean *the structuring of management and non-*
management staff levels; the allocation of strategic, tactical and operational
concerns to within management and non-management staff levels; and hence the
"lines of command": who does what and who reports to whom — administra-
tively and functionally.

A Transportation Management & Organisation Facet — Narrative. In the previous section on support technology we did not describe who or which "ordered"
the change of hub states. We could claim that this might very well be a task for
management.

(We here look aside from such possibilities that the domain being modelled
has some further support technology which advices individual hub controllers as
when to change signals and then into which states. We are interested in finding
an example of a management & organisation facet — and the upcoming one
might do!)

So we think of a 'net hub state management' for a given net. That management
is divided into a number of 'sub-net hub state managements' where the sub-nets
form a partitioning of the whole net. For each sub-net management there are two
kinds management interfaces: one to the overall hub state management, and one

for each of interfacing sub-nets. What these managements do, what traffic state information they monitor, etcetera, you can yourself "dream" up. Our point is this: We have identified a management organisation.

A Transportation Management & Organisation Facet — Formalisation.

type
    HIsLIs = HI-set×LI-set,
    MgtNet′ = HIsLIs×N, MgtNet={|mgtnet:MgtNet′•wf_MgtNet(mgtnet)|}
    Part′ = HIsLIs-set×N, Part={|part:Part′•wf_Part(part)|}
value
    wf_MgtNet: MgtNet′ → **Bool**
    wf_MgtNet((his,lis),n) ≡
        [The his component contains all the hub ids.
         of links identified in lis]
    wf_Part: Part′ → **Bool**
    wf_Part(hisliss,n) ≡
        ∀ (his,lis):HIsLIs •
            (his,lis) ∈ hisliss ⇒ wf_MgtNet((his,lis),n)∧
            [no sub−net overlap and together they "span" n]

Etcetera.

Domain Rules and Regulations

Domain Rules. By a **domain rule** we mean *some text (in the domain) which prescribes how people or equipment are expected to behave when dispatching their duty, respectively when performing their function.*

Domain Regulations. By a **domain regulation** we mean *some text (in the domain) which prescribes what remedial actions are to be taken when it is decided that a rule has not been followed according to its intention.*

A Transportation Rules & Regulations Facet — Narrative. The purpose of maintaining an appropriate set of hub (and link) states may very well be to guide traffic into "smooth sailing" — avoiding traffic accidents etc. But this requires that vehicle drivers obey the hub states, that is, the signals. So there is undoubtedly a rule that says: *Obey traffic signals.*And, in consequence of human nature, overlooking or outright violating signals there is undoubtedly a regulation that says: *Violation of traffic signals is subject to fines and . . . .*

A Transportation Rules & Regulations Facet — Formalisation. We shall, regretfully, not show any formalisation of the above mentioned rule and regulation. To do a proper job at such a formalisation would require that we formalise traffics, say as (a type of) continuous functions from time to pairs of net and vehicle positions, that we define a number of auxiliary (traffic monitoring) functions, including such which test whether from one instance of traffic, say at time t to a "next" instance of time, t', some one or more vehicles have violated the rule[3], etc. The "etcetera" is ominous: It implies modelling traffic wardens (police trying to apprehend the "sinner"), 'etc.' ! We rough-sketch an incomplete formalisation.

type
    T [time], V [vehicle], Rel_Distance = {| f:Rel • 0<f<1 |}
    VPos == VatH(h:H) | VonL(hif:HI,l:L,hit:HI,rel_distance:Rel_Distance)
    Traffic = T → (N × (V $\overrightarrow{m}$ VPos))
value violations: Traffic → (T×T) → V-**set**

Vehicle positions are either at hubs or some fraction f down a link (l) from some hub (hit) towards the connected hub (hit). Traffic maps time into vehicle positions. We omit a lengthy description of traffic well-formedness.

Domain Scripts. By a *domain script* we mean *the structured, almost, if not outright, formally expressed, wording of a rule or a regulation that has legally binding power, that is, which may be contested in a court of law.*

A Transportation Script Facet — Narrative. Regular buses ply the network according to some time table. We consider a train time table to be a script. Let us take the following to be a sufficiency narrative description of a train time table. For every train line, identified by a line number unique to within, say a year of operation, there is a list of train hub visits. A train hub visit informs of the intended arrival and departure times at identified hubs (i.e., train stations) such that "neighbouring" hub visits, (t_{a_i}, h_i, t_{d_i}) and (t_{a_j}, h_j, t_{d_j}), satisfy the obvious that a train cannot depart before it has arrived, and cannot arrive at the next, the "neighbouring" station before it has departed from the previous station, in fact, $t_{a_j} - t_{d_i}$ must be commensurate with the distance between the two stations.

A Transportation Script Facet — Formalisation.

type
    TLin
    HVis = T × HI × T
    Journey' = HVis*, Journey = {|j:Journey'•**len** j≥2|}
    TimTbl' = (TLin $\overrightarrow{m}$ Journey) × N
    TimTbl = {| timtbl:TimTbl' • wf_TimTbl(timtbl) |}
value
    wf_TimTbl: TimTbl' → **Bool**
    wf_TimTbl(tt,n) ≡

> [all hubs designated in tt must be hubs of n]
> [and all journeys must be along feasible links of n]
> [and with commensurate timing net n constraints]

Domain Human Behaviour. By *human behaviour* we mean *any of a quality spectrum of carrying out assigned work: from (i)* **careful, diligent** *and* **accurate,** *via (ii)* **sloppy** *dispatch, and (iii)* **delinquent** *work, to (iv) outright* **criminal** *pursuit.*

Transportation Human Behaviour Facets — Narrative. We have already exemplified aspects of human behaviour in the context of the transportation domain, namely vehicle drivers not obeying hub states. Other example can be given: drivers moving their vehicle along a link in a non-open direction, drivers waving their vehicle off and on the link, etcetera. Whether rules exists that may prohibit this is, perhaps, irrelevant. In any case we can "speak" of such driver behaviours — and then we ought formalise them !

Transportation Human Behaviour Facets — Formalisation. But we decide not to. For the same reason that we skimped proper formalisation of the violation of the *"obey traffic signals"* rule. But, by now, you've seen enough formulas and you ought trust that it can be done.

$$\text{off\_on\_link: Traffic} \to (T \times T) \xrightarrow{\sim} (V \underset{m}{\overrightarrow{\to}} \text{VPos} \times \text{VPos})$$
$$\text{wrong\_direction: Traffic} \to T \xrightarrow{\sim} (V \underset{m}{\overrightarrow{\to}} \text{VPos})$$

3.4 Discussion

We have given a mere glimpse of a domain description. A full description of a reasonably "convincing" domain description will take years to develop and will fill many pages (hundreds, ... (!)).

4 Requirements Engineering

The objective of requirements engineering is to create a requirements prescription: A requirements prescription specifies externally observable properties of entities, functions, events and behaviours of *the machine* such as the requirements stakeholders wish them to be. The *machine* is what is required: that is, the *hardware* and *software* that is to be designed and which are to satisfy the requirements. A *requirements prescription* thus (*putatively*) expresses what there should be. A requirements prescription expresses nothing about the design of the possibly desired (required) software. We shall show how a major part of a requirements prescription can be "derived" from "its" prerequisite domain description.

The Example Requirements. The domain was that of transportation. The requirements is now basically related to the issuance of tickets upon vehicle entry to a toll road net and payment of tickets upon the vehicle leaving the toll road net both issuance and collection/payment of tickets occurring at toll booths which are hubs somehow linked to the toll road net proper. Add to this that vehicle tickets are sensed and updated whenever the vehicle crosses an intermediate toll road intersection.

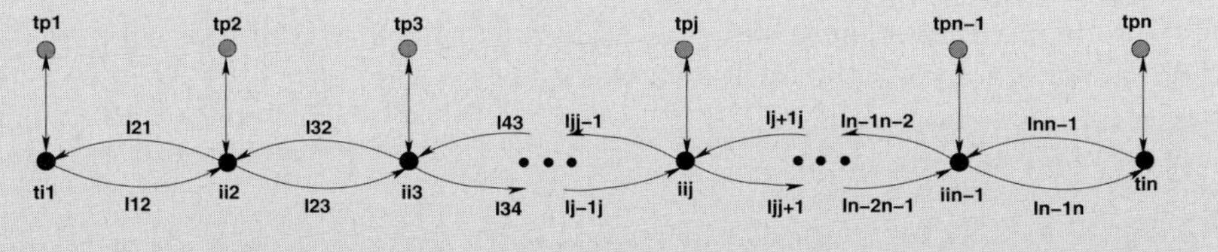

Fig. 1. A simple, linear toll road net: tp_i: toll plaza i, ti_1, ti_n: terminal intersection k, ii_k: intermediate intersection k, $1<k<n$ l_{xy}: tollway link from i_x to i_y, $y=x+1$ or $y=x-1$ and $1\leq x<n$.

4.1   Stages of Requirements Engineering

The following are the stages of requirements engineering: stakeholder identification, *business process re-engineering*, *domain requirements development*, *interface development*, machine requirements development, requirements verification and validation, and requirements satisfiability and feasibility.

The domain requirements development stage consists of a number of steps: projection, instantiation, determination, extension, and fitting.

We shall basically only cover business process re-engineering and   domain requirements development

4.2   Business Process Re-engineering

Business process re-engineering (BPR) re-evaluates the intrinsics, support technologies, management & organisation, rules & regulations, scripts, and human behaviour facets while possibly changing some or all of these, that is, possibly rewriting the corresponding parts of the domain description.

Re-engineering Domain Entities. The net is arranged as a linear sequence of two or more (what we shall call) intersection hubs. Each intersection hub has a single two-way link to (what we shall call) an entry/exit hub (toll plaza); and each intersection hub has either two or four one-way (what we shall call) tollway links: the first and the last intersection hub (in the sequence) has two tollway links and all (what we shall call) intermediate intersections has four tollway links. We introduce a pragmatic notion of net direction: "up" and "down" the net, "from one end to the other". This is enough to give a hint at the re-engineered domain.

Re-engineering Domain Operations. We first briefly sketch the tollgate Operations. Vehicles enter and leave the tollway net only at entry/exit hubs (toll plazas). Vehicles collect and return their tickets from and to tollgate ticket issuing, respectively payment machines. Tollgate ticket-issuing machines respond to sensor pressure from "passing" vehicles or by vehicle drivers pressing ticket-issuing machine buttons. Tollgate payment machines accept credit cards, bank notes or coins in designated currencies as payment and returns any change.

We then briefly introduce and sketch an operation performed when vehicles cross intersections: The vehicle is assumed to possess the ticket issued upon entry (in)to the net (at a tollgate). At the crossing of each intersection, by a vehicle, its ticket is sensed and is updated with the fact that the vehicle crossed the intersection.

The updated domain description section on support technology will detail the exact workings of these tollgate and internal intersection machines and the domain description section on human behaviour will likewise explore the man/machine facet.

Re-engineering Domain Events. The intersections are highway-engineered in such a way as to deter vehicle entry into opposite direction tollway links, yet, one never knows, there might still be (what we shall call ghost) vehicles, that is vehicles which have somehow defied the best intentions, and are observed moving along a tollway link in the wrong direction.

Re-engineering Domain Behaviours. The intended behaviour of a vehicle of the tollway is to enter at an entry hub (collecting a ticket at the toll gate), to move to the associated intersection, to move into, where relevant, either an upward or a downward tollway link, to proceed (i.e., move) along a sequence of one or more tollway links via connecting intersections, until turning into an exit link and leaving the net at an exit hub (toll plaza) while paying the toll.

• • •

This should be enough of a BPR rough sketch for us to meaningfully proceed to requirements prescription proper.

4.3   Domain Requirements Prescription

A domain requirements prescription is that part of the overall requirements prescription which can be expressed solely using terms from the domain description. Thus to construct the domain requirements prescription all we need is collaboration with the requirements stakeholders (who, with the requirements engineers, developed the BPR) and the possibly rewritten (resulting) domain description.

Domain Projection. By a *domain projection* we mean *a subset of the domain description, one which leaves out all those entities, functions, events, and (thus) behaviours that the stakeholders do not wish represented by the machine.*

The resulting document is a *partial domain requirements prescription*.

Domain Projection — Narrative. We copy the domain description and call the copy a 0th version domain requirements prescription. From that document we remove all mention of link insertion and removal functions, to obtain a 1st version domain requirements prescription.

Domain Projection — Formalisation. We do not show the resulting formalisation.

Domain Instantiation. By *domain instantiation* we mean *a refinement of the partial domain requirements prescription, resulting from the projection step, in which the refinements aim at rendering the entities, functions, events, and (thus) behaviours of the partial domain requirements prescription more concrete, more specific. Instantiations usually render these concepts less general.*

Domain Instantiation — Narrative. The 1st version domain requirements prescription is now updated with respect to the properties of the toll way net: We refer to Fig. 1 and the preliminary description given in Sect. 4.2. There are three kinds of hubs: tollgate hubs and intersection hubs: terminal intersection hubs and proper, intermediate intersection hubs. Tollgate hubs have one connecting two way link. linking the tollgate hub to its associated intersection hub. Terminal intersection hubs have three connecting links: (i) one, a two-way link, to a tollgate hub, (ii) one one-way link emanating to a next up (or down) intersection hub, and (iii) one one-way link incident upon this hub from a next up (or down) intersection hub. Proper intersection hubs have five connecting links: one, a two way link, to a tollgate hub, two one way links emanating to next up and down intersection hubs, and two one way links incident upon this hub from next up and down intersection hub. (Much more need be narrated.) As a result we obtain a 2nd version domain requirements prescription.

Domain Instantiation — Formalisation, Toll Way Net.

type
    TN = ((H × L) × (H × L × L))* × H × (L × H)
value
    abs_N: TN → N
    abs_N(tn) ≡ (tn_hubs(tn),tn_hubs(tn))
    pre wf_TN(tn)

    tn_hubs: TN → **H-set**,
    tn_hubs(hll,h,(_,hn)) ≡
        {h,hn} ∪ {thj,hj|((thj,tlj),(hj,lj,lj′)):
            ((H×L)×(H×L×L))•((thj,tlj),(hj,lj,lj′))∈ **elems** hlll}
    tn_links: TN → **L-set**
    tn_links(hll,_,(ln,_)) ≡ ... as above ...

theorem ∀ tn:TN • wf_TN(tn) ⇒ wf_N(abs_N(tn))

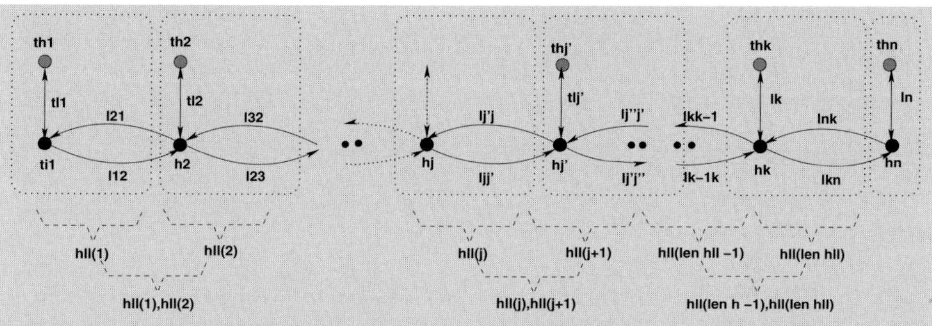

Fig. 2. A simple, linear toll road net: th_i: toll plaza i, h_1, h_n: terminal intersections, h_2, h_j, h'_j, h_k: intermediate intersections, $1<j≤k$, $k=n$-1 l_{xy}, l_{yx}: tollway link from h_x to h_y and from h_y to h_x, $1≤x<n$. l_{x-1x}, l_{xx-1}: tollway link from h_{x-1} to h_x and h_x to h_{x-1}, $1≤x<n$, dashed links are not in formulas.

Domain Instantiation — Formalisation, Well-formedness.

type
    LnkM == plaza | way
value
    wf_TN: TN → **Bool**
    wf_TN(tn:(hll,h,(ln,hn))) ≡

wf_Toll_Lnk(h,ln,hn)(plaza) $\land$ wf_Toll_Ways(hll,h) $\land$
wf_State_Spaces(tn) [to be defined under Determination]

value
   wf_Toll_Ways: $((H{\times}L){\times}(H{\times}L{\times}L))^* \times H \to$ **Bool**
   wf_Toll_Ways(hll,h) $\equiv$
     $\forall$ j:**Nat** $\bullet$ $\{j,j+1\}\subseteq$**inds** hll $\Rightarrow$
      let $((thj,tlj),(hj,ljj',lj'j)) = $ hll(j),
         $(_,(hj',_,_)) = $ hll(j+1) **in**
      wf_Toll_Lnk(thj,tlj,hj)(plaza) $\land$
      wf_Toll_Lnk(hj,ljj',hj')(way) $\land$ wf_Toll_Lnk(hj',lj'j,hj)(way) **end** $\land$
     let $((thk,tlk),(hk,lk,lk')) = $ hll(**len** hll) **in**
      wf_Toll_Lnk(thk,tlk,hk)(plaza) $\land$
      wf_Toll_Lnk(hk,lk,hk')(way) $\land$ wf_Toll_Lnk(hk',lk',hk)(way) **end**

value
   wf_Toll_Lnk: $(H{\times}L{\times}H) \to$ LnkM $\to$ **Bool**
   wf_Toll_Lnk(h,l,h')(m) $\equiv$
     obs_Ps(l)$=\{$(obs_HI(h),obs_LI(l),obs_HI(h')),
           (obs_HI(h'),obs_LI(l),obs_HI(h))$\} \land$
    obs_Σ(l) $= $ **case** m **of**
          plaza $\to$ obs_Ps(l),
          way $\to \{$(obs_HI(h),obs_LI(l),obs_HI(h'))$\}$ **end**

Domain Determination. By *domain determination* we mean a refinement of
*the partial domain requirements prescription, resulting from the instantiation
step, in which the refinements aim at rendering the entities, functions, events,
and (thus) behaviours of the partial domain requirements prescription less non-
determinate, more determinate. Instantiations usually render these concepts less
general.*

Domain Determination — Narrative. We single out only two 'determinations':
The link state spaces. There is only one link state: the set of all paths through
the link, thus any link state space is the singleton set of its only link state. *The
hub state spaces* are the singleton sets of the "current" hub states which allow
these crossings: (i) from terminal link back to terminal link, (ii) from terminal
link to emanating tollway link, (iii) from incident tollway link to terminal link,
and (iv) from incident tollway link to emanating tollway link. Special provision
must be made for expressing the entering from the outside and leaving toll plazas
to the outside.

Domain Determination — Formalisation.

$$wf\_State\_Spaces: TN \rightarrow \textbf{Bool}$$
$$wf\_State\_Spaces(hll,hn,(thn,tln)) \equiv$$
$$\textbf{let } ((th1,tl1),(h1,l12,l21)) = hll(1),$$
$$((thk,ljk),(hk,lkn,lnk)) = hll(\textbf{len } hll) \textbf{ in}$$
$$wf\_Plaza(th1,tl1,h1) \wedge wf\_Plaza(thn,tln,hn) \wedge$$
$$wf\_End(h1,tl1,l12,l21,h2) \wedge wf\_End(hk,tln,lkn,lnk,hn) \wedge$$
$$\forall j:\textbf{Nat} \cdot \{j,j+1,j+2\} \subseteq \textbf{inds } hll \Rightarrow$$
$$\textbf{let } (,(hj,ljj,lj'j)) = hll(j),((thj',tlj'),(hj',ljj',lj'j')) = hll(j+1) \textbf{ in}$$
$$wf\_Plaza(thj',tlj',hj') \wedge wf\_Interm(ljj,lj'j,hj',tlj',ljj',lj'j') \textbf{ end end}$$

$$wf\_Plaza(th,tl,h) \equiv$$
$$obs\_H\Sigma(th) = \{ \ [\,crossings\ at\ toll\ plazas\,]$$
$$("external",obs\_HI(th),obs\_LI(tl)),$$
$$(obs\_LI(tl),obs\_HI(th),"external"),$$
$$(obs\_LI(tl),obs\_HI(th),obs\_LI(tl))\} \wedge$$
$$obs\_H\Omega(th) = \{obs\_H\Sigma(th)\} \wedge obs\_L\Omega(tl) = \{obs\_L\Sigma(tl)\}$$

$$wf\_End(h,tl,l,l') \equiv$$
$$obs\_H\Sigma(h) = \{[\,crossings\ at\ 3-link\ end\ hubs\,]$$
$$(obs\_LI(tl),obs\_HI(h),obs\_LI(tl)),(obs\_LI(tl),obs\_HI(h),obs\_LI(l)),$$
$$(obs\_LI(l'),obs\_HI(h),obs\_LI(tl)),(obs\_LI(l'),obs\_HI(h),obs\_LI(l))\} \wedge$$
$$obs\_H\Omega(h) = \{obs\_H\Sigma(h)\} \wedge$$
$$obs\_L\Omega(l) = \{obs\_L\Sigma(l)\} \wedge obs\_L\Omega(l') = \{obs\_L\Sigma(l')\}$$

$$wf\_Interm(ul\_1,dl\_1,h,tl,ul,dl) \equiv$$
$$obs\_H\Sigma(h) = \{[\,crossings\ at\ properly\ intermediate,\ 5-link\ hubs\,]$$
$$(obs\_LI(tl),obs\_HI(h),obs\_LI(tl)),(obs\_LI(tl),obs\_HI(h),obs\_LI(dl\_1)),$$
$$(obs\_LI(tl),obs\_HI(h),obs\_LI(ul)),(obs\_LI(ul\_1),obs\_HI(h),obs\_LI(tl)),$$
$$(obs\_LI(ul\_1),obs\_HI(h),obs\_LI(ul)),(obs\_LI(ul\_1),obs\_HI(h),obs\_LI(dl\_1)),$$
$$(obs\_LI(dl),obs\_HI(h),obs\_LI(tl)),(obs\_LI(dl),obs\_HI(h),obs\_LI(dl\_1)),$$
$$(obs\_LI(dl),obs\_HI(h),obs\_LI(ul))\} \wedge$$
$$obs\_H\Omega(h) = \{obs\_H\Sigma(h)\} \wedge obs\_L\Omega(tl) = \{obs\_L\Sigma(tl)\} \wedge$$
$$obs\_L\Omega(ul) = \{obs\_L\Sigma(ul)\} \wedge obs\_L\Omega(dl) = \{obs\_L\Sigma(dl)\}$$

Not all determinism issues above have been fully explained. But for now we should — in principle — be satisfied.

Domain Extension. By domain extension we understand the *introduction of domain entities, functions, events and behaviours that were not feasible in the original domain, but for which, with computing and communication, there is the possibility of feasible implementations, and such that what is introduced become part of the emerging domain requirements prescription.*

Domain Extension — Narrative. The domain extension is that of the controlled access of vehicles to and departure from the toll road net: the entry to (and departure from) tollgates from (respectively to) an **"an external"** net — which we do not describe; the new entities of tollgates with all their machinery; the user/machine functions: upon entry: driver pressing entry button, tollgate delivering ticket; upon exit: driver presenting ticket, tollgate requesting payment, driver providing payment, etc.

One added (extended) domain requirements: as vehicles are allowed to cruise the entire net payment is a function of the totality of links traversed, possibly multiple times. This requires, in our case, that tickets be made such as to be sensed somewhat remotely, and that intersections be equipped with sensors which can record and transmit information about vehicle intersection crossings. (When exiting the tollgate machine can then access the exiting vehicles sequence of intersection crossings — based on which a payment fee calculation can be done.)

All this to be described in detail — including all the thinks that can go wrong (in the domain) and how drivers and tollgates are expected to react.

Domain Extension — Formalisation. We suggest only some signatures:

type
Mach, Ticket, Cash, Payment, Map_TN
value
obs_Cash: Mach $\rightarrow$ Cash, obs_Tickets: M $\rightarrow$ Ticket-**set**
obs_Entry, obs_Exit: Ticket $\rightarrow$ HI, obs_Ticket: V $\rightarrow$ (Ticket|nil)

calculate_Payment: (HI$\times$HI) $\rightarrow$ Map_TN $\rightarrow$ Payment

press_Entry: M $\rightarrow$ M $\times$ Ticket   [gate up]
press_Exit: M $\times$ Ticket $\rightarrow$ M $\times$ Payment
payment: M $\times$ Payment $\rightarrow$ M $\times$ Cash   [gate up]

Domain Extension — Formalisation of Support Technology. This example provides a classical requirements engineering setting for embedded, safety critical, real-time systems, requiring, ultimately, the techniques and tools of such things as Petri nets, statecharts, message sequence charts or live sequence charts and temporal logics (DC, TLA+).

Requirements Fitting. The issue of requirements fitting arises when two or more software development projects are based on what appears to be the same domain. The problem then is to harmonise the two or more software development projects by harmonising, if not too late, their requirements developments.

We thus assume that there are n domain requirements developments, d_{r_1}, d_{r_2}, ..., d_{r_n}, being considered, and that these pertain to the same domain — and can hence be assumed covered by a same domain description.

By requirements fitting we mean *a harmonisation of $n > 1$ domain requirements that have overlapping (common) not always consistent parts and which results in n 'modified and partial domain requirements', and m 'common domain requirements' that "fit into" two or more of the 'modified and partial domain requirements'*.

Requirements Fitting — Narrative. We postulate two domain requirements: We have outlined a domain requirements development for software support for a toll road system. We have earlier hinted at domain operations related to insertion of new and removal of existing links and hubs. We can therefore postulate that there are two domain requirements developments, both based on the transport domain: one, $d_{r_{toll}}$, for a toll road computing system monitoring and controlling vehicle flow in and out of toll plazas, and another, $d_{r_{maint.}}$, for a toll link and intersection (i.e., hub) building and maintenance system monitoring and controlling link and hub quality and for development.

The fitting procedure now identifies the shared of awareness of the net by both $d_{r_{toll}}$ and $d_{r_{maint.}}$ of nets (N), hubs (H) and links (L). We conclude from this that we can single out a common requirements for software that manages net, hubs and links. Such software requirements basically amounts to requirements for a database system. A suitable such system, say a relational database management system, DB_{rel}, may already be available with the customer.

In any case, where there before were two requirements $(d_{r_{toll}}, d_{r_{maint.}})$ there are now four: (i) $d'_{r_{toll}}$, a modification of $d_{r_{toll}}$ which omits the description parts pertaining to the net; (ii) $d'_{r_{maint.}}$, a modification of $d_{r_{maint.}}$ which likewise omits the description parts pertaining to the net; (iii) $d_{r_{net}}$, which contains what was basically omitted in $d'_{r_{toll}}$ and $d'_{r_{maint.}}$; and (iv) $d_{r_{db:i/f}}$ (for database interface) which prescribes a mapping between type names of $d_{r_{net}}$ and relation and attribute names of DB_{rel}.

Much more can and should be said, but this suffices as an example in a software engineering methodology paper.

Requirements Fitting — Formalisation. We omit lengthy formalisation.

Domain Requirements Consolidation. After projection, instantiation, determination, extension and fitting, it is time to review, consolidate and possibly restructure (including re-specify) the domain requirements prescription before the next stage of requirements development.

5   Discussion

5.1   An 'Odyssey'

Our 'Odyssey' has ended. A long example has been given.

We have shown that requirements engineering can have an abstraction basis in domain engineering; and we have shown that we do not have to start software development with requirements engineering, but that we can start software development with domain engineering and then proceed to a more orderly requirements engineering phase than witnessed today.

5.2   Claims of Contribution

What is essentially new here is the claim and its partial validation that one can and probably should put far more emphasis on domain modelling, the domain modelling concepts, principles and techniques of business process domain intrinsics, domain support technologies, domain management and organisation, domain rules and regulations, domain scripts and domain human behaviour; the identification of, and the decomposition of the requirements development process into, domain requirements, interface requirements and machine requirements; the domain requirements "derivation" concepts, principles and techniques of projection, instantiation, determination, extension and fitting and the identification of structuring of the interface ground requirements shared entities, shared operations, shared events and shared behaviours.

5.3   Comparison to Other Work

Jackson's Problem Frame approach [4] cleverly alternates between domain analysis, requirements development and software design. For more satisfactory comparisons between our domain engineering approach and past practices and writings on domain analysis we refer to [3].

5.4   A Critique

A major presentation of domain and of requirements engineering is given in [1, Chaps. 8–16 and 17–24]. [3] provides a summary, more complete presentation of domain engineering than the present paper allows, while [2] discusses a set of research issues for domain engineering. Papers, like [3,2], but for requirements engineering, with more a complete presentation, respectively a discussion of research issues for this new kind of requirements engineering might be desirable. The current paper's Sect. 4 provided a slightly revised structuring of the interface requirements engineering.

Some of the development steps within the domain modelling and likewise within the requirements modelling are refinements, and some are extensions. If we ensure that the extensions are what is known as Conservative extensions then all theorems of the source of the extension go through and are also valid in the extension. Although such things are here rather clear much more should be said here about ensuring Conservative extensions. We do not since the current paper is is not aimed at the finer issues of the development but at the domain to requirements "derivation" issues.

5.5   Programming Methodology Versus Software Engineering

The following question has been formulated: *How to make a programming methodological approach like this become everyday software engineering practice ? For example, a small company willing to launch on the market a new idea in a specific domain, needs under this approach to build up a full domain formalization. I fear that this could be felt as a too large burden. On the other hand, using pre-cooked, public, standardized or third-party formalizations of specific domains could end in constraining the imagination of innovators ?*

The programming methodological answer is: Yes, one must build a domain description, informal and, ideally speaking also formal.

The software engineering answer is: how "full" it should be: at least "big" enough to encompass the requirements.

The science and engineering answer is: public universities must experimentally develop and research sufficiently broad (scope) domain theories, while private software houses adapt these to their narrower (span) application domains, thus establishing proprietary, corporate assets.

The research answer is: We must study a programming methodology like the one put forward in this paper. We must do so because the programming methodology appears logical, sound. We cannot abstain from studying this programming methodology just because (even a majority of) software engineers "feels" that it *is too large a burden* to follow this approach "slavishly".

Acknowledgments

I gratefully acknowledge support from Université Henri Poincaré (UHP), Nancy, and from INRIA (l'Institut National de Recherche en Informatique et en Automatique) both of France, for my two month stay at LORIA (Laboratoire Lorrain de Recherche en Informatique et ses Applications), Nancy, in the fall of 2007. I especially and warmly thank Dominique Méry for hosting me. And I thank the organisers of Ugo Montanari's Festschrift, Pierpaolo Degano, Jose Meseguer and Rocco De Nicola, for inviting me — thus forcing me to willingly write this paper.

References

1. Bjørner, D.: Software Engineering: Domains, Requirements and Software Design. Texts in Theoretical Computer Science, the EATCS Series, vol. 3. Springer, Heidelberg (2006)
2. Bjørner, D.: Domain Theory: Practice and Theories, Discussion of Possible Research Topics. In: Jones, C.B., Liu, Z., Woodcock, J. (eds.) ICTAC 2007. LNCS, vol. 4711, pp. 1–17. Springer, Heidelberg (2007)

3. Bjørner, D.: Domain Engineering. In: Boca, P., Bowen, J. (eds.) BCS FACS Seminars, London, UK. Lecture Notes in Computer Science, the BCS FAC Series, pp. 1–42. Springer, Heidelberg (to appear, 2008)
4. Jackson, M.A.: Problem Frames — Analyzing and Structuring Software Development Problems. Addison–Wesley Longman Publishing Co., Inc, Boston, MA (2001)

Laudatio: Ugo 65 Years

There is one thing, rather egoistically expressed, that epitomises my scientific relationship with Ugo:

That in 1980 — it all started at a computer science conference at CNRS in Paris in May 1980 — he was able to bring my groups' interest in developing, professionally, using formal techniques (in the "school" of VDM in those days) an Ada compiler together with an Italian consortium of universities and industry (Olivetti) — and that, together we were able to obtain the European Community funded (multi-annual programme, CEC MAP) Ada Compiler Development project.

For me that project and its follow-on CEC MAP project, the Formal Definition of Ada, also strongly helped on the way by Ugo and his contacts, notably Egidio Astesiano — whose work on that project is legendary — became an important step in my scientific life.

And Ugo was a main source for those two major CEC MAP projects. For that I very grateful to Ugo.

There is another thing, less egoistically expressed, that epitomises my friendship with Ugo:

Our social life together, in Pisa, off work, in Holte, at home, and at conferences around Italy, two of them on Capri (1982 and 1985), on at Taormina (June 2003), one at Pisa (Sept. 2003), and elsewhere, around the world: with Norma Lijtmaer — we shall never forget her.

Thanks, Ugo, for bringing science big steps forward, in Italy and elsewhere; thanks for your trust in me; and thanks for many wonderful times together. All the very best wishes for the very many next years.

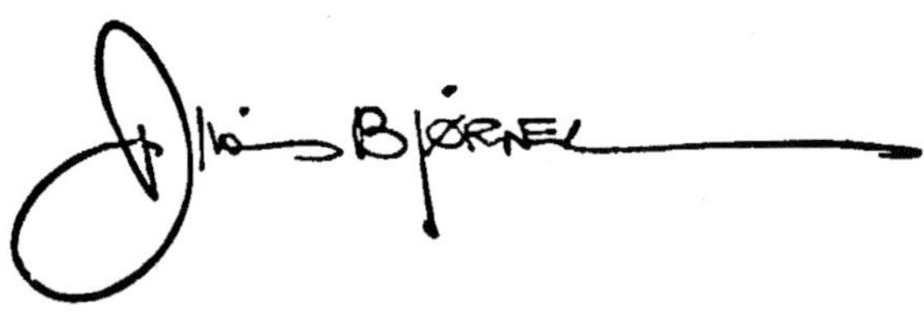

Business Process Modeling for Organizational Knowledge Management

Luca Abeti[1], Paolo Ciancarini[2], and Rocco Moretti[2]

[1] IMT Institute for Advanced Studies, Piazza S.Ponziano, 55100 Lucca, Italy
luca.abeti@imtlucca.it
[2] University of Bologna, CS Dept., Mura Zamboni 7, 40126 Bologna, Italy
ciancarini@cs.unibo.it, moretti@cs.unibo.it

Abstract. The growing complexity of a networked and information-dependent economy requires the innovation of the adopted processes together with their related services. In particular, many Small and Medium-sized Enterprises (SME's) currently base their organizational models in a resource-centric view rather than in a knowledge-based organizational model which is a fundamental bound to their innovation capabilities. This paper presents a framework for organizational knowledge management. Our approach is based on Business Process Modeling (BPM), that is the main modeling practice connecting the management and engineering disciplines in software development. The aim is to present how the software requirements analysis can help in formalizing and sharing the knowledge concerning the business processes. Besides, we show how the service and ontology abstractions can be useful for software development.

1 Introduction

In recent years Ugo Montanari published as sole author or as coauthor a number of papers on some fundamental theoretical issues related to Service-Oriented Architectures, see for instance [8, 9, 10, 13, 22]. Service-Oriented Architectures are software architectures that enable new application scenarios in which small, loosely coupled pieces of functionality are published, consumed, and combined with other functions over a network. A key feature of the SOA's is a two-level model to implement global enterprise systems, where business functions are implemented by individual services and business processes are built as combinations of services.

This technologically-oriented research trend has to confront another, more socio-organizationally-oriented research trend. In fact, the studies about enterprise organizational processes have brought deep changes in the economy and society [11]. The evolution of Business Process Modeling (BPM) has been strongly influenced by its relationships with the new technologies like business process reengineering [16]. Despite the close relationship between the business process view in technology and economy, this concept is considered differently and for different goals by the software engineers and managers [26].

P. Degano et al. (Eds.): Montanari Festschrift, LNCS 5065, pp. 301–311, 2008.

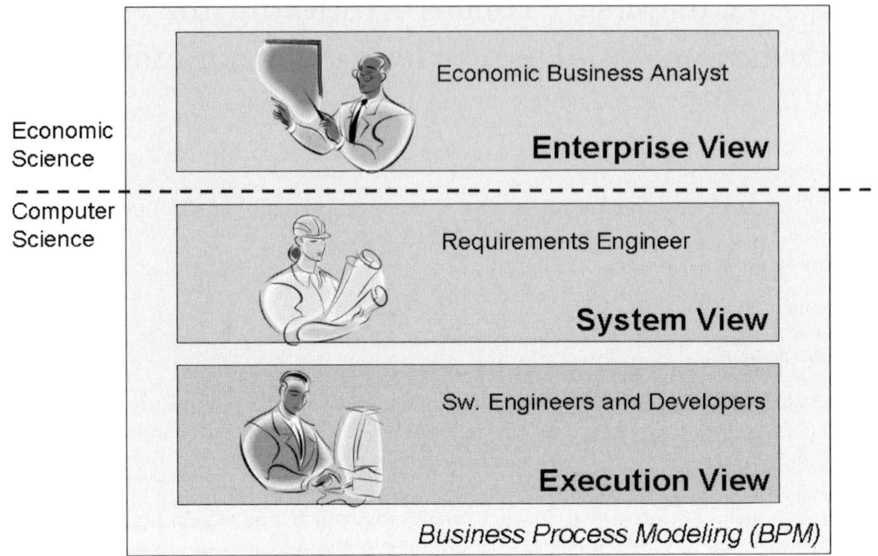

Fig. 1. The three views for BPM

The Business Processes (BP's) modelers of contemporary enterprises have to consider the changes enacted by the new technologies in their processes [29]. On the other hand, the new software engineering technologies influence BP's in the development of service systems [14]. Besides, the emergence of new enterprise models, such as networked and service-oriented enterprises, requires open and interoperable technologies supporting their processes.

Figure 1 identifies the three points of view representing the three main perspectives for BPM:

- **Enterprise view**: its goal is the business improvement. At this level the models deal with the organization, the strategies, the business rules, the business domains, the internal and external BP's, etc. It is the BPM perspective commonly used by the economic business analysts.
- **System view**: at this level the goal is to acquire the early-requirements of a system by means of a business analysis. The models concern the organization, the internal BP's, the business entities, the systems, the architectures, etc. It is the perspective of the software business analysts.
- **Execution view**: at this level the goal is to define an executable model for BP's. It is the software engineers perspective for the system design and implementation. This view of BP's is shared with the developers.

The capability to design a system by means of the service abstraction for its components has important implications in software engineering [22, 30] and can be easily connected to the BP models by means of a design based on the Model-Driven Architecture (MDA) [1]. This issue can be addressed in the *System view* perspective and can be managed by software engineering languages and methods [6].

However, the BP-to-Service mapping can be treated differently in the *System* and *Execution views* because we can distinguish two further perspectives, namely: a *dynamic* and *static* representation of BP's and services. Our approach concerns only the static mapping of the knowledge related to the BP's into services and ontology abstractions. The dynamic behaviors and reconfigurations for services and architectures are considered orthogonal issues for the design of distributed systems by means of the BP abstraction [10, 17].

Currently do not exist methods and tools for BP's development that enable to connect all the three BPM views exposed in Figure 1. Indeed, the current Integrated Development Environments (IDE's) do not fully support the management of both the service abstraction and the related BP technologies evolution.

This paper honors Ugo focussing on the study of the relationship between software engineering and BP modeling. We propose a requirements-driven software development method that considers all the needs and motivations related to the BP's view in order to support the software systems development. Our approach uses the service as intermediate abstraction in order to incrementally model the system starting from its BP's representation. In our approach, the services are considered as autonomous computational entities that can be managed by means of model-driven tools in order to implement the systems.

In this method others modeling languages can be added in order to better support the implementation of BPM. For instance, constraints for service models can be added by means of the SENSORIA Reference Modeling Language (SRML) metamodel [7] in our model-driven tool.

In the next section, we show an overview of our approach. Section 3 presents some IDE and tools supporting the service abstraction management. In Section 4, we present our conclusions.

2   BP's for Organizational Knowledge Management

In this section, we present a method for product knowledge formalization based on software engineering and BPM techniques. This method presents three general phases which correspond to the three BPM views exposed in Section 1. These phases are: *Business Modeling, System Modeling* and *System Implementation.*

In order to define a method able to manage the entire business and to achieve the software knowledge formalization, we assume that the method itself can be tuned for a specific project. Indeed, the BPM process may vary for many reasons:

- BPM can be realized for merely knowledge acquisition or for analysis and requirements identification
- the system development can require a sketch, blueprint or detailed representation
- the development requires more emphasis on structural or behavioral modeling
- the stakeholders, analysts or managers can be not confident with the adopted modeling language

- the reasons of software development could be not well understood
- the methods and processes used inside the company can be inadequate and
  difficult to change

Such reasons are an important aspect of our work because high varying business and problems require to vary also the processes used for BPM [6].

This approach allows to address high-level-requirements in a distributed application and can be exploited in the development of a SOA.

2.1   Business Modeling

The goals of the Business Modeling phase is to wrap the current management with a technology support and to constantly share the knowledge among the economists and the system engineers inside a company.

The focus of the Business Modeling phase is on collecting and maintaining all the knowledge about the company organization, the strategies, the goals, the risks and the management-related issues. In order to face these aspects, this phase is decomposed into two sub-phases: the *informal brainstorming* and the *enterprise knowledge formalization*. The former sub-phase is a collective learning activity [23] trying to realize a shared knowledge-base about the company organization, the strategies and the goals.

Usually different stakeholders assign different meanings to the constructs (e.g., actors, goals, strategies, organizational units) of the organizational knowledge based on their mental models [2]. In this sub-phase, the models are mainly used to structure the problems and the organization. In the latter sub-phase the business process analysts and the managers try to formalize the knowledge in order to check its consistency and to discuss with the stakeholders. Thus, these sub-phases are cyclic Business Modeling sub-phases performed many times taking into account the specific company, project or stakeholders needs. These sub-phases enable to move from a tacit knowledge of the company into a codified and consistent knowledge consisting of BP's for the represented organization.

In the Business Modeling phase the degree of competency, the background and the belief of each participant may vary significantly. Thus, it does not make sense to propose a standard sets of steps and restrictive guidelines for Business Modeling. Besides, the language provided by Si* [21], UML [25] and the Business Process Management Notation (BPMN) [26] can be useful for enterprise knowledge formalization. This set of guidelines and languages is not restrictive because usually the stakeholders and managers are reluctant to spend time on brainstorming, process formalization, learning, training and becoming confident with formal specification languages and methods.

We now briefly discuss the three languages mentioned above. The UML Use Case diagrams have been chosen because of their proved efficacy in stakeholder-analyst collaboration [12]. The Use Case diagrams use intuitive and quick to understand concepts such as: *actor, use case, inclusion, extension, system* *boundary* etc. BPMN is a specification of the Object Management Group (OMG) and the Business Process Management Initiative (BPMI.org) [26]. It synthesizes the best practices of the BPM community and defines a graphical notation for the BP's similar to a flow-chart. Such a notation is both consistent with UML and understandable by the stakeholders, analysts, business users, developers, etc. Moreover, BPMN is widely extensible and can be incrementally adopted. Si* considers a set of primitive concepts such as *actor, goal, task,* and *resource* in order to model a socio-technical system. An *actor* is an active entity having strategic goals and performing actions to achieve these goals. A *goal* is a strategic interest of an actor. A *Soft-goal* is similar to a goal, but the fulfillment condition is not clearly defined. A *task* specifies a sequence of actions that can be executed to achieve a goal. Finally, a *resource* represents a physical or an informational entity.

Goals, tasks, and resources are often related among them in many ways. In particular, three relations have been identified, namely *AND/OR decomposition, means-end,* and *contribution* relations. The AND/OR decomposition combines AND and OR refinements of a root goal into sub-goals. The Means-end relation identifies tasks providing means for achieving a goal and the resources produced or consumed by a task. The contribution relation identifies the impact of the achievement of goals and tasks on the achievement of other goals and tasks. Due to the general meaning of the Si* concepts, we use them without a prescriptive formalism in the informal brainstorming sub-phase. Thus, starting from the beginning of our practice, the tacit knowledge is acquired in terms of the concepts that we will use in the enterprise knowledge formalization sub-phase and in the System Modeling phase.

We distinguish between functional and non-functional specifications. The functional specifications are formalized using the UML Use Case diagrams as regards the static aspects of BP's, and BPMN as regards the dynamic interactions among the BP's concepts. The non-functional specifications closer to tacit knowledge are formalized by using Si* diagrams. The informal brainstorming sub-phase allows to identify an unorganized and not-formalized set of early-requirements that giving a first sketch of the domain and the knowledge concerning the organization. The enterprise knowledge formalization sub-phase is the first phase trying to obtain a formal and analyzable BP representation. The cyclic informal brainstorming sub-phase helps the stakeholders in understanding their implicit knowledge and defining a shared knowledge base of the processes. The enterprise knowledge formalization sub-phase tries to organize the informal knowledge in order to analyze it in the System Modeling phase.

The modeling of goals and strategies (that are not caught by Use Cases and BPMN models) are essential for BPM design and knowledge acquisition. For this purpose, our approach uses the Si* notation. Si* helps to model strategical and operational aspects of the business. By means of the Si* concepts, we are able to connect formally represented systems and actors to, for instance, business units, manager aims, practices and company policies.

2.2 System Modeling

At the beginning of the System Modeling phase, we use the outputs of the Business Modeling phase in order to derive an analysis similar to the Tropos early-requirements analysis [19]. The Business Modeling phase provides three inputs to the System Modeling phase:

1. A formalized knowledge about the business and business processes represented in a set of diagrams.
2. Some indicative choices or purposes to develop new systems and products.
3. A first analysis concerning the technologies that can be used to realize new goals or improve the existing business.

In the System Modeling phase, further technical choices are made in technical brainstormings. In particular, we consider the non-functional requirements defined by means of Si*. Besides, the goals and the soft-goals and their relationships with the system can be deeply analyzed by exploiting the Tropos early-requirements analysis process [15, 19]. The early-requirements analysis concerns with the understanding of a problem by studying an existing organizational setting. The intentions of the stakeholders are modeled as goals and goal dependencies among actors, and analyzed by means some form of goal analysis. The output of this phase is an organizational model including the relevant actors and their respective dependencies for the achievement of the goals and the soft-goals, and for performing or obtaining resources. The System Modeling phase includes six models that can be defined by means of Si*:

- **Actor Model**: allows to identify the actors and their objectives, entitlements, and capabilities. The agents are also described in terms of the roles they play.
- **Social Model**: allows to identify and analyze the social relationships among the system actors and the stakeholders. Trust and distrust relationships between actors are discovered and modeled (i.e., expectations of actors about the capabilities and behaviors of other actors).
- **Goal Model**: allows to model the goals from the point of view of an actor. The impact of the goals on the achievement of other goals is analyzed and modeled in order to refine the requirements models and to elicit new social relations among the actors.
- **Execution Dependency Model**: allows to identify the actors depending on other actors for achieving their goals, executing the tasks and supplying the resources. In this model the assignments of responsibilities among the actors are discovered and modeled.
- **Delegation Model**: allows to model an actor delegating to other actors the achievement of goals, the execution of tasks and the access to the resources. This model enables the transfer of rights among the actors.
- **Task/Resource Model**: allows to elicit and model the tasks and the resources providing means for the achievement of goals. The impact of the tasks on the achievement of goals is analyzed and modeled.

These models enable to define a rigorous representation of the enterprise knowledge by means of the Actor, Social, Execution Dependency and Delegation Models. Besides, they enable to perform some analysis on such a knowledge by means of the Goal model and Task/Resource Model. The analyses that can be carried out are:

- **Means-end analysis**: aimed to identify tasks, goals or resources that provide means for achieving a specific goal.
- **Goal/Resource refinement**: analyzes and decomposes the goals and/or resources in terms of AND/OR decompositions.
- **Contribution Analysis**: studies the impact of the tasks and the achievements of goals on the achievement of other goals.

Starting from the Business Modeling phase, if in the System Modeling phase the need emerges to reengineer BP's or new IT system, the Si* diagrams must be considered together with UML Use Cases and BPMN diagrams. In this way, the IT System requirements can be derived and used in one or more System Implementation phases.

One of the most important activities of the System Modeling phase is the mapping from the business process concepts into the concepts useful in the System Implementation phase. We do not suggest to map BP's directly into the programming languages concepts (e.g., objects and classes), but we try to exploit the ontology and service concepts as intermediate abstractions enabling to move from business processes to implementation paradigms. Starting from the Si* and BPMN models, one or more application service models, depending on the number of the systems to implement, are defined.

The concept of service is considered as a very general abstraction for software development and can be used to represent a wide range of interacting software components [30]. In the same way, the UML Use Cases are used to derive a set of ontological concepts to be represented in one or more ontology models. In order to perform the Use Case-to-Ontology mapping, we can use the Rational Unified Process Business Model-to-System mapping rules [20] to incrementally redefine the Use Case models in a system-centric perspective. The application service and ontology models are both represented in diagrams similar to the UML Class Diagrams [25]. We propose this type of diagrams since it is very intuitive and has an easy to remember semantics. Besides, the service abstraction is useful for a logical division of the software [28]. Such a division is more coarse-grained than the division obtained by components or objects. The selection of the service and ontology abstractions represents a meeting point between the designer and developer requirements in the translation from BP's to services. On the one hand, to derive services from BP's it is useful to realize high level interfaces that are representative of the actual provided business services. On the other hand, the granularity of the services helps to define software components that can be easily changed and reused. The definition of the appropriate level of granularity for the services should consider both cohesion (i.e., the degree of relatedness of service functions) and coupling (i.e., the degree of service independence). The cohesion and coupling information can be derived in the BP's-to-Service mapping. The non-functional requirements can be derived from the Si* models by means of goals and soft-goals dependences analysis.

2.3   System Implementation

The last phase of our practice is the System Implementation phase whose goal is to model the concrete executable support for the BP's. The aim of this phase may change depending on the specific structure of the Business Modeling and System Modeling phases. The System Implementation phase is not necessarily performed because in the System Modeling phase it is possible to decide that no changes have to be made in the current BP's and that no new systems have to be realized.

The outputs of the System Modeling phase are the service and the ontology models. Such models are used to realize a detailed design of the system. Depending on the system nature (e.g., Web applications, centralized systems, Grid systems, agent-based systems, etc.), a specific system implementation abstraction is used in order to model the system (e.g., classes, agents, Web Services, Web pages, CORBA components, Entity Java Beans, etc.). In this context, the service model represents the behavioral aspects of the software, while the ontology model represents the concepts used in the static aspects. For instance, the services in the application service model can be used in the UML behavioral models (e.g., the activity diagrams) or in the execution business process languages (e.g., BPEL4WS [4]). In an analogous way, the ontology models can be used for the UML structural models (e.g., the class diagrams) or to define conceptual and logical models of databases.

3   Tools for the System Modeling Phase

In this section we briefly describe some tools useful in the System Modeling phase described in Section 2.2. In order to support the mapping of the BP's represented by means of Si*, UML Use Cases and BPMN into the service abstraction, we need tools allowing to define, model, and deploy in a simple and platform-independent way the software services. In particular, we focus on Uniframe [5] and the MOdeling TOol for Grid and Agent Services (MOTO-GAS) [1].

Uniframe [5] aims to overcome the platform heterogeneity of distributed systems by using MDA-based service-oriented models. Uniframe defines an abstraction for a unified architecture, relies on MDA in order to design the service models, and uses a formal specification language to define the components and to support their connection. Uniframe investigates many component-based and service-oriented issues and defines an abstraction for a unified architecture. It uses UML to design service models and MDA to realize the Unified Meta-Component Model (UMM) used as a glue putting together different technologies.

MOTO-GAS [1] is a tool that enables to model both stateful and stateless services. A meaningful aspect of MOTO-GAS is the support for the Web Service Resource Framework (WSRF) [27]. Such support is given through the WSRF MetaObject Facility (MOF) [24] metamodel. MOTO-GAS allows to define a platform-independent application service model which can be mapped into an instance of the WSRF MOF metamodel. In this context, the application service model defined in the System Modeling phase represents an application defined by a set of services and resources in a Platform-Independent Model (PIM) described by UML. By means of the MDA-based ATLAS Transformation Language (ATL) [3, 18], it is possible to produce a Platform-Specific Model (PSM) complying with the WSRF MOF metamodel.

MOTO-GAS uses some existing plug-ins implementing the MDA specifications and allowing an effective consistency between the realized models and their respective metamodels. This set of plug-ins allows to define the application service models and to automatically generate: a WSDL definition, the Java service skeleton, the Web Service Deployment Descriptor (WSDD) file, and the Java Naming and Directory Interface (JNDI) deployment file for the service application used in the System Implementation phase.

Both Uniframe and MOTO-GAS are useful to support the System Modeling phase down to the System Implementation phase of our proposed practice. In particular, a model-driven support is considered an essential feature in order to develop platform-independent models of the system derived from the BP models designed in the Business Modeling phase.

4   Conclusions

In this paper we have presented a modeling approach considering a whole business and its organizational issues. It is aimed to obtain a full interaction between technologists and managers involved in the business innovation performed by means of the new technologies.

Our proposed method aims to analyze the use of BPM in software engineering. We have combined three complementary aspects represented by: Si*, UML Use Cases, and BPMN. Even though these aspects overlap, they enable to manage BP's from different points of view.

Considering a software engineering point of view, the service and ontology concepts represent strategic elements enabling to connect the economic and execution view of BP's. In particular, the service abstraction represents the bottleneck for BP's because it allows to map the enterprise BP's into the executive BP's and the system functionalities. Such BP's are mapped into services that are connected in order to create execution BP's and system functionalities.

A framework including both the support for the collaborative work among stakeholders, and the support for model-driven development of the system and ontology abstractions is far from being realized. In this context a further support of the Business Modeling phase in a model-driven IDE may consist in automating the mapping of the Si*, UML Use Case and BPMN models into the service and ontology models.

Acknowledgements

This work is partially supported by the MIUR FIRB project TOCAI.IT URL: http://www.dis.uniroma1.it/~tocai/

References

[1] Abeti, L., Ciancarini, P., Moretti, R.: Service oriented software engineering for modeling agents and services in Grid systems. Multiagent and Grid Systems Journal 2(2), 135–148 (2006)

[2] Adamides, E., Karacapilidis, N.: A knowledge centred framework for collaborative business process modeling. Business Process Management Journal 12(5), 557–575 (2006)

[3] Allilaire, F., Idrissi, T.: ADT: Eclipse development tools for ATL. In: Akehurst, D. (ed.) 2nd European Workshop on Model Driven Architecture (MDA) with an emphasis on Methodologies and Transformations (EWMDA-2), Canterbury, UK, pp. 171–178. The Computing Laboratory (2004)

[4] BEA Systems International Business Machines Corporation and Microsoft Corporation. Business Process Execution Language for Web Services (BPEL4WS) specifications (May 2003)

[5] Benatallah, B., Dijkman, R., Dumas, M., Maamar, Z.: Service Composition: Concepts, Techniques, Tools, and Trends. In: Stojanovic, Z., Dahanayake, A. (eds.) Service-Oriented Software System Engineering: Challenges and Practices, ch. 3, pp. 68–87. Idea Group Publishing, Hershey, PA (2005)

[6] Berztiss, A.T., Bubenko, J.A.: A software process model for business reengineering. In: Proceedings of Information Systems Development for Decentralized Organizations (ISDO 1995), an IFIP 8.1 Working Conference, Norwell, MA, USA, August 1995, pp. 184–200. Chapman & Hall - Kluwer Academic Publishers (1995)

[7] Bruni, R., Lafuente, A., Montanari, U., Tuosto, E.: Service oriented architectural design. In: Barthe, G., Fournet, C. (eds.) TGC 2007. LNCS, vol. 4912, pp. 186–203. Springer, Heidelberg (2008)

[8] Bruni, R., Lafuente, A., Montanari, U., Tuosto, E.: Style based reconfigurations of Software Architectures. Technical Report TR-07-17, Dipartimento di Informatica, University of Pisa (2007)

[9] Bruni, R., Melgratti, H., Montanari, U.: Theoretical foundations for compensations in flow composition languages. In: POPL 2005: Proc. 32nd ACM Symp. on Principles of Programming Languages, vol. 40, pp. 209–220. ACM Press, New York (2005)

[10] Buscemi, M., Montanari, U.: CC-Pi: a constraint-based language for specifying Service Level Agreements. In: De Nicola, R. (ed.) ESOP 2007. LNCS, vol. 4421, pp. 18–32. Springer, Heidelberg (2007)

[11] Davenport, T.: Process Innovation: Reengineering work through information technology. Harvard Business School Press, Boston (1993)

[12] Dobing, B., Parsons, J.: How UML is used. Communications of the ACM 49(5), 109–113 (2006)

[13] Ferrari, G.L., Hirsch, D., Lanese, I., Montanari, U., Tuosto, E.: Synchronised Hyperedge Replacement as a Model for Service Oriented Computing. In: de Boer, F.S., Bonsangue, M.M., Graf, S., de Roever, W.-P. (eds.) FMCO 2005. LNCS, vol. 4111, pp. 22–43. Springer, Heidelberg (2006)

[14] Ghezzi, C.: Software Engineering: Emerging Goals and Lasting Problems. In: Baresi, L., Heckel, R. (eds.) FASE 2006. LNCS, vol. 3922, Springer, Vienna, Austria (2006)

[15] Giorgini, P., Kolp, M., Mylopoulos, J., Pistore, M.: The Tropos methodology: an overview. In: Gleizes, M.P., Bergenti, F., Zambonelli, F. (eds.) Methodologies And Software Engineering For Agent Systems, ch. 5, pp. 89–105. Kluwer Academic Publishing, Norwell, MA, USA (2004)

[16] Hammer, M.: Reengineering Work: Don't Automate, Obliterate. Harvard Business Review 68(4), 104–112 (1990)

[17] Hirsch, D., Montanari, U.: Consistent transformations for software architecture styles of distributed systems. In: Stefanescu, G. (ed.) Workshop on Distributed Systems. ENTCS, vol. 28, p. 4 (1999)

[18] Jouault, F., Kurtev, I.: Transforming models with ATL. In: Bruel, J.-M. (ed.) MoDELS 2005. LNCS, vol. 3844, pp. 128–138. Springer, Heidelberg (2006)

[19] Kolp, M., Giorgini, P., Mylopoulos, J.: Organizational patterns for early requirements analysis. In: Eder, J., Missikoff, M. (eds.) CAiSE 2003. LNCS, vol. 2681, pp. 617–632. Springer, Heidelberg (2003)

[20] Kruchten, P.: The Rational Unified Process: An Introduction, 3rd edn. The Addison-Wesley Object Technology Series. Addison-Wesley Longman Publishing, Boston (2003)

[21] Massacci, F., Mylopoulos, J., Zannone, N.: An ontology for secure socio-technical systems. In: IGI Global (ed.) Handbook of Ontologies for Business Interaction. Information Science Reference, Hershey, PA, USA, vol. 1, p. 469 (December 2007)

[22] Montanari, U.: Web Services and Models of Computation. Electronic Notes in Theoretical Computer Science 105, 5–9 (2004) (invited talk)

[23] Morecroft, J.: Mental models and learning in system dynamics practice. In: Michael (ed.) Systems Modelling: Theory and Practice, ch. 7, pp. 101–126. John Wiley, Hoboken (2004)

[24] OMG. Meta-Object Facility (MOF), v. 1.4 (March 2002),
http://www.omg.org/technology/documents/formal/mof.htm

[25] OMG. Unified Modeling Language (UML) specification v. 2 (2004),
http://www.omg.org/technology/documents/modeling_spec_catalog.htm

[26] OMG. Business Process Modeling Notation (BPMN) specification v. 1.0 (2006),
http://www.bpmn.org/Documents

[27] Organization for the Advancement of Structured Information Standards (OASIS). Web Services Resource, 2005. Working Draft (2005),
http://docs.oasisopen.org/wsrf/2005/03/wsrfWSResource1.2draft03.pdf

[28] Papazoglou, P., Yang, J.: Design methodology for web services and business processes. In: Buchmann, A., Casati, F., Fiege, L., Hsu, M.-C., Shan, M.-C. (eds.) TES 2002. LNCS, vol. 2444, pp. 175–233. Springer, Heidelberg (2002)

[29] Paulson, L.: Services Science: A New Field for Today's Economy. IEEE Computer Magazine 39(8), 18–21 (2006)

[30] Tsai, W.T.: Service-oriented system engineering: A new paradigm. In: Service-Oriented System Engineering, 2005. SOSE 2005. IEEE International Workshop, Washington, DC, USA, October 2005, vol. 0, pp. 3–8. IEEE Computer Society, Los Alamitos (2005)

Event-Based Service Coordination*

Gian-Luigi Ferrari[1], Roberto Guanciale[2], Daniele Strollo[1,2] and Emilio Tuosto[3]

[1] Università degli Studi di Pisa, Dipartimento di Informatica
Largo B. Pontecorvo 3 I-56127, Pisa, Italy
{giangi,strollo}@di.unipi.it
[2] Institute for Advanced Studies IMT Lucca
Piazza S. Ponziano 6, 55100, Lucca, Italy
{roberto.guanciale,daniele.strollo}@imtlucca.it
[3] University of Leicester, Computer Science Department
University Road, LE17RH, Leicester, UK
et52@mcs.le.ac.uk

Abstract. In this paper we tackle the problem of designing and implementing a framework for programming service coordination policies. In particular, we illustrate the design and the prototype implementation of Java Signal Core Layer (JSCL), a coordination middleware for services based on the *event notification* paradigm. We formally motivate the design choices of the JSCL middleware by exploiting a variant of the π-calculus specifically tailored to deal with event notification and distribution. We demonstrate how service coordination polices can be precisely programmed in JSCL by some simple but illustrative case studies.

Keywords: Service Oriented Architectures, Event Notification, Coordination.

1 Introduction

The web service protocol stack (e.g. WSDL, UDDI, SOAP) provides *basic* support for the development of service-oriented architectures by exploiting facilities to publish, discover and invoke network-available services. The service protocol stack has been extremely valuable to highlight the key innovative features of the service oriented computing approach. Most of the current development methodologies are focused on composition of services. Two different approaches can be adopted: *orchestration* and *choreography*. In the orchestration, services are thought as isolated and the main focus relies on their internal behavior. The participants have no acknowledgment of the surrounding network. An intermediate component, the *orchestrator* is responsible to arrange service activities according to the work-flow plan. This strategy provides a *local view* of the participants. From the other hand, the *choreography* model involves all parties and their associated interactions providing a *global view* of the system. Relevant standard technologies have emerged to model coordination policies. Among

* Research supported by the EU FET-GC2 IST-2004-16004 Integrated Project SENSORIA and by the Italian FIRB Project TOCAI.IT.

P. Degano et al. (Eds.): Montanari Festschrift, LNCS 5065, pp. 312–329, 2008.

them, particular relevance is given to the Business Process Execution Language (BPEL4WS) [20], for the orchestration, and Web Service Choreography Description Language (WS-CDL) [28], for the choreography. However, it is not infrequent that such standards have drawbacks. In fact, constructs are often informally specified which usually leads to ambiguities or redundancy. Several research and implementation efforts are currently devoted to provide a clear semantics for their constructs and tools for verification (COWS [21], Global Calculus [8], λ_{req} [1] ORC [24], SCC [3], SOCK [18] to cite a few). However, research is still underway.

A well known paradigm for specifying and programming distributed systems is the *event notification* paradigm (EN, for short), where distributed computational components can act as publishers and/or subscribers. When a component intends to send data to or requests a service from other components, it issues an event that eventually shall trigger a reaction from subscribers that previously subscribed for such kind of events. The EN paradigm seems to provide a suitable framework to deal with *service oriented architectures* (SOAs) that require components to be loosely coupled. Specifically, the EN paradigm features high level coordination mechanisms that allow programmers/designers to decouple components and rely entirely on event handling.

In this paper, we report our experience in using the EN paradigm to design and implement coordination policies for SOA. We designed and implemented a middleware called Java Signal Core Layer [12] (JSCL). A distinguished feature of JSCL consists in the strict interplay among formal semantic foundations, implementation pragmatics and experimental evaluation of the resulting programming constructs. More precisely, all the programming facilities available in JSCL have a clear semantics. Indeed, at the abstract level, the middleware takes the form of the *Signal Calculus* [12, 13, 11] (SC). The SC is a variant of the π-calculus [26] with explicit primitives to deal with event notification and component distribution. At the implementation level, JSCL takes the form of a collection of Java API equipped with a standard development environment (an Eclipse plug-in). The JSCL API's are available at **www.tao4ws.net**.

The SC allows one to define services coordination policies (orchestration and choreography) relying on event notification only. Moreover, it features sessions as a mechanism to synchronize work-flows of distributed and independent components. Remarkably, SC does not assume any centralized mechanism for publishing, subscribing and notifying events. Indeed, each subscriber explicitly defines the class of events it is interested in. In [22] this pattern is referred to as *non brokered*, in contrast with the *brokered* solutions that implements publish/subscribe mechanisms on top of a classification of signals without taking into account the involved components. Basically, brokered solutions rely on global state space e.g. *linda tuple spaces* [15].

All SC notions are reflected in the JSCL API's. Indeed, the design choices underlying the JSCL implementation have been formally motivated in terms of the SC. Hence, SC and JSCL can be regarded as a foundational framework and

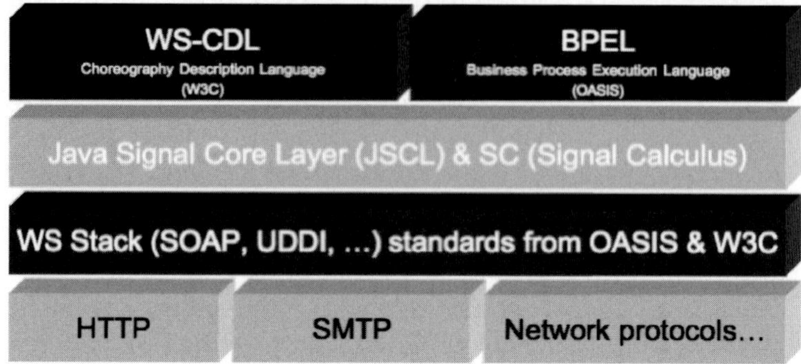

Fig. 1. JSCL-SC technological context

its programming counterpart for specifying, verifying and programming coordination policies of distributed services.

We envisage the impact of our approach on the service oriented computing technologies as follows. Conceptually, the JSCL-SC framework adds a further layer to the basic web service protocol stack (SOAP, UDDI, WSDL). The JSCL-SC layer provides the formal and programming mechanisms to design, verify and program web service coordination policies (e.g. a BPEL4WS orchestrator or a WS-CDL choreography) on top of the basic service protocols. Figure 1 pictorially illustrates the context of our approach.

In this paper we outline the main features of our framework. We refer to [12, 13, 17, 27] for the details. We demonstrate the usefulness of JSCL in the practical programming of coordination policies by some case studies. The focus of our experiments has been on the design and implementation of the work-flow of the coordination, taking into account the possibility of handling long-running transactions in the style of SAGA compensations [14, 5].

This paper is organized as follows. In Section 2 we review the SC process calculus. Section 3 outlines the architectures of the JSCL middleware. Section 4 discusses the case studies. Section 5 yields some concluding remarks.

2   Signal Calculus

In this section, we present a simplified version of SC focusing on session managing only. We assume a set of *topic* names (ranged over by τ), a set of signal variables (ranged over by x) and a set of signal names (ranged over by $s, s_1, s_2...$). In the following the name n ranges over signal names and signal variables. Signal names represent data exchanged among components and should carry additional information even if this feature is not explicitly modeled here. Finally, we assume a set of component names $a, b, ...$ Hereafter, we adopt the notation $\boldsymbol{a}$ to denote a set of component names.

The calculus is centered around the notion of *component*, written as $a[B]_F^R$, that represents a service uniquely identified by a name a, the public address of

$$B ::= \mathbf{out}\langle n : \tau @ \tau'\rangle.B \qquad \textit{(Signal emission)}$$
$$| \quad (\nu\tau)B \qquad\qquad\quad \textit{(Topic restriction)}$$
$$| \quad \mathbf{rupd}\,(R)\,.B \qquad\quad \textit{(Reaction update)}$$
$$| \quad \mathbf{fupd}(F).B \qquad\quad \textit{(Flow update)}$$
$$| \quad B \mid B' \qquad\qquad\quad \textit{(Parallel)}$$
$$| \quad !B \qquad\qquad\qquad \textit{(Bang)}$$
$$| \quad 0 \qquad\qquad\qquad\; \textit{(Empty behavior)}$$

Fig. 2. Behaviors

the service, with internal behavior B and interfaces R and F respectively called *reactions* and *flows*.

Figure 2 displays the syntax of SC behaviors. The *signal emission* $\mathbf{out}\langle n : \tau @ \tau'\rangle.B$ spawns signal n of topic τ over the session τ' and then continues as B. Topics can be generated dynamically by *restriction*. Restriction acts as a binder for the declared topic; namely, the occurrences of τ in $(\nu\tau)B$ are bound. The calculus provides two primitives to allow a component to dynamically change its interface: the *reaction update* $\mathbf{rupd}\,(R)\,.B$ and the *flow update* $\mathbf{fupd}(F).B$. The former installs a new reaction R in the interface part of components and the latter adds F to its flows. Recursive behaviors are obtained via replication $!B$ with the usual interpretation that there are infinitely many copy of B available, namely $!B$ is equivalent to $B \mid !B$. The empty and parallel constructs have the obvious meaning.

Reactions describe how a component reacts upon the reception of a signal and their syntax is given by the following grammar:

$$R ::= 0 \quad | \quad x : \tau @ \lambda\tau' \to B \quad | \quad x : \tau @ \tau' \to B \quad | \quad R|R$$

The *empty reaction* 0 cannot react to any signal. The *lambda reaction* $x : \tau @ \lambda\tau' \to B$ is triggered by signals having topic τ independently from the actual session the event belongs to. The *check reaction* $x : \tau @ \tau' \to B$ reacts only to signals having topic τ issued for the session identified by the topic τ'. Once a reaction has been fired, the behavior B is spawned in the component and will start its executed in parallel with the existing behaviors. Notice that for a lambda reaction the name τ' is bound in the behavior B, while for a check reaction it is a free name. Moreover, in both reactions the variable x acts as a binder for the name of the received signal. *Reaction composition* allows a component to react to different kinds of signal in different ways.

Flows describe the component view of the coordination policies. Their syntax is defined as follows:

$$F ::= 0 \quad | \quad \tau \rightsquigarrow \boldsymbol{a} \quad | \quad F|F.$$

The *empty flow* 0 does not deliver any kind of signals, the *single flow* $\tau \rightsquigarrow \boldsymbol{a}$ delivers signals having topic τ to the components specified in the set $\boldsymbol{a}$ and, finally, flows can be composed in parallel.

Networks describe the component distribution and carry signals exchanged among components (the syntax of networks is given below).

$$N ::= \emptyset \mid a[B]_F^R \mid N \parallel N \mid \langle s : \tau \textcircled{c} \tau' \rangle @a \mid (\nu\tau)N$$

A network can be empty $\emptyset$, a single component $a[B]_F^R$, or the parallel composition of networks $N \parallel N'$. Networks handle signals exchanged among components. The signal emission spawns into the network, for each target component, an "envelope" $\langle s : \tau \textcircled{c} \tau' \rangle @a$ containing the signal and the target component name a. Finally, restriction allows the scope extrusion of freshly generated topics over networks.

2.1 Operational Semantics

We now present the operational semantics of the SC.

The structural congruence over reactions, flows, behaviors and networks is the smallest congruence relation that satisfies the commutative monoidal laws for $(R, |, 0)$, $(F, |, 0)$, $(B, \mid, 0)$ and $(N, \parallel, \emptyset)$. Additionally, the following laws hold:

$$(\nu\tau)0 \equiv 0, \qquad ((\nu\tau)B) \mid B' \equiv (\nu\tau)(B \mid B'), \text{ if } \tau \notin fn(B')$$
$$(\nu\tau)\emptyset \equiv \emptyset, \qquad ((\nu\tau)N) \parallel N' \equiv (\nu\tau)(N \parallel N'), \text{ if } \tau \notin fn(N')$$

and, if $B \equiv B'$,

$$x : \tau \textcircled{c} \lambda\tau' \to B \equiv x : \tau \textcircled{c} \lambda\tau' \to B' \tag{1}$$
$$x : \tau \textcircled{c} \tau' \to B \equiv x : \tau \textcircled{c} \tau' \to B' \tag{2}$$

where x can be alpha converted both in (1) and (2), while τ' only in (1).

Finally, for the structural congruence over networks the following equations hold:

$$a[0]_F^0 \equiv \emptyset, \qquad \frac{F_1 \equiv F_2 \quad B_1 \equiv B_2 \quad R_1 \equiv R_2}{a[B_1]_{F_1}^{R_1} \equiv a[B_2]_{F_2}^{R_2}}, \qquad \frac{\tau \notin fn(R) \cup fn(F) \cup \{a\}}{a[(\nu\tau)B]_F^R \equiv (\nu\tau)a[B]_F^R}.$$

To simplify the definition of the reduction relation over networks, we introduce an auxiliary function on flows. The *flow projection*, $(F)\!\downarrow_\tau$, is inductively defined as follows:

$$(\tau \rightsquigarrow a)\!\downarrow_\tau = a \qquad (\tau \rightsquigarrow a)\!\downarrow_{\tau'} = (0)\!\downarrow_{\tau'} = \emptyset \qquad (F_1|F_2)\!\downarrow_\tau = (F_1)\!\downarrow_\tau \cup (F_2)\!\downarrow_\tau$$

Intuitively, the projection $(F)\!\downarrow_\tau$ takes a flow and a topic and yields the set of names of components which have subscribed for events of topic τ. In other words, when a signal having topic τ occurred, it must be delivered to all the subscribed components, identified by the names in the set $(F)\!\downarrow_\tau$.

The reduction relation $\to$ over networks is defined by the rules depicted in Figure 3.

$$a[\mathbf{rupd}\,(R')\,.B' \mid B]_F^R \to a[B' \mid B]_F^{R\mid R'} \quad \text{(ReactionUpd)}$$

$$a[\mathbf{fupd}(F').B' \mid B]_F^R \to a[B' \mid B]_{F\mid F'}^R \quad \text{(FlowUpd)}$$

$$\frac{(F)\!\downarrow_\tau = \{b_1,\ldots,b_n\}}{a[\mathbf{out}\langle s : \tau\copyright\tau'\rangle.B' \mid B]_F^R \to a[B' \mid B]_F^R \parallel \langle s : \tau\copyright\tau'\rangle@b_1 \parallel \ldots \parallel \langle s : \tau\copyright\tau'\rangle@b_n} \quad \text{(Emit)}$$

$$\langle s : \tau\copyright\tau'\rangle@a \parallel a[B]_F^{x:\tau\copyright\tau'\to B'\mid R} \to a[B\mid\{s/x\}B']_F^R \quad \text{(RCActivation)}$$

$$\langle s : \tau\copyright\tau'\rangle@a \parallel a[B]_F^{x:\tau\copyright\lambda\tau_1\to B'\mid R} \to a[B\mid\{s/x\}\{\tau'/\tau_1\}B']_F^{x:\tau\copyright\lambda\tau_1\to B'\mid R} \quad \text{(RLActivation)}$$

$$\frac{N \to N'}{N \parallel N_1 \to N' \parallel N_1} \text{(NStep)} \qquad \frac{N \equiv N' \quad N \to N_1}{N' \to N_1} \text{(NStruct)}$$

Fig. 3. Operational semantics

Reactions can be added to a component by the rule *ReactionUpd*. The rule extends the interface of the component named a by appending to the set of installed reactions the new reaction. Similarly, the *FlowUpd* extends the flow interface of a component by appending the new flow. The *Emit*, *RCActivation* and *RLActivation* rules define notification dispatching: at emission time, component a spawns into the network a signal targeted to all the components ($c_i \in \boldsymbol{b}$). Once a signal envelope has been spawn into the network, the *RCActivation* or the *RLActivation* rules can be applied in accordance with the kind of the installed reactions. Notice that the application of these rules activates the behavior associated to the reactions applying the suitable variable substitutions. Finally a check reaction has a *stateless* interpretation: after its execution, the reaction is removed by the component interface.

2.2 Joining Events

To explain the SC programming model, we specify a work-flow synchronization mechanism. Let us consider a network consisting of four components: an emitter E, two intermediate components C_1 and C_2, and the join service J. The emitter E starts the communications raising toward C_1 and C_2 two events having different topics. Both components C_1 and C_2 perform an internal computation and then notify their termination by issuing an event to the join service J. The join service J waits for the termination of both components and then executes its internal behavior B. The signals sent to C_1 and C_2 are both related to the same session τ. This session is later used by the component J to synchronize the work-flow. The two intermediate services C_1 and C_2 can concurrently perform their tasks, while the execution of the service J can be triggered only after the completion of their executions.

This work-flow pattern can be specified by the SC network $E \parallel C_1 \parallel C_2 \parallel J$, where:

$$E \triangleq e[(\nu\tau)\mathbf{out}\langle s : \tau_1 \copyright \tau\rangle.\mathbf{out}\langle s : \tau_2 \copyright \tau\rangle.0]^0_{\tau_1 \rightsquigarrow c_1 | \tau_2 \rightsquigarrow c_2}$$

$$C_i \triangleq c_i[0]^{x:\tau_i \copyright \lambda\tau \rightarrow \mathbf{out}\langle x:\tau_i \copyright \tau\rangle.0}_{\tau_i \rightsquigarrow j}, \qquad\qquad i = 1,2$$

$$J \triangleq j[0]^{x:\tau_1 \copyright \lambda\tau \rightarrow \mathbf{rupd}(x':\tau_2 \copyright \tau \rightarrow B).0}_0$$

The join component has only one active reaction installed for signals having topic τ_1. When the two intermediary services forward their signals, the envelope containing the τ_2 event cannot be consumed by the join, and remains pending over the network. The reception of the τ_1 envelope triggers the activation of the join generic reaction. The reaction *reads* the session of the signal τ_1 and creates a new specialized reaction for the signal topic τ_2. This reaction can be triggered only by signals that refer to the session received by the τ_1 signal. Once such kind of signal is received, the behavior B is executed.

3   Java Signal Core Layer

The Signal Calculus has been used to formally drive the prototype implementation of a middleware, called Java Signal Core Layer (JSCL), aimed at programming service coordination policies in an event notification based paradigm.

JSCL has been designed and implemented by a two-level architecture reflecting the structure of the SC. The lower level is called the *Inter Object Communication Layer* (`iocl`). The `iocl` layer provides the primitives for handling network interactions. Indeed, the `iocl` layer abstracts from the actual networking technologies in order to hide the network complexity to the higher layer called *Signal and Component Layer* (SCL). The naming facilities for identifying components, the capabilities for data serialization and message delivering have been implemented by the `iocl` layer. Several instances of the `iocl` may coexist within a JSCL program. The basic idea is that each `iocl` instance acts as the bridge among several network infrastructures (e.g. Web Services, CORBA, remote methods, etc). Intuitively, the `iocl` represents the SC *network* and supports component distribution and the notification/delivery of messages to the distributed components. The SCL layer provides all the facilities to create and handle components, signals, reactions by introducing suitable Java API's. Hence, JSCL supports implementation of event-based coordination policies by enabling Java-like programming techniques.

We do not discuss here the concrete implementation of the JSCL layers (we refer to [27] for the detailed presentation). In order to give a flavor of how JSCL has been implemented, we present the JSCL signal delivery protocol. This protocol is displayed in Figure 4. Once the component $S1$ raises a new event, the resulting notification is delivered to the list of components ($S2$) subscribed for the corresponding signal type. The signal delivering is implemented by demanding the local `iocl` service $iocl_L$ to serialize the message and to contact the remote

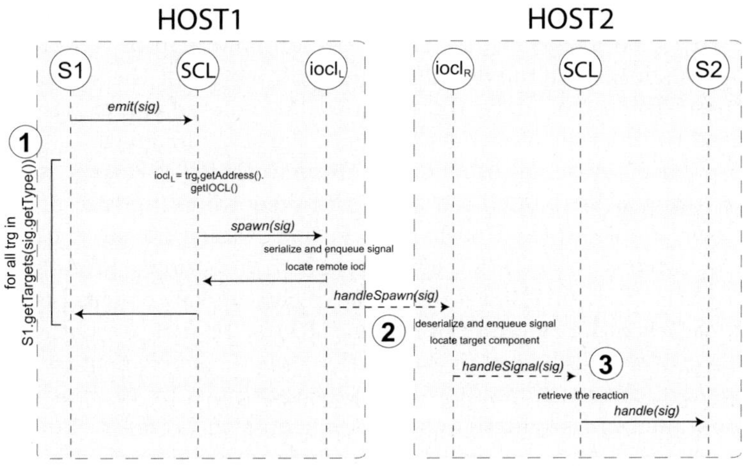

Fig. 4. Signal delivery protocol

`iocl` service $iocl_R$ (Step 1). This step is performed in an asynchronous way (Step 2). Once the message has been received, the remote `iocl` service $iocl_R$ performs the data deserialization and forwards the signal to the component $S2$ (Step 3).

4   Case Studies

To illustrate some of the features and the design-programming facilities made available by our framework, we consider two small case studies. First, we address the problem of composing Web Services in long-running transactional business processes, where compensations must be dealt with appropriately. We illustrate the JSCL implementation of a module which provides suitable primitives for wrapping and invoking Web Services as activities in long-running transactions. The second case study concerns the design and implementation of a car emergency system. We assume a car equipped with a diagnostic system that continuously reports on the status of the vehicle. When the car experiences some major failure (e.g. engine overheating, exhausted battery, flat tires) the in-car emergency service is invoked to select the appropriate tow-truck and garage services.

4.1   Implementing Long Running Transactions in JSCL

One of the emerging issues when aggregating Web Services is constituted by the so-called *long-running transactions* (LRTs), i.e., the possibility of requiring a set of Web Services interactions to be executed atomically. Note that the problem is not just to coordinate the updates of a distributed repository (e.g., a database), since components are loosely coupled and any of them is responsible for maintaining the consistency on local data. In order to achieve atomicity, LRTs may use *compensations*, namely, ad-hoc activities that are responsible for undoing the effects of partial executions when the overall orchestration cannot be completed. In fact, most of the standards proposed for orchestrating Web Services (e.g. BPEL4WS [20]) include primitives for handling LRTs. Noteworthy, all those proposals formalize the orchestration syntax but not the semantics, whose informal description can make the intended behavior of constructs ambiguous and can lead to different implementations of the same language.

Here, we take advantage of a formal framework for isolating and studying LRTs since our goal is to build a framework for coordinating transactional compositions over a solid formal basis. The formal framework we choose is **Naïve Sagas** [5], a process calculus for compensable transactions. From the existing calculi for LRTs [6, 7, 10, 2, 9, 19, 23], we have chosen **Naïve Sagas** because it exposes the orchestration mechanism behind LTRs. In fact, activities in a saga are described at the high level of abstraction, where the elementary actions are not interpreted. Transactional flows are processes built by composing with the standard parallel and sequential composition plus the *compensation pair* construct. Given two actions A and B, the compensation pair $A \div B$ corresponds to a process that uses B as compensation for A. Intuitively, $A \div B$ yields two flows of execution: the *forward flow* and the *backward flow*. During the forward flow, $A \div B$ starts its execution by running A and then, when A finishes: (i) B is "installed" as compensation for A, and (ii) the control is forwardly propagated to the other stages of the transactions. In case of a failure in the rest of the transaction, the backward flow starts so that the effects of executing A must be rolled back. This is achieved by activating the installed compensation B and afterward by propagating the rollback to the activities that were executed before A. Note that B is not installed if A is not executed.

With JSCL the transactional blocks are obtained by suitable wrappers, called *Transactional Components* (TC). To implement the behavior of a transactional component we need three kinds of topics

- SIG_{CMT} is used to notify that the entire work-flow has been successfully completed,
- SIG_{FW} is used to activate the next steps of the chain within a transactional work-flow (forward flow),
- SIG_{RB} is used to activate the compensations (backward flow).

The JSCL implementation of the transactional component (TC) is illustrated in Code 1 in the appendix. A TC is constructed (see lines 3-7) by specifying its address, the main activity (A) to be performed and its compensation (C). The component is initialized by creating the flows for both the SIG_{RB} and SIG_{FW} topics (see lines 13-19). The declaration of reactions for TC is implemented by the method `initReactions`. Notice that TC installs just one reaction for handling requests of SIG_{FW} (lines 23-59). The line 27 declares the activation condition for the installed reaction. The first parameter, FW, declares the topic, while the second parameter is used to specialize reactions on a session (*null* is used for *lambda reactions*).

The block 28-58 contains the declaration of the task to be executed. Once a signal having topic SIG_{FW} is received, the method `handle` is invoked. The components tries to execute the main activity A (line 33) and if a failure happens,

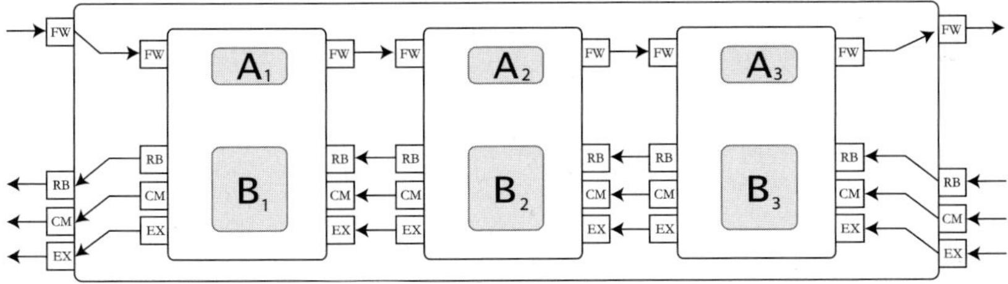

Fig. 5. SAGA compensable process: an example

the rollback signal is sent out (lines 37-38) and the reaction terminates. If A is successfully executed, the compensation for that session flow can be installed (block 45-55). Notice that, in this case the rollback will be activated only for the requests owning to the current session. A check reaction is used to handle this case. The reaction for rollback requests consists of executing the compensation activity C propagating the request along the backward chain.

The JSCL implementation of Naïve Sagas provides components implementing *parallel* and *sequential* structural composition of *transactional gates*. The composition constructs keep the structure of TC and can be reused in further compositions. Figure 5 shows the (intuitive) implementation of the Naïve Sagas compensable process $P \triangleq A_1 \div B_1; A_2 \div B_2; A_3 \div B_3$.

Correctness of the JSCL implementation of Naïve Sagas can be formally stated via a semantic preserving encoding of Naïve Sagas in SC (we refer to [17] for the technical treatment). Here, we simply provide via an example the intuition of the encoding. We assume as given two functions $[\![A \div B]\!](x, \tau_s)$ and $[\![B]\!]_{rb}(x, \tau_s)$ that translate a Naïve Sagas process $A \div B$ and the compensation B to SC internal behaviors. The two functions work on signal named x having session τ_s. We also assume that the first function translates the successful return statements into the signal emission $\mathbf{out}\langle x : fw\textcircled{c}\tau_s\rangle.\mathbf{rupd}\,(x : rb\textcircled{c}\tau_s \to [\![B]\!]_{rb}(x, \tau_s))$ and the exception rising into $\mathbf{out}\langle x : rb\textcircled{c}\tau_s\rangle.0$, and that the second function translates the successful return statements into the signal emission $\mathbf{out}\langle x : rb\textcircled{c}\tau_s\rangle.0$ and the exception rising into $\mathbf{out}\langle x : ex\textcircled{c}\tau_s\rangle.0$. The Naïve Sagas compensable process, previously described, is represented by the SC network $[\![P]\!]$:

$$[\![A_1 \div B_1]\!] \triangleq p_1[0]_{fw \rightsquigarrow p_2}^{x:fw\textcircled{c}\lambda\tau_s \to [\![A_1 \div B_1]\!](x,\tau_s)}$$

$$[\![A_2 \div B_2]\!] \triangleq p_2[0]_{fw \rightsquigarrow p_3 | rb \rightsquigarrow p_1}^{x:fw\textcircled{c}\lambda\tau_s \to [\![A_2 \div B_2]\!](x,\tau_s)}$$

$$[\![A_3 \div B_3]\!] \triangleq p_3[0]_{rb \rightsquigarrow p_2}^{x:fw\textcircled{c}\lambda\tau_s \to [\![A_3 \div B_3]\!](x,\tau_s)}$$

$$[\![P]\!] \quad \triangleq [\![A_1 \div B_1]\!] \mid [\![A_2 \div B_2]\!] \mid [\![A_3 \div B_3]\!]$$

Conceptually, our implementation of Naïve Sagas compensable processes add a further layer to JSCL. This new layer exploits JSCL primitives to define the behavior of transactional constructs according to Naïve Sagas. In other words,

```
1   public class TransactionalComponent {
2    public TransactionalComponent (
3     ComponentAddress addr ,
4     AtomicTask A,
5     AtomicTask C,
6     ComponentAddress [] prev ,
7     ComponentAddress [] next ,
8    )
9    {
10    super (addr ); initReactions (A, C); initFlows (prev , next );
11   }
12
13   public void initFlows (
14    ComponentAddress [] prev ,
15    ComponentAddress [] next)
16   {
17    for (int i =0; i<prev . length ; i++){ createFlow (FW, next [ i ]); }
18    for (int i =0; i<prev . length ; i++){ createFlow (RB, prev [ i ]); }
19   }
20
21   public void initReactions (AtomicTask A, AtomicTask C)
22   {
23    addReaction (
24     new Reaction (
25      // adds a lambda reaction related
26      // to forward topic .
27     new SignalType (FW, null ),
28     new HandlerTask (){
29      public Object handle (Signal signal ){
30       // Retreive the signal session topic as SC lambda binder
31       Object session = (( SignalType ) (signal . getType ())). getSession ();
32      try {
33       A. exec ( signal );
34      } catch (AtomicActionException e){
35       // if A internally fails ,
36       // a signal with type RB is sent back
37       signal . getType (). setTopic (RB);
38       emit ( signal ); return ;
39      }
40      // installs a reaction to handle rollback
41      // coming backward and related to the current
42      // context .
43      // The compensation in installed only if
44      // the main activity has been correctly done
45      addReaction (
46       new Reaction (
47        // adds a check reaction related
48        // to forward topic .
49       new SignalType (RB, session ),
50       new HandlerTask (){
51        public Object handle (Signal signal ){
52         // executes the compensation
53         C. exec ( signal ); emit ( signal );
54        }
55       })
56      );
57     }
58    })
59   );
60   }
61  }
```

Code 1: Transactional Component

Naïve Sagas compensable processes become a specialized variant of JSCL where gates come equipped with few carefully selected signals that are tailored to the treatment of web service transactions. The underlying JSCL layer makes the implementation fully distributed. The prototype implementation of Naïve Sagas compensable process in JSCL has been first presented in [4]. We refer to [27] for a more detailed treatment of the implementation issues.

4.2   The Car Emergency System

In this section we illustrate how the SC can be used for modeling the service coordination issues of the SENSORIA car emergency system [29] where a car manufacturer provides an assistance service to its customers. Once a customer's car breaks down, the system attempts to locate a garage, a tow truck and a rental car service so that the car is towed to the garage and repaired meanwhile the customer may continue his travel. The inter-dependencies between the booking services are summarized as follows:

- the first step is to charge the credit card with a security amount;
- before looking for a tow truck, a garage must be found as it poses additional constraints to the candidate tow trucks;
- if finding a tow truck fails, the garage appointment must be revoked;
- if renting a car succeeds and finding either a tow truck or a garage appointment fails, the car rental must be redirected to the broken down car's actual location;
- if the car rental fails, it should not affect the other services.

To describe the work-flow and the inter-dependencies among services we exploit the standard Business Process Modeling Notation (BPMN [16]) (see Figure 6). Notice that the specification above exploits the transactional and compensation facilities of BPMN.

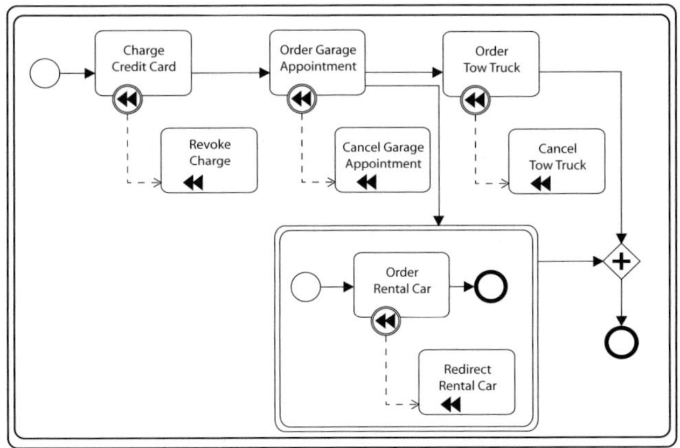

Fig. 6. Car emergency system: the BPMN Specification

The Car Emergency System: The SC specification. We now formally describe the Car Emergency System in SC. Each participant is represented by a SC component making use of the three types of signals below:

- τ_f is used by *forward signals*. These signal propagate the information on the completion of previous activities in the work-flow
- τ_r is used by *rollback signals*. Rollback signals are backwardly propagated by components when their compensations have been executed;
- τ_n is used to notify the current work-flow session to components that have to synchronize several work-flow paths.

A component that represents a BPMN activity is a *transactional component* and can instantiate either a reaction that handles τ_f signals or one that handles τ_r signals. We now specify the basic building block of the SC specification transactional component $TC = TC(a, A, C, \overrightarrow{prev}, \overrightarrow{next})$ where a denotes the location of the component, A the internal behavior and C the compensation. We assume that it propagates τ_f signals to the components in $\overrightarrow{next}$ and τ_r signals to the components in $\overrightarrow{prev}$. The boolean parameter sub states if the component is a sub-transaction. In the following, for the sake of readability, we use $prev$ and $next$ to denote the sets $\overrightarrow{prev}$ and $\overrightarrow{next}$, respectively. Hereafter, we also assume that:

1. if A successfully terminates, then a τ_{ok} signal issued;
2. if A fails, then an exception (a τ_{exc} signal) is raised to inform the component to start the backward flow

$$TC(a, A, C, prev, next, sub) \triangleq a[0]_{\tau_f \leadsto next | \tau_r \leadsto prev | \tau_{exc} \leadsto a | \tau_{ok} \leadsto a}^{R_{TC}(A,C,sub)}$$

where:

$$R_{TC}(A, C, sub) \triangleq \mathbf{rupd} \left(\begin{array}{c} x : \tau_f \textcircled{C} \lambda \tau \rightarrow \\ \left(\begin{array}{c} x : \tau_{ok} \textcircled{C} \tau \rightarrow \\ \left(\mathbf{rupd} \left(\begin{array}{c} x' : \tau_r \textcircled{C} \tau \rightarrow \\ C. \\ \mathbf{out}\langle x' : \tau_r \textcircled{C} \tau \rangle. \\ 0 \end{array} \right) \cdot \right) \\ \mathbf{out}\langle x : \tau_f \textcircled{C} \tau \rangle.0 \\ x : \tau_{exc} \textcircled{C} \tau \rightarrow B_{exc}(x, \tau, sub) \end{array} \right) \\ A \end{array} \right) \Big|$$

and:

$$B_{exc}(x, \tau, sub) \triangleq \left\{ \begin{array}{ll} \mathbf{out}\langle x : \tau_r \textcircled{C} \tau \rangle.0 & if\ sub = false \\[2em] \mathbf{rupd} \left(\begin{array}{c} x' : \tau_r \textcircled{C} \tau \rightarrow \\ \mathbf{out}\langle x' : \tau_r \textcircled{C} \tau \rangle. \\ 0 \end{array} \right) \cdot \\ \mathbf{out}\langle x : \tau_f \textcircled{C} \tau \rangle.0 & otherwise \end{array} \right.$$

In the initial state, the component TC can react only to a τ_f signal. Hence, the session τ starts and A is executed. Concurrently with the activity A, the component installs the reactions to check termination of A (i.e. the emission of signal τ_{ok} or signal τ_{exc}). If the execution of A is successful, a check reaction for a further rollback notification (τ_r) is installed, and the τ_f signal is propagated to the successive stages in the work-flow ($\mathbf{out}\langle x : \tau_f©\tau\rangle.0$). When a τ_r signal for session τ is received, the compensation C is executed and the rollback signal is propagated to previous stages ($\mathbf{out}\langle x' : \tau_r©\tau\rangle.0$). Notice that two handlers (B_{exc}) for the exception of the main activity are provided, according to the *sub* parameter. In the first case, the handler simply starts the backward flow, raising a rollback signal. In the second case, the handler propagates the τ_f signal, since an error of the sub-transaction should not affect the computation of the other components. Moreover the component installs a reaction for τ_r to forward backward flow without executing the compensation.

A sequential work-flow is specified as a chain of transactional components by setting the *next* and *prev* sets suitably. To model the parallel branch, we define the *collector* and *emitter* components as follows:

$$Emitter(a, prev, next, collector) \triangleq$$
$$a[0]^{x:\tau_f©\lambda\tau\to\mathbf{rupd}\left(x':\tau_r©\tau\to\mathbf{rupd}\left(x'':\tau_r©\tau\to\mathbf{out}\langle x'':\tau_r©\tau\rangle.0\right)\right).\mathbf{out}\langle x:\tau_n©\tau\rangle.\mathbf{out}\langle x:\tau_f©\tau\rangle.0}_{\tau_f\rightsquigarrow next|\tau_r\rightsquigarrow prev|\tau_n\rightsquigarrow\{collector\}}$$

$$Collector(a, prev, next) \triangleq$$
$$a[0]^{x:\tau_n©\lambda\tau\to\mathbf{rupd}\left(x':\tau_f©\tau\to\mathbf{rupd}\left(x'':\tau_f©\tau\to\mathbf{rupd}\left(x''':\tau_r©\tau\to\mathbf{out}\langle x''':\tau_r©\tau\rangle.0.\mathbf{out}\langle x'':\tau_f©\tau\rangle.0\right)\right)\right)}_{\tau_f\rightsquigarrow next|\tau_r\rightsquigarrow prev}$$

The emitter represents the entry point of the parallel branch. Basically, it activates the forward flow of *next* components, and synchronizes their backward flows. The synchronization mechanism is implemented by installing two reactions for the topic τ_r and the session τ (through $\mathbf{rupd}\,(x' : \tau_r©\tau \to \mathbf{rupd}\,(x'' : \tau_r©\tau \to ...)))$. After that the synchronization mechanism has been installed, the emitter activates the forward flow ($\mathbf{out}\langle x : \tau_n©\tau\rangle.\mathbf{out}\langle x : \tau_f©\tau\rangle.0$). Notice that the component emits two signals: one having topic τ_f and the other one having topic τ_n. The first signal is delivered to the components representing the parallel activities. The other one is delivered to the collector, informing it of the received session that will be later used by it to implement its synchronization. When the synchronization of the backward flow takes place, the emitter forwards the rollback signal ($\mathbf{out}\langle x'' : \tau_r©\tau\rangle.0$) to the *prev* components.

Similarly, the collector component is responsible to implement the synchronization mechanism for the forward flows and to activate the backward flows of the parallel components when a τ_r signal is received. Notice that the collector exploits a τ_n signal to get information about the session τ.

Summing up, the car emergency system is specified by the following SC network:

$$TC(card, ChargeCredit, RevokeCredit, \{\}, \{garage\}) \parallel$$
$$TC(garage, OrderGarage, CancelGarage, \{card\}, \{e\}) \parallel$$
$$Emitter(e, \{garage\}, \{truck, car\}, \{c\}) \parallel$$
$$TC(truck, OrderTowTruck, CancelTowTruck, \{e, car\}, \{c\}) \parallel$$
$$TC(car, OrderCar, RedirectCar, \{e\}, \{c\}) \parallel$$
$$Collector(c, \{truck, car\}, \{\})$$

The Car Emergency System: the JSCL implementation. The JSCL middleware come equipped with an environment in the form of an Eclipse plug-in that supports the design and development of service coordination policies.The JSCL development environment is composed by three layers: an editor that permits to graphically model service coordination policies, a model transformation that compiles a model into Java code and the JSCL middleware as runtime support. Our development methodology consists of three steps.The first step consists of the graphical definition of the coordination policy of services. The graphical model obtained is then used as input of a compilation facility that generates the JSCL code where the internal logic of each service is rendered through suitable annotations. Finally, the annotations must be finalized to implement the internal behavior of each service.

The JSCL graphical notation captures the sequence of activities via the description of the network topology of components involved in the work-flow. This notation has some similarities with BPMN. The main difference is that BPMN defines directly the flow of the messages exchanged among components exploiting a standard flow-chart notation. Our notation defines the correlation among services, while the message sequence depends by the component internal behavior and is not directly caught at design time. Figure 7 provides the snapshot of the design of the Car Emergency System within the JSCL environment.

Notice that our design exploits a specialized JSCL AndComponent (a generalization of the Join component introduced in Section 2). The first AndComponent is used to synchronize the backward flow before the Garage compensation. The second AndComponent has a twofold role: (i) to synchronize the forward flow, and (ii) to execute the compensation of RentalCar if both the OrderTowTruck fails and the RentalCar main activity has been completed successfully. Notice also that the OrderTowTruck compensation is not executed if the RentalCar fails. The EndPoint component represents the BPMN final state. This component simply forwards the received signals without change their type.

The graphical model is then used to generate the *template* JSCL code where the internal logic of each component has not yet been implemented. Below we show the template code of the ForwardTruck.

```
protected class ForwardTruck extends SignalHandlerTask {
    public Object handle ( Signal  s ){
        try {
            TruckComponent parent = (TruckComponent) getParent();
            //  Program here the internal logic
            parent.state.set(s,ID(), true) ;
```

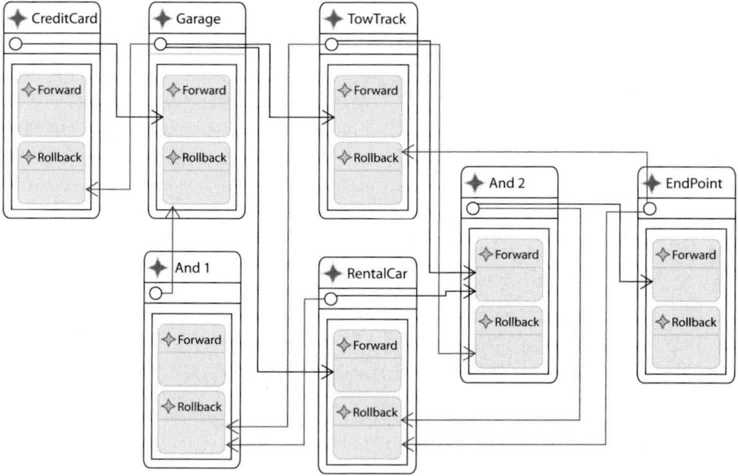

Fig. 7. The JSCL Graphical Design

```
        emit(s) ;
      }
      catch (Exception e ) {
        s.setType (ROLLBACK) ;
        emit ( s ) ;
      }
    }
}
```

Once the template code has been generated, it is possible to implement the
component internal logic using standard Java programming techniques.

5   Concluding Remarks

The SC-JSCL framework has been design to support the specification, the implementation and verification of coordination policies for services oriented applications. Our main goal is to provide general facilities to implement high-level
languages for service oriented architectures (e.g. BPEL4WS [20], BPML [25],WS-
CDL [28]). The strict interplay between SC and JSCL permits to drive and verify
implementation of such languages.

A number of approaches have been introduced to provide the formal foundations of standards for service orchestrations and service choreographies. The
SC-JSCL framework differs from these approaches (COWS [21], Global Calculus [8], λ_{req} [1] ORC [24], SCC [3], SOCK [18] to cite a few), since it focus on a
lower level of abstraction, merging the theoretical formalization with the implementation requirements. Indeed, the emphasis in SC-JSCL is just on designing

general facilities to program coordination patterns on services by exploiting the notion of event notification.

There are a number of directions that we are pursuing for the future development of our framework. In [13], we introduced an algebraic structure over topics. This allows us to implement complex coordination logics directly inside the signal type. Moreover, this provides the foundational description of BPMN-like gateways. We intend to investigate this issue in order to design a BPMN work-flow engine based on the SC/JSCL framework. Furthermore, we plan to extend the SC/JSCL framework with facilities for reasoning and proving properties of coordination policies. On one hand, we are extending the compilation facilities so to generate both the source JSCL code and the SC specification out of the JSCL graphical notation. On the other hand, we plan to integrate in our environment toolkits that provide verification and analysis capabilities for Java programs and other semantic checker (e.g. bisimulation and model checkers) for the SC specification.

References

1. Bartoletti, M., Degano, P., Ferrari, G., Zunino, R.: Secure service orchestration. In: Aldini, A., Gorrieri, R. (eds.) FOSAD 2007. LNCS, vol. 4677. Springer, Heidelberg (2007)
2. Bocchi, L., Laneve, C., Zavattaro, G.: A calculus for long-running transactions. In: Najm, E., Nestmann, U., Stevens, P. (eds.) FMOODS 2003. LNCS, vol. 2884, pp. 124–138. Springer, Heidelberg (2003)
3. Boreale, M., Bruni, R., Caires, L., Nicola, R.D., Lanese, I., Loreti, M., Martins, F., Montanari, U., Ravara, A., Sangiorgi, D., Vasconcelos, V.T., Zavattaro, G.: Scc: A service centered calculus. In: Bravetti, M., Núñez, M., Zavattaro, G. (eds.) WS-FM 2006. LNCS, vol. 4184, pp. 38–57. Springer, Heidelberg (2006)
4. Bruni, R., Ferrari, G., Melgratti, H., Montanari, U., Strollo, D., Tuosto, E.: Java Transactional Web Services: from theory to practice in compensable, distributed long-running transactions. In: Bravetti, M., Kloul, L., Zavattaro, G. (eds.) EPEW/WS-EM 2005. LNCS, vol. 3670, pp. 272–286. Springer, Heidelberg (2005)
5. Bruni, R., Melgratti, H., Montanari, U.: Theoretical Foundations for Compensations in Flow Composition Languages. In: Annual Symposium on Principles of Programming Languages POPL, pp. 209–220. ACM Press, New York (2005)
6. Butler, M., Ferreira, C.: An operational semantics for StAC, a language for modelling long-running business transactions. In: VLBV 2003. LNCS, vol. 2849, pp. 87–104. Springer, Heidelberg (2004)
7. Butler, M., Hoare, T., Ferreira, C.: A trace semantics for long-running transactions. In: Abdallah, A.E., Jones, C.B., Sanders, J.W. (eds.) Communicating Sequential Processes. LNCS, vol. 3525, pp. 133–150. Springer, Heidelberg (2005)
8. Carbone, M., Honda, K., Yoshida, N.: Structured communication-centred programming for web services. In: De Nicola, R. (ed.) ESOP 2007. LNCS, vol. 4421, pp. 2–17. Springer, Heidelberg (2007)
9. Chothia, T., Duggan, D.: An architecture for secure fault-tolerant global applications. Theor. Comput. Sci. 322(3), 567–613 (2004)
10. Danos, V., Krivine, J.: Reversible communicating systems. In: Gardner, P., Yoshida, N. (eds.) CONCUR 2004. LNCS, vol. 3170, pp. 293–307. Springer, Heidelberg (2004)

11. Ferrari, G.L., Guanciale, R., Strollo, D.: Event based service coordination over dynamic and heterogeneous networks. In: Dan, A., Lamersdorf, W. (eds.) ICSOC 2006. LNCS, vol. 4294, pp. 453–458. Springer, Heidelberg (2006)
12. Ferrari, G., Guanciale, R., Strollo, D.: Jscl: A middleware for service coordination. In: Najm, E., Pradat-Peyre, J.-F., Donzeau-Gouge, V.V. (eds.) FORTE 2006. LNCS, vol. 4229, pp. 46–60. Springer, Heidelberg (2006)
13. Ferrari, G., Guanciale, R., Strollo, D., Tuosto, E.: Coordination via types in an event-based framework. In: Derrick, J., Vain, J. (eds.) FORTE 2007. LNCS, vol. 4574, pp. 66–80. Springer, Heidelberg (2007)
14. Garcia-Molina, H., Salem, K.: Sagas. In: Dayal, U., Traiger, I.L. (eds.) SIGMOD Conference, pp. 249–259. ACM Press, New York (1987)
15. Gelernter, D.: Generative communications in Linda. ACM Transactions on Programming Languages and Systems 7(1), 80–112 (1985)
16. Object Management Group. Business process modelling notation. Technical report, http://www.bpmn.org
17. Roberto Guanciale. PhD Thesis. PhD thesis, Institute for Advanced Studies, IMT, Lucca (forthcoming, 2008)
18. Guidi, C., Lucchi, R., Gorrieri, R., Busi, N., Zavattaro, G.: A calculus for service oriented computing. In: Dan, A., Lamersdorf, W. (eds.) ICSOC 2006. LNCS, vol. 4294, pp. 327–338. Springer, Heidelberg (2006)
19. Hosking, A., Jagannathan, S., Vitek, J., Welc, A.: A semantic framework for designer transactions. In: Schmidt, D. (ed.) ESOP 2004. LNCS, vol. 2986, pp. 124–138. Springer, Heidelberg (2004)
20. IBM. Business Process Execution Language (BPEL). Technical report (2005)
21. Lapadula, A., Pugliese, R., Tiezzi, F.: A calculus for orchestration of web services. In: De Nicola, R. (ed.) ESOP 2007. LNCS, vol. 4421, pp. 33–47. Springer, Heidelberg (2007)
22. Liu, Y., Plale, B.: Survey of publish subscribe event systems. Technical Report TR574, Computer Science Department, Indiana University (2003)
23. Mazzara, M., Lucchi, R.: A framework for generic error handling in business processes. In: Proceedings of WS-FM 2004. Elect. Notes in Th. Comput. Sci., vol. 105, pp. 133–145 (2004)
24. Misra, J.: A programming model for the orchestration of web services. In: SEFM, pp. 2–11. IEEE Computer Society Press, Los Alamitos (2004)
25. OMG. Business Process Modeling Language (2002), http://www.bpmi.org
26. Sangiorgi, D., Walker, D.: The π-Calculus: a Theory of Mobile Processes. Cambridge University Press, Cambridge (2002)
27. Strollo, D.: PhD Thesis. PhD thesis, Institute for Advanced Studies, IMT, Lucca (forthcoming, 2008)
28. W3C. Web Services Choreography Description Language (v.1.0). Technical report
29. Wirsing, M., Clark, A., Gilmore, S., Hölzl, M.M., Knapp, A., Koch, N., Schroeder, A.: Semantic-based development of service-oriented systems. In: Najm, E., Pradat-Peyre, J.-F., Donzeau-Gouge, V.V. (eds.) FORTE 2006. LNCS, vol. 4229, pp. 24–45. Springer, Heidelberg (2006)

Dynamically Evolvable Dependable Software:
From Oxymoron to Reality

Carlo Ghezzi[1], Paola Inverardi[2], and Carlo Montangero[3]

[1] Deep-SE Group, DEI-Politecnico di Milano
[2] Dipartimento di Informatica, Università di L'Aquila
[3] Dipartimento di Informatica, Università di Pisa

Abstract. We analyze the main motivations that lead to the present need for supporting continuous software evolution, and discuss some of the reasons for change requirements. Achieving software that is both dynamically evolvable and dependable is our long-term research goal. We do not attempt here to propose a unified solution to dissolve the apparent oximoron, i.e. to reconcile these apparently conflicting goals. Rather, we enlighten different facets of the problem by distilling our experience through three research experience reports. We discuss the lessons learned from the state of the art and practice exemplified by our approaches and outline the directions of possible future research.

1   Introduction

Software evolution has been recognized as a key issue since many years, and several approaches have been identified to tame it. The pioneering work by Belady and Lehman in the late 1970s pointed out that evolution is intrinsic in software: as an application is released for use, the world in which it is situated changes, and therefore new demands arise [26]. Much work was devoted to identifying design techniques that would accommodate future changes [36], and object-oriented design and languages became widely adopted solutions in practical software development. Current progress in the field is documented by the series of specialized workshops, such as IWPSE [1], and working groups sponsored by many organizations, such as [16].

What is new today is the unprecedented degree and speed of change. Software lives in an open world with changing requirements [8]. Software systems never stabilize, but appear to be in a permanent β–*version*. Besides, software evolution, which was traditionally practiced as an off-line activity, nowadays must often be accommodated at run–time. Traditionally, domains where new features or fixes had to be incorporated into a running software were rare, a notable exception being telecommunications. This requirement is now arising for applications in many domains, e.g. those supporting context-aware behaviors.

The challenging problems to be faced to support such high degrees of dynamism and decentralization, are further exacerbated since, more and more often, the applications must also exhibit high degrees of dependability, so that *reliance can be justifiably placed on the services it delivers* [35]. As now commonly

P. Degano et al. (Eds.): Montanari Festschrift, LNCS 5065, pp. 330–353, 2008.

understood, dependability encompasses both functional and nonfunctional software qualities, such as performance, latency (or even guaranteed response time), and availability.

The conjoined need for evolution and dependability is already showing today and is likely to become more and more relevant in the future [17]. Examples can be found in ambient intelligence settings—such as assisted living—and in business processes supporting dynamic enterprise federations. Unfortunately, however, continuous change and dependability are viewed and practiced today as an oxymoron, i.e. as conflicting requirements.

In response to the need for adaptation, software systems have been evolving from closed, static, and centralized architectures to Service Oriented Architectures (SOA), where the components and their connections may change dynamically. So, software systems are increasingly constructed to provide useful services that are exposed for possible use and discoverable by clients through network infrastructures. Services can be composed, recursively, by external parties to provide new added-value services. This emerging scenario is *open*—because new services can appear and disappear—, *dynamic*—because compositions may change dynamically—, and *decentralized*—because no single authority coordinates all developments and their evolution.

In our opinion, to support software evolution and reconcile it with dependability, the entire software development process must be revisited in terms of methods, techniques, languages, and tools. In particular, verification and validation concerns, which in traditional software can be confined within development-time activities, now become perpetual, and extend to software operation.

The paper starts by analyzing the main motivations for supporting continuous change, and classifies the nature of change requirements. It then illustrates three approaches that involved the authors. These contributions describe different attempts to meet dependability requirements for highly evolvable software. Rather than proposing a unified solution to the problem of reconciling the terms of the oxymoron, i.e. of building dynamically evolvable dependable software, which remains our long-term research goal, we decided to enlighten different facets of the problem by distilling our experience around a common case study. We discuss the lessons learned from the state of the art and practice exemplified by our approaches and outline the directions of possible future research.

2   Change Requirements and Research Challenges

In this article, we consider software evolution, as driven by frequent changes in the context embedding the software, which therefore requires continuous dynamic adaptation. That is, we focus on how software running and providing service can deal in-line with changes in its context, differently from what happens in refactoring and reengineering, which as off–line design activities.

Context changes may come from many different sources, and we do not intend here to provide an exhaustive taxonomy. It may be useful, however, to emphasize two relevant classes of, possibly overlapping, contexts and changes, which inspired several of the case-studies we tackled in our research.

Business context. Nowadays, enterprises federate their operations by networking via Web services, to reduce time-to-market. These federations change dynamically according to evolving business goals. For example, a supplier may change its alliance with the distributors and the retailers to improve its sales, or to balance performance and reliability of the overall process. This occurs as the system is operating, according to dynamically gathered business information.

Physical context. In pervasive, ubiquitous applications, user support has to be provided to mobile users, who move across different physical contexts. The change of spatial context, generated as users move, may imply further context changes with respect to available resources, like connectivity, energy, and available software services. Sometimes context changes are generated by sensing physical data other than location, e.g., light or temperature. What is common, however, is that such changes occur while the system is in operation, and they require that system behavior evolves as soon as the change is detected.

In both cases, the underlying notion of change leads to the need for the software to achieve different degrees of *(self-)adaptation/evolution*, and at different levels of granularity, from software architecture to line of code. For instance, in a federated information system, to achieve the required flexibility, the structure of the bindings among services must be defined dynamically as business processes are operational, to maximize the overall quality of service. Similarly, in ambient intelligence applications, often the requirements for context-aware behaviors imply solutions at the software architecture level where the bindings among the components of a distributed software configuration change dynamically because of changes in the physical context. As an example, in an assisted living setting, a request to lower the temperature in a room may be bound to sending a command to the air conditioning system or to sending a command to an actuator that opens the window, depending on the outdoor temperature. In this example, the physical context provides temperature sensing information, which affects the system behavior. The physical context can further change as a consequence of user mobility. For example, if the user moves to a different location, new ways of achieving the same goal of lowering the temperature may become available. The software might be able to discover them and adapt its behavior to maximize the user satisfaction.

The nature of the changes we identified above, and the need to cope with them in real-life systems, challenge our current ability to design software. On the one hand, we need to develop principles and techniques for self-organizing software architectures. On the other, as we observed, dynamic change conflicts with other requirements, most notably with dependability. The main, long-term, open questions are whether dynamic change can coexist with stringent dependability requirements, and how this can be achieved through sound design principles and methods.

3   Approaches to the Development of Adaptable Evolving Software

In this section we present three approaches that address the problem of evolving adaptable services from three different perspectives of the service life cycle, the specification, the deployment and the continuous verification. Then first approach targets business process modelers and final users, i.e. business managers, operators and clients, and provides adaptation by a combination of policies and SOA. Then we present the approach taken in the PLASTIC project to provide a (limited) form of service adaptation at deployment time. Finally, we describe a unified approach to lifelong verification of dynamic service compositions, which encompasses development time—when service compositions are model-checked—and run time—when orchestrated service executions are monitored for compliance with correctness properties.

To clarify our approaches to dependable adaptation and provide concrete grounds for a preliminary discussion on their integration, we use a common running case study, namely the *On Road Assistance* scenario, from SENSORIA [24]. In this scenario, when the diagnostic system of a (very expensive) car of brand CNX reports a severe failure, such that the car is no longer drivable, a notification message is sent to the CNX assistance center, and then the OnRoadAssistance workflow is launched with the accident data, to detect and book appropriate recovery services in the area: garage for repair, tow truck and rental car. After booking, the diagnostic data are automatically transferred to the selected garage, to identify in advance the spare parts needed to perform the repair. Similarly, the GPS data of the stranded vehicle are sent to the towing service. Besides, the driver has to deposit a security payment before being able to order the services.

3.1   An Approach to Business Process Flexibility

The work presented in this section is part of the endevour of the EU IST project SENSORIA (Software Engineering for Service-Oriented Overlay Computers), which aims at a novel comprehensive approach to the development of service oriented software, where foundational theories, techniques and methods are fully integrated in a pragmatic engineering process [43]. A key requirement in this respect is good support for flexibility, to accommodate the variability of the business domains where the systems are deployed, as we discussed in the previous sections. To this end, we propose STPOWLA, a Service–Targeted Policy–Oriented WorkfLow Approach, a novel way to integrate Business Process Management (BPM), Service Oriented Computing (SOC), and Service Oriented Architecture (SOA). STPOWLA introduces a combination of policies and workflows that permits to capture the essential requirements of a business process using a workflow notation and at the same time permits to express variability in a descriptive way by attaching policies to the workflow. From BPM we adopt techniques to model, enact and monitor the business process; SOC and SOA are currently the most promising approach to the design and development of software that can meet the flexibility requirements of the modern enterprises.

The approach. STPOWLA addresses the integration of business processes and Service Oriented Architectures at a high level of abstraction, that is, close to the business goals. STPOWLA exploits this integration to cope with the variety and variability of business requirements, at the same time avoiding that too many details obfuscate the essence of the business workflow. Its driving ideas are that

- a business process is described by a *core* workflow, designed by a *business modeller*, i.e. a professional knowledgeable in the business domain at hand. This model captures the essence of the process as the composition of *tasks* à la BPMN [34], i.e. document transformations that are conceptually elementary in the domain, while accomplishing a meaningful step in the business;
- each task can be characterized along several service level (SL) *dimensions*, with respect to the way it is carried out, i.e. the kind/quality of resources it uses. We use service level in a generalized sense, with respect to the common understanding in the SOA community. Each dimensions is essentially a type of service level, and is used to constrain the admissible implementations of a task. Often dimensions coincide with commonly used service level types, like bandwidth and cost. However, in STPOWLA, the dimensions relate also to domain specific resources, e.g. authorizing roles (simple, double authorization), expertise holders (automatic system or human expert, or both), etc. The dimensions must have a natural, clear interpretation in the domain, since they are the key element in the hands of the stakeholders to adapt a core workflow to their varying requirements;
- the business modeller is in charge of identifying the relevant dimensions, balancing the needs of the business stakeholders and of the IT professionals that implement the tasks;
- the stakeholders (mostly informatically naïf) can adapt the core workflow to their needs by attaching *policies* to various elements of the workflow;
- adaptation occurs either by *refinement* or by *reconfiguration*. In the latter case, the policy prescribes reconfiguration actions, out of a predefined set, which includes, for instance, task insertion and deletion. We have refinement when a policy constrains the admissible characteristics of a task along chosen dimensions;
- the stakeholders can add/delete policies at any time, once the workflow has been installed in the system supporting the business: the changes will affect the enactment of all the instantiations to follow.

Key integration factors. To address its goals, STPOWLA integrates three main ingredients: a graphical workflow notation, a policy language, and SOA. Although intended to be independent of the workflow notation, some high level constructs to build processes out of tasks are presented in [15], and the related UML profile [25]. Here, we can be satisfied with the simplest combinators: sequence, choice, and fork/join, as they are available in UML activity diagrams, for instance.

Currently, we use APPEL [40,42], a general language for expressing policies in a variety of application domains, which was designed to support a clear separation between the *core* language and its specialization for concrete *domains*.

This fosters its use in different business domains. Besides, it has two characteristics that support its use by non technical users: it has a friendly interface to define policies in tabular form, and has been given formal semantics via a mapping to DSTL [32]. An associated technique to detect conflicts when policies are deployed, providing a first early form of dependability [31].

With respect to SOA, StPowla's users, though informatically naïf, should be aware that the business is ultimately carried out by *e–services*, i.e. computational entities that are characterized by *two* sets of parameters: the *invocation* parameters, related to the functionality, and the *service level* parameters. The stakeholders should understand that they can adapt the core workflow by acting on the SL parameters, besides reconfiguring the core.

With respect to implementation, the architecture of the StPowla run-time engine features two execution environments, one for the workflows and one for the policies, interacting via signals that triggers policies from the tasks, and carry back results. The architecture follows naturally from the structure of StPowla specifications, which consist of a workflow part and a policy part.

The policy language. In Appel a *policy* consists of a number of *policy rules*, grouped using a number of operators (**seq**uential, **par**allel, **g**uarded and **u**nguarded choice). A policy rule consists of an optional *trigger*, an optional *condition*, and an *action*. The applicability of a rule depends on whether its conditions are satisfied, once its trigger has occurred. A condition expresses properties of the state of the system and of the trigger parameters. Conditions may be combined with **and**, **or** and **not**, with the expected meaning. Conditions are either domain specific or more generic (e.g. time) predicates; triggers are domain specific. Actions are domain specific: those introduced to support StPowla are discussed below. aslo triggers are domain specific: since in StPowla we associate policies to tasks, we introduce triggers signalling the entry into (`taskEntry`) or the exit from (`taskExit`) a task.

Tasks, policies, and services. Tasks are the units where BPM, SOA and policies converge: the intuitive notion of task is revisited to offer a novel combination of services and policies. To specify tasks, we specialize Appel to deal with software services, by introducing action `req(<Signature>, <InvocationArgs>, <RequiredServiceLevels>)` for service discovery and invocation. The semantics of this action is to *find* a service as described by the first and third arguments, *bind* it, and *invoke* it with the values in the second argument. This dynamic service invocation semantics is inspired by λ^{req}[11] and SRML [28]. With this action, a *default* policy can be associated with each task, so that when the control reaches the task, a service is looked for, bound and invoked, to perform the functionality of the task, with default constraints on any pertinent SL dimension.

Adaptation by refinement occurs when the user overrides the default policy with his own SL constraints, using the composition operators of Appel. SL dimensions are specified in the domain description by their name, set of values, and applicable operators. For instance, dimension `Automation` takes values

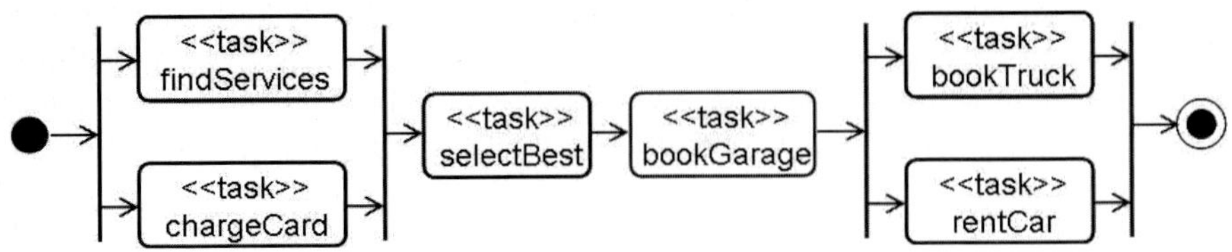

Fig. 1. The On Road Assistance business process

in {automatic, interactive}, where the former excludes the involvement of humans in the fulfillment of the task.

Conditions and arguments to triggers and services can refer to the execution state via *attributes*, i.e. properties of tasks and workflows. There are also global attributes, valid through the workflow of an application. Attributes values are the principal source of information used to adapt the business process: they are used in policy conditions and required SL specifications. APPEL is specialized with actions to set and get the attributes values, with standard syntax.

Attribute types are introduced at different points in the development: a few *predefined* ones, like Integer, etc., come with STPOWLA and are applicable to any task or workflow. Most attributes types are part of the *domain specific* specialization of the APPEL component of STPOWLA, i.e. they come from the ontology of a particular business domain, and often are nothing else but SL dimensions. The set of values of an attribute can grow at operation time, or even utterly new types can be introduced, whenever more choices are offered to the stakeholders by new service implementations.

The running example in StPowla. To attack the *On Road Assistance* scenario in STPOWLA, we first develop the core workflow: the UML4SOA [25] activity diagram in Figure 1 should be self explaining.

Let us see a few policies that adapt the core workflow to different contexts. The first one allows a driver who knows and trusts some of the repairing and towing services of his own town, to choose them directly:

P1: If the car fault happens in the driver's town, then let him select the services to be used. Otherwise choose the services automatically.

Another policy may be introduced to allow the driver to book also a hotel room, if she is far from home, and does not want to go home and then come back to recover her car:

P2: If the car fault happens far from home, also book a hotel room.

Finally, since the whole workflow entails communications with the CNX centre, the driver may be concerned with costs, and choose a low cost connection, as a standard:

P3: Use low cost connections.

The scope of the policies is different: P1 and P2 apply to a single task, while P3 is a global, unconditional policy. P1 and P3 are examples of refinement, since they entails choosing between two implementations of a task, while P2 entails a reconfiguration, since a new task is needed in the workflow. In the following, for sake of space, we will concentrate on the technical facets related to refinements.

To formalize the policies, the modeler defines two attributes: `driverTown`, of type `Town`, bound to the drivers home town, and `crashLocation`, of type `Location`, bound to the car crash location, as detected by the embedded car GPS. Both are workflow attributes. Moreover, `commCost` is a global attribute, of type `CommCost`, a domain dependent SL dimension with values in {`low`, `high`}. We will also use the predefined SL dimension `Automation`.

P1 formalization. This policy applies to task selectBest. It is defined as a sequence, the first argument being the request of an interactive choice and the second one dealing with the automatic case. In APPEL , operator `seq` checks its second argument only if the first one is not applicable, here only if the crash location is different from the driver's town. Hence:

```
P1: appliesTo selectBest
        when taskEntry([])
          if in(thisWorkflow.crashLocation, thisWorkflow.driverTown)
          do req(main, [], [Automation = interactive])
    seq when taskEntry([])
          do req(main, [], [Automation = automatic])
```

P3 formalization. This policy simply sets the global attribute `commCost`:

```
P3: do thisApplication.commCost = low
```

Global policies are executed as soon as they are installed. So, once installed, P3 will influence all subsequent service searches that depend on the value of `commCost`. Note that the user is unaware of which task actually uses the communication services that are affected by this attribute, since they will be buried in the implementation. However, the fact that the attribute is available to the user, has to be seen as a requirement by the implementors. Interesting ways to fulfill such a requirement are presented in the next sections.

Discussion. Of course, it would be possible to define a detailed business model that encapsulates all the options, but it is typical that while the essential process remains the same, the policies change. In the running example, the default installation would include P1-P2, but the driver may discard some: may be, she does not care of P1, since she has no knowledge about garages, neither of P2, since she can stay with friends. The driver might install P3, but in some situations, e.g. if she is scared enough at night, she might override her default choice, as more expensive communications will provide more reliable connections, hence a likely prompter rescue.

More in general, the identification of the SL dimensions is the key aspect of STPOWLA, with respect to both adaptation and dependability. In particular, STPOWLA addresses adaptation to changes in the business context: if an extension of a dimension, or an utterly new dimension, is needed, new services can be developed/searched to take care of the extension, and new policies can be defined and put in place, without the need of intervention on the already existing services and policies.

With respect to dependability, StPowla intends to be a bridge between high level analysis, likely those advocated in [27], which may be performed on the workflows, and the subsequent development phases. As usual in a divide-and–conquer approach, the properties assessed at the workflow level assume that services offering the required service levels are available, and therefore act as requirements for the software processes procuring such services, let it be by in–house development, out–sourcing, buying, exploiting open source projects, and searching registries. The identified SL dimensions, together with the task functional specifications, provide the guiding grid for the procurement activities.

3.2   An Approach to Adaptable SOA

In this section we introduce the approach to adaptation taken in the IST PLASTIC project, whose goal is the rapid and easy development/deployment of self-adapting services for B3G networks [38]. A more extensive presentation of the approach can be found in [4,5].

The PLASTIC development process model relies on model-based solutions to build self-adaptable context-aware services. In PLASTIC a service is equipped with a functional specification that describes behavioral aspects of the modeled service, and with a *Service Level Specification* (SLS) that characterizes the Quality of Service (QoS) that the service can offer to a user in response to her request. Thus the notions of *offered SLS* and *requested SLS* are introduced to address the extra-functional properties that will be used to establish the Service Level Agreement (SLA) between the service provider and the service consumer. The SLA is an entity modeling the conditions on the (possibly negotiated) QoS accepted by both the service consumer and the service provider. SLA represents a kind of contract that is influenced by the service request requirements, the service description and the context where the service has to be provided and consumed (i.e. the union of provider-, network-, and consumer-side context). The contractual procedure may terminate either with or without an agreement.

Accounting for the heterogeneous nature of B3G network-side context, the PLASTIC platform needs to deliver and deploy: (i) provider-side adapted applications to be exposed as services able to offer different SLSs depending on the provider capabilities; (ii) consumer-side adapted applications able to suitably consume the dynamically discovered services so to guarantee the desired service's quality expressed within the SLS requested by the user.

Both consumer- and provider-side applications must correctly run on the respective target devices: this requires the ability to reason on programs and environments in terms of the resources they need and offer, respectively (i.e. resource demand and resource supply), and the ability to suitably adapt the application to the environment that will host it.

PLASTIC's service code (on both consumer and provider side) consists of two parts: the core and the adaptive code. The core code is the frozen portion of the

```
public adaptable class Connection {
  public adaptable void send();
  public adaptable void connect(); ...
}
alternative UMTS adapts Connection {
  public send(Location crashLoc, Data accData) { /*sends via UMTS*/ ... }
  public connect() { /*connects via UMTS*/
    Annotation.resourceAnnotation("UMTS(true)"); ... }
}
alternative WiFi adapts Connection {
  public send(Location crashLoc, Data accData) { /*sends via WiFi*/ ... }
  public connect() { /*connects via WiFi*/
    Annotation.resourceAnnotation("WiFi(true)"); ... }
}
```

Fig. 2. An adaptable class

application and represents the invariant semantics of the service. The adaptive one
is a "generic code" that represents the degree of variability that makes the code
capable to adapt to the execution contexts. This variability based on physical contextual information and consumer needs can be solved hence leading to a set of
alternatives. When a service is invoked, a run-time analysis is performed (on the
available models) and, depending on the analysis results, a new alternative might
be selected among the available ones. The generic code is written by using the
extended version of the Java language supported by the CHAMELEON framework. CHAMELEON allows the development of services that are generic and can
be correctly adapted with respect to a dynamically provided context. The context is characterized in terms of available resources, hardware and/or software.
The approach enables the correct adaptation of the generic code w.r.t. a given
execution context [19,29], and offered and requested SLSs. The framework (fully
implemented in Java) is composed of the following 5 components:

Development Environment. A standard Java development environment that
provides developers with a set of ad-hoc extensions to Java for easily specifying,
in a flexible and declarative way, how the consumer and provider service code
can be adapted. Methods are the smallest building blocks that can be adapted.
The standard Java syntax is enriched by new key-words to specify: *adaptable
classes* that are classes that contain one or more *adaptable methods*; *adaptable
methods* that are the entry-points for a behavior that can be adapted; finally,
adaptation alternatives that specify how one or more *adaptable methods* can
actually be adapted. The output of this step is an extended Java program, i.e. a
generic service code. In Figure 2 we present a snippet of a consumer's adaptable
code. The *adaptable class* Connection contains two *adaptable methods*: send and
connect (introduced by the key-word adaptable). Adaptable methods do not have
a definition in the adaptable class where they are declared but they are defined
within *adaptation alternatives* (see the keywords alternative and adapts). It is
possible to specify more than one alternative for a given adaptable class. The
Connection class has two alternatives: one connects and sends location and data
via the UMTS interface and the other one via the WiFi interface.

```
public class Annotation {
  public static void resourceAnnotation(String ann) {};
  public static void loopAnnotation(int n) {};
  public static void callAnnotation(int n) {};
}
```

Fig. 3. Annotation class

Resource Definition:
```
define Energy as Natural
define Bluetooth as Boolean
define Resolution as {low, medium, high}
```

Resource Demand:
```
{Bluetooth(true),Resolution(high)}
```
Resource Supply:
```
{Bluetooth(false),UMTS(true),Resolution(low)}
```

Fig. 4. A resource definition, a resource demand and a resource supply

Annotations may also add information about particular code instructions (see the keyword **Annotation**). Annotations are specified at the source code level by means of calls to the "do nothing" methods of the **Annotation** class shown in Figure 3. In this way, after compilation, annotations are encoded in the bytecode through well recognizable method calls to allow for easy processing. Annotations can be of three types: (i) resourceAnnotations which directly express a resource demand. For instance, in Figure 2, the method call Annotation.resourceAnnotation ("WiFi(true)") demands for a WiFi interface; (ii) loopAnnotations express an upper bound to the number of loops; (iii) callAnnotations express an upper bound to the number of recursive method calls.

Resource Model. A formal model that permits the characterization of the computational resources needed to consume/provide a service.

Some resources are subject to consumption (e.g. energy, heap space), while others, if present, are never exhausted (e.g. function libraries, network radio interfaces). Thus, we model a *resource* as a typed identifier that can be associated to natural, boolean or enumerated values (left hand-side of Figure 4 shows an example of some resource definitions). Natural values are used for consumable resources whose availability varies during execution. Boolean values define resources that can be present or not (i.e. non-consumable ones). Enumerated values can define non-consumable resources that provide a restricted set of admissible values (e.g. screen resolution, network type). A *resource instance* is an association $res(val)$ where a resource is coupled to its value (e.g. `Bluetooth(true)`). A *resource set* is a set of resource instances with no resource occurring more than once. Centering around the resource model, we also define the notions of *compatibility* between two resource sets to verify if the resource set describing the *resource demand* of an alternative is compatible with the set representing the *resource supply* of an execution environment deciding if the application can there run safely. For instance, in Figure 4, the resource sets are clearly not compatible since the adaptation alternative demands for a bluetooth interface and high screen resolution but the execution environment does not supply any of them. The framework encodes resource sets into XML files that have not been shown for presentation purposes.

1) istore_1 → {CPU(2)}   2) invoke.* → {CPU(4)}
3) .* → {CPU(1), Energy(1)}
4) invokestatic LocalDevice.getLocalDevice() → {Bluetooth(true), Energy(20)}

Fig. 5. A resource consumption profile

Abstract Resource Analyzer (ARA). An interpreter to compute resource consumption. Within our framework, ARA's implementation is based on a transition system which abstracts the JVM w.r.t. resource consumption. In this abstraction we consider bytecode instructions behaviour only by taking into account their effects on resources. ARA inspects the bytecode of the alternatives and is parametric on the *resource consumptions profile* associated to bytecode instructions in a given execution environment.

Resource consumption profiles provide the description of the characteristics of a specific execution environment, in terms of the impact that Java Bytecode instructions have on the resources; these profiles associate resources consumption to particular patterns of Java bytecode instructions specified as *regular expressions*. Since the bytecode is a verbose language[1], this allows to define the resource consumption associated to both basic instructions (e.g., `ipush`, `iload`, etc.) and complex ones, e.g. method calls.

Figure 5 represents an example of a profile over resources defined in Figure 4: the last row states that a call to the `getLocalDevice()` static method of the `LocalDevice` class of the `javax. bluetooth` library requires the presence of Bluetooth on the device (`Bluetooth (true)`), and it causes a consumption of the resource Energy equal to 20 cost units. Note that the expression ".*" matches every bytecode. Through these profiles, ARA can be instantiated w.r.t. the resource-oriented characteristics of a specific execution environment. Basically, ARA takes in a Java program, a profile and returns the program's resource demands.

Execution Environment. Any device that will execute the code. Typically it is provided by PDAs, smart phones, etc. This environment is not strictly part of our framework. However it should provide a declarative description of the resources available to consume/provide the service, plus the consumption profile.

Customizer. It takes in inputs the resource supply and the provided profile, and explores the space of the possible adaptation alternatives. This step delivers consumer/provider standard bytecode. Both consumer and provider can exploit CHAMELEON for selecting/deploying the best suited adaptation alternative for service consumption and provision, respectively.

The PLASTIC service consumption and provision is based on the Web Services (WS) technologies and hence on the WS Interaction Pattern. For PLASTIC adaptation, the WS interaction pattern is slightly modified in order to (possibly) reach the SLA at the end of the discovery phase. Accounting for the different

---

[1] This is particularly true for method invocations where the method is uniquely identified by a fully qualified *id* (base class identifier + name + formal parameters).

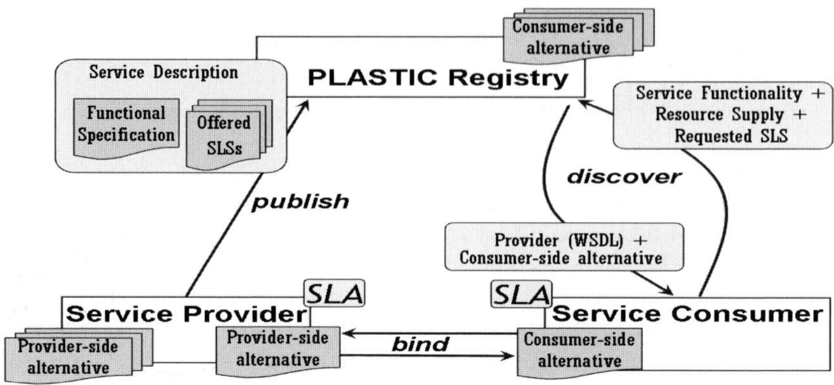

Fig. 6. PLASTIC Services Interaction Pattern

offered and requested SLSs (associated to the different service provision/consumption alternatives, respectively) a "matching" procedure tries to negotiate the SLA. Referring to Figure 6, the steps involved are:

1. The service provider publishes into the PLASTIC Registry the service description in terms of its functional specification and associated SLSs. Each service can be implemented by different *Customizer*-generated alternatives, each one with its own SLS. SLSs are computed on the base of the resource consumption of the alternatives obtained through the quantitative analysis performed by ARA. The WSDL code is coupled with the SLS specifications, which are used by the registry to choose the most suitable provider-side alternative to be used for serving the requests. The registry also stores a set of consumer-side alternatives to be delivered to the consumer and used for suitably consuming the service.
2. The consumer, through a PLASTIC-enabled device, queries the registry for a specific service functionality, additionally providing its resource supply and requested SLSs.
3. The registry searches for a suitable consumer-side alternative that, after delivery, will correctly run on the consumer device with the provided resource supply and will satisfy the requested SLSs by interacting with the suitable provider-side alternative. This means that the SLA can be established only upon existence of (i) a provider-side alternative that has associated a suitable offered SLS and (ii) a suitable consumer-side alternative. If no alternative is able to directly and fully satisfy the request requirements, negotiation is necessary.
4. If the previous phase is successful - i.e., the SLA has been reached - the service consumption can take place under the QoS constraints.

The running example in PLASTIC. Let us now show how PLASTIC may support the implementation of the *On Road Assistance* workflow, as specified in the previous section. Assume that the diagnostic system uses two external radio interfaces - WiFi and UMTS - that, after failure, are no longer functioning. Fortunately, the diagnostic system can back up by connecting to the CNX center through the interfaces provided by any PLASTIC-enabled device connected to it, e.g. through a cable. Let us also assume that the car manufacturer has a preestablished contract with both UMTS and WiFi proprietary network providers that, whenever a backup connection is required, allows the PLASTIC-enabled devices to either connect with low-cost and low-speed UMTS or high-cost and high-speed WiFi.

Note that, even though at production-time the manufacturer may expect that the driver changes her attitude w.r.t. connection cost it cannot predict all the mobile devices that, in the future, might be used as a backup to connect to the CNX center. Thus, the car is equipped with an on-board PLASTIC registry connected to the diagnostic system that stores the two variants of **Connection** (whose meta-code is illustrated in Figure 2) together with their offered SLSs: SLS_{wf} = {Speed=High, Cost=High} and SLS_{umts} = {Speed=Low, Cost=Low} associated to the WiFi and UMTS alternatives, respectively.

When the driver is asked to use a backup connection, she plugs in her device to the diagnostic system. The device searches the registry for a suitable alternative that, after delivery, will be able to correctly run on the device, thus letting the driver specify the requested SLS and connect to the CNX center. Suppose that the device (within its resource supply) specifies that it is equipped with WiFi capabilities only and that the crash location is covered by WiFi hotspots only. The driver also specifies a requested SLS SLS_{req}={Cost=Low}. The only suitable alternative is WiFi but SLS_{wf} does not fully match SLS_{req} since its cost is **High**. A negotiation is necessary and it starts by informing the client that the system can only connect through high-cost WiFi (i.e. the "current version" of the PLASTIC registry can only offer the consumer-side alternative associated to SLS_{wf}). If the driver accepts, the SLA is reached, the WiFi alternative is automatically deployed on her device and the assistance request can take place under the established SLA. Without network coverage, no suitable alternative can be selected, and the workflow fails.

Discussion. The PLASTIC approach to dependable adaptation concerns the variability of both the context and the consumer needs. Adaptation occurs at discovery/deployment– or invocation–time: when the context information becomes available and the consumer preferences have been expressed, CHAMELEON will guarantee an *optimal* alternative, if any.

Within the considered scenario, the change occurs when the UMTS and WiFi interfaces of the diagnostic system break. Adaptation to this change is twofold: on one hand, the diagnostic system reacts by informing the driver of the occurred fault and asking to use a mobile device for a backup connection; on the other hand, adaptation occurs when the suitable alternative (if any) is automatically deployed. These adaptations together present dependability since the ability of the diagnostic system to still provide the CNX service as expected should bring about a considerable degree of trust that the driver has in the system.

Indeed, the framework goes beyond the use we have shown in which we consider only non-consumable resources, i.e. the network interfaces. ARA also permits to infer properties related to consumable resources (e.g. energy consumption, memory foot-print), hence allowing for other dimensions of adaptation.

A form of system evolution can be achieved, by providing the on-board PLASTIC registry with additional adaptation alternatives (through automatic and/or on demand, possibly remote, updates) in order to better fit the driver needs and to extend its scope to newly marketed mobile devices with new features not considered by the current set of alternatives.

3.3   An Approach to Lifelong Verification of SOA

In this section we describe SAVVY (Service Analysis, Verification, and Validation methodologY), which is intended to support the design and operation of dependable Web service compositions. SAVVY has been distilled by research performed in the context of several projects, such as the EU IST SeCSE [39,41] and PLASTIC [38] projects, and the Italian Ministry of Research FIRB project ART DECO [37]. SAVVY is supported by several prototype tools that are currently being integrated in a comprehensive design and execution environment.

The context. SAVVY views software services as components exporting useful functionalities, exposed through the Web and accessible via standard protocols. Services are administered and run by independent parties. To make them usable by others, their specification is published through registries that advertise both their functional and non-functional properties. Their internals, however, are hidden to external users.

Most current research efforts and industrial developments on service technology aim at enabling new business models based on services. In particular, SAVVY's goal is to support the needs of service integrators (also called service brokers) in the definition of new value-adding composite services, through the integration of existing services (or, recursively, composite services). In SAVVY, service composition is achieved by means of a workflow composition language that orchestrates the execution of external remote services. Indeed, Web service compositions leverage remote services to deliver highly dynamic and distributed systems. Since services are administered and run in their own domains, the governance of these systems is intrinsically distributed and services may evolve independently over time. SAVVY focuses on lifelong verification of service compositions, which encompasses design-time and run-time verification.

Dynamic service composition. Most current approaches to Web service compositions assume that the binding between the workflow and the external services is statically defined when the workflow is deployed. SAVVY instead supports dynamic binding. At design time, service integrators may refer to the external services to be composed through their specification, ignoring the actual services that will be orchestrated at run time. They only assume that service implementations which fulfill the required specifications will be available at run time. Any actual services that fulfill the contract encoded by the specification can be selected for use at run time: the binding between a service request and a service provision is therefore established dynamically. Design-time verification checks that composite services deliver the expected functionality and meet the specified quality of service, under the assumption that the external services used in the composition fulfill the contracts defined through their interface.

The motivation and rationale behind this approach is that the marketplace of services is open and continuously evolving. New service providers may join, providing services with different and possibly improved qualities. Service integrators should therefore design their composite services in a way that is as independent as possible of the actual services available at any given time, so that the composition may adapt itself to the dynamic offer available at run time, to optimize some desirable quality of service.

The flexibility of the approach we described above hinders dependability. Because the marketplace is open and decentralized, no central authority is in charge of coordinating it, or even guaranteeing that an exported service honors its contract, i.e., it satisfies its specification. Furthermore, after publishing a service specification in a registry, the service provider may evolve the implementation in a way that breaks the contract, perhaps inadvertently. For these reasons, design-time verification may be threatened by the dynamic bindings established at run time, or by dynamic changes of service implementations. Run-time continuous verification is therefore needed to guarantee the correctness of composite services.

SAVVY considers service compositions described in the BPEL workflow language [3], with correctness statements rendered in ALBERT [7], a temporal logic language suitable for asserting both functional and nonfunctional properties that specify the quality of service (QoS). Furthermore, it assumes the external services to be integrated in the workflow to be stateless. In what follows, we provide a brief introduction to BPEL, ALBERT, design-time verification, and run-time verification in SAVVY.

BPEL. BPEL —Business Process Execution Language for Web services—is a high-level XML-based language for the definition and execution of business processes [3]. It supports the definition of workflows that provide new services, by composing external services in an orchestrated manner. The definition of a workflow contains a set of global variables and the workflow logic is expressed as a composition of *activities*, which include primitives for communicating with other services (*receive, invoke, reply*), executing assignments (*assign*), signaling faults (*throw*), pausing (*wait*), and stopping the execution of the process (*terminate*). Moreover, constructs *sequence, while,* and *switch* provide standard control structures and implement sequences, loops, and branches. Construct *pick* makes the process wait for the occurrence of the first (out of several) incoming message, or for a timeout to expire, after which it executes the activities associated with such an event. The language also support the concurrent execution of activities by means of construct *flow*.

In SAVVY, at design time one can define *abstract workflows*, i.e., the external services orchestrated by the workflow are not bound statically; they are only

Table 1. ALBERT syntax

$$
\begin{aligned}
\phi ::={} & \chi \quad | \quad \neg\phi \quad | \quad \phi \wedge \phi \quad | \quad Becomes(\chi) \quad | \quad Until(\phi,\phi) \quad | \\
& Between(\phi,\phi,K) \quad | \quad Within(\phi,K) \\
\chi ::={} & \psi \text{ relop } \psi \quad | \quad \neg\chi \quad | \quad \chi \wedge \chi \quad | \quad onEvent(\mu) \\
\psi ::={} & \text{var} \quad | \quad \psi \text{ arop } \psi \quad | \quad \text{const} \quad | \quad past(\psi, onEvent(\mu), n) \quad | \\
& count(\phi, K) \quad | \quad count(\phi, onEvent(\mu), K) \quad | \quad \text{fun}(\psi, K) \quad | \\
& \text{fun}(\psi, onEvent(\mu), K) \quad | \quad elapsed(onEvent(\mu)) \\
\text{relop} ::={} & < \quad | \quad \leq \quad | \quad = \quad | \quad \geq \quad | \quad > \\
\text{arop} ::={} & + \quad | \quad - \quad | \quad \times \quad | \quad \div \\
\text{fun} ::={} & sum \quad | \quad avg \quad | \quad min \quad | \quad max \quad | \quad \ldots
\end{aligned}
$$

identified by their ALBERT specification, described hereafter. The binding to specific, concrete service implementations which satisfy their specification only needs to be established at run time.

ALBERT. The ALBERT assertion language [7] is a temporal specification language for stating functional and non-functional properties of BPEL compositions. It is used in SAVVY at design time according to an assume/guarantee specification and proof pattern. Certain ALBERT properties (*AAs–assumed assertions*) specify external services as seen by the workflow: they define the assumptions made on the external services that are composed. Other ALBERT properties (*GAs–guaranteed assertions*) define the properties the BPEL workflow ought to guarantee. GAs precisely state the proof obligations to be honored at design time. AAs precisely state the properties to be verified at run time, when external service invocations are bound to concrete services, to ensure that they behave as expected.

ALBERT formulae predicate over *internal* and *external* variables. The former represent data pertaining to the internal state of the BPEL process in execution. The latter represent data that are used in the verification, but are not part of the process' business logic and must be obtained externally (for example, by invoking other Web services, or by accessing some global, persistent data representing historical information). Given a finite set of variables V and a finite set of natural constants C, an ALBERT formula ϕ is defined as in Table 1, where var $\in V$, const, K, n $\in C$, and *onEvent* is an event predicate. *Becomes, Until, Between* and *Within* are temporal predicates. *count, elapsed, past*, and all the functions derivable from the non-terminal fun are temporal functions of the language. Parameter μ identifies an event: the *start* or the *end* of an *invoke* or *receive* activity, the receipt of a message by a *pick* or an *event handler*, or the execution of any other BPEL activity. The above syntax only defines the language's core constructs. The usual logical derivations are used to define other connectives and temporal operators (e.g. $\vee$, *Always, Eventually, ...*).

As an example, a functional AA on an *invoke* of an external service S can be written as post-condition in the following form:

$$onEvent(end\_S) \rightarrow AA$$

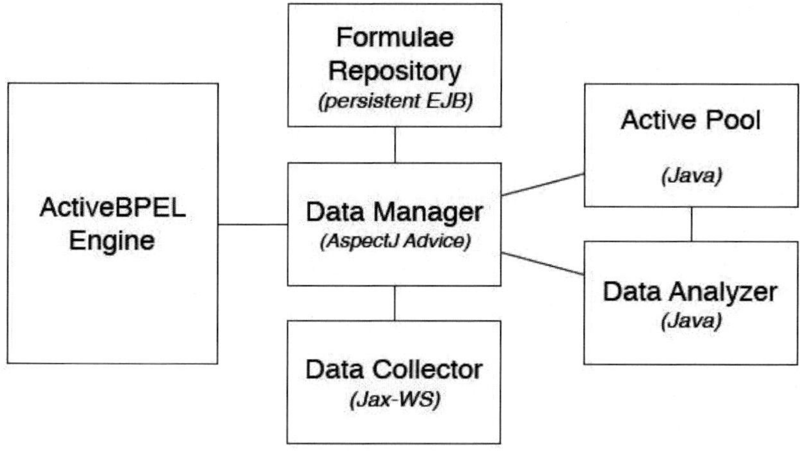

Fig. 7. Monitoring framework

where AA is a predicate on the values computed by service S. It is also possible to express nonfunctional AAs, such as latency in a service response. The following ALBERT formula specifies that the duration of S's invocation should not exceed 1 minute:

$$onEvent(start\_S) \rightarrow Within(onEvent(end\_S), 60)$$

ALBERT can also be used to express GAs. For example, one may state an upper bound to the duration of a certain sequence of activities, which includes external service invocations, performed by a composite BPEL workflow in response to a user input request.

ALBERT semantics is defined in [7] rather conventionally over a *timed state word*, an infinite sequence of states $s = s_1, s_2, \ldots$, where a state s_i is a triple (V_i, I_i, t_i). V_i is a set of $\langle variable, value \rangle$ pairs, I_i is a location of the process and t_i is a time-stamp. States can therefore be considered as snapshots of the process.

Property verification. ALBERT GAs are verified at design time by model checking. SAVVY provides a verification environment based on the Bogor model checker [12] through which verification can be performed. A full implementation of the verification environment is in progress. An implementation (called Dynamo) exists at this stage only for a predecessor of ALBERT, called WSCoL [9], and for an initial temporal extension [6].

ALBERT AAs can then be verified at run-time to ensure compliance of externally invoked services with the properties assumed at design time. This is achieved by the monitoring environment shown in Fig. 7.

The Formulae Repository is a persistent storage for ALBERT formulae to be monitored. The ActiveBPEL Engine [2] is an open-source implementation of a BPEL engine that we have extended with monitoring capabilities using aspect oriented techniques [23]. The **Data Manager** is the aspect that is weaved into the engine, and its main responsibility is to collect the data on which monitoring is to be performed. ALBERT formulae may also refer to external data, which do not belong to the business logic itself, but are retrieved instead from external sources (e.g., context information). This is achieved through **Data Collectors**. All collected data are time-stamped, labeled with the point in the process in which they were extracted, and sent to the **Active Pool**, which stores them to create (bounded) historical sequences. Finally, we have the **Data Analyzer**. This component receives monitoring directives from the **Data Manager** and interacts with the **Active Pool** to execute them. We distinguish between two possible execution models. If the system is critical the monitor has an immediate impact on the process execution: the process is terminated and the error can be dealt with immediately. If, on the other hand, we are monitoring only for collecting information on how the system works, the process continues and all monitoring results are logged for off-line analysis.

A full implementation of the monitoring environment in Fig. 7 is currently in progress. It is based on previous prototypes [10] that were developed in the contexts of SeCSE and PLASTIC. It also inherits features from other approaches, surveyed in [13].

The running example in SAVVY. Let us refer to the *On Road Assistance* scenario introduced earlier, and let us briefly illustrate how this can be handled in SAVVY. It is straightforward to translate the workflow in Figure 1 into BPEL. Each task is implemented as an activity, which invokes an external service. In the following we assume that the interfaces of the external services of the resulting process have the name of the corresponding task.

We now show how to specify sample properties of the *OnRoadAssistance* workflow in ALBERT. We assume for simplicity that external services are activated via synchronous *invoke* activities. In general, $start\_S$ and end_S are events that signal the start and the end of BPEL activity S. We will discuss the meaning of the other variables when we introduce them.

CommunicationBoundResponseTime. Assuming that the *On Road Assistance* workflow runs on the car and that two communication channels are available for use, several external services —namely *FindServices*, *ChargeCard*, and *BookGarage*— reply within 120 seconds when a low-cost communication channel is used, and within 60 seconds if a high-cost communication channel is used. In ALBERT this AA is expressed as a conjunction of formulae, each of which follows the pattern:

$$onEvent(\texttt{start\_S}) \rightarrow$$
$$(\texttt{VCG::getConnection()/cost='low'} \wedge$$
$$Within(onEvent(\texttt{end\_S}), 120) \ \vee$$
$$(\texttt{VCG::getConnection()/cost='high'} \wedge$$
$$Within(onEvent(\texttt{end\_S}), 60))$$

where `VCG::getConnection()/cost` represents an external variable, provided by the `VCG` is the external vehicle communications, which yields contextual information on the communications channel currently in use within the car, and S ranges on the names of the external services listed above.

AssistanceTimeliness. Other external services, namely *BookTruck* and *RentCar*, involve the real word, and are assumed to terminate only when the required assistance has been delivered, e.g. when the tow truck is near the car. As in the previous example, we may assume that an external variable provides contextual information about the quality of service required (and paid for) by the user. Such variable may have values `fast` or `standard`, meaning that the tow truck must arrive in 20 or 50 minutes, respectively. The *RentCar* service follows the same pattern. The ALBERT AAs that formalize these requirements can be expressed similarly to the previous case.

GlobalAssistanceTimeliness. If quality of service `fast` has been selected, the tow truck will be in proximity of the car within 30 minutes from the moment when the credit card is charged. The property can be expressed as follows:

$$onEvent(\texttt{end\_ChargeCard}) \rightarrow$$
$$(\texttt{QOA::getClass()/truckTime='fast'} \rightarrow$$
$$Within(onEvent(\texttt{end\_BookTruck}), 1800)$$

where QOA is an external service that provides information about the quality of assistance to be guaranteed (and subscribed by the user).

The last property is a GA and must be guaranteed to the user by the *OnRoadAssistance* workflow. Its validity can be easily proved by model checking at design time, based on the previous AAs and on structure of the workflow process.

Discussion. SAVVY supports the definition of abstract BPEL workflows, where the orchestrated services are only specified by their required interface, which includes AAs for the semantic aspects. This supports seamless reconfigurations of the bindings to external services at run time. For example, the abstract *FindServices* service may be reified by a default service that is pre-installed in the car, or by an external service available after subscription and payment of a fee. The method ensures, however, that the assumed properties of an orchestrated service's interface will be checked for validity at run time. This ensures that any deviation from a global property of the abstract workflow will be detected in any concrete instantiation.

SAVVY does not specify what should occur if an assumed property is not verified at when run-time verification fails. We are currently investigating different self-healing strategies, which may range from retrying invocation (in the case of transient errors) to re-binding to alternative service candidates to re-planning the workflow.

4 Conclusions

Building dependable dynamically evolvable software is hard and challenging, and no coherent development approach is currently available. We have explored here three research lines that focus on different aspects of the overall problem. The discussion of a common case-study in the three approaches showed that they are complementary, and it might be worthwhile to combine them effectively, as part of future research. In a nutshell, it is possible to see StPowla as an approach to orchestrate PLASTIC services, and provide adaptable business services on a wide variety of devices. SAVVY may be adapted to support design and operation of the service-based system, focusing on property verification.

Indeed, StPowla and PLASTIC work at different conceptual levels, business process and service, respectively, but share the need of a conceptualization of the dimensions along which the service level can be agreed between consumers and providers. Therefore, some of the variability to be resolved by the PLASTIC Customizer, rather than reflecting the choices in the execution platform, could reflect the alternatives available in the business model. In passing, we note that often these alternatives refer to characteristics that can be also seen as pertaining to the execution platform, albeit an abstract one that includes, for instance, the various business roles. Indeed, besides the `Automation` dimension introduced in Section 3.1, often the choice has to do with which business roles (clerk, manager, CEO, ...) are needed to perform or authorize given tasks. Finally, SAVVY does not focus on how changes occur in the system and how they can be managed, but offers an approach to keep them under control, to achieve dependability.

To conclude, we have addressed the problem of dealing with adaptable applications under dependability guarantees. As a concrete and challenging setting we have considered the service oriented paradigm that naturally lends itself to consider the execution dynamic environment intertwined with the static development one. We have presented three separately conceived approaches that well apply at three different but related life cycle stages. At each stage it is possible to express different levels of variability and to assess suitable dependability guarantees on system execution. As we discussed above, the three approaches might be integrated in a unifying framework that permits the specification of the business process, its implementation by means of adaptable services and finally its lifelong verification. The key point of such a unified approach is that dependability can only be achieved by moving software verification and validation activities forward to deployment and execution time. This represents a great research challenge for the models, techniques and methods we have experimented so far. None of them can straightforwardly face the transition and all of them need to be re-elaborated in order to become affordable at execution time.

Laudatio. The core problem of designing and implementing evolving distributed applications that might dynamically re-configure by preserving some kind of dependability in terms of their software architecture and of the module/component type structure is not new. Interestingly enough these kind of problems were already addressed years ago in the scope of the *Progetto Finalizzato Infor-*

matica, in the sub–project *Architettura e Struttura dei Sistemi di Elaborazione*, led by Ugo Montanari, and in which the authors were involved.

For the last two authors, this line of work developed, following a suggestion by Ugo, from initial attempts to provide consistent configuration and reconfiguration capabilities to Ada [18,22] to more developed and advanced proposals [20,21].

The project and Ugo's leadership have also been a source of inspiration for the first author, who was working on modelling concurrency and real-time Ada's constructs [30] at the time, and then developed a research line on verifiable real-time systems [14,33].

In retrospective, that project was an extremely fertile ground for the development of concepts and ideas on (dynamic) software architectures, which years later were addressed in a systematic and comprehensive way by the software engineering community.

Acknowledgments. C. Ghezzi is partially supported by the EU projects SeCSE and PLASTIC and by the Italian FIRB project ART DECO. P. Inverardi is partially supported by the EU project PLASTIC. C. Montangero is partially supported by the EU project SENSORIA IST-2005-16004.

References

1. IWPSE, WEB site, 2007 (2007), http://iwpse2007.inf.unisi.ch/
2. Active Endpoints. Activebpel engine architecture,
   http://www.active-endpoints.com/active-bpel-engine-overview.htm
3. Andrews, T., Curbera, F., Dholakia, H., Goland, Y., Klein, J., Leymann, F., Liu, K., Roller, D., Smith, D., Thatte, S., Trickovic, I., Weerawarana, S.: Business Process Execution Language for Web Services, Version 1.1 (May 2003)
4. Autili, M., Berardinelli, L., Cortellessa, V., Marco, A.D., Ruscio, D.D., Inverardi, P., Tivoli, M.: A development process for self-adapting service oriented applications. In: Krämer, B.J., Lin, K.-J., Narasimhan, P. (eds.) ICSOC 2007. LNCS, vol. 4749, pp. 442–448. Springer, Heidelberg (2007)
5. Autili, M., Cortellessa, V., Benedetto, P.D., Inverardi, P.: On the adptation of context-aware services. In: Krämer, B.J., Lin, K.-J., Narasimhan, P. (eds.) ICSOC 2007. LNCS, vol. 4749, Springer, Heidelberg (2007)
6. Baresi, L., Bianculli, D., Ghezzi, C., Guinea, S., Spoletini, P.: A timed extension of WSCoL. In: Proceedings of the IEEE International Conference on Web Services (ICWS 2007) (July 2007)
7. Baresi, L., Bianculli, D., Ghezzi, C., Guinea, S., Spoletini, P.: Validation of web service compositions. IET Software 1(6), 219–232 (2007)
8. Baresi, L., Di Nitto, E., Ghezzi, C.: Towards Open-World Software. IEEE Computer 39, 36–43 (2006)
9. Baresi, L., Guinea, S.: Towards dynamic monitoring of WS-BPEL processes. In: Benatallah, B., Casati, F., Traverso, P. (eds.) ICSOC 2005. LNCS, vol. 3826, pp. 269–282. Springer, Heidelberg (2005)
10. Baresi, L., Guinea, S.: Dynamo and self-healing bpel compositions. In: ICSE Companion, pp. 69–70. IEEE Computer Society Press, Los Alamitos (2007)

11. Bartoletti, M., Degano, P., Ferrari, G., Zunino, R.: Secure service orchestration. In: Foundations of Security Analysis and Design IV, FOSAD 2006/2007 Tutorial Lectures, pp. 24–74 (2007)
12. Dwyer, M.B., Hatcliff, J., Hoosier, M., Robby,: Building your own software model checker using the Bogor extensible model checking framework. In: Etessami, K., Rajamani, S.K. (eds.) CAV 2005. LNCS, vol. 3576, pp. 148–152. Springer, Heidelberg (2005)
13. Ghezzi, C., Guinea, S.: Run-time monitoring in service-oriented architectures. In: Baresi, L., Di Nitto, E. (eds.) Test and Analysis of Web Services, pp. 237–264. Springer, Heidelberg (2007)
14. Ghezzi, C., Mandrioli, D., Morasca, S., Pezzè, M.: A unified high-level petri net formalism for time-critical systems. IEEE Trans. Softw. Eng. 17(2), 160–172 (1991)
15. Gorton, S., Reiff-Marganiec, S.: Towards a task-oriented, policy-driven business requirements specification for web services. In: Business Process Management, pp. 465–470 (2006)
16. ERCIM Working Group (2007), `http://w3.umh.ac.be/evol/`
17. Inverardi, P.: Software of the future is the future of software? In: Montanari, U., Sannella, D., Bruni, R. (eds.) TGC 2007. LNCS, vol. 4661, pp. 69–85. Springer, Heidelberg (2007)
18. Inverardi, P., Levi, G., Montanari, U., Vallario, G.N.: A distributed KAPSE architecture. Ada Lett. III(2), 55–61 (1983)
19. Inverardi, P., Mancinelli, F., Nesi, M.: A declarative framework for adaptable applications in heterogeneous environments. In: SAC 2004, pp. 1177–1183. ACM Press, New York (2004)
20. Inverardi, P., Martini, S., Montangero, C.: Is type checking practical for system configuration? In: Díaz, J., Orejas, F. (eds.) TAPSOFT 1989 and CCIPL 1989. LNCS, vol. 352, pp. 257–271. Springer, Heidelberg (1989)
21. Inverardi, P., Mazzanti, F.: Experimenting with dynamic linking with ada. Softw., Pract. Exper. 23(1), 1–14 (1993)
22. Inverardi, P., Montanari, U., Vallario, G.N.: How to develop a programming environment almost completely in a compiled language. In: International Computing Symposium 1983 on Application Systems Development, pp. 429–438. Teubner (1983)
23. Kiczales, G., Lamping, J., Mendhekar, A., Maeda, C., Videira Lopes, C., Loingtier, J.M., Irwin, J.: Aspect-oriented programming. In: Aksit, M., Matsuoka, S. (eds.) ECOOP 1997. LNCS, vol. 1241, pp. 220–242. Springer, Heidelberg (1997)
24. Koch, N., Berndl, D.: Requirements modelling and analysis of selected scenarios: Automotive case study. Technical report, SENSORIA EU–IST–016004 (2007), `http://www.pst.informatik.uni-uenchen.de/projekte/Sensoria/del_24/D8.2.a.pdf`
25. Koch, N., Mayer, P., Heckel, R., Gonczy, L., Montangero, C.: UML for service-oriented systems, SENSORIA EU-IST 016004 Deliverable D1.4.a (2007), `http://www.pst.ifi.lmu.de/projekte/Sensoria/del_24/D1.4.a.pdf`
26. Lehman, M.M., Belady, L.A.: Program evolution: processes of software change. Academic Press Professional, Inc., San Diego (1985)
27. Leveson, N.G.: A systems-theoretic approach to safety in software-intensive systems. IEEE Trans. Dependable Sec. Comput. 1(1), 66–86 (2004)
28. Fiadeiro, J.L., Lopes, A., Bocchi, L.: A Formal Approach to Service Component Architecture. Web Services and Formal Methods 4184, 193–213 (2006)

29. Mancinelli, F., Inverardi, P.: Quantitative resource-oriented analysis of java (adaptable) applications. In: WOSP 2007: Proceedings of the 6th international workshop on Software and performance, pp. 15–25. ACM Press, New York (2007)
30. Mandrioli, D., Zicari, R., Ghezzi, C., Tisato, F.: Modeling the ada task system by petri nets. Computer Languages (1985)
31. Montangero, C., Reiff-Marganiec, S., Semini, L.: Logic-based detection of conflicts in APPEL policies. In: Arbab, F., Sirjani, M. (eds.) FSEN 2007. LNCS, vol. 4767, pp. 257–271. Springer, Heidelberg (2007)
32. Montangero, C., Semini, L.: Distributed states logic. In: 9th International Symposium on Temporal Representation and Reasoning (TIME 2002), Manchester, UK, July 2002, IEEE CS Press, Los Alamitos (2002)
33. Morzenti, A., Mandrioli, D., Ghezzi, C.: A model parametric real-time logic. ACM Trans. Program. Lang. Syst. 14(4), 521–573 (1992)
34. OMG. Business process modeling notation. Technical report (February 6, 2006), `http://www.bpmn.org/`
35. IFIP WG 10.4 on Dependable Computing and Fault Tolerance. Dependability: Basic concepts and terminology (October 1990)
36. Parnas, D.L.: On the criteria to be used in decomposing systems into modules. pp. 139–150 (1979)
37. ART DECO Project. Description of Work (2005), `http://artdeco.elet.polimi.it/Artdeco`
38. PLASTIC Project. Description of Work (2005), `http://www.ist-plastic.org`
39. SeCSE Project. Description of Work (2004), `http://secse.eng.it/`
40. Reiff-Marganiec, S., Turner, K.J., Blair, L.: Appel: The accent project policy environment/language. Technical Report TR-161, University of Stirling (December 2005)
41. The SeCSE Team. Designing and deploying service-centric systems: The SeCSE way. In: Proceedings of Service Oriented Computing: a look at the Inside (SOC@Inside'07), workshop colocated with ICSOC 2007 (2007)
42. Turner, K.J., Reiff-Marganiec, S., Blair, L., Pang, J., Gray, T., Perry, P., Ireland, J.: Policy support for call control. Computer Standards and Interfaces 28(6), 635–649 (2006)
43. Wirsing, M., Carizzoni, G., Gilmore, S., Gonczy, L., Koch, N., Mayer, P., Palasciano, C.: Software engineering for service-oriented overlay computers (2007), `http://www.sensoria-ist.eu/files/whitePaper.pdf`

The Temporal Logic of Rewriting:
A Gentle Introduction

José Meseguer

University of Illinois at Urbana-Champaign

Abstract. This paper presents the temporal logic of rewriting TLR^*. Syntactically, TLR^* is a very simple extension of CTL^* which just adds *action atoms*, in the form of spatial action patterns, to CTL^*. Semantically and pragmatically, however, when used together with rewriting logic as a "tandem" of system specification and property specification logics, it has substantially more expressive power than purely state-based logics like CTL^*, or purely action-based logics like $A\text{-}CTL^*$. Furthermore, it avoids the system/property mismatch problem experienced in state-based or action-based logics, which makes many useful properties inexpressible in those frameworks without unnatural changes to a system's specification. The advantages in expresiveness of TLR^* are gained without losing the ability to use existing tools and algorithms to model check its properties: a faithful translation of models and formulas is given that allows verifying TLR^* properties with CTL^* model checkers.

1 Introduction

I feel deeply grateful to Ugo Montanari for our long friendship, and for our equally long scientific collaboration, spanning more than twenty years of our lives. In honoring him on his 65th birthday, I have chosen a topic close to our common interest in algebraic models of *true concurrency*, but emphasizing the interplay between algebraic models and temporal logic properties.

1.1 The System/Property Mismatch Problem

In formal specification there is a natural division of labor between two clear and necessary tasks: (i) to formally specify a system design in a *system specification*; and (ii) to formally specify the requirements that such a system should satisfy in a *property specification*. This leads naturally to a division of labor between two logics: a system specification logic $\mathcal{L}_S$, and a property specification logic $\mathcal{L}_P$. I call such a pair of logics a *tandem*, and use the notation $\mathcal{L}_S/\mathcal{L}_P$. This paper is all about the *RewritingLogic/TLR** tandem, where rewriting logic is used to specify concurrent system, and TLR^* is used to specify their temporal properties.

P. Degano et al. (Eds.): Montanari Festschrift, LNCS 5065, pp. 354–382, 2008.

Some tandems are better than others by being more *expressive*.[1] The formal specification task involves assigning to an informal design *Des* a formal system specification $\mathcal{S}_{Des}$ in $\mathcal{L}_S$, and to an informal requirement *req* a sentence φ_{req} in $\mathcal{L}_P$. The crucial objective is to *faithfully capture* the intended meaning of *Des* and *req*, so that we have in fact an equivalence

$$Des \ \ satisfies \ \ req \qquad \Leftrightarrow \qquad \mathcal{S}_{Des} \models \varphi_{req} \qquad\qquad (\flat)$$

The left side of the equivalence is still in our informal metalanguage, and only the right side is fully formalized: that is the whole point of formal specifications. The *expressiveness problem* for a tandem $\mathcal{L}_S/\mathcal{L}_P$ is, first, whether, assuming we can correctly express *Des* as $\mathcal{S}_{Des}$ in $\mathcal{L}_S$, we can express *req* at all in $\mathcal{L}_P$, and, second, even if we can, how easily and naturally we can do so. What may happen is that we cannot express *req directly* in $\mathcal{L}_P$, that is, there is no φ_{req} to be found such that we have the above equivalence $(\flat)$. However, we may be able to express *req indirectly* by "cooking" the system specification $\mathcal{S}_{Des}$. That is, we may be able to devise a quite different system specification $\widetilde{\mathcal{S}}_{Des}$ that encodes what is inexpressible about *req* in our original specification $\mathcal{S}_{Des}$, and to find a formal property φ_{req} meaningful for $\widetilde{\mathcal{S}}_{Des}$, so that we have an equivalence

$$Des \ \ satisfies \ \ req \qquad \Leftrightarrow \qquad \widetilde{\mathcal{S}}_{Des} \models \varphi_{req} \qquad\qquad (\natural)$$

When situations of this kind, requiring frequent "cooking" of system specifications, arise in a tandem $\mathcal{L}_S/\mathcal{L}_P$, I say that $\mathcal{L}_S/\mathcal{L}_P$ suffers from the *system/property mismatch problem*. The mismatch can sometimes be palliated by finding an alternative tandem $\mathcal{L}'_S/\mathcal{L}'_P$ and a *faithful mapping of tandems*, from $\mathcal{L}'_S/\mathcal{L}'_P$ to our given tandem $\mathcal{L}_S/\mathcal{L}_P$. By this I mean a pair of functions $(K, \gamma) : \mathcal{L}'_S/\mathcal{L}'_P \longrightarrow \mathcal{L}_S/\mathcal{L}_P$, where K maps a system specification $\mathcal{S}'$ in $\mathcal{L}'_S$ to a corresponding system specification $K(\mathcal{S}')$ in $\mathcal{L}_S$, and γ maps a formula φ' in $\mathcal{L}'_P$ to a formula $\gamma(\varphi')$ in $\mathcal{L}_P$ in such a way that we have an equivalence

$$\mathcal{S}' \models' \varphi' \qquad \Leftrightarrow \qquad K(\mathcal{S}') \models \gamma(\varphi') \qquad\qquad (\dagger)$$

where the symbols $\models'$ and $\models$ emphasize that formula satisfaction is taking place in different tandems. A mapping of tandems $(K, \gamma) : \mathcal{L}'_S/\mathcal{L}'_P \longrightarrow \mathcal{L}_S/\mathcal{L}_P$ may allow us to express the design *Des* as a formal specification $\mathcal{S}'_{Des}$ in $\mathcal{L}'_S$, and the requirement *req* as a sentence φ'_{req} in $\mathcal{L}'_S$, so that we have a chain of equivalences

$$Des \ \ satisfies \ \ req \ \ \Leftrightarrow \ \ \mathcal{S}'_{Des} \models' \varphi'_{req} \ \ \Leftrightarrow \ \ K(\mathcal{S}'_{Des}) \models \gamma(\varphi'_{req}) \qquad (\ddagger)$$

This gives us a systematic recipe for "cooking" the system specification $\mathcal{S}_{Des}$ as $K(\mathcal{S}'_{Des})$, and expressing *req* as $\gamma(\varphi'_{req})$.

---

[1] Everything else being equal, being more expressive is clearly an advantage and, as I argue in this paper, the key to avoiding the "system/property mismatch problem." However, greater expressiveness does not always come entirely for free: one may lose some decidability properties, or have higher computational complexity of model checking, or get a harder to understand formalism.

A good example of unending system/property mismatch problems is furnished by tandems typically used in action-based and state-based approaches to the specification of concurrent systems and their properties. At the system specification level, action-based approaches typically adopt the formalisms of labeled transition systems, and state-based approaches that of Kripke structures. At the property specification level, they respectively adopt temporal or modal pure *action logics*, based on the labels of events, and pure *state-based logics*, based on atomic state predicates. For example, the tandem *Kripke/CTL** (or one of its subtandems such as *Kripke/CTL* or *Kripke/LTL*) is widely used in the state-based camp (see [11] for an in-depth discussion of the *Kripke/CTL** tandem). On the other hand, the tandem *LTranSys/A-CTL** proposed by De Nicola and Vaandrager [34], where *A-CTL** is a pure action logic mirror image of *CTL**, is a perfect exponent of a corresponding tandem in the action-based camp. Two important and elegant results in [34] are precisely the proof of correctness of two faithful mappings of tandems, one *Kripke/CTL** $\longrightarrow$ *LTranSys/A-CTL**, and another in the opposite direction *LTranSys/A-CTL** $\longrightarrow$ *Kripke/CTL**. These two mappings palliate mismatch problems caused by the existence of action-based properties that are directly inexpressible in purely state-based logics, and, similarly, of state-based properties that are directly inexpressible in purely action-based logics.

Is this all there is to it? I do not think so. First, the mismatch problems obviated by the above mappings of tandems have to do with properties that are either purely action-based or purely state-based. But as I illustrate with examples in Section 2, many natural properties, by being *mixed* properties involving in an intrinsic way both action-based and state-based aspects, are by their very nature directly *inexpressible* in either *LTranSys/A-CTL** or *Kripke/CTL**. Second, the fact that we can indirectly express a property is helpful, but it is a clear sign of lack of expressiveness in the formalisms involved, and forces one to reason in an indirect, roundabout way about things that should be expressed much more directly and naturally. Third, both labeled transition systems and Kripke structures are quite limited, low-level formalisms. Not only do they each lack what the other has: by assuming unstructured *sets* of elements for their states, both lack support for expressing high-level system structure such as concurrency.

1.2 How *RewritingLogic/TLR** Addresses System/Property Mismatches

I view the *RewritingLogic/TLR** tandem as a good choice of intrinsically more expressive formalisms for both system and property specification. I begin commenting on rewriting logic, and then explain *TLR** as its ideal counterpart at the property specification level. This is certainly neither the first nor the only proposal for combining state-based and action-based features (see, e.g., [37,17,22,36,20,21,8,7,9,1,14,18,23]); however, it is a tandem design proposal with a number of useful new features.

Rewriting logic [30] is a very expressive and general logical framework for concurrent systems [30,25] that includes labeled transition systems and Kripke structures as special cases. Furthermore:

- Labeled transitions are defined *parametrically* and *locally*: the variables in the terms t and t' of a rewrite rule $l : t \longrightarrow t'$ are its *parameters*; and one does not have to specify the *contexts* in such those local transitions happen.
- The locality of rewriting, together with the *algebraic structure* of states, makes possible the intrinsic and explicit expression of a system's *concurrency*, which is directly supported by the logic [30].
- Concurrency is expressed in an "ecumenical" and "nonsectarian" way, without building in any particular concurrency model. Instead, any such models can be easily specified within rewriting logic as specific rewrite theories [30,25].
- Since equational logic is a sublogic, equational theorem proving tools can be used to *combine theorem proving and model checking* (see, e.g., [32,12]).
- Rewrite theories have an *initial model semantics* [30], which supports very useful inductive reasoning principles.
- Another important advantage is *executability*: rewrite theories are *concurrent programs*, which can be both executed and analyzed [12].

What is TLR^*, and what are its key advantages when used in tandem with rewriting logic? From the syntactic point of view, TLR^* is a simple extension of the state-based logic CTL^* (see, e.g., [11]). The extension consists of just adding to CTL^* a new kind of atom. The atoms of CTL^* are state predicates in a set AP of atomic predicates. They are interpreted in a Kripke structure by the structure's labeling function. In TLR^*, besides such state atoms, we can also have *action atoms*, belonging to a set of *spatial action patterns*. In a standard labeled transition system, the only possible action patterns are the labels themselves. Instead, because of the local and parametric nature of rewrite rules, and the algebraic nature of states, we can have much more expressive action atoms that specify not just the fact that an action or event labeled l has happened, but *where*, in the spatial structure of the distributed state, has the action actually happened. The point is that the parametric nature of a rewrite rule $l : t(x_1, \ldots, x_n) \longrightarrow t'(x_1, \ldots, x_n)$ involving variables $x_1, \ldots, x_n$ is shared by its label l, which we can view also as a *parametric label* $l(x_1, \ldots, x_n)$. Furthermore, the *local* nature of a rewrite rule, which allows it to happen in many different contexts, can be constrained by specifying in a spatial action pattern the allowable spatial "shapes" we are interested in having as contexts. Therefore, spatial action patterns have the general form $C[l(t_1, \ldots, t_n)]$, where l is the rule label indicating the type of action, C is a context term indicating the spatial shape of the context in which the action is taking place, and $t_1, \ldots, t_n$ are terms constraining the shape of the parameters the rule can be instantiated with.

What are then the key advantages of TLR^*? First, its simultaneous support for both state-based and action-based properties, and for mixed properties involving both action and state aspects. Second, the expressiveness of action-based patterns to localize actions. Third, that this greater expressiveness is achieved *without* losing the wealth of model checking techniques and tools of the state-based tradition. I show in Section 4.3 that there is a faithful mapping of tandems $RewritingLogic/TLR^* \longrightarrow Kripke/CTL^*$ that makes it possible to model check a TLR^* formula using a CTL^* model checker with an acceptable cost in complexity for the kinds of properties one wishes to verify in practice.

I motivate TLR^* in Section 2 by means of a simple example. Rewrite theories, proof terms, computations, and spatial action patterns are covered in Section 3. TLR^*, its syntax and semantics, and the mapping of tandems making it possible to model check TLR^* properties using CTL^* model checkers, are studied in Section 4. Sections 5 and 6 discuss related work and present some conclusions.

2   A Simple Example

To motivate TLR^* and why temporal logics that are either solely state-based or solely action-based cannot directly express natural properties one would like to specify and verify, I use a simple fault-tolerant client-server communication protocol. There can be many clients and many servers, and each server can serve many clients. For simplicity I assume that each client communicates with a single server. The purpose of the communication is for the client to ask a question from the server and then receive an answer. For simplicity I assume that both the question and the answer are natural numbers. The server S uses a function f, only known to the server itself, that, given a question N from client C, computes the answer $f(S, C, N)$. The communication environment is faulty: messages can arrive out of order, can be duplicated and can be lost. In spite of the faulty environment, the protocol satisfies a very natural property: under appropriate fairness assumptions, any client asking a question does eventually receive the corresponding answer.

The protocol has a simple system specification as a rewrite theory $\mathcal{R} = (\Sigma, E, R)$, with (Σ, E) an equational theory defining the states as elements of the initial algebra $T_{\Sigma/E}$, and R a collection of labeled rewrite rules describing the message-passing asynchronous transitions between clients and servers. The fact that messages can arrive out of order is modeled by making the communication medium a "soup," in which clients, servers, and messages are floating. Algebraically, this soup is represented as a multiset of sort *State*, built up with a multiset union operator, denoted with empty syntax (juxtaposition), satisfying laws of commutativity ($X\ Y = Y\ X$) and associativity ($(X\ Y)\ Z = X\ (Y\ Z)$), and having the empty multiset (denoted *null*) as an identity ($X\ null = X$). The elements floating in this soup are the states of different clients and servers and the messages. Clients and servers have names belonging to a sort *Oid*. Client states are represented as four-tuples of the form $[C, S, N, W]$, where C is the client's name, S is the name of the server it wants to communicate with, N is the natural number representing the question, and W is either a natural number corresponding to the answer if it has already been received, or the *nil* value if the answer has not yet been received. Servers are stateless and are represented as structures $[S]$, with S the server's name. Messages are all represented as pairs of the form $I \lhd CNT$, with I the addressee's name, and CNT the message contents. Such contents are pairs (J, N), with J the sender's name, and N a number. There is also the function f computed by the server, whose specific defining equations need not concern us here. I now explain the rewrite rules R:

$$req : [C, S, N, nil] \longrightarrow [C, S, N, nil] \ S \lhd (C, N)$$
$$reply : \ S \lhd (C, N) \ [S] \longrightarrow [S] \ C \lhd (S, f(S, C, N))$$
$$rec : \ C \lhd (S, M) \ [C, S, N, W] \longrightarrow [C, S, N, M]$$
$$dupl : \ I \lhd CNT \longrightarrow I \lhd CNT \ \ I \lhd CNT$$
$$loss : \ I \lhd CNT \longrightarrow null$$

All variables in the terms of the above rules are written with capital letters and have the sorts already explained above. For example, C and S have sort Oid, N and M have sort Nat, and W has a supersort of Nat, say $DefaultNat$, containing also the *nil* element. Note that all the rules are "soup rewriting" rules. Thanks to the associativity and commutativity of multiset union, rules such as *reply* and *rec* express the asynchronous nature of the message passing communication, since they can be applied as long as the client or server is in a state matching part of the rule's lefthand side pattern and somewhere in the soup there is a message matching the rest of that lefthand side pattern. The meaning of the rules is now quite obvious. Rule *req* means that, as long as the client has not yet received an answer, it can keep resending its request. Rule *reply* means that the server can answer repeated requests from a client. Rule *rec* means that the client can receive the answer from the client and will store the answer in its fourth component. The rules *dupl* and *loss* model the faulty environment and have the obvious meaning: messages can be either duplicated or lost.

This simple protocol satisfies the key property that under suitable fairness assumptions any client asking a question does eventually receive the answer from its server. Fairness assumptions are needed, because without them the *loss* rule could preempt all communication. As I show in what follows, even this very simple property cannot be directly expressed in either a purely state-based logic, or a purely action-based logic without modifying the above system specification.

Consider an initial state with server a and clients b and c of the form

$$[a] \ \ [b, a, 7, nil] \ \ [c, a, 17, nil]$$

and let us assert this property for a and b with the TLR^* formula

$$\mathbf{A}(FairnessAssumptions \ \Rightarrow \ \mathbf{F} \ rec(b))$$

I focus for the moment on the conclusion formula $\mathbf{F} \ rec(b)$ and will explain later the *FairnessAssumptions* subformula. The spatial action $rec(b)$ illustrates a key feature of TLR^*. $rec(b)$ is shorthand notation for $rec(b, S, M, N, W)$. It asserts that the general action[2] $rec(C, S, M, N, W)$ corresponding to applying the *rec*

---

[2] Note that the variables of a given rewrite rule are listed in their textual order of appearance in its lefthand side.

rule has taken place with the rule's variable C *instantiated* to b. Since the remaining variables in the spatial action are "don't care" variables, by convention we can use any of the following three equivalent notations:

$$rec(b, S, M, N, W) = rec(b, \_, \_, \_, \_) = rec(b)$$

Therefore, $rec(b)$ allows us to *localize* the action rec to the client b.

Can we replace $rec(b)$ by a state-based formula? Yes and no. For the above initial state, in which client b has not yet received an answer, it *is* possible to replace $rec(b)$ in the above universally-quantified formula by the state-based formula $\neg(answered.b) \wedge \mathbf{X} \; answered.b$, where the state predicate $answered.b$ holds of a state if there is a client named b whose last component is a natural number, and does not hold if either b is not present or its last component is *nil*. However, *there is no state-based formula* equivalent to $rec(b)$ without changing our protocol specification. To see why not, notice that such a formula should presumably be a path formula ϕ that must be evaluated on a *sequence*, say $\rho = \rho(1) \, \rho(2) \, \ldots \, \rho(n) \, \ldots$ of states. Consider the state

$$[a] \quad [b, a, 7, f(a, b, 7)] \quad a \lhd (b, 7) \quad [c, a, 17, nil]$$

in which b has already received an answer from a, but there is still another copy of its request message in the soup, either because b had sent two requests, or because the *dupl* rule was applied. Consider now the sequence of states ρ with $\rho(0)$ the above state, $\rho(1)$ the state obtained from $\rho(0)$ by removing the message $a \lhd (b, 7)$ from the soup, and $\rho(n+2)$ obtained from $\rho(n+1)$ by adding a fresh new copy of the message $a \lhd (c, 17)$. The intrinsic ambiguity is that in passing from $\rho(0)$ to $\rho(1)$ either the rule rec, with C instantiated to b is applied, so that we have $\rho \models \phi$, or the rule *loss* could have been applied, so that we have $\rho \not\models \phi$, a blatant contradiction. The point is that we have no way to tell from the sequence ρ which rule *was* applied. In particular, the state-based formula $\neg(answered.b) \wedge \mathbf{X} \; answered.b$ is *not* equivalent to $rec(b)$, and in fact does not hold for any state in the above sequence ρ.

Another case in point is the localized spatial action $req(b)$ (which is shorthand for $req(b, S, N)$). It is utterly impossible to define a state-based formula ϕ asserting that the action $req(b)$ *has* taken place. Consider a state such as

$$[a] \quad [b, a, 7, nil] \quad a \lhd (b, 7) \quad [c, a, 17, nil]$$

in which b has already sent its request, and consider the infinite sequence π of states obtained from this state by adding to it one more copy of the message $a \lhd (b, 7)$ each time. Since such a sequence can be obtained by applying the rule req with C instantiated to b at every single step, if such a formula ϕ existed, we should have $\pi \models \phi$. But since the same sequence π can also be obtained by applying the *dupl* rule at every single step, we must conclude that $\pi \not\models \phi$. Again, given the sequence of states π, we have no way to tell whether the action $req(b)$ *has* taken place or not: the only thing we can say for sure is that it *could* have taken place, since the scenario in which $req(b)$ is applied at least once, and the *dupl* rule is applied whenever $req(b)$ is not applied, is consistent with π.

One could adopt a weaker "possibilistic" semantics of the "could have happened" kind by interpreting the holding of an action δ on the pair of subsequent states $\pi(0)$ and $\pi(1)$ of a sequence of states π as the *possibility* of performing an action of the kind $\delta : \pi(0) \longrightarrow \pi(1)$. More precisely, we could define a satisfaction relation $\pi \models_\Diamond \delta$, read "$\delta$ holds possibly in π," by the equivalence

$$\pi \models_\Diamond \delta \quad \Leftrightarrow \quad (\pi(0), \pi(1)) \in Pairs(\delta)$$

where, by definition, $Pairs(\delta)$ denotes the set of pairs of states where we can reach the second state from the first by the spatial action δ. The problem with this possibilistic semantics is that it blurs the distinction between different actions, since in general various other actions δ' could also perform a transition of the form $\delta' : \pi(0) \longrightarrow \pi(1)$; therefore, by asserting $\pi \models_\Diamond \delta$ we would also be committed to asserting $\pi \models_\Diamond \delta'$. For π our example sequence, we would have $\pi \models_\Diamond req(b)$ *and* $\pi \models_\Diamond dupl$, making it impossible to distinguish between these two different actions. Instead, what we would like to have (and TLR^* has) is not a "possibilistic" semantics, but an "actual" semantics, telling us which action has *actually* taken place. In such an actual semantics, we could simultaneously assert that $req(b)$ has taken place *and* $dupl$ has not, even though the pair of states involved could be related by either $req(b)$ or $dupl$. This, of course, is what is impossible in the state-based semantics. As I explain in Section 3.3, the notion of *computation* needed for the actual semantics provided by TLR^* is not just a sequence of states π, but a *pair* (π, γ), where π is a sequence of states, and γ a sequence of rewrite proofs between such states. Such a notion of computation allows us to distinguish which actions have actually taken place; and to give semantics to both action-based and state-based properties. For π our example state sequence, this resolves the above paradox $\pi \models \phi$, and $\pi \not\models \phi$ for ϕ a formula expressing the action $req(b)$, since now we have many different computations of the form (π, γ), (π, γ'), etc. γ could be an infinite sequence of $req(b)$ actions; and γ' could be an infinite sequence of $dupl$ actions, giving us $(\pi, \gamma) \models req(b)$ and $(\pi, \gamma) \not\models dupl$, together with $(\pi, \gamma') \models dupl$ and $(\pi, \gamma') \not\models req(b)$.

The *FairnessAssumptions* subformula is interesting in its own right, because it illustrates two important points: (i) the impossibility of directly expressing a natural property like fairness in either purely state-based or purely action-based logics; and (ii) the flexibility of TLR^* to express not just what might be called standard fairness properties, but the more expressive "localized fairness" properties that I proposed in [29]. Standard notions of fairness typically concern *transitions*, see, e.g., [24]. For example, the strong fairness of a transition labeled l can be expressed in CTL^*-like notation by the formula $Fair(l)$ defined by:

$$Fair(l) = \mathbf{GF}\ enabled(l) \ \Rightarrow\ \mathbf{GF}\ taken(l)$$

Similarly, the weak fairness (justice) property of l can be expressed in CTL^*-like notation by the formula $Just(l)$ defined by:

$$Just(l) = \mathbf{FG}\ enabled(l) \ \Rightarrow\ \mathbf{GF}\ taken(l)$$

Note, however, that the CTL^*-like notation is *misleading*, since it seems to suggest that $Fair(l)$ and $Just(l)$ can be directly expressed as state-based formulas. But this is just what in general may be utterly impossible. Consider again the rule *req* in our example, and the requirement $Just(req)$. The requirement $enabled(req)$ is purely state-based and has a straightforward specification. We can just extend our protocol specification by adding a new sort *Prop* of atomic propositions having a constant *enabled.req*, and a binary function symbol $\_ \models \_ : State \times Prop \longrightarrow Bool$ defined in its positive case by the equation

$$X\ [C, S, N, nil] \models enabled.req = true$$

with X a variable of sort *State*. The requirement $enabled(req)$ is then the atomic state predicate *enabled.req*. The really problematic requirement is $taken(req)$, which in TLR^* is just the spatial action $req = req(C, S, N)$. This requirement is utterly inexpressible in CTL^* *without* cooking the system specification, since assuming a path formula ϕ that could express it, the same sequence of states π considered above for $req(b)$ would also give us $\pi \models \phi$ and $\pi \not\models \phi$.

Let us turn to the formula *FairnessAssumptions*. We need suitable fairness requirements for rules *req*, *reply*, and *rec*. For *req*, it seems clear that some kind of justice requirement will suffice. Here is where TLR^*'s flexibility to *localize* spatial actions becomes crucial. The problem is that $Just(req)$ is *not* what we want as an ingredient in *FairnessAssumptions*. To see why not, notice that there can be multiple clients in our state (in the initial state we had clients b and c). Now notice that a computation in which the action $req(c)$ is taken at every step does indeed satisfy the $Just(req)$ formula; but b is utterly *starved*. What we need is to *localize* to b the justice requirement, which becomes the TLR^* formula

$$Just(req(b)) = \mathbf{FG}\ enabled.req(b)\ \Rightarrow\ \mathbf{GF}\ req(b)$$

where *enabled.req* now becomes a *parametric* atomic proposition operator $enabled.req : Oid \longrightarrow Prop$ defined in its positive case by the equation

$$X\ [C, S, N, nil] \models enabled.req(C) = true.$$

Similarly, $Fair(reply)$ is not what we want: in this case we need to localize it to *both a and b*, since even when localized to a, the server a could be answering questions from c in a fair way, but utterly starving b; so we get the requirement

$$Fair(reply(a, b)) = \mathbf{FG}\ enabled.reply(a, b)\ \Rightarrow\ \mathbf{GF}\ reply(a, b)$$

where *enabled.reply* is defined in its positive case by the equation

$$X\ S \triangleleft (C, N)\ [S] \models enabled.reply(S, C) = true.$$

I hope that the general pattern to define such localized fairness formulas in TLR^* is sufficiently clear from these examples. Our desired fairness requirement for a and b is then the TLR^* formula:

$$FairnessAssumptions = Just(req(b))\ \wedge\ Fair(reply(a, b))\ \wedge\ Fair(rec(b)).$$

Fairness properties as the one above are interesting because, by intrinsically combining the need for *both* state-based predicates and actions, they underscore the lack of expressive power of *both* solely state-based logics like CTL^* and solely action-based logics like $A\text{-}CTL^*$ [34], since in general none of them can *directly* express such fairness properties without cooking the given system specification. A key advantage of TLR^* is that it can directly and naturally express both state-based and action-based properties, as well as *mixed* properties that intrinsically combine state-based and action-based aspects. A further advantage is that, by exploiting the algebraic nature of the state structure supported by rewriting logic specifications, it allows actions that are not just plain labels, but that can be spatially located and can be easily *localized* to the special situations just needed.

3   Rewrite Theories, Computations and Spatial Actions

I explain key aspects of rewriting logic needed to use it in tandem with TLR^*. Since in rewriting logic concurrent computation and logical deduction *coincide*, the notion of *proof term* is crucial. Proof terms are at the heart of the finite and infinite *computations* that we can associate to a rewrite theory to evaluate the truth of TLR^* formulas. Spatial actions are then useful *patterns* that characterize a family of corresponding proof terms as their *instances*.

3.1   Rewrite Theories

A *rewrite theory* is a triple $\mathcal{R} = (\Sigma, E, R)$, with (Σ, E) an equational theory, and R a collection of rewrite rules. There are various possibilities for the equational logic in which the equational theory (Σ, E) is specified: one can choose unsorted, many-sorted, order-sorted, or even membership equational logic. To keep the exposition as simple as possible, I will assume that (Σ, E) is a many-sorted equational theory. The rewrite rules in R are of the form $l : q \longrightarrow r$, with l a label, q and r Σ-terms of the same sort, and such that the set of variables $vars(r)$ appearing in the rule's righthand side is a subset of the variables $vars(q)$ appearing on the lefthand side. Although one can associate a different label to each different rule in R, it is also possible for several rules to *share* the same label. All the ideas in this paper can be extended to the more general rewrite theories in [3,12], which, besides using a more expressive equational logic, can have *conditional* rewrite rules, and where rewriting under some argument positions of a function symbol f in Σ can be forbidden by declaring them *frozen*.

Intuitively, a rewrite theory $\mathcal{R} = (\Sigma, E, R)$ specifies a concurrent system whose states are elements of the initial algebra $T_{\Sigma/E}$ defined by the equational theory (Σ, E), and whose rewrite rules define concurrent transitions between those states. For example, for the rewrite theory $\mathcal{R} = (\Sigma, E, R)$ specifying the fault-tolerant client-server protocol discussed in Section 2, the states are the soups of clients, servers and messages, that is, elements of sort *State* in the initial algebra $T_{\Sigma/E}$, and the rewrite rules specify the protocol transitions that can concurrently take place in such soups. Therefore, mathematically each state is modeled as an E-equivalence class $[t]_E$ of ground terms, and rewriting happens *modulo* E, that is, we rewrite not just terms t but rather E-equivalence classes $[t]_E$ representing states. A *one-step rewrite* $[t]_E \longrightarrow^1_{\mathcal{R}} [t']_E$ exists in $\mathcal{R}$ iff there exists $u \in [t]_E$ such that u can be rewritten to v using some rule $l : q \longrightarrow r$ in R in the standard way, denoted $u \longrightarrow^1_R v$, and we furthermore have $v \in [t']_E$. More precisely, we have $u \longrightarrow^1_R v$ using $l : q \longrightarrow r$ in R iff there is a position[3] p in u and a many-sorted substitution θ such that $u|_p = \theta(q)$ and $v = u[\theta(r)]_p$.

The problem is that for arbitrary E and R, with, say, E and R finite, whether $[t]_E \longrightarrow^1_{\mathcal{R}} [t']_E$ holds is in general *undecidable*. The most useful rewrite theories satisfy additional conditions under which we can decide in a finite number of steps whether $[t]_E \longrightarrow^1_{\mathcal{R}} [t']_E$ holds or not. Call a rewrite theory *computable* if it is of the form $\mathcal{R} = (\Sigma, E \cup A, R)$, with E, A, and R finite and is such that:

1. Equality modulo A is decidable, and there exists a *matching algorithm modulo A* producing a finite number of A-matching substitutions, or failing otherwise, that can implement rewriting in A-equivalence classes. This implies that for a rewrite theory of the form $\mathcal{R}' = (\Sigma, A, Q)$ with Q a finite set of rewrite rules, whether $[t]_A \longrightarrow^1_{\mathcal{R}'} [t']_A$ holds or not is *decidable*.

2. $(\Sigma, E \cup A)$ is *ground terminating and confluent modulo A* [13]. That is: (i) in $\mathcal{R}_{E/A} = (\Sigma, A, E)$ there are no infinite sequences of the form

$$[t_1]_A \longrightarrow^1_{\mathcal{R}_{E/A}} [t_2]_A \cdots [t_n]_A \longrightarrow^1_{\mathcal{R}_{E/A}} [t_{n+1}]_A \cdots$$

where the $[t_i]_A \in T_{\Sigma/A}$ ($\mathcal{R}_{E/A}$-rewriting terminates on ground A-equivalence classes); and (ii) for each $[t]_A \in T_{\Sigma/A}$ there is a *unique* A-equivalence class $[can_{E/A}(t)]_A \in T_{\Sigma/A}$ called the *E-canonical form* of $[t]_A$ modulo A such that any terminating sequence (possibly of length zero) beginning at $[t]_A$,

$$[t]_A \longrightarrow^1_{\mathcal{R}_{E/A}} [t_1]_A \cdots [t_n]_A \longrightarrow^1_{\mathcal{R}_{E/A}} [t_{n+1}]_A$$

where $[t_{n+1}]_A$ cannot be further rewritten with E module A, necessarily has $[t_{n+1}]_A = [can_{E/A}(t)]_A$. The ground confluence and termination assumptions make the mapping $[t]_{E \cup A} \mapsto [can_{E/A}(t)]_A$ *bijective*. Therefore, we can uniquely represent $E \cup A$-equivalence clases as A-equivalence classes in E-canonical form modulo A.

3. In addition to condition (2), the rules R are *ground coherent* relative to the equations E modulo A (in a somewhat stronger sense than in [39]). This precisely means that, if we decompose the rewrite theory $\mathcal{R} = (\Sigma, E \cup A, R)$ into the simpler theories $\mathcal{R}_{E/A} = (\Sigma, A, E)$ and $\mathcal{R}_{R/A} = (\Sigma, A, R)$ (which have decidable relations $\longrightarrow^1_{\mathcal{R}_{E/A}}$ and $\longrightarrow^1_{\mathcal{R}_{R/A}}$ because of (1)), then for each

---

[3] See [13] for basic notation on term rewriting. Positions in a term are denoted as strings of nonzero natural numbers and represent tree positions when the term is parsed as a tree. Two useful notions are that of a subterm of a given term t at a given position p, denoted $t|_p$, and of replacement in t of such a subterm by another term u at position p, denoted $t[u]_p$. For example, in the term $t = x + ((z+0)+y)$, the subterm at position 2.1 is $z + 0$, and the replacement $t[z]_{2.1}$ is the term $x + (z + y)$.

ground A-equivalence class $[t]_A$ such that $[t]_A \longrightarrow^1_{\mathcal{R}_{R/A}} [t']_A$ using a given rewrite rule $l : q \longrightarrow r$ in R we can always find a corresponding rewrite $[can_{E/A}(t)]_A \longrightarrow^1_{\mathcal{R}_{R/A}} [t'']_A$ using the *same* rewrite rule $l : q \longrightarrow r$ in R such that $[can_{E/A}(t')]_A = [can_{E/A}(t'')]_A$.

Conditions (1)–(3) then imply that for each sort s in Σ, $(\rightarrow^1_{\mathcal{R},s})$ is a *computable binary relation* on $T_{\Sigma/E \cup A,s}$: one can decide $[t]_{E \cup A} \rightarrow^1_{\mathcal{R}} [u]_{E \cup A}$ by generating the finite set of all one-step $\mathcal{R}$-rewrites modulo A of $can_{E/A}(t)$ and testing if any of them has the same E-canonical form modulo A as $[can_{E/A}(u)]_A$. The above computability requirements (1)–(3) are quite natural and are typically met in practical rewriting logic specifications. They are assumed by the Maude system as part of the "contract" with the user to make the theory executable in an efficient way. Conditions (2) and (3) can be checked for some equational axioms A using the Maude Church-Rosser, Termination, and Coherence tools [12].

To make the integration of rewriting logic and TLR^* smoother, I define the class $RWTh_0$ as the class of rewrite theories $\mathcal{R}$ satisfying:

- $\mathcal{R}$ is computable and has a sort *State* as its chosen sort of states.
- If $\mathcal{R}$ has a sort named *Prop* (it need not have it in general), then it must also have a sort named *Bool* with constants *true* and *false* and an operator $\_ \models \_ : State \times Prop \longrightarrow Bool$. *Prop* is the designated sort of atomic state predicates, and $\models$ is the function defining whether a given state satisfies a given state predicate. Furthermore, if Σ is the signature of $\mathcal{R}$, then we will define the subsignature $\Pi \subseteq \Sigma$ of its *state predicate symbols* as the set of all operators in Σ of the form $p : A_1 \times \ldots \times A_n \longrightarrow Prop$, with $n \geq 0$. The sorts $A_1, \ldots, A_n$ are called the *parameter sorts* of the atomic state predicate p.
- $\mathcal{R}$ is *deadlock-free*. This means that in $\mathcal{R}$ there are no finite sequences

$$[t_1]_A \longrightarrow^1_{\mathcal{R}} [t_2]_A \ldots [t_n]_A \longrightarrow^1_{\mathcal{R}} [t_{n+1}]_A$$

such that $[t_{n+1}]_A$ cannot be further rewritten (i.e., it is a "deadlock state"). This is not at all a strong restriction, since, as explained in [32,12], any rewrite theory $\mathcal{R}$ whose rules do not have rewrites in their conditions can be transformed into a semantically equivalent theory $\widehat{\mathcal{R}}$ that is deadlock-free.

3.2   Proof Terms

Rewriting logic has inference rules that, given a rewrite theory $\mathcal{R}$, infer all the concurrent computations possible in the system specified by $\mathcal{R}$ [30,3]. That is, given two states $[u], [v] \in T_{\Sigma/E \cup A}$, one can *reach* $[v]$ from $[u]$ by some possibly complex concurrent computation if and only if one can *prove* $\mathcal{R} \vdash [u] \longrightarrow [v]$, where the sequent $[u] \longrightarrow [v]$ does not necessarily denote a single rewrite step, but can instead correspond to a complex combination of: (i) concurrent rewrites (different subterms being rewritten simultaneously, as, for example, client b re-sending a request to server a, and a responding to an earlier copy of that request *simultaneously* in the protocol of Section 2); and (ii) sequential compositions of such concurrent rewrites. In rewriting logic any such complex computation reaching $[v]$ from $[u]$ is witnessed by a *proof term*, say λ, written $\mathcal{R} \vdash \lambda : [u] \longrightarrow [v]$. For example, in our client-server protocol we may have a state $[u]$ of the form

$$[a] \quad [b, a, 7, nil] \quad a \lhd (b, 7)$$

and a state $[v]$ of the form

$$[a] \quad [b, a, 7, nil] \quad a \lhd (b, 7) \quad b \lhd (a, f(a, b, 7))$$

and the proof $\lambda : [u] \longrightarrow [v]$ that $[v]$ can be reached from $[u]$ by the concurrent computation in which b resends the message and, simultaneously, a replies to the message present in $[u]$, is the proof term $\lambda = req(b, a, 7) \; reply(a, b, 7)$. Therefore, a proof term gives us a precise description of how a concurrent computation reaching a state $[v]$ from a state $[u]$ has happened. But several such computations, specified by several proof terms, may in some sense be *equivalent*. For example, we may consider the concurrent request and reply described by our λ above to be in some sense equivalent to both: (i) the sequential computation in which the request happens first and the reply second; and (ii) the one in which first the reply happens and then the request is sent. That is, we have the following equality of proof terms, all of which reach $[v]$ from $[u]$ in an equivalent manner:

$$
\begin{aligned}
&req(b, a, 7) \; reply(a, b, 7) \\
&= (req(b, a, 7) \; [a] \quad a \lhd (b, 7)) \; ; \; (reply(a, b, 7) \; [b, a, 7, nil] \quad a \lhd (b, 7))) \\
&= (reply(a, b, 7) \; [b, a, 7, nil]) \; ; \; (req(b, a, 7) \; [a] \quad b \lhd (a, f(a, b, 7)))
\end{aligned}
$$

where $\_ ; \_$ is the sequential composition operator between proof terms. Note that now each proof subterm in the above sequential compositions describes a *one-step* rewrite of the form $\gamma : [u] \longrightarrow^1_{\mathcal{R}} [w]$, or of the form $\gamma : [w] \longrightarrow^1_{\mathcal{R}} [v]$, for some w. For example, we have $(req(b, a, 7) \; [a] \quad a \lhd (b, 7)) : [u] \longrightarrow^1_{\mathcal{R}} [w]$ for $[w]$ the state

$$[a] \quad [b, a, 7, nil] \quad a \lhd (b, 7) \quad a \lhd (b, 7)$$

Rewriting logic defines a general equivalence relation between proof terms [30,3] which satisfies the property that any proof term λ has (possibly many) equivalent interleaving descriptions as a sequential composition $\gamma_1; \ldots; \gamma_k$ of one-step proof terms γ_i. Such one-step proof terms have a very simple algebraic description: they are all of the form $\gamma = t[l(u_1, \ldots, u_n)]_p$, with $t, u_1, \ldots, u_n$ Σ-terms, and l a rule label, indicating that at position p in t, a rule labeled l and having variables, say, $x_1, \ldots, x_n$ has been applied with substitution $\theta = \{x_1 \mapsto u_1, \ldots, x_n \mapsto u_n\}$. Note the slight technicality that, to make the expression $l(u_1, \ldots, u_n)$ unambiguous, we have to agree on some order of the variables $x_1, \ldots, x_n$. I assume that we use the *textual order* in which they appear in the rule's lefthand side. For example, if the lefthand side is the term $f(z, y, g(y, x, z))$, then the textual order is z, y, x.

One-step proof terms will be very useful to label atomic transitions in the computations on which we will evaluate the truth of path formulas in TLR^*. We just need a few more technical details to make the treatment of such one-step proof terms *canonical* in the context of computable rewrite theories. The point is that, given a computable rewrite theory $\mathcal{R} = (\Sigma, E \cup A, R)$ and given a state $[u]_{E \cup A}$ we can have in general an *infinite* number of one-step proof terms rewriting it, because the equivalence relation between proof terms includes the equivalence modulo $E \cup A$ among its axioms [30]. What we really want is the notion of a *canonical* one-step proof term. This will allow us to have only a *finite* number of such canonical proof terms specifying all the one-step rewrites from a given state. This is made possible by property (3) (ground coherence) of a computable rewrite theory, which gives us a systematic way of associating to an arbitrary one-step proof term a corresponding canonical proof term. Indeed, ground coherence ensures that, given a rewrite proof $t[l(u_1, \ldots, u_n)]_p : [t]_A \longrightarrow^1_{\mathcal{R}_{R/A}} [t']_A$ using a given rewrite rule $l : q \longrightarrow r$ in R we can always find a corresponding rewrite proof $can_{E/A}(t)[l(v_1, \ldots, v_n)]_{p'} : [can_{E/A}(t)]_A \longrightarrow^1_{\mathcal{R}_{R/A}} [t'']_A$ using the *same* rewrite rule $l : q \longrightarrow r$ in R such that $[can_{E/A}(t')]_A = [can_{E/A}(t'')]_A$. Let us call $[can_{E/A}(t)[l(v_1, \ldots, v_n)]_{p'}]_A$ a *canonical* one-step proof term for $\mathcal{R} = (\Sigma, E \cup A, R)$. Since R is finite and there exists an A-matching algorithm, there is only a finite number of such canonical proof terms, representing in a canonical way all one-step rewrites from a given state, and such canonical proof terms can be effectively computed. We can also make canonical the representation of the states reached from a given state by one-step rewrites: we can associate to the rewrite proof $can_{E/A}(t)[l(v_1, \ldots, v_n)]_{p'} : [can_{E/A}(t)]_A \longrightarrow^1_{\mathcal{R}_{R/A}} [t'']_A$ its corresponding *canonical one-step rewrite proof*, namely, $[can_{E/A}(t)[l(v_1, \ldots, v_n)]_{p'}]_A : [can_{E/A}(t)]_A \longrightarrow^1_{\mathcal{R}_{R/A}} [can_{E/A}(t'')]_A$. I denote canonical one-step rewrite proofs

$$[can_{E/A}(t)]_A \xrightarrow{\gamma}{}^1 [can_{E/A}(t'')]_A$$

where $\gamma = [can_{E/A}(t)[l(v_1, \ldots, v_n)]_{p'}]_A$, leaving implicit the reference to R .

3.3   Computations

Canonical one-step rewrite proofs are the key ingredient to arrive at our desired notion of *computation*, on which the truth of TLR^* path formulas will be evaluated. Recall that if $\mathcal{R} = (\Sigma, E \cup A, R) \in RWTh_0$, then $\mathcal{R}$, besides being computable, is assumed to be deadlock-free and to have a sort named *State* corresponding to the states of the system specified by $\mathcal{R}$. We can then define two useful sets. First, the set $(Can_{\Sigma/E,A})_{State}$ of all A-equivalence classes of the form $[can_{E/A}(t)]_A$, where t is a ground Σ-term of sort *State*. Of course, what $(Can_{\Sigma/E,A})_{State}$ describes is the set of all *states* of the system specified by $\mathcal{R}$ in their canonical form representation. Second, we can define the set $CanPTerms^1(\mathcal{R})$ of all one-step canonical proof terms in $\mathcal{R}$. Then, by definition, an *infinite computation* in $\mathcal{R} \in RWTh_0$ is a pair of functions (π, γ), with $\pi : \mathbb{N} \longrightarrow Can_{\Sigma/E,A_{State}}$, and $\gamma : \mathbb{N} \longrightarrow CanPTerms^1(\mathcal{R})$ such that for all $n \in \mathbb{N}$,

$$\pi(n) \xrightarrow{\gamma(n)}{}^1 \pi(n+1)$$

is a canonical one-step rewrite proof in $\mathcal{R}$. Graphically, (π, γ) looks as follows:

$$\pi(0) \xrightarrow[\quad]{\gamma(0)}{}^1 \pi(1) \xrightarrow[\quad]{\gamma(1)}{}^1 \pi(2) \ldots \pi(n) \xrightarrow[\quad]{\gamma(n)}{}^1 \pi(n+1) \ldots$$

$Comp(\mathcal{R})^\infty$ denotes the set of infinite computations in $\mathcal{R}$, and for each $[t] \in Can_{\Sigma/E, A_{State}}$, $Comp(\mathcal{R})^\infty_{[t]}$ denotes the infinite computations starting at $[t]$, that is, those computations (π, γ) such that $\pi(0) = [t]$. Given an infinite computation (π, γ) and a number $i \in \mathbb{N}$, $(\pi, \gamma)^i$ denotes the suffix of (π, γ) beginning at position i, that is, the pair of functions $(\pi \circ s^i, \gamma \circ s^i)$ with s the successor function, s^0 the identity function, and $s^{n+1} = s \circ s^n$.

3.4   Spatial Actions

Spatial actions are the action atoms of TLR^*. They generalize one-step proof terms, which can be thought of as *ground-instantiated spatial actions*. A one-step proof term tells us exactly *what* action takes place, with which *instantiation* of the rule's variables, and at which exact *position* in the state. Spatial actions describe instead *patterns*, that in general specify not just a single one-step proof term, but a possibly infinite set of such proof terms. Roughly speaking, we can think of spatial actions as "one-step proof terms with variables," but they are slightly more general than that, since I also allow rule labels themselves, with no instantiation information, as the most general kind of spatial actions. Given a rewrite theory $\mathcal{R} = (\Sigma, E \cup A, R) \in RWTh_0$, the simplest way to describe its corresponding set $SP(\Omega, L)$ of spatial actions is as a set of terms with variables in a signature associated to $\mathcal{R}$. First of all, note that, since $\mathcal{R}$ is computable, the equations E are ground confluent and terminating modulo A. This means that in the canonical forms $[can_{E/A}(t)]$ not all function symbols in Σ may be present, since some of them may correspond to functions defined by equations in E which are evaluated away when a canonical form is reached. For example, our client-server protocol specification may represent natural numbers in Peano notation with a constant 0, and a successor function s, and can have an addition function $+$ defined by the equations $N + 0 = N$, and $N + s(M) = s(N + M)$. Then, the symbol $+$ will *never* be present in any canonical form of a ground term. Call $\Omega \subset \Sigma$ the subsignature of *constructors* associated to the ground confluent and terminating (modulo A) equational theory $(\Sigma, E \cup A)$, where, by definition, $f \in \Omega$ iff there is a ground term t such that f is a function symbol in $[can_{E/A}(t)]$. In our example, 0 and s, and also $\_\_$, are constructors, but $+$ is not a constructor. As it will become clear in what follows, making the constructor signature Ω explicit is an important technical requirement to make the checking of whether a given canonical 1-step proof term γ is an *instance* of a spatial action pattern δ *decidable*. Note that, given a rewrite theory $\mathcal{R} = (\Sigma, E \cup A, R) \in RWTh_0$, all its canonical proof terms $\gamma \in CanPTerms^1(\mathcal{R})$ must be of the form $\gamma = [t[l(u_1, \ldots, u_n)]_p]_A$, with $t, u_1, \ldots, u_n$, Ω-terms.

Let us now define the set $SP(\Omega, L)$ of $\mathcal{R}$'s spatial action patterns, where Ω is the subsignature of constructors and L is the set of labels labeling rules in R, and where I assume $\Omega \cap L = \emptyset$. The signature $\Omega(L)$ extends Ω by adding:

- a fresh new sort Top
- for each rewrite rule $l : q \longrightarrow r$ in R with q, r of sort B, and with the textually-ordered set of variables in q having respective sorts $B_1, \ldots, B_n$:
  - an operator $l : B_1 \times \ldots \times B_n \longrightarrow B$
  - a constant l of sort B
  - an operator $top : B \longrightarrow Top$

Let X be a many-sorted set of variables with an infinite set of variables for each sort in Ω. Consider the algebras: (i) $T_{\Omega(L)/A}(X)$ of A-equivalence classes of $\Omega(L)$-terms with variables in X; and (ii) $T_{\Omega/A}(X)$ of A-equivalence classes of Ω-terms with variables in X. $SP(\Omega, L)$ is the subset of $T_{\Omega(L)/A}(X)$ defined by:

- for each $l \in L$, $[l]_A, [top(l)]_A \in SP(\Omega, L)$
- $[l(u_1, \ldots, u_n)]_A \in SP(\Omega, L)$ if $l \in L$, $[l(u_1, \ldots, u_n)]_A \in T_{\Omega(L)/A}(X)$, and $u_1, \ldots, u_n \in T_{\Omega/A}(X)$
- $[top(l(u_1, \ldots, u_n))]_A \in SP(\Omega, L)$ if $l \in L$, $[top(l(u_1, \ldots, u_n))]_A \in T_{\Omega(L)/A}(X)$, and $u_1, \ldots, u_n \in T_{\Omega/A}(X)$
- $[v[l(u_1, \ldots, u_n)]_p]_A \in SP(\Omega, L)$ if p is not the empty (top) position, $l \in L$, $[v[l(u_1, \ldots, u_n)]_p]_A \in T_{\Omega(L)/A}(X)$, and $v, u_1, \ldots, u_n \in T_{\Omega/A}(X)$.

Note that $CanPTerms^1(\mathcal{R}) \subseteq SP(\Omega, L)$, so that any canonical one-step proof term is a ground version of some spatial action pattern. The main purpose of spatial action patterns is to endow TLR^* with flexible action patterns with varying degrees of generality: they can range from the most general patterns of the form $[l]_A$ (that only indicate that *some* rule labeled l has been applied *somewhere* in the state structure) to fully instantiated canonical one-step proof terms (that fully specify the position, rule label, and variable instantiation), to anything in between. As explained in Section 2, action patterns $\delta \in SP(\Omega, L)$ can often be *abbreviated* by following a few simple conventions, making their notation more succinct. I will often leave the A-equivalence class $[\delta]_A$ of an action pattern δ implicit, and work directly with a representative δ. An action pattern of the form l describes a rule labeled l that can be applied *anywhere*. In our running protocol example, the action pattern req allows the req rule to be applied to any client anywhere in the soup. An action pattern $l(u_1, \ldots, u_n)$ allows l to also be applied anywhere, but constrains the variable instantiation θ to be itself a further instance of the substitution $x_1 \mapsto u_1, \ldots, x_n \mapsto u_n$. For example, with the pattern $req(b)$, the req rule can only be applied to client b. Action patterns of the form $top(l(u_1, \ldots, u_n))$ are needed to cover the case where l is applied at the top of the term. For example, $top(req(b))$ only allows application of the req rule to states consisting of just one client named b, with no other clients, servers, or messages present in the state. The most fully spatial patterns are those of the form $v[l(u_1, \ldots, u_n)]_p$ with v a nonempty context and p a position. For example, we could constrain the req rule to be applied to states containing only one client, its corresponding server, and no other clients, servers, or messages, by means of the spatial action pattern $[S]\ req(C, S)$, which is obtained from the nonempty context term $[S]\ X$ by replacing X by $req(C, S)$ at position 2.

The last point to be explained is how we can effectively *check* that a given canonical one-step proof term γ is an *instance* of a spatial action pattern $\delta \in SP(\Omega, L)$. This check will be essential when defining the semantics of TLR^*. First of all, notice that, by our assumption that there is an A-matching algorithm, the instance-of relation modulo A between $[u], [v] \in T_{\Omega(L)/A}(X)$, denoted $[u]_A \preceq_A [v]_A$, and defined by $[u]_A \preceq_A [v]_A$ iff there is a many-sorted substitution θ such that $[u]_A = [\theta(v)]_A$, is a *decidable* relation, since it amounts to checking whether u *matches* the pattern v modulo A. The *instance-of* relation between a canonical one-step proof term γ and a spatial action pattern $\delta \in SP(\Omega, L)$, denoted $\gamma \sqsubseteq_A \delta$, is a slight variant of the $\preceq_A$ relation defined as follows:

- $[v[l(u_1, \ldots, u_n)]_p]_A \sqsubseteq_A [l]_A$
- $[v[l(u_1, \ldots, u_n)]_p]_A \sqsubseteq_A [l(v_1, \ldots, v_n)]_A$ iff $[l(u_1, \ldots, u_n)]]_A \preceq_A [l(v_1, \ldots, v_n)]_A$
- $[v[l(u_1, \ldots, u_n)]_p]_A \sqsubseteq_A [w[l(v_1, \ldots, v_n)]_{p'}]_A$ iff $[v[l(u_1, \ldots, u_n)]_p]_A \preceq_A [w[l(v_1, \ldots, v_n)]_{p'}]_A$
- $[l(u_1, \ldots, u_n)]_A \sqsubseteq_A [top(l)]_A$
- $[l(u_1, \ldots, u_n)]_A \sqsubseteq_A [top(l(v_1, \ldots, v_n))]_A$ iff $[l(u_1, \ldots, u_n)]]_A \preceq_A [l(v_1, \ldots, v_n)]_A$.

4   The Temporal Logic of Rewriting

I first introduce the syntax of TLR^*, and then define its satisfaction semantics on a rewrite theory. By restricting the spatial action patterns and/or the atomic state predicates used, one obtains various sublogics of TLR^*, including well-known state-based and action-based sublogics. The generality and expressiveness of the $RewritingLogic/TLR^*$ tandem allow it to unify as special cases both the state-based tandem $Kripke/CTL^*$, based on Kripke structures, and the action-based tandem $DFLTranSys/TLR^*(L)$, based on (deadlock-free) labeled transition systems. I also show that there is a pair of model and formula transformations faithfully mapping the tandem $RewritingLogic/TLR^*$ to the tandem $Kripke/CTL^*$. This makes possible the use of standard CTL^* model checkers to verify TLR^* properties of finite-state systems specified by rewrite theories.

4.1 TLR^* Syntax

TLR^* is a family of logics parameterized by the spatial actions $SP(\Omega, L)$ and the signature of atomic propositions Π. The most general of these logics is TLR^*, a generalization of the state-based CTL^* logic that allows both spatial actions and state predicates in formulas. Similarly, TLR is the sublogic of TLR^* generalizing CTL, and $LTLR$ is the sublogic generalizing LTL.

Everything is parameterized by the spatial actions $SP(\Omega, L)$ and the signature of state predicates Π. For example, TLR^* is the parametric family $TLR^*(SP(\Omega, L), \Pi)$. The key classes I consider are: (i) $TLR^*(SP(\Omega, L), \Pi)$, and $PTLR^*(SP(\Omega, L), \Pi)$, which generalize, respectively, state and path formulas in CTL^*; (ii) $TLR(SP(\Omega, L), \Pi)$, and $PTLR(SP(\Omega, L), \Pi)$, which generalize, respectively, state and path formulas in CTL; (iii) $QFR(SP(\Omega, L), \Pi)$, the (path-) quantifier-free formulas; (iv) $LTLR(SP(\Omega, L), \Pi)$, which generalizes LTL formulas; (v) the spatial actions in positive or negative form $Atom(SP(\Omega, L))$; and (vi) the atomic state predicates in positive or negative form $Atom(\Pi)$. I assume that all state predicate constants and function symbols are constructors, i.e., that there is a subsignature containment $\Pi \subseteq \Omega$, and then define the set $Prop(\Pi)$ of *atomic propositions* as the set of ground terms $Prop(\Pi) = T_{\Omega_{Prop}}$. I use BNF-like notation to characterize the syntax of each of these logics, with variables:

$$\delta : SP(\Omega, L), \; \lambda : Atom(SP(\Omega, L)), \; p : Prop(\Pi), \; \alpha : Atom(\Pi)$$

$$\varphi, \varphi' : TLR^*(SP(\Omega, L), \Pi), \; \phi, \phi' : PTLR^*(SP(\Omega, L), \Pi)$$

$$\zeta, \zeta' : TLR(SP(\Omega, L), \Pi), \; \mu : PTLR(SP(\Omega, L), \Pi), \; \eta, \eta' : QFR(SP(\Omega, L), \Pi).$$

- $Atom(SP(\Omega, L)) : \; \delta \mid \neg\delta$
- $Atom(\Pi) : \; \top \mid \bot \mid p \mid \neg p$
- $TLR^*(SP(\Omega, L), \Pi) : \; \alpha \mid \neg\varphi \mid \varphi \vee \varphi' \mid \varphi \wedge \varphi' \mid \mathbf{A}\phi \mid \mathbf{E}\phi$
- $PTLR^*(SP(\Omega, L), \Pi) : \; \varphi \mid \lambda \mid \neg\phi \mid \phi \vee \phi' \mid \phi \wedge \phi' \mid \mathbf{X}\phi \mid \phi\mathbf{U}\phi' \mid \phi\mathbf{R}\phi' \mid \phi\mathbf{W}\phi' \mid \mathbf{F}\phi \mid \mathbf{G}\phi$
- $TLR(SP(\Omega, L), \Pi) : \; \alpha \mid \neg\zeta \mid \zeta \vee \zeta' \mid \zeta \wedge \zeta' \mid \mathbf{A}\mu \mid \mathbf{E}\mu$
- $PTLR(SP(\Omega, L), \Pi) : \; \zeta \mid \lambda \mid \mathbf{X}\zeta \mid \zeta\mathbf{U}\zeta' \mid \zeta\mathbf{R}\zeta' \mid \mathbf{F}\zeta \mid \mathbf{G}\zeta$
- $QFR(SP(\Omega, L), \Pi) : \; \lambda \mid \alpha \mid \neg\eta \mid \eta \vee \eta' \mid \eta \wedge \eta' \mid \mathbf{X}\eta \mid \eta\mathbf{U}\eta' \mid \eta\mathbf{R}\eta' \mid \eta\mathbf{W}\eta' \mid \mathbf{F}\eta \mid \mathbf{G}\eta$
- $LTLR(SP(\Omega, L), \Pi) : \; \mathbf{A}\eta$

For example, the formula

$$\mathbf{A}(FairnessAssumptions \; \Rightarrow \; \mathbf{F} \; rec(b))$$

specifying the the client-server protocol property already discussed in Section 2 belongs to $LTLR(SP(\Omega, L), \Pi)$, with Ω, resp. Π, the constructor, resp. predicate, subsignature of Σ in our example system specification $\mathcal{R} = (\Sigma, E \cup A, R)$.

Smaller, useful sublogics of $TLR^*(SP(\Omega, L), \Pi)$ can be obtained by restricting the atomic propositions and/or spatial actions allowed. That is, we can define sublogics of TLR^* parameterized by a subset $W \subseteq SP(\Omega, L)$ of spatial actions, and a subset $\Delta \subseteq Prop(\Pi)$ of atomic propositions. Specifically, the sublogic $PTLR^*(W, \Delta) \subseteq PTLR^*(SP(\Omega, L), \Pi)$ is defined by the set-theoretic formula $PTLR^*(W, \Delta) = \{\phi \in PTLR^*(SP(\Omega, L), \Pi) \mid sp(\phi) \subseteq W \; \wedge \; prop(\phi) \subseteq \Delta\}$, where $sp(\phi)$ denotes the set of spatial action subformulas of ϕ, and $prop(\phi)$ denotes the set of atomic proposition subformulas. In exactly the same way we can define all the smaller sublogics, such as $TLR^*(W, \Delta)$, $TLR(W, \Delta)$, $LTLR(W, \Delta)$, and so on. In the particular case where we make just one of these parameters empty without restricting the other, we encounter well-known logics. When $W = \emptyset$, and $\Delta = Prop(\Pi)$, we obtain specializations to the *state-based logics* CTL^*, CTL, and LTL; that is, $CTL^*(\Pi) = TLR^*(\emptyset, \Pi)$, $CTL(\Pi) = TLR(\emptyset, \Pi)$, and $LTL(\Pi) = LTLR(\emptyset, \Pi)$. When $W = SP(\Omega, L)$ and $\Delta = \emptyset$, one obtains *pure action logics*: $TLR^*(SP(\Omega, L)) = TLR^*(SP(\Omega, L), \emptyset)$, $TLR(SP(\Omega, L)) = TLR(SP(\Omega, L), \emptyset)$, and $LTLR(SP(\Omega, L)) = LTLR(SP(\Omega, L), \emptyset)$. These pure action logics can be further restricted by allowing only atomic labels $l \in L$ as actions, i.e., by choosing $W = L$. In this way we obtain $TLR^*(L) = TLR^*(L, \emptyset)$, $TLR(L) = TLR(L, \emptyset)$, and $LTLR(L) = LTLR(L, \emptyset)$. The mixed action-state logic $SE\text{-}LTL$ in [8,9] is also obtained this way as: $SE\text{-}LTL(L, \Pi) = LTLR(L, \Pi)$.

4.2 TLR^* Semantics

The semantics of a state formula $\varphi \in TLR^*(SP(\Omega, L), \Pi)$ is given by the satisfaction relation $\mathcal{R}, [t] \models \varphi$ defined below, where $\mathcal{R} \in RWTh_0$ has subsignatures of constructors Ω and of state predicates Π, and $[t]$ is a state, that is, an A-equivalence class $[t]_A$ in E-canonical form modulo A and of sort $State$, where $E \cup A$ are the equations in $\mathcal{R}$.

Similarly, the semantics of a path formula $\phi \in PTLR^*(SP(\Omega, L), \Pi)$ is given by the satisfaction relation $\mathcal{R}, (\pi, \gamma) \models \phi$, where (π, γ) is an infinite computation in $Comp(\mathcal{R})^\infty$. Since one can express all of $TLR^*(SP(\Omega, L), \Pi)$ and $PTLR^*(SP(\Omega, L), \Pi)$ in terms of $Atom(SP(\Omega, L))$, $Atom(\Pi)$, the basic connectives $\top$, $\neg$, $\vee$, $\mathbf{X}$, $\mathbf{U}$, and the universal path quantifier $\mathbf{A}$, it is enough to define the semantics for the atoms and for those connectives. Since TLR^* generalizes CTL^*, the semantic definitions are entirely similar to those for CTL^* (see, e.g., [11]). The key new addition is the semantics of *spatial actions*. The satisfaction relation is defined inductively as follows:

- $\mathcal{R}, [t] \models \top$
- $\mathcal{R}, [t] \models p \;\Leftrightarrow\; E \cup A \vdash t \models p = \mathit{true}$
- $\mathcal{R}, [t] \models \neg\varphi \;\Leftrightarrow\; \mathcal{R}, [t] \not\models \varphi$
- $\mathcal{R}, [t] \models \varphi \vee \varphi' \;\Leftrightarrow\; \mathcal{R}, [t] \models \varphi$ or $\mathcal{R}, [t] \models \varphi'$
- $\mathcal{R}, [t] \models \mathbf{A}\phi \;\Leftrightarrow\; \forall\, (\pi, \gamma) \in Comp(\mathcal{R})^\infty_{[t]} \;\; \mathcal{R}, (\pi, \gamma) \models \phi$
- $\mathcal{R}, (\pi, \gamma) \models \varphi \;\Leftrightarrow\; \mathcal{R}, \pi(0) \models \varphi$
- $\mathcal{R}, (\pi, \gamma) \models \delta \;\Leftrightarrow\; \gamma(0) \sqsubseteq_A \delta$
- $\mathcal{R}, (\pi, \gamma) \models \neg\phi \;\Leftrightarrow\; \mathcal{R}, (\pi, \gamma) \not\models \phi$
- $\mathcal{R}, (\pi, \gamma) \models \phi \vee \phi' \;\Leftrightarrow\; \mathcal{R}, (\pi, \gamma) \models \phi$ or $\mathcal{R}, (\pi, \gamma) \models \phi'$
- $\mathcal{R}, (\pi, \gamma) \models \mathbf{X}\phi \;\Leftrightarrow\; \mathcal{R}, (\pi, \gamma)^1 \models \phi$
- $\mathcal{R}, (\pi, \gamma) \models \phi\mathbf{U}\phi' \;\Leftrightarrow\; \exists k \in \mathbb{N}\; s.t.\; \mathcal{R}, (\pi, \gamma)^k \models \phi' \wedge \forall 0 \leq i < k \;\; \mathcal{R}, (\pi, \gamma)^i \models \phi$

At the syntactic level, we have already seen that TLR^* contains both the state-based logic CTL^* and the pure action logic $TLR^*(L)$, where L is a parameter denoting the given set L of labels, as sublogics. This generality has a similar counterpart at the semantic level, where both Kripke structures and labeled transition systems can be seen as very simple special cases of rewrite theories (with rewrite rules that rewrite atomic constants instead of rewriting general terms). As explained in [31], we have two faithful embeddings of tandems

$$Kripke/CTL^* \hookrightarrow RewritingLogic/TLR^* \hookleftarrow DFLTranSys/TLR^*(L)$$

where $DFLTranSys$ stands for the class of deadlock-free labeled transition systems. Other pure action logics can be embedded into the $RewritingLogic/TLR^*$ tandem by faithful mappings of tandems. For example, two formula translations $\gamma : HM(L) \longrightarrow TLR^*(L)$, and $\beta : A\text{-}CTL^*(L) \longrightarrow TLR^*(L)$, from, respectively, Hennesy-Milner logic [19], and De Nicola and Vaandrager $A\text{-}CTL^*$ [34], are defined in [31] and shown to yield faithful mappings of tandems

$$DFLTranSys/A\text{-}CTL^*(L) \xrightarrow{(id,\beta)} DFLTranSys/TLR^*(L) \xleftarrow{(id,\gamma)} DFLTranSys/HM(L)$$

4.3   Reduction to State-Based Temporal Logics

Although the example in Section 2 has shown that in general there is no *direct* way of expressing TLR^* properties in state-based logics *without changing the system specification*, there is, however, a systematic *indirect* way of achieving such a reduction to state-based logics, namely, to CTL^*. This requires changing both the system specification, by associating to a rewrite theory a suitable Kripke structure, and translating the TLR^* formula φ into a corresponding CTL^* formula $\widetilde{\varphi}$. That is, we need to define a suitable faithful mapping of tandems. Since any TLR^* formula ϕ only involves a finite set $sp(\phi)$ of spatial actions as subformulas, it is always sufficient to consider formulas in $TLR^*(W, \Pi)$ with W *finite*. The faithful mapping of tandems we seek is a mapping parametric in W

$$(\mathcal{K}_W, (\widetilde{\_})) : RewritingLogic/TLR^*(W, \Pi) \longrightarrow Kripke/CTL^*(\Pi \cup W)$$

with W a finite set of spatial actions in the given rewrite theory $\mathcal{R}$, and Π the subsignature of state predicates in $\mathcal{R}$; and where in the translated CTL^* formula $\widetilde{\varphi}$ a spatial action pattern $\delta \in W$ becomes an additional *state predicate*.

Recall that a *Kripke structure* on a set AP of atomic propositions is a triple $\mathcal{K} = (A, R, \mathcal{L})$, with A a set of states, $R \subseteq A \times A$ a total transition relation, and $\mathcal{L} : A \longrightarrow \mathcal{P}(AP)$ a labeling function assigning to each state $a \in A$ the set $\mathcal{L}(a) \subseteq AP$ of the atomic propositions that hold in a. Given $\mathcal{R} \in RWTh_0$ and a finite $W \subseteq SP(\Omega, L)$, the construction $\mathcal{K}_W$ maps the rewrite theory $\mathcal{R}$ to the following Kripke structure $\mathcal{K}_W(\mathcal{R})$:

- Its set of states is $(Can_{\Sigma/E,A})_{State} \times \mathcal{P}(W)$; that is, states are pairs $([t], U)$, with $[t]$ a state of $\mathcal{R}$, and $U \subseteq W$ a subset of spatial actions in W.
- Its transition relation is defined by the equivalence: $([t], U) \longrightarrow ([t'], V)$ iff there is a canonical one-step rewrite proof $[t] \overset{\gamma}{\longrightarrow}^1 [t']$ in $\mathcal{R}$, and V is the set

$$act_W(\gamma) = \{\delta \in W \mid \gamma \sqsubseteq_A \delta\}.$$

  That is, a transition $([t], U) \longrightarrow ([t'], V)$ is possible iff there exists a one-step rewrite $[t] \overset{\gamma}{\longrightarrow}^1 [t']$ in $\mathcal{R}$, and V is the set of *all* spatial action patterns in W of which γ is an instance (note that V could be empty).
- Its set of atomic propositions is the set $Prop(\Pi) \cup W$, and the labeling function maps a state $([t], U)$ to the set of atomic propositions $\mathcal{L}_{\mathcal{R}}([t]) \cup U$, where, by definition, $\mathcal{L}_{\mathcal{R}}([t]) = \{p \in Prop(\Pi) \mid E \cup A \vdash [t] \models p = true\}$; that is, the atomic propositions holding in a state $([t], U)$ are exactly those holding in $[t]$ *plus* the propositions U.

In typical model-checking applications of this mapping of tandems, we will be interested in using the construction $\mathcal{K}_W(\mathcal{R})$ to model check a TLR^* formula φ using a CTL^* model checker, and we will choose the smallest W possible, namely, $W = sp(\varphi)$. In most practical cases such a set W will furthermore be *unambiguous*, in the sense that different action patterns in W never overlap. For example, if $W \subseteq L$ is just a set of labels, different labels always denote different sets of one-step proof terms and therefore such an W is always unambiguous; but we can have also unambiguous action patterns such as $l(0)$ and $l(s(x))$, where l is parametric on, say, the natural numbers with 0 and successor s, since $l(0)$ and $l(s(x))$ can never have a common ground instance. The interest of having W unambiguous is that then all transitions $([t], U) \longrightarrow ([t'], V)$ in $\mathcal{K}_W(\mathcal{R})$ have U and V either singleton subsets of W, or the empty set. This means that the original state space of reachable states of $\mathcal{R}$ from some initial state, if is finite, is then only increased by a $|W| + 1$ factor. In, [31] a detailed unifiability-based decision procedure is given to determine if a given W is unambiguous.

Given now a formula $\varphi \in TLR^*(W, \Pi)$ (resp. $\phi \in PTLR^*(W, \Pi)$) we can associate to it a formula $\widetilde{\varphi} \in CTL^*(W \cup \Pi)$ (resp. $\widetilde{\varphi} \in PCTL^*(W \cup \Pi)$) by systematically replacing each occurrence of a spatial action $\delta \in W$ in φ by the formula $\mathbf{X}\delta$. The construction $\mathcal{K}_W(\mathcal{R})$, together with the above formula translation $(\widetilde{})$ define a mapping of tandems $(\mathcal{K}_W, (\widetilde{})) : RewritingLogic/TLR^*(W, \Pi) \longrightarrow Kripke/CTL^*(\Pi \cup W)$. This mapping is a *faithful* mapping of tandems preserving the satisfaction relations $\models$ in $TLR^*(W, \Pi)$ and $\models_{CTL^*}$ in $CTL^*(\Pi \cup W)$. This is shown by the following theorem proved in detail in [31].

Theorem 1. *Given a rewrite theory $\mathcal{R} \in RWTh_0$ and a finite $W \subseteq SP(\Omega, L)$, for each state $[t]$ in $\mathcal{R}$, subset $U \subseteq W$, and formula $\varphi \in TLR^*(W, \Pi)$, the following equivalence holds:*

$$\mathcal{R}, [t] \models \varphi \Leftrightarrow \mathcal{K}_W(\mathcal{R}), ([t], U) \models_{CTL^*} \widetilde{\varphi}$$

The above theorem gives us a systematic way of verifying TLR^* properties of a rewrite theory $\mathcal{R}$ by verifying the CTL^* translation of those properties on the Kripke structure $\mathcal{K}_W(\mathcal{R})$. Notice that if $\mathcal{R}$ is finite-state and computable, then so is $\mathcal{K}_W(\mathcal{R})$. Therefore, in such case we can use standard CTL^* model-checking algorithms —or CTL or LTL algorithms if the translated formula falls within those sublogics— to verify the given TLR^* property of our original system. Note, also, that given any formula $\varphi \in TLR^*(SP(\Omega, L), \Pi)$, we can always choose $W = sp(\varphi)$. Therefore, the computational cost to model check $\widetilde{\varphi}$ from an initial state $([t], U)$ can be estimated in terms of the size $|\mathcal{R}_{[t]}|$ of the original rewrite theory $\mathcal{R}$ starting at $[t]$, which we can define as $|\mathcal{R}_{[t]}| = |Reach_{\mathcal{R}}([t])| + | \longrightarrow^1_{[t]} |$, where $Reach_{\mathcal{R}}([t])$ is the set of states reachable from $[t]$, and $\longrightarrow^1_{[t]}$ is the restriction of the one-step rewrite relation to states in $Reach_{\mathcal{R}}([t])$. Similarly, let $\mathcal{K}_W(\mathcal{R})_{([t],U)}$ denote the sub-Kripke structure of $\mathcal{K}_W(\mathcal{R})$ determined by the states reachable from $([t], U)$, let $Reach_{\mathcal{K}_W(\mathcal{R})}([t], U)$ denote the set of such reachable states, and let $\longrightarrow_{(t,U)}$ denote the restriction of the transition relation to states reachable from $([t], U)$. The size of this sub-Kripke structure is then $|\mathcal{K}_W(\mathcal{R})_{([t],U)}| = |Reach_{\mathcal{K}_W(\mathcal{R})}([t], U)| + | \longrightarrow_{(t,U)} |$. We then have the inequalities: $|Reach_{\mathcal{K}_W(\mathcal{R})}([t], U)| \leq |Reach_{\mathcal{R}}([t])| \cdot 2^{|sp(\varphi)|}$, and $| \longrightarrow_{(t,U)} | \leq | \longrightarrow^1_{[t]} | \cdot 2^{2 \cdot |sp(\varphi)|}$. Therefore, we obtain the bound $|\mathcal{K}_W(\mathcal{R})_{([t],U)}| \leq |\mathcal{R}_{[t]}| \cdot 2^{2 \cdot |sp(\varphi)|}$, and for the most common case when W is unambiguous the considerably better bound $|\mathcal{K}_W(\mathcal{R})_{([t],U)}| \leq |\mathcal{R}_{[t]}| \cdot (|sp(\varphi)| + 1)^2$. Therefore, since the complexity of model checking a CTL, resp., LTL or CTL^* formula $\widetilde{\varphi}$ on a Kripke structure

$\mathcal{B} = (B, \rightarrow_\mathcal{B}, \mathcal{L}_\mathcal{B})$ is $O(|\widetilde{\varphi}| \cdot |\mathcal{B}|)$, resp., $|\mathcal{B}| \cdot 2^{O(|\widetilde{\varphi}|)}$ (see [11]), for $\mathcal{B} = \mathcal{K}_W(\mathcal{R})_{([t], U)}$ we conclude that the complexity of model checking $\widetilde{\varphi}$ is $O(|\widetilde{\varphi}| \cdot |\mathcal{R}_{[t]}| \cdot 2^{2 \cdot |sp(\varphi)|})$ (resp. $O(|\widetilde{\varphi}| \cdot |\mathcal{R}_{[t]}| \cdot (|sp(\varphi)| + 1)^2)$ in the unambiguous case) if $\widetilde{\varphi} \in CTL$; and $O(|\mathcal{R}_{[t]}| \cdot 2^{2 \cdot |sp(\varphi)|} \cdot 2^{O(|\widetilde{\varphi}|)})$ (resp. $O(|\mathcal{R}_{[t]}| \cdot (|sp(\varphi)| + 1)^2 \cdot 2^{O(|\widetilde{\varphi}|)})$ in the unambiguous case) if $\widetilde{\varphi}$ is a LTL or CTL^* formula. In all cases we only incur an extra factor $2^{2 \cdot |labels(\varphi)|}$ (resp. $(|sp(\varphi)| + 1)^2$ in the most common, unambiguous case), which in practice should not be too big for a typical φ. Note that in the worse case we have $|\widetilde{\varphi}| = 2 \cdot |\varphi|$. The above are rather *naive* estimates, since by generalizing the state-based model-checking algorithms to "native" TLR^* algorithms that can work directly on $\mathcal{R}$, the above bounds could be greatly improved. For example, for $LTLR$ formulas, the model checking approach in Theorem 3 in [8,9], based on the so-called state-event product of a labeled Kripke structure and a Büchi automaton, does not require any increase in the state space of $\mathcal{R}$.

4.4   The Example Revisited

The general reduction to state-based logics can be illustrated with our client-server protocol example and φ our example $LTLR$ formula in Section 2. The set $sp(\varphi)$ is $sp(\varphi) = \{req(b), reply(a, b), rec(b)\}$. Since different spatial actions in $sp(\varphi)$ have different labels, $sp(\varphi)$ is clearly unambiguous. We can then define a rewrite theory $\mathcal{R}_{sp(\varphi)}$ whose rules are the following modified versions of our original rules:

$req:\ \{X\ [C, S, N, nil] \mid A\} \longrightarrow$ **if** $C == b$ **then**
  $\{X\ [C, S, N, nil]\ \ S \lhd (C, N) \mid req(b)\}$ **else** $\{X\ [C, S, N, nil]\ \ S \lhd (C, N) \mid \tau\}$ **fi**

$reply:\ \{X\ \ S \lhd (C, N)\ [S] \mid A\} \longrightarrow$ **if** $C == b\ and\ S == a$
     then $\{X\ [S]\ \ C \lhd (S, f(S, C, N)) \mid reply(a, b)\}$
     else $\{X\ [S]\ \ C \lhd (S, f(S, C, N)) \mid \tau\}$ **fi**

$rec:\ \{X\ \ C \lhd (S, M)\ [C, S, N, W] \mid A\} \longrightarrow$ **if** $C == b$
     then $\{X\ [C, S, N, M]\ \mid rec(b)\}$
     else $\{X\ [C, S, N, M]\ \mid \tau\}$ **fi**

$dupl:\ \{X\ \ I \lhd CNT \mid A\} \longrightarrow \{X\ \ I \lhd CNT\ \ I \lhd CNT \mid \tau\}$

$loss:\ \{X\ \ I \lhd CNT \mid A\} \longrightarrow \{X \mid \tau\}$

The equational theory $(\Sigma, E \cup A)$ of the original rewrite theory $\mathcal{R}$ for our protocol is extended by: (i) renaming the sort *State* to *OldState*; (ii) adding a new sort, *Action* with a constant τ (denoting the empty set $\emptyset$) and operators $req, rec :$ $Oid \longrightarrow Action$, and $reply : Oid \times Oid \longrightarrow Action$; (iii) adding a new sort *State* together with an operator $\{\_ \mid \_\} : OldState \times Action \longrightarrow State$; that is, now a state is a pair $\{X \mid A\}$, with X an *OldState*, and A an *Action*; and (iv) adding an **if then else fi** operator. The equations $E \cup A$ remain unchanged, except for the obvious new equations for the **if then else fi** operator, and the need to lift to the new state structure the satisfaction equations for localized enabledness state predicates *enabled.req, enabled.rec* : $Oid \longrightarrow Prop$ and *enabled.reply* :

$Oid \times Oid \longrightarrow Prop$. For example, the positive case of the *enabled.req* predicate is defined by

$$\{X \; [C, S, N, nil] \mid A\} \models enabled.req(C) = true$$

This example also illustrates an additional point, namely, that the $\mathcal{K}_W(\mathcal{R})$ construction can be decomposed into *two* simpler constructions: (i) a rewrite theory transformation $\mathcal{R} \mapsto \mathcal{R}_W$, adding to the states of $\mathcal{R}$ a second component containing finite sets of spatial actions in W; and (ii) the standard construction $\mathcal{R} \mapsto \mathcal{K}(\mathcal{R})$ mapping a rewrite theory to its underlying Kripke structure (see [12]), so that we have $\mathcal{K}_W(\mathcal{R}) = \mathcal{K}(\mathcal{R}_W)$. The decomposition $\mathcal{K}_W(\mathcal{R}) = \mathcal{K}(\mathcal{R}_W)$ is particularly useful when model checking $\mathcal{R}, [t] \models \varphi$ with φ a *LTLR* formula: we can model check instead $\mathcal{R}_W, ([t], \emptyset) \models \widetilde{\varphi}$ in the Maude *LTL* model checker, which implicitly performs the $\mathcal{K}(\mathcal{R})$ construction.

Our transformed system, although very simple, has an *infinite* number of reachable states, even for simple initial states such as the one discussed in Section 2. Therefore, we cannot verify the *LTLR* formula $\varphi = \mathbf{A}(FairnessAssumptions \Rightarrow \mathbf{F}\, rec(b))$ by model checking the *LTL* formula $\widetilde{\varphi}$ directly on $\mathcal{R}_{sp(\varphi)}$ using standard algorithms. However, we can verify it by defining a simple finite-state *equational abstraction* $\widehat{\mathcal{R}}_{sp(\varphi)}$ of $\mathcal{R}_{sp(\varphi)}$ [32], and model checking $\widetilde{\varphi}$ on $\widehat{\mathcal{R}}_{sp(\varphi)}$. For abstractions in general (see Chapter 13 and Theorem 16 in [11]), and equational abstractions in particular [32], if a $\mathbf{A}CTL^*$ formula holds for the abstraction, then it holds also for the original system. The equational abstraction $\widehat{\mathcal{R}}_{sp(\varphi)}$ in question is the rewrite theory obtained by adding to $\mathcal{R}_{sp(\varphi)}$ the single equation:

$$I \triangleleft CNT \; I \triangleleft CNT \; = \; I \triangleleft CNT$$

This equation causes a lack of coherence for the rules *reply*, *rec*, and *loss*, so a simple process of "coherence completion" has to be performed by adding to $\mathcal{R}_{sp(\varphi)}$ extra versions of these rules that are coherent. For example, we need to add an extra *loss* rule of the form

$$loss : \{X \; I \triangleleft CNT \mid A\} \longrightarrow \{X \; I \triangleleft CNT \mid \tau\}$$

The *LTL* formula $\widetilde{\varphi}$ can then be model checked using Maude's *LTL* model checker on the initial state *init* defined by the term

$$\{[a] \;\; [b, a, 7, nil] \;\; [c, a, 17, nil] \mid \tau\}$$

which by Theorem 1 is the counterpart for $\mathcal{R}_{sp(\varphi)}$ and $\widehat{\mathcal{R}}_{sp(\varphi)}$ of the initial state for $\mathcal{R}$ in Section 2. Indeed, Maude's *LTL* model checker gives us the answer

```
Maude> red modelCheck(init,tilde(FairnessAssumptions -> <> rec(b))) .
result Bool: true
```

Appendix A in [31] contains the Maude specification of the theories $\mathcal{R}_{sp(\varphi)}$ and $\widehat{\mathcal{R}}_{sp(\varphi)}$, as well as the automation of the mapping $\varphi \mapsto \widetilde{\varphi}$.

5   Related Work

There is much related work on both state-based and action-based logics. This section is *not* a survey, and I do not attempt to cover what is indeed a very vast field. I do however comment on various logics that I view as most closely related to TLR^*. They break down naturally into: (i) state-based logics; (ii) action-based logics; and (iii) mixed logics supporting both actions and state predicates.

The connections to well-known state-based logics such as LTL, CTL, and CTL^* (see, e.g., [24], [11]), to action-based logics such as Hennessy-Milner logic [19], and A-CTL^* [34], and to the mixed action-state logic SE-LTL in [8,9], all of which can be viewed as special cases of the $RewritingLogic/TLR^*$ tandem, have already been discussed in Sections 4.1–4.2.

Regarding other state-based logics, I would like to briefly comment on the Spatial Logic for Concurrency of Caires and Cardelli [5,6], which is primarily a spatial modal logic for process calculi in the π-calculus spirit and has some close relationships to the modal μ-calculus. Caires and Cardelli use spatial features only for *state predicates* and do not distinguish between different actions, which are all handled by a single, unlabeled diamond modality[4]. Instead, TLR^* uses spatial features for *action patterns*, while spatial features are not used for state predicates. If one accepts the equation "*concurrency* $= \pi$-*calculus*," the logic of Caires and Cardelli is indeed a state-based logic for concurrency. From a more ecumenical point of view, it is a logic well suited for π-calculus-like concurrent systems, which are particular kinds of rewrite theories [38].

Regarding various action-based temporal and modal logics, I refer the reader to a survey by Mateescu [27], where, besides Hennessy-Milner logic and A-CTL^*, various other action-based logics are both discussed in detail and compared with each other, including: (i) the μ-A-CTL [15] extension of A-CTL^*, which adds to A-CTL a least fixpoint operator and has the same expressive power as the modal μ-calculus; (ii) Propositional Dynamic Logic (PDL) [37,17]; and (iv) the modal μ-calculus [22] (μL). As already pointed out, the closest analogue of TLR^* among action logics is A-CTL^*, which can be viewed as a special case of TLR^*. The modal μ-calculus is itself more expressive than A-CTL, PDL, and Hennessy-Milner logic (see [27] for a careful comparison). To the work covered in Mateescu's survey, I would also add the $NPATRL$ logic [28] associated to the NRL Protocol Analyzer. Since this tool uses backwards symbolic search to verify cryptographic protocols, the $NPATRL$ logic uses *past* temporal logic operators and event atoms.

The third class of logics to compare TLR^* with are logics supporting both actions and state predicates. First of all, two of the logics in Mateescu's survey, namely PDL and the modal μ-calculus are already in this category, since they support both state predicates and actions. In some sense, the closest analogue to PDL is not TLR^* itself but, instead, the strategy language and the strategy formulas presented in [31], which, although not having the same primitives or expressive power as PDL, can be thought of as some kind of dynamic logic,

---

[4] See however the related [4], which has an action-labeled diamond.

which shares with *PDL* the general idea of using regular expressions and of being interpreted on finite computations. Regarding the modal μ-calculus, it is well-known that it can encode both *CTL* and *CTL** (see, e.g., [11], [2]). Therefore, more than a direct comparison with *TLR**, the best comparison rests on the observation that, just as *TLR** generalizes *CTL**, there is a similar generalization of μL to a *modal μ-calculus of rewriting* μLR, whose syntax I define below. As *TLR**, μLR is likewise parametric on the constructors Ω, labels L, and predicate signature Π of a rewrite theory $\mathcal{R}$. Therefore, we have a parametric family of formulas, where assuming as before that $\Pi \subseteq \Omega$, and the set of atomic propositions $Prop(\Pi) = T_{\Omega_{Prop}}$, and using as before the variables $\delta : SP(\Omega, L)$, and $p : Prop(\Pi)$, and assuming a disjoint set $SVar$ of variables $X, Y, Z, \ldots$, that in the semantics will range over subsets of the set of states, plus $\varphi, \varphi' : \mu LR(SP(\Omega, L), \Pi)$, we have the following BNF-like syntax definition:

$$\mu LR(SP(\Omega, L), \Pi) : \quad p \mid X \mid \varphi \wedge \varphi' \mid [\delta]\varphi \mid \neg\varphi \mid \nu X.\varphi$$

where every free occurrence of X in $\nu X.\varphi$ must occur positively, that is, within the scope of an even number of negations. The semantics of μLR can then be defined along standard lines (see [31]). The ideas in this paper could have been developed for μLR instead than for *TLR**. My preference for *TLR** is motivated by pragmatic reasons of the kind already pointed out in Footnote 1.

Yet another line of work involves several extensions of either *A-CTL** or *A-CTL* supporting both actions and state predicates, including, e.g., [36,18,1,14]. Although almost all these logics use unstructured labels, one noteworthy exception is the logic presented in [14], which shares some common features with *TLR**, since action expressions involving logical variables are used to describe interactions between services in a service-oriented architecture. Three other approaches proposing mixed logics with both state-predicates and actions are: (i) the extension of the *SE-LTL* in [8,9] to a universally path quantified logic involving ω-regular expressions [7]; (ii) the *ESTL* logic of events and states for Petri nets of [21]; and (iii) the Kripke modal transition systems of [20], and their use in the verification of safety and liveness properties in the context of the modal μ-calculus.

The work most closely related to the one presented here, and indeed one of its sources of inspiration, is that on *VLRL*, a Verification Logic for Rewriting Logic developed in [16,26]. Two common similarities are: (i) state-based predicates can be expressed in both logics; and (ii) both *VLRL* and *TLR** support *spatial action patterns*. However, *VLRL* is a Hennessy-Milner-like modal logic without temporal modalities. This is compensated for by providing an interface that extends *VLRL* with some temporal logic operators. In this regard, the *TLR** solution is simpler (no such interface is needed) and, on the temporal logic dimension, more expressive. A mapping from *VLRL* action formulae into *LTL*, similar in spirit to the reduction from *TLR** to *CTL** given in Section 4.3, was presented in [35].

Two other logics that combine actions and state-based formulas are the UNITY logic of Chandy and Misra [10], and Misra's logic for Seuss [33]. In both cases, an action corresponds to a command c in a UNITY or Seuss program, and can be mentioned in a Hoare-like triple $\{p\}\, c\, \{q\}$. Such triples are elegantly deployed in both

[10] and [33] to reason axiomatically about temporal logic properties. One important difference with TLR^* is that, since the meaning of temporal logic formulas is defined by universal or existential quantification over the actions of a program, actions as such do not appear in temporal logic formulas, which remain state-based.

Although Lamport's Temporal Logic of Actions (TLA) [23] and TLR^* have a similar motivation, both methodologically and technically TLA and TLR^* are very different. Methodologically, the division of labor between system and property specification logics supported by the $RewritingLogic/TLR^*$ tandem is flatly rejected in TLA. Lamport proposes TLA in [23] as:

> "a simpler approach in which both the algorithm and the property are specified by formulas in a single logic."

This has important technical consequences. The key one is that TLA builds into the logic a shared variable, imperative programming model. TLA formulas talk about states by making mathematical statements about program variables. To capture state change, the usual technique of using primed versions of the program variables is adopted. Then an *action* in Lamport's sense is a predicate involving both primed and unprimed variables, and therefore describing a set of pairs of states. Although TLA is a logic of actions, its semantics, however, remains state-based, in the precise sense of being defined over sequences of states. It is indeed what I called in Section 2 a "possibilistic" semantics, so that an action $\mathcal{A}$ in Lamport's sense holds true of a sequence of states π, which in my notation I would write $\pi \models_\Diamond \mathcal{A}$, if and only if the pair of states $(\pi(0), \pi(1))$ belongs to the binary relation defined by $\mathcal{A}$. For the reasons explained in Section 2, as any other "possibilistic semantics," TLA cannot distinguish between two different actions that happen to relate the same pair of states: one would need to add some kind of history variables to make such distinctions. For somebody accepting the equation "*concurrent system specification = imperative concurrent algorithm*," TLA is an attractive framework. But of course this is just a specific approach to concurrent system specification with its own advantages and drawbacks.

6   Conclusions

I have explained the "system/property mismatch problem," which plagues both purely action-based and purely state-based formal specification tandems. I have then proposed the $RewritingLogic/TLR^*$ tandem as a way to generalize and unify a wide range of action-based and state-based tandems and avoid many of these mismatch problems. Although this is neither the first nor the only proposal for combining state-based and action-based formalisms, it has a number of unique advantages. At the system specification level, rewrite theories allow much higher level descriptions than either Kripke structures or labeled transition systems, support true concurrency, and allow easy combination of algorithmic and deductive reasoning. At the property specification level, the generalization from CTL^* to TLR^* is so straightforward as not to cause any additional ease-of-use problems. Indeed, the opposite is the case, since action-based properties, requiring complex indirect expression in CTL^* as well as cooking of system specifications, now become trivial to specify with expressive action-based patterns and do not require any such cooking. At the algorithmic level, because of the existence of a faithful mapping of tandems $RewritingLogic/TLR^* \longrightarrow Kripke/CTL^*$, TLR^* properties can be model checked using CTL^* model checkers at reasonable cost for the formulas typically encountered in practice.

Acknowledgments: The ideas presented here in more mature form were first developed during two successive weekends in Santa Fe, New Mexico, and Santa Barbara, California, in February 2007. I most cordially thank my friends Jim and Madelene Jackson in Santa Fe, and Emmett and Jadja Mc Donough in Santa Barbara, for their warm hospitality, which made those weekends delightful both personally and scientifically. As already pointed out, my earlier work with José Fiadeiro, Tom Maibaum, Narciso Martí-Oliet and Isabel Pita on $VLRL$ [16,26] has been a source of inspiration. I thank Stefania Gnesi for very helpful suggestions to improve the paper and to discuss related work, Joe Hendrix, Camilo Rocha and Ralf Sasse for helping in the preparation of [31], Jayadev Misra, Alan Emerson, and Mahesh Viswanathan for very helpful discussions on these ideas, and Santiago Escobar, Michael Katelman, Narciso Martí-Oliet, Peter Ölveczky, Grigore Roşu, Alberto Verdejo, and the anonymous referees for their very helpful suggestions to improve the paper. Finally, I thank the audiences at UIUC, the Universidad Complutense de Madrid, SRI International, the University of Texas at Austin, and the University of Leicester for their very helpful comments and suggestions during seminar talks in which I presented these ideas. This research has been supported in part by ONR Grant N00014-02-1-0715 and NSF Grants CNS-05-24516 and CNS-07-16638.

References

1. Beek, M., Fantechi, A., Gnesi, S., Mazzanti, F.: An action/state-based model-checking approach for the analysis of communication protocols for Service-Oriented Applications. In: Proc. FMICS. LNCS, Springer, Heidelberg (to appear, 2008)
2. Bradfield, J., Stirling, C.: Modal Mu-Calculi. In: Handbook of Modal Logic, vol. 3, Elsevier, Amsterdam (2006)
3. Bruni, R., Meseguer, J.: Semantic foundations for generalized rewrite theories. Theor. Comput. Sci. 360(1-3), 386–414 (2006)
4. Caires, L.: Behavioral and spatial observations in a logic for the pi-calculus. In: Walukiewicz, I. (ed.) FOSSACS 2004. LNCS, vol. 2987, pp. 72–87. Springer, Heidelberg (2004)
5. Caires, L., Cardelli, L.: A spatial logic for concurrency (part I). Inf. Comput. 186(2), 194–235 (2003)
6. Caires, L., Cardelli, L.: A spatial logic for concurrency - II. Theor. Comput. Sci. 322(3), 517–565 (2004)
7. Chaki, S., Clarke, E., Grumberg, O., Ouaknine, J., Sharygina, N., Touili, T., Veith, H.: State/event software verification for branching-time specifications. In: Romijn, J.M.T., Smith, G.P., van de Pol, J. (eds.) IFM 2005. LNCS, vol. 3771, pp. 53–69. Springer, Heidelberg (2005)

8. Chaki, S., Clarke, E.M., Ouaknine, J., Sharygina, N., Sinha, N.: State/event-based software model checking. In: Boiten, E.A., Derrick, J., Smith, G.P. (eds.) IFM 2004. LNCS, vol. 2999, pp. 128–147. Springer, Heidelberg (2004)
9. Chaki, S., Clarke, E., Ouaknine, J., Sharygina, N., Sinha, N.: Concurrent software verification with states, events, and deadlocks. Formal Aspects of Computing 17, 461–483 (2005)
10. Chandy, K.M., Misra, J.: Parallel Program Design: A Foundation. Addison-Wesley, Reading (1988)
11. Clarke, E.M., Grumberg, O., Peled, D.A.: Model Checking. MIT Press, Cambridge (2001)
12. Clavel, M., Durán, F., Eker, S., Meseguer, J., Lincoln, P., Martí-Oliet, N., Talcott, C.: All About Maude - A High-Performance Logical Framework. LNCS, vol. 4350. Springer, Heidelberg (2007)
13. Dershowitz, N., Jouannaud, J.-P.: Rewrite systems. In: van Leeuwen, J. (ed.) Handbook of Theoretical Computer Science, vol. B, pp. 243–320. North-Holland, Amsterdam (1990)
14. Fantechi, A., Gnesi, S., Lapadula, A., Mazzanti, F., Pugliese, R., Tiezzi, F.: A model checking approach for verifying COWS specifications. In: FASE 2008. LNCS, vol. 4961, pp. 230–245. Springer, Heidelberg (2008)
15. Fantechi, A., Gnesi, S., Ristori, G.: From ACTL to Mu-Calculus. In: Proc. ERCIM Workshop on Theory and Practice of Verificiation (Pisa, Italy, December 1992), pp. 3–10. IEI-CNR (1992)
16. Fiadeiro, J., Martí-Oliet, N., Maibaum, T., Meseguer, J., Pita, I.: Towards a verification logic for rewriting logic. In: Bert, D., Choppy, C., Mosses, P.D. (eds.) WADT 1999. LNCS, vol. 1827, pp. 438–458. Springer, Heidelberg (2000)
17. Fischer, M.J., Ladner, R.E.: Propositional dynamic logic of regular programs. J. Comput. Syst. Sci. 18(2), 194–211 (1979)
18. Gnesi, S., Mazzanti, F.: A Model Checking Verification Environment for UML Statecharts. In: Proceedings XLIII AICA Annual Conference, University of Udine - AICA 2005, October 2-5 (2005)
19. Hennessy, M., Milner, R.: Algebraic laws for nondeterminism and concurrency. Journal of the Association for Computing Machinery 32(1), 137–172 (1985)
20. Huth, M., Jagadeesan, R., Schmidt, D.: Modal transition systems: A foundation for three-valued program analysis. In: Sands, D. (ed.) ESOP 2001. LNCS, vol. 2028, pp. 155–169. Springer, Heidelberg (2001)
21. Kindler, E., Vesper, T.: ESTL: A temporal logic for events and states. In: Desel, J., Silva, M. (eds.) ICATPN 1998. LNCS, vol. 1420, pp. 365–384. Springer, Heidelberg (1998)
22. Kozen, D.: Results on the propositional mu-calculus. Theor. Comput. Sci. 27, 333–354 (1983)
23. Lamport, L.: A temporal logic of actions. ACM Trans. on Prog. Lang. and Systems 16(3), 872–923 (1994)
24. Manna, Z., Pnueli, A.: The Temporal Logic of Reactive and Concurrent Systems – Specification. Springer, Heidelberg (1992)
25. Martí-Oliet, N., Meseguer, J.: Rewriting logic: roadmap and bibliography. Theoretical Computer Science 285, 121–154 (2002)
26. Martí-Oliet, N., Pita, I., Fiadeiro, J.L., Meseguer, J., Maibaum, T.S.E.: A verification logic for rewriting logic. J. Log. Comput. 15(3), 317–352 (2005)
27. Mateescu, R.: Logiques temporelles basées sur actions pour la vérification des systèmes asynchrones. Technique et Science Informatiques 22(4), 461–495 (2003); also, INRIA Report 5032 (December 2003)

28. Meadows, C., Syverson, P.F., Cervesato, I.: Formal specification and analysis of the group domain of interpretation protocol using NPATRL and the NRL protocol analyzer. Journal of Computer Security 12(6), 893–931 (2004)
29. Meseguer, J.: Localized fairness: A rewriting semantics. In: Giesl, J. (ed.) RTA 2005. LNCS, vol. 3467, pp. 250–263. Springer, Heidelberg (2005)
30. Meseguer, J.: Conditional rewriting logic as a unified model of concurrency. Theoretical Computer Science 96(1), 73–155 (1992)
31. Meseguer, J.: The temporal logic of rewriting. Technical Report UIUCDCS-R-2007-2815, CS Dept., University of Illinois at Urbana-Champaign (February 2007)
32. Meseguer, J., Palomino, M., Martí-Oliet, N.: Equational abstractions. In: Baader, F. (ed.) CADE 2003. LNCS (LNAI), vol. 2741, Springer, Heidelberg (2003)
33. Misra, J.: A Discipline of Multiprogramming. Springer, Heidelberg (2001)
34. Nicola, R.D., Vaandrager, F.W.: Action versus state based logics for transition systems. In: Guessarian, I. (ed.) LITP 1990. LNCS, vol. 469, pp. 407–419. Springer, Heidelberg (1990)
35. Palomino, M., Pita, I.: Proving VLRL action properties with the Maude model checker. Electr. Notes Theor. Comput. Sci. 117, 113–133 (2005)
36. Pecheur, C., Raimondi, F.: Symbolic model checking of logics with actions. In: Edelkamp, S., Lomuscio, A. (eds.) MoChArt IV. LNCS (LNAI), vol. 4428, pp. 113–128. Springer, Heidelberg (2007)
37. Pratt, V.R.: Semantical considerations on Floyd-Hoare logic. In: FOCS 1976, pp. 109–121. IEEE Computer Society Press, Los Alamitos (1976)
38. Thati, P., Sen, K., Martí-Oliet, N.: An executable specification of asynchronous Pi-Calculus semantics and may testing in Maude 2.0. In: Gadducci, F., Montanari, U. (eds.) Proc. 4th. Intl. Workshop on Rewriting Logic and its Applications. ENTCS, Elsevier, Amsterdam (2002)
39. Viry, P.: Equational rules for rewriting logic. Theoretical Computer Science 285, 487–517 (2002)

A Heterogeneous Approach to UML Semantics[*]

María Victoria Cengarle[1], Alexander Knapp[2], Andrzej Tarlecki[3], and Martin Wirsing[2]

[1] Technische Universität München
cengarle@in.tum.de
[2] Ludwig-Maximilians-Universität München
{knapp,wirsing}@pst.ifi.lmu.de
[3] Uniwersytet Warszawski
tarlecki@mimuw.edu.pl

Abstract. UML models consist of several diagrams of different types describing different views of a software system ranging from specifications of the static system structure to descriptions of system snapshots and dynamic behaviour. In this paper a heterogeneous approach to the semantics of UML is proposed where each diagram type can be described in its "natural" semantics, and the relations between diagram types are expressed by appropriate translations. More formally, the UML family of diagram types is represented as a "heterogeneous institution environment": each diagram type is described as an appropriate institution where typically the data structures occurring in a diagram are represented by signature elements whereas the relationships between data and the dynamic behaviour of objects are captured by sentences; in several cases, the diagrams are themselves the sentences. The relationship between two diagram types is described by a so-called institution comorphism, and in case no institution comorphism exists, by a co-span of such comorphisms. Consistency conditions between different diagrams are derived from the comorphism translations. This heterogeneous semantic approach to UML is illustrated by several example diagram types including class diagrams, OCL, and interaction diagrams.

1   Introduction

Almost exactly 40 years ago, computer programs had become so large and complex that many software development projects failed and maintaining software programs was almost unmanageable. Neither pragmatic methods for software construction nor the scientific foundations of programming were established at the time; the need for systematic software development techniques was so urgent and evident that in 1968 the first software engineering conference was organized in Garmisch [39].

Today, software systems are larger than ever, and software is the central innovating factor of many high-tech products, services, and systems, e.g. in consumer electronics, automotive applications, telecommunications, and business. Our daily life and work depend more and more on such software-intensive systems. Driven by techniques such as

[*] This work has been partially sponsored by the project SENSORIA IST-2005-016004, and the DFG projects InfoZert, MAEWA and rUML.

P. Degano et al. (Eds.): Montanari Festschrift, LNCS 5065, pp. 383–402, 2008.

object-orientation, service-orientation, or model-transformation, practical software engineering methods have considerably evolved and many companies follow well-defined software development processes for constructing larger and larger software systems. A substantial body of theoretical foundations of programming is available and formal modelling and analysis techniques like abstraction and refinement techniques, model checking or theorem proving have undergone a steep development during the last years. However, at the same time we experience that many software systems are error-prone, unstable, have security holes, and do not meet the required quality standards. There is still a large gap between industrial practice and formal approaches: pragmatic modelling languages and techniques lack formal foundations, inhibiting the development of powerful analysis and development tools, while formal approaches are often too difficult to use, do not scale easily to complex software-intensive systems, and are not well-integrated with pragmatic methods. Aspects such as distribution, mobility, heterogeneity, quality of service, security, trust, and dynamically changing infrastructures and environments are not well supported by actual engineering methods.

Bridging this gap and advancing software engineering theory and methods is one of the main aims of our common research with Ugo Montanari during the last several years. In the project AGILE [1] we developed an architectural approach to software mobility which was based on a uniform mathematical framework to support sound methodological principles, formal analysis, and refinement. The aim of SENSORIA [48,50] is to develop a novel comprehensive approach to the engineering of service-oriented software systems where foundational theories, techniques and methods are fully integrated in a pragmatic software development process. With SENSORIA techniques software engineers can model a system in the usual way by using standard high-level visual modelling languages such as UML; but they get additional help for reasoning about functional and non-functional properties of the system by mathematical models which run in the background hidden from the developer.

As a prerequisite for the SENSORIA approach we investigate in this paper the semantics of UML. Models expressed in this language consist of several diagrams of different types describing different views of a software system ranging from specifications of the static system structure to descriptions of system snapshots and dynamic behaviour. For example, UML 1.x offers nine different types of diagrams for describing different static and dynamic aspects of a system. Depending on the modelling purpose one may employ e.g. class diagrams, component diagrams, state diagrams, sequence diagrams, activity diagrams, or instance diagrams. UML 2.0 adds several new diagram types and enhances the expressiveness and semantics of sequence diagrams and activity diagrams considerably.

We propose here a new "heterogeneous approach" to the semantics of UML which concentrates on the comparison and integration of different modelling formalisms, and in which

- each diagram type can be described in its "natural" semantics,
- relations between diagram types are expressed by appropriate translations, and
- consistency conditions can be derived between diagrams of different type.

More formally, we present the UML family of diagram types as a "heterogeneous institution environment". An institution is given by a type of signatures, a type of sentences, a notion of model, and a notion of validity of a sentence in a model. Each diagram type is described as an appropriate institution, and each diagram instance is a specification in the institution. Typically, the data structures occurring in a diagram are named by signature elements whereas the dynamic behaviour of objects and relationships between data are described by sentences; in several cases, the diagrams are themselves the sentences. The semantic concepts involved are captured by the models considered, with the validation between sentences and models determining the semantics of particular diagrams. The relationship between two diagram types is described by so-called institution comorphisms, and in case no institution comorphism exists, by a co-span of such comorphisms. Consistency conditions between different diagrams are derived from the comorphism translations or the co-span construction. We illustrate the heterogeneous semantic approach to UML by several diagram types including class diagrams, OCL, and interaction diagrams; we demonstrate the use of heterogeneous UML institutions by means of the e-learning case study of the SENSORIA project [50].

The remainder of this paper is structured as follows: In Sect. 2 we review the related work to the semantics on modelling languages and, in particular, on UML semantics. Sect. 3 introduces the theory of heterogeneous institution environments. In Sect. 4 we present sketches of the institutions for class diagrams, OCL, and interaction diagrams and in Sect. 5 we sketch how these different languages can be linked by institution comorphisms and co-spans. The implications for consistency conditions between diagrams of these different types are discussed in Sect. 6. Finally, in Sect. 7 we conclude with a short discussion of the results of the paper and an outlook on further research topics.

2   Related Work

Giving an integrated semantics to UML is a difficult task due to the complexity and variety of the different diagram types. The classical approaches to programming language semantics are adequate only for a restricted subset of the specification and modelling tasks of software development. For example, denotational semantics is an elegant framework for compositionally modelling functional behaviour but it is not so appropriate for the dynamic behaviour of concurrent processes; in contrast, structural operational semantics (SOS) is well-suited for the latter but treats data structures only in a syntactic way. Algebraic specifications are appropriate for modelling complex data types and associated operations, but they are not easily usable for specifying reactive systems, in spite of the elegant work started with [2] on axiomatic process algebra.

In order to model static functional aspects as well as dynamic concurrent behaviour, several researchers investigated extensions or combinations of these methods. Broy proposes a denotational "system model" based on stream-processing functions in combination with abstract data types (see e.g. [8,9]) and currently uses this system model for developing a complete UML semantics [5,6,7], where diagrams are taken as predicates that a system model instance has to satisfy. Other approaches propose the combination of CSP and Z [23,47] or a combination of algebraic specifications and labelled

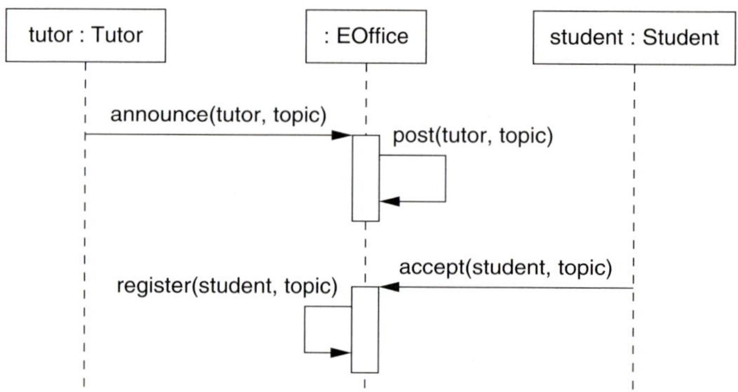

Fig. 4. E-office example: sequence diagram for accepting a topic

like for instance seq, par, loop, etc. (for details see [16]). Well-formed formulas combine atomic ones using conjunction, negation, universal quantification, and equality of variables. As usual, sentences are closed formulas.

For instance the interaction diagram in Fig. 4 declares classes Tutor, Topic, EOffice, and Student, and messages announce(Tutor, Topic), post(Tutor, Topic), accept(Student, Topic), and register(Student, Topic). Given variables $tutor$: Tutor, $eoffice$: EOffice, and $student$: Student, the only sentence represented by the diagram is the atomic formula given by the following pomset:

$$[\{X_1, X_2, X_3, X_4, X_5, X_6, X_7, X_8\},$$
$$\{X_1 < X_2 < X_3 < X_4 < X_6 < X_7 < X_8, \ X_5 < X_6,\},$$
$$\{X_1 \mapsto \mathsf{snd}(tutor, eoffice, \mathsf{announce}(tutor, topic)),$$
$$X_2 \mapsto \mathsf{rcv}(tutor, eoffice, \mathsf{announce}(tutor, topic)),$$
$$X_3 \mapsto \mathsf{snd}(eoffice, eoffice, \mathsf{post}(tutor, topic)),$$
$$X_4 \mapsto \mathsf{rcv}(eoffice, eoffice, \mathsf{post}(tutor, topic)),$$
$$X_5 \mapsto \mathsf{snd}(student, eoffice, \mathsf{accept}(student, topic)),$$
$$X_6 \mapsto \mathsf{rcv}(student, eoffice, \mathsf{accept}(student, topic)),$$
$$X_7 \mapsto \mathsf{snd}(eoffice, eoffice, \mathsf{register}(student, topic)),$$
$$X_8 \mapsto \mathsf{rcv}(eoffice, eoffice, \mathsf{register}(student, topic))\}]$$

In the institution of interactions, an interpretation maps class names to sets of object instances and messages to message instances. Object instances may interchange message instances; we define send events as triples of a sender instance, a receiver instance, and a message instance; receive events are defined similarly. Traces are sequences of send/receive events. Models for interactions are defined to be pairs (P, N) of sets of traces.

In order to grasp the notion of models for interactions and of satisfaction relation of an interaction by a model, recall that interactions depict possible message interchange scenarios, and perhaps forbid (other) scenarios by use of the interaction-building operator neg. In general, hence, interactions do not completely specify a software system. Given an interpretation, given a valuation of variables by object instances (of the correct class), an interaction induces *positive* and *negative* traces of send/receive events. These induced sets, of all positive traces and of all negative traces, need not be disjoint, and their union need not contain every possible trace [16]. Given a valuation, a model (P, N) satisfies an interaction if the negative traces of the interaction diagram under that valuation are contained in N, and the genuine positive ones (i.e., positive and non-negative) are included in P. Equality, negation, conjunction and universal quantification are interpreted as usual.

On the one hand, for example the trace

snd(tutor, eoffice, announce(tutor, topic)) · rcv(tutor, eoffice, announce(tutor, topic)) ·
snd(eoffice, eoffice, post(tutor, topic)) · rcv(eoffice, eoffice, post(tutor, topic)) ·
snd(student, eoffice, accept(student, topic)) · rcv(student, eoffice, accept(student, topic)) ·
snd(eoffice, eoffice, register(student, topic)) · rcv(eoffice, eoffice, register(student, topic))

positively satisfies the interaction diagram of Fig. 4, where tutor, topic, eoffice and student are instances of the corresponding classes, and the message instances coincide with the messages declared in the signature of above. On the other hand, no trace negatively satisfies the interaction, i.e., the set of negative traces of the interaction is empty.

An equivalent abstract representation of the diagram of Fig. 4 is $\mathrm{seq}(S_1, S_2)$, where S_1 is the labelled pomset

$$
\begin{aligned}
[\{X_1, X_2, X_3, X_4\}, \\
\{X_1 < X_2 < X_3 < X_4\}, \\
\{X_1 \mapsto \mathrm{snd}(\textit{tutor}, \textit{eoffice}, \mathrm{announce}(\textit{tutor}, \textit{topic})), \\
X_2 \mapsto \mathrm{rcv}(\textit{tutor}, \textit{eoffice}, \mathrm{announce}(\textit{tutor}, \textit{topic})), \\
X_3 \mapsto \mathrm{snd}(\textit{eoffice}, \textit{eoffice}, \mathrm{post}(\textit{tutor}, \textit{topic})), \\
X_4 \mapsto \mathrm{rcv}(\textit{eoffice}, \textit{eoffice}, \mathrm{post}(\textit{tutor}, \textit{topic})) \quad \}]
\end{aligned}
$$

and S_2 the labelled pomset

$$
\begin{aligned}
[\{Y_1, Y_2, Y_3, Y_4\}, \\
\{Y_1 < Y_2 < Y_3 < Y_4\}, \\
\{Y_1 \mapsto \mathrm{snd}(\textit{student}, \textit{eoffice}, \mathrm{accept}(\textit{student}, \textit{topic})), \\
Y_2 \mapsto \mathrm{rcv}(\textit{student}, \textit{eoffice}, \mathrm{accept}(\textit{student}, \textit{topic})), \\
Y_3 \mapsto \mathrm{snd}(\textit{eoffice}, \textit{eoffice}, \mathrm{register}(\textit{student}, \textit{topic})), \\
Y_4 \mapsto \mathrm{rcv}(\textit{eoffice}, \textit{eoffice}, \mathrm{register}(\textit{student}, \textit{topic})) \quad \}]
\end{aligned}
$$

and seq combines its arguments sequentially in a "instance-wise" manner: For each instance, all events for this instance in the first argument must happen before all events for this instance in the second argument. Hence, we could moreover have split the pomset S_2 into S_2^1 and S_2^2 with S_2^1 representing the sending and reception of the message accept(*student, topic*), and S_2^2 the sending and reception of the message register(*student, topic*), and combine S_2^1 and S_2^2 using the seq interaction-building operator. In other words, the abstract representation of a diagram needs not be unique. These different representations are nevertheless equivalent, i.e., have the same models.

Signature morphisms in the interactions institution work similarly to the signature morphisms in the static structure institution, renaming classes and messages. The translation of sentences along a signature morphism is canonical, and reducts forget all traces mentioning events only expressible over the target signature.

4.3   OCL Institution

OCL signatures declare class names, query names (that correspond to attributes and query methods), and method names. Class names are equipped with a partial order relation representing the inheritance hierarchy. Default (or built-in) types extend these declarations. So for instance the set of class names is closed under application of the type constructors Set and Sequence (which is equivalent to list construction). This extended type system is used to define a (unique) type for each query name and each method name. The inheritance hierarchy together with a built-in subtype relation induce the OCL-subtype relation, that comprises class names as well as query and method names. The sentences defined by an OCL signature are invariants and pre/postconditions, as in the example shown in Fig. 5. The corresponding signature declares, possibly among others, class names EOffice, Student, and Topic, query names topic and student for EOffice, student for Topic, topic for Student, and a method name register for EOffice. OCL presentations consist of an OCL signature and a set of OCL sentences over that signature.

```
context EOffice::accept(student : Student, topic : Topic)
pre: topic.student->empty() and student.topic->empty()
post: topic.student = student and student.topic = topic
```

Fig. 5. E-office example: OCL specification of accept

The OCL interpretations map class names to sets of created objects, and provide a mechanism to retrieve functions that implement query names and method names; see [15]. The former functions do not modify the state of objects, whereas the latter may modify the state. Models of OCL theories are state transition systems, whose states are sets of created objects and whose transitions are labelled by a method invocation and the corresponding return value, so that the target state of a transition is the result of applying the method on the origin state. Moreover, every state of these models observes the invariants of the theory, and any two adjacent states satisfy the pre-/postconditions required for the method that labels the transition connecting these two states.

Signature morphisms, translations and reducts, again, are built similarly to the corresponding notions in the static structure institution.

5   Linking UML Institutions

We study how particular views of a given software system, as represented in different languages designed for their specification, can be linked with each other. In particular, we investigate the natural links between the UML institutions as sketched above.

Following the presentation in Sect. 3, there are two possibilities to link institutions: direct translation from one institution to another via a comorphism, and definition of a new mediating institution to which both institutions are embedded (see definition of *sink* in Sect. 3).

```
context Tutor inv: eoffice->count() = 1
context Tutor inv: eoffice->tutors->includes(self)
context EOffice
    inv: tutors->forall(x | x.eoffice->includes(self))
```

Fig. 6. EOffice example: multiplicity axiom for tuteof translated into OCL

The institution of class diagrams can easily be embedded into the OCL institution. Indeed, class names are mapped to class names, typed attributes to queries, typed method to method names, and role names to set-valued queries (this means, the same class name possibly gets more queries when translated). Sentences of a theory presentation given by a class diagram are translated into OCL invariants. So, for instance, the first *association*-sentence associated with Fig. 2 is translated into the OCL invariants of Fig. 6. In general, for a binary association $a \subseteq r_1 : c_1 \times r_2 : c_2$ with declared multiplicity $association(a, r_1 : c_1 : m_1, r_2 : c_2 : m_2)$, the translation of each of the multiplicities consists of a constraint for navigability and up to two constraints for cardinality: if m_1 is of the form $n_1..n_2$, then

```
context c₁ inv: r₂->forall(x | x.r₁->contains(self))
context c₂ inv: r₁->count() >= n₁ and r₁->count() <= n₂ ;
```

if m_1 is of the form $n..\star$, then

```
context c₁ inv: r₂->forall(x | x.r₁->contains(self))
context c₂ inv: r₁->count() >= n₁ ;
```

and similarly for m_2.

This translation gets somewhat more involved for other than binary associations and the various kinds of multiplicities. Nevertheless, the translation is rather straightforward: the same as above, it takes care of back navigability and cardinalities within the bounds imposed by multiplicities.

Given an OCL model for a signature, we extract from it a model of an embedded class diagram signature by taking the set of states of the OCL model.

The institution of class diagrams cannot so easily be embedded in the institution of interactions. Again, class names could be mapped to class names, and typed methods to sets of messages. But it is not trivial how to embed typed attributes and association names, nor how to translate declarations of multiplicities for associations. Since class diagrams can be embedded into OCL, this matter, however, is not so crucial if the OCL institution and the interactions institution can be linked.

For this an auxiliary institution OCL+I can be devised that contains all the elements of the OCL institution as well as all the elements of the institution of interactions. Signatures declare class names (as OCL signatures and interaction signatures do), query names and method names (as in the OCL institution); query and method names, together with variables typed over declared class names, induce messages (which correspond to messages in the institution of interactions). Sentences are either OCL sentences over class names, query names and method names, or interaction sentences over class names and induced messages. A model is a set of so-called runs; cf. [14,11]. Runs are sequences of pairs, each pair consisting of a set of created objects and a set of events. An

A set of sentences $\Phi_1 \subseteq \mathbf{Sen}_{\mathcal{I}_1}(\Sigma_1)$ is *consistent* with a set of sentences $\Phi_2 \subseteq \mathbf{Sen}_{\mathcal{I}_2}(\Sigma_2)$ with respect to a signature $\Sigma \in |\mathbf{Sign}_{\mathcal{I}_2}|$ and signature morphisms $\sigma_1 : \rho^{Sign}(\Sigma_1) \to \Sigma$ and $\sigma_2 : \Sigma_2 \to \Sigma$ if $\rho^{Sen}_{\Sigma_1}(\Phi_1)$ and Φ_2 are consistent with respect to Σ, σ_1 and σ_2, i.e., if there is a model $M \in \mathbf{Mod}_{\mathcal{I}_2}(\Sigma)$ such that $\sigma_1(\rho^{Sen}_{\Sigma_1}(\Phi_1)) \cup \sigma_2(\Phi_2)$ is satisfied by M, or equivalently, $\rho^{Mod}_{\Sigma_1}(M|_{\sigma_1})$ satisfies Φ_1 and $M|_{\sigma_2}$ satisfies Φ_2.

Generalizing this further, if institutions $\mathcal{I}_1$ and $\mathcal{I}_2$ are linked by a co-span $\mathcal{I}_1 \xrightarrow{\rho_1} \mathcal{I} \xleftarrow{\rho_2} \mathcal{I}_2$, two sets of sentences $\Phi_1 \subseteq \mathbf{Sen}_{\mathcal{I}_1}(\Sigma_1)$ and $\Phi_2 \subseteq \mathbf{Sen}_{\mathcal{I}_2}(\Sigma_2)$ are *consistent* with respect to a signature $\Sigma \in |\mathbf{Sign}_{\mathcal{I}}|$ and signature morphisms $\sigma_1 : \rho_1{}^{Sign}(\Sigma_1) \to \Sigma$ and $\sigma_2 : \rho_2{}^{Sign}(\Sigma_2) \to \Sigma$ if $\rho_1{}^{Sen}_{\Sigma_1}(\Phi_1)$ and $\rho_2{}^{Sen}_{\Sigma_2}(\Phi_2)$ are consistent with respect to Σ, σ_1 and σ_2. This is equivalent to the existence of a model $M \in \mathbf{Mod}_{\mathcal{I}}(\Sigma)$ such that $\rho_1{}^{Mod}_{\Sigma_1}(M|_{\sigma_1}) \in \mathbf{Mod}_{\mathcal{I}_1}(\Sigma_1)$ satisfies Φ_1 and $\rho_2{}^{Mod}_{\Sigma_2}(M|_{\sigma_2}) \in \mathbf{Mod}_{\mathcal{I}_2}(\Sigma_2)$ satisfies Φ_2.

In the heterogeneous institution environment for UML, the institutions for static structures and for OCL are linked by a comorphism. Given a class diagram as a specification in the institution of static structures and a set of invariants and operation specifications, we are interested in their consistency with respect to shared class names, attributes, and translated association ends. For example, when linking the translation of the class diagram in Fig. 2 with the OCL specification in Fig. 5, the conjuncts of the postcondition of `accept` imply each other due to the *association*-axiom for topstu.

As long as both specifications do not require any particular objects of the classes in the class diagram to exist, their consistency can always be witnessed by a model, i.e., a state transition system, whose states are empty sets of object instances. However, for our e-office example, if the OCL specification contains, e.g.,

```
context Tutor
inv: Tutor.allInstances()->count() >= 2
```

a common model must show at least two instances of Tutor, as this symbol is identified with `Tutor`. Thus, the invariants induced in the comorphism translation of class diagrams to OCL, which, in particular, require then at least one e-office to exist, must not contradict the invariants in the extended OCL specification.

The institution OCL+I integrates the semantics of OCL and of interactions via a co-span of institution comorphisms. It does so by presupposing a particular way of linking operation specifications from OCL with sequencing obligations from interactions. Here, we are interested in their consistency with respect to shared class names on the one hand, and identifying query and method calls with messages on the other. Consider, for example, the operation specification for accept in Fig. 5 and the interaction in Fig. 4. Then, in order to be able to obtain a common model in OCL+I, it is required that the precondition of accept does not contradict the postcondition of announce, as the postcondition time of announce is the same as the precondition time of accept; indeed, this is the case in our specification when accept is called only by a student lacking a topic.

The particular integrating institution OCL+I with the co-span of institution co-morphisms presented above is by no means the only possibility to link the OCL and the interactions institution. We could also choose a looser definition of satisfaction of an interaction by a run: a trace is obtained from a run by not only eliminating the set of objects and eliminating the marks but also skipping all those events that happen on objects that are not mentioned in the interaction. In this situation, the postcondition of an operation and the precondition of another operation, albeit the termination of the first and the start of the second happen immediately after each other in an interaction, do not have to show any correlation, as always some event, external to the interaction, could interfere and restore the precondition of the second operation.

7   Conclusions

We have presented a general framework for constructing the semantics of different UML diagram types in a flexible way. The framework is used as a mathematical basis of UML in the SENSORIA development approach for service-oriented systems. It relies on the mathematical theory of institutions and offers a new approach to the semantics of heterogeneous system specifications in a "heterogeneous institution environment". It allows one to describe each diagram type in its "natural" semantics. Different diagram types are integrated via appropriate translations (into each other or into intermediate institutions), and in this way their semantic consistency can be analysed. Another advantage of our approach is that other system models can be easily integrated. For instance, rewriting logic and temporal logic are themselves institutions [12,13].

Institutions provide an elegant and robust framework for programming in the large and in particular for compositionality. Indeed, the satisfaction condition ensures that properties fulfilled by parts of a development do not get lost when putting those parts together. This is also true when, even at different places, those parts are made more specific using a refinement relation based on model-class inclusion. The trade-off is the loss of expressive power regarding some reflective properties like closed-world assumption (cf. the OCL constructs `Type` and `Type.allInstances`). Depending on the application, however, this is a price we are willing to pay, since the compositionality gained applies not only to the development of the software system as such but also to the verification of the whole system, which may proceed by verification of the parts.

In contrast to Mossakowski [36], we keep the different institutions (of the heterogeneous institution environment) separate and do not aim at integrating them into a single (heterogeneous) institution using the so called Grothendieck construction. The latter only puts the institutions side by side and allows at most the sharing of syntactic constructs. As a consequence, from this construction we do not get additional insight like e.g. consistency conditions, and may moreover loose the intuitive separation of the diverse views offered by the individual institutions in a heterogeneous institution environment.

The ideas in this paper present only a first step to a comprehensive heterogeneous approach to system development which will support also model transformations, refinement, and deployment to particular programming languages environments, and provide relationships to "single" system models such as Broy's stream-based system model. Currently we are studying the embedding of service-oriented concepts into our heterogeneous system model approach, ranging from declarative specifications of SCA in SRML [22] and Montanari's Architectural Design Rewriting [10] to process algebraic specifications of the dynamic behaviour of services [4,28,30].

Acknowledgements. We would like to thank an anonymous referee for many valuable suggestions and José Meseguer for fruitful discussions.

References

1. Andrade, L., Baldan, P., Baumeister, H., Bruni, R., Corradini, A., De Nicola, R., Fiadeiro, J.L., Gadducci, F., Gnesi, S., Hoffman, P., Koch, N., Kosiuczenko, P., Lapadula, A., Latella, D., Lopes, A., Loreti, M., Massink, M., Mazzanti, F., Montanari, U., Oliveira, C., Pugliese, R., Tarlecki, A., Wermelinger, M., Wirsing, M., Zawłocki, A.: AGILE: Software architecture for mobility. In: Wirsing, M., Pattinson, D., Hennicker, R. (eds.) WADT 2003. LNCS, vol. 2755. Springer, Heidelberg (2003)
2. Bergstra, J.A., Klop, J.W.: Process Algebra for Synchronous Communication. Information and Control 60(1–3), 109–137 (1984)
3. Bernot, G., Coudert, S., Gall, P.L.: Towards Heterogenous Formal Specifications. In: Wirsing, M., Nivat, M. (eds.) AMAST 1996. LNCS, vol. 1101, pp. 458–472. Springer, Heidelberg (1996)
4. Boreale, M., Bruni, R., Caires, L., De Nicola, R., Lanese, I., Loreti, M., Martins, F., Montanari, U., Ravara, A., Sangiorgi, D., Vasconcelos, V., Zavattaro, G.: SCC: a Service Centered Calculus. In: Bravetti, M., Zavattaro, G. (eds.) Proc. 3rd Int. Wsh. Web Services and Formal Methods (WS-FM 2006). Lect. Notes Comp. Sci., vol. 4184, pp. 38–57. Springer, Heidelberg (2006)
5. Broy, M., Cengarle, M.V., Rumpe, B.: Semantics of UML – Towards a System Model for UML: The Structural Data Model. Technical Report TUM-I0612, Institut für Informatik, Technische Universität München (June 2006)
6. Broy, M., Cengarle, M.V., Rumpe, B.: Semantics of UML – Towards a System Model for UML: The Control Model. Technical Report TUM-I0710, Institut für Informatik, Technische Universität München (February 2007)
7. Broy, M., Cengarle, M.V., Rumpe, B.: Semantics of UML – Towards a System Model for UML: The State Machine Model. Technical Report TUM-I0711, Institut für Informatik, Technische Universität München (February 2007)
8. Broy, M., Stølen, K.: Specification and Development of Interactive Systems: Focus on Streams, Interfaces, and Refinement. Springer, Heidelberg (2001)
9. Broy, M., Wirsing, M.: Algebraic State Machines. In: Rus, T. (ed.) AMAST 2000. LNCS, vol. 1816, pp. 89–118. Springer, Heidelberg (2000)
10. Bruni, R., Lluch-Lafuente, A., Montanari, U., Tuosto, E.: Style-based architectural reconfigurations. Technical Report TR-07-17, Computer Science Department, University of Pisa (2007)
11. Calegari, D.: UML 2.0 Interactions with OCL/RT Constraints. Master's thesis, InCo-PEDECIBA, Technical report 07-17 (2007)
12. Cengarle, M.V.: The Rewriting Logic institution. Technical Report 9801, Ludwig-Maximilians-Universität München, Institut für Informatik (1998)
13. Cengarle, M.V.: The Temporal Logic institution. Technical Report 9805, Ludwig-Maximilians-Universität München, Institut für Informatik (1998)
14. Cengarle, M.V., Knapp, A.: Towards OCL/RT. In: Eriksson, L.-H., Lindsay, P.A. (eds.) FME 2002. LNCS, vol. 2391, pp. 390–409. Springer, Heidelberg (2002)
15. Cengarle, M.V., Knapp, A.: OCL 1.4/1.5 vs. OCL 2.0 Expressions: Formal Semantics and Expressiveness. Softw. Syst. Model. 3(1), 9–30 (2004)
16. Cengarle, M.V., Knapp, A.: UML 2.0 Interactions: Semantics and Refinement. In: Jürjens, J., Fernandez, E.B., France, R., Rumpe, B. (eds.) Proc. 3rd Int. Wsh. Critical Systems Development with UML (CSDUML 2004), pp. 85–99 (2004); Technical Report TUM-I0415, Institut für Informatik, Technische Universität München (2004)
17. Cengarle, M.V., Knapp, A.: An Institution for OCL 2.0. Technical Report 0801, Institut für Informatik, Ludwig-Maximilians-Universität München (2008)

18. Cengarle, M.V., Knapp, A.: An Institution for UML 2.0 Interactions. Technical Report TUM-I0808, Institut für Informatik, Technische Universität München (2008)
19. Cengarle, M.V., Knapp, A.: An Institution for UML 2.0 Static Structures. Technical Report TUM-I0807, Institut für Informatik, Technische Universität München (2008)
20. Diaconescu, R.: Institution-independent Model Theory. Birkhäuser (to appear, 2008)
21. Ehrig, H., Padberg, J., Orejas, F.: From basic views and aspects to integration of specification formalisms. In: Paun, G., Rozenberg, G., Salomaa, A. (eds.) Current Trends in Theoretical Computer Science: Entering the 21th Century, pp. 202–214. World Scientific, Singapore (2001)
22. Fiadeiro, J.L., Lopes, A., Bocchi, L.: A Formal Approach to Service Component Architecture. In: Bravetti, M., Núñez, M., Zavattaro, G. (eds.) WS-FM 2006. LNCS, vol. 4184, pp. 193–213. Springer, Heidelberg (2006)
23. Fischer, C.: CSP-OZ: How to Combine Z with a Process Algebra. In: Bowman, H., Derrick, J. (eds.) Proc. 2nd Int. Conf. Formal Methods for Open Object-Based Distributed Systems (FMOODS 1997), pp. 423–438. Chapman & Hall, Boston (1997)
24. Gadducci, F., Montanari, U.: The Tile Model. In: Plotkin, G., Stirling, C., Tofte, M. (eds.) Proof, Language, and Interaction: Essays in Honour of Robin Milner. Foundations Of Computing Series, pp. 133–166. The MIT Press, Cambridge (2000)
25. Goguen, J.A.: Data, schema, ontology and logic integration. Logic J. IGPL 13(6), 685–715 (2005)
26. Goguen, J.A., Burstall, R.M.: Introducing Institutions. In: Clarke, E., Kozen, D. (eds.) Logics of Programs. LNCS, vol. 164, pp. 221–256. Springer, Heidelberg (1984)
27. Goguen, J.A., Burstall, R.M.: Institutions: Abstract Model Theory for Specification and Programming. J. ACM 39(1), 95–146 (1992)
28. Guidi, C., Lucchi, R., Busi, N., Gorrieri, R., Zavattaro, G.: SOCK: A Calculus for Service-Oriented Computing. In: Dan, A., Lamersdorf, W. (eds.) ICSOC 2006. LNCS, vol. 4294, pp. 327–338. Springer, Heidelberg (2006)
29. Goguen, J.A., Rosu, G.: Institution Morphisms. Form. Asp. Comp. 13(3-5), 274–307 (2002)
30. Lapadula, A., Pugliese, R., Tiezzi, F.: A Calculus for Orchestration of Web Services. In: De Nicola, R. (ed.) ESOP 2007. LNCS, vol. 4421, pp. 33–47. Springer, Heidelberg (2007)
31. MacLane, S.: Categories for the Working Mathematician. Springer, Heidelberg (1971)
32. Mayer, P., Schroeder, A., Koch, N.: A Model-Driven Approach to Service Orchestration. In: Proc. IEEE Int. Conf. Services Computing (SCC 2008) (submitted, 2008)
33. Meseguer, J.: General Logics. In: Ebbinghaus, H.D., Fernandez-Prida, J., Garrido, M., Lascar, D. (eds.) Logic Colloquium 1987, pp. 275–329. North-Holland, Amsterdam (1989)
34. Meseguer, J.: Conditional Rewriting Logic as a Unified Model of Concurrency. Theo. Comp. Sci. 96, 73–155 (1992)
35. Meseguer, J., Montanari, U.: Mapping Tile Logic into Rewriting Logic. In: Parisi-Presicce, F. (ed.) WADT 1997. LNCS, vol. 1376, pp. 62–91. Springer, Heidelberg (1998)
36. Mossakowski, T.: Heterogenous Development Graphs and Heterogeneous Borrowing. In: Nielsen, M., Engberg, U. (eds.) FOSSACS 2002. LNCS, vol. 2303, pp. 326–341. Springer, Berlin (2002)
37. Mossakowski, T.: Heterogeneous Specification and the Heterogeneous Tool Set. Habilitation thesis, Universität Bremen (2005)
38. Mossakowski, T., Maeder, C., Lüttich, K.: The Heterogeneous Tool Set. In: Grumberg, O., Huth, M. (eds.) TACAS 2007. LNCS, vol. 4424, pp. 519–522. Springer, Heidelberg (2007)
39. Naur, P., Randell, B.: Software Engineering — Report on a Conference sponsored by the NATO Science Committee. NATO Sci. Affairs Div., Bruxelles, Garmisch (1969)
40. Pepper, P., Wirsing, M.: A Method for the Development of Correct Software. In: Jähnichen, S., Broy, M. (eds.) KORSO 1995. LNCS, vol. 1009, pp. 27–57. Springer, Heidelberg (1995)

41. Pratt, V.: Modeling Concurrency with Partial Orders. Int. J. Parallel Program. 15(1), 33–71 (1986)
42. Reggio, G., Repetto, L.: CASL-CHART: A Combination of Statecharts and the Algebraic Specification Language CASL. In: Rus, T. (ed.) AMAST 2000. LNCS, vol. 1816, pp. 243–272. Springer, Berlin (2000)
43. Tarlecki, A.: Institution representation. Unpublished note, Dept.of Computer Science, University of Edinburgh (1987)
44. Tarlecki, A.: Moving between Logical Systems. In: Haveraaen, M., Dahl, O.-J., Owe, O. (eds.) Abstract Data Types 1995 and COMPASS 1995. LNCS, vol. 1130, pp. 478–502. Springer, Heidelberg (1996)
45. Tarlecki, A.: Towards heterogeneous specifications. In: Gabbay, D., de Rijke, M. (eds.) Frontiers of Combining Systems. Studies in Logic and Computation, vol. 2, pp. 337–360. Research Studies Press (2000)
46. Tarlecki, A.: Abstract Specification Theory: An Overview. In: Broy, M., Pizka, M. (eds.) Models, Algebras, and Logics of Engineering Software. NATO Science Series — Computer and System Sciences, vol. 191, pp. 43–79. IOS Press, Amsterdam (2003)
47. Wehrheim, H.: Behavioural Subtyping in Object-Oriented Specification Formalisms. Habilitationsschrift, Carl-von-Ossietzky-Universität Oldenburg (2002)
48. Wirsing, M., Clark, A., Gilmore, S., Hölzl, M., Knapp, A., Koch, N., Schroeder, A.: Semantic-Based Development of Service-Oriented Systems. In: Najm, E., Pradat-Peyre, J.-F., Donzeau-Gouge, V.V. (eds.) FORTE 2006. LNCS, vol. 4229, pp. 24–45. Springer, Heidelberg (2006)
49. Wirsing, M., Knapp, A.: View Consistency in Software Development. In: Wirsing, M., Knapp, A., Balsamo, S. (eds.) RISSEF 2002. LNCS, vol. 2941, pp. 341–357. Springer, Heidelberg (2004)
50. Wirsing, M., Nicola, R.D., Gilmore, S., Hölzl, M.M., Lucchi, R., Tribastone, M., Zavattaro, G.: Sensoria process calculi for service-oriented computing. In: Montanari, U., Sannella, D., Bruni, R. (eds.) TGC 2007. LNCS, vol. 4661, pp. 30–50. Springer, Heidelberg (2007)

Ugo Montanari and Concurrency Theory

Roberto Gorrieri

Dipartimento di Scienze dell'Informazione, Università di Bologna
Mura A. Zamboni, 7, 40127 Bologna, Italy
`gorrieri@cs.unibo.it`

1   Introduction

This short note introduces the reader to the contributed papers in the area of concurrency theory that have been written in honour of Ugo Montanari. I also draw the attention to one of the contributions of Ugo in this area that, in my opinion, constitutes a pearl in theoretical computer science and would deserve to be investigated further.

Ugo Montanari is – undoubtedly – one the fathers of the theory of Concurrency. He started his research in this area about 25 years ago and contributed from the very beginning [29,27,25,26] with many stimulating ideas and proposals. In my role of former student of Ugo, I had the chance and the honour to study first, and to follow then, some of these initial achievements. In my opinion, there is a *fil rouge* that links together many of Ugo's results in this area: the goal to find general, parametric, uniform, compositional and, possibly, universal theories that give comprehensive pictures of many subtheories that have been studied in isolation. The impressive list of papers that can be given to substantiate this claim includes at least the following [3,7,8,9,10,15,16,18]. Among the many specific lines of research in this setting, I would like to mention the goal to connect together distributed models of concurrency (such as Petri Nets and graph grammars) with the world of process algebras, whose theories were developed for automata-based models where concurrency is not a primitive concept. Many papers are devoted to reach this goal; among them, we mention at least the following [1,2,4,6,12,13,14,19,23]. In particular, in the last part of this note, I will draw the attention of the reader to one specific contribution of Ugo (and co-workers) that I personally consider a pearl in theoretical computer science: the definition of structural operational semantics for a process algebra that produces Petri nets instead of ordinary labeled transition systems, which is often known under the name of the *DDM approach*, after Degano, De Nicola and Montanari.

2   The Contributed Papers

The six papers that are collected in this section of the volume are valuable contributions that somehow span over Ugo's interests in this area.

- *On the Synthesis of Zero-safe Nets* by Philippe Darondeau is an interesting contribution about a class of Petri nets that Bruni and Montanari have proposed in [11], in order to compensate for the lack of mechanisms to force

P. Degano et al. (Eds.): Montanari Festschrift, LNCS 5065, pp. 403–408, 2008.
© Springer-Verlag Berlin Heidelberg 2008

synchronization of transitions in P/T nets. In zero-safe nets it is indeed natural to define transactional behaviour where many transitions are executed atomically. While Bruni and Montanari have mainly studied the semantics of this class of nets, Darondeau here addresses the so-called synthesis problem: given the specification of transaction systems in the form of linear step automata in which the atomic steps between two stable states form a regular language, find the corresponding implementations producing and consuming resources in the form of bounded and well-behaved zero-safe nets. Darondeau proposes algorithmic techniques for this, resorting to a generalization of the theory of regions for zero-safe nets.

— The paper *Decomposition of Persistent Petri Nets*, by Eike Best, deals with persistent Petri nets, a class of nets characterized by the absence of conflicts in behaviour, while admitting shared input places among distinct transitions. The main result is a theorem stating that, in a bounded persistent net with a strongly connected marking graph, any cycle of the marking graph can be decomposed into a set of disjoint simple cycles. Starting from a known result by Keller, the author develops the proof of the main theorem, showing, along the way, several specific properties of persistent nets.

— In the paper *Secure Data Flow in a Calculus for Context Awareness*, Doina Bucur and Mogens Nielsen present a process calculus for modelling context-aware computing in the setting of ubiquitous computing systems. The process calculus, based on Mobile Ambients, is equipped with new operators in order to capture context provision and context discovery. The basic idea is to introduce macro definitions residing inside ambients and macro calls, that may cross the ambients boundaries. Moreover, since the network is highly reconfigurable, a type system is proposed in order to enforce distributed security policies of the hosting locations. Each ambient has a security policy in the form of a process type, which has to be satisfied by any internal ambient in the ambients hierarchy. The validation of security policies is realized by a combination of static and dynamic type checking.

— Paola Quaglia, a former student of Ugo, in the paper *On Beta-binders communications*, focuses on the Beta-binder calculus, which has been proposed in order to make easier the description and the analysis of biological systems. This kind of language follows the line initially proposed by Fontana and Buss in 1996, and subsequently applied by Regev and Shapiro in 2001 to use process calculi in the biological context. The beta-binders language is close to Ambient-like calculi developed by Cardelli, even if without nesting of ambients, with the distinguishing feature to permit a form of communication in which the matching of input and output is rather loose. The main motivation of the paper is to help understanding the differences between Ambient and Beta-binders. This is done by developing a novel labeled semantics for the beta-binders, which complements the one originally defined as a reduction semantics. Then Quaglia presents a formal relation between the two semantics, stipulating that the two are equivalent.

— The paper titled *On the asynchronous nature of the asynchronous π-calculus*, by R. Beauxis, C. Palamidessi and F.D. Valencia, supports the claim that

the asynchronous π-calculus is really a formalism for asynchronous communication, as commonly intended in the community of distributed systems. To this aim, a new π-based calculus with buffers is presented, whose semantics more clearly reflects the intuition of asynchrony, as all the communication among processes are mediated via buffers. This new calculus is three-fold, as buffers can be either bags, or stacks or queues. The first interesting result is that the bag π-calculus is in strict correspondence with the asynchronous π-calculus in either direction. Such a correspondence is so strict that no doubt can be raised about the fact the asynchronous π-calculus is indeed a calculus for asynchronous communication. On the contrary, the authors show that no similar result can be obtained when buffers are stacks or queue.
- The paper titled *StonyCam: a Formal Framework for Modeling, Analyzing and Regulating Cardiac Myocytes*, by E. Bartocci, F. Corradini, R. Grosu, E. Merelli, O. Riganelli and Scott A. Smolka, is a brief and clear overview of the StonyCam research program, jointly involving the Universities of Stony Brook and Camerino. The problem of cardiac electrical disturbances, such as atrial fibrillation, is introduced along with the need for a concretely useful analysis and prediction tools. Throughout the paper a bird's eye view is offered regarding all the related work and the main techniques used so far. In particular it is described how to model excitable cells using the theory of hybrid automata, obtaining models amenable to formal analysis with a tenfold speedup in simulation time with respect to classical systems of non-linear differential equations. As an example it is also described an interesting method for decting emergent behaviours.

3   One of Ugo's Pearls

Many are the excellent contributions that Ugo has achieved during his brilliant and exciting research life. And I am not old (or good) enough to remember all of them! But for sure one of such results that I had the opportunity to apply in some of my initial papers, as a student exercise [24,20], was the technique to give a Petri net semantics to process algebras, CCS in particular [28,23,21,22]. This technique – Ugo developed in joint work with Pierpaolo Degano and Rocco De Nicola, and for this reason it has been sometimes called the *DDM technique* – attracted a lot of interest, as an attempt to reconcile two mainstreams of concurrency theory that were developed independently and somehow in antagonism: Petri nets and process algebras. Other researchers contributed to the success of this technique, in particular Ernst-Rüdiger Olderog, who made it popular with his significant textbook [17].

The key aspect of this technique is a non-trivial extension of Plotkin' Structual Operational Semantics [30,5] (SOS for short), which was originally conceived in order to produce (labeled) transitions systems as semantic models of programs, to the more elaborate setting of Petri Nets. In the original Plotkin's approach, the terms of an algebra of programs are the states of the transition systems, while the transitions are generated by means of an inference system, defined inductively on the syntax of terms.

The very nice point is that the infinite (in states and transitions) labeled transition system is defined implicitly with two finite, generative structures: a grammar for the states/processes and a finite set of axioms and inference rules for the transitions (that represent the evolution of a global state into another global state). The transition system generated in this way is for the whole language, but of course, one can focus his/her attention only on the part that is reachable from a particular state/process. In the case of CCS, this SOS is particularly concise and elegant.

Degano, De Nicola & Montanari have transposed this idea to Petri nets by defining a grammar for generating the infinite set of CCS-oriented places of the net and a finite set of axioms and inference rules generating the infinite set of CCS-oriented net transitions. The global state (called marking in the Petri net terminology) of the net is represented as a (multi-)set of places and the transitions represent the evolution of a set of places (not necessarily the global state) into another (multi-)set of places. There is a precise mapping dec of CCS processes to markings of the net and the net associated to a CCS process E is the net reachable from the marking $dec(E)$. The resulting Petri net for CCS, on the one hand, offers two main advantages:

- it is now possible to investigate non-interleaving semantics for CCS, based on some notions of causality and conflict [21] that are somehow primitive in Petri nets while completely absent in ordinary transition systems;
- it is possible to apply well-known Petri nets analysis techniques also to CCS processes (e.g., reachability, coverability, net invariants).

On the other hand, this cross-fertilization between the area of Petri nets and process algebra produced benefits also for the area of Petri nets, where some ideas developed in the process algebra community were imported, notably bisimulation equivalence.

If one tries to look with a up-to-date perspective to what has been done, one immediately observes that there is a lot of work that could have been done and that still deserves to be done. In fact, the Petri net for CCS that DDM developed is just one example of a missing general theory of SOS for Petri nets. The theory of transition systems specification via SOS has a large body of results (e.g., formats that ensure that bisimulation is a congruence) that could conceivably approached also for a to-be-developed theory of Petri nets specification via SOS. For instance, it could be interesting to study which formats should have the inference rules to ensure that a Petri net is safe, or of a special class (elementary, rather than Place/Transitions or else), or such that history preserving bisimulation is a congruence. Another aspect is the meaning of negative premises (that might perhaps be better modeled with the use of inhibitors). Summing up, I think that the seminal papers that Ugo and co-workers have produced in this area should be generalized to start a new line of cross-fertilization between the Petri net area and the process algebra community.

4   Conclusion

I think that probably the most impressive distinguishing quality of Ugo is his ability to be a real *maestro*, in following the scientific work of many students, most of which have taken positions in various Italian and international Universities or research insitutions.

I would like to end this short note by quoting (part of) my dedication to Ugo I wrote in my PhD thesis, that somehow reflects this aspect of Ugo's character:

> *Writing a Ph.D thesis under the supervision of Ugo Montanari has been undoubtedly a unique experience. He is a shine researcher full of scientific interests which tries to instill in the mind of his students and collaborators with Socratic maieutic.*

And indeed, Ugo often acts as a new Socrates in guiding the initial steps of his many students. And he is so strongly affectionate to them, that he would never part from them! (And this explains why many of them still live in Pisa ...)

References

1. Baldan, P., Gadducci, F., Montanari, U.: Modelling Calculi with Name Mobility using Graphs with Equivalences. Electr. Notes Theor. Comput. Sci. 176(1), 85–97 (2007)
2. Bruni, R., Melgratti, H.C., Montanari, U.: Event Structure Semantics for Nominal Calculi. In: Baier, C., Hermanns, H. (eds.) CONCUR 2006. LNCS, vol. 4137, pp. 295–309. Springer, Heidelberg (2006)
3. Montanari, U., Pistore, M.: History-Dependent Automata: An Introduction. In: Bernardo, M., Bogliolo, A. (eds.) SFM-Moby 2005. LNCS, vol. 3465, pp. 1–28. Springer, Heidelberg (2005)
4. Bruni, R., Montanari, U.: Concurrent models for Linda with transactions. Mathematical Structures in Computer Science 14(3), 421–468 (2004)
5. Plotkin, G.D.: A structural approach to operational semantics. J. Log. Algebr. Program 60(61), 117–139 (2004)
6. Baldan, P., Corradini, A., Montanari, U.: Bisimulation Equivalences for Graph Grammars. In: Brauer, W., Ehrig, H., Karhumäki, J., Salomaa, A. (eds.) Formal and Natural Computing. LNCS, vol. 2300, pp. 158–190. Springer, Heidelberg (2002)
7. Corradini, A., Heckel, R., Montanari, U.: Compositional SOS and beyond: a coalgebraic view of open systems. Theor. Comput. Sci. 280(1-2), 163–192 (2002)
8. Bruni, R., Meseguer, J., Montanari, U., Sassone, V.: Functorial Models for Petri Nets. Inf. Comput. 170(2), 207–236 (2001)
9. Ferrari, G.L., Montanari, U.: Parameterized Structured Operational Semantics. Fundam. Inform. 34(1-2), 1–31 (1998)
10. Ferrari, G.L., Montanari, U., Mowbray, M.: Structured Transition Systems with Parametric Observations: Observational Congruences and Minimal Realizations. Mathematical Structures in Computer Science 7(3), 241–282 (1997)
11. Bruni, R., Montanari, U.: Zero-safe nets, or transition synchronization made simple. Electr. Notes Theor. Comput. Sci. 7 (1997)
12. Montanari, U., Pistore, M.: Concurrent semantics for the pi-calculus. Electr. Notes Theor. Comput. Sci. 1 (1995)

13. Gorrieri, R., Montanari, U.: On the Implementation of Concurrent Calculi in Net Calculi: Two Case Studies. Theor. Comput. Sci. 141(1& 2), 195–252 (1995)
14. Montanari, U., Yankelevich, D.: Combining CCS and Petri Nets Via Structural Axioms. Fundam. Inform. 20(1/2/3), 193–229 (1994)
15. Degano, P., Nicola, R.D., Montanari, U.: Universal Axioms for Bisimulations. Theor. Comput. Sci. 114(1), 63–91 (1993)
16. Montanari, U., Yankelevich, D.: A Parametric Approach to Localities. In: Kuich, W. (ed.) ICALP 1992. LNCS, vol. 623, pp. 617–628. Springer, Heidelberg (1992)
17. Olderog, E.R.: Nets, Terms and Formulas. In: Cambridge Tracts in Theoretical Computer Science 23, Cambridge University Press, Cambridge (1991)
18. Ferrari, G.L., Montanari, U.: Towards the Unification of Models for Concurrency. In: Arnold, A. (ed.) CAAP 1990. LNCS, vol. 431, pp. 162–176. Springer, Heidelberg (1990)
19. Degano, P., Nicola, R.D., Montanari, U.: A Partial Ordering Semantics for CCS. Theor. Comput. Sci. 75(3), 223–262 (1990)
20. Brogi, A., Gorrieri, R.: A Distributed, Net-Oriented Semantics for Delta Prolog. In: Díaz, J., Orejas, F. (eds.) CAAP 1989 and TAPSOFT 1989. LNCS, vol. 351, pp. 162–177. Springer, Heidelberg (1989)
21. Degano, P., Nicola, R.D., Montanari, U.: On the Consistency of "Truly Concurrent" Operational and Denotational Semantics. In: LICS 1988, pp. 133–141. IEEE CS Press, Los Alamitos (1988)
22. Degano, P., Nicola, R.D., Montanari, U.: Partial orderings descriptions and observations of nondeterministic concurrent processes. In: de Bakker, J.W., de Roever, W.-P., Rozenberg, G. (eds.) Linear Time, Branching Time and Partial Order in Logics and Models for Concurrency. LNCS, vol. 354, pp. 438–466. Springer, Heidelberg (1989)
23. Degano, P., Nicola, R.D., Montanari, U.: A Distributed Operational Semantics for CCS Based on Condition/Event Systems. Acta Inf. 26(1/2), 59–91 (1988)
24. Degano, P., Gorrieri, R., Marchetti, S.: An Exercise in Concurrency: a CSP Process as a Condition/Event System. In: Rozenberg, G. (ed.) APN 1988. LNCS, vol. 340, pp. 85–105. Springer, Heidelberg (1988)
25. Degano, P., Montanari, U.: A model for distributed systems based on graph rewriting. J. ACM 34(2), 411–449 (1987)
26. Degano, P., Montanari, U.: Concurrent Histories: A Basis for Observing Distributed Systems. J. Comput. Syst. Sci. 34(2/3), 422–461 (1987)
27. Degano, P., Montanari, U.: Liveness Properties as Convergence in Metric Spaces. In: STOC 1984, pp. 31–38. ACM Press, New York (1984)
28. Degano, P., Nicola, R.D., Montanari, U.: CCS is an (Augmented) Contact Free C/E System. In: Venturini Zilli, M. (ed.) Mathematical Models for the Semantics of Parallelism. LNCS, vol. 280, pp. 144–165. Springer, Heidelberg (1987)
29. Castellani, I., Montanari, U.: Graph Grammars for distributed systems. In: Ehrig, H., Nagl, M., Rozenberg, G. (eds.) Graph Grammars 1982. LNCS, vol. 153, pp. 20–38. Springer, Heidelberg (1983)
30. Plotkin, G.D.: A structural approach to operational semantics. Technical Report DAIMI FN-19, Aarhus University (1981)

On the Synthesis of Zero-Safe Nets

Philippe Darondeau

IRISA, campus de Beaulieu, F-35042 Rennes Cedex, France
darondeau@irisa.fr

Abstract. Zero-Safe nets are an adequate framework for representing linear step automata. We define a synthesis procedure based on regions for deriving Zero-Safe nets from such automata.

1   Introduction

Zero-Safe nets were introduced by R. Bruni and U. Montanari to compensate for the lack of a mechanism to force synchronization of transitions in PT-nets [1]. A Zero-Safe net is a PT-net with distinguished places called zero-places. The other places are called stable places. A marking in which every zero-place is empty is a stable marking. A stable step is a sequence of transitions from a stable marking to a stable marking, such that tokens produced and fed into stable places during the step are not used until the completion of the step. R. Bruni and U. Montanari provided categorical characterizations of the operational and abstract semantics of Zero-Safe nets for two different token management policies, reflecting the collective, resp. the individual token philosophies [1,2,3,4].

Our purpose is to revert the path from Zero-Safe nets to their semantics. Given an automaton whose atomic transitions are sequences of micro-steps, the problem is to synthesize a Zero-Safe net that behaves according to this specification. It is moreover desirable that the synthesis procedure be effective, so that Zero-Safe net implementations of linear step automata can be computed, hence the accent is put here on algorithms.

A privileged field of application of Zero-Safe nets is Workflows. A Workflow Net [5] is a Petri net with two special places i and o, used to mark the begin and the end of a procedure composed of tasks modelled by transitions. The other places of the net express dependences between tasks. A Workflow Net is *sound* if, whenever it has been initialized with one token put in place i and all other places empty, the procedure terminates eventually with one token in place o and all other places empty. A Workflow Net may be seen as a Zero-Safe net with two stable places i and o. The complete executions of the procedure correspond thus with the stable steps of the Zero-Safe net. If the Workflow Net is sound, any sequence of transitions of the Zero-Safe net may be extended into a stable step. Note that a transition may occur an unbounded number of times in a transaction owing to the possible presence of cycles in Workflow Nets. More general workflow systems in which resources (or documents or goods) are taken and released (or consumed and produced) by transactions may also be represented as Zero-Safe nets, using stable places to hold these resources.

P. Degano et al. (Eds.): Montanari Festschrift, LNCS 5065, pp. 409–426, 2008.
© Springer-Verlag Berlin Heidelberg 2008

We propose in this paper an algorithm that produces Zero-Safe nets labelled injectively on transitions from (finite) linear step automata, in which a transition from q to q' is a sequence of micro-steps in a regular language $L(q, q')$. The synthesized Zero-Safe nets are well-behaved in the sense that any sequence of concurrently enabled transitions from a stable marking may be extended into a stable step. Compensation [7] is therefore not needed in workflows implemented by Zero-Safe net synthesis. As a matter of fact, due to the injective labelling of nets, some linear step automata cannot be realized by any well-behaved Zero-Safe net. Therefore, we propose a decision and synthesis algorithm whose output is either a well-behaved Zero-Safe net, realizing the linear step automaton given in input, or the answer that no such net exists. The algorithm takes time polynomial in the size of step automata.

The synthesis of Zero-Safe nets may be studied with the collective or with the individual token philosophies. Individual tokens are better suited for representing parallel workflow transactions. However, for the sake of simplicity, we have chosen to work with collective tokens.

The rest of the paper is organized as follows. Section 2 recalls the background of Zero-Safe nets. Section 3 defines Zero-Safe regions of linear step automata and proposes regional axioms for characterizing the linear step automata that may be implemented with bounded and well-behaved Zero-Safe nets. Section 4 establishes the axiomatic characterization. Sections 5 and 6 provide a decision and synthesis algorithm.

2 Zero-Safe Nets

This section, which borrows mainly from [4], recalls the definition of Zero-Safe nets and their semantics. An illustration is given at the end of the section.

Definition 1 (PT-net). *A PT-net is a tuple $N = (P, T, F, M_0)$ where P and T are disjoint sets of places and transitions, F maps $(P \times T) \cup (T \times P)$ to $\mathbb{N}$, and $M_0 : P \to \mathbb{N}$ is the initial marking. A marking is a map $M : P \to \mathbb{N}$. Transition t has concession at M if $M(p) \geq F(p, t)$ for every place p, in which case it may be fired. Firing t at M leads to the marking M' defined with $M(p) - F(p, t) = M'(p) - F(t, p)$ for all p (notation: $M[t\rangle M')$. A firing sequence of N is a finite or infinite sequence $M_0[t_1\rangle \dots M_n[t_{n+1}\rangle \dots$.*

Definition 2 (Zero-Safe net). *A Zero-Safe net (or ZS-net) is a PT-net $N = (P, T, F, M_0)$ where $P = S \cup Z$ is the union of two disjoint subsets S and Z, and $M_0(z) = 0$ for all $z \in Z$. Places in S, resp. in Z, are called stable places, resp. zero places. A marking of N is stable if $M(z) = 0$ for every zero place $z \in Z$. Let $\rho = M_0[t_1\rangle \dots M_{n-1}[t_n\rangle M_n$ be a non-empty firing sequence of N, then:*

ρ is a stable step of N if

- *$\sum_{i=1}^{n} F(s, t_i) \leq M_0(s)$ for all $s \in S$ (concurrent enabling)*
- *M_0 and M_n are stable markings (stable fairness)*

ρ is a stable transaction of N if in addition

- *markings $M_1, \ldots, M_{n-1}$ are not stable (atomicity)*
- *$\sum_{i=1}^{n} F(s, t_i) = M_0(s)$ for all $s \in S$ (perfect enabling)*

We write $M\{[t_1 \ldots t_n \rangle M'$ to mean that $M[t_1\rangle \ldots [t_n\rangle M'$ is a stable transaction of the ZS-net (P, T, F, M).

Zero-places can be used to coordinate and synchronize in a single transaction any number of transitions of the net [4]. In this view, unstable markings are dealt with as unobservable and they are abstracted from. Tokens fed into stable places during a stable transaction (also called stable tokens) are frozen until the completion of the transaction. This is compatible with the intuition that stable transactions implement sequences of micro-steps in (linear step) automata where the updating of the current state is done on step completion. Another strong requirement of stable transactions is that every stable token available at the outset of a transaction is consumed during the transaction. This requirement adds to the expressive power of ZS-nets (it is in fact not restrictive since, for modelling idle resources, one can always introduce "idling" transitions with self-loops connecting them with stable places).

A main contribution of the work on ZS-nets presented in [1,2,3,4] was to define and compare two forms of truly-concurrent semantics of these nets, induced from the collective token and from the individual token philosophy, respectively. Both semantics are defined by considering permutations on stable transactions and by mapping each stable transaction to its equivalence class w.r.t. permutations. In this work, we forget about true concurrency and provide ZS-nets with *linear step* semantics expressed by automata as follows.

Definition 3. *A linear step automaton over T is a tuple $A = (Q, L, q_0)$ where Q is the set of states, q_0 is the initial state, and $L : Q \times Q \to \mathcal{P}(T^*)$ defines for all states q and q' a (possibly empty) set $L(q, q')$ of non-empty sequences of micro-steps from q to q'. Two linear step automata $A = (Q, L, q_0)$ and $A' = (Q', L', q_0')$ over T are isomorphic ($A \cong A'$) if there exists a bijection $\phi : Q \to Q'$ such that $\phi(q_0) = q_0'$ and $L(q_1, q_2) = L'(\phi(q_1), \phi(q_2))$ for all $q_1, q_2 \in Q$. A linear step automaton is reachable if any state $q \in Q$ equals q_n for some sequence $q_0 q_1 \ldots q_n$ such that $L(q_i, q_{i+1}) \neq \emptyset$ for all i. A linear step automaton is regular if Q is finite and all languages $L(q, q')$ are regular subsets of T^*. A linear step automaton is trivial if it has a single state q_0 and $L(q_0, q_0) = \emptyset$.*

In general, $L(q, q') \cap L(q, q'') \neq \emptyset$ for $q' \neq q''$ (linear step automata may be non-deterministic), and $L(q, q') L(q', q'')$ may intersect both $L(q, q'')$ and its complement (linear step automata may have non-atomic steps).

A Zero-Safe net implements a linear step automaton if and only if this automaton may be reconstructed (up to isomorphism) as an abstraction of the Zero-Safe net. The abstractions that we consider are the following.

Definition 4 (Reachable Stable State Graph). *Let $N = (S \cup Z, T, F, M_0)$ be a ZS-net. The set of reachable stable markings of N (notation: $RSM(N)$)*

is the least set of stable markings containing M_0 such that $M \in RSM(N)$ and $M\{[w\rangle M'$ entail $M' \in RSM(N)$ for all $w \in T^*$. The reachable stable state graph of N (notation $RSSG(N)$) is the linear step automaton $(RSM(N), L, M_0)$ defined with $L(M, M') = \{w \mid M\{[w\rangle M'\}$.

The problem addressed in the next sections is the following: given a linear step automaton $A = (Q, L, q_0)$, regular and reachable, decide whether $A \cong RSSG(N)$ for some ZS-net $N = (S \cup Z, T, F, M_0)$, well-behaved in the sense given by the following definition.

Definition 5. *A ZS-net N is well-behaved if, for any marking $M \in RSM(N)$, every concurrently enabled sequence $M[t_1\rangle \ldots [t_n\rangle M'$ extends a stable transaction or may be extended to a stable transaction.*

For general ZS-nets, an external control is needed at run time to avoid performing sequences of transitions that could not be extended into sequences of stable transactions. Such a control is problematic since an effective condition recognizing the inadequate transaction segments is not known [4]. This problem does precisely not occur with well-behaved ZS-nets, hence they can be implemented easily, all the more when they are bounded in the sense given by the following definition.

Definition 6. *A ZS-net N is bounded if $RSM(N)$ is finite and there exists an integer B such that, for any marking $M \in RSM(N)$ and for any concurrently enabled sequence $M[t_1\rangle \ldots [t_n\rangle M'$, $M'(z) \leq B$ for every zero-place z.*

The next proposition shows that all well-behaved ZS-nets realizing regular step-automata are bounded, hence their implementation is easy.

Proposition 1. *Let $A = (Q, L, q_0)$ be a linear step automaton, regular and reachable, then any well-behaved ZS-net N such that $A \cong RSSG(N)$ is bounded.*

Proof. Let $N = (S \cup Z, T, F, M_0)$. As $A \cong RSSG(N)$, $RSM(N)$ is finite. It remains to show that there exists an integer B such that, for any $M \in RSM(N)$ and for any concurrently enabled firing sequence $\rho = M[t_1\rangle M_1 \ldots [t_n\rangle M_n$ in which all markings M_i are unstable, $M_n(z) \leq B$ for every zero-place $z \in Z$. As N is well-behaved, such firing sequences ρ may always be extended into transactions $\rho' = M[t_1\rangle M_1 \ldots [t_n\rangle M_n [t_{n+1}\rangle M_{n+1} \ldots [t_{n+k}\rangle M'$ where $M' \in RSM(N)$. Let $u = t_1 \ldots t_n$ and $v = t_{n+1} \ldots t_{n+k}$. Seeing that $RSM(N)$ is finite, in order to prove the proposition, it suffices to show that $M_n(z) \leq B(M, M')$ for some bound $B(M, M')$ depending only on M, M' and z. Now, in view of the isomorphism $A \cong RSSG(N)$, the set of words $w \in T^*$ generated by transactions from M to M' is equal to $L(q, q')$ for some $q, q' \in Q$. Let $\mathcal{A} = (\mathcal{Q}, \delta, q, q')$ be the minimal automaton accepting the regular language $L(q, q')$, with partial transition map $\delta : Q \times T \rightharpoonup Q$, initial state $q \in Q$, and final state $q' \in Q$. We claim that $M_n(z)$ depends only upon $\delta(q, u)$. To show this, let $u' = t'_1 \ldots t'_m$ such that $\delta(q, u) = \delta(q, u')$. As $uv \in L(q, q')$, necessarily $u'v \in L(q, q')$, and there should exist a corresponding transaction $M[t'_1\rangle \ldots [t'_m\rangle M'_m [t_{n+1}\rangle \ldots [t_{n+k}\rangle M'$. Therefore $M_n = M'_m$ and $M_n(z)$ can take at most $|Q|$ different values. $\qquad\square$

Remark 1. For a well-behaved ZS-net $N = (S \cup Z, T, F, M_0)$, let $RM(N)$ be the set of all markings reached from M_0 by sequences of transactions followed by a transaction segment, and let $RSG(N)$ be the graph obtained by connecting the markings in $RM(N)$ with the transitions $M[t > M'$ occurring in transactions. Using the same reasoning as in the above proof, one can show that for any two well-behaved and bounded ZS-nets N and N', $RSSG(N) \cong RSSG(N')$ if and only if $RSG(N) \cong RSG(N')$.

Remark 2. $A \cong RSSG(N)$ cannot hold unless A is deterministic. So we could restrict ourselves to deterministic automata. However, the decision and synthesis algorithm constructed in this paper copes with the general case.

We now illustrate the definitions given so far on a ZS-net that serves as running example in the remaining sections.

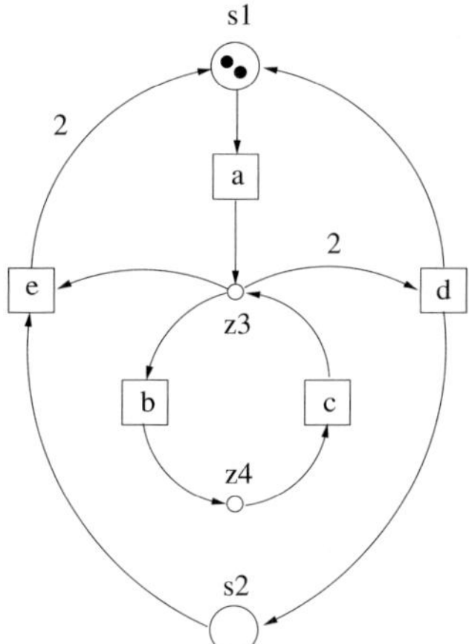

Fig. 1. A ZS-net

Example 1. Consider the ZS-net N shown in Fig. 1. It has 2 stable places $s1$ and $s2$, 2 zero-places $z3$ and $z4$, and 5 transitions a to e. In this net, $F(e, s1) = 2$, $F(z3, d) = 2$, and all other values taken by F are 1 or 0, e.g. $F(a, z3) = 1$ and $F(a, z4) = 0$. In the initial marking M_0, there are two tokens in the stable place $s1$ and all other places are empty. $RSM(N)$ contains exactly two stable markings, namely $M_0 = (2, 0, 0, 0)$ and $M_1 = (1, 1, 0, 0)$. $RSSG(N)$ is the linear step automaton $A = (RSM(N), L, M_0)$ defined with $L(M_0, M1) = (a\,(bc)^* \,\|\, a\,(bc)^*)\,d$ (where $\|$ denotes the shuffle operator), and $L(M_1, M_0) = a\,(bc)^*e$. The sets $L(M_0, M_0)$ and $L(M_1, M_1)$ are empty. The ZS-net N is well-behaved (and it is bounded). $\qquad\square$

3   Zero-Safe Regions

In this section, we define the ZS-regions of a linear step automaton by adapting the concept of regions invented by Ehrenfeucht and Rozenberg for elementary nets [8,9] and extended later on to PT-nets with the step firing rule [10] and to general types of nets in [11]. To begin with, we introduce two monoids.

Definition 7. *Let* $(\mathbb{N} \times \mathbb{N}, +, (0,0))$ *be the monoid with the composition operation* $(m, n) + (m', n') = (m + m', n + n')$, *and let* $(\mathbb{N} \times \mathbb{N}, \cdot, (0,0))$ *be the monoid defined with the alternative composition operation* $(m, n) \cdot (m', n') =$ *if* $n \geq m'$ *then* $(m, n + n' - m')$ *else* $(m + m' - n, n')$ *(the reader may check the associativity of this composition).*

Let $A = RSSG(N)$ be the reachable stable state graph of a ZS-net $N = (S \cup Z, T, F, M_0)$, hence $A = (Q, L, q_0)$ with $Q = RSM(N)$ and $q_0 = M_0$. In this particular case of reachable linear step automata, the following properties may be observed.

PROPERTY S
Each stable place $s \in S$ induces a map $\phi_s : Q \to \mathbb{N}$, defined with $\phi_s(q) = M(s)$ for $q = M \in RSM(N)$. Let $\eta_s : T \to (\mathbb{N} \times \mathbb{N})$ be defined with $\eta_s(t) = (F(s,t), F(t,s))$. Then for all states q, q' and for all words $t_1 t_2 \ldots t_n$ in $L(q, q')$, $(\phi_s(q), \phi_s(q')) = \eta_s(t_1) + \eta_s(t_2) + \ldots + \eta_s(t_n)$. Indeed, let $q = M$ and $q' = M'$ then $L(q, q') = L(M, M')$, and by definition of stable transactions and perfect enabling, $M(s) = F(s, t_1) + \ldots + F(s, t_n)$ and $M'(s) = F(t_1, s) + \ldots + F(t_n, s)$.

PROPERTY Z
For any zero-place $z \in Z$, let $\eta_z : T \to (\mathbb{N} \times \mathbb{N})$ be defined with $\eta_z(t) = (F(z,t), F(t,z))$. The map η_z may be extended from T to T^* by setting $\eta_z(t_1 t_2 \ldots t_n) = \eta_z(t_1) \cdot \eta_z(t_2) \cdot \ldots \cdot \eta_z(t_n)$. Then for all states q and q', for all words $w \in L(q, q')$, and for all prefixes $u = t_1 t_2 \ldots t_n$ of $w = uv$, $\eta_z(u) = (0, j)$ for some $j \geq 0$ and $j = 0$ if $u \in L(q, q')$. To see this, consider a one-place PT-net $N_z = (\{z\}, T, F_z, M_z)$, defined with $F_z(z, t) = F(z, t)$, $F_z(t, z) = F(t, z)$, and some $M_z(z) \geq 0$. If we let $\eta_z(u) = ({}^\bullet u, u^\bullet)$, then the integer ${}^\bullet u$ is the least value of $M_z(z)$ such that $M_z[t_1\rangle \ldots [t_n\rangle M'_z$ for some M'_z, and this M'_z is given by $M'_z(z) = M_z(z) - {}^\bullet u + u^\bullet$. Now $u = t_1 t_2 \ldots t_n$ is a prefix of $w = uv \in L(q, q')$, $q = M$ and $q' = M'$ are two stable markings of N, and $M_z[t_1\rangle \ldots [t_n\rangle M'_z$ for $M_z(z) = M(z) = 0$ by definition of stable steps. Therefore ${}^\bullet u = 0$ and $\eta_z(u) = (0, j)$ for some j. If $u = w \in L(q, q') = L(M, M')$, then $M'_z(z) = M'(z) = 0$ because M' is a stable marking, hence in this case $j = u^\bullet = M'_z(z) - M_z(z) + {}^\bullet u = 0$.

Example 2. Consider the linear step automaton $A = RSSG(N)$ defined in Example 1. Let $\phi_{s1}(M_0) = 2$ and $\phi_{s1}(M_1) = 1$. Let η_{s1} be the map defined with $\eta_{s1}(a) = (1, 0)$, $\eta_{s1}(d) = (0, 1)$, $\eta_{s1}(e) = (0, 2)$, and $\eta_{s1}(x) = (0, 0)$ for other x. Take any word $w \in L(M_0, M_1)$, then w has 2 occurrences of a, 1 occurrence of d and no occurrence of e, hence $\eta_{s1}(w) = (1, 0) + (1, 0) + (0, 1) = (2, 1) = (M_0(s1), M_1(s1))$. Now let η_{z3} be the map defined with $\eta_{z3}(a) = (0, 1)$, $\eta_{z3}(b) = (1, 0)$, $\eta_{z3}(c) = (0, 1)$, $\eta_{z3}(d) = (2, 0)$, and $\eta_{z1}(e) = (1, 0)$. Define $u = aabcb$ and $v = bccd$ thus $w = uv \in L(M_0, M_1)$. Then $\eta_{z3}(u) = (0, 1) \cdot (0, 1) \cdot (1, 0) \cdot (0, 1) \cdot (1, 0)$. Seeing that $(0, 1) \cdot (0, 1) = (0, 2)$ and $(1, 0) \cdot (0, 1) = (1, 1)$, $\eta_{z3}(u) = (0, 2) \cdot (1, 1) \cdot (1, 0) = (0, 2) \cdot (1, 0) = (0, 1)$. Similarly, $\eta_{z3}(v) = (1, 0) \cdot (0, 1) \cdot (0, 1) \cdot (2, 0)$

and seeing that $(0,1) \cdot (2,0) = (1,0)$, $\eta_{z3}(v) = (1,1) \cdot (1,0) = (1,0)$. Therefore, $\eta_{z3}(w) = (0,1) \cdot (1,0) = (0,0)$. $\qquad\square$

Let $A = (Q, L, q_0)$ be now *any* reachable linear step automaton over T. The ZS-regions of A are defined by the following statements that mimick the respective properties S and Z.

Definition 8. *An S-region of A is a pair of maps $\phi_s : Q \to \mathbb{N}$ and $\eta_s : T \to (\mathbb{N} \times \mathbb{N})$ such that, for all states q and q', $(\phi_s(q), \phi_s(q')) = \eta_s(t_1) + \eta_s(t_2) + \ldots + \eta_s(t_n)$ for all words $t_1 t_2 \ldots t_n$ in $L(q, q')$.*

Definition 9. *A Z-region of A is a map $\eta_z : T \to (\mathbb{N} \times \mathbb{N})$ such that for all states q and q', $\eta_z(t_1) \cdot \eta_z(t_2) \cdot \ldots \cdot \eta_z(t_n) = (0, j)$ for all prefixes $u = t_1 \ldots t_n$ of words $uv \in L(q, q')$, and $j = 0$ if $u \in L(q, q')$.*

We use the term ZS-region to mean informally the S-regions or Z-regions of A. In the sequel we use freely the notations $\eta_s(u)$ and $\eta_z(u)$ with $u \in T^*$, to mean the values of the extended maps η_s and η_z ranging over the respective monoids $(\mathbb{N} \times \mathbb{N}, +, (0,0))$ and $(\mathbb{N} \times \mathbb{N}, \cdot, (0,0))$. The goal of the next section is to find necessary and sufficient conditions on the ZS-regions of a reachable linear step automaton A such that $A \cong RSSG(N)$. In order to get an intuition of these conditions, we come back to the particular case where $A = (Q, L, q_0)$ is the reachable stable state graph of a ZS-net $N = (S \cup Z, T, F, M_0)$.

Proposition 2. *If N is well-behaved and bounded, then the linear step automaton $A = RSSG(N)$ is regular.*

Proof. We must show that for all states q and q', the language $L(q, q')$ is regular. Let $q = M$ and $q' = M'$. The transactions $M[t_1\rangle \ldots [t_n\rangle M'$ in N correspond bijectively with the transactions $\mathcal{M}[t_1\rangle \ldots [t_n\rangle \mathcal{M}'$ in the extended ZS-net $\mathcal{N} = (S \cup S' \cup Z, T, \mathcal{F}, \mathcal{M})$ defined, together with the marking $\mathcal{M}'$, as follows:

- each stable place $s \in S$ has a replica $s' \in S'$
- $\mathcal{M}(s) = M(s)$ and $\mathcal{M}(s') = 0$
- $\mathcal{M}'(s) = 0$ and $\mathcal{M}'(s') = M'(s)$
- $\mathcal{F}(s, t) = F(s, t)$, $\mathcal{F}(s', t) = 0$ and $\mathcal{F}(t, s') = F(t, s)$, $\mathcal{F}(t, s) = 0$
- $\mathcal{M}(z) = M(z)$, $\mathcal{F}(z, t) = F(z, t)$ and $\mathcal{F}(t, z) = F(t, z)$

As N is well-behaved and bounded, the set of markings $\mathcal{M}_i$ found in transactions $\rho = \mathcal{M}[t_1\rangle \ldots [t_i\rangle \mathcal{M}_i \ldots [t_n\rangle \mathcal{M}'$ in $\mathcal{N}$ is finite. Let $\mathcal{Q}$ be the set of these markings, and let $\mathcal{A}$ be the automaton over T with the set of states $\mathcal{Q}$, the initial state $\mathcal{M}$, the final state $\mathcal{M}'$, and the labelled transitions $\mathcal{M}_{i-1}[t_i\rangle \mathcal{M}_i$. Clearly, $\mathcal{A}$ accepts $L(q, q')$. As $\mathcal{A}$ is a finite automaton, $L(q, q')$ is regular. The linear step automaton A is therefore regular (in the different sense given by Def. 3). $\qquad\square$

Proposition 3. *For any two distinct states q and q', $\phi_s(q) \neq \phi_s(q')$ for some stable place $s \in S$.*

Proof. Let $q = M$ and $q' = M'$, then $M(z) = M'(z) = 0$ for every zero-place z. Therefore, $M(s) \neq M'(s)$ for some stable place s, i.e. $\phi_s(q) \neq \phi_s(q')$. $\qquad\square$

Proposition 4. *For any state $q \in Q$, word $u \in T^*$, and transition $t \in T$, if $u = \varepsilon$ or $uv \in L(q, q')$ for some $v \in T^+$ and $q' \in Q$, but $utv \notin L(q, q')$ for any $v \in T^*$ and $q' \in Q$, then $\eta_s(ut) = (i, j)$ and $\phi_s(q) < i$ for some stable place $s \in S$, or $\eta_z(u) = (0, j)$, $\eta_z(t) = (i', j')$, and $j < i'$ for some zero-place $z \in Z$.*

Proof. Let $q = M$ and $u = t_1 \ldots t_n$, then $M[t_1\rangle \ldots [t_n\rangle M_n$ for some marking M_n and this firing sequence is concurrently enabled. We claim that, with the hypotheses made in the proposition, there cannot exist any concurrently enabled firing sequence $M[t_1\rangle \ldots [t_n\rangle M_n[t\rangle M_{n+1}$. In order to establish this claim, we assume the opposite and we derive a contradiction. As the ZS-net N is well-behaved, the firing sequence $M[t_1\rangle \ldots [t_n\rangle M_n[t\rangle M_{n+1}$ extends a stable transaction or it may be extended to a stable transaction. If the firing sequence may be extended to a stable transaction, then $utv \in L(q, q')$ for some $v \in T^*$ and $q' \in Q$, contradicting the hypotheses. If the firing sequence may be truncated to a stable transaction $M[t_1\rangle \ldots [t_m\rangle M_m$ for some $m \leq n$, then by the condition of atomicity, $uv \notin L(q, q')$ for any $v \in T^+$ and $q' \in Q$, contradicting again the hypotheses. Therefore, the claim has been established. Seeing that $M[t_1\rangle \ldots [t_n\rangle M_n$ is concurrently enabled, in order that there exists no concurrently enabled firing sequence $M[t_1\rangle \ldots [t_n\rangle M_n[t\rangle M_{n+1}$, it is necessary and sufficient that $\eta_s(t_1 \ldots t_n t) = (i, j)$ and $\phi_s(q) < i$ for some stable place $s \in S$, or $\eta_z(t_1 \ldots t_n) = (0, j)$, $\eta_z(t) = (i', j')$, and $j < i'$ for some zero-place $z \in Z$. $\qquad\square$

Proposition 5. *For any states q and q', if $w = uv \in L(q, q')$ and u and v are both non-empty, then $\eta_z(u) = (0, j)$ with $j \neq 0$ for some $z \in Z$.*

Proof. Let $q = M$ and $q' = M'$, hence $L(q, q') = L(M, M')$. Let $u = t_1 \ldots t_n$ and $v = t_{n+1} \ldots t_k$, then there exists a transaction $M[t_1\rangle M_1 \ldots [t_n\rangle M_n \ldots [t_k\rangle M_k$ ending with $M_k = M'$. In view of the atomicity of transactions, $M_n(z) \neq 0$ for some zero-place z. In view of Property Z, $\eta_z(u) = (0, M_n(z))$. Therefore, the proposition holds for $j = M_n(z)$. $\qquad\square$

Example 3. Consider the linear step automaton $A = RSSG(N)$ defined in Example 1. The language $L(M_0, M1) = (a\,(bc)^* \parallel a\,(bc)^*)\,d$ is regular since the shuffle of two regular languages is regular. As $M_0(s1) = 2$ and $M_1(s1) = 1$, the stable place $s1$ separates all stable markings (illustrating Proposition 3). Let $u = aabcb$ (see Example 2), then ud and ue are not prefixes of any word in $L(M_0, M1)$ or $L(M_0, M_0)$ (the latter set is empty). As regards ud, $\eta_{z3}(u) = (0, 1)$, $\eta_{z3}(d) = (2, 0)$, and $1 < 2$ (illustrating Proposition 4). As regards ue, $\eta_{s2}(u2) = (1, 0)$, $\phi_{s2}(M_0) = M_0(s2) = 0$, and $0 < 1$. Let $v = bccd$ like in Example 2, thus $uv \in L(M_0, M1)$. As we already saw, $\eta_{z3}(u) = (0, 1)$, and $1 \neq 0$ (illustrating Proposition 5). $\qquad\square$

Given two isomorphic reachable linear step automata $A = (Q, L, q_0)$ and $A' = (Q', L', q_0')$, their respective sets of S-regions correspond obviously through the isomorphism $\phi : Q \to Q'$, and their sets of Z-regions are identical. In view of Properties S and Z and Propositions 2 to 5, in order that $A \cong RSSG(N)$ for some bounded and well-behaved ZS-net N, the following conditions on $A = (Q, L, q_0)$ must be satisfied:

- REGULAR REPRESENTATION

  The linear step automaton A is regular.
- SEPARATION

  For any two distinct states q and q', $\phi_s(q) \neq \phi_s(q')$ for some S-region of A.
- FORWARD CLOSURE

  For any state $q \in Q$, word $u \in T^*$, and transition $t \in T$,

  if $u = \varepsilon$ or $uv \in L(q, q')$ for some $v \in T^+$ and $q' \in Q$,

  but $utv \notin L(q, q')$ for any $v \in T^*$ and $q' \in Q$ then

  $\eta_s(ut) = (i, j)$ and $\phi_s(q) < i$ for some S-region of A, or

  $\eta_z(u) = (0, j)$, $\eta_z(t) = (i', j')$, and $j < i'$ for some Z-region of A.
- ATOMICITY

  For any states q and q', if $w = uv \in L(q, q')$ and u and v are both non-empty,
  then $\eta_z(u) = (0, j)$ with $j \neq 0$ for some Z-region of A.

Example 4.

- Let $A = (Q, L, q_0)$ where $Q = \{q_0, q_1\}$, $L(q_0, q_1) = L(q_1, q_0) = a$, and $L(q_0, q_0) = L(q_1, q_1) = \emptyset$, then A does not satisfy the condition of Separation and it cannot be realized by a ZS-net.
- Let $A = (Q, L, q_0)$ where $Q = \{q_0, q_1\}$, $L(q_0, q_1) = abbc + acbb$, and $L(q_1, q_0) = L(q_0, q_0) = L(q_1, q_1) = \emptyset$, then A does not satisfy the Forward Closure condition and it cannot be realized by a ZS-net. Adding $abcb$ to $L(q_0, q_1)$ yields Forward Closure.
- Let $A = (Q, L, q_0)$ where $Q = \{q_0, q_1\}$, $L(q_0, q_1) = aa$, and $L(q_1, q_0) = L(q_0, q_0) = L(q_1, q_1) = \emptyset$, then then A does not satisfy the condition of Atomicity, and it cannot be realized by a ZS-net.

$\square$

We will show in the next section that, for any regular and reachable linear step automaton A, the conditions of Separation, Forward Closure, and Atomicity suffice to guarantee that $A \cong RSSG(N)$ for some well-behaved ZS-net N, providing the desired characterization.

4   Establishing the Axioms for ZS-Net Synthesis

Throughout the section, $A = (Q, L, q_0)$ is a regular and reachable linear step automaton, whereas $\mathcal{A} = (\mathcal{Q}, \delta, q, \mathcal{Q}_F)$ denotes a trim automaton over T, with partial transition map $\delta : Q \times T \rightharpoonup Q$, initial state $q \in Q$, and set of final states $Q_F \subseteq Q$. We recall that an automaton is trim if every state can be reached from the initial state and co-reached from some final state.

In a first stage, we show that when Separation, Forward Closure, and Atomicity hold for A, finite subsets $\mathcal{S}$ and $\mathcal{Z}$ of ZS-regions of A witness for this fact. Since Q is finite, this property is obvious for Separation, but it is less immediate for Forward Closure and for Atomicity.

Lemma 1. *Atomicity, when it holds for A, is witnessed by a finite subset of Z-regions of A.*

Proof. Let $\eta_z : T \to (\mathbb{N} \times \mathbb{N})$ be any Z-region of A. By definition of Z-regions, $\eta_z(w) = (0,0)$ for $w \in L(q, q')$. For fixed states q and q', let $\mathcal{A} = (\mathcal{Q}, \delta, q, \mathcal{Q}_F)$ be a finite automaton accepting the regular language $L(q, q')$. We claim that $\eta_z(u)$ depends only upon $\delta(q, u)$, hence there exists a unique map $\phi_z : \mathcal{Q} \to \mathbb{N}$ such that $\eta_z(u) = (0, \phi_z(\delta(q, u)))$, for all prefixes u of words $w \in L(q, q')$. In order to establish this claim, we show that $\eta_z(u) = \eta_z(u')$ whenever uv and $u'v$ belong to $L(q, q')$ for some v. Let $\eta_z(u) = (0, j)$ and $\eta_z(u') = (0, j')$, thus we want to show $j = j'$, and let $\eta_z(v) = (k, l)$. In order that $\eta_z(uv) = (0, 0)$, it is necessary that $j \geq k$ and $j + l - k = 0$, hence $j = k$ and $l = 0$. Similarly, $\eta_z(u'v) = (0, 0)$ entails $j' = k$. Therefore $j = j'$ and the claim is established. Now, if atomicity holds for A, then there must exist, for any decomposition $w = uv$ of a word $w \in L(q, q')$ into non-empty factors u and v, some Z-region η_z with $\phi_z(\delta(q, u)) > 0$ (hence $\delta(q, u) \neq q$). As $\delta(q, u)$ ranges over $|Q| - 1$ values, it suffices to choose at most $(|\mathcal{Q}| - 1)$ Z-regions of A as witnesses for the condition of Atomicity w.r.t. the fixed pair of states q and q', and there are finitely many such pairs. $\square$

The proofs of Proposition 1 and Lemma 1 establish two specific versions of the general claim stated in the proposition below.

Proposition 6. *Let $q \in Q$, $Q' \subseteq Q$ and let $\mathcal{A} = (\mathcal{Q}, \delta, q, \mathcal{Q}_F)$ be a trim automaton accepting $\cup_{q' \in Q'} L(q, q')$. Then $\delta(q, u) = \delta(q, u')$ entails that $\eta_z(u) = \eta_z(u')$ for any Z-region of A and $\eta_s(u) = \eta_s(u')$ for any S-region of A.*

Proof. Straightforward adaptation of the proof of Lemma 1. $\square$

Lemma 2. *Forward Closure, when it holds for A, is witnessed by a finite subset of ZS-regions of A.*

Proof. For fixed q in A, let $\mathcal{A} = (\mathcal{Q}, \delta, q, \mathcal{Q}_F)$ be a trim automaton accepting the regular language $\mathcal{L}(q) = \cup_{q' \in Q} L(q, q')$. If Forward Closure holds, then for any $u \in T^*$ and $t \in T$ such that $u = \varepsilon$ or $uv \in \mathcal{L}(q)$ for some $v \in T^+$, but $utv \notin \mathcal{L}(q)$ for any $v \in T^*$, there should exist an S-region of A such that $\eta_s(ut) = (i, j)$ and $\phi_s(q) < i$ or a Z-region of A such that $\eta_z(u) = (0, j)$, $\eta_z(t) = (i', j')$, and $j < i'$. In view of Proposition 6, it suffices to choose at most $|\mathcal{Q}|$ ZS-regions of A as witnesses for the condition of Forward Closure w.r.t. the fixed state q (i.e. relatively to $\mathcal{L}(q)$), and there are finitely many states $q \in Q$. $\square$

The next stage, before showing that Separation, Forward Closure, and Atomicity characterize the (reachable) linear step automata that may be realized with (well-behaved) ZS-nets, is to define the ZS-net synthesized from a subset of ZS-regions of $A = (Q, L, q_0)$.

Definition 10. *Given a linear step automaton $A = (Q, L, q_0)$ over T and two finite subsets $\mathcal{S}$ and $\mathcal{Z}$ of S-regions and Z-regions, respectively, of A, the ZS-net synthesized from $\mathcal{S}$ and $\mathcal{Z}$ is the ZS-net $N(\mathcal{S}, \mathcal{Z}) = (S \cup Z, T, F, M_0)$ defined as follows, up to an isomorphism of ZS-nets over T:*

- S is a set in bijection with $\mathcal{S}$, where the bijection maps $s \in S$ to $(\phi_s, \eta_s) \in \mathcal{S}$,
- Z is a set in bijection with $\mathcal{Z}$, where the bijection maps $z \in Z$ to $\eta_z \in \mathcal{Z}$,
- for all $s \in S$, $M_0(s) = \phi_s(q_0)$, and $\eta_s(t) = (F(s,t), F(t,s))$ for all t,
- for all $z \in Z$, $M_0(z) = 0$, and $\eta_z(t) = (F(z,t), F(t,z))$ for all t

We can now state the main theorem we want to prove.

Theorem 1. *Let $A = (Q, L, q_0)$ be a regular and reachable linear step automaton over T, satisfying the conditions of Separation, Forward Closure and Atomicity. Let $\mathcal{S}$ and $\mathcal{Z}$ be finite sets of S-regions and Z-regions, respectively, of A, witnessing for the satisfaction of these conditions. The ZS-net $N(\mathcal{S}, \mathcal{Z})$ is well-behaved, and it realizes A, i.e. $A \cong RSSG(N(\mathcal{S}, \mathcal{Z}))$.*

The proof of Theorem 1 is cut into four propositions. The first three propositions establish the isomorphism $A \cong RSSG(N(\mathcal{S}, \mathcal{Z}))$. The last proposition shows that $N(\mathcal{S}, \mathcal{Z})$ is well-behaved. The following definition of the relation $\sim$ intends to capture the isomorphism $A \cong RSSG(N(\mathcal{S}, \mathcal{Z}))$. In the sequel, we let $N = N(\mathcal{S}, \mathcal{Z})$ for convenience.

Definition 11. *Let $\sim \subseteq Q \times RSM(N)$ be the relation such that $q \sim M$ if and only if $\phi_s(q) = M(s)$ for all $s \in S$.*

Proposition 7. *Let $q \sim M$ and $w \in L(q, q')$, then $M\{[w\rangle M'$ and $q' \sim M'$ for some $M' \in RSM(N)$.*

Proof. For any S-region (ϕ_s, η_s) in $\mathcal{S}$, $\eta_s(w) = (\phi_s(q), \phi_s(q'))$ and $\phi_s(q) = M(s)$. For any Z-region η_z in $\mathcal{Z}$, $\eta_z(w) = (0, 0)$. If we let $w = t_1 \ldots t_n$, then $\eta_s(w) = (\sum_i F(s, t_i), \sum_i F(t_i, s))$, hence $M[t_1\rangle M_1 \ldots [t_n\rangle M_n$, this firing sequence is concurrently and perfectly enabled , and it leads to a stable marking M_n, since $\eta_z(w) = (0, 0)$ for all z. Let $M' = M_n$. For any S-region (ϕ_s, η_s) in $\mathcal{S}$, $M'(s) = M(s) - \sum_i F(s, t_i) + \sum_i F(t_i, s) = \sum_i F(t_i, s)$, hence $(M(s), M'(s)) = \eta_s(w) = (\phi_s(q), \phi_s(q'))$ and thus $q' \sim M'$. Finally, for $i < n$, $\eta_z(t_1 \ldots t_i) = (0, j)$ and $j \neq 0$ for some Z-region η_z in $\mathcal{Z}$ (by the condition of Atomicity), hence all markings M_i $(i < n)$ are unstable. Therefore, $M\{[w\rangle M'$. $\square$

Proposition 8. *Let $q \sim M$ and $M\{[w\rangle M'$ then $w \in L(q, q')$ for some $q' \sim M'$.*

Proof. We first show, by an induction on the prefixes v of w ($\neq \varepsilon$), that w must be a prefix of some word in $\cup_{q' \in Q} L(q, q')$. Suppose this property holds for v, with $w = vtv'$. If it does not hold for vt, then by the Forward Closure condition, the transition t cannot be fired at the marking of $N(\mathcal{S}, \mathcal{Z})$ reached by v, in contradiction with $M\{[w\rangle M'$. We show next that $w \in \cup_{q' \in Q} L(q, q')$. If this is not the case, then $ww' \in \cup_{q' \in Q} L(q, q')$ for some $w' \neq \varepsilon$. By the Atomicity condition, $\eta_z(w) = (0, j)$ and $j \neq 0$ for some Z-region η_z in $\mathcal{Z}$, but this is impossible since on the one hand, $\eta_z(w) = (0, j)$ entails $M'(z) = j$ by definition of the zero-places of $N(\mathcal{S}, \mathcal{Z})$ (Def. 10), and on the other hand, $M'(z) = 0$ because M' is a stable marking. Finally, if $w \in L(q, q')$ then $q' \sim M'$ because $w \in L(q, q')$ entails by Proposition 7 that $M\{[w\rangle M''$ and $q' \sim M''$ for some $M'' \in RSM(N)$, and the firing rule of ZS-nets is deterministic. $\square$

Proposition 9. *The relation $\sim$ is a bijection between Q and $RSM(N)$.*

Proof. Let $q, q' \in Q$ and $M, M' \in RSM(N)$. If $q \sim M$ and $q \sim M'$, then $M(s) = \phi_s(q)$ and $M'(s) = \phi_s(q)$ for every S-region (ϕ_s, η_s) in $\mathcal{S}$, hence $M = M'$ (as M and M' are stable, $M(z) = M'(z) = 0$ for any Z-region η_z in $\mathcal{Z}$). If $q \sim M$ and $q' \sim M$, then $M(s) = \phi_s(q)$ and $M(s) = \phi_s(q')$ for every S-region (ϕ_s, η_s) in $\mathcal{S}$, hence $q = q'$ by Separation. $\qquad\square$

Propositions 7, 8, and 9 entail directly the isomorphism $A \cong RSSG(N(\mathcal{S}, \mathcal{Z}))$.

Proposition 10. $N = N(\mathcal{S}, \mathcal{Z})$ *is well-behaved.*

Proof. Let $M \in RSM(N)$, then $M \sim q$ for some $q \in Q$ because $A \cong RSSG(N)$. Let $\rho = M[t_1\rangle M_1 \ldots [t_n\rangle M_n$ such that $M(s) \geq \sum_i F(s, t_i)$ for every S-region (ϕ_s, η_s) in $\mathcal{S}$. If some prefix $v = t_1 \ldots t_{i-1}$ of $w = t_1 \ldots t_n$ belongs to $\cup_{q' \in Q} L(q, q')$, then by Proposition 7, the initial segment of ρ with length $i - 1$ is a stable transaction, hence ρ extends a stable transaction. In the converse case, we show by induction on $i \leq n$ that vt_i must be a prefix of some word in $\cup_{q' \in Q} L(q, q')$. Assume for a contradiction that $v = \varepsilon$ or v is a prefix of some word in $\cup_{q' \in Q} L(q, q')$ but vt_i is not. By the Forward Closure condition, $\eta_s(vt_i) = (k, j)$ and $\phi_s(q) < k$ for some S-region in $\mathcal{S}$, or $\eta_z(v) = (0, j)$, $\eta_z(t_i) = (k, j')$, and $j < k$ for some Z-region in $\mathcal{Z}$. In both cases, seeing that $\phi_s(q) = M(s)$, $\eta_s(t) = (F(s, t), F(t, s))$ and $\eta_z(t) = (F(z, t), F(t, z))$ for all $s \in S$ and $z \in Z$ by construction of $N(\mathcal{S}, \mathcal{Z})$, $M[t_1\rangle M_1 \ldots M_{i-1}[t_i\rangle$ is not a concurrently enabled firing sequence of N. It follows from this contradiction that N is well-behaved. $\qquad\square$

The proof of Theorem 1 has been completed. In view of Prop. 2, Theorem 1 may also be formulated equivalently as follows.

Theorem 2. *A reachable linear step automaton A over T may be realized by a bounded and well-behaved ZS-net N over T if and only if it is regular (Def 3) and the conditions of Separation, Forward Closure and Atomicity are satisfied. In this case, let $\mathcal{S}$ and $\mathcal{Z}$ be finite sets of S-regions and Z-regions, respectively, of A, witnessing for the satisfaction of these conditions, then the synthesized net $N = N(\mathcal{S}, \mathcal{Z})$ is a solution to the ZS-net realization problem for A.*

The above theorems provide a characterization of the ZS-net realizable linear step automata, but they say nothing about the decision of the realization problem nor about the effective construction of a solution $N = N(\mathcal{S}, \mathcal{Z})$. These aspects are treated in sections 5 and 6.

5   Computing S-Regions and Z-Regions

In this section, we supply a linear algebraic characterization of the S-regions and Z-regions of a regular and reachable linear step automaton $A = (Q, L, q_0)$, which we assume non-trivial. We also suppose that A is provided with auxiliary data as follows. For all states q and q' such that $L(q,q') \neq \emptyset$, $\mathcal{A}(q,q')$ is a trim automaton accepting $L(q,q')$, and $w(q,q')$ is a distinguished word in $L(q,q')$. For all states q, $\mathcal{A}(q)$ is a trim automaton accepting $\cup_{q' \in Q} L(q,q')$. We define the *size* of A as the sum of the sizes of all automata $\mathcal{A}(q,q')$, which is greater than $|Q|$ and in general greater than the sum of the sizes of all automata $\mathcal{A}(q)$.

Notation 3 *Let $\mathcal{A} = (\mathcal{Q}, \delta, q, \mathcal{Q}_F)$ be a trim automaton over T, with initial state $q \in Q$. The set of transitions $q' \xrightarrow{t} \delta(q',t)$ may be divided into arcs and chords such that the arcs form a spanning tree $\mathcal{T}$ with root q (the number of chords is $|T| - |\mathcal{Q}| + 1$ where $|T|$ is the number of transitions of $\mathcal{A}$). Let (u,t,v) denote the chord $\delta(q,u) \xrightarrow{t} \delta(q,v)$ where $\delta(q,ut) = \delta(q,v)$ and u and v are labels of rooted paths in $\mathcal{T}$. $\mathcal{T}(q,q')$ and $\mathcal{T}(q)$ denote arbitrary spanning trees for $\mathcal{A}(q,q')$ and $\mathcal{A}(q)$, respectively.*

Notation 4 *$L(q,q') \neq \emptyset$ is noted $q \to q'$ for short.*

Notation 5 *For any map $\eta : T \to (\mathbb{N} \times \mathbb{N})$ and for any word $w \in T^*$, we let $\eta(w) = (\eta^\circ w, w^\circ \eta)$ in the monoid $((\mathbb{N} \times \mathbb{N}), +, (0,0))$, and $\eta(w) = (\eta^\bullet w, w^\bullet \eta)$ in the monoid $((\mathbb{N} \times \mathbb{N}), \cdot, (0,0))$ - see Def. 7.*

Let (ϕ_s, η_s) be an S-region of A, thus $\phi_s : Q \to \mathbb{N}$ and $\eta_s : T \to (\mathbb{N} \times \mathbb{N})$. By definition of S-regions, $q \to q' \Rightarrow \phi_s(q) = \eta_s^\circ w(q,q')$ and $\phi_s(q') = w(q,q')^\circ \eta_s$, hence $\eta_s : T \to (\mathbb{N} \times \mathbb{N})$ determines $\phi_s : Q \to \mathbb{N}$ (since A is non-trivial). Conversely, in order that a map $\eta_s : T \to (\mathbb{N} \times \mathbb{N})$ determines an S-region of A, it is necessary and sufficient that the following conditions hold for all states q, q', q'' and for all sink states q_f (i.e. states q_f with no successor state w.r.t. the relation $\to$):

- $q_0 \to q$ and $q_0 \to q' \Rightarrow \eta_s^\circ w(q_0,q) = \eta_s^\circ w(q_0,q')$
- $q \to q'$ and $q' \to q'' \Rightarrow w(q,q')^\circ \eta_s = \eta_s^\circ w(q',q'')$
- $q \to q_f$ and $q' \to q_f \Rightarrow w(q,q_f)^\circ \eta_s = w(q',q_f)^\circ \eta_s$
- $\eta_s(u) = \eta_s(w(q,q'))$ for all $u \in L(q,q')$

Now for any $w \in T^*$, $\eta_s^\circ w = \sum_t w[t] \times \eta_s^\circ t$ and $w^\circ \eta_s = \sum_t w[t] \times t^\circ \eta_s$, where $w[t]$ is the number of occurrences of t in w. The first three conditions amount thus to a finite set of linear homogeneous equations in the variables $\eta_s^\circ t$ and $t^\circ \eta_s$ $(t \in T)$. Let $\mathcal{A}(q,q') = (\mathcal{Q}, \delta, q, \mathcal{Q}_F)$. In order that the fourth condition holds, it is necessary and sufficient that first, $\eta_s^\circ u = \eta_s^\circ w(q,q')$ and $u^\circ \eta_s = w(q,q')^\circ \eta_s$ for any word u labelling a path from q to (some state in) $\mathcal{Q}_F$ in the spanning tree $\mathcal{T}(q,q')$, and second, that $\eta_s^\circ ut = \eta_s^\circ v$ and $ut^\circ \eta_s = v^\circ \eta_s$ for any chord (u,t,v) (notation 5). The necessity of the second condition was already stated in Proposition 6. This yields anew a finite set of linear homogeneous equations in the variables $\eta_s^\circ t$ and $t^\circ \eta_s$. Altogether, the S-regions of A are characterized by a finite linear system $\Sigma_S(A)$, made of homogeneous equations in the variables $\eta_s{}^\circ t$ and $t^\circ\eta_s$, $t \in T$. This system has size polynomial in the size of A. Each non-negative integer solution of $\Sigma_S(A)$, seen as a map $\eta_s : T \to (\mathbb{N} \times \mathbb{N})$, viz $\eta_s(t) = (\eta_s{}^\circ t, t^\circ\eta_s)$, induces a unique map $\phi_s : Q \to \mathbb{N}$ such that (ϕ_s, η_s) is an S-region of A.

By definition of Z-regions, a map $\eta_z : T \to (\mathbb{N} \times \mathbb{N})$ is a Z-region of A if and only if $q \to q'$ entails $\eta_z{}^\bullet u = 0$ for every prefix u of a word in $L(q, q')$, and $u^\bullet\eta_z = 0$ for every word $u \in L(q, q')$. When both conditions hold, let $A(q) = (Q, \delta, q, Q_F)$, then $\delta(q, u) = \delta(q, u') \Rightarrow \eta_z(u) = \eta_z(u')$ by Proposition 6. Therefore, a map $\eta_z : T \to (\mathbb{N} \times \mathbb{N})$ defines a Z-region of A if and only if the following conditions hold for all q, u, t, u', t', v', w such that $q \in Q$, u labels a rooted path in the spanning tree $\mathcal{T}(q)$, $t \in T$, (u', t', v') is a chord of $\mathcal{T}(q)$ (notation 5), and w labels a rooted path from q to (some state in) Q_F in $\mathcal{T}(q)$:

- $\delta(\delta(q, u), t))$ defined $\Rightarrow \eta_z{}^\bullet t \leq u^\bullet\eta_z$ (i.e. $\eta_z{}^\bullet ut = \eta_z{}^\bullet u$)
- $u't'^\bullet\eta_z = v'^\bullet\eta_z$
- $w^\bullet\eta_z = 0$

As the first condition should hold for every prefix of u, entailing that $\eta_z{}^\bullet u = 0$, it may be replaced equivalently with: $\delta(\delta(q, u), t))$ defined $\Rightarrow \eta_z{}^\bullet t \leq u^\bullet\eta_z - \eta_z{}^\bullet u$. For $u = t_1 \ldots t_n$, this condition may be rewritten to the linear inequality $\eta_z{}^\bullet t \leq \sum_{i \leq n}(t_i{}^\bullet\eta_z - \eta_z{}^\bullet t_i)$. The second and third conditions may be rewritten similarly to linear equations. Altogether, the Z-regions of A are characterized by a finite linear system $\Sigma_Z(A)$, made of homogeneous equations and inequalities in the variables $\eta_z{}^\bullet t$ and $t^\bullet\eta_z$, $t \in T$. This system has size polynomial in the size of A. Each non-negative integer solution of $\Sigma_Z(A)$ defines a Z-region of A.

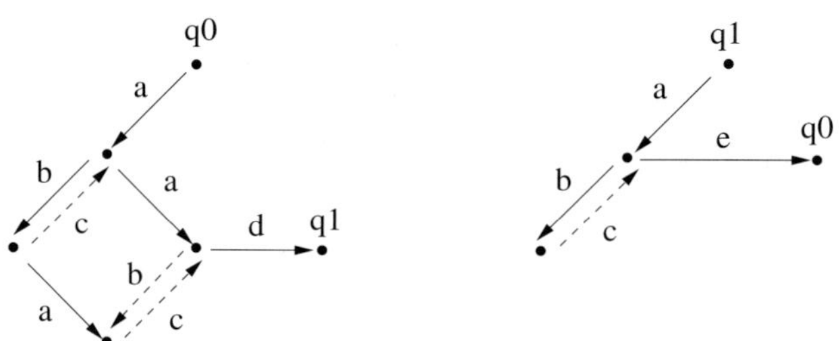

Fig. 2. Spanning Trees (solid arcs) and Chords (dashed)

Example 5. Let $A = (Q, L, q_0)$ where $Q = \{q_0, q_1\}$, $L(q_0, q_1) = (a\,(bc)^* \| a\,(bc)^*)\,d$, $L(q_1, q_0) = a\,(bc)^*e$, and $L(q_0, q_0) = L(q_1, q_1) = \emptyset$ (see Example 1). Let $\mathcal{A}(q_0, q_1)$ and $\mathcal{A}(q_1, q_0)$ be the two automata shown in Figure 2, where the spanning trees are represented with solid arcs while the chords are dashed. Let $w(q_0, q_1) = aad$ and $w(q_1, q_0) = ae$.

The equations in $\Sigma_S(A)$ are:

$$aad^\circ \eta_s = \eta_s{}^\circ ae \qquad \text{(from } q_0 \to q_1 \text{ and } q_1 \to q_0)$$
$$ae^\circ \eta_s = \eta_s{}^\circ aad \qquad \text{(from } q_1 \to q_0 \text{ and } q_0 \to q_1)$$
$$\eta_s(aad) = \eta_s(aad) \qquad (aad \text{ is a path in } \mathcal{T}(q_0, q_1))$$
$$\eta_s(abc) = \eta_s(a) \qquad \text{(first chord } c \text{ of } \mathcal{T}(q_0, q_1))$$
$$\eta_s(aba) = \eta_s(aab) \qquad \text{(chord } b \text{ of } \mathcal{T}(q_0, q_1))$$
$$\eta_s(abac) = \eta_s(aa) \qquad \text{(second chord } c \text{ of } \mathcal{T}(q_0, q_1))$$
$$\eta_s(ae) = \eta_s(ae) \qquad (ae \text{ is a path in } \mathcal{T}(q_1, q_0))$$
$$\eta_s(abc) = \eta_s(a) \qquad \text{(chord } c \text{ of } \mathcal{T}(q_1, q_0))$$

After simplifications, one gets:

$$\eta_s{}^\circ b = b^\circ \eta_s = \eta_s{}^\circ c = c^\circ \eta_s = 0$$
$$2\eta_s{}^\circ a + \eta_s{}^\circ d = a^\circ \eta_s + e^\circ \eta_s$$
$$\eta_s{}^\circ a + \eta_s{}^\circ e = 2a^\circ \eta_s + d^\circ \eta_s$$

The equations in $\Sigma_Z(A)$ induced from $L(q_0, q_1)$ are (after some elimination):

$$\eta_s{}^\bullet a = 0 \qquad (\delta(q_0, a) \text{ is defined})$$
$$\eta_s{}^\bullet b \le a^\bullet \eta_s \qquad (\delta(q_0, ab) \text{ is defined})$$
$$\eta_s{}^\bullet c \le a^\bullet \eta_s + b^\bullet \eta_s - \eta_s{}^\bullet b \qquad (\delta(q_0, abc) \text{ is defined})$$
$$\eta_s{}^\bullet d \le 2a^\bullet \eta_s \qquad (\delta(q_0, aad) \text{ is defined})$$
$$abc^\bullet \eta_s = a^\bullet \eta_s \qquad \text{(first chord } c \text{ of } \mathcal{T}(q_0, q_1))$$
$$aab^\bullet \eta_s = aba^\bullet \eta_s \qquad \text{(chord } b \text{ of } \mathcal{T}(q_0, q_1))$$
$$abac^\bullet \eta_s = aa^\bullet \eta_s \qquad \text{(second chord } c \text{ of } \mathcal{T}(q_0, q_1))$$
$$aad^\bullet \eta_s = 0 \qquad (aad \text{ is a path of } \mathcal{T}(q_0, q_1))$$

After simplifications, one gets:

$$\eta_s{}^\bullet a = 0$$
$$2a^\bullet \eta_s = \eta_s{}^\bullet d$$
$$d^\bullet \eta_s = 0$$
$$\eta_s{}^\bullet b \le a^\bullet \eta_s$$
$$b^\bullet \eta_s + c^\bullet \eta_s = \eta_s{}^\bullet b + \eta_s{}^\bullet c$$

The remaining equations in $\Sigma_Z(A)$ induced from $L(q_1, q_0)$ are:

$$a^\bullet \eta_s = \eta_s{}^\bullet e$$
$$e^\bullet \eta_s = 0 \qquad\qquad\qquad \square$$

The linear systems $\Sigma_S(A)$ and $\Sigma_Z(A)$ are used in section 6 to obtain a decision and synthesis procedure for the problem $? (\exists N)(A \cong RSSG(N))$.

6   A Decision and ZS-Net Synthesis Procedure

Let $A = (Q, L, q_0)$ be a regular and reachable linear step automaton where $q_0 \to q'$ for some q' (possibly equal to q_0). For all $q \in Q$, let $\mathcal{A}(q)$ be a trim automaton accepting $\cup_{q'\in Q}L(q,q')$ and for all q' such that $q \to q'$, let $w(q,q')$ be a distinguished word in $L(q,q')$. By Theorem 2, A may be realized by a well-behaved ZS-net N over T if and only if the conditions of Separation, Forward Closure and Atomicity are satisfied.

Proposition 11. *Separation is decidable.*

Proof. One should decide, for any two distinct states $q, q' \in Q$, whether $\phi_s(q) \neq \phi_s(q')$ for some S-region (ϕ_s, η_s). By definition of S-regions, $q_1 \to q \Rightarrow \phi_s(q) = w(q_1,q)^\circ\eta_s$ and $q \to q_2 \Rightarrow \phi_s(q) = \eta_s^\circ w(q,q_2)$. Suppose, e.g., $q_1 \to q$ and $q' \to q'_2$ then $\phi_s(q) \neq \phi_s(q')$ if and only if $w(q_1,q)^\circ\eta_s \neq \eta_s^\circ w(q',q'_2)$ for some non-negative integer solution η_s of $\Sigma_S(A)$. The existence of such a solution can be decided using time polynomial in the size of this linear homogeneous system.
$\square$

Proposition 12. *Atomicity is decidable.*

Proof. One should decide, for all states $q \in Q$ and for all words $u \in T^+$ such that $uv \in \cup_{q'\in Q}L(q,q')$ for some $v \in T^+$, whether $u^\bullet\eta_z \neq 0$ for some Z-region η_z. Let $A(q) = (Q, \delta, q, Q_F)$ and let $T(q)$ be the associated spanning tree. By Proposition 6, $\delta(q,u) = \delta(q,u')$ entails $\eta_z(u) = \eta_z(u')$ for any Z-region η_z. In order to decide upon the condition of Atomicity, it suffices therefore to check first that $\delta(q_1,t)$ is undefined for all $q_1 \in Q_F$ and $t \in T$, and second that $u^\bullet\eta_z \neq 0$ for any word u labelling a path from q to some $q_2 \notin Q_F$ in $T(q)$. There exist finitely many such words u. Let $u = t_1 \ldots t_n$ then $u^\bullet\eta_z = \sum_{i\leq n}(t_i^\bullet\eta_z - \eta_z^\bullet t_i)$ by definition of Z-regions. Thus $u^\bullet\eta_z > 0$ for some Z-region η_z if and only if $\sum_{i\leq n}(t_i^\bullet\eta_z - \eta_z^\bullet t_i) > 0$ for some non-negative integer solution η_z of $\Sigma_Z(A)$. The existence of such a solution can be decided using time polynomial in the size of $\Sigma_Z(A)$.
$\square$

Proposition 13. *Forward closure is decidable.*

Proof. One should decide, for all states $q \in Q$, for all words $u \in T^*$ such that $u = \varepsilon$ or $uv \in \cup_{q'\in Q}L(q,q')$ for some $v \in T^+$, and for all transitions $t \in T$ such that $utv \notin \cup_{q'\in Q}L(q,q')$ for any $v \in T^*$, whether $\phi_s(q) - \eta_s^\circ u < \eta_s^\circ t$ for some S-region (ϕ_s, η_s) or $u^\bullet\eta_z < \eta^\bullet_z t$ for some Z-region η_z.

Let $A(q) = (Q, \delta, q, Q_F)$. By Proposition 6, $\delta(q,u) = \delta(q,u')$ entails $\eta_s(u) = \eta_s(u')$ for any S-region η_s and $\eta_z(u) = \eta_z(u')$ for any Z-region η_z. It suffices therefore to check the Forward Closure condition for $u = \varepsilon$ and for words $u = t_1 \ldots t_n$ labelling paths in $T(q)$, hence for finitely many words.

The Forward Closure condition, when restricted to Z-regions, writes as the linear inequality $\sum_{i\leq n}(t_i^\bullet\eta_z - \eta_z^\bullet t_i) < \eta_z^\bullet t$. One can decide whether this inequality holds for some Z-region of A using time polynomial in the size of $\Sigma_Z(A)$.

The Forward Closure condition, when restricted to S-regions, expresses in two different forms according to cases. If $q' \to q$ and $u = \varepsilon$, the inequality to check is $w(q',q)^\circ\eta_s < \eta_s^\circ t$. If $q \to q'$, then the inequality to check is $\eta_s^\circ w(q,q') < \eta_s^\circ u + \eta_s^\circ t$. In both cases, one can decide whether the appropriate inequality holds for some S-region of A using time polynomial in the size of $\Sigma_S(A)$.
$\square$

The above three propositions have established the following result.

Theorem 6. *The Zero-Safe net realization problem for regular and reachable linear step automata is decidable. The decision takes time polynomial in the size of linear step automata.*

7   Conclusion

The inventors of the Zero-Safe nets studied mainly their formal semantics. We tried in this paper to show algorithmic techniques for implementing systems of transactions producing and consuming resources in the form of bounded and well-behaved Zero-Safe nets. The specifications of transaction systems which we have considered are linear step automata in which the atomic steps between two stable states form a regular language. We have shown that it is decidable whether a linear step automaton may be implemented by a Zero-Safe net in which any execution started from a stable state can be completed to a stable transaction. A direction in which this work could be extended is transaction mining. In this context, the states of the step automaton are known exactly and they should be implemented up to a bijection, but the sets of atomic steps $L(q, q')$ are meant as partial observations and other atomic steps may be added in the implementation. We have some reasons to believe that the saturated ZS-net $N(\mathcal{S}, \mathcal{Z})$ synthesized from *all* ZS-regions of the specification (linear step automaton) yields in this case the least over-approximation of the specification that may be realized by Zero-Safe nets. It is worth noting that this saturated ZS-net may always be reduced to a finite ZS-subnet with an isomorphic reachable stable state graph (using standard techniques of linear algebra for representing a polyhedral cone by the finite set of extremal rays of this cone).

Acknowledgements. Thanks to the reviewers for their helpful suggestions.

References

1. Bruni, R., Montanari, U.: Zero-Safe nets, or transition synchronization made simple. In: Proc. Express. ENTCS, vol. 7 (1997)
2. Bruni, R., Montanari, U.: Zero-safe nets: The individual token approach. In: Parisi-Presicce, F. (ed.) WADT 1997. LNCS, vol. 1376, pp. 122–140. Springer, Heidelberg (1998)
3. Bruni, R., Montanari, U.: Zero-safe nets: Comparing the collective and individual token approaches. Information and Computation 156, 46–89 (2000)
4. Bruni, R., Montanari, U.: Transactions and Zero-Safe Nets. In: Ehrig, H., Juhás, G., Padberg, J., Rozenberg, G. (eds.) APN 2001. LNCS, vol. 2128, pp. 380–426. Springer, Heidelberg (2001)
5. van der Aalst, W.: Verification of Workflow Nets. In: Azéma, P., Balbo, G. (eds.) ICATPN 1997. LNCS, vol. 1248, pp. 407–426. Springer, Heidelberg (1997)
6. van Hee, K., Serebrenik, A., Sidorova, N., Voorhoeve, M.: Soundness of Resource-Constrained Workflow Nets. In: Ciardo, G., Darondeau, P. (eds.) ICATPN 2005. LNCS, vol. 3536, pp. 250–267. Springer, Heidelberg (2005)

7. Acu, B., Reisig, W.: Compensation in Workflow Nets. In: Donatelli, S., Thiagarajan, P.S. (eds.) ICATPN 2006. LNCS, vol. 4024, pp. 65–83. Springer, Heidelberg (2006)
8. Ehrenfeucht, A., Rozenberg, G.: Partial (Set) 2-Structures; Part I: Basic Notions and the Representation Problem. Acta Informatica 27, 315–342 (1990)
9. Ehrenfeucht, A., Rozenberg, G.: Partial (Set) 2-Structures; Part II: State Spaces of Concurrent Systems. Acta Informatica 27, 343–368 (1990)
10. Mukund, M.: Petri Nets and Step Transition Systems. International Journal of Foundations of Computer Science 3(4), 443–478 (1992)
11. Badouel, E., Darondeau, P.: Theory of Regions. In: Reisig, W., Rozenberg, G. (eds.) APN 1998. LNCS, vol. 1491, pp. 529–586. Springer, Heidelberg (1998)

A Note on Persistent Petri Nets

Eike Best

Parallel Systems, Department of Computing Science
Carl von Ossietzky Universität Oldenburg, D-26111 Oldenburg, Germany
`eike.best@informatik.uni-oldenburg.de`

Prepared for Ugo Montanari's 65th birthday

1   Introduction

Petri nets have traditionally been motivated by their ability to express concurrency. Subclasses of Petri nets with concurrency but without choices or conflicts have been studied extensively. One of the best known – and comparatively restricted – such class are the marked graphs [4]. However, perhaps the largest class of (intuitively) choice-free nets are the persistent nets [7], a class of nets that is significantly larger than marked graphs.

Some early results about persistent nets are Keller's theorem [6], which will be recalled in a later part of this paper, and the famous semilinearity result of Landweber and Robertson [7], which states that the set of reachable markings of a persistent net is semilinear. In this paper, we show that in bounded persistent nets, the smallest cycles of its reachability graph enjoy a uniqueness property.

2   Definitions

A Petri net (S, T, F, M_0) consists of two finite and disjoint sets S (places) and T (transitions), a function $F\colon ((S \times T) \cup (T \times S)) \to \mathbb{N}$ (flow) and a marking M_0 (the initial marking). A marking is a mapping $M\colon S \to \mathbb{N}$.

The incidence matrix C is an $S \times T$ matrix of integers where the entry corresponding to a place s and a transition t is, by definition, equal to the number $F(t, s) - F(s, t)$. A T-invariant J is a vector of integers with index set T satisfying $C \cdot J = \underline{0}$ where $\cdot$ is the inner (scalar) product, and $\underline{0}$ is the vector of zeros with index set S. For a sequence $\sigma \in T^*$ of transitions, its Parikh vector $\Psi(\sigma)$ is a vector of natural numbers with index set T, where $\Psi(\sigma)(t)$ is equal to the number of occurrences of t in σ.

A transition t is enabled (or activated, or firable) in a marking M (denoted by $M[t\rangle$) if, for all places s, $M(s) \geq F(s, t)$. If t is enabled in M, then t can occur (or fire) in M, leading to the marking M' defined by $M'(s) = M(s) + F(t, s) - F(s, t)$ (notation: $M[t\rangle M'$). We apply definitions of enabledness and of the reachability relation to transition (or firing) sequences $\sigma \in T^*$, defined inductively: $M[\varepsilon\rangle$ and $M[\varepsilon\rangle M$ are always true; and $M[\sigma t\rangle$ (or $M[\sigma t\rangle M'$) iff there is some M'' with $M[\sigma\rangle M''$ and $M''[t\rangle$ (or $M''[t\rangle M'$, respectively).

P. Degano et al. (Eds.): Montanari Festschrift, LNCS 5065, pp. 427–438, 2008.

A marking M is reachable (from M_0) if there exists a transition sequence σ such that $M_0[\sigma\rangle M$. The reachability graph of N, with initial marking M_0, is the graph whose vertices are the markings reachable from M_0 and where an edge labelled with t leads from M to M' iff $M[t\rangle M'$. Figure 1 shows an example where on the right-hand side, M_0 denotes the marking shown in the Petri net on the left-hand side.

The marking equation states that if $M[\sigma\rangle M'$, then $M' = M + C \cdot \Psi(\sigma)$.

A Petri net is k-bounded if in any reachable marking M, $M(s) \le k$ holds for every place s, and bounded if there is some k such that it is k-bounded. A finite Petri net (and we consider only such nets in the sequel) is bounded if and only if the set of its reachable markings is finite.

A net $N = (S, T, F, M_0)$ is called a marked graph if $\sum_{t \in T} F(s,t) \le 1$ and $\sum_{t \in T} F(t,s) \le 1$ for all places s.

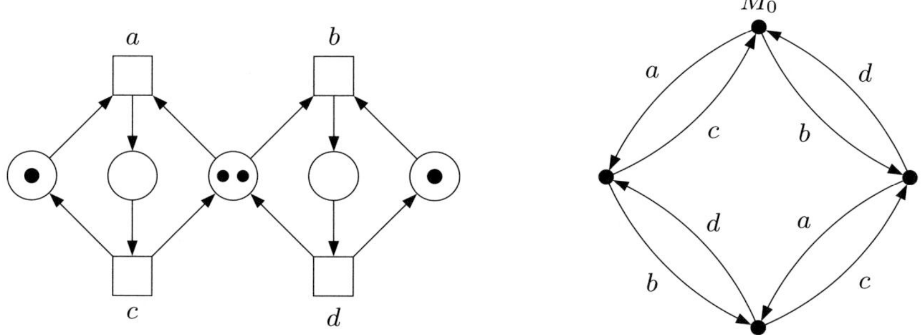

Fig. 1. A persistent Petri net (l.h.s.) and its reachability graph (r.h.s.)

3   Persistent Nets, and Related Notions

A net N, with some initial marking, will be called persistent, if whenever $M[t_1\rangle$ and $M[t_2\rangle$ for a reachable marking M and transitions $t_1 \ne t_2$, then $M[t_1 t_2\rangle$.

Two sequences $M[\sigma\rangle$ and $M[\sigma'\rangle$, firable from M, are said to arise from each other by a single permutation if they are the same, except for the order of an adjacent pair of transitions, thus:

$$\sigma \;=\; t_1 \ldots t_k t t' \ldots t_n \quad \text{and} \quad \sigma' \;=\; t_1 \ldots t_k t' t \ldots t_n.$$

Two sequences $M[\sigma\rangle$ and $M[\sigma'\rangle$ are said to be permutations of each other (from M, written $\sigma \equiv_M \sigma'$) if they are both firable at M and arise out of each other through a (possibly empty) sequence of single permutations.

By $\tau \stackrel{\bullet}{-} \sigma$, we will informally mean the sequence that is obtained when the maximal subsequence of σ that occurs – in some order – inside τ, is erased from τ. Formally, $\tau \stackrel{\bullet}{-} \sigma$ can be defined by induction on the length of σ:

$$\tau \stackrel{\bullet}{-} \varepsilon = \tau$$

$$\tau \stackrel{\bullet}{-} t = \begin{cases} \tau, \text{ if there is no transition } t \text{ in } \tau \\ \text{the sequence obtained by erasing the first } t \text{ in } \tau, \text{ otherwise} \end{cases}$$

$$\tau \stackrel{\bullet}{-} (t\sigma) = (\tau \stackrel{\bullet}{-} t) \stackrel{\bullet}{-} \sigma.$$

By this definition, we have that $\tau \stackrel{\bullet}{-} \sigma$ and $\sigma \stackrel{\bullet}{-} \tau$ contain no common transitions.

The following theorem is the basis for our subsequent considerations. We will provide a new proof, since we will refer to the technical details of this proof in the sequel.

Theorem 1. KELLER [6]
Let N be a persistent net and let τ and σ be two firing sequences starting from some reachable marking M. Then

$$\tau(\sigma \stackrel{\bullet}{-} \tau) \ \text{ and } \ \sigma(\tau \stackrel{\bullet}{-} \sigma)$$

are also firing sequences from M, and what is more, they are permutations of each other, i.e., $\tau(\sigma \stackrel{\bullet}{-} \tau) \equiv_M \sigma(\tau \stackrel{\bullet}{-} \sigma)$. In particular, the marking reached after $\tau(\sigma \stackrel{\bullet}{-} \tau)$ equals the marking reached after $\sigma(\tau \stackrel{\bullet}{-} \sigma)$.

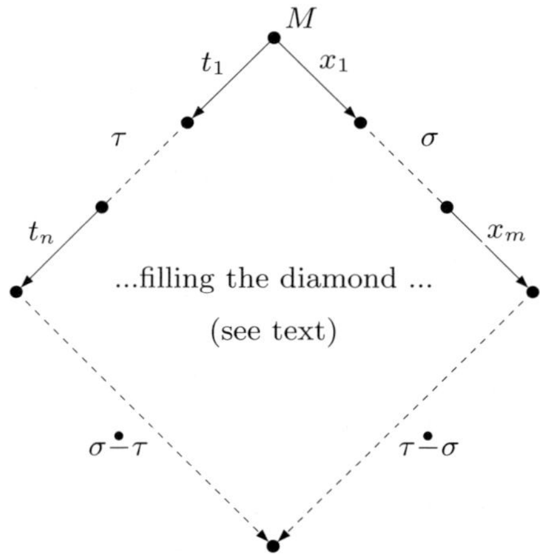

Fig. 2. Outlining the proof of Keller's theorem

Proof:
Let $\tau = t_1 \ldots t_n$ and $\sigma = x_1 \ldots x_m$. Persistency allows us to complete "small diamonds". For instance, if $t_1 \neq x_1$, then both $M[t_1 x_1\rangle$ and $M[x_1 t_1\rangle$. The proof proceeds by "completing big diamonds", such as the one shown in Figure 2. Special care needs to be taken if one of the t_i equals one of the x_j. We proceed systematically through $t_1, \ldots, t_n$, in this order, trying to match as many transitions t_i as possible with transitions x_j from σ.

The starting point are the North, West and East corners of the diamond shown in Figure 2. For each t_i, we will define an i'th line from Northwest to Southeast. Formally, each line is of the form $\widehat{M}[\widehat{\sigma}\rangle$, where $\widehat{M}$ is a marking reachable from M by some subsequence of τ and $\widehat{\sigma}$ is some subsequence of σ. Line 0, i.e. the starting line, is given by σ which leads from the North corner to the East corner; formally, it is defined to be $M[\sigma\rangle$. We distinguish two cases:

- If t_i does not have some transition $x_j = t_i$ on the previous (the $(i{-}1)$'th) line, we draw a new (i'th) line after t_i which is an exact parallel to the previous line, and we cut the resulting parallelogram by small t_i-labelled arcs form Northeast to Southwest. This is justified by the small-diamond completion property of persistent nets.
- If, however, on the $(i{-}1)$'th line, there are some x-transitions that are the same as t_i, we choose the first j for which $x_j = t_i$ holds and we draw an exact parallel only up to the starting point of x_j. The endpoint of this parallel will be merged with the endpoint of x_j, and from that point onwards, the new line is the same as the previous line. Thus, the new line contains one arc (namely, an arc labelled with x_j) less than the previous line. If x_j happens to be the first x-transition on the previous line, this construction corresponds to the special case of merging the endpoints of t_i and x_j. The resulting parallelogramoid is again subdivided by small t_i-arcs, but this time only up to the arc before x_j, if there is any.

It is clear that the last line, from the West corner to the resulting South corner, corresponds to the sequence $\sigma \overset{\bullet}{-} \tau$ of unmatched x-transitions, and it is also not hard to see that the line from the East corner to the South corner corresponds to the sequence $\tau \overset{\bullet}{-} \sigma$ of unmatched t-transitions.   □ 1

Had we started with σ instead with τ, we would possibly have ended up with a different interior of the diamond. However, the borders of the diamond are unique.

Figures 3–5 exhibit some examples with $\tau = t_1 t_2 t_3$ and $\sigma = x_1 x_2 x_3$.

In Figure 3, we assume that the only common transition between $t_1 t_2 t_3$ and $x_1 x_2 x_3$ is $t_2 = x_3$. If, additionally, $t_1 = x_1$, we get the diamond shown in Figure 4. If $t_3 = x_2$ holds further, then the three southmost nodes are merged into a single node, and the t_3- and x_2-arcs leading into them collapse into a single arc labelled t_3 (or x_2), and we arrive at the diamond shown in Figure 5.

Clearly, if two sequences are permutations of each other, then they have the same Parikh vector. In persistent nets, the converse is also true, provided they are both firable from some marking M:

Lemma 1. PARIKH-EQUIVALENCE AND $\equiv$
Let N be a persistent net, M a reachable marking, and $M[\tau\rangle$ and $M[\sigma\rangle$ two transition sequences which are firable at M. If $\Psi(\tau) = \Psi(\sigma)$, then $\tau \equiv_M \sigma$.

Proof: By the definitions of Parikh vector and $\overset{\bullet}{-}$, we have $\Psi(\tau) = \Psi(\sigma)$ if and only if both $\tau \overset{\bullet}{-} \sigma = \varepsilon$ and $\sigma \overset{\bullet}{-} \tau = \varepsilon$. The lemma now follows directly from Theorem 1.   □ 1

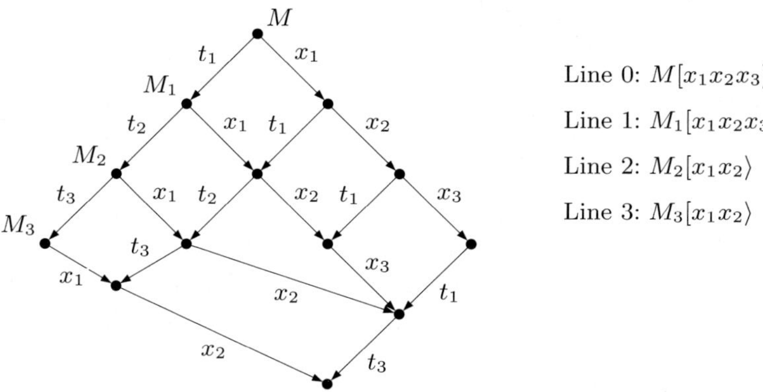

Fig. 3. Diamond for $t_1t_2t_3$ and $x_1x_2x_3$, assuming $t_2{=}x_3$

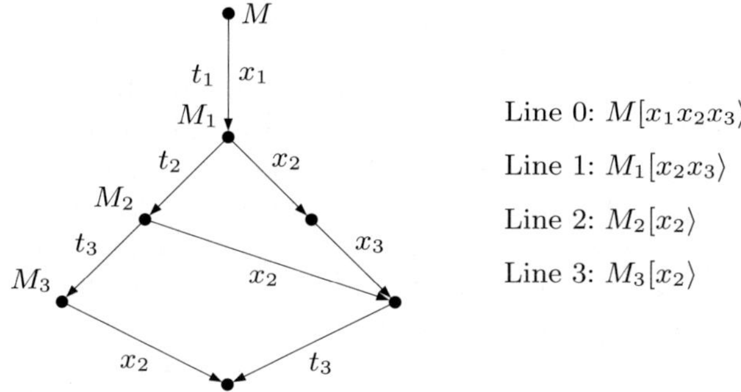

Fig. 4. Diamond for $t_1t_2t_3$ and $x_1x_2x_3$, assuming $t_1{=}x_1{\wedge}t_2{=}x_3$

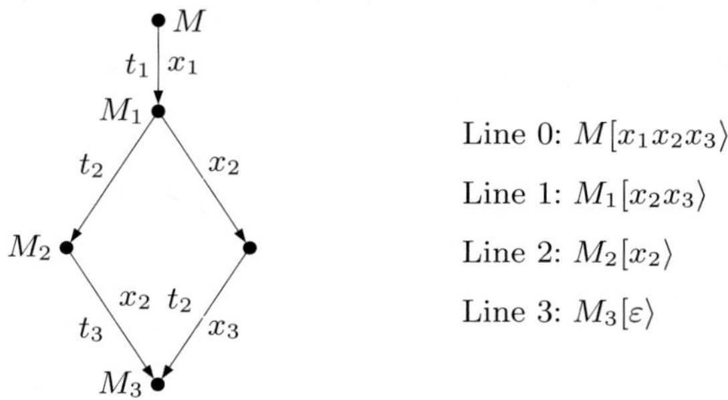

Fig. 5. Diamond for $t_1t_2t_3$ and $x_1x_2x_3$, assuming $t_1{=}x_1{\wedge}t_2{=}x_3{\wedge}t_3{=}x_2$

Lemma 2. A PERMUTATION LEMMA

Let N be persistent and let $M[\gamma\rangle$ and $M[\kappa\gamma\rangle$ be two firing sequences from M, with transition sequences γ and κ.
Then there is a firing sequence $M[\gamma\kappa'\rangle$ such that $\Psi(\kappa) = \Psi(\kappa')$ and $\kappa\gamma \equiv_M \gamma\kappa'$.

That is, κ can as a whole be permuted with γ, albeit, possibly, up to transition re-orderings within κ.

Proof: By Keller's theorem, $M[\gamma\rangle$ and $M[\kappa\gamma\rangle$ imply that $M[\gamma(\kappa\gamma\overset{\bullet}{-}\gamma)\rangle$. Put $\kappa' = \kappa\gamma\overset{\bullet}{-}\gamma$. Clearly, $\Psi(\kappa\gamma\overset{\bullet}{-}\gamma) = \Psi(\kappa)$, and hence $\Psi(\gamma\kappa') = \Psi(\kappa\gamma)$. The fact that $\kappa\gamma \equiv_M \gamma\kappa'$ now follows from Lemma 1, since both sequences are firable from M. $\square$ 2

Lemma 3. DISJOINT SEQUENCES CONTRADICT BOUNDEDNESS

Let N be bounded and persistent and let τ and σ be two firing sequences, both firable at M and leading to the same marking $\widetilde{M}$. Further, suppose that $M \neq \widetilde{M}$. Then there is at least one transition which occurs both in τ and in σ.

Proof: We will prove the lemma by contradiction. First, note that by $M \neq \widetilde{M}$ and because τ and σ lead to $\widetilde{M}$, $\tau \neq \varepsilon \neq \sigma$. Assume that τ and σ are transition-disjoint. Then $\tau\overset{\bullet}{-}\sigma = \tau$ and $\sigma\overset{\bullet}{-}\tau = \sigma$. By (the proof of) Keller's theorem, we get a special case of the diamond in Figure 2, namely Diamond 1 in Figure 6. By assumption, the markings reached after τ and σ are the same, so that the West and East corners of Diamond 1 are the same marking $\widetilde{M}$. Because σ is firable from the West corner, it is also firable from the East corner, and thus we get Diamond 2, again from transition-disjointness. The West and East corners of Diamond 2 are again the same, because both are the marking obtained by firing σ from $\widetilde{M}$. In this way, we get an infinite sequence $M[\sigma\rangle\widetilde{M}[\sigma\rangle\widehat{M}[\sigma\rangle\overline{M}[\sigma\rangle\ldots$ along the Eastern ridge of the diamonds. Moreover, since $M \neq \widetilde{M}$, there is some place s with $M(s) \neq \widetilde{M}(s)$. Since the effect of σ is monotonic, it must be the case that both $\widetilde{M}(s) \neq \widehat{M}(s)$ and $M(s) \neq \widehat{M}(s)$. Thus we also have that $\widehat{M}$ is not in $\{M, \widetilde{M}\}$, and that $\overline{M}$ is not in $\{M, \widetilde{M}, \widehat{M}\}$, and so on. However, this contradicts boundedness.
Hence, our assumption was wrong, and instead, the claim of the lemma is true.
                    $\square$ 3

4   Uniqueness of Simple Cycles in the Reachability Graph

A transition sequence τ is called cyclic if $C \cdot \Psi(\tau) = \underline{0}$, i.e. if $\Psi(\tau)$ is a T-invariant. From the marking equation, τ is cyclic if and only if $M[\tau\rangle M$, for all markings M which activate τ.

Definition 1. DECOMPOSABILITY AND SIMPLICITY

A cyclic transition sequence τ is called decomposable if $\tau = \tau_1\tau_2$ such that τ_1 and τ_2 are cyclic and $\tau_1 \neq \varepsilon \neq \tau_2$.

A cyclic firing sequence $M[\tau\rangle M$ is called simple if there is no decomposable permutation $\tau' \equiv_M \tau$, also starting from M. $\square$ 1

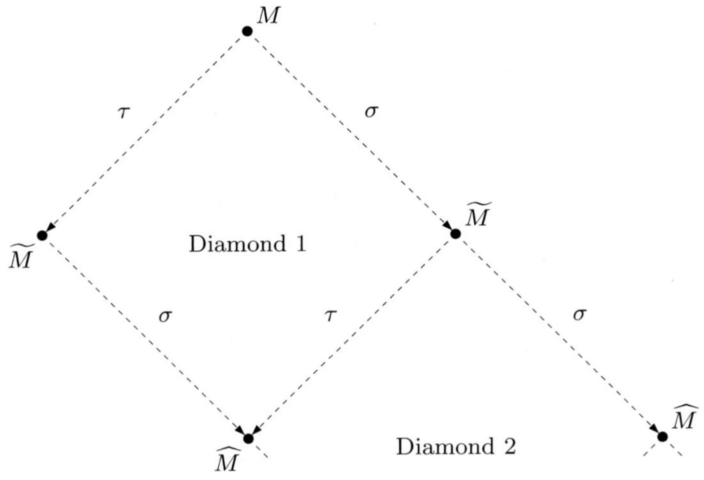

Fig. 6. Completing diamonds, starting from M

In other words, a non-simple sequence can be permuted such that the permuted sequence has a smaller cyclic subsequence leading from some marking back to the same marking. In Figure 1, for example, we have that:

$abcd$ is not decomposable
$acbd$ is decomposable, namely by $\tau_1 = ac$ and $\tau_2 = bd$
$M_0[ac\rangle M_0$ is simple
$M_0[abcd\rangle M_0$ is not simple, because of the permutation $M_0[acbd\rangle M_0$.

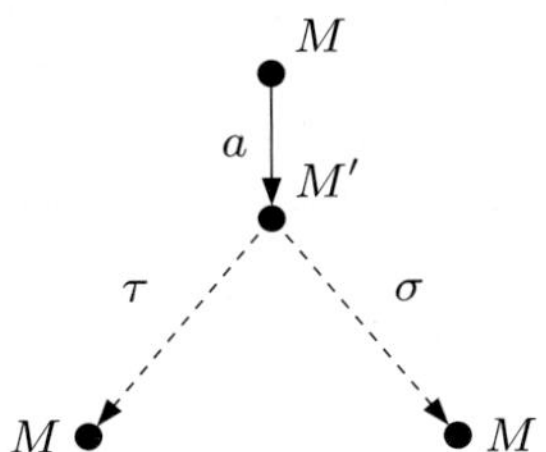

Fig. 7. Two simple cycles with the same initial transition a

Theorem 2. NONDISJOINT SIMPLE CYCLES ARE UNIQUE, UP TO PERMUTATION
Let N, with some initial marking, be bounded and persistent. Let $M[a\tau\rangle M$ and $M[a\sigma\rangle M$, with some transition a and transition sequences τ and σ (see Figure 7). Suppose that both $M[a\tau\rangle M$ and $M[a\sigma\rangle M$ are simple. Then $a\tau \equiv_M a\sigma$.

Proof: First, we prove that $\tau \stackrel{\bullet}{-} \sigma = \varepsilon$ implies $\sigma \stackrel{\bullet}{-} \tau = \varepsilon$. To see this, suppose that we have $\tau \stackrel{\bullet}{-} \sigma = \varepsilon$ and $\sigma \stackrel{\bullet}{-} \tau \neq \varepsilon$; we derive a contradiction.

By Keller's theorem, $a\tau(\sigma \stackrel{\bullet}{-} \tau)$ is firable from M and permutes $a\sigma(\tau \stackrel{\bullet}{-} \sigma)$, from M. Because $a\sigma = a\sigma(\tau \stackrel{\bullet}{-} \sigma)$ reproduces M, so does $a\tau(\sigma \stackrel{\bullet}{-} \tau)$. But because $a\tau$ reproduces M as well, we have

$$M\,[\,\underbrace{a\tau}_{\neq \varepsilon}\,\rangle\, M\,[\,\underbrace{\sigma \stackrel{\bullet}{-} \tau}_{\neq \varepsilon}\,\rangle\, M.$$

But this contradicts the simplicity of $M[a\sigma\rangle M$, since $a\sigma \equiv_M a\tau(\sigma \stackrel{\bullet}{-} \tau)$.

By symmetry, it follows that the proof can be split into two separate cases:

$$\tau \stackrel{\bullet}{-} \sigma = \varepsilon = \sigma \stackrel{\bullet}{-} \tau \quad \text{or} \quad \tau \stackrel{\bullet}{-} \sigma \neq \varepsilon \neq \sigma \stackrel{\bullet}{-} \tau.$$

Case 1: $\tau \stackrel{\bullet}{-} \sigma = \varepsilon = \sigma \stackrel{\bullet}{-} \tau$.
Then we have $\Psi(\tau) = \Psi(\sigma)$, whence also $\Psi(a\tau) = \Psi(a\sigma)$, and the desired result $\tau \equiv_M \sigma$ follows directly from Lemma 1.

Case 2: $\tau \stackrel{\bullet}{-} \sigma \neq \varepsilon \neq \sigma \stackrel{\bullet}{-} \tau$.
We will derive a contradiction, showing that actually only Case 1 remains, and thus proving the theorem. We prove the following in turn (and in increasing order of difficulty):

1. The sequences $\sigma \stackrel{\bullet}{-} \tau$ and $\tau \stackrel{\bullet}{-} \sigma$ are both firable from M.
2. Starting from M, they lead to the same marking, say to $\widetilde{M}$.
3. $\widetilde{M} \neq M$.

When this is proved, we may use boundedness and apply Lemma 3, yielding the result that $\sigma \stackrel{\bullet}{-} \tau$ and $\tau \stackrel{\bullet}{-} \sigma$ must have some transition in common. This is a contradiction to their definition.

Proof of 1.:
Let $M[a\rangle M'$ (cf. Figure 7). From M', both τ and σ are firable, both leading to M. By Keller's theorem, also $\tau(\sigma \stackrel{\bullet}{-} \tau)$ and $\sigma(\tau \stackrel{\bullet}{-} \sigma)$ are firable from M'. Hence both $\sigma \stackrel{\bullet}{-} \tau$ and $\tau \stackrel{\bullet}{-} \sigma$ are firable from M.

Proof of 2.:
Since $a\tau$ and $a\sigma$ lead from M to M, the shorter sequences τ and σ also have the same effect on markings (i.e.: if τ is firable from K and leads to marking K', and if σ is firable from K, then σ also leads to K').

By Part 1., $a\tau(\sigma \stackrel{\bullet}{-} \tau)$ and $a\sigma(\tau \stackrel{\bullet}{-} \sigma)$ are both firable from M and moreover, are in relation $\equiv_M$. It follows that $\sigma \stackrel{\bullet}{-} \tau$ and $\tau \stackrel{\bullet}{-} \sigma$ must also have the same effect on markings, and if they are fired from M, they lead to the same marking, $\widetilde{M}$.

Proof of 3.:
By contradiction. Assume that $\widetilde{M} = M$.

First of all, we have the following reproducing firing sequence:

$$M[\tau \stackrel{\bullet}{-} \sigma\rangle M, \tag{1}$$

simply by Part 2. above, and by $\widetilde{M} = M$. Note that $\tau \stackrel{\bullet}{-} \sigma$ is just the sequence of transitions (in the correct order) from τ that are *not* matchable with transitions in σ, in the sense of (our proof of) Keller's theorem.

Secondly, we will show that there is also a firing sequence

$$M[\rho\rangle M \quad \text{such that } \rho \equiv_M a(\tau \stackrel{\bullet}{-} (\tau \stackrel{\bullet}{-} \sigma)). \tag{2}$$

Note that $\tau \stackrel{\bullet}{-} (\tau \stackrel{\bullet}{-} \sigma)$ is exactly (again in the correct order) the remaining sequence of transitions from τ, i.e. the ones that *are* matchable with transitions from σ.

We now construct the sequence ρ in (2). Note from Part 1. above (i.e., essentially from the proof of Keller's theorem) that $a\tau$ is of the following form:

$$a\, \gamma_0 \chi_1 \gamma_1 \chi_2 \cdots \chi_r \gamma_r, \quad \text{with } \gamma_0 \ldots \gamma_r = \tau \stackrel{\bullet}{-} \sigma \text{ and } \chi_1 \ldots \chi_r = \tau \stackrel{\bullet}{-} (\tau \stackrel{\bullet}{-} \sigma), \tag{3}$$

with suitable $r \geq 0$.

We define a series of markings $M = M_0, M_1, \ldots, M_{r+1} = M$ as follows:

$$M = M_0[\gamma_0\rangle M_1[\gamma_1\rangle M_2 \ \ldots \ M_r[\gamma_r\rangle M_{r+1} = M. \tag{4}$$

This is possible because by Part 1., $\gamma_0 \ldots \gamma_r = \tau \stackrel{\bullet}{-} \sigma$ is firable from $M = M_0$, and by Part 2., we have $M_{r+1} = M$.

Using Lemma 2 repeatedly, we may derive the following series of permutations of the firing sequence $M[a\tau\rangle M$:

1 $\quad M[a\tau\rangle M = M_0[a\gamma_0\chi_1\gamma_1\chi_2\gamma_2\chi_3 \cdots \chi_r\gamma_r\rangle M_0$

2 $\qquad\qquad M_0[\gamma_0\rangle$

3 $\qquad\qquad M_0[\gamma_0 a\chi_1\gamma_1\chi_2\gamma_2\chi_3 \cdots \chi_r\gamma_r\rangle M_0$

4 $\qquad\qquad M_0[\gamma_0\rangle \, M_1[a\chi_1\gamma_1\chi_2\gamma_2\chi_3 \cdots \chi_r\gamma_r\rangle M_0$

5 $\qquad\qquad\qquad M_1[\gamma_1\rangle$

6 $\qquad\qquad\qquad M_1[\gamma_1\rho_1\chi_2\gamma_2\chi_3 \cdots \chi_r\gamma_r\rangle M_0 \text{ with } \Psi(\rho_1) = \Psi(a\chi_1)$

7 $\qquad\qquad\qquad M_1[\gamma_1\rangle \, M_2[\rho_1\chi_2\gamma_2\chi_3 \cdots \chi_r\gamma_r\rangle M_0$

8 $\qquad\qquad\qquad\qquad M_2[\gamma_2\rangle$

9 $\qquad\qquad\qquad\qquad M_2[\gamma_2\rho_2\chi_3 \cdots \chi_r\gamma_r\rangle M_0 \text{ with } \Psi(\rho_2) = \Psi(\rho_1\chi_2)$

$$\vdots$$

Line 1 comes from the original decomposition of $a\tau$, Formula (3). Line 2 comes from (4). Line 3 stems from Lemma 2, applied to $\kappa = a$ and $\gamma = \gamma_0$; line 4 is the same sequence as line 3, where M_1 is inserted. Line 5 comes from (4), and line 6 is another application of Lemma 2 with $\kappa = a\chi_1$ and $\gamma = \gamma_1$. Line 7 introduces M_2 into line 6, line 8 comes from (4), and line 9 is a third application of Lemma 2 with $\kappa = \rho_1\chi_2$ and $\gamma = \gamma_2$.

Continuing in this way, we get a firing sequence

$$M[\gamma_0 \ldots \gamma_r\rangle M[\rho_r\rangle M, \quad \text{with } \Psi(\rho_r) = \Psi(a\chi_1 \ldots \chi_r).$$

The marking reached after this sequence is indeed M, because the Parikh vector of $\gamma_0 \ldots \gamma_r \rho_r$ is the same as that of $a\tau$. Then we may finally put $\rho = \rho_r$, proving (2) because according to (3), $\chi_1 \ldots \chi_r = \tau \overset{\bullet}{-} (\tau \overset{\bullet}{-} \sigma)$, and because of Lemma 1.

Combining (1) and (2), we have the firing sequence

$$M[(\tau \overset{\bullet}{-} \sigma)\rangle M[\rho\rangle M.$$

This sequence is decomposable, because both parts are nonempty: $\tau \overset{\bullet}{-} \sigma \neq \varepsilon$ by assumption, and $\rho \neq \varepsilon$ because a is contained in ρ. But because the sequence $M[(\tau \overset{\bullet}{-} \sigma)\rho\rangle M$ is a permutation of $M[a\tau\rangle M$ by construction and by Lemma 1, we get a contradiction to the simplicity of $M[a\tau\rangle M$.

Hence the assumption was wrong, and $\widetilde{M} \neq M$ must hold true. This ends the proof of Part 3., and thus also the proof of the theorem. $\qquad\square$ 2

5   Concluding Remarks

The presence of transition a which is common to both cycles is relevant in the set of premises of Theorem 2. Otherwise we could not have proved a contradiction to simplicity, and without this precondition, the lemma is indeed not correct. To see this, consider Figure 1 where two disjoint cycles lead through M_0.

The next two examples show that both premises of the theorem are actually necessary.

First, consider Figure 8. The net shown there is bounded, but not persistent. Its reachability graph has two simple cycles, at and ax, which are not transition-disjoint and do not have the same Parikh vector, violating the conclusion of Theorem 2.

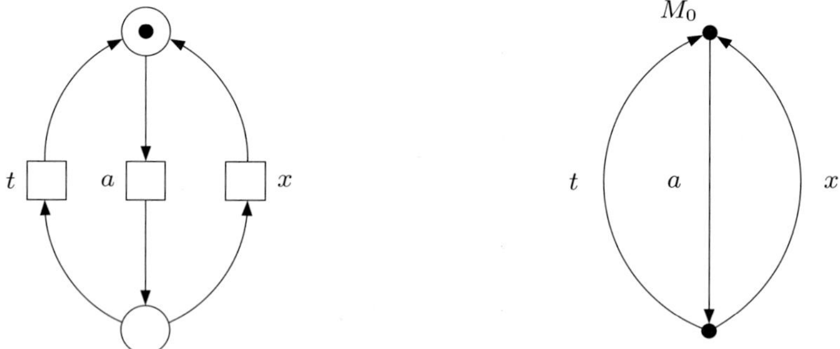

Fig. 8. A non-persistent Petri net (l.h.s.) and its reachability graph (r.h.s.)

Next, consider Figure 9. The net shown there is persistent, but not bounded. Its reachability graph is not shown in this paper because it is infinite. Instead, one its possible converability graphs is shown on the right-hand side of the figure. Again, there are two simple cycles, at and ax, which are not transition-disjoint

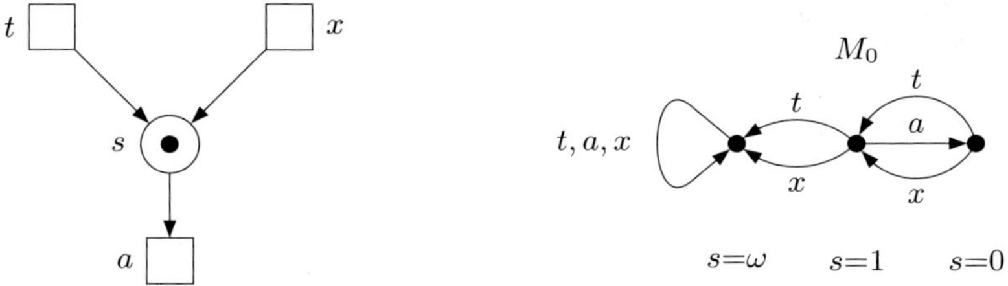

Fig. 9. A non-bounded Petri net (l.h.s.) and a coverability graph (r.h.s.)

and do not have the same Parikh vector, and the conclusion of Theorem 2 is once more violated.

In the special case of marked graphs, Theorem 2 is well-known. If two cycles in the reachability graph of a marked graph start with the same transition a and cannot be decomposed into smaller cycles, then both contain every transition of the connected component of a (in the Petri net) exactly once, and no other transition. In general persistent nets, however, it may be the case that one transition occurs more often than others in a simple cycle; for example, consider Figure 10.

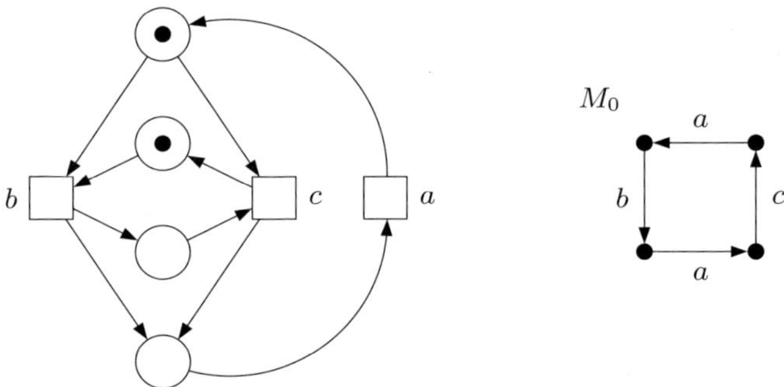

Fig. 10. A bounded persistent Petri net (l.h.s.) and its reachability graph (r.h.s.)

Persistent nets can become quite complex. For example, every free-choice net [5] can canonically be transformed into a persistent net, by superimposing one-token cycles on the output transitions of a place that has more than one output transition (as in Figure 10, which can be seen as a version of Figure 8 with such a cycle).

The result described in this paper has arisen in the context of an attempt to prove two conjectures described in [2], one relating to the conflict-freeness hierarchy defined in that paper, and another one relating to the concept of separability, also described there and investigated more fully in [3]. The first conjecture states that bounded and live persistent nets can essentially be simplified to behaviourally conflict-free nets, where the latter means that any two enabled transitions do not share a common input place. For instance, the net shown in Figure 1 can trivially be simplified in this way, by omitting the place with two tokens. The second conjecture states that bounded and live persistent Petri nets are separable in the following sense: If the initial marking, say M, of such a net, say N, is a k-multiple of another one, then N with initial marking M behaves as k disjoint copies of N with initial marking $(1/k) \cdot M$.

Both conjectures are still open, but Theorem 2 has, in the meantime, been used and generalised in order to obtain a characterisation of the cycles in the reachability graph of a bounded and persistent Petri net [1]. Namely, Theorem 2 can essentially be extended to all non-disjoint simple cycles of the reachability graph, not just ones that start at the same marking or with the same transition.

Acknowledgements

I am much indebted to Harro Wimmel for patiently listening to numerous failed attempts at formulating and/or proving the result described in this paper, and for providing several insightful and helpful suggestions. I would also like to thank Javier Esparza and Philippe Darondeau for comments.

References

1. Best, E., Darondeau, P.: Decomposition Theorems for Bounded Persistent Petri Nets. In: van Hee, K., Valk, R.(eds.) Proc. of ATPN 2008, Xi'An (to appear, 2008)
2. Best, E., Darondeau, P., Wimmel, H.: Making Petri Nets Safe and Free of Internal Transitions. Fundamenta Informaticae 80, 75–90 (2007)
3. Best, E., Esparza, J., Wimmel, H., Wolf, K.: Separability in Conflict-free Petri Nets. In: Virbitskaite, I., Voronkov, A. (eds.) PSI 2006. LNCS, vol. 4378, pp. 1–18. Springer, Heidelberg (2007)
4. Commoner, F., Holt, A.W., Even, S., Pnueli, A.: Marked Directed Graphs. J. Comput. Syst. Sci. 5(5), 511–523 (1971)
5. Desel, J., Esparza, J.: Free Choice Petri Nets. Cambridge Tracts in Theoretical Computer Science, 242 pages (1995) ISBN:0-521-46519-2
6. Keller, R.M.: A Fundamental Theorem of Asynchronous Parallel Computation. In: Tse-Yun, F. (ed.) Parallel Processing. LNCS, vol. 24, pp. 102–112. Springer, Heidelberg (1975)
7. Landweber, L.H., Robertson, E.L.: Properties of Conflict-Free and Persistent Petri Nets. JACM 25(3), 352–364 (1978)

Secure Data Flow in a Calculus for Context Awareness

Doina Bucur and Mogens Nielsen

BRICS, Department of Computer Science
University of Aarhus, Denmark
{doina,mn}@brics.dk

Abstract. We present a Mobile-Ambients-based process calculus to describe context-aware computing in an infrastructure-based Ubiquitous Computing setting. In our calculus, computing agents can provide and discover contextual information and are owners of security policies. Simple access control to contextual information is not sufficient to insure confidentiality in Global Computing, therefore our security policies regulate agents' rights to the provision and discovery of contextual information over distributed flows of actions. A type system enforcing security policies by a combination of static and dynamic checking of mobile agents is provided, together with its type soundness.

Keywords: Ubiquitous Computing, Mobile Ambients, context awareness, security, type system.

1   Introduction

The ubiquitous computing systems encourage a constantly changing execution environment for their computing entities. In such settings, context awareness is a computing paradigm in which schemes for context provision and discovery make applications aware of the changes taking place in their computing context, allowing them to gain advantage from context change, instead of employing a middleware layer for hiding the changes from the application.

On the coordinates of the surveys upon context awareness of Chen, Schilit and Abowd [11,19,1], *contextual information* or computing context is any piece of information used by an application in order to infer knowledge about other interesting computing entities (objects, persons or places) in its environment. *Primary context* can be divided into (possibly overlapping) categories: resources, such as neighbouring printers or the degree of congestion in the network; user context, meaning an object or person's location or status; physical context, such as temperature or moment in time; history of context, meaning the recording of any primary context over time. More complex conclusions about the environment can be drawn by combining several pieces of primary context into what is denoted by *secondary context*: a source of information is indexed by one type of context, after which the result gets indexed by another.

P. Degano et al. (Eds.): Montanari Festschrift, LNCS 5065, pp. 439–456, 2008.

In order for applications to be able to access contextual information in a dynamic network, *context provision* is the dissemination of contextual information from the host entity to a network neighbourhood, to enlarge its visibility scope; *context discovery* is the locating of the provided context by interested applications. In large networks, the design for context provision and discovery is complicated by the network size: network-wide provision is not feasible, and provision to a locality of players is used instead.

According to Schilit [19], the basic way in which applications make use of the surrounding context is by *contextual information and commands*, i.e. requests for either data or actions, which produce different results depending on the specific context in which they are issued. A context-aware application can also involve *context-triggered actions*, i.e. rules which specify what commands to be automatically executed, given certain properties which the context fulfills. Finally, applications can employ *automatic contextual reconfiguration* to add and remove entire software components in response to certain properties of the context.

Ubiquitous systems are numerous, highly dynamic and their contextual information should be made accessible to a selected subset of the players in the network, features which make such systems difficult to enforce security policies upon. Simple access control upon pieces of contextual information cannot prevent their disclosure by agents collaborating on accessing them; static type checking cannot verify, on its own, that a highly dynamic network respects policies, hence dynamic type checking is called to verify the instances in which agents or code move in the network. Given that such systems are selective, privileges are fine-grained, so that users have different rights upon different pieces of contextual information.

Our calculus models infrastructure-based (as opposed to ad hoc) ubiquitous systems, inherently of hierarchical topology. In such systems, logical partitions are imposed over the network and all communication between mobile entities is mediated by the infrastructure; rooms, floors, buildings, and campuses are such logical cells which act as mediators for context provision, context discovery and general communication for the entities in their scope. Early systems in this category include the Xerox ParcTab [20], Active Badge [21], and GUIDE [12].

Computing entities are modelled by mobile ambients enclosing processes and other ambients, and the topology of the network evolves by the ambients' running of in and out movements. Furthermore, we introduce a modelling of contextual information, distributed through the network over scopes of various sizes and expressed by named macros. The context of an ambient is then the collected contextual information hosted by all ambients enclosing it up until the root ambient, at the ambient's current position. Context provision is performed by an ambient through defining the named macro up in the network to a specified ambient destination to increase the macro's scope. Context discovery is performed by an ambient calling a macro name from a specified ambient and having the call replaced with the macro body, if such a macro exists in the current context. The context changes whenever ambients move and provide or consume contextual information.

The focus of this paper is to study the fulfilling of distributed security policies, in such a setting in which contextual information crosses boundaries in a hierarchical topology. The type of a process is a pair composed by a function recording the effects (provision or discovery) which ambients inner to the process have on other ambients, and by a set of ambient names which are allowed to reside in the process. Any ambient can host security policies in the form of process types, which all processes residing under this ambient should fulfill. This gives, at any given moment in the evolution of the network, for any process sitting in a context composed by a line of ancestor ambients (each with a policy of its own), that the process must adhere to the composition of all ancestor policies. Static type checking verifies that an initial state of the network is well-typed, for dynamic type checking to then verify the incremental movements of definitions and ambients in the network. We show well-typedness to be verified over any sequence of reductions in the system, define errors and show that they cannot appear in a well-typed system.

The rest of this paper is organised as follows: Section 2 presents the syntax and the basic semantics of our calculus, Section 3 the type systems, the notion of well-typedness and the subject reduction theorem, and Section 4 the operational semantics and the notion of type soundness. Finally, Section 5 illustrates a case study modelling a ubiquitous computing infrastructure in a hospital, and Section 6 reviews closely related work and concludes.

2   A Calculus for Context Awareness

In this section we briefly present the syntax and the basic, untyped semantics of the calculus. In the complete syntax from Fig. 1, the nil process 0, parallel composition $P \mid P'$, ambient $a[P]$, name restriction $\nu z\, P$ and movement primitives $in\, a.P$ and $out.P$ are all inherited from and have the same meaning as in the Mobile Ambients calculus [10]; as in the Boxed Ambients calculus [7], there is no *open* capability. From Zimmer's calculus for context awareness [22] we borrow the idea of contextual information as macro definitions residing at ambients. A definition of the basic form $def\, f \triangleright Q\, in\, P$ defines macro f as being the process Q in a floating definition $(f \triangleright Q)$ and continues execution with P, and any call f for this macro would be replaced by its body Q; a simplified semantics for a definition and call are:

$$\text{DEF} \quad def\, f \triangleright Q\, in\, P \longrightarrow (f \triangleright Q)\, P \qquad \text{CALL} \quad (f \triangleright Q)\, f \longrightarrow Q$$

The decorations τ and G on ambients and floating definitions from Fig. 1 and Table 1 in the following are security and entry policies, respectively, and their syntax and meaning are detailed in Section 3; they are ignored in this section.

Macro definitions come in two flavours: a *one-shot definition* $f \triangleright Q$ is one which is consumed by that macro being called and is suitable for modelling network packets (if one sees data communication as a simple feature of context); a *permanent definition* $!f \triangleright Q$ is one which can be instantiated by any number of macro calls, and in fact behaves like an infinite set of one-shot definitions.

processes	$P ::= 0$	no process
	$\mid \quad P \mid P'$	parallel composition
	$\mid \quad f^a$	macro call, $f \in \mathcal{F}$
	$\mid \quad def^a\, D\, in\, P$	macro definition
	$\mid \quad a_G^\tau\,[P]$	mobile entity, $a \in \mathcal{A}$
	$\mid \quad E\,P$	public definitions
	$\mid \quad \nu z\, P$	name restriction, $z \in \mathcal{F} \cup \mathcal{A}$
	$\mid \quad in\, a.P$	movement in
	$\mid \quad out.P$	movement out
definitions	$E ::= (D)^a$	floating definition, upward
	$\mid \quad (D)^{a,\tau}$	floating definition, downward
	$D ::= F \mid !F$	
	$F ::= f \triangleright P$	macro definition

Fig. 1. Syntax

Unlike Zimmer's calculus, definitions and calls of contextual information are not only made by processes to and from their enclosing ambient, but cross multiple ambients' boundaries; hence, definitions and calls are tagged with the identity of the destination and source ambient, respectively: process $def^b f \triangleright Q\, in\, P$ publishes macro f at any ambient b in the ancestor ambient line of this process, and process f^b calls macro f from any such ambient. The calls and definitions being tagged with the name of a destination ambient fits the infrastructure setting, in which mobile entities have a degree of knowledge about the identities of servers, gateways and about the protocols in the network, at least such that identities of other points of interest can be provided by calling these.

For this, a process defining macro f to ambient b evolves into a floating definition destined to b:

$$\text{DEF} \quad def^b f \triangleright Q\, in\, P \longrightarrow (f \triangleright Q)^b\, P$$

for the floating definition to travel upwards to destination in fluid movements of the form

$$(f \triangleright Q)^b\, P \mid R \equiv (f \triangleright Q)^b\, (P \mid R)$$

(included among the structural congruence rules in Table 1) and

$$\text{UP} \quad a\,[(f \triangleright Q)^b P] \longrightarrow (f \triangleright Q)^b\, a\,[P]$$

(a semantics rule in Section 4). If permanent, a definition at its destination $b\,[(!f \triangleright Q)^b P]$ expands single instances upon calls, with $(!F) \equiv (F)(!F)$. When called, a single floating definition moves down from its host ambient to the calling process by the inverse of the upward movements above:

$$\text{DOWN} \quad (f \triangleright Q)^b a[P] \longrightarrow a\,[(f \triangleright Q)^b P]$$

for the call to be fulfilled at the calling ambient:

$$\text{CALL} \quad (f \rhd Q)^b \, f^b \longrightarrow Q$$

This scope extension for contextual information follows the idea that context is formed by publishing information from a source to a wider locality of users; it allows for the modelling of context provision and discovery over entire localities of agents, unlike Zimmer's model, which bounds the communication of context within the enclosing ambient of the communicating process. This scheme also effectively models Schilit's contextual information and commands from [19]: a call for a service has a dynamic interpretation varying with context. Furthermore, the intermediate steps a floating definition takes in order to reach a calling ambient gives that either contextual information or its destination can unexpectedly become unreachable with the ambients' changing of location; this fits the profile of highly dynamic networks.

Table 1. Structural congruence is the smallest congruence relation satisfying these rules

$$
\begin{array}{ll}
P|0 \equiv P & \nu z\, 0 \equiv 0 \\
P|Q \equiv Q|P & \nu z\, \nu w\, P \equiv \nu w\, \nu z\, P \\
(P|Q)|R \equiv P|(Q|R) & \nu z\,(P|Q) \equiv P|\nu z\, Q \text{ if } z \notin fn(P) \\
 & \nu z\,(a_G^\tau[P]) \equiv a_G^\tau[\nu z\, P] \text{ if } z \notin fn(a_G^\tau[]) \\
(!F)^{a,\tau} \equiv (F)^{a,\tau}(!F)^{a,\tau} & \nu z\, E\, P \equiv E\, \nu z\, P \text{ if } z \notin fn(E) \\
E_1 E_2 \equiv E_2 E_1 & \\
E\, P|Q \equiv E(P|Q) & \alpha\text{-conversion}
\end{array}
$$

As an example, take a hospital's ubiquitous system supporting collaboration among mobile employees (inspired by [2]); the hospital network infrastructure is ambient hni, and a doctor's personal digital assistant doc currently residing in the operating ward ow updates the network about his location $docloc$ of current value P, so that the tag of any nurse (at any location in the hospital, such as office of) to locate him:

$$hni\left[ow\left[doc\left[def^{hni}!docloc \rhd P\, in\, 0\right]\right] \mid of\left[nurse\left[docloc^{hni}\right]\right]\right]$$

The nurse's code cannot execute without the $docloc$ macro available in its context, but after the definition becomes visible the system is:

$$hni\left[(!docloc \rhd P)^{hni}\left(ow\left[doc\,[]\right] \mid of\left[nurse\left[docloc^{hni}\right]\right]\right)\right]$$

and one instance of the definition travels downwards to meet the request:

$$hni\left[(!docloc \rhd P)^{hni}\left(ow\left[doc\,[]\right] \mid of\left[nurse\left[(docloc \rhd P)^{hni}docloc^{hni}\right]\right]\right)\right]$$

and $(docloc \rhd P)^{hni}docloc^{hni}$ reduces to P.

We use two sets of names in our syntax: macro names f belong to an infinite set $\mathcal{F}$, and ambient names a belong to an infinite set $\mathcal{A}$. A restriction νz can only be made upon any name z in $\mathcal{F} \cup \mathcal{A}$, and α-conversion substitutes names from $\mathcal{F} \cup \mathcal{A}$. We adapt the convention: when considering a process, we assume that the bound names of the process are different from its free names, if necessary, after a number of α-conversion steps.

Crucially, the restriction operator ν is the only binder in our syntax; i.e., a macro name in a macro definition is not a binder. Hence, the bound and free names of a process $bn(P)$, $fn(P)$, the substituted process $P\alpha$ and the extrusion of restrictions are defined as is standard.

3   Type Systems and Well-Typedness

We introduce a typing system which specifies the effects which a process in our calculus is allowed to have. A type is two-fold, as depicted in Fig. 2: on one hand, a security policy τ is a function mapping a source ambient name to a destination ambient name and to a set of effects which the source ambient can have upon the destination ambient. An effect is the ability to run either a definition of or a call for a macro name f. On the other hand, the mere presence of an ambient in a process is a securable feature, thus we also enforce entry policies G, as being subsets from the set of ambient names.

$$
\begin{array}{ll}
\textit{Entry policy} & G \subseteq \mathcal{A} \\
\textit{Effects} & \mathcal{E} = \bigcup_{f \in \mathcal{F}} \{def(f), call(f)\} \\
\textit{Security policy} & \tau = \mathcal{A} \to (\mathcal{A} \to \mathcal{P}(\mathcal{E}))
\end{array}
$$

Fig. 2. Entry policies and security policies

As is natural for Mobile-Ambients-based calculi, policies reside at the ambient membrane. We call a *subambient* of a any ambient enclosed by a either directly, or at any inner level. Then, we write $a_G^\tau[P]$ to specify that P should satisfy policies τ and G. If $b \notin G$, then P should not have a subambient b at all. Furthermore, if $\tau(b)(c) \not\ni def(f)$, then P should not have a subambient b directly enclosing a process $def^c f \triangleright Q \, in \, R$; similarly for macro calls.

As an example, the policy τ of the hospital network infrastructure hni_G^τ is such that only employees have access to the patient record of a certain VIP, accessible through the protocol (i.e., macro name) vip directed at hni; thus, for any visitor vis the policy states that $\tau(vis)(hni) \not\ni call(vip)$ and applies throughout the hospital system, without the need of it being multiplied (for example, at the lower, ward level). Also, supposing that one floor $floor3_C^\gamma$ of the hospital is closed to anybody but the hospital's employees, $vis \notin C$, so that no further policies upon vis are needed inside $floor3$.

A fully-permissive entry policy is $\mathcal{A}$, and a fully-permissive security policy is denoted by ω, with

$$
\forall b, c \in \mathcal{A} \quad \forall f \in \mathcal{F} \quad \omega(b)(c) \ni def(f) \text{ and } \omega(b)(c) \ni call(f).
$$

Two policies can be composed to denote a policy which satisfies both original policies; the composition of entry policies G and H is $G \cap H$, while the composition of security policies τ and σ is the policy $\tau \cap \sigma$, with

$$(\tau \cap \sigma)(b)(c) = \tau(b)(c) \cap \sigma(b)(c).$$

Composition is both commutative and associative, and ω is the identity element, $\omega \cap \tau = \tau$, for all τ. We denote the set of all security policies by $\mathcal{T}$.

We do not detail the matter of defining a particular syntax for policies. We only assume such a syntax to infer a notion of *free names* of policies ${}^{\tau}_{G}$, written $fn\left({}^{\tau}_{G}\right)$. For all non-free combination of names, the policies are implicitly either fully-permissive or fully-dismissive; then, for the free names, policies specify explicitly either restrictions or allowances, in both cases finitely many.

This notion of free names only needs to satisfy that the set of free ambient names, $fn\left({}^{\tau}_{G}\right) \cap \mathcal{A}$, and the set of free macro names, $fn\left({}^{\tau}_{G}\right) \cap \mathcal{F}$, are both finite sets, and that the following conditions are satisfied:

- for entry policies, the set of allowed names of subambients G is either finite or cofinite, relative to the finite set of free ambient names $fn\left({}^{\tau}_{G}\right) \cap \mathcal{A}$;
- for security policies, any set of allowed definitions and calls $\tau(a)(b)$, $\forall a, b \in \mathcal{A}$ is either finite or cofinite relative to the finite set of free macro names $fn\left({}^{\tau}_{G}\right) \cap \mathcal{F}$, and this former set is uniformly defined for all non-free ambient names.

Given our hierarchical network topology of mobile agents, each hosting a policy, a process sitting in the scope of a set of ambients should comply with the collected policies of those ambients. The initial state of a system is statically checked for compliance with all the system's policies, and then only individual moves of ambients are checked in the operational semantics described in Section 4.

3.1   A Type System for Active Code

There are two interconnected type systems; the one for *active* processes (i.e. all processes not a macro definition body), with type statements of the form $a^{\tau}_{G} \vdash P$ stating that P running at ambient a complies with the type τ, is depicted in Fig. 3. The one for *inactive* processes (i.e. definition bodies, which travel over ambient boundaries before being executed) is discussed subsequently. Finally, Def. 1 at the end of the section defines a process to be well-typed if it is both actively and inactively well-typed.

For the compactness of our typing expressions, we write $?f$ to stand for both permanent $!f$ and one-shot f definitions; similarly, we write $\sigma?$ to stand for both σ and no policy.

The interesting rule is AMB: in a top-down checking fashion, for a subambient $b^{\sigma}_{H}[P]$ sitting in an ambient a to be actively typed, process P has to be typed with the composed security policy and the intersected entry policy given by the policies of b and a; hence, in any statement of the form $a^{\tau}_{G} \vdash P$, P sits in ambient

$$\text{NULL} \quad a_G^\tau \vdash 0 \qquad \text{PAR} \quad \frac{a_G^\tau \vdash P \quad a_G^\tau \vdash P'}{a_G^\tau \vdash P|P'} \qquad \text{CALL} \quad \frac{\tau(a)(b) \ni call(f)}{a_G^\tau \vdash f^b}$$

$$\text{DEF} \quad \frac{\tau(a)(b) \ni def(f) \quad a_G^\tau \vdash P}{a_G^\tau \vdash def^b\, ?f \triangleright Q\, in\, P}$$

$$\text{AMB} \quad \frac{b \in G \quad b_{G \cap H}^{\tau \cap \sigma} \vdash P}{a_G^\tau \vdash b_H^\sigma[P]}$$

$$\text{MSG} \quad \frac{a_G^\tau \vdash P}{a_G^\tau \vdash (D)^{b,\sigma?}P} \qquad \text{RES} \quad \frac{a_G^\tau \vdash P}{a_G^\tau \vdash \nu z\, P}\, z \notin fn\left(_G^\tau\right)$$

$$\text{IN} \quad \frac{a_G^\tau \vdash P}{a_G^\tau \vdash in\, b.P} \qquad \text{OUT} \quad \frac{a_G^\tau \vdash P}{a_G^\tau \vdash out.P}$$

Fig. 3. Type system for active code

a, τ is the security policy composed of all security policies belonging to ambients ancestor to P, and G is the intersection of their entry policies.

Then, for a process with a call effect f^b sitting in a_G^τ to be actively typed, the condition in the CALL rule is $\tau(a)(b) \ni call(f)$, for the effect to be allowed by the composed policy τ. The same goes for definition effects in the DEF rule, with a further check on the continuation process P following the definition. Entry policies are checked in the AMB rule, in which a subambient b is allowed to run at a_G^τ if $b \in G$.

In the RES rule, $\nu z\, P$ is typed with respect to a_G^τ if z is not one of the free names of policies $_G^\tau$, i.e. those names upon which the policies explicitly specify restrictions or allowances. On the other hand, if z is such a name, z is α-converted to a fresh name. Thus, if e.g. a fresh ambient is created using the restriction operator, at type checking the ambient name is non-free and is checked against implicit policies (e.g. fully-restrictive).

Movement capabilities in and out are not type checked themselves, but the operational semantics will impose well-typedness conditions for moves, as shown in Section 4. Furthermore, we assume that a system starts in a state without floating definitions, which allows us to only check statically definition bodies only once in their defining process (DEF), and not when floating (MSG).

3.2 A Type System for Inactive Code

For the type checking of definition bodies in Fig. 4, given that definitions travel over ambient boundaries, the check is more complex. Intuitively, a defining process $def^b\, ?f \triangleright Q\, in\, P$ has as result the definition $(?f \triangleright Q)^b$ travelling upwards to ambient b and then downwards to any calling ambient; in the inactive type system below, it is ensured that when having arrived at b, the body process Q complies with the composed policies of b and its ancestors. Subsequently in

Section 4, the remaining down movements will be checked dynamically, to ensure that Q complies with all the policies imposed within b down to the calling ambient.

The *context function* $\mathcal{C}$ records the context, in terms of security and entry policies, with which a definition body should comply when travelling upwards, in order to maintain the well-typedness of the system. The function is totally defined on the ambient names set $\mathcal{A}$ and takes values in pairs of security policies (from the set $\mathcal{T}$) and entry policies (from $\mathcal{P}(\mathcal{A})$). Formally, the type of this function is (written here using a nonstandard notation reflecting our notation for decorating ambients):

$$\mathcal{C} : \mathcal{A} \longrightarrow {}^{\mathcal{T}}_{\mathcal{P}(\mathcal{A})}.$$

Thus, whenever $\mathcal{C}(b) = {}^{\tau}_{G}$, we have $a\mathcal{C}(b) = a^{\tau}_{G}$.

The intuition is that, given a definition $def^b\, f \triangleright Q\, in\, P$, its policy-wise context $\mathcal{C}(b)$ returns the collected policies of ambients above b, so that in order for Q to be run under b, $b\mathcal{C}(b)[Q]$ needs to be well-typed (i.e., both actively and inactively typed against the collected policies $\mathcal{C}(b)$).

As with both security and entry policies, $\mathcal{C}$ has a fully-permissive instantiation Ω with:

$$\Omega(\mathcal{A}) = {}^{\omega}_{\mathcal{A}}.$$

We frequently abuse the notation and write $\mathcal{C}(H) = {}^{\tau}_{G}$ to state that the value of $\mathcal{C}$ for all elements in set H is ${}^{\tau}_{G}$. Also, we frequently define an instantiation of such a context function $\mathcal{C}$ by writing updates upon Ω, of the form $\Omega\,[H \rightarrow {}^{\tau}_{G}]$.

$$\text{N\scriptsize ULL}\quad \mathcal{C}, a^{\tau}_{G} \vdash_{\circ} 0 \qquad \text{P\scriptsize AR}\quad \frac{\mathcal{C}, a^{\tau}_{G} \vdash_{\circ} P \quad \mathcal{C}, a^{\tau}_{G} \vdash_{\circ} P'}{\mathcal{C}, a^{\tau}_{G} \vdash_{\circ} P|P'} \qquad \text{C\scriptsize ALL}\quad \mathcal{C}, a^{\tau}_{G} \vdash_{\circ} f^b$$

$$\text{D\scriptsize EF}\quad \frac{\mathcal{C}, a^{\tau}_{G} \vdash_{\circ} P \quad a^{\tau}_{G}[Q]\ \text{well-typed}}{\mathcal{C}, a^{\tau}_{G} \vdash_{\circ} def^a\,?f \triangleright Q\, in\, P} \qquad \frac{\mathcal{C}, a^{\tau}_{G} \vdash_{\circ} P \quad b\mathcal{C}(b)[Q]\ \text{well-typed}}{\mathcal{C}, a^{\tau}_{G} \vdash_{\circ} def^b\,?f \triangleright Q\, in\, P}$$

$$\text{A\scriptsize MB}\quad \frac{\mathcal{C}\,[\{a\} \cup G \rightarrow {}^{\tau}_{G}], b^{\tau \cap \sigma}_{G \cap H} \vdash_{\circ} P}{\mathcal{C}, a^{\tau}_{G} \vdash_{\circ} b^{\sigma}_{H}[P]}$$

$$\text{M\scriptsize SG}\quad \frac{\mathcal{C}, a^{\tau}_{G} \vdash_{\circ} P}{\mathcal{C}, a^{\tau}_{G} \vdash_{\circ} (D)^{b,\sigma?}P} \qquad \text{R\scriptsize ES}\quad \frac{\mathcal{C}, a^{\tau}_{G} \vdash_{\circ} P}{\mathcal{C}, a^{\tau}_{G} \vdash_{\circ} \nu z\, P}\ z \notin fn\,(\mathcal{C}) \cup fn\,({}^{\tau}_{G})$$

$$\text{I\scriptsize N}\quad \frac{\mathcal{C}, a^{\tau}_{G} \vdash_{\circ} P}{\mathcal{C}, a^{\tau}_{G} \vdash_{\circ} in\, b.P} \qquad \text{O\scriptsize UT}\quad \frac{\mathcal{C}, a^{\tau}_{G} \vdash_{\circ} P}{\mathcal{C}, a^{\tau}_{G} \vdash_{\circ} out.P}$$

Fig. 4. Type system for inactive code

Inactive type statements have the form $\mathcal{C}, a^{\tau}_{G} \vdash_{\circ} P$. This intuitively means that all the definition bodies Q waiting to be activated in processes of the form $def^b f \triangleright Q\, in\, R$ inside P sitting in ambient a will have to be able to run inside the destination ambient: if the destination is the host ambient a itself, then the definition bodies have to comply with a_G^τ, in which, as in the case of the active type system, τ and G are composed by or intersected from all ambients above. On the other hand, for a destination $b \neq a$, the context function $\mathcal{C}(b)$ returns the pair of collected policies from the set of ambients ancestor to b (if b is in P's context) or from the maximal set of ambients which can ever exist above b, were P to move under b.

For consistency, we assume that if one of our systems is not of the form $a_G^\tau[P]$, there exists a unique ambient $world_A^\omega$ with full permissions at the root of the system. The type checking then proceeds top-down, while also collecting contextual policies in a context function $\mathcal{C}$: at top level, there exists no restricting context other than the enclosing root ambient a_G^τ, and a typing statement looks like $\Omega, a_G^\tau \vdash\circ P$; when another ambient is encountered, $\Omega, a_G^\tau \vdash\circ b_H^\sigma[Q]$ if $\Omega[\{a\} \cup G \rightarrow {}_G^\tau], b_{G \cap H}^{\tau \cap \sigma} \vdash\circ Q$ and the context function for ambient b is the updated $\Omega[\{a\} \cup G \rightarrow {}_G^\tau]$, meaning that if a definition in Q is destined to b, it should comply with the composed policies of a and b; if destined to a or any other ambient allowed inside a, it should comply with the policies of a. This last fact is to ensure such a well-typedness, that the dynamic checks at in movements are simple, as will be detailed in Section 4.

We then define well-typedness as being the dual, active and inactive, checking of processes.

Definition 1 (Well-typedness). *A system $a_G^\tau[P]$ is well-typed if $a_G^\tau \vdash P$ and $\Omega, a_G^\tau \vdash\circ P$. A system P without a root ambient is well-typed if $world_A^\omega \vdash P$ and $\Omega, world_A^\omega \vdash\circ P$.*

4   Operational Semantics and Type Soundness

Our operational semantics from Fig. 5 preserves the flavour of reduction rules IN, OUT, STRUCT and CONTEXT from standard Mobile Ambients. Moreover, rules UP and DOWN depict the crossing of ambient borders by definitions floating from their origin to the destination or from the latter to a calling ambient; if these movements are successful, a pair composed by a called definition adjacent to its call reduces to the definition body. In Fig. 5, the *active contexts* **C** are the processes with one hole in active locations (i.e, locations in which a process can suffer a reduction immediately), as in the following:

$$\mathbf{C} ::= [\cdot] \mid \mathbf{C}|P \mid a_G^\tau[\mathbf{C}] \mid E\,\mathbf{C} \mid \nu z\,\mathbf{C}$$

Static checking, discussed in Section 3, verifies that an initial state of the system is well-typed. It includes, in the inactive type checking from Fig. 4, a scheme to verify that even the bodies of those definitions destined to ambients which are not currently in the context will safely run at those ambients, whenever they become present. That verifies, for example, that in the setting

$$c_K^\rho\,[a_G^\tau[P] \mid b_H^\sigma\,[in\,a.Q \mid R]],$$

given that $a \in K$, definitions from $Q \mid R$ destined to a are statically checked against the policies of c_K^ρ. Thus, when b performs the in movement, it is sufficient

$$\text{UP} \quad a_G^\tau[(D)^b P] \longrightarrow (D)^b a_G^\tau[P] \qquad a_G^\tau[(D)^a P] \longrightarrow a_G^\tau[(D)^{a,\tau} P]$$

$$\text{DOWN} \quad \frac{a_G^{\sigma \cap \tau}[Q] \text{ well-typed}}{(f \triangleright Q)^{b,\sigma} a_G^\tau[P] \longrightarrow a_G^\tau[(f \triangleright Q)^{b,\sigma \cap \tau} P]}$$

$$\text{DEF} \quad def^a\, D\, in\, P \longrightarrow (D)^a P \qquad \text{CALL} \quad (f \triangleright P)^{a,\tau} f^a \longrightarrow P$$

$$\text{IN} \quad \frac{b \in G \quad b_G^\tau \vdash Q \mid R \quad \Omega\,[\{a\} \cup G \rightarrow\ _G^\tau]\,, b_G^\tau \vdash\!\circ\ Q \mid R}{a_G^\tau[P] \mid b_H^\sigma[in\, a.Q \mid R] \longrightarrow a_G^\tau[P \mid b_H^\sigma[Q \mid R]]}$$

$$\text{OUT} \quad a_G^\tau[P \mid b_H^\sigma[out.Q \mid R]] \longrightarrow a_G^\tau[P] \mid b_H^\sigma[Q \mid R]$$

$$\text{STRUCT} \quad \frac{P \equiv P' \quad P \longrightarrow Q \quad Q \equiv Q'}{P' \longrightarrow Q'} \qquad \text{CONTEXT} \quad \frac{P \longrightarrow Q}{\mathbf{C}[P] \longrightarrow \mathbf{C}[Q]}$$

Fig. 5. Operational semantics

to dynamically check that $Q \mid R$ is well-typed against a's policies, as shown by rule IN.

Furthermore, since static checking insures that the body of a definition is safe to be run in the destination ambient it reached after moving upwards, all that is left for the DOWN movement to dynamically check is the compliance with the collected policies of each ambient down until the calling one. This is ensured by decorating the floating definition $(f \triangleright Q)^a$, when it has reached ambient a_G^τ, with a *signature* equal to τ; with every down movement, the signature is composed with the policy of the newly-crossed ambient, and Q is checked against its signature and the crossed ambient's group policy. This scheme has the flavour of firewall-carrying code.

As an observation, the same static checking scheme, together with the dynamic checking at the DOWN movement, have the side effect that some breakings of policy could be caught by either of these checks (and are caught by the earliest, static checking), if the case is such that the ambient tree in which the definition moves up before reaching the destination also includes the calling ambient. On the other hand, if the upward and downward trees are disjunct, only the dynamic check can verify the definition body.

Furthermore, the reader may have wondered about our less standard approach of modelling the availability of contextual information partially through reduction (rules UP, DOWN in Fig. 5), instead of structural congruence. Our choice is motivated by the fact that we want to model explicitly the operational behaviour of firewalls in terms of filtering capabilities, in the event of contextual information crossing firewalls during downward moves.

We can now state the subject reduction property, as follows.

Theorem 1 (Subject Reduction). *If P is well-typed and $P \xrightarrow{*} Q$, then Q is well-typed.*

In the behaviour of a system it is considered an error, $P \rightarrow err$, if an effect (be it a definition or a call) breaks the security policy of any ambient in its context, or if an ambient's presence breaks the entry policy of any ambient in its context, as in Fig. 6. The superscript τ and the subscript G decorating a context $\mathbf{C}$ are the composition of all security policies above $[\cdot]$, and the intersection of all entry policies above $[\cdot]$, respectively. The composed τ and G for a context $\mathbf{C}$ are defined inductively in Table 2.

$$\text{ERR\_DEF} \quad \frac{(\sigma \cap \tau)(a)(b) \not\ni def(f)}{\mathbf{C}_H^\sigma \left[a_G^\tau \left[def^b\, f \triangleright Q\, in\, P \mid R \right] \right] \longrightarrow err}$$

$$\text{ERR\_CALL} \quad \frac{(\sigma \cap \tau)(a)(b) \not\ni call(f)}{\mathbf{C}_H^\sigma \left[a_G^\tau \left[f^b \mid P \right] \right] \longrightarrow err}$$

$$\text{ERR\_IN} \quad \frac{H \not\ni a}{\mathbf{C}_H^\sigma \left[a_G^\tau \left[P \right] \right] \longrightarrow err} \qquad \text{ERR\_STR} \quad \frac{P \equiv P' \quad P' \longrightarrow err}{P \longrightarrow err}$$

Fig. 6. Errors

Table 2. The policies of contexts

$\mathbf{C}_G^\tau$	τ	G
$[\cdot]$	ω	A
$\mathbf{C}_H^\sigma \mid P$	σ	H
$c_K^\rho [\mathbf{C}_H^\sigma]$	$\sigma \cap \rho$	$K \cap H$
$E\, \mathbf{C}_H^\sigma$	σ	H
$(\nu z)\mathbf{C}_H^\sigma$	σ	H

We then state that if a system is well-typed, it can never display an error.

Theorem 2 (Type Soundness). *If P is well-typed then $P \not\xrightarrow{*} err$.*

5   Case Study: Ubiquitous Computing in a Hospital

Consider a ubiquitous computing infrastructure in a hospital, inspired by the AWARE project [2], as in Fig. 7. The hospital network infrastructure hni maintains the patient records and keeps track of the location of doctors, nurses, patient and visitors, all of them carrying PDAs, by having their PDAs announce their presence periodically.

The patient records are read and updated by nurses and doctors using either their tabs or the terminals in the operation ward and patients' ward. Patients and visitors have lower degrees of rights upon accessing–with their tabs or at the terminals–patient records and employee locations.

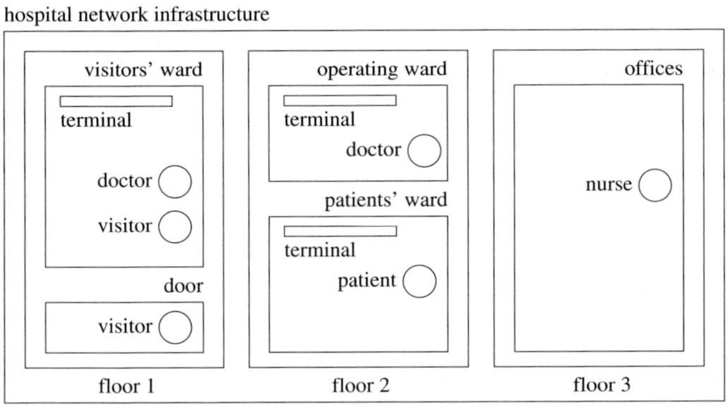

hospital network infrastructure

visitors' ward   operating ward   offices

terminal   terminal

doctor

doctor

visitor   patients' ward

door   terminal

patient

visitor   nurse

floor 1   floor 2   floor 3

Fig. 7. The hospital system

5.1   The Guessing Visitor

The policy τ of the hospital network infrastructure hni_G^τ is such that only employees have access to the patient record P of a certain VIP, accessible through the protocol (i.e., macro name) vip directed at hni; for any visitor vis the policy states that

$$\tau(vis)(hni) \not\ni call(vip).$$

An overly-curious visitor vis_V^ω who guessed the protocol for accessing the VIP's record would enter through the door $door_{\mathcal{A}}^\omega$ (i.e., a new ambient vis_V^ω would be run as a subambient of $door$, after being sprouted from a banged definition resident at $door$). The initial, static type checking determines that the system is not well-typed. As an observation, expression $\nu h \left(def^{door}\,!h \triangleright Q \mid h^{door}\,in\,h^{door}\right)$ effectively models $!Q$.

The system, only focused on the door, is formalised as

$$hni_G^\tau[\,(vip \triangleright P)^{hni}$$
$$door_{\mathcal{A}}^\omega\,\left[\nu h\,\left(def^{door}\,!h \triangleright vis_V^\omega\,\left[vip^{hni}\right] \mid h^{door}\,in\,h^{door}\right)\right]].$$

We show that its well-typedness depends on at least the policy condition which is not met, $\tau(vis)(hni) \not\ni call(vip)$; the other conditions for its well-typedness, possibly satisfied, are not depicted. The system is well-typed if it is also inactively well-typed, as in the derivation tree in Fig. 8. Given that the well-typedness of the system depends on a condition which is false, the error is raised at this stage.

5.2   The Conspiring Nurse

Have a nurse $nurse_N^\omega$ who agreed to conspire with the overly-curious visitor vis_V^ω. They plan on an indirect access scheme to the record vip^{hni}, for the visitor. The visitor's actions will be inconspicuous: the visitor residing in the visitors' ward vw and the nurse in her office of agree upon a new, private service (i.e., macro name)

$$\cfrac{h \notin fn\left(\tfrac{\tau}{G}\right) \quad \cfrac{\cfrac{\cfrac{\cfrac{\cfrac{\cfrac{\cfrac{\Omega, hni_G^\tau \vdash\!\circ (vip \triangleright P)^{hni} door_A^\omega \left[\nu h \left(def^{door} !h \triangleright vis_V^\omega \left[vip^{hni}\right] \mid h^{door} in\, h^{door}\right)\right]}{\Omega, hni_G^\tau \vdash\!\circ door_A^\omega \left[\nu h \left(def^{door} !h \triangleright vis_V^\omega \left[vip^{hni}\right] \mid h^{door} in\, h^{door}\right)\right]}}{\Omega\left[\{hni\} \cup G \to \tfrac{\tau}{G}\right], door_G^\tau \vdash\!\circ \nu h \left(def^{door} !h \triangleright vis_V^\omega \left[vip^{hni}\right] \mid h^{door} in\, h^{door}\right)}}{\Omega\left[\{hni\} \cup G \to \tfrac{\tau}{G}\right], door_G^\tau \vdash\!\circ def^{door} !h \triangleright vis_V^\omega \left[vip^{hni}\right] \mid h^{door} in\, h^{door}}}{door_G^\tau \left[vis_V^\omega \left[vip^{hni}\right]\right] \text{ well-typed}}}{door_G^\tau \vdash vis_V^\omega \left[vip^{hni}\right]}}{vis_{G \cap V}^\tau \vdash vip^{hni}}}{\tau(vis)(hni) \ni call(vip)}$$

Fig. 8. The derivation tree for the guessing visitor scenario

key, assuming that the nurse is allowed to define key, $key \notin fn(\tau)$. The nurse makes key give access to the VIP record's service name: $def^{hni} key \triangleright vip^{hni} in\, 0$, for then the visitor to call key^{hni}.

The hospital system is now:

$$hni_H^\tau \big[(!vip \triangleright P)^{hni} \, \nu\, key\,\big($$
$$floor1_A^\alpha \left[vw_A^\sigma \left[vis_V^\omega \left[key^{hni}\right]\right]\right] \mid$$
$$floor3_C^\gamma \left[of_O^\pi \left[nurse_N^\omega \left[def^{hni} key \triangleright vip^{hni} in\, 0\right]\right]\right]\big)\big]$$

Static checking on this state of the system passes without errors. The dynamic checking at runtime raises the error the moment in which the nurse's definition is about to enter the visitor's ambient. The reduction steps, up until the raising of the error, are:

$$\xrightarrow{*} \nu\, key\, hni_H^\tau \big[\left(key \triangleright vip^{hni}\right)^{hni} (!vip \triangleright P)^{hni}$$
$$floor1_A^\alpha \left[vw_A^\sigma \left[vis_V^\omega \left[key^{hni}\right]\right]\right] \mid$$
$$floor3_C^\gamma \left[of_O^\pi \left[nurse_N^\omega \left[\,\right]\right]\right]$$

$$\xrightarrow{*} hni_H^\tau \big[(!vip \triangleright P)^{hni}$$
$$floor1_A^\alpha \left[vw_A^\sigma \left[\nu\, key \left(\left(key \triangleright vip^{hni}\right)^{hni, \tau \cap \alpha \cap \sigma} vis_V^\omega \left[key^{hni}\right]\right)\right]\right] \mid$$
$$floor3_C^\gamma \left[of_O^\pi \left[nurse_N^\omega \left[\,\right]\right]\right]$$

in which the down movement

$$\left(key \triangleright vip^{hni}\right)^{hni, \tau \cap \alpha \cap \sigma} vis_V^\omega \left[key^{hni}\right] \longrightarrow vis_V^\omega \left[\left(key \triangleright vip^{hni}\right)^{hni, \tau \cap \alpha \cap \sigma} key^{hni}\right]$$

is allowed only if $vis_V^{\tau \cap \alpha \cap \sigma} \left[vip^{hni}\right]$ is well-typed, a condition which depends on $\tau(vis)(hni) \ni call(vip)$:

$$\cfrac{\cfrac{vis_V^{\tau \cap \alpha \cap \sigma} \left[vip^{hni}\right] \text{ well-typed}}{vis_V^{\alpha \cap \tau} \vdash vip^{hni}}}{\tau(vis)(hni) \ni call(vip)}$$

5.3   The Wandering Visitor

Have the hospital policies devise a scheme to limit visitors from loading their own services (say, named key) throughout the hospital. For this, the first floor $floor1_{\mathcal{A}}^{\alpha}$, enclosing the door and the visitors' ward, has $vis \in A$ and α allowing visitors to publish services, but only up to the floor's level: $\alpha(vis)(floor1) \ni def(key)$, but $\alpha(vis)(hni) \not\ni def(key)$. The second floor still allows patients to be present during visiting hours, but imposes that they shouldn't publish anything while there, to any destination: both $\beta(vis)(floor2) \not\ni def(key)$ and $\beta(vis)(hni) \not\ni def(key)$. There should never be a visitor on the third floor, $vis \notin C$, hence γ poses no further restrictions upon the visitor's actions.

Have the visitor vis_V^{ω} planning to tour the hospital's three floors in search of spots to load the system with his key service. A try at publishing key at hni from the visitor's ward:

$$hni_H^{\tau}[floor1_{\mathcal{A}}^{\alpha}\,[vw_{\mathcal{A}}^{\sigma}\,[vis_V^{\omega}\,[def^{hni}key \triangleright P\,in\,0]]]]$$

is signalled at static checking, since the active well-typedness of vis_V^{ω} in this context of security policies depends on a security condition which doesn't hold:

$$\frac{vis_{H\cap V}^{\tau\cap\alpha\cap\sigma} \vdash def^{hni}key \triangleright P\,in\,0}{\alpha(vis)(hni) \ni def(key)}$$

The visitor would, however, be able to load the first floor (which acts like a sandbox for his definitions) with his service, if performing $def^{floor1}key \triangleright P\,in\,0$.

If still undeterred in his trials, he could walk to the second floor having in mind to try defining $def^{floor2}key \triangleright P\,in\,0$:

$$hni_H^{\tau}[floor1_{\mathcal{A}}^{\alpha}\,[vw_{\mathcal{A}}^{\sigma}\,[]]\,\mid$$
$$vis_V^{\omega}\,[in\,floor2.def^{floor2}key \triangleright P\,in\,0]\,\mid$$
$$floor2_B^{\beta}\,[]]$$

In this case, the operational semantics for the in movement raises the error when dynamically checking that:

$$\frac{vis_B^{\beta} \vdash def^{floor2}key \triangleright P\,in\,0}{\beta(vis)(floor2) \ni def(key)}$$

and the same happens if moving to the third floor, upon the condition $vis \notin C$ at the dynamic checking for in.

6   Related Work, Conclusions and Future Work

There are few direct formal models of context awareness in the presence of mobility. Among these, Birkedal et al. [3] propose a complex model of context awareness able to model both the usual reconfigurations of the context, and queries upon context; noting that context queries cannot be naturally modelled with one bigraphical reactive system (BRS), it proposes a solution (called a Plato-graphical model), such that its expressivity suits well sophisticated real-world context-aware systems. Braione [5] builds contextual reactive systems (CRS) upon reactive systems, RS. The difference between a CRS and a RS is the presence of a function which captures an association between elementary rules and their allowed reaction contexts, so that a CRS can express inhibitor and enabler factors for interaction. Both models are yet to achieve full results in studying behavioural equivalences and proving program properties.

Roman, Julien and Payton [16,18] build a language-based model of context-aware systems under mobility, an interesting feature of which is the fact that agents have as context the exposed (not private) variables of other agents. The work also has a limited associated proof logic, with program properties being expressed as predicate relations whose validity can be derived. Kjærgaard and Bunde-Pedersen's Conawa calculus [17] models context using several context trees, one for each category of context information (e.g., one for location information, one for activities and one for printers). An ambient entity will have a pointer-like presence in one or more trees, with the usual in/out ambient capabilities extended for mobility in multiple contexts.

Furthermore, we found inspiration in work on securing information flow in programming languages, such as Boudol's typing of information flow [4] in a multi-threaded ML-like language, when declassifying information for legal users (a stronger type system with flow policies, also guarding against termination leaks).

A number of type systems were introduced for Mobile Ambients: Cardelli, Ghelli and Gordon [9], Coppo et al. [13], Gorla, Hennessy and Sassone [14], Bugliesi, Castagna and Crafa [8], among others, introduce message exchange types, the typing of capabilities and actions, type-level groups of ambient names (effectively giving policies for crossing, opening, exchanging messages), and various security policies as membranes at ambient boundaries. Of particular interest is Gorla and Pugliese's enforcing of security policies via types in μKLAIM [15] for its fine-grained security features, assigning different privileges to users over different resources in a flat topology of networks.

Our calculus aims at capturing a notion of context awareness in infrastructure-based ubiquitous systems over a well-understood and applicable formalisation such as Mobile Ambients; we also put to use our background in designing protocols for ubiquitous systems (Bucur and Bardram, [6]) to ground this work in practice. Unlike the standard ambient communication scheme, we model context and its communication by exposing and calling named macros across ambient boundaries and policies, extending Zimmer's [22] flat context communication model. We design for a model of dynamic context with contextual information being macros distributed over varying network scopes; we follow the previous work in the field of designing for security with Mobile Ambients and apply security policies at ambients' membranes, limiting the capabilities enclosed processes can exhibit, to then type processes in regard to the hierarchy of policies enclosing them. Also, we keep the fine-grained policies on the lines of μKLAIM, only applied over our hierarchical topology of locations.

Our calculus is applicable for modelling and reasoning upon a fraction of the aspects of context-aware computing, in systems deployed over cell-based, hierarchical topologies. The model can aid the understanding of the workings of context in such mobile systems, and guide the implementation of an added software layer for security. Moreover, we feel that the basic ideas of representing contextual information, its communication and its use are applicable to a greater extent; thus, as part of the future work we intend to focus on modelling context awareness in ad hoc ubiquitous systems.

Acknowledgements. The authors wish to thank the anonymous reviewers for their comments on improving this paper.

References

1. Abowd, G.D., Dey, A.K., Brown, P.J., Davies, N., Smith, M., Steggles, P.: Towards a better understanding of context and context-awareness. In: Gellersen, H.-W. (ed.) HUC 1999. LNCS, vol. 1707, pp. 304–307. Springer, Heidelberg (1999)
2. Bardram, J.E., Hansen, T.R.: The AWARE architecture: supporting context-mediated social awareness in mobile cooperation. In: CSCW 2004: Proceedings of the 2004 ACM conference on Computer supported cooperative work, pp. 192–201. ACM Press, New York (2004)
3. Birkedal, L., Debois, S., Elsborg, E., Hildebrandt, T., Niss, H.: Bigraphical models of context-aware systems. In: Aceto, L., Ingólfsdóttir, A. (eds.) FOSSACS 2006. LNCS, vol. 3921. Springer, Heidelberg (2006)
4. Boudol, G.: On typing information flow. In: Van Hung, D., Wirsing, M. (eds.) ICTAC 2005. LNCS, vol. 3722, pp. 366–380. Springer, Heidelberg (2005)
5. Braione, P.: Operational congruences for contextual reactive systems. Technical report, 33, DEI, Politecnico di Milano (2004)
6. Bucur, D., Bardram, J.E.: Resource discovery in activity-based sensor networks. Mobile Networks and Applications (MONET) 12(2-3), 129–142 (2007)
7. Bugliesi, M., Castagna, G., Crafa, S.: Boxed ambients. In: Kobayashi, N., Pierce, B.C. (eds.) TACS 2001. LNCS, vol. 2215, pp. 38–63. Springer, Heidelberg (2001)
8. Bugliesi, M., Castagna, G., Crafa, S.: Reasoning about security in mobile ambients. In: Larsen, K.G., Nielsen, M. (eds.) CONCUR 2001. LNCS, vol. 2154, pp. 102–120. Springer, Heidelberg (2001)
9. Cardelli, L., Ghelli, G., Gordon, A.D.: Types for the ambient calculus. Inf. Comput. 177(2), 160–194 (2002)
10. Cardelli, L., Gordon, A.D.: Mobile ambients. In: Nivat, M. (ed.) FOSSACS 1998. LNCS, vol. 1378. Springer, Heidelberg (1998)
11. Chen, G., Kotz, D.: A Survey of Context-Aware Mobile Computing Research. Dartmouth Computer Science Technical Report TR2000-381 (2000)
12. Cheverst, K., Davies, N., Mitchell, K., Friday, A.: The role of connectivity in supporting context-sensitive applications. In: Gellersen, H.-W. (ed.) HUC 1999. LNCS, vol. 1707, pp. 193–207. Springer, Heidelberg (1999)

13. Coppo, M., Dezani-Ciancaglini, M., Giovannetti, E., Salvo, I.: Mobility types for mobile processes in mobile ambients. Electronic Notes in Theoretical Computer Science 78 (2002)
14. Gorla, D., Hennessy, M., Sassone, V.: Security policies as membranes in systems for global computing. Logical Methods in Computer Science 1(1:3), 1–23 (2005)
15. Gorla, D., Pugliese, R.: Enforcing security policies via types. In: Hutter, D., Müller, G., Stephan, W., Ullmann, M. (eds.) Security in Pervasive Computing. LNCS, vol. 2802, pp. 86–100. Springer, Heidelberg (2004)
16. Julien, C., Payton, J., Roman, G.-C.: Reasoning about context-awareness in the presence of mobility. Electr. Notes Theor. Comput. Sci. 97, 259–276 (2004)
17. Kjærgaard, M.B., Bunde-Pedersen, J.: Towards a formal model of context awareness. In: First International Workshop on Combining Theory and Systems Building in Pervasive Computing 2006 (CTSB 2006) (2006)
18. Roman, G.C., Julien, C., Payton, J.: A formal treatment of context-awareness. In: Wermelinger, M., Margaria-Steffen, T. (eds.) FASE 2004. LNCS, vol. 2984, pp. 12–36. Springer, Heidelberg (2004)
19. Schilit, B., Adams, N., Want, R.: Context-aware computing applications. In: IEEE Workshop on Mobile Computing Systems and Applications (1994)
20. Schilit, B.N.: A System Architecture for Context-Aware Mobile Computing. PhD thesis (1995)
21. Want, R., Hopper, A., Falcão, V., Gibbons, J.: The active badge location system. Technical Report 92.1, Olivetti Research Ltd. (ORL) (1992)
22. Zimmer, P.: A calculus for context-awareness. Technical Report BRICS Report Series RS-05-27

On Beta-Binders Communications[*]

Paola Quaglia

Dipartimento di Ingegneria e Scienza dell'Informazione, Università di Trento, Italy

Abstract. Beta-binders is a bio-inspired formalism with a formal reduction semantics in the process calculi style. The terms of the language are boxes with an internal processing engine and provided with interfaces for interactions with the other boxes in the environment.

Although Beta-binders shares some features with Ambient-like calculi, it exhibits a quite distinctive communication paradigm which might show to be relevant to modelling scenarios other than the biological ones. In the perspective of developing a formal theory for the language, and hence deepening the understanding of such a communication paradigm, here we define a labelled semantics for Beta-binders and show its correspondence to the original reduction semantics.

1   Introduction

Beta-binders [13] is one of the formalisms that have been recently defined to allow the linguistic description of biological scenarios and to reason about them by exploiting the methods and techniques developed over the last couple of decades for specifying and analyzing the behaviour of distributed systems (see, e.g., [7,9,2]). Other examples of bio-inspired process languages are BioAmbients [14], and Brane Calculi [1]. They both provide primitives to model wrappers of entities, as well as primitives for interacting with entities external to the local wrapper. This is in line with the observation that biological entities typically have an internal processing unit, as well as a sort of surrounding border through which the internal unit can receive and communicate signals. Beta-binders shares this view with the above mentioned formalisms, and indeed the language describes boxes equipped with sites for interaction with other boxes.

The Beta-binders communication paradigm, however, significantly departs from the paradigms adopted by either typical process calculi or other bio-inspired languages. The main difference is due to the fact that communication between boxes is driven by a notion of *compatibility* of their sites, rather than by some notion of site *complementarity*. This point is better introduced by resorting to the prototypical example of reactive system: the vending machine. Writing in CCS [7], the specification of a simple vending machine could be the following:

$$Ven = 35cents.\overline{tea}.\ Ven + 50cents.\overline{coffee}.\ Ven.$$

---

[*] This work has been partially sponsored by the PRIN 2006 Project BISCA - *Sistemi e calcoli di ispirazione biologica e loro applicazioni*.

P. Degano et al. (Eds.): Montanari Festschrift, LNCS 5065, pp. 457–472, 2008.

458   P. Quaglia

In order to get a tea or a coffee from such machine, the user has to precisely coordinate with its actions. He has to insert in the machine enough coins to be entitled to collect his preferred drink (i.e., he has to execute either $\overline{35cents}$, the action complementary to $35cents$, or $\overline{50cents}$, complementary to $50cents$), then he can push the appropriate button (i.e., he can perform either tea or $coffee$). Let us now imagine that the user has to interact with a much more obscure vending machine: he just knows that he can insert coins and that, depending on the amount of money, the machine will deliver some drink. A possible specification of such machine and of a user willing to spend 40 cents would look like the following:

$$Ven' = [35 - 49]cents.\,\overline{tea}.\ Ven' + [50 - 69]cents.\,\overline{coffee}.\ Ven'$$

$$Usr = \overline{40cents}.\,drink.\,nil.$$

where $[35 - 49]cents$ means "upon receiving from 35 to 49 cents", and $[50 - 69]cents$ has an analogous meaning. Now $\overline{40cents}$ is complementary to neither $[35 - 49]cents$ nor $[50 - 69]cents$, and $drink$ is complementary to neither $\overline{tea}$ nor $\overline{coffee}$. Nonetheless, 40 cents is enough for a tea. Can Usr interact with Ven' and get a tea?

Biological systems behave in some respect as obscure vending machines: the same high level event may be triggered by possibly different, and sometimes not well-known, low level interactions. Beta-binders builts on the intuition that, in order to underpin predictive modelling or simply infer behavioural models from incomplete information, abstraction and non-determinism might be not enough. A starting point for a suitable modelling language could be leaving some room to "under-determinedness". In Beta-binders, this is done by partially relaxing the idea that interactions should always be based on the strict matching of complementary actions, be them pairs of input and output, or pairs of request to enter an ambient and permission to do so, or else. In other terms, Beta-binders moves from the idea that Usr above should be entitled to interact with Ven' and get a tea from the machine.

More specifically, Beta-binders encapsulates (extended) π-calculus processes [8,16] into boxes with *typed interaction sites*. For example, the following is a graphical representation of a system given by three parallel components.

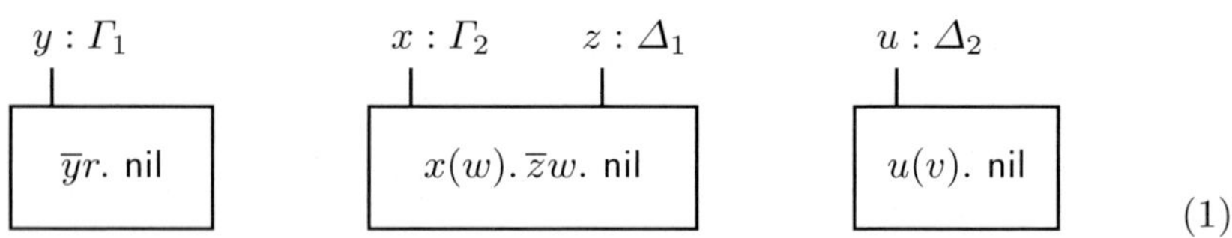

$$(1)$$

Each of the pairs $y : \Gamma_1$, $x : \Gamma_2$, $z : \Delta_1$, and $u : \Delta_2$ denotes the name and the type of the corresponding interaction site. The definition of the operational semantics of the language is parametric w.r.t. both the domain of types and a binary relation of type compatibility. Interaction within a box is ruled out as it is in π-calculus. Interaction between a box B_1 and a box B_2 can take place only if the processes encapsulated in B_1 and B_2 are ready to perform complementary actions (input/output) over one of their interfaces, and if the types of these interfaces are compatible. For instance, assuming Γ_1 and Γ_2 be compatible, in (1) the leftmost box can communicate with the central one. This involves matching the output action $\overline{y}r$ over $y : \Gamma_1$, and the input action $x(w)$ over $x : \Gamma_2$.

The system illustrated in (1) also highlights that communication between boxes can substantially modify the future interaction potentials of boxes. Suppose for instance that (Γ_1, Γ_2) and (Δ_1, Δ_2) are pairwise compatible. Then the communication between the first two boxes also triggers a communication between the second and the third box. This feature, which is particularly useful in modelling pathways (see, e.g. [6]), is yet another source of the desired under-determinedness.

In Beta-binders, different typing policies and notions of compatibility may be adopted correspondingly to distinct specification needs. Choosing the appropriate typing may reveal essential to model interaction paradigms that hardly fit into the standard action/co-action view of communication. This was the case, e.g., in rendering the the so-called *shape spaces theory* [10], a mathematical model of the interactions among components of the immune system. In the corresponding Beta-binders model [11], types of interaction sites were taken to be strings of 0s and 1s encoding both geometric information (e.g., size and shape of motifs) and physical characteristics of molecular determinants (e.g., charge and ability to form hydrogen bonds). Compatibility between types, in turn, was based on a notion of distance between strings.

A range of biological scenarios have been modelled in Beta-binders, and many of them have also been analyzed using a simulator [15] based on the stochastic extension of the language [3]. Beta-binders, indeed, is mainly inspired by bio-chemical interactions. The primitives of the language, though, are quite abstract to suggest that the formalism could be applied to modelling scenarios coming from fields other than life sciences. Boxes, for instance, could be nodes of a network and the actual typing of interaction sites might implement constraints on the communications between nodes (based, e.g., on belief or trust level). Beta-binders, however, still lacks a formal theory which would help understanding the potentials of its communication paradigm.

The original semantics of the language is given in reduction style, namely reasoning on behaviours up to syntactic restructuring of processes. As always, this eases the presentation of the operational semantics by pulling out some intricacies due to the syntactic structure of terms. At the same time, reduction semantics partially limits syntax-driven reasoning and hence makes mathematical treatment harder than it could be. Here we provide an alternative characterization of Beta-binders semantics in terms of a labelled transition system and show the operational correspondence between the labelled and the reduction semantics. This is, we believe, a first step towards deeper investigations on the formalism, and hence towards a better understanding of the applicability of its communication paradigm. Not least, the simplification of formal developments can promote a shift of attention to possible typing disciplines for interaction sites which is where the potentials of the language seem mainly to come from.

The rest of the paper is organized as follows. Section 2 contains a short overview of Beta-binders. Section 3 presents the definition of the labelled semantics for the language. The results about the operational correspondence of the two transition systems are reported in Section 4. The paper ends with some concluding remarks in Section 5.

2   Beta-Binders Overview

This section reviews the definition of Beta-binders, a bio-inspired language whose terms are (essentially) π-calculus processes wrapped into boxes with interaction capabilities. As in the π-calculus, the existence of a countably infinite set of names (ranged over by $x, y, z, \ldots$) is assumed. Furthermore, a special class of binders, called *beta binders*, is introduced. Each binder characterizes an interaction site by means of an identifier and an associated type. The domain of types is left unspecified. It can be arbitrarily instantiated under the proviso that it is decidable whether types are pairwise *compatible* or not.

Processes are generated by the following grammar.

$$B ::= \mathsf{Nil} \ \Big| \ \mathbb{B}[\,P\,] \ \Big| \ B \parallel B$$

$$\mathbb{B} ::= \beta(x,\,\Gamma) \ \Big| \ \beta^h(x,\,\Gamma) \ \Big| \ \beta(x,\,\Gamma)\,\mathbb{B} \ \Big| \ \beta^h(x,\,\Gamma)\,\mathbb{B}$$

$$P ::= \mathsf{nil} \ \Big| \ x(w).\,P \ \Big| \ \overline{x}y.\,P \ \Big| \ P \mid P \ \Big| \ \nu y\,P \ \Big| \ !\,P \ \Big|$$
$$\mathsf{expose}(x,\,\Gamma).\,P \ \Big| \ \mathsf{hide}(x).\,P \ \Big| \ \mathsf{unhide}(x).\,P$$

where Nil is the deadlocked process, $\mathbb{B}[\,P\,]$ denotes the process P enclosed in a box with interaction capabilities $\mathbb{B}$, and $B_1 \parallel B_2$ is the parallel composition of B_1 and B_2.

Interaction capabilities $\mathbb{B}$ are represented by sequences of elements of either the shape $\beta(x,\,\Gamma)$ or the shape $\beta^h(x,\,\Gamma)$, the so-called *elementary beta binders*. Intuitively, the binder $\beta(x,\,\Gamma)$ represents an active, i.e. potentially interacting, site of the box. Beta binders can be (or be made) hidden to prevent possible interactions through them. A hidden binder is denoted by $\beta^h(x,\,\Gamma)$. In either $\beta(x,\,\Gamma)$ or $\beta^h(x,\,\Gamma)$, the name x identifies the interaction site and is called the *subject* of the binder, while Γ is the *type* of x. No requirement is set over the domain of types $\mathcal{T}$, but for assuming that each of its possible instances goes along with the definition of a *symmetric compatibility relation*, and that the predicate $\mathsf{comp} : \mathcal{T} \times \mathcal{T} \to \{\mathsf{true}, \mathsf{false}\}$, which returns true iff its argument types are compatible, is decidable.

The grammar for P generates extended π-calculus processes. The deadlocked nil, as well as the input and output prefixes, and the operators for parallel composition, restriction and replication, have the same meaning as in π-calculus. The added prefixes expose, hide, and unhide are directives for changing the interaction capabilities of the enclosing box by, respectively, adding a new interaction

Table 1. Structural congruences $\equiv$ and $\equiv_b$

$P_1 \equiv P_2$ if $P_1 \equiv_\alpha P_2$

$P \mid \mathsf{nil} \equiv P, \quad P_1 \mid P_2 \equiv P_2 \mid P_1, \quad P_1 \mid (P_2 \mid P_3) \equiv (P_1 \mid P_2) \mid P_3$

$\nu z\,\mathsf{nil} \equiv \mathsf{nil}, \quad \nu z\,\nu w\,P \equiv \nu w\,\nu z\,P, \quad \nu z\,(P_1 \mid P_2) \equiv P_1 \mid \nu z\,P_2$ provided $z \notin \mathsf{fn}(P_1)$

$!P \equiv P \mid {!P}$

$\mathbb{B}\mathbb{B}'[\,P_1\,] \equiv_b \mathbb{B}'\mathbb{B}[\,P_2\,]$ provided $P_1 \equiv P_2$

$B \equiv_b B'$ if $(B = \beta^+(x : \Delta)\mathbb{B}[\,P\,]$ and $B' = \beta^+(y : \Delta)\mathbb{B}[\,P\{y/x\}\,])$ or

$\qquad\qquad (B' = \beta^+(x : \Delta)\mathbb{B}[\,P\,]$ and $B = \beta^+(y : \Delta)\mathbb{B}[\,P\{y/x\}\,])$

$\qquad\qquad$ with y fresh in P and in $\mathsf{sub}(\mathbb{B})$ and where β^+ stays for either β or β^h

$B \parallel \mathsf{Nil} \equiv_b B, \quad B_1 \parallel B_2 \equiv_b B_2 \parallel B_1, \quad B_1 \parallel (B_2 \parallel B_3) \equiv_b (B_1 \parallel B_2) \parallel B_3$

---

site, hiding an existing visible site, making visible a hidden site. The usual definitions of *free names* $\mathsf{fn}(\_)$, of *bound names* $\mathsf{bn}(\_)$, of *names* $\mathsf{n}(\_)$, and of *name substitution* are extended by stipulating that $\mathsf{expose}(x, \Gamma)\,.\,P$ is a binder for x in P.

Notational conventions. We write $\tilde{u}$ as a shorthand for the tuple $u_1 \ldots u_n$ of names, and use $\nu\tilde{u}$ for $\nu u_1 \ldots \nu u_n$. Also, with a slight abuse of notation, we sometime read tuples as sets.

The set of the subjects of all the elementary beta binders in $\mathbb{B}$ is denoted by $\mathsf{sub}(\mathbb{B})$, and we write $\mathbb{B} = \mathbb{B}_1\mathbb{B}_2$ to mean that $\mathbb{B}$ is the beta binder given by the juxtaposition of $\mathbb{B}_1$ and $\mathbb{B}_2$. A binder $\mathbb{B}$ is said to be *well-formed* when the subjects of its elementary components are all distinct. We use $\Delta, \Delta_1, \ldots, \Gamma, \Gamma_1, \ldots$ to range over site types, and $B, B_1, \ldots$ to range of Beta-binders processes. When all the binders $\mathbb{B}_1, \mathbb{B}_2, \ldots$ occurring in B are well-formed, B itself is said to be *well-formed.*

The meta-variables $\mathbb{B}_1, \mathbb{B}_2, \ldots$ are overloaded to stay for either a beta binder or the empty string. For instance, we write $\beta(x, \Gamma)\,\mathbb{B}_1[\,P\,]$ to mean a process that could have no other interface besides x. Notice, however, that by definition each process has to have at least one (possibly hidden) interface. So, for example, $\mathbb{B}_1$ is meant to be different from the empty string in $\mathbb{B}_1[\,P\,]$. $\qquad\qquad\square$

The operational semantics for Beta-binders makes use of both a structural congruence over pi-processes and a structural congruence over boxes. The needed congruences are the smallest relations satisfying the laws in Table 1. The laws of structural congruence over pi-processes are the typical π-calculus axioms, where $\equiv_\alpha$ is used to denote α-equivalence that is extended to deal with expose binders in the natural way. The first axiom for boxes serves two purposes. It declares that the actual ordering of elementary beta binders within a composite binder is irrelevant, and states that the structural congruence of internal pi-processes is reflected at the level of boxes. The second law is a sort of α-conversion axiom for boxes. It states that the subject of elementary beta binders can be refreshed under the proviso that name clashes in the internal process are avoided and that well-formedness of binders is preserved. The latest laws are just the monoidal axioms for the parallel composition of boxes.

Table 2. Beta-binders reduction semantics

$$(\text{intra}) \quad \frac{P \equiv \nu\tilde{u}\,(x(w).\,P_1 \mid \overline{x}z.\,P_2 \mid P_3)}{\mathbb{B}[\,P\,] \twoheadrightarrow \mathbb{B}[\,\nu\tilde{u}\,(P_1\{z\!/\!w\} \mid P_2 \mid P_3)\,]}$$

$$(\text{inter}) \quad \frac{P \equiv \nu\tilde{u}\,(x(w).\,P_1 \mid P_2) \qquad\qquad Q \equiv \nu\tilde{v}\,(\overline{y}z.\,Q_1 \mid Q_2)}{\beta(x,\,\Gamma)\,\mathbb{B}_1[\,P\,] \parallel \beta(y,\,\Delta)\,\mathbb{B}_2[\,Q\,] \twoheadrightarrow \beta(x,\,\Gamma)\,\mathbb{B}_1[\,P'\,] \parallel \beta(y,\,\Delta)\,\mathbb{B}_2[\,Q'\,]}$$

where $P' = \nu\tilde{u}\,(P_1\{z\!/\!w\} \mid P_2)$ and $Q' = \nu\tilde{v}\,(Q_1 \mid Q_2)$
provided $\text{comp}(\Gamma, \Delta)$ and $x, z \notin \tilde{u}$ and $y, z \notin \tilde{v}$

$$(\text{expose}) \quad \frac{P \equiv \nu\tilde{u}\,(\text{expose}(x,\,\Gamma).\,P_1 \mid P_2)}{\mathbb{B}[\,P\,] \twoheadrightarrow \mathbb{B}\,\beta(y,\,\Gamma)[\,\nu\tilde{u}\,(P_1\{y\!/\!x\} \mid P_2)\,]}$$

provided $y \notin \tilde{u} \cup \text{sub}(\mathbb{B}) \cup \text{fn}(P_2)$

$$(\text{hide}) \quad \frac{P \equiv \nu\tilde{u}\,(\text{hide}(x).\,P_1 \mid P_2)}{\mathbb{B}\,\beta(x,\,\Gamma)[\,P\,] \twoheadrightarrow \mathbb{B}\,\beta^h(x,\,\Gamma)[\,\nu\tilde{u}\,(P_1 \mid P_2)\,]}$$

provided $x \notin \tilde{u}$

$$(\text{unhide}) \quad \frac{P \equiv \nu\tilde{u}\,(\text{unhide}(x).\,P_1 \mid P_2)}{\mathbb{B}\,\beta^h(x,\,\Gamma)[\,P\,] \twoheadrightarrow \mathbb{B}\,\beta(x,\,\Gamma)[\,\nu\tilde{u}\,(P_1 \mid P_2)\,]}$$

provided $x \notin \tilde{u}$

$$(\text{redex}) \quad \frac{B \twoheadrightarrow B'}{B \parallel B'' \twoheadrightarrow B' \parallel B''} \qquad\qquad (\text{struct}) \quad \frac{B \equiv_b B_1 \qquad B_1 \twoheadrightarrow B_2 \qquad B_2 \equiv_b B'}{B \twoheadrightarrow B'}$$

The reduction relation describing the operational semantics of Beta-binders is defined by the axioms and rules collected in Table 2. Here we actually present a subset of the language. The original Beta-binders semantics also provides ways to *join* boxes together and to *split* a box in two. This is achieved by relaying on the definition of (one or more instances of) computable functions f_{join} and f_{split} that are used both to check whether the conditions for joining or splitting boxes are met, and to assess the structure of boxes resulting from either the merging or the splitting. To leave the user free to choose different strategies for aggregation and disaggregation of boxes, the operational semantics is parametric w.r.t. the definition of those functions. It would not be particularly difficult to accommodate in the present framework the management of specific definitions of f_{join} and f_{split}. For these reasons, in this work we stick to the main backbone of Beta-binders semantics, which is given by the rules reported in Table 2.

The axiom intra concerns communications between pi-processes within the same box. If the internal process is structurally equivalent to $\nu\tilde{u}\,(x(w).\,P_1 \mid \overline{x}z.\,P_2 \mid P_3)$, then the box can perform a reduction leading to a process with unchanged external interface and with internal process $\nu\tilde{u}\,(P_1\{z\!/\!w\} \mid P_2 \mid P_3)$.

The axiom inter describes possible interactions between boxes, and shows how compatibility of types is used to match complementary actions performed by parallel Beta-binders processes. Notice that, whichever notion of type compatibility is assumed, the communication ability is only determined by the types of the involved beta binders rather than by their subjects. Information passes from the box containing the process which exhibits the output prefix to the box enclosing the process that performs the input action. Also observe that the communicated name z is required to be free in the sending process. This definition of the axiom inter corresponds to considering the borders of the box as the farthest limit that restricted names can reach. This, in turn, is in line with the design principles of the language which considers boxes as first class scope delimiters.

The axiom expose is used to add a new binder to a box. The name x declared in the prefix expose(x, Γ) is a placeholder which can be renamed to avoid clashes with both the subjects of the other binders of the containing box, and the free names of processes outside the scope of the binding expose prefix.

The axiom hide forces a binder to become hidden, and the unhide prefix, dual to hide, makes visible a hidden binder.

As usual for reduction semantics, the rules redex and struct are meant, respectively, to interpret the reduction of a parallel subcomponent as a reduction of the global process, and to infer a reduction after a proper structural shuffling of the process at hand.

A final observation about the semantics is that it preserves well-formedness of processes.

Proposition 1. *If B is well-formed and $B \rightarrow B'$ then B' is well-formed.*

Proof. See [12]. $\qquad\square$

3   Labelled Semantics

This section presents the labelled semantics for Beta-binders. To ease the presentation, binders are grouped into a list. Also, the operator $\_ * \_$ is added to the grammar for π-processes, so that $\mathbb{B}[\,P\,]$ is rendered by $L * P$ where L represents $[\![\mathbb{B}]\!]$ after the following definition of the translation function $[\![\_]\!]$:

$$[\![\mathbb{B}]\!] = [\,] \quad \text{if } \mathbb{B} \text{ is empty}$$

$$[\![\beta(x,\,\Gamma)\,\mathbb{B}']\!] = (x,\Gamma) :: [\![\mathbb{B}']\!] \qquad [\![\beta^h(x,\,\Gamma)\,\mathbb{B}']\!] = (x^h,\Gamma) :: [\![\mathbb{B}']\!]$$

where $[\,]$ denotes the empty list, and "::" is the usual cons operator. Also, we simply use the notation x^h, instead of adopting triples with a component playing as a flag, to denote that the binder named x is hidden. The full mapping of Beta-binders terms into processes of the updated language is obtained by extending the definition of the translation function with the following clauses:

$$[\![\text{Nil}]\!] = \text{Nil} \qquad\qquad [\![\mathbb{B}[\,P\,]]\!] = [\![\mathbb{B}]\!] * P \qquad\qquad [\![B_1 \parallel B_2]\!] = [\![B_1]\!] \parallel [\![B_2]\!] \,.$$

Table 3. Structural congruence $\equiv_l$

$L_1 * P_1 \equiv_l L_2 * P_2$ provided $((x, \Gamma)$ is in L_1 iff (x, Γ) is in $L_2)$ and $P_1 \equiv P_2$

$M \equiv_l M'$ if $(M = (x^+, \Delta) :: L * P$ and $M' = (y^+, \Delta) :: L * P\{y/x\})$ or
$\qquad\quad (M' = (x^+, \Delta) :: L * P$ and $M = (y^+, \Delta) :: L * P\{y/x\})$
$\qquad$ with y fresh in P and in firsts(L) and
$\qquad\quad$ where x^+, y^+ stay for either x, y or x^h, y^h

$M \parallel \mathsf{Nil} \equiv_l M, \quad M_1 \parallel M_2 \equiv_l M_2 \parallel M_1, \quad M_1 \parallel (M_2 \parallel M_3) \equiv_l (M_1 \parallel M_2) \parallel M_3$

We let lists of beta binders be ranged over by $L, L', \ldots$, and let M, N, M', $N', \ldots$ stay for translated processes. The following functions are assumed to be defined on lists of pairs:

- firsts(L) returns the set of the first components of the pairs in L;
- ch(L, z, w) returns a list which is the same as L but for the fact that $w \in$ firsts(L) is changed into z.

Also, we write $(x, \Gamma) \in L$ if Γ is the type paired with $x \in$ firsts(L). Applying ch$(L, _, _)$ is the only way to change the elements of firsts(L). Indeed we stipulate that name substitution does not affect lists, namely $(L * P)\{z/x\} = L * (P\{z/x\})$ for all z and x. The following mathematical development requires adapting the structural congruence $\equiv_b$ to deal with translated processes. To this end $\equiv_b$ is refreshed into $\equiv_l$, which is defined to be the smallest congruence satisfying the laws in Table 3.

In the labelled semantics, transition labels α are given by:

$$\alpha ::= \tau \mid x(w) \mid \overline{x}z \mid \overline{x}(w) \mid \mathrm{hd}\langle x \rangle \mid \mathrm{uh}\langle x \rangle \mid \mathrm{ex}(w : \Gamma) \mid x : \Gamma(w) \mid \overline{x : \Gamma}\langle z \rangle$$

The input and output actions $x(w)$, $\overline{x}z$, and $\overline{x}(w)$ have the same informal interpretation as in the late π-calculus semantics. The actions $\mathrm{hd}\langle x \rangle$, $\mathrm{uh}\langle x \rangle$, and $\mathrm{ex}(w : \Gamma)$ correspond, respectively, to the execution of a hiding, unhiding, and expose directive. The informal meaning of $x : \Gamma(w)$ is that the process can execute an input over the beta binder named x and typed by Γ. Dually, the action $\overline{x : \Gamma}\langle z \rangle$ denotes the transmission of z over the binder x with type Γ. We refer to these latest kind of actions as to *global input* and to *global output*, respectively.

The usual convention about fn(α) and bn(α) apply to input labels and to either free or bound output labels. For the other actions we assume that fn$(\mathrm{hd}\langle x \rangle) = $ fn$(\mathrm{uh}\langle x \rangle) = $ fn$(x : \Gamma(w)) = \{x\}$, and fn$(\overline{x : \Gamma}\langle z \rangle) = \{x, z\}$, and bn$(\mathrm{ex}(w : \Gamma)) = $ bn$(x : \Gamma(w)) = \{w\}$. Also, fn$(L * P) = $ fn(P).

Table 4 reports the labelled operational semantics of Beta-binders processes. As common, the symmetric rules for symmetric operators (here $\mid$ and $\parallel$) are omitted. The rules in the upper portion of the table, namely those named (i-_) are the usual rules of the late π-calculus transition system [9] augmented with rules for dealing with the new prefixes for hiding, unhiding, and exposing a binder. The kind of directive that is being executed is recorded in the label.

Table 4. Labelled semantics

(i-inp) $x(y).P \overset{x(w)}{\rightarrow} P\{w/y\}$ $w \notin \mathrm{fn}(\nu y\, P)$ (i-out) $\overline{x}y.P \overset{\overline{x}y}{\rightarrow} P$

(i-exp) $\mathrm{expose}(y,\Gamma).P \overset{\mathrm{ex}(w:\Gamma)}{\rightarrow} P\{w/y\}$ $w \notin \mathrm{fn}(\nu y\, P)$

(i-hd) $\mathrm{hide}(x).P \overset{\mathrm{hd}\langle x \rangle}{\rightarrow} P$ (i-uh) $\mathrm{unhide}(x).P \overset{\mathrm{uh}\langle x \rangle}{\rightarrow} P$

(i-par) $\dfrac{P \overset{\alpha}{\rightarrow} P'}{P \mid Q \overset{\alpha}{\rightarrow} P' \mid Q}$ $\mathrm{bn}(\alpha)\cap\mathrm{fn}(Q)=\emptyset$ (i-com) $\dfrac{P \overset{\overline{x}y}{\rightarrow} P' \quad Q \overset{x(w)}{\rightarrow} Q'}{P \mid Q \overset{\tau}{\rightarrow} P' \mid Q'\{y/w\}}$

(i-close) $\dfrac{P \overset{x(w)}{\rightarrow} P' \quad Q \overset{\overline{x}(w)}{\rightarrow} Q'}{P \mid Q \overset{\tau}{\rightarrow} \nu w\,(P' \mid Q')}$ (i-open) $\dfrac{P \overset{\overline{x}y}{\rightarrow} P'}{\nu y\, P \overset{\overline{x}(w)}{\rightarrow} P'\{w/y\}}$ $y \neq x,\ w \notin \mathrm{fn}(\nu y\, P')$

(i-res) $\dfrac{P \overset{\alpha}{\rightarrow} P'}{\nu y\, P \overset{\alpha}{\rightarrow} \nu y\, P'}$ $y \notin \mathrm{n}(\alpha)$ (i-bang) $\dfrac{P \mid !P \overset{\alpha}{\rightarrow} P'}{!P \overset{\alpha}{\rightarrow} P'}$

(inp) $\dfrac{P \overset{x(w)}{\rightarrow} P'}{L * P \overset{x:\Delta(w)}{\rightarrow} L * P'}$ $(x,\Delta)\in L$ (out) $\dfrac{P \overset{\overline{y}z}{\rightarrow} P'}{L * P \overset{\overline{y:\Gamma}\langle z \rangle}{\rightarrow} L * P'}$ $(y,\Gamma)\in L$

(hd) $\dfrac{P \overset{\mathrm{hd}\langle x \rangle}{\rightarrow} P'}{L * P \overset{\tau}{\rightarrow} \mathrm{ch}(L, x^h, x) * P'}$ $x \in \mathrm{firsts}(L)$ (uh) $\dfrac{P \overset{\mathrm{uh}\langle x \rangle}{\rightarrow} P'}{L * P \overset{\tau}{\rightarrow} \mathrm{ch}(L, x, x^h) * P'}$ $x^h \in \mathrm{firsts}(L)$

(tau) $\dfrac{P \overset{\tau}{\rightarrow} P'}{L * P \overset{\tau}{\rightarrow} L * P'}$ (exp) $\dfrac{P \overset{\mathrm{ex}(w:\Gamma)}{\rightarrow} P'}{L * P \overset{\tau}{\rightarrow} (w,\Gamma) :: L * P'}$ $w \notin \mathrm{firsts}(L)$

(par) $\dfrac{M \overset{\alpha}{\rightarrow} M'}{M \parallel N \overset{\alpha}{\rightarrow} M' \parallel N}$ $\mathrm{bn}(\alpha)\cap\mathrm{fn}(N)=\emptyset$ (com) $\dfrac{M \overset{x:\Delta(w)}{\rightarrow} M' \quad N \overset{\overline{y:\Gamma}\langle z \rangle}{\rightarrow} N'}{M \parallel N \overset{\tau}{\rightarrow} M'\{z/w\} \parallel N'}$ $\mathrm{comp}(\Delta,\Gamma)$

The single interesting point to observe is about (i-exp). Analogously to what happens for input actions in (i-inp), the transition label shows a bound name which, but for being fresh in the residual process, can be arbitrarily chosen in a set of infinitely many names. This fact guarantees two main properties of the transition system. The first one is relative to the interplay between (i-exp) and (exp). The role of this latest rule is to add a new binder to the list L in $L * P$ while avoiding possible clashes of binder names. To this end, the definition of (i-exp) ensures that it is always possible to choose a name w which is fresh enough to let the action $\text{ex}(w : \Gamma)$ pass the (exp) rule. The second issue is relative to the behaviour of $\text{expose}(y, \Gamma) . P$ in parallel compositions. By the (i-par) rule, a process like $R = \text{expose}(y, \Gamma) . P \mid Q$ can actually execute the action $\text{ex}(w : \Gamma)$ only if w is not free in Q. This guarantees that possible future interactions over the newly exposed binder will not affect Q which was actually outside the scope of $\text{expose}(y, \Gamma)$ in the original process R.

The lower portion of Table 4 shows the rules needed to infer the behaviour of processes of either the form $L * P$ or of the form $M \parallel N$. Rule (inp) states that, if P can execute the input action $x(w)$ and x is a binder paired with Δ in L, then $L * P$ performs the special global input action $x : \Delta(w)$. Dually, rule (out) states how the global output action $\overline{y : \Gamma}\langle z \rangle$ can be performed by $L * P$ as an immediate result of the execution of $\overline{y}z$ by P. Under the proviso that Δ and Γ are compatible, global input and global output actions are matched together by the (com) rule to give rise to a communication between boxes.

Rule (hd), applicable only if the binder x is listed in its unhidden form in L, transforms the internal action $\text{hd}\langle x \rangle$ into a τ action by $L * P$ upon the appropriate updating of L. The interpretation of (uh) is analogous.

Possible τ actions performed by the internal process P as result of communications (either closing or not) are also executed by $L * P$ (rule (tau)). Eventually, rule (par) is the upper level analogue of (i-par).

4   Operational Correspondence

The operational correspondence between the reduction semantics and the labelled semantics is reported below, together with a few intermediate results which are needed to show the main ones.

The next three lemmas are auxiliary to showing how labelled transitions relate to reductions (Theorem 1).

Lemma 1. *If $P \xrightarrow{\alpha} P'$ then one of the following holds for P, α, and P':*

1. $\alpha = x(w)$ *and* $P \equiv \nu\tilde{u}\,(x(y).\,P_1 \mid P_2)$ *and* $P' \equiv \nu\tilde{u}\,(P_1\{w/y\} \mid P_2)$ *for some* $x, w, \tilde{u}, y, P_1$, *and* P_2 *such that* $x \notin \tilde{u}$ *and* $w \notin \text{fn}(P_2) \cup \tilde{u}$;
2. $\alpha = \overline{x}y$ *and* $P \equiv \nu\tilde{u}\,(\overline{x}y.\,P_1 \mid P_2)$ *and* $P' \equiv \nu\tilde{u}\,(P_1 \mid P_2)$ *for some* $x, y, \tilde{u}, P_1$, *and* P_2 *such that* $x, y \notin \tilde{u}$;
3. $\alpha = \text{ex}(w : \Gamma)$ *and* $P \equiv \nu\tilde{u}\,(\text{expose}(y, \Gamma).\,P_1 \mid P_2)$ *and* $P' \equiv \nu\tilde{u}\,(P_1\{w/y\} \mid P_2)$ *for some* $w, \Gamma, \tilde{u}, y, P_1$, *and* P_2 *such that* $w \notin \text{fn}(P_2) \cup \tilde{u}$;
4. $\alpha = \text{hd}\langle x \rangle$ *and* $P \equiv \nu\tilde{u}\,(\text{hide}(x).\,P_1 \mid P_2)$ *and* $P' \equiv \nu\tilde{u}\,(P_1 \mid P_2)$ *for some* $x, \tilde{u}, P_1$, *and* P_2 *such that* $x \notin \tilde{u}$;
5. $\alpha = \text{uh}\langle x \rangle$ *and* $P \equiv \nu\tilde{u}\,(\text{unhide}(x).\,P_1 \mid P_2)$ *and* $P' \equiv \nu\tilde{u}\,(P_1 \mid P_2)$ *for some* $x, \tilde{u}, P_1$, *and* P_2 *such that* $x \notin \tilde{u}$;
6. $\alpha = \overline{x}(w)$ *and* $P \equiv \nu z\,\nu\tilde{u}\,(\overline{x}z.\,P_1 \mid P_2)$ *and* $P' \equiv \nu\tilde{u}\,((P_1 \mid P_2)\{w/z\})$ *for any* $z \notin \text{n}(P)$ *and for some* $x, w, \tilde{u}, P_1$, *and* P_2 *such that* $x \neq z$ *and* $x, z, w \notin \tilde{u}$;
7. $\alpha = \tau$ *and* $P \equiv \nu\tilde{u}\,(x(w).\,P_1 \mid \overline{x}z.\,P_2 \mid P_3)$ *and* $P' \equiv \nu\tilde{u}\,(P_1\{z/w\} \mid P_2 \mid P_3)$ *for some* $x, w, z, \tilde{u}, P_1, P_2$, *and* P_3.

Proof. By induction on depth of inference. For the inductive base the five axioms (i-inp), (i-out), (i-exp), (i-hd), and (i-uh) are considered. In each of these cases process P_2 in the thesis is nil, and $\tilde{u}$ is the empty sequence of names. For the inductive step, we only show the argument for the most interesting case.

Suppose that the last rule applied to infer $P \xrightarrow{\alpha} P'$ is (i-close). By definition of the rule, for some x, w, Q, Q', R and R', it holds that $P = Q \mid R$, $\alpha = \tau$, and $P' = \nu w \, (Q' \mid R')$ with $Q \xrightarrow{x(w)} Q'$ and with $R \xrightarrow{\overline{x}(w)} R'$.

Then by inductive hypothesis:

$$Q \equiv Q_r = \nu\tilde{v}\,(x(y).\,Q_1 \mid Q_2) \qquad\qquad R \equiv R_r = \nu z\,\nu\tilde{u}\,(\overline{x}z.\,R_1 \mid R_2)$$

$$Q' \equiv Q'_r = \nu\tilde{v}\,(Q_1\{w/y\} \mid Q_2) \qquad\qquad R' \equiv R'_r = \nu\tilde{u}\,((R_1 \mid R_2)\{w/z\})$$

with

$$x \notin \tilde{v} \cup \{z\} \cup \tilde{u}, \text{ and } w \notin \mathrm{fn}(Q_2) \cup \tilde{v}, \text{ and } z, w \notin \tilde{u} \,. \tag{2}$$

Also, we can safely assume that

$$\tilde{v} \notin \mathrm{fn}(R_r) \cup \mathrm{fn}(R'_r), \text{ and } \tilde{v} \cap (\tilde{u} \cup \{z\}) = \emptyset, \text{ and } z, \tilde{u} \notin \mathrm{fn}(Q_r) \cup \mathrm{fn}(Q'_r) \,. \tag{3}$$

Indeed the involved terms could be α-converted to simultaneously meet the conditions corresponding, for fresh $\tilde{u}', \tilde{v}', z'$, to those listed in both (2) and (3). Simply notice that any possible α-conversion would affect neither x, which by (2) is free in both Q_r and R_r, nor w, which again by (2) is free in both Q'_r and R'_r.

Now observe that $z \notin \mathrm{fn}(x(y).\,Q_1 \mid Q_2)$, by $z \notin \mathrm{fn}(Q_r)$ and $z \notin \tilde{v}$. Then, for w' fresh in P, P', Q_r, R_r, Q'_r, and R'_r:

$$\begin{aligned}
P = \ & Q \mid R \\
\equiv \ & \nu\tilde{v}\,(x(y).\,Q_1 \mid Q_2) \mid \nu z\,\nu\tilde{u}\,(\overline{x}z.\,R_1 \mid R_2) \\
\equiv \ & \nu z\,\nu\tilde{u}\,\nu\tilde{v}\,(x(y).\,Q_1 \mid Q_2 \mid \overline{x}z.\,R_1 \mid R_2) \\
\equiv_\alpha \ & \nu w'\,\nu\tilde{u}\,\nu\tilde{v}\,(x(y).\,Q_1 \mid Q_2 \mid \overline{x}w'.\,R_1\{w'/z\} \mid R_2\{w'/z\})
\end{aligned}$$

$$\begin{aligned}
P' = \ & \nu w\,(Q' \mid R') \\
\equiv \ & \nu w\,(\nu\tilde{v}\,(Q_1\{w/y\} \mid Q_2) \mid \nu\tilde{u}\,(R_1\{w/z\} \mid R_2\{w/z\})) \\
\equiv \ & \nu w\,\nu\tilde{u}\,\nu\tilde{v}\,(Q_1\{w/y\} \mid Q_2 \mid R_1\{w/z\} \mid R_2\{w/z\}) \\
\equiv_\alpha \ & \nu w'\,\nu\tilde{u}\,\nu\tilde{v}\,(Q_1\{w/y\}\{w'/w\} \mid Q_2\{w'/w\} \mid R_1\{w/z\}\{w'/w\} \mid R_2\{w/z\}\{w'/w\}) \\
\equiv_\alpha \ & \nu w'\,\nu\tilde{u}\,\nu\tilde{v}\,(Q_1\{w'/y\} \mid Q_2 \mid R_1\{w'/z\} \mid R_2\{w'/z\})
\end{aligned}$$

The above argument shows the fundamental role of α-conversion in carrying out the proof, and also serves to clarify the choice of the inductive handle adopted for bound actions. Above, we do not simply refresh z to w but rather resort to a fresh name w' which is used in both the development for P and in that for P'. Notice that refreshing z to w would be the same as simplifying the inductive handle by just taking $z = w$ in the sixth item of the statement. This simplified handle would actually work in the present setting. In fact we could prove that

"If $P \xrightarrow{\overline{x}y} P'$ then $\mathrm{fn}(P) = \mathrm{fn}(P') \cup \{x, y\}$", and this would be the key issue in showing that "If $P \xrightarrow{\overline{x}(w)} P'$ then $P \equiv \nu w \, \nu\tilde{u} \, (\overline{x}w.P_1 \mid P_2)$ and $P' \equiv \nu\tilde{u} \, (P_1 \mid P_2)$ with $x \neq w$ and $x, w \notin \tilde{u}$". If the language were extended with the choice operator, however, the latest properties would not hold any longer (think, e.g., of $P_1 = (\overline{x}y. \, \mathsf{nil} + \overline{x}w. \, \mathsf{nil}) \xrightarrow{\overline{x}y} \mathsf{nil}$ and of $P_2 = \nu y \, P_1 \xrightarrow{\overline{x}(w)} \mathsf{nil}$). So we preferred to state and carry out the most general argument. $\qquad\square$

Lemma 2. *If B is well-formed and $[\![B]\!] \xrightarrow{\alpha} M$ then either $\alpha = \tau$ or $\alpha \in \{x : \Delta(w), \overline{x} : \overline{\Delta}\langle w\rangle\}$ for some x, Δ, and w.*

Proof. The proof is by induction on the inference of $[\![B]\!] \xrightarrow{\alpha} M$. The base of the induction is relative to the case when $[\![B]\!]$ has the shape $L * P$, and the inductive step to the case when $[\![B]\!]$ has the form $M \parallel N$. [Base] The last rule applied in the inference is either (inp) or (out) or (hd) or (uh) or (tau) or (exp), hence the thesis. [Step] By the definition of (com), the definition of (par), and the inductive hypothesis. $\qquad\square$

Lemma 3. *If B is well-formed and $[\![B]\!] \xrightarrow{\alpha} M'$ with $\alpha \neq \tau$ then $[\![B]\!] \equiv_l L * P \parallel M_1$ and $M' \equiv_l L * P' \parallel M_1$ for some L, P, M_1 and P' such that one of the following holds for α, L, P and P':*

1. *$\alpha = x : \Delta(w)$ and $P \equiv \nu\tilde{u} \, (x(w). Q_1 \mid Q_2)$ and $P' \equiv \nu\tilde{u} \, (Q_1\{{}^z\!/\!_w\} \mid Q_2)$ for some $x, \Delta, w, \tilde{u}, Q_1$ and Q_2 such that $(x, \Delta) \in L$ and $x \notin \tilde{u}$ and $w \notin \mathrm{fn}(Q_2) \cup \tilde{u}$;*
2. *$\alpha = \overline{x} : \overline{\Delta}\langle z\rangle$ and $P \equiv \nu\tilde{u} \, (\overline{x}z. Q_1 \mid Q_2)$ and $P' \equiv \nu\tilde{u} \, (Q_1 \mid Q_2)$ for some $x, \Delta, z, \tilde{u}, Q_1$ and Q_2 such that $x, z \notin \tilde{u}$.*

Proof. The proof is by induction on the inference of $[\![B]\!] \xrightarrow{\alpha} M'$. By Lemma 2 and the hypothesis $\alpha \neq \tau$, the action performed by $[\![B]\!]$ can only be a global input or a global output. Then the induction base is limited to the analysis of the rules (inp) and (out), while the rule (par) is relevant for the induction step. In the first two cases mentioned above, the thesis comes by appealing to Lemma 1. $\qquad\square$

Theorem 1. *If B is well-formed and $[\![B]\!] \xrightarrow{\tau} M'$ then $B \twoheadrightarrow B'$ with $[\![B']\!] \equiv_l M'$.*

Proof. By induction on the inference of $[\![B]\!] \xrightarrow{\tau} M'$. The base of the induction is relative to the case when $[\![B]\!]$ has the shape $L * P$, and the inductive step to the case when $[\![B]\!]$ has the form $M \parallel N$.

So, for the base case, we assume that either (hd) or (uh) or (tau) or (exp) is in turn the last rule applied in the inference. In each case, by appealing to Lemma 1, we can deduce both the structure of P in $L * P = [\![B]\!]$ and the structure of the derivative process M'. Then the thesis comes by observing that, respectively, either the axiom (hide) or (unhide) or (intra) or (expose) of the reduction semantics in Table 2 can be applied to get $B \twoheadrightarrow B'$ for B' such that $[\![B']\!] \equiv_l M'$.

The relevant rules to analyze for the inductive step are the rules (par) and (com). In the first case, the thesis comes by the inductive hypothesis and the application of the (redex) reduction rule. When instead the last rule applied in the inference is (com), we appeal to Lemma 3 and the thesis comes by applying to B the (inter) reduction rule. $\qquad\square$

As expected, relating reductions to labelled transitions is harder than proving Theorem 1. Lemma 4 is a useful intermediate result to achieve the desired goal.

Lemma 4. *If B is well-formed and $B \twoheadrightarrow B'$ then there exist $\mathbb{B}_1, \mathbb{B}_2, \tilde{u}, \tilde{v}, x, y, z, w, \Delta, \Gamma, P_1, P_2, P_3, Q_1, Q_2,$ and B_1 such that one of the following holds:*

1. $B \equiv_b \mathbb{B}_1[\,\nu\tilde{u}\,(x(w).\,P_1 \mid \overline{x}z.\,P_2 \mid P_3)\,] \parallel B_1$ *and*
   $B' \equiv_b \mathbb{B}_1[\,\nu\tilde{u}\,(P_1\{z\!/\!w\} \mid P_2 \mid P_3)\,] \parallel B_1$;
2. $B \equiv_b \beta(x,\,\Gamma)\,\mathbb{B}_1[\,\nu\tilde{u}\,(x(w).\,P_1 \mid P_2)\,] \parallel \beta(y,\,\Delta)\,\mathbb{B}_2[\,\nu\tilde{v}\,(\overline{y}z.\,Q_1 \mid Q_2)\,] \parallel B_1$
   and $B' \equiv_b \beta(x,\,\Gamma)\,\mathbb{B}_1[\,\nu\tilde{u}\,(P_1\{z\!/\!w\} \mid P_2)\,] \parallel \beta(y,\,\Delta)\,\mathbb{B}_2[\,\nu\tilde{v}\,(Q_1 \mid Q_2)\,] \parallel B_1$
   with comp(Γ, Δ) *and* $x, z \notin \tilde{u}$ *and* $y, z \notin \tilde{v}$;
3. $B \equiv_b \mathbb{B}_1[\,\nu\tilde{u}\,(\mathsf{expose}(x,\,\Gamma).\,P_1 \mid P_2)\,] \parallel B_1$ *and*
   $B' \equiv_b \mathbb{B}_1\,\beta(y,\,\Gamma)[\,\nu\tilde{u}\,(P_1\{y\!/\!x\} \mid P_2)\,] \parallel B_1$ *with* $y \notin \tilde{u} \cup \mathsf{sub}(\mathbb{B}_1) \cup \mathrm{fn}(P_2)$;
4. $B \equiv_b \mathbb{B}_1\,\beta(x,\,\Gamma)[\,\nu\tilde{u}\,(\mathsf{hide}(x).\,P_1 \mid P_2)\,] \parallel B_1$ *and*
   $B' \equiv_b \mathbb{B}_1\,\beta^h(x,\,\Gamma)[\,\nu\tilde{u}\,(P_1 \mid P_2)\,] \parallel B_1$ *with* $x \notin \tilde{u}$;
5. $B \equiv_b \mathbb{B}_1\,\beta^h(x,\,\Gamma)[\,\nu\tilde{u}\,(\mathsf{unhide}(x).\,P_1 \mid P_2)\,] \parallel B_1$ *and*
   $B' \equiv_b \mathbb{B}_1\,\beta(x,\,\Gamma)[\,\nu\tilde{u}\,(P_1 \mid P_2)\,] \parallel B_1$ *with* $x \notin \tilde{u}$.

Proof. First we show that if $B \twoheadrightarrow B'$ then there exists a derivation of $B \twoheadrightarrow B'$ in *normal form*, namely a derivation where the (struct) rule is used exactly once and exactly as the very last rule of the derivation. This statement can be proven by showing the following.

- An instance of the (struct) rule can always be added at the end of a derivation. This is simply achieved by using the reflexivity of $\equiv_b$.
- An occurrence of the (struct) rule followed by an occurrence of the (redex) rule can be converted into an instance of (redex) followed by an instance of (struct). This is obtained by re-arranging derivations as shown below.

$$\frac{\dfrac{\vdots}{\dfrac{B \equiv_b B_1 \quad B_1 \twoheadrightarrow B_2 \quad B_2 \equiv_b B'}{B \twoheadrightarrow B'}\text{(struct)}}}{B \parallel B'' \twoheadrightarrow B' \parallel B''}\text{(redex)}$$

$$\frac{B \parallel B'' \equiv_b B_1 \parallel B'' \quad \dfrac{\dfrac{\vdots}{B_1 \twoheadrightarrow B_2}}{B_1 \parallel B'' \twoheadrightarrow B_2 \parallel B''}\text{(redex)} \quad B_2 \parallel B'' \equiv_b B' \parallel B''}{B \parallel B'' \twoheadrightarrow B' \parallel B''}\text{(struct)}$$

- Two consecutive occurrences of the (struct) rule can be condensed into one. This is achieved by exploiting the transitivity of $\equiv_b$ and re-arranging derivations as it is done below.

$$\frac{B \equiv_b B_1 \quad \dfrac{B_1 \equiv_b B_3 \quad B_3 \twoheadrightarrow B_4 \quad B_4 \equiv_b B_2}{B_1 \twoheadrightarrow B_2}\text{(struct)} \quad B_2 \equiv_b B'}{B \twoheadrightarrow B'}\text{(struct)}$$

$$\frac{B \equiv_b B_1 \equiv_b B_3 \quad B_3 \twoheadrightarrow B_4 \quad B_4 \equiv_b B_2 \equiv_b B'}{B \twoheadrightarrow B'}\text{(struct)}$$

From the above, a derivation of $B \twoheadrightarrow B'$ in normal form is such that an axiom is first applied, then zero or more (redex) rules come, and finally (struct) is applied. Then the desired thesis, where each of the listed cases corresponds to one of the possible distinct axioms driving the derivation of $B \twoheadrightarrow B'$ in normal form. $\qquad\square$

Theorem 2. *If B is well-formed and $B \twoheadrightarrow B'$ then $[\![B]\!] \xrightarrow{\tau}\equiv_l [\![B']\!]$.*

Proof. By $B \twoheadrightarrow B'$ and Lemma 4 we know which is the possible structure of both B and B'. The proof is by case analysis of the five distinct possible shapes of B. Here we only report the most interesting case, corresponding to the second item in Lemma 4. This is the case when, for some x, Γ, $\mathbb{B}_1$, $\tilde{u}$, w, P_1, P_2, y, Δ, $\mathbb{B}_2$, $\tilde{v}$, z, Q_1, Q_2, and B_3, the two processes B and B' are such that:

$$B \equiv_b \beta(x, \Gamma)\,\mathbb{B}_1[\,\nu\tilde{u}\,(x(w).\,P_1 \mid P_2)\,] \parallel \beta(y, \Delta)\,\mathbb{B}_2[\,\nu\tilde{v}\,(\overline{y}z.\,Q_1 \mid Q_2)\,] \parallel B_3$$

$$B' \equiv_b \beta(x, \Gamma)\,\mathbb{B}_1[\,\nu\tilde{u}\,(P_1\{z/w\} \mid P_2)\,] \parallel \beta(y, \Delta)\,\mathbb{B}_2[\,\nu\tilde{v}\,(Q_1 \mid Q_2)\,] \parallel B_3$$

with $\mathrm{comp}(\Gamma, \Delta)$ and $x, z \notin \tilde{u}$ and $y, z \notin \tilde{v}$.

Then $[\![B]\!] \equiv_l M_1 \parallel M_2 \parallel M_3$ where M_1 and M_2 are such that:

$$M_1 = L_1 * \nu\tilde{u}_3\,(\nu\tilde{u}_1\,x(w).\,P_1 \mid \nu\tilde{u}_2\,P_2)$$

$$M_2 = L_2 * \nu\tilde{v}_3\,(\nu\tilde{v}_1\,\overline{y}z.\,Q_1 \mid \nu\tilde{v}_2\,Q_2)$$

with $L_1 = [\![\beta(x, \Gamma)\,\mathbb{B}_1]\!]$ and $L_2 = [\![\beta(y, \Delta)\,\mathbb{B}_2]\!]$, and where $\tilde{u} = \tilde{u}_1\tilde{u}_2\tilde{u}_3$ and $\tilde{v} = \tilde{v}_1\tilde{v}_2\tilde{v}_3$ with $\tilde{u}_i$ such that $\mathrm{fn}(P_1) \cap \mathrm{fn}(P_2) \cap \tilde{u} \subseteq \tilde{u}_3$, $\mathrm{fn}(P_2) \cap \tilde{u}_1 = \emptyset$, and $\mathrm{fn}(P_1) \cap \tilde{u}_2 = \emptyset$.

Then, by $x \notin \tilde{u}$,

$$M_1 \xrightarrow{x:\Gamma(w')} M_1' = L_1 * \nu\tilde{u}_3\,(\nu\tilde{u}_1\,P_1\{w'/w\} \mid \nu\tilde{u}_2\,P_2)$$

for some w' such that $w' \notin \mathrm{fn}(\nu\tilde{u}_2\,P_2)$ by (i-par) and $w' \notin \tilde{u}_1\tilde{u}_3$ by (i-res). Here notice that if M_1 is composed in parallel with some N different from M_2 (which happens, e.g., for $M_1 \parallel M_3 \parallel M_2 \equiv_l M_1 \parallel M_2 \parallel M_3$), then

$$M_1 \parallel N \xrightarrow{x:\Gamma(w')} M_1' \parallel N$$

and w' is also such that $w' \notin \mathrm{fn}(N)$ by (par). So $N\{z/w'\} = N$ for any z. Also, by $y, z \notin \tilde{v}$,

$$M_2 \xrightarrow{\overline{y:\Delta\langle z\rangle}} M_2' = L_2 * \nu\tilde{v}_3 \left(\nu\tilde{v}_1\, Q_1 \mid \nu\tilde{v}_2\, Q_2\right) \equiv_l L_2 * \nu\tilde{v}\left(Q_1 \mid Q_2\right)$$

using (i-out), (i-res), (i-par), and (out). Then, by (par) and (com),

$$M_1 \parallel M_2 \parallel M_3 \xrightarrow{\tau} M_1'\{z/w'\} \parallel M_2' \parallel M_3$$

where:

$$
\begin{aligned}
& M_1'\{z/w'\} \\
=\ & L_1 * (\nu\tilde{u}_3\,(\nu\tilde{u}_1\, P_1\{w'/w\} \mid \nu\tilde{u}_2\, P_2))\{z/w'\} && \text{by } z \notin \tilde{u}_3 \\
=\ & L_1 * \nu\tilde{u}_3\,((\nu\tilde{u}_1\, P_1\{w'/w\} \mid \nu\tilde{u}_2\, P_2)\{z/w'\}) && \text{by } w' \notin \mathrm{fn}(\nu\tilde{u}_2\, P_2) \\
=\ & L_1 * \nu\tilde{u}_3\,((\nu\tilde{u}_1\, P_1\{w'/w\})\{z/w'\} \mid \nu\tilde{u}_2\, P_2) && \text{by } z \notin \tilde{u}_1 \\
=\ & L_1 * \nu\tilde{u}_3\,(\nu\tilde{u}_1\,(P_1\{w'/w\}\{z/w'\}) \mid \nu\tilde{u}_2\, P_2) && \\
\equiv_\alpha\ & L_1 * \nu\tilde{u}_3\,(\nu\tilde{u}_1\,(P_1\{z/w\}) \mid \nu\tilde{u}_2\, P_2) && \text{by } \mathrm{fn}(P_1) \cap \tilde{u}_2 = \emptyset,\ z \notin \tilde{u}_2 \\
\equiv_l\ & L_1 * \nu\tilde{u}_3\tilde{u}_2\,(\nu\tilde{u}_1\,(P_1\{z/w\}) \mid P_2) && \text{by } \mathrm{fn}(P_2) \cap \tilde{u}_1 = \emptyset \\
\equiv_l\ & L_1 * \nu\tilde{u}\,(P_1\{z/w\} \mid P_2) &&
\end{aligned}
$$

Hence the thesis, by $\llbracket M_1'\{z/w'\} \parallel M_2' \parallel M_3 \rrbracket \equiv_l \llbracket B' \rrbracket$. $\qquad\square$

5 Concluding Remarks

We reviewed Beta-binders, a bio-inspired formalism which interprets processes as boxes with typed interaction sites. The language shows a particular communication paradigm which was originally defined to try to deal with the modelling of uncertain behaviours. The paradigm is based on both the notion of complementarity of actions, and the notion of compatibility of the types associated with box interaction sites. Boxes are essentially closed worlds. Provided (and modulo) the compatibility of sites, they can however communicate the one with the other much in the π-calculus style.

The original Beta-binders semantics is given in reduction style, which, by its own nature, limits the applicability of syntax-driven techniques and hence typically makes it difficult to mathematically deal with behavioural properties. In the perspective of developing a formal theory for the language, here we defined a labelled semantics for Beta-binders and showed its correspondence with the original semantics. The labelled semantics could undergo a stochastic extension along the lines of the extension presented in [3]. From the stochastic point of view, the crucial role played by the (redex) rule in [3] would be delegated to (par) and (com), so to get an operational correspondence between the two extensions.

The alternative semantics defined in this paper is still infinitely branching. In its case, though, the impact factors are those common to labelled semantics for calculi with naming: the unguarded behaviour of the operator for recursion, and α-equivalence. Once unguarded recursion is banned, α-equivalence can be tackled in very many symbolic ways (see, e.g., [5,4,17]).

References

1. Cardelli, L.: Brane Calculi. In: Danos, V., Schachter, V. (eds.) CMSB 2004. LNCS (LNBI), vol. 3082. Springer, Heidelberg (2005)
2. Cardelli, L., Gordon, A.D.: Mobile Ambients. Theoretical Computer Science 240(1), 177–213 (2000)
3. Degano, P., Prandi, D., Priami, C., Quaglia, P.: Beta-binders for Biological Quantitative Experiments. In: Proc. 4th Workshop on Quantitative Aspects of Programming Languages, QAPL 2006. ENTCS, vol. 164, pp. 101–117 (2006)
4. Ferrari, G.-L., Montanari, U., Quaglia, P.: A π-calculus with Explicit Substitutions. Theoretical Computer Science 168(1), 53–103 (1996)
5. Hennessy, M., Lin, H.: Symbolic Bisimulations. Theoretical Computer Science 138, 353–389 (1995)
6. Larcher, R., Ihekwaba, A., Priami, C.: A BetaWB model for the NFkB pathway. Technical Report TR-25-2007, The Microsoft Research - University of Trento Centre for Computational and Systems Biology (2007)
7. Milner, R.: Communication and Concurrency. International Series in Computer Science. Prentice-Hall, Englewood Cliffs (1989)
8. Milner, R.: Communicating and mobile systems: the π-calculus. Cambridge University Press, Cambridge (1999)
9. Milner, R., Parrow, J., Walker, D.: A Calculus of Mobile Processes, Part I and II. Information and Computation 100(1), 1–77 (1992)
10. Perelson, A.S., Oster, G.F.: Theoretical studies of clonal selection: minimal antibody repertoire size and reliability of self-non-self discrimination. J. Theoretical Biology 81(4) (1979)
11. Prandi, D., Priami, C., Quaglia, P.: Shape spaces in formal interactions. ComPlexUs 2(3-4), 128–139 (2006)
12. Prandi, D., Priami, C., Quaglia, P.: Communicating by Compatibility. Journal of Logic and Algebraic Programming (to appear, 2007)
13. Priami, C., Quaglia, P.: Beta Binders for Biological Interactions. In: Danos, V., Schachter, V. (eds.) CMSB 2004. LNCS (LNBI), vol. 3082, pp. 21–34. Springer, Heidelberg (2005)
14. Regev, A., Panina, E.M., Silverman, W., Cardelli, L., Shapiro, E.Y.: Bioambients: an abstraction for biological compartments. Theoretical Computer Science 325(1), 141–167 (2004)
15. Romanel, A., Dematté, L., Priami, C.: The Beta Workbench. Technical Report TR-03-2007, The Microsoft Research - University of Trento Centre for Computational and Systems Biology (2007)
16. Sangiorgi, D., Walker, D.: The π-calculus: a Theory of Mobile Processes. Cambridge University Press, Cambridge (2001)
17. Shinwell, M.R.: The Fresh Approach: functional programming with names and binders. PhD thesis, University of Cambridge Computer Laboratory (2005)

On the Asynchronous Nature of the Asynchronous π-Calculus[*]

Romain Beauxis, Catuscia Palamidessi, and Frank D. Valencia

INRIA Saclay and LIX, École Polytechnique

Abstract. We address the question of what kind of asynchronous communication is exactly modeled by the asynchronous π-calculus (π_a). To this purpose we define a calculus $\pi_\mathfrak{B}$ where channels are represented explicitly as special buffer processes. The base language for $\pi_\mathfrak{B}$ is the (synchronous) π-calculus, except that ordinary processes communicate only via buffers. Then we compare this calculus with π_a. It turns out that there is a strong correspondence between π_a and $\pi_\mathfrak{B}$ in the case that buffers are bags: we can indeed encode each π_a process into a strongly asynchronous bisimilar $\pi_\mathfrak{B}$ process, and each $\pi_\mathfrak{B}$ process into a weakly asynchronous bisimilar π_a process. In case the buffers are queues or stacks, on the contrary, the correspondence does not hold. We show indeed that it is not possible to translate a stack or a queue into a weakly asynchronous bisimilar π_a process. Actually, for stacks we show an even stronger result, namely that they cannot be encoded into weakly (asynchronous) bisimilar processes in a π-calculus without mixed choice.

1 Introduction

In the community of Concurrency Theory the asynchronous π-calculus (π_a) [14,5] is considered, as its name suggests, a formalism for *asynchronous communication*. The reason is that this calculus satisfies some basic properties which are associated to the abstract concept of asynchrony, like, for example, the fact that a send action is non-blocking and that two send actions on different channels can always be swapped (see, for instance, [24,22]).

In other communities, like Distributed Computing, however, the concept of asynchronous communication is much more concrete, and it is based on the assumption that the messages to be exchanged are placed in some communication means while they travel from the sender to the receiver. We will call such communication devices *buffers*. In general, it is also assumed that the action of placing a message in the buffer and the action of receiving the message from the buffer do not take place at the same time, i.e. the exchange is not instantaneous.

A frequent question that people then ask about the asynchronous π-calculus, is "In what sense is this a model of asynchronous communication"? Often they

---

[*] This work has been partially supported by the INRIA ARC project ProNoBiS and by the INRIA DREI Équipe Associée PRINTEMPS.

P. Degano et al. (Eds.): Montanari Festschrift, LNCS 5065, pp. 473–492, 2008.

are puzzled by the communication rule of π_a, which is literally the same as the one of the (synchronous) π-calculus [18]:

$$\frac{P \xrightarrow{\bar{x}y} P' \quad Q \xrightarrow{xy} Q'}{P \mid Q \xrightarrow{\tau} P' \mid Q'} \tag{1}$$

where $\bar{x}y$ represents the action of sending a message y on channel x, and xy represents the action of receiving a message y from channel x. The rule suggests that these two actions take place simultaneously, in a handshaking fashion.

To our experience the most convincing explanation of the asynchrony of π_a is that, because of the lack of output prefix in π_a, in rule (1) P must be of the form $\bar{x}y \mid P'$, and the transition $P \xrightarrow{\bar{x}y} P'$ does not really represent the event of sending. Originally, P must come from a process of the form $C[\bar{x}y \mid R]$ for some prefix context $C[\ ]$ and some process R, and the "real" event of sending takes place at some arbitrary time between the moment $\bar{x}y$ gets at the top-level (i.e. when it is no more preceded by a prefix) and the event of receiving y, the latter being what the rule (1) really represents. In the interval between the two events, R evolves asynchronously into P'. Of course, at this point another question arises: "What happens to the message y in the meanwhile?" The best way to see it is that y is placed in some buffer labeled with x. But then the legitimate question follows on what kind of buffer this is, since it is well known that a distributed system can behave very differently depending on whether the channels are bags, queues, or stacks, for instance. In this paper we address precisely this latter question.

Our approach is to define a calculus π_T where buffers are represented explicitly, like it was done, for instance, in [3,9]. The symbol T stands for $\mathfrak{B}$, $\mathfrak{Q}$, and $\mathfrak{S}$, in the case of bags, queues, and stacks respectively. The actions of receiving from and sending to a given buffer are represented by input and output transitions, respectively. The object of the action is the name being transmitted and subject is the name (or type) of the source buffer if the action is an output, or the destination buffer if the action is an input. The base language for π_T is the (synchronous) π-calculus with guarded choice, except that processes communicate only via buffers. Then we compare this calculus with π_a. It turns out that there is a strong correspondence between π_a and $\pi_{\mathfrak{B}}$ (the case of bags). More precisely, if we interpret the π_a send process $\bar{x}y$ as a bag of type x which contains the single message y, then there is a strong asynchronous bisimulation between a π_a process P and its translation $[\![P]\!]$ in $\pi_{\mathfrak{B}}$ (notation $P \sim [\![P]\!]$). This result reflects the intuitive explanation of the asynchronous nature of π_a given above. On the other hand, we can encode (although in a more involved way) each $\pi_{\mathfrak{B}}$ process P into a π_a process $[\![P]\!]^1$, equivalent to P modulo weak asynchronous bisimilarity (notation $P \approx [\![P]\!]$).

We would like to point out that the the case of bags represents a feature of communication in distributed systems, namely the fact that the order in which

---

[1] We use the same notation $[\![\cdot]\!]$ to indicate both the encoding from π_a into $\pi_{\mathfrak{B}}$ and the one from $\pi_{\mathfrak{B}}$ into π_a. It will be clear from the context which is the intended one.

messages are sent is not guaranteed to be preserved in the order of arrival. In the case of the π-calculus, it is sufficient to allow the messages to be made available in any order to the receiver (which is exactly the property which characterize the bags as a data structure), because there are no primitives that are able to make a "snapshot" of the system, and in particular to detect the absence of a message. For languages which contain such a kind of primitive however (for instance Linda, with the `imp` construct) the abstraction from the order is not sufficient, and the faithful representation of distributed communication would require a more sophisticated model. A proposal for such model is the *unordered semantics* of [7]. In that paper the authors argue, convincingly, that the unordered semantics is the most "asynchronous" semantics and the "right" one for distributed systems.

In case the buffers are queues or stacks, on the contrary, the correspondence between π_T and π_a does not hold. We show indeed that there is no encoding of stacks or queues, represented as described above, into π_a modulo weak asynchronous bisimulation. By "encoding modulo $\mathcal{R}$" we mean an encoding that translates P into a process that is in relation $\mathcal{R}$ with P. Actually for stacks we prove a stronger result: they cannot be translated, modulo weak bisimilarity, even into π_{sc}, the fragment of the (synchronous) π-calculus where the mixed guarded choice operator is replaced by separate-choice, i.e. a choice construct that can contain either input guards, or output guards, but not both. In other words, the least we need to encode stacks is a mixed-choice construct where both input and output guards (aka prefixes) are present.

The above result does not mean, obviously, that queues and stacks cannot be simulated in π_a: we will indeed discuss a possible way to simulate them by encoding the send and receive actions on buffers into more complicated protocols. The meaning of our negative result is only that a queue (respectively a stack) and any translation of it in π_a (respectively in π_{sc}) cannot be related by a relation like weak (asynchronous) bisimilarity, which requires a strict correspondence between transitions.

1.1   Justifying the Choice of the Languages

The results presented in this paper would hold also if we had considered the asynchronous version of CCS [4] instead than the asynchronous π-calculus. The reasons why we have chosen the latter are the following. The asynchronous π-calculus was the first process calculus to represent asynchronous communication by using a send primitive with no continuation (*asynchronous send*), and in the concurrency community it has become paradigmatic of this particular approach to asynchrony. Moreover, the expressive power of the asynchronous π-calculus has been widely investigated, especially in relation to other asynchronous calculi, and in comparison with synchronous communication.

We have chosen the π-calculus with (mixed) guarded choice as a base language for π_T because in the π-calculus the main expressive gap between synchronous and asynchronous communication lies exactly in between mixed choice and separate choice [20,19]. In other words, the π-calculus with mixed choice cannot be encoded in the asynchronous π-calculus in any "reasonable" way, while the π-calculus with separate choice can. The choice of a synchronous language as the basis for π_T is motivated by the fact that it allows a precise control of the communication mechanism: The processes communicate with each other via the buffers, but the interaction between a process and a buffer is synchronous. So the buffers are the only source of asynchrony in π_T, which makes the encoding from the asynchronous π-calculus into π_T more interesting. Furthermore this model of asynchronous communication is very close to the concrete implementation of distributed systems [21].

1.2   Justifying the Criteria for the Encodings

As we stated above, our main positive result is the correspondence between π_a and $\pi_\mathfrak{B}$, expressed in one direction by one encoding

$$[\![\cdot]\!] : \pi_a \to \pi_\mathfrak{B} \quad \text{with } P \sim [\![P]\!] \text{ for all } P \text{ in } \pi_a$$

and, in the other direction, by another encoding

$$[\![\cdot]\!] : \pi_\mathfrak{B} \to \pi_a \quad \text{with } P \approx [\![P]\!] \text{ for all } P \text{ in } \pi_\mathfrak{B}$$

We consider the above properties of the encodings as quite strong, and therefore supporting the claim of a strict correspondence between π_a and $\pi_\mathfrak{B}$. They imply for instance the condition of *operational correspondence*, which is one of the properties of a "good" encoding according to Gorla ([10,11]).

One may question why we did not rather prove the existence of a *fully abstract* encoding between π_a and $\pi_\mathfrak{B}$. We recall that, given a language $\mathcal{L}_1$ equipped with an equivalence relation $\sim_1$, and a language $\mathcal{L}_2$ equipped with an equivalence relation $\sim_2$, an encoding $[\![\cdot]\!] : \mathcal{L}_1 \to \mathcal{L}_2$ is called fully abstract if and only if

$$\text{for every } P, Q \text{ in } \mathcal{L}_1, \quad P \sim_1 Q \;\Leftrightarrow\; [\![P]\!] \sim_2 [\![Q]\!] \text{ holds}$$

Full abstraction has been adopted sometimes in literature as a criterion for expressiveness. We do not endorse this approach: In our opinion, full abstraction can be useful to transfer the theory of a language to another language, but it is not a good criterion for expressiveness. The reason is that it can be, at the same time, both too strong and too weak a requirement. Let us explain why.

Too strong: Consider the asynchronous π-calculus π_a, equipped with its "natural" notion of equivalence, the weak asynchronous bisimilarity, and the π-calculus π, equipped with weak bisimilarity. Let $[\![\cdot]\!]$ be the standard encoding from π_a to π defined as

$$[\![\bar{x}y]\!] = \bar{x}y.0$$

and homomorphic on the other operators. Most people would agree that this is a pretty straightforward, natural encoding, showing that π is at least as powerful as π_a. Still, it does not satisfy the full abstraction criterion. In fact, there are weakly asynchronous bisimilar processes P, Q such that $[\![P]\!]$ and $[\![Q]\!]$ are not weakly bisimilar. Take, for example, $P = 0$ and $Q = x(y).\bar{x}y$.

Too weak: Consider an enumeration of Turing machines, $\{T_n\}_n$, and an enumeration of minimal finite automata $\{A_n\}_n$, with their standard language-equivalence $\equiv$. Consider the following encoding of Turing machines into (minimal) finite automata:

$$\llbracket T_m \rrbracket = \llbracket T_k \rrbracket \text{ if } k < m \text{ and } T_k \equiv T_m$$
$$\llbracket T_m \rrbracket = A_n \quad \text{otherwise}$$

where n is the minimum number such that A_n has not been used to encode any T_k with $k < m$. By definition, we have that $\forall m, n\ T_m \equiv T_n \ \Leftrightarrow\ \llbracket T_m \rrbracket \equiv \llbracket T_n \rrbracket$, but this certainly does not prove that finite automata are as powerful as Turing machines!

Note that the second encoding, from Turing machines to finite automata, is non-effective. This is fine for our purpose, which is simply to show that full abstraction alone, i.e. without extra conditions on the encoding, is not a very meaningful notion. Of course, it would be even more interesting to exhibit an *effective* and fully abstract encoding between some L_1 and L_2, while most people would agree that L_2 is strictly less powerful than L_1. But this is out of the scope of this paper, and we leave it as an open problem for the interested reader.

1.3   Plan of the Paper

In the next section we recall some standard definitions. In Section 3, we introduce the notion of buffer and the different types of buffer we will consider. Then in Section 4, we define a π-Calculus communicating through bags. In Section 5, we study the correspondence between the π-Calculus with bags and the π_a-calculus. The main bisimilarity results are established there. In Section 6, we use the properties from Section 3 to prove the impossibility results for stacks and queues. Section 7 discusses related work. Finally, Section 8 concludes and outlines some directions of future research.

2   Preliminaries

2.1   The Asynchronous π-Calculus: π_a

We assume a countable set of *names*, ranged over by $x, y, \ldots$, and for each name x, a *co-name* $\bar{x}$. In the asynchronous π-Calculus, henceforth denoted as π_a, *processes* are given by the following syntax:

$$P, Q, \ldots \ := \ 0 \ \Big|\ \bar{x}z \ \Big|\ x(y).P \ \Big|\ \nu x P \ \Big|\ P \mid Q \ \Big|\ !P$$

The 0 represents an empty (or terminated) process. Intuitively, an *output* $\bar{x}z$ represents a particle in an implicit medium tagged with a name x indicating that it can be received by an *input process* $x(y).P$ which behaves, upon receiving z, as $P[z/y]$, namely, the process where every free occurence of y is replaced by z.

Table 1. Transition rules for the π_a-calculus

$$(in) \ \frac{}{x(y).P \xrightarrow{xz} P[z/y]} \qquad\qquad (out) \ \frac{}{\overline{x}z \xrightarrow{\overline{x}z} 0}$$

$$(sync) \ \frac{P \xrightarrow{\overline{x}y} P', \ Q \xrightarrow{xy} Q'}{P\,|\,Q \xrightarrow{\tau} P'\,|\,Q'} \qquad (\nu) \ \frac{P \xrightarrow{\alpha} P', \ a \notin fn(\alpha)}{\nu a P \xrightarrow{\alpha} \nu a P'}$$

$$(open) \ \frac{P \xrightarrow{\overline{x}y} P', \ x \neq y}{\nu y P \xrightarrow{\overline{x}(y)} P'} \qquad (close) \ \frac{P \xrightarrow{\overline{x}(y)} P', \ Q \xrightarrow{xy} Q'}{P\,|\,Q \xrightarrow{\tau} \nu y(P'\,|\,Q')}$$

$$(comp) \ \frac{P \xrightarrow{\alpha} P', \ bn(\alpha) \cap fn(Q) = \emptyset}{P\,|\,Q \xrightarrow{\alpha} P'\,|\,Q} \qquad (bang) \ \frac{P\,|\,!P \xrightarrow{\alpha} R}{!P \xrightarrow{\alpha} R}$$

$$(cong) \ \frac{P \equiv P' \ P' \xrightarrow{\alpha} Q' \ Q' \equiv Q}{P \xrightarrow{\alpha} Q}$$

We assume that P is α-renamed before applying $[z/y]$ so to avoid name-capture. A substitution σ causes name-capture in P if it replaces a name y by a name z for which one or more free occurences of y in P are in the scope of a binder for z. Furthermore, $x(y).P$ binds y in P. The other binder is the *restriction* $\nu x P$ which declares a name x private to P. The *parallel composition* $P\,|\,Q$ means P and Q running in parallel. The *replication* $!P$ means $P\,|\,P\,|\,\ldots$, i.e. an unbounded number of copies of P.

We use the standard notations $bn(Q)$ for the *bound names* in Q, and $fn(Q)$ for the *free names* in Q, and write $\nu x_1 \ldots x_n P$ to denote $\nu x_1 \ldots \nu x_n P$.

The (early) transition semantics of π_a is given in terms of the relation $\xrightarrow{\alpha}$ in Table 1. The label α represents an action which can be of the form τ (silent), $\overline{x}z$ (free output), $\overline{x}(y)$ (bound output) or xz (free input). Transitions are quotiented by the *structural congruence* relation $\equiv$ below.

Definition 1. *The relation $\equiv$ is the smallest congruence over processes satisfying α-conversion and the commutative monoid laws for parallel composition with 0 as identity.*

3 Buffers

A buffer in this paper is basically a data structure that accepts messages and resends them later. We consider different types of buffers, depending on the policy used for outputting a previously received message. We focus on the following policies, that can be considered the most common:

- Bag, or unordered policy: any message previously received (and not yet sent) can be sent next.
- Queue, or FIFO policy: only the oldest message received (and not yet sent) can be sent next.

- Stack, or LIFO policy: only the last message received (and not yet sent) can be sent next.

Let us now formally define these three types of buffer. We need to keep the information about the order of reception to decide which message can be sent next. This will be achieved using a common core definition for all kinds of buffers. We will use $M \in \mathcal{M}$ to denote a message that the buffers can accept.

Definition 2 (Buffer). *A buffer is a finite sequence of messages:*

$$B = M_1 * ... * M_k, \ k \geq 0, \ M_i \in \mathcal{M} \ \text{(B is the empty sequence if } k = 0\text{)}.$$

$*$ is a wild card symbol for the three types of buffers. Then, we will use the notation $M_1 \diamond ... \diamond M_k$ for a bag, $M_1 \triangleleft ... \triangleleft M_k$ for a queue, $M_1 \triangleright ... \triangleright M_k$ for a stack.

A reception on a buffer is the same for all kinds of policies:

Definition 3 (Reception on a buffer). *Let $B = M_1 * ... * M_k$. We write $B \xrightarrow{M} B'$ to represent the fact that B receives the message M, becoming $B' = M * B = M * M_1 * ... * M_k$.*

The emission of a message is different for the three types of buffers:

Definition 4 (Sending from a buffer). *Let $B = M_1 * ... * M_k$. We write $B \xrightarrow{\overline{M}} B'$ to represent the fact that B sends the message M, becoming B', where:*

- *If $* = \diamond$ (bag case) then $M = M_i$ for some $i \in \{1, ..., k\}$ and $B' = M_1 \diamond ... \diamond M_{i-1} \diamond M_{i+1} \diamond ... \diamond M_k$.*
- *If $* = \triangleleft$ (queue case) then $M = M_k$ and $B' = M_1 \triangleleft ... \triangleleft M_{k-1}$.*
- *If $* = \triangleright$ (stack case) then $M = M_1$ and $B' = M_2 \triangleright ... \triangleright M_k$.*

Finally, we introduce here the notion of *buffer's content* and *sendable items*.

Definition 5 (Buffer's content). *A buffer's content is the multiset of messages that the buffer has received and has not yet sent:*

$$C(M_1 * ... * M_k) = \{M_1, ..., M_k\}$$

Definition 6 (Buffer's sendable items). *A buffer's sendable items is the multiset of messages that can be sent immediately:*

$$S(M_1 \diamond M_2 \diamond \cdots \diamond M_k) = \{M_1, M_2, \ldots, M_k\}$$
$$S(M_1 \triangleleft M_2 \triangleleft \cdots \triangleleft M_k) = \{M_k\}$$
$$S(M_1 \triangleright M_2 \triangleright \cdots \triangleright M_k) = \{M_1\}$$

Note that $S(B)$ is empty iff $C(B)$ is empty. Furthermore, if B is a bag, then $C(B) = S(B)$.

Remark 1. If B is a buffer such that $B \xrightarrow{\overline{M_1}}$ and $B \xrightarrow{\overline{M_2}}$ with $M_1 \neq M_2$ then B must be a bag, i.e. B cannot be a stack or a queue.

4 A π-Calculus with Bags

In this section, we define a calculus for asynchronous communications obtained by enriching the synchronous π-Calculus with bags, and forcing the communications to take place only between (standard) processes and bags.

We decree that the bag's messages are names. Each bag is able to send and receive on a single channel only, and we write B_x for a bag on the channel x. We use $\emptyset_x$ to denote an empty bag on channel x, and $\{y\}_x$ for the bag on channel x, containing a single message, y.

Definition 7. *The $\pi_{\mathfrak{B}}$-calculus is the set of processes defined by the grammar:*

$$P, Q ::= \sum_{i \in I} \alpha_i.P_i \,\Big|\, P \mid Q \,\Big|\, \nu x\, P \,\Big|\, !P \,\Big|\, B_x$$

where B_x is a bag, I is a finite indexing set and each α_i can be of the form $x(y)$ or $\overline{x}z$. If $|I| = 0$ the sum can be written as 0 and if $|I| = 1$ the symbol "$\sum_{i \in I}$" can be omitted.

The early transition semantics is obtained by redefining the rules *in* and *out* in Table 1 and by adding the rules in_{bag} and out_{bag} for bag communication as defined in Table 3. Note that they are basically the rules for the (synchronous) π-Calculus except that communication can take place only between (standard) processes and bags. In fact, the rule *out* guarantees that a process can only output to a bag. Furthermore the only rule that generates an output transition is out_{bag}, hence a process can only input, via *sync* and *close*, from a bag.

The structural equivalence $\equiv$ consists of the standard rules of Definition 1, plus scope extrusion, plus $P \equiv P|\emptyset_x$. This last rule allows any process to have access to a buffer even if the process itself is blocked by a binder. A typical example would be $P = \nu x(\overline{x}y.x(z).Q)$, which could not execute any action without this rule. Thanks to the rule, we have:

$$P \equiv \nu x(\overline{x}y.x(z).Q \mid \emptyset_x) \rightarrow \nu x(x(z).Q \mid \{y\}_x) \rightarrow \nu x(Q[y/z] \mid \emptyset_x)$$

Note that we could restrict the application of $P \equiv P|\emptyset_x$ to the case in which P is a pure process (not containing a bag already), i.e. a term of the asynchornous π-calculus). We do not impose this constraint here because it is not necessary, and also because we believe that allowing multiple bags on the same channel name and for the same process is a more natural representation of the concept of channel in distributed systems. Later in this paper, when dealing with stacks and queues, we will have to adopt this restriction in order to be consistent with the nature of stacks and queues.

A consequence of the rule $P \equiv P|\emptyset_x$ is that every process P is always *input-enabled*. This property is in line with other standard models of asynchronous communication, for example the Input/output automata (see, for instance, [15]), the input-buffered agents of Selinger [24] and the Honda-Tokoro original version of the asynchronous π-calculus [14].

The scope extrusion equivalence – $(\nu z\, P) \mid Q \equiv \nu z\, (P \mid Q)$ if $z \notin fn(Q)$ – has been added even though the *open* and *close* rules are present. This is to allow

Table 2. Structural congruence for the π-Calculus with bags

$$P \equiv Q \quad \text{if } P \text{ and } Q \text{ are } \alpha\text{-convertible}$$
$$P \mid Q \equiv Q \mid P$$
$$(P \mid Q) \mid R \equiv P \mid (Q \mid R)$$
$$(\nu z\, P) \mid Q \equiv \nu z\, (P \mid Q) \quad \text{if } z \notin fn(Q)$$
$$P \equiv P \mid \emptyset_x \quad \text{for all possible } x$$

Table 3. Transition rules for the π-Calculus with bags

$$(in) \quad \frac{\alpha_j = xy}{\sum_{i \in I} \alpha_i.P_i \xrightarrow{xz} P_j[z/y]} \qquad\qquad (out) \quad \frac{B_x \xrightarrow{xy} B'_x}{(\overline{x}y.P + \sum_{i \in I} \alpha_i.P_i) \mid B_x \xrightarrow{\tau} P \mid B'_x}$$

$$(in_{bag}) \quad \frac{B_x \xrightarrow{y} B'_x}{B_x \xrightarrow{xy} B'_x} \qquad\qquad (out_{bag}) \quad \frac{B_x \xrightarrow{\overline{y}} B'_x}{B_x \xrightarrow{\overline{x}y} B'_x}$$

$$(sync) \quad \frac{P \xrightarrow{\overline{x}y} P', \; Q \xrightarrow{xy} Q'}{P \mid Q \xrightarrow{\tau} P' \mid Q'} \qquad\qquad (\nu) \quad \frac{P \xrightarrow{\alpha} P', \; a \notin fn(\alpha)}{\nu a P \xrightarrow{\alpha} \nu a P'}$$

$$(open) \quad \frac{P \xrightarrow{\overline{x}y} P', \; x \neq y}{\nu y P \xrightarrow{\overline{x}(y)} P'} \qquad\qquad (close) \quad \frac{P \xrightarrow{\overline{x}(y)} P', \; Q \xrightarrow{xy} Q', \; y \notin fn(Q)}{P \mid Q \xrightarrow{\tau} \nu y(P' \mid Q')}$$

$$(comp) \quad \frac{P \xrightarrow{\alpha} P', \; bn(\alpha) \cap fn(Q) = \emptyset}{P \mid Q \xrightarrow{\alpha} P' \mid Q} \qquad\qquad (bang) \quad \frac{P \mid !P \xrightarrow{\alpha} R}{!P \xrightarrow{\alpha} R}$$

$$(cong) \quad \frac{P \equiv P' \; P' \xrightarrow{\alpha} Q' \; Q' \equiv Q}{P \xrightarrow{\alpha} Q}$$

scope extrusion to apply also in some particular case where those rules would not help. A good example is $\nu x\, \overline{x}y$: only a buffer can make an output action, so this process would not be able to use the *open* rule.

The basic input and output transitions for bags given by in_{bag} and out_{bag} are defined in terms of receive and send transitions on buffers in Definition 3 and 4. The following remark follows trivially from the rules in Table 3.

Remark 2. Let B_x a bag process. Then $B_x \xrightarrow{y} B'_x$ iff $B_x \xrightarrow{xy} B'_x$. Similarly, $B_x \xrightarrow{\overline{y}} B'_x$ iff $B_x \xrightarrow{\overline{x}y} B'_x$.

The notions of *free names* and *bound names* for ordinary processes are defined as usual. For bags, we define them as follows. Recall that C gives the content of a buffer (see Definition 5).

Definition 8 (Bag's free and bound names). *Let B_x be a bag with content $C(B_x) = \{y_1, \ldots, y_k\}$. The free variables fn and the bound variables bn of B_x are defined as $fn(B_x) = \{x, y_1, \ldots, y_k\}$ and $bn(B_x) = \emptyset$.*

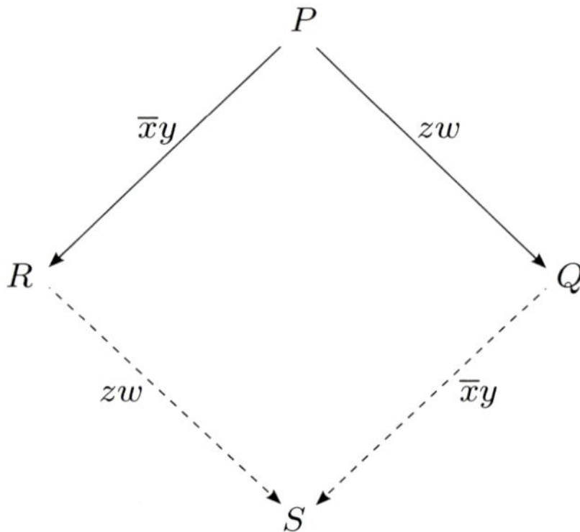

Fig. 3. Lemma of confluence

Lemma 3 (Confluence with τ). *Let $P \in \pi_a$. Assume that $P \xrightarrow{\tau} R$ and $P \xrightarrow{\overline{x}y} Q$. Then, either*

1. $P \xrightarrow{xz}$ for any z, or

2. there exists $S \in \pi_{sc}$ such that: $Q \xrightarrow{\tau} S$ and $R \xrightarrow{\overline{x}y} S$.

Proof. We have to consider the possibility that the transition $P \xrightarrow{\tau} R$ is the result of a synchronization between $P_1 \xrightarrow{xy} Q_1$ and $P_2 \xrightarrow{\overline{x}y} Q_2$, where P_1 and P_2 are parallel subprocesses in P, and the latter transition is the one which induces $P \xrightarrow{\overline{x}y} Q$. If this is the case, then $P \xrightarrow{xz}$ for any z (note that x cannot be bound in P because $P \xrightarrow{\overline{x}y}$). On the other hand, if $P \xrightarrow{\tau} R$ does not involve the transition that induces $P \xrightarrow{\overline{x}y} Q$, then the proof is the same as for Lemma 2 (see [20], Lemma 4.1).

Theorem 4. *Let $[\![\ ]\!]$ be an encoding from $\pi_{\ominus}$ into π_{sc}. Then there exists $P \in \pi_{\ominus}$ such that $P \not\approx [\![P]\!]$.*

Proof. Let P be a stack B_x of the form $y \triangleright \ldots$. Assume by contradiction that $B_x \approx [\![B_x]\!]$ (i.e. B_x is *weakly* asynchronously bisimilar to $[\![B_x]\!]$) . Then B_x must be weakly bisimilar to $[\![B_x]\!]$. In fact, if $B_x \xrightarrow{xz} B'_x \approx [\![B_x]\!] | \overline{x}z \approx B_x | \overline{x}z$, then we would have both $B_x \xrightarrow{\overline{x}y}$ and $B_x \xrightarrow{\overline{x}z}$, which by Remark 1 is not possible.

Let $z \neq y$. We have $B_x \xrightarrow{\overline{x}y} B'_x$ and $B_x \xrightarrow{xz}$. Since B_x is weakly bisimilar to $[\![B_x]\!]$, we have, for some P, $[\![B_x]\!] \xrightarrow{\tau}^* P \xrightarrow{\overline{x}y} \xrightarrow{\tau}^*$ and $P \xrightarrow{\tau}^* \xrightarrow{xz} \xrightarrow{\tau}^*$.

Let us assume that the number of τ steps before P inputs xz is not zero. That is to say, $P \xrightarrow{\tau} P' \xrightarrow{\tau}^* \xrightarrow{xz}$. From Lemma 3, we have that either $P \xrightarrow{xz}$ for any z, or $P' \xrightarrow{\overline{x}y}$. Then, by re-applying this reasoning to each sequence of τ transitions before the input of xz, we eventually get $P \xrightarrow{\overline{x}y}$ and $P \xrightarrow{xz}$. By applying Lemma 2 we have $P \xrightarrow{\overline{x}y} \xrightarrow{zz} P'$ and $P \xrightarrow{zz} \xrightarrow{\overline{x}y} P'$. From the fact that

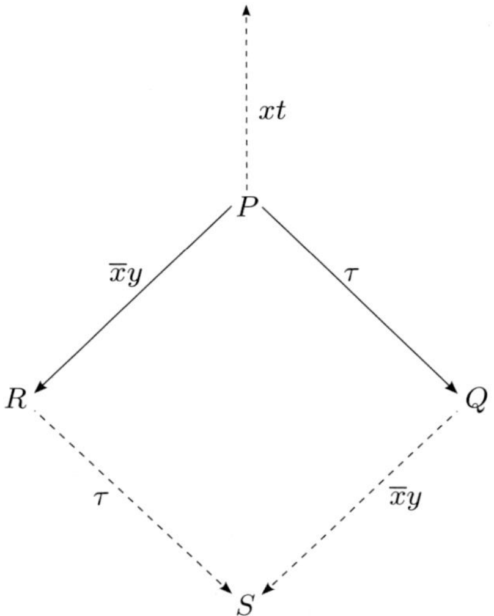

Fig. 4. Confluence with τ

B_x and $[\![B_x]\!]$ are weakly bisimilar, we get $B_x \overset{\overline{x}y}{\Longrightarrow}\overset{xz}{\Longrightarrow}$ and $B_x \overset{xz}{\Longrightarrow}\overset{\overline{x}y}{\Longrightarrow}$. Finally, we observe that the last sequence is not possible, because after the input action xz a stack can only perform an output of the form $\overline{x}z$.

Figure 5 illustrates the fact that the encoded process must have a point where confluence occurs, which is used in this proof.

Remark 4. Also in this case we could give a stronger result, namely that for any encoding $[\![\ ]\!] : \pi_{\mathfrak{G}} \to \pi_{sc}$ and any stack B_x, $[\![B_x]\!] \not\approx B_x$. Again the idea is that if B_x is not in the right form (i.e. it is empty), then we can make an input step so to get a stack with one element.

7   Related Work

The first process calculi proposed in literature (CSP [6,13], CCS [16,17], ACP [2]) were all based on synchronous communication primitives. This is because synchronous communication was considered somewhat more basic, while asynchronous communication was considered a derived concept that could be expressed using buffers (see, for instance, [13]). Some early proposals of calculi based purely on asynchronous communication were based on forcing the interaction between processes to be always mediated by buffers [3,9]. This is basically the same principle that we use in this paper for the definition of $\pi_{\mathfrak{B}}$.

At the beginning of the 90's, asynchronous communication became much more popular thanks to the diffusion of the Internet and the consequent increased interest for widely distributed systems. The elegant mechanism for asynchronous communication (the asynchronous send) proposed in the asynchronous π-calculus [14,5] was very successful, probably because of its simple and basic

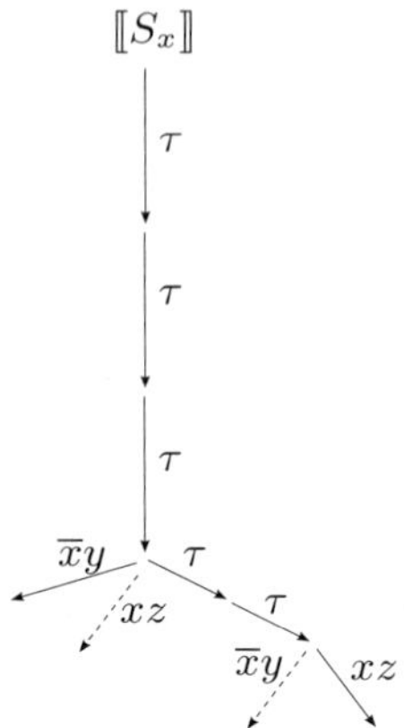

Fig. 5. Impossibility to encode a stack

nature, in line with the tradition of process calculi. Thus it rapidly became the standard approach to asynchrony in the community of process calculi, and it was adopted, for instance, also in Mobile Ambients [8]. A communication primitive (*tell*) similar to the asynchronous send was also proposed, independently, within the community of Concurrent Constraint Programming [23].

We are not aware of any attempt to compare the two approaches to asynchrony (the one with explicit buffers and the one with the asynchronous send). However, our negative results concerning the non-encodability of stacks and queues in π_a use some properties of the asynchronous π-calculus that had been already presented in literature [20,22]. Similar properties were also investigated in [12,24] with the purpose of characterizing the nature of asynchronous communication.

An interesting study of various levels of asynchrony in communication for Linda-like languages has been carried out in [7]. In this paper, the authors investigate three different semantics of the output operation: the *instantaneous*, the *ordered*, and the *unordered* semantics. The first two essentially correspond to the semantics defined in this paper for π_a and $\pi_\mathcal{B}$, respectively. The third one corresponds to the semantics for $\pi_\mathcal{B}$ with the additional possibility of temporal reordering of messages between their sending and their arrival. As argued in the introduction, the last two cases should coincide in languages which do not have the possibility of detecting the absence of a message, so these three calculi (π_a, $\pi_\mathcal{B}$, and $\pi_\mathcal{B}$ with unordered semantics) should be equivalent (up to weak asynchronous bisimilarity)[3].

8   Conclusion and Future Work

In this paper we have investigated the relation between the asynchronous π-calculus and a calculus $\pi_\mathcal{B}$ where asynchronous communication is achieved via

---

[3] This conjecture is due to Roberto Gorrieri.

the explicit use of buffers. We have proved that there is a tight correspondence when the buffers are bags, namely we have exhibited encodings in both directions correct with respect to asynchronous bisimulation. For queues and stacks, on the contrary, we have proved an impossibility result, namely that they cannot be translated into asynchronously bisimilar processes belonging to the asynchronous π-calculus.

We aim at applying these results for modeling and verifying (using the tools developed for the asynchronous π-calculus) widely distributed systems with asynchronous and bag-like communication.

Another line of research is to develop variants of the asynchronous π-calculus in which communication is based on stack-like and queue-like disciplines, and investigate their theories. The motivation is to model and verify distributed systems with the corresponding kind of communication.

Acknowledgment

This work originates from a suggestion of John Mitchell and Andre Scedrov. They remarked that outsiders to Concurrency Theory do not see so clearly why the asynchronous π-calculus represents a model of asynchronous communication. So they recommended to give a justification in terms of a model (buffers) which looks probably more natural to a wide audience, as this could make the asynchronous π-Calculus more widely appreciated.

We also wish to thank Roberto Gorrieri, who helped us with very insightful comments, and the anonymous reviewers.

References

1. Amadio, R.M., Castellani, I., Sangiorgi, D.: On bisimulations for the asynchronous π-calculus. Theoretical Computer Science 195(2), 291–324 (1998); An extended abstract appeared in Sassone, V., Montanari, U. (eds.): CONCUR 1996. LNCS, vol. 1119, pp. 147–162. Springer, Heidelberg (1996)
2. Bergstra, J., Klop, J.: Process algebra for synchronous communication. Information and Control 60(1,3), 109–137 (1984)
3. Bergstra, J.A., Klop, J.W., Tucker, J.V.: Process algebra with asynchronous communication mechanisms. In: Brookes, S.D., Winskel, G., Roscoe, A.W. (eds.) Seminar on Concurrency. LNCS, vol. 197, pp. 76–95. Springer, Heidelberg (1985)
4. Boreale, M., Nicola, R.D., Pugliese, R.: Trace and testing equivalence on asynchronous processes. Information and Computation 172(2), 139–164 (2002)
5. Boudol, G.: Asynchrony and the π-calculus (note). Rapport de Recherche 1702, INRIA, Sophia-Antipolis (1992)
6. Brookes, S., Hoare, C., Roscoe, A.: A theory of communicating sequential processes. J. ACM 31(3), 560–599 (1984)
7. Busi, N., Gorrieri, R., Zavattaro, G.: Comparing three semantics for linda-like languages. Theoretical Computer Science 240(1), 49–90 (2000)
8. Cardelli, L., Gordon, A.D.: Mobile ambients. Theoretical Computer Science (TCS) 240(1), 177–213 (2000)

9. de Boer, F.S., Klop, J.W., Palamidessi, C.: Asynchronous communication in process algebra. In: Scedrov, A. (ed.) Proceedings of the 7th Annual IEEE Symposium on Logic in Computer Science, Santa Cruz, CA, June 1992, pp. 137–147. IEEE Computer Society Press, Los Alamitos (1992)
10. Gorla, D.: On the relative expressive power of asynchronous communication primitives. In: Aceto, L., Ingólfsdóttir, A. (eds.) FOSSACS 2006. LNCS, vol. 3921, pp. 47–62. Springer, Heidelberg (2006)
11. Gorla, D.: On the criteria for a 'good' encoding: a new approach to encodability and separation results. Technical report, Università di Roma "La Sapienza" (2007)
12. He, J., Josephs, M.B., Hoare, C.A.R.: A theory of synchrony and asynchrony. In: Broy, M., Jones, C.B. (eds.) Programming Concepts and Methods, Proc. of the IFIP WG 2.2/2.3, Working Conf. on Programming Concepts and Methods, pp. 459–478. North-Holland, Amsterdam (1990)
13. Hoare, C.A.R.: Communicating Sequential Processes. Prentice-Hall, Englewood Cliffs (1985)
14. Honda, K., Tokoro, M.: An object calculus for asynchronous communication. In: America, P. (ed.) ECOOP 1991. LNCS, vol. 512, pp. 133–147. Springer, Heidelberg (1991)
15. Lynch, N.: Distributed Algorithms. Morgan Kaufmann, San Francisco (1996)
16. Milner, R.: A Calculus of Communication Systems. LNCS, vol. 92. Springer, Heidelberg (1980)
17. Milner, R.: Communication and Concurrency. International Series in Computer Science. Prentice-Hall, Englewood Cliffs (1989)
18. Milner, R., Parrow, J., Walker, D.: A calculus of mobile processes, I and II. Information and Computation 100(1), 1–40 & 41–77 (1992); A preliminary version appeared as Technical Reports ECF-LFCS-89-85 and -86, University of Edinburgh (1989)
19. Nestmann, U.: What is a 'good' encoding of guarded choice? Journal of Information and Computation 156, 287–319 (2000); An extended abstract appeared in the Proceedings of EXPRESS 1997, vol. 7, ENTCS (1997)
20. Palamidessi, C.: Comparing the expressive power of the synchronous and the asynchronous pi-calculus. Mathematical Structures in Computer Science 13(5), 685–719 (2003); A short version of this paper appeared in POPL 1997
21. Pierce, B.C., Turner, D.N.: Pict: A programming language based on the pi-calculus. In: Plotkin, G., Stirling, C., Tofte, M. (eds.) Proof, Language and Interaction: Essays in Honour of Robin Milner, pp. 455–494. MIT Press, Cambridge (1998)
22. Sangiorgi, D., Walker, D.: The π-calculus: a Theory of Mobile Processes. Cambridge University Press, Cambridge (2001)
23. Saraswat, V.A., Rinard, M., Panangaden, P.: Semantic foundations of concurrent constraint programming. In: Conference Record of the Eighteenth Annual ACM Symposium on Principles of Programming Languages, pp. 333–352. ACM Press, New York (1991)
24. Selinger, P.: First-order axioms for asynchrony. In: Mazurkiewicz, A., Winkowski, J. (eds.) CONCUR 1997. LNCS, vol. 1243, pp. 376–390. Springer, Heidelberg (1997)

StonyCam: A Formal Framework for Modeling, Analyzing and Regulating Cardiac Myocytes[*]

Ezio Bartocci[1,2], Flavio Corradini[1], Radu Grosu[2], Emanuela Merelli[1], Oliviero Riganelli[1,2], and Scott A. Smolka[2]

[1] Dipartimento di Matematica e Informatica, Università di Camerino
Camerino I-62032, Italy
[2] Department of Computer Science, Stony Brook University
NY 11794-4400, USA

The authors warmly thank Ugo for providing a constant source of inspiration for their research. Ugo's way of doing research has inspired and will continue to inspire past, present and future generations of computer scientists.

Abstract. This paper presents a formal framework, experimental infrastructure, and computational environment for modeling, analyzing and regulating the behavior of cardiac tissues. Based on the theory of hybrid automata, we aim at providing suitable tools to be used in devising strategies for the pharmacological or other forms of treatment of cardiac electrical disturbances.

1   Introduction

Atrial fibrillation (Afib) is an abnormal rhythm originating in the upper chambers of the heart afflicting 2-3 million Americans and whose incidence rises with increasing age. Due to the "graying" of our population, 12-16 million Americans may be affected by 2050. Not only is its incidence of epidemic proportions, its morbidity is also significant. Among possible sequelae of the disease are thrombi in the fibrillating atria and emboli released to the pulmonic and systemic circulations. Although its importance to public health cannot be questioned, therapies remain problematic. Persistence of the abnormal rhythm results in electrical remodeling of the atria reinforcing its existence. Drugs are frequently ineffective and because of their lack of selectivity can induce arrhythmias themselves. Frequently, electrical cardioversion is tried which is not uniformly successful. Finally, for intractable Afib, the abnormal reentrant pathways are mapped and the tissue is radiofrequency ablated, which may result in a non-functional atrium.

This is a humbling observation from an engineering point of view, highlighting the complexity of the heart and the need for reliable analysis and prediction *in-silico* tools for cardiac behavior. Such tools would be of great use in devising rational strategies for pharmacological or other intervention in cardiac electrical disturbances such as Afib. During the last two years, we have worked (see

---

[*] This work was supported by the Investment Funds for Basic Research (FIRB) project "LITBIO: Laboratory of Interdisciplinary Technologies in Bioinformatics".

P. Degano et al. (Eds.): Montanari Festschrift, LNCS 5065, pp. 493–502, 2008.

Figure 1) towards a formal framework, experimental infrastructure, and computational environment for modeling, analyzing and controlling the behavior of excitable cells such as cardiac myocytes.

In this paper we provide a brief overview of the results obtained so far and discuss directions for future work. We start in Section 2 by describing how we modeled the behavior of excitable tissue using networks of hybrid automata (HA). With respect to the classical approach which uses systems of non-linear ordinary differential equations HA models, by combining discrete and continuous processes, are able to successfully capture the morphology of the excitation event (action potential) of cardiac cells [22]. In section 3, we show how this approach also enhances the analysis capabilities of this biological phenomena. In particular it renders possible large-scale simulation of cardiac-cell networks and the detection of emergent behavior such as fibrillation [9]. Once such a behavior has been identified, one could use electrical therapy in order to restore normal physiological function. This means that any time a critical behavior is predicted, depending on the type of spatial pattern, the right repair strategy is performed either at low-level, e.g. by introducing an artificial inhibitor or catalyst agent to regulate the ion channels of cell membranes, or at higher-level, e.g. by resetting the behavior of a group of myocytes forcing a global correct behavior. We are investigating solutions from the area of the networks of dynamical elements, where distributed synchronization is obtained by dividing the whole network into groups or regions of fully synchronized elements [18] while elements from different groups are not necessarily synchronized and can be of entirely different dynamics [25]. Section 4 offers our concluding remarks and directions for future work.

2   Modeling Excitable Cells Using Hybrid Automata

An excitable cell has the ability to propagate an electrical signal, known at the cellular level as the *Action Potential* (AP), to neighboring cells. An AP corresponds to a difference in electrostatic potential between the inside and outside of a cell, and is caused by the flow of ions across the cell membrane. The major ion species involved in this process are sodium, potassium and calcium; they flow through multiple voltage-gated ion channels as pore-forming proteins in the cell membrane. Excitation disturbances can occur in the behavior of these ion channels at the cell level, or in the propagation of the electrical waves at the cell-network level.

Generally, an AP is an externally triggered event: a cell fires an action potential as an "all-or-nothing" response to a supra-threshold stimulus, and each AP follows the same sequence of phases and maintains the same magnitude regardless of the applied stimulus. During an AP, generally no re-excitation can occur. The early portion of an AP is known as "absolute refractory period" due to its non-responsiveness to further stimulation. The "relative refractory period" is the interval immediately following during which an altered secondary excitation event is possible if the stimulation strength or duration is raised. Examples of excitable cells are neurons, cardiac myocytes and skeletal muscle cells.

Fig. 1. StonyCam Group from left to right: Ezio Bartocci, Flavio Corradini, Emanuela Merelli, Scott Smolka, Oliviero Riganelli, Radu Grosu

Despite differences in AP duration, morphology and underlying ion currents, the following major AP phases can be identified across different species of excitable cells: *resting, rapid upstroke, early repolarization phase, plateau and late repolarization*, and *final repolarization* (identical to the resting phase due to the cyclic nature of an AP). The resting state features a constant transmembrane potential (difference between the inside and outside potential of the cell) that is about -80 mV for most species of cardiac cells; i.e. the membrane is polarized at rest. During the AP upstroke, the transmembrane potential rapidly changes, from negative to positive; i.e. the membrane depolarizes. This is followed by an early repolarization phase. A slower, plateau phase is present in most mammalian action potentials, during which calcium influx facilitates muscle contraction. After this phase, a faster final repolarization brings the potential back to the resting state.

The classical mathematical model [4,14,11] of excitable cell involve complex systems of nonlinear differential equations. Such models not only impair formal analysis but also impose high computational demands on simulations, especially in large-scale 2D and 3D cell networks. To address this state of affairs, we have developed *Cycle-Linear Hybrid Automata* (CLHA) models (see Figure 2). The CLHA formalism was designed to be both abstract enough to admit formal analysis and efficient simulation and expressive enough to capture the AP morphology and restitution properties exhibited by classical non linear excitable-cell models. The basic idea behind the CLHA model is the observation that, during an AP, an excitable cell cycles through four basic modes of operations - resting, stimulated,

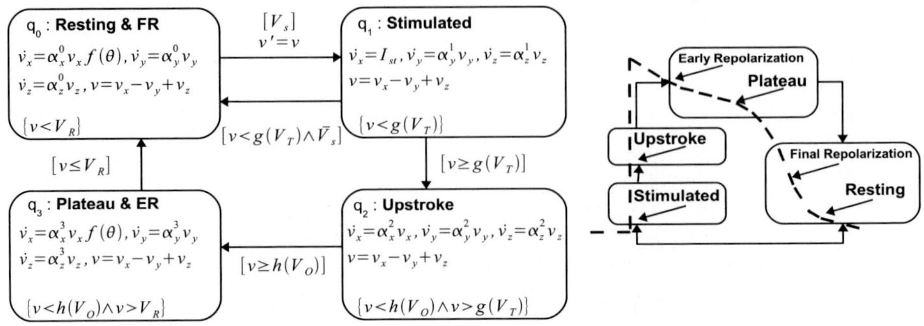

Fig. 2. CLHA and the corresponding Action Potential

upstroke, early repolarization, plateau final repolarization - and the dynamics of each mode is essentially linear and time-invariant. To capture possible non linear, frequency-dependent properties such as restitution, the CLHA model is equipped with one-cycle memory of the cells voltage and per-mode parameters of the current cycle's linear time invariant system of differential equations are updated according to this voltage. Consequently, the models behavior is linear in any one cycle but appropriately non linear overall. For more details on CLHA, we refer the reader to [22]. A CLHA approximates AP and other bio-electrical properties of several representative excitable-cell types, with reasonable accuracy [21,22,10]. This derivation was first performed manually [21,22]. In [10], we showed that it is possible to automatically learn a much simpler cycle-linear hybrid automaton for cardiac myocytes, which describes their action potential up to a specified error margin. Moreover, as we have shown in [2,3], one can use a variant of this model [21,20,23,24,22] to efficiently (up to an order of magnitude faster) and accurately simulate the behavior of myocyte networks, and, in particular, induce spirals and fibrillation. The term *Cycle-Linear* is used to highlight the cyclic structure of CLHA, and the fact that while in each cycle they exhibit linear dynamics, the coefficients of the corresponding linear equations and mode-transition guards may vary in interesting ways from cycle to cycle. These CLHA models were found to capture essential cell features, are amenable to formal analysis, and exhibit, respect to the classical models, a nearly ten-fold speedup in a simulation of 400x400 cell network.

3 Simulation and Analysis of Networks of Cardiac Myocytes

3.1 Simulation

In order to simulate the emergent behavior of cardiac tissue, we have developed CELLEXCITE [3], a CLHA-based simulation environment for excitable-cell networks. CELLEXCITE allows the user to sketch a tissue of excitable cells, plan the

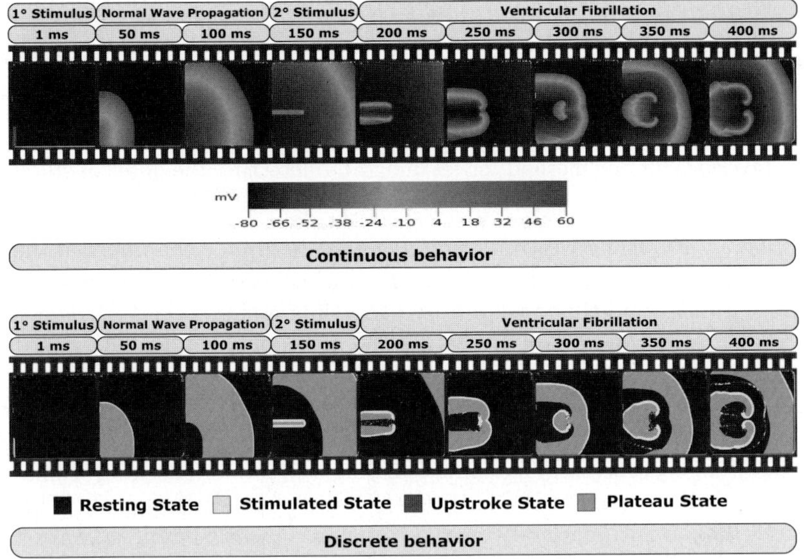

Fig. 3. Simulation of continuous and discrete behavior of CLHA network

stimuli to be applied during simulation, and customize the arrangement of cells by selecting the appropriate lattice. Figure 3 presents our simulation results for a 400×400 CLHA network. The network was stimulated twice during simulation, at different regions. The results we obtain demonstrate the feasibility of using CLHA networks to capture and mimic different spatiotemporal behavior of wave propagation in 2D isotropic cardiac tissue, including normal wave propagation (1-150 ms); the creation of spirals, a precursor to fibrillation (200-250 ms); and the break-up of such spirals into more complex spatiotemporal patterns, signaling the transition to ventricular fibrillation (250-400 ms).

As can be clearly seen in Figure 3, a particular form of discrete abstraction, in which the action potential value of each CLHA in the network is *discretely abstracted* to its corresponding mode, faithfully preserves the network's waveform and other spatial characteristics. Hence, for the purpose of learning and detecting spirals within CLHA networks, we can exploit discrete mode-abstraction to dramatically reduce the system state space.

3.2 Detecting Emergent Behavior

One of the most important and intriguing questions in systems biology is how to formally specify *emergent behavior in biological tissue*, and how to efficiently predict and detect its onset. A prominent example of such behavior is electrical *spiral waves* in spatial networks of cardiac myocytes (heart cells). Spiral waves of this kind are a precursor to a variety of cardiac disturbances, including *atrial*

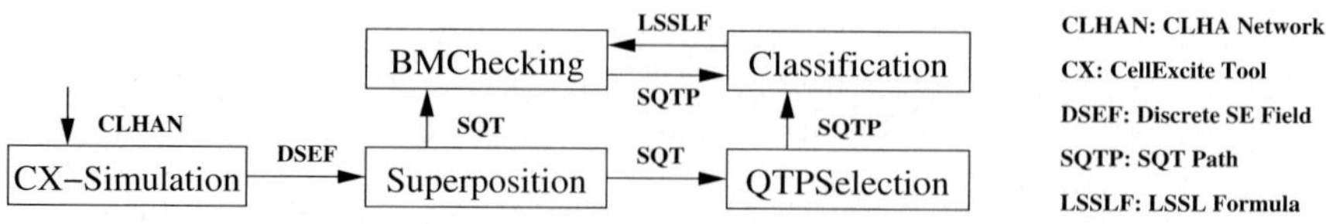

Fig. 4. Overview of our method for learning and detecting the onset of spiral waves

fibrillation, an abnormal rhythm originating in the upper chambers of the heart. Moreover, the likelihood of developing atrial fibrillation increases with age.

In [9], we addressed this question by proposing a simple and efficient method for learning and automatically detecting the onset of spiral waves in cardiac tissue (see Figure 4). Underlying our method is a *linear spatial-superposition logic* (LSSL), which we have developed for specifying properties of spatial networks. LSSL is discussed in greater detail below. Our method also builds upon hybrid-automata, image-processing, machine-learning and model-checking techniques to first learn an LSSL formula (LSSLF) that characterizes such spirals. The resulting LSSLF is then automatically checked against a *quadtree* representation [17] of the scalar electric (SE) field, produced by simulating a hybrid automata network modeling the myocytes, at each discrete time step. The quadtree representation is obtained via hybrid abstraction [19] and hierarchical superposition of the elementary units within the field.

A key observation concerning our simulations (see Figure 3) is that a particular form of hybrid abstraction, in which the action potential value of each CLHA in the network is *discretely abstracted* to its corresponding mode, faithfully preserves a spiral's topological characteristic; i.e. its shape. Hence, for the purpose of learning and detecting the onset of spirals within CLHA networks, we can exploit hybrid abstraction to dramatically reduce the system state space. A similar hybrid abstraction is possible for voltage recordings in live cell networks, but this is outside the scope of this paper.

The state space of a 400×400 CLHA network is still prohibitively large even after applying the above-described hybrid abstraction: it contains $4^{160,000}$ states, as each CLHA has four modes! To combat this state explosion, we use a spatial abstraction inspired by [12]: we regard the state of each automaton as a probability distribution and define the *superposition* of a set of states as the probability that an arbitrary state in this set has a particular mode. By successively applying superposition to the network, we obtain a tree structure, the root of which is the state-superposition of the entire CLHA network, and the leaves of which are the states of the individual CLHA. The particular superposition tree structure we employ, quadtrees, is inspired by image-processing techniques [17]. We shall refer to quadtrees obtained in this manner as *superposition-quadtrees* (SQT).

Our LSSL is an appropriate logic for reasoning about *paths* in superposition-quadtrees, and the spatial properties of a CLHA network in which we are interested, including spirals, can be cast in LSSL. For example, we have observed that the presence of a spiral can be formulated in LSSL as follows: Given an

SQT, is there a path from its root leading to the core of a spiral? Based on this observation, we build a machine-learning classifier, the training-set records for which correspond to the probability distributions of the nodes along such paths. Each node distribution corresponds to an attribute of a training-set record, with the number of attributes bounded by the depth of the SQT. An additional attribute is used to classify the record as either spiral or non-spiral. For spiral-free SQTs, we simply record the path of maximum distribution.

For training purposes, we use the CELLEXCITE simulator [2,3] to generate, upon successive time steps, snapshots of a 400×400 CLHA networks and their hybrid abstraction; see Figures 4,3. Training data for the classifier is then generated by converting the hybrid-abstracted snapshots into SQTs and selecting paths leading to the core of a spiral (if present). The resulting table is input to the decision-tree algorithm of the Weka machine-learning tool suite [8], which produces a classifier in the form of a predicate over the node-distribution attributes.

The syntax of LSSL is similar to that of linear temporal logic, with LSSL's Next operator corresponding to *concretization* (anti-superposition). Moreover, a (finite) sequence of LSSL Next operators corresponds to a path through an SQT. The classifier produced by Weka can therefore be regarded as an LSSL formula. The meaning of such a path is that of a magnifying glass, which starting from the root, produces an increasingly detailed but more focused view of the image. This effect is analogous to *concept hierarchy* in data mining [13] and arguably similar to the way the brain organizes knowledge: a human can recognize a word or a picture without having to look at all of the characters in the word or all of the details in the picture, respectively.

We are now in a position to view spiral detection as a bounded-model-checking problem [5]: Given the SQT Q generated from the discrete scalar electric field of a CLHA network and an LSSL formula φ learned through classification, is there a finite path $\pi \in Q$ satisfying the LSSL formula φ, i.e. $\pi \models \varphi$? We use this observation to check in real time, i.e. at each discrete simulation time step, whether or not a spiral has been created. More precisely, the LSSL formula we use states that no spiral is present, and we thus obtain as a counterexample one or all the paths leading to the core of a spiral. In the latter case, we can identify the number of spirals in the scalar field and their actual position.

4   Conclusion

The StonyCam collaboration has been a highly fruitful one to date, resulting in the development of HA-based models of complex networks of excitable cells, the CELLEXCITE simulator for such networks, and techniques for learning and detecting emergent behavior (spirals) in cardiac tissue. Much work remains to be done, especially in the engineering of distributed coordination and control algorithms for myocyte networks.

In other ongoing and future work, analyzing large-scale networks of cardiac myocytes requires a flexible and powerful simulation environment. Along these lines, we are investigating the use of multiagent systems (MAS) and graphical processing units (GPU). MASs would offer us increased flexibility while GPUs would offer us increased computational power. A MAS is an autonomous software entity able to perceive and react to the changes of the surrounding environment. A MAS consists of a collection of interactive agents and a set of coordination rules. It constitutes a suitable benchmark for simulating the actions and interactions of autonomous real entities in a network to assess their effects on the system as a whole. This programming paradigm allows to easily add new entities and to modify the behavior of existing ones even in a zooming-in and zooming-out approach [6]. Following [7,1], we would like to investigate a distributed coordination model based on simulation-time model checking for the online prediction of critical behaviors in cardiac tissue.

Regarding GPUs, they implement a number of graphical primitive operations in a very efficient manner. The use of graphics hardware has recently shown promising results in massive simulations of complex behavioral models [15] and in general-purpose stream computations [16]. We would like to explore their computational power as well in our simulation environment for cardiac tissue.

Acknowledgement

StonyCam is a joint research program involving the Departments of Computer Science at Stony Brook University and Camerino University. We would like to thank Maria Rita Di Berardini, Emilia Entcheva, Anita Wasilewska, and Pei Ye for joint work and personal discussions.

References

1. Angeletti, M., Baldoncini, A., Cannata, N., Corradini, F., Culmone, R., Forcato, C., Mattioni, M., Merelli, E., Piergallini, R.: Orion: a spatial multiagent system framework for computational cellular dynamics of metabolic pathways. In: Bioinformatics Italian Society Meeting, BITS 2006 (2007)
2. Bartocci, E., Corradini, F., Di Berardini, M., Entcheva, E., Grosu, R., Smolka, S.: Spatial Networks of Hybrid I/O Automata for Modeling Excitable Tissue. In: The proceedings of FBTC 2007, pp. 86–102 (2007)
3. Bartocci, E., Corradini, F., Entcheva, E., Grosu, R., Smolka, S.A.: Cellexcite: An efficient simulation environment for excitable cells. BMC Bioinformatics 9 (2007)
4. Beeler, G., Reuter, H.: Recostruction of the action potential of ventricular mycardial fibres. J Physiol 268, 177–210 (1977)
5. Biere, A., Cimatti, A., Clarke, E., Strichman, O., Zhu, Y.: Bounded model checking. In: Zelkowitz, M. (ed.) Advances in Computers: Highly Dependable Software, vol. 58, Academic Press, London (2003)
6. Cannata, N., Corradini, F., Merelli, E., Omicini, A., Ricci, A.: An agent-oriented conceptual framework for systems biology. Transactions on Computational Systems Biology III 3737, 105–122 (2005)

7. Corradini, F., Merelli, E.: Hermes: Agent-based middleware for mobile computing. In: Bernardo, M., Bogliolo, A. (eds.) SFM-Moby 2005. LNCS, vol. 3465, pp. 234–270. Springer, Heidelberg (2005)

8. Frank, E., Hall, M.A., Holmes, G., Kirkby, R., Pfahringer, B., Witten, I.H., Trigg, L.: WEKA – a machine learning workbench for data mining. In: The Data Mining and Knowledge Discovery Handbook, pp. 1305–1314. Springer, Heidelberg (2005)

9. Grosu, R., Bartocci, E., Corradini, F., Entcheva, E., Smolka, S.A., Wasilewska, A.: Learning and detecting emergent behavior in networks of cardiac myocytes. In: Proc. of HSCC 2008, the 11th International Conference on Hybrid Systems: Computation and Control (2008)

10. Grosu, R., Mitra, S., Ye, P., Entcheva, E., Ramakrishnan, I., Smolka, S.: Learning cycle-linear hybrid automata for excitable cells. In: Bemporad, A., Bicchi, A., Buttazzo, G. (eds.) HSCC 2007. LNCS, vol. 4416, pp. 245–258. Springer, Heidelberg (2007)

11. Hodgkin, A.L., Huxley, A.F.: A quantitative description of membrane currents and its application to conduction and excitation in nerve. J Physiol 117, 500–544 (1952)

12. Kwon, Y., Agha, G.: Scalable modeling and performance evaluation of wireless sensor networks. In: IEEE Real Time Technology and Applications Symposium, pp. 49–58 (2006)

13. Lu, Y.: Concept hierarchy in data mining: Specification, generation and implementation. Master's thesis, Simon Fraser University (1997)

14. Luo, C.H., Rudy, Y.: A dynamic model of the cardiac ventricular action potential. i. simulations of ionic currents and concentration changes. Circ. Res. 74(6), 1071–1096 (1994)

15. Miyazaki, R., Yoshida, S., Nishita, T., Dobashi, Y.: A method for modeling clouds based on atmospheric fluid dynamics. In: Pacific Conference on Computer Graphics and Applications, pp. 363–373 (2001)

16. Owens, J.D.: A survey of general-purpose computation on graphics hardware. Computer Graphics Forum. 26(34), 80–113 (2007)

17. Shusterman, E., Feder, M.: Image compression via improved quadtree decomposition algorithms. IEEE Transactions on Image Processing 3(2), 207–215 (1994)

18. Strogatz, S.H.: From kuramoto to crawford: exploring the onset of synchronization in populations of coupled oscillators. Phys. D 143(1-4), 1–20 (2000)

19. Tabuada, P., Pappas, G.J.: Hybrid abstractions that preserve timed languages. In: Di Benedetto, M.D., Sangiovanni-Vincentelli, A.L. (eds.) HSCC 2001. LNCS, vol. 2034, pp. 501–514. Springer, Heidelberg (2001)

20. True, M., Entcheva, E., Smolka, S., Ye, P., Grosu, R.: Efficient event-driven simulation of excitable hybrid automata. In: Proc. of EMBS 2006, the 28th IEEE International Conference of the Engineering in Medicine and Biology Society, pp. 3150–3153. IEEE Computer Society Press, Los Alamitos (2006)

21. Ye, P., Entcheva, E., Grosu, R., Smolka, S.: Efficient modeling of excitable cells using hybrid automata. In: Proc. of CMSB 2005, the 3rd Workshop on Computational Methods in Systems Biology, pp. 216–227 (2005)

22. Ye, P., Entcheva, E., Smolka, S., Grosu, R.: A cycle-linear hybrid-automata model for excitable cells. IET Journal of Systems Biology (SYB) (2007)

23. Ye, P., Entcheva, E., True, M.R., Smolka, S.A., Grosu, R.: A cycle-linear approach to modeling action potentials. In: Proc. of EMBS 2006, the 28th IEEE International Conference of the Engineering in Medicine and Biology Society, pp. 3931–3934. IEEE Press, Los Alamitos (2006)

24. Ye, P., Entcheva, E., True, M., Smolka, S., Grosu, R.: Hybrid automata as a unifying framework for modeling cardiac cells. In: Proc. of EMBS 2006, the 28th IEEE International Conference of the Engineering in Medicine and Biology Society, pp. 4151–4154. IEEE Press, Los Alamitos (2006)
25. Zhang, Y., Hu, G., Cerdeira, H.A., Chen, S., Braun, T., Yao, Y.: Partial synchronization and spontaneous spatial ordering in coupled chaotic systems. Physical Review E 63(2) (2001)

Models of Computation:
A Tribute to Ugo Montanari's Vision

Roberto Bruni[1] and Vladimiro Sassone[2]

[1] Dipartimento di Informatica, Università di Pisa, Italia
[2] ECS, University of Southampton, UK

Ugo Montanari's Models of Computation

Ugo's research activity in the area of Models of Computation (MoC, for short) has been prominent, influential and broadly scoped. Ugo's trademark is that undefinable ability to understand and distill computational aspects into new models as if you were reading them out of some evident connection between well-know models: only, most often, that connection is really visible only after Ugo shows the way. Like experienced sailors have trusted compasses and sextants to help them find the best routes to harbour, Ugo relies on a bag of favourite tools which he has used along the years to deliver a variety of contributions to the MoC area. To mention just three (in alphabetic order): algebraic techniques, concurrency theory, and unification mechanisms.

In this introductory contribution we would like to recall some of the influential MoC models put forward by Ugo which cut across the three approaches. Before doing that, it is worth devoting some space to discuss the three aspects separately. Notably, the use of category theory is a pervasive common trait.

Algebraic techniques. By algebraic techniques we refer broadly to the use of universal algebras and initial model semantics; of universal coalgebras and final semantics; and of bialgebras. Many interesting papers witness Ugo's leading role in exploiting algebraic techniques during his entire scientific career. Indeed, his contributions are too many to mention all in the space allocated to this overview; we shall therefore attempt to convey the sense of Ugo's broad-spectrum contribution by recapping only a few key results.

Reference [43] is the first paper on final, observational semantics in abstract data types, and the main reference for one of the MoC contributed papers in this volume. It presented several key insights in software specification and development for the first time, like the separation between given sorts and newly specified ones, whereby the given sorts lay the ground to define the observable behaviour for the new sorts. Another key suggestion is that the specification of new data types is often partial —in the sense that it may include *"don't care"* cases— and that many realisations can exist that exhibit equivalent observable behaviour but are not isomorphic. In fact, [43] shows that the isomorphism classes of observably equivalent algebras conforming to the partial specification form a complete lattice, yielding a so-called loose semantics.

Possibly the best known of Ugo's papers, [52] exposes the underlying monoidal structure of the category of Petri net computations. The title itself is revealing: *Petri nets are monoids*. Besides doing what it says on the tin, this paper opened a long-lasting and fruitful collaboration with José Meseguer, and a research line on the initial semantics of

P. Degano et al. (Eds.): Montanari Festschrift, LNCS 5065, pp. 503–509, 2008.
© Springer-Verlag Berlin Heidelberg 2008

computational models in which, after [33], we got deeply involved ourselves [55,56,17,24,18]. The key insight is that by lifting the algebraic structure of (machine) states to the level of computations via a so-called free construction, one can gain a deeper understanding of the axioms which regulate equivalent computations (or processes), an idea that is also the basis of Meseguer's Rewriting Logic [51]. This theme of lifting the algebraic structure of states to the level of concurrent computations has motivated the study of Structured Transition Systems [30,39,37], while finding axiomatisations of computational structures has been reconsidered in [49,16,27,9,11].

A more recent result in coalgebraic semantics [57] has paved the way to the efficient verification techniques for the π-calculus [35], and to the bialgebraic semantics of fusion calculus [25]. The technique proposed in [57] addresses the issue of finding a suitable setting to develop a coalgebraic semantics for the π-calculus, so as to characterise minimal process realisations. The key difficulty is the proper handling of fresh names, tackled by exploiting a category of name-permutation algebras which underpins the coalgebraic treatment of the π-calculus operational semantics.

Concurrency theory. Concurrency theory encompasses many different techniques and approaches, ranging from bisimilarity and contextual equivalences to event structure semantics. It is harder here to make a representative selection of a few seminal papers, because of the quality and volume of Ugo's work in the area of concurrency models.

Given our previous lives in 'Petri-land,' we cannot help but mention the work on unfolding semantics that generalised Winskel's approach from the class of safe nets to a wide class of place/transition nets [54]. An unfolding semantics accounts for a full fledged view of the admissible computations, including concurrency, causality and conflict aspects: the so-called "truly concurrent" semantics. Exploiting mathematical tools from category theory, the main result establishes that a chain of adjunctions (a suitable categorical notion indicating that the corresponding construction is as good as possible) leads from the category of Petri nets to the category of prime event structures, which is equivalent to the category of coherent finitary prime algebraic domains (because of this, the unfolding approach is sometimes referred to as a denotational semantics).

More recently, it was shown that such event structure semantics can be extended to a more sophisticated setting of contextual nets and graph transformation systems, where e.g., multiple concurrent read accesses to the same resource and inhibiting conditions for the occurrence of certain events can be accounted for. The price to pay was the introduction of more complex event structures [3,28,2,4,5]. Significantly related is also [1]. The extension has made it possible to provide event structure semantics to mobile calculi for free by encoding them in graph transformation systems [14,15].

The paper [21] presents a mathematical setting building upon some analogies in the representation of names, locations and causal links as shared entities. Such a uniform treatment of different concepts opens the way to the definition of a general-purpose meta-model to be instantiated to several cases of interest. The main result shows that the framework can be applied to the basic parallel processes with weak synchronisation, by defining an operational semantics that accounts for concurrency aspects and a causal abstract semantics and showing it equivalent with bisimilarity via "causal trees" [32].

It is finally worth to mention Ugo's work on transactional extensions of concurrent frameworks [44,20,12,13,7].

Unification mechanisms. Logic programming and its extensions, in particular with concurrent constraints, are one of the long-term research interests of Ugo's. In logic programming, resolution steps are based on the notion of unification between the head of a logic clause and a selected atomic sub-goal. The Martelli-Montanari algorithm [50] is arguably the best known unification algorithm for constructing the "most general unifier" (mgu) between a sub-goal and the head of a clause. Since those brilliant beginnings, the view of unification as an elegant coordination mechanism has been a recurrent source of inspiration in Ugo's work on MoC. We mention here three cases.

Reference [30] builds on the view of mgu as a categorical "equaliser:" clauses are seen as rewrite rules whose variables can be further instantiated freely, and the computational model of a logic program is a suitable 2-category. The interesting point is that the 2-cells of the 2-category are equipped with an algebraic structure that captures some concurrency aspects. If there exists a refutation for the goal G with computed answer substitution θ, then in the 2-category model we can find a refutation for the goal $G\theta$ but not necessarily one for G. This situation is improved in [22], where the mgu is expressed as a categorical "pullback" square and double-categories are considered instead of 2-categories. This setting can account for the dynamic creation of fresh variables and deal with the computed answer substitutions instead of just the correct answer substitutions.

The ideas in [22] are further developed in [6], where logic programming "resolution rule" is generalised to MoC tailored to the needs of the general server-to-client bindings required by the service oriented applications. When a new service is discovered, not only it must adapt to the client, e.g., accepting a list of parameters, but vice versa the client too must sometimes adapt to the server in order to establish the connection. Then, the mgu represents the minimal possible adaptation that should be sought in order to minimise the possible degradation.

Combined approaches. Much of Ugo's scientific thinking can be characterised as the aspiration to combine modelling elements so that the combination of the parts is more expressive and flexible than their mere sum. Below we point out some examples.

The CHARM [31], Concurrency and Hiding in an Abstract Rewriting Machine, is an abstract machine that combines algebraic techniques typical of process calculi with the experience in constraint logic programming and graph transformation systems. Characteristic of the CHARM is the ability to capture the essence of concurrent computations in systems composed by a global, shared part and locally distributed resources.

GDS [26,34], Grammars for Distributed Systems, combines distributed computation based on Hoare synchronisation with concurrent histories. This model later evolved in Synchronized Hyperedge Replacement, SHR [45,46,40,36,48], where different synchronisation mechanisms are considered together with node merging and splitting.

HD-automata [57] (see also the section on Software Verification in the present volume), for History Dependent Automata, are an extension of ordinary automata aimed to endow them with name handling features: states and transition labels may contain names which can represent, e.g., communication channels or locations in distributed systems. Each transition establishes a correspondence between the names in the source state, those in the label and those in the target state. HD-automata permit an adequate representation of the behavior of calculi with name mobility, as names can be garbage-collected and reused to identify 'verification-friendly' processes semantics.

The Tile Model [41,53,58,19,29,38,8,47,16,42,10,23] combines the modularity of Structured Transition Systems with Meseguer's Rewriting Logic approach. While rewrite rules in Rewriting Logic can be applied in any context and with any actual parameters, the Tile Model allows rewritings to be inhibited under certain contexts. In category theory, this correspond to move from 2-categories to double-categories. Moreover, as tiles have been designed around concurrent systems, it is common to consider a monoidal structure of states that gives raise to a monoidal double-category of computations. Thanks to these features, the Tile Model offers a framework where the specification of process calculi with name passing, causality and locality becomes uniform and several important results can be accounted for at the meta-theoretical level.

Papers on Models of Computation in This Volume

The six contributed papers in this section of the present volume cover several of Ugo's favourite topics; other papers on models of computation are included in other chapters dealing with more specific contexts and applications.

Martín Abadi: Automatic mutual exclusion and atomicity checks. This contribution presents a calculus for studying the Automatic Mutual Exclusion (AME) programming model. Roughly, the AME calculus consists of a concurrent lambda calculus with references, extended with constructs for thread spawning, yielding, blocking and atomic execution. A type system ensures that atomic blocks are not violated through yield executions. The main results show soundness and progress theorems.

Samson Abramsky: Petri nets, discrete physics, and distributed quantum computation. This inspired paper builds interesting connections between separate fields, and does so by building upon some of Ugo's best known work. In fact, it describes analogies between Petri Nets, monoidal categories with additional structure, and quantum mechanics (in particular quantum information).

Filippo Bonchi, Maria Grazia Buscemi, Vincenzo Ciancia and Fabio Gadducci: A category of explicit fusions. The paper introduces a suitable category E of equivalence relations and shows it suitable to represent (abstract) syntax and semantics (via an endofunctor B on Set^E) of the calculus of explicit fusions. The main result gives a bijection between inside-outside bisimulations and coalgebraic bisimulations for B.

José Luiz Fiadeiro: What do semantics matter when the meat is overcooked? This paper presents a model for configuration management of service-oriented applications modelled with the language developed by the EU funded Sensoria project. The model makes use of various of Ugo's favourite ingredients: roughly, business configurations are represented as graphs; constraint systems play the role of business policies; a module requiring a set of services is seen as a clause in logic programming style; the reconfiguration that happens when a service is called for instantiation (via the usual service-oriented mechanism of discovery, selection, and binding) is modelled by a sort of resolution.

Nicoletta Sabadini and Robert Walters: Calculating Colimits Compositionally. Recent years witnessed a renewed interest in exploring the dichotomy between the algebraic and the graphical presentations of a system, a topics to which Ugo has also contributed. Along these lines, the paper gives an algebraic description for finite colimits in a category based on the cospan construction, whence the graphical counterpart.

Donald Sannella and Andrzej Tarlecki: Observability concepts in abstract data type specification, 30 years later. Last but not least, the paper is ideal for closing our overview, because it presents in a modern fashion the pioneering ideas of Ugo on abstract data type specification [43], commenting upon which we opened this contribution.

References

1. Baldan, P., Bruni, R., Montanari, U.: Pre-nets, read arcs and unfolding: A functorial presentation. In: Wirsing, M., Pattinson, D., Hennicker, R. (eds.) WADT 2003. LNCS, vol. 2755, pp. 145–164. Springer, Heidelberg (2003)
2. Baldan, P., Corradini, A., Montanari, U.: Unfolding and event structure semantics for graph grammars. In: Thomas, W. (ed.) FOSSACS 1999. LNCS, vol. 1578, pp. 73–89. Springer, Heidelberg (1999)
3. Baldan, P., Corradini, A., Montanari, U.: Contextual Petri nets, asymmetric event structures, and processes. Inform. and Comput. 171(1), 1–49 (2001)
4. Baldan, P., Corradini, A., Montanari, U.: Relating SPO and DPO graph rewriting with Petri nets having read, inhibitor and reset arcs. Elect. Notes in Th. Comput. Sci. 127(2), 5–28 (2005)
5. Baldan, P., Corradini, A., Montanari, U., Ribeiro, L.: Unfolding semantics of graph transformation. Inform. and Comput. 205(5), 733–782 (2007)
6. Bonchi, F., König, B., Montanari, U.: Saturated semantics for reactive systems. In: Proc. of LICS 2006, pp. 69–80. IEEE Computer Society Press, Los Alamitos (2006)
7. Bruni, R., Butler, M.J., Ferreira, C., Hoare, C.A.R., Melgratti, H.C., Montanari, U.: Comparing two approaches to compensable flow composition. In: Abadi, M., de Alfaro, L. (eds.) CONCUR 2005. LNCS, vol. 3653, pp. 383–397. Springer, Heidelberg (2005)
8. Bruni, R., de Frutos-Escrig, D., Martí-Oliet, N., Montanari, U.: Bisimilarity congruences for open terms and term graphs via tile logic. In: Palamidessi, C. (ed.) CONCUR 2000. LNCS, vol. 1877, pp. 259–274. Springer, Heidelberg (2000)
9. Bruni, R., Gadducci, F., Montanari, U.: Normal forms for algebras of connections. Theoret. Comput. Sci. 286(2), 247–292 (2002)
10. Bruni, R., Gadducci, F., Montanari, U., Sobocinski, P.: Deriving weak bisimulation congruences from reduction systems. In: Abadi, M., de Alfaro, L. (eds.) CONCUR 2005. LNCS, vol. 3653, pp. 293–307. Springer, Heidelberg (2005)
11. Bruni, R., Lanese, I., Montanari, U.: A basic algebra of stateless connectors. Theoret. Comput. Sci. 366(1-2), 98–120 (2006)
12. Bruni, R., Melgratti, H.C., Montanari, U.: Extending the zero-safe approach to coloured, reconfigurable and dynamic nets. In: Desel, J., Reisig, W., Rozenberg, G. (eds.) ACPN 2003. LNCS, vol. 3098, pp. 291–327. Springer, Heidelberg (2004)
13. Bruni, R., Melgratti, H.C., Montanari, U.: Nested commits for mobile calculi: Extending join. In: Proc. of IFIP TCS 2004, pp. 563–576. Kluwer Academic Publishers, Dordrecht (2004)
14. Bruni, R., Melgratti, H.C., Montanari, U.: Event structure semantics for nominal calculi. In: Baier, C., Hermanns, H. (eds.) CONCUR 2006. LNCS, vol. 4137, pp. 295–309. Springer, Heidelberg (2006)
15. Bruni, R., Melgratti, H.C., Montanari, U.: Event structure semantics for dynamic graph grammars. In: Proc. of PNGT 2006. Elect. Communic. of the EASST, vol. 2, EASST (2007)
16. Bruni, R., Meseguer, J., Montanari, U.: Symmetric monoidal and cartesian double categories as a semantic framework for tile logic. Math. Struct. in Comput. Sci. 12(1), 53–90 (2002)
17. Bruni, R., Meseguer, J., Montanari, U., Sassone, V.: Functorial models for Petri nets. Inform. and Comput. 170(2), 207–236 (2001)

18. Bruni, R., Meseguer, J., Montanari, U., Sassone, V.: Algebraic theories for contextual pre-nets. In: Blundo, C., Laneve, C. (eds.) ICTCS 2003. LNCS, vol. 2841, pp. 256–270. Springer, Heidelberg (2003)
19. Bruni, R., Montanari, U.: Cartesian closed double categories, their lambda-notation, and the pi-calculus. In: Proc. of LICS 1999, pp. 246–265. IEEE Computer Society Press, Los Alamitos (1999)
20. Bruni, R., Montanari, U.: Zero-safe nets: Comparing the collective and individual token approaches. Inform. and Comput. 156(1-2), 46–89 (2000)
21. Bruni, R., Montanari, U.: Dynamic connectors for concurrency. Theoret. Comput. Sci. 281(1–2), 131–176 (2002)
22. Bruni, R., Montanari, U., Rossi, F.: An interactive semantics of logic programming. Theory and Practice of Logic Programming 1(6), 647–690 (2001)
23. Bruni, R., Montanari, U., Sassone, V.: Observational congruences for dynamically reconfigurable tile systems. Theoret. Comput. Sci. 335(2-3), 331–372 (2005)
24. Bruni, R., Sassone, V.: Algebraic models for contextual nets. In: Welzl, E., Montanari, U., Rolim, J.D.P. (eds.) ICALP 2000. LNCS, vol. 1853, pp. 175–186. Springer, Heidelberg (2000)
25. Buscemi, M.G., Montanari, U.: A compositional coalgebraic model of fusion calculus. J. Log. Algebr. Program 72(1), 78–97 (2007)
26. Castellani, I., Montanari, U.: Graph grammars for distributed systems. In: Ehrig, H., Nagl, M., Rozenberg, G. (eds.) Graph Grammars 1982. LNCS, vol. 153, pp. 20–38. Springer, Heidelberg (1983)
27. Coccia, M., Gadducci, F., Montanari, U.: GS-lambda theories: A syntax for higher-order graphs. Elect. Notes in Th. Comput. Sci. 69 (2002)
28. Corradini, A., Ehrig, H., Löwe, M., Montanari, U., Rossi, F.: An event structure semantics for safe graph grammars. In: Pro. of PROCOMET 1994. IFIP Transactions, vol. A-56, pp. 423–444. North-Holland, Amsterdam (1994)
29. Corradini, A., Heckel, R., Montanari, U.: Tile transition systems as structured coalgebras. In: Ciobanu, G., Păun, G. (eds.) FCT 1999. LNCS, vol. 1684, pp. 13–38. Springer, Heidelberg (1999)
30. Corradini, A., Montanari, U.: An algebraic semantics for structured transition systems and its application to logic programs. Theoret. Comput. Sci. 103, 51–106 (1992)
31. Corradini, A., Montanari, U., Rossi, F.: An abstract machine for concurrent modular systems: CHARM. Theoret. Comput. Sci. 122(1–2), 165–200 (1994)
32. Darondeau, P., Degano, P.: Causal trees. In: Ronchi Della Rocca, S., Ausiello, G., Dezani-Ciancaglini, M. (eds.) ICALP 1989. LNCS, vol. 372, pp. 234–248. Springer, Heidelberg (1989)
33. Degano, P., Meseguer, J., Montanari, U.: Axiomatizing the algebra of net computations and processes. Acta Inform. 33(7), 641–667 (1996)
34. Degano, P., Montanari, U.: A model for distributed systems based on graph rewriting. Journal of the ACM 34(2), 411–449 (1987)
35. Ferrari, G.L., Gnesi, S., Montanari, U., Pistore, M.: A model-checking verification environment for mobile processes. ACM Transactions on Software Engineering and Methodology 12(4), 440–473 (2003)
36. Ferrari, G.L., Hirsch, D., Lanese, I., Montanari, U., Tuosto, E.: Synchronised hyperedge replacement as a model for service oriented computing. In: de Boer, F.S., Bonsangue, M.M., Graf, S., de Roever, W.-P. (eds.) FMCO 2005. LNCS, vol. 4111, pp. 22–43. Springer, Heidelberg (2006)
37. Ferrari, G.L., Montanari, U.: Parameterized structured operational semantics. Fundam. Inform 34(1-2), 1–31 (1998)
38. Ferrari, G.L., Montanari, U.: Tile formats for located and mobile systems. Inform. and Comput. 156(1/2), 173–235 (2000)

39. Ferrari, G.L., Montanari, U., Mowbray, M.: Structured transition systems with parametric observations: observational congruences and minimal realizations. Math. Struct. in Comput. Sci. 7(3), 241–282 (1997)

40. Ferrari, G.L., Montanari, U., Tuosto, E.: A LTS semantics of ambients via graph synchronization with mobility. In: Restivo, A., Ronchi Della Rocca, S., Roversi, L. (eds.) ICTCS 2001. LNCS, vol. 2202, pp. 1–16. Springer, Heidelberg (2001)

41. Gadducci, F., Montanari, U.: The tile model. In: Proof, Language and Interaction: Essays in Honour of Robin Milner, pp. 133–166. MIT Press, Cambridge (2000)

42. Gadducci, F., Montanari, U.: Comparing logics for rewriting: rewriting logic, action calculi and tile logic. Theoret. Comput. Sci. 285(2), 319–358 (2002)

43. Giarratana, V., Gimona, F., Montanari, U.: Observability concepts in abstract data type specifications. In: Mazurkiewicz, A. (ed.) MFCS 1976. LNCS, vol. 45, pp. 576–587. Springer, Heidelberg (1976)

44. Gorrieri, R., Marchetti, S., Montanari, U.: A2CCS: Atomic actions for CCS. Theoret. Comput. Sci. 72(2&3), 203–223 (1990)

45. Hirsch, D., Inverardi, P., Montanari, U.: Reconfiguration of software architecture styles with name mobility. In: Porto, A., Roman, G.-C. (eds.) COORDINATION 2000. LNCS, vol. 1906, pp. 148–163. Springer, Heidelberg (2000)

46. Hirsch, D., Montanari, U.: Synchronized hyperedge replacement with name mobility. In: Larsen, K.G., Nielsen, M. (eds.) CONCUR 2001. LNCS, vol. 2154, pp. 121–136. Springer, Heidelberg (2001)

47. König, B., Montanari, U.: Observational equivalence for synchronized graph rewriting with mobility. In: Kobayashi, N., Pierce, B.C. (eds.) TACS 2001. LNCS, vol. 2215, pp. 145–164. Springer, Heidelberg (2001)

48. Lanese, I., Montanari, U.: Hoare vs Milner: Comparing synchronizations in a graphical framework with mobility. Elect. Notes in Th. Comput. Sci. 154(2), 55–72 (2006)

49. Laneve, C., Montanari, U.: Axiomatizing permutation equivalence. Math. Struct. in Comput. Sci. 6(3), 219–249 (1996)

50. Martelli, A., Montanari, U.: An efficient unification algorithm. ACM Transactions on Programming Languages and Systems 4, 258–282 (1982)

51. Meseguer, J.: Conditional rewriting logic as a unified model of concurrency. Theoret. Comput. Sci. 96, 73–155 (1992)

52. Meseguer, J., Montanari, U.: Petri nets are monoids. Inform. and Comput. 88, 105–155 (1990)

53. Meseguer, J., Montanari, U.: Mapping tile logic into rewriting logic. In: Parisi-Presicce, F. (ed.) WADT 1997. LNCS, vol. 1376, pp. 62–91. Springer, Heidelberg (1998)

54. Meseguer, J., Montanari, U., Sassone, V.: Process versus unfolding semantics for place/transition Petri nets. Theoretical Computer Science 153(1–2), 171–210 (1996)

55. Meseguer, J., Montanari, U., Sassone, V.: On the semantics of place/transition Petri nets. Math. Struct. in Comput. Sci. 7(4), 359–397 (1997)

56. Meseguer, J., Montanari, U., Sassone, V.: Representation theorems for Petri nets. In: Freksa, C., Jantzen, M., Valk, R. (eds.) Foundations of Computer Science. LNCS, vol. 1337, pp. 239–249. Springer, Heidelberg (1997)

57. Montanari, U., Pistore, M.: Structured coalgebras and minimal HD-automata for the pi-calculus. Theoret. Comput. Sci. 340(3), 539–576 (2005)

58. Rossi, F., Montanari, U.: Graph rewriting, constraint solving and tiles for coordinating distributed systems. Applied Categorical Structures 7(4), 333–370 (1999)

Automatic Mutual Exclusion
and Atomicity Checks

Martín Abadi[1,2]

[1] Microsoft Research
[2] University of California, Santa Cruz

Abstract. This paper provides an introduction to the Automatic Mutual Exclusion (AME) programming model and to its formal study, through the AME calculus. AME resembles cooperative multithreading; in the intended implementations, however, software transactional memory supports the concurrent execution of atomic fragments. This paper also studies simple dynamic and static mechanisms for atomicity checks in AME.

1   Introduction

Transactions promise a practical mechanism for synchronization that should facilitate the design and coding of a wide range of concurrent systems. In particular, in shared-memory concurrency, systems based on transactions may achieve the efficiency of fine-grained locking while reducing the opportunities for deadlocks, race conditions, and other bugs. For these benefits to be realized, however, advances in low-level implementations of transactions do not suffice. Also needed are corresponding languages and programming techniques (e.g., [9,10,6,3]).

The principle "Lo bueno, si breve, dos veces bueno" does not necessarily apply to transactions. Although long-running transactions can lead to excessive conflicts and may complicate hardware-based implementation strategies, they also support a conservative style of programming in which transactions, with their guarantees, are the default. This style is embodied, in particular, in the Automatic Mutual Exclusion (AME) model [12,1], which is the focus of this paper.

AME can be seen as cooperative multithreading on top of software transactional memory (STM) [14]. In the spirit of cooperative multithreading, calls to the construct `yield` delimit atomic fragments of computations. STM allows multiple sequential code fragments to execute at the same time, each within a transaction.

Yielding requires care. For instance, consider a call to a library method made from within a transaction. As long as the execution remains within the same transaction, the caller need not be concerned with concurrent calls to the library or any other concurrent activity. On the other hand, the library method may decide to interrupt the transaction by yielding, perhaps in order to interact with the outside world. In this case, the caller may need to consider interleavings of other computations, restoring invariants if necessary. In this paper we explore a mechanism for asserting that, dynamically, yielding should not happen in a particular piece of code. Yielding can be turned into (caught) run-time errors,

P. Degano et al. (Eds.): Montanari Festschrift, LNCS 5065, pp. 510–526, 2008.

and transactional recovery may optionally mask those errors altogether. We also define and study a simple static type system that indicates whether yielding is possible in a piece of code. A practical version of this type system has been implemented for an extension of C# on Bartok-STM [11].

In sum, the goals of this paper are to provide an introduction to AME and to its formal study (largely as a review of recent work [12,1]), and also to advance a specific aspect of AME and its theory. Section 2 describes AME, informally. Sections 3 and 4 define the AME calculus and its high-level formal semantics. Section 5 and 6 concern dynamic atomicity checks and the static type system, respectively. Section 7 establishes the soundness of the static type system with respect to the dynamic atomicity checks. Section 8 concludes by mentioning some further work. An appendix contains proofs.

Similar themes have been explored in other projects. For instance, in the Mianjin language, type annotations distinguish routines that may perform communication [13]. More recently (independently from the AME work), the model Transactions with Isolation and Cooperation (TIC) includes a type system for atomicity [15]. In both Mianjin and TIC, the type systems are defined semiformally. Further, other research on types for atomicity offers powerful analyses that apply to Java and similar languages [7]. While some of their ideas may be useful in implementations of AME, they may be less necessary at the AME source level, because of the reliance on cooperation and transactions. In another direction, research on sagas explores techniques that reconcile atomicity and responsiveness for long-lived transactions, with sophisticated treatments of nesting, parallelism, and compensation (which are beyond the scope of the present paper) [8,5]. Finally, research on cooperative multithreading includes techniques for proving that yielding must eventually happen, guaranteeing fairness in single-threaded implementations [4].

2 Automatic Mutual Exclusion

AME encourages programmers to use transactions: code is executed in transactions by default. The intent is that the pervasive use of transactions will lead to clearer programs with fewer synchronization bugs. However, for interactions with legacy components and other computations that should not be placed in transactions, code can be marked explicitly as "unprotected".

In AME, running a program consists of executing a set of asynchronous method calls. The semantics of AME guarantees that the program execution is equivalent to executing each of these calls (or their fragments, as explained below) in some serial order. An asynchronous method call is created by an invocation `async MethodName(<args>)`. The caller continues immediately after this invocation. AME achieves concurrency by executing asynchronous method calls in transactions, overlapping the execution of multiple calls, with roll-backs when conflicts occur. If a transaction initiates other asynchronous method calls, their execution is deferred until the initiating transaction commits, and they are discarded if the initiating transaction aborts.

$$
\begin{aligned}
V \in \textit{Value} \ &= c \mid x \mid \lambda x.\, e \\
c \in \textit{Const} \ &= \texttt{unit} \mid \texttt{false} \mid \texttt{true} \\
x, y \in \textit{Var} \ & \\
e, f \in \textit{Exp} \ \ &= V \\
&\mid e\, f \\
&\mid \texttt{ref}\ e \mid\ !e \mid e := f \\
&\mid \texttt{async}\ e \\
&\mid \texttt{yield} \\
&\mid \texttt{blockUntil}\ e
\end{aligned}
$$

Fig. 1. Syntax of the AME calculus (without unprotected sections)

An asynchronous method call may also invoke `yield`. A `yield` call breaks a method into multiple atomic fragments, implemented by committing one transaction and starting a new one. These atomic fragments are delimited dynamically by the calls to `yield`, not statically scoped like explicit atomic blocks [7,9]. AME thus avoids some of the pitfalls of pure event-based programming models (in particular, "stack ripping" [2].). With this addition, the overall execution of a program is guaranteed to be a serialization of its atomic fragments.

An atomic fragment may include any number of guards, each of the form `blockUntil(<predicate>)`. An atomic fragment executes to completion only if all the guards encountered in the course of the execution have predicates that evaluate to true. The implementation of `blockUntil` does nothing if the predicate holds, but otherwise it aborts the current atomic fragment and re-executes it later (at a time when it is likely to succeed).

As indicated above, AME provides block-structured **unprotected** sections. We omit them here, for simplicity. It is straightforward to extend the results of this paper to them, although the semantics of **unprotected** sections can be delicate.

3   The AME Calculus

In our formal study of AME, we focus on a small but expressive language that we call the AME calculus. This calculus includes constructs for AME, higher-order functions, and imperative features.

In Figure 1 we define the syntax of the AME calculus, omitting unprotected sections. This syntax is untyped; we define a type system in Section 6. The syntax introduces syntactic categories of values, constants, variables, and expressions. The values are constants, variables, and lambda abstractions ($\lambda x.\, e$). In addition to values and to expressions of the forms `async` e, `blockUntil` e, and `yield`, the expressions include notations for function application (ef), allocation (`ref` e, which allocates a new reference location and returns it after initializing it to the value of e), dereferencing (!e, which returns the contents in the reference location that is the value of e), and assignment ($e := f$, which sets the reference location that is the value of e to the value of f).

We write $\mathtt{let}\ x = e\ \mathtt{in}\ e'$ for $(\lambda x.\ e')\ e$, and also write $e; e'$ for $\mathtt{let}\ x = e\ \mathtt{in}\ e'$ when x does not occur free in e'. Including standard control structures and other common constructs (directly or by encodings) is routine.

As a small example, let us consider the following code fragment:

```
blockUntil !r₀;
r₁ := e₁;
r₂ := e₂;
async (r₃ := e₃);
yield
```

in which r_0, r_1, r_2, and r_3 are variables that represent reference locations, and e_1, e_2, and e_3 are arbitrary expressions. This code fragment blocks until r_0 holds $\mathtt{true}$, then it performs assignments to r_1 and r_2, forks an expression that will perform an assignment to r_3, and finally yields.

Intuitively, a programmer may expect that the assignments to r_1 and r_2 (but not r_3) happen within the same transaction, and this property will indeed hold if e_1 and e_2 are simple values. However, in general, the evaluations of e_1 and e_2 may trigger calls to $\mathtt{yield}$, so the assignments may happen in different transactions. For instance, e_2 might be a call to a function with body $\mathtt{yield}; (\mathtt{blockUntil}\ !r_4); !r_5$, which yields, waits until the value in r_4 is true, and then returns the value in r_5. In that case, some other thread may execute between the assignments, may observe inconsistent values in r_1 and r_2, and may misbehave as a result. Therefore, it is useful to have dynamic or static means of guaranteeing that expressions such as e_1 and e_2 do not yield. Sections 5 and 6 address this goal.

4   High-Level Semantics

This section presents a semantics for the AME calculus. This semantics is intended to provide a clear, high-level model, rather than a description of possible underlying implementation techniques. Accordingly, the semantics does not model optimistic concurrency, conflict detection, roll-back, and other important low-level features. In [1] we consider richer and weaker semantics that add these features. Those weaker semantics implement the high-level semantics—though under non-trivial assumptions that restrict the sharing of data between transactions and unprotected code.

4.1   States

As described in Figure 2, a state $\langle \sigma, T, e \rangle$ consists of the following components:

- a reference store σ,
- a collection of expressions T, which we call the pool,
- a distinguished active expression e.

$$\begin{aligned}
S &\in \quad State = RefStore \times ExpSeq \times Exp \\
\sigma &\in RefStore = RefLoc \rightharpoonup Value \\
r &\in \quad RefLoc \subset Var \\
T &\in \quad ExpSeq = Exp^*
\end{aligned}$$

Fig. 2. State space

A reference store σ is a finite mapping of reference locations to values. Formally, reference locations are special kinds of variables that can be bound only by a reference store. We write $RefLoc$ for the set of reference locations. We assume that $RefLoc$ is infinite, so $RefLoc - dom(\sigma)$ is never empty. For every state $\langle \sigma, T, e \rangle$, we require that if $r \in RefLoc$ occurs free in $\sigma(r')$, in T, or in e, then $r \in dom(\sigma)$. This condition will be assumed for initial states and will be preserved by computation steps.

4.2   Steps

A transition relation takes an execution from one state to the next. According to this transition relation, when the active expression is `unit`, an expression from the pool becomes the active expression. It is then evaluated as such until it produces `unit` or until it yields. No other computation is interleaved with this evaluation. Each evaluation step produces a new state. Unless the active expression is `unit`, this new state is obtained by decomposing the active expression into an evaluation context and a subexpression that describes an operation (for instance, a function application or an allocation).

As usual, a context is an expression with a hole $[\,]$, and an evaluation context is a context of a particular kind. Given a context $\mathcal{C}$ and an expression e, we write $\mathcal{C}[\,e\,]$ for the result of placing e in the hole in $\mathcal{C}$. We use the evaluation contexts defined by the grammar:

$$\mathcal{P} = [\,]\ |\ \mathcal{P}\ e\ |\ V\ \mathcal{P}\ |\ \texttt{ref}\ \mathcal{P}\ |\ !\mathcal{P}\ |\ \mathcal{P} := e\ |\ r := \mathcal{P}\ |\ \texttt{blockUntil}\ \mathcal{P}$$

Figure 3 gives rules that specify the transition relation. The string "Trans" in the names of the rules refers to "transition" rules, not to "transaction". In these rules, we write $e[V/x]$ for the result of the capture-free substitution of V for x in e, and write $\sigma[r \mapsto V]$ for the store that agrees with σ except at r, which is mapped to V.

(Trans Activate) applies when the active expression is `unit` and the pool is not empty; it takes an expression from the pool as the new active expression. In all other rules, a subexpression in an evaluation context in the active expression determines a possible next operation. For instance, in (Trans Appl), the subexpression is a function application $(\lambda x.\, e)\, V$, so the next operation is beta reduction, and the result $e[V/x]$ of this beta reduction replaces $(\lambda x.\, e)\, V$ in the evaluation context. Similarly, in (Trans Yield), the subexpression is `yield`, so

$$\langle \sigma, T, \mathcal{P}[\ (\lambda x.\, e)\ V\ ]\rangle \quad\longmapsto\ \langle \sigma, T, \mathcal{P}[\ e[V/x]\ ]\rangle \qquad \text{(Trans Appl)}$$

$$\langle \sigma, T, \mathcal{P}[\ \texttt{ref}\ V\ ]\rangle \quad\longmapsto\ \langle \sigma[r \mapsto V], T, \mathcal{P}[\ r\ ]\rangle \qquad \text{(Trans Ref)}$$
$$\text{if } r \in \mathit{RefLoc} - \mathit{dom}(\sigma)$$

$$\langle \sigma, T, \mathcal{P}[\ !r\ ]\rangle \quad\longmapsto\ \langle \sigma, T, \mathcal{P}[\ V\ ]\rangle \qquad \text{(Trans Deref)}$$
$$\text{if } \sigma(r) = V$$

$$\langle \sigma, T, \mathcal{P}[\ r := V\ ]\rangle \quad\longmapsto\ \langle \sigma[r \mapsto V], T, \mathcal{P}[\ \texttt{unit}\ ]\rangle \quad \text{(Trans Set)}$$

$$\langle \sigma, T, \mathcal{P}[\ \texttt{async}\ e\ ]\rangle \quad\longmapsto\ \langle \sigma, T.e, \mathcal{P}[\ \texttt{unit}\ ]\rangle \qquad \text{(Trans Async)}$$

$$\langle \sigma, T, \mathcal{P}[\ \texttt{blockUntil true}\ ]\rangle \longmapsto \langle \sigma, T, \mathcal{P}[\ \texttt{unit}\ ]\rangle \qquad \text{(Trans Block)}$$

$$\langle \sigma, T, \mathcal{P}[\ \texttt{yield}\ ]\rangle \quad\longmapsto\ \langle \sigma, T.\mathcal{P}[\ \texttt{unit}\ ], \texttt{unit}\rangle \qquad \text{(Trans Yield)}$$

$$\langle \sigma, T.e.T', \texttt{unit}\rangle \quad\longmapsto\ \langle \sigma, T.T', e\rangle \qquad \text{(Trans Activate)}$$

Fig. 3. Transition rules of the abstract machine

---

unit replaces yield, the active expression is moved to the pool, and the new active expression is unit. No rule applies in some cases, for instance when the active expression is blockUntil false. Lower-level semantics may abort and roll-back in such cases [1].

These rules are more compact than previous ones, simply because of the omission of unprotected computations. Further variants are possible. In particular, we may consider adding the rule:

$$\langle \sigma, T, \mathcal{P}[\ \texttt{yield}\ ]\rangle \longmapsto \langle \sigma, T, \mathcal{P}[\ \texttt{unit}\ ]\rangle$$

This rule represents a short-cut: it can be derived by composing (Trans Yield) and (Trans Activate).

5   Dynamic Atomicity Checks

We extend the calculus with a construct that asserts the absence of yielding in a computation. We focus on the high-level semantics of Section 4, though similar extensions and corresponding results can be obtained for other semantics.

The extension goes as follows:

- We extend the syntax of the language with terms of the form $\langle e \rangle$. Informally, $\langle e \rangle$ means that there should be no yield in the course of the evaluation of e. (This notation is inspired by Lamport's angle brackets, which also indicate atomicity.)

- We also extend the evaluation contexts, so that evaluation can proceed under $\langle \cdot \rangle$. Their grammar becomes:

$$\mathcal{P} = [\,] \mid \mathcal{P}\ e \mid V\ \mathcal{P} \mid \texttt{ref}\ \mathcal{P} \mid !\mathcal{P} \mid \mathcal{P} := e \mid r := \mathcal{P} \mid \texttt{blockUntil}\ \mathcal{P} \mid \langle \mathcal{P} \rangle$$

- We extend all the rules of the operational semantics to these terms and these evaluation contexts, and also add a rule to the operational semantics:

$$\langle \sigma, T, \mathcal{P}[\ \langle V \rangle\ ]\rangle \longmapsto \langle \sigma, T, \mathcal{P}[\ V\ ]\rangle \quad \text{(Trans Assert)}$$

Given that $\langle e \rangle$ asserts that there is no yield in the course of the evaluation of e, this rule says that the assertion can be dismissed when e is a value V (not subject to further evaluation).

These extensions are conservative, in the sense that they affect neither the operational semantics nor the typing (in the type system of Section 6) of expressions without assertions. Therefore, some of the main results below (Theorems 1 and 3) apply also without the extensions.

Consider a transition that is an instance of (Trans Yield), so this transition is of the form:

$$\langle \sigma, T, \mathcal{P}[\ \texttt{yield}\ ]\rangle \longmapsto \langle \sigma, T.\mathcal{P}[\ \texttt{unit}\ ], \texttt{unit}\rangle$$

for some σ, T, and $\mathcal{P}$. We say that this transition is an atomicity violation if $\mathcal{P}$ is of the form $\mathcal{P}'[\ \langle \mathcal{P}'' \rangle\ ]$, for instance if $\mathcal{P}[\ \texttt{yield}\ ]$ is $\langle \texttt{yield} \rangle$ or $!\langle \texttt{ref yield} \rangle$.

What should we do with an atomicity violation? There are at least three distinct possibilities:

1. Continue the computation despite the atomicity violation; in this case, the main use of $\langle \cdot \rangle$ is as a marker that allows us to explain what went wrong. The present definition of the operational semantics embodies this possibility. Accordingly, the results below concern this possibility as well.
2. Stop the computation, allowing for recovery.
   Formally, it would suffice to remove the transitions that constitute atomicity violations, with the understanding that any computation that has not committed may be rolled back, and perhaps retried later. Specifically, we would restrict (Trans Yield) to:

$$\langle \sigma, T, \mathcal{P}[\ \texttt{yield}\ ]\rangle \longmapsto \langle \sigma, T.\mathcal{P}[\ \texttt{unit}\ ], \texttt{unit}\rangle$$
$$\text{if } \mathcal{P} \text{ is not of the form } \mathcal{P}'[\ \langle \mathcal{P}'' \rangle\ ]$$

   Thus, $\langle \mathcal{P}''[\texttt{yield}] \rangle$ would be analogous to $\texttt{blockUntil false}$.
3. Stop the computation with a fatal error.
   Formally, we could add a special state $\textbf{wrong}$ that would represent errors, and change the operational semantics for producing errors instead of allowing atomicity violations. Specifically, we would restrict (Trans Yield), as above, and add:

$$\langle \sigma, T, \mathcal{P}'[\ \langle \mathcal{P}''[\texttt{yield}] \rangle\ ]\rangle \longmapsto \textbf{wrong}$$

With all these options, it is attractive to prove that, for some class of good programs, atomicity violations are not possible. The next section provides a type system for this purpose.

$$s, t \in \mathit{Type} \;=\; \texttt{Unit}$$
$$\mid\; \texttt{Bool}$$
$$\mid\; s \rightarrow^p t$$
$$\mid\; \texttt{Ref}\; t$$
$$p, q \in \{\texttt{Yields}, \texttt{NoYields}\}$$

Fig. 4. Types for yielding

6   Static Atomicity Checks

This section defines a simple type system for atomicity checking. This type system can be seen as an alternative to the dynamic approach described above in Section 5. However, the two approaches may be combined; moreover, the dynamic approach is useful for formulating the correctness of the static approach (in Section 7).

The type system is based on the syntax of types of Figure 4, and is defined in terms of formal judgments:

$$E \vdash \diamond \qquad\qquad E \text{ is a well-formed typing environment}$$
$$E\,; p \vdash e : t \qquad e \text{ is a well-typed expression of type } t \text{ in } E \text{ with } p$$

The typing rules of Figure 5 operate on these judgments.

The type of an expression depends on a typing environment E, which maps variables to types. The typing environment is organized as a sequence of bindings, and we use $\emptyset$ to denote the empty environment:

$$E ::= \emptyset \mid E, x : t$$

The core of the type system is the set of rules for the judgment $E\,; p \vdash e : t$ (read "e is a well-typed expression of type t in typing environment E with effect p"). The intent is that, if this judgment holds, then e yields values of type t with effect p, and the free variables of e are given bindings consistent with the typing environment E. When p is $\texttt{Yields}$, this means that the evaluation of e may yield; when p is $\texttt{NoYields}$, this means that the evaluation of e definitely does not yield. We require that $\langle \cdot \rangle$ appears only around expressions with effect $\texttt{NoYields}$. We write $q <: p$ for $p = q$ or $p = \texttt{Yields}$. We say that e is well-typed when there exist E, p, and t such that $E\,; p \vdash e : t$.

As a design choice, we arrange that every expression that can be typed with effect $\texttt{NoYields}$ can also be typed with effect $\texttt{Yields}$. For instance, we allow giving the effect $\texttt{Yields}$ to the constant $\texttt{true}$, although the evaluation of $\texttt{true}$ will obviously never yield. This property ensures that effects are not invalidated by computation. For example, consider the expression $\texttt{yield}; \texttt{true}$, which has effect $\texttt{Yields}$ and produces the result $\texttt{true}$. Because $\texttt{true}$ has effects $\texttt{NoYields}$ and also $\texttt{Yields}$, the effect of $\texttt{yield}; \texttt{true}$ continues to be derivable after reduction to $\texttt{true}$.

$$\emptyset \vdash \diamond \qquad\qquad (\text{Env } \emptyset)$$

$$\frac{E \vdash \diamond \quad x \notin dom(E)}{E, x : t \vdash \diamond} \qquad\qquad (\text{Env } x)$$

$$\frac{E \vdash \diamond}{E\,;p \vdash \texttt{unit} : \texttt{Unit}} \qquad\qquad (\text{Exp Unit})$$

$$\frac{E \vdash \diamond}{E\,;p \vdash \texttt{false} : \texttt{Bool}} \qquad\qquad (\text{Exp Bool }\texttt{false})$$

$$\frac{E \vdash \diamond}{E\,;p \vdash \texttt{true} : \texttt{Bool}} \qquad\qquad (\text{Exp Bool }\texttt{true})$$

$$\frac{E, x : t, E' \vdash \diamond}{E, x : t, E'\,;p \vdash x : t} \qquad\qquad (\text{Exp } x)$$

$$\frac{E, x : s\,;p \vdash e : t}{E\,;q \vdash \lambda x.\, e : s \to^{p} t} \qquad\qquad (\text{Exp Fun})$$

$$\frac{E\,;p \vdash e_1 : s \to^{q} t \quad E\,;p \vdash e_2 : s \quad q <: p}{E\,;p \vdash e_1\ e_2 : t} \qquad\qquad (\text{Exp Appl})$$

$$\frac{E\,;p \vdash e : t}{E\,;p \vdash \texttt{ref}\ e : \texttt{Ref}\ t} \qquad\qquad (\text{Exp Ref})$$

$$\frac{E\,;p \vdash e : \texttt{Ref}\ t}{E\,;p \vdash {!}e : t} \qquad\qquad (\text{Exp Deref})$$

$$\frac{E\,;p \vdash e_1 : \texttt{Ref}\ t \quad E\,;p \vdash e_2 : t}{E\,;p \vdash e_1 := e_2 : \texttt{Unit}} \qquad\qquad (\text{Exp Set})$$

$$\frac{E\,;p \vdash e : \texttt{Unit}}{E\,;q \vdash \texttt{async}\ e : \texttt{Unit}} \qquad\qquad (\text{Exp Async})$$

$$\frac{E\,;p \vdash e : \texttt{Bool}}{E\,;p \vdash \texttt{blockUntil}\ e : \texttt{Unit}} \qquad\qquad (\text{Exp Block})$$

$$\frac{E \vdash \diamond}{E\,;\texttt{Yields} \vdash \texttt{yield} : \texttt{Unit}} \qquad\qquad (\text{Exp Yield})$$

$$\frac{E\,;\texttt{NoYields} \vdash e : t}{E\,;p \vdash \langle e \rangle : t} \qquad\qquad (\text{Exp Assert})$$

Fig. 5. Rules of the first-order type system for yielding

There are alternative methods for achieving the same effect. These include the use of a system with subtyping, which would also provide more flexibility at function types. The present method is simpler and enables us to focus on the core system. Undoubtedly richer type disciplines are possible.

7   Soundness

Intuitively, the correctness of the type system is the property that says that if an expression has effect `NoYields` statically then it does not yield at run-time. However, in the course of evaluation, the expression may change, and that should not be an excuse for yielding. So it is convenient to tag the expression, and to keep the tag on the expression even if the expression changes until its evaluation completes. The angle brackets of Section 5 serve as such a tag.

As a first step in the soundness proof, we generalize the type system to states $\langle \sigma, T, e \rangle$. We write

$$E \, ; p_1. \cdots .p_n, p \vdash \langle \sigma, e_1. \cdots .e_n, e \rangle$$

if

- $dom(\sigma) = dom(E) \cap RefLoc$,
- for all $r \in dom(\sigma)$, there exists t such that $E(r) = \mathtt{Ref}\ t$ and $E \, ; \mathtt{NoYields} \vdash \sigma(r) : t$,
- $E \, ; p_i \vdash e_i : \mathtt{Unit}$ for all $i = 1..n$,
- $E \, ; p \vdash e : \mathtt{Unit}$.

The first condition relates the domains of σ and E. The second one says that E assigns types of the appropriate form to reference locations, and that σ maps these reference locations to expressions of appropriate types, with effect `NoYields` (because these expressions must be values). The remaining conditions require typing the expressions e_1, ..., e_n, and e.

We say that $\langle \sigma, e_1. \cdots .e_n, e \rangle$ is well-typed if there exist E and p_1, ..., p_n, p such that $E \, ; p_1. \cdots .p_n, p \vdash \langle \sigma, e_1. \cdots .e_n, e \rangle$.

We obtain that typability is preserved by computation:

Theorem 1 (Preservation of Typability). *If $\langle \sigma, T, e \rangle \longmapsto^* \langle \sigma', T', e' \rangle$ and $\langle \sigma, T, e \rangle$ is well-typed, then so is $\langle \sigma', T', e' \rangle$.*

Partly as a corollary, we also obtain a result that expresses the correctness of `NoYields`:

Theorem 2 (Atomicity Soundness). *If $\langle \sigma, T, e \rangle \longmapsto^* \langle \sigma', T', e' \rangle$ and $\langle \sigma, T, e \rangle$ is well-typed, then none of the transitions in $\langle \sigma, T, e \rangle \longmapsto^* \langle \sigma', T', e' \rangle$ is an atomicity violation.*

Moreover, we obtain a progress result, which characterizes when a computation may stop and implies that computations do not get stuck in unexpected ways:

Theorem 3 (Progress). *If $\langle \sigma, T, e \rangle$ is well-typed, the only free variables in $\langle \sigma, T, e \rangle$ are reference locations, and $\langle \sigma, T, e \rangle \longmapsto^* \langle \sigma', T', e' \rangle$, then:*

1. *there exists* $\langle \sigma'', T'', e'' \rangle$ *such that* $\langle \sigma', T', e' \rangle \longmapsto \langle \sigma'', T'', e'' \rangle$; *or*
2. e' *is of the form* $\mathcal{P}[$ `blockUntil false` $]$; *or*
3. e' *is* `unit` *and* T' *is empty.*

The proofs of these three theorems are in an appendix.

8   Further Work

This paper provides an introduction to the AME programming model and advances one aspect of its development and formal study. We conclude with a brief description of other recent and ongoing work on this model.

To date, we have only limited experience in programming in the AME model. While this experience is rather encouraging, further experience may conceivably lead to refinements in the constructs for AME. For instance, we have briefly considered expressive generalizations of `yield`. In any case, it seems likely that the need for atomicity checking will persist.

The semantics presented in this paper is a high-level description of the intended meanings of the AME constructs. Lower-level semantics embody various strategies for the implementation of these constructs. For instance, those lower-level semantics can include optimistic concurrent execution of transactions, with in-place updates to memory, conflict detection, and roll-backs [11]. In particular, the implementation of AME for C# on Bartok-STM relies on these features. Such strategies may have great advantages in performance and responsiveness, but they can lead to surprising results. We have therefore worked on describing those strategies precisely and on analyzing their properties in detail [1]. The correctness of these strategies require substantial assumptions which say, roughly, that transactional and non-transactional computations do not share data directly. Several versions of these assumptions lead to correctness results, though with different specifics. Some of these versions, and the corresponding trade-offs, are the subject of ongoing work.

Acknowledgements

This paper is based on the original work on AME by Michael Isard and Andrew Birrell, and on further, ongoing joint work with Tim Harris and Johnson Hsieh. Dan Grossman made useful comments on the type system of Section 6. I am grateful to all of them.

References

1. Abadi, M., Birrell, A., Harris, T., Isard, M.: Semantics of transactional memory and automatic mutual exclusion. In: Proc. 35th ACM SIGPLAN-SIGACT Symposium on Principles of Programming Languages, pp. 63–74 (2008)
2. Adya, A., Howell, J., Theimer, M., Bolosky, W.J., Douceur, J.R.: Cooperative task management without manual stack management. In: Proc. 2002 USENIX Annual Technical Conference, pp. 289–302 (2002)

3. Allen, E., Chase, D., Hallett, J., Luchangco, V., Maessen, J.-W., Ryu, S., Steele, G.L., Jr., Tobin-Hochstadt, S.: The Fortress language specification, v1.0β. Technical report, Sun Microsystems (March 2007)

4. Boudol, G.: Fair cooperative multithreading. In: Caires, L., Vasconcelos, V.T. (eds.) CONCUR 2007. LNCS, vol. 4703, pp. 272–286. Springer, Heidelberg (2007)

5. Bruni, R., Melgratti, H.C., Montanari, U.: Theoretical foundations for compensations in flow composition languages. In: Proc. 32nd ACM SIGPLAN-SIGACT Symposium on Principles of Programming Languages, pp. 209–220 (2005)

6. Carlstrom, B.D., McDonald, A., Chafi, H., Chung, J., Minh, C.C., Kozyrakis, C., Olukotun, K.: The Atomos transactional programming language. In: PLDI 2006: Proc. 2006 ACM SIGPLAN Conference on Programming Language Design and Implementation, pp. 1–13 (2006)

7. Flanagan, C., Qadeer, S.: A type and effect system for atomicity. In: Proc. 2003 ACM SIGPLAN Conference on Programming Language Design and Implementation, pp. 338–349 (2003)

8. Garcia-Molina, H., Salem, K.: Sagas. In: Proc. ACM SIGMOD 1987 Annual Conference, pp. 249–259 (1987)

9. Harris, T., Fraser, K.: Language support for lightweight transactions. In: Proc. 18th Annual ACM SIGPLAN Conference on Object-Oriented Programming, Systems, Languages, and Applications, pp. 388–402 (2003)

10. Harris, T., Marlow, S., Peyton-Jones, S., Herlihy, M.: Composable memory transactions. In: PPoPP 2005: Proc. 10th ACM SIGPLAN Symposium on Principles and Practice of Parallel Programming, pp. 48–60 (2005)

11. Harris, T., Plesko, M., Shinnar, A., Tarditi, D.: Optimizing memory transactions. In: Proc. 2006 ACM SIGPLAN Conference on Programming Language Design and Implementation, pp. 14–25 (2006)

12. Isard, M., Birrell, A.: Automatic mutual exclusion. In: Proc. 11th Workshop on Hot Topics in Operating Systems (2007)

13. Roe, P., Szyperski, C.A.: Mianjin: A parallel language with a type system that governs global system behaviour. In: Weck, W., Gutknecht, J. (eds.) JMLC 2000. LNCS, vol. 1897, pp. 38–50. Springer, Heidelberg (2000)

14. Shavit, N., Touitou, D.: Software transactional memory. In: Proc. 14th Annual ACM Symposium on Principles of Distributed Computing, pp. 204–213 (1995)

15. Smaragdakis, Y., Kay, A., Behrends, R., Young, M.: Transactions with isolation and cooperation. In: Proc. 22nd Annual ACM SIGPLAN Conference on Object-Oriented Programming, Systems, Languages, and Applications, pp. 191–210 (2007)

16. Andrew, K.: Wright and Matthias Felleisen. A syntactic approach to type soundness 115(1), 38–94 (1994)

Appendix: Proofs

Auxiliary Results. We rely on a few auxiliary results. Several of them are routine, and we omit the corresponding proofs. These include a replacement lemma (in the style of Wright and Felleisen [16]), a substitution lemma, and a lemma that deals with updates to the state.

Lemma 1 (Replacement). *Consider a derivation $\mathcal{D}$ of $E\,;p \vdash \mathcal{P}[\ e_0\ ] : t$. Assume that this derivation includes, as a subderivation, a proof $\mathcal{D}_0$ of the judgment $E\,;p_0 \vdash e_0 : t_0$ for the occurrence of e_0 in $\mathcal{P}[\ \cdot\ ]$. Assume that we also* *have a derivation $\mathcal{D}'_0$ of $E\,;p_0 \vdash e'_0 : t_0$ for some e'_0. Let $\mathcal{D}'$ be obtained from $\mathcal{D}$ by replacing $\mathcal{D}_0$ with $\mathcal{D}'_0$, and e_0 with e'_0 in $\mathcal{P}$. Then $\mathcal{D}'$ is a derivation of $E\,;p \vdash \mathcal{P}[\,e'_0\,] : t$.*

Lemma 2 (Substitution). *If $E, x : s, E'\,;p \vdash e : t$ and $E\,;\mathtt{NoYields} \vdash e' : s$ then $E, E'\,;p \vdash e[e'/x] : t$.*

Lemma 3 (Update). *Assume that $r \in dom(\sigma)$ and $E(r) = \mathtt{Ref}\ t_0$. If $E\,;p_1.$ $\cdots .p_n, p \vdash \langle \sigma, e_1. \cdots .e_n, e \rangle$ and $E\,;\mathtt{NoYields} \vdash V : t_0$, then $E\,;p_1. \cdots .p_n, p \vdash \langle \sigma[r \mapsto V], e_1. \cdots .e_n, e \rangle$.*

The remaining lemmas are more specific to our study, so we outline their proofs. They say that values can be typed as not yielding, if they can be typed at all; that expressions that do not yield may be seen as yielding; and that `yield` can never appear in an evaluation context when the type system does not indicate yielding. They also provide an analysis of the possible forms of well-typed expressions.

Lemma 4. *If $E\,;p \vdash V : t$ then $E\,;\mathtt{NoYields} \vdash V : t$.*

This lemma holds simply because, in all the rules that can be used as the last one for typing a value ((Exp `unit`), (Exp `false`), (Exp `true`), (Exp x), and (Exp Fun)), the type system leaves the choice of effect completely unconstrained.

Lemma 5. *If $E\,;\mathtt{NoYields} \vdash e : t$ then $E\,;\mathtt{Yields} \vdash e : t$.*

The proof of Lemma 5 is by induction on the derivation of $E\,;\mathtt{NoYields} \vdash e : t$, with a case analysis on which rule is applied last. No rule forces a conclusion with `NoYields`: some rules where the conclusion may have effect `NoYields` (like (Exp Async) and (Exp Assert)) leave the choice of effect unconstrained, while others (like (Exp Appl) and (Exp Ref)) propagate the effect used in the hypotheses of the rule application. In the latter case, `Yields` can be used instead of `NoYields` also in the hypotheses of the rule application, by induction hypothesis and, in the case of (Exp Appl), because $q <: \mathtt{Yields}$ always holds.

Lemma 6. *It is never the case that $E\,;\mathtt{NoYields} \vdash \mathcal{P}[\,\mathtt{yield}\,] : t$.*

The proof of Lemma 6 is by induction on typing derivations, with a case analysis on which rule is applied last.

- The cases of (Exp Unit), (Exp Bool `false`), (Exp Bool `true`), (Exp x), and (Exp Fun) are trivial, since the expressions typed there are values and cannot be the one in question.
- The case for (Exp Yield) is trivial because it gives an effect `Yields`.
- The cases of (Exp Appl), (Exp Ref), (Exp Deref), (Exp Set), and (Exp Block) are all by applications of the induction hypothesis, which are possible because the effects in the hypotheses of the rules are the same as the effects in their conclusions.
- The case for (Exp Async) is excluded because a context $\mathcal{P}$ cannot be of the form `async` $\mathcal{P}'$, so in this case $\mathcal{P}$ must be [ ], and `async` · cannot match `yield`.

- The case for (Exp Assert) is by application of the induction hypothesis, since the effects in the hypothesis of the rule is `NoYields`.

Lemma 7. *Suppose that e is a well-typed expression in which the only free variables are reference locations (with types of the form* `Ref` t*). Then e is a value or an expression of the form* $\mathcal{P}[\, f \,]$*, where f has one of the forms* $(\lambda x.\, e')\, V$*,* `ref` V*,* $!r$*, $r := V$,* `async` e'*,* `blockUntil true`*,* `blockUntil false`*,* `yield`*, and* $\langle V \rangle$*.*

The proof of Lemma 7 is by induction on the typing of e, with a case analysis on the last rule in the typing derivation.

- In the cases of (Exp Unit), (Exp Bool `false`), (Exp Bool `true`), (Exp x), and (Exp Fun), e is a value.
- In the case of (Exp Appl), e cannot be a value. If $e_1\, e_2$ is well-typed, then e_1 and e_2 must be well-typed, and we apply the induction hypothesis to them. Suppose first that e_1 is a value. Because the type of e_1 must be a function type, e_1 must be of the form $\lambda x.\, e'$. (It cannot be a variable because reference locations do not have function types.) If e_2 is also a value V, we obtain that e is of the required form, with $[\,]$ for $\mathcal{P}$. If e_2 is not a value, then it is of the form $\mathcal{P}'[\, f \,]$, for an appropriate f, and we let $\mathcal{P}$ be $e_1\, \mathcal{P}'$. If e_1 is not a value, then it is of the form $\mathcal{P}'[\, f \,]$, for an appropriate f, and we let $\mathcal{P}$ be $\mathcal{P}'\, e_2$.
- In the case of (Exp Ref), e cannot be a value. If `ref` e_1 is well-typed, then e_1 must be well-typed, and we apply the induction hypothesis to it. Suppose first that e_1 is a value. We obtain that e is of the required form, with $[\,]$ for $\mathcal{P}$. If e_1 is not a value, then it is of the form $\mathcal{P}'[\, f \,]$, for an appropriate f, and we let $\mathcal{P}$ be `ref` $\mathcal{P}'$.
- In the case of (Exp Deref), e cannot be a value. If $!e_1$ is well-typed, then e_1 must be well-typed, and we apply the induction hypothesis to it. Suppose first that e_1 is a value. Because the type of e_1 must be a reference type, e_1 must be a reference location r. We obtain that e is of the required form, with $[\,]$ for $\mathcal{P}$. If e_1 is not a value, then it is of the form $\mathcal{P}'[\, f \,]$, for an appropriate f, and we let $\mathcal{P}$ be $!\mathcal{P}'$.
- In the case of (Exp Set), e cannot be a value. If $e_1 := e_2$ is well-typed, then e_1 and e_2 must be well-typed, and we apply the induction hypothesis to them. Suppose first that e_1 is a value. Because the type of e_1 must be a reference type, e_1 must be a reference location r. If e_2 is also a value V, we obtain that e is of the required form, with $[\,]$ for $\mathcal{P}$. If e_2 is not a value, then it is of the form $\mathcal{P}'[\, f \,]$, for an appropriate f, and we let $\mathcal{P}$ be $r := \mathcal{P}'$. If e_1 is not a value, then it is of the form $\mathcal{P}'[\, f \,]$, for an appropriate f, and we let $\mathcal{P}$ be $\mathcal{P}' := e_2$.
- The cases of (Exp Async) and (Exp Yield) are immediate, using the context $[\,]$.
- In the case of (Exp Block), e cannot be a value. If `blockUntil` e_1 is well-typed, then e_1 must be well-typed, and we apply the induction hypothesis to it. Suppose first that e_1 is a value; according to the typing rules, it can be only `false` and `true`. (It cannot be a variable because reference locations do not have type `Bool`.) We obtain that e is of the required form, with $[\,]$ for $\mathcal{P}$. If e_1 is not a value, then it is of the form $\mathcal{P}'[\, f \,]$, for an appropriate f, and we let $\mathcal{P}$ be `blockUntil` $\mathcal{P}'$.

– In the case of (Exp Assert), e cannot be a value. If $\langle e_1 \rangle$ is well-typed, then e_1 must be well-typed, and we apply the induction hypothesis to it. Suppose first that e_1 is a value. We obtain that e is of the required form, with $[\ ]$ for $\mathcal{P}$. If e_1 is not a value, then it is of the form $\mathcal{P}'[\ f\ ]$, for an appropriate f, and we let $\mathcal{P}$ be $\langle \mathcal{P}' \rangle$.

Proof of Theorem 1. We prove that if $\langle \sigma, e_1.\cdots.e_n, e \rangle \longmapsto \langle \sigma', e'_1.\cdots.e'_{n'}, e' \rangle$ and $\langle \sigma, e_1.\cdots.e_n, e \rangle$ is well-typed then so is $\langle \sigma', e'_1.\cdots.e'_{n'}, e' \rangle$. The theorem follows immediately by induction.

The proof is by cases on the operational-semantics rule being applied. In each case, we show that if

$$E\,;\,p_1.\cdots.p_n, p \vdash \langle \sigma, e_1.\cdots.e_n, e \rangle$$

then

$$E'\,;\,p'_1.\cdots.p'_{n'}, p' \vdash \langle \sigma', e'_1.\cdots.e'_{n'}, e' \rangle$$

where, unless indicated otherwise, $E' = E$, $n' = n$, and $p'_i = p_i$ for $i = 1..n$. In several cases, we consider the typings of certain subexpressions that occur in evaluation contexts; those typings are with respect to E, since the holes in the contexts are never under binders.

– (Trans Appl): The typing of $\langle \sigma, T, \mathcal{P}[\ (\lambda x.\ e)\ V\ ] \rangle$ must rely on (Exp Appl) and (Exp Fun). Specifically, we must have $E\,;\,p_0 \vdash (\lambda x.\ e)\ V : t_0$ for some t_0 and p_0, and therefore $E\,;\,p_0 \vdash \lambda x.\ e : t_1 \to^{q_0} t_0$ for some $q_0 <: p_0$ and $E\,;\,p_0 \vdash V : t_1$ for some t_1, and therefore $E, x : t_1\,;\,q_0 \vdash e : t_0$. By Lemma 5, $E, x : t_1\,;\,q_0 \vdash e : t_0$ and $q_0 <: p_0$ imply $E, x : t_1\,;\,p_0 \vdash e : t_0$. By Lemma 2, we obtain $E\,;\,p_0 \vdash e[V/x] : t_0$. By Lemma 1, we obtain a typing of $\langle \sigma, T, \mathcal{P}[\ e[V/x]\ ] \rangle$.
– (Trans Ref): The typing of $\langle \sigma, T, \mathcal{P}[\ \mathtt{ref}\ V\ ] \rangle$ must rely on (Exp Ref). Specifically, we must have $E\,;\,p_0 \vdash \mathtt{ref}\ V : \mathtt{Ref}\ t_0$ for some t_0 and p_0, and therefore $E\,;\,p_0 \vdash V : t_0$. By Lemma 4, we obtain $E\,;\,\mathtt{NoYields} \vdash V : t_0$. We extend E with $r : \mathtt{Ref}\ t_0$. We can do this extension because $r \in \mathit{RefLoc} - \mathit{dom}(\sigma)$, hence $r \notin \mathit{dom}(E)$. By a weakening (adding $r : \mathtt{Ref}\ t_0$ to E for typing $\langle \sigma, T, \mathcal{P}[\ \mathtt{ref}\ V\ ] \rangle$) and Lemma 1, we obtain a typing of $\langle \sigma, T, \mathcal{P}[\ r\ ] \rangle$.
– (Trans Deref): The typing of $\langle \sigma, T, \mathcal{P}[\ !r\ ] \rangle$ must rely on (Exp Deref). Specifically, we must have $E\,;\,p_0 \vdash !r : t_0$ for some t_0 and p_0, and therefore $E\,;\,p_0 \vdash r : \mathtt{Ref}\ t_0$. Since r is a variable, its type must come from the environment E, so by hypothesis $E\,;\,\mathtt{NoYields} \vdash V : t_0$ where $V = \sigma(r)$. By Lemma 5, we also have $E\,;\,\mathtt{Yields} \vdash V : t_0$, which is useful in case p_0 is $\mathtt{Yields}$. By Lemma 1, we obtain a typing for $\langle \sigma, T, \mathcal{P}[\ V\ ] \rangle$.
– (Trans Set): The typing of $\langle \sigma, T, \mathcal{P}[\ r := V\ ] \rangle$ must rely on (Exp Set). Specifically, we must have $E\,;\,p_0 \vdash r := V : \mathtt{Unit}$ for some p_0, and therefore $E\,;\,p_0 \vdash V : t_0$ and $E\,;\,p_0 \vdash r : \mathtt{Ref}\ t_0$ for some p_0. By Lemma 4, $E\,;\,p_0 \vdash V : t_0$ implies $E\,;\,\mathtt{NoYields} \vdash V : t_0$. Since r is a variable, its type must come from the environment E. By Lemma 1, we can transform a typing of $\langle \sigma, T, \mathcal{P}[\ r := V\ ] \rangle$ into a typing of $\langle \sigma, T, \mathcal{P}[\ \mathtt{unit}\ ] \rangle$, and since $E\,;\,\mathtt{NoYields} \vdash V : t_0$ and $E(r) = \mathtt{Ref}\ t_0$, we also obtain a typing of $\langle \sigma[r \mapsto V], T, \mathcal{P}[\ \mathtt{unit}\ ] \rangle$ by Lemma 3.

- (Trans Async): The typing of $\langle \sigma, T, \mathcal{P}[\ \texttt{async}\ e\ ]\rangle$ must rely on (Exp Async). Specifically, we must have $E\,; p_0 \vdash \texttt{async}\ e : \texttt{Unit}$ for some p_0, and therefore that $E\,; q_0 \vdash e : \texttt{Unit}$ for some q_0. By Lemma 1, we can transform a typing of $\langle \sigma, T, \mathcal{P}[\ \texttt{async}\ e\ ]\rangle$ into a typing of $\mathcal{P}[\ \texttt{unit}\ ]$, and then into a typing of $\langle \sigma, T.e, \mathcal{P}[\ \texttt{unit}\ ]\rangle$ by letting $n' = n + 1$ and adding q_0 to the sequence of effects.
- (Trans Block): The typing of $\langle \sigma, T, \mathcal{P}[\ \texttt{blockUntil true}\ ]\rangle$ must rely on (Exp Block), specifically on a derivation of $E\,; p_0 \vdash \texttt{blockUntil true} : \texttt{Unit}$ for some p_0. By Lemma 1, we obtain a typing of $\langle \sigma, T, \mathcal{P}[\ \texttt{unit}\ ]\rangle$.
- (Trans Yield): This case requires a trivial rearrangement in the effects: $n' = n + 1$, $p'_{n+1} = p$, and $p' = \texttt{NoYields}$.
- (Trans Activate): This case requires a trivial rearrangement in the effects: $n' = n - 1$, and the effect p_i that corresponds to the expression e is skipped in $p'_1.\cdots.p'_{n'}$, and becomes p'.
- (Trans Assert): The typing of $\langle \sigma, T, \mathcal{P}[\ \langle V \rangle\ ]\rangle$ must rely on (Exp Assert). Specifically, we must have $E\,; p_0 \vdash \langle V \rangle : t_0$ for some t_0 and p_0, and $E\,;$ $\texttt{NoYields} \vdash V : t_0$, so $E\,; p_0 \vdash V : t_0$ by Lemma 5. By Lemma 1, we obtain a typing of $\mathcal{P}[\ V\ ]$ and then of $\langle \sigma, T, \mathcal{P}[\ V\ ]\rangle$.

Proof of Theorem 2. By Theorem 1, if $\langle \sigma, T, e \rangle$ is well-typed then so are all the states reached in the computation $\langle \sigma, T, e \rangle \longmapsto^* \langle \sigma', T', e' \rangle$. Therefore, it suffices to prove that if $\langle \sigma, T, e \rangle$ is well-typed and $\langle \sigma, T, e \rangle \longmapsto \langle \sigma', T', e' \rangle$, then this transition is not an atomicity violation. The claim in the theorem then follows by induction.

So suppose that $\langle \sigma, T, e \rangle$ is well-typed and $\langle \sigma, T, e \rangle \longmapsto \langle \sigma', T', e' \rangle$. This transition could be an atomicity violation only if e is of the form $\mathcal{P}'[\ \langle \mathcal{P}''[\ \texttt{yield}\ ]\rangle\ ]$ for some $\mathcal{P}'$ and $\mathcal{P}''$. If $\langle \sigma, T, e \rangle$ is well-typed, then so is e, and therefore also $\langle \mathcal{P}''[\ \texttt{yield}\ ]\rangle$, because a state can be well-typed only if all its components and their subexpressions are well-typed. By the typing rule for assertions, the fact that $\langle \mathcal{P}''[\ \texttt{yield}\ ]\rangle$ is well-typed implies that $E'\,; \texttt{NoYields} \vdash \mathcal{P}''[\ \texttt{yield}\ ] : t'$ for some E' and t'. We conclude by Lemma 6.

Proof of Theorem 3. According to Theorem 1, the state $\langle \sigma', T', e' \rangle$ is well-typed. Since the rules of the operational semantics do not introduce free variables other than reference locations, the only free variables in e' are reference locations. The desired conclusion follows from Lemma 8, given next.

Lemma 8. *If $\langle \sigma, T, e \rangle$ is well-typed, and the only free variables in e are reference locations, then:*

1. *there exists $\langle \sigma', T', e' \rangle$ such that $\langle \sigma, T, e \rangle \longmapsto \langle \sigma', T', e' \rangle$; or*
2. *e is of the form $\mathcal{P}[\ \texttt{blockUntil false}\ ]$; or*
3. *e is $\texttt{unit}$ and T is empty.*

In order to prove Lemma 8, we apply Lemma 7 to e.

- If e is a value, then it must be $\texttt{unit}$ because $\langle \sigma, T, e \rangle$ is well-typed and reference locations do not have type $\texttt{Unit}$. If T is empty, then we are in the third case. Otherwise, rule (Trans Activate) applies, and we are in the first case.

- If e is of the form $\mathcal{P}[\,$ blockUntil false $]$, then we are immediately in the second case.
- If e is of the form $\mathcal{P}[\,f\,]$ where f is one of $(\lambda x.\, e')\ V$, ref V, $!r$, $r := V$, async e', blockUntil true, yield, and $\langle V \rangle$, then (Trans Appl), (Trans Ref), (Trans Deref), (Trans Set), (Trans Async), (Trans Block), (Trans Yield), or (Trans Assert) apply, respectively, and we are in the first case again. In the case of (Trans Ref), we use that $RefLoc - dom(\sigma)$ is never empty. In the case of (Trans Deref), we rely on the condition that if $r \in RefLoc$ occurs free in e then $r \in dom(\sigma)$, and on the fact that σ maps reference locations to values.

Petri Nets, Discrete Physics, and Distributed Quantum Computation

Samson Abramsky

Oxford University Computing Laboratory

This paper is dedicated to Ugo Montanari on the occasion of his 65th birthday.

Abstract. We shall describe connections between Petri nets, quantum physics and category theory. The view of Net theory as a kind of discrete physics has been consistently emphasized by Carl-Adam Petri. The connections between Petri nets and monoidal categories were illuminated in pioneering work by Ugo Montanari and José Meseguer. Recent work by the author and Bob Coecke has shown how monoidal categories with certain additional structure (dagger compactness) can be used as the setting for an effective axiomatization of quantum mechanics, with striking applications to quantum information. This additional structure matches the extension of the Montanari-Meseguer approach by Marti-Oliet and Meseguer, motivated by linear logic.

1   Introduction

In this paper, we shall be concerned with links between three, *prima facie* very different, areas:

- Models of concurrent computation, especially Petri nets.
- Physics, especially quantum mechanics and quantum information.
- Monoidal categories with additional structure (e.g. compact closure [16]).

In particular, we are motivated by the following previous work:

- Petri's seminal work, which has always emphasized links between his Net Theory and Physics [24,25,26].
- The pioneering work by Ugo Montanari and José Meseguer [21] using monoidal categories as a setting for Net Theory, further extended by Marti-Oliet and Meseguer [20].
- Our own work with Bob Coecke [4,5], using monoidal categories as a setting for a novel axiomatization of quantum mechanics, with applications to quantum information.

Thus the situation can be depicted as follows:

P. Degano et al. (Eds.): Montanari Festschrift, LNCS 5065, pp. 527–543, 2008.
© Springer-Verlag Berlin Heidelberg 2008

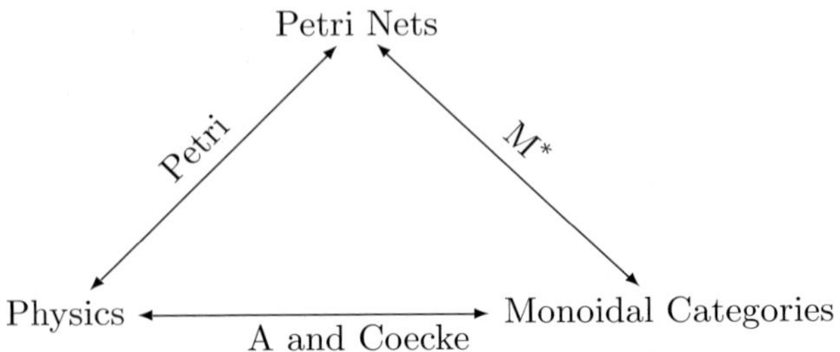

2   Petri Nets as Discrete Physics

An important quality of Petri's conception of concurrency, as compared with "linguistic" approaches such as process calculi, is that it seeks to explain fundamental concepts: **causality**, **concurrency**, **process**, etc. in a syntax-independent, "geometric" fashion. Another important point, which may originally have seemed merely eccentric, but now looks rather ahead of its time, is the extent to which Petri's thinking was explicitly influenced by physics (see e.g. [25]). As one example, note that K-density comes from one of Carnap's axiomatizations of relativity [11]. To a large extent, and by design, Net Theory can be seen as a kind of **discrete physics**: **lines** are time-like causal flows, **cuts** are space-like regions, **process unfoldings** of a **marked net** are like the solution trajectories of a differential equation.

This acquires new significance today, when the consequences of the idea that "Information is Physical" [17] are being explored in the rapidly developing field of quantum informatics. Moreover, the need to recognize the spatial structure of distributed systems has become apparent, and is made explicit in formalisms such as the Ambient calculus [10], and Milner's bigraphs [23].

We shall illustrate these points with some quotations from [25].

> "This paper attempts to provide a common basis for physical and computational ways of thinking. ... If this approach should turn out to be a small, but definite step towards the remote (perhaps illusory) goal of founding technology and natural sciences on a theory of information flow, the author would feel rewarded beyond merit."

The paper discusses **four levels of description** for processes and systems. We concentrate on the first two.

Level 0: Concurrency Structure

> "Concurrency is short for "the binary relation of cotemporality of world points". Here we follow closely the axiomatizations of relativistic space-time ...
>
> For individuals, we take the time layers of **signals**, the smallest propagators of physical effects. Some signals are particles, others propagate

... through interaction. The history of each signal is a "world line" and consists of world points. Let x and y be individuals; we write $x < y$ if $x \neq y$ and a signal passes from x to y. We define

$$x \text{ co } y \Leftrightarrow \text{neither } x < y \text{ nor } y < x$$

$$x \text{ li } y \Leftrightarrow x < y \text{ or } y < x \text{ or } x = y$$

Let

$$\text{Co}(x) := \{z \mid x \text{ co } z\} \qquad \text{Li}(x) := \{z \mid x \text{ li } z\}$$

If $\text{Co}(x) = \text{Co}(y)$ or $\text{Li}(x) = \text{Li}(y)$, we shall collect x and y into a cluster; such clusters are equivalence classes of world points, and will be the individuals of Level 1."

Level 1: Occurrence Nets

"We shall now describe the structure of the set X of all occurrences, and its partition into a set S of state elements, and a set T of transition elements. ...

A subset $l \subset X$ will be called a **Line** iff it is a maximal set of occurrences which are pairwise in relation li.

A subset $c \subset X$ will be called a **Cut** iff it is a maximal set of occurrences which are pairwise in relation co.

The old physical postulate that every Cut represents a spatial distribution ... can now be written as 'every Cut meets every Line', i.e. as **K-density**."

2.1   Causal Sets and Other Roads

Quite independently, physicists have recently been thinking along strikingly similar lines, in one of the radical current approaches to **quantum gravity**, which is being developed by Raphael Sorkin and his collaborators [28,9].

Following intuitions going back to Riemann and Einstein, the aim with Causal Sets is to build a theory of space-time which is ultimately (at the "Planck scale") **discrete**. A causal set is just a **locally finite poset**. The elements are events, the ordering is causality. The aim is to build everything back from these ingredients, under the slogan

$$\text{Order} + \text{Number} = \text{Geometry.}$$

"Number" here refers to counting the events which have occurred in a given region of spacetime; this is meaningful by local finiteness, and leads to a notion of "volume". There is a "dynamics" which comes from the **growth** of a poset. Large-scale structural properties of space-time should emerge from stochastic properties of such growth.

One may also note the popular book by Lee Smolin [27]: the discussion in the first few chapters, especially of the relational view of spacetime, is very much in the same spirit.

We also mention the striking recent work by Keye Martin and Prakash Panangaden [19], which builds back the spacetime manifold from the causal order, using ideas from domain theory.

3   Interlude: Symmetric Monoidal Categories

We briefly recall and motivate the basic setting of symmetric monoidal categories. For further details, we refer to standard texts such as [18].

3.1   Categories

A category $\mathcal{C}$ has **objects** (types) A, B, C, $\ldots$, and for each pair of objects A, B a set of **morphisms** $\mathcal{C}(A, B)$. (Notation: $f : A \to B$). It also has identities $\mathrm{id}_A : A \to A$, and composition $g \circ f$ when types match:

$$A \xrightarrow{\ f\ } B \xrightarrow{\ g\ } C$$

Categories allow us to deal explictly with **typed processes**, e.g.

Logic	Programming	Computation
Propositions	Data Types	States
Proofs	Programs	Transitions

3.2   Symmetric Monoidal Categories

A **symmetric monoidal category** comes equipped with an associative operation $\otimes$, the "tensor product", which acts on **both** objects **and** morphisms — a **bifunctor**:

$$A \otimes B \qquad f_1 \otimes f_2 : A_1 \otimes A_2 \longrightarrow B_1 \otimes B_2$$

There is also a symmetry operation

$$\sigma_{A,B} : A \otimes B \longrightarrow B \otimes A$$

which satisfies some 'obvious' rules, e.g. naturality:

$$
\begin{array}{ccc}
A_1 \otimes A_2 & \xrightarrow{\ f_1 \otimes f_2\ } & B_1 \otimes B_2 \\
\Big\downarrow{\sigma_{A_1,A_2}} & & \Big\downarrow{\sigma_{B_1,B_2}} \\
A_2 \otimes A_1 & \xrightarrow{\ f_1 \otimes f_2\ } & B_2 \otimes B_1
\end{array}
$$

The Logic of Tensor Product Tensor can express **independent** or **concurrent** actions (mathematically: bifunctoriality):

$$
\begin{array}{ccc}
A_1 \otimes A_2 & \xrightarrow{\; f_1 \otimes \mathrm{id}\;} & B_1 \otimes A_2 \\[2pt]
{\scriptstyle \mathrm{id} \otimes f_2} \Big\downarrow & & \Big\downarrow {\scriptstyle \mathrm{id} \otimes f_2} \\[2pt]
A_1 \otimes B_2 & \xrightarrow[\; f_1 \otimes \mathrm{id}\;]{} & B_1 \otimes B_2
\end{array}
$$

But tensor is **not** a 'cartesian' or categorical product, in the sense that **we cannot reconstruct an 'element' of the tensor from its components**.

This turns out to comprise the absence of natural diagonals (copying) and projections (deleting):

$$
A \xrightarrow{\;\Delta\;} A \otimes A \qquad\qquad A_1 \otimes A_2 \xrightarrow{\;\pi_i\;} A_i
$$

$$
\mathrm{Cf.}\quad A \vdash A \wedge A \qquad\qquad A_1 \wedge A_2 \vdash A_i.
$$

Hence there is a direct connection to "resource-sensitive" logics such as linear logic [14].

A basic example, familiar to Computer Scientists, is given by **Rel**, the category with sets as objects and **relations** as arrows. Here the usual cartesian product of sets gives a tensor product, but **not** the categorical product. In particular, although we can define diagonals and projections, they are not **natural**. For diagonals, this means that the diagram

$$
\begin{array}{ccc}
X & \xrightarrow{\;\Delta_X\;} & X \times X \\[2pt]
{\scriptstyle R} \Big\downarrow & & \Big\downarrow {\scriptstyle R \times R} \\[2pt]
Y & \xrightarrow[\;\Delta_Y\;]{} & Y \times Y
\end{array}
$$

does not commute in general, where $\Delta_X = \{(x,(x,x)) \mid x \in X\}$ is the usual diagonal, and $R \subseteq X \times Y$ can be any relation.

A fundamental example for Quantum Mechanics and Quantum Information is **FdHilb**, the category of finite-dimensional complex Hilbert spaces and linear maps, with the standard concrete tensor product of linear algebra [4].

4   Petri Nets and Monoidal Categories

In the late 1980's there was a brief flowering of work relating Petri nets with monoidal categories and Linear logic [29,13,21,20]. This work does not seem to have had much lasting impact on the Petri net community, but the work by Ugo Montanari and José Meseguer in particular has been influential on wider developments in concurrency and graph rewriting, e.g. [22]. We shall briefly summarize their approach.

4.1   The Meseguer and Montanari Approach

Petri Nets are defined as

$$N = (S^{\otimes}, T, \delta_0, \delta_1)$$

where

- $S^{\otimes}$ is a free commutative monoid of states
- T is the set of transitions
- $\delta_0, \delta_1 : T \longrightarrow S^{\otimes}$ give the **source** and **target** of each transition.

A multiset of S-elements is just another way of thinking of a distribution of tokens.

Example

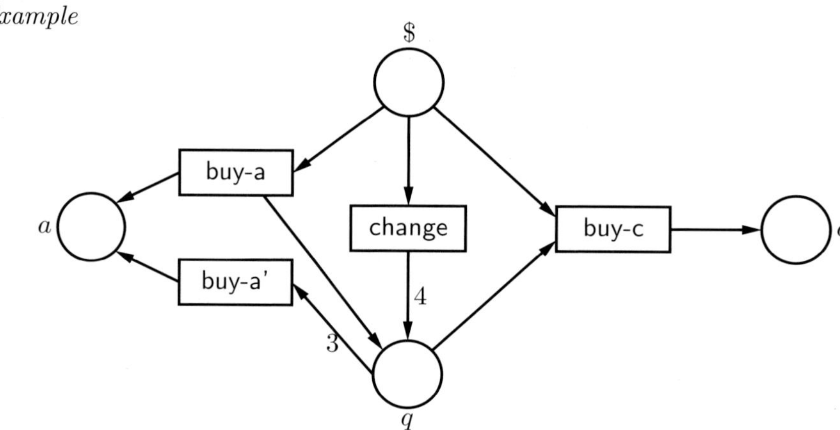

The transitions (axioms, arrows) are:

$$\text{buy-c} : \$ \otimes q \longrightarrow c \quad \text{buy-a} : \$ \longrightarrow a \otimes q$$

$$\text{buy-a'} : q^3 \longrightarrow a \quad \text{change} : \$ \longrightarrow q^4$$

Petri Categories By **closing up** the transitions under sequential and parallel composition, a general notion of **process** is obtained:

- Given $\alpha : A \longrightarrow B$ and $\beta : B \longrightarrow C$, form $\alpha; \beta : A \longrightarrow C$. Taking sequential composition to be associative, and adding idle transitions $1_A : A \longrightarrow A$ which are identities for sequential composition, this gives a **category**.
- Given $\alpha : A_1 \longrightarrow B_1$ and $\beta : A_2 \longrightarrow B_2$, form the parallel composition $\alpha \otimes \beta : A_1 \otimes A_2 \longrightarrow B_1 \otimes B_2$. If we assume the key **(bi)functoriality axiom**

$$(\alpha_1 \otimes \alpha_2); (\beta_1 \otimes \beta_2) = (\alpha_1; \beta_1) \otimes (\alpha_2; \beta_2)$$

this gives a **symmetric monoidal category**.

Meseguer and Montanari identify the particular kind of symmetric monoidal categories which arise in this way — the **Petri categories**. They show that the Best-Devillers theory of sequential and concurrent behaviours [8] can be recaptured in a systematic way in this framework.

5   Processes in Monoidal Categories: A General Perspective

We have seen that Petri Nets can be seen as **particular examples of symmetric monoidal categories**. But why not turn this around? Why not see **any** symmetric monoidal category as **a setting for describing computational processes in a resource sensitive way, closed under sequential and parallel composition?**

There is a natural objection to this, that we would then lose the additional **concrete, combinatorial structure** of Petri nets, and the corresponding graphical formalism, which is so much a part of how they are used.

But this objection does not really hold water! Monoidal categories, quite generally, admit a beautiful **graphical calculus** or **diagrammatic notation** which makes equational proofs perspicuous, and is sound and complete for equational reasoning in monoidal categories [15]. It also supports links with Logic (e.g. Proof Nets) and with Geometry (Knots, Braids, Temperley-Lieb algebra etc.)[1,3].

5.1   Outline of the Graphical Calculus

In the graphical calculus we depict processes by boxes, and we label the inputs and outputs of these boxes by **types** which indicate the kind of system on which these processes act:

Algebraically, these correspond to:

$$1_A : A \to A, \quad f : A \to B, \quad g \circ f, \quad 1_A \otimes 1_B, \quad f \otimes 1_C, \quad f \otimes g, \quad (f \otimes g) \circ h$$

respectively. (The convention in these diagrams is that the 'upward' vertical direction represents progress of time.)

Kets, Bras and Scalars: A special role is played by boxes with either no input or no output, *i.e.* arrows of the form $I \longrightarrow A$ or $A \longrightarrow I$ respectively, where I is the unit of the tensor. In the setting of **FdHilb** and Quantum Mechanics, they correspond to **states** and **costates** respectively (cf. Dirac's kets and bras [12]), which we depict by triangles. **Scalars** then arise naturally by composing these elements (cf. inner-product or Dirac's bra-ket):

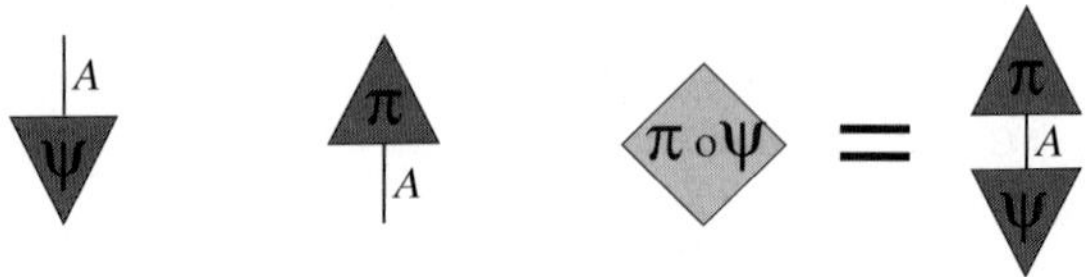

Formally, scalars are arrows of the form $I \longrightarrow I$. In the physical context, they provide numbers ("probability amplitudes" etc.). For example, in **FdHilb**, the tensor unit is $\mathbb{C}$, the complex numbers, and a linear map $s : \mathbb{C} \longrightarrow \mathbb{C}$ is determined by a single number, $s(1)$. In **Rel**, the scalars are the boolean semiring $\{0, 1\}$.

This graphical notation can be seen as a substantial two-dimensional generalization of **Dirac notation** [12]:

$$\langle \phi \mid \qquad \mid \psi \rangle \qquad \langle \phi \mid \psi \rangle$$

Note how the geometry of the plane absorbs functoriality and naturality conditions, e.g.:

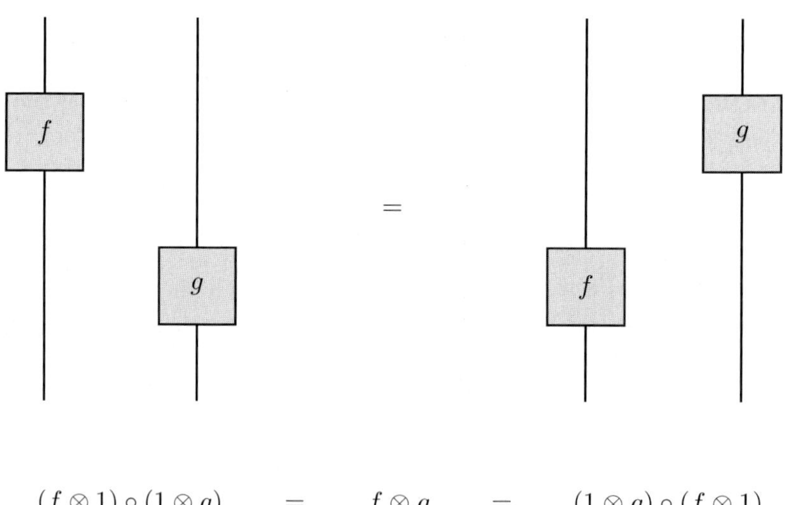

$$(f \otimes 1) \circ (1 \otimes g) \quad = \quad f \otimes g \quad = \quad (1 \otimes g) \circ (f \otimes 1)$$

6   Deficits and Cancellation

We shall now consider an extension of the Meseguer–Montanari approach, due to Marti-Oliet and Meseguer [20]. For initial motivation, consider the following example:

Example

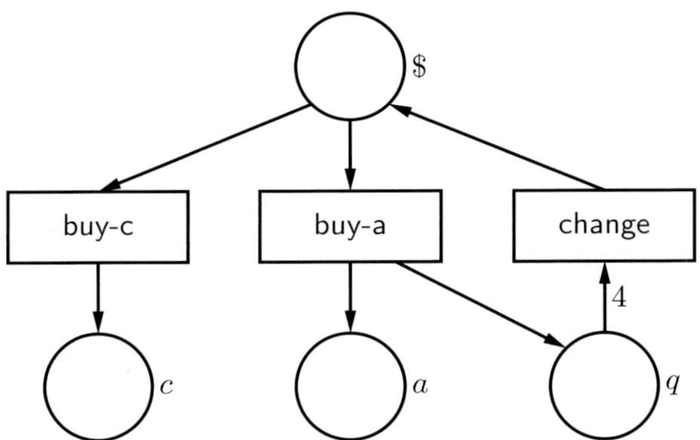

Transitions:

$$\text{buy-c} : \$ \longrightarrow c \qquad \text{buy-a} : \$ \longrightarrow a \otimes q \qquad \text{change} : q^4 \longrightarrow \$$$

Note that there is no way of getting an apple if we start with three quarters!

6.1   The Financial Game

Marti-Oliet and Meseguer introduce the following modification of the usual token game, motivated by the aim of extending the connection between Petri Nets and monoidal structures to the whole of Multiplicative Linear Logic [14].

- **Negative** or **dual** tokens a^* etc. are introduced. We can use these to represent situations involving **deficits** as well as the usual presence of resources.
- As well as the usual firing rules, we now have the opportunity to "borrow" resources, creating **both** the resource, **and** the corresponding deficit.

- Conversely, given a resource and a corresponding deficit, we can **cancel them**, removing both:

Deficits and Cancellation   The basic transitions we need are:

$$I \longrightarrow a^* \otimes a \qquad\qquad a \otimes a^* \longrightarrow I$$

where I is the unit of the monoidal structure, creating and cancelling a deficit.

   We can now produce a computation in the above example to get an apple from three quarters:

$$q^3 \longrightarrow q^3 \otimes I \longrightarrow q^3 \otimes q^* \otimes q \longrightarrow \$ \otimes q^* \longrightarrow a \otimes q \otimes q^* \longrightarrow a$$

This is a well known idea in Category theory: it takes us from symmetric monoidal to **compact closed** categories [16]. How does this look in our graphical calculus for monoidal categories?

Cups and Caps

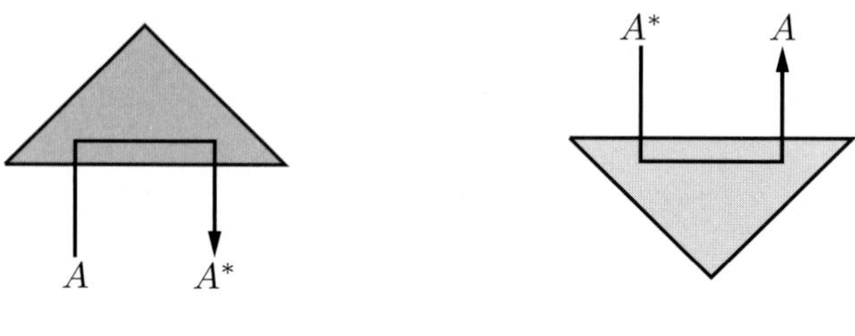

$$\epsilon_A : A \otimes A^* \longrightarrow I \qquad\qquad \eta_A : I \longrightarrow A^* \otimes A.$$

$$\text{Caps} = \text{Cancellations}; \text{Cups} = \text{Deficits}.$$

6.2   Graphical Calculus for Information Flow

Compact Closure : The basic algebraic laws for units and counits.

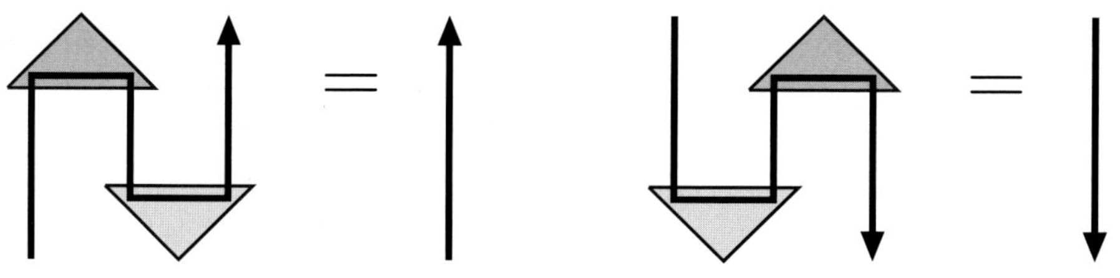

$$(\epsilon_A \otimes 1_A) \circ (1_A \otimes \eta_A) = 1_A \qquad\qquad (1_{A^*} \otimes \epsilon_A) \circ (\eta_A \otimes 1_{A^*}) = 1_{A^*}$$

In terms of deficits and cancellations:

Names and Conames in the Graphical Calculus The units and counits are powerful; they allow us to define a **closed structure** on the category. In particular, we can form the **name** $\ulcorner f \urcorner$ of any arrow $f : A \rightarrow B$, as a special case of λ-abstraction, and dually the **coname** $\llcorner f \lrcorner$:

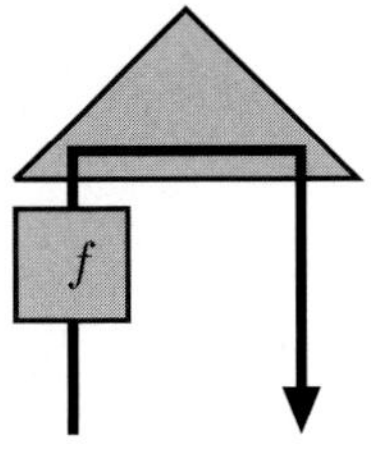 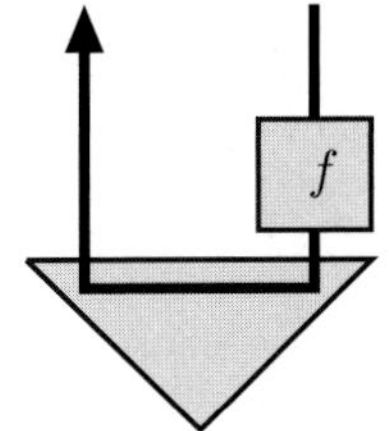

$$\llcorner f \lrcorner : A \otimes B^* \to I \qquad\qquad \ulcorner f \urcorner : I \to A^* \otimes B$$

This is the general form of Map-State duality:

$$\mathcal{C}(A \otimes B^*, I) \simeq \mathcal{C}(A, B) \simeq \mathcal{C}(I, A^* \otimes B).$$

7 Monoidal Categories and Physics

Having related physics to Petri nets, and Petri nets to monoidal categories, we shall now show how to close the circle by relating (quantum) physics to monoidal categories, following [4]. Moreover, a key role will be played by the compact closed structure which was described in the previous section, as the abstract form of the "negative flows" of deficits and cancellations, introduced by Marti-Oliet and Meseguer to correspond to the logical structure of linear negation and implication. Here the same structure arises with a **physical** motivation, and plays a crucial rôle in explicating the information flows arising from **quantum entanglement**.

7.1 Bits and Qubits

Classical Bits:

- have two values 0, 1
- are freely readable and duplicable
- admit arbitrary data transformations

Qubits:

- have a 'sphere' of values spanned by $|0\rangle$, $|1\rangle$

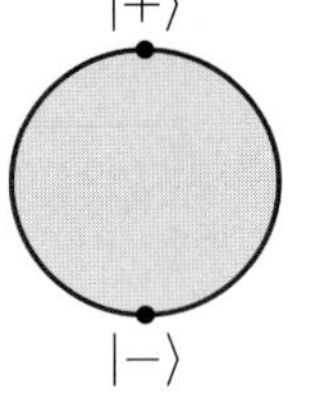

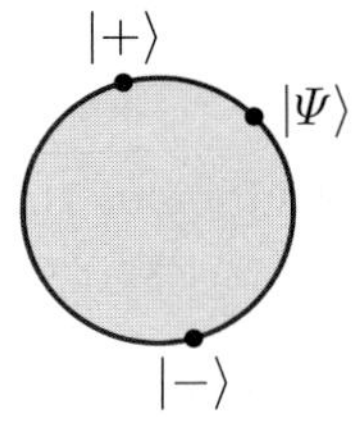

- measurements of qubits
  - have two outcomes $|-\rangle$, $|+\rangle$
  - change the value $|\psi\rangle$
- admit **unitary transformations**, *i.e.* antipodes and angles are preserved.

Formally, a qubit is a vector[1] in the two-dimensional complex Hilbert space $\mathbb{C}^2$. This space allows for two degrees of freedom when we measure the qubit in a given basis; we get one of two possible answers, conventionally '0' or '1' in the standard basis. **Which** of these answers we get is in general uncertain; the state of the qubit tells us only the **probability** with which we will get each of the two possible answers. Moreover, measurement has an **effect** on the system being measured; it "collapses" to the basis state corresponding to the outcome of the measurement.

7.2   Quantum Entanglement

We consider for illustration two standard examples of two-qubit entangled states, the Bell state:

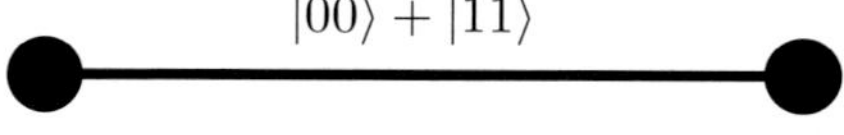

and the EPR state:

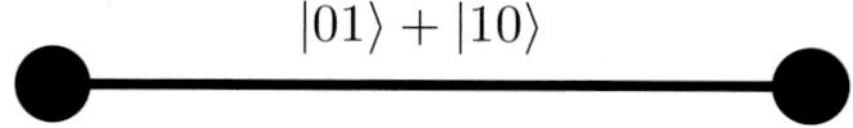

In quantum mechanics, compound systems are represented by the **tensor product** of Hilbert spaces: $\mathcal{H}_1 \otimes \mathcal{H}_2$. A typical element of the tensor product has the form:

$$\sum_i \lambda_i \cdot \phi_i \otimes \psi_i$$

where ϕ_i, ψ_i range over basis vectors, and the coefficients λ_i are complex numbers. **Superposition** encodes **correlation**: in the Bell state, the off-diagonal elements have zero coefficients. This gives rise to Einstein's "spooky action at a distance". Even if the particles are spatially separated, measuring one has an effect on the state of the other. In the Bell state, for example, when we measure one of the two qubits we may get either 0 or 1, but once this result has been obtained, it is certain that the result of measuring the other qubit will be the same.

This leads to Bell's famous theorem [6]: QM is **essentially non-local**, in the sense that the correlations it predicts exceed those of any "local realistic theory".

---

[1] Really by a one-dimensional subspace, or **ray**.

From 'paradox' to 'feature': Teleportation

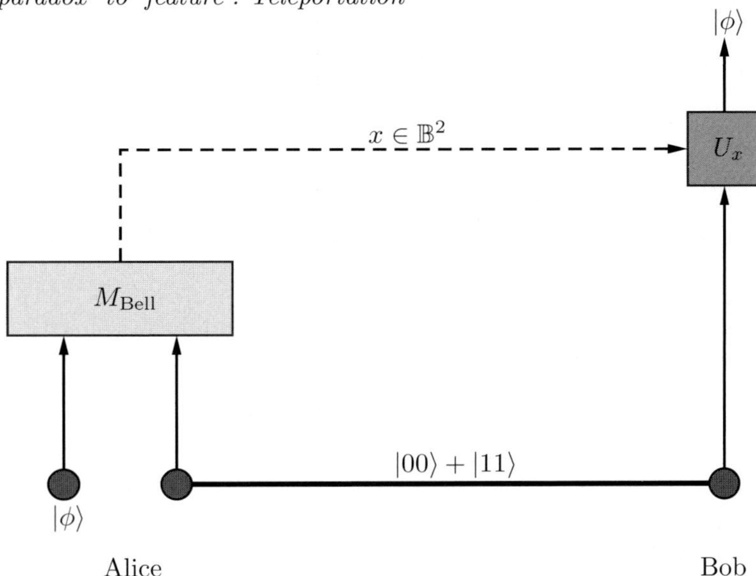

In the teleportation protocol [7], Alice sends an unknown qubit ϕ to Bob, using a shared Bell pair as a "quantum channel". By performing a measurement in the Bell basis on ϕ and her half of the entangled pair, a collapse is induced on Bob's qubit. Once the result x of Alice's measurement is transmitted by classical communication to Bob (there are four possible measurement outcomes, hence this requires two classical bits), Bob can perform a corresponding unitary correction U_x on his qubit, after which it will be in the state ϕ.

7.3 Categorical Quantum Mechanics and Diagrammatics

We now outline the categorical approach to quantum mechanics developed in [4,5]. The **same** graphical calculus and underlying algebraic structure which we have seen in the previous section has been applied to quantum information and computation, yielding an incisive analysis of **quantum information flow**, and powerful and illuminating methods for reasoning about quantum informatic processes and protocols [4].

Bell States and Costates: The cups and caps we have already seen in the guise of deficit and cancellation operations, now take on the rôle of **Bell states and costates** (or preparation and test of Bell states), the fundamental building blocks of quantum entanglement. (Mathematically, they arise as the transpose and co-transpose of the identity, which exist in any finite-dimensional Hilbert space by "map-state duality").

540   S. Abramsky

The formation of **names** and **conames** of arrows (*i.e.* map-state and map-costate duality) is conveniently depicted thus:

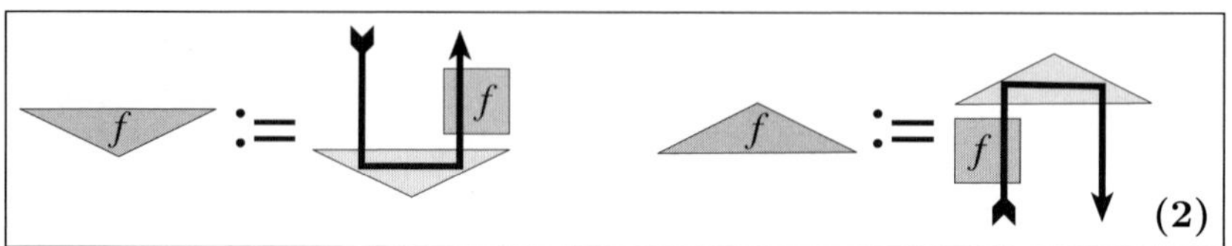

$$(2)$$

The key lemma in exposing the quantum information flow in (bipartite) entangled quantum systems can be formulated diagrammatically as follows:

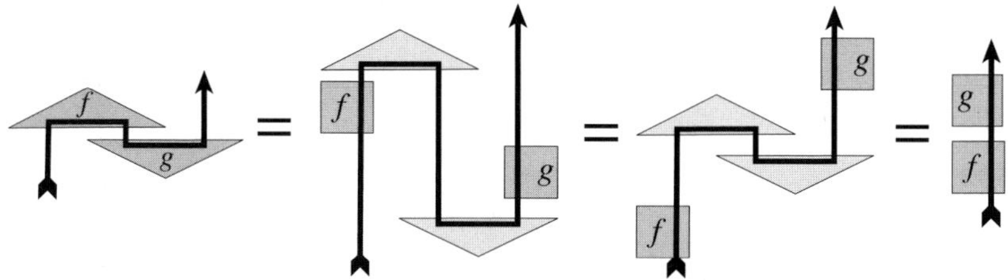

Note in particular the interesting phenomenon of "apparent reversal of the causal order" . While on the left, physically, we first prepare the state labeled g and then apply the costate labeled f, the global effect is *as if* we first applied f itself first, and only then g. This corresponds to the apparent reversal of flow of computations in the token game on Petri nets achieved with deficits and cancellations.

Derivation of quantum teleportation. This is the most basic application of compositionality in action. We can read off the basic quantum mechanical potential for teleportation immediately from the geometry of Bell states and costates:

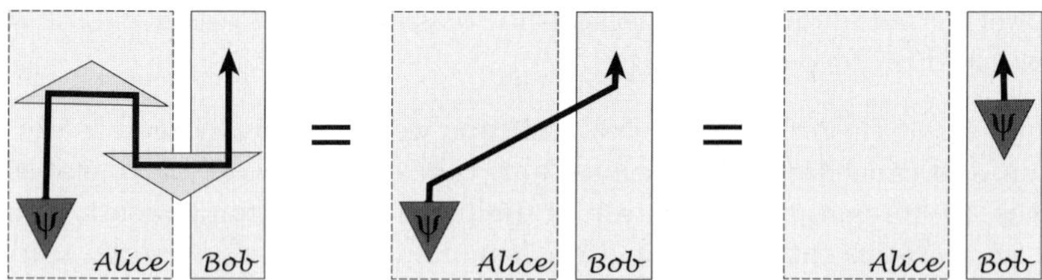

The Bell state forming the shared channel between Alice and Bob appears as the downwards triangle in the diagram; the Bell costate forming one of the possible measurement branches is the upwards triangle. The information flow of the input qubit from Alice to Bob is then immediately evident from the diagrammatics.

This is not quite the whole story, because of the non-deterministic nature of measurements. But in fact, allowing for this shows the underlying **design principle** for the teleporation protocol. Namely, we find a measurement basis such that each possible branch i through the measurement is labelled, under map-state duality, with a unitary map f_i. The corresponding correction is then just the inverse map f_i^{-1}. Using our lemma, the full description of teleportation becomes:

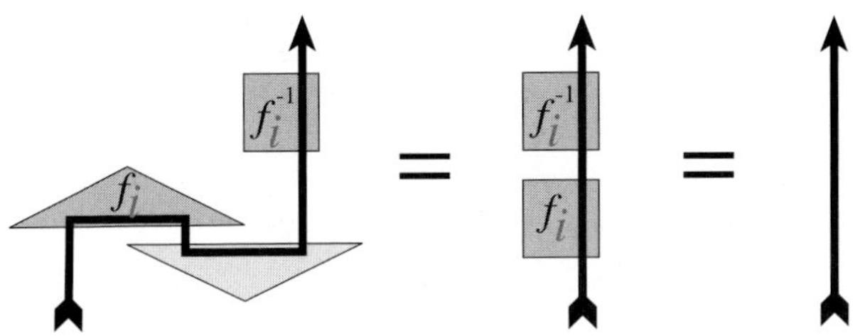

8   Conclusions

We have described a striking nexus of ideas arising from several different sources. Conceptually, the most interesting feature has been the need for "negative information flow", which has arisen from several sources:

- In logic, with the need to account for negative polarities, as created by connectives such as (linear) negation or implication, leading to Marti-Oliet and Meseguer's proposal of deficits and cancellations as a computational correlate in the setting of Petri net dynamics.
- Physically, e.g. in the desciption of quantum teleportation, these negative flows run counter to the normal flow of time and causality, and form part of the enigma of quantum mechanics.
- Mathematically, we are led to dualities and closed structure.
- Geometrically, we go in the direction of loops, tangles and knots.

Standard Petri net theory, with its careful enforcement of local causality and information flow, explicitly influenced by relativity theory, does not suffice to capture the non-local features of quantum mechanics, which are exploited in quantum information. Perhaps it is not too fanciful to see here in microcosm some of the foundational obstacles to formulating an adequate theory of quantum gravity!

The further elaboration and deepening of this nexus of ideas offers many interesting challenges. In particular, a full analysis of distributed quantum computation, e.g. quantum security protocols, which should take into account both the quantum and the classical ingredients of the protocols, and also reflect enough of the spatio-temporal structure to capture the salient distributed features, will require a deeper understanding of the ingredients we have assembled here, and of how they may be related and combined.

References

1. Abramsky, S.: Abstract Scalars, Loops, and Free Traced and Strongly Compact Closed Categories. In: Fiadeiro, J.L., Harman, N.A., Roggenbach, M., Rutten, J. (eds.) CALCO 2005. LNCS, vol. 3629, pp. 1–31. Springer, Heidelberg (2005)
2. Abramsky, S.: What are the fundamental structures of concurrency? We still don't know! In: Proceedings of the Workshop Essays on Algebraic Process Calculi (APC 25). Electronic Notes in Theoretical Computer Science, vol. 162, pp. 37–41 (2006)
3. Abramsky, S.: Temperley-Lieb algebra: from knot theory to logic and computation via quantum mechanics. In: Chen, G., Kauffman, L., Lomonaco, S. (eds.) Mathematics of Quantum Computation and Quantum Technology, pp. 515–558. Taylor and Francis, Abington (2007)
4. Abramsky, S., Coecke, B.: A categorical semantics of quantum protocols. In: Proceedings of the 19th Annual IEEE Symposium on Logic in Computer Science, pp. 415–425 (2004), arXiv:quant-ph/0402130
5. Abramsky, S., Coecke, B.: Abstract physical traces. Theory and Applications of Categories 14, 111–124 (2005)
6. Bell, J.S.: On the Problem of Hidden Variables in Quantum Mechanics. Reviews of Modern Physics (1966)
7. Bennet, C.H., Brassard, G., Crépeau, C., Jozsa, R., Peres, A., Wooters, W.K.: Teleporting an unknown quantum state via dual classical and Einstein-Podolsky-Rosen channels. Physical Review Letters 70, 1895–1899 (1993)
8. Best, E., Devillers, R.: Sequential and concurrent behaviour in Petri net theory. Theoretical Computer Science (1987)
9. Bombelli, L., Lee, J., Meyer, D., Sorkin, R.D.: Spacetime as a causal set. Phys. Rev. Lett. 59, 521–524 (1987)
10. Cardelli, L., Gordon, A.D.: Mobile Ambients. In: Nivat, M. (ed.) FOSSACS 1998. LNCS, vol. 1378, pp. 140–155. Springer, Heidelberg (1998)
11. Carnap, R.: Introduction to Symbolic Logic with Applications. Dover Books (1958)
12. Dirac, P.A.M.: The Principles of Quantum Mechanics, 3rd edn. Oxford University Press, Oxford (1947)
13. Gehlot, V., Gunter, C.: Normal process representatives. In: Proceedings LiCS (1990)
14. Girard, J.-Y.: Linear Logic. Theoretical Computer Science 50(1), 1–102 (1987)
15. Joyal, A., Street, R.: The geometry of tensor calculus I. Advances in Mathematics 88, 55–112 (1991)
16. Kelly, G.M., Laplaza, M.L.: Coherence for compact closed categories. Journal of Pure and Applied Algebra 19, 193–213 (1980)
17. Landauer, R.: Information is physical. Physics Today 44, 23–29 (1991)
18. Mac Lane, S.: Categories for the Working Mathematician, 2nd edn. Springer, Heidelberg (1998)
19. Martin, K., Panangaden, P.: A Domain of spacetime intervals for General Relativity. Communications of Mathematical Physics (November 2006)
20. Marti-Oliet, N., Meseguer, J.: From Petri Nets To Linear Logic Through Categories. IJFCS 2(4), 297–399 (1991)
21. Meseguer, J., Montanari, U.: Petri Nets Are Monoids. Information and Computation 88, 105–155 (1990)
22. Milner, R.: Calculi for Interaction. Acta Informatica 33 (1996)
23. Milner, R.: Pure bigraphs: Structure and dynamics. Inf. Comput. 204(1), 60–122 (2006)

24. Petri, C.A.: Fundamentals of a Theory of Asynchronous Information Flow. IFIP Congress, 386–390 (1962)
25. Petri, C.-A.: State-Transition Structures in Physics and in Computation. International Journal of Theoretical Physics 21(12), 979–993 (1982)
26. Petri, C.A.: Nets, Time and Space. Theor. Comput. Sci. 153(1–2), 3–48 (1996)
27. Smolin, L.: Three Roads to Quantum Gravity. Phoenix (2000)
28. Sorkin, R.D.: A Specimen of Theory Construction from Quantum Gravity. In: Leplin, J. (ed.) The Creation of Ideas in Physics: Studies for a Methodology of Theory Construction (Proceedings of the Thirteenth Annual Symposium in Philosophy, held Greensboro, North Carolina, March, 1989, pp. 167–179. Kluwer Academic Publishers, Dordrecht (1995)
29. Winskel, G.: Petri nets, algebras, morphisms and compositionality. Information and Computation 72, 197–238 (1987)

A Category of Explicit Fusions

Filippo Bonchi[1], Maria Grazia Buscemi[2],
Vincenzo Ciancia[1], and Fabio Gadducci[1]

[1] Dipartimento di Informatica, University of Pisa, Italy
{fibonchi,ciancia, gadducci}@di.unipi.it
[2] IMT Lucca Institute for Advanced Studies, Italy
m.buscemi@imtlucca.it

Abstract. Name passing calculi are nowadays an established field on
its own. Besides their practical relevance, they offered an intriguing challenge, since the standard operational, denotational and logical methods
often proved inadequate to reason about these formalisms. A domain
which has been successfully employed for languages with asymmetric
communication, like the π-calculus, are presheaf categories based on (injective) relabelings, such as $Set^{\mathbb{I}}$. Calculi with *symmetric* binding, in the
spirit of the *fusion calculus*, give rise to new research problems. In this
work we examine the *calculus of explicit fusions*, and propose to model
its syntax and semantics using the presheaf category $Set^{\mathbb{E}}$, where $\mathbb{E}$ is the
category of equivalence relations and equivalence preserving morphisms.

1 Introduction

Among the many research themes Ugo Montanari considered in his carrier, he
was always concerned with the semantics of interactive systems. In our work we
consider a few topics, related to such a large area, that always interested Ugo,
namely final semantics, coalgebras, names and name fusions, and constraints.

Denotational Semantics *via* Final Object. In [35] Scott and Strachey introduced denotational semantics as a way of formalizing the meaning of programming languages: to each expression of the language a *denotation* is assigned, i.e.,
an object in a mathematical domain. In their original proposal, each program denotes a continuous function on a partially ordered set, mapping each input of the
program into the corresponding output. Despite its expressiveness, the approach
is less adequate in modelling the semantics of *interactive systems*: indeed, in this
case the non-deterministic behaviour of a program is more important than the
function it computes: thus, these systems can not be simply denoted as if they
were input-output functions.

An important tenet of denotational semantics is that it should be *compositional*, i.e, the denotation of a program expression has to be constructed by
the denotation of its sub-expressions. This property allows one to reason inductively on the program structure, providing a general methodology for proving
properties of programs.

P. Degano et al. (Eds.): Montanari Festschrift, LNCS 5065, pp. 544–562, 2008.

A lot of effort has been spent to give compositional denotational semantics to concurrent programming languages. Usually, this has been accomplished by restricting the focus to simple computational models exhibiting fundamental aspects of concurrent computations. Amongst the various proposals, process calculi are one of the most successful: a set of basic operators defines the syntax of the calculus (more formally, the set of operators is a signature and the expressions of the language are the elements of the *initial algebra* associated to such a signature) and for each operator there is a set of (*SOS* [32]) rules describing the behaviour of the composite expression in terms of the behaviours of its sub-expressions. The resulting operational semantics is usually a *labeled transition system* (LTS) where labels represent interactions amongst the various sub-expressions (representing components) of an expression (a system).

Moreover, Universal Coalgebra [33] provides a good categorical framework for the denotation of process calculi. Given a behavioural endo-functor B, one can define the category of B-coalgebras and B-cohomomorphisms. By choosing a certain endo-functor on Set (i.e., the category of sets and functions), we get the category of all labeled transition systems and "zig-zag" morphisms (i.e., morphisms that both respect and reflect transitions). This category has a *final object* F, i.e., from every LTS there is a unique morphism to F.

In this setting, one can easily define the denotation of an LTS as its image through this unique morphism. This idea, that nowadays is called *final semantics*, was originally proposed for abstract data types by Giarratana, Gimona and Montanari in [20] and it is still central in Ugo's work (see e.g. [11,2]).

Compositional Denotational Semantics *via* Initial and Final Object. This kind of representation is not completely satisfactory, because the intrinsic algebraic structure of states is lost, and compositionality of denotational semantics is not reflected in this model. In [37], Turi and Plotkin provide a solution by means of *bialgebras*. These are pairs composed of a Σ-algebra and a B-coalgebra for Σ and B two endo-functors on Set related by a distributive law λ. Roughly, they have shown that providing a set of SOS rules (in some well-behaved format) corresponds to defining λ; the syntax of the formalism is the initial algebra for Σ and the semantics domain is the final coalgebra for B. This uniquely induces a (bialgebraic) morphism (representing the denotational semantics) that maps each element of the initial Σ-algebra (i.e., each term of the syntax) into the final B-coalgebra (representing the denotation of the terms). Since morphisms also respect the operations of Σ, the denotational semantics is also compositional.

This approach is general enough to allow one, by simply modifying either the base category or the associated endo-functors, also to handle sophisticated process calculi having complex variable binding (like the value passing CCS [25] and the π-calculus [26]). More precisely, in order to represent both syntax and semantics, one has to consider proper endo-functors not on Set, but on categories Set^C of *covariant presheaves* over some C. These are functors from a generic category C of interfaces and contexts to Set. Intuitively, a presheaf maps each object i of C to the set of states having i as interface, and each arrow $c : i \to j$ to a function turning states with interface i into states with interface j.

As an example, abstract syntax with variable binding has been tackled as a category of endo-functors Σ over $Set^{\mathbb{F}}$ [14,21], for $\mathbb{F}$ the category of finite cardinals (i.e., sets of variables) and all functions (i.e., variable substitutions).

Models for Names. The index category C can be made up of a single object. This is the case of presheaves in the category $Set^{\mathbb{G}}$ with $\mathbb{G}$ the group of permutations over natural numbers. This category is essentially the same as the *nominal sets* of Pitts and Gabbay [16], used to model abstract syntax with variable binding, and of *permutation algebras*, that have been exploited to give a final semantics of the π-calculus [28,6]. This corresponds to an *untyped* view of interfaces, where the actual element of interest is the action of a presheaf on the *arrows* of the one-object category.

Having more objects in the index category corresponds instead to a *typed* framework. One of the most widely investigated cases is $\mathbb{I}$, that is, the full subcategory of $\mathbb{F}$ containing only injective morphisms. Both the early and late semantics of the π-calculus [26] can be characterized by proper endo-functors on $Set^{\mathbb{I}}$ [36,13,15]. Intuitively, any object i of $\mathbb{I}$ is mapped to a set of processes whose free names belong to i, and an arrow is mapped into an injective renaming. In this perspective, it is easy to understand the roots of a problem which is common to the typed and untyped case: interfaces of π-processes can always be enlarged (an operation that corresponds to allocate new names) but never contracted (so that two names can never be coalesced).

A solution to this problem was proposed by Montanari and Pistore by introducing *history dependent automata* [27,31], where at any step of the execution a set of names can be junked away, obtaining an operational model where finite state verification of name passing calculi can be carried out in interesting cases. A coalgebraic formulation of HD-automata was given in [10,11], by employing the category of *named sets*.

An equivalence result between the categories of named sets, permutation algebras, $Set^{\mathbb{G}}$ and the full subcategory of $Set^{\mathbb{I}}$ of pullback preserving functors (also known as *Schanuel topos*) has been given by the fourth author, Marino Miculan and Ugo in [17]. An extension of this result to the associated categories of coalgebras has recently been proposed in [9], by the third author and Ugo. There, a garbage-collecting functor for name abstraction is defined on named sets, allowing the coalgebraic framework of HD-automata to be generalised to calculi different than the π-calculus.

In the previous proposals, arrows of the index category are just injections of names, and thus names can not be fused. In order to tackle non injective substitutions of names, one can consider different index categories. The open semantics of the π-calculus [34], for example, can be defined using an endo-functor on $Set^{\mathbb{D}}$, for $\mathbb{D}$ the category of irreflexive relations and relation-preserving morphisms [19,24]. Roughly, every object is a set of names equipped with an irreflexive relation such that two related names are considered distinguished. Morphisms are functions between sets of names that preserve the relation and thus they can not coalesce those names that are considered distinct.

Explicit Fusions. In this work we introduce the category $\mathbb{E}$, of equivalence relations, in order to properly model *explicit fusions* [39], i.e., processes that equate two names, allowing all the processes running in parallel with them to use one name in place of the other. Our interest for the explicit fusion calculus comes from the fact that it lies half-way between the fusion calculus [30] (to which Ugo has devoted much attention in the last years [4,22,18,12,8]) and the CC-Pi [7]. Indeed, explicit fusions are just instances of the more general concept of *named c-semiring* introduced in [7] by the second author and Ugo.

Each object of $\mathbb{E}$ is an equivalence relation over a set of names. Analogously to $\mathbb{I}$, morphisms preserve names (i.e., names can not be junked away), but equivalence classes can be enlarged, thus obtaining semantical fusion of names without loosing any syntactic name. We prove that $\mathbb{E}$ is suitable for providing both syntax and semantics to the *calculus of explicit fusions* [39] by showing both a syntactic endo-functor Σ and a behavioural endo-functor B on $Set^{\mathbb{E}}$.

Unfortunately, it seems hard to recover the Turi-Plotkin framework, since the operational behaviour of explicit fusion calculus is not compositional with respect to the semantical name fusions provided by the arrows of $\mathbb{E}$. This might suggest that $\mathbb{E}$ is not adequate for our purposes, but the main result states that *inside-outside bisimulations* [38] (i.e., the standard bisimulation of explicit fusion) are in one to one correspondence with coalgebraic bisimulations for the endo-functor B. Moreover, considering a proper *saturated semantics* [29,1] in $Set^{\mathbb{E}}$ might bring to a compositional operational semantics, and thus to a bialgebraic representation.

Thus the paper presents no conclusive result for the denotational semantics of explicit fusion, but just a solid base for further studies. Moving beyond explicit fusion, $\mathbb{E}$ is the first step toward a deeper understanding of named c-semirings and then for a Turi-Plotkin characterization of CC-Pi [7]. Moreover, by slightly modifying $\mathbb{E}$, thanks to its definition as a comma category, one can obtain a category of equivalence relations and distinctions that is suitable for open π-calculus [34] and D-fusion [4]. Finally, these formalisms share a symbolic semantics that can not be naively tackled through coalgebras. Giving a presheaf semantics for them will hopefully allow these symbolic semantics to be characterized through *normalized coalgebras* as shown by the first author and Ugo in [2,3].

Synopsis. In § 2 we give a brief overview of the explicit fusion calculus. In § 3 we introduce the category $\mathbb{E}$ of equivalence relations. In § 4 we show how to use $\mathbb{E}$ to characterise an abstract syntax of explicit fusion calculus as an initial algebra. In § 5 we define a behavioural endo-functor B on $Set^{\mathbb{E}}$ and we state a correspondence between inside-outside bisimulations and B-bisimulations. In § 6 we draw some conclusions and provide directions for future work.

2 Background on the Explicit Fusion Calculus

The explicit fusion calculus is a variant of the π-calculus that aims at guaranteeing asynchronous broadcasting of name equivalences to the environment. In order to ease its introduction, this section presents the explicit fusion calculus in the standard π-calculus fashion rather than in the "commitment" style of [39].

Assume a set of *names* $\mathcal{N}$, ranged over by $x, y, \dots$ The *(explicit) fusion processes*, ranged over by $P, Q, \dots$ are defined by the syntax in Fig. 1(a). The τ prefix stands for a silent action, while the complementary output $\overline{x}\langle y \rangle$ and input $x\langle y \rangle$ prefixes are used for communications. Unlike π-calculus, the input prefix is not a binder, hence input and output operations are fully symmetric. As usual, $\mathbf{0}$ stands for the inert process, $P \,|\, P$ for the parallel composition, $!P$ for the replication operator and $(x)P$ for the process that makes the name x local in P. An *explicit fusion* $x = y$ is a process that exists concurrently with the rest of the system and that enables to use the names x and y interchangeably.

We now define the *structural* congruences $\equiv_1$ and $\equiv_2$ as the least congruences over processes closed with respect to α-conversion and satisfying the axioms in Fig. 1(b) and Fig. 1(c), respectively. Finally, we define the structural congruence $\equiv$ as the transitive closure $(\equiv_1 \cup \equiv_2)^*$. With respect to the standard structural congruence of fusion processes, here we changed a few axioms. In particular, we weakened the axiom called *reflexivity*, equating $x = x$ and $\mathbf{0}$: in our case, we offer a name-preserving version, equating $x = x \,|\, P$ with P, whenever P contains x free. Moreover, instead of the *subtraction* axiom, that equates the processes $(\nu x)(x = y)$ and $\mathbf{0}$ (thus allowing us to remove names from an equivalence relation via restriction), we prefer to add an explicit α-conversion law for our processes, which is apparently the reason underlying the introduction of the subtraction axiom in [39]. Summing up, we considered a set of axioms guaranteeing that any parallel composition of fusions is closed with respect to the composition with any fusion, when that fusion occurs in the induced equivalence relation plus three axioms. These axioms (the left-most of Fig. 1(c)) ensure what is called small-step substitution in Fig. 2 of [39]: this is further made explicit by Proposition 1, taking into account the equivalence relation $\mathsf{Eq}(P)$ (Fig 1(d)), specifying the name equivalences induced by a process P: it properly generalizes [39, Lemma 5].

Proposition 1 (decomposition). *Let P, Q be processes such that $P = C[Q]$ for a unary context $C[-]$ that does not bind the names in $\mathrm{fn}(P)$ (i.e., such that the placeholder $-$ does not fall in the scope of a restriction operator (x), for any $x \in \mathrm{fn}(P)$). Moreover, let E_P be the process*

$$\prod_{\{x,y \in \mathrm{fn}(P) \,|\, (x,y) \in \mathsf{Eq}(P)\}} x = y$$

(where $\prod$ denotes the multiple parallel composition). Then, $P \equiv_2 C[E_P \,|\, Q]$.

The rules of the operational semantics, as depicted in Fig. 1(e), recall the rules of the π-calculus. We assume a set of labels

$$\mathcal{L} = \{\tau, z\langle y \rangle, \overline{z}\langle y \rangle, z(w), \overline{z}(w)\}$$

where z, y are free and w is bound. We let $\lambda, \lambda_1, \dots$ range over $\mathcal{L}$. By $\mathsf{Eq}(P) \vdash \lambda_1 = \lambda_2$ we mean that P contains enough fusions to interchange λ_1 and λ_2. Formally, we have

$$\mathsf{Eq}(P) \vdash \tau = \tau$$

$$\mathsf{Eq}(P) \vdash z\langle y\rangle = x\langle w\rangle \quad \text{if } (z,x),(y,w) \in \mathsf{Eq}(P) \qquad \text{(similarly for output)}$$

$$\mathsf{Eq}(P) \vdash z(v) = x(v) \quad \text{if } (z,x) \in \mathsf{Eq}(P) \qquad \text{(similarly for output)}$$

Unlike the π-calculus, the synchronization of two complementary processes $x\langle y\rangle.P$ and $\overline{x}\langle z\rangle.Q$ yields an explicit fusion $y = z$ rather than binding y to z. According to rule (FUS), a process can undergo any transition up to interchanging names that are fused. For instance, a process $x = y \,|\, \overline{x}\langle z\rangle.P$ can make both actions $\overline{x}\langle z\rangle$ and $\overline{y}\langle z\rangle$.

In [38] several bisimulations have been proposed for the explicit fusion calculus: they were all proved to coincide, and in fact to be congruences. For convenience, as a reference semantics we consider the *inside-outside* bisimulation that is much like π-calculus open bisimulation.

Definition 1 (inside-outside bisimulation). *Let R be a symmetric relation on fusion processes. We say that R is an inside-outside bisimulation if whenever $P \, R \, Q$ holds then*

1. *$\mathsf{Eq}(P) = \mathsf{Eq}(Q)$;*
2. *If $P \xrightarrow{\lambda} P'$ with $\mathrm{bn}(\lambda) \cap \mathrm{fn}(Q) = \emptyset$ then $Q \xrightarrow{\lambda} Q'$ and $P' \, R \, Q'$;*
3. *$P \,|\, x = y \, R \, Q \,|\, x = y$, for all fusions $x = y$.*

We let $\sim^{io}$ denote the largest such bisimulation, the inside-outside bisimilarity.

According to clause *1*, two processes $x = y$ and $\mathbf{0}$ are not bisimilar, since $\mathsf{Eq}(x = y)$ and $\mathsf{Eq}(\mathbf{0})$ differ. However, bisimilarity is not name preserving, so that $\mathbf{0}$ is indeed bisimilar to $x = x$. Finally, note that clause *3* allows bisimilarity to distinguish the following processes (inspired by an example described by Boreale and Sangiorgi [5] for the π-calculus)

$$P = \,!\overline{y}\langle\rangle.x\langle\rangle.\tau.z\langle\rangle \,|\, !x\langle\rangle.\overline{y}\langle\rangle.\tau.z\langle\rangle \qquad Q = \,!(w)(\overline{y}\langle\rangle.\overline{w}\langle\rangle \,|\, x\langle\rangle.w\langle\rangle.z\langle\rangle)$$

When inserted into the context $\_ \,|\, x = y$, Q can perform an action $\xrightarrow{z\langle\rangle}$ after two steps (synchronizing $\overline{y}\langle\rangle$ with $x\langle\rangle$), while P at least after three steps.

3 A Category of Name Equivalences

The paper aims at extending the presheaf approach in order to tackle the calculus of explicit fusions. To this end, we now define a category of equivalence relations $\mathbb{E}$ that will be used to represent (the sets of) fusion processes as a presheaf

$$\pi ::= \tau \mid \overline{x}\langle y\rangle \mid x\langle y\rangle \qquad\qquad P ::= \mathbf{0} \mid x = y \mid \pi.P \mid P \mid P \mid (x)P \mid {!}P$$

(a) syntax

$$P \mid \mathbf{0} \equiv_1 P \qquad\qquad P \mid Q \equiv_1 Q \mid P \qquad\qquad (P \mid Q) \mid R \equiv_1 P \mid (Q \mid R)$$

$${!}P \equiv_1 P \mid {!}P \qquad (x)(y)P \equiv_1 (y)(x)P \qquad P \mid (x)Q \equiv_1 (x)(P \mid Q) \quad \text{if } x \notin \mathrm{fn}(P)$$

(b) structural congruence, I

$$x = y \mid \pi.P \equiv_2 x = y \mid \pi.(x = y \mid P) \qquad\qquad x = x \mid P \equiv_2 P \quad \text{if } x \in \mathrm{fn}(P)$$

$$x = y \mid (z)P \equiv_2 x = y \mid (z)(x = y \mid P) \quad \text{if } z \notin \{x, y\} \qquad\qquad x = y \equiv_2 y = x$$

$$x = y \mid {!}P \equiv_2 {!}(x = y \mid P) \qquad\qquad x = y \mid y = z \equiv_2 x = z \mid y = z$$

(c) structural congruence, II

$$\mathsf{Eq}(\mathbf{0}) = \mathsf{Eq}(\pi.P) = \mathsf{Id} \qquad\qquad \text{the identity relation}$$

$$\mathsf{Eq}(x = y) = \{(x, y), (y, x)\} \cup \mathsf{Id} \qquad\qquad \text{smallest equivalence including } (x, y)$$

$$\mathsf{Eq}(P \mid Q) = (\mathsf{Eq}(P) \cup \mathsf{Eq}(Q))^* \qquad\qquad \text{transitively-closed union}$$

$$\mathsf{Eq}((x)\,P) = \mathsf{Eq}(P) \setminus \{(y, z) \mid x \in \{y, z\}\} \qquad\qquad \text{removing name from equivalence classes}$$

$$\mathsf{Eq}({!}\,P) = \mathsf{Eq}(P) \qquad\qquad \text{removing replication operator}$$

(d) equivalence relation $\mathsf{Eq}(P)$

(PREF)
$$\pi.P \xrightarrow{\pi} P$$

(COMM)
$$\frac{P \xrightarrow{x\langle y\rangle} P' \quad Q \xrightarrow{\overline{x}\langle w\rangle} Q'}{P \mid Q \xrightarrow{\tau} P' \mid Q' \mid y = w}$$

(PAR)
$$\frac{P \xrightarrow{\lambda} P' \quad \mathrm{bn}(\lambda) \cap \mathrm{fn}(Q) = \emptyset}{P \mid Q \xrightarrow{\lambda} P' \mid Q}$$

(RES)
$$\frac{P \xrightarrow{\lambda} P' \quad x \notin \mathrm{n}(\lambda)}{(x)\,P \xrightarrow{\lambda} (x)\,P'}$$

(OPEN-I)
$$\frac{P \xrightarrow{z\langle x\rangle} P' \quad (x, z) \notin \mathsf{Eq}(P)}{(x)\,P \xrightarrow{z(x)} P'}$$

(FUS)
$$\frac{P \xrightarrow{\lambda} Q \quad \mathsf{Eq}(P) \vdash \lambda = \lambda'}{P \xrightarrow{\lambda'} Q}$$

(STRUCT)
$$\frac{P \equiv P' \xrightarrow{\lambda} Q' \equiv Q}{P \xrightarrow{\lambda} Q}$$

(e) operational semantics (omitting rule (OPEN-O) for output)

Fig. 1. Explicit fusion calculus

$\mathbb{E} \to \textit{Set}$. The objects of $\mathbb{E}$ are basically surjective functions on sets and the arrows of $\mathbb{E}$ are defined by taking suitable injective functions.

Definition 2 (The category $\mathbb{E}$). *The category $\mathbb{E}$ of equivalence relations is the category whose objects are surjective functions $h : s \twoheadrightarrow t$ in Set and whose arrows $\sigma : h \to h'$ are pairs $\langle \sigma_1, \sigma_2 \rangle$ of functions in Set such that $\sigma_1 : s \hookrightarrow s'$ is injective and the diagram below commutes.*

$$
\begin{array}{ccc}
s & \xrightarrow{\ \sigma_1\ } & s' \\
\Big\downarrow{\scriptstyle h} & & \Big\downarrow{\scriptstyle h'} \\
t & \xrightarrow[\ \sigma_2\ ]{} & t'
\end{array}
$$

The domain s of the function h specifies a set of names while the codomain represents a set of equivalence classes involving the names of s. Note that every s' has at least the same arity of s. Furthermore, the equivalence classes in t must be preserved in t': in other words, the equivalence classes of the names in s, as represented in t and carried along σ_1 to s', may be either left unchanged or further merged in t', but they can not be broken.

Note that all arrows in $\mathbb{E}$ are monomorphims. We let $\mathbb{E}(h, h')$ denote the set of all arrows in $\mathbb{E}$ with source h and target h'. Each object h of $\mathbb{E}$ defines a functor $\mathbb{E}(h, \_) : \mathbb{E} \to Set$ as follows. Every object h' of $\mathbb{E}$ is mapped into the set $\mathbb{E}(h, h')$. An arrow $\sigma : h' \to h''$ of $\mathbb{E}$ is mapped in the function $\mathbb{E}(h, \sigma) : \mathbb{E}(h, h') \to \mathbb{E}(h, h'')$, defined by post-composition: for any $\rho \in \mathbb{E}(h, h')$, $\mathbb{E}(h, \sigma)(\rho) = \rho; \sigma$.

Let $\{\star\}$ denote the one element set, graphically represented as $\bullet$, and let us denote by 1 and E the following two objects of $\mathbb{E}$ (corresponding to $id_{\{\star\}}$ and $[id_{\{\star\}}, id_{\{\star\}}]$, respectively, for the uniquely-induced arrow)

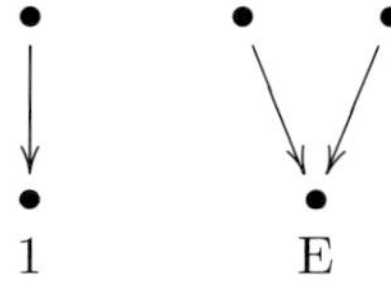

1   E

Then, $\mathbb{E}(1, h)$ is the set of all monomorphisms that map $\star$ to an element of the domain of h: hence, it is isomorphic to it. Similarly, $\mathbb{E}(E, h)$ abstractly represents the set of explicit fusions $x = y$ (for x, y different names) that hold in h. Hereafter, we denote the set $\mathbb{E}(1, \_)$ by *Names* and the set $\mathbb{E}(E, \_)$ by *Fus*.

On the structure of $\mathbb{E}$ Consider the comma category $ID_{Set} \downarrow ID_{Set}$, i.e. the category whose objects are triples $\langle s, t, h : s \to t \rangle$ and whose arrows are pairs $\sigma_1 : s \to s'$, $\sigma_2 : t \to t'$ such that the diagram below commutes.

$$
\begin{array}{ccc}
s & \xrightarrow{\ \sigma_1\ } & s' \\
\Big\downarrow & & \Big\downarrow \\
t & \xrightarrow[\ \sigma_2\ ]{} & t'
\end{array}
$$

Let $IN_{\mathbb{I}} : \mathbb{I} \to Set$ be the functor that injects the category $\mathbb{I}$ into Set. Consider the category $IN_{\mathbb{I}} \downarrow ID_{Set}$, where $\mathbb{I}$ is the category of sets and injective functions: in the definition above, it simply means that σ_1 must always be a monomorphism. Now, $\mathbb{E}$ is the full subcategory of $IN_{\mathbb{I}} \downarrow ID_{Set}$, including only those objects $\langle s, t, h : s \twoheadrightarrow t \rangle$ such that h is surjective.

All limits do exist in $IN_{\mathbb{I}} \downarrow ID_{Set}$: they are simply computed point-wise. This is not the case instead in $\mathbb{E}$. Let us consider e.g. pullbacks: given any two arrows $\langle \sigma_1, \sigma_2 \rangle : h \to h^*$, and $\langle \sigma_1', \sigma_2' \rangle : h' \to h^*$, their pullback exists only if either σ_2 or σ_2' is mono. Note that if both components of an arrow σ are mono, then σ is a *regular* monomorphism in $IN_{\mathbb{I}} \downarrow ID_{Set}$ (and in $\mathbb{E}$ as well), meaning that it can be computed as the equalizer of two arrows.

In the following, we sometimes consider functors from $\mathbb{E}$ to Set that are pullback-preserving: this fact which basically implies that all regular monomorphisms in $\mathbb{E}$ are mapped into injective functions in Set.

4 Abstract Syntax

In this section we consider the category $Set^{\mathbb{E}}$ of functors from $\mathbb{E}$ to Set (called *presheaves* over $\mathbb{E}^{op}$) and natural transformations. This category can be used for both the syntax and the semantics of the explicit fusion calculus.

In the syntax, objects of the so-called *index* category $\mathbb{E}$ can be viewed as *types* representing the equivalence classes of process names. The presheaf for the syntax gives, for each index $h : s \twoheadrightarrow t$, the set of those processes that can be typed with h, i.e., processes whose set of free names coincides with s, and whose equivalence class is the one induced by h.

In Section 4.1 we provide an account of the concrete syntax of the explicit fusion calculus as a presheaf, exploiting the previously defined equivalence relation Eq. Then, in Section 4.3 we explain how this syntax can be described as an initial algebra in $Set^{\mathbb{E}}$ for a suitable endo-functor, preserving types. For the purpose, in Section 4.2 we introduce a number of basic constructions that are used as a meta-language in $Set^{\mathbb{E}}$, and are distinguishing features of this category.

4.1 Typing the Concrete Syntax

For the definition of the syntactic presheaf, for each arrow $h : s \twoheadrightarrow t$ we introduce the notation Ker_h, representing the equivalence relation induced by h (obviously defined), and the process P_h, containing all the pairs in an equivalence relation h, viewed as fusions (analogous to the process E_P induced by $\mathsf{Eq}(P)$)

$$P_h = \prod_{\{x,y \in s \mid h(x)=h(y)\}} x = y.$$

Consider the functor $Syn\colon \mathbb{E} \to Set$, defined on objects as

$$Syn\,(h : s \twoheadrightarrow t) = \{P \mid \mathrm{fn}(P) = s \wedge \mathsf{Eq}(P) = Ker_h \cup \mathsf{Id}\} \,/ \equiv_2 .$$

The intuition behind this definition is that if an agent P is in $Syn\,(h : s \twoheadrightarrow t)$, then its free names are the elements of s, and its equivalence classes $\mathsf{Eq}(P)$ are exactly those described by h, modulo adding the pairs (x, x) for $x \notin \mathrm{fn}(P)$.

The action of the functor on arrows must injectively relabel processes, whilst it syntactically adds as many fusions as needed (actually, all of them) to put a process in the equivalence class chosen as destination

$$Syn\,(\langle \sigma_1 : s \hookrightarrow s', \sigma_2 : t \to t' \rangle : h \to h')(P) = P\sigma_1 \mid P_{h'}$$

4.2   A Basic Metalanguage in $Set^{\mathbb{E}}$

Besides the usual constructors for the polynomial endo-functors (namely constants, finite sums and products), $Set^{\mathbb{E}}$ actually allows us to define additional constructors that are well-suited for handling fusions.

Bottom operator $F^{\perp}$. For any $s \in Set$, let $s_{\perp}$ be $s + \{\star\}$ and for any $h : s \twoheadrightarrow t$, let $h_{\perp} : s_{\perp} \twoheadrightarrow t_{\perp}$ be the arrow $[h, id_{\{\star\}}]$. The bottom operator $F^{\perp}$ is an endo-functor on $Set^{\mathbb{E}}$ defined as $F^{\perp}(h) = F(h_{\perp})$ and $F^{\perp}(\langle \sigma_1, \sigma_2 \rangle) = F(\langle (\sigma_1)_{\perp}, (\sigma_2)_{\perp} \rangle)$.

Box operator $F^{\square}$. The *box operator* $F^{\square}$ is the endo-functor on $Set^{\mathbb{E}}$ defined on objects and arrows as below. Let $h : s \twoheadrightarrow t$ be an object of $\mathbb{E}$. Then

$$F^{\square}(h) = \sum_{p} F(h; p)$$

where $p : t \twoheadrightarrow p(t)$ is a (identity preserving) morphism from t to a partition $p(t)$ of t. Intuitively, p is an epimorphism further merging the names in s. Choosing $p(t)$ as a target of p simply amounts to choose a canonical representative for each isomorphic merging of names.

$$
\begin{array}{ccc}
s & \xrightarrow{\;id_s\;} & s \\
h \downarrow & & \downarrow h;p \\
t & \xrightarrow{\;p\;} & p(t)
\end{array}
$$

Let $h : s \twoheadrightarrow t$ and $h' : s' \twoheadrightarrow t'$ be objects of $\mathbb{E}$ and $\sigma = \langle \sigma_1, \sigma_2 \rangle : h \to h'$ be an arrow of $\mathbb{E}$. Then

$$F^{\square}(\sigma) = \sum_{p} F(\langle \sigma_1, \sigma_2^* \rangle)$$

where σ_2^* is uniquely induced by the pushout depicted in the diagram below for each p, noting that $\langle \sigma_1, \sigma_2^* \rangle : h; p \to h'; p^*$.

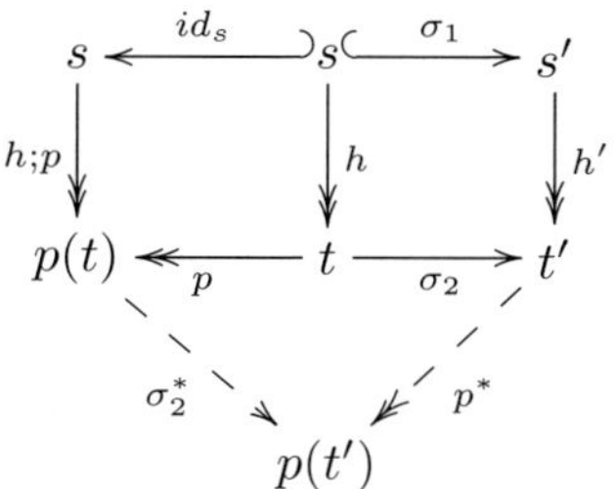

Roughly, $F^\square$ may add any possible name equivalence to an equivalence relation on s. The box operator is used in Section 4.3 to define the functor for prefixes of the explicit fusion. Indeed, a fusion process with a prefix $\pi.P$ has no equivalence relation on its names, while P may have any equivalence.

Shift operator F^δ. The *shift operator* is the functor defined on objects and arrows as follows. Let $h : s \twoheadrightarrow t$ be an object of $\mathbb{E}$. Then

$$F^\delta(h) = F^\perp(h) + \sum_e F([h, e])$$

where $e : \{\star\} \to t$ is a function mapping $\star$ in some element of t and $[h, e] : s + \{\star\} \twoheadrightarrow t$ is the function uniquely induced by the coproduct, mapping all elements of s in $h(s)$ and $\star$ in $e(\star)$.

Let $h : s \twoheadrightarrow t$ and $h' : s' \twoheadrightarrow t'$ be objects of $\mathbb{E}$ and $\sigma = \langle \sigma_1, \sigma_2 \rangle : h \to h'$ be an arrow of $\mathbb{E}$. Then

$$F^\delta(\sigma) = F^\perp(\langle \sigma_1, \sigma_2 \rangle) + \sum_e F(\langle (\sigma_1)_\perp, \sigma_2 \rangle)$$

noting that $e; \sigma_2 : \{\star\} \to t'$ and $\langle (\sigma_1)_\perp, \sigma_2 \rangle : [h, e] \to [h', e; \sigma_2]$.

The shift operator is used to define the functor for the restriction operation in the explicit fusion. In fact, F^δ is a variant of the functor for name generation used to model restriction in the π-calculus. The main point here is that objects are now equivalence relations. Hence, when generating a new name x, it is necessary to specify whether x is equivalent to any other name already occurring in the process, or it belongs to its own equivalence class.

4.3   Explicit Fusion Syntax as an Initial Algebra

An abstract syntax for the explicit fusion calculus is captured by the initial algebra of the endo-functor $\Sigma : Set^{\mathbb{E}} \to Set^{\mathbb{E}}$ that is defined below.

$$
\begin{aligned}
\Sigma F = \ &1 &&\text{(inert process)}\\
+\ &F &&\text{(replication)}\\
+\ &F \times F &&\text{(parallel composition)}\\
+\ &F^\square &&\text{(tau prefix)}\\
+\ &Names \times Names \times F^\square &&\text{(input prefix)}\\
+\ &Names \times Names \times F^\square &&\text{(output prefix)}\\
+\ &F^\delta &&\text{(restriction)}
\end{aligned}
$$

We briefly comment on the component functors. For the purpose, we introduce the notation $!_s$ representing the unique morphism of type $s \twoheadrightarrow \{\star\}$ from s to the final object in Set, viewed as an object of $\mathbb{E}$ (noting that $!_{\{\star\}} = id_{\{\star\}}$). Moreover, we say that an object $h : s \twoheadrightarrow t$ of $\mathbb{E}$ is *included* in an object $h_1 : s_1 \twoheadrightarrow t_1$, and we write $h \sqsubseteq h_1$, if there exists an arrow $\sigma : h \to h_1$.

The functor for the inert process just returns a singleton at each stage; whilst the functor for replication just returns an additional copy of the processes at each stage, as if prefixing them with a suitable operator.

Table 1. Inference rules for Abstract Syntax

$$\mathbf{0} \in F(id_\emptyset) \qquad \frac{p \in F(h) \quad h \sqsubseteq h_1}{p \in F(h_1)} \qquad \frac{p \in F(h)}{!p \in F(h)}$$

$$\frac{p \in F^\square(h)}{\tau.p \in F(h)} \qquad \frac{p \in F^\delta(h)}{(\star)p \in F(h)} \qquad \frac{p \in F(h) \quad q \in F(h)}{p \,|\, q \in F(h)}$$

$$\frac{a \in Names(h) \quad b \in Names(h) \quad p \in F^\square(h)}{\overline{a}\langle b\rangle.p \in F(h)}$$

$$\frac{a \in Names(h) \quad b \in Names(h) \quad p \in F^\square(h)}{a\langle b\rangle.p \in F(h)}$$

The three functors for prefixing are similar, the idea being of "hiding" equivalences of a process P when prefixed as $\pi.P$, similarly to what is done in the definition of $\mathsf{Eq}(P)$. To this purpose, we use the functor $F^\square$: it returns, at stage $h : s \twoheadrightarrow t$, all the processes occurring in $F(h')$ for all h' such that $h \sqsubseteq h' \sqsubseteq !_s$.

The parallel composition of two elements $a \in F(h_1)$ and $b \in F(h_2)$ can only be found at a stage h such that $h_1 \sqsubseteq h$ and $h_2 \sqsubseteq h$. Indeed, being a presheaf, the product $F \times F$ is a functor of type $\mathbb{E} \to Set$ defined as $(F \times F)(h) = F(h) \times F(h)$. Hence, we can not find the product of two elements h_1 and h_2 belonging to *different* stages, but we have to "inject" them into a common stage h, where both can be found. Note that, since arrows can never remove fusions from objects of $\mathbb{E}$, this requirement is equivalent to stating that $\mathsf{Eq}(P \,|\, Q) = (\mathsf{Eq}(P) \cup \mathsf{Eq}(Q))^*$, carrying on the intuition that the stage of an element always includes its fusions.

Next, we consider F^δ. We can exemplify its meaning using the process $(x)(x = y)$. This process belongs to $F^\delta(id_{\{y\}})$, since its set of free names is indeed included in $\{y\}$. However, it is obtained from the process $x = y$, which does not belong to $F(id_{\{x,y\}})$, but rather to $F(!_{\{x,y\}})$. Hence, to capture the process $(x)(x = y)$ in the abstract syntax, $F(!_{\{y,\star\}})$ has to be a subset of $F^\delta(id_{\{y\}})$.

The previous discussion is summed up by Table 1. The different stages h account for the equivalences. Note e.g. that $\mathbf{0}$ belongs to $F(id_\emptyset)$, for $id_\emptyset$ the identity of the empty set. It thus belongs also to $F(id_{\{x\}})$ and to $F(!_{\{x,y\}})$, for any name x, y. However, in the latter stage it represents the process $\mathbf{0} \,|\, x = y$.

4.4   Including Fusions, Syntactically

We could have introduced explicitly the fusions in our syntactical functor, simply considering the functor Fus to represent them

$$\begin{aligned} \Sigma'F \;=\; & \Sigma F && \text{(fusionless calculus)} \\ & + \; Fus && \text{(explicit fusions)} \end{aligned}$$

This ensures, by definition, that a process $x = y$ (for x, y different names) can not be found in $F(h)$ unless $(x,y) \in Ker_h$, i.e. that the syntax reflects the equivalences of the objects in the index category. These constraints propagate to the parallel composition of fusions with arbitrary processes, and they induce exactly

Proof. Consider the requirements of $\mathbb{E}$-transition system. The first condition just states that we are working with a presheaf. The second condition imposes the correct type. The third and fourth requirement just impose that the transition structure is a natural transformation between F and $B(F)$, for each functor F.

$\square$

We now move to introduce a suitable notion of $\mathbb{E}$-bisimulation.

Definition 4. *Let $(F, \rightarrow)$ and $(G, \rightarrow)$ be two $\mathbb{E}$-transition systems. Let $R \subseteq \bigoplus_{h \in \mathbb{E}} F(h) \times G(h)$ be an $\mathbb{E}$-sorted family of symmetric relations. We say that R is an $\mathbb{E}$-bisimulation if whenever $f R_h g$ then*

- *if $\sigma : h \rightarrow i$, then $F(\sigma)(f) R_i G(\sigma)(g)$,*
- *if $h \vdash f \xrightarrow{\alpha} h' \vdash f'$ then $h \vdash g \xrightarrow{\alpha} h' \vdash g'$ and $f' R_{h'} g'$.*

We finally state the main result of the paper.

Proposition 3. *$\mathbb{E}$-bisimulations are in one to one correspondence with coalgebraic bisimulations for the endo-functor $B : Set^{\mathbb{E}} \rightarrow Set^{\mathbb{E}}$.*

Proof. Let us now consider the definition of the coalgebraic bisimulation for the endo-functor $B : Set^{\mathbb{E}} \rightarrow Set^{\mathbb{E}}$.

A presheaf R and two natural transformations $a : R \rightarrow F$ and $b : R \rightarrow G$ form a bisimulation (R, a, b) between (F, α) and (G, β) if (R, a, b) is a monic span[1] in $Set^{\mathbb{E}}$ and there exists in $Set^{\mathbb{E}}$ a natural transformation $\gamma : R \rightarrow B(R)$ such that the following commutes.

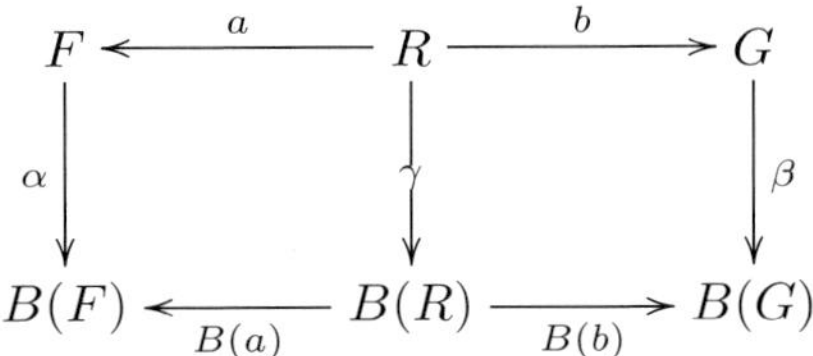

First of all note that in any category C with binary product, (R, a, b) is a monic span if and only if the induced morphism $\langle a, b \rangle : R \rightarrow F \times G$ is a monomorphism. In Definition 4 we require that $R \subseteq \bigoplus_{h \in \mathbb{E}} F(h) \times G(h)$, and this is equivalent to restricting to those $\langle a, b \rangle$ that are injections of R into $F \times G$, instead of simply monomorphisms. Therefore, the one to one correspondence holds only up to isomorphism, as it is standard in the theory of coalgebras.

Then, the first requirement of Definition 4 coincides with the fact that R is a presheaf and a, b are natural transformations. Indeed, if $f R_h g$ (i.e., $(f, g) \in R(h)$) and $\sigma : h \rightarrow i$, then $R(\sigma)(f, g) \in R(i)$, because R is a functor $\mathbb{E} \rightarrow Set$. Notice that $R(\sigma)(f, g) = (F(\sigma)(f), G(\sigma)(g))$, when considering $\langle a, b \rangle$ as the injection of R into $F \times G$. Indeed, since $a : R \rightarrow F$ is a natural transformation, then $a_i(R(\sigma)(f, g)) = F(\sigma)a_h(f, g) = F(\sigma)(f)$ (because $\langle a, b \rangle$ is the injection and not simply a mono). Similarly $b_i(R(\sigma)(f, g)) = G(\sigma)b_h(f, g) = G(\sigma)(g)$.

---

[1] The triple (R, a, b) is a monic span (or monic pair) in a category C if for all $h, i \in C$ it holds that $h = i$ whenever $h; a = i; a$ and $h; b = i; b$ do.

The second requirement of Definition 4 (together with the fact that R is symmetric) coincides with requiring the existence of a γ making the above diagram commute. This is a standard reasoning in the theory of coalgebras. The interested reader is referred to [33, Example 2.1]. $\qquad\square$

The correspondence between inside-outside bisimulations and B-bisimulatios is now evident. Both definitions require that the same equations must hold in the compared processes (this is implicitly expressed in $\mathbb{E}$-bisimulations by requiring that the relation is indexed over $\mathbb{E}$). Moreover, both definitions require that the relation is closed with respect to name fusions (this is equivalent to requiring that the relation is closed under all arrows of $\mathbb{E}$). And finally, the cases of bound input and bound output (expressed in inside-outside bisimulation by bn $(\alpha) \cap$ fn $(Q) = \emptyset$) is safely tackled by considering as the bound name a new name for both processes.

5.2   A Further Abstraction

We round up the section by showing an alternative definition of the behavioural functor. The definition would allow one to observe directly the classes of name equivalences, instead of the names themselves.

Let $Equiv : \mathbb{E} \to Set$ be the functor defined as $Equiv(h : s \twoheadrightarrow t) = t$ and $Equiv(\langle \sigma_1, \sigma_2 \rangle) = \sigma_2$. The behavioural endofunctor $B : Set^{\mathbb{E}} \to Set^{\mathbb{E}}$ is formally defined as

$$
\begin{aligned}
BF \;=\; \mathcal{P}_f(\\
F^{\square} & \qquad\qquad \text{(tau action)} \\
+\; Equiv \times Equiv \times F^{\square} & \qquad\qquad \text{(input action)} \\
+\; Equiv \times Equiv \times F^{\square} & \qquad\qquad \text{(output action)} \\
+\; Equiv \times F^{\perp^{\square}} & \qquad\qquad \text{(bound input action)} \\
+\; Equiv \times F^{\perp^{\square}} & \qquad\qquad \text{(bound output action)} \\
)
\end{aligned}
$$

¿From one side, this implicitly mimics the rule FUS, by forcing all processes to perform actions with equivalent names. On the other hand, this could be useful as an efficient characterization, since there would only be one transition for any two actions from a process P, as long as they are identified by the equivalence class $\mathsf{Eq}(P)$. We leave the exploration of this functor as future work.

6   Conclusions and Further Work

In this paper, we introduced the category $\mathbb{E}$ of equivalence classes, and we started the study of the presheaf category $Set^{\mathbb{E}}$. Some preliminary results, summed up in the last section, show that the category $Set^{\mathbb{E}}$ seems the right universe for providing denotational models for the fusion calculus.

Much work remains to be done. First of all, our propositions just proved that $Set^{\mathbb{E}}$ is the right category for discussing inside-outside bisimulation, but it is yet to be proved that the operational semantics induced by the rules in

Fig. 1(c) coincides with the unique morphism from the initial Σ-algebra T_Σ to $B(T_\Sigma)$ of our construction. The main problem concerns the fact that the operational semantics of explicit fusion calculus is not compositional with respect to name fusion, and thus it does not correspond to any arrow in $Set^{\mathbb{E}}$ (natural transformation). A possible way out of the empasse could be to consider the category of functors from $\mathbb{E}$ to Set and *lax* natural transformations. Another solution might consist in considering the *context transition system* in the spirit of [29,23]: for all $\sigma \in \mathbb{E}$, $p \xrightarrow{\sigma,l} p'$ if and only if $\sigma(p) \xrightarrow{l} p'$. It is evident that such an operational semantics is clearly compositional with respect to all fusions, and thus it is trivially an arrow in $Set^{\mathbb{E}}$.

The latter solution could lead us to a coalgebraic characterization of the so called *efficient bisimulation* [38, Definition 9]. There, as in the case of *open bisimulation* [34], instead of considering transitions $p \xrightarrow{\sigma,l} p'$ for all possible σ, only the *minimal* σ's are considered. This is also similar to *reactive systems*, as proposed by Leifer and Milner [23]. The exact correspondence with this approach is shown in [3] for the case of open π-calculus, and can be trivially extended to explicit fusion calculus. Unfortunately, the definition of efficient bisimulation is asymmetric and thus it seems hard to characterize it through canonical coalgebras. Probably, *normalized coalgebras* [2] can be fruitfully employed for this aim.

In more general terms, we would like to have a better understanding of the properties of $Set^{\mathbb{E}}$. In particular, we plan to check if alternative characterisations exist, mimicking the correspondence between nominal sets and Schanuel topos holding for the subcategory of $Set^{\mathbb{I}}$ of pullback-preserving functors. As a start, we noticed that in $\mathbb{E}$ pullbacks exist only along regular monomorphims.

As a next step, we would like to address those calculi featuring *distinctions*, such as D-Fusion [4] and the open semantics for π-calculus [34]. A suitable denotational model could be obtained by considering the category $\mathbb{D}$ of irreflexive graphs [24]. Given the injection $IN_{\mathbb{D}} : \mathbb{D} \rightarrow Set$, we should then study the comma category $IN_{\mathbb{I}} \downarrow IN_{\mathbb{D}}$, thus equipping each equivalence class (and each pair of names belonging to them) with a suitable irreflexive relation.

References

1. Bonchi, F., König, B., Montanari, U.: Saturated semantics for reactive systems. In: Proc. of LICS, pp. 69–80. IEEE, Los Alamitos (2006)
2. Bonchi, F., Montanari, U.: Coalgebraic models for reactive systems. In: Kok, J.N., Koronacki, J., Lopez de Mantaras, R., Matwin, S., Mladenič, D., Skowron, A. (eds.) ECML 2007. LNCS (LNAI), vol. 4701, pp. 364–380. Springer, Heidelberg (2007)
3. Bonchi, F., Montanari, U.: Symbolic semantics revisited. In: Amadio, R. (ed.) FOSSACS 2008. LNCS, vol. 4962, pp. 395–412. Springer, Heidelberg (2008)
4. Boreale, M., Buscemi, M.G., Montanari, U.: D-fusion: A distinctive fusion calculus. In: Chin, W.-N. (ed.) APLAS 2004. LNCS, vol. 3302, pp. 296–310. Springer, Heidelberg (2004)
5. Boreale, M., Sangiorgi, D.: Some congruence properties for pi-calculus bisimilarities. Theoret. Comput. Sci. 198(1-2), 159–176 (1998)

6. Buscemi, M.G., Montanari, U.: A first order coalgebraic model of pi-calculus early observational equivalence. In: Brim, L., Jančar, P., Křetínský, M., Kucera, A. (eds.) CONCUR 2002. LNCS, vol. 2421, pp. 449–465. Springer, Heidelberg (2002)

7. Buscemi, M.G., Montanari, U.: Cc-pi: A constraint-based language for specifying service level agreements. In: De Nicola, R. (ed.) ESOP 2007. LNCS, vol. 4421, pp. 18–32. Springer, Heidelberg (2007)

8. Buscemi, M.G., Montanari, U.: A compositional coalgebraic model of fusion calculus. J. Log. Algebr. Program. 72(1), 78–97 (2007)

9. Ciancia, V., Montanari, U.: A name abstraction functor for named sets. In: Proc. of CMCS. Elect. Notes in Th. Comput. Sci (to appear, 2008)

10. Ferrari, G.L., Montanari, U., Pistore, M.: Minimizing transition systems for name passing calculi: A co-algebraic formulation. In: Nielsen, M., Engberg, U. (eds.) FOSSACS 2002. LNCS, vol. 2303, pp. 129–158. Springer, Heidelberg (2002)

11. Ferrari, G.L., Montanari, U., Tuosto, E.: Coalgebraic minimization of HD-automata for the pi-calculus using polymorphic types. Theoret. Comput. Sci. 331(2-3), 325–365 (2005)

12. Ferrari, G.L., Montanari, U., Tuosto, E., Victor, B., Yemane, K.: Modelling fusion calculus using HD-automata. In: Fiadeiro, J.L., Harman, N.A., Roggenbach, M., Rutten, J. (eds.) CALCO 2005. LNCS, vol. 3629, pp. 142–156. Springer, Heidelberg (2005)

13. Fiore, M.P., Moggi, E., Sangiorgi, D.: A fully abstract model for the π-calculus. Inf. Comput. 179(1), 76–117 (2002)

14. Fiore, M.P., Plotkin, G.D., Turi, D.: Abstract syntax and variable binding. In: Proc. of LICS, pp. 193–202 (1999)

15. Fiore, M.P., Turi, D.: Semantics of name and value passing. In: Proc. of LICS, pp. 93–104. IEEE, Los Alamitos (2001)

16. Gabbay, M., Pitts, A.M.: A new approach to abstract syntax with variable binding. Formal Asp. Comput. 13(3-5), 341–363 (2002)

17. Gadducci, F., Miculan, M., Montanari, U.: About permutation algebras (pre)sheaves and named sets. Higher-Order and Symbolic Computation 19(2-3), 283–304 (2006)

18. Gadducci, F., Montanari, U.: Graph processes with fusions: Concurrency by colimits, again. In: Kreowski, H.-J., Montanari, U., Orejas, F., Rozenberg, G., Taentzer, G. (eds.) Formal Methods in Software and Systems Modeling. LNCS, vol. 3393, pp. 84–100. Springer, Heidelberg (2005)

19. Ghani, N., Yemane, K., Victor, B.: Relationally staged computations in calculi of mobile processes. In: The Programming Language Ada. LNCS, vol. 106, pp. 105–120. Springer, Heidelberg (1981)

20. Giarratana, V., Gimona, F., Montanari, U.: Observability concepts in abstract data type specifications. In: Mazurkiewicz, A. (ed.) MFCS 1976. LNCS, vol. 45, pp. 576–587. Springer, Heidelberg (1976)

21. Hofmann, M.: Semantical analysis of higher-order abstract syntax. In: Proc. of LICS, pp. 204–213. IEEE, Los Alamitos (1999)

22. Lanese, I., Montanari, U.: Mapping fusion and synchronized hyperedge replacement into logic programming. Theory Pract. Log. Program. 7(1-2), 123–151 (2007)

23. Leifer, J.J., Milner, R.: Deriving bisimulation congruences for reactive systems. In: Palamidessi, C. (ed.) CONCUR 2000. LNCS, vol. 1877, pp. 243–258. Springer, Heidelberg (2000)

24. Miculan, M., Yemane, K.: A unifying model of variables and names. In: Sassone, V. (ed.) FOSSACS 2005. LNCS, vol. 3441, pp. 170–186. Springer, Heidelberg (2005)

25. Milner, R.: Communication and Concurrency. Prentice-Hall, Englewood Cliffs (1989)
26. Milner, R., Parrow, J., Walker, D.: A calculus of mobile processes, I and II. Inform. and Comput. 100(1), 1–40 (1992)
27. Montanari, U., Pistore, M.: An introduction to history dependent automata. In: Proc. of HOOTS. Elect. Notes in Th. Comput. Sci, vol. 10, pp. 170–188 (1997)
28. Montanari, U., Pistore, M.: Structured coalgebras and minimal HD-automata for the pi-calculus. Theoret. Comput. Sci. 340(3), 539–576 (2005)
29. Montanari, U., Sassone, V.: Dynamic congruence vs. progressing bisimulation for CCS. Fundamenta Informaticae 16(1), 171–199 (1992)
30. Parrow, J., Victor, B.: The fusion calculus: Expressiveness and symmetry in mobile processes. In: Proc. of LICS, pp. 176–185. IEEE, Los Alamitos (1998)
31. Pistore, M.: History Dependent Automata. PhD thesis, Università di Pisa, Dipartimento di Informatica, Available at University of Pisa as PhD. Thesis TD-5/99 (1999)
32. Plotkin, G.: A structural approach to operational semantics. Technical Report DAIMI FN-19, Aarhus University, Computer Science Department (1981)
33. Rutten, J.J.M.M.: Universal coalgebra: a theory of systems. Theoret. Comput. Sci. 249(1), 3–80 (2000)
34. Sangiorgi, D.: A theory of bisimulation for the π-calculus. Acta Inform. 33(1), 69–97 (1996)
35. Scott, D., Strachey, C.: Toward a mathematical semantics for computer languages. In: Programming Research Group Technical Monograph, Oxford University, Computing Laboratory, vol. PRG-6 (1971)
36. Stark, I.: A fully abstract domain model for the π-calculus. In: Proc. of LICS, pp. 36–42. IEEE, Los Alamitos (1996)
37. Turi, D., Plotkin, G.D.: Towards a mathematical operational semantics. In: Proc. of LICS, pp. 280–291. IEEE, Los Alamitos (1997)
38. Wischik, L., Gardner, P.: Strong bisimulation for the explicit fusion calculus. In: Walukiewicz, I. (ed.) FOSSACS 2004. LNCS, vol. 2987, pp. 484–498. Springer, Heidelberg (2004)
39. Wischik, L., Gardner, P.: Explicit fusions. Theoret. Comput. Sci. 340(3), 606–630 (2005)

What Do Semantics Matter
When the Meat Is Overcooked?

José Luiz Fiadeiro

Department of Computer Science, University of Leicester
University Road, Leicester LE1 7RH, UK
`jose@mcs.le.ac.uk`

Abstract. We develop an abstract operational model for configuration management under service-oriented computing. This semantics is based on a graph-based representation of the configuration of global computers and an operational model of service-oriented dynamic reconfiguration based on a resolution-like mechanism similar to concurrent constraint programming. A resolution step involves a goal executed by a business activity and a clause that corresponds to a complex service. Unification captures service discovery, ranking and selection based on SLA-constraint optimisation and interpretations between specifications of conversations expected by the goal and provided by the discovered service. The resolvent is a reconfiguration of the original business activity that results from binding the goal with the discovered service.

1 Introduction

Given the breadth of Ugo Montanari's interests and expertise, it would not have been too difficult to contribute a paper in an area that he has touched. To match the depth of his 'touch' is, however, a much more difficult challenge. Not many people have made such profound contributions to what in computer science we usually call 'semantics', i.e. the definition of mathematical structures that explain given computational phenomena. Something that is particularly difficult in this area is to make sure that we do not obfuscate the subject of study. The title of this paper is a quote from Jan Moir, a British food critic (it is debatable whether the British know more about semantics or overcooked meat): her protest is (quite rightly) directed to sophisticated elaborations that end up destroying ingredients that would have deserved a much lighter touch. To some of us, Italian cuisine is precisely about simplicity and attention to what the products being cooked require to bring the best in them. Not surprisingly, Ugo excels in this tradition, and I am very fortunate to have been exposed to his culinary skills as a guest at a most memorable dinner that he cooked in May 2006.

In this paper, I have tried to pay tribute to Ugo by 'cooking' a 'dish' using some of the ingredients that he has cultivated (and earned fame). I should say immediately that this not a ribollita, quite the contrary: it is very much work in progress within the SENSORIA project, one of many to which Ugo has contributed during his long career. The dish is called "semantics of service-oriented configuration management".

Service-oriented computing (SOC) is a new paradigm in which interactions are no longer based on the exchange of products with specific parties – what is known as

P. Degano et al. (Eds.): Montanari Festschrift, LNCS 5065, pp. 563–580, 2008.

clientship in object-oriented programming – but on the provisioning of services that are procured through a process of discovery and negotiation that takes place at run-time, establishing a service-level agreement between the two parties. While it is recognised that specialised programming language primitives are needed that address the challenges raised by this new paradigm, we are still lacking models that are abstract enough to understand the foundations of the paradigm independently of the way services are programmed, namely from their current Web manifestation [3].

In previous papers (e.g. [2,12]), we have reported on the static aspects of SRML, a modelling language for service-oriented computing that we are defining within the SENSORIA project [20]. More precisely, we have focused on an algebraic approach for service description at the higher level of 'business modelling', and on techniques through which (simple) services can be assembled, at design-time, to create more complex services. Our contributions include language primitives for orchestrating interactions and a logic for specifying properties of conversations. Both the language and the associated logic are 'technology agnostic' in the sense that they are based on a semantic model that abstracts away from the languages in which services are programmed and the middleware that supports the coordination of interactions [1,13]. They are also expressive enough to accommodate orchestrations programmed in languages such as BPEL [7].

In this paper, we address the run-time aspects that are concerned with the way configurations of global computers change as services are discovered, selected, instantiated and bound to the applications that procured them. Once again, our aim is to provide an operational model of service-oriented configuration management that is independent of the technologies that provide the middleware infrastructure over which services can be deployed, published and discovered. For this purpose, we propose an approach inspired by (soft) concurrent constraint programming [6,19]: the process of reconfiguration is formalised in a resolution-style operational semantics that builds on the declarative algebraic semantics of SRML modelling primitives; the process of discovery, matching, ranking and selection involves unification/matching mechanisms based on c-semiring based techniques for constraint satisfaction and optimisation [5]. Familiarity with concurrent constraint programming is not strictly required as the analogy is used only for putting in context the different aspects of the operational model and its declarative semantics.

In Section 2, we lay the table by making precise what we mean by a configuration. Sections 3 and 4 address the static architectural aspects: how configurations can be structured in terms of business activities and services. Sections 5 and 6 address the dynamic aspects, i.e. how configurations change as business activities discover and bind to services. Throughout the paper, we make use of methods and techniques developed by Ugo and his colleagues. In Section 7, we say how we would like to continue doing so.

2 Configurations of Global Computers

Graphs are one of the most important commodities for any researcher working in computer science, a bit like hot (preferably boiling) water for cooks: they are not so much ingredients (i.e. they are not food as such) but enablers or domains over which one can cook a semantics. Ugo has excelled in the development and use of graph-based techniques. Our paper will also use graphs galore.

Ugo has been involved in pioneering work on the use of graphs for modelling software architectures (e.g. [15]). Because our aim is to develop a semantic domain for the way configurations of global computers are redefined as applications execute and get bound to other applications that offer required services, we choose to view configurations as graphs constituted of components (applications deployed over a given execution platform) and wires (interconnections between components over a given communication network) in a given state of execution (as in Fig. 1).

We denote by **COMP** and **WIRE** the set of all components and wires, respectively. Every component $c \in$ **COMP** and wire $w \in$ **WIRE** may be in a number of states (e.g. valuations of local state variables), the set of which is denoted by **STATE**$_c$ and **STATE**$_w$, respectively. We denote by **STATE** the corresponding indexed family of sets of states. The precise nature of these local states is of no particular importance for this paper.

A state configuration *SF* is defined to consist of:

- A simple graph G, i.e. a set *nodes(SF)* and a set *edges(SF)*; each edge *e* is associated with one and only one (unordered) pair $n \leftrightarrow m$ of nodes. We take *nodes(SF)* $\subseteq$ **COMP** (i.e. nodes are components) and *edges(SF)* $\subseteq$ **WIRE** (i.e. edges are wires).

- A (configuration) state S, i.e. an assignment of a state $S(c) \in$ **STATE**$_c$ to every $c \in nodes(SF)$ and $S(w) \in$ **STATE**$_w$ to every $w \in edges(SF)$.

A state configuration <G,S> can change because either the state function S or the graph G change. We treat these two kinds of changes separately: a computation step (state change) may trigger a reconfiguration step, which needs to complete before the next computation step is performed.

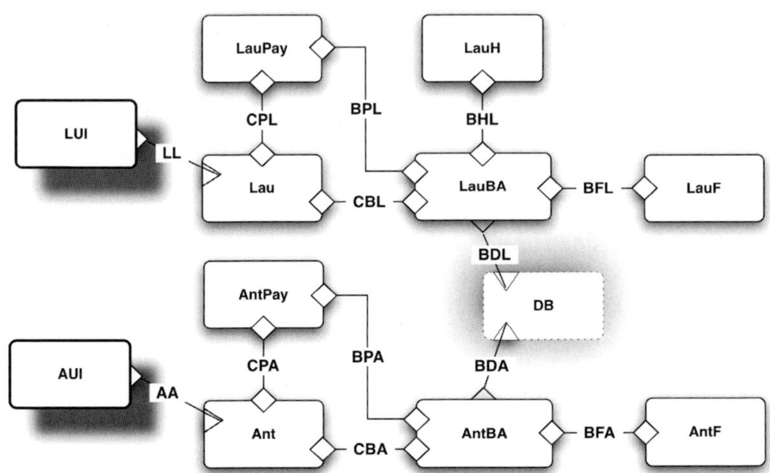

Fig. 1. The graph of a state configuration with 12 components and 13 wires

Changes to the state result from the computations executed by components and the coordination activities performed by the wires that connect them. The computational model that we are defining for SRML is explained in detail in [1]; besides providing local states for components and wires, configuration states include information on which events are pending; state transitions account for the effects of executing events on the local states and publication events. Because the configuration model that we discuss in this paper is largely independent of the computational one (except for what we call internal configuration policies below), we refrain from giving a detailed definition of that model and restrict ourselves to the aspects that are essential for understanding the way computation steps lead to reconfiguration ones. A computation step relates two state configurations $\langle G,src\rangle \rightarrow \langle G,trg\rangle$ by changing the state and keeping the graph invariant; reconfiguration steps, which change the graph but not the state, are considered later in the paper.

3   Services as Architectural Units

Our goal is to provide a semantics for changes in the configuration of a system, i.e. its graph, as resulting from a service-oriented architecture. More precisely, we are going to present a semantics for services as the basic units of configuration management. In this model, changes in the configuration graph – in the components that are active and the wires that connect them – result from the fact that state changes can trigger the discovery, ranking and selection of services that give rise to the addition of new components and wires that connect them to the rest of the configuration.

For this purpose, we need an architectural model of services. The basic elements of the architectural model of SRML are called modules. Together with some of his colleagues, Ugo gave a complete formalisation of the static aspects of this model in [9]. In this paper, we recall some of its essential parts and extend it to the dynamic aspects, i.e. service discovery and binding.

In Fig. 2 we present the structure of a module that defines a service provided through an interface **CR** of type *Customer* for booking a flight and a hotel for a given itinerary and dates. The service relies on a component **BA** of type *BookingAgent* that orchestrates interactions with a service **FA** of type *FlightAgent* (for booking flights), a service **HA** of type *HotelAgent* (for booking hotel rooms), a service **PA** of type *PayAgent* (for handling payments), and an external component **DB** of type *UsrDB* (that stores information about registered users).

Modules are also defined as graphs. Although we use the same icons for state configurations as for modules, the nodes of modules are not components and the edges are not wires: modules involve abstract models, i.e. the labels of the graph are types, not instances. The types abstract from the components the business roles that they play in the activity performed by the service and, from the wires, the connectors that are responsible for coordinating the way the components interact.

Some of the nodes of a module may consist of interfaces to a pool of shared components: this is the case of **DB** of type *UsrDB*, i.e. a database of users. Several such "uses-interfaces" can be included in a module. Other nodes – **PA**, **HA**, and **FA** – consist of "requires-interfaces" to external services that may need to be discovered for the service to fulfil its business goal. This goal is captured in a "provides-interface" – the node **CR** of type *Customer*.

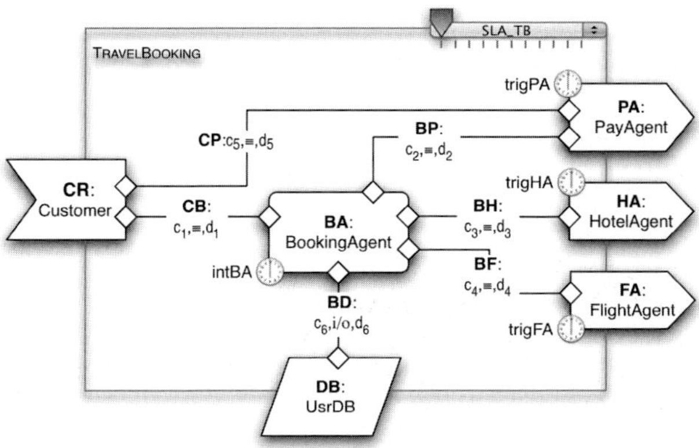

Fig. 2. The structure of a module defining the booking service of a travel agency

This notion of service module was inspired by concepts proposed in the Service Component Architecture (SCA) [21]: they are abstractions of (composite) services whose execution involves a number of interactions among coarse-grained components that perform tasks according to the underlying business logic, as well as external entities that also play a role in the business domain. These external parties are not explicitly identified in the module but only implicitly through what we have called external-interfaces. External interfaces are more than syntactic declarations: they are typed by business protocols – abstract specifications of the conversations in which the parties are required to be involved – or by layer protocols in the case of uses-interfaces – abstract specifications of the remote interactions supported with the external party. Likewise, the components themselves are not explicitly identified in the module. Instead, the module includes semantic interfaces – business roles – that model the way interactions are orchestrated by the components.

The operational model that we wish to present for configuration management is independent of the formalisms used for defining business roles, business protocols, layer protocols and connectors. Therefore, we will not discuss these formalisms in the paper; we assume instead that we have available sets **_BROL_**, **_BUSP_** and **_LAYP_** of specifications of business roles, business protocols and layer protocols, respectively (see [12] for an overview of the formalisms used in SRML).

The difference between business roles and protocols is that components that correspond to the business roles are created and bound to their interfaces when the module is instantiated (i.e. when a new session of the service is initiated) whereas the external services that correspond to the business protocols are bound to the require interfaces at run-time after a process of discovery, ranking and selection triggered according to the internal configuration policy of the module.

The difference with respect to layer protocols is that these bind to shared components that persist independently of the activities performed by the services, whereas business roles bind to components that are created when the session starts and have no persistency beyond that session. Hence, in the case of the *TravelBooking* service, a new instance of *BookingAgent* is generated for each new session whereas all sessions will share the same component that binds to *UsrDB* (i.e. they all share the same database of users).

Connectors, which label wires, are triples $<\mu_A,P,\mu_B>$ where:

- P is an "interaction protocol." Every interaction protocol has two roles $roleA_P$ and $roleB_P$, and a glue $glue_P$. The glue is a description of the coordination mechanisms enforced by the protocol, which we assume to be given in a formalism *IGLU*.
- μ_A and μ_B are 'attachments' that connect the roles of the protocol to the entities (business roles or protocols) being interconnected.

Interaction protocols are often just straight connections between ports identifying which interactions in *roleA* correspond to which interactions in *roleB*. In many cases, the interaction glue may include the routing of events, encryption/decryption of messages, or transforming sent data to the format expected by the receiver. In a connector, the interaction protocol is bound to the parties via attachments: these are mappings from the roles to the signatures of the parties identifying which interactions of the parties perform which roles in the protocol. We use *CNCT* to designate the set of connectors. See [2] for a more detailed account of how connectors are formalised in SRML. In software architecture, one can define connectors that involve an arbitrary number of roles, but service-oriented architectures involve only interactions between two partners.

In addition to a graph, a module identifies two important aspects related to the way a service can change a configuration:

- An internal configuration policy (indicated by the symbol ⊕) that identifies the triggers of the external service discovery process, and the initialisation and termination conditions of the components.
- An external configuration policy (indicated by the symbol ▽▭SLA▭) that consists of the variables and constraints that determine the quality profile to which the discovered services need to adhere.

The configuration policies (both internal and external) are discussed below together with a formal definition of the notion of module.

4   Business Configurations

As already explained, we approach the operational aspects of SOC from the point of view of the execution of business processes: our aim is to see state configurations as a result of the joint execution of a number of activities that can trigger the discovery and binding of external services. In the previous section, we discussed how services define architectural units. In this section, we discuss how the configuration itself is structured so that these units can be plugged together. This configuration structure is given by what we call business activities.

We take business activities to be characterised, in every configuration, by

- A sub-configuration, i.e. a subset of the components, and the wires between them, that execute as part of the activity.
- A workflow that implements the "business logic" of the activity.

For instance, we would like to recognise two activities in Fig. 1 whose sub-configurations are as depicted in Fig. 3. Intuitively, both correspond to two instances of the same business logic (two costumers booking their travel) but at different stages of their workflow: one (launched by *LUI*) is already connected to a flight and a hotel agent (*LauF* and *LauH*, respectively) but the other (launched by *AUI*) is also connected to a (different) flight agent (*AntF*) still has to find a hotel agent. Both share a database *DB* (of users), which is a persistent component.

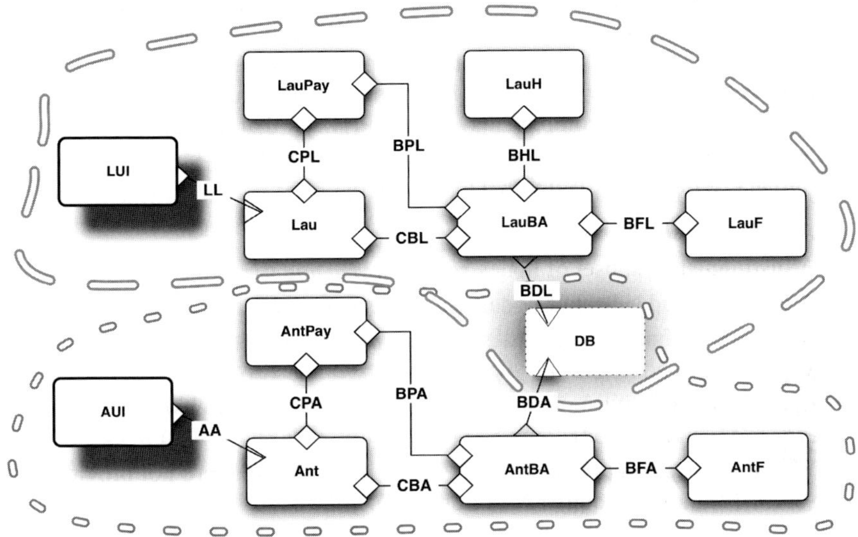

Fig. 3. The sub-configurations corresponding to two business activities

What we are calling 'business workflow' is formally captured by typing the sub-configuration of the activity by what we call an 'activity module'. For instance, the activity module depicted in Fig. 4 types some of the components of the configuration depicted in Fig. 1 with business roles and some of its wires with connectors.

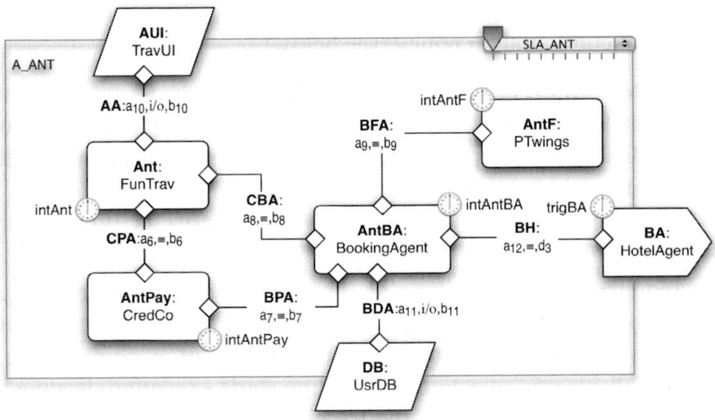

Fig. 4. An activity module

A business configuration consists of:

- A state configuration *SF*.
- A partial mapping B that assigns a module B(a) to the activities $a \in$ A that are active in *SF* – the workflow being executed by a in the configuration *SF*.
- A mapping C that assigns an homomorphism C(a) of graphs *body*(B(a))$\rightarrow$*SF* to every activity $a \in$ A that is active in *SF*. This homomorphism types the nodes of the activity with business roles or layer protocols – i.e. C$(a)(n)$:*label*$_{B(a)}(n)$ for every node n – and the edges with connectors – i.e. C$(a)(e)$: *label*$_{B(a)}(e)$ for every edge e of the body of the activity.

The homomorphism labels the components and wires of the state configuration with the business elements (business roles and the connectors) that they fulfil in the activity.

5   Services as Clauses

The operational semantics that we wish to put forward for service-oriented reconfiguration is inspired by another of Ugo's contributions to computer science, this time to concurrent constraint programming (CCP) in its 'soft' version [6]. More precisely, our approach is not CCP *sensu stricto*: it borrows aspects and techniques from CCP but it also adds a few (interesting) new ingredients.

The analogy starts with the identification of every service module with a 'clause':

$$P \xleftarrow[body(M)]{} R_1, \ldots, R_n$$

where *label*$_M$(*provides*(M))=P and *label*$_M$(*requires*(M))=$\{R_1,\ldots,R_n\}$. In logic programming "speak", the clause states that, to obtain P, one has to find $R_1,\ldots,R_n$ and execute *body(M)*. The execution of the body corresponds, in a sense, to the computation that, in the execution of a Horn clause, is performed to provide an 'answer' – what in logic programming corresponds to a substitution.

Using this representation, a business activity a is of the form:

$$\xleftarrow[B(a)]{} R_1, \ldots, R_n$$

where $B(a)$ is *body*(B(a)). That is, a business activity corresponds to a goal clause: finding $R_1,\ldots,R_n$ and providing an answer through the execution of $B(a)$.

Like in concurrent logic programming (CLP), we are not interested in the "don't know" ("angelic") non-determinism that results from exploring, through backtracking, all possible alternative matches to the R_i: if we are not happy with the chosen service provider, we cannot go back in time and restart with another provider! That is, we 'commit' to the choice of service provider. We deal instead with what is sometimes called "indeterminism" (or "don't care" non-determinism), which results from the existence of a choice of service provider. In CLP, this choice is controlled by the 'guard' assigned to each clause – a sequence of goals that appears before the body of the clause, which need to be executed successfully for the clause to be chosen and the body to be executed. Among all clauses that have satisfiable guards, one of them is chosen and the execution 'commits' to it.

In CCP [19], one works in a more general setting in which all processes executing can interact by means of a shared set of constraints to which they can add new constraints ('tell') or check if they entail a given constraint ('ask'). Soft CCP [6] generalises these mechanisms even further by working over a c-semiring as used in Section 0 for external configuration policies: one can then choose among all satisfiable clauses that maximise the degree of satisfaction relative to the set of constraints. In our setting, this means that we work instead with clauses of the form:

$$P \xleftarrow[body(M)]{} sla(M) \mid R_1, ..., R_n$$

where the set $sla(M)$ acts as a 'soft-guard'.

As explained later in the paper, when the discovery of a requires-interface R_i of an activity

$$\xleftarrow[B(a)]{} sla(a) \mid R_1, ..., R_n$$

is triggered, the matching process identifies service modules (clauses) M

$$P \xleftarrow[body(M)]{} sla(M) \mid T_1, ..., T_m$$

and a 'unifier'-morphism ρ that is an interpretation between R_i and P, and makes the combination $sla(B(a)) \oplus_{R,\rho} sla(M)$ of the sets of constraints of $B(a)$ and M consistent.

The selection of the clause (service provider) is made among those that maximise the degree of satisfaction of the combined set of constraints. The resolvent is another goal clause corresponding to the reconfiguration of the business activity a:

$$\xleftarrow[B'(a)]{} sla'(a) \mid R_1, ..., R_{i-1}, T_1, ..., T_m, R_{i+1}, ..., R_n$$

where $B'(a)=body(B(a) \oplus_{R,\rho} M)$ and $sla'(a)=contract(sla(B(a)) \oplus_{R,\rho} sla(M))$ as defined below. That is, $B'(a)$ is the body of the new workflow of the activity a that results from the binding with the discovered service and $sla'(a)$ is the contract negotiated between a and M, which extends the amalgamated set of constraints of both a and M. That is, from the point of view of CCP, new constraints are added to the current set of the activity (each activity has its own set of constraints and its execution interferes with other activities only through the shared persistent components).

When a state is reached in which the activity a is an empty clause of the form

$$\xleftarrow[B(a)]{} sla(a)$$

the resolution process for that activity will have ended, meaning that the activity does not need any external services and will continue executing according to the same workflow until completion (though one may simplify the configuration by removing components as they finish executing). By then, all relevant quality-of-service variables will have been instantiated according to $sla(a)$.

However, one may not need to discharge all the requires-interfaces (i.e. bind them to service providers). The resolution step does not spawn immediately all the body goals in parallel; instead, we wait for the triggers declared for each goal T_i (in the example above) to become true in order to launch the corresponding discovery, ranking and selection process. Notice that the evaluation of the triggers will change as execution proceeds. This is because the body $B(a)$ of the activity will be executing and changing the state over which the triggers are evaluated, and $sla(a)$ will itself change as new constraints are added. From the point of view of concurrent programming, this means that the goals are guarded by their triggers, i.e. they are of the form $(ask(trigger(n_i))\rightarrow R_i)$ where n_i is the node labelled by R_i and $ask(c)$ checks if condition c is entailed by the current state and set of constraints.

Because the state may change, what matters is not so much the consistency of the trigger with the current state and sla, but the fact that it may become true in a future state. However, contrarily to CCP, we do not take non-satisfiability as failure. As seen in Section 0, the internal configuration policy contains a termination condition for each component interface that determines when the execution of the instances should stop. For any $(ask(trigger(n_i))\rightarrow R_i)$ still outstanding when all components have terminated, the condition $trigger(n_i)$ will not be satisfiable in time, which we do not consider to be a failure. Therefore, the components that are delivering the service may finish executing their (distributed) workflow without having triggered all the conditions in the service internal configuration policy.

6 Reconfiguration as Resolution

In logic programming, different strategies may be adopted for choosing the next query to be processed, which in our case means the next service to be discovered. In our model, this choice is given by the occurrence of triggers. As mentioned in Section 0, every module declares, as part of its internal configuration policy, the triggering conditions that apply to their requires-interfaces. Given a business configuration $BC=\langle <G,S>,B,C\rangle$ and an activity a, each condition $trigger(r)$ is evaluated over the state S. If the condition $trigger(R)$ for a given requires-interface R holds in BC, the "unification" process is launched, which should return a service that "best" fits the business protocol $label_{B(a)}(R)$ and the external configuration policy of $B(a)$.

In our setting, this unification process involves three steps, which we can outline as follows:

- Discovery. This step consists in finding the services – among those that are able to guarantee the properties of the business protocol $label_{B(a)}(R)$ associated with R – with which it is possible to reach a service-level agreement.
- Ranking. For each service M discovered in the previous step, we calculate the most favourable service-level agreement that can be achieved – the contract that will be established between the two parties if M is selected. This calculation uses a notion of satisfaction that takes into account the preferences of the activity a and the service M.
- Selection. Select one of the services that maximises the level of satisfaction offered by the corresponding contract.

We are now going to define each of these steps in more detail, though most of the technical aspects need to be consulted in [5] and [13,14]. Consider a business configuration $BC=\langle SF,B,C\rangle$ and let R be a requires-interface of a business activity a such that $trigger(R)$ holds in SF. The discovery phase returns all the service modules M that satisfy the following properties:

- There is a specification morphism $\rho\!: label_{B(a)}(R)\rightarrow label_M(provides(M))$, i.e. the behavioural properties offered by the provides interface of the candidate service module entail the properties required by the requires-interface of the activity up to a suitable translation between the languages of both.

- The constraint system $cs(M)$ of the external policy of M is compatible with that of $cs(B(a))$. This means that we can extend the mapping ρ in such a way that, for every variable v in $cs(B(a))$:
- if owner(v)=R, there exists $\rho(v)$ in cs(M) such that type(v)=type'($\rho(v)$) and owner'($\rho(v)$)=provides(M);
- if *owner(v)* is a wire $i\leftrightarrow R$ then, for every wire w' in M of the form *provides(M)$\leftrightarrow j$*, there is a variable $\rho(v,w')$ in $cs(M)$ s.t. *owner'($\rho(v,w')$)=w'* and *type(v)=type'($\rho(v,w')$))*.
- The combination $sla(B(a))\oplus_{R,\rho}sla(M)$ of the sets of constraints of $B(a)$ and M is consistent (as defined below).

Intuitively, compatibility means that each discovered service needs to support the properties required by the activity through the business protocol associated with R and the negotiation of the configuration parameters associated with R, i.e. those configuration parameters that belong to R or to the wires that connect R to the components of the activity module. The first condition (entailment of properties) is handled through the logic that is adopted for specifying business protocols (see [12] for a flavour of the logic used in SRML). The second condition ensures that is indeed possible to achieve a service-level agreement between the activity and the service module. Compatibility of the constraint systems of $B(a)$ and M relative to R ensures that they can be combined, which gives rise to another constraint system.

The combined constraint system $cs(B(a))\oplus_{R,\rho}cs(M)$ is defined as follows:

- Its domain D'' is the union $D\cup D'$ of the domains of $cs(B(a))$ and $cs(M)$.
- Its set of variables V is the disjoint union of $cs(B(a))$ and $cs(M)$ except for all pairs $v|\rho(v)$ and $v|\rho(v,w')$, which give rise to variables (those involved in the negotiation). Notice that, if *owner(v)* is a wire $i\leftrightarrow R$, then we may end up with several "aliases" $v|\rho(v,w')$, one for each wire w' in M of the form *provides(M)$\leftrightarrow j$*. We denote by $neg(R,\rho)$ the set of such variables.

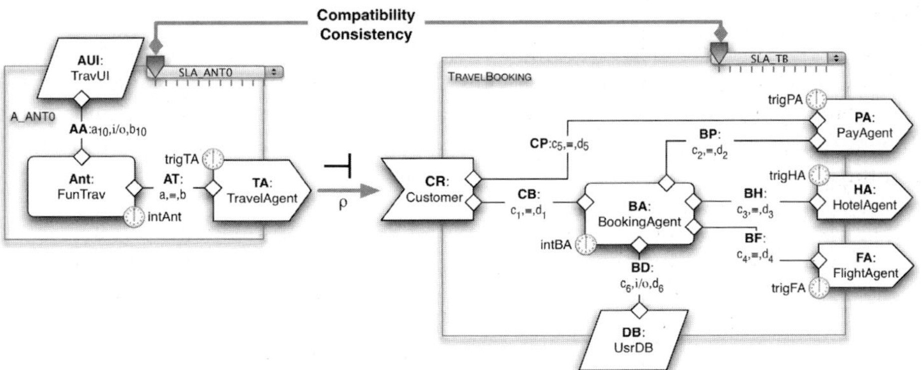

Fig. 5. The elements involved in unification

The combined set of constraints $sla(B(a)) \oplus_{R,\rho} sla(M)$ is defined by 'lifting' the constraints of $sla(B(a))$ and $sla(M)$ to the new constraint system.

In order to illustrate these constructions, consider that, at a certain point of the execution of the workflow of the business activity A_ANT0 in **Fig. 5**, the condition $trigTA$ becomes true and triggers the unification process for TA. For the service module $T_{RAVEL}B_{OOKING}$ to be discovered, we would need to

- Establish a specification morphism between $TravelAgent$ (the business protocol that types TA) and $Customer$ (the business protocol that types the provides interface CR of $T_{RAVEL}B_{OOKING}$) showing that the properties required in $TravelAgent$ are entailed by those of $Customer$.
- Check that the constraint systems of A_ANT0 and $T_{RAVEL}B_{OOKING}$ are compatible.

Finally, we can discuss how contracts are established. Together with the set $neg(R,\rho)$ of the variables being negotiated (those in the domain of ρ), the set of constraints $sla(B(a)) \oplus_{R,\rho} sla(M)$ defines a constraint problem. In the c-semiring framework, the solution of this constraint problem is again a constraint and, hence, it assigns a degree of satisfaction to each possible tuple of values for the variables in $neg(R,\rho)$. Ranking a discovered service M in our framework consists in finding an assignment that maximizes the degree of satisfaction. The constraint that results from the negotiation is denoted by $contract(B(a) \oplus_{R,\rho} M)$. The selected service is one with maximal rank.

It remains to define the new business configuration that results from the process of discovery, ranking and selection – what we could call the 'resolution step' using the analogy with logic programming. This includes the new state configuration that results from instantiating the selected service over the current configuration and binding it to the business activity a that triggered the process, and the typing of the business activity with a new module.

We start by defining the activity module that will type a in the new business configuration. Consider that a service module M is returned by the selection process upon the occurrence of $trigger(R)$ where R is a requires-interface of $B(a)$. The binding of R with an instance of M involves the assembly of modules $B(a)$ and M, giving rise to a new module that corresponds to the new execution plan of a. This new module is the composition $B(a) \oplus_{R,\rho} M$ (depicted in Fig. 6 for A_ANT0 and $T_{RAVEL}B_{OOKING}$) defined as follows:

- The graph of $B(a) \oplus_{R,\rho} M$ is obtained from the sum (disjoint union) of the graphs of $B(a)$ and M by eliminating the nodes R and $provides(M)$, and adding an edge $i \leftrightarrow j$ between any two nodes i and j such that $i \leftrightarrow R$ is an edge of $B(a)$ and $provides(M) \leftrightarrow j$ is an edge of M. The requires-interfaces are those of $B(a)$, except for R, and those of M. Given that $provides(M)$ has been eliminated, there are no provides-interfaces; we obtain an activity module B that defines the new execution plan of the activity a.
- The labels of the resulting graph are inherited from the graphs of $B(a)$ and M, except for the new edges $i \leftrightarrow j$ that result from the binding of R and $provides(M)$

through the morphism ρ. These are calculated by composing the connectors that label $i{\leftrightarrow}R$ and $provides(M){\leftrightarrow}j$. This process of composition is detailed in [13]: basically, we need to compose the glues of the connectors through the roles that have been bound through the signature morphism.

- The external configuration policy is $contract(B(a)\oplus_{R,\rho}M)$ and the triggers, initialisation and termination conditions of the internal configuration policy are all inherited from $B(a)$ and M.

We take this module to provide the reconfigured execution plan of the business activity a. We can now define the new state and business configurations that result from the discovery and binding processes. The current state configuration is modified as follows:

- New components (nodes) are added to the service layer, which are typed by the business roles of $components(M)$.
- New wires (edges) are added that are typed with the connectors that link together the new components introduced in the previous step.
- New wires are added between the new components and the ones that were already present in the configuration, which are typed by the composed connectors that result from the bindings.
- New wires are added that bind the new service components to the shared persistent components, which are typed by the layer protocols of $uses(M)$. Notice that we do not create new shared persistent components (instances) in this process: such components are used, not created by services.
- The new components and wires are initialised so as to satisfy the internal configuration policy of M.

The new business configuration B' is the same as B except for activity a for which $B'(a)$ is $B(a)\oplus_{R,\rho}M$. The homomorphism is as defined by the typing of nodes and wires discussed above.

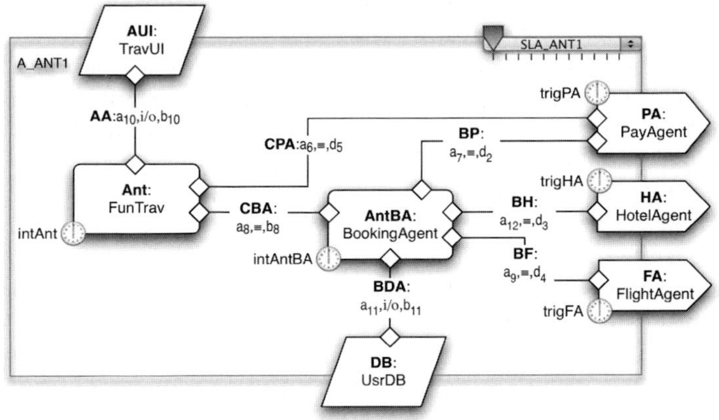

Fig. 6. A new session of *TravelBooking* starts and reconfigures the workflow of *ANT*

7 Final Tasting

We cooked a dish using some of the ingredients that Ugo has given us during his career. As any good amateur of Italian cuisine would have done, the chosen ingredients are of top quality. As for the dish, "provate per credere"…

We certainly hope not to have 'overcooked the meat'. In our opinion, widely shared within the SENSORIA project, there is a lot of research that needs to be done towards a methodological and mathematical characterisation of the service-oriented computing paradigm [20]. Our approach differs from other work on Web Services (e.g. [3]) and SOC in general (e.g. [21]) in that we address not the middleware architectural layers or low-level design issues, but what we call the 'business level'. That is, we view SOC as operating over configurations of global computers that are typed by business activities, which may need to discover and bind to external services as they execute and, therefore, reconfigure their activity.

This emphasis on the business dimension is well apparent in the semantic model that we proposed in the sense that it separates the reconfiguration (business level) from the computation dimension (state level). More specifically, our model makes only minimal assumptions about the computational aspects that account for state changes and interactions, as well as the languages and formalisms that are used for specifying the workflows executed by components, the interaction protocols established through the wires, and the properties that describe the properties of services. The specific formalisms used in the SENSORIA Reference Modelling Language (SRML) are presented in [2,12]. Other popular formalisms for modelling (web) services are those also adopted for business workflows [17,18], as well different kinds of process calculi (e.g. [8,10,16]). However, the workflow-oriented formalisms tend not to address dynamic reconfiguration and the process calculi tend not to address it separately from computation. As far as we know, SRML is the first service-modelling language to separate these two concerns.

The semantics of the actual reconfiguration operated during a resolution step was given based on algebraic, graph-based techniques [13,14]. The notion of configuration and module were formalised in terms of graphs and their labelling with different kinds of components, connectors, specifications and specification morphisms. In this context, another interesting semantics of the reconfiguration process that we would like to explore is the use of graph transformations, for instance as in [9] where the architectural style of SRML has been defined by Ugo and some of his colleagues.

Another aspect worth investigating is the "interleaving" of the synthesis/resolution process with the execution of the activity whose workflow is being synthesised. Having offered separate models for these two processes, we intend to investigate how the reconfiguration process can be analysed in conjunction with the computations that are being performed by components and the coordination mechanisms on the interactions performed by the wires. For this purpose, we will rely on calculi (e.g. [16]) and logics (e.g. [4]) that are being developed within SENSORIA. Another avenue that we would like to explore in this respect is the use of graphs as a computational model, for instance as developed in [11], once more with Ugo's contribution.

Acknowledgments

The reviewers kindly suggested many improvements and extensions, which unfortunately would not fit in the space available for the paper. Most of the work reported in this paper was developed in collaboration with Antónia Lopes, Laura Bocchi, and João Abreu. I would like to thank our colleagues in the SENSORIA project for many useful discussions on the topics covered in this paper, which includes not only Ugo but also many of Ugo's "children" (and "grandchildren"). This is the last contribution to computer science that I would like to thank Ugo for: to have populated the area with so many brilliant scientists (though their performance as cooks still needs to be properly tested…).

References

1. Abreu, J., Fiadeiro, J.: A coordination model for service-oriented interactions. In: Lea, D., Zavattaro, G. (eds.) COORDINATION 2008. LNCS, vol. 5052, pp. 1–16. Springer, Heidelberg (2008)
2. Abreu, J., Bocchi, L., Fiadeiro, J.L., Lopes, A.: Specifying and composing interaction protocols for service-oriented system modelling. In: Derrick, J., Vain, J. (eds.) FORTE 2007. LNCS, vol. 4574, pp. 358–373. Springer, Heidelberg (2007)
3. Alonso, G., Casati, F., Kuno, H., Machiraju, V.: Web Services. Springer, Heidelberg (2004)
4. ter Beek, M., Fantechi, A., Gnesi, S., Mazzanti, F.: An action/state-based model checking approach for the analysis of communication protocols for Service-Oriented Applications. In: Formal Methods for Industrial Critical Systems. LNCS, Springer, Heidelberg (to appear)
5. Bistarelli, S., Montanari, U., Rossi, F.: Semiring-based constraint satisfaction and optimization. Journal of the ACM 44(2), 201–236 (1997)
6. Bistarelli, S., Montanari, U., Rossi, F.: Soft concurrent constraint programming. ACM Transactions on Computational Logic 7(3), 563–589 (2006)
7. Bocchi, L., Hong, Y., Lopes, A., Fiadeiro, J.: From BPEL to SRML: a formal transformational approach. In: Dumas, M., Heckel, R. (eds.) Web Services and Formal Methods. LNCS, vol. 4937, pp. 92–107. Springer, Berlin, Heidelberg, New York (2008)
8. Boreale, M., et al.: SCC: a service centered calculus. In: Bravetti, M., Núñez, M., Zavattaro, G. (eds.) WS-FM 2006. LNCS, vol. 4184, pp. 38–57. Springer, Berlin, Heidelberg, New York (2006)
9. Bruni, R., Lluch Lafuente, A., Montanari, U., Tuosto, E.: Service oriented architectural design. In: Trustworthy Global Computing, Springer, Berlin, Heidelberg, New York (to appear, 2007)
10. Buscemi, M., Montanari, U.: CC-Pi: A constraint-based language for specifying service level agreements. In: De Nicola, R. (ed.) ESOP 2007. LNCS, vol. 4421, pp. 18–32. Springer, Berlin, Heidelberg, New York (2007)
11. Ferrari, G.F., Hirsch, D., Lanese, I., Montanari, U., Tuosto, E.: Synchronised hyperedge replacement as a model for service oriented computing. In: de Boer, F.S., Bonsangue, M.M., Graf, S., de Roever, W.-P. (eds.) FMCO 2005. LNCS, vol. 4111, pp. 22–43. Springer, Berlin, Heidelberg, New York (2006)

12. Fiadeiro, J.L., Lopes, A., Bocchi, L.: A formal approach to service-oriented architecture. In: Bravetti, M., Núñez, M., Zavattaro, G. (eds.) WS-FM 2006. LNCS, vol. 4184, pp. 193–213. Springer, Berlin, Heidelberg, New York (2006)
13. Fiadeiro, J.L., Lopes, A., Bocchi, L.: Algebraic semantics of service component modules. In: Fiadeiro, J.L., Schobbens, P.-Y. (eds.) WADT 2006. LNCS, vol. 4409, pp. 37–55. Springer, Berlin, Heidelberg, New York (2007)
14. Fiadeiro, J.L., Schmitt, V.: Structured co-spans: an algebra of interaction protocols. In: Mossakowski, T., Montanari, U., Haveraaen, M. (eds.) CALCO 2007. LNCS, vol. 4624, pp. 194–200. Springer, Berlin, Heidelberg, New York (2007)
15. Hirsch, D., Montanari, U.: Two graph-based techniques for software architecture reconfiguration. Electronic Notes in Theoretical Computer Science 51, 177–190 (2001)
16. Lapadula, A., Pugliese, R., Tiezzi, F.: Calculus for orchestration of web services. In: De Nicola, R. (ed.) ESOP 2007. LNCS, vol. 4421, pp. 33–47. Springer, Berlin, Heidelberg, New York (2007)
17. Ouyang, C., Verbeek, E., van del Aalst, W.M.P., Dumas, M., ter Hofstede, A.H.M.: Formal semantics and analysis of control flow in WS-BPEL. Science of Computer Programming 67(2-3), 162–198 (2007)
18. Reisig, W.: Modeling and analysis techniques for web services and business processes. In: Steffen, M., Zavattaro, G. (eds.) FMOODS 2005. LNCS, vol. 3535, pp. 243–258. Springer, Berlin, Heidelberg, New York (2005)
19. Saraswat, V.A.: Concurrent Constraint Programming. MIT Press, Cambridge, Massachusetts (1993)
20. SENSORIA consortium (2007), http://www.sensoria-ist.eu/files/whitePaper.pdf
21. The Open Service Oriented Architecture collaboration, http://www.osoa.org

Calculating Colimits Compositionally

Robert Rosebrugh[1], Nicoletta Sabadini[2], and Robert F.C. Walters[2,*]

[1] Department of Mathematics and Statistics
Mt. Allison University, Sackville,
New Brunswick, Canada
[2] Dipartimento di Scienze delle Cultura, Politiche e dell'Informazione,
Università dell'Insubria, Italy

Abstract. We show how finite limits and colimits can be calculated compositionally using the algebras of spans and cospans, and give as an application a proof of the Kleene Theorem on regular languages.

1   Introduction

In Computer Science:
The state spaces of systems are often described by finite limits or colimits in a category $\mathbf{E}$ parametrized by a graph G which describes the underlying geometry of the system. It is desirable that there is also an algebraic description, so that the limit or colimit is described by an expression, rather than geometrically.

This goes back to the beginnings of computer science, where (i) a program may be described either by a flow chart (goto's), or program text (while) (Böhm-Jacopini), (ii) a language may be specified by an automaton or an expression (Kleene). And of course it is present in innumerable areas of computer science (Petri nets versus process algebras, wysiwyg versus markup, graph versus term rewriting, etc.) and mathematics.

In Category Theory:
Finite limits and colimits are parametrized by graphs; that is, geometrically. We show that they can also be described by expressions in an algebra. As an application we prove Kleene's theorem.

Perhaps the first proposal for a strict relation between graphic and algebraic/categorical descriptions arose in the work of R. Penrose [20] in his graphical description of the tensor calculus in 1971. C.C. Elgot [6] began the algebraicization of flowcharts and circuits introducing a categorical algebra which contained three basic operations; in circuit terminology - series and parallel composition, and feedback. In subsequent work this algebra has been intensively developed by S.L. Bloom and Z. Esik, with the state of progress being recorded in their monograph [2].

---

* The authors gratefully acknowledge financial support from the Universitá dell'Insubria, the Italian Government PRIN project ART (*Analisi di sistemi di Riduzione mediante sistemi di Transizione*), and the Canadian NSERC.

P. Degano et al. (Eds.): Montanari Festschrift, LNCS 5065, pp. 581–592, 2008.

In a parallel development G.M. Kelly and M.L. Laplaza [16] gave a graphical account of compact closed monoidal categories. Following [25], A. Joyal and R.H. Street [9] discovered the notion of braided monoidal category, a major impulse towards the study of geometry and higher dimensional categories. Among the many categorical developments was the discovery in [4] of the Frobenius equation and the recognition by Joyal of its importance in 2-dimensional cobordism theory [17].

Ugo Montanari and his collaborators have played a fundamental role in related developments in computer science, beginning with the fundamental paper [19] relating Petri nets and monoidal categories. Gadducci and Heckel [7],[8] discovered that a category of cospans of graphs is a free symmetric monoidal category with appropriate structure and axioms (one being the Frobenius axiom (see also [22])), and gave an algebraic formulation of double-pushout graph rewriting.

In the last ten years there has been a flowering of applications of monoidal categories in geometry, physics and computer science. We cite just three directions in computer science [10], [3], [1].

The algebra in which finite limits and colimits in $\mathbf{E}$ may be expressed compositionally is an appropriate structure on spans and cospans in the category. This fact is a partial explanation for the algebra of spans and cospans introduced in [12],[14] and developed in various papers, such as [13],[15],[21].

This note is an expanded version of a lecture [23] to Category Theory 2007, Carvoeiro, Portugal, 18th June 2007. A more detailed version with full proofs is in preparation [24].

2   What Algebra?

Assume now $\mathbf{E}$ is a category with finite colimits. What is the algebra in which finite colimits in $\mathbf{E}$ can be described by expressions?

It is $cospan(\mathbf{E})$, considered as a symmetric monoidal category in which each object has a commutative separable algebra structure. We call a category with such a structure $wscc$ (well-supported compact closed [26]). To be precise:

Definition 1. *A commutative separable algebra [4] in a symmetric monoidal category is an object A equipped with four arrows*

$$! : I \longrightarrow A, \quad \nabla : A \otimes A \longrightarrow A, \quad \text{¡} : A \longrightarrow I, \quad \Delta : A \longrightarrow A \otimes A$$

such that $(A, \nabla, !)$ forms a commutative monoid, $(A, \Delta, \text{¡})$ forms a cocommutative comonoid, and the following three axioms hold

$$(1_A \otimes \nabla)(\Delta \otimes 1_A) = \Delta\nabla = (\nabla \otimes 1_A)(1_A \otimes \Delta),$$

$$\nabla\Delta = 1_A.$$

We can draw a picture of the last three extra axioms, namely:

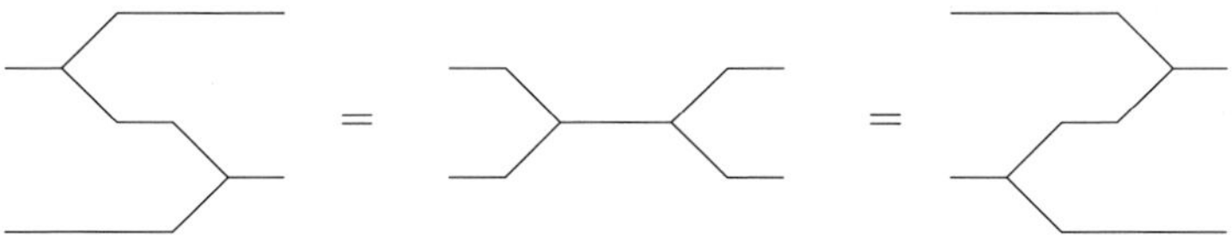

The wscc structure induces a self-dual compact closed structure on the category, and we denote the units and counits of this structure as

$$\eta_A : I \to A \otimes A \ (= \Delta \cdot \,!), \ \varepsilon_A : A \otimes A \to I \ (= \, \textrm{¡} \cdot \nabla).$$

For some background to these axioms see also [17].

2.1   The wscc Structure on Span and Cospan Categories

We will describe the wscc structure on $cospan(\mathbf{E})$ for $\mathbf{E}$ a finitely cocomplete category – the dual structure on $span(\mathbf{E})$ will then be clear.

An object of $cospan(\mathbf{E})$ is an object of $\mathbf{E}$; an arrow of $cospan(\mathbf{E})$ from A to B is an isomorphism class of cospans from A to B; that is, of pairs of arrows

$$\alpha_1, \alpha_2 : A \to R \leftarrow B.$$

We will use the notation $\alpha_1, \alpha_2; A \longleftrightarrow B$ to distinguish cospans from arrows in $\mathbf{E}$. However given any arrow $f : A \to B$ there are special cospans denoted $\boldsymbol{f} = f, 1_B : A \longleftrightarrow B$ and $f^o = 1_B, f : B \longleftrightarrow A$. Composition of cospans is by pushout. Now to describe the wscc structure of $cospan(\mathbf{E})$. The monoidal structure is sum. The special arrows

$$! : I \longrightarrow A, \ \nabla : A \otimes A \longrightarrow A, \ \textrm{¡} : A \longrightarrow I, \ \Delta : A \longrightarrow A \otimes A$$

are (using ∇ both for the codiagonal in $\mathbf{E}$ and the structure in $cospan(\mathbf{E})$, and similarly overloading the symbol !)

$$! = \,! : 0 \longleftrightarrow A, \ \nabla = \boldsymbol{\nabla} : A{+}A \longleftrightarrow A, \ \textrm{¡} =\,!^o : A \longleftrightarrow 0, \ \Delta = \nabla^o : A \longleftrightarrow A{+}A.$$

2.2 $Cspn(Graph/|\mathbf{E}|)$

Let $Graph$ be the category of finite graphs, let $|\mathbf{E}|$ be the underlying graph (possibly infinite) of $\mathbf{E}$. Consider $Graph/|\mathbf{E}|$, the category with objects $diagrams$ in $\mathbf{E}$, and morphisms $compatible$ $graph$ $morphisms$. Then $Cspn(Graph/|\mathbf{E}|)$ is the full subcategory of $cospan(Graph/|\mathbf{E}|)$ whose objects are $discrete$ diagrams in $\mathbf{E}$.

Notice that colimits in this category are calculated as in $Graph$, and are unrelated to colimits in $\mathbf{E}$.

We will denote diagrams using set-theoretical notation; for example

$$\{A \xrightarrow[\;g\;]{\;f\;} B\}$$

denotes the diagram with two parallel arrows.

It will be useful to introduce a way of picturing arrows in $Cspn(Graph/|\mathbf{E}|)$ (engineering notation). Represent the objects in the centre graph of the cospan as points, and arrows in the centre as components with one input (to the left) and one output (to the right) joined to those points which are the domain and codomain of the arrow. Represent the graph morphisms of the cospan as input and output wires of the whole picture.

Example 1. Consider the following cospan of diagrams

$$\{A\} \longrightarrow \{A \underset{g}{\overset{f}{\rightleftarrows}}\, B \xrightarrow{\;k\;} C\} \longleftarrow \{C\}\ .$$

This cospan could be pictured as

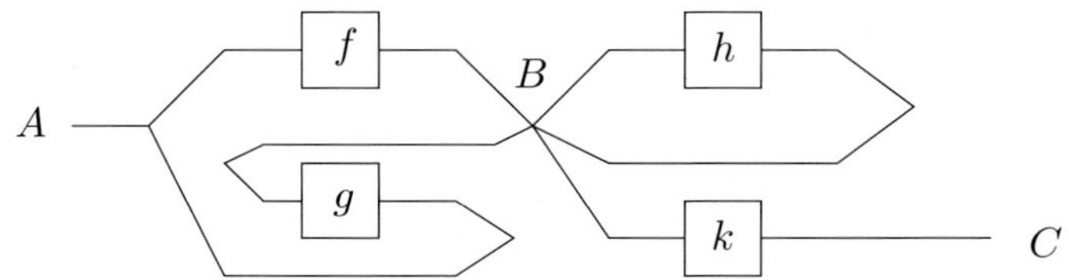

3   The Theorem

Taking colimit of diagrams in $\mathbf{E}$ induces a functor

$$colim : Graph/|\mathbf{E}| \to \mathbf{E}.$$

Theorem 1. *The functor $colim : Graph/|\mathbf{E}| \to \mathbf{E}$ extends to a functor*

$$colim : Cspn(Graph/|\mathbf{E}|) \to cospan(\mathbf{E})$$

which preserves the wscc structure.

The definition of the extended colimit is just applying colimit to cospans. It is straightforward that this colim preserves the constants of wscc structure of $Cspn(Graph/|\mathbf{E}|)$, and that colim of the cospan of the diagram

$$\{A\} \longrightarrow \{A \xrightarrow{\;f\;} B\} \longleftarrow \{B\}$$

is $f : A \longleftrightarrow B$. The fact that colim preserves the tensor is also clear. What remains to prove is the fact that colim is a functor – we outline the proof below.

Another special case of colim is worth remarking. Consider a cospan in which the centre diagram is also discrete, so that we may consider the cospan to be of the form

$$\{A_i\}_{(i\in I)} \xrightarrow{\phi:I\to J} \{B_j\}_{(j\in J)} \xleftarrow{\psi:J\leftarrow K} \{C_k\}_{(k\in K)} \ .$$

Then colimit applied to this cospan is

$$\Sigma_{i\in I}A_i \xrightarrow{colim(\phi)} \Sigma_{j\in J}B_j \xleftarrow{colim(\psi)} \Sigma_{k\in K}C_k \ ,$$

where $colim(\phi)\cdot inj_i = inj_{\phi(i)}$ $(i\in I)$ and $colim(\psi)\cdot inj_k = inj_{\psi(k)}$ $(k\in K)$

Remark 1. $Cspn(Graph/|\mathbf{E}|)$ is the result of freely adding wscc category structure to the graph $|\mathbf{E}|$ (a special case of this result was proved in [22]). This means that diagrams in $|\mathbf{E}|$ may be written as expressions in the wscc structure of $Cspn(Graph/|\mathbf{E}|)$ with constants being the cospans of the form $\boldsymbol{f}$ for arrows f of $|\mathbf{E}|$.

Then colim preserves wscc expressions, so the colimit of any diagram may be written as an expression in $cospan(\mathbf{E})$. *This is the compositionality of the calculation of colimits, mentioned in the title.*

3.1   The Example of Coequalizers

Consider the following cospan of diagrams in $\mathbf{E}$:

$$\{A\} \longrightarrow \{A \mathrel{\substack{\xrightarrow{f}\\[-0.3em] \xrightarrow{g}}} B\} \longleftarrow \{B\} \ .$$

The cospan of diagrams may be pictured, as described above, as

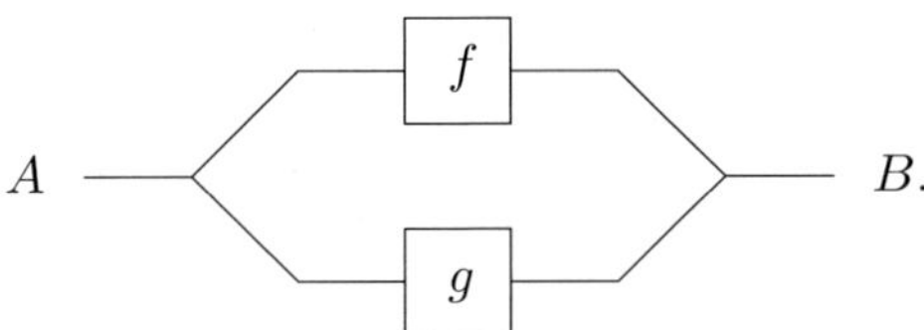

It is clear from the picture that the cospan may be expressed as the following composite in $Cspn(Graph/|\mathbf{E}|)$:

$$\{A\} \xleftarrow{\Delta} \{A\}+\{A\} \xrightarrow{\{f\}+\{g\}} \{B\}+\{B\} \xleftarrow{\nabla} \{B\}.$$

Applying colimit we see that the coequalizer of f and g may be expressed as the composite in $cospan(\mathbf{E})$

$$A \xleftarrow{\Delta} A+A \xrightarrow{f+g} B+B \xleftarrow{\nabla} B.$$

The composite of these three cospans is the pushout Q of the following diagram in E

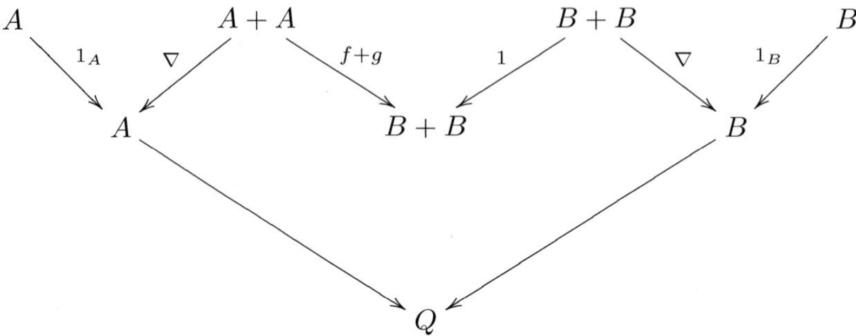

It is easy to verify directly that Q so defined is the coequalizer of f and g.

Example 2. By the same kind of reasoning the colimit of the diagram in example 1 may be given by the expression of cospans in **E**

$$(\epsilon_B + C) \cdot (\boldsymbol{h} + B + \boldsymbol{k}) \cdot (\varDelta + B) \cdot \varDelta \cdot (\nabla + \epsilon_A) \cdot (\boldsymbol{f} + B + \boldsymbol{g} + A) \cdot (A + \eta_B + A) \cdot \varDelta.$$

Remark 2. Theorem 1 has an analogue for limits. If **E** has finite limits the functor $lim : (Graph/|\mathbf{E}|)^{op} \to \mathbf{E}$ extends to a functor

$$lim : Cspn(Graph/|\mathbf{E}|) \to span(\mathbf{E})$$

which preserves the wscc structure. This permits the compositional calculation of finite *limits* in **E**. In fact the equalizer of two arrows $A \underset{g}{\overset{f}{\rightrightarrows}} B$ may be calculated by the same expression as that of the coequalizer above but evaluated in $span(\mathbf{E})$ rather than $cospan(\mathbf{E})$, since in both cases the expression is determined by the wscc structure of $Cspn(Graph/|\mathbf{E}|)$.

3.2　Sketch of Proof of Theorem

The main point to check in showing that colim is a monoidal functor is (a special case of) the following:

Consider a diagram D of diagrams in **E** parametrized by a graph G; that is, a graph morphism $D : G \to Graph/|\mathbf{E}|$. We can do two things.

(1) Calculate first the colimit of D in $Graph/|\mathbf{E}|$ to obtain a diagram in **E** of which we may then take the colimit in **E**, that is calculate

$$colim_{\mathbf{E}}(colim_{Graph/|\mathbf{E}|}(D)).$$

(2) Calculate the colimit of

$$G \overset{D}{\longrightarrow} Graph/\mathbf{E} \overset{colim}{\longrightarrow} \mathbf{E}$$

that is, calculate $colim_{\mathbf{E}}(colim_{\mathbf{E}} \cdot D)$.

Lemma $colim_{\mathbf{E}}(colim_{Graph/|\mathbf{E}|}(D)) \cong colim_{\mathbf{E}}(colim_{\mathbf{E}} \cdot D)$.
Sketch of proof.
It suffices to show for any $X \in \mathbf{E}$ a bijection between cocones

$$colim_{Graph/|\mathbf{E}|}(D) \longrightarrow X$$

and cocones

$$colim_{\mathbf{E}} \cdot D \longrightarrow X$$

But it is not hard to show that both of these are equivalent to a "compatible family" of cocones

$$D(g) \longrightarrow X \quad (g \in G).$$

A Very Special Case. Consider a diagram D of diagrams in $\mathbf{E}$, namely

$$D = \{\{A\}\ \{B\ C\}\}.$$

Then

$$colim_{\mathbf{E}}(colim_{Graph/|\mathbf{E}|}(D)) = colim_{\mathbf{E}}(\{A\ B\ C\}) = A + B + C,$$

whereas

$$colim_{\mathbf{E}}(colim_{\mathbf{E}} \cdot D) = colim_{\mathbf{E}}(\{A\ B + C\}) = A + (B + C).$$

The lemma says exactly that the triple sum may be formed by repeated double sums, which has as a consequence the associative law for sums. It is clear that the general form of the lemma implies many further "associative laws" - any two wscc expressions which yield the same diagram evaluate to the same object in $cospan(\mathbf{E})$.

3.3   Example of Theorem

A general cospan in $Cspn(Graph/|\mathbf{E}|)$ from $\varnothing$ to $\varnothing$ with centre D may be constructed by taking the disjoint sum of all the arrows, and then equating vertices appropriately. This yields a formula for the general colimit of a finite diagram as follows. Let Σdom denote the graph $\sum_{\alpha \in D}\{dom(\alpha)\}$ and Σcod denote the graph $\sum_{\alpha \in D}\{codom(\alpha)\}$. Let $\Sigma\alpha$ denote the graph $\sum_{\alpha \in D}\{\alpha\}$. Let Σobj denote the diagam consisting of all the objects in the D. Finally, let i_{dom} and i_{cod} denote the discrete cospans corresponding to the domain and codomain functions on the arrows of the graph parametrizing D. Then the cospan may be written

$$\{\} \xleftarrow{\ \ \eta\ \ } \Sigma dom + \Sigma dom \xleftarrow{\ i_{cod} \cdot (\sum \alpha) + i_{dom}\ } \Sigma obj + \Sigma obj \xleftarrow{\ \ \epsilon\ \ } \{\}\ .$$

Evaluating this formula instead in $cospan(\mathbf{E})$ gives the classical formula for colimits in terms of the coequalizer of two arrows from $\sum_{\alpha \in D} dom(\alpha)$ to $\sum_{A \in D} A$ (α arrow in D, A object in D).

4   Limits and Colimits of Monoidal Diagrams

Systems in computer science are not usually constructed from parts with one input and one output, like arrows in a graph. Components have multiple inputs and outputs; that is, they are arrows in a monoidal graph.

Definition 2. *A monoidal graph (A, V, d_0, d_1) consists of a set V of vertices, and a set A of arcs and two functions $d_0, d_1 : A \longrightarrow V^*$ (V^* the free monoid on V). A morphism of monoidal graphs $\phi = (\phi_0, \phi_1)$ from (A, V, d_0, d_1) to (B, W, d_0, d_1) consists of two functions $\phi_1 : A \to B$ and $\phi_0 : V \to W$ such that $\phi_0^* d_0 = d_0 \phi_1, \phi_0^* d_1 = d_1 \phi_1$. We denote the category (actually a presheaf category) of monoidal graphs as $MonGraph$. There is an obvious notion then of a* monoidal diagram *in a monoidal category since any monoidal category has an underlying monoidal graph.*

Definition 3. *Let $\mathbf{E}$ be a category with finite colimits, regarded as a monoidal category with sum as tensor. A cocone q of a monoidal diagram D to an object X is a family of arrows $(q_i : A_i \longrightarrow X)$ (A_i objects of the diagram D) such that for any arrow $f : A_{i_1} + A_{i_2} + \cdots + A_{i_m} \to A_{j_1} + A_{j_1} + \cdots + A_{j_n}$ in the diagram*

$$(q_{j_1} | q_{j_2} | q_{j_n} | \cdots | q_{j_n}) \cdot f = (q_{i_1} | q_{i_2} | q_{i_3} \cdots | q_{i_m}).$$

A colimit of monoidal diagram D is an object C with a cocone q from D which is univeral; that is, any cocone to an object X factors uniquely through q.

4.1 $Cspn(MonGraph/|\mathbf{E}|)$

Let $\mathbf{E}$ be a category with finite colimits, regarded as a monoidal category with sum as tensor, and let $|\mathbf{E}|$ denote the underlying monoidal graph of $\mathbf{E}$. Then $Cspn(MonGraph/|\mathbf{E}|)$ denotes the full subcategory of $cospan(MonGraph/|\mathbf{E}|)$ whose objects are *discrete* diagrams in $\mathbf{E}$. Just as with $Cspn(Graph/|\mathbf{E}|)$ we may picture arrows in $Cspn(MonGraph/|\mathbf{E}|)$, the only difference being that components may have several input and output wires. Monoidal colimits may also be calculated compositionally, in the algebra $cospan(\mathbf{E})$, by a result analogous to Theorem 1. Taking the monoidal colimit of diagrams in $\mathbf{E}$ induces a functor

$$moncolim : MonGraph/|\mathbf{E}| \to \mathbf{E}.$$

Theorem 2. *The functor moncolim $: MonGraph/|\mathbf{E}| \to \mathbf{E}$ extends to a functor*
$$moncolim : Cspn(MonGraph/|\mathbf{E}|) \to cospan(\mathbf{E})$$
which preserves the wscc structure.

We look at one example only.

4.2   Example

Consider the following cospan of monoidal diagrams D in $\mathbf{E}$: the centre is the diagram with three objects A, B, C, and one arrow $f : A + C \to B + C$; the left hand side is the diagram $\{A\}$; the right hand side is the diagram $\{B\}$. Pictured, the cospan is

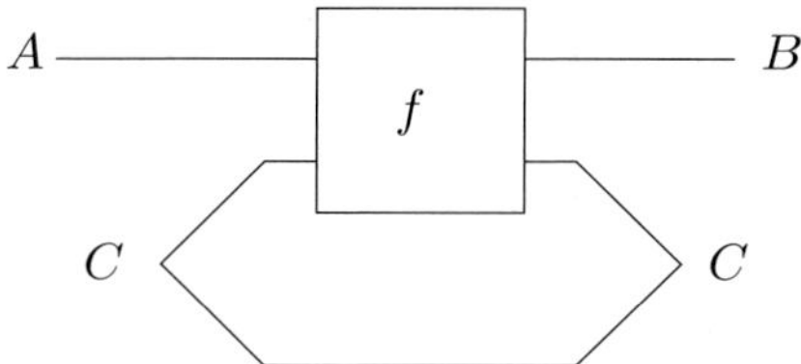

From the picture we see immediately that this cospan of diagrams is expressible as a composite in $Cspn(MonGraph/|\mathbf{E}|)$, namely

$$\{A\} \xleftarrow{\{A\}+\eta_{\{C\}}} \{A\} + \{C\} + \{C\} \xleftarrow{\{f\}+\{C\}} \{B\} + \{C\} + \{C\} \xleftarrow{\{B\}+\epsilon_{\{C\}}} \{B\} .$$

Applying monoidal colimit yields the fact that the monoidal colimit of the original diagram is the following composite in $cospan(\mathbf{E})$:

$$A \xleftarrow{A+\eta_C} A + C + C \xleftarrow{f+C} B + C + C \xleftarrow{B+\epsilon_C} B .$$

Hence the colimit of the original diagram can be calculated as the pushout below.

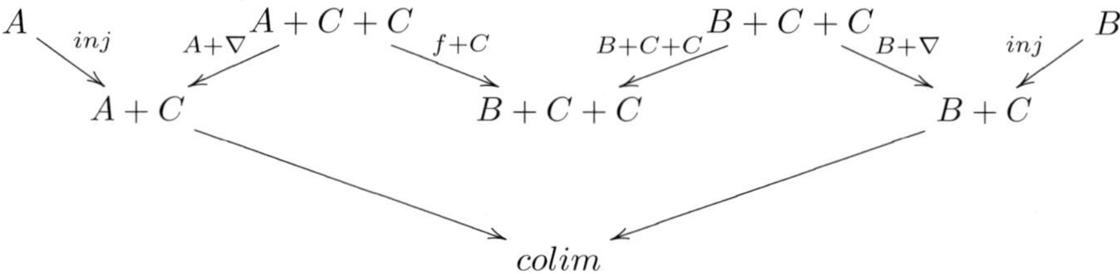

The colimit consists of orbits of $A + B + C$ under f. The pullback of the resulting cospan is the partial function obtained by iterating f.

5   The Kleene Theorem

Theorem 3. (Kleene)
The languages recognized by finite state automata are the closure of singletons under union, concatenation and iteration.

To prove this classical theorem the category $\mathbf{E}$ we consider is $\wp(\Sigma^*)\text{-}Cat$, categories enriched in Σ-languages. There is a composite of wscc functors

$$Cspn(Graph/\Sigma) \xrightarrow{\Phi_1} Cspn(Graph/\wp(\Sigma^*)) \xrightarrow{\Phi_2} cospan(\wp(\Sigma^*)\text{-}Cat).$$

which takes a labelled graph (with input and output states) to the $\wp(\Sigma^*)$-category whose homs are the languages traced out from the domain to the codomain. The existence of wscc functor Φ_1 is implied by [22], and of Φ_2 (=colim) by Theorem 1.

This is already a Kleene-type theorem since, conceptually, the Kleene Theorem says that behaviour is an operation-preserving morphism from an algebra of systems to an algebra of possible behaviours, which implies that the perceived behaviours are the smallest class of possible behaviours closed under operations. In this case, the algebra of systems, that is the left-hand side, is generated as a wscc category by single labelled edges, and hence the image on the right-hand side is also generated as a wscc category by singleton languages.

However it is not the classical Kleene theorem, since the right-hand side does not consist of single languages and the wscc operations of $cospan(\wp(\Sigma^*)\text{-}Cat)$ are not the Kleene operations. Further the functor does not lose internal states.

To obtain a theorem closer to the classical Kleene theorem we consider *corelations* between $\wp(\Sigma^*)$ categories, by which we mean cospans which are jointly bijective on objects. Then we compose the above wscc functor $\Phi_2\Phi_1$ with a further wscc functor

$$\Phi_3 : cospan(\wp(\Sigma^*)\text{-}Cat) \longrightarrow corel(\wp(\Sigma^*)\text{-}Cat)$$

which uses the bijective-on-objects fully-faithful factorization to obtain from a cospan of $\wp(\Sigma^*)$-categories a corelation of $\wp(\Sigma^*)$-categories.

The final composite

$$\Phi_3\Phi_2\Phi_1 : Cspn(Graph/\Sigma) \longrightarrow corel(\wp(\Sigma^*)\text{-}Cat)$$

takes a labelled graph with initial and final states to the category with objects only the initial and final states, and whose homs are the languages traced out.

To finish a proof of the classical Kleene theorem we need to show that the wscc operations in $corel(\wp(\Sigma^*)\text{-}Cat)$, at the level of languages (homs), may be expressed in terms of union, concatenation and $(\ )^*$.

Clearly the operation of tensor of corelations does not change the languages which occur as homs. The problem is the composition. But in a wscc category the composition of two arrows may be expressed in terms of the tensor and composition with the constants of the compact closed structure; pictured, this is the fact that:

So the general operation of composition in $corel(\wp(\Sigma^*)\text{-}Cat)$ may be reduced to the very special case of the colimit identifying two objects in a category.

Consider a $\wp(\Sigma^*)$-category X containing two objects x and y. The colimit category $X' = X/(x = y)$ has hom $X'(z, w)$ equal to

$$X(z,w) \cup (X(z,x) \cup X(z,y)) \cdot (X(x,x) \cup X(x,y) \cup X(y,x) \cup X(y,y))^* \cdot (X(x,w) \cup X(y,w)),$$

expressible using only Kleene operations. Hence the result.

This proof is very close to one of the usual proofs of Kleene (if you strip the superstructure). Notice that the passage Φ_1 permits the introduction of ε moves; that is, homs which consist of the empty word. The superstructure has the advantage of suggesting needed generalizations, for example, to parallelism.

6   Comments

The theorem we have described concerns calculating colimits as objects, not as functors. We have not shown the compositionality of morphisms between colimits. We believe that this is connected with the algebra of span and cospan as symmetric monoidal bicategories, rather than as categories. We have made initial progress in understanding this question in [18], by considering a very special case, where we identify the role of 2-separable object.

There is a more precise relation between this work and the paper of Ugo Montanari and José Meseguer [19]. Monoidal graphs may be thought of as Petri nets without markings. Processes of a net G are arrows in the free symmetric monoidal category on G. But in [5] we show that the free symmetric monoidal category on G is an easily identifiable subcategory of $Cspn(MonGraph/G)$, with the same objects as $Cspn(MonGraph/G)$, but with arrows being cospans of "linear monoidal diagrams", that is, those parametrized by monoidal graphs without loops, forking or merging. Linear monoidal diagrams are a generalization of paths in a graph. (Other classes of free categories are similarly obtainable by specializing the type of graphs in the cospans).

References

1. Abramsky, S., Coecke, B.: A Categorical Semantics of Quantum Protocols. In: Proceedings of the 19th Annual IEEE Symposium on Logic in Computer Science: LICS 2004, pp. 415–425. IEEE Computer Society, Los Alamitos (2004)
2. Bloom, S.L., Esik, Z.: Iteration Theories: the equational logic of iterative processes. EATCS Monographs in Theoretical Computer Science. Springer, Heidelberg (1993)
3. Bruni, R., Gadducci, F., Montanari, U.: Normal forms for algebras of connection. Theor. Comput. Sci. 286(2), 247–292 (2002)
4. Carboni, A., Walters, R.F.C.: Cartesian bicategories I. Journal of Pure and Applied Algebra 49, 11–32 (1987)
5. de Francesco Albasini, L., Rosebrugh, R., Sabadini, N., Walters, R.F.C.: Cospans and free symmetric monoidal categories (in preparation)
6. Elgot, C.C.: Monadic computation and iterative algebraic theories, Logoc Colloquium 1973, Studies in Logic 80, pp. 175–230. North Holland, Amsterdam (1975)

7. Gadducci, F., Heckel, R.: An inductive view of graph transformation. In: Parisi-Presicce, F. (ed.) WADT 1997. LNCS, vol. 1376, pp. 223–237. Springer, Heidelberg (1998)
8. Gadducci, F., Heckel, R., Llabrés, M.: A bi-categorical axiomatisation of concurrent graph rewriting. In: Proc. CTCS 1999, Category Theory and Computer Science. Electronic Notes in Theoretical Computer Science, vol. 29, Elsevier Sciences, Amsterdam (1999)
9. Joyal, A., Street, R.H.: The geometry of tensor calculus I. Advances in Math. 88, 55–112 (1991)
10. Joyal, A., Street, R., Verity, D.: Traced monoidal categories. Mathematical Proceedings of the Cambridge Philosophical Society 119(3), 447–468 (1996)
11. Katis, P., Sabadini, N., Walters, R.F.C.: Bicategories of processes. Journal of Pure and Applied Algebra 115, 141–178 (1997)
12. Katis, P., Sabadini, N., Walters, R.F.C.: Span(Graph): A categorical algebra of transition systems. In: Johnson, M. (ed.) AMAST 1997. LNCS, vol. 1349, pp. 307–321. Springer, Heidelberg (1997)
13. Katis, P., Sabadini, N., Walters, R.F.C.: On the algebra of systems with feedback and boundary. Rendiconti del Circolo Matematico di Palermo Serie II Suppl. 63, 123–156 (2000)
14. Katis, P., Sabadini, N., Walters, R.F.C.: A formalisation of the IWIM Model. In: Porto, A., Roman, G.-C. (eds.) COORDINATION 2000. LNCS, vol. 1906, pp. 267–283. Springer, Heidelberg (2000)
15. Katis, P., Sabadini, N., Walters, R.F.C.: Feedback, trace and fixed-point semantics. Theoret. Informatics Appl. 36, 181–194 (2002)
16. Kelly, G.M., Laplaza, M.L.: Coherence for compact closed categories. J. Pure Appl. Algebra 19, 193–213 (1980)
17. Kock, J.: Frobenius algebras and 2D topological Quantum Field Theories. Cambridge University Press, Cambridge (2004)
18. Menni, M., Sabadini, N., Walters, R.F.C.: A universal property of the monoidal 2-category of cospans of ordinals and surjections. Theory and Applications of Categories 18(19), 631–653 (2007)
19. Meseguer, J., Montanari, U.: Petri Nets Are Monoids. Information and Computation 88, 105–155 (1990)
20. Penrose, R.: Applications of negative dimensional tensors. In: Combinatorial Mathematics and its Applications, p. 221. Academic Press, London (1971)
21. Rosebrugh, R., Sabadini, N., Walters, R.F.C.: Minimization and minimal realization in Span(Graph). Mathematical Structures in Computer Science 14, 685–714 (2004)
22. Rosebrugh, R., Sabadini, N., Walters, R.F.C.: Generic commutative separable algebras and cospans of graphs. Theory and Applications of Categories 15(6), 264–277 (2005)
23. Rosebrugh, R., Sabadini, N., Walters, R.F.C.: Calculating colimits and limits compositionally. Category Theory 2007, Carvoeiro, Portugal, 18th (June 2007)
24. Rosebrugh, R., Sabadini, N., Walters, R.F.C.: Calculating colimits and limits compositionally (in preparation)
25. Walters, R.F.C.: Lecture to the Sydney Category Seminar (January 26, 1983)
26. Walters, R.F.C.: The tensor product of matrices, Lecture. In: International Conference on Category Theory, Louvain-la-Neuve (1987)

Observability Concepts in Abstract Data Type Specification, 30 Years Later[*]

Donald Sannella[1] and Andrzej Tarlecki[2,3]

[1] Laboratory for Foundations of Computer Science, University of Edinburgh
[2] Institute of Informatics, Warsaw University
[3] Institute of Computer Science, Polish Academy of Sciences

Abstract. We recall the contribution of Montanari's paper [GGM76] and sketch a framework for observable behaviour specification that blends some of these early ideas, seen from a more modern perspective, with our own approach.

1   Introduction

The starting point for this work is a brief paper [GGM76] coauthored by Ugo Montanari and published in 1976. This appears to be the first of many papers to study observational aspects of the algebraic approach to software specification and development, where the overall idea is that one should regard a specification of a system as constraining its observable behaviour, and nothing more. Such a view is required to cope with many examples. However, it adds significant technical complexities to the simple and elegant algebraic approach. Some of these remain unresolved today, even after 30 years of research.

[GGM76] starts by challenging the initial algebra approach to specifications of abstract data types, then recently introduced by early versions of [GTW78]. Most importantly, [GGM76] points out that not all sorts of data in a data type play the same role: one should separate the given, "old" sorts from the "new" ones, to be specified and implemented. What really matters then is the behaviour of the data type as viewed via these old sorts only; the implementation details of the new sorts play a secondary role. Such *observable behaviour* is captured by the evaluation function restricted to terms that are of old sorts, but in general use the new operations and involve new sorts internally. Another crucial insight in [GGM76] is that in general there are many non-isomorphic algebras that display the same observable behaviour. They show that the set of isomorphism classes of such algebras (limited to the ones generated by the old sorts) forms a complete lattice — a nice technical result which, however, is not used to insist that any such specific algebra is always chosen (as in the initial [GTW78] or final [Wan79] algebra approaches) since all of them are equally adequate implementations of the given observable behaviour. Such behaviours are specified in [GGM76] by

---

[*] This work has been partially supported by European projects IST-2005-015905 MO-BIUS (DS, AT) and IST-2005-016004 SENSORIA (AT).

P. Degano et al. (Eds.): Montanari Festschrift, LNCS 5065, pp. 593–617, 2008.

giving a partial evaluation function, which assigns values to some terms of old sorts only, marking the others as "don't care" cases (indicated by assigning to them a special "value" α, a notation that we will maintain here). The latter captures the situation where the specifier permits the behaviour to be chosen arbitrarily (but consistently with other choices) in any particular implementation. Particular implementations for such a behaviour specification in [GGM76] are captured as (generated) algebras that conform to the specification in the obvious sense.

Quite a few points made in [GGM76] were very insightful in their historical context. This is the first place we know of where several key ideas appear, including some that underlie most of our own contributions to the area. First, the stress on the need for loose specifications, which need not determine behaviour unambiguously (up to isomorphism) was of key importance. The results on the lattice properties of the class of models for a given observable behaviour initiated a line of research in this direction, including a debate on the issue of initial vs. final interpretation of algebraic specifications. One aspect which disappeared in later work was the method of presenting specifications by using an explicitly given set of data on which the data type is based, with behaviour specified by indicating the results of evaluation of some terms, while explicitly marking others as "don't care" cases. The authors' techniques turn out to be very close to "abstract model specifications" in the style of VDM [Jon80]. The main contribution though is the idea of limiting specifications to observable parts of behaviour only, thus introducing observability aspects to algebraic specification.

The pioneering role of [GGM76] is underlined by the fact that it cites just 12 references, some of them unpublished, with only a few concerning algebraic specifications. Hardly any other papers in the field could have been mentioned then: at the time, this is essentially all that there was! This has to be contrasted with the outburst of work in the area in the following years, as for instance summarised in the bibliography [BKL+91] some 15 years later, or in the overview presentations of the field in [Wir90] or the more recent [AKKB99]. One important line of activity concerned observability aspects, with an extensive literature of its own, including [Rei81] and numerous papers presenting further developments in various directions, at diverse levels of abstraction. This includes for instance the popular *hidden algebra* framework [GM00] and our own work [ST87] aimed at bringing this closer to logical characterisation via elementary equivalence, with [BH06] offering a recent elegant approach benefiting from all this experience.

We reiterate some of the ideas presented in [GGM76] here, looking back at more than 30 years of work on algebraic specification, and trying to blend what happened with these ideas with our current personal perspective. We sketch a framework for observable behaviour specification and development, reconsidering some of the work presented earlier [ST88b, BST02, BST08] in a different technical setting. It is reassuring that, after shifting to quite a different specification technology, inspired by [GGM76], our basic ideas on system specification, architectural design and development under an observational view of specifications still stand.

2   Algebraic Preliminaries

Signatures and signature morphisms are as usual, except for the treatment of the distinguished sort *bool*.

Definition 2.1. *A signature $\Sigma = \langle S, \Omega \rangle$ consists of a set S of sort names and an $S^* \times S$-indexed set Ω of operation names, where $f \in \Omega_{\langle s_1 \cdots s_n, s \rangle}$ is written $f \colon s_1 \times \cdots \times s_n \to s$. We require that bool $\in S$ and that no operations in Ω take arguments in bool. If A is an S-sorted set, then $\Sigma(A)$ denotes the signature obtained from Σ by adding the elements of A to Ω as constants.*

If $\Sigma' = \langle S', \Omega' \rangle$, then a signature morphism $\sigma \colon \Sigma \to \Sigma'$ consists of a mapping of sort names, $\sigma \colon S \to S'$, and an $S^* \times S$-sorted mapping of operation names, with $\sigma_{\langle s_1 \cdots s_n, s \rangle} \colon \Omega_{\langle s_1 \cdots s_n, s \rangle} \to \Omega'_{\langle \sigma(s_1) \cdots \sigma(s_n), \sigma(s) \rangle}$, such that $\sigma(s) = bool$ iff s is bool.

We regard *bool* as the sort of logical meta-values, where operations that deliver results in *bool* are like predicates. Forbidding operations taking arguments in *bool* corresponds to the fact that applying a predicate to a tuple of terms would normally yield an atomic formula, not a term. We treat predicates here as operations, with this restriction, rather than as relations, for the sake of technical convenience. Observations (see Sect. 4) will be terms of sort *bool*.

Algebras and their homomorphisms are defined as usual, except that we fix the interpretation of the distinguished sort *bool* to be the set $\mathbb{B} = \{true, false\}$.

Definition 2.2. *Given a signature $\Sigma = \langle S, \Omega \rangle$, a Σ-algebra $\mathbf{A}$ consists of an S-sorted carrier set A and, for each operation name $f \colon s_1 \times \cdots \times s_n \to s$, a function $f_{\mathbf{A}} \colon A_{s_1} \times \cdots \times A_{s_n} \to A_s$. We require that $A_{bool} = \mathbb{B}$.*

A Σ-homomorphism $m : \mathbf{A} \to \mathbf{B}$ between Σ-algebras $\mathbf{A}$ and $\mathbf{B}$ is an S-sorted family of functions $m_s : A_s \to B_s$, $s \in S$, that preserve the values of operations, as usual. We require that m_{bool} is the identity on $\mathbb{B}$.

Given a signature morphism $\sigma \colon \Sigma \to \Sigma'$, for any Σ'-algebra $\mathbf{A}'$, its σ-reduct is the Σ-algebra $\mathbf{A} = \mathbf{A}'|_\sigma$ given by $A_s = A'_{\sigma(s)}$ for $s \in S$, and $f_{\mathbf{A}} = \sigma(f)_{\mathbf{A}'}$ for $f \in \Omega$. Reducts of Σ'-homomorphisms and of S'-sorted sets as well as of $(S'$-sorted) functions and relations between them are defined analogously.

Signatures and their morphisms form a category, which is cocomplete. Σ-algebras and homomorphisms between them form a category (which is also cocomplete, with the algebra $\mathbf{T_\Sigma}$ of ground Σ-terms as the initial object). For any signature morphism $\sigma \colon \Sigma \to \Sigma'$, σ-reduct is a functor. Moreover, the assignments of the categories of algebras to signatures, and of reduct functors to signature morphisms form a (contravariant) functor from the category of signatures to the category of "all" categories. This functor is continuous, so that in particular the following amalgamation lemma holds:

Lemma 2.3. *Consider a pushout in the category of signatures.*

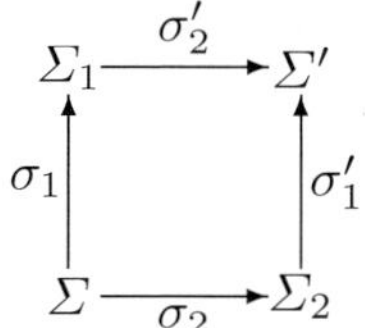

Then for any Σ_1-algebra $\mathbf{A_1}$ and Σ_2-algebra $\mathbf{A_2}$ with common Σ-reduct $\mathbf{A_1}|_{\sigma_1} = \mathbf{A_2}|_{\sigma_2}$, there exists a unique Σ'-algebra $\mathbf{A}'$ such that $\mathbf{A}'|_{\sigma_2'} = \mathbf{A_1}$ and $\mathbf{A}'|_{\sigma_1'} = \mathbf{A_2}$; and similarly for homomorphisms.

Given a signature $\Sigma = \langle S, \Omega \rangle$ and an S-sorted set A, we will consider the set $T_\Sigma(A)$ of Σ-terms with "variables" in A (when A is empty, we write T_Σ for $T_\Sigma(A)$). Equivalently we could take $T_{\Sigma(A)}$, where elements of A are considered as additional constants. The distinction will be disregarded whenever convenient. Terms that are in A will be referred to as *data*; the others as *non-data*.

Given a signature morphism $\sigma \colon \Sigma \to \Sigma'$, S-sorted set A and S'-sorted set A' such that $A \subseteq A'|_\sigma$, σ induces the translation $\sigma : T_\Sigma(A) \to T_{\Sigma'}(A')$ in the obvious way: $\sigma(a) = a$ for data terms $a \in A$, and extending this to non-data terms by replacing each Σ-operation name f with Σ'-operation name $\sigma(f)$.

By a $\Sigma(A)$-context we mean any Σ-term t that in addition to the operation names from Σ and data from A may contain an occurrence of a special variable $\square$. Then, for any term $t' \in T_\Sigma(A)$, we write $t(t')$ for the term in $T_\Sigma(A)$ obtained by substituting t' for $\square$ in t.[1]

3 Behaviours and Behaviour Specifications

Inspired by the notion of a (complete) specification in [GGM76] as a "black-box" view of models, we will not deal explicitly with algebras here, but rather concentrate on the study of their *behaviours*. Let $\Sigma = \langle S, \Omega \rangle$.

Definition 3.1. *A Σ-behaviour is an S-sorted carrier set A together with an S-sorted evaluation function $ev \colon T_\Sigma(A) \to A$ such that: $ev(a) = a$ for all data $a \in A$; if $ev(t') = a'$ then $ev(t(a')) = ev(t(t'))$ for all terms $t' \in T_\Sigma(A)$ and $\Sigma(A)$-contexts t; and $A_{bool} = \mathbb{B}$. We will use evaluation functions ev to refer to behaviours, with carriers left implicit.*

A knowledgeable reader will recognise the notion of an algebra for the monad T_Σ. In this definition, it suffices to consider contexts t of the form $f(a_1, \ldots, \square, \ldots, a_n)$ for $a_1, \ldots, a_n \in A$. Examples for this and other definitions will come in Sect. 5.

Definition 3.2. *Given Σ-behaviours $ev_1 \colon T_\Sigma(A_1) \to A_1$ and $ev_2 \colon T_\Sigma(A_2) \to A_2$, a homomorphism $m \colon ev_1 \to ev_2$ is an S-sorted function $m \colon A_1 \to A_2$, such that m_{bool} is the identity function on $\mathbb{B}$, and if $m(ev_1(t_1)) = ev_2(t_2)$ then $m(ev_1(t(t_1))) = ev_2(\hat{t}(t_2))$ for all $\Sigma(A_1)$-contexts t, $\Sigma(A_1)$-terms t_1 and $\Sigma(A_2)$-terms t_2, where $\hat{t}$ results from t by replacing data $a \in A_1$ by $m(a) \in A_2$.*

In this definition, it is once again sufficient to consider contexts t of the form $f(a_1, \ldots, \square, \ldots, a_n)$ for $a_1, \ldots, a_n \in A_1$, and t_1 and t_2 that are data terms.

As usual, semantics (behaviours) can be translated along signature morphisms in the opposite direction to the translation of syntax (terms):

Definition 3.3. *Consider a signature morphism $\sigma \colon \Sigma \to \Sigma'$ and Σ'-behaviour $ev' \colon T_{\Sigma'}(A') \to A'$. The σ-reduct of ev' is the Σ-behaviour $ev'|_\sigma \colon T_\Sigma(A) \to A$ where $A = A'|_\sigma$ and for $t \in T_\Sigma(A)$, $(ev'|_\sigma)(t) = ev'(\sigma(t))$.*

---

[1] To be precise, this requires a careful identification of the sort for the variable $\square$ and the term t' — whenever convenient, we will continue omitting such details here.

Proposition 3.4. *Given a signature morphism* $\sigma\colon \Sigma \to \Sigma'$, *the* σ-*reduct* $m|_\sigma$ *of any homomorphism* $m\colon ev'_1 \to ev'_2$ *between* Σ'-*behaviours is a homomorphism between their* σ-*reducts,* $m|_\sigma\colon ev'_1|_\sigma \to ev'_2|_\sigma$.

There is a 1–1 correspondence between Σ-behaviours and Σ-algebras, and between homomorphisms as above and ordinary homomorphisms on algebras, as recalled in Sect. 2. This gives an isomorphism between the category of Σ-behaviours and the category of Σ-algebras, and carries over to the reduct functors determined by signature morphisms.

Σ-behaviours are specified by indicating what the values of certain terms should be, while explicitly indicating that the values of other terms are not constrained. We use α for the latter "don't care" case, following [GGM76].

Definition 3.5. *A* Σ-*behaviour specification is an* S-*sorted carrier set* A *together with an* S-*sorted function* $h\colon T_\Sigma(A) \to A \cup \{\alpha\}$ *such that:* $h(a) = a$ *for all data* $a \in A$; *if* $h(t') = a'$ *then* $h(t(a')) = h(t(t'))$ *for all terms* $t' \in T_\Sigma(A)$ *and* $\Sigma(A)$-*contexts* t; *and* $A_{bool} = \mathbb{B}$. *We will use functions* h *to refer to behaviour specifications, leaving their carriers implicit.*

It is important to understand that α ("don't care") does not mean that any choice of value will do; as we will see, the choice taken needs to respect the values of those terms that are specified as non-α.

There are two natural orderings on behaviour specifications, both reflecting the degree to which behaviour is constrained.

Definition 3.6. *Let* $h_1\colon T_\Sigma(A_1) \to A_1 \cup \{\alpha\}$ *and* $h_2\colon T_\Sigma(A_2) \to A_2 \cup \{\alpha\}$ *be* Σ-*behaviour specifications, and let* $ev\colon T_\Sigma(A) \to A$ *be a* Σ-*behaviour.*

1. h_2 *refines* h_1, *written* $h_1 \rightsquigarrow h_2$, *if* $A_1 \subseteq A_2$ *and* h_2 *conforms to* h_1, *that is: for each* $t \in T_\Sigma(A_1)$, $h_2(t) = h_1(t)$ *whenever* $h_1(t) \neq \alpha$. *Then* ev *satisfies* h_1 *if* ev *(viewed as a behaviour specification) refines* h_1. *We write* $Mod(h_1)$ *for the class of all* Σ-*behaviours that satisfy* h_1.
2. h_2 *strongly refines* h_1 *if* h_2 *refines* h_1 *and* $A_1 = A_2$. *Then* $ev\colon T_\Sigma(A) \to A$ *strongly satisfies* h_1 *if* ev *strongly refines* h_1, *which requires* $A_1 = A$.

For any Σ-behaviour specification $h\colon T_\Sigma(A) \to A \cup \{\alpha\}$ we define its *free extension* by adding a new value for each of the "don't care" cases. It corresponds to the "initial symbolic representation" of h from [GGM76] (with all sorts viewed as "old") and is constructed as follows. Let I extend A by all non-data terms $t \in T_\Sigma(A)$ such that all subterms t' of t with $h(t') \neq \alpha$ are data. This not only requires that $h(t) = \alpha$, but also that all subterms of t are evaluated as far as determined by h. Then, the free extension of h is the only function $ev_I\colon T_\Sigma(I) \to I$ such that $ev_I(t) = h(t)$ for all terms $t \in T_\Sigma(A)$ with $h(t) \neq \alpha$, $ev_I(t) = t$ for all terms t in $I \setminus A$, and $ev_I(t(t')) = ev_I(t(ev_I(t')))$ for all contexts t and terms t'. Now, ev_I is not necessarily a Σ-behaviour, because I_{bool} need not be $\mathbb{B}$. However, we can obtain a Σ-behaviour from ev_I by choosing a function $f\colon I_{bool} \to \mathbb{B}$ that extends identity on $\mathbb{B}$. Let then $\hat{f}$ extend f to an S-sorted function that is the identity on all sorts other than *bool*, and let $\hat{I}$ be an S-sorted set such that

$\hat{I}_s = I_s$ for all sorts $s \neq bool$ and $\hat{I}_{bool} = \mathbb{B}$. Then $ev_h^f \colon T_\Sigma(\hat{I}) \to \hat{I}$, such that $ev_h^f(t) = \hat{f}(ev_I(t))$ for $t \in T_\Sigma(\hat{I})$, is a Σ-behaviour that satisfies h. This proves the following:

Lemma 3.7. *Every Σ-behaviour specification $h \colon T_\Sigma(A) \to A \cup \{\alpha\}$ is satisfiable.*

It is not the case that any behaviour specification h is *strongly* satisfiable, unless we add the requirement that whenever $h(t_1) = \alpha$ then there exists some $a \in A$ such that if $h(t(t_1)) \neq \alpha$ then $h(t(a)) = h(t(t_1))$. (See condition (c) in the definition of specification in [GGM76].)

Lemma 3.8. *For Σ-behaviour specifications h_1 and h_2, $h_1 \rightsquigarrow h_2$ iff $Mod(h_2) \subseteq Mod(h_1)$.*

For technical convenience, let us introduce an additional special specification $\varnothing$ that is not satisfied by any behaviour, $Mod(\varnothing) = \varnothing$. We extend our definition of refinement to cover $\varnothing$ in the natural way, so that Lemma 3.8 still holds.

Complex behaviour specifications can be built in a structured way from simpler specifications, using standard specification-building operations such as derive (reduct), translate, union, defined in terms of their model classes in [ST88a]. For behaviour specifications of the form considered here, these kernel specification-building operations may be defined "internally".

Definition 3.9. *Reduct, translation and union of behaviour specifications are defined as follows:*

1. *Given a signature morphism $\sigma : \Sigma \to \Sigma'$ and Σ'-behaviour specification $h' : T_\Sigma(A') \to A' \cup \{\alpha\}$, its σ-reduct $h'|_\sigma$ is the Σ-behaviour specification $h : T_\Sigma(A|_\sigma) \to A|_\sigma \cup \{\alpha\}$ defined by $h(t) = h'(\sigma(t))$ for all $t \in T_\Sigma(A|_\sigma)$.*
2. *Given a signature morphism $\sigma : \Sigma \to \Sigma'$ and a Σ-behaviour specification $h : T_\Sigma(A) \to A \cup \{\alpha\}$, let $A'_{s'} = \bigcup\{A_s \mid \sigma(s) = s'\}$ for all sorts s' in Σ'. First, we define an auxiliary relation between terms $t' \in T_{\Sigma'}(A')$ and data $a' \in A'$, written h via σ forces t' to a', as the least relation such that*
   (a) if $h(t) = a$ then h via σ forces $\sigma(t)$ to a, and
   (b) if h via σ forces t'' to a'' and h via σ forces $t'(a'')$ to a' then h via σ forces $t'(t'')$ to a'.
   Then the σ-translation $\sigma(h)$ of h is $\varnothing$ if for some term $t' \in T_{\Sigma'}(A')$ and two distinct data $a', a'' \in A'$, h via σ forces t' to a' and h via σ forces t' to a''. Otherwise, $\sigma(h)$ is the Σ'-behaviour specification $h' : T_\Sigma(A') \to A' \cup \{\alpha\}$ such that for each term $t' \in T_{\Sigma'}(A')$, $h'(t') = a'$ if h via σ forces t' to a' and $h'(t') = \alpha$ if such $a' \in A'$ does not exist.
3. *Given Σ-behaviour specifications $h_1 : T_\Sigma(A_1) \to A_1 \cup \{\alpha\}$ and $h_2 : T_\Sigma(A_2) \to A_2 \cup \{\alpha\}$, first define an auxiliary relation between terms $t \in T_\Sigma(A_1 \cup A_2)$ and data $a \in (A_1 \cup A_2)$, written $\{h_1, h_2\}$ forces t to a, as the least relation such that*
   (a) if $h_1(t) = a$ or $h_2(t) = a$ then $\{h_1, h_2\}$ forces t to a and
   (b) if $\{h_1, h_2\}$ forces t_0 to a_0 and $\{h_1, h_2\}$ forces $t(a_0)$ to a then $\{h_1, h_2\}$ forces $t(t_0)$ to a.

Then, the union $h_1 + h_2$ *of* h_1 *and* h_2 *is* $\emptyset$ *if for some term* $t \in T_\Sigma(A_1 \cup A_2)$ *and two distinct data* $a, a' \in (A_1 \cup A_2)$, $\{h_1, h_2\}$ *forces* t *to* a *and* $\{h_1, h_2\}$ *forces* t *to* a'. *Otherwise,* $h_1 + h_2$ *is the* Σ-*behaviour specification* $h : T_\Sigma(A_1 \cup A_2) \to (A_1 \cup A_2) \cup \{\alpha\}$ *such that for each term* $t \in T_\Sigma(A_1 \cup A_2)$, $h(t) = a$ *if* $\{h_1, h_2\}$ *forces* t *to* a *and* $h(t) = \alpha$ *if such* $a \in (A_1 \cup A_2)$ *does not exist.*

Theorem 3.10. *1. Given a signature morphism* $\sigma : \Sigma \to \Sigma'$ *and* Σ'-*behaviour specification* h', $Mod(h'|_\sigma) \supseteq Mod(h')|_\sigma$, *where* $Mod(h')|_\sigma$ *is the class of all* σ-*reducts of* Σ'-*behaviours in* $Mod(h')$.

   2. Given a signature morphism $\sigma : \Sigma \to \Sigma'$ *and* Σ-*behaviour specification* h, $Mod(\sigma(h)) = Mod(h)|_\sigma^{-1}$, *where* $Mod(h)|_\sigma^{-1}$ *is the class of all* Σ'-*behaviours with* σ-*reducts in* $Mod(h)$.

   3. Given two Σ-*behaviour specifications* h_1 *and* h_2, *we have* $Mod(h_1 + h_2) = Mod(h_1) \cap Mod(h_2)$.

Note that for some (even injective) signature morphisms $\sigma : \Sigma \to \Sigma'$ and Σ'-behaviour specifications h' there may be no Σ-behaviour specification h such that $Mod(h) = Mod(h')|_\sigma$.

As usual, the above are just "kernel" operations on specifications, which underly more complex ones that are closer to what will be used in practical examples. For instance, if h is a Σ-behaviour specification, then we can write behaviour specifications over signatures extending Σ as follows (also permitting self-explanatory notational variants whenever convenient):

h **then** *signature-extension* **with** *sort-definitions* **and** *behaviour-definition*

Here, *signature-extension* is an extension of Σ, possibly contributing new sort names S' and operation names over $(S \cup S')^* \times (S \cup S')$, resulting in a new signature Σ', *sort-definitions* provides carrier definitions for the sorts in S', and *behaviour-definition* defines the functions h_s for all sorts $s \in S \cup S'$ excluding those in S for which no new terms arise. In all the examples below, it will be the case that the resulting specification is equivalent to a behaviour specification $h' : T_{\Sigma'}(A') \to A'$ where for each sort s in Σ', A'_s and h'_s are either inherited from h or defined explicitly in *sort-definitions* and *behaviour-definition*, respectively, and moreover, if h'_s is defined in *behaviour-definition* then it extends h_s. In fact, such a behaviour specification may be defined explicitly referring to the operations introduced by Def. 3.9, similarly as the standard enrich operation is defined in terms of translation and union [ST88a].

4   Observable Behaviour

We now begin to focus on the main theme of this work, and look at what happens when we consider only *observable* behaviour. As usual, we regard values of some sorts as directly observable while the remaining sorts are treated as internal, with properties of their elements made visible only via observations, which are terms producing a result of observable sort. However, the technical means used to achieve this are somewhat different, at least superficially, from much previous work in this area. Without loss of generality, we take *bool* to be the only observable sort. In view of our earlier discussion on the role of *bool*, this means that we observe only the results of predicate applications, and take none of the ordinary "data" sorts as observable. This departs from standard approaches (a recent exception being [BST08]), where choosing a non-empty set of observable data sorts is crucial to have any observations at all and it is appropriate for this set to vary in the process of modular development (e.g. the parameter sorts in specifications of local constructions must be locally considered as observable). The former is taken care of by assuming that appropriate predicates are introduced into the specifications considered; the latter will be achieved in a technically different way here, see Def. 7.7 below.

Definition 4.1. *An* observable Σ-behaviour *is a function* $ev\colon (T_\Sigma)_{bool} \to \mathbb{B}$. *The* observable part *of a Σ-behaviour* $ev\colon T_\Sigma(A) \to A$ *is the restriction of* ev_{bool} *to* $(T_\Sigma)_{bool}$, *written as* $Obs(ev)\colon (T_\Sigma)_{bool} \to \mathbb{B}$.

The domain of the observable behaviour $Obs(ev)$ is properly included in the domain of the *bool* component of the behaviour ev: the latter also includes observations on non-ground terms, as well as the constants *true* and *false*. Including *true* and *false* would do no harm, but it is inappropriate to regard observations on unreachable values of non-observable sort as relevant to observable behaviour.

When producing a specification, we are actually interested in specifying *observable* behaviour; what happens with non-observable components is of no interest, except insofar as they affect the observable behaviour. The definition of observable behaviour specification is analogous to the definition of behaviour specification above and is inspired by the notion of "specification" in [GGM76], which is highlighted as their key definition.

Definition 4.2. *An* observable Σ-behaviour specification *is a function* $h\colon (T_\Sigma)_{bool} \to \mathbb{B} \cup \{\alpha\}$. *An observable Σ-behaviour* $ev\colon (T_\Sigma)_{bool} \to \mathbb{B}$ *satisfies* h *if* $ev(t) = h(t)$ *whenever* $h(t) \neq \alpha$. *A Σ-behaviour* $ev\colon T_\Sigma(A) \to A$ *(observationally) satisfies* h *if its observable part* $Obs(ev)\colon (T_\Sigma)_{bool} \to \mathbb{B}$ *satisfies* h. *We write* $Mod_{Obs}(h)$ *for the class of all such behaviours. Finally, the* observable part *of a Σ-behaviour specification* $h\colon T_\Sigma(A) \to A \cup \{\alpha\}$ *is the restriction of* h *to the set* $(T_\Sigma)_{bool}$; *we write this as* $Obs(h)\colon (T_\Sigma)_{bool} \to \mathbb{B} \cup \{\alpha\}$.

Lemma 4.3. *Let* $h\colon T_\Sigma(A) \to A \cup \{\alpha\}$ *be a Σ-behaviour specification. Then* $Mod(h) \subseteq Mod_{Obs}(Obs(h))$, *that is, for all Σ-behaviours* ev, *if* ev *satisfies* h *then* ev *observationally satisfies* $Obs(h)$.

Any observable Σ-behaviour specification $h\colon (T_\Sigma)_{bool} \to \mathbb{B} \cup \{\alpha\}$ may be equivalently considered as the behaviour specification $h^+\colon T_\Sigma(\mathbb{B}^+) \to \mathbb{B}^+ \cup \{\alpha\}$ that adds empty carriers for all sorts other than *bool* and maps all terms of these sorts to α.

Lemma 4.4. *Let* $h\colon (T_\Sigma)_{bool} \to \mathbb{B} \cup \{\alpha\}$ *be an observable Σ-behaviour specification. Then* $Mod_{Obs}(h) = Mod(h^+)$, *that is, for all Σ-behaviours* ev, ev *observationally satisfies* h *iff* ev *satisfies* h^+.

Any notion of observable behaviour gives rise to an equivalence between behaviours, whereby two behaviours are equivalent iff their observable parts coincide. This equivalence, methods for proving it, and conditions under which it is preserved by constructions on behaviours, is central to the study of observability concepts in specifications.

Definition 4.5. *Two Σ-behaviours, $ev\colon T_\Sigma(A) \to A$ and $ev'\colon T_\Sigma(A') \to A'$, are* observationally equivalent, *written $ev \equiv ev'$, if $Obs(ev) = Obs(ev')$.*

The following definition, and its use in the sequel, is derived from Schoett's notion of correspondence for Σ-algebras in [Sch87].

Definition 4.6. *Let $ev\colon T_\Sigma(A) \to A$ and $ev'\colon T_\Sigma(A') \to A'$ be Σ-behaviours. A Σ-correspondence $\rho\colon ev \bowtie ev'$ is an S-sorted relation $\rho \subseteq A \times A'$ such that ρ_{bool} is the identity relation on $\mathbb{B}$, and ρ is preserved by operations: for any operation name $f\colon s_1 \times \cdots \times s_n \to s$ and terms $t_1, \ldots, t_n \in T_\Sigma(A)$ and $t'_1, \ldots, t'_n \in T_\Sigma(A')$ of the respective argument sorts, if $ev(t_1) \ \rho_{s_1} \ ev'(t'_1)$ and $\cdots$ and $ev(t_n) \ \rho_{s_n} \ ev'(t'_n)$ then $ev(f(t_1, \ldots, t_n)) \ \rho_s \ ev'(f(t'_1, \ldots, t'_n))$.*

In fact, it is enough to consider here $t_1, \ldots, t_n$ and $t'_1, \ldots, t'_n$ to be constants in A and A' respectively.

Σ-correspondences can also be presented as spans of Σ-behaviour homomorphisms, see e.g. [BST08].

Proposition 4.7. *Let $ev\colon T_\Sigma(A) \to A$ and $ev'\colon T_\Sigma(A') \to A'$ be Σ-behaviours. Then $ev \equiv ev'$ iff there is a Σ-correspondence $\rho\colon ev \bowtie ev'$.*

Proposition 4.8. *Consider any Σ-behaviour specification $h\colon T_\Sigma(A) \to A \cup \{\alpha\}$ and its observable part $Obs(h)\colon (T_\Sigma)_{bool} \to \mathbb{B} \cup \{\alpha\}$. For every Σ-behaviour $ev\colon T_\Sigma(B) \to B$ that observationally satisfies $Obs(h)$, there exists a Σ-behaviour $ev'\colon T_\Sigma(C) \to C$ such that ev' satisfies h, and $ev \equiv ev'$.*

5   Examples

The following examples illustrate the definitions above. Examples in the sequel will build on these and will provide further illustrations.

We give an observable behaviour specification of symbol tables for a programming language with block structure and local variable declarations. This builds on the following specification of identifiers as strings.

> IDENT =
>> **eqsort** *ident = string*
>> **opns**   "a", "b", ..., "any string you like", ... : *string*

The notion of *eqsort* is borrowed from Standard ML's "eqtypes" (equality types). Since *ident* is an eqsort, it comes with an implicit operation $=\colon ident \times ident \to bool$ such that $h_{bool}^{\mathrm{IDENT}}(i = j) = true$ iff i and j are identical strings. We allow ourselves to write $i = j$ below in place of $h_{bool}^{\mathrm{IDENT}}(i = j) = true$ for brevity.

We use an informal notation for extending observable behaviour specifications, with a meaning that is analogous to that defined above for the case of behaviour specifications. According to the following specification, a symbol table records identifiers without associating any information to them. Identifiers that are added after entering a block are forgotten once the end of the block is reached, since identifiers within the block are no longer in scope at that point.

$$\text{SYMTAB} = \text{IDENT } \textbf{then}$$
$$\textbf{sort } \; symtab$$
$$\textbf{opns } \; empty \colon symtab$$
$$add \colon ident \times symtab \to symtab$$
$$enter \colon symtab \to symtab$$
$$leave \colon symtab \to symtab$$
$$isin \colon ident \times symtab \to bool$$

with

$$h^{\text{SYMTAB}}_{bool}(isin(i, empty)) = false$$
$$h^{\text{SYMTAB}}_{bool}(isin(i, add(i', t))) = \alpha \textbf{ if } t \text{ is unbalanced}$$
$$\textbf{else } true \textbf{ if } i = i'$$
$$\textbf{else } h^{\text{SYMTAB}}_{bool}(isin(i, t))$$
$$h^{\text{SYMTAB}}_{bool}(isin(i, enter(t))) = \alpha \textbf{ if } t \text{ is unbalanced}$$
$$\textbf{else } h^{\text{SYMTAB}}_{bool}(isin(i, t))$$
$$h^{\text{SYMTAB}}_{bool}(isin(i, leave(t))) = \alpha \textbf{ if } leave(t) \text{ is unbalanced}$$
$$\textbf{else } h^{\text{SYMTAB}}_{bool}(isin(i, L(t)))$$

where a (ground) term t is *unbalanced* if there is a subterm of t containing more occurrences of *leave* than *enter*. Informally, such a t is erroneous in the sense that it indicates an attempt to exit a block that has not been entered. Otherwise we say that t is *well-balanced*. Then L is an auxiliary function that for each t such that $leave(t)$ is well-balanced yields the term immediately inside the first use of *enter* that is not matched by a preceding *leave* (formally: $L(add(i, t)) = L(t)$, $L(enter(t)) = t$, and $L(leave(t)) = L(L(t))$). The meaning of an equation like $h^{\text{SYMTAB}}_{bool}(isin(i, empty)) = false$ above is that it holds for all of its ground instances. (That is the reason why we needed constants in IDENT.)

SYMTAB is then an observable behaviour specification over the indicated signature (i.e., the signature of IDENT, including $=$, together with the sort $symtab$ and the operations listed above).

A refinement of SYMTAB is SYMTAB$'$, given by the following function, which specifies choices for the cases that SYMTAB leaves open. It makes no use of α, and so it determines a single observable behaviour; all of the behaviours that satisfy it are observationally equivalent. We have decided here that for unbalanced terms t, $isin(i, t)$ yields true for any identifier i.

$$h_{bool}^{\text{Symtab}'}(isin(i, empty)) = false$$
$$h_{bool}^{\text{Symtab}'}(isin(i, add(i', t))) = true \textbf{ if } i = i'$$
$$\textbf{else } h_{bool}^{\text{Symtab}'}(isin(i, t))$$
$$h_{bool}^{\text{Symtab}'}(isin(i, enter(t))) = h_{bool}^{\text{Symtab}'}(isin(i, t))$$
$$h_{bool}^{\text{Symtab}'}(isin(i, leave(t))) = true \textbf{ if } leave(t) \text{ is unbalanced}$$
$$\textbf{else } h_{bool}^{\text{Symtab}'}(isin(i, L(t)))$$

Examples of behaviours over the signature of Symtab are given below in
the form of SML structures. When no partiality, exceptions, polymorphism etc.
arises, as below, this amounts to a definition of an algebra and therefore of a
behaviour in our sense. Given such a structure definition Str, we will write
Str.s for its carrier of sort s, and ev^{Str} for its behaviour function.

The definitions below build on the definition of *ident* as the eqsort (or in SML,
eqtype) *string* with constants as above.

```
structure LST =
  struct
    type symtab = (ident list) list
    val empty = [[]]  : (ident list) list
    fun add(i,[]) = []
      | add(i,l::st) = (i::l)::st
    fun enter [] = []
      | enter st = []::st
    fun leave [] = []
      | leave(l::st) = st
    fun isin(i,[]) = false
      | isin(i,[]::st) = isin(i,st)
      | isin(i,(j::l)::st) = (i=j) orelse isin(i,l::st)
  end
```

Here, symbol tables are represented as lists of lists of identifiers. A list of identifiers represents the set of identifiers declared in a given block. A list of these
lists is used to record block structure; this works because of the way that blocks
can be nested. The behaviour determined by Lst satisfies Symtab but does not
satisfy Symtab$'$.

A different behaviour Lst$'$ is obtained by making the *isin* function yield *true*
for unbalanced symbol tables. In the code for Lst, we just replace the definition
of *isin* by

```
fun isin(i,[]) = true
  | isin(i,[[]]) = false
  | isin(i,[]::st) = isin(i,st)
  | isin(i,(j::l)::st) = (i=j) orelse isin(i,l::st)
```

(Note that the order of clauses matters in SML: the third clause only applies to
non-empty st.) Lst$'$ satisfies Symtab$'$ and so also satisfies Symtab.

A different behaviour with the same observable part as Lst is given by the
following structure, in which symbol tables are represented using functions. The

set of identifiers within a given block is represented by its characteristic function, of type $ident \to bool$, and a stack of these (represented as an "array" of sets, $int \to (ident \to bool)$ together with an integer "pointer" to the top of the stack) is used to record block structure.

```
structure SST =
  struct
    type symtab = int -> (ident -> bool) * int
    val empty = (fn n => if n=0 then (fn i => false)
                              else (fn i => true) ,
                         0)
    fun add(i,(st,m)) =
          (fn n => if n=m
                      then fn j => (i=j) orelse st(n)(j)
                      else st(n) ,
              m)
    fun enter(st,m) =
          if m<0 then (st,m)
          else (fn n => if n>m then fn j=>false else st(n) ,
              m+1)
    fun leave(st,m) = (st,m-1)
    fun isin(i,(st,m)) = if m<0 then false
                              else if m=0 then st(m)(i)
                                  else st(m)(i) orelse isin(i,(st,m-1))
  end
```

Now, LST and SST are observationally equivalent but there is no homomorphism from either to the other, even if we restrict their carriers to the values of ground terms. However, there are correspondences $\rho\colon ev^{\text{SST}} \bowtie ev^{\text{LST}}$ that witness the behavioural equivalence. One such correspondence relates all pairs $\langle st, m\rangle$ for $m < 0$ with the empty list, and then for $m \geq 0$ it relates $\langle st, m\rangle$ with all lists $[l_0, \ldots, l_m]$ such that $st(i)(j) = true$ iff j occurs at least once in l_i.

6   Implementations

We write specifications because we are interested in developing programs that implement them. One way of proceeding is top-down, by stepwise refinement: we refine the original specification of requirements to another one that is easier to implement by filling in design decisions such as choosing between the options of behaviour left open in "don't care" cases.

The issue of implementing specifications by programs is not mentioned in [GGM76]. An elegant approach to this issue has been developed in the years since then, coping with both observational and non-observational views of specifications. In this section and the next one we adapt this existing approach to the present framework, using our own work [ST88b, BST02, BST08] as a basis.

To produce a program from a specification, we proceed in stages by reducing the problem to a simpler one. At each stage, we postulate a solution to the simpler problem, and show by construction how to turn such a solution into a solution to the overall problem.

Definition 6.1. *Let Σ and Σ' be signatures. A* construction *from Σ to Σ', written $\kappa\colon \Sigma \Rightarrow \Sigma'$, is a function mapping any Σ-behaviour to a Σ'-behaviour.[2] Given a Σ-behaviour specification $h\colon T_\Sigma(A) \to A \cup \{\alpha\}$ and a Σ'-behaviour specification $h'\colon T_{\Sigma'}(A') \to A' \cup \{\alpha\}$, we say that h* implements h' via κ, *written $h' \overset{}{\underset{\kappa}{\rightsquigarrow}} h$, if κ maps each Σ-behaviour satisfying h to a Σ'-behaviour satisfying h', i.e. $\kappa(Mod(h)) \subseteq Mod(h')$, where $\kappa(Mod(h))$ denotes the image of $Mod(h)$ under κ. When we want to emphasize correctness of κ in relating h and h' rather than the relationship between h and h', we say that κ is* correct w.r.t. h *and* h'.

Although the definition says that a construction $\kappa\colon \Sigma \Rightarrow \Sigma'$ is a mathematical function, it is best viewed as the semantic function underlying a *parameterised program* [Gog96], or in SML terms a *functor*, which produces the components (sorts and operations) required by Σ' when supplied with the components required by Σ. Then $h' \overset{}{\underset{\kappa}{\rightsquigarrow}} h$ amounts to a reduction of the task of implementing h' to the task of implementing h, where κ supplies code to fill in the gap.

We can easily compose successive implementations. Then, once we have reduced the problem to one we have already solved, we obtain a solution to the original problem.

Proposition 6.2. *If $h_1 \overset{}{\underset{\kappa_1}{\rightsquigarrow}} h_2 \overset{}{\underset{\kappa_2}{\rightsquigarrow}} h_3$ then $h_1 \overset{}{\underset{\kappa_2;\kappa_1}{\rightsquigarrow}} h_3$. Thus, if $h_1 \overset{}{\underset{\kappa_1}{\rightsquigarrow}} \cdots \overset{}{\underset{\kappa_{n-1}}{\rightsquigarrow}} h_n$ and ev_n satisfies h_n, then $\kappa_1(\cdots(\kappa_{n-1}(ev_n))\cdots)$ satisfies h_1.*

If we regard each construction as supplying some code, then composing a chain of constructions combines all of these program fragments into a single program.

This picture can be considerably enhanced to accommodate architectural system design [AG97] using multi-argument constructions to combine smaller components into a larger system — see [SST92, BST02].

A more sophisticated version of Def. 6.1 is needed to deal with the distinction between ordinary and observable behaviours. Since only observable aspects should determine correctness of implementations, in an implementation step $h' \overset{}{\underset{\kappa}{\rightsquigarrow}} h$ it is too restrictive to require κ to deliver behaviours that "strictly" satisfy h'. We weaken this to satisfaction of only the observable part of h'.

Definition 6.3. *Given a Σ-behaviour specification $h\colon T_\Sigma(A) \to A \cup \{\alpha\}$ and a Σ'-behaviour specification $h'\colon T_{\Sigma'}(A') \to A' \cup \{\alpha\}$, we say that h* observationally implements h' via κ, *written $h' \overset{Obs}{\underset{\kappa}{\rightsquigarrow}} h$, if κ maps each Σ-behaviour satisfying h to a Σ'-behaviour satisfying $Obs(h')$, i.e. $\kappa(Mod(h)) \subseteq Mod_{Obs}(Obs(h'))$. Again, when we want to emphasize correctness of κ in relating h and h', we say that κ is* observationally correct w.r.t. h *and* h'.

This definition may be phrased in terms of observational equivalence: $h' \overset{Obs}{\underset{\kappa}{\rightsquigarrow}} h$ if for every Σ-behaviour ev satisfying h, there is a Σ'-behaviour ev' satisfying h' such that $\kappa(ev) \equiv ev'$. By Lemma 4.3, if $h' \overset{}{\underset{\kappa}{\rightsquigarrow}} h$ then $h' \overset{Obs}{\underset{\kappa}{\rightsquigarrow}} h$.

---

[2] Constructions involved in practical examples may turn out to be *partial functions*; this may be dealt with similarly as in [BST02, BST08], so we disregard this issue here for the sake of simplicity.

But now composition of correctness is not so straightforward! Since observationally correct constructions build results that observationally satisfy the result specification given arguments that satisfy the argument specification "strictly", the following additional property is required to ensure that no problems arise when constructions are composed.

Definition 6.4. *A construction* $\kappa\colon \Sigma \Rightarrow \Sigma'$ *is* stable *if it preserves observational equivalence, that is, for all Σ-behaviours ev_1 and ev_2, $ev_1 \equiv ev_2$ implies* $\kappa(ev_1) \equiv \kappa(ev_2)$.

Proposition 6.5. *If* $h_1 \overset{Obs}{\underset{\kappa_1}{\rightsquigarrow}} h_2 \overset{Obs}{\underset{\kappa_2}{\rightsquigarrow}} h_3$ *and κ_1 is stable then* $h_1 \overset{Obs}{\underset{\kappa_2;\kappa_1}{\rightsquigarrow}} h_3$.

This suggests that in order to compose observational implementations, we must check that constructions are stable as well as checking that the implementing specification of one matches the implemented specification of the other. The definition of stability of $\kappa\colon \Sigma \Rightarrow \Sigma'$ involves quantification over all pairs of Σ-behaviours, so that could be difficult. But when constructions are determined by parameterised programs in a programming language, e.g. functors in SML, then it is possible to shift the burden of proof to the programming language designers by requiring that all *expressible* constructions be stable. This is entirely reasonable since it corresponds to requiring that parameterised programs respect abstraction boundaries: κ may freely use the components of its parameter that are listed in Σ, but may not take advantage of their particular internal properties.

6.1   Examples

A possible implementation of symbol tables as specified in Sect. 5, in terms of identifiers, proceeds in three steps. We implement SYMTAB by LBUNCH, then LBUNCH by BUNCH, and finally BUNCH by IDENT. The intermediate specifications BUNCH and LBUNCH are as follows.

> BUNCH = IDENT **then**
>> **sort** *bunch*
>> **opns** *emptybunch* : *bunch*
>>> *defaultbunch* : *bunch*
>>> *addid* : *ident* × *bunch* → *bunch*
>>> *isinbunch* : *ident* × *bunch* → *bool*
>> **with**
>>> *bunch* = *ident list*

and

$$h^{\text{BUNCH}}_{bunch}(emptybunch) = []$$
$$h^{\text{BUNCH}}_{bunch}(defaultbunch) = \alpha$$
$$h^{\text{BUNCH}}_{bunch}(addid(i,b)) = \alpha \ \textbf{if} \ h^{\text{BUNCH}}_{bunch}(b) = \alpha$$
$$\textbf{else} \ i::h^{\text{BUNCH}}_{bunch}(b)$$
$$h^{\text{BUNCH}}_{bool}(isinbunch(i,b)) = \alpha \ \textbf{if} \ h^{\text{BUNCH}}_{bunch}(b) = \alpha$$
$$\textbf{else case} \ h^{\text{BUNCH}}_{bunch}(b) \ \textbf{of} \ [] \quad \Rightarrow \mathit{false}$$
$$j::b' \Rightarrow \mathit{true} \ \textbf{if} \ i = j$$
$$\textbf{else} \ isinbunch(i,b')$$

$$\text{LBunch} = \text{Bunch } \textbf{then}$$

$$\textbf{sort } lb$$

$$\textbf{opns } emptylb : lb$$
$$addbunch : bunch * lb \rightarrow lb$$
$$popbunch : lb \rightarrow lb$$
$$topbunch : lb \rightarrow bunch$$
$$isemptylb : lb \rightarrow bool$$

$$\textbf{with}$$
$$lb = bunch\ list$$

and

$$h_{lb}^{\text{LBunch}}(emptylb) = \texttt{[]}$$
$$h_{lb}^{\text{LBunch}}(addbunch(b,l)) = \alpha \ \textbf{if } h_{lb}^{\text{LBunch}}(l) = \alpha \ \textbf{or } h_{bunch}^{\text{LBunch}}(b) = \alpha$$
$$\textbf{else } b\!::\!h_{lb}^{\text{LBunch}}(l)$$
$$h_{lb}^{\text{LBunch}}(popbunch(l)) = \alpha \ \textbf{if } h_{lb}^{\text{LBunch}}(l) = \alpha$$
$$\textbf{else case } h_{lb}^{\text{LBunch}}(l) \textbf{ of } \texttt{[]} \ \Rightarrow \alpha$$
$$b\!::\!l' \Rightarrow l'$$
$$h_{bunch}^{\text{LBunch}}(topbunch(l)) = \alpha \ \textbf{if } h_{lb}^{\text{LBunch}}(l) = \alpha$$
$$\textbf{else case } h_{lb}^{\text{LBunch}}(l) \textbf{ of } \texttt{[]} \ \Rightarrow \alpha$$
$$b\!::\!l' \Rightarrow b$$
$$h_{bool}^{\text{LBunch}}(isemptylb(l)) = \alpha \ \textbf{if } h_{lb}^{\text{LBunch}}(l) = \alpha$$
$$\textbf{else case } h_{lb}^{\text{LBunch}}(l) \textbf{ of } \texttt{[]} \ \Rightarrow true$$
$$b\!::\!l' \Rightarrow false$$
$$h_{bunch}^{\text{LBunch}}(addid(i,b)) = \alpha \ \textbf{if } h_{bunch}^{\text{LBunch}}(b) = \alpha$$
$$\textbf{else } i\!::\!h_{bunch}^{\text{LBunch}}(b)$$
$$h_{bool}^{\text{LBunch}}(isinbunch(i,b)) = \alpha \ \textbf{if } h_{bunch}^{\text{LBunch}}(b) = \alpha$$
$$\textbf{else case } h_{bunch}^{\text{LBunch}}(b)$$
$$\textbf{of } \texttt{[]} \ \Rightarrow false$$
$$j\!::\!b' \Rightarrow true \ \textbf{if } i = j$$
$$\textbf{else } isinbunch(i,b')$$

Bunch describes an abstract data type for a collection of identifiers with membership. LBunch, short for "layered bunch", comes from the recognition that a stack-like structure is relevant to dealing with entering and leaving blocks.

Note that the last two clauses of LBunch need to be included, although they are essentially repeated from Bunch: here b ranges over a larger set of terms. A full-blown specification language for writing specifications in this style would include notational conventions for circumventing this kind of boring and error-prone repetition, as well as other inconveniences of the notation above.

Now we give the constructions, which are functions from behaviours to behaviours, in a number of variants, as SML functors. We proceed bottom-up, starting with two variants of the implementation of Bunch by Ident.

```
functor FB(structure I : IDENT) : BUNCH =
  struct
    open I
    type bunch = ident list
```

```
          val emptybunch = []
          val defaultbunch = []
          fun addid(i,b) = i::b
          fun isinbunch(i,[]) = false
            | isinbunch(i,j::b) = (i=j) orelse isinbunch(i,b)
       end
```

Clearly, for any behaviour $\textrm{ID}$ that satisfies $\textrm{IDENT}$, $\textrm{FB}(\textrm{ID})$ yields a behaviour that satisfies $\textrm{BUNCH}$, and so $\textrm{FB}$ (the semantic function underlying the functor FB) is correct w.r.t. $\textrm{IDENT}$ and $\textrm{BUNCH}$, i.e. $\textrm{BUNCH} \underset{\textrm{FB}}{\rightsquigarrow} \textrm{IDENT}$.

```
       functor FB'(structure I : IDENT) : BUNCH =
         struct
           open I
           type bunch = ident -> bool
           val emptybunch = fn i => false
           val defaultbunch = fn i => true
           fun addid(i,f) = fn j => (i=j) orelse f(j)
           fun isinbunch(i,f) = f(i)
         end
```

Given any behaviour $\textrm{ID}$ that satisfies $\textrm{IDENT}$, $\textrm{FB}'(\textrm{ID})$ yields a behaviour that does not satisfy $\textrm{BUNCH}$. However, it does observationally satisfy the observable part of $\textrm{BUNCH}$ and so $\textrm{FB}'$ is observationally correct w.r.t. $\textrm{IDENT}$ and $\textrm{BUNCH}$, i.e. $\textrm{BUNCH} \underset{\textrm{FB}'}{\overset{Obs}{\rightsquigarrow}} \textrm{IDENT}$.

Now we consider two ways of implementing $\textrm{LBUNCH}$ by $\textrm{BUNCH}$.

```
       functor FLB(structure B : BUNCH) : LBUNCH =
         struct
           open B
           type lb = bunch list
           val emptylb = []
           fun addbunch(b,l) = b::l
           fun popbunch [] = []
             | popbunch(b::l) = l
           fun topbunch [] = defaultbunch
             | topbunch(b::l) = b
           fun isemptylb [] = true
             | isemptylb(b::l) = false
         end
```

For any behaviour B that satisfies $\textrm{BUNCH}$, $\textrm{FLB}(B)$ yields a behaviour that satisfies $\textrm{LBUNCH}$, and thus $\textrm{LBUNCH} \underset{\textrm{FLB}}{\rightsquigarrow} \textrm{BUNCH}$.

```
       functor FLB'(structure B : BUNCH) : LBUNCH =
         struct
           open B
           type lb = (int -> bunch) * int
           val emptylb = (fn n => defaultbunch , ~1)
           fun addbunch(b,(f,m)) = (fn n => if n>m then b else f(n) , m+1)
```

```
    fun popbunch(f,m) = (f,m-1)
    fun topbunch(f,m) = f(m)
    fun isemptylb(f,m) = m<0
  end
```

For any behaviour B that satisfies BUNCH, FLB$'(B)$ yields a behaviour that
satisfies the observable part of LBUNCH, and so LBUNCH $\underset{\text{FLB}'}{\overset{Obs}{\rightsquigarrow}}$ BUNCH.
  Finally, we give two ways of implementing SYMTAB by LBUNCH.

```
  functor FST(structure LB : LBUNCH) : SYMTAB =
    struct
      eqtype ident = LB.ident
      val "a" = LB."a"   val "b" = LB."b"   ...
      type symtab = LB.lb
      val empty = LB.addbunch(LB.emptybunch,LB.emptylb)
      fun add(i,st) =
          if LB.isemptylb(st) then st
          else LB.addbunch(LB.addid(i,LB.topbunch(st)),LB.popbunch(st))
      fun enter(st) = if LB.isemptylb(st) then st
                      else LB.addbunch(LB.emptybunch,st)
      val leave = LB.popbunch
      fun isin(i,st) =
          if LB.isemptylb(st) then false
          else if LB.isemptylb(LB.popbunch(st))
              then LB.isinbunch(i,LB.topbunch(st))
              else LB.isinbunch(i,LB.topbunch(st))
                  orelse LB.isin(i,LB.popbunch(st))
    end

  functor FST'(structure LB : LBUNCH) : SYMTAB =
    struct
      eqtype ident = LB.ident
      val "a" = LB."a"   val "b" = LB."b"   ...
      type symtab = LB.lb
      val empty = LB.addbunch(LB.emptybunch,LB.emptylb)
      fun add(i,st) =
          if LB.isemptylb(st) then st
          else LB.addbunch(LB.addid(i,LB.topbunch(st)),LB.popbunch(st))
      fun enter(st) =
          if LB.isemptylb(st) then st
          else LB.addbunch(LB.emptybunch,st)
      val leave = LB.popbunch
      fun isin(i,st) =
          LB.isemptylb(st) orelse
          LB.isinbunch(i,LB.topbunch(st)) orelse
          LB.isin(i,LB.popbunch(st))
    end
```

For any behaviour LB that satisfies LBUNCH, each of FST(LB) and FST$'$(LB)
yields a behaviour that satisfies SYMTAB, and thus we have SYMTAB $\underset{\text{FST}}{\rightsquigarrow}$

LBUNCH and SYMTAB $\xrightarrow[\text{FST}']{}$ LBUNCH. And so, for each of these, we also have an observational implementation, SYMTAB $\xrightarrow[\text{FST}]{Obs}$ LBUNCH etc.

All of these constructions are stable; since they are coded as closed SML functors, under the very plausible conjecture that all closed SML-expressible functors are stable. (Actually proving this result would be very tedious, as the proof would need to consider the entire definition of SML. But it would only need to be done once, thereafter freeing implementors from the obligation to check stability case by case.) Consequently, the obvious compositions of these constructions yield behaviours that observationally satisfy SYMTAB. In particular, for any behaviour ID that satisfies IDENT, $\text{FST}(\text{FLB}(\text{FB}(\text{ID})))$ corresponds to LST, $\text{FST}'(\text{FLB}(\text{FB}(\text{ID})))$ corresponds to LST$'$ and $\text{FST}(\text{FLB}'(\text{FB}'(\text{ID})))$ corresponds to SST. Other combinations, like $\text{FST}(\text{FLB}'(\text{FB}(\text{ID})))$ and $\text{FST}'(\text{FLB}(\text{FB}'(\text{ID})))$, yield still different structures.

Now consider the following modification to `FST`:

```
functor FSTBAD(structure LB : LBUNCH) : SYMTAB =
  struct
    eqtype ident = LB.ident
    ...
    fun isin(i,st) =
         if (st = LB.emptylb) then false
         else if LB.popbunch(st) = LB.emptylb
             then LB.isinbunch(i,LB.topbunch(st))
             else LB.isinbunch(i,LB.topbunch(st))
                   orelse LB.isin(i,LB.popbunch(st))
  end
```

(This is not valid in SML: the type `LB.lb` is not required to admit equality and so `st = LB.emptylb` does not typecheck.)

FSTBAD is still observationally correct: for every behaviour that satisfies LBUNCH, FSTBAD builds a behaviour that observationally satisfies SYMTAB (since for each LB satisfying LBUNCH, for all well-balanced terms t of type $symtab$, the values of $isin(i, t)$ in FSTBAD(LB) and in FST(LB) coincide). But FSTBAD is not stable. So, if we take $\text{FLB}'(\text{FB}'(\text{ID}))$, which does not satisfy LBUNCH (even though it satisfies the observable part of it), and then apply FSTBAD to build BAD $= \text{FSTBAD}(\text{FLB}'(\text{FB}'(\text{ID})))$, BAD need not satisfy SYMTAB. Indeed, it does not: evaluating the term $isin(j, leave(add(i, enter(empty))))$ in BAD gives $true$, and this is incorrect according to SYMTAB.

7   Local Constructions in Global Contexts

Very informally, constructions as discussed in Sect. 6 were considered at the "global" level of the entire system under development: they build a behaviour that implements the overall requirements specification given any argument behaviour satisfying a specification of the part of the system yet to be implemented. But in each development step the construction typically uses only a relatively small part of its argument to add a new part of the result, and just passes most of the argument over to the result without using or touching it in any way.

We will now have a closer look at one aspect of modular development, namely at the use of *local* constructions, which take as argument only as much of the behaviour as is necessary to build some new part of the result. Such local constructions give rise to constructions at the "global" level, whereby the required argument is cut out of the global context and the result combined with the same global context, thus extending it by new parts and contributing to the overall system implementation. We capture this idea using the pushout technique and amalgamation, as is standard in algebraic specifications [EM85].

A technical tool we need is an extension of Lemma 2.3 to behaviours and correspondences.

Lemma 7.1. *Consider a pushout in the category of signatures.*

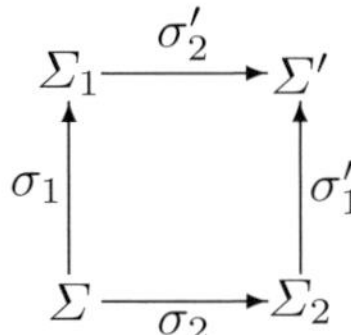

Then for any Σ_1-behaviour ev_1 and Σ_2-behaviour ev_2 with common Σ-reduct $ev_1|_{\sigma_1} = ev_2|_{\sigma_2}$, there exists a unique Σ'-behaviour ev' such that $ev'|_{\sigma_2'} = ev_1$ and $ev'|_{\sigma_1'} = ev_2$. Similarly for correspondences: for any Σ_1-correspondence $\rho_1\colon ev_{1,1} \bowtie ev_{1,2}$ and Σ_2-correspondence $\rho_2\colon ev_{2,1} \bowtie ev_{2,2}$ with common Σ-reducts $ev_{1,1}|_{\sigma_1} = ev_{2,1}|_{\sigma_2}$, $ev_{1,2}|_{\sigma_1} = ev_{2,2}|_{\sigma_2}$, and $\rho_1|_{\sigma_1} = \rho_2|_{\sigma_2}$, there exists a unique Σ'-correspondence $\rho'\colon ev_1' \bowtie ev_2'$ such that $\rho'|_{\sigma_2'} = \rho_1$ and $\rho'|_{\sigma_1'} = \rho_2$, where $ev_1'|_{\sigma_2'} = ev_{1,1}$, $ev_1'|_{\sigma_1'} = ev_{2,1}$, $ev_2'|_{\sigma_2'} = ev_{1,2}$, $ev_2'|_{\sigma_1'} = ev_{2,2}$.

We are ready now to state the main definitions for this section:

Definition 7.2. *Given a signature morphism $\iota\colon \Sigma \to \Sigma'$, a (local) construction along ι is a function κ that maps any Σ-behaviour ev to a Σ'-behaviour $\kappa(ev)$ such that $\kappa(ev)|_\iota = ev$.*

Then, given a ("global context") signature Σ_G and a ("fitting") morphism $\gamma\colon \Sigma \to \Sigma_G$, the construction κ^γ along $\iota'\colon \Sigma_G \to \Sigma_G'$ induced by κ via γ (or the γ-lifting κ^γ of κ) is defined for any Σ_G-behaviour ev_G so that $\kappa^\gamma(ev_G)$ is the Σ_G'-behaviour such that $\kappa^\gamma(ev_G)|_{\iota'} = ev_G$ and $\kappa^\gamma(ev_G)|_{\gamma'} = \kappa(ev_G|_\gamma)$, where 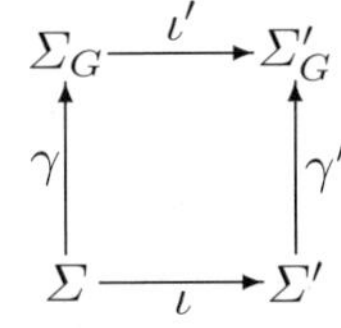 *Σ_G' and morphisms $\iota'\colon \Sigma_G \to \Sigma_G'$, $\gamma'\colon \Sigma' \to \Sigma_G'$ are given by a pushout of $\iota\colon \Sigma \to \Sigma'$ and $\gamma\colon \Sigma \to \Sigma_G$.*

As argued in Sect. 6, the key property of (global) constructions is stability. Unfortunately, since enlarging the context of use typically expands observable behaviour, stability of a local construction will not ensure that its lifting along a fitting morphism is stable as well. Following [Sch87] we introduce a stronger property, which is preserved by lifting along any morphism.

Definition 7.3. *A construction κ along a signature morphism $\iota\colon \Sigma \to \Sigma'$ is locally stable if it extends correspondences, that is, for any Σ-behaviours ev_1 and ev_2, any Σ-correspondence $\rho\colon ev_1 \bowtie ev_2$ extends to a Σ'-correspondence $\rho'\colon \kappa(ev_1) \bowtie \kappa(ev_2)$ such that $\rho'|_\iota = \rho$.*

Theorem 7.4. *A construction κ along a signature morphism $\iota\colon \Sigma \to \Sigma'$ is locally stable if and only if for all signatures Σ_G and fitting morphisms $\gamma\colon \Sigma \to \Sigma_G$, the γ-lifting κ^γ of κ is stable.*

The "only if" part of this theorem is what we need: local stability of a local construction ensures its stability in any context of use; the "if" part shows that no weaker condition can be given.

We now turn to the issue of specifying local constructions and using their specifications to justify correctness of global implementation steps. The standard approach would be that to specify local constructions along $\iota'\colon \Sigma \to \Sigma'$, one gives a Σ-behaviour specification that determines the arguments intended for the construction, and a Σ'-behaviour specification that describes the results built. Unfortunately, this doesn't work in the framework discussed in this paper. Any behaviour specification constrains the behaviour only on elements that are in the specification's carrier, since behaviours which satisfy that specification are required to conform on such elements only, not on additional elements that may also be present in their carriers. When a local construction is used in a global context, the set of data for which the behaviour is of importance may grow beyond what we can take explicit account of when writing a single specification. We accommodate this by making the result specification depend on the carriers of the argument supplied.[3]

Definition 7.5. *A specification of local constructions along $\iota\colon \Sigma \to \Sigma'$ consists of a Σ-behaviour specification $h\colon T_\Sigma(A) \to A \cup \{\alpha\}$ and a function that maps any set $X \supseteq A$ to a Σ'-behaviour specification $h'_X\colon T_\Sigma(X') \to X' \cup \{\alpha\}$. We write such a specification as $\Pi X\colon h{\to}h'_X$.*

A local construction κ along $\iota\colon \Sigma \to \Sigma'$ strictly satisfies (or, is strictly correct w.r.t.) $\Pi X\colon h{\to}h'_X$ if for any Σ-behaviour $ev\colon T_\Sigma(X) \to X$ that satisfies h, $\kappa(ev)$ is a Σ'-behaviour that satisfies h'_X.

The conditions to ensure that a strictly correct local construction lifts to a strictly correct construction in a global context are now rather natural:

Theorem 7.6. *Let κ be a local construction along $\iota\colon \Sigma \to \Sigma'$ that strictly satisfies $\Pi X\colon h{\to}h'_X$.*

Consider a signature Σ_G with a fitting morphism $\gamma\colon \Sigma \to \Sigma_G$ and the usual pushout. Let $h_G\colon T_{\Sigma_G}(A_G) \to A_G \cup \{\alpha\}$ and $h'_G\colon T_{\Sigma'_G}(A'_G) \to A'_G \cup \{\alpha\}$ be behaviour specifications. If $\gamma(h)$ refines h_G, and for each $X \supseteq A_G|_\gamma$, h'_G refines $\iota'(h) + \gamma'(h'_X)$ then the γ-lifting κ^γ of κ is strictly correct w.r.t. h_G and h'_G.

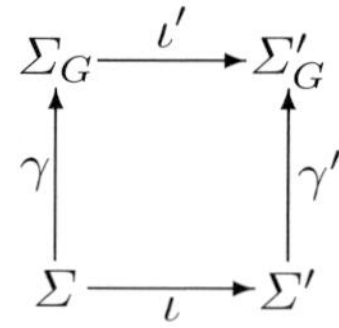

Let us now turn to observational correctness. As with stability, observational correctness for (global) constructions would be too weak for local constructions:

---

[3] Allowing, more generally, the result specification to depend on the entire argument behaviour, not just its carriers, should not cause extra technical difficulties.

when such a local construction is used in a global context, where more observations are available, correctness would be lost in general. The following definition strengthens the requirements appropriately:

Definition 7.7. *A local construction κ along $\iota\colon \Sigma \to \Sigma'$ locally satisfies (or, is locally correct w.r.t.) $\Pi X{:}h{\to}h'_X$ if for any Σ-behaviour $ev\colon T_\Sigma(X) \to X$ that satisfies h, there is a Σ'-behaviour ev' that satisfies h'_X and a Σ'-correspondence $\rho'\colon ev' \bowtie \kappa(ev)$ such that $\rho'|_\iota$ is the identity (which implies $ev'|_\iota = ev$).*

Local correctness is stronger than observational correctness of κ, which would just state that $\kappa(ev)$ observationally satisfies h'_X. By requiring that this is "witnessed" by a correspondence that is identity on the argument sorts, we "locally" fix the argument sorts as observable, thus allowing arbitrary observations for them to be added in the context of use, as the following key theorem shows.

Theorem 7.8. *Let κ be a local construction along $\iota\colon \Sigma \to \Sigma'$ that is locally stable and locally satisfies $\Pi X{:}h{\to}h'_X$.*

Consider a signature Σ_G with a fitting morphism $\gamma\colon \Sigma \to \Sigma_G$ and the usual pushout. Let $h_G\colon T_{\Sigma_G}(A_G) \to A_G \cup \{\alpha\}$ and $h'_G\colon T_{\Sigma'_G}(A'_G) \to A'_G \cup \{\alpha\}$ be behaviour specifications. If $Mod(h_G) \subseteq Mod_{Obs}(Obs(h_G + \gamma(h)))$, and for each $X \supseteq A_G|_\gamma$, $Mod(\iota'(h) + \gamma'(h'_X)) \subseteq Mod_{Obs}(Obs(h'_G))$ then the γ-lifting κ^γ of κ is observably correct w.r.t. h_G and h'_G.

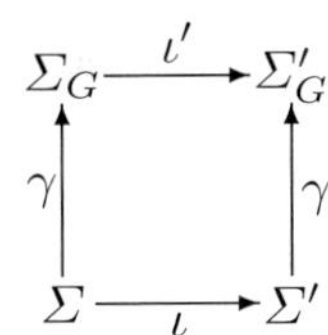

7.1   Examples

Recall constructions $\textsc{Flb}$ and $\textsc{Flb}'$ as defined in Sect. 6. Any argument behaviour B (which has to satisfy $\textsc{Bunch}$) is inherited by the result behaviours $\textsc{Flb}(B)$ and $\textsc{Flb}(B)$, but it is passed there untouched and otherwise (apart from the type *bunch* and the constant *defaultbunch*) is not used by $\textsc{Flb}$ and $\textsc{Flb}'$. Therefore the essence of either of the two constructions can be captured over a smaller signature as a local construction as follows.

We start with a simple argument specification:

$\textsc{Elem} =$
>   **sort** *elem*
>   **opns** *default*: *elem*
>   **with**
>   *elem* = *unit*
>   **and**
> $h_{elem}^{\textsc{Elem}}(default) = \alpha$

The following function yields specifications for the result given a carrier for sort *elem*; note that all these result specifications have a common signature.

$$\text{STACK}(\textbf{sort}\ elem) =$$
$$\textbf{sort}\ \ stack$$
$$\textbf{opns}\ emptystack : stack$$
$$push : elem \times stack \rightarrow stack$$
$$pop : stack \rightarrow stack$$
$$top : stack \rightarrow elem$$
$$isempty : stack \rightarrow bool$$
$$\textbf{with}$$
$$stack = elem\ \texttt{list}$$

and

$$h_{stack}^{\text{STACK}}(emptystack) = \texttt{[]}$$
$$h_{stack}^{\text{STACK}}(push(b,l)) = \alpha\ \textbf{if}\ h_{stack}^{\text{STACK}}(l) = \alpha\ \textbf{or}\ h_{elem}^{\text{STACK}}(b) = \alpha$$
$$\textbf{else}\ b\!::\!h_{stack}^{\text{STACK}}(l)$$
$$h_{stack}^{\text{STACK}}(pop(l)) = \alpha\ \textbf{if}\ h_{stack}^{\text{STACK}}(l) = \alpha$$
$$\textbf{else case}\ h_{stack}^{\text{STACK}}(l)\ \textbf{of}\ \texttt{[]}\quad \Rightarrow \alpha$$
$$b\!::\!l' \Rightarrow l'$$
$$h_{elem}^{\text{STACK}}(top(l)) = \alpha\ \textbf{if}\ h_{stack}^{\text{STACK}}(l) = \alpha$$
$$\textbf{else case}\ h_{stack}^{\text{STACK}}(l)\ \textbf{of}\ \texttt{[]}\quad \Rightarrow \alpha$$
$$b\!::\!l' \Rightarrow b$$
$$h_{bool}^{\text{STACK}}(isempty(l)) = \alpha\ \textbf{if}\ h_{stack}^{\text{STACK}}(l) = \alpha$$
$$\textbf{else case}\ h_{stack}^{\text{STACK}}(l)\ \textbf{of}\ \texttt{[]}\quad \Rightarrow true$$
$$b\!::\!l' \Rightarrow false$$

Here are two local constructions along the obvious inclusion of the signature of ELEM into the signature of STACK($elem$) (for any carrier for $elem$).

```
functor FSTACK(structure E : ELEM) : STACK(E.elem) =
  struct
    open E
    type stack = elem list
    val emptystack = []
    fun push(b,l) = b::l
    fun pop [] = []
      | pop(b::l) = l
    fun top [] = default
      | top(b::l) = b
    fun isempty [] = true
      | isempty(b::l) = false
  end

functor FSTACK'(structure E : ELEM) : STACK(E.elem) =
  struct
    open E
    type stack = (int -> elem) * int
    val emptystack = (fn n => default , -1)
    fun push(b,(f,m)) = (fn n => if n>m then b else f(n), m+1)
    fun pop(f,m) = (f,m-1)
```

```
    fun top(f,m) = f(m)
    fun isempty(f,m) = m<0
  end
```

FSTACK and FSTACK$'$ are locally stable. As in Sect. 6.1, this should follow from the fact that they are coded as SML functors. FSTACK and FSTACK$'$ are also locally correct.

FLB and FLB$'$ arise as global constructions induced by FSTACK and FSTACK$'$, respectively, via the fitting morphism from the signature of ELEM to the signature of BUNCH which maps *elem* to *bunch* and *default* to *defaultbunch*. This is, of course, assuming the appropriate choice of names in the pushout signature, where we need to rename *stack* to *lb*, *emptystack* to *emptylb*, etc. Stability and observational correctness of FLB and FLB$'$ follow from local stability and local correctness of FSTACK and FSTACK$'$, respectively, using Thms. 7.4 and 7.8.

8   Conclusion

This is a new look at some of our previous work on observational interpretation of specifications and its role in software specification and development [ST88b, BST02, BST08], presented in a new framework inspired by the ideas in [GGM76]. We have focused on a key idea in [GGM76], that of viewing a system via its behaviour, given by the evaluation function for terms of sort *bool*, and consequently, that of specifying behaviour by indicating the results of evaluating some terms and leaving others as "don't care" cases. The resulting framework and its observational aspects are sketched in Sects. 3 and 4; our work on systematic software development is then adapted to this framework, concentrating on the use of local constructions in a global context, in Sects. 6 and 7.

The specifications considered in this new framework are considerably more restrictive than axiomatic specifications. Even the simple constraint that the values of two terms coincide cannot be captured without giving their common value explicitly. On the other hand, such specifications offer a visible link to so-called "abstract model specifications" [Jon80] with mechanisms for making looseness in specifications explicit, via the designation of "don't care" cases, rather than implicit, by simply omitting axioms from specifications that constrain required behaviour. The ramifications of this link remains to be investigated. It is possible to add axiomatic specifications to this framework, for which there are two approaches: one uses an observational interpretation of the axioms, where equality refers to indistinguishability via the available operations; the other uses the standard interpretation, but closes model classes of specifications under behavioural equivalence.

Other important parts of the story are not covered here for lack of space. One major issue concerns proof: to show that a behaviour satisfies a specification, or that one specification implements another, requires formal proofs about behaviours and specifications. An observational view of specifications complicates this task, although again requiring stability helps considerably. But this is still

has started, the values sent and received will be of the appropriate type, thus avoiding runtime errors, and no process can get stuck forever.

The contribution "The pairing of contracts and session types" by Cosimo Laneve and Luca Padovani presents a thorough comparison of session types and contracts, two formalisms aiming at specifying high-level structured patterns of communication between distributed services. The paper provides a nice synthesis of previous works on session types and contracts and sheds light on the relationship between them in the context of service-oriented computing.

The contribution "Specifying and Analysing SOC Applications with COWS' by Alessandro Lapadula, Rosario Pugliese and Francesco Tiezzi summarizes some results related to the COWS process calculus and its usage as a model in service oriented computing. The paper illustrates how to use COWS for service orchestration and discovery, and then focuses on two formal methods for ensuring correctness of certain aspects of the computation: a static type system that guarantees secrecy of data, and a verification framework based on modal logics to verify functional properties of systems specified using COWS.

The last contribution "Approximating Behaviors in Embedded System Design" by Roberto Passerone and Alberto Sangiovanni Vincentelli deals with verification of embedded systems. Embedded systems are electronic devices that operate in the context of a physical environment, by sensing and reacting to a set of stimuli. The paper reviews a formal methodology (recently proposed by the authors) for approximating behaviours in embedded system design and verification. The scenario is where one has an abstract description of the system (the specification), a more concrete one (the refinement, implementation) and abstract versions of both (for verification purposes). The approach is not limited to such a "square": in general one can have several refinement levels. The paper focuses on three behavioral models based on trace algebras (each tailored to a different level of abstraction). It is shown how to map concrete traces into more abstract ones via trace algebra homomorphisms that induce conservative approximations.

References

1. Abadi, M., Gordon, A.D.: A calculus for cryptographic protocols: The spi calculus. Information and Computation 148(1), 1–70 (1999)
2. Boreale, M., Bruni, R., Caires, L., Nicola, R.D., Lanese, I., Loreti, M., Martins, F., Montanari, U., Ravara, A., Sangiorgi, D., Vasconcelos, V.T., Zavattaro, G.: SCC: A service centered calculus. In: Bravetti, M., Núñez, M., Zavattaro, G. (eds.) WS-FM 2006. LNCS, vol. 4184, Springer, Heidelberg (2006)
3. Bouali, A., Gnesi, S., Larosa, S.: The integration project for the jack environment. EATCS Bulletin, Centrum voor Wiskunde en Informatica (CWI) 54, 207–223 (1994)
4. Bouali, A., Ressouche, A., Roy, V., de Simone, R.: The FC2TOOLS set. In: Alur, R., Henzinger, T.A. (eds.) CAV 1996. LNCS, vol. 1102, pp. 441–445. Springer, Heidelberg (1996)

5. Ciancia, V., Montanari, U.: A name abstraction functor for named sets. In: Coalgebraic Methods in Computer Science 2008 (to appear, 2008)
6. Clarke, E.M., Wing, J.M.: Formal methods: State of the art and future directions. ACM Computing Surveys 28(4), 626–643 (1996)
7. Cleaveland, R., Parrow, J., Steffen, B.: The Concurrency Workbench: A Semantics-Based Tool for the Verification of Concurrent Systems. ACM Transactions on Programming Languages and Systems 15(1), 36–72 (1993)
8. Ferrari, G.L., Ferro, G., Gnesi, S., Montanari, U., Pistore, M., Ristori, G.: An Automata Based Verification Environment for Mobile Processes. In: Brinksma, E. (ed.) TACAS 1997. LNCS, vol. 1217, pp. 275–289. Springer, Heidelberg (1997)
9. Ferrari, G.L., Gnesi, S., Montanari, U., Pistore, M.: A model-checking verification environment for mobile processes. ACM Trans. Softw. Eng. Methodol. 12(4), 440–473 (2003)
10. Ferrari, G.L., Gnesi, S., Montanari, U., Pistore, M., Ristori, G.: Verifying Mobile Processes in the HAL Environment. In: Y. Vardi, M. (ed.) CAV 1998. LNCS, vol. 1427, pp. 511–515. Springer, Heidelberg (1998)
11. Ferrari, G.L., Montanari, U., Pistore, M.: Minimizing Transition Systems for Name Passing Calculi: A Co-algebraic Formulation. In: Nielsen, M., Engberg, U. (eds.) FOSSACS 2002. LNCS, vol. 2303, pp. 129–143. Springer, Heidelberg (2002)
12. Ferrari, G.L., Montanari, U., Tuosto, E.: Coalgebraic minimization of HD-automata for the pi-calculus using polymorphic types. Theor. Comput. Sci. 331(2-3), 325–365 (2005)
13. Ferrari, G.L., Montanari, U., Tuosto, E., Victor, B., Yemane, K.: Modelling fusion calculus using HD-automata. In: Fiadeiro, J.L., Harman, N.A., Roggenbach, M., Rutten, J. (eds.) CALCO 2005. LNCS, vol. 3629, pp. 142–156. Springer, Heidelberg (2005)
14. Focardi, R., Gorrieri, R.: A classification of security properties. Journal of Computer Security 3(1) (1995)
15. Gabbay, M., Pitts, A.M.: A new approach to abstract syntax with variable binding. Formal Asp. Comput. 13(3-5), 341–363 (2002)
16. Gadducci, F., Miculan, M., Montanari, U.: About permutation algebras (pre)sheaves and named sets. Higher-Order and Symbolic Computation 19(2-3), 283–304 (2006)
17. Milner, R., Parrow, J., Walker, D.: A Calculus of Mobile Processes, I and II. Information and Computation 100(1), 1–40,41–77 (1992)
18. Montanari, U., Buscemi, M.: A First Order Coalgebraic Model of π-Calculus Early Observational Equivalence. In: Brim, L., Jančar, P., Křetínský, M., Kucera, A. (eds.) CONCUR 2002. LNCS, vol. 2421, pp. 449–465. Springer, Heidelberg (2002)
19. Montanari, U., Pistore, M.: Checking Bisimilarity for Finitary π-Calculus. In: Lee, I., Smolka, S.A. (eds.) CONCUR 1995. LNCS, vol. 962, pp. 42–56. Springer, Heidelberg (1995)
20. Montanari, U., Pistore, M.: Finite state verification for the asynchronous pi-calculus. In: Cleaveland, W.R. (ed.) TACAS 1999. LNCS, vol. 1579, pp. 255–269. Springer, Heidelberg (1999)
21. Montanari, U., Pistore, M.: Structured coalgebras and minimal HD-automata for the pi-calculus. Theor. Comput. Sci. 340(3), 539–576 (2005)
22. Montanari, U., Pistore, M., Yankelevich, D.: Efficient minimization up to location equivalence. In: Riis Nielson, H. (ed.) ESOP 1996. LNCS, vol. 1058, pp. 265–279. Springer, Heidelberg (1996)

23. Park, D.: Concurrency and Automata on Infinite Sequences. In: Deussen, P. (ed.) GI-TCS 1981. LNCS, vol. 104, pp. 167–183. Springer, Heidelberg (1981)
24. Pistore, M. History Dependent Automata. PhD Thesis, Dipartimento di Informatica, Università di Pisa (1999)
25. Pitts, A.M.: Nominal logic, a first order theory of names and binding. Inf. Comput. 186(2), 165–193 (2003)
26. Sangiorgi, D., Walker, D.: The π-Calculus: a Theory of Mobile Processes. Cambridge University Press, Cambridge (2002)

History Dependent Automata for Service Compatibility[*]

Vincenzo Ciancia[1], Gian-Luigi Ferrari[1], Marco Pistore[2], and Emilio Tuosto[3]

[1] Department of Computer Science, University of Pisa
{ciancia,giangi}@di.unipi.it
[2] Center for Information Technology - IRST, Fondazione Bruno Kessler
pistore@fbk.eu
[3] Department of Computer Science, University of Leicester
et52@mcs.le.ac.uk

Abstract. We use *History Dependent Automata* (HD-automata) as a syntax-indepentend formalism to check compatibility of services at binding time in *Service-Oriented Computing*.

Informally speaking, service requests are modelled as pairs of HD-automata $\langle C_o, C_r \rangle$; C_r describes the (abstract) behaviour of the searched service and C_o the (abstract) behaviour guaranteed by the invoker. Symmetrically, service publication consists of a pair of HD-automata $\langle S_o, S_r \rangle$ such that S_o provides an (abstraction of) of the behaviour guaranteed by the service and S_r yields the requirement imposed to invokers. An invocation $\langle C_o, C_r \rangle$ matches a published interface $\langle S_o, S_r \rangle$ when C_o simulates S_r and S_o simulates C_r.

1   Introduction

Over the last few years nominal calculi have been envisaged as a suitable model for *Service-Oriented Computing* (SOC). As a matter of fact, *names* provide a uniform mechanism for abstracting a variety of different concepts like addresses, links, continuations, distributed objects, localities, causal dependencies, cryptographic keys and session identifiers. Also, the dynamicity issues usually arising in distributed computing (e.g., network reconfiguration, link mobility) can benefit from the sophisticated linguistic mechanisms of nominal calculi such as binding and scope extrusion. The π-calculus [12, 22] is a small but illustrative example of nominal calculus. Many of the concepts outlined above can be formally described and explained in terms of the π-calculus.

In the nineties, Montanari and Pistore [14, 15, 20] introduced *History Dependent automata* (HD-automata) as a "syntax-independent" automata based model amenable to represent the behaviour of a whole spectrum of formalisms that stress the role of names to refer to suitable semantics concepts. For instance, CCS with causality and localities and some dialects of the π-calculus have been

---

[*] Research partially supported by the EU FP6-IST IP 16004 SENSORIA and by the UK project HIDEA4SOC.

P. Degano et al. (Eds.): Montanari Festschrift, LNCS 5065, pp. 625–641, 2008.

semantically described by HD-automata. Indeed, different versions of HD-automata have been defined. The simplest version can easily be translated to finite-state automata, but possibly with a larger number of states.

A more sophisticated variant consists of HD-automata *with symmetries* where states are equipped with name symmetries (i.e., groups of name permutations). HD-automata with symmetries yield two main benefits: a faithful representation of linguistic mechanisms like scope extrusion and a re-adaptation of the partition refinement algorithm [18] for semantic minimisation of HD-automata. Basically, semantic minimisation provides a "garbage collection" mechanism of names so that states can be eliminated if their behaviour is mimicked by other states, possibly "re-using" their names with different meanings. Noteworthy, a theory based on coalgebras in a category of "named sets" can be developed for this kind of HD-automata, which extends the applicability of the approach to other nominal calculi and guarantees the existence of the minimal automaton within the same bisimilarity class [16, 6].

Also, the coalgebraic theory constitutes the formal basis upon which several verification toolkits have been defined and implemented. In fact, on the one hand, the front-end towards the π-calculus and the translation algorithm for the simplest version of HD-automata have been implemented in the HAL tool [4, 5], which relies on the JACK verification environment [1] for handling semantic verification via standard finite-state automata (e.g., model checking). And, on the other hand, the minimisation algorithm, naturally suggested by the coalgebraic framework, has been implemented in the Mihda toolkit [7, 8].

Finally, other variations of HD-automata have been defined by introducing different algebraic operations [9], and are based on a algebraic-coalgebraic theory [13]. Moreover, HD-automata have been considered with respect to bisimulation-like behavioural equivalences (e.g., π-calculus bisimulation [12] or fusion calculus hyperbisimulation [19]).

In this paper, we introduce a semantic framework for HD-automata based on *simulation* and propose to use it as a mechanism for dealing with semantic-based discovery of services within the service-oriented context. Our main goal is to define a foundational framework to express *behaviour*-based service discovery. Current standards for service discovery (i.e., UDDI and WSDL) provide purely syntactic techniques. As a consequence, composing services only on the basis of syntactic WSDL interfaces may lead to composite services that fall short in meeting their requirements.

In our approach, service descriptions are annotated with HD-automata abstracting the behaviour of the service. Informally speaking, service requests are modelled as pairs of HD-automata $\langle C_o, C_r \rangle$; C_r describes the (abstract) behaviour of the searched service and C_o the (abstract) behaviour guaranteed by the invoker. Symmetrically, service publication consists of a pair of HD-automata $\langle S_o, S_r \rangle$ where S_o provides an (abstraction of) of the behaviour guaranteed by the service and S_r yields the requirement imposed to invokers. Operationally, a service invocation $\langle C_o, C_r \rangle$ *matches* a published interface $\langle S_o, S_r \rangle$ when C_o simulates S_r and S_o simulates C_r. Hence, in our approach the operation of service discovery becomes a semantic-based operation: the service registry is searched for a service matching the semantic abstractions.

In this paper we report our preliminary results in the exploitation of HD-automata as intermediate language to represent semantic-based discovery of services.

Structure of the Paper. § 2 fixes the notations, revisits the main notions underlying HD-automata, and gives the definition of HD-automata. § 3 recasts the definition of *simulation* relation given in [20] in our context. § 4 gives an example applying the framework to the problem of semantical versioning of protocols. § 5 summarises our work, draws conclusions and sketches some research directions.

2   Background

This section collects the main notations and some basic concepts used throughout the paper.

Let X, Y, Z, ... be sets. Then:

- $\mathcal{P}(X)$ (resp. $\mathcal{P}_{fin}(X)$) is the set of subsets (resp. finite subsets) of X;
- $[X \to Y]$ is the set of maps f with domain $\mathrm{dom}\, f \stackrel{\mathrm{def}}{=} X$ and codomain $\mathrm{cod}\, f \stackrel{\mathrm{def}}{=} Y$; $\mathrm{Im}\, f \stackrel{\mathrm{def}}{=} \{f(x) \in Y \mid x \in X\}$ is the image of f;
- $[X \stackrel{inj}{\to} Y]$ is the set of injective maps from X to Y;
- $\iota : X \hookrightarrow Y$ is the inclusion map from X to Y (implicitely assuming that $X \in \mathcal{P}(Y)$) and, if $f \in [Y \to Z]$ then $f|_X = f \circ \iota$ is the restriction of f to X;
- $Aut(X) = \{f \in [X \to X] \mid f \text{ bijective}\}$ is the set of *automorphisms* (or *permutations*) of X.

To avoid cumbersome parenthesis, $\stackrel{inj}{\to}$ has precedence over $\in$ and sometimes square brackets will be omitted from the denotation of functional domains.

We use P and R to range over (π-calculus) processes built on a countable set of *names* ω. Elements of $[\omega \to \omega]$ are name substitutions and $P\sigma$ denotes the agent obtained by applying the substitution $\sigma \in [\omega \to \omega]$ to P.

2.1   Automata as Coalgebras

We will define HD-automata as *coalgebras* for a functor on the category of **NSet** (§ 2.2). To make the presentation more clear, we first summarise how classical automata can be specified as coalgebras for which very basic notions from category theory have to be introduced. (The interested reader is referred to, e.g., [21].)

Recall that a *category* is a collection of *objects* a, b, ... and *morphisms* $f : a \to b$ from a to b ($\mathrm{dom}\, f \stackrel{\mathrm{def}}{=} a$ and $\mathrm{cod}\, f \stackrel{\mathrm{def}}{=} b$ resp. are the *domain* and *codomain* of f). A category is subject to the following axioms:

- if f and g are morphisms for which $\mathrm{cod}\, f = \mathrm{dom}\, g$, the *composition* of f and g, written $g \circ f$, is a morphism of the category and $\mathrm{dom}\, g \circ f = \mathrm{dom}\, f$ and $\mathrm{cod}\, g \circ f = \mathrm{cod}\, g$;

- morphism composition is associative: if f, g and h can be composed, $(h \circ g) \circ f = h \circ (g \circ f)$;
- for each object a there is the *identity morphism* $id_a : a \to a$; identities are such that $f \circ id_{\mathrm{dom}\, f} = f = id_{\mathrm{cod}\, f} \circ f$, for any morphism f.

A *functor* $\mathcal{F}$ *from* $\mathcal{A}$ *to* $\mathcal{B}$ (written $\mathcal{F} : \mathcal{A} \to \mathcal{B}$) trasforms objects and morphisms of the category $\mathcal{A}$ resp. into objects and morphisms of the category $\mathcal{B}$, preserving identities and composition. Formally, any object a in $\mathcal{A}$ is mapped to an object $\mathcal{F}a$ in $\mathcal{B}$ and any morphism $f : a \to b$ in $\mathcal{A}$ is sent to a morphism $\mathcal{F}f : \mathcal{F}a \to \mathcal{F}b$ in $\mathcal{B}$ such that:

$$\mathcal{F}id_a = id_{\mathcal{F}a}, \qquad \mathcal{F}(g \circ f) = (\mathcal{F}g) \circ (\mathcal{F}f).$$

For simplicity we limit ourselves to $\mathcal{S}et$, the category of sets and total functions with the usual function composition. The singleton set $\{\star\}$ is indicated as $\mathbf{1}$ and the disjoint union of $\mathbf{1}$ with a set X is denoted by $X + \mathbf{1}$.

Definition 1 ($\mathcal{F}$-Coalgebra). *Let $\mathcal{F} : \mathcal{S}et \to \mathcal{S}et$ be a fixed endofunctor on $\mathcal{S}et$ (i.e., a functor from $\mathcal{S}et$ to $\mathcal{S}et$). A pair (X, α) is a coalgebra for $\mathcal{F}$ (or $\mathcal{F}$-coalgebra) iff X is an object of $\mathcal{S}et$ (i.e., a set) and $\alpha : X \to \mathcal{F}X$ is an arrow of $\mathcal{S}et$ (i.e., a total function from X to $\mathcal{X}$).*

Given two $\mathcal{F}$-coalgebras (X, α) and (Y, β), a function $f : X \to Y$ in $\mathcal{S}et$ is an $\mathcal{F}$-coalgebra *(homo)morphism* iff

$$
\begin{array}{ccc}
X & \xrightarrow{\ f\ } & Y \\
\alpha \downarrow & & \downarrow \beta \\
\mathcal{F}X & \xrightarrow[\mathcal{F}f]{} & \mathcal{F}Y
\end{array}
\qquad \text{commutes, namely } \beta \circ f = \mathcal{F}f \circ \alpha.
$$

It is immediate to see that coalgebras for $\mathcal{F} = \mathcal{P}(L \times \_)$ coincide with transition systems labelled by L; in fact, if (X, α) is a $\mathcal{P}(L \times \_)$-coalgebra then X is the set of states and, for $x \in X$, $\alpha(x)$ are the outgoing transitions of x. *Vice versa*, given a L-labelled transition system T whose set of states is X one can define the coalgebra (X, α) by letting $\alpha : x \mapsto \{\langle l, y \rangle \mid \langle x, l, y \rangle$ is a transition in $T\}$ for each $x \in X$. Similarly, finitely branching L-labelled transition systems correspond to $\mathcal{P}_{fin}(L \times \_)$-coalgebras. Also, and more importantly, coalgebra morphisms enable the definition of *coalgebraic bisimulation* that nicely corresponds to the familiar notion of bisimulation in labelled transitions systems. As usual, we let $\pi_1 : X \times Y \to X$ and $\pi_2 : X \times Y \to Y$ be the projections on the cartesian product of X and Y.

Definition 2 (Coalgebraic bisimulation). *A bisimulation between two $\mathcal{F}$-coalgebras (X, α) and (Y, β) is a set $B \subseteq X \times Y$ such that there is $\theta : B \to \mathcal{F}B$ making $\pi_1' = \pi_1 \circ \iota$ and $\pi_2' = \pi_2 \circ \iota$ two homomorphism, where $\iota : B \hookrightarrow X \times Y$.*

In other words,

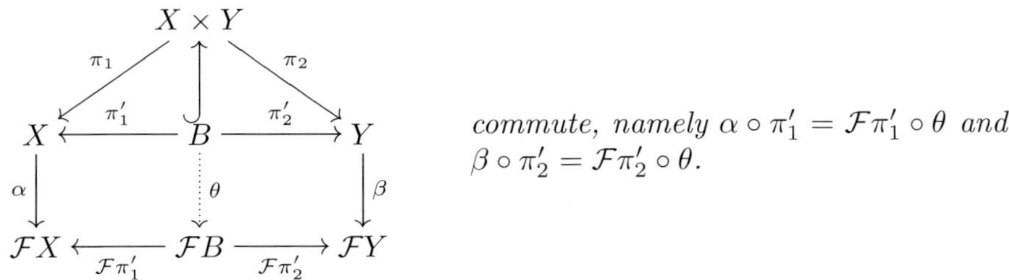

commute, namely $\alpha \circ \pi_1' = \mathcal{F}\pi_1' \circ \theta$ *and* $\beta \circ \pi_2' = \mathcal{F}\pi_2' \circ \theta$.

The correspondence between coalgebra homomorphisms and bisimulations is made precise by the following theorem, that holds under mild conditions on the functor $\mathcal{F}$ (namely, preservation of *weak pullbacks*):

Theorem 1 ([21]). *Morphism* $f : X \to Y$ *is an homomorphisms between two* $\mathcal{F}$*-coalgebras* (X, α) *and* (Y, β) *iff* $\{(x, f(x)) \mid x \in X\}$ *is a bisimulation between* (X, α) *and* (Y, β).

Similar results have been proved for HD-automata and in the next sections we recast the coalgebraic framework for HD-automata.

2.2   Named Sets and Named Functions

The definition of HD-automata relies on the notion of *named sets* (to represent the states) and *named functions* (to represent transitions).

Definition 3 (Named set). *A* named set *consists of a triple* $\langle Q, \|\_\|, G \rangle$ *where*

- Q *is a set of* states;
- $\|\_\| : Q \to \mathcal{P}_{fin}(\omega)$ *maps each state* $q \in Q$ *to the set of names of* $\|q\|$;
- $G : Q \to \mathcal{P}(Aut(\omega))$ *maps each state* $q \in Q$ *to* $G(q)$ *which is a subgroup of* $Aut(\|q\|)$ *called* symmetry of q.

We let $L, M, N, \ldots$ range over named sets; states, set of names and symmetry maps of a named set N are written as Q_N, $\|\_\|_N$ and G_N, resp. (subscripts will be removed if clear from the context).

By Definition 3, each state $q \in Q_N$ of a named set N is equipped with a finite set of names $\|q\|_N$ (called *support*) together with a group of permutations over such names. Intuitively, q represents a set of states "using" names in $\|q\|_N$ (the names of q) that are "indistinguishable" (according to a suitable equality of states) under the permutations in $G_N(q)$.

Example 1. The set of π-calculus agents can be given a named set structure by taking as elements sets q of agents and setting $G(q)$ to be the symmetry made of all permutations in $Aut(\omega)$ that applied to agents in q yield a structurally congruent agent still in q. Namely, if $P, R \in q$ then $P \equiv R\rho$ for a $\rho \in G(q)$. Observe that all the agents in q have the same set of free names which is actually the support of q, $\|q\|$.

A key feature of HD-automata is that names do not have a global meaning. In fact, names are deemed *local* to states and transitions. This makes possible garbage collection of unused names which is usually absent in ordinary transition systems. For instance, in the ordinary semantics of the π-calculus, the agent $R(x) = (\nu y)\bar{x}y.R(y)$ reaches an infinite number of agents because all $R(z)$ with fresh z are different. Instead, in the HD-automata representation of $R(x)$ it is the "history" of computation that establishes the freshness of names; hence, all $R(z)$ with fresh z collapse on a single state. *Named functions* yield the "history" of computations.

Definition 4 (Named function). *A* named function *F between two named sets N and M is a pair of functions $\langle h, \Sigma \rangle$ such that*

- $h : Q_N \to Q_M$;
- $\Sigma : Q_N \to \mathcal{P}(\omega \overset{inj}{\to} \omega)$ *such that for all $q \in Q_N$, $\Sigma(q) \in \mathcal{P}_{fin}(\|h(q)\|_M \overset{inj}{\to} \|q\|_N)$ and, for all $\sigma \in \Sigma(q)$*

$$\sigma \circ G_M(h(q)) = \Sigma(q) \tag{1}$$

$$G_N(q) \circ \sigma \subseteq \Sigma_F(q). \tag{2}$$

We write $F : N \to M$ to denote a named function from N to M and, if $F = \langle h, \Sigma \rangle$, h_F (resp. Σ_F) denotes h (resp. Σ).

Condition (1) intuitively states that a function in Σ_F traces the history of names of q when mapped via h_F. Remarkably, Σ_F contains in general many injective functions: one for each permutation in the symmetry of $h_F(q)$. In other words, Σ_F is obtained by saturating an injective function with $G_M(q)$. This avoids the possibility to have two different mappings, that only differ for a permutation which is already in the symmetry of an element.

Condition (2) might look a bit obscure at a first glance, however it can be explained as follows: if we interpret σ as the representative of the history of names along a transition, condition (2) states that permutations in the symmetry of q respect such a history, namely they do not "represent" a transition which is not encompassed by $\Sigma(q)$.

We conclude this section by defining identity and composition of named functions.

Definition 5 (Identity and composition). *Let L, M, N be named sets. The identity named function on N is $id_N = \langle id_{Q_N}, \lambda q.G_N(q) \rangle$. If $F : L \to M$ and $H : M \to N$, the composition of F and H is $\langle h_H \circ h_F, \Sigma \rangle$ where*

$$\Sigma(q) = \{\sigma \circ \sigma' \mid \sigma \in \Sigma_F(q) \wedge \sigma' \in \Sigma_H(h_F(q))\}.$$

The composition of F and H is denoted as $H \circ F$.

Summing up, named sets and named functions form the category **NSet** [6, 3]. In [11], it is shown how **NSet** is categorially equivalent to the category of *nominal sets* [10], which is the same as algebras over the permutation signature [17].

2.3 HD-Automata

We define HD-automata as labelled transition systems where a set of *local* names is associated to each state. Locality of names is a key aspect of HD-automata and, intuitively, asserts that identity of names used in a given state is not related to that of names used in other states. This, combined with symmetries associated to states, enables the minimisation of HD-automata [17, 3]. Roughly speaking, the symmetry of a state q specifies how names of q can be interchanged without affecting its observable behaviour. This also allows the size of the state space of HD-automata to be reduced yielding a bisimulation checking algorithm. Here we neglect some technical details in favour of a simpler presentation. Specifically, we do not consider the *normalisation* operation (cf. [3, 6]).

We fix the named set $L = \langle Q_L, \|\text{-}\|_L, G_L \rangle$ of *labels* where

- Q_L is the *set* of labels (ranged over by l);
- $\|l\|_L \in \mathcal{P}_{fin}(\omega)$ are the names "exposed" in a transition labelled by l;
- for each label $l \in Q_L$, $G_L(l)$ is usually the trivial group $\{id_{\|l\|_L}\}$.

Example 2. A label representing the π-calculus output can be given by a label $\textbf{out} \in Q_L$ for which $\|\textbf{out}\|_L = \{x, y\}$, yielding the *subject* and the *object* names of the output. Notice that an output where subject and object coincide can be represented by a label $\textbf{out1} \in Q_L$ such that $\|\textbf{out1}\|_L$ is a singleton. Similarly, a *bound output* (where the object is a private name extruded to the environment) can be represented by a label $\textbf{bout}$ having a single name for the subject.

HD-automata can be defined as coalgebras for the functor $\mathcal{T}$ (given below) that equips labelled transition systems with the notion of *local names* and *binding* emerging from the base category **NSet**.

Let $\widehat{N}$, the set of *(L-)transitions* on a named set N, be defined as

$$\widehat{N} = \left\{ \langle q, l, n, \pi, \vartheta \rangle \mid q \in Q_N, l \in Q_L, n \in \mathcal{P}_{fin}(\omega), \pi : \|l\|_L \xrightarrow{inj} n, \vartheta : \|q\|_N \xrightarrow{inj} n + 1 \right\}.$$

Each transition consists of a destination state q, a label l, a set of names n (representing the observable names of the whole transition), and two injections π and ϑ. The former maps the observable names of the label l into n, while ϑ provides the history of the names of the destination state along the transition, by mapping them to n. Remarkably, $\text{Im } \vartheta \subseteq n + 1$ accounts for the generation of a fresh name. Specifically, if a name of q is mapped on $\star \in 1$ then it is fresh (for simplicity we assume that at most one fresh name can be generated).

Transitions on N form a named set $Tr[N] = \langle \widehat{N}, \|\text{-}\|_{\widehat{N}}, G_{\widehat{N}} \rangle$ where

- $\|\langle q, l, n, \pi, \vartheta \rangle\|_{\widehat{N}} = n$;
- $G_{\widehat{N}}(\langle q, l, n, \pi, \vartheta \rangle) = \{\rho \in Aut(n) \mid \rho|_{\text{Im } \pi} = id_{\text{Im } \pi} \wedge \rho^* \circ \vartheta \circ G_N(q) = \vartheta \circ G_N(q)\}$, where $\rho^* \in Aut(n + 1)$ is the map $\rho + id_1$ (i.e., $\rho^*|_n = \rho$ and $\rho^*|_1 = id_1$).

The symmetry of a transition is given by the permutations that preserves the mappings $\vartheta \circ G_N(q)$, that is the meaning of the names as given by the symmetry of the target state q (and the label l).

Proposition 1. *If N is named set, then $Tr[N]$ is a named set.*

The functor $\mathcal{T} : \mathbf{NSet} \to \mathbf{NSet}$ is specified by describing its action on objects (i.e., named sets) and morphisms (i.e., named functions) of $\mathbf{NSet}$. The action of the functor $\mathcal{T}$ on an object N is

$$\mathcal{T}N = \langle \mathcal{P}_{fin}(Q_{Tr[N]}), \|-\|_{\mathcal{T}N}, G_{\mathcal{T}N} \rangle$$

where the set of names of $T \in \mathcal{P}_{fin}(Q_{Tr[N]})$ is the union of the set of names of all transitions in T, and the symmetry of T is the set of permutations for which transitions in T "are preserved". Formally,

$$\|T\|_{\mathcal{T}N} = \bigcup_{\langle q,l,n,\pi,\vartheta \rangle \in T} n, \qquad G_{\mathcal{T}N}(T) = \bigcap_{t \in T} G_{\widehat{N}}(t).$$

The action of $\mathcal{T}F = \langle h, \Sigma \rangle$ on a named function $F : N \to M$ is given by

$$h(T) = \bigcup_{\langle q,l,n,\pi,\vartheta \rangle \in T} \{ \langle h_F(q), l, (\mathrm{Im}\ \vartheta \circ \sigma \cup \mathrm{Im}\ \pi) \setminus \mathbf{1}, \pi, \vartheta \circ \sigma \rangle \mid \sigma \in \Sigma_F(q) \}$$

$$\Sigma(T) = \iota \circ G_{\mathcal{T}N}(T), \text{ where } \iota : \|h(T)\|_M \hookrightarrow \|T\|_N$$

Namely, $\mathcal{T}$ maps transitions on N to transitions on M replacing q with $h_F(q)$ in each $\langle q, l, \mathcal{N}, \pi, \vartheta \rangle$ and relating names of $h_F(q)$ to names of q via $\Sigma_F(q)$. It is worth noticing that the names of the transitions T are a superset of the names $h(T)$ because they are the union of the images of $\vartheta \circ \sigma$ (and π) of transitions in $h(T)$. In fact, $\mathrm{Im}\ \vartheta \subseteq n$ by definition of ϑ and the composition with the function σ can only restrict the image of $\vartheta \circ \sigma$ (which happens when some name of q is discarded by F).

Proposition 2. *If F is a named function, then $\mathcal{T}F$ is a named function.*

Definition 6. *An HD-automata is a named function $H : N \to \mathcal{T}N$, namely it is a coalgebra for the functor $\mathcal{T}$ on $\mathbf{NSet}$.*

3   Simulation for HD-Automata

This section recasts the definition of the *simulation* relation originally presented in [20] (Definition 7.11, Chapter 7) in our context.

Definition 7 (HD-Simulation). *Let $H : N \to \mathcal{T}N$ and $K : M \to \mathcal{T}M$ be two HD-automata. A relation $\mathfrak{S} \subseteq Q_N \times Aut(\omega) \times Q_M$ is an HD-simulation iff whenever $(q, \delta, q') \in \mathfrak{S}$ for any $\langle q_1, l, n_1, \pi_1, \vartheta_1 \rangle \in h_H(q)$ there are $\langle q_2, l, n_2, \pi_2, \vartheta_2 \rangle \in h_K(q')$ and $\delta' \in Aut(\omega)$ such that*

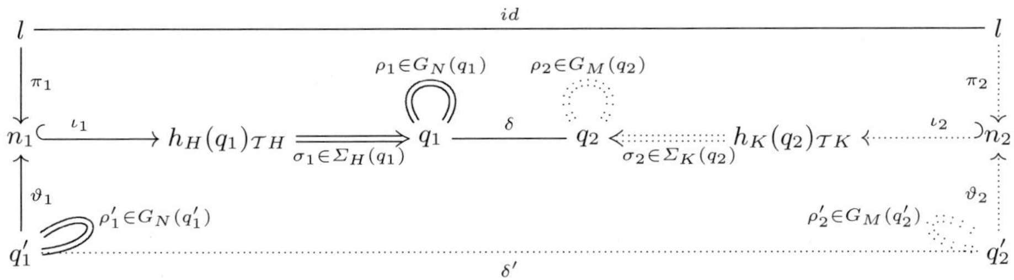

Fig. 1. HD-automata simulation

- *for all $\rho_1 \in G_N(q_1)$, $\sigma_1 \in \Sigma_H(q_1)$ and $\rho_1' \in G_{TN}(q_1')$ there are $\rho_2 \in G_M(q_2)$, $\sigma_2 \in \Sigma_K(q_2)$ and $\rho_2' \in G_{TM}(q_2)$ such that*
    1. *$\delta \circ \rho_1 \circ \sigma_1|_{n_1} \circ \vartheta_1 \circ \rho_1' = \rho_2 \circ \sigma_2|_{n_2} \circ \vartheta_2 \circ \rho_1' \circ \delta'$ on the names x of q_1 for which $\vartheta_1(\rho_1'(x)) \notin \mathbf{1}$;*
    2. *$\vartheta_1(\rho_1'(x)) \in \mathbf{1} \iff \vartheta_2(\rho_2(\delta'(x))) \in \mathbf{1}$;*
    3. *$\delta \circ \rho_1 \circ \sigma_1|_{n_1} \circ \pi_1 = \rho_2 \circ \sigma_2|_{n_2} \circ \pi_2$;*
- *$(q_1', \delta', q_2') \in \mathfrak{S}$.*

where δ (resp. δ') is used as map from the names of q_1 (resp. q_1') to the names of q_2 (resp. q_2') in the compositions above.

Definition 7 can be explained using the diagram in Figure 1 where

- bijections are represented by lines,
- elements of sets of maps Σ are represented by double lines,
- universally (resp. existentially) quantified maps are represented by solid (resp. dotted) lines and
- all the arrows involving each of the states q_1, q_1', q_2 and q_2' are meant to be maps from/to the names of the state.

For instance, ρ_2 is a double dotted line because it is an existentially quantified permutation in the symmetry of q_2.

In Figure 1 the sub-diagram consisting of π_1, ϑ_1 and $\sigma_1|_{n_1} = \sigma_1 \circ \iota_1$ describes how the coalgebra H maps the names of the transition $\langle q_1', l, \pi, \vartheta_1 \rangle \in h_H(q_1)$ on the names of q_1 through the injective maps in $\Sigma_H(q_1)$ (and similarly for the sub-diagram consisting of π_2, ϑ_2 and $\sigma_2|_{n_2} = \sigma_2 \circ \iota_2$).

Intuitively, Definition 7 states that any transition from q_1 is matched by a transition from q_2. However, names of such transitions and their relationship with names of q_1 and q_2 are of concern. As a matter of fact, symmetries may yield several different representations of equivalent states and transitions. Therefore, the conditions on δ state that names of q_1 and q_2 must be related so that the simulation is independent of their symmetries. More precisely, for any possible representation ρ_1 of q_1 and meaning σ_1 of the names in the derivative of q_1 (in the coalgebra H) it is possible to find corresponding representation ρ_2 and σ_2 for q_2 (in the coalgebra K) in such a way that:

- the history of names in transitions are compatible;
- freshness of names is preserved;
- observed names are the same up to permutations of the symmetries.

Notice that the conditions above resp. correspond to conditions 1, 2 and 3 of Definition 7.

Given an HD-simulation $\mathfrak{S}$ and $\langle q, \delta, q' \rangle \in \mathfrak{S}$, Definition 7 demands δ to be an automorphism of ω. In practice, name correspondences δ can be maps in $[\|q\| \to \|q'\|]$ such that $G(\delta) \stackrel{\text{def}}{=} \{(x, \delta(x)) \mid x \in \|q\|\}$ is a *partial bijection* in $\|q\| \times \|q'\|$, namely for all $(x', y') \in G(\delta)$, $x = x' \iff y = y'$.

Proposition 3. *For each HD-simulation $\mathfrak{S}$ there is an HD-simulation $\mathfrak{S}'$ such that for all $\langle q, \delta, q' \rangle \in \mathfrak{S}$ there is $\langle q, \delta', q' \rangle \in \mathfrak{S}'$ such that $\delta' \in [\|q\| \to \|q'\|]$ is a partial bijection.*

Moreover, since each partial bijection on ω can be extended to an automorphism, the following proposition holds.

Proposition 4. *Let $\mathfrak{S}$ be a set of triples $\langle q, \delta, q' \rangle$ where q and q' are states of two HD-automata and $\delta \in [\|q\| \to \|q'\|]$ is a partial bijection. If $\mathfrak{S}$ satisfies the conditions of Definition 3 then*

$$\mathfrak{S}' \stackrel{\text{def}}{=} \{\langle q, \hat{\delta}, q' \rangle \mid \langle q, \delta, q' \rangle \in \mathfrak{S} \wedge \hat{\delta} \in Aut(\omega) \wedge \hat{\delta}|_{\|q\|} = \delta\}$$

is an HD-simulation.

4   A Motivating Example

This section gives an example illustrating the main features of our approach, namely *symmetrical* and behavioural service matching with explicit handling of names.

Typically, in a distributed setting such as SOC, services and invokers agree on the adopted communication protocol during their very first interaction, by selecting a version identifier that is uniquely associated to a known protocol. In other words, an exact matching of the identifiers would establish a sort of "contract" about the communication protocol, avoiding unexpected requests or replies.

However, distributed protocols commonly evolve from one version to another of a service, making the usage of version identifiers a very fragile mechanism. In our framework, version identifiers are replaced by *behaviours* and simulation is used to establish the matching. This allows new deployed versions of services to slightly deviate from communication protocols in certain cases (e.g., by adding functionality to certain stages).

4.1   The Scenario

Consider a service offering a "forwarding" mechanism whereby invokers send an address a and a message m to be sent to a. The service sends m to a after checking some conditions that we neglect for simplicity. Also, suppose that the service handles *sessions* through *cookies*. For instance, at the beginning of the interaction, a unique cookie is assigned to invokers. (Note that session handling is a requirement imposed by the service to invokers, and it is a completely separate issue from the effective service offered.) At any time during the evolution of the interaction, the service may require to check the invoker's cookie. Let this version of the service be called `protocol-A`. A different version, called `protocol-B`, allows the service to *refresh* cookies (i.e., to send a fresh cookie which should replace the old one) at any stage of the protocol.

Since cookies are expected to be fresh (in practice, being randomly generated with a very low probability of collision), it is necessary to explicitly handle fresh resource generation. The absence of garbage collection of *old* session cookies would yield infinite state systems, since an unbound number of (unused) cookies should be maintained. Remarkably, since clients must record at least the *last* received cookie (which is always fresh), it is not possible to model either of the protocols without memory at all. Therefore HD-automata are a reasonable choice, being close to ordinary labelled transition systems, with the added benefit of name handling and garbage collection.

In order to make the presentation clearer, in the following:

- states and labels are written together with their set of local names; for instance, $\mathbf{q}\{x,y\}$ or $\mathbf{addr}\{a\}$ denote a state q with two names x and y or a label $\mathbf{addr}$ whose observed name is a;
- the HD-automata transition

$$\underset{X}{q_0} \xrightarrow{\; 1\; Z,\, \vartheta\;} \underset{Y}{q_1}$$

  represents a transition from $q\ X$ to $q'\ Y$ with label $1\ Z$ and mapping θ from the names Y of q' to those X of q such that π is the identity map on Z and ϑ behaves as the identity of $Y \cap X$; moreover, θ is omitted if it is the identity;
- we write

$$\underset{X}{q_0} \xrightarrow{\; \substack{1_1\ Z_1,\,\theta_1/\ldots/ \\ 1_n\ Z_n,\,\theta_n}\;} \underset{Y}{q_1}$$

  for a set of transitions from $q\ X$ to $q\ Y$ with labels $1_1\ Z_1, \vartheta_1,\ldots, 1_n\ Z_n, \vartheta_n$.

4.2   Binding Services Using HD-Automata Simulation

We assume that a service publishes (together with its signature) a pair of HD-automata $\langle S_o, S_r \rangle$ built on top of a fixed named set of labels. Intuitively, S_o yields the (abstracted) behaviour of the service (its *offer*) and S_r is the requirement that the service imposes to invokers to allow them to bind and make invocations.

We show here how this approach works for the scenario discussed in § 4.1 for which we define the named set of labels $\langle Lab, \|\_\|, G \rangle$ as follows:

- *Lab* is the set $\{\textbf{setC}, \textbf{readC}, \textbf{addr}, \textbf{msg}, \textbf{quit}\}$ (where **setC** and **readC** are for setting a new cookie or reading a new one, **addr** is for communicating an address, **msg** is for communicating a message and **quit** determines the end of the protocol);
- $\|\_\|$ maps **setC** and **quit** to the empty set while **addr** and **msg** are mapped to a singleton;
- as assumed for named sets of labels, G maps each label to the trivial symmetry containing only the identity.

The HD-automaton representing the service requirements is S_r below:

$$
\begin{array}{c}
\textbf{readC}\{c'\}/ \\
\textbf{setC}\{\}, \vartheta_{c'}/ \\
\textbf{addr}\{a'\} \\
\circlearrowright
\end{array}
$$

$$
\underset{\{a',m'\}}{sr_0} \xrightarrow{\ \textbf{setC}\{\},\vartheta_{c'}\ } \underset{\{a',m',c'\}}{sr_1} \xrightarrow{\ \textbf{msg}\{m'\}\ } \underset{\{a',m',c'\}}{sr_2} \xrightarrow{\ \textbf{quit}\{\}\ } \underset{\{\}}{sr_3}
$$

where $\vartheta_{c'}$ maps c' to $\star$ and a' (resp. m') to a' (resp. m'). HD-automaton S_r formalises `protocol-B` service requirements and requires invokers to accept a cookie and to be ready to reset it (**setC**) or provide it (**readC**) at any time during the protocol before the message is sent. After sending the message, the invoker has to quit.

Remark 1. It is a simple observation that requirements for `protocol-A` can be obtained by removing the transition labelled **setC**$\{\}$ from sr_1 in S_r.

The service offers the behaviour S_o below to invokers that fulfill S_r:

$$
\underset{\{a_1,a_2,c'\}}{\overset{\textbf{setC}\{\},\vartheta_{c'}\,\circlearrowright}{so_0}} \xrightarrow[\textbf{addr}\{a_2\}]{\textbf{addr}\{a_1\}/} \underset{\{a_1,a_2,c'\}}{\overset{\begin{array}{c}\textbf{setC}\{\}, \vartheta_{c'}/ \\ \textbf{readC}\{c'\}/ \\ \textbf{addr}\{a_1\}/ \\ \textbf{addr}\{a_2\} \\ \circlearrowright\end{array}}{so_1}} \xrightarrow{\ \textbf{msg}\{m'\}\ } \underset{\{a_1,a_2,m',c'\}}{so_2} \xrightarrow{\ \textbf{quit}\{\}\ } \underset{\{\}}{so_3}
$$

where the cookie can be refreshed any number of times before and after getting an (states so_0 and so_1). Once an address is sent, the service moves in state so_1 where cookies can be refreshed or required for checking an unbound number of times. In so_1 the service is also keen to accept addresses (either a_1 or a_2) before sending the message. Finally, if the message is sent, the service halts.

Remark 2. Noteworthy, so_1 could be replaced by a state so_1' equipped with the symmetry $\{id_{a_1,a_2}, (a_1\ a_2)\}$ (where $(a_1\ a_2)$ is the transposition of a_1 and a_2). However, for simplicity we prefer to stick with the current more explicit representation.

Remark 3. Again, removing the transition **setC{}** from so_1 in S_o yields the HD-automaton for `protocol-A`.

Symmetrically to service publication, service invocations have to specify a pair $\langle C_r, C_o \rangle$ of HD-automata so that C_r describes the behaviour required by the invoker to the service and C_o yields the offered guarantees. An invocation matching $\langle S_r, S_o \rangle$ is represented by $\langle C_o, C_r \rangle$ where C_o is:

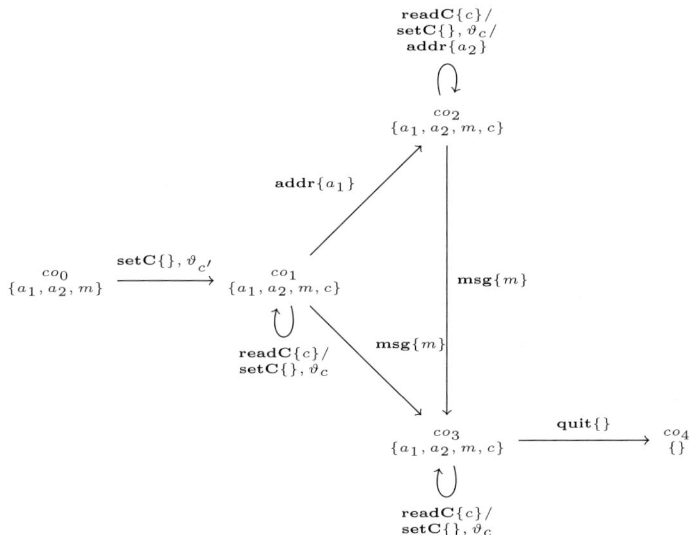

and C_r is

C_r simply requires to the service the capacity of executing the sequence of transitions setting the cookie, receiving the addresses and forwarding. (Notice that the invocation does not require the service to stop.)

C_o guarantees that the client accepts a request to set the cookie, then in each state is capable to provide the previously set cookie upon request, or to refresh it, thus respecting the protocol imposed by the service provider.

While it is immediate to see that S_o simulates C_r, it is less obvious that C_o simulates S_r. To show this, we build an explicit simulation between S_r and C_o represented in the following figure.

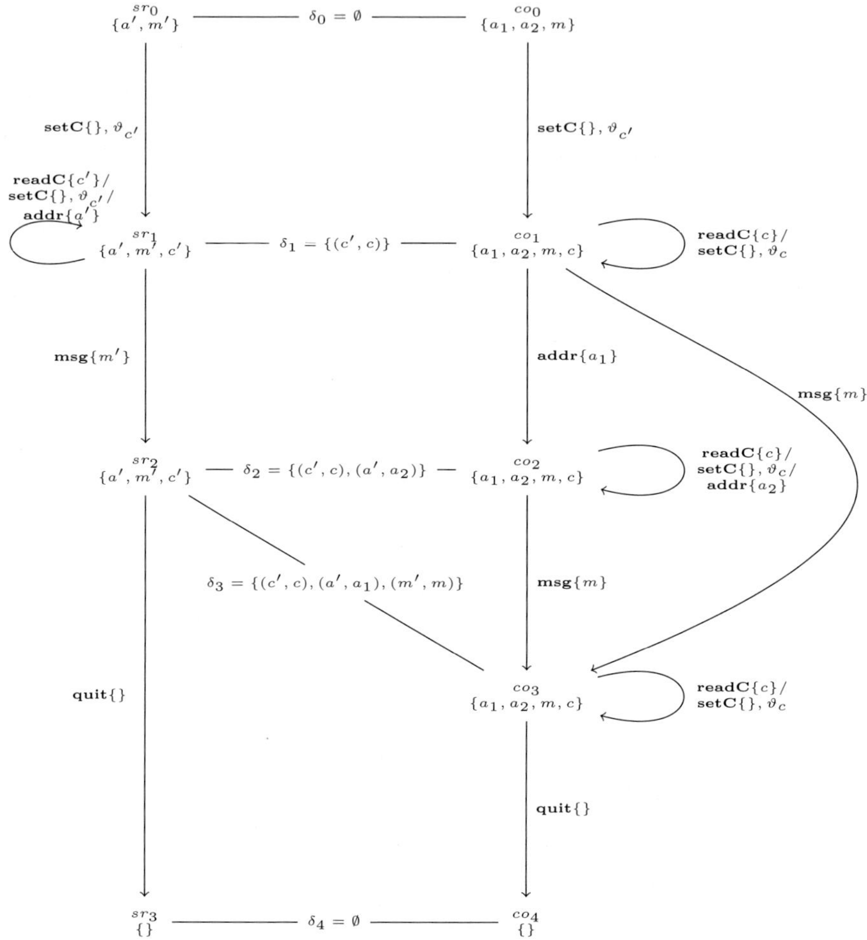

A simple check shows that

$$\mathfrak{S} = \{\langle sr_0, \delta_0, co_0 \rangle, \langle sr_1, \delta_1, co_1 \rangle, \langle sr_2, \delta_2, co_2 \rangle, \langle sr_2, \delta_3, co_3 \rangle, \langle sr_3, \delta_4, co_4 \rangle\}$$

yields an HD-simulation (by Proposition 4).

5   Conclusions and Future Work

We have introduced the foundations of a notion of *behavioural* matching of services, that keeps in account resource generation in a *finitistic* way employing

HD-automata. Our framework inherits the algorithmic properties of HD-automata and it can support effective usages in design environments for SOC.

As a a case study here we considered the versioning problem of protocols. In the usual approach, versioning relies on version identifiers. Local deviations to a protocol, hence, either lead to incompatibility between clients and servers, or to servers mimicking a version identifier without giving real guarantees on the fact that the protocol is respected. We tackled this problem exploiting the semantic HD-automata machinery for the fresh generation of names.

A pair of matching interfaces (i.e., $\langle C_o, C_r \rangle$ and $\langle S_o, S_r \rangle$ where C_o simulates S_r and S_o simulates C_r) determines a contract. A possible research direction concerns the synthesis of a *monitor* out of a contract so that the fulfillment of the contract can be enforced during the execution of the communication protocol. This is an interesting research direction as, during the execution of the communication protocol, only traces can be observed (while at service binding time the interfaces of two services can be represented by abstractions of their interactive behaviours). We argue that a monitor can be automatically synthesised from contracts.

Laudatio

HD-automata appeared for the first time in CONCUR'95, in an article by Ugo Montanari and the third author of this paper [14]. At that time, HD-automata were called π-automata, and they were simply seen as an efficient structure in an algorithm for detecting names that were syntactically present in π-calculus agents but that did not play any semantic role.

Since then, HD-automata have evolved into a reference model for nominal calculi, both from a theoretical and from a practical point of view. HD-automata can be defined by extending to structures with names several frameworks initially proposed for classical automata, ranging from the the *categorical* setting of Nielsen and Winskel (see [20]), to Rutten's *coalgebraic* setting (see this paper, [3] and the recent work [2]), to the *bialgebraic* setting of Turi and Plotkin (see [11, 17]). The semantic framework of HD-automata has been used to model and analyse concepts such as locality, causality, link mobility and cryptography [5]; more recently, they have also been used to capture concepts typical of SOC, as shown in this paper. Moreover, HD-automata have been exploited as the formal basis for verification toolkits such as HAL and Mihda [4, 7].

Along the years, the research on HD-automata has involved several scientists, including PhD students of Ugo, of whom we would like to remember Gioia Ristori, who will unfortunately not be able to celebrate Ugo's 65th birthday. If HD-automata have become a reference model for nominal calculi, it is thanks to the work of all these scientists. However, it is also and primarily thanks to Ugo, to his intuition that HD-automata were much more than the efficient data structure of CONCUR'95, to his idea that being able to manage names in a syntax-independent way is the key to understand and analyse nominal calculi, and to his capability to guide the research and to prove that this intuition was true.

References

[1] Bouali, A., Gnesi, S., Larosa, S.: The Integration Project for the JACK Environment. Bulletin of the EATCS 54, 207–223 (1994)

[2] Ciancia, V., Montanari, U.: A name abstraction functor for named sets. In: Coalgebraic Methods in Computer Science 2008 (to appear, 2008)

[3] Ferrari, G.L., Montanari, U., Tuosto, E.: Coalgebraic minimization of hd-automata for the pi-calculus using polymorphic types. Theoretical Computer Science 331(2-3), 325–365 (2005)

[4] Ferrari, G., Ferro, G., Gnesi, S., Montanari, U., Pistore, M., Ristori, G.: An Automata Based Verification Environment for Mobile Processes. In: Brinksma, E. (ed.) TACAS 1997. LNCS, vol. 1217, pp. 275–289. Springer, Heidelberg (1997)

[5] Ferrari, G., Gnesi, S., Montanari, U., Pistore, M.: A Model Checking Verification Environment for Mobile Processes. ACM Transactions on Software Engineering and Methodology 12(4), 440–473 (2003)

[6] Ferrari, G., Montanari, U., Pistore, M.: Minimizing Transition Systems for Name Passing Calculi: A Co-algebraic Formulation. In: Nielsen, M., Engberg, U. (eds.) FOSSACS 2002. LNCS, vol. 2303, pp. 129–143. Springer, Heidelberg (2002)

[7] Ferrari, G., Montanari, U., Tuosto, E.: From Co-algebraic Specifications to Implementation: The Mihda toolkit. In: de Boer, F.S., Bonsangue, M.M., Graf, S., de Roever, W.-P. (eds.) FMCO 2002. LNCS, vol. 2852, pp. 319–338. Springer, Heidelberg (2003)

[8] Ferrari, G., Montanari, U., Tuosto, E.: Coalgebraic Minimisation of HD-automata for the π-Calculus in a Polymorphic λ-Calculus. Theoretical Computer Science 331, 325–365 (2005)

[9] Ferrari, G., Montanari, U., Tuosto, E., Victor, B., Yemane, K.: Modelling and Minimising the Fusion Calculus Using HD-Automata. In: Fiadeiro, J.L., Harman, N.A., Roggenbach, M., Rutten, J. (eds.) CALCO 2005. LNCS, vol. 3629, pp. 142–156. Springer, Heidelberg (2005)

[10] Gabbay, M., Pitts, A.M.: A new approach to abstract syntax with variable binding. Formal Aspects of Computing 13(3-5), 341–363 (2002)

[11] Gadducci, F., Miculan, M., Montanari, U.: About permutation algebras (pre)sheaves and named sets. Higher-Order and Symbolic Computation 19(2-3), 283–304 (2006)

[12] Milner, R., Parrow, J., Walker, D.: A Calculus of Mobile Processes, I and II. Information and Computation 100(1), 1–40,41–77 (1992)

[13] Montanari, U., Buscemi, M.: A First Order Coalgebraic Model of π-Calculus Early Observational Equivalence. In: Brim, L., Jančar, P., Křetínský, M., Kucera, A. (eds.) CONCUR 2002. LNCS, vol. 2421, pp. 449–465. Springer, Heidelberg (2002)

[14] Montanari, U., Pistore, M.: Checking Bisimilarity for Finitary π-Calculus. In: Lee, I., Smolka, S.A. (eds.) CONCUR 1995. LNCS, vol. 962, pp. 42–56. Springer, Heidelberg (1995)

[15] Montanari, U., Pistore, M.: History Dependent Automata. Technical report, Dipartimento di Informatica, Università di Pisa, TR-11-98 (1998)

[16] Montanari, U., Pistore, M.: π-Calculus, Structured Coalgebras, and Minimal HD-Automata. In: Leung, K.-S., Chan, L., Meng, H. (eds.) IDEAL 2000. LNCS, vol. 1983, Springer, Heidelberg (2000); An extended version will be published on Theoretical Computer Science

[17] Montanari, U., Pistore, M.: Structured coalgebras and minimal hd-automata for the π-calculus. Theoretical Computer Science 340, 539–576 (2005)

[18] Paige, R., Tarjan, R.: Three Partition Refinement Algorithms. SIAM Journal on Computing 16(6), 973–989 (1987)

[19] Parrow, J., Victor, B.: The Fusion Calculus: Expressiveness and Symmetry in Mobile Processes. In: Annual Symposium on Logic in Computer Science. IEEE Computer Society Press, Los Alamitos (1998)

[20] Pistore, M.: History Dependent Automata. PhD thesis, Università di Pisa, Dipartimento di Informatica, available at University of Pisa as PhD Thesis TD-5/99 (1999)

[21] Rutten, J.J.M.M.: Universal coalgebra: a theory of systems. Theoretical Computer Science 249(1), 3–80 (2000)

[22] Sangiorgi, D., Walker, D.: The π-Calculus: a Theory of Mobile Processes. Cambridge University Press, Cambridge (2002)

A Type System for Client Progress
in a Service-Oriented Calculus[*]

Lucia Acciai and Michele Boreale

Dipartimento di Sistemi e Informatica, Università di Firenze.
`{lacciai,boreale}@dsi.unifi.it`

Abstract. We introduce a type system providing a guarantee of *client progress* for a fragment of CaSPiS, a recently proposed process calculus for service-oriented applications. The interplay of sessioning and data-orchestration primitives makes the design of a type system for CaSPiS challenging. Our main result states that in a well-typed CaSPiS system, and in absence of divergence, any client invoking a service is guaranteed not to get stuck during the execution of a conversation protocol because of inadequate service communication capabilities.

Keywords: process calculi, service-oriented computing, pi-calculus, type systems.

1   Introduction

Recent years have seen the emergence of web-based applications composed by several loosely coupled components, often referred to as web services, relying on message-passing as the sole means of cooperation. This technological shift has in turn led to the formulation of a new computational paradigm underpinning the construction of such applications and known as *Service Oriented Computing (SOC)*. Equipping SOC with rigorous semantic foundations is the subject of a very active research area. We just mention here the SENSORIA project [17], a large, EU-funded research initiative aiming at the development of a comprehensive approach to the engineering of SOC software systems, starting from rigorous methodological foundations.

CaSPiS (*Calculus of Sessions and Pipelines*, [2]) is a language currently being considered in SENSORIA as a candidate core calculus for SOC programming. CaSPiS design, influenced both by Cook and Misra's Orc [8] and by the pi-calculus [16], is centered around the notions of *session* and of *pipeline*. In CaSPiS, these concepts, and the related programming primitives, are viewed as natural tools for structuring client-service interaction and orchestration, the following description of CaSPiS is partly adopted from [2].

In CaSPiS, service definitions and invocations are written like (nullary) input and output prefixes in CCS, thus we have:

$$s.P \quad \text{and} \quad \overline{s}.Q$$

---

[*] Research supported by the Project FET-GC II IST-2005-16004 SENSORIA.

P. Degano et al. (Eds.): Montanari Festschrift, LNCS 5065, pp. 642–658, 2008.

where s is the name of the service. There is an important difference, though, as the bodies P and Q are not quite continuations, but rather protocols that, within a session, govern interaction between (instances of) the client and the server. As an example

$$currency\_converter.(x).\langle x * r \rangle \quad \text{and} \quad \overline{currency\_converter}.\langle amount \rangle.(y).\langle y \rangle^{\uparrow}$$

are respectively: a service that once called waits for an amount expressed in euros and then sends back its counter-value in US dollars, computed according to an exchange rate r; and a client that passes argument $amount$ to the service, then waits for the counter-value and returns this value as a result.

A session is generated as the result of a service invocation and represents an ongoing conversation between a client and a service. In the variant of CaSPiS we consider, a session is written $[P \| Q]$, with P and Q the communicating protocols running at the client and at the service side, respectively. For instance, synchronization of the client and of the service described above triggers a new session

$$[\langle amount \rangle.(y).\langle y \rangle^{\uparrow} \ \| \ (x).\langle x * r \rangle] \ .$$

Here, after one reduction step, the counter-value $x * r$ will be computed by the service protocol and then sent to the client:

$$[(y).\langle y \rangle^{\uparrow} \ \| \ \langle amount * r \rangle] \ \rightarrow \ [\langle amount * r \rangle^{\uparrow} \ \| \ \mathbf{0}] \ .$$

The remaining activity will be performed by the client side, which will emit $amount * r$ outside the session. In fact, values can be returned outside a session to the enclosing environment using the return operator, $\langle \cdot \rangle^{\uparrow}$. These values can be used to start new activities. To orchestrate flows of data arising from different sessions, CaSPiS provides the programmers with a *pipe* operator, written $P > Q$. As an example, pipes allow to pass the results produced by one service invocation in P onto the next service Q in a given chain of invocations; or to wait for the results produced by two concurrent invocations before invoking a third service. For instance, what follows is a client that invokes the service *currency_converter* and then checks if the amount is available on his bank account:

$$\overline{currency\_converter}.\langle amount \rangle.(y).\langle y \rangle^{\uparrow} > (z).\overline{check\_bank\_availability}.\langle z \rangle \ .$$

Very often, client-service interactions in a SOC scenario comprise not only the exchange of messages between the two main parties, but also invocation of subsidiary services. The results produced by these subsidiary invocations are used in the main (top level) session. For this reason, CaSPiS allows service invocations to be placed inside sessions, hence giving rise to hierarchies of invocations and nested sessions. As an example, suppose the exchange rate from euros to US dollars in the example above is not fixed and that service *currency_converter* calls service *exchange_rates* for obtaining the up to date rate as described below.

$$currency\_converter.\Big((x).\overline{exchange\_rates}.(\langle \text{``€/\$''} \rangle.(z)\langle z \rangle^{\uparrow}) > (r).\langle x * r \rangle \Big) \ .$$

Interaction of the client above and the new version of the *currency_converter* service will lead to

$$\left[(y).\langle y\rangle^{\uparrow}\,\|\!|\,\left([\langle\text{``}\text{€/\$}\text{''}\rangle.(z)\langle z\rangle^{\uparrow})\,\|\!|\,R] > (r).\langle amount * r\rangle\right)\right]$$

where R is the interaction protocol of service *exchange_rates*. Once R gets "€/\$" message, it provides the up to date exchange rate *rate*, the innermost session passes this value through the pipeline and the whole process reduces to

$$\left[(y).\langle y\rangle^{\uparrow}\,\|\!|\,\left([\mathbf{0}\,\|\!|\,R'] > \langle amount * rate\rangle\right)\right]$$

and then to

$$\left[\langle amount * rate\rangle^{\uparrow}\,\|\!|\,\left([\mathbf{0}\,\|\!|\,R'] > \mathbf{0}\right)\right].$$

The presence of pipes and nested sessions makes the dynamics of a CaSPiS session quite complex: it is substantially different from simple type-regulated interactions as found in the pi-like languages of, e.g. [11,10], or in the finite-state contract languages of [5,4,6].

The present paper is a contribution towards developing programming techniques for safe client-service interaction in a SOC scenario. Technically, we offer a type system for CaSPiS that provides guarantees of *client progress*. In practice, this means that in a well-typed CaSPiS system, and in absence of divergence, any client invoking a service is guaranteed, during the execution of a conversation protocol, not to get stuck because of inadequate service communication capabilities. More generally, we hope that some of the concepts we discuss here may be further developed and find broader applications in the future.

There are three key aspects involved in the design of our type system. A first aspect concerns *abstraction*: types focus on flows of I/O value-types and ignore the rest (actual values, service calls, …). Specifically, types take the form of CCS-like terms describing I/O flows of processes. In fact, a tiny fragment of CCS, with no synchronization and restriction, is employed, where the role of atomic actions is played by basic types. A second aspect concerns *compliance* of client protocols with service protocols, which is essential to avoid deadlocks. In the type system, the operational abstractions provided by types are employed to effectively check client-service compliance. To this purpose, types are required to account for process I/O behaviour quite precisely. Indeed, approximation might easily result into ignoring potential client-service deadlocks. A final aspect concerns the nesting of sessions. A session at a lower level can exercise *effects* on the upper level, say the level of any enclosing session. To describe this phenomenon, we follow [3,14] and keep track, in the type system, of the behaviour at both the current level and at the level of a (fictitious) enclosing session. This results in type judgments of the form $P : [\mathsf{S}]\mathsf{T}$, where S is the the current-level type and T is the upper-level effect of P. Note that the distinction between types and effects we make here is somehow reminiscent of the type-and-effects systems of [18], with the difference that our effects are very simple (sequences of outputs) and are exercised on an upper level of activity rather than on a shared memory.

The version of CaSPiS considered in this paper differs from the "official" one in [2] in one important respect: we restrict our attention to the case where values can be returned outside a session only on the client side (the same restriction applies to the language considered in [3]). The theoretical reasons for doing so will be discussed in the concluding section. From a practical point of view, this limitation means that, once a session is started, for the service there will be no "feedback" of sort as to what is going on inside the session. This is somehow consistent with the idea that services should be stateless entities.

Related work. Our work is mainly related to Bruni and Mezzina's [3] and to Lanese et al.'s [14]. In these papers, type systems for languages affine to CaSPiS are put forward. In particular, the language considered in [3] is essentially CaSPiS with the restriction discussed above. The language of [14], SSCC, differs from CaSPiS essentially because streams, rather than pipes, are provided for data orchestration of different activities. We share with [3,14] the two-level types technique. In some important aspects, though, our system differs from theirs, resulting into a gain of simplicity and generality. These aspects we discuss below. First, we take advantage of the restriction that values can be returned only to the client and adopt a new syntax and operational semantics for sessions that spares us the necessity of explicit session names and the annoying "balancing" conditions on them (see also [2]). Second, our type system does not suffer from certain heavy restrictions of [3,14], like for example, forcing either of the two components in a parallel composition to have a null effect. Also, the client-service compliance relation we adopt is more flexible than the bare complementarity relation inherited from session-types disciplines employed in [3,14]. Finally, our client-progress theorem is an immediate consequences of two natural properties of the system (subject reduction, type safety). In particular, we do not have to resort to a complex system of obligations and capabilities *a l* Kobayashi [13,12], like [3] does. This is a benefit partly of a precise operational correspondence between processes and types and partly of our new syntax for sessions. Note that, in [14], synchronization problems related to data streams prevent achieving a deadlock-freeness result.

Both CaSPiS and SSCC evolved from *SCC* (*Serviced Centered Calculus*) [1], a language that arose from a joint effort of various partners involved in the SENSORIA consortium. The original proposal turned out later to be unsatisfactory in some important respects. In particular, SCC had no dedicated mechanisms for data orchestration and came equipped with no type system. These problems motivated the proposal of a few evolutions of SCC. As mentioned above, SSCC is stream-oriented, in that values produced by sessions are stored into dedicated queues, accessible by their names, while CaSPiS relies solely on pipes. Another evolution of SCC is the language in [19], featuring message-passing primitives for communication in all directions (within a session, from inside to outside and vice-versa).

Structure of the paper. The rest of the paper is organized as follows. In Section 2 we present CaSPiS⁻, the variant of CaSPiS we will consider. *Client progress*, the property we wish to capture with our system, is also defined. A language of types is introduced in Section 3, while a type system is presented in Section 4. Results about the type system are discussed in Section 5, culminating in Corollary 1, which asserts that well-typed processes enjoy the client progress property. We conclude with a few remarks concerning the limitation of our system and further work in Section 6.

Table 1. Syntax of CaSPiS$^-$

$$P, Q ::= \sum_{i \in I} \pi_i.P_i \quad \text{Guarded Summation} \qquad \pi ::= (x : b) \quad \text{Input Prefix}$$

$$\mid \langle u \rangle^\uparrow \qquad\qquad \text{Return} \qquad\qquad\qquad\quad \mid \langle u \rangle \qquad \text{Output Prefix}$$

$$\mid s.P \qquad\qquad\quad \text{Service Definition}$$

$$\mid \overline{u}.P \qquad\qquad\quad \text{Service Invocation}$$

$$\mid [P \| Q] \qquad\qquad \text{Session}$$

$$\mid P > Q \qquad\qquad \text{Pipeline}$$

$$\mid P | Q \qquad\qquad\quad \text{Parallel Composition}$$

$$\mid (\nu s) P \qquad\qquad \text{Restriction}$$

$$\mid *P \qquad\qquad\quad \text{Replication}$$

2 Processes

2.1 Syntax and Semantics

We introduce below the variant of CaSPiS, which we christen CaSPiS$^-$, that we have chosen as a target language for our type system.

Syntax. We presuppose the following disjoint sets: a set $\mathcal{B}$ of *base values*, a countable set $\mathcal{N}$ of *service names* ranged over by $n, s, \ldots$ and a countable set $\mathcal{V}$ of *variables*, ranged over by $x, y, \ldots$. In the following, we let u be a generic element of $\mathcal{N} \cup \mathcal{B} \cup \mathcal{V}$ and v be a generic element of $\mathcal{N} \cup \mathcal{B}$. We presuppose a set $\mathcal{B}t$ of *base types*, b, b$'$, $\ldots$ which include name *sorts* $\mathcal{S}, \mathcal{S}', \ldots$. We finally presuppose a generic base-typing relation, mapping base values and service names to base types, written $v : $ b, with the obvious proviso that service names are mapped to sorts and base values are mapped to the remaining base types.

The syntax of the calculus is reported in Table 1. Input prefixes are annotated with types b, which are associated to input variables. In service definitions and invocations, $s.P$ and $\overline{s}.Q$, processes P and Q are the protocols followed respectively by the service and client side. As in [2], the grammar defined in Table 1 should be considered as a run-time syntax. In particular sessions $[P \| Q]$ can be generated at run-time, upon service invocation, but a programmer is not expected to explicitly use them. In $[P \| Q]$ processes P and Q represent respectively the rest of the client and the service protocol to be executed. The free and bound names and variables of a term are defined as expected. In the following, we suppose each bound name in a process different from free, and we identify terms up to alpha-equivalence. We denote by $\text{fn}(P)$, resp. $\text{fv}(P)$, the set of free names, resp. variables, of P, and indicate with $\mathcal{P}$ the set of closed terms, that is, the set of process terms with no free variables. In what follows, we abbreviate the empty summation by **0**.

CaSPiS$^-$ is essentially the close-free fragment of the calculus in [2], but for a major difference: in CaSPiS$^-$ sessions are one-sided. In particular, sessions are executed on the client side and all returned values are available only at this side. This simplification allows us to dispense with session names and balancing conditions on them – see [2] – which are necessary when the two sides of a sessions are distinct and far apart. Practically, this limitation means that services in $\mathsf{CaSPiS^-}$ cannot return values and are stateless. Another, minor difference from [2] is that here returns are asynchronous. Finally, for the sake of simplicity we do not consider structured values and expressions, which can be easily accommodated.

Semantics. The operational semantics of the calculus is given in terms of a *labelled transition relation*, $\xrightarrow{\lambda}$ defined as the least relation generated by axioms in Table 2. Labels λ can be of the following form: input (v), output $\langle v \rangle$ or $(\nu\hat{s})\langle s \rangle$ – where $(\nu\hat{s})$ indicates that the restriction (νs) may or may not be present –, return $\langle v \rangle^{\uparrow}$ or $(\nu\hat{s})\langle s \rangle^{\uparrow}$, service definition $(\nu\tilde{n})s\langle R \rangle$, service invocation $\overline{s}(R)$ and synchronization τ. It is worth noticing that service definitions are persistent, (DEF), and only (synchronous) in-session value-passing is allowed, (S-COM$_l$) and (S-COM$_r$). As already stated, sessions are one-sided, (CALL), and possible returns arising from the service protocol Q are ignored – there is no symmetric rule of (S-RET). Note that in (P-PASS) we have used an optional restriction $(\nu\hat{n})$ to indicate that the passed value might be a bound service name. Finally, note the run-time type check in (IN), which avoids type mismatch between the received object and the expected base type. From a computational point of view, this rule should not be particularly worrying, since we are only considering checks on base values. Note that in pi-like process calculi, static checks on channels are often sufficient to avoid such type mismatches – see e.g. the sorting system of [15]. In $\mathsf{CaSPiS^-}$, this solution is not viable as communication takes place freely inside sessions. In fact, an alternative to run-time checks would be assigning "tags" to I/O actions, to regulate data-exchange inside sessions, which would essentially amount to re-introducing a channel-based discipline, which is not our main concern here. Note that this issue does not arise in [3], because, as discussed in the Introduction, their type system discards the parallel composition of two or more outputs inside sessions.

We shall often refer to a silent move $P \xrightarrow{\tau} P'$ as a *reduction*; $P \Rightarrow P'$ and $P \overset{\lambda}{\Rightarrow} P'$ mean respectively $P \xrightarrow{\tau}{}^{*} P'$ and $P \xrightarrow{\tau}{}^{*} \xrightarrow{\lambda} \xrightarrow{\tau}{}^{*} P'$.

2.2   Client Progress Property

The client progress property will be defined in terms of an error predicate. Informally, an error occurs when the client protocol of an active session tries to send to or receive a value from the service side, but the session as a whole is blocked. This is formalized by the predicate $\rightarrow_{\mathrm{ERR}}$ defined below. In the definition, we rely on two standard notions, structural congruence and contexts, briefly introduced below. *Structural congruence*, $\equiv$, is defined as the least congruence over (open) processes preserved by substitutions and satisfying the axioms in Table 3. In the vein of [2], the laws in Table 3 comprise the structural rules for parallel composition and restriction from the pi-calculus, plus some extra scope extension laws for pipelines and sessions.

Contexts, $C[\cdot], C'[\cdot], \ldots$, are process terms with a hole; we shall indicate with $C[P]$ the process obtained by replacing the hole with P. The notion of context can be generalized to n-holes contexts as expected. We say a context is *static* if its hole is not under

Table 2. Labeled Semantics

$$\text{(IN)}\ \frac{v:\mathsf{b}}{(x:\mathsf{b}).P \xrightarrow{(v)} P[v/x]} \qquad \text{(OUT)}\ \frac{}{\langle v\rangle.P \xrightarrow{\langle v\rangle} P} \qquad \text{(RET)}\ \frac{}{\langle v\rangle^\uparrow \xrightarrow{\langle v\rangle^\uparrow} \mathbf{0}}$$

$$\text{(REP)}\ \frac{P\,|\,{*}P \xrightarrow{\lambda} P'}{{*}P \xrightarrow{\lambda} P'} \qquad \text{(DEF)}\ \frac{}{s.P \xrightarrow{s\langle P\rangle} s.P} \qquad \text{(CALL)}\ \frac{}{\overline{s}.P \xrightarrow{\overline{s}(Q)} [P\|Q]}$$

$$\text{(SYNC}_l)\ \frac{P \xrightarrow{(\nu\tilde{n})s\langle R\rangle} P' \quad Q \xrightarrow{\overline{s}(R)} Q'}{P|Q \xrightarrow{\tau} (\nu\tilde{n})(P'|Q')} \qquad \text{(S-RET)}\ \frac{P \xrightarrow{(\nu\hat{v})\langle v\rangle^\uparrow} P'}{[P\|Q] \xrightarrow{(\nu\hat{v})\langle v\rangle} [P'\|Q]}$$

$$\text{(S-PASS}_l)\ \frac{P \xrightarrow{\lambda} P' \quad \lambda ::= \tau \mid \overline{s}(Q) \mid (\nu\tilde{n})s\langle Q\rangle}{[P\|Q] \xrightarrow{\lambda} [P'\|Q]} \qquad \text{(S-COM}_l)\ \frac{P \xrightarrow{(v)} P' \quad Q \xrightarrow{(\nu\hat{v})\langle v\rangle} Q'}{[P\|Q] \xrightarrow{\tau} (\nu\hat{v})[P'\|Q']}$$

$$\text{(S-SYNC}_l)\ \frac{P \xrightarrow{(\nu\tilde{n})s\langle R\rangle} P' \quad Q \xrightarrow{\overline{s}(R)} Q'}{[P\|Q] \xrightarrow{\tau} (\nu\tilde{n})[P'\|Q']} \qquad \text{(SUM)}\ \frac{\pi_j.P_j \xrightarrow{\lambda} P_j \quad j\in I \quad |I|>1}{\sum_{i\in I}\pi_i.P_i \xrightarrow{\lambda} P_j}$$

$$\text{(P-PASS)}\ \frac{P \xrightarrow{\lambda} P' \quad \lambda \neq (\nu\hat{v})\langle v\rangle}{P > Q \xrightarrow{\lambda} P' > Q} \qquad \text{(P-SYNC)}\ \frac{P \xrightarrow{(\nu\hat{v})\langle v\rangle} P' \quad Q \xrightarrow{(v)} Q'}{P > Q \xrightarrow{\tau} (\nu\hat{v})(P' > Q|Q')}$$

$$\text{(PAR}_l)\ \frac{P \xrightarrow{\lambda} P' \quad \mathrm{fn}(Q)\cap \mathrm{bn}(\lambda)=\emptyset}{P|Q \xrightarrow{\lambda} P'|Q} \qquad \text{(R-PASS)}\ \frac{P \xrightarrow{\lambda} P' \quad n\notin \mathrm{n}(\lambda)}{(\nu n)P \xrightarrow{\lambda} (\nu n)P'}$$

$$\text{(OPEN)}\ \frac{P \xrightarrow{\lambda} P' \quad \lambda ::= (\nu\tilde{a})s\langle R\rangle \mid (\nu\tilde{a})\langle n\rangle \mid (\nu\tilde{a})\langle n\rangle^\uparrow \quad s\neq n \quad n\in \mathrm{fn}(\lambda)}{(\nu n)P \xrightarrow{(\nu n)\lambda} P'}$$

Symmetric versions of (S-SYNC$_l$), (PAR$_l$), (S-PASS$_l$), (S-SYNC$_l$) and (S-COM$_l$) are not displayed.

Table 3. Structural Congruence

$$(P|Q)|R \equiv P|(Q|R) \qquad\qquad P|Q \equiv Q|P \qquad\qquad P|\mathbf{0} \equiv P$$
$$(\nu n)(\nu m)P \equiv (\nu m)(\nu n)P \qquad\qquad {*}P \equiv {*}P|P$$
$$(\nu n)P|Q \equiv (\nu n)(P|Q) \qquad (\nu n)P > Q \equiv (\nu n)(P > Q) \qquad \text{if } n\notin \mathrm{fn}(Q)$$
$$[Q\|((\nu n)P)] \equiv (\nu n)[Q\|P] \qquad [((\nu n)P)\|Q] \equiv (\nu n)[P\|Q] \qquad \text{if } n\notin \mathrm{fn}(Q)$$

the scope of a dynamic operator (input and output prefixes, replication, service definitions and invocations, and the right-hand side of a pipeline). In essence, active subterms in a process P are those surrounded by a static context.

Definition 1 (error). $P \to_{\mathrm{ERR}}$ *if and only if whenever* $P \equiv C[[Q\|R]]$, *with* $C[\cdot]$ *static, and* $Q \xrightarrow{\lambda}$, *with* $\lambda ::= (v) \mid (\nu\hat{v})\langle v\rangle$, *then* $[Q\|R] \xrightarrow{\lambda'} \!\!\!\!/\;$, *with* $\lambda' ::= \tau \mid \overline{s}(P')$.

A process guarantees client progress if it is error-free at run-time.

Definition 2 (client progress). *Let be $P \in \mathcal{P}$. We say P guarantees client progress if and only if whenever $P \Rightarrow P'$ then $P' \not\to_{\text{ERR}}$.*

The above definition of error may seem too liberal, as absence of error does not actually guarantee progress of the session if $[Q \| R] \xrightarrow{\overline{s}(P')}$ and service s is not available. In fact, we are interested in processes where such situations do not arise: we call these processes well-formed, and define them formally below. First, we need a notion of s-receptive process, a process where a service definition for the service name s is available under a static context, hence is active.

Definition 3 (s-receptive process). *Let be P an (open) process. P is s-receptive if $s \in \mathrm{fn}(P)$ and $P \equiv C[s.R]$ for some static $C[\cdot]$ not binding s.*

Definition 4 (well-formed process). *Let be $P \in \mathcal{P}$. P is* well-formed *if and only if for each $s \in \mathrm{fn}(P)$ P is s-receptive and whenever $P \equiv C[(\nu s)Q]$ process Q is s-receptive.*

Well-formedness is preserved by reductions.

Lemma 1. *Let P be well-formed. $P \Rightarrow P'$ implies P' is well-formed.*

The following lemma ensures that, in well-formed processes, each active service call can be immediately served, thus substantiating our previous claim that our definition of error is adequate for well-formed processes.

Lemma 2. *Let $P \in \mathcal{P}$ be well-formed. If $P \equiv C[\overline{s}.Q]$, with $C[\cdot]$ static, then either $C[\cdot]{=}C_0[\,C_1[\cdot]\,|\,C_2[s.R]\,]$, $C[\cdot]{=}C_0[\,[C_1[\cdot]\|\|C_2[s.R]]\,]$ or $C[\cdot]{=}C_0[\,[C_1[s.R]\|\|C_2[\cdot]]\,]$, for some static contexts $C_0[\cdot]$, $C_1[\cdot]$ and $C_2[\cdot]$, and for some process R.*

3   Types

In this section we introduce syntax and semantics of types, essentially a fragment of CCS corresponding to BPP processes [7].

The set $\mathcal{T}$ of types is defined by the grammar in Table 4. Recall that b, b′, ... range over base types in $\mathcal{B}t$, including sorts. Notice that, like in [3], we need not nested session types in our system, because in order to check session safety it is sufficient to check local, in-session communications. In what follows we abbreviate with 0 the empty summation type.

The semantics of types is described in terms of a labelled transition relation, $\xrightarrow{\alpha}$, derived from the axioms in Table 5. It is worth noticing that input and output prefixes, ?b and !b, cannot synchronize with each other – we only have interleaving in this fragment of CCS.

The basic requirement for ensuring client progress is *type compliance* between client and service protocols involved in sessions, defined below. In the following, we indicate with $\overline{\alpha}$ the *coaction* of α: $\overline{?b} = !b$ and $\overline{!b} = ?b$. This notation is extended to sets of actions as expected. Moreover, we indicate with $I(\mathsf{S})$ the set of initial actions S can perform: $I(\mathsf{S}) = \{\alpha \mid \exists \mathsf{S}' : \mathsf{S} \xrightarrow{\alpha} \mathsf{S}'\}$. Type compliance is defined co-inductively and guarantees that, given two compliant types S and T, either S is stuck, or there is at least one action from S matched by a coaction from T. Formally:

Table 4. Syntax of types

$$\mathsf{T}, \mathsf{S}, \mathsf{U}, \mathsf{V} ::= \sum_i \alpha_i.\mathsf{T}_i \quad \text{Guarded Summation} \qquad \alpha ::= \;!b \quad \text{Output Prefix}$$
$$\mid \mathsf{T} \mid \mathsf{T} \qquad\quad \text{Interleaving} \qquad\qquad\quad \mid\; ?b \quad \text{Input Prefix}$$
$$\mid *\mathsf{T} \qquad\qquad \text{Replication}$$

Table 5. Labeled transition system of types

$$(\textsc{Sum-t}) \;\; \frac{j \in I}{\displaystyle\sum_{i \in I} \alpha_i.\mathsf{T}_i \xrightarrow{\alpha_j} \mathsf{T}_j} \qquad\qquad (\textsc{Par-t}_l) \;\; \frac{\mathsf{T} \xrightarrow{\alpha} \mathsf{T}'}{\mathsf{T}|\mathsf{S} \xrightarrow{\alpha} \mathsf{T}'|\mathsf{S}}$$

$$(\textsc{Par-t}_r) \;\; \frac{\mathsf{S} \xrightarrow{\alpha} \mathsf{S}'}{\mathsf{T}|\mathsf{S} \xrightarrow{\alpha} \mathsf{T}|\mathsf{S}'} \qquad\qquad (\textsc{Rep-t}) \;\; \frac{\mathsf{T}| *\mathsf{T} \xrightarrow{\alpha} \mathsf{T}'}{*\mathsf{T} \xrightarrow{\alpha} \mathsf{T}'}$$

Definition 5 (type compliance). *Let be* $\mathsf{S}, \mathsf{T} \in \mathcal{T}$. *Type compliance is the largest relation on types such that whenever* S *is compliant with* T, *written* $\mathsf{S} \propto \mathsf{T}$, *it holds that either* $I(\mathsf{S}) = \emptyset$ *or* $K = I(\mathsf{S}) \cap \overline{I(\mathsf{T})} \neq \emptyset$ *and for each* $\alpha \in K$ *and each* S' *and* T' *such that* $\mathsf{S} \xrightarrow{\alpha} \mathsf{S}'$ *and* $\mathsf{T} \xrightarrow{\overline{\alpha}} \mathsf{T}'$, *it holds that* $\mathsf{S}' \propto \mathsf{T}'$.

4 A Type System for Client Progress

In this section we introduce a type system that ensures client progress, that is, ensures that sessions cannot block as long as the client's protocol is willing to do some action.

The type system is along the lines of those in [3,14] and is reported in Table 6. We presuppose a mapping ob from sorts $\{\mathcal{S}, \mathcal{S}', \ldots\}$ to types $\mathcal{T}$, with the intended meaning that if $ob(\mathcal{S}) = \mathsf{T}$ then names of sort $\mathcal{S}$ represent services whose abstract protocol is T. We take $s : \mathsf{T}$ as an abbreviation of $s : \mathcal{S}$ and $ob(\mathcal{S}) = \mathsf{T}$ for some $\mathcal{S}$. A *context* Γ is a finite partial mapping from types to variables. For u a service name, a base value or a variable, we take

$$\Gamma \vdash u : \mathsf{T}$$

to mean either that $u = s : \mathsf{T}$, or $u = v : b$ or $u = x \in \mathrm{dom}(\Gamma)$ and $\Gamma(x) = \mathsf{T}$. Type judgments are of the form $\Gamma \vdash P : [\mathsf{S}]\mathsf{T}$, where Γ is a context, P is a possibly open process with $\mathrm{fv}(P) \subseteq \mathrm{dom}(\Gamma)$ and S and T are types. Informally, S and T represent respectively the *in-session*, or *internal*, and the *external* types of P. The first one describes inputs and outputs P can perform at the current session level, while the second one represents the outputs P can perform at the parent level – which correspond to P's returns. As already discussed in the Introduction, the external type T describes the effects produced outside the enclosing session, that is the effects visible one level up.

Rule (T-DEF) checks that the internal type of the service protocol corresponds to the type expected by the sorting system; moreover, the rule requires the absence of external effects, hence, as discussed earlier, no returns are allowed on the service protocol. Concerning rules (T-CALL) and (T-SESS), it is worth noticing that the premises ensure compliance between client and service internal types. Rule (T-SUM) requires that each summand exposes the same external type: intuitively, sums are resolved as internal

Table 6. Rules of the type system

$$(\text{T-Out}) \quad \frac{\Gamma \vdash P : [\mathsf{S}]\mathsf{T} \quad \Gamma \vdash u : \mathsf{b}}{\Gamma \vdash \langle u \rangle.P : [!\mathsf{b}.\mathsf{S}]\mathsf{T}} \qquad (\text{T-Res}) \quad \frac{\Gamma \vdash P : [\mathsf{S}]\mathsf{T}}{\Gamma \vdash (\nu a)P : [\mathsf{S}]\mathsf{T}}$$

$$(\text{T-Inp}) \quad \frac{\Gamma, x : \mathsf{b} \vdash P : [\mathsf{S}]\mathsf{T}}{\Gamma \vdash (x : \mathsf{b}).P : [?\mathsf{b}.\mathsf{S}]\mathsf{T}} \qquad (\text{T-Par}) \quad \frac{\Gamma \vdash P : [\mathsf{S}_1]\mathsf{T}_1 \quad \Gamma \vdash Q : [\mathsf{S}_2]\mathsf{T}_2}{\Gamma \vdash P|Q : [\mathsf{S}_1|\mathsf{S}_2](\mathsf{T}_1|\mathsf{T}_2)}$$

$$(\text{T-Ret}) \quad \frac{\Gamma \vdash u : \mathsf{b}}{\Gamma \vdash \langle u \rangle^\uparrow : [0]!\mathsf{b}} \qquad (\text{T-Sum}) \quad \frac{\forall i \in I : \ \Gamma \vdash \pi_i.P_i : [\alpha_i.\mathsf{S}_i]\mathsf{T} \quad |I| \neq 1}{\Gamma \vdash \sum_{i \in I} \pi_i.P_i : [\sum_{i \in I} \alpha_i.\mathsf{S}_i]\mathsf{T}}$$

$$(\text{T-Def}) \quad \frac{s : \mathsf{V} \quad \Gamma \vdash P : [\mathsf{V}]0}{\Gamma \vdash s.P : [0]0} \qquad (\text{T-Call}) \quad \frac{\Gamma \vdash u : \mathsf{V} \quad \Gamma \vdash P : [\mathsf{S}]\mathsf{T} \quad \mathsf{S} \propto \mathsf{V}}{\Gamma \vdash \overline{u}.P : [\mathsf{T}]0}$$

$$(\text{T-Rep}) \quad \frac{\Gamma \vdash P : [\mathsf{S}]\mathsf{T}}{\Gamma \vdash *P : [*\mathsf{S}] * \mathsf{T}} \qquad (\text{T-Sess}) \quad \frac{\Gamma \vdash P : [\mathsf{S}]\mathsf{U} \quad \Gamma \vdash Q : [\mathsf{T}]0 \quad \mathsf{S} \propto \mathsf{T}}{\Gamma \vdash [P \| Q] : [\mathsf{U}]0}$$

$$(\text{T-Pipe}) \quad \frac{\Gamma \vdash P : [\mathsf{S}]\mathsf{T} \quad \Gamma \vdash Q : [?\mathsf{b}.\mathsf{U}]\mathsf{V} \quad \mathrm{monomf}(\mathsf{S}, \mathsf{b}) \quad \mathrm{NoSum}(\mathsf{S})}{\Gamma \vdash P > Q : [\mathsf{S} \bowtie \mathsf{U}](\mathsf{T}|\mathsf{S} @ \mathsf{V})}$$

choices from the point of view of an enclosing session, hence which branch is chosen should not matter as for the external effect. Finally, rule (T-Pipe) deserves some explanations. We put some limitations on the types of the pipeline operands. First, the right-hand process Q is a single, input-prefixed process of type $?\mathsf{b}.\mathsf{U}$, ready to receive a value. Second, we make sure, through predicate $\mathrm{NoSum}(\mathsf{S})$, that the left-hand P's type does not contain any summation. Third, we make sure, through predicate $\mathrm{monomf}(\mathsf{S}, \mathsf{b})$, that the type of the left-hand side of a pipeline is "monomorphic", that is, contains only outputs of the given type b. Formal definition of $\mathrm{NoSum}(\mathsf{S})$ and $\mathrm{monomf}(\mathsf{S}, \mathsf{b})$ are obvious and omitted. We will come back to these restrictions in Remark 1.

The auxiliary functions $\bowtie$ and $@$ are used to build respectively the internal and the external type of $P > Q$ starting from the types of P and Q. In essence, both $\mathsf{S} \bowtie \mathsf{U}$ and $\mathsf{S} @ \mathsf{V}$ spawn a new copy of type U and V, respectively, in correspondence of each output prefix in S. The main difference is that in $@$ inputs in S are discarded, while in $\bowtie$ they are preserved. This because S is an internal type, hence its actions cannot be observed from the external viewpoint. Formally, $\mathsf{S} \bowtie \mathsf{U}$ and $\mathsf{S} @ \mathsf{V}$ are inductively defined on the structure of S as follows.

$$
\begin{aligned}
!\mathsf{b}.\mathsf{S} \bowtie \mathsf{U} &= \mathsf{U}|(\mathsf{S} \bowtie \mathsf{U}) & \qquad !\mathsf{b}.\mathsf{S} @ \mathsf{U} &= \mathsf{U}|(\mathsf{S} @ \mathsf{U}) \\
?\mathsf{b}.\mathsf{S} \bowtie \mathsf{U} &= ?\mathsf{b}.(\mathsf{S} \bowtie \mathsf{U}) & \qquad ?\mathsf{b}.\mathsf{S} @ \mathsf{U} &= \mathsf{S} @ \mathsf{U} \\
*\mathsf{S} \bowtie \mathsf{U} &= *(\mathsf{S} \bowtie \mathsf{U}) & \qquad *\mathsf{S} @ \mathsf{U} &= *(\mathsf{S} @ \mathsf{U}) \\
(\mathsf{S}_1|\mathsf{S}_2) \bowtie \mathsf{U} &= (\mathsf{S}_1 \bowtie \mathsf{U})|(\mathsf{S}_2 \bowtie \mathsf{U}) & \qquad (\mathsf{S}_1|\mathsf{S}_2) @ \mathsf{U} &= (\mathsf{S}_1 @ \mathsf{U})|(\mathsf{S}_2 @ \mathsf{U})
\end{aligned}
$$

Note that $\mathrm{NoSum}(\mathsf{S})$ ensures the absence of summations on the internal type S, hence we intentionally omit definitions of $\bowtie$ and $@$ for this case.

Example 1 (pipelines). Consider the process P below, which calls two services *ansa* and *bbc*, supposed to reply by sending a newspage of type news, returns an acknowledgment of type ack, sends the received news by e-mail to address a and outputs an acknowledgment:

$$P \stackrel{\triangle}{=} (\nu callnews)\Big(\big(callnews.(\overline{ansa}.(y : \mathsf{news}).\langle y\rangle^\uparrow \mid \overline{bbc}.(y : \mathsf{news}).\langle y\rangle^\uparrow)$$
$$\mid \overline{callnews}.(w : \mathsf{news}).(z : \mathsf{news}).\langle(w \cdot z)\rangle^\uparrow)$$
$$> (x : (\mathsf{news} \times \mathsf{news})).(\langle \mathsf{ack}\rangle^\uparrow \mid \overline{Email}.\langle(x, a)\rangle.(y : \mathsf{ack}).\langle y\rangle^\uparrow)\Big)$$

where we suppose service $Email$ is defined elsewhere with associated protocol of the expected type ?(news $\times$ news $\times$ eAddr).!ack.

Suppose $callnews$:!news | !news. Then the left hand side of the pipeline is of type [!(news $\times$ news)]0 and the right one of type [?(news $\times$ news).!ack]!ack. Hence, given that !(news $\times$ news) $\bowtie$!ack $=$!ack and !(news $\times$ news) @ !ack $=$!ack, the whole process P has associated type [!ack]!ack.

Remark 1 (summations and pipelines). We discuss here the necessity of banning summations on both side of pipelines. Suppose summations on the left hand side are allowed and consider e.g. the following process

$$P \stackrel{\triangle}{=} \big((x : \mathsf{int}).(\langle x\rangle \mid \langle x\rangle \mid \langle x\rangle^\uparrow) + (y : \mathsf{int}).(\langle y\rangle \mid \langle y\rangle^\uparrow)\big) > (w : \mathsf{int}).\langle w\rangle^\uparrow \ .$$

It is clear that

$$(x : \mathsf{int}).(\langle x\rangle \mid \langle x\rangle \mid \langle x\rangle^\uparrow) + (y : \mathsf{int}).(\langle y\rangle \mid \langle y\rangle^\uparrow) : [?\mathsf{int}.(!\mathsf{int} \mid !\mathsf{int})+?\mathsf{int}.!\mathsf{int}]!\mathsf{int}$$
$$(w : \mathsf{int}).\langle w\rangle^\uparrow : [?\mathsf{int}]!\mathsf{int} \ .$$

And by definition of $\bowtie$ and @, $P : [\mathsf{S}]\mathsf{T}$ with

$$\mathsf{S} \stackrel{\triangle}{=} (?\mathsf{int}.(!\mathsf{int} \mid !\mathsf{int})+?\mathsf{int}.!\mathsf{int}) \bowtie 0 \qquad\qquad = 0$$
$$\mathsf{T} \stackrel{\triangle}{=} !\mathsf{int} \mid \big((?\mathsf{int}.(!\mathsf{int} \mid !\mathsf{int})+?\mathsf{int}.!\mathsf{int}) @!\mathsf{int}\big) = !\mathsf{int} \mid \big((!\mathsf{int} \mid !\mathsf{int})+!\mathsf{int}\big) \ .$$

But T contains a non-guarded summation, hence $\mathsf{T} \notin \mathcal{T}$.

Similarly, suppose that summations at top level are allowed on the right-hand side of pipelines, like in

$$Q \stackrel{\triangle}{=} \langle 1\rangle > \big((x : \mathsf{int}).((z : \mathsf{int}).R_1 \mid (w : \mathsf{int}).R_2) + (y : \mathsf{int}).R_3\big)$$

the internal type associated to Q is

$$!\mathsf{int} \bowtie \big((?\mathsf{int}.\mathsf{T}_{R_1} \mid ?\mathsf{int}.\mathsf{T}_{R_2}) + \mathsf{T}_{R_3}\big) \ = \ (?\mathsf{int}.\mathsf{T}_{R_1} \mid ?\mathsf{int}.\mathsf{T}_{R_2}) + \mathsf{T}_{R_3}$$

which contains a non-guarded summation. In fact, we might type sums with distinct input prefixes (external determinism only). In such a manner, each output performed by the left-hand side must be deterministically associated to one choice on the right one and no summation would arise by $\bowtie$. We have preferred to restrict our attention to pipelines where the right-hand side does not contain summations at top level for the sake of simplicity.

Let us now discuss some important differences with [3], relative to how pipelines and parallel compositions are managed. In typing a pipeline, Bruni and Mezzina require that the left-hand side be a single output if the right-hand side contains more than a single input (or the vice-versa). As discussed, we only require absence of summations on the left-hand side. E.g. in [3] the process $(x).\big(*\langle x\rangle\big) > (y).\bar{s}.\langle y\rangle.(z).\langle z\rangle^{\uparrow}$, which receives a value and uses it to call service s an unbounded number of times, is not well-typed, while it is in our system. Concerning parallel composition, they require that either of the two components has a null type. This means that, e.g. a process invoking two services in parallel, and then return something, like in Example 1, are not well typed in their system. Concerning sessions, in [3] the authors decide to keep the two-sided structure of the original calculus, but ignore all effects on the service side. From the point of view of expressiveness, this is essentially equivalent to using one-sided sessions, like we do.

5 Results

The first step towards proving that well-typed processes guarantee client progress is establishing the usual subject reduction property (Proposition 1). Next, we prove that if a type is not stuck, the associated process is not stuck either (Proposition 2). Finally, type safety (Theorem 1), stating that a well typed process cannot immediately generate an error, is sufficient to conclude.

In the following, we denote by $\Gamma \vdash^n P : [S]T$ a type judgment whose derivation from the rules in Table 6 has depth n. Moreover, we abbreviate $\emptyset \vdash P : [S]T$ as $P : [S]T$. Finally, we say $P \in \mathcal{P}$ is *well-typed* if $P : [S]T$ for some S and T.

Lemma 3 (substitution). *If* $\Gamma, x : b \vdash^n P : [S]T$ *and* $v : b$ *then* $\Gamma \vdash^m P[v/x] : [S]T$, *with* $m \leq n$.

Lemma 4.
1. *Whenever* $S \xrightarrow{?b} S'$ *then* $S \bowtie T \xrightarrow{?b} S' \bowtie T$.
2. *Whenever* $S \xrightarrow{!b} S'$ *then* $S \bowtie T = T | S' \bowtie T$.
3. *Whenever* $S \bowtie T \xrightarrow{?b} V$ *then either* $S \xrightarrow{?b} S'$ *and* $V = S' \bowtie T$ *or* $S \xrightarrow{!b'} S'$, $T \xrightarrow{?b} T'$ *and* $V = T' | S' \bowtie T$.
4. *Whenever* $S \bowtie T \xrightarrow{!b} V$ *then* $S \xrightarrow{!b'} S'$, $T \xrightarrow{!b} T'$ *and* $V = T' | S' \bowtie T$.

Lemma 5.
1. *Whenever* $S \xrightarrow{?b} S'$ *then* $S @ T = S' @ T$.
2. *Whenever* $S \xrightarrow{!b} S'$ *then* $S @ T = T | S' @ T$.
3. *Whenever* $S @ T \xrightarrow{!b} V$ *then* $S \xrightarrow{!b'} S'$, $T \xrightarrow{!b} T'$ *and* $V = T' | S' @ T$.

Proposition 1 (subject reduction). *Suppose* $P : [S]T$. *Then*

1. *whenever* $P \xrightarrow{(v)} P'$, *for some* $v : b$, *then* $S \xrightarrow{?b} S'$ *and* $P' : [S']T$;
2. *whenever* $P \xrightarrow{(v\hat{v})\langle v\rangle} P'$, *for some* $v : b$, *then* $S \xrightarrow{!b} S'$ *and* $P' : [S']T$;
3. *whenever* $P \xrightarrow{\langle v\rangle^{\uparrow}} P'$, *with* $v : b$, *then* $T \xrightarrow{!b} T'$ *and* $P' : [S]T'$;
4. *whenever* $P \xrightarrow{(v\tilde{n})s\langle Q\rangle} P'$ *then* $P' : [S]T$;
5. *whenever* $P \xrightarrow{\bar{s}(Q)} P'$, *with* $s : U$ *and* $Q : [U]0$, *then* $P' : [S]T$;
6. *whenever* $P \xrightarrow{\tau} P'$ *then* $P' : [S]T$.

Proof. The proof is straightforward by induction on the derivation of $P : [\mathsf{S}]\mathsf{T}$ and proceeds by distinguishing the last tying rule applied. The most interesting case is (T-SESS), which we examine below (concerning other cases, note that case (T-INP) relies on Lemma 3 and (T-PIPE) relies on Lemma 3, 4 and 5).

(T-SESS): by $[P\|Q] : [\mathsf{S}]\mathsf{0}$ and the premises of the rule, we get $P : [\mathsf{T}]\mathsf{S}, Q : [\mathsf{U}]\mathsf{0}$ and $\mathsf{T} \propto \mathsf{U}$. We distinguish various cases, depending on the rule applied for deducing $[P\|Q] \xrightarrow{\lambda}$.

   (S-RET): $\lambda = (\nu\hat{v})\langle v\rangle$ and by the premises of the rule and (S-RET), it must be $P \xrightarrow{\langle v\rangle^{\uparrow}} P'$. Suppose $v : \mathsf{b}$. Hence, by applying the inductive hypothesis to P, we get $P' : [\mathsf{T}]\mathsf{S}'$, with $\mathsf{S} \xrightarrow{!\mathsf{b}} \mathsf{S}'$, and $[P'\|Q] : [\mathsf{S}']\mathsf{0}$, by (T-SESS).

   (S-PASS$_l$): by the premises of the rule, we get $P \xrightarrow{\lambda} P'$, and by applying the inductive hypothesis to P, we get $P' : [\mathsf{T}]\mathsf{S}$. Therefore, $[P'\|Q] : [\mathsf{S}]\mathsf{0}$, by (T-SESS).

   (S-COM$_l$): $\lambda = \tau$ and by the premises of the rule, we get $P \xrightarrow{(v)} P'$ and $Q \xrightarrow{\langle v\rangle} Q'$. Suppose that $v : \mathsf{b}$. By applying the inductive hypothesis to both P and Q, we get $\mathsf{T} \xrightarrow{?\mathsf{b}} \mathsf{T}', \mathsf{U} \xrightarrow{!\mathsf{b}} \mathsf{U}', P : [\mathsf{T}']\mathsf{S}$ and $Q : [\mathsf{U}']\mathsf{0}$. Moreover, by definition of $\propto$ it must be $\mathsf{T}' \propto \mathsf{U}'$. Hence, by (T-SESS), $[P'\|Q'] : [\mathsf{S}]\mathsf{0}$.

   (S-SYNC$_l$): $\lambda = \tau$ and by the premises of the rule, we get $P \xrightarrow{(\nu\tilde{n})s\langle R\rangle} P', Q \xrightarrow{\overline{s}(R)} Q', P' : [\mathsf{T}]\mathsf{S}$ and $Q' : [\mathsf{U}]\mathsf{0}$. Therefore, by (T-SESS), we get $[P\|Q] : [\mathsf{S}]\mathsf{0}$.

   (S-PASS$_r$), (S-COM$_r$), (S-SYNC$_r$): the proof proceeds in a similar way.

Proposition 2. *Suppose* $P : [\mathsf{S}]\mathsf{T}$. *Then:*

1. *whenever* $\mathsf{S} \xrightarrow{\alpha}$ *then* $P \xrightarrow{\lambda}$ *with* $\lambda ::= \tau \mid \overline{s}(Q) \mid \lambda'$ *and either* $\lambda' = (v)$, *if* $\alpha =?\mathsf{b}$, *or* $\lambda' = (\nu\hat{v})\langle v\rangle$, *if* $\alpha =!\mathsf{b}$, *for some* $v : \mathsf{b}$;
2. *whenever* $\mathsf{T} \xrightarrow{!\mathsf{b}}$ *then* $P \xrightarrow{\lambda}$ *with* $\lambda ::= \tau \mid \langle v\rangle^{\uparrow} \mid (v') \mid (\nu\hat{v}')\langle v'\rangle \mid \overline{s}(Q)$, *for some* $v : \mathsf{b}$.

Proof. The proof is straightforward by induction on the derivation of $P : [\mathsf{S}]\mathsf{T}$ and proceeds by distinguishing the last typing rule applied. For the first result, the most interesting cases are (T-SESS) and (T-PIPE).

(T-SESS): by $[P\|Q] : [\mathsf{S}]\mathsf{0}$ and the premises of the rule, we get $P : [\mathsf{T}]\mathsf{S}, Q : [\mathsf{U}]\mathsf{0}$ and $\mathsf{S} \propto \mathsf{U}$.

By applying the inductive hypothesis to P, given that it must be $\alpha =!\mathsf{b}$ for some b, we get $P \xrightarrow{\lambda}$ with $\lambda ::= \tau \mid \langle v\rangle^{\uparrow} \mid (v') \mid (\nu\hat{v}')\langle v'\rangle \mid \overline{s}(Q)$, for some $v : \mathsf{b}$.

If $P \xrightarrow{\tau}$, then $[P\|Q] \xrightarrow{\tau}$, by (S-PASS$_l$). Similarly, if $P \xrightarrow{\overline{s}(Q)}$, then $[P\|Q] \xrightarrow{\overline{s}(Q)}$, by (S-PASS$_l$).

If $P \xrightarrow{\langle v\rangle^{\uparrow}}$ then $[P\|Q] \xrightarrow{\langle v\rangle}$ with $v : \mathsf{b}$, by (S-RET).

Otherwise, by Proposition 1 (subject reduction) and $\lambda ::= (v') \mid (\nu\hat{v}')\langle v'\rangle$ for some $v' : \mathsf{b}'$, we get $\mathsf{T} \xrightarrow{\alpha}$, with $\alpha ::=?\mathsf{b}' \mid !\mathsf{b}'$. Hence, by $\propto$, $\mathsf{U} \xrightarrow{\overline{\alpha}}$ and by applying the inductive hypothesis to Q we get either $Q \xrightarrow{\overline{\lambda}}, Q \xrightarrow{\tau}$, or $Q \xrightarrow{\overline{s}(Q)}$. In the first case, either (S-COM_l) or (S-COM_r) can be applied for deducing $[P\|Q] \xrightarrow{\tau}$. In both the second and the third case, rule (S-PASS_r) can be applied for deducing either $[P\|Q] \xrightarrow{\tau}$ or $[P\|Q] \xrightarrow{\bar{s}(Q)}$.

(T-PIPE): by $P > Q : [S \bowtie U]T|S @ V$ and the premises of the rule, we get $P : [S]T$, $Q : [?b'.U]V$, $\text{NoSum}(S)$ and $\text{monomf}(S, b')$.

Suppose $\alpha =?b$. By Lemma 4, $S \bowtie U \xrightarrow{?b}$ implies either $S \xrightarrow{?b}$ or $S \xrightarrow{!b'}$ and $U \xrightarrow{?b}$. In both cases, by applying the inductive hypothesis to P we get $P \xrightarrow{\lambda}$, with $\lambda ::= \tau \mid (v) \mid (\nu\hat{v}')\langle v'\rangle \mid \bar{s}(Q)$, for some $v : b$ and $v' : b'$. Therefore, either $P > Q \xrightarrow{\tau}$, $P > Q \xrightarrow{(v)}$ or $P > Q \xrightarrow{\bar{s}(Q)}$, by (P-PASS) and (P-SYNC).

Suppose $\alpha =!b$. By Lemma 4, $S \xrightarrow{!b'}$, $S \bowtie U = S' \bowtie U|U$ and $U \xrightarrow{!b}$. Hence, again by applying the inductive hypothesis to P, we get either $P > Q \xrightarrow{\tau}$ or $P > Q \xrightarrow{\bar{s}(Q)}$, by either (P-PASS) or (P-SYNC).

Concerning the second result, the most interesting case is (T-PIPE) and proceeds by applying Lemma 5 instead of Lemma 4 as shown for the previous case.

The following theorem is the main result of the paper.

Theorem 1 (type safety). *Suppose P is well typed. Then $P \not\rightarrow_{\text{ERR}}$.*

Proof. Suppose by contradiction that $P \rightarrow_{\text{ERR}}$. This means that $P \equiv C[[P_1\|P_2]]$, $P_1 \xrightarrow{\lambda}$, with $\lambda ::= (v) \mid (\nu\hat{v})\langle v\rangle$ and $[P_1\|P_2] \not\xrightarrow{\lambda'}$, with $\lambda' ::= \tau \mid \bar{s}(R)$.

Given that P is well typed, by induction on $C[\cdot]$ we can prove that $[P_1\|P_2]$ is well-typed too, hence there are suitable S, T and U such that $P_1 : [S]T$, $P_2 : [U]0$ and $S \propto U$.

Now, by $P_1 \xrightarrow{\lambda}$, for some λ, and by Proposition 1 (subject reduction), we deduce that there is a suitable $\alpha ::=?b \mid !b$ such that $S \xrightarrow{\alpha}$. Hence $I(S) \neq \emptyset$. By definition of $\propto$, we get $I(S) \cap \overline{I(U)} \neq \emptyset$. That is, there is at least one α such that $S \xrightarrow{\alpha}$ and $U \xrightarrow{\bar{\alpha}}$.

Suppose $\alpha =?b$ (the case when $\alpha =!b$ is similar). By Proposition 2, we have $P_1 \xrightarrow{\lambda}$ and $P_2 \xrightarrow{\lambda'}$, with $\lambda ::= (v) \mid \tau \mid \bar{s}(Q)$ and $\lambda' ::= (\nu\hat{v})\langle v\rangle \mid \tau \mid \bar{s}(Q')$, for a suitable $v : b$. Now, if either λ or λ' is a τ or a service call, we get a contradiction, because we would get a transition for $[P_1\|P_2]$ violating $P \rightarrow_{\text{ERR}}$. The only possibility we are left with is $\lambda = (v)$ and $\lambda' = (\nu\hat{v})\langle v\rangle$, but this would imply $[P_1\|P_2] \xrightarrow{\tau}$, contradicting again $P \rightarrow_{\text{ERR}}$.

Corollary 1 (client progress). *Suppose P is well typed. Then P guarantees client progress.*

Proof. By Proposition 1 and Theorem 1.

Example 2. Consider the system Sys below, composed by:

- a directory of services D, which upon invocation offers a set of services $\tilde{s}_i$. We suppose that each service definition $s_i.P_i$ is well typed;
- a client C that asks S service to compute the summation of two integers and outputs its value;

– a service S, that, upon invocation: (1) asks D for the name of an available service of type $\mathcal{S}_{sum}$, with $ob(\mathcal{S}_{sum}) = ?int.?int.!int$ – that is a service capable of receiving two integers and computing and outputting their sum; (2) invokes the received service and gets the result of the computation; and, (3) passes this value to its client.

$$D \triangleq (\nu\tilde{s}_i)(dir.\textstyle\sum_i \langle s_i\rangle | \prod_{s_i} s_i.P_i)$$
$$C \triangleq \overline{sum}.\langle 2\rangle.\langle 3\rangle.(w : int)\langle w\rangle^{\uparrow}$$
$$S \triangleq sum.(z : int).(y : int).$$
$$\left(\overline{dir}.((x : \mathcal{S}_{sum}).\langle x\rangle^{\uparrow}) > (y : \mathcal{S}_{sum}).\overline{y}.(\langle z\rangle.\langle y\rangle.(w' : int).\langle w'\rangle^{\uparrow})\right)$$
$$Sys \triangleq C|(\nu\, dir)(S|D) .$$

The whole system is well typed assuming $sum :?int.?int.!int$ and $dir : \sum_i !\mathcal{S}_i$, with $s_i : \mathcal{S}_i$ for each i ($Sys : [!int]0$) and, as expected,

$$Sys \Rightarrow\equiv [\langle 5\rangle^{\uparrow}\||\mathbf{0}] \mid (\nu\, dir)(S|D) .$$

Example 3 (divergence). Let us show a simple example of a process that is well typed but diverges. Let s be a service with associated type $!b$ and let be $Q = \overline{s}.(x : b).\langle x\rangle^{\uparrow}$.

It is easy to see that $(x : b).\langle x\rangle^{\uparrow} : [?b]!b$, (T-RET) and (T-IN), $?b \propto !b$ and $\Gamma \vdash \overline{s}.(x : b).\langle b\rangle^{\uparrow} : [!b]0$, (T-CALL).

Note also that by (SYNC):

$$Q|s.Q \rightarrow [((x : b).\langle x\rangle^{\uparrow})\||Q]|s.Q \rightarrow [((x : b).\langle x\rangle^{\uparrow})\|| ([((x : b).\langle x\rangle^{\uparrow})\||Q])]|s.Q \rightarrow \cdots$$

hence $Q \mid s.Q$ diverges.

More complex types are needed for avoiding such kind of divergences. E.g. types extended with service calls, and an extended type system for ensuring termination and livelock freedom, in the style of [9,13]. We let these extensions as future works.

6   Conclusion

We have presented a type system ensuring client progress for well typed CaSPiS^{-} processes. While capturing an interesting class of services, the system we propose suffers from an important limitation with respect the language in [2]: CaSPiS^{-} does not allow values produced inside a session to be returned to the service. Overcoming this limitation would imply allowing non-null effects in the body P of a service definition $s.P$, at the same time labelling those effects as "potential" – as they are to be exercised only if and when s is invoked.

It would also be important to account in a type-theoretic framework another for another important feature offered by the language in [2]: the possibility of explicitly closing sessions and handling the corresponding compensation actions.

Although the compliance relation we make use of already offers some flexibility on the client side, it would be interesting to extend the type system with subtyping on service protocols. This would imply, in the first place, understanding when two service protocols in CaSPiS can be considered as *conformant*, that is, equivalent from the point of view of any client. To this purpose, a good starting point is represented by the theories of [5,6], which provide notions of conformance for contracts, that is, service protocols.

References

1. Boreale, M., Bruni, R., Caires, L., De Nicola, R., Lanese, I., Loreti, M., Martins, F., Montanari, U., Ravara, A., Sangiorgi, D., Vasconcelos, V.T., Zavattaro, G.: SCC: A Service Centered Calculus. In: Bravetti, M., Núñez, M., Zavattaro, G. (eds.) WS-FM 2006. LNCS, vol. 4184, pp. 38–57. Springer, Heidelberg (2006)
2. Boreale, M., Bruni, R., De Nicola, R., Loreti, M.: Sessions and Pipelines for Structured Service Programming. FMOODS 2008 (to appear, 2008), http://rap.dsi.unifi.it/sensoria/
3. Bruni, R., Mezzina, L.G.: A deadlock free type system for a calculus of services and sessions (submitted, 2007)
4. Carpineti, S., Laneve, C.: A Basic Contract Language for Web Services. In: Sestoft, P. (ed.) ESOP 2006. LNCS, vol. 3924, pp. 197–213. Springer, Vienna (2006)
5. Carpineti, S., Castagna, G., Laneve, C., Padovani, L.: A formal account of contracts for web services. In: Bravetti, M., Núñez, M., Zavattaro, G. (eds.) WS-FM 2006. LNCS, vol. 4184, pp. 148–162. Springer, Vienna (2006)
6. Castagna, G., Gesbert, N., Padovani, L.: A theory of contracts for web services. In: POPL 2008, pp. 261–272. ACM Press, San Francisco (2008)
7. Christensen, S., Hirshfeld, Y., Moller, F.: Bisimulation equivalence is decidable for basic parallel processes. In: Best, E. (ed.) CONCUR 1993. LNCS, vol. 715, pp. 143–157. Springer, Heidelberg (1993)
8. Cook, W.R., Misra, J.: Computation Orchestration: A Basis for Wide-Area Computing. Journal of Software and Systems Modeling (2006), http://www.cs.utexas.edu/~wcook/projects/orc/
9. Deng, Y., Sangiorgi, D.: Ensuring Termination by Typability. Information and Computation 204(7), 1045–1082 (2006)
10. Gay, S., Hole, M.: Subtyping for session types in the pi-calculus. Acta Informatica 42(2), 191–225 (2005)
11. Honda, K., Vasconcelos, V.T., Kubo, M.: Language primitives and type disciplines for structured communication-based programming. In: Hankin, C. (ed.) ESOP 1998. LNCS, vol. 1381, pp. 22–138. Springer, Heidelberg (1998)
12. Kobayashi, N.: A New Type System for deadlock-Free Processes. In: Baier, C., Hermanns, H. (eds.) CONCUR 2006. LNCS, vol. 4137, pp. 233–247. Springer, Heidelberg (2006)
13. Kobayashi, N.: A type system for lock-free processes. Information and Computation 177(2), 122–159 (2002)
14. Lanese, I., Martins, F., Vasconcelos, V.T., Ravara, A.: Disciplining Orchestration and Conversation in Service-Oriented Computing. In: Fifth IEEE International Conference on Software Engineering and Formal Methods (SEFM 2007), pp. 305–314. IEEE Press, London (2007)
15. Milner, R.: The polyadic π-calculus: a tutorial. Logic and Algebra of Specification, pp. 203–246. Springer, Heidelberg (1993)

16. Milner, R., Parrow, J., Walker, D.: A calculus of Mobile Processes, part I and II. Information and Computation 100, 1–40 and 41–78 (1992)
17. SENSORIA: Software Engineering for Service-Oriented Overlay Computers. FET-GC II IST-2005-16004 EU Project, http://www.sensoria-ist.eu/
18. Talpin, J.P., Jouvelot, P.: The Type and Effect Discipline. Information and Computation 111(2), 245–296 (1994)
19. Vieira, H.T., Caires, L., Seco, J.C.: The Conversation Calculus: a Model of Service Oriented Computation. In: Drossopoulou, S. (ed.) ESOP 2008. LNCS, vol. 4960, pp. 269–283. Springer, Heidelberg (2008)

Session and Union Types for Object Oriented Programming[*]

Lorenzo Bettini[1], Sara Capecchi[1], Mariangiola Dezani-Ciancaglini[1],
Elena Giachino[1], and Betti Venneri[2]

[1] Dipartimento di Informatica, Università di Torino
[2] Dipartimento di Sistemi e Informatica, Università di Firenze

Dedicated to Ugo Montanari on the Occasion of his 65th Birthday

Abstract. In network applications it is crucial to have a mechanism to guarantee that communications evolve correctly according to the agreed protocol. Session types offer a method for abstracting and validating structured communication sequences (sessions). In this paper we propose union types for refining and enhancing the flexibility of session types in the context of communication centred and object oriented programming. We demonstrate our ideas through an example and a calculus formalising the main issues of the present approach. The type system garantees that, in well-typed executable programs, after a session has started, the values sent and received will be of the appropriate type, and no process can get stuck forever.

Keywords: Sessions, Object Oriented Programming, Session Types, Union Types.

1 Introduction

Writing safe communication protocols has become a central issue in the theory and practice of concurrent and distributed computing. The actual standards still leave to the programmer much of the responsibility in guaranteeing that the communication will evolve as agreed by all the involved agents.

Session types [26, 27] offer a method for abstracting and validating structured communication sequences (*sessions*). This is achieved by giving types to communication channels, in terms of the types of values sent or received, e.g., the type `?int.!bool` expresses that an integer will be received and then a boolean value will be sent. A session involves channels of dual session type, thus guaranteeing that, after a session has started, the values sent and received will be of the appropriate type. Since the specification of a session is a type, the conformance test of programs with respect to specifications becomes type checking.

The popularity of class-based *object oriented* languages justifies the interest in searching for class definitions which naturally include communication primitives. For

---

[*] This work has been partially supported by MIUR project EOS DUE and by EU Project Software Engineering for Service-Oriented Overlay Computers (SENSORIA, contract IST-3-016004-IP-09).

P. Degano et al. (Eds.): Montanari Festschrift, LNCS 5065, pp. 659–680, 2008.

this reason, an amalgamation of communication centred and object oriented programming has been first proposed in [18], where methods are unified with sessions and choices are based on the classes of exchanged objects.

Union types have been shown useful for enhancing the flexibility of subtyping in various settings [1, 21, 11, 10, 29]. For example a bank can answer **yes** or **not** according to the balance between an account and an item price. If **yes** and **not** are objects of classes **OK** and **NoMoney** respectively, then the class of the object **answer** is naturally the union of the two classes **OK** and **NoMoney**, *i.e.* **OK** ∨ **NoMoney**. Without union types typing **answer** would require a superclass of both **OK** and **NoMoney** to be already defined, and this superclass could include unwanted objects. With union types we can express communications between parties which manipulate heterogeneous objects just by sending and receiving objects which belong to subclasses of one of the classes in the union. In this way the flexibility of object-oriented depth-subtyping is enhanced, by strongly improving the expressiveness of choices based on the classes of sent/received objects.

The aim of the present paper is to discuss and formalise the use of union types for session-centred communications in a core object-oriented calculus. A preliminary version of the basic calculus, without union types, is defined in [18]. In the present paper, the calculus of [18] is formally revised, so that typing and semantics are rather cleaner and simpler. Furthermore, the extension to union types, which is the main novelty of the present proposal, poses specific problems in formulating reduction and typing rules to ensure that communications are safe while flexible.

We first present an example which illustrates the main features of our approach and then we formalise these features through a featherweight representation. We call SAM$^\vee$ (*Sessions Amalgamated with Methods plus union types*) the language of the example and $\mathscr{F}$SAM$^\vee$ the formalising calculus.

SAM$^\vee$ *Overview.* SAM$^\vee$, as the language of [18], is concerned with the amalgamation of the object oriented features with the session part, but it is agnostic w.r.t. to the remaining features of the language, such as whether the language is distributed or concurrent, and the features for synchronisation.

In SAM$^\vee$, sessions and methods are "amalgamated": invocation takes place on an object and the execution takes place immediately and concurrently with the requesting thread (indeed, SAM$^\vee$ is multi-threaded and the communication is asynchronous). Thus, it keeps the method-like invocation mechanism while involving two threads, typical of session based communication mechanisms. The body is determined by the class of the receiving object (avoiding in this way the usual branch/select primitives [27]), and any number of communications interleaved with computation is possible. Sessions are defined in a class, which can have also fields. We believe that the above amalgamated model of session naturally reflects our intuition of services. Furthermore, it can neatly encode "standard" methods.

A thread can make a session request through e.s$\{e'\}$, where e is an expression denoting an object, s is the name of a session defined in the object's class; then, e is evaluated to an object o, and the session body of s in o's class is executed concurrently with e', introducing a new pair of fresh channels k and $\tilde{k}$ (one for each communication direction) to perform communications between the session body and e'^1. Notice that channels are implicit, and are not written by the programmer. At every step, in each thread, there is only one single active channel on which communications are performed.

The expressions $\mathsf{send}(e)$ and $\mathsf{rec}(x)$ send and receive objects on the active channel, respectively. The expression $\mathsf{sendC}(e)\{C_1 \Rightarrow e_1 \, [] \ldots [] \, C_n \Rightarrow e_n\}$ (where C means Case) evaluates e to an object and sends it on the active channel, and then continues with e_i, where C_i is the class that best fits the class of the object sent. The counter part of sendC is the expression $\mathsf{recC}(x)\{C_1 \Rightarrow e_1 \, [] \ldots [] \, C_n \Rightarrow e_n\}$, where the choice is based on the class of the object received. The expression $\mathsf{sendW}(e)\{C_1 \Rightarrow e_1 \, [] \ldots [] \, C_n \Rightarrow e_n\}$ (where W means While) is similar to $\mathsf{sendC}(e)\{C_1 \Rightarrow e_1 \, [] \ldots [] \, C_n \Rightarrow e_n\}$, except that it allows for enclosed cont, which continues the execution at the nearest enclosing sendW. The expression $\mathsf{recW}(x)\{C_1 \Rightarrow e_1 \, [] \ldots [] \, C_n \Rightarrow e_n\}$ has the obvious meaning. Finally, $e \bullet s \{\}$ delegates the current session to the object resulting from the evaluation of e; the body of the session s in the class of that object is executed concurrently, using the current session. At the end, the final value of the body is passed to the current thread.

Related Papers. We describe $\mathscr{F}\mathsf{SAM}^\vee$ following *Featherweight Java* [30], which today has become a standard for class based object calculi.

Session Types have been first introduced to model communication protocols between π-calculus processes [26, 32, 27]. They have been made more expressive by enriching them with correspondence assertions [3], subtyping [24], bounded polymorphism [23] and safer by assuring deadlock-freedom [14]. More recently session types have been extended to multi-party communications [2, 9].

Session types have been developed also for CORBA [33], for functional languages [25, 34], for boxed ambients [22], for the W3C standard description language for Web Services CDL [8, 35, 31, 28], for operating systems [19], and for object oriented programming languages [17, 16, 13, 15, 18, 6]. In [6] generic types are added to a language/-calculus based on the approach of [18]; independently from the different typing extensions, $\mathscr{F}\mathsf{SAM}^\vee$ improves the definition of the calculi of [18] and [6], both in syntax and in operational semantics.

Union types have been proved useful for functional languages [1, 11], for object-oriented languages [29], for languages manipulating semi-structured data [21] and for the π-calculus [10]. We will tell more on the relations between the present paper and [29] at the end of Subsection 6.1.

There are many concurrent object-oriented languages and calculi in the literature; for this topic we refer to the related work section of [20].

Paper Structure. In Section 2 we describe $\mathsf{SAM}^\vee$ in terms of an example. We then proceed by formalising the calculus $\mathscr{F}\mathsf{SAM}^\vee$, its typing and semantics. Section 7 draws some future work directions.

2   An Example

In this section, we describe $\mathsf{SAM}^\vee$ through an example, which expresses a typical collaboration pattern, *c.f.* [35, 7, 8], and which refines the example of [18]. This simple

---

[1] k and $\tilde{\mathsf{k}}$ play for channels a role similar to that of this for objects.

662   L. Bettini et al.

```
1   sessiontype Shopping_ST =
2     !Item.?Money.μα.!{ OK  ⇒ !AccntNr.!Money.?{OK ⇒ ?Date, NoMoney ⇒ ε },
3                       NoDeal  ⇒ ε ,
4                       MakeAnOffer  ⇒ ? { Money ⇒ α, NoDeal ⇒ ε } }
5
6   sessiontype ExaminePrice_ST = ?Money.!Ok∨NoDeal∨MakeAnOffer
7
8   sessiontype Sell_ST =
9     ?Item.!Money.μα.?{ OK  ⇒ ?AccntNr.?Money.!{ OK⇒!Date, NoMoney ⇒ ε },
10                      NoDeal  ⇒ ε,
11                      MakeAnOffer  ⇒ ! { Money ⇒ α, NoDeal ⇒ ε } }
12
13  sessiontype CalDelDate_ST = ?Item.!Date
14
15  sessiontype CalNewPrice_ST = ?Money.!Money∨NoDeal
16
17  sessiontype CreditCheck_ST = ?AccntNr.?Money
```

Fig. 1. Session types for the buyer-seller example

protocol contains essential features which demonstrate the expressiveness of the id-
ioms of $SAM^\vee$. The buyer negotiates a price from a seller, and if and when they have
reached agreement, he sends his bank account number so that it gets verified that he
has enough money. If he has enough money, he receives the delivery date, otherwise the
deal falls through. The seller *delegates* to a bank the part of the session that checks the
money in the account. Such delegation has traditionally been expressed through higher
order sessions; instead, $SAM^\vee$ can delegate the current session through a session call as
in [18,6]. The negotiation allows several rounds: the buyer may either accept the price,
or break the negotiation, or require a better deal by sending different kinds of answers;
in the latter case, the seller might respond by sending a better price, or might break the
negotiation sending a negative answer. Thus, branch selection in control structures is
based on the dynamic class of an object sent.

The session types `Shopping_ST` and `Sell_ST` (see Figure 1) describe the communication pattern between the `Buyer` and the `Seller`.

The session type `Shopping_ST` describes the above protocol from the point of view
of the buyer. The part `!Item.?Money` indicates sending an `Item` followed by receipt
of a `Money`. The recursive type $\mu\alpha.$`! { OK ⇒ ..., NoDeal ⇒ ..., MakeOnOffer ⇒`
`... }` describes the negotiation part, whereby an object is sent, and then, depending on
whether the actual object sent belongs to class `OK`, `NoDeal`, or `MakeAnOffer`, the first,
second or third branch is taken. In the first branch, the account number and the price is
sent; then, either `OK` followed by a `Date`, or a `NoMoney` is received. In the third branch,
a further object is received, and if that object is a `Money`, then the negotiation resumes
on the basis of it, whereas if it is a `NoDeal`, the negotiation ends.

Note that in both `Shopping_ST` and `Sell_ST` the recursion variable is nested inside
multiple choices, so that this behaviour could not have been expressed using regular
expressions as in [15]. The use of recursive types has also other advantages, like that
of allowing iterative expressions with multiple exit points and multiple recursions. In

```
1   class Buyer {
2    AccntNr accnt; Seller seller;
3
4    String∨Date Shopping_ST shopping
5        { Item prodId := ....;
6          seller.sell{
7             send(prodID);
8             Money price := rec;
9             examinePrice{send(price);
10                      OK∨NoDeal∨MakeAnOffer resp := rec};
11          sendW(resp){
12             OK ⇒ { send(accnt); send(price);
13                    recC(x) { OK ⇒ Date delivDate:=rec; []
14                              NoMoney ⇒ new String("no money"); } }[]
15          NoDeal ⇒ new String("refusing proposed price") []
16          MakeAnOffer ⇒ {
17                    recC(x) { Money ⇒ examinePrice{send(x);
18                                          resp := rec};
19                                          cont; []
20                              NoDeal ⇒ new String("offer refused")}
21                          }
22             } //end of session call sell
23           } //end of session shopping
24
25   Object ExaminePrice_ST examinePrice
26       { Money price := rec
27         ... //code for decision
28        send(resp)
29       } // end of session examinePrice
30   }
```

Fig. 2. The class Buyer

Figure 2 we show the implementation of the class Buyer. It has the fields accnt and seller, which will contain the account number and the seller used to buy products.

The class Buyer supports two sessions called shopping and examinePrice. Session shopping has session type Shopping_ST and return type String∨Date. The union type String∨Date describes the possible results of the negotiation: in case of success the session ends returning the date of the delivery of the item; in case of failure it returns a string describing the reason. In the body of this session the desired product is determined and stored in prodId (line 5). Then, a session request is made to the seller to run session sell (line 6). Thus, the seller will run the body of sell in parallel with the remaining part of the session body of shopping, and a connection will be created between the two threads. On this connection, the buyer will send an Item (line 7), receive a Money and store it in price (line 8). Based on its value the buyer will calculate his response calling session examinePrice which returns the answer in resp (lines 9 and 10). Let us notice that resp's type is the union of all the possible answers' type. On line 11, the buyer enters a loop with sendW, where he sends resp, and branches according to its class. If resp is OK, indicating acceptance of the price,

```
1    class Seller {
2       Bank bank;
3
4       String∨Item Sell_ST sell
5          { Item prodID := rec;
6            Money price=...;
7            send(price);
8            recW( x ){
9                OK ⇒ { sendC( bank•check{ } )
10                          {OK ⇒ calDelDate{send(prodID);
11                                           Date date := rec};
12                              send(date); prodID []
13                          NoMoney ⇒ new String("failed bank transaction");
14                          } } []
15           NoDeal ⇒ new String("proposal refused by buyer") []
16           MakeAnOffer ⇒ { calNewPrice{send(price);
17                               Money∨NoDeal resp := rec} ;
18                               sendC(resp){ Money ⇒ cont []
19                                       NoDeal ⇒ new String("refusing
                                               proposed price"); } }
19           }
20        } //end of session sell
21
22     Object CalDelDate_ST calDelDate
23        { Item item := rec
24          ... //code for calculate date
25          send(date)
26        } // end of session calDelDate
27
28     Object CalNewPrice_ST calNewPrice
29        { Money price := rec
30          ... //code for response (if the answer is positive then the field
31             //price is updated with the new price)
32          sendC(resp)
33        } // end of session calNewPrice
34     }
```

Fig. 3. The class `Seller`

then the buyer will send his account number, and price (line 12); and will receive an object which may be OK, in which case he will receive a `Date` and store it in `delivDate` (line 13), or will receive a `NoMoney` (line 14). In this case the reason of failure is stored in the string `failure`. If `resp` is `NoDeal`, indicating that the price is unacceptable, then the session terminates. If the response is `MakeAnOffer`, inviting the `seller` to make a better offer, then the rest depends on the other party's response, indeed Line 17 contains `recC` indicating that a value will be received, and the remaining steps will be determined by its class. If the value received is a `Money` then session `examinePrice` is called, which returns the buyer's reaction in `resp`, and the recursion will continue (line 19). If the value received is `NoDeal`, then the loop will be abandoned.

```
1  class Bank {
2      Ok∨NoMoney CreditCheck_ST check
3         { AccntNr accnt := rec;
4           Money amt := rec;
5           .... // code for check
6           If(response) then new Ok else new NoMoney;
7         }// end of session check
8      }
```

Fig. 4. The class Bank

Notice that in order to get an arbitrary number of repetitions, it is crucial to allow objects of different classes to be sent in the different iterations of while loops.

The session type `Sell_ST` describes the protocol from the point of view of the `seller`, and is "dual" to `Shopping_ST`. We now consider the class `Seller`, from Figure 3. The session body for `sell` starts by receiving the description of an `Item`, calculating and sending its price. Then, in line 8, it enters a `recW` loop, which is the counterpart to the `sendW` loop from `shopping` and performs all the `seller`'s negotiation. The interesting feature shown here is *delegation*, on line 9, whereby, the `bank` is requested to continue the session, using the current connection, and by application of the session body `check`. At the end of the execution of `check` the session will continue according to the bank answer (`OK` or `NoMoney`).

The session type for `check` from class `Bank` in Figure 4 is the receipt of a `AccntNr` and a `Money` followed by sending either `OK`, or a `NoMoney` object. Note that the session body for `check` is *not aware* whether it will be called through a session request, or through delegation. The return type of `check` is the union of the types of the possible answers, *i.e.* `OK∨NoMoney`.

Notice that the sessions `examinePrice`, `calNewDate` and `calNewPrice` are examples of the implementation in $\text{SAM}^\vee$ of methods, since they start by receiving arguments and after elaborating them send a result.

3   Syntax

This section presents the syntax of $\mathscr{F}\text{SAM}^\vee$ (Figure 5), a minimal concurrent and imperative core calculus, based on *Featherweight Java* [30] (abbreviated with FJ). $\mathscr{F}\text{SAM}^\vee$ supports the basic object-oriented features and session request, session delegation, branching sending/receiving and loops. In details, $\mathscr{F}\text{SAM}^\vee$ encompasses the following linguistic features: basic object oriented expressions, session bodies and communication constructs that combine send/receive with branching and loops.

We use grey to indicate expressions that are produced during the reduction process, but do not occur in the source code of a program. We also use the standard convention of denoting with $\overline{\xi}$ a sequence of elements $\xi_1, ..., \xi_n$.

Union types are defined as in [29]: they are built out of class names by the union operator (denoted by $\vee$).

Programs are defined from a collection of classes. The metavariables C and D, possibly with subscripts, range over class names. Each class has a name, a list of *fields* of the

$$
\begin{array}{llll}
\text{(union type)} & \mathsf{T} & ::= & \mathsf{C} \mid \mathsf{T} \vee \mathsf{T} \\
\text{(class)} & \mathsf{L} & ::= & \mathsf{class}\ \mathsf{C} \lhd \mathsf{C}\ \{\ \overline{\mathsf{T}\,\mathsf{f}};\ \overline{\mathsf{S}}\ \} \\
\text{(session)} & \mathsf{S} & ::= & \mathsf{T}\,\mathsf{t}\,\mathsf{s}\,\{\,\mathsf{e}\,\} \\
\text{(expression)} & \mathsf{e} & ::= & \mathsf{x} \mid \mathsf{this} \mid \mathsf{cont} \mid \boxed{\mathsf{o}} \mid \mathsf{e};\mathsf{e} \mid \mathsf{e.f}:=\mathsf{e} \mid \mathsf{e.f} \mid \mathsf{new}\,\mathsf{C}(\overline{\mathsf{e}}) \\
& & \mid & \mathsf{e.s}\,\{\mathsf{e}\} \mid \mathsf{e} \bullet \mathsf{s}\,\{\,\boxed{\mathsf{k}}\,\} \\
& & \mid & \boxed{\mathsf{k.}}\,\mathsf{sendC}(\mathsf{e})\{\mathsf{C} \Rightarrow \mathsf{e} \,[]\, \mathsf{C} \Rightarrow \mathsf{e}\} \\
& & \mid & \boxed{\mathsf{k.}}\,\mathsf{recC}(\mathsf{x})\{\mathsf{C} \Rightarrow \mathsf{e} \,[]\, \mathsf{C} \Rightarrow \mathsf{e}\} \\
& & \mid & \boxed{\mathsf{k.}}\,\mathsf{sendW}(\mathsf{e})\{\mathsf{C} \Rightarrow \mathsf{e} \,[]\, \mathsf{C} \Rightarrow \mathsf{e}\} \\
& & \mid & \boxed{\mathsf{k.}}\,\mathsf{recW}(\mathsf{x})\{\mathsf{C} \Rightarrow \mathsf{e} \,[]\, \mathsf{C} \Rightarrow \mathsf{e}\} \\
\text{(parallel threads)} & P & ::= & \boxed{\mathsf{e} \mid P \parallel P}
\end{array}
$$

Fig. 5. Syntax, where syntax occurring only at runtime appears shaded. Syntax for session types t is in Figure 9.

form $\overline{\mathsf{T}\,\mathsf{f}}$, where f represents the field name and T its type, and a list of *sessions* of the form $\mathsf{T}\,\mathsf{t}\,\mathsf{s}\,\{\,\mathsf{e}\,\}$, where T is the return type, t the session type, s the session name, and e the session body. For the sake of conciseness the symbol $\lhd$ represents class extension, as in [30]. All classes are defined as extensions of the topmost class Object.

Expressions include variables, that are both standard term variables x and the special variables this and cont. The variable this is considered implicitly bound in any session declaration. Instead, sendW and recW are the only binders for cont, that represents the continuation by recursive computation. Let us notice that free occurrences of cont in e are not bound in the expression $\mathsf{sendW}(\mathsf{e})\{\dots\}$: actually no occurrence of cont can appear in e if this expression is typable (see rule SENDW-T in Figure 11).

In a *session request* $\mathsf{e.s}\,\{\mathsf{e}'\}$ we call the expression e' the *co-body* of the request (since it will be evaluated concurrently with the body of requested session).

In the *session delegation* expression, $\mathsf{e} \bullet \mathsf{s}\,\{\mathsf{k}\}$, the channel k is added by the operational semantics in order to keep track of the channel to pass to the delegated session.

Channels are implicit in the source language syntax. At runtime, communication channels k are introduced at each new session request. We denote the dual with $\tilde{\,\cdot\,}$, where $\tilde{\mathsf{k}}$ is again a runtime channel, and where $\tilde{\,\cdot\,}$ is an involution: $\tilde{\tilde{\mathsf{k}}} = \mathsf{k}$. Whenever a thread uses a channel k, the other participant in the communication uses its dual $\tilde{\mathsf{k}}$. The operational semantics associates to k and $\tilde{\mathsf{k}}$ two different queues of messages; when a thread, which uses the channel k, wants to receive a message it will inspect the queue associated to k, while, when it sends a message it will add it to the queue associated to $\tilde{\mathsf{k}}$ (see Section 5).

The body of a *communication expression* is a pair of alternatives $\{\mathsf{C}_1 \Rightarrow \mathsf{e}_1 \,[]\, \mathsf{C}_2 \Rightarrow \mathsf{e}_2\}$, whose choice depends on the class of the object that is sent or received. In particular, in case of sendW and recW the expressions e_i can contain cont, representing recursive computations.

With respect to $\mathsf{SAM}^\vee$ we only have binary choices as bodies of communication expressions in $\mathscr{F}\mathsf{SAM}^\vee$, since the other forms can be encoded with them. First of all, a unary choice $\{\mathsf{C} \Rightarrow \mathsf{e}\}$ can be simply encoded as $\{\mathsf{C} \Rightarrow \mathsf{e} \,[]\, \mathsf{C} \Rightarrow \mathsf{new}\,\mathsf{Object}()\}$, since sending or receiving an object of class C will always choose the first alternative. With unary choices we can encode the two communication constructs used in the example, i.e., send for sending, and rec for receiving, that we omit from $\mathscr{F}\mathsf{SAM}^\vee$: the expression $\mathsf{sendC(e)\{Object} \Rightarrow \mathsf{new\ Object()\}}$ encodes $\mathsf{send(e)}$ and in a similar way, $\mathsf{recC(x)\{Object} \Rightarrow \mathsf{x\}}$ encodes rec.

With binary choices we can also encode n-ary choices for $n > 2$, getting in this way the constructs used in the example of Section 2. The informal idea of such encoding, given the set of classes to be used in the choices, is to take the first (in the left to right order) relative minimum from this set and use it to define the first branch; the second branch will use Object and we iterate this procedure to write the expression of this second branch, until we remain with only two choices. Thus, for instance, consider the choice $\{C_1 \Rightarrow e_1 \,[\!] \, C_2 \Rightarrow e_2 \,[\!] \, C_3 \Rightarrow e_3\}$, where C_1 is not related to the other classes and C_2 is a subclass of C_3; we can encode this choice with the (nested) binary choice: $\{C_1 \Rightarrow e_1 \,[\!]$ $\mathsf{Object} \Rightarrow \{C_2 \Rightarrow e_2 \,[\!] \, C_3 \Rightarrow e_3\}\}$. Notice that this encoding is correct also if $C_2 = C_3$.

The types used for selecting branches in a choice are class names. This simplifies the formal treatment and the proofs, but, again, we can encode choices with arbitrary union types by n-ary choices in a straightforward way; e.g., $\{C_1 \vee C_2 \Rightarrow e \,[\!] \, C_3 \Rightarrow e'\}$ is encoded as $\{C_1 \Rightarrow e \,[\!] \, C_2 \Rightarrow e \,[\!] \, C_3 \Rightarrow e'\}$.

A *runtime expression* is either a user expression (i.e. an expression in Figure 5 without shaded syntax) or an expression containing channels and/or object indentifiers. Furthermore, threads of runtime expressions can occur at runtime (see the operational semantics). Parallel threads are ranged over by P. Fully evaluated objects will be represented by object identifiers denoted by o.

The main novelty of $\mathscr{F}\mathsf{SAM}^\vee$ w.r.t. FJ is that session invocation can involve the creation of concurrent and communicating threads. Other minor differences are: we do not have cast and overriding, which are orthogonal to our approach; we do not have explicit constructors, then in the object instantiation expression $\mathsf{new}\ C(\overline{\mathsf{e}})$, the values $\overline{\mathsf{o}}$ to which $\overline{\mathsf{e}}$ reduce are the initial values of the fields.

Notice that standard methods can be seen as special cases of sessions. In fact, a method declaration can be (informally) encoded as a session with nested recCs (one for each parameter) and with one sendC returning the method body. Similarly, method calls are special cases of session requests: the passing of arguments is encoded as nested sendCs (one for each argument) and the object returned by the method body is retrieved via one recC. This encoding will use unary choices that can be rendered with binary choices as explained above.

4   Auxiliary Functions

As in FJ, a class table CT is a mapping from class names to class declarations with domain $\mathscr{D}(CT)$. Then a program is a pair (CT, e) of a class table (containing all the class definitions of the program) and an expression e (an expression belonging to the source language representing the program's main entry point). The class Object has no fields and its declaration does not appear in CT. As in FJ, from any CT we can read off the subtype relation between classes, as the transitive closure of $\lhd$ clause; moreover this relation is extended in order to relate types built out of union (Figure 6). As usual considering union types modulo the equivalence relation induced by $<$: we get the commutativity and associativity of $\vee$. Therefore each union type can be written

$$T <: T \qquad \frac{T <: T' \quad T' <: T''}{T <: T''} \qquad \frac{\text{class } C \vartriangleleft D \{ \overline{T\,f}; \, \overline{S} \} \in CT}{C <: D}$$

$$T <: T \vee T' \qquad T' <: T \vee T' \qquad \frac{T' <: T \quad T'' <: T}{T' \vee T'' <: T}$$

$$\textit{fields}(\texttt{Object}) = \bullet \qquad \frac{\textit{fields}(D) = \overline{T'\,f'} \quad \text{class } C \vartriangleleft D \{ \overline{T\,f}; \, \overline{S} \}}{\textit{fields}(C) = \overline{T\,f}, \overline{T'\,f'}}$$

$$\frac{\textit{fields}(C) = \overline{T\,f}}{\textit{ftype}_w(f_i, C) = \textit{ftype}_r(f_i, C) = T_i}$$

$$\textit{ftype}_w(f, T_1 \vee T_2) = \begin{cases} \textit{ftype}_w(f, T_1) & \text{if } \textit{ftype}_w(f, T_1) <: \textit{ftype}_w(f, T_2), \\ \textit{ftype}_w(f, T_2) & \text{if } \textit{ftype}_w(f, T_2) <: \textit{ftype}_w(f, T_1), \\ \bot & \text{otherwise.} \end{cases}$$

$$\textit{ftype}_r(f, T_1 \vee T_2) = \textit{ftype}_r(f, T_1) \vee \textit{ftype}_r(f, T_2)$$

$$\frac{\text{class } C \vartriangleleft D \{ \overline{T\,f}; \, \overline{S} \} \quad T\,t\,s\,\{\,e\,\} \in \overline{S}}{\textit{stype}(s, C) = \{t\}} \qquad \frac{\text{class } C \vartriangleleft D \{ \overline{T\,f}; \, \overline{S} \} \quad s \notin \overline{S}}{\textit{stype}(s, C) = \textit{stype}(s, D)}$$

$$\textit{stype}(s, T_1 \vee T_2) = \textit{stype}(s, T_1) \cup \textit{stype}(s, T_2)$$

$$\frac{\text{class } C \vartriangleleft D \{ \overline{T\,f}; \, \overline{S} \} \quad T\,t\,s\,\{\,e\,\} \in \overline{S}}{\textit{rtype}(s, C) = T} \qquad \frac{\text{class } C \vartriangleleft D \{ \overline{T\,f}; \, \overline{S} \} \quad s \notin \overline{S}}{\textit{rtype}(s, C) = \textit{rtype}(s, D)}$$

$$\textit{rtype}(s, T_1 \vee T_2) = \textit{rtype}(s, T_1) \vee \textit{rtype}(s, T_2)$$

$$\frac{\text{class } C \vartriangleleft D \{ \overline{T\,f}; \, \overline{S} \} \quad T\,t\,s\,\{\,e\,\} \in \overline{S}}{\textit{sbody}(s, C) = e} \qquad \frac{\text{class } C \vartriangleleft D \{ \overline{T\,f}; \, \overline{S} \} \quad s \notin \overline{S}}{\textit{sbody}(s, C) = \textit{sbody}(s, D)}$$

Fig. 6. Subtyping and Lookup Functions

as $C_1 \vee \ldots \vee C_n$ for $n \geq 1$: we say that the classes $C_1, \ldots, C_n$ *build* the union type $C_1 \vee \ldots \vee C_n$. A union type $C_1 \vee \ldots \vee C_n$ is *proper* if $n > 1$.

We assume a fixed *CT* that satisfies some usual sanity conditions as in FJ [30]. Thus, in the following, instead of writing $CT(C) = \texttt{class} \ \ldots$ we will simply write `class C ...`.

We define auxiliary functions (see Figure 6) to lookup fields and sessions from *CT*; these functions are used in the typing rules and in the operational semantics. The *fields* lookup function is as in FJ. As for field type lookup we distinguish between the contexts where the field is used for reading (*ftype$_r$*) from those where it is for writing (*ftype$_w$*). The *stype* and *rtype* return a set of session types and the return type of a session, respectively, while *sbody* returns the body of a session. As in FJ these functions may have to inspect the class hierarchy in case the required element is not present in the current class.

Notice that the type lookup functions take a type as argument (not simply a class name) because the receiver expression of a field/session access may be of a proper union type. As for field type lookup, when the field is used in read mode, in case of a proper union type, we simply return the union type of the result of *ftype$_r$* invoked on the argument types (if both retrievals succeed). On the contrary, when a field is updated, due to the contravariance relation, in case of a proper union type we must

$$
e\langle k\rangle = \begin{cases}
e_1\langle k\rangle ; e_2\langle k\rangle & \text{if } e = e_1 ; e_2, \\
e_1\langle k\rangle.f & \text{if } e = e_1.f, \\
e_1\langle k\rangle.f := e_2\langle k\rangle & \text{if } e = e_1.f := e_2, \\
e_1\langle k\rangle.s\{e_2\} & \text{if } e = e_1.s\{e_2\}, \\
e_1\langle k\rangle \bullet s\{k\} & \text{if } e = e_1 \bullet s\{\ \}, \\
k.\mathsf{sendC}(e_0\langle k\rangle)\{\overline{C \Rightarrow e\langle k\rangle}\} & \text{if } e = \mathsf{sendC}(e_0)\{\overline{C \Rightarrow e}\}, \\
k.\mathsf{recC}(x)\{\overline{C \Rightarrow e\langle k\rangle}\} & \text{if } e = \mathsf{recC}(x)\{\overline{C \Rightarrow e}\}, \\
k.\mathsf{sendW}(e)\{\overline{C \Rightarrow e\langle k\rangle}\} & \text{if } e = \mathsf{sendW}(e)\{\overline{C \Rightarrow e}\}, \\
k.\mathsf{recW}(x)\{\overline{C \Rightarrow e\langle k\rangle}\} & \text{if } e = \mathsf{recW}(x)\{\overline{C \Rightarrow e}\}, \\
e & \text{otherwise.}
\end{cases}
$$

$$
e\lfloor e'/\mathsf{cont}\rfloor = \begin{cases}
e_1\lfloor e'/\mathsf{cont}\rfloor ; e_2\lfloor e'/\mathsf{cont}\rfloor & \text{if } e = e_1 ; e_2, \\
e_1\lfloor e'/\mathsf{cont}\rfloor.f & \text{if } e = e_1.f, \\
e_1\lfloor e'/\mathsf{cont}\rfloor.f := e_2\lfloor e'/\mathsf{cont}\rfloor & \text{if } e = e_1.f := e_2, \\
e_1\lfloor e'/\mathsf{cont}\rfloor.s\{e_2\} & \text{if } e = e_1.s\{e_2\}, \\
e_1\lfloor e'/\mathsf{cont}\rfloor \bullet s\{k\} & \text{if } e = e_1 \bullet s\{k\}, \\
k.\mathsf{sendC}(e_0)\{\overline{C \Rightarrow e\lfloor e'/\mathsf{cont}\rfloor}\} & \text{if } e = k.\mathsf{sendC}(e_0)\{\overline{C \Rightarrow e}\}, \\
k.\mathsf{recC}(x)\{\overline{C \Rightarrow e\lfloor e'/\mathsf{cont}\rfloor}\} & \text{if } e = k.\mathsf{recC}(x)\{\overline{C \Rightarrow e}\}, \\
e' & \text{if } e = \mathsf{cont}, \\
e & \text{otherwise.}
\end{cases}
$$

Fig. 7. Channel Addition and Continuation Replacement

return the intersection of the result of *ftype*$_w$ on the arguments; however, in the absence of multiple inheritance, the only possible cases are those listed in Figure 6, thus we can avoid introducing intersection types.

As for the *stype* lookup function, it returns a set of session types; in case it is invoked with a class name as argument, it will return a singleton. The interesting case is when it is invoked with a proper union type: it will return the union of the sets corresponding to the argument types, so that we have all the session types of the classes that build the union type (see how it is used in the typing system, Figures 11 and 13). The *rtype* lookup function behaves in a covariant way since the resulting object cannot be used in writing mode. We notice that *sbody* is only invoked with a class name as type argument, since we invoke sessions on objects only, and all objects have a class name as type.

It is easy to verify that all lookup functions applied to equivalent union types return either equivalent union types or the same sets of session types, whenever they are defined.

5   Operational Semantics

Objects passed in asynchronous communications are stored in a *heap*. A heap h is a finite mapping with domain consisting of objects and channel names. Its syntax is given by:

$$
h ::= [\ ] \mid o \mapsto (C, \overline{f : o}) \mid k \mapsto \overline{o} \mid h :: h
$$

where :: denotes heap concatenation.

During evaluation, any expression $\text{new } C(\bar{o})$ will be replaced by a new object identifier o. The heap will then maps the object identifier o to the pair $(C, \overline{f : o})$ of its class name C and the list of its fields with corresponding objects $\bar{o}$; this mapping is denoted by $o \mapsto (C, \overline{f : o})$.

The form $h[o \mapsto h(o)[f \mapsto o']]$ denotes the update of the field f of the object o with the object o'.

Channel names are mapped to queues of objects: $k \mapsto \bar{o}$. The heap produced by $h[k \mapsto \bar{o}]$ maps the channel k to the queue $\bar{o}$. With some abuse of notation we write $o :: \bar{o}$ and $\bar{o} :: o$ to denote the queue whose first and last element is o, respectively.

Heap membership for object identifiers and channels is checked using standard set notation, by identifying h with its domain, we can also write $o \in h$, and $k \in h$.

The values that can result from normal termination are parallel threads of fully evaluated objects.

In the reduction rules we make use of the special *channel addition* operation $\wr \ldots \wr$, and of the *continuation replacement* operation $\lfloor \ldots / \text{cont} \rfloor$ (their formal definitions are in Figure 7, where $\{C \Rightarrow e\}$ is short for $\{C_1 \Rightarrow e_1 \,[\!]\, C_2 \Rightarrow e_2\}$). We denote by $e \wr k \wr$ the source expression e in which all occurrences of receive, send, and delegation expressions which *are not* within the co-body of a session request are extended, so that they explicitly mention the channel k they will use (remember that channel names are not written by the programmer). Also, we denote by $e \lfloor e' / \text{cont} \rfloor$ the expression e in which all occurrences of cont, that *are not* within the co-body of a session request or within the body of a send/receive loop, are replaced by e', thus preserving the correct nested structure of while expressions. For example $\text{recC}(x)\{C_1 \Rightarrow x \,[\!]\, C_2 \Rightarrow \text{cont}\} \wr k \wr \lfloor e' / \text{cont} \rfloor = k.\text{recC}(x)\{C_1 \Rightarrow x \,[\!]\, C_2 \Rightarrow e'\}$.

The reduction is a relation between pairs of threads and heaps:

$$P, h \longrightarrow P', h'$$

Reduction rules use evaluation contexts (based on runtime syntax) that capture the notion of the "next subexpression to be reduced":

$$\mathscr{E} ::= [-] \mid \mathscr{E}; e \mid \mathscr{E}.f \mid \mathscr{E}.f := e \mid o.f := \mathscr{E} \mid \mathscr{E}.s\{e\} \mid$$
$$\mathscr{E} \bullet s\{k\} \mid k.\text{sendC}(\mathscr{E})\{C_1 \Rightarrow e_1 \,[\!]\, C_2 \Rightarrow e_2\}$$

The explicit mention of the evaluation context is needed in rule SESSREQ-R (Figure 8), in which a new thread is generated in parallel with the evaluation context.

Reduction rules are in Figure 8. Rule PAR-R models the execution of parallel threads. In this rule parallel composition is considered modulo structural equivalence. As usual, we define structural equivalence rules asserting that parallel composition is associative and commutative:

$$P \,||\, P_1 \equiv P_1 \,||\, P \quad P \,||\, (P_1 \,||\, P_2) \equiv (P \,||\, P_1) \,||\, P_2 \quad P \equiv P' \Rightarrow P \,||\, P_1 \equiv P' \,||\, P_1$$

The successive four rules define the execution of standard object-oriented constructions.

Rule SESSREQ-R models the connection between the co-body e of a session request $o.s\{e\}$ and the body e' of the session s, in the class of the object o. This connection is

PAR-R
$$\frac{e,h \longrightarrow P,h'}{e \parallel P_1,h \longrightarrow P \parallel P_1,h'}$$

SEQ-R
$$\mathcal{E}[o;e],h \longrightarrow \mathcal{E}[e],h$$

FLD-R
$$\frac{h(o) = (C,\overline{f:o})}{\mathcal{E}[o.f_i],h \longrightarrow \mathcal{E}[o_i],h}$$

NEWC-R
$$\frac{\mathit{fields}(C) = \overline{T\,f} \qquad o \notin h}{\mathcal{E}[\text{new}\,C(\overline{o})],h \longrightarrow \mathcal{E}[o],h::[o \mapsto (C,\overline{f:o})]}$$

FLDASS-R
$$\mathcal{E}[o.f := o'],h \longrightarrow \mathcal{E}[o'],h[o \mapsto h(o)[f \mapsto o']]$$

SESSREQ-R
$$\frac{h(o) = (C,\_) \qquad \mathit{sbody}(s,C) = e' \qquad k,\tilde{k} \notin h}{\mathcal{E}[o.s\{e\}],h \longrightarrow \mathcal{E}[e\langle k\rangle] \parallel [o/\texttt{this}]e'\langle\tilde{k}\rangle,h[k,\tilde{k} \mapsto ()]}$$

SESSDEL-R
$$\frac{h(o) = (C,\_) \qquad \mathit{sbody}(s,C) = e}{\mathcal{E}[o \bullet s\{k\}],h \longrightarrow \mathcal{E}[[o/\texttt{this}]e\langle k\rangle],h}$$

SENDCASE-R
$$\frac{h(\tilde{k}) = \overline{o} \qquad h(o) = (C,\_) \qquad C \Downarrow \{C_1,C_2\} = C_i}{\mathcal{E}[k.\text{send}C(o)\{C_1 \Rightarrow e_1 \,[\!]\, C_2 \Rightarrow e_2\}],h \longrightarrow \mathcal{E}[e_i],h[\tilde{k} \mapsto \overline{o} :: o]}$$

RECEIVECASE-R
$$\frac{h(k) = o :: \overline{o} \qquad h(o) = (C,\_) \qquad C \Downarrow \{C_1,C_2\} = C_i}{\mathcal{E}[k.\text{rec}C(x)\{C_1 \Rightarrow e_1 \,[\!]\, C_2 \Rightarrow e_2\}],h \longrightarrow \mathcal{E}[[o/x]e_i],h[k \mapsto \overline{o}]}$$

SENDWHILE-R
$$\mathcal{E}[k.\text{send}W(e)\{C_1 \Rightarrow e_1 \,[\!]\, C_2 \Rightarrow e_2\}],h \longrightarrow \mathcal{E}[k.\text{send}C(e)\{C_1 \Rightarrow e'_1 \,[\!]\, C_2 \Rightarrow e'_2\}],h$$

$$\text{where} \quad e'_i = \lfloor e_i/\texttt{cont}\rfloor k.\text{send}W(e)\{C_1 \Rightarrow e_1 \,[\!]\, C_2 \Rightarrow e_2\}/\texttt{cont}$$

RECEIVEWHILE-R
$$\mathcal{E}[k.\text{rec}W(x)\{C_1 \Rightarrow e_1 \,[\!]\, C_2 \Rightarrow e_2\}],h \longrightarrow \mathcal{E}[k.\text{rec}C(x)\{C_1 \Rightarrow e'_1 \,[\!]\, C_2 \Rightarrow e'_2\}],h$$

$$\text{where} \quad e'_i = \lfloor e_i/\texttt{cont}\rfloor k.\text{rec}W(x)\{C_1 \Rightarrow e_1 \,[\!]\, C_2 \Rightarrow e_2\}/\texttt{cont}$$

Fig. 8. Reduction Rules

established through a pair of fresh channels $k,\tilde{k}$. For this purpose the expression $o.s\{e\}$ reduces, in the same context, to its own co-body $e\langle k\rangle$ and in parallel, outside the context, it spawns the body $[o/\texttt{this}]e'\langle\tilde{k}\rangle$ of the called session. The explicit substitution of k in e and of $\tilde{k}$ in e' ensures that the communication is on the fresh dual channels k and $\tilde{k}$. Thus, an object can serve *any number* of session requests. For example,

$$o.s\{\text{send}C(5)\{C_1 \Rightarrow e_1 \,[\!]\, C_2 \Rightarrow e_2\}\};\text{new}\,C(\,) \longrightarrow$$
$$k.\text{send}C(5)\{C_1 \Rightarrow e_1\langle k\rangle \,[\!]\, C_2 \Rightarrow e_2\langle k\rangle\};\text{new}\,C(\,) \parallel$$
$$\tilde{k}.\text{rec}C(x)\{C'_1 \Rightarrow [o/\texttt{this}]e'_1\langle\tilde{k}\rangle \,[\!]\, C'_2 \Rightarrow [o/\texttt{this}]e'_2\langle\tilde{k}\rangle\}$$

if $\text{rec}C(x)\{C'_1 \Rightarrow e'_1 \,[\!]\, C'_2 \Rightarrow e'_2\}$ is the body of session s in the class of the object o. Notice that there is no ambiguity in this rule, since

$$(k.\text{send}C(5)\{C_1 \Rightarrow e_1\langle k\rangle \,[\!]\, C_2 \Rightarrow e_2\langle k\rangle\} \parallel$$
$$\tilde{k}.\text{rec}C(x)\{C'_1 \Rightarrow [o/\texttt{this}]e'_1\langle\tilde{k}\rangle \,[\!]\, C'_2 \Rightarrow [o/\texttt{this}]e'_2\langle\tilde{k}\rangle\});\text{new}\,C(\,)$$

is not a thread according to the syntax of $\mathcal{F}\text{SAM}^\vee$.

Rule SESSDEL-R replaces the session delegation $o \bullet s\{k\}$ by $[o/\texttt{this}]e\langle k\rangle$, where e is the body of the session s, in the class of the object o. This allows a part of the communication to be delegated via the channel k to the object o: this delegation is transparent for the thread using the dual channel $\tilde{k}$. When the delegated job is over, the original thread can resume the communication via the channel k. For example

$$o \bullet s\{k\} \longrightarrow k.\text{recC}(x)\{C_1 \Rightarrow [^o/\text{this}]e_1 \langle k\rangle \mathbin{[\!]} C_2 \Rightarrow [^o/\text{this}]e_2 \langle k\rangle\}$$

if $\text{recC}(x)\{C_1 \Rightarrow e_1 \mathbin{[\!]} C_2 \Rightarrow e_2\}$ is the body of session s in the class of the object o.

The communication rule for sendC, SENDCASE-R, puts the object o, *i.e.* the result of evaluating the expression e, in the queue associated to the dual channel $\tilde{k}$ of the communication channel k. The computation then proceeds with the expression e_i, if $C_1 \neq C_2$ and C_i is the smallest class in $\{C_1, C_2\}$ to which the object o belongs. Otherwise, if $C_1 = C_2$ and o belongs both to C_1 and to C_2, then the computation proceeds with e_1[2]. This is given by the condition $h(o) = (C, \_)$ and by the following definition of $C \Downarrow \{C_1, C_2\} = C_i$, using the subtyping relation (Figure 6):

$$C \Downarrow \{C_1, C_2\} = \begin{cases} C_i & \text{if } C <: C_i \text{ and } C <: C_j \text{ with } i \neq j \text{ imply } C_j \not<: C_i, \\ C_1 & \text{if } C <: C_1 = C_2, \\ \bot & \text{otherwise.} \end{cases}$$

Dually the receive communication rule takes an object o from the queue associated to channel k and returns the expression $[o/x]e_i$, if $h(o) = (C, \_)$ and $C \Downarrow \{C_1, C_2\} = C_i$.

In rules SENDCASE-R and RECEIVECASE-R it is understood that the transition cannot fire if $C \Downarrow \{C_1, C_2\} = \bot$. However we will see that $C \Downarrow \{C_1, C_2\}$ is always defined in well-typed expressions.

Rules SENDWHILE-R and RECEIVEWHILE-R simply realize the repetition using the case communication expressions. Note that $\text{sendW}(\mathscr{E})\{e_1 \Rightarrow C_1 \mathbin{[\!]} e_2 \Rightarrow C_2\}$ is not an evaluation context, since we do not want to reduce the expression which controls the loop before the application of rule SENDWHILE-R, in which the sendW expression is unfolded.

Only communication and delegation expressions containing explicit channels can be reduced. So, for example, $\text{sendC}(o)\{e\}$ and $o \bullet s\{\}$ are stuck; however, as we will see in Subsection 6.1, the latter cannot be typed and the former is not an initial expression (type soundness is only guaranteed for initial expressions).

6 Typing

Session types, ranged over by t, describe the communications that take place during a session. The syntax of session types is in Figure 9, where we use $\dagger$ as a symbol that stands for either ! or ?. By ε we denote the *empty* communication, and the *concatenation* $t_1.t_2$ expresses the communications in t_1 followed by those in t_2. The session type ε is the neutral element of concatenation, so that $\varepsilon.t = t = t.\varepsilon$ for all t.

The types $!\{C_1 \Rightarrow t_1 \mathbin{[\!]} C_2 \Rightarrow t_2\}$ and $?\{C_1 \Rightarrow t_1 \mathbin{[\!]} C_2 \Rightarrow t_2\}$ express the sending and the receiving of an object, respectively: depending on the class of this object the communication will proceed with one of the t_i. In $\mu\alpha.\dagger\{C_1 \Rightarrow t_1 \mathbin{[\!]} C_2 \Rightarrow t_2\}$ the *session type variable* α can occur inside t_i with the usual meaning of representing the whole session type. We consider recursive session types modulo fold/unfold: i.e., $\mu\alpha.t =$

---

[2] In this particular case, there is no other motivation for selecting the smallest index but to avoid introducing non-deterministic choices. From this point of view, alternative solutions could be just as sound: for instance, the selection of the greatest index or linguistic restrictions on the expressions e_i, e.g., the condition $e_1 = e_2$ whenever $C_1 = C_2$.

$$\dagger ::= \; ! \; | \; ? \hspace{6cm} \text{direction}$$
$$t ::= \; \varepsilon \mid \alpha \mid \odot \mid \dagger\{C \Rightarrow t \,[\!]\, C \Rightarrow t\} \mid \mu\alpha.\dagger\{C \Rightarrow t \,[\!]\, C \Rightarrow t\} \mid t.t \hspace{1.5cm} \text{session type}$$

Fig. 9. Syntax of Session Types

$[\mu\alpha.t/\alpha]t$. So we equate $\mu\alpha.\dagger\{C_1 \Rightarrow t_1 \,[\!]\, C_2 \Rightarrow t_2\}$ to $\dagger\{C_1 \Rightarrow t_1 \,[\!]\, C_2 \Rightarrow t_2\}$ when α does not occur in $\dagger\{C_1 \Rightarrow t_1 \,[\!]\, C_2 \Rightarrow t_2\}$.

The type $\odot$ is used only as session type for the command $\mathtt{cont}$: it plays the role of a place holder which will be replaced by a type variable when the while expression is completed (see rules SENDW-T and RECEIVEW-T in Figure 11).

We say that a session type is *closed* if it does not contain occurrences of free session type variables and of $\odot$. Therefore, each closed session type has one of the following shapes:

- ε;
- $\mu\alpha.\dagger\{C_1 \Rightarrow t_1 \,[\!]\, C_2 \Rightarrow t_2\}$ or $\dagger\{C_1 \Rightarrow t_1 \,[\!]\, C_2 \Rightarrow t_2\}$;

or a concatenation of the session types above. For simplicity we will use in definitions unfolded recursive types whenever possible.

6.1   Typing of Channel Free Expressions

In this subsection we define typing for user expressions, in which communication channels are implicit. For technical reasons it is useful to consider also expressions with occurrences of object identifiers, which are not directly expressible in user syntax. We call these expressions *channel free expressions*. The term environments therefore will contain also type assignments to object identifiers. This permits a simpler formulation of the runtime typing rules, as we will see in next subsection.

The typing judgement has the shape

$$\Gamma \vdash e : T \,\mathbf{;}\, t$$

where Γ is a term environment, which maps $\mathtt{this}$, $\mathtt{cont}$, variables and objects to types, and t represents the session type of the (implicit) active channel.

In order to allow (possible) multiple occurrences of a variable or $\mathtt{cont}$ with different types inside an expression, we define the following "update" operation on Γ (z ranges over $\mathtt{this}$, $\mathtt{cont}$, term variables, and object identifiers):

$$\Gamma(z : T)(z') = \begin{cases} T & \text{if } z' = z \\ \Gamma(z') & \text{otherwise.} \end{cases}$$

Thus, the operation $\Gamma(z : T)$ has the effect of adding $z : T$ to Γ, but after deleting a declaration of z from Γ (if there is one). This will avoid checking well-formedness of term environments and does not require an explicit weakening rule to add an assumption on the $\mathtt{cont}$ variable (when typing nested while communication expressions).

To assure a safe communication between two threads we must require their session types to be *dual*, i.e., that each send will correspond to a receive and vice versa. The duality is then the symmetric relation generated by the rules of Figure 10, in which we consider folded recursive types, otherwise the definition would not be well-founded.

$$\varepsilon \bowtie \varepsilon \qquad\qquad \alpha \bowtie \alpha \qquad\qquad \frac{t_1 \bowtie t_1' \quad t_2 \bowtie t_2'}{t_1.t_2 \bowtie t_1'.t_2'}$$

$$\frac{C_1 \vee C_2 <: C_1' \vee C_2' \quad C_i \Downarrow \{C_1',C_2'\} = C_j' \Rightarrow t_i \bowtie t_j' \quad C_l' \Downarrow \{C_1,C_2\} = C_k \Rightarrow t_k \bowtie t_l'}{\mu\alpha.!\{C_1 \Rightarrow t_1 \,[\!]\, C_2 \Rightarrow t_2\} \bowtie \mu\alpha.?\{C'_1 \Rightarrow t'_1 \,[\!]\, C'_2 \Rightarrow t'_2\}}$$

Fig. 10. Duality Relation

The exchanged values must also be of one of the classes expected by the receiver. All possible choices on the basis of the class of the exchanged value must continue with session types which are dual of each other. For this reason we have to perform checks on the type of the exchanged values in both directions:

- for any sent value of type C_i such that $C_i \Downarrow \{C_1',C_2'\} = C_j'$ for some $1 \leq j \leq 2$ we require $t_i \bowtie t_j'$;
- for any received value of type C_l' such that $C_l' \Downarrow \{C_1,C_2\} = C_k$ for some $1 \leq k \leq 2$ we require $t_k \bowtie t_l'$.

For instance, let us consider the session types $!\{\texttt{Shape} \Rightarrow t_1 \,[\!]\, \texttt{String} \Rightarrow t_2\}$ and $?\{\texttt{Triangle} \Rightarrow t_3 \,[\!]\, \texttt{Object} \Rightarrow t_4\}$ where $\texttt{Triangle} <: \texttt{Shape}$. At run time a $\texttt{Triangle}$ can be sent as a $\texttt{Shape}$, thus the types t_1 and t_3 have to be dual. Notice that, thanks to the absence of generics we can be more flexible w.r.t. [6]: the types used in the choices (actually their union) of the send can be subtypes of the ones expected (in the dual receive).

Typing rules for channel free expressions are in Figure 11. For the sake of simplicity in rule NEWC-T we require that the initialisation of an object does not involve communications. Notice that in rule SEQ-T we use session type concatenation to represent that first the communications in e_1 and then those in e_2 are performed.

The rule for session request SESSREQ-T relies on the duality relation (Figure 10) to assure that all the bodies of the session s in the classes which build the union type T and the co-body e' of the request will communicate properly. In typing session delegation (rule SESSDEL-T) we take into account that the whole expression will be replaced by the session body defined in the class of the expression to which the session is delegated (cf. the reduction rule SESSDEL-R, Figure 8). Notice that the condition $stype(s, T)=\{t'\}$ does not imply T be one class, but only that all definitions of s in the classes which build T have the same session types. If a session has session type ε, then it is meaningless to use it in a delegation, while it is sensible to use it in a request. For this reason we require $t' \neq \varepsilon$ in rule SESSDEL-T.

Rules SENDC-T and RECEIVEC-T require all possible alternative expressions to have the same type T, but they can implement different communication sequences t_i. Rule SENDC-T prescribes that the class type of e is the union type of the classes used in the choice. Without union types the typing rule for the same construct in [18] was much more demanding and less clear. The typing rules for the while communication expressions are similar, but they also discharge the assumption on cont and replace the occurrences of $\circledcirc$ in session types by a fresh variable α which will be bound by μ. In rule SENDW-T typing e with session type ε prevents e from containing occurrences

AXIOM-T
$$\Gamma \vdash \mathsf{z} : \Gamma(\mathsf{z}) \,\natural\, \varepsilon \qquad \mathsf{z} \neq \mathsf{cont}$$

CONT-T
$$\Gamma \vdash \mathsf{cont} : \Gamma(\mathsf{cont}) \,\natural\, \circledcirc$$

SUB-T
$$\dfrac{\Gamma \vdash \mathsf{e} : \mathsf{T} \,\natural\, \mathsf{t} \qquad \mathsf{T} <: \mathsf{T}'}{\Gamma \vdash \mathsf{e} : \mathsf{T}' \,\natural\, \mathsf{t}}$$

NEWC-T
$$\dfrac{\mathit{fields}(\mathsf{C}) = \overline{\mathsf{T}\,\mathsf{f}} \qquad \Gamma \vdash \mathsf{e}_i : \mathsf{T}_i \,\natural\, \varepsilon}{\Gamma \vdash \mathsf{new}\,\mathsf{C}(\overline{\mathsf{e}}) : \mathsf{C} \,\natural\, \varepsilon}$$

FLD-T
$$\dfrac{\Gamma \vdash \mathsf{e} : \mathsf{T} \,\natural\, \mathsf{t}}{\Gamma \vdash \mathsf{e}.\mathsf{f} : \mathit{ftype}_r(\mathsf{f},\mathsf{T}) \,\natural\, \mathsf{t}}$$

SEQ-T
$$\dfrac{\Gamma \vdash \mathsf{e} : \mathsf{T} \,\natural\, \mathsf{t} \qquad \Gamma \vdash \mathsf{e}' : \mathsf{T}' \,\natural\, \mathsf{t}'}{\Gamma \vdash \mathsf{e};\mathsf{e}' : \mathsf{T}' \,\natural\, \mathsf{t}.\mathsf{t}'}$$

FLDASS-T
$$\dfrac{\Gamma \vdash \mathsf{e} : \mathsf{T} \,\natural\, \mathsf{t} \qquad \Gamma \vdash \mathsf{e}' : \mathit{ftype}_w(\mathsf{f},\mathsf{T}) \,\natural\, \mathsf{t}'}{\Gamma \vdash \mathsf{e}.\mathsf{f} := \mathsf{e}' : \mathit{ftype}_w(\mathsf{f},\mathsf{T}) \,\natural\, \mathsf{t}.\mathsf{t}'}$$

SESSREQ-T
$$\dfrac{\Gamma \vdash \mathsf{e} : \mathsf{T} \,\natural\, \mathsf{t} \qquad \Gamma \vdash \mathsf{e}' : \mathsf{T}' \,\natural\, \mathsf{t}' \qquad \mathsf{t}' \bowtie \mathsf{t}'' \;\; \forall \mathsf{t}'' \in \mathit{stype}(\mathsf{s},\mathsf{T})}{\Gamma \vdash \mathsf{e}.\mathsf{s}\{\mathsf{e}'\} : \mathsf{T}' \,\natural\, \mathsf{t}}$$

SESSDEL-T
$$\dfrac{\Gamma \vdash \mathsf{e} : \mathsf{T} \,\natural\, \mathsf{t} \qquad \mathit{stype}(\mathsf{s},\mathsf{T}) = \{\mathsf{t}'\} \qquad \mathsf{t}' \neq \varepsilon \qquad \mathit{rtype}(\mathsf{s},\mathsf{T}) = \mathsf{T}'}{\Gamma \vdash \mathsf{e} \bullet \mathsf{s}\{\} : \mathsf{T}' \,\natural\, \mathsf{t}.\mathsf{t}'}$$

SENDC-T
$$\dfrac{\Gamma \vdash \mathsf{e} : \mathsf{C}_1 \vee \mathsf{C}_2 \,\natural\, \mathsf{t} \qquad \Gamma \vdash \mathsf{e}_i : \mathsf{T} \,\natural\, \mathsf{t}_i}{\Gamma \vdash \mathsf{sendC}(\mathsf{e})\{\mathsf{C}_1 \Rightarrow \mathsf{e}_1 \,[\!]\, \mathsf{C}_2 \Rightarrow \mathsf{e}_2\} : \mathsf{T} \,\natural\, \mathsf{t}.!\{\mathsf{C}_1 \Rightarrow \mathsf{t}_1 \,[\!]\, \mathsf{C}_2 \Rightarrow \mathsf{t}_2\}}$$

RECEIVEC-T
$$\dfrac{\Gamma(\mathsf{x} : \mathsf{C}_i) \vdash \mathsf{e}_i : \mathsf{T} \,\natural\, \mathsf{t}_i}{\Gamma \vdash \mathsf{recC}(\mathsf{x})\{\mathsf{C}_1 \Rightarrow \mathsf{e}_1 \,[\!]\, \mathsf{C}_2 \Rightarrow \mathsf{e}_2\} : \mathsf{T} \,\natural\, ?\{\mathsf{C}_1 \Rightarrow \mathsf{t}_1 \,[\!]\, \mathsf{C}_2 \Rightarrow \mathsf{t}_2\}}$$

SENDW-T
$$\dfrac{\Gamma \vdash \mathsf{e} : \mathsf{C}_1 \vee \mathsf{C}_2 \,\natural\, \varepsilon \qquad \Gamma(\mathsf{cont} : \mathsf{T}) \vdash \mathsf{e}_i : \mathsf{T} \,\natural\, \mathsf{t}_i \qquad \alpha \text{ fresh in } \mathsf{t}_1, \mathsf{t}_2}{\Gamma \vdash \mathsf{sendW}(\mathsf{e})\{\mathsf{C}_1 \Rightarrow \mathsf{e}_1 \,[\!]\, \mathsf{C}_2 \Rightarrow \mathsf{e}_2\} : \mathsf{T} \,\natural\, \mu\alpha.!\{\mathsf{C}_1 \Rightarrow [\alpha/\circledcirc]\mathsf{t}_1 \,[\!]\, \mathsf{C}_2 \Rightarrow [\alpha/\circledcirc]\mathsf{t}_2\}}$$

RECEIVEW-T
$$\dfrac{\Gamma(\mathsf{cont} : \mathsf{T})(\mathsf{x} : \mathsf{C}_i) \vdash \mathsf{e}_i : \mathsf{T} \,\natural\, \mathsf{t}_i \qquad \alpha \text{ fresh in } \mathsf{t}_1, \mathsf{t}_2}{\Gamma \vdash \mathsf{recW}(\mathsf{x})\{\mathsf{C}_1 \Rightarrow \mathsf{e}_1 \,[\!]\, \mathsf{C}_2 \Rightarrow \mathsf{e}_2\} : \mathsf{T} \,\natural\, \mu\alpha.?\{\mathsf{C}_1 \Rightarrow [\alpha/\circledcirc]\mathsf{t}_1 \,[\!]\, \mathsf{C}_2 \Rightarrow [\alpha/\circledcirc]\mathsf{t}_2\}}$$

Fig. 11. Typing Rules for Channel Free Expressions

of communications and cont[3]. Notice that requiring that both branch expressions in a choice operation have the same union type T does not imply that we require them to be of the same class: in fact, T can be a proper union type. For instance, we may have that $\Gamma \vdash \mathsf{e}_1 : \mathsf{T}_1 \,\natural\, \mathsf{t}_1$ and $\Gamma \vdash \mathsf{e}_2 : \mathsf{T}_2 \,\natural\, \mathsf{t}_2$; by subsumption (rule SUB-T) we also have that $\Gamma \vdash \mathsf{e}_1 : \mathsf{T}_1 \vee \mathsf{T}_2 \,\natural\, \mathsf{t}_1$ and $\Gamma \vdash \mathsf{e}_2 : \mathsf{T}_1 \vee \mathsf{T}_2 \,\natural\, \mathsf{t}_2$. Then, $\mathsf{T} = \mathsf{T}_1 \vee \mathsf{T}_2$.

Figure 12 defines well-formed class tables. Rule SESS-WF type checks the session bodies with respect to the current class C taking as term environment the association between this and C. Notice that $\circledcirc$ has no dual type, so sessions whose bodies would be typed with types containing $\circledcirc$ would be useless. This justifies the condition that t must be closed in rule SESS-WF.

A last remark is that, since no typing rule generates free session type variables, then all session types in typing judgements are closed unless they contain occurrences of $\circledcirc$.

---

[3] Note that this typing allows e to contain session requests, since the execution of these requests will use different channels to communicate.

$$\frac{\text{S{\small ESS}-WF}}{\{\texttt{this}:\texttt{C}\} \vdash \texttt{e}:\texttt{T}\,\S\,\texttt{t} \qquad \texttt{t is closed}}{\texttt{T t s}\{\texttt{e}\}\ \texttt{ok in C}} \qquad \frac{\text{C{\small LASS}-WF}}{\texttt{D ok} \qquad \overline{\texttt{S}}\ \texttt{ok in C}}{\texttt{class C} \lhd \texttt{D}\,\{\,\overline{\texttt{T f}};\ \overline{\texttt{S}}\,\}\ \texttt{ok}}$$

Fig. 12. Well-formed Class Tables

The rules, presented in this section, define how the type system checks that the declarative part of the program and the main part are well-typed, also with respect to session types used in declarations of sessions. However, when considering well-typed executable programs, we require that they are closed with respect to term variables and that all communication expressions occur inside session co-bodies, that is, that they are typed in the empty type environment with an empty session type. Namely, an *initial* expression $\texttt{e}$ is such that $\emptyset \vdash \texttt{e}:\texttt{T}\,\S\,\varepsilon$ for some $\texttt{T}$. It is easy to verify that the set of initial expressions is the set of closed and well-typed user expression. For example, let us consider the stuck expressions $\texttt{sendC(o)}\{\texttt{e}\}$ and $\texttt{o}\bullet\texttt{s}\{\}$:

- $\texttt{sendC(o)}\{\texttt{e}\}$ is well-typed but its session type is not empty,
- $\texttt{o}\bullet\texttt{s}\{\}$ is not well-typed.

We conclude this subsection by comparing $\mathscr{F}\mathsf{SAM}^\vee$ with FJ $\vee$, an extension of FJ with union types, proposed by Igarashi and Nagira in [29]. They define union types as in the present paper: the essential difference is that they have traditional methods instead of sessions.

The method signatures are of the shape $\overline{\texttt{T}} \to \texttt{T}$, where both the parameter types $\overline{\texttt{T}}$ and the return type $\texttt{T}$ are union types. The method type lookup function applied to a method name $\texttt{m}$ and to a union type $\texttt{T}$ gives a set of method signatures, *i.e.* all the signatures which $\texttt{m}$ has in the classes which build $\texttt{T}$. This is similar to our *stype* function, which returns a set of session types.

The rule of method call checks that the types of the parameters agree with all the signatures found by the method type lookup function for the union type of the object. Also our rule S{\small ESS}R{\small EQ}-T requires the session type of the co-body be dual to all the session types returned by the *stype* function.

It is easy to check that the encoding of methods by sessions sketched at the end of Section 3 extends without changes to methods with union types.

6.2   Typing of Runtime Expressions

During evaluation of well-typed programs, channel names are made explicit in send and receive expressions, as well as in session delegation. Thus, in order to show how well-typedness is preserved under evaluation, we need to define new typing rules for runtime expressions. Furthermore, in typing runtime expressions, we must take into account the session types of more than one channel: runtime expressions contain explicit channel names (used for communication) thus session types must be associated with channel names in an appropriate way. Then judgements have the form

$$\Gamma \vdash_{\texttt{r}} \texttt{e}:\texttt{T}\,\S\,\Sigma$$

where Σ denotes a *session environment* which maps channels to session types.

AXIOM-RT
$$\Gamma \vdash_r z : \Gamma(z) \,;\, \emptyset \qquad z \neq \text{cont}$$

FLD-RT
$$\frac{\Gamma \vdash_r e : T \,;\, \Sigma}{\Gamma \vdash_r e.f : ftype_r(f,T) \,;\, \Sigma}$$

SUB-RT
$$\frac{\Gamma \vdash_r e : T \,;\, \Sigma \qquad T <: T'}{\Gamma \vdash_r e : T' \,;\, \Sigma}$$

NEWC-RT
$$\frac{fields(C) = \overline{T\,f} \qquad \Gamma \vdash_r e_i : T_i \,;\, \emptyset}{\Gamma \vdash_r \text{new}\,C(\overline{e}) : C \,;\, \emptyset}$$

CONT-RT
$$\Gamma(\text{cont} : T) \vdash_r \text{cont} : T \,;\, \{k : \circledcirc\}$$

SEQ-RT
$$\frac{\Gamma \vdash_r e : T \,;\, \Sigma \qquad \Gamma \vdash_r e' : T' \,;\, \Sigma'}{\Gamma \vdash_r e\,;e' : T' \,;\, \Sigma.\Sigma'}$$

FLDASS-RT
$$\frac{\Gamma \vdash_r e : T \,;\, \Sigma \qquad \Gamma \vdash_r e' : ftype_w(f,T) \,;\, \Sigma'}{\Gamma \vdash_r e.f := e' : ftype_w(f,T) \,;\, \Sigma.\Sigma'}$$

SESSREQ-RT
$$\frac{\Gamma \vdash_r e : T \,;\, \Sigma \qquad \Gamma \vdash e' : T' \,;\, t' \qquad t' \bowtie t'' \ \forall t'' \in stype(s,T)}{\Gamma \vdash_r e.s\{e'\} : T' \,;\, \Sigma}$$

SESSDEL-RT
$$\frac{\Gamma \vdash_r e : T \,;\, \Sigma \qquad stype(s,T) = \{t\} \qquad t \neq \varepsilon \qquad rtype(s,T) = T'}{\Gamma \vdash_r e \bullet s\{k\} : T' \,;\, \Sigma.\{k : t\}}$$

SENDC-RT
$$\frac{\Gamma \vdash_r e : C_1 \vee C_2 \,;\, \Sigma \qquad \Gamma \vdash_r e_i : T \,;\, \{k : t_i\}}{\Gamma \vdash_r k.\text{sendC}(e)\{C_1 \Rightarrow e_1 \,[\!]\, C_2 \Rightarrow e_2\} : T \,;\, \Sigma.\{k :!\{C_1 \Rightarrow t_1 \,[\!]\, C_2 \Rightarrow t_2\}\}}$$

RECEIVEC-RT
$$\frac{\Gamma(x : C_i) \vdash_r e_i : T \,;\, \{k : t_i\}}{\Gamma \vdash_r k.\text{recC}(x)\{C_1 \Rightarrow e_1 \,[\!]\, C_2 \Rightarrow e_2\} : T \,;\, \{k :?\{C_1 \Rightarrow t_1 \,[\!]\, C_2 \Rightarrow t_2\}\}}$$

SENDW-RT
$$\frac{\Gamma \vdash_r e : C_1 \vee C_2 \,;\, \emptyset \qquad \Gamma(\text{cont} : T) \vdash_r e_i : T \,;\, \{k : t_i\} \qquad \alpha \text{ fresh in } t_1, t_2}{\Gamma \vdash_r k.\text{sendW}(e)\{C_1 \Rightarrow e_1 \,[\!]\, C_2 \Rightarrow e_2\} : T \,;\, \{k : \mu\alpha.!\{C_1 \Rightarrow [\alpha/\circledcirc]t_1 \,[\!]\, C_2 \Rightarrow [\alpha/\circledcirc]t_2\}\}}$$

RECEIVEW
$$\frac{\Gamma(\text{cont} : T)(x : C_i) \vdash_r e_i : T \,;\, \{k : t_i\} \qquad \alpha \text{ fresh in } t_1, t_2}{\Gamma \vdash_r k.\text{recW}(x)\{C_1 \Rightarrow e_1 \,[\!]\, C_2 \Rightarrow e_2\} : T \,;\, \{k : \mu\alpha.?\{C_1 \Rightarrow [\alpha/\circledcirc]t_1 \,[\!]\, C_2 \Rightarrow [\alpha/\circledcirc]t_2\}\}}$$

Fig. 13. Typing Rules for Runtime Expressions

A session environment maps only a finite set of channels to session types different from ε, and all the remaining to ε. We can then represent one session environment with an infinite number of finite sets which give all the meaningful associations and some of the others. For example $\{k : t\}$ and $\{k : t, k' : \varepsilon\}$ represent the same environment. This choice avoids an explicit weakening rule for session environments. Figure 13 gives the typing rules for runtime expressions, which differ from those for channel free expressions for having session environments instead of a unique session type. For this reason we extend the *concatenation* of session types to *session environments* as follows:

$$\Sigma.\Sigma'(k) = \Sigma(k).\Sigma'(k)$$

Notice that in rule SESSREQ-RT we are making use of the judgement $\Gamma \vdash e' : T \,;\, t'$, where the expression e' does not contain channels, but it can contain object identifiers. This justifies our choice of considering channel free expressions instead of user expressions in the typing rules of previous subsection. Notice also that the session environments of the branches in the communication expressions only contain the current channel as subject, since these expressions will never be reduced before the selection has been done. In rule SENDW-RT we assume $\Gamma \vdash_r e : C_1 \vee C_2 \, \S \, \emptyset$, since the evaluation of e cannot start before the sendW expression has been unfolded to a sendC.

The typing rules for runtime expressions differ from the ones for user expressions only in assigning the session type to explicit channels, not in the union type.

6.3   Type Soundness

Our type system enjoys subject reduction and assures progress ($\longrightarrow^*$ is the reflexive and transitive closure or $\longrightarrow$):

If $\emptyset \vdash e_0 : T_0 \, \S \, \varepsilon$ and $e_0, [\,] \longrightarrow^* P, h$, where $P \equiv e_1 \parallel \ldots \parallel e_n$, then:

- for each e_i we get $\Gamma \vdash_r e_i : T_i \, \S \, \Sigma_i$ for some Γ, T_i, Σ_i $(1 \leq i \leq n)$, and there is j $(1 \leq j \leq n)$ such that $T_j = T_0$, and;
- either $P, h \longrightarrow P', h'$ for some P', h', or for all i $(1 \leq i \leq n)$ e_i is an object identifier.

  The runtime errors which our type system prevents are:
  1. the selection of a field and the request of a session which do not belong to the class of the current object;
  2. the creation of a pair of dual channels whose communication sequences do not perfectly match.

Proofs, more examples and discussions can be found in the extended version of this paper, available at http://www.di.unito.it/~dezani/papers /bcdgvfull.pdf.

7   Conclusion

In the present paper we showed, through the language $\mathsf{SAM}^\vee$, how the addition of union types to an object oriented language with session types enhances flexibility.

The amalgamation of communication centred and object oriented programming, as it has been developed in [18,6] and in the present paper, can be extended in various directions. In particular we plan to integrate this approach with multi-party session communication [9] and with safe failure recovery [4].

Moreover, we want to study the extension of union and intersection to session types, following the intuition given by union/intersection of contracts in [12].

Lastly it would be interesting to integrate session primitives with name constraints as introduced in [5] in order to allow specification of Quality of Service requirements.

Acknowledgements. We thank the referees for their helpful comments. The final version of the paper improved due to their suggestions.

References

1. Barbanera, F., Dezani-Ciancaglini, M., de'Liguoro, U.: Intersection and Union Types: Syntax and Semantics. Information and Computation 119, 202–230 (1995)
2. Bonelli, E., Compagnoni, A.: Multipoint Session Types for a Distributed Calculus. In: Barthe, G., Fournet, C. (eds.) TGC 2007. LNCS, vol. 4912, pp. 240–256. Springer, Heidelberg (to appear, 2008)
3. Bonelli, E., Compagnoni, A., Gunter, E.: Correspondence Assertions for Process Synchronization in Concurrent Communications. Journal of Functional Programming 15(2), 219–248 (2005)

4. Boreale, M., Bruni, R., Caires, L., Nicola, R.D., Lanese, I., Loreti, M., Martins, F., Montanari, U., Ravara, A., Sangiorgi, D., Vasconcelos, V., Zavattaro, G.: SCC: a Service Centered Calculus. In: Bravetti, M., Núñez, M., Zavattaro, G. (eds.) WS-FM 2006. LNCS, vol. 4184, pp. 38–57. Springer, Heidelberg (2006)
5. Buscemi, M., Montanari, U.: CC-Pi: A Constraint-Based Language for Specifying Service Level Agreements. In: De Nicola, R. (ed.) ESOP 2007. LNCS, vol. 4421, pp. 18–32. Springer, Heidelberg (2007)
6. Capecchi, S., Coppo, M., Dezani-Ciancaglini, M., Drossopoulou, S., Giachino, E.: Amalgamating Sessions and Methods in Object Oriented Languages with Generics (submitted, 2007)
7. Carbone, M., Honda, K., Yoshida, N.: A Calculus of Global Interaction Based on Session Types. In: Fernández, M., Kirchner, C. (eds.) SecReT 2006. ENTCS, vol. 171(3), pp. 127–151. Elsevier, Amsterdam (2007)
8. Carbone, M., Honda, K., Yoshida, N.: Structured Communication-Centred Programming for Web Services. In: De Nicola, R. (ed.) ESOP 2007. LNCS, vol. 4421, pp. 2–17. Springer, Heidelberg (2007)
9. Carbone, M., Honda, K., Yoshida, N.: Multiparty Asynchronous Session Types. In: Necula, G.C., Wadler, P. (eds.) POPL 2008, pp. 273–284. ACM Press, New York (2008)
10. Castagna, G., De Nicola, R., Varacca, D.: Semantic Subtyping for the π-calculus. In: Panangaden, P. (ed.) LICS 2005, pp. 92–101. IEEE Computer Society Press, Los Alamitos (2005)
11. Castagna, G., Frisch, A.: A Gentle Introduction to Semantic Subtyping. In: Barahona, P., Felty, A.P. (eds.) PPDP 2005, pp. 198–199. ACM Press, New York (2005)
12. Castagna, G., Gesbert, N., Padovani, L.: A Theory of Contracts for Web Services. In: Necula, G.C., Wadler, P. (eds.) POPL 2008, pp. 261–272. ACM Press, New York (2008)
13. Coppo, M., Dezani-Ciancaglini, M., Yoshida, N.: Asynchronous Session Types and Progress for Object-Oriented Languages. In: Bonsangue, M.M., Johnsen, E.B. (eds.) FMOODS 2007. LNCS, vol. 4468, pp. 1–31. Springer, Heidelberg (2007)
14. Dezani-Ciancaglini, M., de' Liguoro, U., Yoshida, N.: On Progress for Structured Communications. In: Barthe, G., Fournet, C. (eds.) TGC 2007. LNCS, vol. 4912, pp. 257–275. Springer, Heidelberg (2008)
15. Dezani-Ciancaglini, M., Drossopoulou, S., Giachino, E., Yoshida, N.: Bounded Session Types for Object-Oriented Languages. In: de Boer, F., Bonsangue, M. (eds.) FMCO 2006. LNCS, vol. 4709, pp. 207–245. Springer, Heidelberg (2007)
16. Dezani-Ciancaglini, M., Mostrous, D., Yoshida, N., Drossopoulou, S.: Session Types for Object-Oriented Languages. In: Thomas, D. (ed.) ECOOP 2006. LNCS, vol. 4067, pp. 328–352. Springer, Heidelberg (2006)
17. Dezani-Ciancaglini, M., Yoshida, N., Ahern, A., Drossopoulou, S.: A Distributed Object Oriented Language with Session Types. In: De Nicola, R., Sangiorgi, D. (eds.) TGC 2005. LNCS, vol. 3705, pp. 299–318. Springer, Heidelberg (2005)
18. Drossopoulou, S., Dezani-Ciancaglini, M., Coppo, M.: Amalgamating the Session Types and the Object Oriented Programming Paradigms. In: MPOOL 2007 (2007),
   http://homepages.fh-regensburg.de/mpool/mpool07/programme.html
19. Fähndrich, M., Aiken, M., Hawblitzel, C., Hodson, O., Hunt, G.C., Larus, J.R., Levi, S.: Language Support for Fast and Reliable Message-based Communication in Singularity OS. In: Zwaenepoel, W. (ed.) EuroSys 2006. ACM SIGOPS, pp. 177–190. ACM Press, New York (2006)
20. Fournet, C., Laneve, C., Maranget, L., Rémy, D.: Inheritance in the Join Calculus. In: Kapoor, S., Prasad, S. (eds.) FST TCS 2000. LNCS, vol. 1974, pp. 397–408. Springer, Heidelberg (2000)
21. Gapeyev, V., Pierce, B.C.: Regular Object Types. In: Cardelli, L. (ed.) ECOOP 2003. LNCS, vol. 2743, pp. 151–175. Springer, Heidelberg (2003)

22. Garralda, P., Compagnoni, A., Dezani-Ciancaglini, M.: BASS: Boxed Ambients with Safe Sessions. In: Maher, M. (ed.) PPDP 2006, pp. 61–72. ACM Press, New York (2006)
23. Gay, S.: Bounded Polymorphism in Session Types. In: MSCS (to appear, 2008)
24. Gay, S., Hole, M.: Subtyping for Session Types in the Pi-Calculus. Acta Informatica 42(2/3), 191–225 (2005)
25. Gay, S., Vasconcelos, V.T., Ravara, A.: Session Types for Inter-Process Communication. TR 2003–133, Department of Computing, University of Glasgow (2003)
26. Honda, K.: Types for Dyadic Interaction. In: Best, E. (ed.) CONCUR 1993. LNCS, vol. 715, pp. 509–523. Springer, Heidelberg (1993)
27. Honda, K., Vasconcelos, V.T., Kubo, M.: Language Primitives and Type Disciplines for Structured Communication-based Programming. In: Hankin, C. (ed.) ESOP 1998. LNCS, vol. 1381, pp. 22–138. Springer, Heidelberg (1998)
28. Honda, K., Yoshida, N., Carbone, M.: Web Services, Mobile Processes and Types. EATCS Bulletin 91, 160–188 (2007)
29. Igarashi, A., Nagira, H.: Union Types for Object Oriented Programming. Journal of Object Technology 6(2), 31–52 (2007)
30. Igarashi, A., Pierce, B.C., Wadler, P.: Featherweight Java: a Minimal Core Calculus for Java and GJ. ACM TOPLAS 23(3), 396–450 (2001)
31. Sparkes, S.: Conversation with Steve Ross-Talbot. ACM Queue 4(2), 14–23 (2006)
32. Takeuchi, K., Honda, K., Kubo, M.: An Interaction-based Language and its Typing System. In: Halatsis, C., Philokyprou, G., Maritsas, D., Theodoridis, S. (eds.) PARLE 1994. LNCS, vol. 817, pp. 398–413. Springer, Heidelberg (1994)
33. Vallecillo, A., Vasconcelos, V.T., Ravara, A.: Typing the Behavior of Objects and Components using Session Types. In: Brogi, A., Jacquet, J.-M. (eds.) FOCLASA 2002. ENTCS, vol. 68(3), pp. 439–456. Elsevier, Amsterdam (2002)
34. Vasconcelos, V.T., Gay, S., Ravara, A.: Typechecking a Multithreaded Functional Language with Session Types. Theorical Computer Science 368(1-2), 64–87 (2006)
35. Web Services Choreography Working Group. Web Services Choreography Description Language (2002), http://www.w3.org/2002/ws/chor/

The Pairing of Contracts and Session Types

Cosimo Laneve[1] and Luca Padovani[2]

[1] Department of Computer Science, University of Bologna
[2] Information Science and Technology Institute, University of Urbino

Dedicated to Ugo Montanari in occasion of his 65th birthday

Abstract. We pair session types and contracts using two encodings. The encoding of session types accommodates width and depth subtyping, two properties that partially hold in contracts. The encoding of contracts accommodates complex synchronization patterns, since session types own a simple control protocol. The encodings allow one to use the two formalisms interchangeably, within the context of dyadic interactions.

1 Introduction

Service Oriented technologies and Web Services have been recently proposed as a new way of distributing and organizing complex applications across the Internet. The success of these technologies has fostered the development of formal methods for statically analyzing and verifying the behavior of concurrent and distributed systems. Two such methods – session types [14,11,10] and contracts [3,15,4] – aim at describing the communication protocol implemented by services. These methods provide a foundation for statically checking that a process implements a given communication protocol and formally characterize *compliance* (when a client interacts successfully with a service) and *safe replacement* (when it is possible to replace a process with another one).

Session types and contracts find their origins in two different domains: the former ones derive from the domain of type theory and type systems, whereas contracts are more related to the study of behavioral equivalences, such as bisimulation and testing equivalence [7,13]. Both languages are equipped with similar constructors. For example, the session type

$$S \stackrel{\text{def}}{=} \mu x.\&\langle \texttt{Login} : \oplus\langle \texttt{Wrong} : x; \texttt{Ok} : \&\langle \texttt{VoteA} : \texttt{end}; \texttt{VoteB} : \texttt{end}\rangle; \texttt{Cheat} : \texttt{end}\rangle\rangle$$

and the contract

$$\texttt{I}[\sigma] \stackrel{\text{def}}{=} \texttt{I}[\texttt{rec } x.\texttt{Login}.(\overline{\texttt{Wrong}}.x \oplus \overline{\texttt{Ok}}.(\texttt{VoteA} + \texttt{VoteB}) \oplus \overline{\texttt{Cheat}})]$$

where $\texttt{I} = \{\texttt{Login}, \overline{\texttt{Wrong}}, \overline{\texttt{Ok}}, \overline{\texttt{Cheat}}, \texttt{VoteA}, \texttt{VoteB}\}$, represent a simple service for an online ballot between two candidates A and B. Before a client is allowed to vote, he must provide a valid login token that the system uses for ensuring that preferences are expressed at most once, for otherwise the voter is identified as a cheater.

In contracts, the actions such as $\texttt{Login}$ and $\overline{\texttt{Cheat}}$ represent atomic communications between the voter and the service and we have two binary operators for encoding alternatives: a $+$ indicates an *external choice* (the voter chooses the candidate) whereas

P. Degano et al. (Eds.): Montanari Festschrift, LNCS 5065, pp. 681–700, 2008.

a $\oplus$ indicates an *internal choice* (the service decides whether the login is valid or not). In session types we have *branches* $\&\langle\cdots\rangle$ and *choices* $\oplus\langle\cdots\rangle$ that play the same roles played by $+$ and $\oplus$. However, branches indicate the receipt of a label expressing the decision of the voter, whereas choices indicate that the service emits a label expressing the alternative it has chosen. Hence, the direction of the exchanged messages is encoded in the type of alternative, rather than in the labels themselves. Also, while actions in contracts may encode the exchange of data between the voter and the service, labels in session types are only meant to implement the control part of the protocol. In fact, typical presentations of session types admit additional constructs for representing the exchange of data, which we omit in this paper as they are irrelevant for the results that follow.

The relationship between session types and contracts also regards the semantics, although, in this case, their different origin is evident because the relations are mostly the opposite. For example, the session type

$$T \stackrel{\text{def}}{=} \&\langle \text{Login} : \oplus\langle \text{Ok} : \&\langle \text{VoteA} : \text{end}; \text{VoteB} : \text{end}\rangle\rangle\rangle$$

and the contract

$$\text{I}[\tau] \stackrel{\text{def}}{=} \text{I}[\text{Login}.\overline{\text{Ok}}.(\text{VoteA} + \text{VoteB})]$$

both describe a ballot service that does not care about cheaters and always accepts login tokens regardless their validity. It turns out that T is a *subtype* of S (notation $T \leq S$) and $\text{I}[\sigma]$ is a *subcontract* of $\text{I}[\tau]$ (notation $\text{I}[\sigma] \preceq \text{I}[\tau]$). That is, the subtype relation $T \leq S$ embodies the notion of safe substitutability: every term having type S may be replaced by a term having type T without affecting the context in a sensible way. The subcontract relation $\text{I}[\sigma] \preceq \text{I}[\tau]$ embodies the notion of successful interaction (called *compliance*): the set of clients succeeding in interacting with a service of contract $\text{I}[\sigma]$ also succeed with a service of contract $\text{I}[\tau]$.

There are also some differences between the theory of session types and the one of contracts. In one direction, these differences mainly regard the so-called *width subtyping*. Session types, much alike object-oriented type disciplines, enjoy the property $\&\langle \text{VoteA} : \text{end}; \text{VoteB} : \text{end}\rangle \leq \&\langle \text{VoteA} : \text{end}\rangle$ (the ballot service can be extended with more candidates without invalidating former voters), namely width extensions of capabilities is always possible. In contracts this is not the case. To discuss the point, consider the contracts $\text{I}[\sigma_1] = \{\text{VoteA}, \overline{\text{VoteB}}\}[\text{VoteA}]$ and $\text{I}[\sigma_2] = \{\text{VoteA}, \overline{\text{VoteB}}\}[\text{VoteA} + \text{VoteB}]$ and the client $\text{K}[\rho] = \{\overline{\text{VoteA}}, \overline{\text{VoteB}}, e\}[\overline{\text{VoteA}}.e + \overline{\text{VoteB}}.\overline{\text{VoteB}}.e]$. Such client tries to vote for candidate A once or for B twice. It is easy to verify that $\text{K}[\rho]$ successfully interacts with $\text{I}[\sigma_1]$ whilst the interaction may fail with $\text{I}[\sigma_2]$, therefore $\text{I}[\sigma_1] \npreceq \text{I}[\sigma_2]$. In contracts width subtyping is admitted provided the additional capabilities are not present in the interface of the smaller contract. For example $\{\text{VoteA}\}[\text{VoteA}] \preceq \{\text{VoteA},\text{VoteB}\}[\text{VoteA}+\text{VoteB}]$ (the client $\text{K}[\rho]$ is not a valid client for $\{\text{VoteA}\}[\text{VoteA}]$ because it has a larger interface). In the other direction, the differences follow by the fact that session types embody the property that sessions are supposed to be completed *symmetrically* by both parties, whilst contracts are biased towards clients, which are free to interrupt the interaction any time they please. As a consequence, the resulting theory of contracts admits *depth subtyping*, namely the replacement of a service with another one providing a *longer* communication protocol, whereas this is not possible in session types. Another difference is that contracts describe a more abstract synchronization pattern than session types do. For example, in the theory of contracts, a client such as

$$\{\overline{\text{Login}}, \text{Ok}, \overline{\text{VoteA}}, \overline{\text{VoteB}}, e\}[\overline{\text{Login}}.\text{Ok}.(\overline{\text{VoteA}}.e + \overline{\text{VoteB}}.e)]$$

may successfully interact with a service as $\textsc{i}[\tau]$, while, in session types, a branch cannot be matched by another branch. That is, session types describe a communication in which no handshaking between the interacting parties ever occurs. There is always exactly one party having control, and this party has to explicitly notify the other one about the (internal) choices it has made.

In this contribution we undertake a thorough comparison between session types and contracts for assessing a precise relationship between the two formalisms. We define two encodings, one from session types to contracts, and the other from contracts to session types. These encodings allows one to use the two formalisms interchangeably, without losing any relevant information. Therefore it is possible to argue about session types by means of the subcontract relation and, conversely, about contracts by using the deductive system of the subtyping relation. However, because of the differences between the two theories, the two encodings are not one the converse of the other. Let us discuss this issue with few examples. We encode the session type

$$T' \overset{\text{def}}{=} \&\langle \text{Login} : \oplus\langle \text{Ok} : \text{end}; \text{Cheat} : \text{end}\rangle\rangle$$

into the contract

$$\textsc{i}'[\tau'] \overset{\text{def}}{=} [\text{Login}, \overline{\text{Ok}}, \overline{\text{Cheat}}](\text{Login}.(\overline{\text{Ok}} \oplus \overline{\text{Cheat}}) + \overline{\text{Ok}}.\Omega + \overline{\text{Cheat}}.\Omega)$$

where the terms $\overline{\text{Ok}}.\Omega$ and $\overline{\text{Cheat}}.\Omega$ have been added in order to enforce width subtyping in the contract. If a client of $\textsc{i}'[\tau']$ attempts actions that are not explicitly allowed by T' then a catastrophic state – Ω – is reached, meaning that, in practice, such actions are *not* guaranteed.

To illustrate the encoding of contracts into session types we discuss the encoding of $\{\text{VoteA}, \text{VoteB}\}[\text{VoteA} \oplus \text{VoteB}]$, which eventually generates the session type

$$\begin{aligned}
\oplus\langle &\{\text{VoteA}\} : \&\langle \emptyset : \text{end}; \{\text{VoteA}\} : \text{end}\rangle; \\
&\{\text{VoteB}\} : \&\langle \emptyset : \text{end}; \{\text{VoteB}\} : \text{end}\rangle; \\
&\{\text{VoteA}, \text{VoteB}\} : \&\langle \emptyset : \text{end}; \{\text{VoteA}\} : \text{end}; \{\text{VoteB}\} : \text{end}\rangle\rangle
\end{aligned}$$

where we use sets of actions as labels. This type manifests a blow up of the input contract that is needed for compiling the complex synchronization patterns of contracts. Indeed, in the theory of contracts, $\{\text{VoteA}, \text{VoteB}\}[\text{VoteA} \oplus \text{VoteB}]$ is equivalent to $\{\text{VoteA}, \text{VoteB}\}[\text{VoteA} \oplus \text{VoteB} \oplus (\text{VoteA} + \text{VoteB})]$. That is, an internal choice between two alternatives means that one *or possibly both* are available. The encoding of contracts has to model explicitly which alternative is taken by sending a notification to the partner.

Related work. The research on contracts was inspired by "CCS without τ's" [8] and by Hennessy's model of acceptance trees [12,13]. Contracts are an alternative representation of acceptance trees. The relation $\preceq$ was first introduced in [3], albeit it suffered from the lack of a clean semantic characterization and from the fact that it was not transitive. The version of $\preceq$ used in this paper is the same as the one introduced in [15]. In fact, $\preceq$ resembles the must preorder (and it reduces to the must preorder when the interfaces are large enough), but it arises from a notion of compliance that significantly differs from the notion of "passing a test" in the testing framework [7] and that more realistically describes well-behaved clients of Web services. The version of $\preceq$ we work with is actually a stricter version of a more powerful subcontract relation that has been investigated in [4].

Session types have been originally proposed in [14] and subsequently extended for dealing with functional languages [16], asynchrony, object-orientation [10]. In this paper we take [11] as the main reference for session types because it focuses on the subtyping relation. It is worth to notice that we restrict our analysis on the control aspects of session types, whilst other features such as first-class sessions and name passing, which are described in [11], have not been investigated in the framework of contracts yet.

Structure of the paper. We present the formal syntax and semantics of contracts in Section 2 and of session types in Section 3. Sections 4 and 5 present the encoding from session types to contracts and from contracts to session types, respectively. Section 6 concludes by summarizing the main similarities and differences between contracts and session types.

2   Contracts

The syntax of contracts uses an infinite set of *names* $\mathcal{N}$ ranged over by $a, b, c, \ldots$, and a disjoint set of *co-names* $\overline{\mathcal{N}}$ ranged over by $\overline{a}, \overline{b}, \overline{c}, \ldots$. Names and co-names are generically called *actions*. We let $\overline{\overline{a}} = a$ and use $\alpha, \beta, \ldots$ to range over actions; we let $\mathrm{I}, \mathrm{J}, \mathrm{K}, \ldots$ and $\mathrm{R}, \mathrm{S}, \ldots$ to range over (finite) sets of actions and we extend the operation $\overline{\cdot}$ to sets of actions so that $\overline{\mathrm{R}} = \{\overline{\alpha} \mid \alpha \in \mathrm{R}\}$. An infinite set of variables is also used, which is ranged over by $x, y, z, \ldots$.

Contracts are pairs $\mathrm{I}[\sigma]$ where I is a finite subset of $\mathcal{N} \cup \overline{\mathcal{N}}$ representing the *static interface* of the contract (all the actions occurring in σ must also occur in I), whereas σ, called *behavior*, is defined by the grammar:

$$\sigma \quad ::= \quad \mathbf{0} \quad | \quad \alpha.\sigma \quad | \quad \sigma \oplus \sigma \quad | \quad \sigma + \sigma \quad | \quad x \quad | \quad \mathtt{rec}\, x.\sigma$$

Informally, $\mathbf{0}$ describes the inactive behavior; $\alpha.\sigma$ describes the behavior that performs an action α and then behaves like σ; $\sigma \oplus \tau$ describes the behavior that autonomously decides whether to behave as σ or as τ; $\sigma + \tau$ describes the behavior that lets the environment choose whether it should behave as σ or as τ; finally, $\mathtt{rec}\, x.\sigma$ describes a recursive behavior that is equivalent to $\sigma\{\mathtt{rec}\, x.\sigma/x\}$. In the following we write Ω for the behavior $\mathtt{rec}\, x.x$.

Behaviors retain a *transition relation* that is inductively defined by the rules

$$\alpha.\sigma \xrightarrow{\alpha} \sigma \qquad \sigma \oplus \tau \longrightarrow \sigma \qquad \frac{\sigma \xrightarrow{\alpha} \sigma'}{\sigma + \tau \xrightarrow{\alpha} \sigma'} \qquad \frac{\sigma \longrightarrow \sigma'}{\sigma + \tau \longrightarrow \sigma' + \tau}$$

$$\mathtt{rec}\, x.\sigma \longrightarrow \sigma\{\mathtt{rec}\, x.\sigma/x\}$$

plus the symmetric rules for $\oplus$ and $+$. We write $\Longrightarrow$ for the reflexive and transitive closure of $\longrightarrow$; $\sigma \xRightarrow{\alpha} \sigma'$ for $\sigma \Longrightarrow \xrightarrow{\alpha} \Longrightarrow \sigma'$; $\sigma \xRightarrow{\alpha}$ if there exists σ' such that $\sigma \xRightarrow{\alpha} \sigma'$. We write $\sigma\uparrow$ if σ has an infinite internal computation $\sigma = \sigma_0 \longrightarrow \sigma_1 \longrightarrow \sigma_2 \longrightarrow \cdots$ and $\sigma\downarrow$ if not $\sigma\uparrow$. We write $\sigma\downarrow\alpha_1 \cdots \alpha_n$ if $\sigma\downarrow$ and, if $\sigma \xRightarrow{\alpha_1} \sigma'$ implies $\sigma'\downarrow\alpha_2 \cdots \alpha_n$; we write $\sigma\uparrow\varphi$ otherwise. For example $\Omega\uparrow$, $\mathtt{rec}\ x.a + x\uparrow$, and $\mathtt{rec}\ x.(a.x + b.x)\downarrow\varphi$ for every $\varphi \in \{a,b\}^*$. We let $\mathtt{init}(\sigma)$ be $\{\alpha \mid \sigma \xRightarrow{\alpha}\}$.

A basic use of contracts is to verify whether a client protocol is *compliant with* a service protocol. This compliance is possible if, independently of the internal choices of both client and service, the client successfully completes every interaction with the service. We now formalize the notions of "interaction" and of "successful completion":

- The interaction of a client and a service is defined by the relation $\longrightarrow$ over pairs of behaviors as follows:

$$\frac{\rho \longrightarrow \rho'}{\rho \parallel \sigma \longrightarrow \rho' \parallel \sigma} \qquad \frac{\sigma \longrightarrow \sigma'}{\rho \parallel \sigma \longrightarrow \rho \parallel \sigma'} \qquad \frac{\rho \xrightarrow{\alpha} \rho' \quad \sigma \xrightarrow{\bar{\alpha}} \sigma'}{\rho \parallel \sigma \longrightarrow \rho' \parallel \sigma'}$$

where we assume that ρ is a client contract and σ is a service contract. As usual we write $\Longrightarrow$ for the reflexive and transitive closure of $\longrightarrow$.
- The successful completion of the client is modeled using a special name e. The client has successfully completed the interaction with the service if no further synchronization with the service is possible and the client can emit an e action. We assume that behaviors never manifest co-names $\bar{\mathsf{e}}$.

Definition 1 (Compliance). *The (client) contract* $\mathrm{K}[\rho]$ *is compliant with the (service) contract* $\mathrm{I}[\sigma]$, *written* $\mathrm{K}[\rho] \dashv \mathrm{I}[\sigma]$, *if* $\overline{\mathrm{K} \setminus \{\mathsf{e}\}} \subseteq \mathrm{I}$ *and* $\rho \parallel \sigma \Longrightarrow \rho' \parallel \sigma'$ *implies*

1. *if* $\rho' \parallel \sigma' \not\longrightarrow$, *then* $\{\mathsf{e}\} \subseteq \mathtt{init}(\rho')$;
2. *if* $\sigma'\uparrow$, *then* $\{\mathsf{e}\} = \mathtt{init}(\rho')$.

According to the notion of behavioral compliance, if a client $\mathrm{K}[\rho]$ is compliant with a service $\mathrm{I}[\sigma]$ then it should never attempt to perform actions that are not allowed by the interface of the service it is interacting with. If the client-service conversation terminates, then the client is in a successful state (it will emit e). For example, $a.\mathsf{e} + b.\mathsf{e} \dashv \bar{a} \oplus \bar{b}$ and $a.\mathsf{e} \oplus b.\mathsf{e} \dashv \bar{a} + \bar{b}$ but $a.\mathsf{e} \oplus b.\mathsf{e} \not\dashv \bar{a} \oplus \bar{b}$ because of the computation $a.\mathsf{e} \oplus b.\mathsf{e} \parallel \bar{a} \oplus \bar{b} \Longrightarrow a.\mathsf{e} \parallel \bar{b} \not\longrightarrow$ where the client waits for an interaction on a in vain. Similarly, the client must reach a successful state if the conversation does not terminate but the divergence is due to the service. In this case, however, the client cannot rely on any signal from the service, not even an end-of-connection one, so it is required to do nothing but terminate.

Following De Nicola and Hennessy's approach to process semantics [7], this test induces a preorder on services on the basis of the set of clients that comply with a given service.

Definition 2 (Subcontract). *A contract* $\mathrm{I}[\sigma]$ *is a* subcontract *of* $\mathrm{J}[\tau]$, *written* $\mathrm{I}[\sigma] \preceq \mathrm{J}[\tau]$, *if and only if, for every* $\mathrm{K}[\rho]$, *we have* $\mathrm{K}[\rho] \dashv \mathrm{I}[\sigma]$ *implies* $\mathrm{K}[\rho] \dashv \mathrm{J}[\tau]$. *We let* $\mathrm{I}[\sigma] \simeq \mathrm{J}[\tau]$ *if both* $\mathrm{I}[\sigma] \preceq \mathrm{J}[\tau]$ *and* $\mathrm{J}[\tau] \preceq \mathrm{I}[\sigma]$.

That is, if a client is compliant with a service $I[\sigma]$ and $I[\sigma] \preceq J[\tau]$, then the same client is also compliant with $J[\tau]$. Hence, the service $J[\tau]$ can be safely used where $I[\sigma]$ is expected. As usual it is easier to figure out inequalities: $\{a,b\}[a] \not\preceq \{a,b\}[a.b]$ because $\{\overline{a},\overline{b},\mathsf{e}\}[\overline{a}.(\mathsf{e}+\overline{b})] \dashv \{a,b\}[a]$ but $\{\overline{a},\overline{b},\mathsf{e}\}[\overline{a}.(\mathsf{e}+\overline{b})] \not\dashv \{a,b\}[a.b]$; $\{a,b\}[a] \not\preceq \{a,b\}[a+b]$ because $\{\overline{b},\mathsf{e}\}[\mathsf{e}+\overline{b}] \dashv \{a,b\}[a]$ but $\{\overline{b},\mathsf{e}\}[\mathsf{e}+\overline{b}] \not\dashv \{a,b\}[a+b]$.

Since the set of clients compliant with a given service is usually infinite, Definition 2 gives little insight on the properties of $\preceq$. This calls for a direct, coinductive characterization of $\preceq$, which also happens to be easier to work with in the proofs of the results that follow (a further coinductive characterization of the testing preorder appears in [5], but is somehow less direct as it is based on a generalized prebisimulation preorder over a suitably transformed transition system).

Definition 3 (Coinductive subcontract). *Let* $\sigma \Downarrow R$ *(read* σ *has* ready set R*) if and only if* $\sigma \Longrightarrow \sigma'$ *and* $R = \mathtt{init}(\sigma')$. *The relation* $\mathscr{R}$ *is a* coinductive subcontract *if* $I[\sigma] \mathscr{R} J[\tau]$ *implies* $I \subseteq J$ *and whenever* $\sigma\downarrow$ *then*

1. *$\tau\downarrow$, and*
2. *$\tau \Downarrow R$ implies $\sigma \Downarrow R'$ and $R' \subseteq R$, and*
3. *$\alpha \in I$ and $\tau \overset{\alpha}{\Longrightarrow} \tau'$ implies that there exist $\sigma_1,\ldots,\sigma_n$ such that $\sigma \overset{\alpha}{\Longrightarrow} \sigma_i$ for every $1 \le i \le n$ and $I[\bigoplus_{1 \le i \le n} \sigma_i] \mathscr{R} J[\tau']$.*

By this definition, a contract $I[\sigma]$ such that $\sigma\uparrow$ is the smallest one with interface I. When $\sigma\downarrow$, condition 1 constrains the larger contract $J[\tau]$ to converge as well, since clients might rely on the convergence of σ to complete successfully. Condition 2 states that $J[\tau]$ must exhibit a more deterministic behavior: the smaller the number of ready sets is, the more deterministic the contract is. Furthermore, $J[\tau]$ should expose *at least* the same capabilities as the smaller one ($R' \subseteq R$). Condition 3 is perhaps the most subtle one, as it deals with all the possible derivatives of the smaller contract. The point is that $\{a,b,c\}[a.b + a.c] \simeq \{a,b,c\}[a.(b \oplus c)]$ since, after interacting on a, a client of the service on the left side of $\simeq$ is not aware of which state the service is in (it can be either b or c). Hence, we have to consider all of the possible derivatives after a, thus reducing to verifying $\{a,b,c\}[(b+c) \oplus b \oplus c] \mathscr{R} \{a,b,c\}[b \oplus c]$ which trivially holds.

The set-theoretic and the coinductive versions of the subcontract relation do coincide, as the following proposition states (a proof can be found in the full version of [15]).

Proposition 1. *$\preceq$ is the largest coinductive subcontract.*

The theory of $\preceq$ has been thoroughly studied in [15]; in particular it has been put in correspondence with a well-known equivalence – the *must testing* [13]. A useful property relating $\preceq$ and $\longrightarrow$ is in order.

Proposition 2. *If $\sigma \Longrightarrow \sigma'$, then $I[\sigma] \preceq I[\sigma']$.*

The generality of contracts (allowing for arbitrary mixtures of unguarded internal and external choices) makes them hard to compare directly with session types, in which choices and branches are always guarded and syntactically layered. It is thus useful to introduce a normal form for contracts that is more amenable to such a comparison and is also necessary in the encoding presented in Section 5. For every $I[\sigma]$ there exists σ'

in normal form such that $\mathrm{I}[\sigma] \simeq \mathrm{I}[\sigma']$. This result is slightly more general than those in [6,7,13], where normal forms are used to demonstrate the completeness of an axiomatization and are defined for recursion-free terms only. The normal form uses particular families of sets of actions – the *acceptance set*.

Definition 4 (acceptance set [13]). *The* acceptance set *of a behavior σ, denoted by $\mathscr{A}(\sigma)$, is defined as* $\mathscr{A}(\sigma) \stackrel{\text{def}}{=} \{\mathrm{R} \subseteq \texttt{init}(\sigma) \mid \exists \mathrm{S} \subseteq \mathrm{R} : \sigma \Downarrow \mathrm{S}\}$. *We let $\mathscr{A}, \mathscr{A}', \ldots$ range over acceptance sets.*

Definition 5 (normal form). *A behavior σ is in* normal form *if either $\sigma = \Omega$ or $\sigma = x$ or $\sigma = \bigoplus_{\mathrm{R}\in\mathscr{A}(\sigma)} \sum_{\alpha\in\mathrm{R}} \alpha.\sigma_\alpha$ or $\sigma = \texttt{rec}\ x.\bigoplus_{\mathrm{R}\in\mathscr{A}(\sigma)} \sum_{\alpha\in\mathrm{R}} \alpha.\sigma_\alpha$ and each σ_α is in normal form.*

Let $\sigma(\alpha) \stackrel{\text{def}}{=} \bigoplus_{\sigma\Longrightarrow\stackrel{\alpha}{\longrightarrow}\sigma'} \sigma'$ and let $\texttt{nf}(\sigma)$ be a family of behavior names defined as follows

$$\texttt{nf}(\sigma) \stackrel{\text{def}}{=} \begin{cases} \Omega & \text{if } \sigma\uparrow \\ \bigoplus_{\mathrm{R}\in\mathscr{A}(\sigma)} \sum_{\alpha\in\mathrm{R}} \alpha.\texttt{nf}(\sigma(\alpha)) & \text{otherwise} \end{cases}$$

It is not obvious that $\sigma(\alpha)$ is well defined and that behavior names $\texttt{nf}(\sigma)$ may be folded into a finite behavior (by using recursion and variables). In order to prove these facts, we represent behaviors as syntax trees where variables are pointers to the corresponding binder [1]. For example, the figure on the right shows the syntax tree of $\sigma = \texttt{rec}\ x.a.\texttt{rec}\ y.b.(x+y)$.

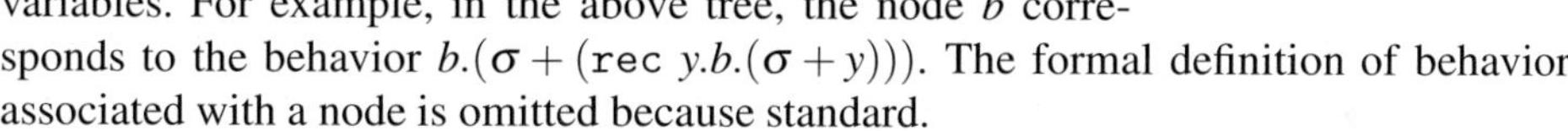

As usual, every node in syntax trees corresponds to a closed term (i.e. a behavior) that is unique up to the name of bound variables. For example, in the above tree, the node b corresponds to the behavior $b.(\sigma + (\texttt{rec}\ y.b.(\sigma + y)))$. The formal definition of behavior associated with a node is omitted because standard.

Lemma 1. *Let $\lfloor\sigma\rfloor \stackrel{\text{def}}{=} \{\tau \mid \exists\alpha : \alpha.\tau \text{ occurs in the syntax tree of } \sigma\}$. If $\sigma \stackrel{\varphi}{\Longrightarrow} \sigma'$, then $\lfloor\sigma'\rfloor \subseteq \lfloor\sigma\rfloor$.*

Proof. We prove that the property holds by induction on the derivation of $\sigma \longrightarrow \sigma'$ or $\sigma \stackrel{\alpha}{\longrightarrow} \sigma'$ (symmetric cases are omitted). The lemma follows by a straightforward induction on the length of the derivation $\sigma \stackrel{\varphi}{\Longrightarrow} \sigma'$.

$(\sigma = \alpha.\sigma' \stackrel{\alpha}{\longrightarrow} \sigma')$ We distinguish two subcases: if σ does occur in σ', then $\lfloor\sigma'\rfloor = \lfloor\sigma\rfloor$; if σ does *not* occur in σ', then $\lfloor\sigma'\rfloor = \lfloor\sigma\rfloor \setminus \{\sigma'\} \subset \lfloor\sigma\rfloor$.

$(\sigma = \sigma' \oplus \tau \longrightarrow \sigma')$ We conclude immediately $\lfloor\sigma'\rfloor \subseteq \lfloor\sigma' \oplus \tau\rfloor = \lfloor\sigma\rfloor$.

$(\sigma = \tau + \tau', \tau \stackrel{\alpha}{\longrightarrow} \sigma')$ By induction hypothesis we have $\lfloor\sigma'\rfloor \subseteq \lfloor\tau\rfloor$ hence $\lfloor\sigma'\rfloor \subseteq \lfloor\tau\rfloor \subseteq \lfloor\tau + \tau'\rfloor = \lfloor\sigma\rfloor$.

$(\sigma = \tau + \tau'', \tau \longrightarrow \tau', \sigma' = \tau' + \tau'')$ By induction hypothesis we have $\lfloor\tau'\rfloor \subseteq \lfloor\tau\rfloor$, hence $\lfloor\sigma'\rfloor = \lfloor\tau' + \tau''\rfloor \subseteq \lfloor\tau + \tau''\rfloor = \lfloor\sigma\rfloor$.

$(\sigma = \texttt{rec}\ x.\tau, \sigma \longrightarrow \tau\{\sigma/x\} = \sigma')$ We conclude immediately $\lfloor\sigma'\rfloor = \lfloor\sigma\rfloor$. $\qquad\square$

It is worth to notice that Lemma 1 does not hold if behaviors are extended with a parallel operator "$|$". For example if $\sigma = \texttt{rec}\ x.a \mid (b.x + 0)$, then $\sigma(b) = a \mid \sigma$, which is not a subtree in the syntax tree of σ.

Lemma 2. *Let $D(\sigma) \stackrel{\text{def}}{=} \{\sigma(\varphi) \mid \sigma \stackrel{\varphi}{\Longrightarrow}\}$. Then $D(\sigma)$ is finite.*

Proof. Let $D_0, D_1, \ldots$ be the family of sets defined as follows:

$$D_0 \stackrel{\text{def}}{=} \{\sigma\} \qquad D_{i+1} \stackrel{\text{def}}{=} D_i \cup \{\sigma' \mid \exists \sigma \in D_i : \exists \alpha : \sigma \stackrel{\alpha}{\Longrightarrow} \sigma'\}$$

We only need to show that there exists n such that $D_n = D_{n+1}$, because each $\sigma(\varphi)$ is obtained by joining some continuations σ', for some subtrees $\alpha.\sigma'$ in σ. Let n be the cardinality of $\lfloor \sigma \rfloor$ (n is the number of *distinct* subtrees in the syntax tree of σ and it exists because σ is finite). By contradiction, assume that there exists $\sigma_{n+1} \in D_{n+1} \setminus D_n$. Then there exist $\sigma_1, \ldots, \sigma_n$ and $\alpha_1, \ldots, \alpha_n, \alpha_{n+1}$ such that

$$\sigma \stackrel{\alpha_1}{\Longrightarrow} \sigma_1 \stackrel{\alpha_2}{\Longrightarrow} \cdots \stackrel{\alpha_n}{\Longrightarrow} \sigma_n \stackrel{\alpha_{n+1}}{\Longrightarrow} \sigma_{n+1}$$

and $\sigma_i \in D_i \setminus D_{i-1}$ for every $1 \leq i \leq n$. We have $\sigma_{n+1} \in \lfloor \sigma_n \rfloor$. By Lemma 1 we deduce $\sigma_{n+1} \in \lfloor \sigma \rfloor$. Since $\lfloor \sigma \rfloor$ has only n distinct subtrees, σ_{n+1} is reachable from σ with at most n reductions, thus $\sigma_{n+1} \in D_n$, which contradicts $\sigma_{n+1} \in D_{n+1} \setminus D_n$. $\qquad \square$

Theorem 1. *For every $\mathrm{I}[\sigma]$ there exists σ' in normal form such that $\mathrm{I}[\sigma] \simeq \mathrm{I}[\sigma']$. The behavior σ' is the folding of $\mathtt{nf}(\sigma)$.*

Proof. By Lemma 2 there are finitely many $\sigma(\varphi)$. These $\sigma(\varphi)$ are in one-to-one correspondence with the behavior names that may be recursively invoked by $\mathtt{nf}(\sigma)$. Therefore the set of such behavior names is finite and $\mathtt{nf}(\sigma)$ may be folded into a behavior with recursion and variables.

To prove the subcontract equivalence, let $\mathscr{R}$ be the least relation containing $\preceq$ and such that

1. if $\mathrm{I}[\sigma] \preceq \mathrm{J}[\tau]$ and τ' is the normal form of τ, then $\mathrm{I}[\sigma] \mathrel{\mathscr{R}} \mathrm{J}[\tau']$;
2. if $\mathrm{I}[\sigma] \preceq \mathrm{J}[\tau]$ and σ' is the normal form of σ, then $\mathrm{I}[\sigma'] \mathrel{\mathscr{R}} \mathrm{J}[\tau]$.

We prove that $\mathscr{R}$ is a subcontract relation; the theorem follows directly.

Let $\mathrm{I}[\sigma] \mathrel{\mathscr{R}} \mathrm{J}[\tau']$, where τ' is the normal form of τ. Then $\mathrm{I}[\sigma] \preceq \mathrm{J}[\tau]$. As regards condition 1 in the definition of coinductive subcontract, notice that from $\mathrm{I}[\sigma] \preceq \mathrm{J}[\tau]$ we have $\sigma\!\downarrow$ implies $\tau\!\downarrow$, hence $\tau'\!\downarrow$ from the definition of normal form. As regards condition 2, let $\tau' \Downarrow \mathrm{R}$, then $\mathrm{R} \in \mathscr{A}(\tau)$, hence $\tau \Downarrow \mathrm{R}'$ and $\mathrm{R}' \subseteq \mathrm{R}$ by definition of $\mathscr{A}(\tau)$. From $\mathrm{I}[\sigma] \preceq \mathrm{J}[\tau]$ we obtain $\sigma \Downarrow \mathrm{S}$ and $\mathrm{S} \subseteq \mathrm{R}'$, hence we conclude $\mathrm{S} \subseteq \mathrm{R}$. As regards condition 3, let $\tau' \stackrel{\alpha}{\Longrightarrow} \tau''$. Then $\mathrm{J}[\tau'(\alpha)] \preceq \mathrm{J}[\tau'']$ and, by definition, $\tau'(\alpha)$ is in normal form. Then $\tau \stackrel{\alpha}{\Longrightarrow}$ and, by $\mathrm{I}[\sigma] \preceq \mathrm{J}[\tau]$, $\sigma \stackrel{\alpha}{\Longrightarrow}$. Therefore $\mathrm{I}[\sigma(\alpha)] \preceq \mathrm{J}[\tau(\alpha)]$, hence by definition of $\mathscr{R}$, $\mathrm{I}[\sigma(\alpha)] \mathrel{\mathscr{R}} \mathrm{J}[\tau''']$, where τ''' is the normal form of $\tau(\alpha)$. By definition of normal form $\tau''' = \tau'(\alpha)$ and we conclude.

Let $\mathrm{I}[\sigma'] \mathrel{\mathscr{R}} \mathrm{J}[\tau]$, where σ' is the normal form of σ. Then $\mathrm{I}[\sigma] \mathrel{\mathscr{R}} \mathrm{J}[\tau]$. As regards condition 1 in the definition of coinductive subcontract, $\sigma'\!\downarrow$ implies $\sigma\!\downarrow$, hence $\tau\!\downarrow$. As regards condition 2, let $\tau \Downarrow \mathrm{R}$. From $\mathrm{I}[\sigma] \preceq \mathrm{J}[\tau]$ we have $\sigma \Downarrow \mathrm{S}$ and $\mathrm{S} \subseteq \mathrm{R}$ and we conclude by observing that $\sigma' \Downarrow \mathrm{S}$. As regards condition 3, let $\tau \stackrel{\alpha}{\Longrightarrow} \tau'$. Then there exist $\sigma_1, \ldots, \sigma_n$ such that $\mathrm{I}[\bigoplus_{1 \leq i \leq n} \sigma_i] \preceq \mathrm{J}[\tau']$. Since $\mathrm{I}[\sigma(\alpha)] \preceq \mathrm{I}[\bigoplus_{1 \leq i \leq n} \sigma_i]$ and the normal form of $\sigma(\alpha)$ is equal to $\sigma'(\alpha)$ we conclude $\mathrm{I}[\sigma'(\alpha)] \mathrel{\mathscr{R}} \mathrm{J}[\tau']$ by definition of $\mathscr{R}$. $\qquad \square$

By Theorem 1, from now on we let $\mathtt{nf}(\sigma)$ be the normal form of σ.

3   Session Types

Similarly to a contract, a session type is meant to describe the type of a communication as a sequence of actions and branching points. The syntax of session types uses an infinite set of labels $\mathscr{L}$ ranged over by $\ell, \ell', \ldots$, and an infinite set of variables ranged over by $x, y, z, \ldots$. Session types are defined by the grammar:

$$S \quad ::= \quad \text{end} \quad | \quad \oplus\langle \ell_i : S_i^{i \in I} \rangle \quad | \quad \&\langle \ell_i : S_i^{i \in I} \rangle \quad | \quad x \quad | \quad \mu x.S$$

The session type end describes a completed communication; the session type $\oplus\langle \ell_i : S_i^{i \in I} \rangle$ describes a *choice* where a process autonomously decides to proceed according to one of the continuations S_i's. Before doing so, the process notifies the partner of the communication by sending a label ℓ_i. The session type $\&\langle \ell_i : S_i^{i \in I} \rangle$ describes a *branch* where a process is ready to proceed according to any of the continuations S_i's. Before doing so, the process waits for a label ℓ_i from the process it is interacting with. The label uniquely identifies the continuation. The session type $\mu x.S$ describes a recursive type, much like a recursive contract.

Following [11], session types are taken *contractive*, namely every occurrence of a variable is guarded by at least a choice or a branch. Unlike [11], however, we omit session types describing the communications of actual data (not merely labels) during a session. Typically, such communications are represented by session types of the form $!t.S$ (for "send a message of type t on the channel and then continue as S") or $?t.S$ (for "receive a message of type t from the channel and then continue as S"). However, as far as the comparison of session types and contracts is concerned, the presence of such actions is not particularly relevant, since we are going to focus on the control part of an interaction. The addition of $!t$ and $?t$ actions to the above grammar does not pose any particular problem for all of the results that follow, but at the same time it does not contribute any significant insight in the comparison.

The semantics of session types is defined in terms of a subtyping relation $\leq$, which is presented by means of a deductive system or as a coinductive definition [11]. In this contribution we stick to this latter presentation that turns out to be easier to compare with the subcontract relation defined in the previous section. More precisely, $S \leq T$, whenever every process with type S can be used where a process with type T is expected (in this respect, $\leq$ differs from $\preceq$ as it is oriented as a real subtyping relation). Because of the structure of session types, $\leq$ is quite straightforward to define: basically every choice in S must be defined on a subset of the labels occurring in the corresponding choice in T, whereas any branch in S must be defined on a superset of the labels occurring in the corresponding branch in T. Let

$$\text{unfold}(S) \stackrel{\text{def}}{=} \begin{cases} \text{unfold}(T\{S/x\}) & \text{if } S = \mu x.T \\ S & \text{otherwise} \end{cases}$$

That is, $\text{unfold}(\cdot)$ removes topmost recursions until it reveals a branch or a choice. It is worth to notice that $\text{unfold}(\cdot)$ is always defined because session types are contractive.

Definition 6 (Subtyping). *The relation $\mathscr{R}$ is a* coinductive subtyping *if $S \mathscr{R} T$ implies*

1. if $\text{unfold}(S) = \text{end}$, then $\text{unfold}(T) = \text{end}$,

2. *if* $\mathtt{unfold}(S) = \oplus\langle \ell_i : S_i^{i\in I}\rangle$, *then* $\mathtt{unfold}(T) = \oplus\langle \ell_j : T_j^{j\in J}\rangle$ *and* $I \subseteq J$ *and for every* $i \in I$ *we have* $S_i \mathcal{R} T_i$,
3. *if* $\mathtt{unfold}(S) = \&\langle \ell_i : S_i^{i\in I}\rangle$, *then* $\mathtt{unfold}(T) = \&\langle \ell_j : T_j^{j\in J}\rangle$ *and* $J \subseteq I$ *and for every* $j \in J$ *we have* $S_j \mathcal{R} T_j$.

Let $\leq$ be the largest coinductive subtyping.

For example, we have $\oplus\langle \ell : S\rangle \leq \oplus\langle \ell : S; \ell' : T\rangle$ since the smaller session type represents a process that behaves more deterministically than the one represented by the larger session type. Dually we have $\&\langle \ell : S; \ell' : T\rangle \leq \&\langle \ell : S\rangle$ since the process represented by the smaller session type offers a wider range of continuations (it is ready to accept a superset of the labels accepted by the process represented by the larger session type). Unlike contracts, however, we have $\&\langle \ell : S\rangle \not\leq \mathtt{end}$ and $\oplus\langle \ell : S\rangle \not\leq \mathtt{end}$. In the first case the process *waits for* a label that never arrives, in the second case it *sends* a label that the matching party is not ready to receive.

Session types are naturally equipped with a $\mathtt{dual}(\cdot)$ function computing the dual type.

$$\mathtt{dual}(\mathtt{end}) = \mathtt{end}$$
$$\mathtt{dual}(\oplus\langle \ell_i : S_i{}^{i\in I}\rangle) = \&\langle \ell_i : \mathtt{dual}(S_i){}^{i\in I}\rangle$$
$$\mathtt{dual}(\&\langle \ell_i : S_i{}^{i\in I}\rangle) = \oplus\langle \ell_i : \mathtt{dual}(S_i){}^{i\in I}\rangle$$
$$\mathtt{dual}(x) = x$$
$$\mathtt{dual}(\mu x.S) = \mu x.\mathtt{dual}(S)$$

As for $\mathtt{unfold}(\cdot)$, $\mathtt{dual}(\cdot)$ is always defined because session types are contractive. If S is the session type representing the behavior of a process, then $\mathtt{dual}(S)$ represents the behavior of a "canonical" process that interacts successfully with the first one. The theory of session types does not formalize the notion of successful interaction, but we will be able to reason about it by means of one of our encodings in Section 5.

The duality between choices and branches in session types is formalized by the following proposition. Let $\mathtt{in}(S)$ (respectively, $\mathtt{out}(S)$) be the set of labels occurring in branches (respectively, choices) of S. Then subtyping induces a relation on labels $\mathtt{in}(\cdot)$ and $\mathtt{out}(\cdot)$.

Proposition 3. *If $S \leq T$ then (1) $\mathtt{out}(S) \subseteq \mathtt{out}(T)$, (2) $\mathtt{in}(T) \subseteq \mathtt{in}(S)$, and (3) $\mathtt{dual}(T) \leq \mathtt{dual}(S)$.*

Notice that from the previous proposition and from the fact that $\mathtt{dual}(\cdot)$ is the inverse of itself ($\mathtt{dual}(\mathtt{dual}(S)) = S$), we have $S \leq T$ if and only if $\mathtt{dual}(T) \leq \mathtt{dual}(S)$.

4   Encoding Session Types into Contracts

In order to encode session types into contracts we preliminarily need to set a correspondence between labels and names. For simplicity we let $\mathscr{L} = \mathscr{N}$. Therefore, in the following, we will not distinguish between labels and names and we will address them with $\ell, a, \dots$.

The encoding $[\![S]\!]_I$ of a session type S is defined with respect to an interface I (Table 1). The terminal behavior σ_e is e if the session type being encoded regards a client,

Table 1. Encoding of session types into contracts

$$[\![\text{end}]\!]_\text{I} = \sigma_e$$
$$[\![\oplus\langle \ell_i : S_i{}^{i\in 1..n}\rangle]\!]_\text{I} = \bigoplus_{i\in 1..n} \overline{\ell_i}.[\![S_i]\!]_\text{I}$$
$$[\![\&\langle \ell_i : S_i{}^{i\in 1..n}\rangle]\!]_\text{I} = \Sigma_{i\in 1..n}\, \ell_i.[\![S_i]\!]_\text{I} + \Sigma_{\alpha\in\text{I}\setminus\{\ell_1,\dots,\ell_n\}}\, \alpha.\Omega$$
$$[\![x]\!]_\text{I} = x$$
$$[\![\mu x.S]\!]_\text{I} = \texttt{rec } x.[\![S]\!]_\text{I}$$

or $\mathbf{0}$ if it regards a service. The only case that deserves discussion is branch. The contract $[\![\&\langle \ell_i : S_i{}^{i\in 1..n}\rangle]\!]_\text{I}$ has as many initial actions as are allowed by the interface, irrespective of the labels of the branches. This enforces width subtyping in the resulting contracts, which is otherwise false, in general. In facts, $\{a,b\}[a] \not\preceq \{a,b\}[a+b]$ because the client $\{\overline{a},\overline{b}\}[\overline{a}.\text{e}+\overline{b}]$ is compliant with the first service, but not with the second one. By translating $S = \&\langle a : S'\rangle$ as simply $\{a,b\}[a.[\![S']\!]]$ (the action b is in the interface because it occurs in S'), then a client such as $\{\overline{a},\overline{b}\}[\overline{a}.\text{e}+\overline{b}.\rho]$ could fail when one replaces $[\![S]\!]_{\{a,b\}}$ with $[\![T]\!]_{\{a,b\}}$ and $T \leq S$, for instance when $T = \&\langle a : S'; b : T'\rangle$. To avoid these cases, one has to exclude $\{\overline{a},\overline{b}\}[\overline{a}.\text{e}+\overline{b}.\rho]$ from those clients that are compliant with $[\![S]\!]$ because it attempts to do an action ($\overline{b}$) that is not explicitly allowed by S. This exclusion follows by translating a branch into an external choice "$+$" with a summand for every action in the interface: the actions that do not appear as labels of the branch are given a continuation Ω. Therefore, the contract $[\![S]\!]_{\{a,b\}}$ does not admit, in general, a client $\{\overline{a},\overline{b}\}[\overline{a}.\text{e}+\overline{b}.\rho]$ but only a client $\{\overline{a},\overline{b}\}[\overline{a}.\text{e}+\overline{b}.\text{e}]$ (this client has to terminate after an interaction on b, without requiring any further capability to the service). The contract Ω is the smallest one with a given interface: a client compliant with Ω will also comply with any other service, in particular with those resulting from the encoding of smaller session types.

Example 1. According to the encoding, we have

- $[\![\&\langle a : \text{end}\rangle]\!]_{\{a\}} = a;$
- $[\![\&\langle a : \text{end}\rangle]\!]_{\{a,b\}} = a + b.\Omega;$
- $[\![\&\langle a : \text{end}; b : \text{end}\rangle]\!]_{\{a,b\}} = a + b;$
- $[\![\oplus\langle a : \text{end}\rangle]\!]_{\{a\}} = [\![\oplus\langle a : \text{end}\rangle]\!]_{\{a,b\}} = \overline{a};$
- $[\![\oplus\langle a : \text{end}; b : \text{end}\rangle]\!]_{\{a,b\}} = \overline{a} \oplus \overline{b}.$

We notice that $\&\langle a : \text{end}; b : \text{end}\rangle \leq \&\langle a : \text{end}\rangle$ and $\{a\}[a] \preceq \{a,b\}[a + b.\Omega] \preceq \{a,b\}[a+b]$. Similarly, $\oplus\langle a : \text{end}\rangle \leq \oplus\langle a : \text{end}; b : \text{end}\rangle$ and $\{a,b\}[\overline{a} \oplus \overline{b}] \preceq \{a,b\}[\overline{a}]$. ∎

More generally, the encoding of Table 1 makes the subtyping relation and the subcontract relation agree.

Theorem 2. *Let* $\text{J} = \text{in}(T)\cup\overline{\text{out}(T)}$ *and* $\text{I} = \text{in}(S)\cup\overline{\text{out}(T)}$. *Then* $S \leq T$ *if and only if* $\text{J}[[\![T]\!]_\text{J}] \preceq \text{I}[[\![S]\!]_\text{I}]$.

Proof. ("only if" part) Let $\mathscr{R}$ be the least relation such that

1. if $\mathrm{J} \subseteq \mathrm{I}$ then $\mathrm{J}[\Omega] \ \mathscr{R} \ \mathrm{I}[\sigma]$;
2. if $S \leq T$ and $\mathrm{J} = \mathtt{in}(T) \cup \overline{\mathtt{out}(T)}$ and $\mathrm{I} = \mathtt{in}(S) \cup \overline{\mathtt{out}(T)}$, then $\mathrm{J}[[\![T]\!]_{\mathrm{J}}] \ \mathscr{R} \ \mathrm{I}[[\![S]\!]_{\mathrm{I}}]$.

We prove that $\mathscr{R}$ is a coinductive subcontract. Let $\mathrm{J}[[\![T]\!]_{\mathrm{J}}] \ \mathscr{R} \ \mathrm{I}[[\![S]\!]_{\mathrm{I}}]$. By Proposition 3, $\mathrm{J} \subseteq \mathrm{I}$. By definition, $[\![\cdot]\!]$. always yields behaviors σ such that $\sigma \downarrow$. Therefore condition 1 in Definition 3 is trivially satisfied. As regards conditions 2 and 3, we reason by cases on S. By definition of $\leq$, $S \leq T$ if and only if $\mathtt{unfold}(S) \leq \mathtt{unfold}(T)$. Therefore we may restrict to cases where neither S nor T are recursive types.

$(S = \mathtt{end})$ Then $T = \mathtt{end}$. Condition 2 follows immediately; condition 3 is vacuous.

$(S = \oplus\langle \ell_h : S_h \ ^{h \in H}\rangle)$ Then $T = \oplus\langle \ell_k : T_k \ ^{k \in K}\rangle$ and $H \subseteq K$ and $S_h \leq T_h$, for every $h \in H$. As regards condition 2, if $[\![S]\!]_{\mathrm{I}} \Downarrow \mathrm{R}$ then $\overline{\ell_h} \in \mathrm{R}$ for some $h \in H$. Since $H \subseteq K$, $[\![T]\!]_{\mathrm{J}} \Downarrow \{\overline{\ell_h}\}$ and $\{\overline{\ell_h}\} \subseteq \mathrm{R}$. As regards conditions 3, by Proposition 2, it suffices to consider the transitions $[\![S]\!]_{\mathrm{I}} \xrightarrow{\overline{\ell_h}} [\![S_h]\!]_{\mathrm{I}}$ for every $h \in H$. Since $H \subseteq K$, $[\![T]\!]_{\mathrm{J}} \xrightarrow{\overline{\ell_h}} [\![T_h]\!]_{\mathrm{J}}$ and $[\![T_h]\!]_{\mathrm{J}} \ \mathscr{R} \ [\![S_h]\!]_{\mathrm{I}}$ follows by $S_h \leq T_h$;

$(S = \&\langle \ell_k : S_k \ ^{k \in K}\rangle)$ Then $T = \&\langle \ell_h : T_h \ ^{h \in H}\rangle$ with $H \subseteq K$. Condition 2 follows because $[\![T]\!]_{\mathrm{J}} \Downarrow \{\ell_h \mid h \in H\}$ and $[\![S]\!]_{\mathrm{I}} \Downarrow \{\ell_k \mid k \in K\}$ and $\{\ell_h \mid h \in H\} \subseteq \{\ell_k \mid k \in K\}$. As regards condition 3, by Proposition 2, it suffices to consider $[\![S]\!]_{\mathrm{I}} \xrightarrow{\ell_k} [\![S_k]\!]_{\mathrm{I}}$ with $k \in \mathrm{J}$. There are two subcases: either (i) $\ell_k \notin \{\ell_h \mid h \in H\}$ or (ii) $\ell_k \in \{\ell_h \mid h \in H\}$. In subcase (i), $[\![T]\!]_{\mathrm{J}} \xrightarrow{\ell_k} \Omega$ and $\mathrm{J}[\Omega] \ \mathscr{R} \ \mathrm{I}[[\![S_k]\!]_{\mathrm{I}}])$ by definition of $\mathscr{R}$. In subcase (ii), $[\![T]\!]_{\mathrm{J}} \xrightarrow{\ell_k} [\![T_k]\!]_{\mathrm{J}}$. From $S_k \leq T_k$ we derive $\mathrm{J}[[\![T_k]\!]_{\mathrm{J}}] \ \mathscr{R} \ \mathrm{I}[[\![S_k]\!]_{\mathrm{I}}]$ by definition of $\mathscr{R}$.

("if" part) We prove that if $S \not\leq T$, then $\mathrm{J}[[\![T]\!]_{\mathrm{J}}] \not\preceq \mathrm{I}[[\![S]\!]_{\mathrm{I}}]$. We reason by cases on the shape of S and T. It is sufficient to consider those cases in which none of the types begins with a recursion and $S \leq T$ holds directly, without looking at the session types in the choices and branches possibly found in S and T.

$(S = \mathtt{end})$ It must be $T \neq \mathtt{end}$. Then $[\![S]\!]_{\mathrm{I}} \Downarrow \{\mathtt{e}\}$ whereas $[\![T]\!]_{\mathrm{J}} \Downarrow \mathrm{R}$ implies $\mathrm{R} \neq \emptyset$ and $\mathtt{e} \notin \mathrm{R}$;

$(S = \oplus\langle \ell_i : S_i \ ^{i \in I}\rangle)$ If $T = \mathtt{end}$, then we can reason as for the case $S = \mathtt{end}$ and conclude that $\mathrm{J}[[\![T]\!]_{\mathrm{J}}] \not\preceq \mathrm{I}[[\![S]\!]_{\mathrm{I}}]$. If $T = \&\langle \ell_j : T_j \ ^{j \in J}\rangle$, then $[\![T]\!]_{\mathrm{J}}$ has only one ready set $\mathrm{J} = \mathtt{in}(T) \cup \overline{\mathtt{out}(T)}$, which contains at least two actions and hence cannot be smaller than $\{\ell_i\}$, for every $i \in I$. If $T = \oplus\langle \ell_j : T_j \ ^{j \in J}\rangle$ and $I \not\subseteq J$, then there exists $i \in I$ such that $i \notin J$. So $[\![S]\!]_{\mathrm{I}} \Downarrow \{\overline{\ell_i}\}$ whereas $[\![T]\!]_{\mathrm{J}} \Downarrow \mathrm{R}$ implies $\mathrm{R} \neq \emptyset$ and $\overline{\ell_i} \notin \mathrm{R}$;

$(S = \&\langle \ell_i : S_i \ ^{i \in I}\rangle)$ If $T = \mathtt{end}$ we conclude immediately since no ℓ_i is equal to $\mathtt{e}$. If $T = \oplus\langle \ell_j : T_j \ ^{j \in J}\rangle$, then from $\mathrm{I} = \mathtt{in}(S) \cup \overline{\mathtt{out}(T)}$ we have that $j \in J$ implies $\overline{\ell_j} \in \mathrm{I}$. Now we have $[\![S]\!]_{\mathrm{I}} \xrightarrow{\overline{\ell_j}} \Omega$, whereas $[\![T]\!]_{\mathrm{J}} \xrightarrow{\overline{\ell_i}} \sigma$ implies $\sigma \downarrow$. If $T = \&\langle \ell_j : T_j \ ^{j \in J}\rangle$ and $J \not\subseteq I$, then there exists $j \in J$ such that $j \notin I$. If $\mathrm{J} \not\subseteq \mathrm{I}$ there is nothing to prove. If $\mathrm{J} \subseteq \mathrm{I}$, then $[\![S]\!]_{\mathrm{I}} \xrightarrow{\overline{\ell_j}} \Omega$, whereas $[\![T]\!]_{\mathrm{J}} \xrightarrow{\overline{\ell_j}} \sigma$ implies $\sigma \downarrow$. $\square$

The following proposition shows that internal moves in the encoding of a session type S correspond to reducing its choices and determining a session type $S' \leq S$.

Proposition 4. *If $[\![S]\!]_I \Longrightarrow \sigma$, then $\sigma = [\![S']\!]_I$ and $S' \leq S$.*

Proof. It is sufficient to consider the case $[\![S]\!]_I \longrightarrow \sigma$, then the lemma follows directly. We reason by cases on the structure of S, there are only two possibilities:

$(S = \oplus\langle \ell_i : S_i \ ^{i \in K}\rangle)$ Then $[\![S]\!]_I = \oplus_{i \in I} \overline{\ell_i}.[\![S_i]\!]_I \longrightarrow \oplus_{i \in J} \overline{\ell_j}.[\![S_j]\!]_I = [\![\oplus\langle \ell_j : S_j \ ^{j \in J}\rangle]\!]_I$ with $J \subseteq I$. We conclude by observing that $\oplus\langle \ell_j : S_j \ ^{j \in J}\rangle \leq S$.

$(S = \mu x.S')$ Then $[\![S]\!]_I = \mathtt{rec}\ x.[\![S']\!]_I \longrightarrow [\![S']\!]_I\{\mathtt{rec}\ x.[\![S']\!]_I/x\} = [\![S'\{S/x\}]\!]_I$ and this is the only possible reduction. We conclude because $S'\{S/x\} \leq S$. $\qquad\square$

The encoding in Table 1 allows us to relate session types and their duals. The relation is exactly the compliance of Definition 1.

Theorem 3. *Let $I = \mathtt{in}(S) \cup \overline{\mathtt{out}(S)}$ and $J = \mathtt{in}(T) \cup \overline{\mathtt{out}(T)}$ and $S \leq \mathtt{dual}(T)$. Then $[\![S]\!]_I \dashv [\![T]\!]_J$.*

Proof. Let $\mathscr{R}$ be the least relation such that if $S \leq \mathtt{dual}(T)$ and $\mathtt{in}(S) \cup \overline{\mathtt{out}(S)} \subseteq I$ and $\mathtt{in}(T) \cup \overline{\mathtt{out}(T)} \subseteq J$, then $[\![S]\!]_I \mathscr{R} [\![T]\!]_J$. We prove that $\mathscr{R}$ is a compliance relation. Let $[\![S]\!]_I \mathscr{R} [\![T]\!]_J$. By Proposition 4, if $[\![S]\!]_I \longrightarrow \rho$, then $\rho = [\![S']\!]_I$ and $S' \leq S \leq \mathtt{dual}(T)$. Symmetrically, if $[\![T]\!]_J \longrightarrow \sigma$, then $\sigma = [\![T']\!]_J$ and $T' \leq T$, hence, by Proposition 3(3), $S \leq \mathtt{dual}(T) \leq \mathtt{dual}(T')$. It follows that we may restrict our analysis to cases where neither the encoding of S nor the encoding of T can perform internal moves:

$(S = \mathtt{end})$ Then $T = [\![\mathtt{end}]\!]_J$. Therefore $[\![S]\!]_I \parallel [\![T]\!]_J \nrightarrow$ and $[\![S]\!]_I \xrightarrow{\ e\ };$

$(S = \oplus\langle \ell : S'\rangle)$ Then $T = \&\langle \ell_i : T_i \ ^{i \in K}\rangle_J$ and $\ell = \ell_i$ for some $i \in K$ and $S' \leq \mathtt{dual}(T_i)$.

     Now $[\![S]\!]_I \xrightarrow{\ \overline{\ell}\ }$ and $[\![T]\!]_J \xrightarrow{\ \ell\ }$, hence $[\![S]\!]_I \parallel [\![T]\!]_J \longrightarrow [\![S']\!]_I \parallel [\![T_i]\!]_J$ and we conclude $[\![S']\!]_I \mathscr{R} [\![T_i]\!]_J$ by definition of $\mathscr{R}$.

$(S = \&\langle \ell_i : S_i \ ^{i \in I}\rangle)$ Dual of the previous case. $\qquad\square$

5   Encoding Contracts into Session Types

The encoding of contracts to session types is partially defined. In particular, we will restrict the encoding to behaviors that are convergent. The reason for this restriction derives from the fact that session types describe communications whose end is agreed by both parties having type end. The behavior Ω, on the other hand, represents a service that may not pay any attention to the communication it is involved in, and it has no corresponding session type.

A behavior σ is *strongly convergent* if, for every sequence φ, $\sigma \downarrow \varphi$. The following encodings are defined on contracts with *strongly convergent* behaviors. It is also convenient to restrict the encoding of client contracts to those whose behavior never leads to $\mathbf{0}$ without emitting e. For example, the behavior $a.\mathtt{e} + b.\mathbf{0}$ describes a client that succeeds if the service proposes $\overline{a}$, but that fails if the service proposes $\overline{b}$. Such a behavior has no counterpart in session types, where unacceptable labels are not explicitly specified. In general, if a client is unable to handle a particular action, like b in the example, it should simply omit that action from its behavior. We say that a (client) contract $\mathtt{I}[\rho]$ is *canonical* if, whenever $\rho \xrightarrow{\ \varphi\ } \rho'$ is maximal, then $\varphi = \varphi'\mathtt{e}$ and $\mathtt{e} \notin \mathtt{names}(\varphi')$. For example

$\{a\}[a.e]$, $\{a\}[\texttt{rec}\ x.a.x]$, and $\emptyset[\Omega]$ are canonical; $\{a,b\}[a.e+b.\mathbf{0}]$ and $\{a\}[\texttt{rec}\ x.a+x]$ are not canonical. For strongly convergent, canonical client behaviors it is the case that the subcontract relation can be safely used for replacing equivalent clients still preserving compliance. This result is fundamental in the encoding of client contracts, as the encoding relies on the normal form.

Proposition 5. *Let* $\textsc{k}[\rho]$ *be a strongly convergent, canonical client contract such that* $\textsc{k}[\rho] \dashv \textsc{i}[\sigma]$ *and* $\textsc{k}[\rho] \simeq \textsc{k}[\rho']$. *Then* $\textsc{k}[\rho'] \dashv \textsc{i}[\sigma]$.

Proof. First of all note that if $\textsc{k}[\rho] \simeq \textsc{k}[\rho']$ and ρ is strongly convergent, so is ρ'. Let $\rho' \parallel \sigma \Longrightarrow \rho'' \parallel \sigma'$ be an interaction of ρ' and σ. Then $\rho' \overset{\varphi}{\Longrightarrow} \rho''$ and $\sigma \overset{\overline{\varphi}}{\Longrightarrow} \sigma'$ for some sequence φ of actions. From $\textsc{k}[\rho] \simeq \textsc{k}[\rho']$ and the fact that ρ and ρ' are strongly convergent we derive that $\rho \overset{\varphi}{\Longrightarrow} \rho''' \nrightarrow$ for some ρ''' such that $\texttt{init}(\rho''') \subseteq \texttt{init}(\rho'')$, hence $\rho \parallel \sigma \Longrightarrow \rho''' \parallel \sigma'$. Assume $\rho'' \parallel \sigma' \nrightarrow$. Then $\rho''' \parallel \sigma' \nrightarrow$, hence $\{e\} \subseteq \texttt{init}(\rho''') \subseteq \texttt{init}(\rho'')$. Assume $\sigma'\uparrow$. From $\textsc{k}[\rho] \dashv \textsc{i}[\sigma]$ and the fact that ρ is canonical we derive that $\{e\} = \texttt{init}(\rho''') \subseteq \texttt{init}(\rho'')$. From $\textsc{k}[\rho] \simeq \textsc{k}[\rho']$ we derive $\textsc{k}[\rho(\varphi)] \simeq \textsc{k}[\rho'(\varphi)]$ and $\rho(\varphi) \Longrightarrow \rho'''$ and $\rho'(\varphi) \Longrightarrow \rho''$. Hence $\texttt{init}(\rho'') \subseteq \texttt{init}(\rho(\varphi)) = \{e\}$, namely $\{e\} = \texttt{init}(\rho''')$. In both cases we conclude $\textsc{k}[\rho'] \dashv \textsc{i}[\sigma]$. $\qquad\square$

The encoding of contracts into session types is reported in Table 2. We write $\textsc{s} \bowtie \mathscr{A}$ ($\mathscr{A}$ is an acceptance set) if for every $\textsc{r} \in \mathscr{A}$, either $e \in \textsc{r}$ or $\textsc{s} \cap \textsc{r} \neq \emptyset$. We use an injective map $\widehat{\cdot}$ from sets of actions to labels that we leave unspecified.

Table 2. Encoding of contracts into session types

Encoding of client contracts/behaviors:

$$\mathscr{C}[\![\textsc{k}[\rho]]\!] = \oplus\langle\widehat{\textsc{k}\setminus\{e\}} : \mathscr{C}[\![\texttt{nf}(\rho)]\!]_{\textsc{k}\setminus\{e\}}\rangle$$

$$\mathscr{C}[\![\textstyle\bigoplus_{\textsc{r}\in\mathscr{A}}\sum_{\alpha\in\textsc{r}}\alpha.\rho_\alpha]\!]_\textsc{k} = \&\langle\widehat{\overline{\textsc{s}}} : \oplus\langle\underbrace{\widehat{\emptyset} : \texttt{end};}_{\text{if } e \in \textsc{r}}\ \widehat{\{\overline{\alpha}\}} : \mathscr{C}[\![\rho_\alpha]\!]_\textsc{k}^{\textsc{r}\in\mathscr{A},\alpha\in\textsc{s}\cap\textsc{r}}\rangle^{\textsc{s}\subseteq\overline{\textsc{k}},\textsc{s}\bowtie\mathscr{A}}\rangle$$

$$\mathscr{C}[\![\texttt{rec}\ x.\rho]\!]_\textsc{k} = \mu x.\mathscr{C}[\![\rho]\!]_\textsc{k}$$
$$\mathscr{C}[\![x]\!]_\textsc{k} = x$$

Encoding of service contracts/behaviors:

$$\mathscr{S}[\![\textsc{i}[\sigma]]\!] = \&\langle\widehat{\textsc{k}} : \mathscr{S}[\![\texttt{nf}(\sigma)]\!]_\textsc{k}^{\textsc{k}\subseteq\textsc{i}}\rangle$$

$$\mathscr{S}[\![\textstyle\bigoplus_{\textsc{r}\in\mathscr{A}}\sum_{\alpha\in\textsc{r}}\alpha.\sigma_\alpha]\!]_\textsc{k} = \oplus\langle\widehat{\textsc{r}\cap\textsc{k}} : \&\langle\widehat{\emptyset} : \texttt{end};\ \widehat{\{\alpha\}} : \mathscr{S}[\![\sigma_\alpha]\!]_\textsc{k}^{\alpha\in\textsc{r}\cap\textsc{k}}\rangle^{\textsc{r}\in\mathscr{A}}\rangle$$

$$\mathscr{S}[\![\texttt{rec}\ x.\rho]\!]_\textsc{k} = \mu x.\mathscr{S}[\![\rho]\!]_\textsc{k}$$
$$\mathscr{S}[\![x]\!]_\textsc{k} = x$$

The encoding distinguishes between client and service contracts. One reason is that, unlike sessions, which are supposed to be completed *symmetrically* by both parties, the theory of contracts is biased towards clients, which are free to interrupt the interaction any time they please. The other reason is that contracts describe a more abstract synchronization pattern than session types do, and the encoding of contracts into session types has to render this synchronization pattern, which amounts to a little handshaking protocol.

The session types corresponding to the client contract $\kappa[\rho]$ and the service contract $\iota[\sigma]$ are denoted by $\mathscr{C}[\![\kappa[\rho]]\!]$ and $\mathscr{S}[\![\iota[\sigma]]\!]$, respectively. The labels in the types represent finite sets of actions in the source contracts. Let us discuss the encoding of a service contract. The first step in the generation of the session type is the offering of an interface to the client. Any client that asks no more capabilities than those offered by the service can connect. Hence, the service offers as many interfaces as the number of the subsets of its own interface. For every offered interface, the continuation session type is a specialized encoding of the service's contract, restricted to the offered interface: $\mathscr{S}[\![\sigma]\!]_\kappa$. For the proper encoding of the behavior, we resort to the normal form (Definition 5). A behavior of the form $\bigoplus_{R \in \mathscr{A}} \sum_{\alpha \in R} \alpha.\sigma_\alpha$ is one in which the service can be in as many states as the cardinality of $\mathscr{A}$. In each state $R \in \mathscr{A}$, the service is ready to perform any of the actions in R. So, the service begins by communicating to the client the state it is in (restricted to the interface of the connected client). Then, the service accepts a singleton action, among those that are available, indicating the choice of the client, or the special action $\widehat{\emptyset}$ denoting the fact that the client has decided to terminate at this stage. We notice that the term

$$\oplus \langle \widehat{R \cap K} : \& \langle \widehat{\emptyset} : \mathrm{end}; \widehat{\{\alpha\}} : \mathscr{S}[\![\sigma_\alpha]\!]_K{}^{\alpha \in R \cap K} \rangle^{R \in \mathscr{A}} \rangle$$

is well formed: every label of a branch or a choice has exactly one continuation because the behavior $\bigoplus_{R \in \mathscr{A}} \sum_{\alpha \in R} \alpha.\sigma_\alpha$ is in normal form.

In some sense the encoding of a service contract is "kind" as it tries to accommodate the largest number of clients (not only those connecting with exactly the same interface as the service, but also with smaller ones). Conversely, the encoding of a client contract is selfish in that it only encodes the client behavior, whose only purpose is to successfully achieve its task. The first action of the client is the selection of the service interface that matches with its own (containing the co-actions of the client's interface). Then, at each interaction, the client must be ready to accept a set $\overline{S}$ of actions from the service, representing the state the service is in. However, not all such states are suitable for the client. In particular, let $\mathscr{A}$ be the acceptance set of the client; the client will only accept those states $\overline{S}$ such that, for every $R \in \mathscr{A}$, either $e \in R$ (the client is ready to terminate) or $S \cap R \neq \emptyset$ (the client and the service can synchronize). Said otherwise, the fewer sets $\overline{S}$ the client can accept from the service, the more demanding it is. Once a set has been accepted, the client may choose the special label $\widehat{\emptyset}$, signaling its intention to terminate the interaction, or it may choose a label $\widehat{\{\alpha\}}$, signaling the intention of synchronizing on α. Because of the way the sets $\overline{S}$ are accepted by the client, it is always possible to pursue at least one of such possibilities.

There are discretions in the two encodings $\mathscr{C}[\![\cdot]\!]$ and $\mathscr{S}[\![\cdot]\!]$: the alternation between $\oplus$ and $\&$ may be reversed without changing the following correctness results.

Example 2. Consider the service contracts

$$\iota[\sigma] \stackrel{\mathrm{def}}{=} \{a,b\}[a \oplus b] \qquad \text{and} \qquad \iota[\tau] \stackrel{\mathrm{def}}{=} \{a,b\}[a+b]$$

and notice that $\iota[\sigma] \preceq \iota[\tau]$. Their respective encodings are

$$\mathscr{S}[\![\mathrm{I}[\sigma]]\!] = \&\langle\widehat{\emptyset} : \oplus\langle\widehat{\emptyset} : \&\langle\widehat{\emptyset} : \mathrm{end}\rangle\rangle;$$
$$\widehat{\{a\}} : \oplus\langle\widehat{\emptyset} : \&\langle\widehat{\emptyset} : \mathrm{end}\rangle; \widehat{\{a\}} : \&\langle\widehat{\emptyset} : \mathrm{end}; \widehat{\{a\}} : \mathrm{end}\rangle\rangle;$$
$$\widehat{\{b\}} : \oplus\langle\widehat{\emptyset} : \&\langle\widehat{\emptyset} : \mathrm{end}\rangle; \widehat{\{b\}} : \&\langle\widehat{\emptyset} : \mathrm{end}; \widehat{\{b\}} : \mathrm{end}\rangle\rangle;$$
$$\widehat{\{a,b\}} : \oplus\langle\widehat{\{a\}} : \&\langle\widehat{\emptyset} : \mathrm{end}; \widehat{\{a\}} : \mathrm{end}\rangle;$$
$$\widehat{\{b\}} : \&\langle\widehat{\emptyset} : \mathrm{end}; \widehat{\{b\}} : \mathrm{end}\rangle;$$
$$\widehat{\{a,b\}} : \&\langle\widehat{\emptyset} : \mathrm{end}; \widehat{\{a\}} : \mathrm{end}; \widehat{\{b\}} : \mathrm{end}\rangle\rangle\rangle$$
$$\mathscr{S}[\![\mathrm{I}[\tau]]\!] = \&\langle\widehat{\emptyset} : \oplus\langle\widehat{\emptyset} : \&\langle\widehat{\emptyset} : \mathrm{end}\rangle\rangle;$$
$$\widehat{\{a\}} : \oplus\langle\widehat{\{a\}} : \&\langle\widehat{\emptyset} : \mathrm{end}; \widehat{\{a\}} : \mathrm{end}\rangle\rangle;$$
$$\widehat{\{b\}} : \oplus\langle\widehat{\{b\}} : \&\langle\widehat{\emptyset} : \mathrm{end}; \widehat{\{b\}} : \mathrm{end}\rangle\rangle;$$
$$\widehat{\{a,b\}} : \oplus\langle\widehat{\{a,b\}} : \&\langle\widehat{\emptyset} : \mathrm{end}; \widehat{\{a\}} : \mathrm{end}; \widehat{\{b\}} : \mathrm{end}\rangle\rangle\rangle$$

and now we have $\mathscr{S}[\![\mathrm{I}[\tau]]\!] \leq \mathscr{S}[\![\mathrm{I}[\sigma]]\!]$. Notice that, in the encoding of $\mathrm{I}[\sigma]$, we have $\mathscr{A}(\sigma) = \{\{a\}, \{b\}, \{a,b\}\}$. ∎

More generally, the encodings of related contracts are related session types, except that the direction of the preorder is reversed.

Theorem 4. *Let σ and τ strongly convergent. Then $\mathrm{I}[\sigma] \preceq \mathrm{J}[\tau]$ if and only if $\mathscr{S}[\![\mathrm{J}[\tau]]\!] \leq \mathscr{S}[\![\mathrm{I}[\sigma]]\!]$.*

Proof. ("only if" part) Let $\mathscr{R}$ be the least relation such that

- if $\mathrm{I}[\sigma] \preceq \mathrm{J}[\tau]$ and σ and τ are strongly convergent, then $\mathscr{S}[\![\mathrm{J}[\tau]]\!] \mathscr{R} \mathscr{S}[\![\mathrm{I}[\sigma]]\!]$; additionally, if $\mathrm{K} \subseteq \mathrm{I}$ then $\mathscr{S}[\![\mathrm{nf}(\tau)]\!]_\mathrm{K} \mathscr{R} \mathscr{S}[\![\mathrm{nf}(\sigma)]\!]_\mathrm{K}$.

We prove that $\mathscr{R}$ is a coinductive subtyping. Let $S \mathscr{R} T$. We have two possibilities, according to the definition of $\mathscr{R}$.

1. $(S = \mathscr{S}[\![\mathrm{J}[\tau]]\!], T = \mathscr{S}[\![\mathrm{I}[\sigma]]\!],$ and $\mathrm{I}[\sigma] \preceq \mathrm{J}[\tau])$ By definition of $\mathscr{S}[\![\cdot]\!]$ we have $S = \&\langle\widehat{\mathrm{K}} : \mathscr{S}[\![\mathrm{nf}(\tau)]\!]_\mathrm{K} \overset{\mathrm{K}\subseteq\mathrm{J}}{}\rangle$ and $T = \&\langle\widehat{\mathrm{K}} : \mathscr{S}[\![\mathrm{nf}(\sigma)]\!]_\mathrm{K} \overset{\mathrm{K}\subseteq\mathrm{I}}{}\rangle$. From $\mathrm{I} \subseteq \mathrm{J}$ we have $\{\widehat{\mathrm{K}} \mid \mathrm{K} \subseteq \mathrm{I}\} \subseteq \{\widehat{\mathrm{K}} \mid \mathrm{K} \subseteq \mathrm{J}\}$, hence each label in the topmost branch of T also occurs as a label in the topmost branch of S. Now take $\mathrm{K} \subseteq \mathrm{I}$. We have to show that $\mathscr{S}[\![\mathrm{nf}(\tau)]\!]_\mathrm{K} \mathscr{R} \mathscr{S}[\![\mathrm{nf}(\sigma)]\!]_\mathrm{K})$, but this follows immediately from the definition of $\mathscr{R}$ and from $\mathrm{I}[\sigma] \preceq \mathrm{J}[\tau]$.
2. $(S = \mathscr{S}[\![\mathrm{nf}(\tau)]\!]_\mathrm{K}, T = \mathscr{S}[\![\mathrm{nf}(\sigma)]\!]_\mathrm{K}, \mathrm{K} \subseteq \mathrm{I},$ and $\mathrm{I}[\sigma] \preceq \mathrm{J}[\tau])$ We have several subcases depending on the shape of the normal form of σ and τ. Assume that $\mathrm{nf}(\sigma) = \bigoplus_{R\in\mathscr{A}(\sigma)}\Sigma_{\alpha\in R}\,\alpha.\sigma_\alpha$ and $\mathrm{nf}(\tau) = \bigoplus_{S\in\mathscr{A}(\tau)}\Sigma_{\beta\in S}\,\beta.\tau_\beta$. Then $S = \oplus\langle\widehat{\mathrm{S}\cap\mathrm{K}} : \&\langle\widehat{\emptyset} : \mathrm{end}; \widehat{\{\beta\}} : \mathscr{S}[\![\tau_\beta]\!]_\mathrm{K} \overset{\beta\in\mathrm{S}\cap\mathrm{K}}{}\rangle \overset{\mathrm{S}\in\mathscr{A}(\tau)}{}\rangle$ and $T = \oplus\langle\widehat{\mathrm{R}\cap\mathrm{K}} : \&\langle\widehat{\emptyset} : \mathrm{end}; \widehat{\{\alpha\}} : \mathscr{S}[\![\sigma_\alpha]\!]_\mathrm{K} \overset{\alpha\in\mathrm{R}\cap\mathrm{K}}{}\rangle \overset{\mathrm{R}\in\mathscr{A}(\sigma)}{}\rangle$. From $\mathrm{I}[\sigma] \preceq \mathrm{J}[\tau]$ we have that $\{\mathrm{S}\cap\mathrm{K} \mid \mathrm{S} \in \mathscr{A}(\tau)\} \subseteq \{\mathrm{R}\cap\mathrm{K} \mid \mathrm{R} \in \mathscr{A}(\sigma)\}$, hence each label in the topmost choice of S also occurs as a label in the topmost choice of T. Take $\mathrm{S} \in \mathscr{A}(\tau)$. The label $\widehat{\mathrm{S}\cap\mathrm{K}}$ occurs in both S and T, hence the corresponding branches have exactly the same set of labels $\{\widehat{\emptyset}\} \cup \{\widehat{\{\alpha\}} \mid \alpha \in \mathrm{S}\cap\mathrm{K}\}$. Let $S' = \mathscr{S}[\![\tau_\alpha]\!]_\mathrm{K}$ and $T' = \mathscr{S}[\![\sigma_\alpha]\!]_\mathrm{K}$. From $\mathrm{I}[\mathrm{nf}(\sigma)] \preceq \mathrm{J}[\mathrm{nf}(\tau)]$ and $\alpha \in \mathrm{I}$ we have that $\mathrm{nf}(\tau) \overset{\alpha}{\Longrightarrow}$ implies $\mathrm{nf}(\sigma) \overset{\alpha}{\Longrightarrow}$ and $\mathrm{I}[\sigma_\alpha] \preceq \mathrm{J}[\tau_\alpha]$, so we conclude $S' \mathscr{R} T'$ by definition of $\mathscr{R}$.

Assume that $\mathrm{nf}(\sigma) = \mathrm{rec}\ x.\sigma'$, where σ' is in normal form. Then $\mathrm{I}[\sigma] \simeq \mathrm{I}[\sigma'\{\mathrm{nf}(\sigma)/_x\}]$ and $\sigma'\{\mathrm{nf}(\sigma)/_x\}$ is itself in normal form. Since σ is strongly convergent, we eventually reach some σ'' in normal form such that σ'' does not begin with a recursion and $\mathrm{I}[\sigma] \simeq \mathrm{I}[\sigma'']$. Similarly for τ. Hence, we easily reduce to the previous subcase.

("if" part) We prove that $\mathrm{I}[\sigma] \not\preceq \mathrm{J}[\tau]$ implies $\mathscr{S}[\![\mathrm{J}[\tau]]\!] \not\leq \mathscr{S}[\![\mathrm{I}[\sigma]]\!]$. It suffices to consider all the possibilities by which $\mathrm{I}[\sigma] \not\preceq \mathrm{J}[\tau]$ directly:

1. if $\mathrm{I} \not\subseteq \mathrm{J}$, then there exists $\alpha \in \mathrm{I}$ such that $\alpha \notin \mathrm{J}$. Then $\widehat{\{\alpha\}}$ is a label occurring in the topmost branch of $\mathscr{S}[\![\mathrm{I}[\sigma]]\!]$ but not occurring in the topmost branch of $\mathscr{S}[\![\mathrm{J}[\tau]]\!]$, hence $\mathscr{S}[\![\mathrm{J}[\tau]]\!] \not\leq \mathscr{S}[\![\mathrm{I}[\sigma]]\!]$;

2. let $\mathrm{R}_1,\ldots,\mathrm{R}_n$ be the ready sets of σ and assume that there exists R such that $\tau \Downarrow \mathrm{R}$ and for every $1 \leq i \leq n$ we have $\mathrm{R}_i \not\subseteq \mathrm{R}$, that is for every $1 \leq i \leq n$ there exists $\alpha_i \in \mathrm{R}_i \cap \mathrm{I}$ and $\alpha_i \notin \mathrm{R}$. From $\mathrm{I} \subseteq \mathrm{J}$ we know that both $\mathscr{S}[\![\mathrm{I}[\sigma]]\!]$ and $\mathscr{S}[\![\mathrm{J}[\tau]]\!]$ have the label $\widehat{\mathrm{I}}$ in their corresponding topmost branches with continuations $\mathscr{S}[\![\mathrm{nf}(\sigma)]\!]_{\mathrm{I}}$ and $\mathscr{S}[\![\mathrm{nf}(\tau)]\!]_{\mathrm{I}}$ respectively. Then, by the fact that $\widehat{\cdot}$ is injective, it follows that $\widehat{\mathrm{R} \cap \mathrm{I}} \notin \{\widehat{\mathrm{R}_i \cap \mathrm{I}} \mid \mathrm{R}_i \in \mathscr{A}(\sigma)\}$, hence $\widehat{\mathrm{R} \cap \mathrm{I}}$ is a label occurring in the topmost choice of $\mathscr{S}[\![\mathrm{nf}(\tau)]\!]_{\mathrm{I}}$ but not occurring in the topmost choice of $\mathscr{S}[\![\mathrm{nf}(\sigma)]\!]_{\mathrm{I}}$, so we conclude $\mathscr{S}[\![\mathrm{J}[\tau]]\!] \not\leq \mathscr{S}[\![\mathrm{I}[\sigma]]\!]$;

3. assume that there exists $\alpha \in \mathrm{I}$ such that $\tau \overset{\alpha}{\Longrightarrow}$ and $\sigma \overset{\alpha}{\not\Longrightarrow}$. Then $\tau \Downarrow \mathrm{S}$ where $\alpha \in \mathrm{S}$ whereas $\sigma \Downarrow \mathrm{R}$ implies $\alpha \notin \mathrm{R}$. Hence we can reason as for the previous case and conclude $\mathscr{S}[\![\mathrm{J}[\tau]]\!] \not\leq \mathscr{S}[\![\mathrm{I}[\sigma]]\!]$. $\qquad\square$

The strict correspondence between $\leq$ and $\preceq$ allows one to use them interchangeably, as discussed in the example below.

Example 3. Consider the client contracts

$$\mathrm{K}[\rho] \overset{\mathrm{def}}{=} \{\bar{a},\bar{b},\mathrm{e}\}[\bar{a}.\mathrm{e} + \bar{b}.\mathrm{e}] \qquad \text{and} \qquad \mathrm{K}[\rho'] \overset{\mathrm{def}}{=} \{\bar{a},\bar{b},\mathrm{e}\}[\bar{a}.\mathrm{e} \oplus \bar{b}.\mathrm{e}]$$

and notice that $\mathrm{K}[\rho] \dashv \mathrm{I}[\sigma]$ whereas $\mathrm{K}[\rho'] \not\dashv \mathrm{I}[\sigma]$, where $\mathrm{I}[\sigma]$ is the service contract defined in Example 2. The respective encodings of these two client contracts are

$$\mathscr{C}[\![\mathrm{K}[\rho]]\!] = \oplus\langle\widehat{\{a,b\}} : \&\langle\widehat{\{a\}} : \oplus\langle\widehat{\{a\}} : \mathrm{end}\rangle;$$
$$\widehat{\{b\}} : \oplus\langle\widehat{\{b\}} : \mathrm{end}\rangle;$$
$$\widehat{\{a,b\}} : \oplus\langle\widehat{\{a\}} : \mathrm{end}; \widehat{\{b\}} : \mathrm{end}\rangle\rangle\rangle$$
$$\mathscr{C}[\![\mathrm{K}[\rho']]\!] = \oplus\langle\widehat{\{a,b\}} : \&\langle\widehat{\{a,b\}} : \oplus\langle\widehat{\{a\}} : \mathrm{end}; \widehat{\{b\}} : \mathrm{end}\rangle\rangle\rangle$$

and now we notice that $\mathscr{S}[\![\mathrm{I}[\sigma]]\!] \leq \mathrm{dual}(\mathscr{C}[\![\mathrm{K}[\rho]]\!])$, namely it is safe to use $\mathscr{S}[\![\mathrm{I}[\sigma]]\!]$ to interact successfully with $\mathscr{C}[\![\mathrm{K}[\rho]]\!]$. On the other hand $\mathscr{S}[\![\mathrm{I}[\sigma]]\!] \not\leq \mathrm{dual}(\mathscr{C}[\![\mathrm{K}[\rho']]\!])$ because $\mathscr{S}[\![\mathrm{I}[\sigma]]\!]$ may be in a state where only a or only b are available, whilst the client $\mathrm{K}[\rho']$ autonomously decides which of the two actions to execute. Notice however that $\mathscr{S}[\![\mathrm{I}[\tau]]\!] \leq \mathrm{dual}(\mathscr{C}[\![\mathrm{K}[\rho']]\!])$. $\qquad\blacksquare$

More generally, if $\mathrm{K}[\rho] \dashv \mathrm{I}[\sigma]$ holds then it is safe to use the service $\mathscr{S}[\![\mathrm{I}[\sigma]]\!]$ to interact with the client $\mathscr{C}[\![\mathrm{K}[\rho]]\!]$ and in fact $\mathrm{dual}(\mathscr{C}[\![\mathrm{K}[\rho]]\!])$ is the *principal service type* that interacts successfully with $\mathscr{C}[\![\mathrm{K}[\rho]]\!]$.

Theorem 5. $\text{K}[\rho] \dashv \text{I}[\sigma]$ *if and only if* $\mathscr{S}[\![\text{I}[\sigma]]\!] \leq \text{dual}(\mathscr{C}[\![\text{K}[\rho]]\!])$.

Proof. ("only if" part) Let $\mathscr{R}$ be the least relation such that

- if $\text{K}[\rho] \dashv \text{I}[\sigma]$ and ρ and σ are strongly convergent, then $\mathscr{S}[\![\text{I}[\sigma]]\!] \mathscr{R} \text{ dual}(\mathscr{C}[\![\text{K}[\rho]]\!])$ and $\mathscr{S}[\![\text{nf}(\sigma)]\!]_{\overline{\text{K}\setminus\{\text{e}\}}} \mathscr{R} \text{ dual}(\mathscr{C}[\![\text{nf}(\rho)]\!]_{\text{K}\setminus\{\text{e}\}})$.

It is sufficient to show that $\mathscr{R}$ is a coinductive subtyping. Let $S \mathscr{R} T$. We have two possibilities.

1. $(S = \mathscr{S}[\![\text{I}[\sigma]]\!], \ T = \text{dual}(\mathscr{C}[\![\text{K}[\rho]]\!]), \text{ and } \text{K}[\rho] \dashv \text{I}[\sigma])$ Then $S = \&\langle\widehat{\text{H}} : \mathscr{S}[\![\text{nf}(\sigma)]\!]_{\text{H}} {}^{\text{H}\subseteq\text{I}}\rangle$ and $T = \&\langle\text{K}\setminus\{\text{e}\} : \text{dual}(\mathscr{C}[\![\text{nf}(\rho)]\!]_{\text{K}\setminus\{\text{e}\}})\rangle$. From $\overline{\text{K}\setminus\{\text{e}\}} \subseteq \text{I}$ we have that the label in the topmost branch of T also occurs as a label in the topmost label of S. We must show that $\mathscr{S}[\![\text{nf}(\sigma)]\!]_{\text{H}} \mathscr{R} \text{ dual}(\mathscr{C}[\![\text{nf}(\rho)]\!]_{\text{K}\setminus\{\text{e}\}})$, but this is obvious by definition of $\mathscr{R}$.

2. $(S = \mathscr{S}[\![\text{nf}(\sigma)]\!]_{\overline{\text{K}}}, \ T = \text{dual}(\mathscr{C}[\![\text{nf}(\rho)]\!]_{\text{K}}), \text{ and } (\text{K}\cup\{\text{e}\})[\rho] \dashv \text{I}[\sigma])$ Let $\text{nf}(\sigma) = \bigoplus_{\text{R}\in\mathscr{A}(\sigma)} \sum_{\alpha\in\text{R}} \alpha.\sigma_\alpha$ and $\text{nf}(\rho) = \bigoplus_{\text{R}\in\mathscr{A}(\rho)} \sum_{\alpha\in\text{R}} \alpha.\rho_\alpha$. Then

$$S = \oplus\langle\widehat{\text{R}\cap\overline{\text{K}}} : \&\langle\widehat{\emptyset} : \text{end}; \widehat{\{\alpha\}} : \mathscr{S}[\![\text{nf}(\sigma_\alpha)]\!]_{\overline{\text{K}}} {}^{\alpha\in\text{R}\cap\overline{\text{K}}}\rangle {}^{\text{R}\in\mathscr{A}(\sigma)}\rangle$$
$$T = \oplus\langle\widehat{\text{s}} : \&\langle\widehat{\emptyset} : \text{end}; \widehat{\{\overline{\alpha}\}} : \text{dual}(\mathscr{C}[\![\text{nf}(\rho_\alpha)]\!]_{\text{K}}) {}^{\text{R}\in\mathscr{A}(\rho),\alpha\in\text{S}\cap\text{R}}\rangle {}^{\text{S}\subseteq\text{K},\text{S}\bowtie\mathscr{A}(\rho)}\rangle$$

Let $\text{R} \in \mathscr{A}(\sigma)$ be a ready set of the service. Then $\overline{\text{R}\cap\overline{\text{K}}} = \overline{\text{R}}\cap\text{K} \subseteq \text{K}$. Furthermore, by definition of $\bowtie$, we have that $\overline{\text{R}\cap\overline{\text{K}}} \bowtie \mathscr{A}(\rho)$ if and only if for every $\text{R}' \in \mathscr{A}(\rho)$ we have either $\text{e} \in \text{R}'$ or $\overline{\text{R}\cap\overline{\text{K}}}\cap\text{R}' \neq \emptyset$. But $\text{R}' \subseteq \text{K}$, hence $\overline{\text{R}\cap\overline{\text{K}}}\cap\text{R}' = \overline{\text{R}}\cap\text{R}'$. So $\overline{\text{R}\cap\overline{\text{K}}} \bowtie \mathscr{A}(\rho)$ is a direct consequence of the hypothesis $(\text{K}\cup\{\text{e}\})[\rho] \dashv \text{I}[\sigma]$. Furthermore, from the same hypothesis and from $\text{nf}(\rho) \stackrel{\alpha}{\Longrightarrow}$ and $\text{nf}(\sigma) \stackrel{\overline{\alpha}}{\Longrightarrow}$ it follows that $(\text{K}\cup\{\text{e}\})[\rho_\alpha] \dashv \text{I}[\sigma_\alpha]$, hence we conclude $\mathscr{S}[\![\text{nf}(\sigma_\alpha)]\!]_{\overline{\text{K}}} \mathscr{R} \text{ dual}(\mathscr{C}[\![\text{nf}(\rho_\alpha)]\!]_{\text{K}})$ by definition of $\mathscr{R}$.

("if" part) Trivial by the definitions of the encoding, the details are left to the reader. $\qquad\square$

By combining Theorem 5 and Proposition 3(3) we derive an interesting dual result, by which $\text{dual}(\mathscr{S}[\![\text{I}[\sigma]]\!])$ is the *principal client type* that interacts successfully with $\mathscr{S}[\![\text{I}[\sigma]]\!]$.

Corollary 1. $\text{K}[\rho] \dashv \text{I}[\sigma]$ *if and only if* $\mathscr{C}[\![\text{K}[\rho]]\!] \leq \text{dual}(\mathscr{S}[\![\text{I}[\sigma]]\!])$.

6   Discussion

The encoding of session types into contracts (Section 4) is simple and almost homomorphic with respect to the operations. The branch is the only operation whose encoding requires some care, by adding extra $\ell.\Omega$ subterms in the generated contract for those labels ℓ not mentioned in the session type. That is, the encoding renders clearly that processes involved in a session may only perform actions that are explicitly allowed by the type of the partner process. In other words, session types describe communications where at any time exactly one of the two interacting parties has control and no hand-shaking occurs. Interestingly, by interpreting Ω as a catastrophic state of a process and by considering a slightly weaker notion of subcontract relation called *safe must* in [2], we have $\alpha.\Omega \preceq \mathbf{0}$, meaning that, in practice, action α is *not* guaranteed.

The encoding of contracts into session types (Section 5) manifests an exponential blow up of the encoded contract. This indicates that contracts are more abstract than session types and are capable of expressing more complex synchronization scenarios. Part of this added expressiveness derives from the operators $+$ and $\oplus$, which can be used for composing arbitrary behaviors in a liberal way. As we have anticipated in the introduction, the encoding of $\{a,b\}[a \oplus b]$ explicitly notifies the client if only a is available, or if only b is available, or if both a and b are available. This is possible only if the service has centralized control over its own resources and can decide about the availability of a and b. If, on the other hand, the service is a collection of possibly distributed processes (as in service choreographies) it may be impractical or impossible to provide such information to the client.

The lesson we learn is that there is a strict correspondence between contracts and session types that allows one to use them interchangeably *within the context of dyadic interactions*, where both parties provide centralized control on the communication protocol. Contracts go beyond session types in that they permit to characterize *arbitrary processes* in addition to *sessions*. For instance, the process $\overline{a} \mid \overline{b}$, which stands for the parallel composition of two smaller processes sending messages $\overline{a}$ and $\overline{b}$, can be seen as having two unrelated sessions with type $\oplus\langle a : \mathsf{end}\rangle$ and $\oplus\langle b : \mathsf{end}\rangle$. On the other hand, the *whole* process can be typed according to one of the following behaviors

$$\Omega \qquad \overline{a}.\Omega + \overline{b}.\Omega \qquad \overline{a}.\overline{b} + \overline{b}.\Omega \qquad \overline{a}.\Omega + \overline{b}.\overline{a} \qquad \overline{a}.\overline{b} + \overline{b}.\overline{a}$$

representing regular, increasingly accurate approximations of the process behavior.

Recently type systems with sessions have been extended so as to guarantee stronger progress properties [9]. Such type systems must necessarily consider a broader perspective that takes into account the mutual dependencies and interactions between sessions. In this respect, we plan to investigate whether the additional expressiveness of contracts already provides the required machinery for dealing with these issues.

References

1. Asperti, A., Laneve, C.: Interaction systems I: The theory of optimal reductions. Mathematical Structures in Computer Science 4(4), 457–504 (1994)
2. Boreale, M., De Nicola, R., Pugliese, R.: Basic observables for processes. Information and Computation 149(1), 77–98 (1999)
3. Carpineti, S., Castagna, G., Laneve, C., Padovani, L.: A formal account of contracts for Web Services. In: Bravetti, M., Núñez, M., Zavattaro, G. (eds.) WS-FM 2006. LNCS, vol. 4184, pp. 148–162. Springer, Heidelberg (2006)
4. Castagna, G., Gesbert, N., Padovani, L.: A theory of contracts for web services. In: POPL 2008: Proceedings of the 35th annual ACM SIGPLAN-SIGACT symposium on Principles of programming languages, pp. 261–272. ACM, New York (2008)
5. Cleaveland, R., Hennessy, M.: Testing equivalence as a bisimulation equivalence. Formal Aspects of Computing 5(1), 1–20 (1993)

6. De Nicola, R.: Two complete axiom systems for a theory of communicating sequential processes. Information and Control 64(1–3), 136–172 (1985)
7. De Nicola, R., Hennessy, M.: Testing equivalences for processes. Theoretical Computer Science 34, 83–133 (1984)
8. De Nicola, R., Hennessy, M.: CCS without τ's. In: Ehrig, H., Levi, G., Montanari, U. (eds.) CAAP 1987 and TAPSOFT 1987. LNCS, vol. 249, pp. 138–152. Springer, Heidelberg (1987)
9. Dezani-Ciancaglini, M., de' Liguoro, U., Yoshida, N.: On Progress for Structured Communications. In: Barthe, G., Fournet, C. (eds.) TGC 2007. LNCS, vol. 4912, pp. 257–275. Springer, Heidelberg (2008)
10. Dezani-Ciancaglini, M., Mostrous, D., Yoshida, N., Drossopoulou, S.: Session Types for Object-Oriented Languages. In: Thomas, D. (ed.) ECOOP 2006. LNCS, vol. 4067, pp. 328–352. Springer, Heidelberg (2006)
11. Gay, S., Hole, M.: Subtyping for session types in the π-calculus. Acta Informatica 42(2-3), 191–225 (2005)
12. Hennessy, M.: Acceptance trees. JACM: Journal of the ACM 32(4), 896–928 (1985)
13. Hennessy, M.: Algebraic Theory of Processes. Foundation of Computing. MIT Press, Cambridge (1988)
14. Honda, K.: Types for dyadic interaction. In: Best, E. (ed.) CONCUR 1993. LNCS, vol. 715, pp. 509–523. Springer, Heidelberg (1993)
15. Laneve, C., Padovani, L.: The *must* preorder revisited – an algebraic theory for web services contracts. In: Caires, L., Vasconcelos, V.T. (eds.) CONCUR. LNCS, vol. 4703, pp. 212–225. Springer, Heidelberg (2007)
16. Vasconcelos, V., Gay, S., Ravara, A.: Type checking a multithreaded functional language with session types. Theoretical Computer Science 368 (2006)

Specifying and Analysing SOC Applications with COWS[*]

Alessandro Lapadula, Rosario Pugliese, and Francesco Tiezzi

Dipartimento di Sistemi e Informatica Università degli Studi di Firenze

Dedicated to Ugo Montanari on the occasion of his 65th birthday

Abstract. COWS is a recently defined process calculus for specifying and combining service-oriented applications, while modelling their dynamic behaviour. Since its introduction, a number of methods and tools have been devised to analyse COWS specifications, like e.g. a type system to check confidentiality properties, a logic and a model checker to express and check functional properties of services. In this paper, by means of a case study in the area of automotive systems, we demonstrate that COWS, with some mild linguistic additions, can model all the phases of the life cycle of service-oriented applications, such as publication, discovery, negotiation, orchestration, deployment, reconfiguration and execution. We also provide a flavour of the properties that can be analysed by using the tools mentioned above.

1   Introduction

In recent years, the increasing success of e-business, e-learning, e-government, and other similar emerging models, has led the World Wide Web, initially thought of as a system for human use, to evolve towards an architecture for *service-oriented computing* (SOC) supporting automated use. SOC advocates the use of loosely coupled 'services', to be understood as autonomous, platform-independent, computational entities that can be described, published, discovered, and assembled, as the basic blocks for building interoperable and evolvable systems and applications. While early examples of technologies that are at least partly service-oriented date back to CORBA, DCOM, J2EE and IBM WebSphere, the most successful instantiation of the SOC paradigm are probably the more recent *web services*. These are sets of operations that can be published, located and invoked through the Web via XML messages complying with given standard formats. To support the web service approach, several new languages and technologies have been designed and many international companies have invested a lot of efforts.

Current software engineering technologies for SOC, however, remain at the descriptive level and lack rigorous formal foundations. We are still experiencing a gap between practice (programming) and theory (formal methods and analysis techniques) in the design of SOC applications. The challenges come from the necessity of dealing at once with issues like communication, co-operation, resource usage, security, failures, etc. in a setting where demands and guarantees can be very different for the many different components. Many researchers have hence put forward the idea of using *process*

---

[*] This work has been supported by the EU project SENSORIA, IST-2005-016004.

P. Degano et al. (Eds.): Montanari Festschrift, LNCS 5065, pp. 701–720, 2008.

calculi, a cornerstone of current foundational research on specification and analysis of concurrent, distributed and mobile systems through mathematical — mainly algebraic and logical — tools. Thus, many process calculi have been designed, addressing one aspect or another of SOC and aiming at assessing the adequacy of diverse sets of primitives w.r.t. modelling, combining and analysing service-oriented applications.

Due to their algebraic nature, process calculi convey in a distilled form the compositional programming style of SOC. Thus, for example, many well-known problems related to services composition (e.g., messages not received, race conditions, deadlocks, incompatible behaviours) could be investigated through an adequate and sufficiently expressive process calculus. A major benefit of using process calculi is that they enjoy a rich repertoire of elegant meta-theories, proof techniques and analytical tools that can be likely tailored to the needs of service-based applications. It has been already argued that type systems, model checking and (bi)simulation analysis provide adequate tools to address topics relevant to the web services technology (see e.g. [20,24]). This 'proof technology' can eventually pave the way for the development of automatic property validation tools. Therefore, process calculi might play a central role in laying rigorous methodological foundations for specification and validation of SOC applications.

By taking inspiration from well-known process calculi and from the standard language for orchestration of web services WS-BPEL [22], in [15] we have designed COWS (*Calculus for Orchestration of Web Services*), a process calculus for specifying and combining service-oriented applications, while modelling their dynamic behaviour. We have shown that COWS can model different and typical features of web services, such as, e.g., multiple start activities, receive conflicts, routing of correlated messages, service instances and interactions among them. Since its definition, some linguistic extensions have been introduced to model timed activities [17] and dynamic service discovery and negotiation [19], thus obtaining a linguistic formalism capable of modelling all the phases of the life cycle of service-oriented applications. A number of methods and tools have also been devised to analyse COWS specifications, such as the stochastic extension defined in [23] to enable quantitative reasoning on service behaviours, the type system introduced in [18] to check confidentiality properties, and the logic and model checker presented in [9] to express and check functional properties of services. In this paper, by means of the 'on road assistance scenario', a case study in the area of automotive systems defined and analysed within the EU project SENSORIA [2], we provide a flavour of COWS main features and specification style, and illustrate the classes of properties that can be analysed by using some of the tools mentioned above.

The rest of the paper is organized as follows. Section 2 introduces the scenario that will be used throughout the paper for illustration purposes. Section 3 presents syntax and main features of COWS; this is done in a step-by-step fashion while modelling some services within the scenario and their orchestration. Section 4 shows that also service discovery and negotiation can be naturally modelled in COWS by exploiting some mild linguistic additions, i.e. timed activities, constraints and operations on them. Section 5 sums up a type-based approach for expressing and enforcing confidentiality properties. Section 6 illustrates a logical verification framework including the logic SocL for expressing functional properties of services and the on-the-fly model checker CMC for verifying them. Section 7 concludes the paper with some final remarks.

2   On Road Assistance Scenario

The 'on road assistance scenario' [13] is one of the scenarios in the area of automotive systems defined and analysed within the EU project SENSORIA [2] and describes some functionalities that will be likely available in the near future. The scenario involves a number of services that are discovered and bound at run-time according to levels of service specified at design time, so as to deliver the best available functionalities at agreed levels of quality. A brief description follows.

> The in-vehicle *diagnostic* system reports a severe failure when the car is no longer drivable. The car's *discovery* system then identifies garages, car rentals and towing truck services in the car's vicinity. At this point, the car's *reasoner* system selects a set of adequate services taking into account personalised policies and preferences of the driver, e.g. balancing cost and delay, and tries to order them. Before being enable to order services, the owner of the car has to deposit a security payment, that will be given back if ordering the services fails. Other components of the in-vehicle service platform involved in this assistance activity are a *GPS* service, providing the car's current location, and an *orchestrator*, coordinating all the described services.

An UML-like activity diagram of the orchestration of services is shown in Figure 1. For simplicity, we assume that the orchestration is only triggered either by an 'engine failure' or by a 'low oil level' sensor signal. The process starts with a request from the orchestrator to the bank to charge the driver's credit card with the security deposit payment. This is modelled by the UML action requestCardCharge for charging the credit card whose number is provided as an output parameter of the action call. In parallel to the interaction with the bank, the orchestrator requests the current location of the car from the car's internal GPS service. The current location is modelled as an input to the requestLocation action and subsequently used by the findServices interaction which retrieves a list of services. If no service can be found, an action to compensate the credit card charge will be launched. For the selection of services, the orchestrator synchronises with the reasoner service to obtain the most appropriate (best) services.

Service ordering is modelled by the UML actions orderGarage, orderTowTruck and orderRentalCar. When the orchestrator makes an appointment with the garage, the diagnostic data are automatically transferred to the garage, which could then be able, e.g., to identify the spare parts needed to perform the repair. Then, the orchestrator makes an appointment with the towing service, providing the GPS data of the stranded vehicle and of the garage, to tow the vehicle to the garage. Concurrently, the orchestrator makes an appointment with the rental service, by indicating the location where the car will be handed over to the driver.

The workflow described in Figure 1 models the overall behaviour of the system. Besides interactions among services, it also includes activities using concepts developed for long running business transactions (e.g. in [11,22]). These activities entail fault and compensation handling, kind of specific activities attempting to reverse the effects of previously committed activities, that are an important aspect of SOC applications. Specifically, in the considered scenario, the security deposit payment charged to

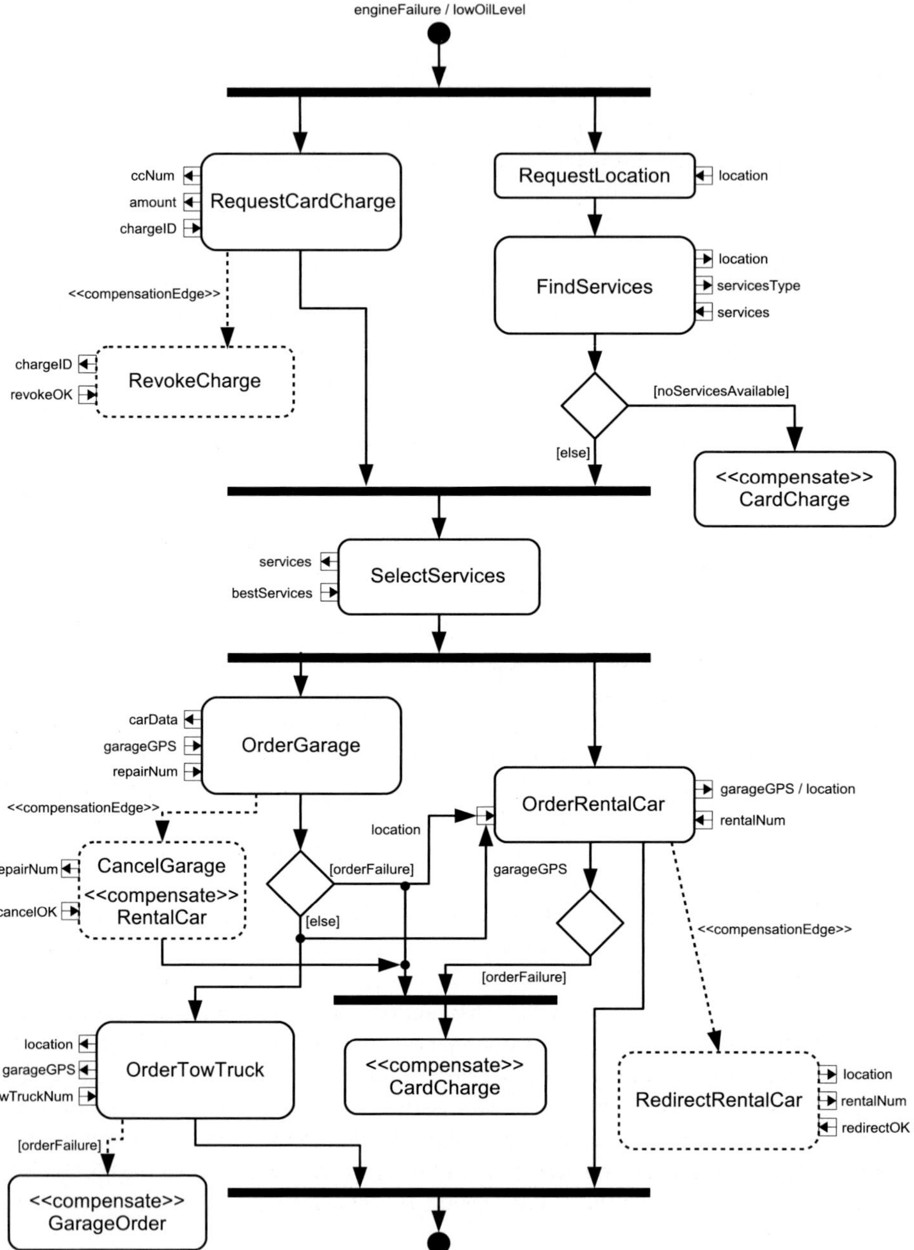

Fig. 1. Orchestration in the on road assistance scenario

Table 1. COWS syntax

$$
\begin{array}{llll}
s ::= & \mathbf{kill}(k) & | \quad u \cdot u'!\bar{e} & \text{(kill, invoke)} \\
& | \quad \sum_{i=0}^{l} p_i \cdot o_i?\bar{w}_i.s_i & | \quad s \,|\, s & \text{(receive-guarded choice, parallel)} \\
& | \quad \{\!|s|\!\} & | \quad [d]\, s \quad | \quad * s & \text{(protection, delimitation, replication)}
\end{array}
$$

the driver's credit card must be revoked if either the discovery phase does not succeed or ordering the services fails, i.e. both garage/tow truck and car rental services reject the requests. Moreover, if ordering a tow truck fails, the garage appointment has to be cancelled and the rental car delivery has to be redirected to the stranded car's actual location. Instead, if ordering the car rental fails, the overall process may not fail, as the activity is enclosed in a sub-transaction.

3   COWS: A Calculus for Orchestration of Web Services

In this section, we report the syntax of COWS and explain the semantics of its primitives in a step-by-step fashion while modelling the on road assistance scenario (the complete specification can be found in [16]). Due to lack of space, here we only provide an informal account of the semantics of COWS and refer the interested reader to [15,14] for a formal presentation, for examples illustrating peculiarities and expressiveness of the language, and for comparisons with other process-based and orchestration formalisms. To get accustomed to using the language one can also use CMC [1], a tool supporting the automated derivation of all computations originating from a COWS term.

3.1   Syntax

The syntax of COWS, given in Table 1, is parameterized by three countable and pairwise disjoint sets: the set of *(killer) labels* (ranged over by $k, k', \ldots$), the set of *values* (ranged over by $v, v', \ldots$) and the set of 'write once' *variables* (ranged over by $x, y, \ldots$). The set of values is left unspecified, but we assume that it includes the set of *names*, ranged over by $m, n, o, p, \ldots$, mainly used to represent partners and operations. The language is also parameterized by a set of *expressions*, ranged over by e, whose exact syntax is deliberately omitted. We just assume that expressions contain, at least, values and variables, and do not include killer labels (that, hence, are *not* communicable values). Partner and operation names can be combined to designate communication *endpoints*; e.g. $p \cdot o$ denotes the endpoint composed of the partner p and the operation o. Being values, partner and operation names can be exchanged in communication, but dynamically received names can only be used to designate endpoints for service invocation. Indeed, endpoints of receive activities are identified statically because their syntax only allows using names and not variables.

We use w to range over values and variables, u to range over names and variables, and d to range over killer labels, names and variables. Notation $\bar{}$ stands for tuples of objects, e.g. $\bar{x}$ is a compact notation for denoting the tuple of variables $\langle x_1, \ldots, x_n \rangle$ (with $n \geq 0$).

In the sequel, we shall use **0** to denote empty choice and + to abbreviate binary choice. We will omit trailing occurrences of **0**, writing e.g. $p \cdot o?\bar{w}$ instead of $p \cdot o?\bar{w}.\mathbf{0}$, and write $[d_1, \ldots, d_n]\, s$ in place of $[d_1] \ldots [d_n]\, s$. We will write $Z \triangleq W$ to assign a symbolic name Z to the term W.

The only *binding* construct is delimitation: $[d]\, s$ binds d in the scope s (the notions of *bound* and *free* occurrences of a name/variable/label are defined accordingly). In fact, differently from most process calculi, receive activities in COWS bind neither names nor variables. This enables e.g. easily modelling and updating the shared state of concurrent threads within each service instance. Delimitation can be used to generate fresh names whose scope can later dynamically change because of taking place of communication. This is exactly as in the π-calculus [21]. However, delimitation is more general than the restriction of the π-calculus since it can be also used to declare variables (thus regulating the range of application of the substitutions generated by communications) and to delimit the field of action of kill activities. Notably, killer labels are dealt with differently from names and variables since, being not communicable values, their scope is statically determined by the corresponding delimitation and can never change.

3.2   Basic Operators for Service Orchestration

The COWS term representing the 'orchestration' in Figure 1 is

$$[p_{car}]\,(Orchestrator \mid GPS \mid Discovery \mid Reasoner \mid SensorsMonitor\,)$$
$$\mid Bank \mid OnRoadRepairServices$$

The services above are composed by using the *parallel composition* operator $\_ \mid \_$ that allows the different components to be concurrently executed and to interact with each other. The *delimitation* operator $[\_]\_$ is used here to declare that p_{car} is a (partner) name known to all services of the in-vehicle platform, i.e. *Orchestrator*, *GPS*, *Discovery*, *Reasoner* and *SensorsMonitor*, and only to them.

Orchestrator, the most important component of the in-vehicle platform, is

$$[x_{carData}]\,(\,p_{car} \cdot o_{engfail}?\langle x_{carData}\rangle.s_{engfail} \;+\; p_{car} \cdot o_{lowoil}?\langle x_{carData}\rangle.s_{lowoil}\,)$$

This term uses the *choice* operator $\_ + \_$ to pick one of those alternative 'recovery' behaviours whose execution can start immediately. The *receive-guarded prefix* operator $p_{car} \cdot o_i?\langle x_{carData}\rangle.\_$ expresses that each recovery behaviour starts with a *receive* activity of the form $p_{car} \cdot o_i?\langle x_{carData}\rangle$ corresponding to reception of a request emitted, when a failure arises, by *SensorsMonitor* (a term representing the behaviour of the 'low level vehicle platform'). *Receives*, together with *invokes*, written as $p \cdot o!\langle e_1, \ldots, e_m\rangle$, are the basic communication activities provided by COWS. Besides input parameters and sent values, they indicate the endpoint $p \cdot o$ through which the communication should occur. $p \cdot o$ can be interpreted as a specific implementation of operation o provided by the service identified by the logic name p. This naming mechanism allows a same service to be identified by means of different logic names, i.e. to play more than one partner role as in WS-BPEL. An inter-service communication takes place when the arguments of a receive and of a concurrent invoke along the same endpoint do match[1],

---

[1] The pattern-matching mechanism permits correlating messages logically forming a same interaction 'session' by means of their same contents. We refer to [15,14] for further details.

and causes replacement of the variables arguments of the receive with the corresponding values arguments of the invoke (within the scope of variables declarations). For example, variable $x_{carData}$, declared local to *Orchestrator* by means of the delimitation operator, is initialized by the receive leading the recovery activity with data provided by *SensorsMonitor*. Notice that, while executing a recovery behaviour, *Orchestrator* does not accept other recovery requests. We are also assuming that it is restarted at the end of the recovery task.

The recovery behaviour $s_{engfail}$ executed when an engine failure occurs is

$$[p_e, o_e, x_{loc}, x_{list}]$$
$$(\,(\,RequestCardCharge \mid RequestLocation.\,FindServices\,)$$
$$\mid\ p_e \cdot o_e?\langle\rangle.\ p_e \cdot o_e?\langle\rangle.\ SelectServices.$$
$$[x_{garageGPS}]\,(\,OrderGarage.\,OrderTowTruck \mid OrderRentalCar\,)\,)$$

$p_e \cdot o_e$ is a scoped endpoint along which successful termination signals (i.e. communications that carry no data) are exchanged to orchestrate execution of the different components. Variables x_{loc}, x_{list} and $x_{garageGPS}$ are used to store the value of the car's current location, the list of closer on road services discovered and the garage's GPS location, respectively. To present the specification of $s_{engfail}$ in terms of the UML actions of Figure 1, we have used an auxiliary 'sequence' notation. Thus, e.g., *RequestLocation.FindServices* indicates that execution of *RequestLocation* terminates before execution of *FindServices* starts. Indeed, *RequestLocation.FindServices* actually stands for the COWS term

$$p_{car} \cdot o_{reqLoc}!\langle\rangle \mid p_{car} \cdot o_{respLoc}?\langle x_{loc}\rangle.$$
$$(\,p_{car} \cdot o_{find}!\langle x_{loc}, ServicesType\rangle$$
$$\mid\ p_{car} \cdot o_{servicesFound}?\langle x_{list}\rangle.\ p_e \cdot o_e!\langle\rangle\ +\ p_{car} \cdot o_{servicesNotFound}?\langle\rangle\,)$$

where *RequestLocation* and *FindServices* are

$$RequestLocation\ \triangleq\ p_{car} \cdot o_{reqLoc}!\langle\rangle \mid p_{car} \cdot o_{respLoc}?\langle x_{loc}\rangle$$

$$FindServices\ \triangleq\ p_{car} \cdot o_{find}!\langle x_{loc}, ServicesType\rangle$$
$$\mid\ p_{car} \cdot o_{servicesFound}?\langle x_{list}\rangle.\ p_e \cdot o_e!\langle\rangle$$

Endpoints of service invocations can also contain variables as, e.g., in the term

$$OrderGarage\ \triangleq\ x_{garage} \cdot o_{order}!\langle p_{car}, x_{carData}\rangle\ \mid$$
$$[x_{repairNum}]\,(p_{car} \cdot o_{garageFail}?\langle\rangle\ +\ p_{car} \cdot o_{garageOK}?\langle x_{repairNum}, x_{garageGPS}\rangle)$$

Here, variable x_{garage} is used to invoke a garage service whose partner name is unknown at design time. This garage will be selected dynamically by activity *SelectService* that, through a communication, replaces x_{garage} with the actual partner name of the garage. Indeed, in COWS dynamic binding of discovered services and service reconfiguration rely on the exchange of partner and operation names in communication.

Bank, the last service we show in this section, can serve multiple requests simultaneously. This behaviour is modelled by exploiting the *replication* operator $*\_$ to spawn in parallel as many copies of its argument term as necessary. The definition of *Bank* is

$$* [x_{cust}, x_{cc}, x_{amount}, o_{checkOK}, o_{checkFail}]$$
$$p_{bank} \cdot o_{charge}?\langle x_{cust}, x_{cc}, x_{amount}\rangle.$$
$$(< \texttt{perform some checks and reply on } o_{checkOK} \text{ or } o_{checkFail}>$$
$$| \ p_{bank} \cdot o_{checkFail}?\langle\rangle. \ x_{cust} \cdot o_{chargeFail}!\langle\rangle$$
$$+ \ p_{bank} \cdot o_{checkOK}?\langle\rangle. \ [chargeID] \ (\ x_{cust} \cdot o_{chargeOK}!\langle chargeID\rangle$$
$$| \ p_{bank} \cdot o_{revoke}?\langle chargeID\rangle.$$
$$< \texttt{revoke } chargeID>. \ x_{cust} \cdot o_{revokeOK}!\langle\rangle \ ) \ )$$

Once prompted by a request, differently from *Orchestrator, Bank* creates one specific instance to serve that request and is immediately ready to concurrently serve other requests. Notably, each instance exploits communication on 'internal' operations $o_{checkOK}$ and $o_{checkFail}$ to model a conditional choice, and creates a new 'charge identifier' by means of the delimitation operator (that acts here as the restriction operator of the π-calculus). Thus, if after some invocations the service receives a message along the endpoint $p_{bank} \cdot o_{revoke}$ to revoke a request, a certain number of service instances could be able to accept it. However, the message is routed to the proper instance by exploiting, as a correlation value, a unique identifier (that is named *chargeID* in the term above) characterizing the instance.

3.3   Fault and Compensation Handling

We now show how to modify the specification described in the previous section for adding the fault and compensation activities depicted in Figure 1. For improving readability, these activities are highlighted by a gray background to distinguish them from 'normal behaviour'. For example, the term modelling the garage ordering is:

$$OrderGarage \ \triangleq \ x_{garage} \cdot o_{order}!\langle p_{car}, x_{carData}\rangle$$
$$| \ [x_{repairNum}] \ p_{car} \cdot o_{garageFail}?\langle\rangle.$$
$$(\ p_{car} \cdot o_{undo}!\langle cc\rangle$$
$$| \ [p, o] \ (p \cdot o!\langle x_{loc}\rangle \ | \ p \cdot o?\langle x_{garageGPS}\rangle) \ )$$
$$+ \ p_{car} \cdot o_{garageOK}?\langle x_{repairNum}, x_{garageGPS}\rangle.$$
$$p_{car} \cdot o_{undo}?\langle garage\rangle.$$
$$(\ x_{garage} \cdot o_{cancel}!\langle x_{repairNum}\rangle$$
$$| \ p_{car} \cdot o_{cancelOK}?\langle\rangle$$
$$| \ p_{car} \cdot o_{undo}!\langle cc\rangle \ | \ p_{car} \cdot o_{undo}!\langle rentalCar\rangle)$$

Thus, if ordering a garage fails, the compensation of the credit card charge is invoked by sending a signal cc (abbreviation of 'card charge') along the endpoint $p_{car} \cdot o_{undo}$ and the rental car delivery is redirected by assigning the car's current location x_{loc} to the variable $x_{garageGPS}$ (this assignment is modelled by means of communication along the private endpoint $p \cdot o$). Otherwise, a compensation handler is installed that is invoked whenever tow truck ordering fails and, in that case, attempts to cancel the garage order and to compensate the credit card charge and the rental car order.

To model fault handling and compensation behaviours, the term *OrderGarage* exploits interactions along the endpoint $p_{car} \cdot o_{undo}$. However, to better support the specification of these aspects, COWS provides two further constructs. *Kill* activities of the form **kill**(k), where k is a killer label, can be used to force termination of all unprotected parallel terms inside the enclosing $[k]$, that stops the killing effect. Kill activities are executed *eagerly* with respect to the other parallel activities but critical code, such as e.g. fault/compensation signals and handlers, can be protected from the effect of a forced termination by using the *protection* operator $\{\_\}$. By exploiting these new features, the recovery behaviour $s_{engfail}$ becomes

$$[\,p_e, o_e, x_{loc}, x_{list}, o_{undo}, k\,]$$
$$(\,(\,RequestCardCharge \mid RequestLocation.\,FindServices\,)$$
$$\mid\ p_e \cdot o_e?\langle\rangle.\ p_e \cdot o_e?\langle\rangle.\ SelectServices.$$
$$[x_{garageGPS}]\,(\,OrderGarage.OrderTowTruck \mid OrderRentalCar)\,)$$

where *RequestCardCharge* and *FindServices* are defined as

$$FindServices \triangleq p_{car} \cdot o_{find}!\langle x_{loc}, ServicesType\rangle$$
$$\mid p_{car} \cdot o_{servicesFound}?\langle x_{list}\rangle.\ p_e \cdot o_e!\langle\rangle$$
$$+\ p_{car} \cdot o_{servicesNotFound}?\langle\rangle.$$
$$(\,\textbf{kill}(k) \mid \{p_{car} \cdot o_{undo}!\langle cc\rangle \mid p_{car} \cdot o_{undo}!\langle cc\rangle\}\,)$$

$$RequestCardCharge \triangleq p_{bank} \cdot o_{charge}!\langle p_{car}, ccNum, amount\rangle$$
$$\mid\ \{\,[x_{chargeID}]\,p_{car} \cdot o_{chargeFail}?\langle\rangle.\ \textbf{kill}(k)$$
$$+\ p_{car} \cdot o_{chargeOK}?\langle x_{chargeID}\rangle.$$
$$(\,p_e \cdot o_e!\langle\rangle \mid p_{car} \cdot o_{undo}?\langle cc\rangle.p_{car} \cdot o_{undo}?\langle cc\rangle.$$
$$(\,p_{bank} \cdot o_{revoke}!\langle x_{chargeID}\rangle$$
$$\mid p_{car} \cdot o_{revokeOK}?\langle\rangle)\,)\,\}$$

Thus, whenever services finding fails, *FindServices* terminates the whole recovery behaviour and sends two signals cc along the endpoint $p_{car} \cdot o_{undo}$. Similarly, if charging the credit card fails, then *RequestCardCharge* terminates the whole recovery behaviour $s_{engfail}$. Otherwise, it installs a compensation handler that takes care of revoking the credit card charge. Activation of this compensation activity requires two signals cc along $p_{car} \cdot o_{undo}$ and, thus, takes place either whenever *FindService* fails or whenever both *OrderGarage* and *OrderRentalCar* (not shown here) fail.

4   Service Publication, Discovery and Negotiation

We have demonstrated so far that COWS can model service specification, orchestration, reconfiguration, and execution. Now, we focus on other important phases of the life cycle of service-oriented applications. In fact, in a service-oriented architecture, services can play essentially three different roles: the provider, the requester and the registry. Providers offer functionalities and publish machine-readable service descriptions on registries to enable automated discovery and invocation by requesters. In addition to the function that the service performs, service descriptions also include non-functional properties, such as e.g., response time, availability, reliability, security, and performance, that jointly represent the *quality of the service* (QoS). Some of these properties could depend on the current run-time configuration of the system (e.g. the maximum allowed bandwidth might depend on the actual load of the server), thus a *dynamic discovery* process is often needed to find a provider that meets the requesters' requirements. Moreover, since services are often developed and run by different organizations, a key issue of the discovery process is to define a flexible *negotiation* mechanism that allows two or more parties to reach a joint agreement about cost and quality of a service, prior to service execution. The outcome of the negotiation phase is a *Service Level Agreement* (SLA), i.e. a contract among the involved parties that sets out both type and bounds on various performance metrics of the service to be provided, and the remedial actions to be performed if these are not met.

We want now to demonstrate that service publication, discovery and SLA negotiation can be naturally modelled in COWS by exploiting the additions of 'timed' activities and 'constraints'. Timed activities have been introduced in [17], by adding specific rules for modelling time passing to the COWS operational semantics, since it is not known to what extent timed computation can be reduced to untimed forms of computation [25]. Specifically, COWS is extended with a WS-BPEL-like *wait* activity of the form $\oplus_e$, that suspends the execution of the invoking service until the time interval whose duration is specified as an argument has elapsed and can be used as a guard for the choice operator. Constraints have been introduced in [19], by exploiting the fact that COWS' definition is parameterised with respect to a few sets of objects, namely the set of values and that of expressions that operate on them. Notably, we do not take a definite standing on which of the many kinds of constraints one should use. For example, one could use *crisp* constraints, that can only be satisfied or violated, or *soft* constraints, that instead can be satisfied with multiple consistency levels (these are usually expressed by means of *c-semirings* [4] and interpreted as levels of preference or importance). From time to time, the appropriate kind of constraints to work with should be chosen depending on what one intends to model.

Still in [19] we argue that the concurrent constraint computing model can be easily mimicked in COWS. This model of computation is based on a shared store of constraints that provides partial information about possible values that program variables can assume. In COWS the store of constraints is represented by the following service:

$$store_C \triangleq [p, o]\, (\, p \bullet o!\langle C \rangle \mid *[x]\, p \bullet o?\langle x \rangle.(\, p_s \bullet o_{get}!\langle x \rangle \mid [y]\, p_s \bullet o_{set}?\langle y \rangle.p \bullet o!\langle y \rangle\,)\,)$$

where C is the multiset of constraints currently in the store, while p_s is a distinguished partner, and o_{get} and o_{set} are distinguished operations. Other services can interact with the store service in mutual exclusion, by acquiring the lock (and, at the same time, the stored value) with a receive along $p_s \bullet o_{get}$ and by releasing the lock (providing the new stored value) with an invoke along $p_s \bullet o_{set}$. The programs running in parallel with the store can act on it by performing operations for adding/removing constraints to/from the store (`tell` and `retract`, respectively), and for checking entailment/consistency of a constraint by/with the store (`ask` and `check`, respectively). For example, the service `tell` $c.s$ willing to perform operation `tell` c and then to continue as service s can be rendered in COWS as follows:

$$[p, o]\, (\, p \bullet o!\langle c \rangle \mid [y]\, p \bullet o?\langle y \rangle.[x]\, p_s \bullet o_{get}?\langle\langle y, x \rangle\rangle.(\{\!\mid p_s \bullet o_{set}!\langle x \uplus \{y\} \rangle \!\mid\!\} \mid s)\,)$$

Due to lack of space, we refer the interested reader to [19] for the implementation of the other operations and further details.

Now, like in cc-pi [5], service descriptions and SLA requirements can be expressed as constraints that can be dynamically generated and composed, and that can be used by the involved parties both for service publication and discovery, and for the SLA negotiation process. Consistency of the set of constraints resulting from negotiation means that the agreement has been reached. Timed activities can be exploited to allow services not to get stuck forever waiting on a receive.

We use the on road assistance scenario to illustrate all such features and to put the related mechanisms to work. Initially, each on road service has to publish its service description on a service registry. For example, assume that a garage service description consists of: a string identifying the kind of provided service, the provider's partner name, and a constraint that defines the garage location. Now, by assuming that the registry provides the operation o_{pub} by means of the partner name p_{reg}, a garage service can request the publication of its description as follows:

$$p_{reg} \cdot o_{pub}! \langle \text{``garage''}, p_{garage}, \mathsf{gps} = (4348.1143N, 1114.7206E) \rangle$$

gps is what we call a *constraint variable*. In fact, it is a specific name and, hence, is not affected by substitution application. Constraint variables are used to avoid that taking place of communication can make the store inconsistent. Indeed, suppose constraints in the store may contain variables and consider the following example:

$$[x]\,(\,store_{\emptyset} \mid \mathtt{tell}(x \leq 5).\,(p \cdot o! \langle 6 \rangle \mid p \cdot o? \langle x \rangle)\,)$$

After action $\mathtt{tell}$ has added the constraint $x \leq 5$ to the store, communication along the endpoint $p \cdot o$ can modify the constraint in $6 \leq 5$, thus making the store inconsistent. To distinguish constraint variables from COWS (true) variables, the formers are written in the typewriter style (e.g. x, y, ...). The service registry can be defined as

$$[o_{DB}]\,(\, * \,[x_{type}, x_p, x_c]\, p_{reg} \cdot o_{pub}? \langle x_{type}, x_p, x_c \rangle . p_{reg} \cdot o_{DB}! \langle x_{type}, x_p, x_c \rangle \ \mid \ R^{search} \,)$$

For each publication request received along the endpoint $p_{reg} \cdot o_{pub}$ from a provider service, the registry service outputs a service description along the private endpoint $p_{reg} \cdot o_{DB}$. The parallel composition of all these outputs represents the database of the registry. The subservice R^{search}, serving the searching requests, is defined as

$$
\begin{aligned}
R^{search} \ \triangleq \ & * \,[x_{type}, x_{client}, x_c, o_{addToList}, o_{askList}] \\
& p_{reg} \cdot o_{search}? \langle x_{type}, x_{client}, x_c \rangle . \,[p_s]\,(\, store_{\emptyset} \mid \mathtt{tell}\ x_c . R' \mid List \,)
\end{aligned}
$$

$$
\begin{aligned}
R' \ \triangleq \ & [k]\,(\, * \,[x_p, x_{const}]\, p_{reg} \cdot o_{DB}? \langle x_{type}, x_p, x_{const} \rangle . \\
& \quad (\,\{\!| p_{reg} \cdot o_{pub}! \langle x_{type}, x_p, x_{const} \rangle |\!\} \mid \mathtt{check}\ x_{const}.\ p_{reg} \cdot o_{addToList}! \langle x_p \rangle \,) \\
& \mid \odot_{\delta}.\,(\,\mathbf{kill}(k) \mid \{\!| \,[x_{list}]\, p_{reg} \cdot o_{askList}? \langle x_{list} \rangle . x_{client} \cdot o_{resp}! \langle x_{list} \rangle |\!\} \,)\,)
\end{aligned}
$$

When a searching request is received along $p_{reg} \cdot o_{search}$, the registry service initializes a new local store (delimitation $[p_s]$ makes $store_{\emptyset}$ inaccessible outside of service R^{search}) by adding the constraint within the query message. Then, it cyclically reads a description (whose first field is the string specified by the client) from the internal database, checks if the provider constraints are consistent with the store and, in case of success, adds the provider's partner name to a list (by exploiting an internal service *List*, that provides operations $o_{addToList}$ and $o_{askList}$). After δ time units from the initialization of the local store, the loop is terminated by executing a kill activity and the current list of providers for service type x_{type} is sent to the client. Notably, reading a description in the database, in this case, consists of an input along $p_{reg} \cdot o_{DB}$ followed by an output along $p_{reg} \cdot o_{pub}$; this way we are guaranteed that, after being consumed, the description is correctly added to the database. It is worth noticing that service descriptions are non-deterministically retrieved, thus the same provider can occur in the returned list many times. This could be avoided by refining the specification, e.g. by tagging each service description with an index (stored in an additional field) that is then exploited to read the descriptions in an ordered way. Moreover, since our notion of time does not rely on the so-called 'maximal progress assumption', i.e. communication does not prevent the execution of timed transitions, there is no guarantee that any service at all is retrieved.

After the user's car breaks down and *Orchestrator* is triggered, the service *Discovery* of the in-vehicle platform will receive from *Orchestrator* a request containing the GPS data of the car, that it stores in x_{loc}, and a string identifying the kind of the required services (see the specification in Section 3.2). By exploiting the latter information, it will know that it has to search a garage, a tow truck and a rental car service. For example, the component taking care of discovering a garage service can be

$$p_{reg} \cdot o_{search}!\langle \text{``garage''}, p_{car}, dist(x_{loc}, \mathsf{gps}) < 20 \rangle \mid [x_{garageList}] \, p_{car} \cdot o_{resp}?\langle x_{garageList} \rangle$$

where the constraint $dist(x_{loc}, \mathsf{gps}) < 20$ means that the required garages must be less than 20 km far from the stranded car's actual location.

Once the discovery phase terminates and *Reasoner* communicates the best garage service to *Orchestrator*, the latter and the selected garage engage in a negotiation phase in order to sign an SLA. First, *Orchestrator* invokes the operation o_{order} provided by the selected garage (see *OrderGarage* definition at page 708); then, it starts the negotiation by performing an operation `tell` that adds *Orchestrator*'s local constraints (i.e. constraints with restricted constraint variables) to the shared global store; finally, it synchronizes with the garage service, by invoking o_{sync}, for sharing its local constraints with it.

```
[cost, duration]
tell ((cost < 1500 ∧ duration < 48) ∨ (cost < 800 ∧ duration ⩾ 48)).
```
$$(x_{garage} \cdot o_{sync}!\langle \mathsf{cost}, \mathsf{duration} \rangle$$
$$\mid [x_{repairNum}] \, p_{car} \cdot o_{garageOK}?\langle x_{reapairNum} \rangle. \cdots + p_{car} \cdot o_{garageFail}?\langle \rangle. \cdots)$$

In our example, the constraints state that for a repair in less than two days the driver is disposed to spend up to 1500 Euros, otherwise he is ready to spend less than 800 Euros.

After the synchronization with *Orchestrator*, the selected garage service tries to impose its first-rate constraint $c = ((\mathsf{cost}' > 2000 \wedge 6 < \mathsf{duration}' < 24) \vee (\mathsf{cost}' > 1500 \wedge \mathsf{duration}' \geqslant 24))$ and, if it fails to reach an agreement within δ' time units, weakens the requirements and retries with the constraint $c' = ((\mathsf{cost}' > 1700 \wedge 6 < \mathsf{duration}' < 24) \vee (\mathsf{cost}' > 1200 \wedge \mathsf{duration}' \geqslant 24))$. Both constraints are specifically generated by the garage service for the occurred engine failure, by exploiting the transmitted diagnostic data. After δ'' time units, if also the second attempt fails, it gives up the negotiation. This negotiation task is modelled as follows:

$$[x_{cost}, x_{duration}, \texttt{cost}', \texttt{duration}']$$
$$p_{garage} \bullet o_{sync}?\langle x_{cost}, x_{duration}\rangle.\,\texttt{tell}\,(x_{cost} = \texttt{cost}' \wedge x_{duration} = \texttt{duration}').$$
$$(\,\texttt{tell}\,c.\,x_{client} \bullet o_{garageOK}!\langle repairNum\rangle$$
$$+\,\odot_{\delta'}.\,(\,\texttt{tell}\,c'.\,x_{client} \bullet o_{garageOK}!\langle repairNum\rangle$$
$$+\,\odot_{\delta''}.\,x_{client} \bullet o_{garageFail}!\langle\rangle\,)\,)$$

Notably, operations $\texttt{tell}$ cannot be used as guards for the choice operator. Thus, a term like $\texttt{tell}\,c.\,s + \odot_e.\,s'$ should be considered as an abbreviation for

$$[p, q, o]\,(\,\texttt{check}\,c.\,(p \bullet o!\langle\rangle \mid q \bullet o?\langle\rangle.\,\texttt{tell}\,c.\,s) \mid \odot_e.\,s' + p \bullet o?\langle\rangle.\,q \bullet o!\langle\rangle\,)$$

Intuitively, if the constraint c is consistent with the store, the timer can be stopped (i.e. communication along $p \bullet o$ makes a choice and removes the wait activity); afterward, the constraint can be added to the store, provided that other interactions that took place in the meantime do not lead to inconsistency. Otherwise, if the timeout expires, the constraint cannot be added to the store.

5   A Type System for Checking Confidentiality Properties

The type system for COWS introduced in [18] permits expressing and forcing policies regulating the exchange of data among interacting services and ensuring that, in that respect, services do not manifest unexpected behaviours. This enables us to check confidentiality properties, e.g., that critical data such as credit card information are shared only with authorized partners. The type system has been obtained by tailoring to COWS the type-based approach for protecting data in distributed systems put forward in [12], in the context of a higher-order functional programming language, and drawn on in [6], in that of languages for global computing.

The types express the policies for data exchange in terms of *regions*, i.e. sets of service partner names attachable to each single datum. Service programmers can thus settle the partners usable to exchange any given datum (and, then, the services that can share it), thus avoiding the datum be accessed (by unwanted services) through unauthorized partners. Then, a type inference system (statically) performs some coherence checks (e.g. the service used in an invocation must belong to the regions of all data occurring in the argument of the invocation) and derives the minimal region annotations for variable declarations that ensure consistency of services initial configuration. COWS operational semantics uses these annotations in very efficient checks (i.e. subset inclusions) to authorise or block transitions, in order to guarantee that computations proceed according to them. This property, called *soundness*, can be stated as follows: a service s is *sound* if, for any datum v in s associated to region r and for all evolutions of s, it holds that v can be exchanged only by using services in r. As a consequence of the type soundness of the language, it follows that well-typed services always comply with the policies regulating the exchange of data among interacting services. In fact, it is also possible to move all dynamic checks to the static phase. This would require a static analysis that gathers information about all the values that each variable can assume at runtime and uses these information to verify the compliance with the specified policies. At the price of a more complex static phase, this approach, on the one hand, would alleviate the runtime checks but, on the other hand, could discard terms that at runtime would behave soundly since statically they cannot guarantee to comply with their policies. We are currently evaluating and implementing the two approaches.

We illustrate now some relevant properties for the on road assistance scenario. Firstly, a driver in trouble must be assured that information about his credit card and his location cannot become available to unauthorized users. Thus, for example, the credit card identifier *ccNum*, communicated by activity *RequestCardCharge* to service *Bank*, gets annotated with the policy $\{p_{bank}\}$, that allows *Bank* to receive the datum but prevents it from transmitting the datum to other services. Other non-critical data, e.g. *amount*, can be transmitted without an attached policy. The typed version of *RequestCardCharge* (where irrelevant fault/compensation details are omitted) is defined as follows

$$p_{bank} \cdot o_{charge}!\langle p_{car}, \{ccNum\}_{\{p_{bank}\}}, amount \rangle$$
$$| \ [x_{chargeID}] \ p_{car} \cdot o_{chargeFail}?\langle\rangle + \ p_{car} \cdot o_{chargeOK}?\langle x_{chargeID} \rangle . p_e \cdot o_e!\langle\rangle$$

Notably, the annotations set by programmers are written as a subscript of the datum to which they refer to. Instead, the annotations put by the type inference, to better distinguish them from those put by the programmers, are written as a superscript of the variable declaration to which they refer to. Thus, the syntax of variable delimitation becomes $[\{x\}^r] \ s$, which means that the datum that dynamically will replace x will be used in s at most by the partners belonging to the region r. Hence, once the type inference phase ends, *Bank* gets annotated as follows

$$* \ [\{x_{cust}\}^{\{p_{bank}\}}, \{x_{cc}\}^{\{p_{bank}\}}, \{x_{amount}\}^{\{p_{bank}\}}, o_{checkOK}, o_{checkFail}]$$
$$p_{bank} \cdot o_{charge}?\langle x_{cust}, x_{cc}, x_{amount} \rangle .$$
$$(\ \texttt{<\ perform\ some\ checks\ and\ reply\ on}\ o_{checkOK}\ \texttt{or}\ o_{checkFail}\texttt{>}$$
$$|\ p_{bank} \cdot o_{checkFail}?\langle\rangle . \ x_{cust} \cdot o_{chargeFail}!\langle\rangle$$
$$+\ p_{bank} \cdot o_{checkOK}?\langle\rangle .$$
$$[chargeID]\ (\ x_{cust} \cdot o_{chargeOK}!\langle chargeID \rangle$$
$$|\ p_{bank} \cdot o_{revoke}?\langle chargeID \rangle .$$
$$\texttt{<\ revoke}\ chargeID\texttt{>} . \ x_{cust} \cdot o_{revokeOK}!\langle\rangle\ ) \ )$$

Indeed, the annotations inferred for variables x_{cust}, x_{cc} and x_{amount} are derived from the use of these variables made by *Bank*. Thus, they are assigned region $\{p_{bank}\}$ because they are only used in the receive along $p_{bank} \cdot o_{charge}$ and, of course, the partner name of the endpoint must belong to the region of the variables.

Suppose instead that service *Bank* (accidentally or maliciously) attempts to reveal the credit card number through some 'internal' operation such as $p_{int} \cdot o!\langle\{x_{cc}\}_r\rangle$, for some region r. For *Bank* to successfully complete the type inference phase, we should have $p_{int} \in r$. Then, as result of the inference, we would get the annotated variable declaration $[\{x_{cc}\}^{r'}]$, for some region r' with $r \subseteq r'$. Now, the interaction between the typed terms *RequestCardCharge* and *Bank* would be blocked by the runtime checks because the datum sent by *RequestCardCharge* would be annotated as $\{ccNum\}_{\{p_{bank}\}}$ while the region r' of the receiving variable x_{cc} is such that $p_{int} \in r \subseteq r' \not\subseteq \{p_{bank}\}$.

When delivering a datum, we can specify different policies according to the invoked service. For example, when sending the car's current location stored in x_{loc} to services *OrderTowTruck* and *OrderRentalCar*, we annotate it with the regions $\{x_{towTruck}\}$ and $\{x_{rentalCar}\}$, respectively. This means that the corresponding service invocations get annotated as follows:

$$x_{towTruck} \bullet o_{order}!\langle p_{car}, \{x_{loc}\}_{\{x_{towTruck}\}}, x_{garageGPS} \rangle$$

$$x_{rentalCar} \bullet o_{redirect}!\langle x_{rentalNum}, \{x_{loc}\}_{\{x_{rentalCar}\}} \rangle$$

Notably, the used policies are not fixed at design time, but *depend* on the partner variables $x_{towTruck}$ and $x_{rentalCar}$, and, thus, will be determined by the services that these variables will be bound to as computation proceeds. For example, consider a towing truck service annotated as follows:

$$TowTruck \triangleq * [\{x_{client}\}^{r_1}, \{x_{carLoc}\}^{r_2}, \{x_{garageLoc}\}^{r_3}, o_{checkOK}, o_{checkFail}]$$
$$p_{towTruck} \bullet o_{order}?\langle x_{client}, x_{carLoc}, x_{garageLoc} \rangle.$$
$$(< \texttt{perform some checks and reply on } o_{checkOK} \text{ or } o_{checkFail} >$$
$$| \ p_{towTruck} \bullet o_{checkFail}?\langle\rangle. \ x_{client} \bullet o_{towTruckFail}!\langle\rangle$$
$$+ \ p_{towTruck} \bullet o_{checkOK}?\langle\rangle.$$
$$[towTruckNum] \ x_{client} \bullet o_{towTruckOK}!\langle towTruckNum \rangle \ )$$

Now, the car's current location can be communicated to the towing truck if, and only if, the region of the variable x_{carLoc} that, after communication, will store the datum and the region of x_{loc} do comply, i.e. $r_2 \subseteq \{p_{towTruck}\}$.

As a final example, the on road services could want to guarantee that critical data sent to the in vehicle services, such as cost and quality of the service supplied, are not disclosed to competitors. For example, suppose that the towing truck services, like *TowTruck* before, must send the estimated travel time (*ETT*) to clients. To prevent this datum from being sent to competitor services, *ETT* is communicated with an attached policy that only authorizes the client partner to access it, as in the following activity

$$x_{client} \bullet o_{towTruckOK}!\langle towTruckNum, \{ETT\}_{\{x_{client}\}} \rangle$$

6   A Logical Framework for Verifying Functional Properties

The logical verification framework introduced in [9] permits checking functional properties of services by abstracting away from the computational contexts in which they are operating. Specifically, services are abstractly considered as entities capable of accepting requests, delivering corresponding responses and, on-demand, cancelling requests, over specified interactions. The 'abstract' service actions are the following: $request(i, c)$, $response(i, c)$, $cancel(i, c)$ and $fail(i, c)$, where the name i indicates the interaction to which the corresponding 'concrete' action (i.e. the action occurring in the COWS specification) belongs, and c denotes a tuple of correlation values that identifies a particular invocation of the action. For example, $request(i, c)$ indicates that the corresponding concrete action represents the initial request of the interaction i and its invocation is identified by the correlation tuple c; similarly, $response(i, c)$, $cancel(i, c)$ and $fail(i, c)$ characterise actions that correspond to a response, a cancellation and a failure notification, respectively, of the interaction i. The name of the interaction or the correlation tuple will be omitted whenever they are not relevant. The correspondence between concrete actions used in the specifications and the abstract actions above must be defined from time to time by the user through appropriate abstraction rules.

Our abstract notion of services can be modelled by Doubly Labelled Transition Systems (L^2TSs, [7]) in a very natural way. Thus, to formalize functional properties of

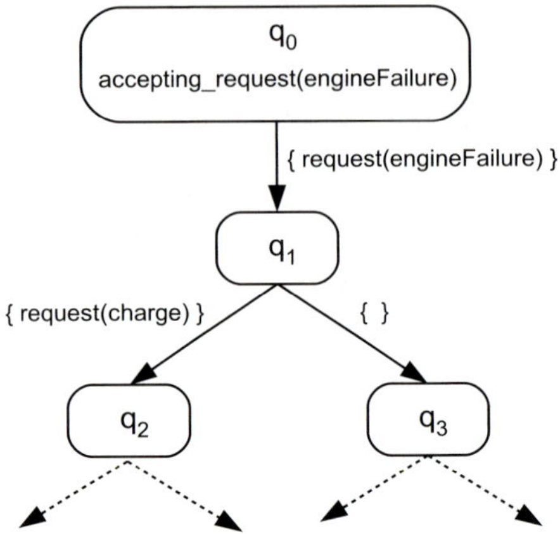

Fig. 3. Excerpt of the abstract L^2TS for the on road assistance scenario

that replace the concrete labels on the transitions with abstract actions (i.e. $request(i, c)$, $response(i, c)$, $cancel(i, c)$ and $fail(i, c)$) and the concrete labels on the states with atomic propositions (such as, e.g., $accepting\_request(i)$). The transformation only involves the concrete actions we want to observe; the concrete actions that are not replaced by their abstract counterparts may not be observed. Thus, the application of the abstraction rules transforms the concrete L^2TS into an 'abstract' one. For example, the abstract L^2TS of the on road assistance scenario shown in Figure 3, is obtained by applying to the concrete L^2TS of Figure 2 the following abstraction rules:

$$
\begin{aligned}
Action: \quad & o_{engfail} \;\rightarrow\; request(engineFailure) \\
Action: \quad & o_{charge} \;\rightarrow\; request(charge) \\
Action: \quad & o_{chargeOK} \;\rightarrow\; response(charge) \\
Action: \quad & p_{garage\_1} \cdot o_{order}\langle \$1, * \rangle \;\rightarrow\; request(garage, \langle \$1 \rangle) \\
Action: \quad & p_{garage\_2} \cdot o_{order}\langle \$1, * \rangle \;\rightarrow\; request(garage, \langle \$1 \rangle) \\
Action: \quad & \$1 \cdot o_{garageOK} \;\rightarrow\; response(garage, \langle \$1 \rangle) \\
Action: \quad & \$1 \cdot o_{garageFail} \;\rightarrow\; fail(garage, \langle \$1 \rangle) \\
Action: \quad & o_{cancel} \;\rightarrow\; cancel(garage) \\
& \ldots \\
Action: \quad & o_{rentalCarOK}\langle \$1 \rangle \;\rightarrow\; response(rentalCar, \langle \$1 \rangle) \\
Action: \quad & o_{redirect}\langle \$1, * \rangle \;\rightarrow\; cancel(rentalCar, \langle \$1 \rangle) \\
State: \quad & o_{engfail} \;\rightarrow\; accepting\_request(engineFailure)
\end{aligned}
$$

Most of the rules are self-explicative, we comment on the remaining ones. Variables "$\$n$" (with n natural number) are used to define parametric abstraction rules. Also the wildcard " $*$ " is used for increasing flexibility. The fourth and fifth rules prescribe that whenever an action over the endpoints $p_{garage\_1} \cdot o_{order}$ or $p_{garage\_2} \cdot o_{order}$ with sent data $\langle p_{car}, data \rangle$ (that match $\langle \$1, * \rangle$) occurs in the label of a transition, then it is replaced by the abstract action $request(garage, \langle p_{car} \rangle)$. This way, the car partner name p_{car} can be used to correlate responses from the contacted garage service. Similarly, the second-last rule prescribes that whenever an action over the operation $o_{redirect}$ with sent data $\langle rentalNum, gps \rangle$ occurs in the label of a transition, then it is replaced by *cancel(rentalCar, $\langle rentalNum \rangle$)*. The last rule works similarly, but it applies to labels of states rather than to labels of transitions.

Finally, the SocL formulae are checked over the abstract L^2TS. To assist the verification process, one can use CMC [1], that is a model checker for SocL formulae over L^2TS, other than an interpreter for COWS terms. One can thus verify that, as expected, all the abstract properties we introduced before do hold for the COWS specification of on road assistance scenario, but the first property, because *Orchestrator* is not a persistent service capable of accepting and serving multiple requests (indeed, as we noted in Section 3.2, it can only perform one recovery task at a time).

7   Concluding Remarks

COWS falls within a main line of research that aims at developing process calculi capable of capturing the basic aspects of service-oriented systems and, possibly, of supporting the analysis of qualitative and quantitative properties of services. We have demonstrated that one can use COWS to model all the phases of the life cycle of SOC applications such as publication, discovery, negotiation, orchestration, deployment, reconfiguration and execution. We believe that the methods and tools we have described for expressing and checking properties of services are already an important added value of using COWS as a modelling language.

The fact that several relevant aspects of SOC systems can be suitably addressed and dealt with in an homogeneous and direct way by using a single linguistic low-level formalism somehow suggests that COWS could serve as a common and convenient basis to enable analysis of service-oriented applications by translation from higher level languages. As further steps in this direction, we are currently studying translations from the service orchestration language WS-BPEL [22] and the SENSORIA Reference Modelling Language SRML [10] into COWS. A short-term goal of this activity is to define, via translation in COWS, an operational semantics for these two high level languages. A long-term goal is to enable using proof techniques and analytical tools developed for COWS, such as the type system and the logical verification framework summed up in this paper, and the stochastic extension defined in [23], to analyse service-oriented applications programmed in WS-BPEL or modelled in SRML.

Acknowledgements. We thank Alessandro Fantechi, Stefania Gnesi and Franco Mazzanti for their contribution on the development of SocL and CMC. The anonymous referees provided helpful suggestions for improving the presentation.

References

1. CMS: an on-the-fly model checker and interpreter for COWS, http://fmt.isti.cnr.it/cmc/
2. Software Engineering for Service-Oriented Overlay Computers (SENSORIA) (2005), http://sensoria.fast.de/
3. ter Beek, M.H., Fantechi, A., Gnesi, S., Mazzanti, F.: An action/state-based model-checking approach for the analysis of communication protocols for Service-Oriented Applications. In: Leue, S., Merino, P. (eds.) FMICS 2007. LNCS, vol. 4916, pp. 133–148. Springer, Heidelberg (2008)

4. Bistarelli, S., Montanari, U., Rossi, F.: Semiring-based constraint satisfaction and optimization. Journal of the ACM 44(2), 201–236 (1997)
5. Buscemi, M., Montanari, U.: CC-Pi: A Constraint-Based Language for Specifying Service Level Agreements. In: De Nicola, R. (ed.) ESOP 2007. LNCS, vol. 4421, pp. 18–32. Springer, Heidelberg (2007)
6. De Nicola, R., Gorla, D., Pugliese, R.: Confining data and processes in global computing applications. Science of Computer Programming 63, 57–87 (2006)
7. De Nicola, R., Vaandrager, F.: Three logics for branching bisimulation. J. ACM 42(2), 458–487 (1995)
8. De Nicola, R., Vaandrager, F.W.: Action versus state based logics for transition systems. In: Guessarian, I. (ed.) LITP 1990. LNCS, vol. 469, pp. 407–419. Springer, Heidelberg (1990)
9. Fantechi, A., Gnesi, S., Lapadula, A., Mazzanti, F., Pugliese, R., Tiezzi, F.: A model checking approach for verifying COWS specifications. In: Fiadeiro, J., Inverardi, P. (eds.) FASE 2008. LNCS, vol. 4961, pp. 230–245. Springer, Heidelberg (2008)
10. Fiadeiro, J., Lopes, A., Bocchi, L.: A formal approach to service component architecture. In: Bravetti, M., Núñez, M., Zavattaro, G. (eds.) WS-FM 2006. LNCS, vol. 4184, pp. 193–213. Springer, Heidelberg (2006)
11. Garcia-Molina, H., Salem, K.: Sagas. In: SIGMOD, pp. 249–259. ACM Press, New York (1987)
12. Kirli, Z.D.: Confined mobile functions. In: 14th IEEE Computer Security Foundations Workshop (CSFW 2001), pp. 283–294. IEEE Computer Society, Los Alamitos (2001)
13. Koch, N.: Automotive case study: UML specification of on road assistance scenario. Technical Report 1, FAST (2007), `http://rap.dsi.unifi.it/sensoria/files/FAST_report_1_2007_ACS_UML.pdf`
14. Lapadula, A., Pugliese, R., Tiezzi, F.: A Calculus for Orchestration of Web Services (full version). Technical report, Dipartimento di Sistemi e Informatica, Univ. Firenze (2006), `http://rap.dsi.unifi.it/cows`
15. Lapadula, A., Pugliese, R., Tiezzi, F.: A Calculus for Orchestration of Web Services. In: De Nicola, R. (ed.) ESOP 2007. LNCS, vol. 4421, pp. 33–47. Springer, Heidelberg (2007)
16. Lapadula, A., Pugliese, R., Tiezzi, F.: COWS specification of the on road assistance scenario. Technical report, Dipartimento di Sistemi e Informatica, Univ. Firenze (2007), `http://rap.dsi.unifi.it/cows`
17. Lapadula, A., Pugliese, R., Tiezzi, F.: C⊕WS: A timed service-oriented calculus. In: Jones, C.B., Liu, Z., Woodcock, J. (eds.) ICTAC 2007. LNCS, vol. 4711, pp. 275–290. Springer, Heidelberg (2007)
18. Lapadula, A., Pugliese, R., Tiezzi, F.: Regulating data exchange in service oriented applications. In: Arbab, F., Sirjani, M. (eds.) FSEN 2007. LNCS, vol. 4767, pp. 223–239. Springer, Heidelberg (2007)
19. Lapadula, A., Pugliese, R., Tiezzi, F.: Service discovery and negotiation with COWS. In: WWV. ENTCS, Elsevier, Amsterdam (to appear, 2007)
20. Meredith, L.G., Bjorg, S.: Contracts and types. Commun. ACM 46(10), 41–47 (2003)
21. Milner, R., Parrow, J., Walker, D.: A calculus of mobile processes, I and II. Inf. Comput. 100(1), 1–40, 41–77 (1992)
22. OASIS WSBPEL TC. Web Services Business Process Execution Language Version 2.0. Technical report, OASIS, April (2007), `http://docs.oasis-open.org/wsbpel/2.0/OS/wsbpel-v2.0-OS.html`
23. Prandi, D., Quaglia, P.: Stochastic COWS. In: Krämer, B.J., Lin, K.-J., Narasimhan, P. (eds.) ICSOC 2007. LNCS, vol. 4749, pp. 245–256. Springer, Heidelberg (2007)
24. van Breugel, F., Koshkina, M.: Models and verification of bpel. Technical report (2006), `http://www.cse.yorku.ca/~franck/research/drafts/tutorial.pdf`
25. van Glabbeek, R.J.: On specifying timeouts. ENTCS 162, 173–175 (2006)

Approximating Behaviors in Embedded System Design

Roberto Passerone[1] and Alberto L. Sangiovanni-Vincentelli[2]

[1] Dipartimento di Ingegneria e Scienza dell'Informazione,
University of Trento, Trento, Italy
[2] Department of Electrical Engineering and Computer Sciences,
University of California, Berkeley, CA

Abstract. Embedded systems are electronic devices that function in the context of a physical environment, by sensing and reacting to a set of stimuli. To simplify the design of embedded systems, different parts are best described using different notations and analyze with different techniques, i.e., the system is said to be *heterogeneous*. We informally refer to the notation and the rules that are used to specify and verify the elements of heterogeneous systems and their collective behavior as a *model of computation*. In this paper, the use of *conservative approximations* (recently introduced by the authors) is reviewed to establish relationships between different models of computation in a design. After presenting the basic definitions, we propose three different models at different levels of abstraction for describing a system and the progression towards its implementation. Then, we derive associated conservative approximations starting from simple homomorphisms between sets of behaviors of the different models.

1   Introduction

Embedded systems are electronic devices that function in the context of a physical environment, by sensing and reacting to a set of stimuli. To simplify the design of an embedded system, its different parts are best described using different notations and analyzed with different techniques. In this case, we say that the system is *heterogeneous*. For example, the model of the software application that runs on a distributed collection of nodes in a sensor network is often concerned only with the initial and final state of the behavior of a reaction. In contrast, the particular sequence of actions of the reaction could be relevant to the design of one instance of a node. Likewise, the notation used in reasoning about a resource management subsystem is often incompatible with the handling of real time deadlines, typical of communication protocols. This form of heterogeneity is also reflected in the structure of the design teams, which increasingly consist of highly specialized groups that focus on the solution of a particular task, under the direction of system architects.

Designers benefit from this separation. First, the system is naturally partitioned into smaller and more manageable parts. Secondly, and more importantly, designers are free to select for each subsystem the rules that are used to specify its behavior as a hierarchical collection of modules (*composition*), and to verify that such behavior conforms to a specification (*refinement verification*). These rules vary widely across different modeling domains, such as the ones outlined above. The restrictions and the intrinsic properties of these rules, which we collectively refer to as a *model of computation*, are the

P. Degano et al. (Eds.): Montanari Festschrift, LNCS 5065, pp. 721–742, 2008.

basis of domain specific techniques that can be used to guarantee the correctness of the implementation in an easier way.

While specified separately, subsystems must eventually interact to form the system behavior, and will in fact do so in the physical implementation. However, system designers are typically interested in not waiting until the final stages of the implementation to validate the system functionality and performance metrics, because the cost of fixing design and specification errors increases dramatically in the later phases of the design flow as amply documented for electronic systems, software and integrated circuits. The costs associated with late discovery of errors and, in particular, of integration errors, have risen to a point that they are no longer sustainable. To witness, consider the recent recalls by Mercedes-Benz of 1.5 million cars for problems with the braking subsystem. Consequently, the ability of the system designer to specify, manage, and verify the functionality and performance of *concurrent behaviors*, within *and across* heterogeneous boundaries, is essential. Most design methodologies that address these problems are based on the processes of abstraction and refinement, that is, of the application of maps that convert and relate different models of computation. However, crossing the boundaries between abstraction levels by abstracting and refining a specification is often not trivial. The most common pitfalls include mishandling of corner cases and inadvertently misinterpreting changes in the communication semantics.

These problems arise because of the poor understanding and the lack of a precise definition of the abstraction and refinement maps used in the flow, which are therefore likely to provide little, if any, guarantee of satisfying a given set of constraints and specifications, without resorting to extensive simulation or tests on prototypes. However, in the face of growing complexity, this approach will have to yield to more rigorous methods. In addition, abstraction and refinement should be designed to preserve, whenever possible, the properties of the design that have already been established. This is essential to increase the value of early, high level models and to guarantee a speedier path to implementation.

In this paper we review abstraction and refinement relationships in the form of *conservative approximations* [3,17,18] introduced by the authors to approach the problem of abstraction and refinement from a formal standpoint. Conservative approximations are closely related to abstract interpretations, and, in addition, preserve refinement verification from an abstract to a concrete model while avoiding the occurrence of false positive results. This property of an abstraction is useful because, presumably, refinement verification is more efficient at the abstract level than it is at the concrete. In this paper we show how to derive models of computation and the corresponding abstraction and refinement maps starting from simple models of behavior. We focus in particular on models that include both continuous and discrete behaviors, and are therefore appropriate for the design of hybrid systems [4].

The rest of the paper is structured as follows. Section 2 gives an overview of our methodology and formal framework and introduces the basic terminology. A set of different agent models for embedded systems are presented in Section 3. Then, we construct relationships between these models and give a general recipe for deriving conservative approximations in Section 4. Section 5 surveys related work and discusses other forms of abstraction. In all cases, the specific abstraction is either an instance of an abstract interpretation (and is therefore unsound for refinement verification), or is a particular case of conservative approximations. Finally, Section 6 provides directions for our future research.

2   Methodology Overview

Before we can investigate the notion of an abstraction, we must provide a way to describe its domain and range, namely the models of computation. In general, an abstraction transforms a block of computation in one model into a block of computation in another model. For example, it may transform a module written in a discrete event language (such as Verilog or VHDL) into a transaction level module that ignores the precise time at which events occur, such as a dataflow language. We therefore represent models of computation at the granularity of the module, or block. In other words, a model of computation is simply the set of blocks that can be expressed in the model. For instance, we represent a model of computation based on finite state machines as the set of finite state machines, or a dataflow model of computation as the set of dataflow actors. However, the representation of the blocks need not be in the form of a programming language. In fact, to simplify the task of defining abstraction functions, we typically represent blocks as the *set of behaviors*, or *traces* that they can exhibit. The nature of these traces obviously depends on the particular model of computation: for instance, they may consist of sequences of values (as in the case of synchronous models), functions of real variables (for more accurate continuous time models), or sets of values representing certain performance metrics (power models, constraints). Because we use traces, we will refer to blocks in any model of computation generically as *agents*, and they will be denoted by the letters p and q. Traces often refer to the externally visible features of agents: their actions, signals, state variables, etc. We do not distinguish among the different types, and we refer to them collectively as a set of *signals W*. Each trace and each agent is then associated with an *alphabet $A \subseteq W$* of the signals it uses.

We make a distinction between two different kinds of traces: *complete* traces and *partial* traces. A complete trace has no endpoint. A partial trace has an endpoint; it can be a prefix of a complete trace or of another partial trace. Every complete trace has partial traces that are prefixes of it; every partial trace is a prefix of some complete trace. The distinction between a complete trace and a partial trace has only to do with the length of the trace (that is, whether or not it has an endpoint), not with what is happening during the trace. For example, a finite string can represent a complete behavior with a finite number of actions, or it can represent a partial behavior.

In our framework, the first step in defining a model of computation is to construct an algebra of traces C. The trace algebra contains the universe of partial traces and the universe of complete traces for the model of computation. The algebra also includes three operations on traces: *projection, renaming* and *concatenation*. Intuitively, these operations correspond to encapsulation, instantiation and sequential composition, respectively. Projection removes all references to a specified set of signals in a trace, hiding them from an external observer, while renaming is used to change the names of the signals, emulating the replacement of actual for formal parameters in function instantiation. Concatenation, which joins two behaviors at their ends, can be used to define the notion of a prefix of a trace. We say that a trace x is a prefix of a trace z if there exists a trace y such that z is equal to x concatenated with y.

Agents in a model of computation do not exist in isolation. For instance, agents can be combined with the operation of parallel composition (denoted by the symbol $\parallel$) to yield a new agent in the *same* model of computation. An *agent algebra* Q is used to define these operators, together with their set of agents. For trace-based agent models, an agent algebra is constructed in a fixed way from the algebra of the corresponding traces, where agents are simply sets of traces. For composition, the new agent combines the behaviors of the original components in such a way that the new behaviors are consistent with those of the components when projected onto their alphabets. Other operations are derived by simply extending to sets of traces the operators of the trace algebra. We will show examples of these derivations later in Section 3.

Different means can often be used to achieve the same goal. Likewise, agents with different behavior may sometimes yield the same result when applied to a particular context. In particular, if an agent p can always be replaced for an agent q in any context without materially changing the outcome of the composition, then we say that p *refines* q. In the rest of this paper we use the symbol $\preceq$ to denote this refinement relationship, and we write $p \preceq q$ whenever p refines q. We also refer to q as a *specification*, and to p as an *implementation* of q. For trace-based models, refinement can be reduced to checking containment between the trace set of agents, and is therefore analogous to verification methods based on language containment.

2.1 Refinement Preserving Abstractions

The choice of levels of abstraction, or models, in a heterogeneous design methodology is obviously very important. Each model must in fact be capable of supporting the desired techniques, and must be detailed enough to provide answers to the specific questions under consideration for the particular subsystems it applies to. An equally important choice has to do with the way the levels of abstraction are connected, or, in other words, with the abstraction and refinement functions that are used to relate the models. In general, many forms of abstraction and refinement are possible. In practice, only those that preserve certain properties of interest are useful. In particular, we are interested in abstractions that preserve the refinement relationship $\preceq$ when moving from a more abstract model to a more concrete one. More formally, assume p and q are agents in a model Q, and that p' and q' are the corresponding abstractions in a model Q'. Then we say that the abstraction preserves the refinement relationship from the abstract to the concrete model if $p' \preceq q'$ implies that also $p \preceq q$.

This property is useful for several reasons. First, refinement verification can be used to establish that an agent satisfies some requirement by comparing it to a specification. It would therefore be at best inconvenient if the result of this verification were lost during a refinement step of the methodology. In the worst case, it could lead to incorrect designs. A second advantage has to do with the efficiency of refinement verification. The process is in fact potentially more efficient at the abstract level because of the lesser amount of information included in the model. An abstraction that preserves the refinement relationship can thus be used to translate a complex verification problem at the concrete level to a simpler problem at the abstract level. This translation is conservative: while the loss of information may make it impossible to establish a refinement relationship between the abstracted agents, it ensures that when the relationship is indeed established it also holds at the concrete level. In other words, false positive results are ruled out.

Conservative approximations are constructed using two abstraction functions, instead of just one. The first function, usually denoted Ψ_u, is applied to the implementation p, while the second function, denoted Ψ_l, is applied to the specification q. The pair (Ψ_l, Ψ_u) forms a conservative approximation whenever $\Psi_u(p) \preceq \Psi_l(q)$ implies $p \preceq q$. Thus, by definition, a conservative approximation always preserves the refinement relationship from the abstract to the concrete model. In the rest of this paper, we first introduce a number of models of interest for the development of embedded systems, and then show how to relate them using conservative approximations, and their inverses, obtained from simple homomorphisms on traces.

The notion of a conservative approximation is independent of the use of traces as an underlying agent model [18]. In particular, it could be used in other contexts, such as branching-time logics, where refinement and equivalence are expressed in terms of simulations. Our motivations for developing a trace-based model is the ease with which conservative approximations can be derived starting from simple homomorphic functions on behaviors.

3   Models of Embedded System Behavior

In this section we will present three models at progressively higher levels of abstraction, by defining a trace algebra and a corresponding agent algebra. We develop our models in the context of hybrid systems [4], a particular kind of heterogeneous systems that combine behaviors expressed as a continuous evolution with the occurrence of instantaneous discrete events. These two aspects of a behavior are often called the *flows* and the *jumps* of the system. Hybrid formalisms are particularly useful when designing embedded control systems, which require modeling the physical behavior of environments that undergo sudden mode changes. The hybrid model, in addition, is necessary for an accurate evaluation of a control strategy based on discrete computations.

The first model that we present, called *metric time*, is intended to represent exactly the evolutions (the flows and the jumps) of a system as a function of global real time. With the second we abstract away the metric while maintaining the total order of occurrence of events. This model is used to define the untimed semantics of embedded applications. Finally, the third trace algebra further abstracts away the information on the event occurrences by only retaining initial and final states and removing the intermediate steps. This simpler model can be used to describe the semantics of some programming language constructs. Later, we will use homomorphisms on trace sets to derive conservative approximations.

3.1   Metric Time

A typical semantics for hybrid embedded systems includes continuous *flows* that represent the continuous dynamics of the system, and discrete *jumps* that represent instantaneous changes of the operating conditions. The system is modeled by its state variables. In our formalization, the evolution of the state variables takes the form of a single piece-wise continuous function over real-valued time, where the continuous segments represent the flows, while the discontinuities between the segments model the jumps. In this paper we assume that the variables of the system take only real or integer values. Real-valued variables are used, for instance, to model quantities such as position and speeds, while integer variables are more appropriate for modalities and other discrete quantities. The sets of real-valued and integer valued variables for a given trace are called $V_{\mathbb{R}}$ and $V_{\mathbb{Z}}$, respectively. Traces may also contain actions, which are discrete events that can occur at any time. Actions do not carry data values. For a given trace, the set of input actions is M_I and the set of output actions is M_O. Actions could be, for example, the commands issued by a user, or signals generated by an embedded controller.

Each trace has a signature γ which is a 4-tuple of the above sets of signals:

$$\gamma = (V_{\mathbb{R}}, V_{\mathbb{Z}}, M_I, M_O).$$

The sets of signals may be empty, but we assume they are disjoint. The *alphabet* of γ is

$$A = V_{\mathbb{R}} \cup V_{\mathbb{Z}} \cup M_I \cup M_O.$$

The set of partial traces for a signature γ is $B_P(\gamma)$. Each element of $B_P(\gamma)$ is a triple $x = (\gamma, \delta, f)$. The non-negative real number δ is the *duration* (in time) of the partial trace. The function f has domain A. For $v \in V_{\mathbb{R}}$, $f(v)$ is a function in $[0, \delta] \to \mathbb{R}$, where $\mathbb{R}$ is the set of real numbers and the closed interval $[0, \delta]$ is the set of real numbers between 0 and δ, inclusive. This function must be piece-wise continuous and right-hand limits must exist at all points. Analogously, for $v \in V_{\mathbb{Z}}$, $f(v)$ is a piece-wise constant function in $[0, \delta] \to \mathbb{Z}$, where $\mathbb{Z}$ is the set of integers. For $a \in M_I \cup M_O$, $f(a)$ is a function in $[0, \delta] \to \{0, 1\}$, where $f(a)(t) = 1$ iff action a occurs at time t in the trace.

The set of complete traces for a signature γ is $B_C(\gamma)$. Each element of $B_C(\gamma)$ is a pair $x = (\gamma, f)$. The function f is defined as for partial traces, except that each occurrence of $[0, \delta]$ in the definition is replaced by $\mathbb{R}^{\not\hspace{0.1em}}$, the set of non-negative real numbers.

To complete the definition of this trace algebra, we must define the operations of projection, renaming and concatenation on traces. The projection operation $proj(B)(x)$ is defined iff $M_I \subseteq B \subseteq A$, where B is the set of signals that must be *retained*. The trace that results is the same as x except that the domain of f is restricted to B. The renaming operation $x' = rename(r)(x)$ is defined iff r is a one-to-one function from A to some $A' \subseteq W$. If x is a partial trace, then $x' = (\gamma', \delta, f')$ where γ' results from using r to rename the elements of γ and $f' = r \circ f$.

The definition of the concatenation operator $x_3 = x_1 \cdot x_2$, where x_1 is a partial trace and x_2 is either a partial or a complete trace, is more complicated. If x_2 is a partial trace, then x_3 is defined iff $\gamma_1 = \gamma_2$ and for all $a \in A$,

$$f_1(a)(\delta_1) = f_2(a)(0),$$

(note that δ_1, δ_2, etc., are components of x_1 and x_2 in the obvious way). Concatenation is defined only when the end points of the two traces match. By doing so, jumps must be modeled explicitly by a trace, and do not arise as a byproduct of concatenation. When defined, $x_3 = (\gamma_1, \delta_3, f_3)$ is such that $\delta_3 = \delta_1 + \delta_2$ and for all $a \in A$,

$$f_3(a)(\delta) = f_1(a)(\delta) \quad \text{for } 0 \leq \delta \leq \delta_1,$$
$$f_3(a)(\delta) = f_2(a)(\delta - \delta_1) \quad \text{for } \delta_1 \leq \delta \leq \delta_3.$$

The concatenation of a partial trace with a complete trace yields a complete trace with a similar definition. If $x_3 = x_1 \cdot x_2$, then x_1 is a *prefix* of x_3.

3.2   Non-metric Time

In the definition of this trace algebra we are concerned with the order in which events occur in the system, but not in their absolute distance or position. This is useful if we want to describe the semantics of a programming language for embedded systems that abstracts from a particular real time implementation. Although we want to remove real time, we want to retain the global ordering on events induced by time. In particular, to simplify the abstraction from metric time to non-metric time described below, we would like to support the case of an uncountable number of events[1]. Sequences are clearly inadequate given our requirements. Instead we use a more general notion of a partially ordered multiset to represent the trace. We quote the original definition given by Pratt [19], and due to Gischer, which begins with the definition of a labeled partial order. We then specialize this notion to our needs.

Definition 1 (Labeled partial order, Pratt [19]). *A* labeled partial order *(lpo) is a 4-tuple* $L = (V, \Sigma, \leq, \mu)$ *consisting of*

1. *a* vertex set V, *typically modeling* events;
2. *an* alphabet Σ *(for symbol set), typically modeling* actions *such as the arrival of integer 3 at port Q, the transition of pin 13 of IC-7 to 4.5 volts, or the disappearance of the 14.3 MHz component of a signal;*
3. *a* partial order $\leq$ *on V, with* $e \leq f$ *typically being interpreted as event e necessarily preceding event f in time; and*
4. *a* labeling function $\mu : V \to \Sigma$ *assigning symbols to vertices, each labeled event representing an* occurrence *of the action labeling it, with the same action possibly having multiple occurrence, that is,* μ *need not be injective.*

A *pomset* (partially ordered multiset) is then the isomorphism class of an lpo, denoted $[V, \Sigma, \leq, \mu]$. By taking lpo's up to isomorphism we confer on pomsets a degree of abstraction equivalent to that enjoyed by strings (regarded as finite linearly ordered labeled sets up to isomorphism), ordinals (regarded as well-ordered sets up to isomorphism), and cardinals (regarded as sets up to isomorphism).

This representation is suitable for the above mentioned infinite behaviors: the underlying vertex set may be based on an uncountable total order that suits our needs. For our application, we do not need the full generality of pomsets. Instead, we restrict ourselves to pomsets where the partial order is total, which we call *tomsets*.

---

[1] In theory, such Zeno-like behavior is possible, for example, for an infinite loop whose execution time halves with every iteration.

Traces have the same form of signature as in metric time:

$$\gamma = (V_{\mathbb{R}}, V_{\mathbb{Z}}, M_I, M_O).$$

Both partial and complete traces are of the form $x = (\gamma, L)$ where L is a tomset. When describing the tomset L of a trace, we will in fact describe a particular lpo, with the understanding that L is the isomorphism class of that lpo. An action $\sigma \in \Sigma$ of the lpo is a function with domain A such that for all $v \in V_{\mathbb{R}}$, $\sigma(v)$ is a real number (the value of variable v resulting from the action σ); for all $v \in V_{\mathbb{Z}}$, $\sigma(v)$ is an integer; and for all $a \in M_I \cup M_O$, $\sigma(v)$ is 0 or 1. The underlying vertex set V, together with its total order, provides the notion of time, a space that need not contain a metric. For both partial and complete traces, there must exist a unique minimal element $min(V)$. The action $\mu(min(V))$ that labels $min(V)$ should be thought of as giving the initial state of the variables in $V_{\mathbb{R}}$ and $V_{\mathbb{Z}}$. For each partial trace, there must exist a unique maximal element $max(V)$ (which may be identical to $min(V)$). As defined above, the set of partial traces and the set of complete traces are not disjoint. It is convenient, in fact, to extend the definitions so that traces are labeled with a bit that distinguishes partial traces from complete traces, although we omit the details.

By analogy with the metric time case, it is straightforward to define projection and renaming on actions $\sigma \in \Sigma$. This definition can be easily extended to lpo's and, thereby, traces. The concatenation operation $x_3 = x_1 \cdot x_2$ is defined iff x_1 is a partial trace, $\gamma_1 = \gamma_2$ and $\mu_1(max(V_1)) = \mu_2(min(V_2))$. When defined, the vertex set V_3 of x_3 is a disjoint union:

$$V_3 = V_1 \uplus (V_2 - min(V2)),$$

ordered such that the orders of V_1 and V_2 are preserved and such that all elements of V_1 are less than all elements of V_2. The labeling function is such that for all $v \in V_3$

$$\mu_3(v) = \mu_1(v) \text{ for } min(V_1) \le v \le max(V_1),$$
$$\mu_3(v) = \mu_2(v) \text{ for } max(V_1) \le v.$$

3.3 Pre-post Time

The third and last trace algebra is concerned with modeling non-interactive constructs of a programming language. In this case we are interested only in an agents possible final states given an initial state. This semantic domain could therefore be considered as a denotational representation of an axiomatic semantics.

We cannot model communication actions at this level of abstraction, so signatures are of the form $\gamma = (V_{\mathbb{R}}, V_{\mathbb{Z}})$ and the alphabet of γ is $A = V_{\mathbb{R}} \cup V_{\mathbb{Z}}$. A non-degenerate state s is a function with domain A such that for all $v \in V_{\mathbb{R}}$, $s(v)$ is a real number (the value of variable v in state s); and for all $v \in V_{\mathbb{Z}}$, $s(v)$ is an integer. We also have a degenerate, undefined state $\perp_*$.

A partial trace $B_P(\gamma)$ is a triple (γ, s_i, s_f), where s_i and s_f are the initial and final states. A complete trace $B_C(\gamma)$ is of the form $(\gamma, s_i, \perp_\omega)$, where $\perp_\omega$ indicates non-termination. This trace algebra is primarily intended for modeling terminating behaviors, which explains why so little information is included on the traces that model non-terminating behaviors.

The operations of projection and renaming are built up from the obvious definitions of projection and renaming on states. The concatenation operation $x_3 = x_1 \cdot x_2$ is defined iff x_1 is a partial trace, $\gamma_1 = \gamma_2$ and the final state of x_1 is identical to the initial state of x_2. As expected, when defined, x_3 contains the initial state of x_1 and the final state of x_2.

3.4   Construction of Agent Models

Our models of agent are constructed in a fixed way from models of traces by considering the set of behaviors that an agent exhibits. An agent over a given trace algebra is a pair (γ, P), where γ is a signature and P is a subset of the traces for that signature. The set P represents the set of possible behaviors of an agent.

An agent algebra has a set of agents over a given trace algebra as its domain. Operations of projection, renaming, parallel composition and serial composition on agents are defined using the operations of the trace algebra, as follows.

Projection and renaming are the simplest operations to define. When they are defined depends on the signature of the agent in the same way that definedness for the corresponding trace algebra operations depends on the signature of the traces. The signature of the result is also analogous. Finally, the set of traces of the result is defined by naturally extending the trace algebra operations to sets of traces. For instance, if $p = (\gamma, P)$ is an agent, then $proj(B)(p) = (\gamma', proj(B)(P))$, where γ' is obtained by γ by retaining only the elements of B. Sequential composition is defined in terms of concatenation in an analogous way. The only difference from projection and renaming is that sequential composition requires two agents as arguments, and concatenation requires two traces as arguments.

Parallel composition of two agents is defined only when all the traces in the agents are complete traces, and the set of output actions of the two agents are disjoint. Let the agent p'' be the parallel composition of p and p'. Then the components of p'' are as follows (M_I and M_O are omitted in pre-post traces):

$$V_{\mathbb{R}}'' = V_{\mathbb{R}} \cup V_{\mathbb{R}}'$$
$$V_{\mathbb{Z}}'' = V_{\mathbb{Z}} \cup V_{\mathbb{Z}}'$$
$$M_O'' = M_O \cup M_O'$$
$$M_I'' = (M_I \cup M_I') - M_O''$$
$$P'' = \{x \in \mathcal{B}_C(\gamma'') : proj(A)(x) \in P \wedge proj(A')(x) \in P'\}.$$

Here, the variables of the composite p'' are the union of the variables of the components p and p'. The actions of the composite are also the union of the actions of the components. An action is regarded as an output of the composite if it is an output of either component. However, an action is an input of the composite if it is an input of one of the components, and it is not at the same time an output of the other component, so that an input can only be connected to one output. The definition of P'' ensures that the behaviors of the composite are all and only the behaviors consistent with the components.

4 Relations Between Models

The three trace algebras defined above cover a wide range of levels of abstraction. The first step in formalizing the relationships between those levels is to define homomorphims between the trace algebras. Trace algebra homomorphisms induce corresponding conservative approximations between the agent algebras, as we shall see.

4.1 Homomorphisms

From metric to non-metric time. A homomorphism from metric time to non-metric time should abstract away detailed timing information. This requires characterizing events in metric time and mapping those events into a non-metric time domain. Since metric time trace algebra is, in part, value based, some additional definitions are required to characterize events at that level of abstraction.

Let x be a metric trace with signature γ and alphabet A such that

$$\gamma = (V_{\mathbb{R}}, V_{\mathbb{Z}}, M_I, M_O),$$
$$A = V_{\mathbb{R}} \cup V_{\mathbb{Z}} \cup M_I \cup M_O.$$

We define the homomorphism h by defining a non-metric time trace $y = h(x)$. This requires building a vertex set V and a labeling function μ to construct an lpo. The trace y is the isomorphism class of this lpo. For the vertex set we take all reals such that an event occurs in the trace x, where the notion of event is formalized in the next several definitions.

Definition 2 (Stable function). *Let f be a function over a real interval to $\mathbb{R}$ or $\mathbb{Z}$. The function is stable at t iff there exists an $\epsilon > 0$ such that f is constant on the interval $(t - \epsilon, t]$.*

Definition 3 (Stable trace). *A metric time trace x is stable at t iff for all $v \in V_{\mathbb{R}} \cup V_{\mathbb{Z}}$ the function $f(v)$ is stable at t; and for all $a \in M_I \cup M_O$, $f(a)(t) = 0$.*

In other words, a trace is stable at a time t if it is possible to find a left neighborhood of t (i.e., an interval $(t - \epsilon, t]$ for $\epsilon > 0$) where the trace is constant and no action occurs. When a trace is not stable at t, then we say that the trace has an event at t.

Definition 4 (Event). *A metric time trace x has an event at $t > 0$ if it is not stable at t. Because a metric time trace doesn't have a left neighborhood at $t = 0$, we always assume the presence of an event at the beginning of the trace. If x has an event at t, the action label σ for that event is a function with domain A such that for all $v \in A$, $\sigma(a) = f(a)(t)$, where f is a component of x as described in the definition of metric time traces.*

Now we construct the vertex set V and labeling function μ necessary to define y and, thereby, the homomorphism h. The vertex set V is the set of reals t such that x has an event at t. While it is convenient to make V a subset of the reals, remember that the tomset that results is an isomorphism class. Hence the metric defined on the set of reals is lost. The labeling function μ is such that for each element $t \in V$, $\mu(t)$ is the action label for the event at t in x.

Note that if we start from a partial trace in the metric trace we obtain a trace in the non-metric trace that has an initial and final event. It has an initial event by definition. It has a final event because the metric trace either has an event at δ (the function is not constant), or the function is constant at δ, and then there must be an event that brought the function to that constant value (which, in case of identically constant functions, is the initial event itself).

To show that h does indeed abstract away information, consider the following situation. Let x_1 be a metric time trace. Let x_2 be same trace where time has been "stretched" by a factor of two (i.e., for all $v \in A_1$, $x_1(a)(t) = x_2(a)(2t)$). The vertex sets generated by the above process are isomorphic (the order of the events is preserved), therefore $h(x_1) = h(x_2)$.

From non-metric to pre-post time. The homomorphism h from the non-metric time traces to pre-post traces requires that the signature of the agent be changed by removing M_I and M_O. Let $y = h(x)$. The initial state of y is formed by restricting $\mu(min(V))$ (the initial state of x) to $V_{\mathbb{R}} \cup V_{\mathbb{Z}}$. If x is a complete trace, then the final state of y is $\perp_\omega$. If x is a partial trace, and there exists $a \in M_I \cup M_O$ and time t such that $f(a)(t) = 1$, the final state of y is $\perp_*$. Otherwise, the final state of y is formed by restricting $\mu(max(V))$.

4.2   Conservative Approximations

As discussed in the introduction, we are interested in relating different models that describe systems at different levels of abstraction. We can accomplish this by deriving a conservative approximation from a homomorphism between trace algebras. Consider two trace algebras C and C'. Intuitively, if $h(x) = x'$, the trace x' is an abstraction of any trace y such that $h(y) = x'$. Thus, x' can be thought of as representing the set of all such y. For instance, a non-metric time trace x' can be thought of the abstraction of all possible stretched versions y in the metric time model. This is easily extended to sets of traces, and therefore to agents. Hence, if $\mathcal{Q}$ and $\mathcal{Q}'$ are agent algebras over C and C' respectively, we use the function Ψ_u that maps an agent $p = (\gamma, P)$ in $\mathcal{Q}$ into the agent $(\gamma, h(P))$ in $\mathcal{Q}'$ as the upper bound in a conservative approximation. A sufficient condition for a corresponding lower bound is: if $x \notin P$, then $h(x)$ is not in the set of possible traces of $\Psi_l(p)$. This leads to the definition of a function $\Psi_l(p)$ that maps P into the set $h(P) - h(\mathcal{B}(\gamma) - P)$, where $\mathcal{B}(\gamma)$ is the set of all traces with alphabet γ. For instance, the lower bound of a metric time agent p into non-metric time includes a trace x' if and only if p contains all its possible concretizations (time stretched versions). The conservative approximation $\Psi = (\Psi_l, \Psi_u)$ is an example of a *conservative approximation induced by h*. A slightly tighter lower bound is also possible (see [3]).

It is straightforward to take the general notion of a conservative approximation induced by a homomorphism, and apply it to specific models. Simply construct trace algebras C and C', and a homomorphism h from C to C'. Recall that these trace algebras act as models of individual behaviors. One can construct the agent algebras $\mathcal{Q}$ over C and $\mathcal{Q}'$ over C', and a conservative approximation Ψ induced by h. Thus, one need only construct two models of individual behaviors and a homomorphism between them to obtain two agent models along with a conservative approximation between the individual agents of the models.

This same approach can be applied to the three trace algebras, and the two homomorphisms between them, that were defined in Section 3, giving conservative approximations between process models at three different levels of abstraction. The application of the upper bound is straightforward, since it is a natural extension to sets of the homomorphism on behaviors. The lower bound, on the other hand, provides complementary information. For instance, the lower bound of a metric-time agent contains all those behaviors for which the agent has an analogous behavior for any possible "stretching" of the time-axis. Thus, the lower bound identifies those behaviors in an agent that are in a sense *speed independent*. Similarly, the conservative approximation from non-metric time to pre-post traces identifies the subset of behaviors of an agent which depend exclusively on the initial and final state of the computation.

4.3   Inverse Approximations

As we have discussed, if $\Psi = (\Psi_l, \Psi_u)$ is a conservative approximation from Q to Q', then $p' = \Psi_u(p)$ represents a kind of upper bound on p. It is instructive to investigate whether there is an agent in Q that is represented exactly by p' rather than just being bounded by p'. If no agent in Q can be represented exactly, then Ψ is abstracting away too much information to be of much use for verification. If every agent in Q can be represented exactly, then Ψ_l and Ψ_u are equal and are isomorphisms from Q to Q'. These extreme cases illustrate that the amount of abstraction in Ψ is related to what agents p are represented exactly by $\Psi_u(p)$ and $\Psi_l(p)$.

To formalize what it means to be represented exactly in the context of conservative approximations, we define the inverse Ψ_{inv} of the conservative approximation Ψ. The inverse of an approximation is a function from the abstract model Q' to the concrete model Q that, as we shall see in this section, completes the relationships between Q and Q' by establishing a refinement map across the models. Normal notions of the inverse of a function are not adequate for constructing the inverse of a conservative approximation Ψ, since Ψ is a pair of functions. Our notion of an inverse is thus based on the following result.

Lemma 1. *Let Q and Q' be models of computation, and let (Ψ_l, Ψ_u) be a conservative approximation from Q to Q'. For all p_1 and p_2 in Q, if $\Psi_l(p_1) = \Psi_u(p_1) = p'$ and $\Psi_l(p_2) = \Psi_u(p_2) = p'$, then $p_1 = p_2$.*

Lemma 1 shows that when the upper and the lower bound coincide for a particular agent p, then, intuitively, the abstraction p' is an exact representation of p. To put it another way, p does not use any of the additional information provided by the concrete level, since it can be determined uniquely from its abstraction p'. It is therefore natural to define $\Psi_{inv}(p') = p$, where p is the agent in Q such that $\Psi_u(p) = \Psi_l(p) = p'$. If $\Psi_l(p) \neq \Psi_u(p)$, then p is not represented exactly in Q'. In this case, p is not in the image of Ψ_{inv}.

Definition 5 (Inverse of a Conservative Approximation). *Let $\Psi = (\Psi_l, \Psi_u)$ be a conservative approximation from Q to Q'. For $p' \in Q'$, the inverse $\Psi_{inv}(p')$ is defined and is equal to p if and only if $\Psi_l(p) = \Psi_u(p) = p'$.*

It follows from the definition that, when Ψ_{inv} is defined, the following identity holds:

$$\Psi_l(\Psi_{inv}(p')) = \Psi_u(\Psi_{inv}(p')).$$

The function Ψ_{inv} need not be defined for all p'. This may happen, for example, if the model $\mathcal{Q}'$ includes information that cannot be expressed exactly in $\mathcal{Q}$. In that case, Ψ_{inv} is a partial function and is only defined for the agents that have an exact representation in both models. When an agent has an exact representation in $\mathcal{Q}$ and $\mathcal{Q}'$, we say that it can be used *indifferently* in the two models, or that it is *polymorphic*. This is because the agent makes no assumption regarding its behavior based on information that can be expressed exclusively in either model of computation. However, the representation of the agent in $\mathcal{Q}$ and $\mathcal{Q}'$ is, in general, different. Thus, this notion extends our ability to reuse agents across models that employ different representations.

For our examples, the inverse of the approximation from metric to non-metric time agent is always defined, and translates a non-metric time agent to a corresponding metric time agent which non-deterministically chooses a given timing for any of its behaviors. This non-determinism is typical of our approach, and is useful to expose the degrees of freedom that are available in a design-by-refinement methodology. Similarly, the inverse of the conservative approximation that goes from pre-post to non-metric time agents builds a concretization where each pair of initial and final states is non-deterministically computed by reordering actions along the time axis.

A conservative approximation thus induces its own inverse in the form of a (possibly partial) refinement map. The inverse is uniquely determined, and, because of the defining properties of a conservative approximation, Ψ_{inv} is one-to-one, and, when restricted to the image of Ψ_{inv}, the functions Ψ_l and Ψ_u are equal and are the inverse of Ψ_{inv}. In addition, when defined, Ψ_{inv} is always monotonic and, if either Ψ_l or Ψ_u is also monotonic, it preserves the ordering of the agents in both directions. Hence, the inverse embeds the abstract model of computation (or at least the part of it where it is defined) into the more concrete model, in a way that is consistent with the chosen abstractions. Different conservative approximations between the same models may therefore induce different embeddings. This is again an indication of the importance of choosing the right abstraction for the problem at hand. The nature of the embedding, in fact, determines how one model is interpreted in terms of the other, and quantifies the amount of information lost during the abstraction.

4.4 Modeling Constructs in Embedded Software

Using Pre-Post Traces. One of the fundamental features of embedded software is that it interacts with the physical world. Conventional axiomatic or denotational semantics of sequential programming languages only model initial and final states of terminating programs. Thus, these semantics are inadequate to fully model embedded software. However, much of the code in an embedded application does computation or internal communication, rather than interacting with the physical world. Such code can be adequately modeled using conventional semantics, as long as the model can be integrated with the more detailed semantics necessary for modeling interaction. Pre-post agents are quite similar to conventional semantics. As described earlier, we can also embed pre-post agents into more detailed models. Thus, we can model the non-interactive parts of an embedded application at a high level of abstraction that is simpler and more natural, while also being able to integrate accurate models of interaction, real-time constraints and continuous dynamics.

As an example we consider the problem of developing software to control an engine in the cutoff region [2]. Here, the behaviors of an automobile engine are divided into regions of operation, each characterized by appropriate control actions to achieve a desired result. The cutoff region is entered when the driver releases the accelerator pedal, thereby requesting that no torque be generated by the engine. In order to minimize power train oscillations that result from suddenly reducing torque, a closed loop control damps the oscillations using carefully timed injections of fuel. The control problem is therefore hybrid, consisting of a discrete (the fuel injection) and a continuous (the power train behavior) systems tightly linked.

```
01. void control_algorithm( void ) {
02.     // state definition
03.     struct state { double x1; double x2; double omega_c; } current_state;
04.     // Init the past three injections (assume injection before cutoff)
05.     double u1, u2, u3 = 1.0;
06.
07.     loop forever {
08.         await( action_request );
09.         read_current_state( current_state );
10.         compute_sigmas( sigma_m, sigma_0, current_state, u1, u2, u3 );
11.         // update past injections
12.         u1 = u2; u2 = u3;
14.         // compute next injection signal
15.         if ( sigma_m < sigma_0 ) {
16.             action_injection( );
17.             u3 = 1.0;
18.         } else {
19.             action_no_injection( );
20.             u3 = 0.0;
21.         }
22.     }
23. }
```

Fig. 1. An embedded control algorithm

Figure 1 shows the top level routine of the control algorithm. Although we use a C-like syntax, the semantics are simplified, as described later. The controller is activated by a request for an injection decision (this happens every full engine cycle). The algorithm first reads the current state of the system (as provided by the sensors on the power train), predicts the effect of injecting or not injecting on the future behavior of the system, and finally controls whether injection occurs. The prediction uses the value of the past three decisions to estimate the position of the future state. The control algorithm involves solving a differential equation, which is done in the call to `compute_sigmas` (see [2] for more details). A nearly optimal solution can be achieved without injecting intermediate amounts of fuel (i.e., either inject no fuel or inject the maximum amount). Thus, the only control inputs to the system are the actions `action_injection` (maximum injection) and `action_no_injection` (zero injection).

The semantics of each statement of the programming language is given by an agent. To simplify the semantics, we assume that inter-process communication is done through shared actions rather than shared variables. A pre-post agent has a signature γ of the form $(V_{\mathbb{R}}, V_{\mathbb{Z}})$. For the semantics of a programming language statement, γ indicates the variables accessible in the scope where the statement appears. For a block that declares local variables, the agent for the statement in the block includes in its signature the local variables. The agent for the block is formed by projecting away the local variables from the agent of the statement.

The sequential composition of two statements is defined as the concatenation of the corresponding agents: the definition of concatenation ensures that the two statements agree on the intermediate state. The traces in the agent for an assignment to variable v are of the form (γ, s_i, s_f), where s_i is an arbitrary initial state, and s_f is identical to s_i except that the value of v is equal to the value of the right-hand side of the assignment statement evaluated in state s_i (we assume the evaluation is side-effect free).

The semantics of a procedure definition is given by an agent with an alphabet $\{v_1, \ldots, v_r\}$ where v_k is the k-th argument of the procedure (these signal names do not necessarily correspond to the names of the formal variables). We omit the details of how this agent is constructed from the text of the procedure definition. More relevant for our control algorithm example, the semantics of a procedure call `proc(a, b)` is the result of renaming $v_1 \rightarrow a$ and $v_2 \rightarrow b$ on the agent for the definition of `proc`. The parameter passing semantics that results is *value-result* (i.e., no aliasing or references) with the restriction that no parameter can be used for both a value and result. More realistic (and more complicated) parameter passing semantics can also be modeled.

To define the semantics of `if-then-else` and `while` loops we define a function $init(x, c)$ to be true if and only if the predicate c is true in the initial state of trace x. The formal definition depends on the particular trace algebra being used. In particular, for pre-post traces, $init(x, c)$ is false for all c if x has $\perp_*$ as its initial state.

For the semantics of `if-then-else`, let c be the conditional expression and let P_T and P_E be the sets of possible traces of the `then` and `else` clauses, respectively. The set of possible traces of the `if-then-else` is

$$P = \{x \in P_T : init(x, c)\} \cup \{x \in P_E : \neg init(x, c)\}$$

Notice that this definition can be used for any trace algebra where $init(x, c)$ has been defined, and that it ignores any effects of the evaluation of c not being atomic.

In the case of `while` loops we first define a set of traces E such that for all $x \in E$ and traces y, if $x \cdot y$ is defined then $x \cdot y = y$. For pre-post traces, E is the set of all traces with identical initial and final states. If c is the condition of the loop, and P_B the set of possible traces of the body, we define $P_{T,k}$ and $P_{N,k}$ to be the set of terminating and non-terminating traces, respectively, for iteration k, as follows:

$$P_{T,0} = \{x \in E : \neg init(x, c)\}$$
$$P_{N,0} = \{x \in E : init(x, c)\}$$
$$P_{T,k+1} = P_{N,k} \cdot P_B \cdot P_{T,0}$$
$$P_{N,k+1} = P_{N,k} \cdot P_B \cdot P_{N,0}$$

The concatenation of $P_{T,0}$ and $P_{N,0}$ at the end of the definition ensures that the final state of a terminating trace does not satisfy the condition c, while that of a non-terminating trace does. Clearly the semantics of the loop should include all the terminating traces. For non-terminating traces, we need to introduce some additional notation. A sequence $Z = < z_0, \ldots >$ is a non-terminating execution sequence of a loop if, for all k, $z_k \in P_{N,k}$ and $z_{k+1} \in z_k \cdot P_B$. This sequence is a chain in the prefix ordering. The initial state of Z is defined to be the initial state of z_0. For pre-post traces, we define $P_{N,\omega}$ to be all traces of the form $(\gamma, s, \perp_\omega)$ where s is the initial state of some non-terminating execution sequence Z of the loop. The set of possible traces of the loop is therefore

$$P = \left(\bigcup_k P_{T,k} \right) \cup P_{N,\omega}.$$

Using Non-Metric Time Traces. Using an inverse conservative approximation, as described earlier, the pre-post trace semantics outlined above can be embedded into the non-metric time agent model. However, this is not adequate for two of the constructs used in Figure 1: `await` and the non-terminating loop. These constructs must be describe directly at the lower level of abstraction provided by non-metric time traces.

As used used in Figure 1, the `await(a)` simply delays until the external action a occurs. Thus, the possible partial traces of `await` are those where the values of the state variables remain unchanged and the action a occurs exactly once, at the endpoint of the trace. The possible complete traces are similar, except that the action a must never occur.

To give a more detailed semantics for non-terminating loops, we define the set of extensions of a non-terminating execution sequence Z to be the set $ext(Z) = \{x \in \mathcal{B}(\gamma) : \forall k[z_k \in pref(x)]\}$. For any non-terminating sequence Z, we require that $ext(Z)$ be non-empty, and have a unique maximal lower bound contained in $ext(Z)$, which we denote $\lim(Z)$. In the above definition of the possible traces of a loop, we modify the definition of the set of non-terminating behaviors $P_{N,\omega}$ to be the set of $\lim(Z)$ for all non-terminating execution sequences Z.

Using Metric Time Traces. Analogous to the embedding discussed in the previous subsection, non-metric time agents can be embedded into the metric-time agent model. Here continuous dynamics can be represented, as well as timing assumptions about programming language statements. Also, timing constraints that a system must satisfy can be represented, so that the system can be verified against those constraints.

5 Related Work

Abstract interpretations [6,7] are a widely used means of relating different domains of computation for the purpose of facilitating the analysis of a system. An abstract interpretation between two domains of computation consists of an abstraction function and of a concretization function that form a Galois connection. The distinguishing feature of an abstract interpretation is that the concretization of the evaluation of an expression using the operators of the abstract domain of computation is guaranteed to be an upper bound of the corresponding evaluation of the same expression using the operators of the concrete domain. Hence, a conservative evaluation can be carried out at the more abstract level, where it is potentially computationally more efficient. Refinement verification, however, is unsound: a positive refinement verification result at the abstract level does not guarantee a corresponding refinement verification result at the concrete level. Conservative approximations overcome this problem because they employ two separate abstraction functions, one for the implementation and one for the specification. Our study shows that this is a necessary condition for the preservation of refinement, and one that is not satisfied by a Galois connection [18]. Conservative approximations and abstract interpretations are however strongly related, in that a pair of Galois connections can be used to construct a conservative approximation [18]. This result is important because it extends the rich field of abstract interpretations to refinement verification.

The study of heterogeneous systems is also a central theme of both the Metropolis [1] and the Ptolemy [11] projects. In Metropolis, a system is composed of processes that communicate over media expressed in a meta-model of computation. Their combination, and their relationships, implicitly determine the interaction semantics. Because of its generality, the underlying meta-model fabrics can be used to promote reuse of diverse components. The communication media, however, must be carefully designed to resolve possible incompatibilities. Our work can be thought of as the theory base for the use of the meta-model to represent heterogeneous systems. In addition, conservative approximations have been used to make the process of platform-based design advocated in the Metropolis project precise, and their application in this area is part of our current work [17].

Similarly, Ptolemy consists of several executable domains of computation that can be mixed in a hierarchy controlled by a global scheduler. Ptolemy does not currently provide a notion of abstraction between the different models in the system. However, an important innovative concept in the design of the Ptolemy II infrastructure is the notion of *domain polymorphism* [12]. An actor (agent) is domain polymorphic if it can be used indifferently, i.e., without modification, in several domains of computation. To check whether an actor can be used in a particular domain, the authors set up a type system based on state machines, which is used to describe the assumptions of each model and each actor relative to an abstract semantics.

Conservative approximations offer a formal way of defining a similar concept of polymorphism, even though they do not rely upon a common underlying semantics, as in the case of Ptolemy. A distinctive feature of conservative approximations is their ability to determine which parts of the models are unaffected by the application of the abstraction. This information is useful because it identifies the elements of the models that can be expressed indifferently under the interpretation of either model, without changing their meaning. Our interpretation of this notion is, however, broader than that introduced in Ptolemy II. In particular, an actor (agent) is polymorphic in our framework when it makes no assumption regarding its behavior based on information that cannot be expressed in the other model. When this is the case, reuse of subsystems can be extended across the boundaries of heterogeneous models. This leads to the notion of the *inverse* of a conservative approximation, which is a refinement map that is used to embed one model into another. The embedding provides us with an interpretation of agents across different models which is consistent with the corresponding abstraction. *An agent is polymorphic precisely when this interpretation is exact.* This has the advantage of making the process of abstraction and refinement of an agent explicit. Elements that do not fall in the range of the inverse can only be approximated by the other model.

Another example of approximation is the *homomorphic reduction* proposed by Kurshan [9,10]. This technique can be applied to models of behavior that consist of languages (sets of sequences) that are recognized by a class of ω automata called L-automata, which are able to express both safety and fairness constraints. Here, each automaton P constructed over a set of symbols L (an L-automaton) accepts a language $\mathcal{L}(P) \subseteq L^\omega$, where L^ω denotes the set of all infinite sequences of symbols from L. Verification in this context is the process of determining whether the language $\mathcal{L}(P)$ recognized by an implementation automaton P is contained in the language $\mathcal{L}(T)$ accepted by the specification automaton T, i.e., that $\mathcal{L}(P) \subseteq \mathcal{L}(T)$. This problem can be reduced to a more abstract language L' by verifying that $\mathcal{L}(P') \subseteq \mathcal{L}(T')$, for appropriate abstract L'-automata P' and T'. The main result[2] states that $\mathcal{L}(P') \subseteq \mathcal{L}(T')$ implies $\mathcal{L}(P) \subseteq \mathcal{L}(T)$ provided there exists a *language homomorphism* $\Phi : L^\omega \mapsto L'^\omega$ such that $\Phi(\mathcal{L}(P)) \subseteq \mathcal{L}(P')$ and $\Phi(\mathcal{L}(T^\#)) \subseteq \mathcal{L}(T'^\#)$. In this case, Φ is said to be *co-linear*[3] for $(P, T; P', T')$. In the co-linearity condition above, the notation $T^\#$ denotes the *dual automaton*[4] of T, which is closely related to language complementation.

We have argued before that one function on languages is not sufficient to guarantee the preservation of such verification result. The apparent contradiction with the use of just one language homomorphism Φ can be reconciled by accounting for the use of the dual automaton in the co-linearity condition. Effectively, if Φ is co-linear for $(P, T; P', T')$, then it can be shown that not only is $\Phi(\mathcal{L}(P)) \subseteq \mathcal{L}(P')$, but also that $\mathcal{L}(T') \subseteq \Phi(\overline{\mathcal{L}(T)})$, where the overline bar denotes language complementation. Hence, the language of the specification T is transformed according to a different abstraction functions, namely $\Theta(\mathcal{L}(T)) = \overline{\Phi(\overline{\mathcal{L}(T)})}$. Interestingly, Φ and Θ form the upper and lower bound of a conservative approximation that is closely related (and under certain conditions equal) to the conservative approximation induced by a homomorphism (see Section 4). Co-linearity of Φ thus simply ensures that $\mathcal{L}(P')$ and $\mathcal{L}(T')$ provide looser bounds, a condition that still guarantees soundness in the verification. Conservative approximations generalize the technique of homomorphic reduction to arbitrary agent models, and can therefore be applied to models that are not described by automata.

Model checking techniques based on abstraction/refinement is also a well studied related field of application for abstraction mappings [5], and is a typical application of the framework of abstract interpretations. The technique consists of first deriving an over-approximation of a state-based model using, for instance, predicate abstraction. The abstract model is constructed in a way that ensures that the property to be verified can be represented exactly (by, for example, an appropriate choice of the predicates). Therefore, if the property is verified in the abstract domain, it is also verified in the

---

[2] Theorem 8.5.2 in [9].

[3] Definition 8.5.1 in [9].

[4] Definitions 6.2.19 and 6.2.26 in [9].

concrete domain. If not, a counter-example is generated and used to refine the abstract domain until the satisfaction of the property is determined. The approach based on conservative approximations differs because, as explained, it allows non-trivial abstraction of the specification, as well as of the implementation. Model checking techniques also exist that use under-approximations, rather than over-approximations, to derive an abstracted model [16]. This is similar to our use of the lower bound function. However, unlike our use of the lower bound, the under-approximation is applied to the implementation, rather than to the specification. This corresponds again to using a Galois connection, one that goes in the reverse direction. By doing this, if the abstract model violates the property under verification, then it can be concluded that also the concrete model violates the property. Instead, if the abstract model satisfies the property, the verification is inconclusive and the abstract model must be refined until the property is proved incorrect, or the abstraction becomes exact. This approach may be useful when the interest lies in finding true counter-examples and bug traces.

Another formalization of abstraction is based on *theory interpretations* [14]. Here, an abstract architecture description and a concrete architecture description are both translated to theories in a logical language (typically first-order logic). The concrete architecture is correct relative to the abstract architecture if there is a theory interpretation I from the abstract theory Θ to the concrete theory Θ'; that is, for every formula F, $F \in \Theta \Rightarrow I(F) \in \Theta'$. In addition, it may be required that I be faithful: $F \notin \Theta \Rightarrow I(F) \notin \Theta'$. Our approach does not interpret architectures, or other agents, as logical theories. Instead, they are directly modeled as mathematical objects. This can be thought of as a model based approach, as opposed to a theory based approach. In a model based approach, within a given model of computation, the refinement relation is just a binary relation on objects in the model. This notion of refinement is easier to reason about than theory interpretations, but it is less flexible for comparing agents in different models of computation. This can be addressed by introducing abstract interpretations or conservative approximations.

Process Spaces [15] is a very general class of concurrency models, and it compares quite closely to trace-based agent models [17]. Given a set of executions $\mathcal{E}$, a Process Space $\mathcal{S}_{\mathcal{E}}$ consists of the set of all the processes (X, Y), where X and Y are subsets of $\mathcal{E}$ such that $X \cup Y = \mathcal{E}$. The sets of executions X and Y of a process are not necessarily disjoint, and they represent the assumptions (Y) and the guarantees (X) of the process with respect to its environment. As in trace-based agent models, executions are abstract objects. Different sets of abstract executions $\mathcal{E}_1$ and $\mathcal{E}_2$ induce different Process Spaces $\mathcal{S}_{\mathcal{E}_1}$ and $\mathcal{S}_{\mathcal{E}_2}$. The notion of process abstraction from $\mathcal{S}_{\mathcal{E}_1}$ to $\mathcal{S}_{\mathcal{E}_2}$ in Process Spaces is related to the notion of conservative approximation. In particular, process abstractions are defined as the Galois connections between process spaces that are derived from a relation on the set of abstract executions. The connections are obtained as axialities [8]. A process abstraction is classified as optimistic or pessimistic according to whether it preserves certain verification results from the concrete to the abstract or from the abstract to the concrete model. These two kinds of abstraction can be used in combination to preserve verification results both ways. However, in that case, the two models are isomorphic since there is effectively no loss information. Optimistic and pessimistic process abstractions roughly correspond to the two abstraction functions of conservative approximations. However, our use of these functions is significantly different, since we apply them in combination (one for the specification, the other for the implementation). Consequently, our models need not be isomorphic, so that we obtain stronger preservation results without sacrificing the abstraction.

Winskel et al. [20] propose a framework based on category theory that is related to ours. In their formalism, each model of computation is turned into a category where the objects are the agents, and the morphisms represent a refinement relationship based on *simulations* between the agents. The authors study a variety of different models that are obtained by selecting arbitrary combinations of three parameters: behavior vs. system (e.g., traces vs. state machines), interleaving vs. non-interleaving (e.g., state machines vs. event structures) and linear vs. branching time. The common operations in a model are derived as universal constructions in the category. Relationships can be constructed by relating the categories corresponding to different models by means of functors, which are homomorphisms of categories that preserve morphisms and their compositions. When categories represent models of computation, functors establish connections between the models in a way similar to abstraction maps and semantic functions. In particular, when the morphisms in the category are interpreted as refinement, functors become essentially monotonic functions between the models, since preserving morphisms is equivalent to preserving the refinement relationship.

In [20], the authors thoroughly study the relationships between the eight different models of concurrency above by relating the corresponding categories through functors. In addition, these functors are shown to be components of *reflections* or *co-reflections*. These are particular kinds of adjoints, which are pairs of functors that go in opposite directions and enjoy properties that are similar to the order preservation of the abstraction and concretization maps of a Galois connection. When the morphisms are interpreted as refinement, reflections and co-reflections generalize the concept of Galois connection to preorders. In fact, the relationships between categories based on adjoints are similar in nature to the abstractions and refinements obtained by abstract interpretations and conservative approximations. However, as described above for abstract interpretations, conservative approximations use independent abstractions for the implementation and the specification in order to derive a stronger result in terms of preservation of the refinement relation, and avoidance of false positive verification results. Indeed, we require two Galois connections, instead of one, to determine a single conservative approximation. In the work presented in [20], this translates in two adjoints per pair of categories.

6 Conclusions

We presented the use of abstraction and refinement functions between models of computation for the verification and design of heterogeneous systems. We compared conservative approximations to abstract interpretations and we showed that, unlike abstract interpretations, conservative approximations always preserve refinement verification results from an abstract to a concrete model, while avoiding false positives. Therefore, conservative approximations are better suited for heterogeneous design methodologies, i.e., methodologies that use several models of computation. In particular, because they always guarantee correctness, conservative approximations provide more flexibility in choosing the verification strategy and the hierarchy of models used in the design flow. We then described how to construct models of computation suitable for the design of embedded systems, and how conservative approximations can be derived for these models starting from simple functions (homomorphisms) over their set of behaviors. In addition, the inverse of a conservative approximation has been shown to identify components that can be used indifferently in several models, thus enabling reuse across domains of computation. The resulting theory can be used as the basis of frameworks that support heterogeneous modeling.

Our current work focuses on extending techniques that make it easier to construct conservative approximations between agent models. The axialities of homomorphisms on behaviors described in this paper is one such example. However, homomorphisms are usually defined to preserve the alphabet of behaviors, so that the induced conservative approximations, too, must preserve the alphabet of agents. More interesting conservative approximations can be constructed by letting the homomorphism change the alphabet of a behavior, for example by hiding certain signals, like clocks and activation signals, that have no meaning in a more abstract model. This is also appropriate for converting a detailed protocol specification into a more abstract, transaction-based, specification. Arbitrary changes of the alphabet are also possible. In this case, however, the homomorphism must not only be applied to the behaviors, but also to the operators, in order to correctly translate expressions. In this case the homomorphism becomes similar to a functor between categories, where a category has behaviors as objects and the operators as morphisms.

A model that uses behaviors as its underlying structure may impose restrictions on the kind of agents that can be constructed. For example, only receptive (or progressive, or input enabled) agents might be allowed. The axialities of a homomorphism, however, may not necessarily yield agents that satisfy such conditions. A promising avenue of future research consists therefore in identifying the agent that most faithfully approximates the missing abstraction, while satisfying the constraints imposed by the model, and while still functioning as the bound of a conservative approximation. This would constitute a generalization of the technique proposed by Loiseaux et al. [13] on property-preserving abstractions in the context of transition systems.

References

1. Balarin, F., Lavagno, L., Passerone, C., Sangiovanni-Vincentelli, A., Watanabe, Y., Yang, G.: Concurrent execution semantics and sequential simulation algorithms for the metropolis meta-model. In: Proceedings of the Tenth International Symposium on Hardware/Software Codesign, Estes Park, CO (May 2002)
2. Balluchi, A., Benedetto, M.D., Pinello, C., Rossi, C., Sangiovanni-Vincentelli, A.: Cut-off in engine control: a hybrid system approach. In: IEEE Conf. on Decision and Control (1997)
3. Burch, J.R.: Trace Algebra for Automatic Verification of Real-Time Concurrent Systems. PhD thesis, School of Computer Science, Carnegie Mellon University (Aug 1992)
4. Carloni, L.P., Passerone, R., Pinto, A., Sangiovanni-Vincentelli, A.L.: Languages and Tools for Hybrid Systems Design. Foundations and Trends in Electronic Design Automation, vol. 1. Now Publishers (2006)
5. Clarke, E.M., Grumberg, O., Peled, D.: Model Checking, 2nd edn. The MIT Press, Cambridge (1999)

6. Cousot, P., Cousot, R.: Abstract interpretation: a unified lattice model for static analysis of programs by construction or approximation of fixpoints. In: Conference Record of the Fourth Annual ACM SIGPLAN-SIGACT Symposium on Principles of Programming Languages, Los Angeles, California, pp. 238–252. ACM Press, New York (1977)
7. Cousot, P., Cousot, R.: Comparing the Galois connection and widening/narrowing approaches to abstract interpretation, invited paper. In: Bruynooghe, M., Wirsing, M. (eds.) PLILP 1992. LNCS, vol. 631, pp. 269–295. Springer, Heidelberg (1992)
8. Erné, M., Koslowski, J., Melton, A., Strecker, G.E.: A primer on galois connections. In: The Design of an Extendible Graph Editor. Ann. New Yosk Acad. Sci, vol. 704, pp. 103–125 (1993)
9. Kurshan, R.P.: Computer-Aided Verification of Coordinating Processes: The Automata-Theoretic Approach. Princeton University Press, Princeton (1995)
10. Kurshan, R.P., McMillan, K.L.: Analysis of digital circuits through symbolic reduction. IEEE Trans. Comput.-Aided Design Integrated Circuits 10(11), 1356–1371 (1991)
11. Lee, E.A.: Overview of the Ptolemy project. Technical Memorandum UCB/ERL M03/25, University of California, Berkeley (July 2003)
12. Lee, E.A., Xiong, Y.: System-level types for component-based design. In: Henzinger, T.A., Kirsch, C.M. (eds.) EMSOFT 2001. LNCS, vol. 2211, Springer, Heidelberg (2001)
13. Loiseaux, C., Graf, S., Sifakis, J., Bouajjani, A., Bensalem, S.: Property preserving abstractions for the verification of concurrent systems. Formal Methods in System Design 6, 1–35 (1995)
14. Moriconi, M., Qian, X., Riemenschneider, R.A.: Correct architecture refinement. IEEE Transactions on Software Engineering 21(4), 356–372 (1995)
15. Negulescu, R.: Process Spaces and the Formal Verification of Asynchronous Circuits. PhD thesis, University of Waterloo, Canada (1998)
16. Pasareanu, C., Pelánek, R., Visser, W.: Concrete model checking with abstract matching and refinement. In: Etessami, K., Rajamani, S.K. (eds.) CAV 2005. LNCS, vol. 3576, Springer, Heidelberg (2005)
17. Passerone, R.: Semantic Foundations for Heterogeneous Systems. PhD thesis, Department of EECS, University of California at Berkeley (May 2004)
18. Passerone, R., Burch, J.R., Sangiovanni-Vincentelli, A.L.: Refinement preserving approximations for the design and verification of heterogeneous systems. Formal Methods in System Design 31(1), 1–33 (2007)
19. Pratt, V.R.: Modelling concurrency with partial orders. International Journal of Parallel Programming 15(1), 33–71 (1986)
20. Sassone, V., Nielsen, M., Winskel, G.: Models for concurrency: Towards a classification. Theoretical Computer Science 170, 297–348 (1996)

Ugo Montanari and Friends

Fabio Gadducci

Dipartimento di Informatica, Università di Pisa, Italy
gadducci@di.unipi.it

Last, but by no means least... This section covers the most important facet of Ugo's personality, namely, his tireless efforts in building national and international relationships, in order to foster the advance of research. Indeed, a large part of Ugo's time is spent in casting a web of links among people as well as institutions, fostering research collaborations and exchanges of ideas. A facet of his personality Ugo is aware of, as he witnesses in his recent book-interview [11].

Consider e.g. the first three contributions of this section, and let us start chauvinistically with the Italian colleague, Angelo Raffaele Meo. Ugo often talks about those five years "lost to his research", where he actively contributed to the shaping of the governmental guidelines for ITC. In the late Seventies, Meo was the chair of the National Program in Computer Sciences for the National Research Council (CNR); Ugo headed one of the three sections of that Program, devoted to Computer Industry. The major subproject of the Program was Campus Net (CNet) Project, lead by Norma Lijtmaer and running from 1979 to 1985. Ugo, Norma and Meo were the driving force behind CNet: the project represented a major focus for Italian ITC, maybe one of the most influential and ambitious among those projects combined into the first Progetto Finalizzato in Informatica sponsored by CNR. Research issues focused on developing the functional specification and implementing prototype versions of distributed systems on local networks. The technical accomplishments were initially spread in a series of technical reports, and the main achievements were collected in a two-volume proceedings of the final symposium [7], held in Pisa in June 24-28, 1985. The presence of a large group of industrial partners, among them Olivetti, had consequences far beyond the scientific assessments of the projects itself, in a booming period for the national industry. Moreover, the project was likely the defining moment for Ugo's interests in concurrency, and one of the key steps in the rise of the whole area in Italy. Some of the outcomes of the project (such as [6], originally a CNet technical report) were later considered pivotal by key people like Robin Milner in the development of nominal calculi. The topic tackled by Meo in his article is maybe far from Ugo's current interests. Nevertheless, let us not forget that Ugo was the coauthor of the first Italian textbook on formal languages and computability [5].

But for Ugo, the research landscape has always been a global one. As a student, you were warmly encouraged to reading articles and submitting your own output to conferences around the world, in order to learn from the discussions and the exchange of ideas, standing indeed on the shoulders of giants. Likewise, also the scope of Ugo's connections has been international. Such a span is well-represented by the many distinguished authors who contributed to the overview

P. Degano et al. (Eds.): Montanari Festschrift, LNCS 5065, pp. 743–746, 2008.

of the Esprit Basic Research Action CEDISYS, a "fruitful project which brought together researchers at the meeting point of true concurrency and process algebra, in the period 1988-1991". A report by Ugo, the project leader, can be found in [10]. I leave to the contributors of the memoir the task of introducing the reader to the scientific broad aims, as well as to the far reaching influences and technical achievements, brought forth by the project, one of the many that Ugo was part of in the latest two decades. As a chronicler, I will limit myself just to unveil the name of the "younger members of the project [who] had been both Ugo Montanari's Master students in Pisa and Matthew Hennessy's PhD students in Edinburgh" referred to in the article, namely, Ilaria Castellani and Rocco De Nicola; and to point out the third leg of the Pisa unit, besides Ugo and Rocco, that is, Pierpaolo Degano.

Many of the participants of the project agree that the presence of CEDYSIS and of similar initiatives was a key factor in paving the way for the CONCUR series of conferences. It is a well-known fact that the concurrency theory forum was established as an off-shot of the similarly named Basic Research Action, lead by Jos C.M. Baeten (indeed, the subtitle in the original call of papers read "Theories of Concurrency: Unification and Extension", thus echoing the name of the BRA; and the conference was heralded as "the first in a series of conferences that will be organized by the ESPRIT project no.3006 CONCUR"). The first edition of the conference was chaired by Jan Willem Klop, and the proceedings were edited by him together with Baeten. Ugo was part of that Program Committee, and since its inception Ugo has been part of the Steering Committee, and one of the main force behind the development of the concurrency community. The wide range of his interests is powerfully reflected in the variety of works he has presented: indeed, so far he is the most significant contributor, with 16 papers, to the conference. The flurry of activities on concurrency left many traces, either in the concurrency mailing list [3] or in more specific venues such as the reports on Strategic Directions in Computing Research by the Concurrency Working Group [2], chaired by Scott Smolka. Ugo and Baeten have been since then strong advocates of the use of process algebras as a perfect tool for the unambiguous specification of systems, and the textbook contribution by Baeten in this section would surely fit in the general mood of that period.

The remaining contributions of the volume are *laudatio* by colleagues, or memories of working experience from former colleagues and students. This time, let us first consider the international side. Turing-laureate Robin Milner discusses the influence of the seminal paper "Petri nets are monoids", coauthored by Ugo with his long-time friend José Meseguer [9]. The timeline of that article is recalled in the general introduction to this volume. As for its contents, it is fair to say that since then Ugo always put monoidal categories at the center of his reflections on concurrency models (see e.g. page 33 in [11]). Probably, Ugo's influence on the community is best rendered by the words of Milner: "[Ugo] is one who deserves credit for bringing to our notice the categorical ideas that can help us think this subject more clearly". The 1973 meeting Milner refers to actually took place on March, 1-3; its proceedings are collected in [4]. It was the first Italian conference devoted to theoretical computer science, and the gathering represented a pivotal step for the "coming of age" of the community. However, Milner's connection with Ugo was in fact established and strengthened in the Nineties by the Europe-funded Basic Research Action (later on, Esprit Working Group) CONFER (for "CONcurrency and Functions: Evaluation and Reduction"), coordinated by Jean Jacques Lévy from 1994 to 2000.

The red trail of European projects leads to Jan Rutten, one of the driving forces behind the renewed interest of computer scientists towards coalgebras, a topic pursued vigorously by Ugo in latest years. His first collaboration with Ugo dates back to the two European Science Project MASK ("Mathematical Structures for Concurrency"), led by Jaco W. De Bakker from 1992 to 1996. Rutten's whimsical paper witnesses the impression made by the appearance of Ugo: this impression is shared by his students, but it can be found in many oral reminiscences of Ugo's peers, since his first appearances in the Seventies...

The Seventies are warmly remembered by Alberto Martelli. Both he and Ugo came from Politecnico di Milano, brought to Pisa by Antonio Grasselli, one of the founding fathers of the Computer Science curriculum in Pisa. Martelli wrote a string of joint papers with Ugo in the decade, on a large variety of subjects, as well as being finally part of the Pisa unit in CNet, before moving to Turin. The relevance of the unification algorithm coauthored by him and Ugo, reported in [8], has been already stressed by the editors of this volume: it is however enlightening to see here also the intellectual environment surrounding such a scientific achievement, and its recasting of its origin in the scientific stimuli from artificial intelligence, so actively pursued by Ugo at the time.

Franco Turini was one of the first students of Ugo in the by then just formed Corso di Laurea in Scienze dell'Informazione in Pisa. He was thus also among the first students to be awarded the Laurea degree in Italy, since the curriculum in the Tuscan town, established in 1969, was at the time the only one of its kind in Italy. He was involved in the Pisa unit of the CNet project, and still called himself a theoretician in his earlier research years. Franco rightly draws the attention of the readers to Ugo's teaching ability. Indeed, Ugo takes great pride in being identified as a *nurturer* (see again his book-interview), and this is reflected in the attention he pays to the didactic exploitation of current research trends, pursuing his teaching excellence "by looking for more elegant proofs of the theorems and a smoother flowing of the results".

Finally, the carrier of Daniel Yankalevich witnesses the tireless effervescence of Ugo we all know quite well. "Dani" has been one of the first students to come out of ESLAI, the Escuela Superior Latinoamericana de Informática (besides the general introduction of this volume, see also the corresponding wikipedia entry [1]). The institute for higher education was founded in 1986 by the Argentinian government under its first Democratic president after the dictatorship, Raul Alfonsín (and unfortunately closed in 1990 by his successor, Carlos Menem). The school "had a considerable impact on informatics teaching and research in Argentina and South America", as the editors of this volume say, and many of his pupils have reached position in universities and industries around the world.

Ugo played a major role in its establishment, together with his late-wife Norma, of Argentinian origin, and one of the many exiles during the military ruling: the experience is warmly remembered by Ugo in his book-interview, and, also as a witness of the applicative side of research, always foremost in Ugo's mind, Dani's memories appear to me as the appropriate closure for the whole volume.

References

1. Escuela Superior Latinoamericana de Informática. Wikipedia, The Free Encyclopedia (2008),
   `en.wikipedia.org/wiki/Escuela_Superior_Latinoamericana_de_Informatica`
2. SDCR Concurrency Working Group Materials (1996),
   `www.cs.sunysb.edu/~sas/sdcr/`
3. The CONCURRENCY Mailing List (2008), `www.cwi.nl/~bertl/concurrency/`
4. Convegno di Informatica Teorica. Editrice Tecnico Scientifica, Pisa (1973)
5. Aiello, M., Albano, A., Attardi, G., Montanari, U.: Elementi di Teoria della Computabilità, Logica, Teoria dei Linguaggi Formali. Editrice Tecnico Scientifica, Pisa (1976)
6. Astesiano, E., Zucca, E.: Parametric channels via label expressions in CCS. Theoretical Computer Science 33, 45–63 (1984)
7. Lijtmaer, N. (ed.): Distributed Systems on Local Networks. Vols I and II. Editrice Tecnico Scientifica, Pisa (1985)
8. Martelli, A., Montanari, U.: An efficient unification algorithm. ACM Transactions on Programming Languages and Systems 4(2), 258–282 (1982)
9. Meseguer, J., Montanari, U.: Petri nets are monoids. Information and Computation 88(2), 105–155 (1990)
10. Montanari, U.: CEDISYS: Compositional distributed systems - State of the art, research goals, references. In: Rozenberg, G. (ed.) APN 1989. LNCS, vol. 424, pp. 507–524. Springer, Heidelberg (1990)
11. Montanari, U.: Idee per diventare informatico. Dalle schede perforate al futuro di internet. Zanichelli, Bologna (2007)

Calculating with Automata

Jos C.M. Baeten

Division of Computer Science, Technische Universiteit Eindhoven

Abstract. Some elements are presented of a forthcoming textbook on automata theory and formal languages, that puts more emphasis on equational reasoning. Some advantages of such an approach are discussed. This paper is dedicated to Ugo Montanari, who has contributed such a lot to concurrency theory and the theory of computational models.

1 Introduction

The theory of automata and formal languages on the one hand, and concurrency theory on the other hand, both present a model of computation. These theories have a lot in common, for instance they share the concept of a transition system or an automaton. Both theories can also stand to gain by adopting notation, methods and techniques developed in the other theory. Here we have a look at an approach to automata and formal languages that uses equational reasoning as has been developed in process algebra.

2 Grammar as a Recursive Specification

We present a few examples of an equational approach to automata theory. We limit ourselves to examples in the class of regular languages, but applications in the setting of context-free languages and Turing machines can be added easily. Crucial to this approach is presenting a grammar as a recursive specification. Consider a simple right-linear grammar.

$$S \rightarrow aT \mid aW$$
$$T \rightarrow aU \mid bW$$
$$U \rightarrow bV \mid bR$$
$$W \rightarrow aR$$
$$R \rightarrow bW \mid \lambda.$$

Consider a simple process algebraic language with the following signature Σ:

- There is a constant **0** denoting unsuccessful termination or deadlock;
- There is a constant **1** denoting successful termination or skip;
- For each element a of the alphabet $\mathcal{A}$, there is a unary operator a_- denoting action prefix;

P. Degano et al. (Eds.): Montanari Festschrift, LNCS 5065, pp. 747–756, 2008.

– There is a binary operator $\_ + \_$ denoting alternative composition or choice.

Now the grammar above can be presented as a recursive specification over a set of variables $\{S, T, U, V, W, R\}$ as follows. This specification is linear, as every summand on the right-hand side is $\mathbf{0}, \mathbf{1}$ or an action prefix of a variable.

$$S = aT + aW$$
$$T = aU + bW$$
$$U = bV + bR$$
$$V = \mathbf{0}$$
$$W = aR$$
$$R = bW + \mathbf{1}.$$

Such a presentation of a grammar did occur in literature, see e.g. [5], [7] or [6], but was not exploited in order to do calculations.

Every closed term over Σ denotes an automaton by means of *Structural Operational Semantics*, SOS for short (see e.g. [1]). The rules in Table 1 define when a closed term x is a final state in an automaton $(x \downarrow)$ and when a closed term x has an a-labeled transition to term x' $(x \xrightarrow{a} x')$. Thus, the two axioms at the top say that every term of the form ax has an a-step to x, and that $\mathbf{1}$ is a final state. The four rules below are the rules for choice: a step from $x + y$ must be a step from x or a step from y (discarding the other component), and $x + y$ is a final state exactly when x or y is.

Starting from a term x, each reachable state is a subterm of x, and so each automaton generated by these rules has finitely many states. Thus, the language of such an automaton is a regular language.

Next, we add a recursive specification over a finite set of variables $\mathcal{V}$, so for each $P \in \mathcal{V}$ there is exactly one equation $P = t_P$, where t_P is a term over Σ and variables $\mathcal{V}$. For a recursive specification, the following rules need to be added, see Table 2. For each term over Σ with only variables from $\mathcal{V}$, an automaton can be constructed. The states in this automaton are again terms over Σ and $\mathcal{V}$, and transitions and termination are given by the rules.

Table 1. Operational rules $(a \in \mathcal{A})$

$$\frac{}{ax \xrightarrow{a} x} \qquad\qquad \frac{}{\mathbf{1} \downarrow}$$

$$\frac{x \xrightarrow{a} x'}{(x + y) \xrightarrow{a} x'} \qquad \frac{y \xrightarrow{a} y'}{(x + y) \xrightarrow{a} y'} \qquad \frac{x \downarrow}{(x + y) \downarrow} \qquad \frac{y \downarrow}{(x + y) \downarrow}$$

Table 2. Operational rules for recursion

$$\frac{t_P \xrightarrow{a} x \quad P = t_P}{P \xrightarrow{a} x} \qquad \frac{t_P \downarrow \quad P = t_P}{P \downarrow}$$

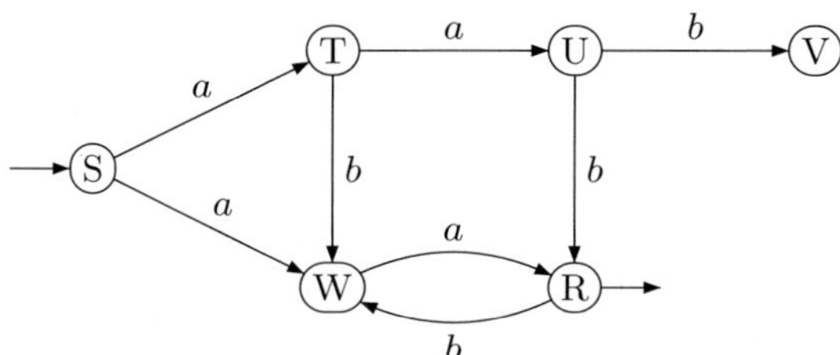

Fig. 1. Example automaton

As an example, for the given recursive specification, using the rules we get exactly the automaton in Figure 1.

Now let us consider the laws that hold in this minimal algebra MA. The laws in Table 3 give the commutativity, associativity and idempotency of alternative composition, and state that **0** is a unit element for alternative composition. In each case, the automata generated by the two sides of the equation are exactly the same, apart from the name of the initial state. Thus, they denote isomorphic automata.

Table 3. Laws of MA

$$
\begin{aligned}
x + y &= y + x \\
(x + y) + z &= x + (y + z) \\
x + x &= x \\
x + \mathbf{0} &= x
\end{aligned}
$$

The laws in Table 4 give the distributivity of action prefix over alternative composition and the zero law for action prefix. The automata generated by the two sides of the equation are not isomorphic, but have the same language.

Both language equivalence and isomorphism are equivalence relations on automata. Language equivalence is also a congruence relation, but isomorphism is

Table 4. Two more laws of MA ($a \in \mathcal{A}$)

$$ax + ay = a(x + y)$$
$$a\mathbf{0} = \mathbf{0}$$

not: the automaton of $\mathbf{1}$ is isomorphic to the automaton of $\mathbf{1} + \mathbf{0}$. However, the automaton of $a(\mathbf{1} + \mathbf{0}) + a\mathbf{1}$ has three states (the other states are $\mathbf{1} + \mathbf{0}$ and $\mathbf{1}$), and the automaton of $a\mathbf{1} + a\mathbf{1}$ has only two states (the other state is $\mathbf{1}$), so they cannot be isomorphic.

If we want to turn isomorphism into a congruence relation, it is needed to relate several nodes with the same behavior in one automaton with several nodes with the same behavior in another automaton. In this way, *bisimulation* is obtained: bisimulation is the most distinguishing congruence containing isomorphism. This means the following: if there is any equivalence on automata that is a congruence for the operators of Σ and that relates any two isomorphic automata, then necessarily any two bisimilar automata must be related.

Now a string is an element of the language of a recursive specification exactly when it is a summand of the initial variable. If we define $x \leq y$ as $x + y = y$, then we have that string w is an element of the language generated by a recursive specification with initial variable S exactly when $S \geq w\mathbf{1}$. In our example recursive specification, the language contains string $aaba$:

$$S = aT + aW \geq aW = aaR = aa(bW + \mathbf{1}) = aabW + aa\mathbf{1} \geq aabW =$$

$$= aabaR = aaba(bW + \mathbf{1}) = aababW + aaba\mathbf{1} \geq aaba\mathbf{1}.$$

We find that recursive specifications over Σ generate exactly the family of regular languages. Recursive specifications over Σ plus sequential composition give the family of context-free languages. Finally, recursive specifications over Σ plus parallel composition with communication and abstraction yields the family of all computable languages (see [2]).

3   Deterministic Automata

A well-known theorem from automata theory states that for each non-deterministic automaton, there is a deterministic automaton that accepts the same language. This theorem is usually proved using the powerset construction: a state of the deterministic automaton consists of a set of states of the non-deterministic automaton (see e.g. [10]). Students find this abstract construction often difficult to grasp. Moreover, it is not always clear to them which subsets are needed in the deterministic automaton, and which are not needed. In the equational approach, all that is needed is the application of the distributive law of Table 4.

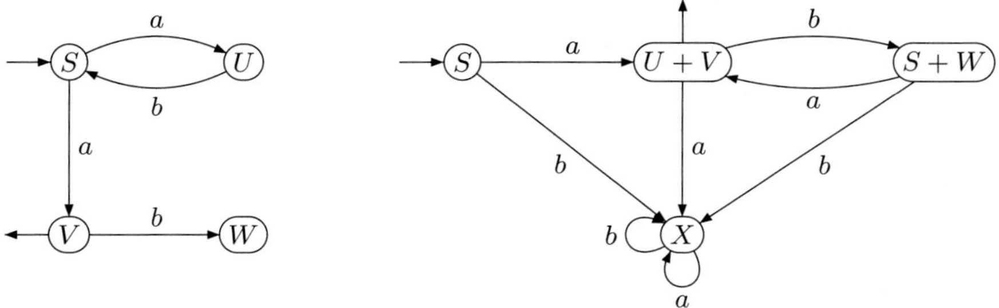

Fig. 2. Finding a deterministic automaton

To illustrate, we give an example.

Example 1. Given is the automaton on the left-hand side of Figure 2, with state variables $\mathcal{S} = \{S, U, V, W\}$. As a recursive specification, we get

$$S = aU + aV$$
$$U = bS$$
$$V = bW + \mathbf{1}$$
$$W = \mathbf{0}.$$

Application of the distributive law yields the following set of equations.

$$S = aU + aV = a(U + V)$$
$$U + V = bS + bW + \mathbf{1} = b(S + W) + \mathbf{1}$$
$$S + W = aU + aV + \mathbf{0} = a(U + V)$$

Commonly, a deterministic automaton has a total transition function, and thus in any state for any label there must be an outgoing transition. In automata theory, a so-called *trap state* is added, that shows every possible behaviour but can never lead to a final state. Equationally, this amounts to the addition of a Chaos process X (see [8]). As this process can never terminate, adding summands leading to it does not change the language generated. The resulting recursive specification becomes

$$S = a(U + V) + bX$$
$$U + V = b(S + W) + \mathbf{1} + aX$$
$$S + W = a(U + V) + bX$$
$$X = aX + bX.$$

The resulting automaton is displayed on the right-hand side of Figure 2.

4   Automata with ε-Transitions

We consider automata with transitions not consuming input. These transitions are usually called ε-transitions or λ-transitions. Taking an example from process algebra, I call these transitions τ-transitions, and add τ-prefixing to the language.

A well-known theorem from automata theory states that for each automaton with ε-transitions, there is an automaton without such transitions accepting the same language. To prove this, again the powerset construction is used (see e.g. [9]). In our case, removing τ-transitions just boils down to the application of the law $\tau x = x$ that is valid in language equivalence.

Example 2. Consider the automaton at the left-hand side of Figure 3. It yields the recursive specification

$$S = aW + \tau T$$
$$T = aU$$
$$U = \tau U + bV + \tau T$$
$$V = \mathbf{0}$$
$$W = \mathbf{1}$$

We calculate:

$$S = aW + \tau T = aW + T = aW + aU$$
$$U = \tau U + bV + \tau T = U + bV + T = U + bV + aU$$
$$V = \mathbf{0}$$
$$W = \mathbf{1}$$

As T is not a reachable state in the resulting automaton, it can be left out. We show the resulting automaton on the right-hand side of Figure 3. Note that the τ-loop on U is just omitted.

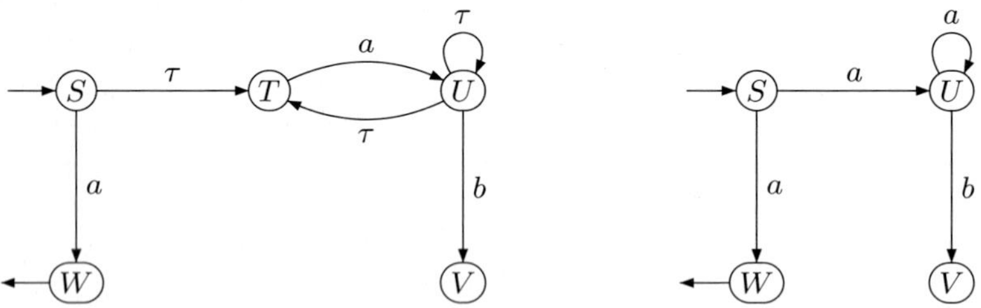

Fig. 3. Finding an automaton without silent steps

5   Regular Expressions

The family of regular languages is exactly the family of languages given by regular expressions. We will see that expressing a regular language by means of a regular expression boils down to solving the recursive equations. Two extra operators are needed: sequential composition and Kleene star or iteration.

The operational rules for the additional operators are given in Table 5. We call the resulting algebra Iteration Algebra. Thus, a sequential composition can start with a step from the first component. If the first component is in a final state, a step from the second component can be taken. A sequential composition is in a final state only if both components are in a final state. An iteration can take a step from the body, entering into the loop. When the loop is finished, the process can be repeated. Any iteration can be exited immediately, not executing the body at all.

The set of reachable states of an automaton with initial state $x \cdot y$ consists of all $x' \cdot y$ where x' is reachable from x, plus, in case the automaton of x has a final state, all y' reachable from y. The set of states reachable from x^* are the terms $x' \cdot x^*$ where x' is reachable from x. In both cases, the set of reachable states is finite, and so we find that every automaton generated by the SOS-rules of Iteration Algebra is finite, so its language is regular.

Table 5. Additional operational rules for sequential composition and Kleene star ($a \in \mathcal{A}$)

$$\frac{x \xrightarrow{a} x'}{x \cdot y \xrightarrow{a} x' \cdot y} \qquad \frac{x \downarrow \quad y \xrightarrow{a} y'}{x \cdot y \xrightarrow{a} y'} \qquad \frac{x \downarrow \quad y \downarrow}{x \cdot y \downarrow}$$

$$\frac{x \xrightarrow{a} x'}{x^* \xrightarrow{a} x' \cdot x^*} \qquad \frac{}{x^* \downarrow}$$

Table 6 shows laws for sequential composition and Kleene star, where in each case the automata generated by both sides of an equation are isomorphic. Sequential composition is associative, and action prefix can be seen as a restricted form of sequential composition. Sequential composition distributes over alternative composition from the right. **1** is a unit element of sequential composition, and **0** is a left zero element. The first law of Kleene star says that the iterated term can be repeated zero times (summand **1**) or at least one time ($x \cdot x^*$). Termination inside an iteration is disregarded, only steps count, as is shown in the first operational rule for Kleene star. Finally, if the iterated term is just **0**, only termination can occur.

Table 6. Isomorphism laws for Iteration Algebra ($a \in \mathcal{A}$)

$$
\begin{aligned}
(ax) \cdot y &= a(x \cdot y) \\
(x \cdot y) \cdot z &= x \cdot (y \cdot z) \\
(x + y) \cdot z &= x \cdot z + y \cdot z \\
\mathbf{1} \cdot x &= x \\
x \cdot \mathbf{1} &= x \\
\mathbf{0} \cdot x &= \mathbf{0} \\
x^* &= x \cdot x^* + \mathbf{1} \\
(x + \mathbf{1})^* &= x^* \\
\mathbf{0}^* &= \mathbf{1}
\end{aligned}
$$

Next, Table 7 shows laws that do transform the automaton, but preserve language equivalence. They say that $\mathbf{0}$ is also a right zero, and sequential composition distributes over alternative composition also from the left.

Table 7. Language equivalence laws for Iteration Algebra ($a \in \mathcal{A}$)

$$
\begin{aligned}
x \cdot \mathbf{0} &= \mathbf{0} \\
x \cdot (y + z) &= x \cdot y + x \cdot z
\end{aligned}
$$

In addition, we need a rule that we call RSP* (see [3]; this is a special case of the Recursive Specification Principle of ACP, see e.g. [4]):

$$
x = y \cdot x + z, \quad y \not\downarrow \quad \implies \quad x = y^* \cdot z.
$$

We provide a not so simple example.

Example 3. Consider a specification over 3 variables where every variable has an outgoing step to every variable.

$$
\begin{aligned}
S &= aS + bT + cU \\
T &= dS + eT + fU \\
U &= gS + hT + iU + \mathbf{1}
\end{aligned}
$$

In order to solve for S, calculate as follows:

$$
S = aS + bT + cU = a\mathbf{1} \cdot S + bT + cU = (a\mathbf{1})^* \cdot (bT + cU) =
$$

$$
= (a\mathbf{1})^* \cdot bT + (a\mathbf{1})^* \cdot cU,
$$

and S has been eliminated. Now use this in the other two equations. They become:

$$
T = dS + eT + fU = d(a\mathbf{1})^* \cdot bT + d(a\mathbf{1})^* \cdot cU + eT + fU =
$$

$$= (d(a1)^* \cdot b1 + e1) \cdot T + (d(a1)^* \cdot c1 + f1) \cdot U$$

and

$$U = gS + hT + iU + 1 = g(a1)^* \cdot bT + g(a1)^* \cdot cU + hT + iU + 1 =$$

$$= (g(a1)^* \cdot b1 + h1) \cdot T + (g(a1)^* \cdot c1 + i1) \cdot U + 1.$$

Next, eliminate T from the equation of T:

$$T = (d(a1)^* \cdot b1 + e1)^* \cdot (d(a1)^* \cdot c1 + f1) \cdot U,$$

and substitute this in the equation of U:

$$U = (g(a1)^* \cdot b1 + h1) \cdot (d(a1)^* \cdot b1 + e1)^* \cdot (d(a1)^* \cdot c1 + f1) \cdot U +$$

$$+ (g(a1)^* \cdot c1 + i1) \cdot U + 1 =$$

$$= \{(g(a1)^* \cdot b1 + h1) \cdot (d(a1)^* \cdot b1 + e1)^* \cdot (d(a1)^* \cdot c1 + f1) + (g(a1)^* \cdot c1 + i1)\}^*.$$

This result can be substituted again in the equation of T to obtain a regular expression for T:

$$T = (d(a1)^* \cdot b1 + e1)^* \cdot (d(a1)^* \cdot c1 + f1) \cdot$$

$$\cdot \{(g(a1)^* \cdot b1 + h1) \cdot (d(a1)^* \cdot b1 + e1)^* \cdot (d(a1)^* \cdot c1 + f1) + (g(a1)^* \cdot c1 + i1)\}^*.$$

Finally, both the expressions for T and U can be substituted in the equation of S:

$$S = ((a1)^* \cdot c1 + (a1)^* \cdot b(d(a1)^* \cdot b1 + e1)^* \cdot (d(a1)^* \cdot c1 + f1)) \cdot$$

$$\cdot \{(g(a1)^* \cdot b1 + h1) \cdot (d(a1)^* \cdot b1 + e1)^* \cdot (d(a1)^* \cdot c1 + f1) + (g(a1)^* \cdot c1 + i1)\}^*.$$

In this example, we converted a finite automaton into a regular expression without the need of a generalized automaton where the edges are labeled by regular expressions.

6   Conclusion

Every undergraduate curriculum in computer science contains a course on automata theory and formal languages. On the other hand, an introduction to concurrency theory is usually not given in the undergraduate program. Both theories as basic models of computation are part of the foundations of computer science. Automata theory and formal languages provide a model of computation where interaction is not taken into account, so a computer is considered as a stand-alone device executing batch processes. On the other hand, concurrency theory provides a model of computation where interaction is taken into account. Concurrency theory is sometimes called the theory of reactive processes.

Both theories can be integrated into one course in the undergraduate curriculum, providing students with the foundation of computing. I teach such a course at my university in the first half of 2008. In this way, both theories have a lot to gain from such an integrated presentation. As an example, it is shown how an equational approach can be useful in automata theory. On the other hand, by working from automata theory, some standardization can be brought to concurrency theory. For instance, it can be achieved that everyone writes **0** instead of $0, nil, \delta, STOP$, and **1** instead of $\epsilon, SKIP$.

References

1. Aceto, L., Fokkink, W.J., Verhoef, C.: Structural operational semantics. In: Bergstra, J.A., Ponse, A., Smolka, S.A. (eds.) Handbook of Process Algebra, pp. 197–292. North-Holland, Amsterdam (2001)
2. Baeten, J.C.M., Bergstra, J.A., Klop, J.W.: On the consistency of Koomen's fair abstraction rule. Theoretical Computer Science 51, 129–176 (1987)
3. Baeten, J.C.M., Corradini, F., Grabmayer, C.A.: A characterization of regular expressions under bisimulation. Journal of the ACM 54(2), 6.1–28 (2007)
4. Baeten, J.C.M., Weijland, W.P.: Process Algebra. In: Number 18 in Cambridge Tracts in Theoretical Computer Science, Cambridge University Press, Cambridge (1990)
5. Brzozowski, J.A.: Derivatives of regular expressions. Journal of the ACM 11(4), 481–494 (1964)
6. Conway, J.H.: Regular Algebra and Finite Machines. Chapman and Hall, Boca Raton (1971)
7. Ginzburg, A.: Algebraic Theory of Automata. Academic Press, London (1968)
8. Hoare, C.A.R.: Communicating Sequential Processes. Prentice-Hall, Englewood Cliffs (1985)
9. Hopcroft, J.E., Motwani, R., Ullman, J.D.: Introduction to Automata Theory, Languages, and Computation. Pearson, London (2006)
10. Linz, P.: An Introduction to Formal Languages and Automata. Jones and Bartlett (2001)

Twenty Years on: Reflections on the CEDISYS Project. Combining True Concurrency with Process Algebra

Gérard Boudol[1], Ilaria Castellani[1], Matthew Hennessy[2], Mogens Nielsen[3], and Glynn Winskel[4]

[1] INRIA, 2004 route des Lucioles, 06902 Sophia Antipolis Cedex, France
[2] Department of Computer Science, O'Reilly Institute, Trinity College, Dublin 2, Ireland
[3] Department of Computer Science, University of Aarhus, IT-parken, Aabogade 34, DK-8200 Aarhus N, Denmark
[4] Computer Laboratory, University of Cambridge, William Gates Building, JJ Thomson Avenue, Cambridge CB3 0FD, UK

Abstract. We recall some memories of the Esprit Basic Research Action CEDISYS, a small, well-focussed and fruitful project which brought together researchers at the meeting point of true concurrency and process algebra, in the period 1988-1991. The project was initiated and effectively animated by Ugo Montanari, a passionate and long-time advocate of both these approaches to the semantics of concurrency.

1 Genesis of the Project

Twenty years after its start, we revisit some of the results of the CEDISYS project, trying to place them in their historical context and to emphasise their impact on later work. The Basic Research Action CEDISYS (Compositional Distributed Systems) was funded by the European Community under the Esprit programme, between January 1988 and December 1991. The project involved four institutions: the three Computer Science Departments of the Universities of Aarhus (Denmark), Pisa (Italy), and Sussex (United Kingdom), and the INRIA Research Unit of Sophia Antipolis (France). The action was coordinated by Ugo Montanari.

The CEDISYS project did not come about by chance. It emerged quite naturally from pre-existing interactions among the partners, and from their increasing convergence of interests and approaches. The Aarhus partners were experts in domain theory and true concurrency models and had been among the founders of Event Structures, a nonsequential model of computation closely related to domains. The Sussex and Sophia partners were process calculi specialists, with a strong interest in true concurrency semantics. As for the Pisa site, Ugo Montanari had been from the very start an enthusiastic advocate of both Milner's calculus CCS and of true concurrency models, whereas other members of the group had been active researchers in the process calculi community. Finally, two

P. Degano et al. (Eds.): Montanari Festschrift, LNCS 5065, pp. 757–777, 2008.

younger members of the project had been both Ugo Montanari's Master students in Pisa and Matthew Hennessy's PhD students in Edinburgh (this tradition of mobility among the sites was to be maintained during and after the project).

It was undoubtedly Ugo's merit to have detected the potential of bringing together these particular partners, and to propose a well-focussed project whose objectives could be easily shared by all. In retrospect, it is clear that the initial conditions were there for the project to become operational very quickly. Indeed, in the three years of its activity, the project turned out to be both lively and cohesive, and, in our view, also very productive.

In the remaining sections we shall try to highlight some of the project's results, without aiming in any way at an exhaustive account (the interested reader is referred for more details to the project's final report [87]).

2   Goal and Context

In this section, we recall the goal of the project and point out its connection to previous and contemporary research. In the final CEDISYS report [87], the overall objective of the project is described as follows:

"developing a fundamental understanding of the nature of concurrency and providing a formal framework useful for describing concurrent and distributed systems. This formal framework should support the specification and development of such systems and should lead to methodologies for proving systems correct and more generally for deriving their properties".

More specifically, the project aimed at *"a theory of concurrency where the distributed nature of processes is properly taken into account"*. The plan was to *"compare existing formalisms, develop models, languages and logics with compositionality and abstraction capabilities"*, as well as to *"experiment with techniques and tools for supporting the implementation of the proposed formalisms"*.

At the time when the project was proposed, two main approaches to the theory of concurrency had been investigated:

- The *true concurrency* approach. Here concurrent processes were represented by means of event-based nonsequential models, among which the most established ones were Petri nets [100,108] and Event Structures [110,94][1]. These models had pleasant characteristics: they were expressive enough to represent the new phenomena arising in concurrent computations, and they were sufficiently concrete to serve both as system models and as execution models. In some sense, they seemed to be able to play the roles of "denotational" and "operational" models at the same time. Moreover, a formal result established that Event Structures were concrete representations of particular

---

[1] Other proposed models were Mazurkiewicz traces [83], Grabowski's partial words [70] and Gischer and Pratt's pomsets [55,103,104]. However, these appeared to be more suited as execution models than as system models.

domains: as such, they appeared as a natural candidate for denoting concurrent processes. However, both Petri nets and Event Structures lacked two important properties of denotational semantics: abstraction, and, up to the mid-eighties, also compositionality. Indeed, the first studies on Petri nets had concentrated on analysis techniques, rather than on techniques for the synthesis and extension of nets. In connection with the advent of process calculi in the early eighties, combinators for Petri nets and Event Structures had started to be investigated, mainly by Winskel [111,113] but also by other researchers [37,38]. Still, these models appeared too intensional to offer a proper denotational semantics for concurrency.

– The *process algebraic* approach. This was a relatively new but very active line of research, initiated by Milner in 1980 with his proposal of a Calculus of Communicating Systems (CCS) [85]. Here concurrent processes were defined as terms of an algebra and interpreted as labelled transition systems, using Plotkin's structural operational semantics (SOS) [102]. The abstract semantics of processes was to be recovered in a second step, by quotienting processes with respect to an observational equivalence, and notably with respect to bisimulation equivalence [99,86]. This equivalence could furthermore be characterised by a set of axioms. The simplicity and elegance of CCS, which made it formally close to the lambda-calculus, together with the powerful mathematical tools that accompanied it (SOS semantics, bisimulation proof method, algebraic laws), immediately turned CCS, and the family of process calculi which rapidly grew around it, into an appealing field of experimentation for a large community of researchers. Indeed, the introduction of CCS and SOS semantics resulted in a renewed interest in operational semantics for concurrency, after the difficult quest for a denotational semantics in the late 70's (conducted by Milner himself and by other researchers). However, the emphasis of the process algebraic approach was on compositionality rather than on abstraction. Abstraction was only recovered a posteriori through observational equivalences, and thus the resulting semantic model amounted to a (syntactic) term model. In spite of this, CCS had many of the qualities of a denotational model. In particular, the axiomatisations of bisimulation and other behavioural equivalences (e.g. testing equivalence [41]) made it possible to reason about processes and reduce them to normal forms. Another important property of process calculi was their ability to describe both complete systems and their abstract specifications, so that the satisfaction relation of the latter by the former could be simply implemented via observational equivalence. Labelled transition systems were also the prime models for a simple modal logic called Hennessy-Milner logic [74], which precisely characterised bisimulation equivalence.

The two approaches just described had somewhat complementary advantages and drawbacks:

– Process calculi enjoyed the properties of compositionality and substitutivity, by virtue of their algebraic syntax and of the existence of behavioural congruences. On the other hand, they used an *interleaving semantics*, simulating concurrency by a nondeterministic choice of sequential interleavings. The characteristic semantic equation of process calculi, as formally expressed by Milner's expansion law, could informally be rendered as: concurrency = sequentiality + nondeterminism. Hence the notions of causality and distribution, which were present in the syntax, were lost in the semantics.

– Petri nets and Event Structures allowed for a clear distinction between concurrency, causality and conflict, which were all primitive relations on events. In contrast with the interleaving equation, they embodied the true concurrency inequation: concurrency $\neq$ causality + conflict. The main drawback of these models was their lack of abstract semantics.

In the light of this situation, two natural questions came to mind: 1) How to formally relate models? and 2) How to bridge the gap between the two approaches, so as to get the best of each world?

Initial answers to these questions had already started to be provided, both by CEDISYS members (to be) and by other researchers. For instance, a comparison of different concurrency models, related by adjunctions within a common categorical framework, had been carried out by Winskel in [114]. Attempts at narrowing the gap between process calculi and models had also started to be made, either by interpreting process combinators in the models, or by increasing the "observational power" of the operational semantics of calculi so as to capture (some amount of) concurrency and causality.

Among the early interpretations of process calculi (CCS, CSP [75], TCSP [21], CCSP [98]) into nonsequential models, we could cite:

– CCS: interpretations into Petri nets by De Cindio et al. [38], Goltz and Mycroft [63], Winskel [113], Nielsen [92], and Vaandrager and van Glabbeek [62]; interpretations into Event Structures by Winskel [111] and (for a subset of CCS) by Boudol and Castellani [13];

– CSP and TCSP: interpretations into Petri nets by De Cindio et al. [37], by Goltz and Reisig [64] and by Goltz and Loogen [82];

– CCSP (a mixture of CCS and TCSP): interpretation into Petri nets by Olderog [98].

A few proposals for enriching the operational semantics of CCS (sometimes in relation to the interpretations above) had also been put forward. The general idea here was to retain more of the syntax of processes in their semantics, by adding "structure" to the labelled transition system of CCS. This structure could be introduced either via structured actions (generalising atomic actions to composite actions) or via structured states (somehow departing from an extensional view of transition systems), or combining both. We could cite here:

– Structured labels: "pomsets" in Boudol and Castellani's semantics for a subset of CCS [13], and "concurrent histories" in the semantics by Degano, De Nicola and Montanari [42];

– Structured states: pairs of local and concurrent residuals in the distributed bisimulation semantics of Castellani and Hennessy [24,27], "grapes" in the partial order semantics of [42].

To conclude this introductory section, we could say that in the course of the eighties, a shift had been made from denotational towards operational semantics: thanks to Plotkin's structural approach and to the rapid development of process calculi, operational semantics had acquired new dignity and had gradually become the touchstone for models of concurrency.

One of the central questions of the CEDISYS project was then: how to enrich the operational semantics of CCS so as to take into account concurrency and distribution, and how to relate the new operational semantics to interpretations of CCS into Petri nets or Event Structures?

3   Some Achievements of the Project

The project's work on noninterleaving semantics for concurrency was structured along five axes: 1) comparison of existing formalisms, 2) models, 3) languages, 4) logics and proof systems, and 5) implementation. It is not our intention here to give an exhaustive account of the work accomplished. Neither shall we try to give a well-balanced description of the project's results. Rather, we aim at giving an idea of the project's collaborative work by picking some representative results which benefitted from the interaction among the partners within axes (1) – (3). We shall try to reconstruct the history of these results within the "4-player ping-pong game" of CEDISYS, and to analyse their influence on subsequent research. As will become evident, Ugo's role in the project was a very prominent and pervasive one: not only was he the inspired instigator and supportive coach of the game (which terminated with four winners and no loser), but he was also personally involved in most of the themes of the project. If the term "pervasive computing" had not yet appeared at that time, the term "pervasive researching" could well have been coined on purpose for Ugo!

As it should be apparent from the discussion in the previous section, there was already a strong convergence of ideas and interests within the consortium, even before the project started. We wish now to show how this convergence turned into a fruitful "cross-fertilization" among the sites in the course of the project: how some existing ideas came to their full development, and how new ideas emerged from the project's "chemical soup" (in the sense of the CHAM model [10] described below...). In doing so, we shall focus on two main strands: 1) noninterleaving operational semantics for CCS, and 2) nonsequential models and abstract semantics.

3.1   Noninterleaving Operational Semantics for CCS

Process calculi include a parallel composition operator and a form of sequential composition. These operators specify respectively a concurrency and a causality relation between the actions of their components. However, these relations are "forgotten" when passing from the syntax to the standard labelled transition semantics of the calculus. To define a noninterleaving operational semantics for CCS the idea was then, quite naturally, to add structure to its labelled transition system so as to retain in the semantics some of the information about concurrency and causality which was present in the syntax. As mentioned in Section 2, some proposals for enriching the operational semantics of CCS along these lines, by adding structure to the actions or to the states of transitions, had already been presented by some of the project's partners. However, these proposals did not completely fulfill the objective, either because they were defined only on subsets of CCS, or because they did not fully agree with existing interpretations of CCS into Event Structures and Petri nets. Indeed, the initial criterion for these noninterleaving operational semantics for CCS was their agreement with interpretations of the calculus into event based semantic domains.

This line of work, aiming at obtaining noninterleaving operational semantics for CCS by enriching its labelled transition system, was intensively pursued in the project. Notable advances were made, both on structured actions and on structured states:

1. **Structured actions.** In the standard labelled transition system of CCS, transitions are labelled by atomic actions. Two ways of generalising such actions were investigated:

   • *Composite actions.* The idea here was to relax the atomicity constraint to allow transitions to be labelled by whole nonsequential computations. In the early work by Boudol and Castellani [13,14], transitions labelled by *pomsets* (partially ordered multisets) had been defined for a simple CCS-like language with parallel composition but no communication. In that case, pomset transitions could be directly constructed by means of new transition rules, whose effect was to transfer the constructs of sequential and parallel composition from the processes to the actions labelling their transitions. However, it was not clear how to extend this direct pomset semantics to the full language CCS, since the parallel composition of two pomsets does not in general yield a pomset when communication is allowed. This led the authors to propose, in [16,15], an indirect construction of a pomset transition starting from a class of permutation-equivalent transition sequences. This construction "a posteriori" of a partial order from an equivalence class of sequential computations was similar to that used for defining Mazurkiewicz traces [84]. The novelty of [16,15] was the use of enriched transitions for CCS called *proved transitions* (because they were labelled with a representation of their proof in the inference system of CCS), from which the concurrency relation between transitions could be derived by syntactic means. A different construction of partial order computations for CCS, starting from enriched sequential computations called *concurrent histories*, obtained by concatenating *atomic histories* (corresponding to single steps) and recording their causality relation, was proposed by Degano, De Nicola and Montanari in [42,47,46]. In that case, the construction exploited structured states rather than structured actions, and made use of the causality rather than the concurrency relation. This semantics will be discussed in more detail later, under the heading "structured states". Note that, while in the direct pomset semantics of [13,14] atomic actions were replaced with composite actions in the transition system itself, in both the proved transition semantics [16,15] and the concurrent history semantics [45,44] actions remained atomic in the basic transition system but were decorated with additional information (they were instances of the "decorated atomic transitions" discussed below). Partial order transitions were only retrieved in a second step.

Clearly related to the generalisation of atomic actions to composite actions was the issue of *action refinement*, which was thoroughly studied within the project. However, the motivation behind action refinement was not that of modelling nonsequential computations, bur rather of managing different levels of abstraction in the description of computations (sequential or not). The project's work on action refinement will be described in Section 3.2.

• *Decorated (atomic) actions.* In this case the idea was to remain with atomic actions but to annotate them with syntactic information so as to identify the component responsible for them. Several kinds of decorated actions were proposed:

– *Actions with localities* [19,20,4,89,91]. Localities were introduced to distinguish different parallel components: they could either be assigned statically to processes (*static localities* [4,91,25]) and then transmitted to their actions, or be dynamically created and associated with actions, and then recorded in the residual process (*dynamic localities* [19,20]). The notion of dynamic locality proposed by Boudol, Castellani, Hennessy and Kiehn was already implicitly present in the distributed bisimulation semantics of [27], embodied in the idea of splitting the residual of a transition among multiple observers; indeed, the idea of associating localities with actions and processes emerged after some insightful (but unsuccessful) attempts by Kiehn [78] at extending to full CCS the distributed bisimulation semantics of [27]. As it were, in the presence of global scope operators like restriction, the idea of splitting the residual of a transition among multiple observers could only be implemented by inserting the observers themselves in the residual, which exactly amounted to the introduction of dynamic localities. Similarly, the notion of static locality was already implicitly contained in both that of "atomic history" by Degano and Montanari [42,49], and in that of "proof" by Boudol and Castellani [16,15]. In this sense, Montanari and Yankelevic's paper on parametric localities [89] may be seen as an accomplishment of both the concept of atomic history and of the notion of locality (and particularly of static locality [4]) which had been meanwhile developed in the project.

- *Actions with causes.* In the model of *causal trees* by Darondeau and Degano [34,35], the actions labelling the branches were annotated with their set of (global) causes in the computation given by the path from the root to the action. In Kiehn's subsequent papers [79,80], this causal semantics was compared with the dynamic locality semantics within a common framework (the "local-global cause transition system"), which also allowed a stronger semantics to be defined, based on the conjunction of local and global causes.

- *Actions with proofs.* In the proved transition system of Boudol and Castellani [16,15], labels go all the way from simple atomic actions to complete "proofs" of transitions. As remarked by the authors in [18], the notion of proof could be weakened in various ways to obtain more specific semantics (for instance, the static locality semantics). Again, the notion of "atomic history" [42] could be seen as a precursor to that of "proof" (although it did not record choices, and hence did not support the definition of a concurrency relation between transitions).

- *Actions with past.* Both actions with dynamic localities and actions with causes are instances of actions with portions of their "computational past" or "global history". In contrast, actions with a "static past" or "local history" were considered in the *event transition system* proposed for CCS by Boudol and Castellani in [18], as an intermediate among three different interpretations of the calculus (into proved transition systems, flow event structures and flow nets).

2. **Structured states.** New forms of structured states were explored, often in connection with the decorated actions described above:

   - *Grapes.* In their early model of *concurrent histories* [49], Degano and Montanari had defined concrete computations with structured initial and final states. These structured states, called *grapes* [42], were essentially the sets of sequential components of a process, each of them being annotated with its "position" within the process. This annotation allowed the computations to be concatenated while recording the causality relation among their actions. The constituent blocks of such computations, called *atomic histories*, could be directly derived by operational rules. From a concurrent history, a partial ordering of actions could be abstracted away. This model was used by Degano, De Nicola and Montanari to define a partial ordering semantics for CCS in [42,47].

   - *Processes with past.* Whenever actions with past are used, it is also necessary to record the past in the states, so that the computational history can be incremented at each step and propagated with subsequent actions. Thus, corresponding to actions with dynamic localities and actions with causes, *processes with localities* and *processes with causes* were introduced. A more general notion of process with past or "marked process" was considered in the event transition system of [18]. A marked process contains all the information of the original process, but additionally specifies which part of the process has been executed. It corresponds to an unfolded Petri net with a marking, or to an event structure with a configuration.

Although not expressly tailored for true concurrency, but reflecting in some sense the notions of distribution and mobility associated with concurrency, a new model for describing the operational semantics of parallel processes, called the *Chemical Abstract Machine* (CHAM), was proposed by Berry and Boudol in [10]. In this model, the state of a system of concurrent agents is viewed as a "chemical solution", in which floating "molecules" can interact with each other according to "reaction rules". The CHAM model introduced a new, simpler way of expressing a reduction semantics in a concurrent scenario, which was to become quite popular in later research on process calculi (notably on the π-calculus [86], the join-calculus [54] and the ambient calculus [22]).

3.2   Abstract Noninterleaving Semantics

Most of the noninterleaving semantics for CCS described in the previous section were compared with interpretations of CCS into nonsequential models such as Event Structures and Petri nets, or used as a basis for defining behavioural equivalences on processes, in order to obtain more abstract semantics.

Nonsequential Models. As mentioned earlier, the initial criterion for assessing the new noninterleaving operational semantics for CCS was their agreement with "denotational semantics", given by interpretations of CCS into semantic domains like Event Structures and Petri nets. Thus, for instance, in [45,44] Degano, De Nicola and Montanari showed the consistency of their operational semantics of [43] with interpretations of CCS into *prime event structures* [94] and *condition-event systems* (a class of Petri nets). In [16], Boudol and Castellani showed that their proved transition semantics agreed with an interpretation of CCS (and SCCS) into *flow event structures*, a class of event structures lying between prime and stable event structures [114], thus generalising their previous result of [13,14]. In a joint paper between Aarhus and Sophia [28], it was shown that parallel composition of flow event structures can be characterized as a categorical product, and that flow event structures obtained as interpretations of CCS terms satisfy a particular structural property (which generalises that given in [13] for a subset of CCS). In [18], the results of [14,16] were extended to *flow nets* [12], a class of "stable" Petri nets, and to *asynchronous transition systems*, a model introduced by Bednarczyk in [9]. Indeed, the need for interpreting process algebras gave a new impulse to the study of semantic domains. Besides the above-mentioned flow event structures and flow nets, which were designed to fit typical process combinators, we could also cite here the models of *bundle event structures* [81] (developed outside the project but shown to correspond to safe flow nets in [17]), *Δ-free event structures* [36], and later on, *transition systems with independence* [117].

On a more abstract level, a line of research that was actively pursued within the project was the comparison of different semantic models within an algebraic or categorical framework. A web of strong formal connections, expressed as adjunctions among various transition-based models presented as categories, had already been established by Winskel in [112,115]. A notable contribution was made by Nielsen, Rozenberg and Thiagarajan with the introduction of the notion of *region*, which allowed the above connections to be extended to a class of structured transition systems [95,96,97]. The idea of region was to be further used by Winskel and Nielsen in [117] to give an adjunction between Petri nets and asynchronous transition systems [9].

The unification of models of concurrency into a common algebraic framework had also been a central concern of Ferrari and Montanari's work [51,53], where the permutation semantics of [14] was lifted to a more abstract setting, which could then be specialised in various ways. For instance, by introducing axioms to equate permutation-equivalent computations, one could obtain partial ordering computations. Also, by considering categories having CCS models as objects, one could use morphisms to represent a specification-implementation relation between two CCS models. In the paper [52], the same authors proposed a generalisation of Milner's expansion theorem to a language called CCCS, a variant of CCS with an "observation prefixing" operator. Observations were elements of an algebra (in fact, they were "proofs" in the sense of [14]). By adding axioms to this algebra, one could instantiate the *extended expansion theorem* to characterise different bisimulation equivalences, such as standard bisimulation and pomset bisimulation. As it were, the extended expansion theorem was a generalisation of that proposed in [13] for pomset bisimulation (where a notation for "pomset prefixing" had similarly been introduced for axiomatisation purposes). In the same spirit, a parametric approach for axiomatising bisimulation equivalences was later proposed in [48]: in this case, the underlying structures, called "observation structures", were node-labelled graphs where the labels represented computations. This framework allowed for the characterisation of a larger number of equivalences. The last model we wish to mention is Gorrieri and Montanari's SCONE calculus [67], a calculus of Petri nets generated by a set of combinators, which was used in particular to implement CCS.

As might be apparent from the above discussion, there were two main "angles of attack" in the project. In general, Aarhus and Pisa partners had a strong concern in generality and parametricity, and tended to privilege a "uniform" or "parametric" approach to the study of nonsequential models. By contrast, Sophia and Sussex partners preferred to focus on particular calculi or models, adopting a more "language-specific" or "model-specific" approach.

Noninterleaving equivalence notions. Based on the various concrete operational semantics described in Section 3.1, several noninterleaving behavioural equivalences and preorders were proposed for CCS. For some of them, axiomatisations or logical characterisations were also provided. Let us briefly recall the main proposals:

- *NMS-equivalence.* In [43], Degano, De Nicola and Montanari had proposed a partial ordering semantics based on a tree-model called Nondeterministic

Measurement Systems (NMS). By applying "observation functions" to NMS's, different equivalence notions could be obtained, among which a partial ordering equivalence called NMS-equivalence.

- *Pomset bisimulation.* The pomset transition system of [13] and its extension in [16] gave rise to the notion of pomset bisimulation. Pomset bisimulation equivalence was unrelated to NMS-equivalence. An axiomatisation of pomset bisimulation for a subset of CCS was given in [13].

- *Mixed-ordering equivalence.* Based on the concurrent history semantics [49], Degano, De Nicola and Montanari proposed in [45,44] an equivalence called "mixed-ordering equivalence", which combined the observation of the temporal ordering with that of the causal ordering of actions. This equivalence was stronger than pomset bisimulation, and was shown by Aceto [2] to coincide with an equivalence proposed by Goltz and Van Glabbeek to deal with action refinement, called *history-preserving bisimulation* [57].

- *Distributed bisimulation.* Some advances were made on distributed bisimulation equivalence. In his PhD thesis [31], Christensen presented a logical characterisation and a decidability result for distributed bisimulation (in its initial formulation with local and concurrent residuals [24]). In [78], Kiehn showed the limits of the original definition of distributed bisimulation, which appeared to be incompatible with the CCS restriction operator. Indeed, in the light of Kiehn's observations [78] and Aceto's results [3], the "right" extension of distributed bisimulation to full CCS turned out to be dynamic location equivalence (see below, and previous discussion at page 763).

- *Location equivalence and preorder.* The static [4] and dynamic [19,20] locality semantics supported two different definitions of "location equivalence" and "location preorder". For the dynamic versions, an axiomatisation was also provided in [20]. In [4], Aceto established the coincidence of the static and dynamic notions for a subset of CCS with only top-level parallelism (this work was later to be extended to full CCS by Castellani [25] and by Mukund and Nielsen [91]). Since static and dynamic localities do not have the same meaning - static localities represent sites, while dynamic localities represent sets of local causes - this result was not obvious. It amounted to proving that observing distribution and observing local causality were essentially the same thing. Dynamic location equivalence was shown to coincide with distributed bisimulation on the subset of CCS without restriction, where the latter had been originally defined. On the smaller subset of CCS with parallel composition but no communication, it was finer than pomset bisimulation (because it recorded how computations were locally extended). As soon as communication was introduced, it became incomparable with pomset bisimulation.

- *Local mixed-ordering equivalence.* A transition system for CCS labelled with static localities, called "spatial transition system", was studied by Montanari and Yankelevich in [118,90]. In this case, transitions with localities were not directly used to define an equivalence, but rather to build a second transition system, labelled by "local mixed partial orders" (where the ordering was a mixture of temporal ordering and local causality). This new transition system was then used to define a behavioural equivalence called local mixed-ordering equivalence. This equivalence, a variant of the mixed-ordering equivalence discussed above, was shown in [118] to coincide with dynamic location equivalence.

— *Causal bisimulation.* The bisimulation equivalence associated with Darondeau and Degano's causal trees, called causal bisimulation [34], was also investigated in relation to action refinement in [36]. It was shown to coincide with history-preserving bisimulation, and hence also with mixed-ordering equivalence. A complete axiomatisation for causal bisimulation was given in [34].

— *Local-global cause equivalence.* A "local-global cause transition system" for CCS, where actions were decorated with both their local and global causes, was proposed by Kiehn in [79] and [80] (see also [77]). By disregarding the set of global causes, respectively local causes, of actions, one could then obtain two different equivalences called "local cause equivalence" and "global cause equivalence", which were shown to coincide respectively with dynamic location equivalence and causal bisimulation. This semantics also supported a stronger equivalence, called local-global cause equivalence, arising from the joint observation of the two sets of causes.

An interesting and somewhat unexpected outcome of these studies on noninterleaving equivalences was that distributed and causal semantics, exemplified respectively by location equivalence and causal bisimulation, were in general incomparable: they only coincided on the finite fragment of CCS without communication (essentially because in that case all causality is local). Another observation was that, while both location equivalence and causal bisimulation are "history-preserving" equivalences (recording respectively the local and the global history), other equivalences like pomset bisimulation were instead "forgetful". A detailed comparison of noninterleaving equivalences was given by Aceto in [2,3]. A more recent comparison, including other equivalences, may be found in [26].

Action refinement. Another question that was thoroughly studied within the project was that of action refinement. The motivation here was to allow for a specification of systems at different levels of abstraction. This was to be used as a conceptual underpinning for the so called "top-down design" of distributed systems, whereby a system is developed by successive refinements of atomic actions into more complex behaviours. Usually an "implementation relation" between two successive descriptions was introduced, in order to guarantee the correctness of the refinement process.

Approaches to action refinement may be classified according to various criteria:

— *atomic* versus *non-atomic* refinement: in the first case the refining behaviour is required to be executed atomically, while in the latter it may be freely interleaved with other behaviours;
— *horizontal* versus *vertical* refinement: in the first case the abstract and concrete descriptions are given in the same formalism, while in the latter case

they are given in different formalisms, and an encoding from the first to the
second formalism has to be provided;
- *syntactic* versus *semantic* refinement: in the first case action refinement is
  introduced in a language by means of a specific operator, and implemented as
  syntactic substitution; in the second case it is defined on a semantic model.

In some early work on atomicity by Boudol and Castellani [14,11], as well as
in the paper [66] by Gorrieri, Marchetti and Montanari, methods for treating
computations as *atomic* (that is, both *recoverable* and *interference-free*, in the
terminology of [11]) had been considered. In both cases, an auxiliary labelled
transition system was introduced for describing atomic computations. However,
these proposals were not explicitly concerned with the issue of action refinement,
but rather with that of atomicity of behaviours. Subsequently, all the project's
work on action refinement was essentially concerned with non-atomic refinement,
both syntactic and semantic, and mostly of the horizontal type. Indeed, non-
atomic refinement was the most popular approach at that time.

Although the motivation behind action refinement was quite different from
that of noninterleaving semantics, a connection between the two questions was
deemed to exist for some time, and resulted in a concurrent development of their
theories. Let us briefly recall the research context at the time.

When dealing with syntactic action refinement, a new refinement operator is
introduced in the language. Then a natural question to address is whether this
operator preserves behavioural equivalences. In a short note [29] published in
1987, Castellano, De Michelis and Pomello showed that standard bisimulation
equivalence was not preserved by action refinement, and conjectured that the
recourse to a noninterleaving semantics was necessary to cope with refinement.
Robustness with respect to action refinement became a new criterion to assess
equivalence notions, and spurred the search for new noninterleaving equivalences.

Hennessy's unpublished manuscript "On the relationship between time and
interleaving" [71], later to become "Axiomatising finite processes" [72], was probably the first attempt to consider actions with duration without explicitly introducing time, by splitting atomic actions into a beginning subaction and an ending
subaction. Strong bisimulation equivalence with respect to these subactions was
called *timed equivalence* in [6], where it was shown to be preserved by action
refinement in the simple setting of CCS without communication and restriction.
The *owl example* from [56] shows that this is no longer the case when the expressiveness of the language is increased to allow communication and some form
of hiding. A strengthening of *timed equivalence* called *refine equivalence* is obtained by requiring, in the bisimulation game, that the matching of begin actions
should determine the matching of the corresponding end actions; the identity of
the original action which gives rise to the begin and end actions is remembered
in the bisimulation game. In [7], the weak version of this equivalence, called *weak
refine equivalence*, was shown to be preserved by refinement in a more expressive
language with communication. Although somewhat complicated to define, the
latter equivalence seems more natural than timed equivalence. It is also technically easier to handle; for example the refinement theorem in [6], in the simple

language, was actually proved for the strong version of refine equivalence, and the result was only obtained indirectly for timed equivalence by showing that the two equivalences coincided on the simple language.

Similar ideas were pursued outside the project by van Glabbeek and Vaandrager [62,56], in the setting of non-interleaving models such as Event Structures and Petri Nets. Their equivalence, called *ST-bisimulation equivalence*, was very close in spirit to refine equivalence, although the decision of which end actions are to be matched was determined by the previous history of associations between actions. In [62], van Glabbeek and Vaandrager showed that ST-bisimulation was robust for action refinement in the setting of Event Structures. Since ST-bisimulation is not a partial order equivalence (it was conceived for real-time and disregarded some causality) this result invalidated the conjecture of [29].

Meanwhile Goltz and van Glabbeek [57] had shown that, except for pomset trace equivalence, none of the partial order equivalences existing at the time was preserved by refinement. In particular pomset bisimulation and NMS equivalence were not robust with respect to refinement. In the same paper, it was shown that a stronger partial order equivalence called *history preserving bisimulation* (already used by Rabinovitch and Trakhtenbrot under the name "behaviour structure equivalence" [107], but reformulated on Event Structures by Goltz and van Glabbeek), was preserved by refinement on prime event structures. This result was to be strengthened in subsequent papers, by considering more general forms of refinement and more general classes of Event Structures [58,59,60].

The above-mentioned work by Aceto and Hennessy mostly concentrated on syntactic refinement. In Aceto's thesis [1], on the other hand, both syntactic and semantic refinement were considered. In a similar vein, Nielsen, Engberg and Larsen presented in [93] four different fully abstract models, based on sets of pomsets, for a simple CCS-like process calculus (essentially the same as in [13]) equipped with a refinement construct. These results were generalised by Aceto and Engberg to a model based on *pomsets failures* (pomsets with refusal sets) in [5]. Semantic refinement was studied by Gorrieri and Montanari in [67,68], as well as in Gorrieri's thesis [65]; here, vertical refinement was privileged and the use of implementation relations, parameterised by refinement functions, was investigated. More specifically, in [67] Gorrieri and Montanari proposed an implementation of the subset of CCS without restriction and relabelling into the net calculus SCONE. This result was generalised in [65] to full CCS and to SCONE$^+$, an extension of SCONE. In [68] the same authors put forward a general methodology for "atomic linear refinement", where each refined process could be proved to be a correct atomic implementation of its source process.

In [36], Darondeau and Degano explored the issue of syntactic and semantic action refinement in their model of causal trees [34]. To this purpose, they proposed an adaptation of prime event structures called "free event structures", which was closed with respect to event refinement. Causal trees were already known to correspond to event structures up to history-preserving bisimulation. Refinement operators were then introduced for causal trees and the agreement of syntactic and semantic refinement was established.

4   Conclusion

In this section, we attempt a short, and certainly incomplete, analysis of the impact of the CEDISYS project on contemporary and subsequent research. As can be gathered from the discussion in the previous sections, the project was the theatre of intense "cross-fertilization". Many examples of mutual influences among the partners have already been cited. Let us point out some others, and mention some later influences on other researchers.

The work on *localities* may be viewed as carrying the project's "trademark". As explained earlier, the notion of dynamic locality stemmed from previous work on distributed bisimulation, which in turn had been inspired by *abstraction homomorphisms* [23,88] (or rather their extension to true concurrency), and by some earlier work by Degano, De Nicola and Montanari on partial order semantics for CCS [42]. The later work by Corradini and De Nicola on localities [32] was also very much in the spirit of CEDISYS. Localities are now an intrinsic component of the wide range of calculi used to study "network aware" computing, with varying characteristics reflecting the different facets of network computing; typical examples include the ambient calculus [22], the distributed π-calculus [73], the join-calculus [54], and the language KLAIM (locality-based LINDA) [39,40].

The essential characteristics of the *chemical abstract machine* [10] have been widely adopted, and now form part of the natural framework for designing and studying computational calculi. In particular they provide a very convenient and useful methodology for defining the reduction semantics of such calculi.

The notion of *proved transition system* has also been very influential. Apart from its use within the project, let us cite the work by Badouel and Darondeau, who established in [8] a correspondence between (a variant of) the proved transition semantics and an interpretation of CCS into Stark's trace automata [109]. In [50], Degano and Priami studied *proved trees*, a subset of proved transition systems which was well-suited for the comparison with causal trees. Subsequently, proved transition systems were taken up by Priami [105], under the name *enhanced transition systems*, and used for many different purposes; a particularly interesting application area is that of *Computational Systems Biology*, [106]. The study of reversible computing [33,101] also benefitted from this idea; for example, the model used in [101] is very close to that of *event transition systems* [18].

The project's work on *refinement* was both inspired and used by Goltz and van Glabbeek [60,59] as well as by Rensink [69]. These authors also took up the model of *flow event structures*, which turned out to be well-suited for certain forms of refinement. In particular, in [61], Goltz and van Glabbeek proposed a subclass of flow event structures which was both closed under action refinement and well-behaved with respect to parallel composition, thus generalising the work by Castellani and Zhang [28].

The notion of *open map bisimulation* [76,30], introduced a few years after the end of the project, was partly inspired by that of abstraction homomorphism. The general issue of abstraction, namely how to make models such as event structures and Petri nets and their morphisms abstract enough to support a fully fledged "domain theory for concurrency", continues to be pursued, one recent line being through the introduction of a formal treatment of symmetry on such models [116].

We could not close this recollection of the CEDISYS project without a special praise for "Ugo's management style", which led the project to meet its commitments with a Swiss clock punctuality, while being experienced by all its members as a most stimulating and enjoyable collaboration.

Acknowledgements. We would like to thank Luca Aceto, Rocco De Nicola and the anonymous reviewer for helpful comments.

References

1. Aceto, L.: Action-refinement in process algebras. In: Distinguished Dissertations in Computer Science. Cambridge University Press, Cambridge (1992)
2. Aceto, L.: History preserving, causal and mixed-ordering equivalence over stable event structures (Note). Fundamenta Informaticae 17(4), 319–331 (1992)
3. Aceto, L.: Relating distributed, temporal and causal observations of simple processes. Fundamenta Informaticae 17(4), 369–397 (1992)
4. Aceto, L.: A static view of localities. Formal Aspects of Computing 6, 201–222 (1994); Previously appeared as INRIA Research Report n. 1483 (1991)
5. Aceto, L., Engberg, U.: Failure semantics for a simple process language with refinement. In: Biswas, S., Nori, K.V. (eds.) FSTTCS 1991. LNCS, vol. 560, pp. 89–108. Springer, Heidelberg (1991)
6. Aceto, L., Hennessy, M.: Towards action-refinement in process algebras. Information and Computation 103(2), 204–269 (1993); Extended abstract. In: Proceedings LICS 1989, IEEE Computer Society Press (1989)
7. Aceto, L., Hennessy, M.: Adding action refinement to a finite process algebra. Information and Computation 115(2), 179–247 (1994); Extended abstract. In: Leach Albert, J., Monien, B., Rodríguez-Artalejo, M. (eds.) ICALP 1991. LNCS, vol. 510, Springer, Heidelberg (1991)
8. Badouel, E., Darondeau, P.: Structural operational specifications and trace automata. In: Cleaveland, W.R. (ed.) CONCUR 1992. LNCS, vol. 630. Springer, Heidelberg (1992)
9. Bednarczyk, M.: Categories of Asynchronous Systems. PhD thesis, University of Sussex (1988)
10. Berry, G., Boudol, G.: The chemical abstract machine. Theoretical Computer Science 96, 217–248 (1992); Extended abstract. In: Proceedings POPL 1990, pp. 81–94 (1990)
11. Boudol, G.: Atomic actions (Note). Bulletin of the European Association for Theoretical Computer Science 38, 136–144 (1989)
12. Boudol, G.: Flow event structures and flow nets. In: Guessarian, I. (ed.) LITP School 1990. LNCS, vol. 469, pp. 62–95. Springer, Heidelberg (1990)
13. Boudol, G., Castellani, I.: On the semantics of concurrency: partial orders and transition systems. In: Ehrig, H., Levi, G., Montanari, U. (eds.) CAAP 1987 and TAPSOFT 1987. LNCS, vol. 249, pp. 123–137. Springer, Heidelberg (1987)
14. Boudol, G., Castellani, I.: Concurrency and atomicity. Theoretical Computer Science 59, 25–84 (1988)
15. Boudol, G., Castellani, I.: A non-interleaving semantics for CCS based on proved transitions. Fundamenta Informaticae 11(4), 433–452 (1988)

16. Boudol, G., Castellani, I.: Permutation of transitions: an event structure semantics for CCS and SCCS. In: de Bakker, J.W., de Roever, W.-P., Rozenberg, G. (eds.) REX Workshop. Linear Time, Branching Time and Partial Order in Logics and Models for Concurrency. LNCS, vol. 354, Springer, Heidelberg (1989)
17. Boudol, G., Castellani, I.: Flow models of distributed computations: event structures and nets. Research Report 1482, INRIA (1991); Previous version by G. Boudol. In: Guessarian, I.(ed.) LITP School 1990. LNCS, vol. 469, Springer, Heidelberg (1990)
18. Boudol, G., Castellani, I.: Flow models of distributed computations: three equivalent semantics for CCS. Information and Computation 114(2), 247–314 (1994); Extended abstract. In: Guessarian, I. (ed.) LITP School 1990. LNCS, vol. 469, Springer, Heidelberg (1990)
19. Boudol, G., Castellani, I., Hennessy, M., Kiehn, A.: Observing localities. Theoretical Computer Science 114, 31–61 (1993); Extended abstract. In: Tarlecki, A. (ed.) MFCS 1991. LNCS, vol. 520. Springer, Heidelberg (1991)
20. Boudol, G., Castellani, I., Hennessy, M., Kiehn, A.: A theory of processes with localities. Formal Aspects of Computing 6, 165–200 (1994); Extended abstract. In: Cleaveland, W.R. (ed.) CONCUR 1992. LNCS, vol. 630, pp. 165–200. Springer, Heidelberg (1992)
21. Brookes, S., Hoare, C.A.R., Roscoe, A.: A theory of communicating sequential processes. Journal of ACM 31(3), 560–599 (1984)
22. Cardelli, L., Gordon, A.D.: Mobile Ambients. In: Nivat, M. (ed.) ETAPS 1998 and FOSSACS 1998. LNCS, vol. 1378. Springer, Heidelberg (1998)
23. Castellani, I.: Bisimulations and abstraction homomorphisms. Journal of Computer and System Sciences 34, 210–235 (1987)
24. Castellani, I.: Bisimulations for Concurrency. PhD thesis, University of Edinburgh (1988)
25. Castellani, I.: Observing distribution in processes: static and dynamic localities. Int. Journal of Foundations of Computer Science 4(6), 353–393 (1995); Extended abstract. In: Borzyszkowski, A.M., Sokolowski, S. (eds.) MFCS 1993. LNCS, vol. 711. Springer, Heidelberg (1993)
26. Castellani, I.: Process algebras with localities. In: Bergstra, J., Ponse, A., Smolka, S. (eds.) Handbook of Process Algebra, pp. 945–1045. North-Holland, Amsterdam (2001)
27. Castellani, I., Hennessy, M.: Distributed bisimulations. JACM 36(4), 887–911 (1989)
28. Castellani, I., Zhang, G.Q.: Parallel product of event structures. Theoretical Computer Science 179(1-2), 203–215 (1997); Previously appeared as DAIMI Research Report PB-301, 1989, and as INRIA Research Report 1078 (1989)
29. Castellano, L., De Michelis, G., Pomello, L.: Concurrency vs interleaving: an instructive example. Bulletin of the EATCS 31, 12–15 (1987)
30. Cattani, G., Winskel, G.: Profunctors, open maps and bisimulation. MSCS 15(3), 553–614 (2005)
31. Christensen, S.: Distributed bisimilarity is decidable for a class of infinite state-space systems. In: Cleaveland, W.R. (ed.) CONCUR 1992. LNCS, vol. 630. Springer, Heidelberg (1992)
32. Corradini, F., De Nicola, R.: Locality based semantics for process algebras. Acta Informatica 34, 291–324 (1997)
33. Danos, V., Krivine, J.: Reversible communicating systems. In: Gardner, P., Yoshida, N. (eds.) CONCUR 2004. LNCS, vol. 3170, pp. 292–307. Springer, Heidelberg (2004)

34. Darondeau, P., Degano, P.: Causal trees. In: ICALP 1989. LNCS, vol. 372. Springer, Heidelberg (1989)
35. Darondeau, P., Degano, P.: Causal trees: interleaving + causality. In: Guessarian, I. (ed.) LITP School 1990. LNCS, vol. 469. Springer, Heidelberg (1990)
36. Darondeau, P., Degano, P.: Refinement of actions in event structures and causal trees. Theoretical Computer Science 118(1), 21–48 (1993); Extended abstract. In: Rovan, B. (ed.) MFCS 1990. LNCS, vol. 452. Springer, Heidelberg (1990)
37. De Cindio, F., De Michelis, G., Pomello, L., Simone, C.: A Petri net model for CSP. In: Proceedings CIL 1981, Barcelona (1981)
38. De Cindio, F., De Michelis, G., Pomello, L., Simone, C.: Milner's Communicating Systems and Petri Nets. In: European Workshop on Applications and Theory of Petri Nets, pp. 40–59 (1982)
39. De Nicola, R., Ferrari, G., Pugliese, R.: Locality based LINDA: programming with explicit localities. In: Bidoit, M., Dauchet, M. (eds.) CAAP 1997, FASE 1997, and TAPSOFT 1997. LNCS, vol. 1214, Springer, Heidelberg (1997)
40. De Nicola, R., Ferrari, G., Pugliese, R.: KLAIM: a kernel language for agents interaction and mobility. IEEE Trans. on Software Engineering 24(5), 315–330 (1998)
41. De Nicola, R., Hennessy, M.: Testing equivalences for processes. Theoretical Computer Science 43, 83–133 (1984)
42. Degano, P., De Nicola, R., Montanari, U.: Partial ordering derivations for CCS. In: Budach, L. (ed.) FCT 1985. LNCS, vol. 199, pp. 520–533. Springer, Heidelberg (1985)
43. Degano, P., De Nicola, R., Montanari, U.: Observational equivalences for concurrency models. In: Proceedings 3rd IFIP WG 2.2 Working Conference, Ebberup 1986, North-Holland, Amsterdam (1987)
44. Degano, P., De Nicola, R., Montanari, U.: A distributed operational semantics for CCS based on condition/event systems. Acta Informatica 26(1/2), 59–91 (1988)
45. Degano, P., De Nicola, R., Montanari, U.: On the consistency of truly concurrent operational and denotational semantics. In: Proceedings LICS 1988, IEEE Computer Society Press, Los Alamitos (1988)
46. Degano, P., De Nicola, R., Montanari, U.: Partial orderings descriptions and observations of nondeterministic concurrent processes. In: de Bakker, J.W., de Roever, W.-P., Rozenberg, G. (eds.) REX Workshop. Linear Time, Branching Time and Partial Order in Logics and Models for Concurrency. LNCS, vol. 354, Springer, Heidelberg (1989)
47. Degano, P., De Nicola, R., Montanari, U.: A partial ordering semantics for CCS. Theoretical Computer Science 75(3), 223–262 (1990)
48. Degano, P., De Nicola, R., Montanari, U.: Universal axioms for bisimulations. Theoretical Computer Science 114, 63–91 (1993); Extended abstract. In: Proceedings 3rd Workshop on Concurrency and Compositionality, GMD-Studien Nr. 191 (1991)
49. Degano, P., Montanari, U.: Concurrent histories: A basis for observing distributed systems. Journal of Computer and System Sciences 34(2/3), 422–461 (1987)
50. Degano, P., Priami, C.: Proved trees. In: Kuich, W. (ed.) ICALP 1992. LNCS, vol. 623, Springer, Heidelberg (1992)
51. Ferrari, G.: Unifying Models of Concurrency. PhD thesis, University of Pisa (1990)
52. Ferrari, G., Gorrieri, R., Montanari, U.: An extended expansion theorem. In: Abramsky, S. (ed.) TAPSOFT 1991. LNCS, vol. 494. Springer, Heidelberg (1991)
53. Ferrari, G., Montanari, U.: Towards the unification of models of concurrency. In: Arnold, A. (ed.) CAAP 1990. LNCS, vol. 431, Springer, Heidelberg (1990)

54. Fournet, C., Gonthier, G.: The reflexive chemical abstract machine and the join-calculus. In: Proceedings POPL 1996, pp. 372–385 (1996)
55. Gischer, J.L.: Partial orders and the axiomatic theory of shuffle. PhD thesis, Stanford University (1984)
56. van Glabbeek, R.J.: The refinement theorem for ST-bisimulation semantics. In: Proceedings IFIP TC2 Working Conference on Programming Concepts and Methods, pp. 27–52. North-Holland, Amsterdam (1990)
57. van Glabbeek, R.J., Goltz, U.: Equivalence notions for concurrent systems and refinement of actions. In: Kreczmar, A., Mirkowska, G. (eds.) MFCS 1989. LNCS, vol. 379. Springer, Heidelberg (1989)
58. van Glabbeek, R.J., Goltz, U.: Equivalences and refinement. In: Guessarian, I. (ed.) LITP School 1990. LNCS, vol. 469. Springer, Heidelberg (1990)
59. van Glabbeek, R.J., Goltz, U.: Refinement of actions in causality based models. In: de Bakker, J.W., de Roever, W.-P., Rozenberg, G. (eds.) REX Workshop 1989. LNCS, vol. 430. Springer, Heidelberg (1990)
60. van Glabbeek, R.J., Goltz, U.: Refinement of actions and equivalence notions for concurrent systems. Acta Informatica 37(4/5), 229–327 (2001); Previously appeared as Research Report 6/98, University of Hildesheim (1998)
61. van Glabbeek, R.J., Goltz, U.: Well-behaved flow event structures for parallel composition and action refinement. Theoretical Computer Science 311(1-3), 463–478 (2004)
62. van Glabbeek, R.J., Vaandrager, F.W.: Petri net models for algebraic theories of concurrency. In: de Bakker, J.W., Nijman, A.J., Treleaven, P.C. (eds.) PARLE 1987. LNCS, vol. 259. Springer, Heidelberg (1987)
63. Goltz, U., Mycroft, A.: On the relationship of CCS and Petri nets. In: Paredaens, J. (ed.) ICALP 1984. LNCS, vol. 172. Springer, Heidelberg (1984)
64. Goltz, U., Reisig, W.: CSP programs as nets with individual tokens. In: Rozenberg, G. (ed.) APN 1984. LNCS, vol. 188. pp. 169–196. Springer, Heidelberg (1985)
65. Gorrieri, R.: Refinement, Atomicity and Transactions for Process Description Languages. PhD thesis, University of Pisa (1991)
66. Gorrieri, R., Marchetti, S., Montanari, U.: A^2CCS: Atomic actions for CCS. Theoretical Computer Science 72, 203–223 (1990); Extended abstract. In: Dauchet, M., Nivat, M. (eds.) CAAP 1988. LNCS, vol. 299. Springer, Heidelberg (1988)
67. Gorrieri, R., Montanari, U.: SCONE: A simple calculus of nets. In: Baeten, J.C.M., Klop, J.W. (eds.) CONCUR 1990. LNCS, vol. 458. Springer, Heidelberg (1990)
68. Gorrieri, R., Montanari, U.: Towards hierarchical description of systems: a proof system for strong prefixing. Int. Journal of Foundations of Computer Science 1(3), 277–293 (1990)
69. Gorrieri, R., Rensink, A.: Action refinement. In: Bergstra, J., Ponse, A., Smolka, S. (eds.) Handbook of Process Algebra, pp. 1047–1147. North-Holland, Amsterdam (2001)
70. Grabowski, J.: On partial languages. Fundamenta Informaticae 4(1), 427–498 (1981)
71. Hennessy, M.: On the relationship between time and interleaving. Sophia-Antipolis (1980) (unpublished draft)
72. Hennessy, M.: Axiomatising finite concurrent processes. SIAM Journal of Computing 17(5), 997–1017 (1988)
73. Hennessy, M.: A Distributed Pi-calculus. Cambridge University Press, Cambridge (2007)

74. Hennessy, M., Milner, R.: Algebraic laws for nondeterminism and concurrency. JACM 32 (1985)
75. Hoare, C.A.R.: Communicating Sequential Processes. Prentice-Hall, Englewood Cliffs (1985)
76. Joyal, A., Nielsen, M., Winskel, G.: Bisimulation and open maps. In: Proceedings LICS 1993, pp. 418–427. IEEE Computer Society Press, Los Alamitos (1993)
77. Kiehn, A.: Concurrency in process algebras. Habilitation Thesis, Technische Universität München (1999)
78. Kiehn, A.: Distributed bisimulations for finite CCS. Report 7/89, University of Sussex (1989)
79. Kiehn, A.: Local and global causes. Technical Report 342/23/91, Technische Universität München (1991)
80. Kiehn, A.: Comparing locality and causality based equivalences. Acta Informatica 31, 697–718 (1994)
81. Langerak, R.: Bundle event structures: a non-interleaving semantics for LOTOS. In: Proceedings IFIP TC6/WG6.1 Fifth International Conference on Formal Description Techniques for Distributed Systems and Communication Protocols (1992); Previously appeared as University of Twente Research Report (1991)
82. Loogen, R., Goltz, U.: Modelling nondeterministic concurrent processes with event structures. Fundamenta Informaticae XIV(1), 39–74 (1991)
83. Mazurkiewicz, A.: Concurrent program schemes and their interpretation. Report DAIMI PB-78, Aarhus University (1977)
84. Mazurkiewicz, A.: Trace theory. In: Brauer, W., Reisig, W., Rozenberg, G. (eds.) APN 1986. LNCS, vol. 255. Springer, Heidelberg (1987)
85. Milner, R.: A Calculus of Communication Systems. LNCS, vol. 92. Springer, Heidelberg (1980)
86. Milner, R.: Communication and Concurrency. Prentice-Hall, Englewood Cliffs (1989)
87. Montanari, U., Boudol, G., Castellani, I., Ferrari, G., Hennessy, M., Nielsen, M., Winskel, G.: ESPRIT Basic Research Action n. 3011 CEDISYS - Final Report (1992), `http://www-sop.inria.fr/mimosa/personnel/Ilaria.Castellani/CEDISYS-final-report.pdf`
88. Montanari, U., Sgamma, M.: Canonical representatives for observational equivalence. In: Proceedings Colloquium on the Resolution of Equations in Algebraic Structures, pp. 292–319. Academic Press, London (1989)
89. Montanari, U., Yankelevich, D.: A parametric approach to localities. In: Kuich, W. (ed.) ICALP 1992. LNCS, vol. 623. Springer, Heidelberg (1992)
90. Montanari, U., Yankelevich, D.: Location equivalence in a parametric setting. Theoretical Computer Science 149. 299–332 (1995)
91. Mukund, M., Nielsen, M.: CCS, locations and asynchronous transition systems. In: Shyamasundar, R.K. (ed.) FSTTCS 1992. LNCS, vol. 652. Springer, Heidelberg (1992)
92. Nielsen, M.: CCS and its relationship to net theory. In: Brauer, W., Reisig, W., Rozenberg, G. (eds.) APN 1986. LNCS, vol. 255, Springer, Heidelberg (1987)
93. Nielsen, M., Engberg, U.H., Larsen, K.S.: Fully abstract models for a process language with refinement. In: de Bakker, J.W., de Roever, W.-P., Rozenberg, G. (eds.) Linear Time, Branching Time and Partial Order in Logics and Models for Concurrency. LNCS, vol. 354. Springer, Heidelberg (1989)
94. Nielsen, M., Plotkin, G., Winskel, G.: Petri nets, event structures and domains, Part I. Theoretical Computer Science 13(1), 85–108 (1981)

95. Nielsen, M., Rozenberg, G., Thiagarajan, P.S.: Behavioural notions for elementary net systems. Distributed Computing 4, 45–57 (1990)
96. Nielsen, M., Rozenberg, G., Thiagarajan, P.S.: Elementary transition systems. Theoretical Computer Science 96(1), 3–33 (1992)
97. Nielsen, M., Rozenberg, G., Thiagarajan, P.S.: Transition systems, event structures and unfoldings. Information and Computation 118(2), 191–207 (1995)
98. Olderog, E.-R.: Operational Petri net semantics for CCSP. In: Rozenberg, G. (ed.) APN 1987. LNCS, vol. 266. Springer, Heidelberg (1987)
99. Park, D.: Concurrency and automata on infinite sequences. In: Deussen, P. (ed.) GI-TCS 1981. LNCS, vol. 104. Springer, Heidelberg (1981)
100. Petri, C.A.: Nonsequential processes. Research Report 77-05, GMD, Sankt Augustin (1977)
101. Phillips, I., Ulidowski, I.: Reversibility and models for concurrency. Electron. Notes Theor. Comput. Sci. 192(1), 93–108 (2007)
102. Plotkin, G.D.: A structural approach to operational semantics. Report DAIMI FN-19, Computer Science Department, Aarhus University (1981)
103. Pratt, V.R.: On the composition of processes. In: Proceedings POPL 1982 (1982)
104. Pratt, V.R.: Modelling concurrency with partial orders. Journal of Parallel Programming 15(1) (1986)
105. Priami, C.: Enhanced Operational Semantics for Concurrency. PhD thesis, University of Pisa (1996)
106. Priami, C., Plotkin, G. (eds.): Transactions on Computational Systems Biology VI. LNCS, vol. 4220. Springer, Heidelberg (2006)
107. Rabinovich, A., Trakhtenbrot, B.A.: Behavior structures and nets. Fundamenta Informaticae XI(4), 357–404 (1988)
108. Reisig, W.: Petri Nets. EATCS Monographs on Theoretical Computer Science (1985)
109. Stark, E.W.: Connections between a concrete and an abstract model of concurrent systems. In: Schmidt, D.A., Main, M.G., Melton, A.C., Mislove, M.W. (eds.) MFPS 1989. LNCS, vol. 442, pp. 53–79. Springer, Heidelberg (1990)
110. Winskel, G.: Events in Computation. PhD thesis, University of Edinburgh (1980)
111. Winskel, G.: Event structure semantics for CCS and related languages. In: Nielsen, M., Schmidt, E.M. (eds.) ICALP 1982. LNCS, vol. 140. Springer, Heidelberg (1982)
112. Winskel, G.: Categories of models for concurrency. In: Brookes, S.D., Winskel, G., Roscoe, A.W. (eds.) Seminar on Concurrency. LNCS, vol. 197, Springer, Heidelberg (1985)
113. Winskel, G.: A new definition of morphism on Petri nets. In: Fontet, M., Mehlhorn, K. (eds.) STACS 1984. LNCS, vol. 166, pp. 140–150. Springer, Heidelberg (1984)
114. Winskel, G.: Event structures. In: Brauer, W., Reisig, W., Rozenberg, G. (eds.) APN 1986. LNCS, vol. 255, pp. 325–392. Springer, Heidelberg (1987)
115. Winskel, G.: Petri nets, algebras, morphisms and compositionality. Information and Control 72, 197–238 (1987)
116. Winskel, G.: Symmetry and concurrency. In: Mossakowski, T., Montanari, U., Haveraaen, M. (eds.) CALCO 2007. LNCS, vol. 4624, pp. 40–64. Springer, Heidelberg (2007)
117. Winskel, G., Nielsen, M.: Models for concurrency. In: Handbook of Logic in Computer Science, Oxford, vol. 4, pp. 1–148 (1995)
118. Yankelevich, D.: Parametric Views of Process Description Languages. PhD thesis, University of Pisa (1993)

Some Theorems Concerning the Core Function

Angelo Raffaele Meo

Politecnico di Torino

Abstract. In a preceding paper an NP-complete problem has been discussed pertaining to a function, called "core function", which plays an important role in the well known Boolean satisfiability problem (see the first item in the references list). In this paper, some theorems concerning the minimal Boolean implementation of the core function are proved.

1   A Theorem of Boolean Isotonic Functions

Let $\mathbf{f(x_1,x_2, ..., x_n)}$ be an isotonic Boolean function, that is a Boolean function which can be implemented with only **AND** and **OR** gates.

Let $\mathbf{I}_{min}$ be one of its minimum cost implementations among all the implementations of $\mathbf{f}$, the cost being defined as the total number of **AND**, **OR** or **NOT** gates. Let $\mathbf{C}_{min}$ be the cost of $\mathbf{I}_{min}$.

Theorem 1. *There exists always an implementation J of f containing only **AND** and **OR** gates, applied to both complemented and uncomplemented input variables, such that cost(J) $\leq$ 2 * C_{min}*

This theorem is nearly obvious (a proof is reported in Appendix A). It is will be used in the analysis of core function.

2   The Question "P = NP?"

The simple theorem proved in the preceding section will be applied to the analysis of a well known problem, namely the question "**P = NP?**".

A brief description of the definitions and properties well known among the scientists of modern computational complexity theory which will be made reference to, is presented in this section.

P denotes the class of all the decision problems which can be solved in polynomial time.

NP denotes the class of all the decision problems **f** satisfying the property that the function **check(f)** analyzing a witness of the decision problem is polynomial time decidable.

"**P=NP?**", or, in other terms, "Is **P** a proper subset of **NP**?", is one of the most important open questions in modern computational complexity theory.

A decision problem **C** in **NP** is **NP**-complete if it is in **NP** and if every other problem **L** in **NP** is reducible to it in the sense that there is a polynomial time

P. Degano et al. (Eds.): Montanari Festschrift, LNCS 5065, pp. 778–796, 2008.

algorithm which transforms instances of $\mathbf{L}$ into instances of $\mathbf{C}$ producing the same values.

The importance of $\mathbf{NP}$-completeness derives from the fact that, if we find a polynomial time algorithm for just one $\mathbf{NP}$-complete problem, then we can construct polynomial time algorithms for all the problems in $\mathbf{NP}$ and, conversely, if any single $\mathbf{NP}$-complete problem does not have a polynomial time algorithm, than no $\mathbf{NP}$-complete problem has a polynomial time solution.

The analysis discussed in this paper will be based on a well known $\mathbf{NP}$-complete problem which is called "satisfiability problem" or $\mathbf{SAT}$.

Given a Boolean expression containing only the names of a set of variables, the operators $\mathbf{AND}$, $\mathbf{OR}$ and $\mathbf{NOT}$, and parentheses, is there an assignment of $\mathbf{TRUE}$ and $\mathbf{FALSE}$ values to the variables which makes the entire expression $\mathbf{TRUE}$?

It is well known that the problem remains $\mathbf{NP}$-complete also when all the expressions are written in "conjunctive normal form" with $\mathbf{3}$ variables per clause. In this case, the analyzed expressions will be of the type

$$(\mathbf{x}_{11} \ \mathbf{OR} \ \mathbf{x}_{12} \ \mathbf{OR} \ \mathbf{x}_{13}) \ \mathbf{AND}$$
$$(\mathbf{x}_{21} \ \mathbf{OR} \ \mathbf{x}_{22} \ \mathbf{OR} \ \mathbf{x}_{23}) \ \mathbf{AND} \tag{1}$$
$$\cdots \cdots \cdots \cdots \cdots$$
$$(\mathbf{x}_{t1} \ \mathbf{OR} \ \mathbf{x}_{t2} \ \mathbf{OR} \ \mathbf{x}_{t3})$$

where

- $\mathbf{t}$ is the number of clauses or triplets;
- each $\mathbf{x}_{ij}$ is a variable with or without a $\mathbf{NOT}$ operator in front of it;
- each variable can appear multiple times in the expression.

If the deterministic Turing machine is assumed as the computational model, with $\{\mathbf{0,1,b}\}$ as its set of input symbols, the input data appearing on the tape at the beginning of computation can represent the data of expression $(\mathbf{1})$ in the following way

$$\ldots \mathbf{b} \ \mathbf{b} \ \langle\mathbf{binary \ code \ of \ number \ of \ variables}\rangle \ \langle\mathbf{separator}\rangle$$
$$\mathbf{s}_{11} \ \mathbf{n}_{111} \ \mathbf{n}_{112} \cdots \mathbf{n}_{11m}$$
$$\mathbf{s}_{12} \ \mathbf{n}_{121} \ \mathbf{n}_{122} \cdots \mathbf{n}_{12m}$$
$$\mathbf{s}_{13}\mathbf{n}_{131} \ \mathbf{n}_{132}\cdots \mathbf{n}_{13m} \tag{2}$$
$$\mathbf{s}_{21} \ \mathbf{n}_{211} \ \mathbf{n}_{212} \cdots \mathbf{n}_{21m}$$
$$\cdots \cdots \cdots \cdots \cdots \cdots \cdots \cdots \cdots \cdots \cdots \cdots$$
$$\mathbf{s}_{t3} \ \mathbf{n}_{t31} \ \mathbf{n}_{t32} \cdots \mathbf{n}_{t3m} \ \mathbf{b} \ \mathbf{b} \ \ldots$$

where

- $\mathbf{b}$ is the blank symbol;
- $\mathbf{t}$ is the number of triplets;
- $\mathbf{s}_{ij}$ denotes the sign of variable $\mathbf{x}_{ij}$ (with $\mathbf{s}_{ij} = \mathbf{1}$ denoting that $\mathbf{x}_{ij}$ is preceded by operator $\mathbf{NOT}$);

- n_{ijk} denotes the **k**-th component of the binary code $<n_{ij1}\ n_{ij2}\ \ldots\ n_{ijm}>$ representing the name of variable x_{ij};
- the binary code of the number n_v of variables is needed in order to determine the number **m** of binary digits necessary to represent the names of variables according the rule **m**=minimun integer not less than $\log_2 n_v$;
- the binary code of the separator must not be included in the binary code of the number of variables.

Notice that, by neglecting the bits of the binary code of the number of variables and the bits of the separator, the number of input bits on the tape will be

$$\mathbf{n = t * 3 * (1 + \text{minimum integer not less than } \log_2 3^*t)} \qquad (3)$$

since the maximun value of the number of variables is **t*3**.

The properties of Turing machines processing the bit string described by (**2**) will be analyzed in this paper with reference to a family $\{C_n\}$ of Boolean circuits. C_n has **n** binary inputs and produces the same binary output as the corresponding Turing machine.

The equivalence between a deterministic Turing machine **M** processing some input **x** belonging to $\{0,1\}^n$ and an **n**-input Boolean circuit C_n is well known. It is also known that the number of gates, or **AND, OR, NOT** operators, appearing in circuit C_n, is polynomial in the running time $T_M(n)$ of the corresponding Turing machine. We shall prove here that the number of gates of the circuit C_n increases exponentially with **n**; it follows that also $T_M(n)$ increases exponentially.

3   The Core Function

In the case of satisfiability problem with **3** variables for clause, Boolean circuit C_n has **n** inputs which the binary data described in (**2**) are applied to. (Of course, the binary code of the number of variables and the separator are not needed). The output of C_n will take the value **TRUE** when, and only when, there is an assignment of values **TRUE** and **FALSE** to variables making expression (**1**) **TRUE**.

In order to simplify analysis, circuit C_n will be decomposed into two processing layers as shown in Fig. **1**.

A variable **j** of triplet **i** will be defined as "compatible" with variable **k** of triplet **h** when, and only when, either the sign s_{ij} of the former variable is equal to the sign s_{hk} of the latter or the name $<n_{ij1}\ n_{ij2}\ \ldots\ n_{ijm}>$ of the former is different from the name $<n_{hk1}\ n_{hk2}\ldots n_{hkm}>$ of the latter. From that definition it follows that two not compatible variables have different signs and the same name; therefore, their **AND** is identically **FALSE**.

The compatibility layer is composed of **3*t*(3*t-3)/2** identical cells, one for each pair of variables belonging to different triplets.

As shown in Fig. **2**, the inputs of a cell will be the sign s_{ij} and the name $<n_{ij1}\ n_{ij2}\ \ldots n_{ijm}>$ of variable **j** of triplet **i**, and the sign s_{hk} and the name $<n_{hk1}\ n_{hk2}\ldots n_{hkm}>$ of variable **k** of triplet **h**. The ouput of the same cell c_{ijhk} will be **TRUE** when, and only when, the two variables are compatible between themselves.

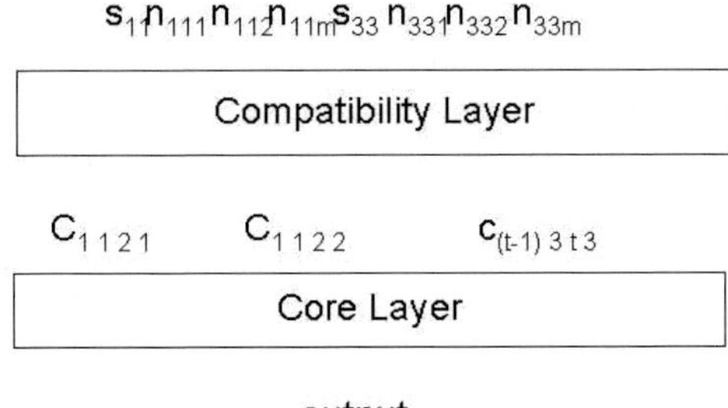

Fig. 1. Decomposition of Boolean circuit $\mathbf{C}_n$ into the compatibility layer and core layer

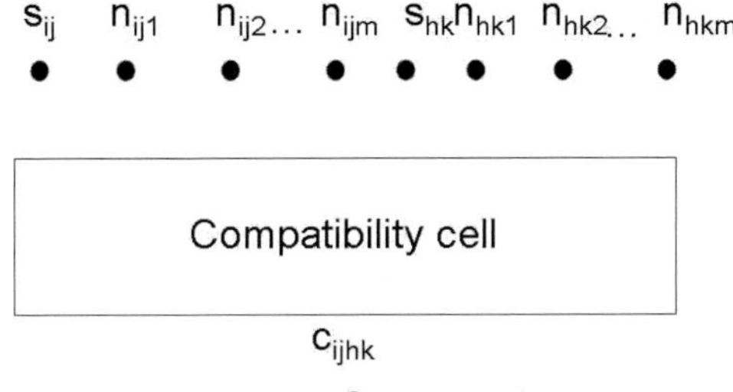

Fig. 2. Compatibility cell

Variable $\mathbf{c}_{ijhk}$ will be called a compatibility variable or simply a compatibility.

The core layer processes only the $\mathbf{3*t*(3*t\text{-}3)/2}$ compatibility variables $\mathbf{c}_{ijhk}$ and produces the global result of computation.

As the circuit $\mathbf{C}_n$, also the global Boolean function implemented by $\mathbf{C}_n$ may be decomposed into two layers of functions. At the compatibility layer, the function implemented by a cell may be written as follows (by using the symbols $\mathbf{*}$, $\mathbf{+}$, and $\mathbf{!}$ for representing $\mathbf{AND, OR}$ and $\mathbf{NOT}$ operators):

$$
\begin{aligned}
\mathbf{c}_{ijhk} = \ & \mathbf{s}_{ij} * \mathbf{s}_{hk} + !\mathbf{s}_{ij} * !\mathbf{s}_{hk} + \\
& + \mathbf{n}_{ij1} * !\mathbf{n}_{hk1} + !\mathbf{n}_{ij1} * \mathbf{n}_{hk1} + \\
& + \mathbf{n}_{ij2} * !\mathbf{n}_{hk2} + !\mathbf{n}_{ij2} * \mathbf{n}_{hk2} + \\
& \cdots\cdots\cdots\cdots\cdots\cdots\cdots\cdots \\
& + \mathbf{n}_{ijm} * !\mathbf{n}_{hkm} + !\mathbf{n}_{ijm} * \mathbf{n}_{hkm}
\end{aligned}
\tag{4}
$$

The Boolean function implemented by the core layer will be called the "Core Function" of order $\mathbf{t}$, where $\mathbf{t}$ is the number of triplets. It will be denoted with the symbol $\mathbf{CF(t)}$. The core function can be determined by proceeding as follows.

Consider one selection of variables appearing in $(\mathbf{1})$, one and only one for each triplet, for all the triplets. Let

$$
\mathbf{1i}_1, \ \mathbf{2i}_2, \ \ldots, \ \mathbf{ti}_t
\tag{5}
$$

(with $\mathbf{i_1}$, $\mathbf{i_2}$, ..., $\mathbf{i_t}$ ∈ $\{\mathbf{1,2,3}\}$) be the indexes <number of triplet, number of variable in the triplet> of the selected variables. Let Π be the product of all the compatibility variables relative to all the pairs of the selected variables

$$\mathbf{P} = \mathbf{c}_{1i12i2} * \mathbf{c}_{1i13i3} * \ldots * \mathbf{c}_{t-1it-1tit} \tag{6}$$

The core function can be defined as the sum

$$\sum \Pi \tag{7}$$

of the products (**6**) relative to all the selections (**5**).

For example, in the case of **3** triplets, the core function can be defined as follows:

$$
\begin{aligned}
\mathbf{CF(3)} &= \mathbf{c}_{1121} * \mathbf{c}_{1131} * \mathbf{c}_{2131} + \\
\mathbf{c}_{1121} &* \mathbf{c}_{1132} * \mathbf{c}_{2132} + \\
\mathbf{c}_{1121} &* \mathbf{c}_{1133} * \mathbf{c}_{2133} + \\
\mathbf{c}_{1122} &* \mathbf{c}_{1131} * \mathbf{c}_{2231} + \\
&...\textbf{(other 22 products)}... \\
\mathbf{c}_{1323} &* \mathbf{c}_{1333} * \mathbf{c}_{2333}
\end{aligned}
\tag{8}
$$

It is easy to prove that Eq. (**1**) is **TRUE** when, and only when, the core function takes the value **TRUE**.

Notice that the processing work of the cell of Fig. **2** increases with the logarithm of the number of the variables since such is the increment of the code of the name of a variable. Therefore, the total processing work of the compatibility layer can be written as:

$$\mathbf{K} * \mathbf{3{*}t} * (\mathbf{3{*}t} - \mathbf{3}) * \log_2(\mathbf{3{*}t})$$

where $\mathbf{3{*}t} * (\mathbf{3{*}t} - \mathbf{3})/\mathbf{2}$ is the total number of the compatibility cells.

Besides, the problem solved by the core layer is clearly in **NP**, because it is easy to verify a witness solution. It follows that, since the compatibility layer polinomially reduces an **NP**-complete problem (**3SAT**) to the problem solved by the core layer, the core function describes a new **NP**-complete problem. This will be the problem discussed in the following sessions.

4   Properties of the Core Function

In a previous paper [**1**] the following properties of a core function have been proved.

Lemma 1 (Property 1). *The core function CF is totally isotonic.*

Lemma 2 (Property 2). *Any product (6) is a prime implicant of the core function (that is, a product of variables which is included by no other).*

Lemma 3 (Property 3). *Since the different selections of each of variables (5) are 3, the number of prime implicants of the core function is equal to 3^t. Each of these prime implicants is essential and is the product of $t*(t-1)/2$ compatibilities.*

In the next section, reference will be made to the following definitions.

Definition 1 (spurious term). *A pair of compatibility variables $\{c_{hklm}, c_{pqrs}\}$ is defined as a spurious pair if*

- *$h = p$ and $k \neq q$; or*
- *$h = r$ and $k \neq s$; or*
- *$l = p$ and $m \neq q$; or*
- *$l = r$ and $m \neq s$.*

For example, the pair $\{c_{1121}, c_{1231}\}$ is a spurious pair since the triplet **1** is associated to two different indexes of variables (**1** and **2**).

A spurious term is a product of compatibility variables countaining the elements of one or more than one spurious pair.

For example, the term

$$c_{1121} * c_{1231} * c_{2131}$$

is a spurious term since it contains the elements of the spurious pair $\{c_{1121}, c_{1231}\}$.

Definition 2 (impure term). *A term T of the core function CF (namely, a product of literals implying CF), which contains one or more complemented variables, will be defined as an impure term. The product of all the uncomplemented variables of T will be defined as the core of T.*

Definition 3 (mark). *Consider a not spurious subset of the variables listed in (6) satisfying the property that each of the indexes of triplet appears at least once in some variable. The product of the variables of such a subset will defined as a "mark" or a "not spurious mark" of the prime implicant which it is part of.*

For example, in the case of CF (4), the product

$$M = c_{1a2b} * c_{1a3c} * c_{1a4d} \tag{9}$$

(where **a, b, c, d** are elements of $\{1,2,3\}$) is a "not spurious mark" of the prime implicant

$$P = c_{1a2b} * c_{1a3c} * c_{1a4d} * c_{2b3c} * c_{2b4d} * c_{3c4d} \tag{10}$$

since all the indexes of triplet appear at least once in (9).

A spurious term in which all the indexes of triplet appear at least once will be called a "spurious mark". Notice that a spurious mark may be the mark of more than one prime implicant. For the example, in the case of $CF(3)$,

$$c_{1121} * c_{1131} * c_{1122}$$

is a spurious mark of both the prime implicants

$$c_{1121} * c_{1131} * c_{2131}$$

and

$$c_{1122} * c_{1131} * c_{2231}.$$

An impure term whose core is a (possibly spurious) mark will be a defined as a (possibly spurious) impure mark.

Definition 4 (extended prime implicants). *A term T of the core function which contains all the uncomplemented literals of a prime implicant will be defined as an "extended prime implicant".*

It may be a spurious prime implicant or an impure prime implicant or both a spurious and impure prime implicant.

Notice that an extended prime implicant can be viewed as a (possibly spurious or impure) mark.

Definition 5 (remainder). *A term which is neither a possibly spurious or impure mark nor an extended prime implicant will be called a "remainder". Generally, a remainder can be associated to one or more prime implicants.*

For example,
$$\mathbf{R} = \mathbf{c}_{2b3c} * \mathbf{c}_{2b4d} * \mathbf{c}_{3c4d} \tag{11}$$
is a remainder of the prime implicant (**10**). As will be shown shortly, a neither spurious or impure remainder $\mathbf{R}$ "belongs" to more than one prime implicant, in the sense that
$$\mathbf{R} \geq \mathbf{P}_i$$
for a set $\{\mathbf{P}_i\}$ of prime implicants of the core function.

For example, in the case of $\mathbf{CF(3)}$, $\mathbf{c}_{2131}$ is a remainder of the prime implicants
$$\mathbf{P}_1 = \mathbf{c}_{1121} * \mathbf{c}_{1131} * \mathbf{c}_{2131}$$
$$\mathbf{P}_2 = \mathbf{c}_{1221} * \mathbf{c}_{1231} * \mathbf{c}_{2131} \tag{12}$$
$$\mathbf{P}_3 = \mathbf{c}_{1321} * \mathbf{c}_{1331} * \mathbf{c}_{2131}$$
On the definitions of mark and remainder the following properties are based.

Lemma 4 (Property 4). *A not spurious mark M specifies a corresponding prime implicant P uniquely. Indeed, if all the indexes of triplet appear in M, the product (6) is completely defined.*

We shall write
$$\mathbf{P} = \mathbf{I(M)}$$
to state that $\mathbf{P}$ is the prime implicant specified by $\mathbf{M}$.

The same equation will be written to state that $\mathbf{P}$ is one of the prime implicants specified by the spurious mark $\mathbf{M}$.

As already mentioned, a remainder $\mathbf{R}$ does not specify a corresponding prime implicant uniquely. In the example relative to $\mathbf{CF(3)}$ above described, three prime implicants correspond to $\mathbf{c}_{2131}$, as shown by (**12**), since a single index of triplet is missing in that remainder. In general, if $\mathbf{z}$ triplets are not involved in $\mathbf{R}$, there are $\mathbf{3}^z$ different ways of involving the missing triplets. Hence the following property follows.

Lemma 5 (Property 5). *A not spurious remainder R in which the indexes of z triplets are missing corresponds to 3^z different prime implicants.*

Finally, the following property can be proved.

Lemma 6 (Property 6). *Let P_1 and P_2 be two products of variables such that $P_1 * P_2$ is equal to a prime implicant P of a core function. Either P_1 or P_2 is a mark of P.*

5   The Reference Architecture

The theorem proved in Appendix A makes it possible to simplify the proof of the fact that the cost of the core function increases exponentially with the number of variables.

Indeed, we shall consider the implementations of the core function with **AND** and **OR** gates only (isotonic implementations). We shall prove that no isotonic implementation exists which increases not exponentially with the number of variables; hence, we shall derive that no implementation of general type (with **NOT** gates also) exists which increases with the number of variables exponentially.

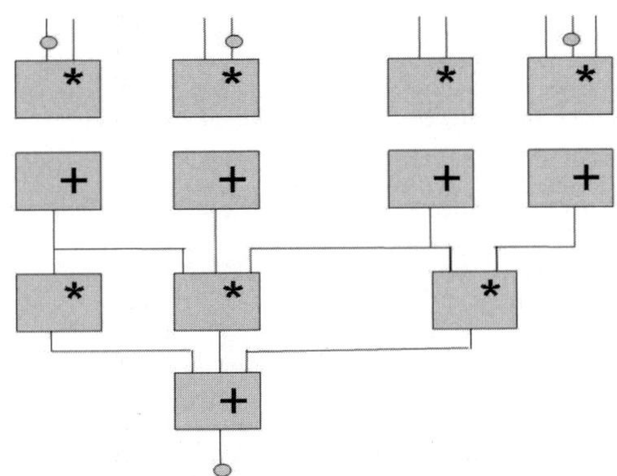

Fig. 3. The reference architecture

Fig. **3** shows the related architecture characterized by a number of subnetworks each of which has the structure shown by Fig. **4**. As an alternative, the network of Fig. **3** might be composed by a single network of the type of Fig. **4**.

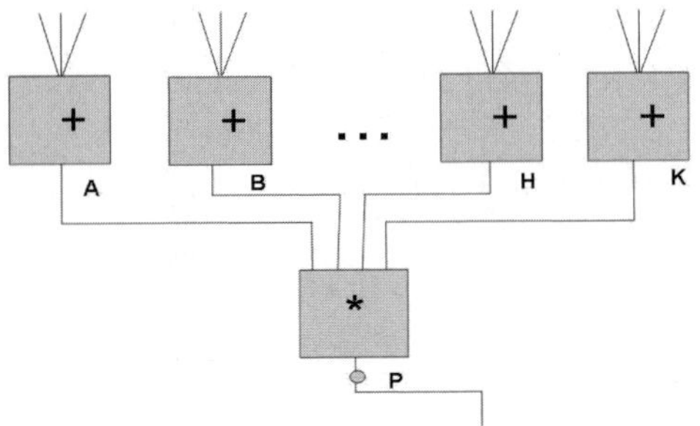

Fig. 4. The structure of a PCA

Each of the circuits presented in Fig. **4** will be called a "primary composite addendum (**PCA**)" and each of the **OR** gates together with its input gates will be called a "primary factor of a primary composite addendum (**PCAF**)" (Fig. **5**).

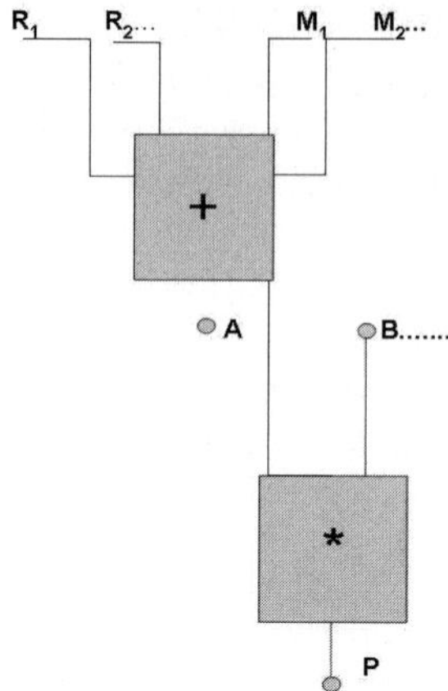

Fig. 5. A structure of a PCAF

Consider the Boolean values of the functions implemented by the circuits preceding nodes **A, B, H, K, P,**
> **val (A)**
> **val (B)**
> ....
> **val (H)**
> **val (P)** = **val (A)** * **val (B)** *...* **val (H)** * **val (K)**

Each of these Boolean values can be expressed as the sum of its prime implicants and, therefore, as a sum of terms of the type:
> $(\mathbf{M}_1 + \mathbf{M}_2 + \mathbf{M}_3 + ... + \mathbf{R}_1 + \mathbf{R}_2 + \mathbf{R}_3 + ...)$

where the $\mathbf{M}_i$'s are marks (or prime implicants) and the $\mathbf{R}_i$'s are remainders.

It is easy to prove that the following four different cases may occur

1. **val (A)** * **val (B)** *...***val (H)** = $\mathbf{M}_{11} + \mathbf{M}_{12} + \mathbf{M}_{13} + ...$
   and
   val (K) = $\mathbf{M}_{21} + \mathbf{M}_{22} + \mathbf{M}_{23} + ... + \mathbf{R}_{21} + \mathbf{R}_{22} + \mathbf{R}_{23} + ...$
2. **val (A)** * **val (B)** *...* **val (H)** = $\mathbf{M}_{11} + \mathbf{M}_{12} + \mathbf{M}_{13} + ...$
   and
   val (K) = $\mathbf{M}_{21} + \mathbf{M}_{22} + \mathbf{M}_{23} + ...$
3. **val (A)** * **val (B)** *...***val (H)** = $\mathbf{M}_{11} + \mathbf{M}_{12} + \mathbf{M}_{13} + ... + \mathbf{R}_{11} + \mathbf{R}_{12} + \mathbf{R}_{13} + ...$
   and
   val (K) = $\mathbf{M}_{21} + \mathbf{M}_{22} + \mathbf{M}_{23} + ...$
4. **val (A)** * **val (B)** *...* **val (H)** = $\mathbf{M}_{11} + \mathbf{M}_{12} + \mathbf{M}_{13} + ...\mathbf{R}_{11} + \mathbf{R}_{12} + \mathbf{R}_{13} + ...$
   and
   val (K) = $\mathbf{M}_{21} + \mathbf{M}_{22} + \mathbf{M}_{23} + ... + \mathbf{R}_{21} + \mathbf{R}_{22} + \mathbf{R}_{23} + ...$
   but, for any **i, j,**
   $\mathbf{R}_{1i} * \mathbf{R}_{2j} = \mathbf{0}.$

Indeed, if any $\mathbf{R}_{1j}$ * $\mathbf{R}_{2j}$ were different fron **0** identically, val (**P**) could not the core function.

For the sake of brevity, we shall restrict analysis to case **1**, leaving the reader the task of analyzing the other three cases by the same arguments we shall apply to case **1**.

Besides, a product $\mathbf{M}_{1j}$ * $\mathbf{M}_{2j}$ must generate a possibly extended prime implicant $\mathbf{I}(\mathbf{M}_{1j})$. We shall assume that for any i

$\mathbf{M}_{1i}$ * $\mathbf{M}_{zj} = \mathbf{I}(\mathbf{M}_{1j})$.

The extension to the general case has no conceptual relevance and is left to the reader for the sake of brevity.

As mentioned, a prime implicant can be interpreted as a mark; so, in the preceding relationships some mark $\mathbf{M}_{ij}$ might be a prime implicant.

6   The Merit and Cost of a PCA

The Boolean function implemented by a **PCA** is the sum of prime implicants, some of which are extended.

An extended prime implicant contains only a fraction of the minterms contained in the corresponding not extended prime implicant. Therefore, we shall define the merit of a not extended prime implicant as the value 1 and the merit of an extended prime implicant as the fraction of minterms of the corresponding prime implicants it covers.

Finally, the merit of a **PCA** will be defined as the sum of the merits of all its extendended and not extended prime implicants.

Generally, the cost of a **PCA** will be defined as the number of **OR** or **AND** gates it contains. However, if a gate or a subnetwork of gates may be used in another **PCA**, its cost will be divided by the number of different prime implicants it might contribute to implementing.

The definitions of merit and cost will be used in the evaluation of the total cost of an implementation of core function.

7   The Structure of PCA's and PCAF's

Any product $\mathbf{M}_{1i}$ * $\mathbf{M}_{2j}$ or $\mathbf{M}_{1i}$* $\mathbf{R}_{2j}$ must coincide with a prime implicant (or with an extended prime implicant) of the core function.

First consider the products of the type $\mathbf{M}_{1i}$ * $\mathbf{R}_{2j}$. It is easy to prove that any product of this family must coincide with one of the prime implicants $\mathbf{I}(\mathbf{M}_{1j})$ generated by $\mathbf{M}_{1i}$.

Notice that a spurious mark might generate a term of the core function which is the extension of two or more prime implicants. However, this case will not be considered here since the number of minterms contained in such an extension decreases very quickly with the number of variables of core function.

A mark might be a spurious term. However, we shall assume that the number of literals making it a spurious term is less than e**N** where e is arbitrarily small and **N** is the total number of compatibilities.

In what follows we shall need a new definition.

Definition 6 (complete and nearly complete marks). *A mark* **M** *will be defined as "complete in a given variable of a given triplet" if it contains all the compatibilities involving that variable.*

For example the mark

$$c_{1121} \quad * \quad c_{1131}$$

of **CF(3)** *is complete in nodo 11.*

A mark **M** *will be defined as "nearly complete in a given variable of a given triplet" if it contains nearly all the compatibilities involving that variable, the exceptions being at maximum equal to* **e*N** *where* **e** *is arbitrarily small and* **N** *the total number of compatibilities.*

On this definition the following theorem is based.

Theorem 2. *Consider two products of the following type:*

$$M_{11} \; * \; R_{21} = I(M_{11})$$
$$M_{12} \; * \; R_{22} = I(M_{12})$$

If **I(M$_{11}$)** *is different from* **I(M$_{12}$)**, *then* **M$_{11}$** *and* **M$_{12}$** *must be nearly complete in two different variables.*

Proof. For the sake of brevity we shall present here only an informal proof. More formal definitions and proofs concerning nearly complete marks and terms can be found in Appendix B.

Assume that M_{11} is complete in a given variable of a given triplet and M_{12} is nearly complete in *another variable* with the exception of a compatibiblity **c**.

Since also $M_{12} * R_{21}$ must be equal to $I(M_{12})$, then R_{21} must contain that compatibility **c**. As a consequence the product $M_{11}* R_{21}$ will contain **c** and, therefore, it will be spurious.

Its merit will be **1/2.**

Assume now that M_{11} is complete in a given variable and M_{12} is nearly complete in another variable with the exceptions of two compatibilities c_1 and c_2. In this case R_{21} must contain both c_1 and c_2, and its merit will be 1/4.

As a synthesis, the merit decreases very quickly with **e*N**.

The previous theorem can be extended to the products of the type $M_{1i} * M_{2j}$ according the following theorem.

Theorem 3. *Consider the products*

$$M_{11} \; * \; M_{21} \text{ and } M_{12} \; * \; M_{22}$$

Then, **M$_{11}$** * **M$_{21}$** *and* **M$_{12}$** * **M$_{22}$** *must be equal to the prime implicants specified by* **M$_{11}$** *and* **M$_{12}$**, *respectively. Besides,* **M$_{11}$** *and* **M$_{12}$** *must be nearly complete in two different variables.*

As a consequence of the preceding theorems we can state the following theorem.

Theorem 4. *If condition 1 of Section 4 is satisfied and the hypothesis is assumed that no complemented variables occur,* **val (A) * val (B)* ... *val (H)** *(Fig. 4) can be expressed as a sum of nearly complete marks or prime implicants.*

As mentioned, the hypothesis has been assumed here that, for any $\mathbf{i}$, $\mathbf{M}_{1i}{}^* \mathbf{M}_{2i}$ is equal to $\mathbf{I}(\mathbf{M}_{1i})$. In general, for some $\mathbf{i}$

$$\mathbf{M}_{1i} * \mathbf{M}_{2i} = \mathbf{I}(\mathbf{M}_{1i})$$

while for other $\mathbf{i}$

$$\mathbf{M}_{1i} * \mathbf{M}_{2i} = \mathbf{I}(\mathbf{M}_{2i}).$$

In this general case the same type of analysis which is going to be presented here can be applied.

8 The Decomposition of a PCA in Factors

Consider again the function

val(A) * val(B) * ...*val(H)

under the hypothesis that it is the sum of nearly complete marks and no complemented variable occur. We are interested now in determing **val(H)**.

Assume

val(A) * val(B)*... = (a+b+c+...)

val(H) = (m+n+o+...)

where $\mathbf{a}$, $\mathbf{b}$, $\mathbf{c}$, ..., $\mathbf{m}$, $\mathbf{n}$, $\mathbf{o}$, ... are the prime implicants of the Boolean functions they represent.

Of course,

(a+b+c+...) * (m+n+o+...) = $\mathbf{M}_{11}$ + $\mathbf{M}_{12}$ + ...

Let

a * m = $\mathbf{M}_{11}$

b * n = $\mathbf{M}_{12}$

Consider

a * n

Since $\mathbf{a}^*\mathbf{n}$ will be one of addenda of $(\mathbf{M}_{11} + \mathbf{M}_{12}+...)$ each of which must be nearly complete in one of its variables, only one of the two following conditions is possible:

1.1 $\mathbf{a}$ is nearly complete in the variables characteristic of $\mathbf{M}_{11}$;
1.2 $\mathbf{n}$ is nearly complete in the variables characterist of $\mathbf{M}_{12}$.

Now consider the product **b * m.**

Since also $\mathbf{b}\ ^*\mathbf{m}$ will be one of the addenda of $(\mathbf{M}_{11} + \mathbf{M}_{12} +...)$, only one of the two following conditions is possible:

2.1 $\mathbf{b}$ is nearly complete in the variables characteristic of $\mathbf{M}_{12}$;
2.2 $\mathbf{m}$ is nearly complete in the variables characteristic of $\mathbf{M}_{11}$.

The combinations of the cases 1.1, 1.2, 2.1, 2.2 produce the following conclusive alternatives:

3.1 $\mathbf{a}$ is nearly complete as $\mathbf{M}_{11}$ and $\mathbf{b}$ is nearly complete as $\mathbf{M}_{12}$;
3.2 $\mathbf{a}$ and $\mathbf{m}$ are nearly complete as $\mathbf{M}_{11}$;
3.3 $\mathbf{b}$ and $\mathbf{n}$ are nearly complete as $\mathbf{M}_{12}$;
3.4 $\mathbf{m}$ is nearly complete as $\mathbf{M}_{11}$ and $\mathbf{n}$ is nearly complete as $\mathbf{M}_{12}$.

In the cases **3.1** and **3.4** (**a** + **b**+...) and (**m** + **n** +...) have the same type of structure as (**M**$_{11}$ + **M**$_{12}$ +...); in the cases **3.2** and **3.3**, (**a** + **b** +...) and (**m** + **n** +...) contain a number of nearly complete terms less by a unit than the number of near complete marks contained in (**M**$_{11}$ + **M**$_{12}$ +...).

The same line of reasoning apply also to the decomposition in factors of (**a** + **b** +...) or (**m** + **n** +...). Therefore, we can state the following theorem.

Theorem 5. *In the decomposition of a **PCA** in factors of the type shown in Fig. 3 and Fig.4 the number of the generated factors is equal to or larger than the number of prime implicants produced by that **PCA**. As a consequence also the merit of a **PCA** is less than the number of the corresponding prime implicants.*

9   The Decomposition of a PCAF in Addenda

Let **val (A)** be the Boolean function implemented by the subnetwork whose output is node **A** of the circuit of Fig. **5**. Node **A** is a **PCAF** resulting from $\mathbf{n}_1$ decomposition operations of **PCA P** in factors. Assume that **val (A)** is equal to

$\mathbf{R}_1$+ $\mathbf{R}_2$ +...+$\mathbf{R}_p$+ $\mathbf{M}_1$ + $\mathbf{M}_2$ + ...+$\mathbf{M}_q$

and, therefore, it contains **q** nearly complete marks.

A might be the output of an **OR** gate whose inputs take the values $\mathbf{R}_1$, $\mathbf{R}_2$,..., $\mathbf{M}_1$, $\mathbf{M}_2$.., as shown in Fig. **5**.

In this case each of the value $\mathbf{M}_1$, $\mathbf{M}_2$, ... can be implemented with an **AND** gate having a subset of the input variable $\mathbf{x}_1$, $\mathbf{x}_2$,... of **CF** as its inputs. It is apparent that the total number of gates implementing **PCA P** is larger than $\mathbf{n}_1$ + **q** and, therefore, it is larger than the number of prime implicants implemented by **P**.

Alternatively A might be the output of an **OR** gate whose inputs take values of the type $\sum \mathbf{R} + \sum \mathbf{M}$ as shown by the example of Fig. **6**. Also in this case, an addendum as ($\mathbf{R}_1$ + $\mathbf{M}_1$ + $\mathbf{M}_2$) can be decomposed in factors by using a number of gates larger than **2**, the number of nearly complete marks. Also in this case the total number of gates implementing **PCA P** is larger than the number of prime implicants implemented by **P**.

10   The Role of Complemented Variables

So far the hypothesis has been assumed that complemented variables play no important role in marks or remainders. Such a simple hypothesis will be removed in this section.

Let us return to the first decomposition in factors shown in Fig. **4** and Fig. **5**. The relationships:

val(A) * **val(B)** *....***val(H)** = **M**$_{11}$ + **M**$_{12}$ +...

val(K) = **M**$_{21}$ + **M**$_{22}$ + ...+ **R**$_{21}$ + **R**$_{22}$ +....

presented in Section **4** are still valid, but now marks and remainders may contain complemented variables.

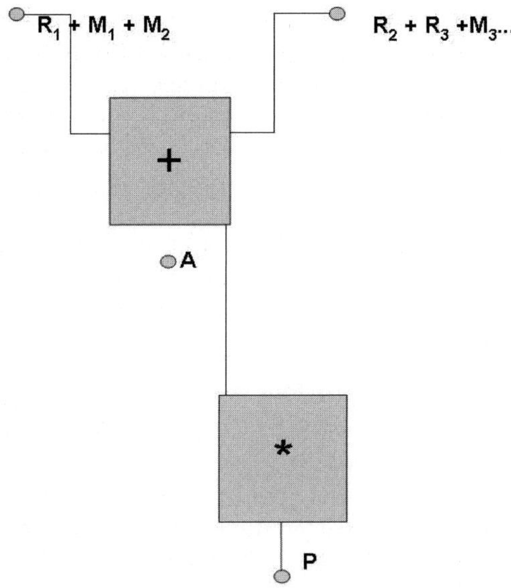

Fig. 6. A second type of structure of a PCAF

Subdivide the set of marks $\mathbf{M}_{1i}$ into the following subsets:
marks containing no complemented variables;
marks containing one complemented variable $!\mathbf{c}_{11}$;
marks containing one complemented variable $!\mathbf{c}_{12}$;

....
marks containing two complemented variables $!\mathbf{c}_{11}\ !\mathbf{c}_{12}$;
marks containing two complemented variables $!\mathbf{c}_{11}!\mathbf{c}_{13}$;

....
marks containing three complemented variables $!\mathbf{c}_{11}\ !\mathbf{c}_{12}\ !\mathbf{c}_{13}$;
and so on.
Of course, some of these subsets may be empty.
Let $\mathbf{m}$ be the number of elements contained in the largest subset.

The evaluations discussed in previous sections apply to first subset and, with minor changes, to each of the other subsets. So, a single decomposition can reduce the number of marks of all the subsets by one unit. It follows that the number of gates contained in a **PCA** will be

cost $\geq$ m

Now we can evaluate the merit of the considered **PCA**. The merit of a prime implicant generated by the marks of the first subset which contain no complemented variables may be equal to **1**, but the merit of each of the prime implicants generated by the marks of the other subsets decreases very quickly with the number of complemented variables contained in the generated mark. Indeed, the merit of each of the prime implicants generated by a mark containing one complemented variable is less than (or equal to) 1/2; the merit of each of the prime implicants generated by a mark containing two complemented variables is less than (or equal to) 1/4; and so on. If **n** is the number of complemented variables used to subdivide the set of marks into disjoint subsets, the total merit of a **PCA** will be less than the following limit:

$$\mathbf{m} * 1 + \mathbf{m} * \mathbf{n} * 1/2 +$$
$$+ \mathbf{m} * \mathbf{bin(n,2)} * (1/2)^2 +$$
$$+ \mathbf{m} * \mathbf{bin(n,3)} * (1/2)^3 +$$

....

where **bin (n,i)** denotes the number of combination of **n** elements **i** by **i**.

11   Conclusion

From the analysis presented in previous section we can deduce an upper bound for the merit of a gate, which is specified by the following value;

$$1 + \mathbf{n}/2 + \mathbf{bin(n,2)} * (1/2)^2 +$$
$$+ \mathbf{bin(n,3)} * (1/2)^3 + ...$$

But **n** is less than **N**, the total number of compatibilities, and $\mathbf{3}^N$ is the number of prime implicants of the core function.

It follows that the number of gates necessary to implement the core function, represented by $\mathbf{3}^N$ divided by the above specified merit of a gate, increases with **N** exponentially.

References

1. Meo, A.R.: On the minimization of core function Acc. Sc. Torino – Memorie Sc. Fis. 29, 155–178, 7 ff. (2005)
2. Sipser, M.: Introduction to the theory of computation, PWS Publ.Comp. (1997)
3. Smale, S.: Mathematical problems for the next century. Mathematical Intelligencer 20(2), 7–15 (1998)
4. Garey, M.R., Johnson, D.S.: Computers and intractability. W.H.Freeman and Co., New York (1979)
5. Asser, G.: Das Reprasentantenproblem im Pradikatenkalkul der ersten Stufe mit Identitat, Zietschr.f.Mathematisches Logik u. Grundlagen der Math. 1, 252–263 (1955)
6. Cook, S.A.: The complexity of theorem proving procedures. In: Proc. 3rd Annual ACM Symp. on Theory of Computing, pp. 151–158. ACM Press, New York (1971)
7. Cook, S.A.: Computational complexity of higher type functions. In: Sarake, I. (ed.) Proceeding of the International Congress of Mathematicians, Kyoto, Japan, pp. 55–60. Springer, Heidelberg (1991)
8. Godel, K.: A letter to J.von Neumann, in Sipser's article listed above. In: Clote, P., Krajicek, J. (eds.) Arithmetic, Proof Theory and Computational Complexity, Oxford University Press, Oxford (1993)
9. Karp, R.M.: Reducibility Among Combinatorial Problems, Complexity of Computer Computations, pp. 85–103. Plenum Press (1972)
10. Levin, L.: Universal Search Problems. Problemy Peredachi Informatsii (= Problems of Information Transmission) 9(3), 265–266 (in Russian) (1973); A partial English transalation, In: Trakhtenbrot, B.A.: A Survey of Russian Approches to Perebor (Brute-force Search) Algorithms. Annals of the History of Computing 6(4), 384–400 (1984)

11. Scholz, H.: Problem #1: Ein ungelostes Problem in der symbolisches Logik. Symbolic Logic 17, 160 (1952)
12. Sipser, M.: The history and status of the P versus NP question. In: Proceedings of the 24th Annual ACM Symposium on Theory of Computing, pp. 603–618. ACM Press, New York (1992)
13. Baker, T., Gill, J., Solovay, R.: Relativizations of the P =? NP question. IAM J. of Computing 4, 431–442 (1975)
14. Boppana, R., Sipser, M.: Complexity of finite functions. In: van Leeuwen, J. (ed.) Handbook of heoretical Computer Science, pp. 758–804 (1990)
15. Krajicek, J.: Bounded arithmetic, propositional logic, and omplexity theory, Encyclopedia of Mathematics and Its Applications, vol. 60. Cambridge University Press, Cambridge (1995)
16. Pudlak, P.: The lengths of proofs. In: Buss, S.R. (ed.) Handbook of Proof Theory, North-Holland, Amsterdam (1998)

Appendix A

Theorem 1. *There exists always an implementation* J *of* f *containing only* **AND** *and* **OR** *gates, applied to both complemented and uncomplemented input variables, such that* $cost(J) \leq 2 * C_{min}$

Proof. Let us divide the gates of implementation I_{min} of f into different levels.

At level **1** we place the gates all inputs of which coincide with the complemented or uncomplemented input variables x_i or $!x_i$ (where $!x_i$ denotes the complement of variable x_i).

Level **2** contains the gates whose inputs coincide with input variables or outputs of level **1** gates.

In general terms, level **q** contains the gates whose inputs coincide with input variables or outputs of levels less than **q**.

We can transform I_{min} into J by deleting **NOT** gates and adding new **AND** or **OR** gates as follows.

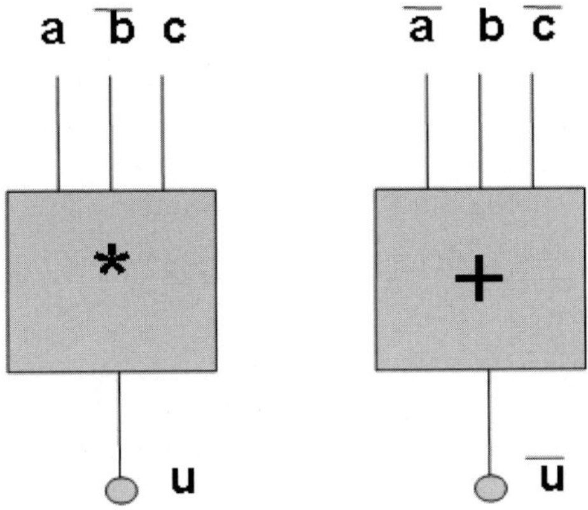

Fig. 7. The transformatin of gates of level 1

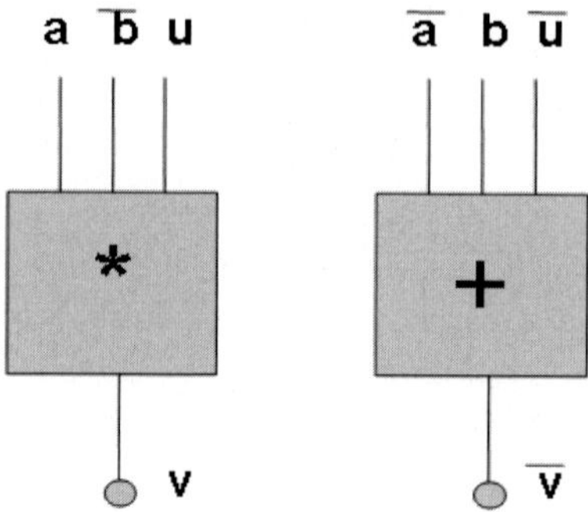

Fig. 8. The transformation of the gates of level 1

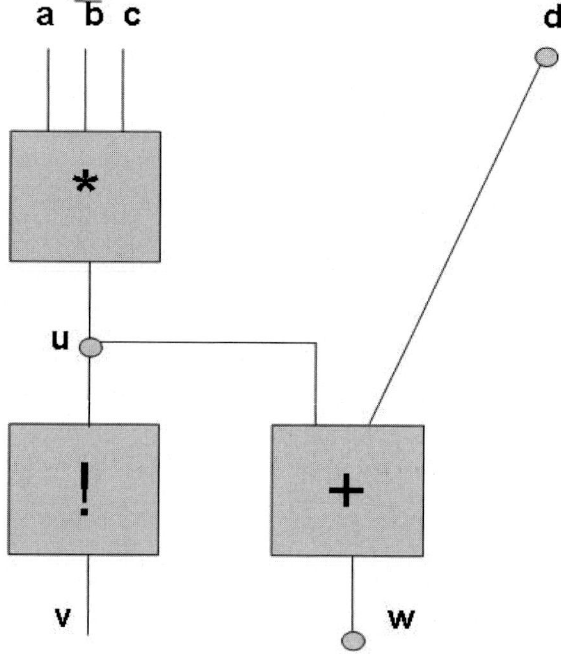

Fig. 9. A two level subnetwork

We start from level **1**.

For any level **1 AND** gate we join an **OR** gate whose inputs are the complements of the inputs of the considered **AND** gate (Fig. **7**). Similarly, for any level **1 OR** gate we join an **AND** gate whose inputs are the complements of the corresponding **OR** gate.

By virtue of such operations, for any output **u** of the level **1** gates a new node will be available in the new circuit we are generating whose value will be !**u**.

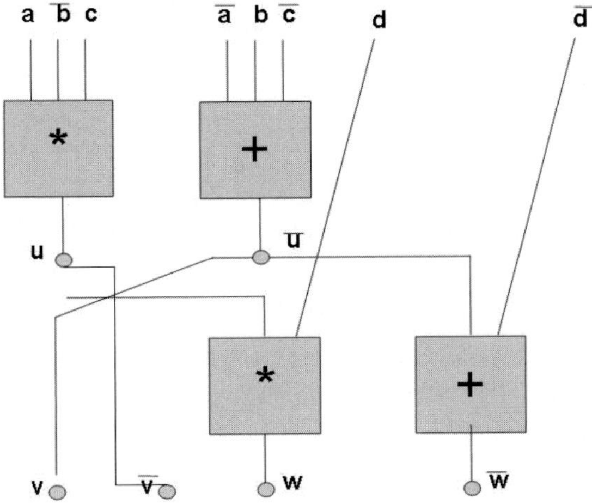

Fig. 10. The transformation of subnetwork of fig. 9

As a second step of processing, for any level **2 AND** gate of implementation $\mathbf{I}_{min}$ we shall join an **OR** gate whose inputs are the complements of the inputs of the corresponding **AND** gate, in both the cases when these inputs coincide with input variables of **f** or with outputs of level **1** gates (Fig. **8**).

A similar transformation will be applied to all level **2 OR** gates.

As an example, the two level subnetwork of Fig. **9** will be transformed into the subnetwork of Fig. **10**. Notice that at the outputs of **J** not only the outputs **u** and **w** of $\mathbf{I}_{min}$ will be available, but also their complements !u and !v.

The preceding operations will be applied to all the levels of implementation $\mathbf{I}_{min}$, in the order of increasing levels. It is apparent that the number of gates of **J** is less than twice the number of gates of $\mathbf{I}_{min}$.

Appendix B

Let **NPCA** be the number of **PCA**'s contained in the minimum cost implementation of core function and **NPCA(N)** the corresponding function of the number **N** of compatibilities.

Let us assume that **NPCA(N)** is an unknown polynomial function **P(N)**. Indeed, if **NPCA(N)** were an exponential function, the cost of the minimum cost implementation would be exponential.

Consider again the product
$$\mathbf{val(P)} = (\mathbf{M}_{11} + \mathbf{M}_{12} + \mathbf{M}_{13} + ...) * (\mathbf{M}_{21} + \mathbf{M}_{22} + \mathbf{M}_{23} + ...)$$
and the following definitions

Definition 7. *An elementary product*
$$\boldsymbol{M}_{1j} * \boldsymbol{M}_{2j} = \boldsymbol{J}_{ij}$$

(which is an extended prime implicant $I(M_{1i})$ or $I(M_{2j})$) is defined as a spurious prime implicant of order s if it can be transformed into a not spurious prime implicant by removing s of its literals.

Definition 8. *The merit of I_{ij} is defined as*
$$merit\ (I_{ij}) = 2^{-s}$$

Definition 9. *An elementary product I_{ij} is considered as "useful" if its merit is larger than $1/P(N)$.*

Indeed, if
$$\mathbf{NPCA} * merit\ (I_{ij})\quad < 1$$
the elementary contributions of an extended prime implicant for any **PCA** are not sufficient to generate a not extended prime implicant.

From the preceding Definitions 8 and 9 the following simple theorem follows.

Theorem 6. *The extended prime implicant is I_{ij} useful if it is spurious of order s and*
$$s < log_2\ P(N)$$

Definition 10. *A spurious (of order s) mark M_{1j} is useful if*
$$s < log_2\ P(N)$$

Indeed, if M_{1j} is not useful, neither a product involving M_{1j} can be useful.

Definition 11. *A mark M or a term T will be defined as "nearly complete in a given variable of a given triplet" if it contains all the compatibilities involving that variable with the exception of a number of them less than $log_2\ P(N)$.*

The theorems of Section 8 are true also by making reference to the preceding Definition 11. The proof is left to the reader.

The Seventies

Alberto Martelli

Dipartimento di Informatica, Università di Torino, Torino
`mrt@di.unito.it`

I arrived in Pisa at the beginning of 1968, after getting my degree from the Politecnico of Milan. At that time Antonio Grasselli was moving from the Politecnico of Milan to organize the new curriculum in Computer Science at the University of Pisa, and wanted to create a research group there. The group was established at the Istituto di Elaborazione dell'Informazione (IEI) of the National Research Council, and included initially Giorgio Levi, Ugo Montanari, Franco Sirovich and myself.

The initial research activities of the group were in the area of image processing, but, after a couple of years, the group, which in the meantime had been joined by Gigina Carlucci Aiello, decided to modify its main research topic. Our conclusion was that Artificial Intelligence was a new and more challenging area, allowing to combine theoretical aspects with problems of practical interest, and we started to redirect our research towards the topic. In those years, all of us spent long periods in the USA, and this gave us an up-to-date view of the main research themes, and allowed to establish international contacts.

The early research activities of Ugo can be placed mainly in this context. Since the beginning Ugo showed a great autonomy in selecting important research topics, and carried out high level research both in Pisa and while visiting US institutions like University of Maryland, Stanford University and Carnegie Mellon. His first papers deal with picture processing, graphics, graph grammars, networks of constraints. These papers, most of which with Ugo as a single author, have been published in famous journals [7,10,9,8,11,12]. Most of these papers were later considered as pioneering papers in the above mentioned areas, and are still widely cited.

My first joint paper with Ugo was about optimal smoothing in picture processing, presented at the IFIP Congress 1971 [2]. The paper describes the use of optimization techniques based on dynamic programming for smoothing pictures, and was the first of a series of papers where we studied the foundations of dynamic programming. In other papers, we approached optimization problems using artificial intelligence methods, consisting in finding minimal cost solutions in ordinary or AND/OR graphs [3,5], and we explored the relationships between dynamic programming and heuristic search methods [4].

In the second half of the seventies, the group had grown, and we became interested also on different topics. In particular, we worked on a project aimed at developing a programming environment based on a symbolic interpreter. I would like also to point out a paper by Rocco De Nicola, Ugo and myself on a formal comparison between message passing and shared memory, which is perhaps the first of Ugo's papers on the theory of concurrent computations [1].

P. Degano et al. (Eds.): Montanari Festschrift, LNCS 5065, pp. 797–798, 2008.

My last paper with Ugo was *An Efficient Unification Algorithm*, published in the TOPLAS in 1982 [6]. Unification in first-order logic is the central step of the inference rule called resolution, used, in particular, by interpreters of the language Prolog, which had appeared in the seventies. The unification algorithm, as originally proposed, could be extremely inefficient; therefore, various researchers started to analyze the complexity of the problem and studied more efficient algorithms. We became interested in this topic in 1976, and developed a linear algorithm, which was presented in a Technical Report in 1976. Unfortunately, a similar result had independently been proposed by Paterson and Wegman, and thus we reformulated our result as an efficient, but not linear, unification algorithm. Our paper represents the unification problem as the solution of a set of equations and presents a nondeterministic algorithm, from which the efficient algorithm is derived. It is interesting to remark that many authors who have cited our paper consider the "Martelli and Montanari algorithm" to be the nondeterministic one, rather than its efficient version.

At the end of the seventies I moved to the University of Torino, but I still remember with great pleasure the years spent in Pisa, when we carried out exciting scientific work in a pleasant friendly atmosphere.

Happy birthday, Ugo.

References

1. De Nicola, R., Martelli, A., Montanari, U.: Communication through message passing or shared memory: A formal comparison. In: ICDCS, pp. 513–522 (1981)
2. Martelli, A., Montanari, U.: Optimal smoothing in picture processing: An application to fingerprints. In: IFIP Congress (1), pp. 173–178 (1971)
3. Martelli, A., Montanari, U.: Additive and/or graphs. In: IJCAI, pp. 1–11 (1973)
4. Martelli, A., Montanari, U.: Form dynamic programming to search algorithms with functional costs. In: IJCAI, pp. 345–350 (1975)
5. Martelli, A., Montanari, U.: Optimizing decision trees through heuristically guided search. Commun. ACM 21(12), 1025–1039 (1978)
6. Martelli, A., Montanari, U.: An efficient unification algorithm. ACM Trans. Program. Lang. Syst. 4(2), 258–282 (1982)
7. Montanari, U.: A method for obtaining skeletons using a quasi-euclidean distance. J. ACM 15(4), 600–624 (1968)
8. Montanari, U.: Heuristically guided search and chromosome matching. Artif. Intell. 1(4), 227–245 (1970)
9. Montanari, U.: A note on minimal length polygongal approximation to a digitized contour. Commun. ACM 13(1), 41–47 (1970)
10. Montanari, U.: Separable graphs, planar graphs and web grammars. Information and Control 16(3), 243–267 (1970)
11. Montanari, U.: On the optimal detection of curves in noisy pictures. Commun. ACM 14(5), 335–345 (1971)
12. Montanari, U.: Networks of constraints: Fundamental properties and applications to picture processing. Inf. Sci. 7, 95–132 (1974)

Categories, Software and Meaning

Robin Milner

University of Cambridge, Computer Laboratory

I am delighted to be able to share in celebrating Ugo's birthday, if not with a formal paper then at least with some loosely-knit philosophical ideas.

I have worked on ideas similar to Ugo's for most of our careers. Ugo had much to do with the stream of expert Italians, many now well-known, who travelled from the warmth of Pisa to the romantic but cooler climate of Edinburgh, often to launch their careers with a PhD there. Going further back, I remember with excitement the meeting on parallel processes at Pisa in 1973, organised I believe by Ugo, the first concurrency conference I ever attended.

I have a growing regard for Ugo's creativity in fundamental notions, and his boldness in formulating and tackling the problems that arise from them. In one particular case, his paper with Pepe Meseguer entitled *Petri nets are monoids*, he opened up an avenue of thought that goes very deep. The paper gives a prominent place to causality, and to causal independence, and thus is a part of the conceptual repertoire not only of Petri nets but also of event structures. But independence arises not only with respect to causality, but also with respect to connectivity and locality. We can call actions independent by virtue of their occurring on different channels, or in different places; these kinds of independence are not the same as causal independence.

Whatever kind of independence we entertain, the idea of a monoidal category arises almost automatically. This is because any notion of independence presupposes its opposite: the dependence of one arrow on another, naturally exhibited by possibility of composing two to make a third. For example, we may compose two actions to form a third if the output state of one is the input state of the other; this is causal dependence. Equally we may compose one substitution with another to form a third if the range of one is the domain of the other; thinking if names as links — or pointers — this is dependence of linking. In either case, if dependence is absent then we can juxtapose the actions to form a third. A monoidal category distils the way that composition and juxtaposition of — respectively — dependent and independent arrows relate to each other; we see it arising in many guises in concurrency. In recent work Ugo, and many others including myself, are using models based upon graphs to exploit the way in which (in)dependence of linking and (in)dependence of placing (i.e. spatial arrangement) co-exist, and so help us to explain the behaviour of mobile interactive systems.

But I propose that we respond at a higher level to the inspiration from monoidal categories. Let us use them not only as models, but also to organise the wide range of models that computer scientist use, in such a way that we can join our science more readily to software engineering. We may not — and perhaps should not — begin with a rigorous treatment. One way to begin is to

P. Degano et al. (Eds.): Montanari Festschrift, LNCS 5065, pp. 799–801, 2008.

note that we surely build models at different levels of abstraction, and we relate them to each other; so let us ask what insight we gain into this relationship, if we entertain the idea that these models are themselves the objects of a category? Of course, we can immediately ask what the arrows would be, and how to compose them; and we can ask what it could mean to juxtapose independent models, and then to juxtapose a pair of arrows (whatever they are) between pairwise independent objects.

I have played with these ideas, and found some sense. Those who are interested may like to read my short essay *The tower of informatic models*, which will be published by Cambridge University Press in a volume entitled *From Semantics to Computer Science: Essays in honour of Gilles Kahn*.

In that essay I propose that a closer link can be formed between computation theory and software engineering by considering every model to have two parts: (a) a set of entities, and (b) their meaning. These phrases are deliberately informal and generous in application. For example, any programming language LAN is a model, once the meaning of programs is supplied either formally or — as in the Algol 60 report — informally. Equally, predicate logic LOG together with the truth-valuation of its formulae forms a model. There is a conflict here with the logicians' normal use of the term 'model'. They use it to mean a semantic space in which formulae are evaluated, whereas here I use it to denote both the logic and that semantic space.

Now, in our putative category of models, we can define an arrow between these two models. Call it $Hoare : LOG \rightarrow LAN$; it consists of the validation (perhaps by Hoare logic) of LAN programs against pre- and post-conditions drawn from LOG. Let us call such an arrow an *explanation*; it demonstrates that the meanings of entities in the more concrete model (in this case LAN) *realises* the meanings of corresponding entities in the more abstract model (LOG). There may also be other kinds of arrow.

Now consider independence. Relevant to the flight of the European Airbus are at least three independent models, informatic, electro-mechanical and meteorological; only the first of these is computational. Each of the three models has both entities and meaning, determined by its own science. The tensor product of these models is then a (partial) explanation of a coordinated model that describes the flight of a whole aircraft. Thus the product model 'explains' the aircraft by means of a product of explanatory arrows, expressed in different sciences. In this example the informatic explanation has indeed been carried out by abstract interpretation.

It is accurate, at least sometimes, to say that the opposite (in the true categorical sense) of an explanation is a 'realisation' of one model by another. In the extreme we have the aircraft itself, realising its models; so we can think of real phenomena — the flight of actual aircraft — as extremal models.

Other examples of explanation and realisation abound. For example, a compiler is a realisation of one language by another. With logics, languages and abstraction one can easily present much of our work as a tower of models built in a monoidal category. Note that explanations are often achieved by what is conveniently called model-checking.

To make these ideas precise will take some effort, and it may be better first to find a way of discussing them informally with software engineers, because then we may make them precise in a way that is acceptable in engineering practice. This effort would be timely; many members of the software engineering community are committed to some form of model-driven engineering (MDE), though they may diverge in their understanding of this term just as we may diverge on what should be a rigorous notion of model. I suggest tht both communities are more likely to succeed if they reach convergence not merely within their communities but also between them.

I hope Ugo will forgive me for this superficial introduction of a subject which I believe to be deep. I don't doubt that if I had discussed it with him first it would be better. But in my opinion he is one who deserves credit for bringing to our notice the categorical ideas that can help us think this subject more clearly.

A Roman Senator

Jan Rutten

CWI and Vrije Universiteit Amsterdam

Somehow I have always viewed Ugo Montanari as a Roman Senator. I do not remember when this started but it must have been rather soon after we first met. Whenever I listen to one of his innumerable lectures — but even without him actually being present — I picture Ugo in a white toga, and imagine him addressing the Roman Senate. No doubt the basis for this image is his physical stature: formidable and noble, with a large head and straight nose. But of course, it is because of the combination of this with his great worthiness, his undoubted authorithy, his spirituality and his conviction, that this image of Ugo as a Roman Senator comes to mind.

As a small sign of my great appreciation for all that Ugo has meant for the scientific community in general, for computer science and coalgebra, more specifically, and for myself in the form of many personal contacts in the course of the last twenty years, I include this little picture[1]:

---

[1] See http://www.flickr.com/photos/diagonals/370704326 for a larger version.

P. Degano et al. (Eds.): Montanari Festschrift, LNCS 5065, pp. 802–803, 2008.

The picture was taken in Roma, City of cities, home of the Senate and its Senators, at the Piazza del Campidoglio, a beautiful square that was conceived by Michelangelo, near the Forum Romanum. The statue is part of the the Fontana della Dea Roma and, alas, does not represent a Senator. Instead it was used by Michelangelo as a representation of the river Tigris, or actually its Roman counterpart the Tevere (which explains the presence of the wolf with Romolo and Remo, on the left). And it comes somehow very close to the image I've been trying to convey above.

Who, one may wonder, is the little mortal sitting on the right, at the feet and in the shadow of the statue? We don't know. It could be Ugo himself, as a young man, still without a beard. It could be (an Italian version of) myself. Could it be you?!

The Semantics of Ugo Montanari

Franco Turini

Dipartimento di Informatica, University of Pisa
`F.Turini@di.unipi.it`

The first time I met Ugo Montanari, he had no beard. It was 1970 and I was a student attending the third year of a master degree in computer science at the University of Pisa ("Laurea in Scienze dell'informazione") and he had just come back from the States and had been given the job of teaching a course on System theory. The subject was very oriented towards classical theoretical engineering, not very much to do with computer science and programming. I thought his teaching style was great, as was the way he was able to demystify complex concepts. However, I was not particularly impressed by the subject. The following year was my fourth and last year, and, when I looked at the list of courses available, I discovered that Ugo Montanari would be teaching the course on Methods in Information Processing ("Metodi per il trattamento della informazione"). Montanari had planned to teach the most important theoretical foundations of computer science: formal languages, computability theory, and the semantics of programming languages. In the US, formal languages and computability theory were already an integral part of any computer science course. But not so in Italy. In fact my course in Pisa was the first of its kind. And the idea of including semantics of programming languages as the third leg of the theoretical corpus of computer science was really a leap into the future, almost visionary. I remember that the learning material was a paper by Johnston [2] presenting the so-called contour model, which was an initial attempt to formalize the operational semantics of a programming language. That course impressed me more than any other in the entire curriculum. Needless to say, I asked Montanari to be my supervisor for my Master's thesis. I wrote a thesis on the semantics of nondeterministic languages, mostly inspired by the contour model, in which I tried to extend the stack-based computation model to handle nondeterministic computations and the need to handle efficiently state saving and recovery. The first ideas developed at that time have subsequently kept me busy working for many years [3,4,5]. Imagine my pleasure when in 1990 I was given the possibility of teaching a course on Methods in information processing in parallel with a course taught by Montanari. There were still three subjects: formal languages, computability and formal semantics of programming languages. In the meanwhile much research had been done in the third area. Scott and Strachey had laid down the foundations of denotational semantics, and a few books on this subject had been published. Montanari and I decided to follow Gordon's book [1] which went down very well with our students. A few years later the teaching of computer science in Italy was redesigned. Courses with a coverage as large as Methods in Information Processing were replaced by shorter courses. A course on Semantics of programming languages was introduced, and Montanari and I

P. Degano et al. (Eds.): Montanari Festschrift, LNCS 5065, pp. 804–805, 2008.

taught it in two parallel courses. We taught both operational and denotational semantics and their relationships, and even some elements of semantics of parallel programming languages. We had a good English book [6] translated for us to use. However I was again impressed by Ugo's desire to improve the teaching, by looking for more elegant proofs of the theorems and a smoother flowing of the results. Well, today Ugo Monatanari and I teach different courses, but I am ready to start tomorrow to design and prepare a new course with him. As has always been in the past, I bet that it would be another highly rewarding experience for me. Finally, let me thank the editors of this book for offering me the opportunity to express my appreciation to someone who really deserves the title of "maestro".

References

1. Gordon, M.: The denotational description of Programming Languages. Springer, Heidelberg (1979)
2. Johnston, J.B.: The contour model of block structured processes, In: Tou, J., Wegner, P. (eds.) Sigplan Notices - Proc. Symp. on Data Structures in Prog. Lang., February 1971, vol. 6(2), pp. 55–82 (1971)
3. Montangero, C., Pacini, G., Turini, F.: MAGMA-LISP: a machine language for Artificial Intelligence. In: Advanced Papers of the Fourth International Joint Conference on Artificial Intelligence, pp. 556–561 (1975)
4. Montangero, C., Pacini, G., Turini, F.: Two-level control structures for Nondeterministic Programming. Communications of ACM 20(10), 725–730 (1977)
5. Turini, F.: MAGMA2: a language oriented to experiments in Control. ACM Trans. Program. Lang. Syst. 6(4), 468–486 (1984)
6. Winskel, G.: The formal semantics of programming languages: an introduction. MIT Press, Cambridge (1993)

Abstraction for a Career in Industry:
A Praise for Ugo's 65 Years

Daniel Yankelevich

Pragma Consultores

I arrived in Italy to join Ugo's group after getting my MsC degree on December 1989, on a Friday night, after a 15 hours trip crossing the ocean. Saturday morning, at 7:30, I received a call at the apart hotel where I was staying. Ugo had the theory that it was better to get over jet lags as soon as possible. At 8:30 AM we had our first meeting to discuss categories, models for concurrent computation, and process algebras.

After that, we had a lot of meetings. Sometimes the discussions were captured in papers that we published, but as very often happens in academia (too often for my taste) most of the knowledge was lost.

I am not sure whether we ever had the discussion that follows or it was just a dream. But it sounds quite real. I've came into Ugo's office at 7:40 AM.

Ugo: - *So, what do you mean when you say you are tired of CCS? CCS is just a process algebra - there can be many. Let's concentrate on the concepts, not on the syntax.*

Me: - Ciao, Ugo, it's nice to see you again.

Ugo: - *However, you will agree that it is not by chance that CCS was designed as it was. You can capture most of the problems of cooperation and concurrency, and if you want them solved, you should first solve them in a simple model, isn't it? (actually: "non è vero?")*

Me: - Well, I have a doubt precisely about that. My point is: when you talk about computable functions, there is an agreement that the function itself (the relation input/output) is the object of interest, and this agreement is needed in order to have a Church Thesis. In the case of concurrent systems, and even interactive systems, there is not such agreement at all. In every conference there are at least two proposals of what a computation should be, what an observation is, or different equivalences

Ugo: - *Any process which could naturally be called an effective procedure can be realized by a Turing machine.*

Me: - Well, if the process is to digitalize input from a digital camera and produce a distorted version of it, this would be modeled too abstractly by a Turing Machine. Probably, the computation required after the input is codified and before the output is printed can be modeled, but for an external observer that is nothing. What is the right level of abstraction?

Ugo: - *May be there is not such thing as a right level of abstraction. Why should someone tell you at what level you must observe the computation? The point is what are you going to do with the computations. I am not a mathematician, and after you understood the model, you want to solve a practical problem. You have to apply the model to build better software, always remember that.*

P. Degano et al. (Eds.): Montanari Festschrift, LNCS 5065, pp. 806–808, 2008.
© Springer-Verlag Berlin Heidelberg 2008

I have always had problems describing my job. I am the founder, and since last year the Chief Innovation Officer, of a technology firm with branches in Latinoamerica and Spain. During the last fifteen years I worked in software-intensive projects and systems. And I think my job, in the last 15 years, was more related to semantics than it could be suspected: I was always translating technology in business terms and sometimes vice-versa.

Many times I have asked myself whether or not the skills I have learned from Ugo helped me in doing my job. The answer is yes. I have not only learned semantics of concurrent languages or process algebras from Ugo. Actually, I have learned from him (and from the other people I have worked with during the same period) three skills that were key to my career. First, the main skill that helped me and I think it is essential for computer scientists, software engineers and related disciplines, is abstraction. This is probably not new: the SWEBOK and many specialists mention abstraction as a key skill for computer professionals. To learn how to deal with abstractions is a continuous process that begins in our early childhood and probably continues all our lives. Ugo masters abstraction as few people I have met. I have learned from him how it is possible to work with abstractions in a systematic way.

The other two skills are also related to the work with abstractions. One is modeling and formalization: how to construct models to represent abstractions, and how to reason and make explicit properties using those models. This is incredibly useful for anyone working with projects and technology. Traditionally, this skill was learned from Physics, but engineers know about the usefulness of models. And Ugo always remembers that he is an Engineer.

The last skill could be named "symbolic transformations", and it consists in transforming representations from a model to a different one. In my PhD thesis I worked with Ugo on the idea of having a very detailed description of a system and then to transform it (using abstractions and formal techniques) in different views that were coherent by construction. More than 15 years later, I have been working with my team on a model for Project Portfolio Management, abstracting away details from the process in order to build three views of the same model. This allowed us to understand what techniques and technology could help us. Even though these two topics (project portfolio management and concurrent programming languages) are very different, both reasoning processes were very similar in their nature.

Besides those skills, Ugo transmits an incredible passion for what he does. When I managed to get in touch with that passion inside myself, I found a greater joy in my job and my career. I think this is not a skill, is something else, but it is invaluable. Abstraction, Formalization and Symbolic Transformations play very important roles in a professional's education. These skills are way more important than particular techniques for specific problems. Everything is important, but you must choose when resources -in particular time- are limited. In the limit, and provocatively, one might talk about a tradeoff between training and education. Training is used to acquire specific techniques to solve specific problems. Education implies a wider formation that allows one to apply the same concepts to new problems.

I have had the luck of having a real education with the group I have worked with at Pisa, much of it achieved through "Socratic discussions" with Ugo. I really recommend this way of learning both to teachers and to trainees - it lasts a lifetime.

Author Index

Printing: Mercedes-Druck, Berlin
Binding: Stein+Lehmann, Berlin

Lecture Notes in Computer Science

Sublibrary 1: Theoretical Computer Science and General Issues

For information about Vols. 1– 4711
please contact your bookseller or Springer

Vol. 5065: P. Degano, R. De Nicola, J. Meseguer (Eds.), Concurrency, Graphs and Models. XV, 810 pages. 2008.

Vol. 5062: K.M. van Hee, R. Valk (Eds.), Applications and Theory of Petri Nets. XIII, 429 pages. 2008.

Vol. 5059: F.P. Preparata, X. Wu, J. Yin (Eds.), Frontiers in Algorithmics. XI, 350 pages. 2008.

Vol. 5050: J.M. Zurada, G.G. Yen, J. Wang (Eds.), Computational Intelligence: Research Frontiers. XVI, 389 pages. 2008.

Vol. 5038: C.C. McGeoch (Ed.), Experimental Algorithms. X, 363 pages. 2008.

Vol. 5036: S. Wu, L.T. Yang, T.L. Xu (Eds.), Advances in Grid and Pervasive Computing. XV, 518 pages. 2008.

Vol. 5035: A. Lodi, A. Panconesi, G. Rinaldi (Eds.), Integer Programming and Combinatorial Optimization. XI, 477 pages. 2008.

Vol. 5029: P. Ferragina, G.M. Landau (Eds.), Combinatorial Pattern Matching. XIII, 317 pages. 2008.

Vol. 5018: M. Grohe, R. Niedermeier (Eds.), Parameterized and Exact Computation. X, 227 pages. 2008.

Vol. 5015: L. Perron, M.A. Trick (Eds.), Integration of AI and OR Techniques in Constraint Programming for Combinatorial Optimization Problems. XII, 394 pages. 2008.

Vol. 5011: A.J. van der Poorten, A. Stein (Eds.), Algorithmic Number Theory. IX, 455 pages. 2008.

Vol. 5010: E.A. Hirsch, A.A. Razborov, A. Semenov, A. Slissenko (Eds.), Computer Science – Theory and Applications. XIII, 411 pages. 2008.

Vol. 5008: A. Gasteratos, M. Vincze, J.K. Tsotsos (Eds.), Computer Vision Systems. XV, 560 pages. 2008.

Vol. 5004: R. Eigenmann, B.R. de Supinski (Eds.), OpenMP in a New Era of Parallelism. X, 191 pages. 2008.

Vol. 4996: H. Kleine Büning, X. Zhao (Eds.), Theory and Applications of Satisfiability Testing – SAT 2008. X, 305 pages. 2008.

Vol. 4988: R. Berghammer, B. Möller, G. Struth (Eds.), Relations and Kleene Algebra in Computer Science. X, 397 pages. 2008.

Vol. 4981: M. Egerstedt, B. Mishra (Eds.), Hybrid Systems: Computation and Control. XV, 680 pages. 2008.

Vol. 4978: M. Agrawal, D. Du, Z. Duan, A. Li (Eds.), Theory and Applications of Models of Computation. XII, 598 pages. 2008.

Vol. 4975: F. Chen, B. Jüttler (Eds.), Advances in Geometric Modeling and Processing. XV, 606 pages. 2008.

Vol. 4974: M. Giacobini, A. Brabazon, S. Cagnoni, G.A. Di Caro, R. Drechsler, A. Ekárt, A.I. Esparcia-Alcázar, M. Farooq, A. Fink, J. McCormack, M. O'Neill, J. Romero, F. Rothlauf, G. Squillero, A.Ş. Uyar, S. Yang (Eds.), Applications of Evolutionary Computing. XXV, 701 pages. 2008.

Vol. 4973: E. Marchiori, J.H. Moore (Eds.), Evolutionary Computation, Machine Learning and Data Mining in Bioinformatics. X, 213 pages. 2008.

Vol. 4972: J. van Hemert, C. Cotta (Eds.), Evolutionary Computation in Combinatorial Optimization. XII, 289 pages. 2008.

Vol. 4971: M. O'Neill, L. Vanneschi, S. Gustafson, A.I. Esparcia Alcázar, I. De Falco, A. Della Cioppa, E. Tarantino (Eds.), Genetic Programming. XI, 375 pages. 2008.

Vol. 4967: R. Wyrzykowski, J. Dongarra, K. Karczewski, J. Wasniewski (Eds.), Parallel Processing and Applied Mathematics. XXIII, 1414 pages. 2008.

Vol. 4963: C.R. Ramakrishnan, J. Rehof (Eds.), Tools and Algorithms for the Construction and Analysis of Systems. XVI, 518 pages. 2008.

Vol. 4962: R. Amadio (Ed.), Foundations of Software Science and Computational Structures. XV, 505 pages. 2008.

Vol. 4961: J.L. Fiadeiro, P. Inverardi (Eds.), Fundamental Approaches to Software Engineering. XIII, 430 pages. 2008.

Vol. 4960: S. Drossopoulou (Ed.), Programming Languages and Systems. XIII, 399 pages. 2008.

Vol. 4959: L. Hendren (Ed.), Compiler Construction. XII, 307 pages. 2008.

Vol. 4957: E.S. Laber, C. Bornstein, L.T. Nogueira, L. Faria (Eds.), LATIN 2008: Theoretical Informatics. XVII, 794 pages. 2008.

Vol. 4943: R. Woods, K. Compton, C. Bouganis, P.C. Diniz (Eds.), Reconfigurable Computing: Architectures, Tools and Applications. XIV, 344 pages. 2008.

Vol. 4942: E. Frachtenberg, U. Schwiegelshohn (Eds.), Job Scheduling Strategies for Parallel Processing. VII, 189 pages. 2008.

Vol. 4941: M. Miculan, I. Scagnetto, F. Honsell (Eds.), Types for Proofs and Programs. VII, 203 pages. 2008.

Vol. 4935: (Ed.), A Practical Programming Model for the Multi-Core Era. 6, 208 pages. 2008.

Vol. 4935: B. Chapman, W. Zheng, G.R. Gao, M. Sato, E. Ayguadé, D. Wang (Eds.), A Practical Programming Model for the Multi-Core Era. 6, 208 pages. 2008.

Vol. 4934: U. Brinkschulte, T. Ungerer, C. Hochberger, R.G. Spallek (Eds.), Architecture of Computing Systems – ARCS 2008. XI, 287 pages. 2008.

Vol. 4927: C. Kaklamanis, M. Skutella (Eds.), Approximation and Online Algorithms. X, 289 pages. 2008.

Vol. 4926: N. Monmarché, E.-G. Talbi, P. Collet, M. Schoenauer, E. Lutton (Eds.), Artificial Evolution. XIII, 327 pages. 2008.

Vol. 4921: S.-i. Nakano, M.. S. Rahman (Eds.), WALCOM: Algorithms and Computation. XII, 241 pages. 2008.

Vol. 4919: A. Gelbukh (Ed.), Computational Linguistics and Intelligent Text Processing. XVIII, 666 pages. 2008.

Vol. 4917: P. Stenström, M. Dubois, M. Katevenis, R. Gupta, T. Ungerer (Eds.), High Performance Embedded Architectures and Compilers. XIII, 400 pages. 2008.

Vol. 4915: A. King (Ed.), Logic-Based Program Synthesis and Transformation. X, 219 pages. 2008.

Vol. 4912: G. Barthe, C. Fournet (Eds.), Trustworthy Global Computing. XI, 401 pages. 2008.

Vol. 4910: V. Geffert, J. Karhumäki, A. Bertoni, B. Preneel, P. Návrat, M. Bieliková (Eds.), SOFSEM 2008: Theory and Practice of Computer Science. XV, 792 pages. 2008.

Vol. 4905: F. Logozzo, D.A. Peled, L.D. Zuck (Eds.), Verification, Model Checking, and Abstract Interpretation. X, 325 pages. 2008.

Vol. 4904: S. Rao, M. Chatterjee, P. Jayanti, C.S.R. Murthy, S.K. Saha (Eds.), Distributed Computing and Networking. XVIII, 588 pages. 2007.

Vol. 4878: E. Tovar, P. Tsigas, H. Fouchal (Eds.), Principles of Distributed Systems. XIII, 457 pages. 2007.

Vol. 4875: S.-H. Hong, T. Nishizeki, W. Quan (Eds.), Graph Drawing. XIII, 402 pages. 2008.

Vol. 4873: S. Aluru, M. Parashar, R. Badrinath, V.K. Prasanna (Eds.), High Performance Computing – HiPC 2007. XXIV, 663 pages. 2007.

Vol. 4863: A. Bonato, F.R.K. Chung (Eds.), Algorithms and Models for the Web-Graph. X, 217 pages. 2007.

Vol. 4860: G. Eleftherakis, P. Kefalas, G. Păun, G. Rozenberg, A. Salomaa (Eds.), Membrane Computing. IX, 453 pages. 2007.

Vol. 4855: V. Arvind, S. Prasad (Eds.), FSTTCS 2007: Foundations of Software Technology and Theoretical Computer Science. XIV, 558 pages. 2007.

Vol. 4854: L. Bougé, M. Forsell, J.L. Träff, A. Streit, W. Ziegler, M. Alexander, S. Childs (Eds.), Euro-Par 2007 Workshops: Parallel Processing. XVII, 236 pages. 2008.

Vol. 4851: S. Boztaş, H.-F.(F.) Lu (Eds.), Applied Algebra, Algebraic Algorithms and Error-Correcting Codes. XII, 368 pages. 2007.

Vol. 4848: M.H. Garzon, H. Yan (Eds.), DNA Computing. XI, 292 pages. 2008.

Vol. 4847: M. Xu, Y. Zhan, J. Cao, Y. Liu (Eds.), Advanced Parallel Processing Technologies. XIX, 767 pages. 2007.

Vol. 4846: I. Cervesato (Ed.), Advances in Computer Science – ASIAN 2007. XI, 313 pages. 2007.

Vol. 4838: T. Masuzawa, S. Tixeuil (Eds.), Stabilization, Safety, and Security of Distributed Systems. XIII, 409 pages. 2007.

Vol. 4835: T. Tokuyama (Ed.), Algorithms and Computation. XVII, 929 pages. 2007.

Vol. 4818: I. Lirkov, S. Margenov, J. Waśniewski (Eds.), Large-Scale Scientific Computing. XIV, 755 pages. 2008.

Vol. 4800: A. Avron, N. Dershowitz, A. Rabinovich (Eds.), Pillars of Computer Science. XXI, 683 pages. 2008.

Vol. 4783: J. Holub, J. Žďárek (Eds.), Implementation and Application of Automata. XIII, 324 pages. 2007.

Vol. 4782: R. Perrott, B.M. Chapman, J. Subhlok, R.F. de Mello, L.T. Yang (Eds.), High Performance Computing and Communications. XIX, 823 pages. 2007.

Vol. 4771: T. Bartz-Beielstein, M.J. Blesa Aguilera, C. Blum, B. Naujoks, A. Roli, G. Rudolph, M. Sampels (Eds.), Hybrid Metaheuristics. X, 202 pages. 2007.

Vol. 4770: V.G. Ganzha, E.W. Mayr, E.V. Vorozhtsov (Eds.), Computer Algebra in Scientific Computing. XIII, 460 pages. 2007.

Vol. 4769: A. Brandstädt, D. Kratsch, H. Müller (Eds.), Graph-Theoretic Concepts in Computer Science. XIII, 341 pages. 2007.

Vol. 4763: J.-F. Raskin, P.S. Thiagarajan (Eds.), Formal Modeling and Analysis of Timed Systems. X, 369 pages. 2007.

Vol. 4759: J. Labarta, K. Joe, T. Sato (Eds.), High-Performance Computing. XV, 524 pages. 2008.

Vol. 4750: M.L. Gavrilova, C.J.K. Tan (Eds.), Transactions on Computational Science I. XI, 181 pages. 2008.

Vol. 4746: A. Bondavalli, F. Brasileiro, S. Rajsbaum (Eds.), Dependable Computing. XV, 239 pages. 2007.

Vol. 4743: P. Thulasiraman, X. He, T.L. Xu, M.K. Denko, R.K. Thulasiram, L.T. Yang (Eds.), Frontiers of High Performance Computing and Networking ISPA 2007 Workshops. XXIX, 536 pages. 2007.

Vol. 4742: I. Stojmenovic, R.K. Thulasiram, L.T. Yang, W. Jia, M. Guo, R.F. de Mello (Eds.), Parallel and Distributed Processing and Applications. XX, 995 pages. 2007.

Vol. 4739: R. Moreno Díaz, F. Pichler, A. Quesada Arencibia (Eds.), Computer Aided Systems Theory – EUROCAST 2007. XIX, 1233 pages. 2007.

Vol. 4736: S. Winter, M. Duckham, L. Kulik, B. Kuipers (Eds.), Spatial Information Theory. XV, 455 pages. 2007.

Vol. 4732: K. Schneider, J. Brandt (Eds.), Theorem Proving in Higher Order Logics. IX, 401 pages. 2007.

Vol. 4731: A. Pelc (Ed.), Distributed Computing. XVI, 510 pages. 2007.

Vol. 4728: S. Bozapalidis, G. Rahonis (Eds.), Algebraic Informatics. VIII, 291 pages. 2007.

Vol. 4726: N. Ziviani, R. Baeza-Yates (Eds.), String Processing and Information Retrieval. XII, 311 pages. 2007.

Vol. 4719: R. Backhouse, J. Gibbons, R. Hinze, J. Jeuring (Eds.), Datatype-Generic Programming. XI, 369 pages. 2007.